Key Ideas in Green Chemistry

1. It is better to **prevent waste** than to treat or clean up waste after it is formed.

2. It is better to **minimize the amount of materials used** in the production of a product.

3. It is better to **use and generate substances that are not toxic**.

4. It is better to **use less energy**.

5. It is better to **use renewable materials** when it makes technical and economic sense.

6. It is better to **design materials that degrade** into innocuous products **at the end of their useful life**.

Source: The Twelve Principles of Green Chemistry by Paul Anastas and John Warner.

2024 Release

Chemistry in Context

Applying Chemistry to Society

Bradley D. Fahlman
Central Michigan University

Nicole Bouvier-Brown
Loyola Marymount University

John S. Kirk
Carthage College

Caroline C. Sloan
University of Kansas

Resa M. Kelly
San Jose State University

Stephanie Ryan
Ryan Education Consulting LLC

ACS
Chemistry for Life®

Mc
Graw
Hill

A Project of the American Chemical Society

CHEMISTRY IN CONTEXT

Published by McGraw Hill LLC, 1325 Avenue of the Americas, New York, NY 10019. Copyright ©2024 by McGraw Hill LLC. All rights reserved. Printed in the United States of America. No part of this publication may be reproduced or distributed in any form or by any means, or stored in a database or retrieval system, without the prior written consent of McGraw Hill LLC, including, but not limited to, in any network or other electronic storage or transmission, or broadcast for distance learning.

Some ancillaries, including electronic and print components, may not be available to customers outside the United States.

This book is printed on acid-free paper.

1 2 3 4 5 6 7 8 9 LWI 29 28 27 26 25 24

ISBN 978-1-266-86702-6
MHID 1-266-86702-3

Cover Images: *Jigsaw puzzle and globe: Purestock/Getty Images; Spider web: William Leaman/Alamy Stock Photo; Batteries: Reed Richards/Alamy Stock Photo; Wireless communications store: Graham Oliver/Echo/ Cultura/Getty Images; Bottle of pills: Stock Footage, Inc./Getty Images; Recycling: asiseeit/iStock/Getty Images*

mheducation.com/highered

Brief Contents

Appendices

Contents

Luiza Dutkiewicz/Shutterstock

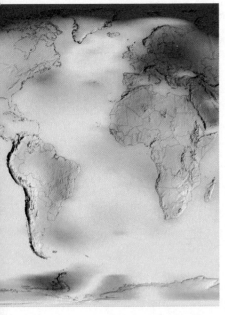

NASA's Scientific Visualization
Studio/Kathryn Mersmann

fizkes/Shutterstock

Stock Footage, Inc./Getty Images

Preface

Climate change. Water contamination. Air pollution. Food shortages. These and other global issues are regularly featured in the media. However, did you know that chemistry plays a crucial role in addressing these challenges? In addition, a fundamental knowledge of chemistry is essential for the design and fabrication of new technologies that improve the quality of our lives. For instance, faster electronic devices, stronger plastics, and more effective medicines and vaccines all rely on the innovations of chemists throughout the world. With our world so dependent on chemistry, it is unfortunate that most chemistry textbooks do not provide significant details regarding real-world applications. Enter *Chemistry in Context*—"the book that broke the mold." Since its inception in 1993, *Chemistry in Context* has focused on the presentation of chemistry fundamentals within a contextual framework.

So, what is "context," and how will this make your study of chemistry more interesting and relevant?

Context! This word is derived from the Latin word meaning "to weave." Hence, *Chemistry in Context* weaves together connections between chemistry and society. In the absence of social issues, there could be no *Chemistry in Context*. Similarly, without teachers and students who are willing (and brave enough) to engage in these issues, there could be no *Chemistry in Context*. As the "Central Science," chemistry is woven into the fabric of practically every issue that our society faces today.

Context! Do you enjoy good stories about the world in which you live? If so, look inside this book for stories that intrigue, challenge, and possibly even motivate you to act in new or different ways. In almost all contexts—local, regional, and global—parts of these stories are still unfolding. The ways in which you and others make choices today will determine the nature of the stories told in the future.

Context! Are you aware that using a real-world context to engage people is a high-impact practice backed up by research on how people learn? *Chemistry in Context* offers real-world contexts to engage learners on multiple levels: personal, social, and global. Given the rapidly changing nature of these contexts, *Chemistry in Context* also offers teachers the opportunity to become learners alongside their students.

Sustainability—The Ultimate Context

Global sustainability is not just a challenge. Rather, it is *the* defining challenge of our century. Accordingly, this release of *Chemistry in Context* continues to focus on this challenge, both as a topic worth studying and as a problem worth solving. As a topic, sustainability provides an important source of content. For example, the tragedy of the commons (Section 2.12), the Triple Bottom Line (Section 6.16), and the concept of cradle-to-cradle (Section 1.9) are all part of this essential content. As a problem worth solving, sustainability generates new questions to ask—ones that can help you imagine and achieve a sustainable future. For example, you will find many questions about the risks and benefits of acting (or not acting) to reduce emissions of greenhouse gases.

Incorporating sustainability requires more than a casual rethinking of the curriculum. Unlike most general chemistry texts, *Chemistry in Context* is context rich. In essence, you can think of our coverage as a "Citizens First" approach that is context-driven, rather than the content-driven "Atoms First" approach used in many general chemistry curricula. Thus, unlike many other textbooks, this book provides interesting real-world scenarios about energy, materials, food, water, and health to convey essential chemistry content alongside the key concepts of sustainability.

Green chemistry, a means to sustainability, continues to be an important theme in *Chemistry in Context*. As with earlier editions, we have continued our strong focus on green chemistry concepts, which are woven throughout this release. This contextual coverage offers a better sense of the need for, and importance of, greening our chemical processes.

Updates to Existing Content

As with all previous editions, this release of *Chemistry in Context* describes how chemistry affects your life and that of your students through discussions of real-world issues and applications. However, you may have noticed that we no longer refer to this revision as the "11th edition," but rather by the "2024 release." *Chemistry in Context* is moving to an Evergreen delivery model, which means it has content, tools, and technology that is updated and relevant, with updates delivered directly to your existing McGraw Hill Connect® course. This means that you will now be able to engage students and freshen up assignments with up-to-date coverage of select topics and assessments, all without having to switch editions or build a new course. This is especially important for this title due to the time-sensitive nature of our content that aims to reflect the most recent developments of our ever-changing world.

In this revision, we have provided a significant update to the breadth of digital assets provided throughout the book. Each chapter contains a variety of new features such as videos and diverse types of interactive activities—each designed to keep the reader engaged and assist with mastery of content. Although these digital assets are embedded directly in the e-book, print users can access all of the multimedia assets by visiting http://www.acs.org/cic.

Similar to the previous edition, digital assets and activities are woven throughout each chapter that direct students to search the Internet to find appropriate data or reports to draw their own conclusions regarding current worldwide issues. However, you will find even more diverse opportunities in this release.

All chapters have also been extensively revised to improve the flow of topics while incorporating new scientific developments, changes in policies, energy trends, and current world events.

Each chapter begins with a video that introduces the context, with a "Reflect" activity for students to ponder before reading the chapter. This is immediately followed by "Compelling Questions," which identify the main questions that are addressed in each section of the chapter.

In this release, we have been more purposeful regarding learning objectives, which are now provided under the heading of every section title. Each chapter then concludes with a "Learning Outcomes," section that lists all of these important concepts, with links to their particular sections.

Just as in the previous edition, the final capstone chapter of the textbook (Chapter 14), is written as a "whodunit" storyline. Concepts from all of the previous 13 chapters are woven into the story, which takes students through the process of investigating crime scenes and employing appropriate techniques for evidence collection and analyses.

Teaching and Learning in Context

This new release of *Chemistry in Context* continues with the organizational scheme used previously. Each chapter delves into a real-world theme that provides a foundation of chemistry concepts that are built upon in later chapters. As shown on pages xv–xvi, a variety of learning resources consisting of videos, and interactive illustrations and simulations, are woven throughout each chapter. Assets that are video-based such as Lightboard tutorials and ACS Reactions series are denoted as "Reflect." A new type of activity is included in this release, denoted as "@Home," which feature lab-based activities that may be performed at home using simple household items. Since our world is filled with controversial topics, we have also included "Let's Debate" activities throughout each chapter. These are meant to foster fruitful student-instructor dialogue that goes beyond the textbook descriptions. Lastly, we have introduced "Real Talk" interviews throughout the book, which describe some real-world applications of chemistry as told by leading scientists and students.

This release also features a variety of embedded in-chapter "Your Turn" activities and end-of-chapter questions, grouped by three types—"Emphasizing Essentials" (basic review, more traditional), "Concentrating on Concepts" (critical thinking or simulation-based), and "Exploring Extensions" (analytical reasoning, often requiring the use of resources outside the textbook). These activities are plentiful and varied. They range from simple practice exercises focusing on traditional chemical principles to those requiring more thorough analysis and integration of applications. In this release, we have more thoughtfully placed these activities throughout each chapter so they do not interrupt the flow, but rather aid in learning for small group work, class discussions, or individual projects to explore interests, as time permits, beyond the core topics.

Chemistry in Context, 2024 Release— A Team Effort

Once again, we have the pleasure of offering our readers a new release of *Chemistry in Context*. But the work is not done by just one individual; rather, it is the work of a talented team. The 2024 Release builds on the legacy of prior author teams led by Cathy Middlecamp, A. Truman Schwartz, Conrad L. Stanitski, and Lucy Pryde Eubanks—all leaders in the chemical education community.

This new release was prepared by myself and co-authors Nicole Bouvier-Brown, John Kirk, Caroline Sloan, Resa Kelly, and Stephanie Ryan. Each author brought their own experiences and expertise to the project, which helped to expand the depth and breadth of the contexts to reach a variety of audiences.

At the American Chemical Society, leadership was provided by LaTrease Garrison, Chief Operating Officer. She supported the writing team, cheering on its efforts to "connect the dots" between chemistry contexts and the underlying fundamental chemistry content. Terri Chambers, Senior Director of the Education Division of the American Chemical Society, provided support and direction throughout the project, with great insights regarding the effective use of *CiC* in the classroom. ACS Program Manager Amanda Koenig reviewed manuscripts, providing thoughtful insights to optimize learning and flow. Society & Media Relations Manager Emily Abbott provided guidance on new content and approaches to multimedia. Technical Manager Karen Trimmer and Program Manager Natalia Martin supported the development of the interactives designed to promote student engagement. Rebecca Presgraves Principal Technologist of QA led the team to meet WCAG (Web Content Accessibility Guidelines) accessibility standards. The ACS Productions group created the Reactions videos that are pointed to throughout the text. The team also edited and added animations to videos created by the author team that are part of the book's digital collection.

The web site housing various simulations and multimedia assets pointed to from the book was designed and executed through collaboration between many members of the ACS Web Strategy and ACS Web Services and Communications teams led by Mark Carpenter, Director, and Lorinda Bullock, Senior Manager, respectively.

The McGraw Hill team was superb in all aspects of this project, with special thanks to Katie Peterson (Senior Product Developer) and Jane Mohr (Content Project Manager) for shepherding the project to the finish line. We also gratefully acknowledge the following individuals at McGraw Hill for their support: Ian Townsend (Executive Portfolio Manager), Hannah Downing (Portfolio Manager), Rose Koos (Senior Director of Product Development), Shirley Hino (Director of Digital Content Development), Tami Hodge (Executive Marketing Manager), Ron Nelms (Assessment Content Project Manager), Lori Hancock and Melissa Homer (Content Licensing Specialists), David Hash (Designer), Robin Reed (Senior Product Developer Manager), and John Murdzek (Copyeditor).

The author team truly benefited from the expertise of a wider community. We would like to thank the following individuals who assisted me in writing and/or reviewing learning-goal-oriented content for **SmartBook**:

Stephanie Ryan, *Ryan Education Consulting LLC*
David Jones, *St. David's School in Raleigh, NC*
Barbara Pappas, *The Ohio State University*

We are very excited by the new content and digital features provided in this release. As you explore the various contexts, we hope that your study of the underlying fundamental chemistry concepts will become more relevant in your life. We believe that the chemistry contexts and content provided herein, alongside the interactive and thought-provoking activities embedded throughout, will make you think differently about the world around you and the challenges we face. The solutions to current and future global problems will require an interdisciplinary approach. Whether you decide to continue your studies in chemistry, or transition to other fields of study, we believe that the critical thinking skills fostered in *Chemistry in Context* will be of value to all of your future endeavors.

Sincerely, on behalf of the author team,

Bradley D. Fahlman
Editor-in-Chief
December 2023

Active Learning Resources

The 2024 release features a variety of interactive features to engage the reader and foster critical thinking skills. Particular styles and icons highlight the placement of these features.

Compelling Questions & Learning Outcomes

At the start of each chapter, you'll find **Compelling Questions** to consider as you journey through the chapter and learn about the real-world applications of chemistry. **Learning Outcomes** are provided at the beginning of each section and summarized at the end of each chapter, which address the key chemistry concepts posed by the Compelling Questions.

COMPELLING QUESTIONS

In this chapter, we will answer the following questions:

1. What is respiration and why is it essential to life?
2. What are the regions of the atmosphere?
3. How do we measure the concentration of pollutants in air?
4. How can we visualize the components present in air?
5. How do we name the substances present in air?
6. What are the health implications of inhaling air pollutants?
7. How do we determine if the air is safe to breathe?
8. How do interpret reported air quality measurements?
9. What are some chemical reactions that create air pollutants?
10. What are combustion reactions and how do these lead to air pollution?
11. How do coal-fired power plants and vehicle use generate air pollutants?
12. How is ground-level ozone produced?
13. What are some harmful components of indoor air?
14. In what ways can we prevent or limit contaminants from polluting our atmosphere?

LEARNING OUTCOMES

The numbers in parentheses indicate the sections within the chapter where these outcomes were discussed.

Having studied this chapter, you should now be able to:

- describe and illustrate the process of respiration (2.1)
- categorize the regions of the atmosphere and compare the compositions of inhaled and exhaled air (2.2)
- describe the concentrations of substances in the air (2.3)
- use macroscopic, molecular, and symbolic representations to describe matter (2.4)
- write formulas and names for molecular compounds, including hydrocarbons (2.5)
- identify and explain the health effects of common air pollutants (2.6)
- assess the risk and toxicity of different air pollutants by exposure (2.7)

- analyze local air quality for air pollutants and make comparisons to national and global trends (2.8)
- describe chemical reactions in words and write their balanced chemical equations (2.9)
- describe the process of complete combustion and its effect on air quality (2.10)
- describe direct sources and formation reactions of air pollutants (2.11)
- describe the formation of tropospheric ozone (2.12)
- define and identify sources and potential health impacts of some indoor pollutants (2.13)
- explain the role of "green chemistry" practices in reducing pollution (2.14)

Videos

Each chapter features an introductory video and associated activity that foster instructor–student dialogue within the context of the real-world application. In addition, an assortment of instructional videos are woven throughout each chapter to assist the reader in grasping fundamental content, as well as understanding the broader applications of topics.

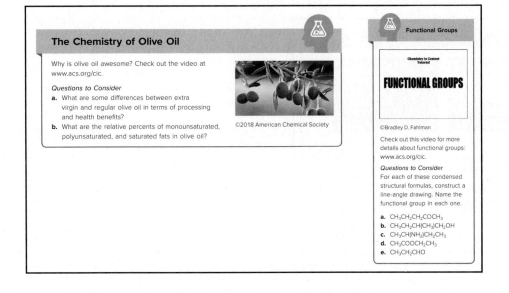

The Chemistry of Olive Oil

Why is olive oil awesome? Check out the video at www.acs.org/cic.

Questions to Consider
a. What are some differences between extra virgin and regular olive oil in terms of processing and health benefits?
b. What are the relative percents of monounsaturated, polyunsaturated, and saturated fats in olive oil?

©2018 American Chemical Society

Functional Groups

Chemistry in Context Tutorial

FUNCTIONAL GROUPS

©Bradley D. Fahlman

Check out this video for more details about functional groups: www.acs.org/cic.

Questions to Consider
For each of these condensed structural formulas, construct a line-angle drawing. Name the functional group in each one.

a. $CH_3CH_2CH_2COCH_3$
b. $CH_3CH_2CH(CH_3)CH_2OH$
c. $CH_3CH(NH_2)CH_2CH_3$
d. $CH_3COOCH_2CH_3$
e. CH_3CH_2CHO

Laboratory Experiments

A variety of laboratory experiments and demos are placed throughout each chapter, as videos or experimental procedures, both with follow-up questions for reflection. Many of these experiments are denoted as "@HOME" activities, which may be done at home using common household items.

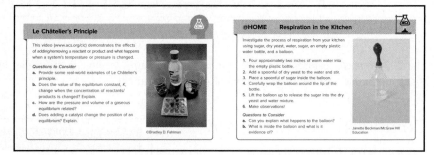

Simulations and Interactive Figures

You'll find numerous interactive figures and PhET simulations referenced throughout chapters to foster student engagement and hands-on learning. Additionally, we provide MolView 3D representations for many chemical structures throughout the textbook.

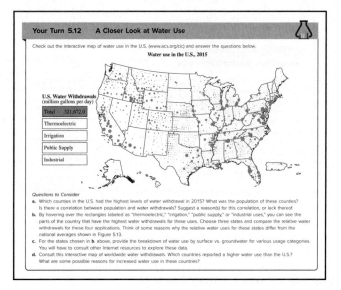

Let's Debate!

A variety of topics for debates are now included in each chapter. These opportunities offer an effective way to increase student motivation for the topics being discussed in lecture and foster student-peer and student-instructor interactions.

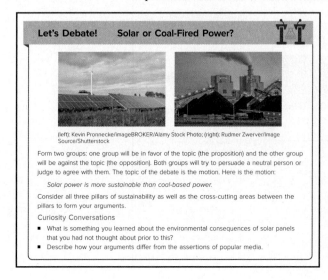

Real Talk

This release includes interviews with scientists and students who are working in fields that involve real-world applications of chemistry. These video-based activities offer a rare glimpse into the life of a scientist and provide a tangible context for the chemistry fundamentals being discussed.

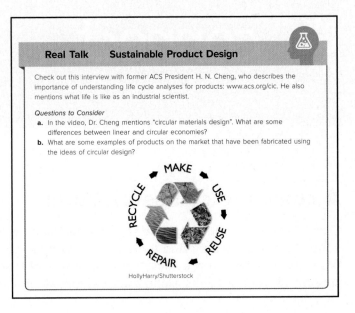

Real Talk   Sustainable Product Design

Check out this interview with former ACS President H. N. Cheng, who describes the importance of understanding life cycle analyses for products: www.acs.org/cic. He also mentions what life is like as an industrial scientist.

Questions to Consider
a. In the video, Dr. Cheng mentions "circular materials design". What are some differences between linear and circular economies?
b. What are some examples of products on the market that have been fabricated using the ideas of circular design?

HollyHarry/Shutterstock

Your Turn Activities: Explorations Beyond The Textbook

Each chapter includes a variety of thought-provoking activities that are woven into the text. There are three types, clearly indicated by color boxes and icons, and they parallel the types of end-of-chapter problem sets.

i) **Emphasizing Essentials**: questions that give the opportunity to practice fundamental skills. To aid with problem-solving skills, we have provided sample questions and answers for many of these activities.

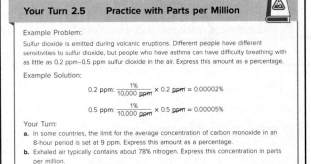

Your Turn 2.5   Practice with Parts per Million

Example Problem:
Sulfur dioxide is emitted during volcanic eruptions. Different people have different sensitivities to sulfur dioxide, but people who have asthma can have difficulty breathing with as little as 0.2 ppm–0.5 ppm sulfur dioxide in the air. Express this amount as a percentage.

Example Solution:

$$0.2 \text{ ppm: } \frac{1\%}{10,000 \text{ ppm}} \times 0.2 \text{ ppm} = 0.00002\%$$

$$0.5 \text{ ppm: } \frac{1\%}{10,000 \text{ ppm}} \times 0.5 \text{ ppm} = 0.00005\%$$

Your Turn:
a. In some countries, the limit for the average concentration of carbon monoxide in an 8-hour period is set at 9 ppm. Express this amount as a percentage.
b. Exhaled air typically contains about 78% nitrogen. Express this concentration in parts per million.

ii) **Concentrating on Concepts**: questions that go beyond the content to highlight a real-world application or social issue.

Your Turn 4.26   Local Temperature Changes

Check out www.acs.org/cic to find data on the temperature changes over the last 100 years in the area where you live. Has the temperature changed much? Now search for a map of the world that shows temperature change. Which areas have changed the most in the last 100 years? Which have changed the least? How does this compare to your area?

iii) **Exploring Extensions**: questions that challenge students to apply their knowledge to question policies, make decisions, and design solutions to global issues.

Your Turn 8.9   Yes, Tons of CO$_2$!

Could an automobile *really* emit 7 tons of carbon dioxide in a year? Do a calculation to verify this. State all the assumptions that you make. If you were to improve your car's gas mileage by 5 miles per gallon, how much CO$_2$ would you save over a year? Aside from buying a hybrid car, what are some ways people can reduce transportation-related carbon dioxide emissions?

A complete course platform

Connect enables you to build deeper connections with your students through cohesive digital content and tools, creating engaging learning experiences. We are committed to providing you with the right resources and tools to support all your students along their personal learning journeys.

65%
Less Time Grading

Laptop: Getty Images: Woman/dog: George Doyle/Getty Images

Every learner is unique

In Connect, instructors can assign an adaptive reading experience with SmartBook® 2.0. Rooted in advanced learning science principles, SmartBook 2.0 delivers each student a personalized experience, focusing students on their learning gaps, ensuring that the time they spend studying is time well spent. **mheducation.com/highered/connect/smartbook**

Study anytime, anywhere

Encourage your students to download the free ReadAnywhere® app so they can access their online eBook, SmartBook® 2.0, or Adaptive Learning Assignments when it's convenient, even when they're offline. And since the app automatically syncs with their Connect account, all of their work is available every time they open it. Find out more at **mheducation.com/readanywhere**

"I really liked this app— it made it easy to study when you don't have your textbook in front of you."

Jordan Cunningham, a student at *Eastern Washington University*

Effective tools for efficient studying

Connect is designed to help students be more productive with simple, flexible, intuitive tools that maximize study time and meet students' individual learning needs. Get learning that works for everyone with Connect.

Education for all

McGraw Hill works directly with Accessibility Services departments and faculty to meet the learning needs of all students. Please contact your Accessibility Services Office, and ask them to email **accessibility@mheducation.com**, or visit **mheducation.com/about/accessibility** for more information.

Affordable solutions, added value

Make technology work for you with LMS integration for single sign-on access, mobile access to the digital textbook, and reports to quickly show you how each of your students is doing. And with our Inclusive Access program, you can provide all these tools at the lowest available market price to your students. Ask your McGraw Hill representative for more information.

Solutions for your challenges

A product isn't a solution. Real solutions are affordable, reliable, and come with training and ongoing support when you need it and how you want it. Visit **supportateverystep.com** for videos and resources both you and your students can use throughout the term.

Updated and relevant content

Our new Evergreen delivery model provides the most current and relevant content for your course, hassle-free. Content, tools, and technology updates are delivered directly to your existing McGraw Hill Connect® course. Engage students and freshen up assignments with up-to-date coverage of select topics and assessments, all without having to switch editions or build a new course.

1

Portable Electronics: The Periodic Table in the Palm of Your Hand

Banner credit above: 279photo Studio/Shutterstock; LifestyleVideoFootage/Shutterstock

WHAT'S IN YOUR CELL PHONE?

Watch the Chapter 1 opening video at www.acs.org/cic to get a glimpse of how chemistry plays a central role in controlling the properties of electronic devices.

a. List some desirable attributes of a cell phone, and some that you would like to see in the future.

b. Cite two elements that combine to form a substance important to your cell phone.

c. What is the expected lifespan of your cell phone?

In this chapter, you will explore the following questions:

1. How do touchscreens work?
2. What are the different components of your portable electronic device made from?
3. How are different properties of materials created from a limited number of elements?
4. How small are atoms?
5. What is it about atoms that may make a material electrically conductive?
6. What are rocks, and how do we isolate and purify metals from these natural sources?
7. How is ordinary sand converted into silicon—the fundamental component of processor chips?
8. How is sand converted into glass, and how can its structure be modified for crack-resistant screens?
9. What is the life cycle of your portable electronic device?
10. What are the environmental implications of recycling your portable electronic device?

Introduction

Email, phone calls, texts, and social media. Our modern society demands constant contact during busy days filled with meetings, classes, travel, and social activities. The tablet or cell phone you hold in your hand is a combination of a variety of materials that have been carefully crafted to give you special capabilities you can't imagine living without.

In order to satisfy the ever-rigorous demands of today's consumer, the latest portable electronics must be lightweight, thin, durable, multifunctional, and easily synced with computers and next-generation wearable devices. Such complex designs are possible only by putting together the elements of the periodic table in many different ways to form materials with the above physical properties that we need or desire.

In this chapter, you will learn about the various components that make up your cell phone, tablet, or other portable electronic devices. Perhaps most importantly, you will discover where these components came from and what happens to them after their lifetime is finished.

1.1 | How Do Touchscreens Work?

Learning Objective: Classify, compare, and visualize the states of matter

It's wintertime, and you need to respond to an urgent text on your smartphone. You touch the screen with a gloved finger and get no response. The hassle of removing your gloves and risking frostbite, just to operate your cell phone or tablet, is an all-too-common occurrence for those who live in cold climates. However, a variety of commercially available gloves use a special thread or have pads sewn into them, allowing users to seamlessly control their touchscreen devices. Most smartphones and tablets will also respond to a stylus. Nevertheless, this begs the question: Why are touchscreens so restrictive in responding to only a small number of stimuli?

Your Turn 1.1   Touchscreen Response

Taking care not to damage your screen, use a variety of materials to touch the screen of your portable electronic device. In addition to your finger, items that may be used include: a paper clip, a plastic pen, a key, a battery, fabrics, pencil lead, a sponge (wet and dry), a pencil eraser, a coin, a glass marble, paper, cardboard, or any other items. Make a list of things the screen responded to and a list that the screen did not. Are there any trends in the types of materials on each list?

How Do Touchscreens Work?

nenetus/123RF

Watch a video for more details on how touchscreens work: www.acs.org/cic.

As you saw in the previous activity, touchscreens respond only to electrically conductive objects. If you have experienced a shock by touching a metal object after sliding your feet across a carpet, you realize that the human body is a conductor of electricity. Other examples of electrically conductive materials are metals such as copper, silver, and aluminum. On the other hand, materials such as concrete, wood, and most plastics do not allow electricity to flow and are called electrically *insulating*. You may be wondering why metallic objects like paper clips and keys, which are also electrically conductive, didn't give a touchscreen response. This is due to touchscreen controls that ignore contact points much smaller than your finger to avoid giving false signals.

The properties of a device are governed by what it is made of—its **composition**. What compositions are required for a touchscreen to be transparent, crack-resistant, and touch-sensitive? This is no minor feat and requires scientists to constantly explore the world around them to select the most appropriate constituents.

Everything around you—the air you breathe, the water you drink, and the mobile device in your hand—is defined as **matter**. Matter is considered to be anything that occupies space and has a mass. However, most relevant to this textbook, the discipline of **chemistry** is the branch of science that focuses on the composition, structure, properties, and changes of matter.

Let's begin our investigation by looking at three phases of matter that are commonly present on Earth: solids, liquids, and gases (Figure 1.1). These phases play a critical role in our daily lives but are often taken for granted. For instance, we breathe in gases on a daily basis; the components in air will be described in Chapter 2. We also drink liquids regularly in the form of water, soft drinks, or coffee, and we eat many solids such as candy, french fries, or potato chips. We are inundated with these phases every day, but what are their defining principles? Let's find out by examining their properties.

@HOME   Soup or Salad?

(left): MaraZe/Shutterstock; (right): Tatjana Baibakova/Shutterstock

Part I. The Properties of Solids and Liquids

Investigate solids and liquids around your house and consider how the properties of solids and liquids differ.

1. Gather the following solids: sugar, baking soda, ice, and lettuce. Gather the following liquids: water, vinegar, tomato soup, and cooking oil.
2. Make a list of ways in which solids differ from one another and a list of ways in which solids are alike.
3. Repeat step 2 for the liquids you have gathered.

Question to Consider

A student claims that liquids are pourable and able to fill a container, but you note that you can actually pour sugar and baking soda to fill a container. Should sugar and baking soda be reclassified as liquids? Why or why not?

Part II. Making Ice Cubes

1. Fill an ice cube tray with water (for dramatic effect add a drop of food coloring)
2. Place the filled tray in the freezer and make observations over time.
3. Empty the contents of the tray into a bowl and make observations over time.

—(continued)

Questions to Consider
a. Was heat added or removed in each scenario?
b. As we will define later, individual water particles are known as "molecules". How did each step affect the motion of the water molecules? How do you know the motion of the water molecules was affected since you could not actually see the water molecules?

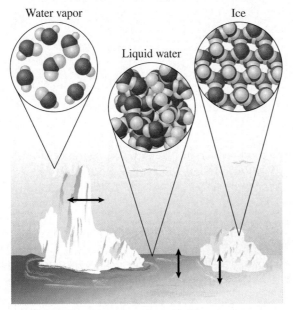

Figure 1.1

Arrangements of water molecules in their different states.

Your Turn 1.2   Macroscopic Properties of Solids, Liquids, and Gases

Check out an interactive simulation (www.acs.org/cic) of atoms and molecules in different states and answer the following questions for solids, liquids, and gases. Provide an example to support each of your answers.

a. Does the phase have a definite volume?
b. Does the phase have a definite shape?
c. Will the phase take the shape of its container?
d. Will the phase completely fill its container?

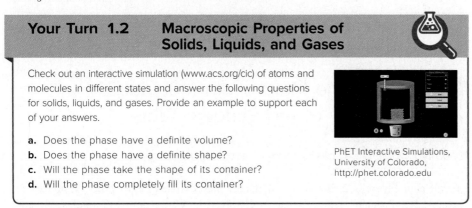

PhET Interactive Simulations, University of Colorado, http://phet.colorado.edu

Based on your answers from Your Turn 1.2, we can create a table (Table 1.1) describing the macroscopic properties of solids, liquids, and gases.

Table 1.1	Macroscopic Properties of Solids, Liquids, and Gases			
Phase	Takes the shape of its container?	Completely fills its container?	Definite volume?	Definite shape?
Solid	No	No	Yes	Yes
Liquid	Yes	No	Yes	No
Gas	Yes	Yes	No	No

The behavior of substances can give us insight into the arrangement and attraction of their constituent atoms and/or molecules. Considering Table 1.1, we can predict the attractions and relative spacing of the particles that compose a substance. In general, the closer atoms and molecules are to one another, the more attracted they are, which leads to a more rigid material such as a solid. As you will explore in the next activity, on average, the atoms or particles of gases are much farther apart than in solids or liquids.

Your Turn 1.3   Visualizing the Phases of Matter

Explore the following simulations and activities for solids, liquids, and gases: www.acs.org/cic. You will use these depictions and the definitions you created in this section to create an even more descriptive model of these phases at the particulate, atomic, or molecular level.

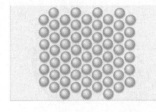
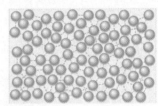
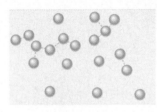
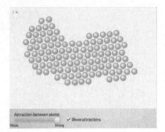

[The Concord Consortium]

What conclusions can you draw from these simulations? Do these representations follow the properties outlined in Table 1.1? Create a representation for 15 atoms of a solid, a liquid, and a gas in a container of your choice.

Hint: Ensure your representation follows all the statements in Table 1.1. For example, your solid should not fill the container, nor take the shape of the container.

1.2 | What's the Matter with Materials? A Survey of the Periodic Table

Learning Objective: Classify matter according to composition

As shown in Figure 1.2, the states of matter may exist as either pure substances or mixtures. For instance, before sugar is dissolved in water, both the solid sugar and liquid water are considered pure substances—each is composed of a single substance. Mixing these separate pure substances together results in a **homogeneous mixture**, which is uniform in composition throughout. Quite often, a homogeneous mixture is referred to as a **solution**. In contrast, if you dig up a handful of soil, you will discover a complicated mixture of sand, particles of varying shapes and colors, liquid water within the pores, and perhaps even some resident earthworms. This is known as a **heterogeneous mixture**, because it is not uniform in composition throughout. That is, the relative amounts of sand, dirt, or rocks will vary from one handful to the next.

As we will describe later, the smallest building blocks of matter are known as **atoms**. An **element** is composed of many atoms of the same type. Every day, we take for granted the use of pure elements such as copper in household pipes, aluminum in home exteriors, lithium in batteries, and carbon in pencil nibs. In contrast, a **compound** is a pure substance made up of two or more different types of atoms in a fixed, characteristic chemical combination.

A **chemical formula** is a symbolic way to represent the elementary composition of a substance. It reveals both the elements present (by chemical symbols—described below) and the atomic ratio of those elements (by the subscripts). For example, in the compound CO_2, the elements carbon (C) and oxygen (O) are present in a ratio of one carbon atom for every two oxygen atoms. Similarly, H_2O indicates two hydrogen atoms

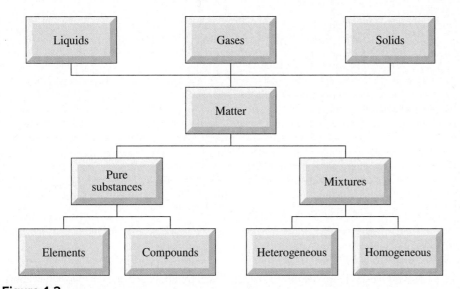

Figure 1.2

A classification scheme for matter.

for each oxygen atom. Note that when an atom occurs only once, such as the O in H_2O or the C in CO_2, the subscript "1" is omitted.

Revisiting a sugar solution, water (H_2O) is a compound consisting of oxygen and hydrogen atoms. Sugar ($C_{12}H_{22}O_{11}$) is also a compound, but instead contains carbon, hydrogen, and oxygen atoms. Even though the types of atoms in compounds and elements are identical, they are bonded to one another differently within each substance. For instance, the oxygen atoms in sugar are identical to the oxygen atoms that comprise elemental oxygen gas (O_2). However, it would take a chemical reaction to break apart the atoms within sugar to return the oxygen atoms to their elemental form—gaseous oxygen.

Classification of Matter

Watch a video to further familiarize yourself with classifying matter: www.acs.org/cic. Then use the classification scheme shown in Figure 1.2 to categorize the following:

a. Your cell phone
b. Aluminum foil
c. Red wine
d. Chlorine gas
e. Stainless steel

McGraw Hill

@HOME   Classifying Matter Around Your Home

Classify matter using things in your home. You will need a soft drink, sugar, table salt, a microwave-safe cup or glass, water, and ice cubes.

1. Pour a soft drink into a microwave-safe cup or glass and make observations. Classify the soft drink in its type of matter.
2. Add ice cubes to the soft drink in the cup and make observations. Classify the type of matter that you are observing. Then, place the cup into the microwave for 30 seconds and make observations after the time is over. Repeat for 30 seconds more and make observations again. Classify the type of matter in the cup.

Steve Cukrov/Shutterstock

—(continued)

3. Measure a spoonful of salt and make observations. Then, add it to a cup of water and make observations. Using only your observations, what classifications are possible for the salt itself? How would the salt water be classified? Keep pouring more salt in it until you start to see salt collecting on the bottom. How would you classify this now? Now, place the cup into the microwave for 30 seconds and make observations after the time is over. Repeat for 30 seconds and make observations again. Classify the type of matter in the cup.

4. Measure a spoonful of sugar and make observations. Then, add it to a cup of water and make observations. Using only your observations, what classifications are possible for the sugar itself? How would the sugar water be classified? Keep pouring more sugar in it until you start to see the sugar collecting in the bottom. How would you classify this now? Now, place the cup into the microwave for 30 seconds and make observations after the time is over. Repeat for 30 seconds and make observations again. Classify the type of matter in the cup.

The Race to Invent the Periodic Table

©2018 American Chemical Society

Check out this video for more details about the origin of the periodic table: www.acs.org/cic.

Questions to Consider

a. One of the elements Mendeleev predicted, he named eka-aluminum. After the element was discovered, this element was named gallium. Another element Mendeleev predicted he called eka-silicon. What is the modern name of eka-silicon?

b. Mendeleev and other chemists at the time organized their tables by increasing atomic mass. The modern-day periodic table is organized by increasing atomic number. The periodic table in Figure 1.3 has the atomic number above each elemental symbol. The atomic mass is the decimal number below the atomic symbol. Looking at a periodic table, identify any elements where the mass decreases with an increase in the atomic number.

As you have already begun to see, chemical symbols are one- or two-letter abbreviations for the elements. These symbols, established by international agreement, are used throughout the world. Some make immediate sense to those who speak English or related languages. For example, oxygen is O, aluminum is Al, lithium is Li, and silicon is Si. However, a number of symbols have their origin in other languages, such as some metals that were discovered by ancient civilizations and given Latin names long ago. For example, argentum (Ag) is silver, ferrum (Fe) is iron, plumbum (Pb) is lead, and hydrargyrum (Hg) is mercury.

Elements have been named for properties, planets, places, and people. Hydrogen (H) means "water former" because hydrogen and oxygen gases burn in a flame to form the compound water (H_2O). Neptunium (Np) and plutonium (Pu) were named after two planets in our solar system. Berkelium (Bk) and californium (Cf) honor the University of California, Berkeley lab in which they were first produced. Two newer elements, flerovium (Fl) and livermorium (Lv), were both named after the laboratories in which they were discovered.

It is fitting that Russian chemist Dmitri Mendeleev (1834–1907) has his own element (mendelevium, Md) because the most common way of arranging the elements— the periodic table—reflects the system he developed. This is an orderly arrangement of all the elements based on similarities in their reactivities and properties.

About 90 elemental substances occur naturally on Earth and, as far as we know, elsewhere in the universe. The other two dozen or so elements, including those most recently discovered, have been created from existing elements through nuclear reactions. Plutonium is probably the best-known synthetic element, although it does occur in trace amounts in nature.

Among all known elements, the vast majority are solids at room temperature. At room temperature, nitrogen (N_2), oxygen (O_2), argon (Ar), and eight other elements are gases; in contrast, only bromine (Br_2) and mercury (Hg) are liquids.

The modern periodic table shown in Figure 1.3 lists each element by number. The green shading indicates the *metals*, which represent most of the periodic table. These elements are usually solid at room temperature, shiny in appearance, may be permanently deformed without breaking or cracking, and are conductors of electricity and heat. Ancient civilizations used some metallic elements (iron, copper, tin, lead, gold, and silver) for weaponry, currency, and decoration. Today, the cases of portable electronic devices sometimes employ the metal aluminum, and the circuitry that powers the device utilizes metals such as gold, copper, and tin.

Far fewer in number are the *nonmetals*—elements that may be in gaseous, liquid, or solid states at room temperature. The nonmetals are characterized by poor conductivity of heat or electricity, and those in the solid state cannot be deformed without cracking or breaking. A mere eight elements fall into a category known as *metalloids*— elements that lie between metals and nonmetals in the periodic table; and whose properties do not fall cleanly into either category. As a reflection of their intermediate electrical conductivity relative to metals and nonmetals, the metalloids are also often

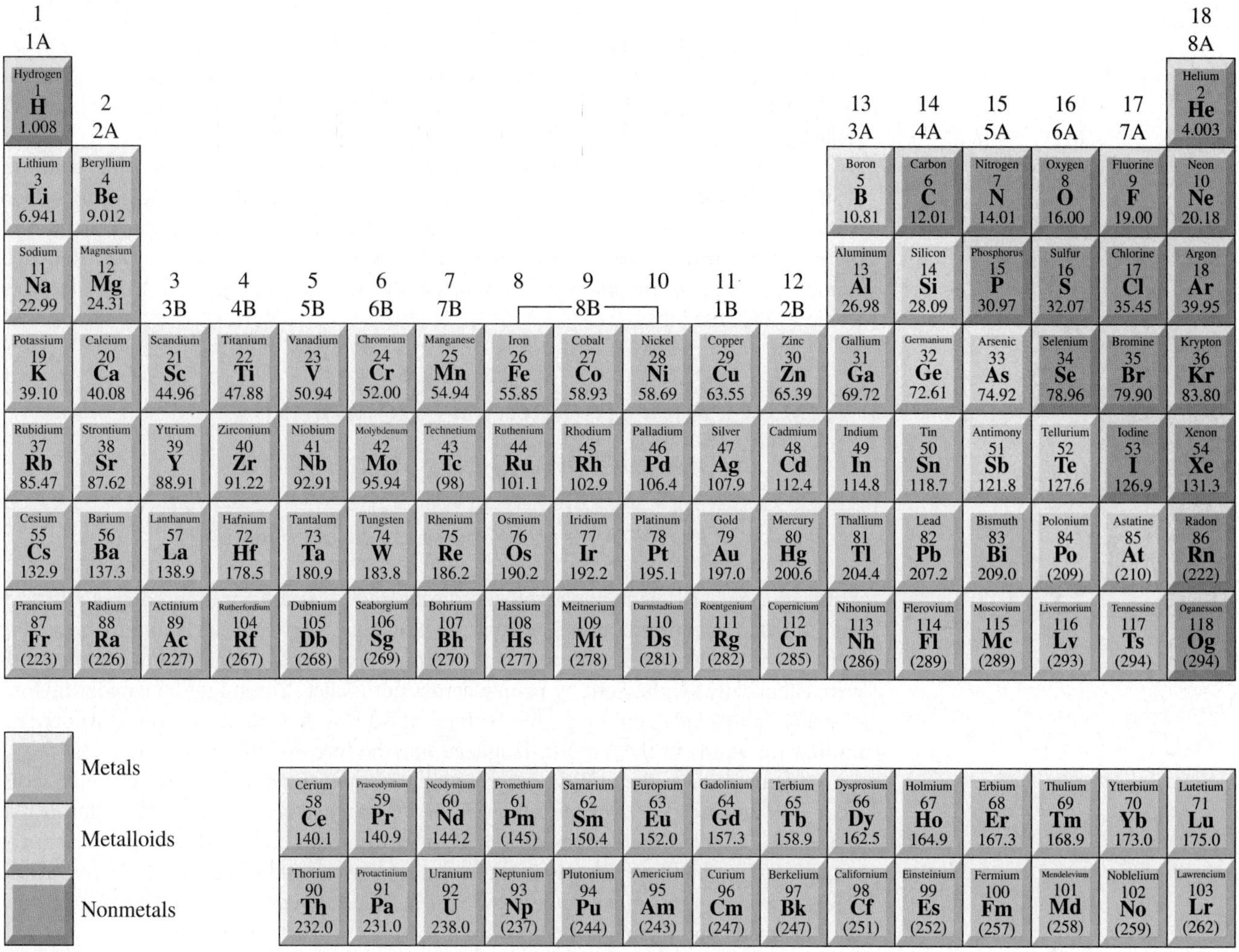

1 1A																		18 8A
Hydrogen 1 **H** 1.008	2 2A											13 3A	14 4A	15 5A	16 6A	17 7A		Helium 2 **He** 4.003
Lithium 3 **Li** 6.941	Beryllium 4 **Be** 9.012											Boron 5 **B** 10.81	Carbon 6 **C** 12.01	Nitrogen 7 **N** 14.01	Oxygen 8 **O** 16.00	Fluorine 9 **F** 19.00		Neon 10 **Ne** 20.18
Sodium 11 **Na** 22.99	Magnesium 12 **Mg** 24.31	3 3B	4 4B	5 5B	6 6B	7 7B	8	9 8B	10	11 1B	12 2B	Aluminum 13 **Al** 26.98	Silicon 14 **Si** 28.09	Phosphorus 15 **P** 30.97	Sulfur 16 **S** 32.07	Chlorine 17 **Cl** 35.45		Argon 18 **Ar** 39.95
Potassium 19 **K** 39.10	Calcium 20 **Ca** 40.08	Scandium 21 **Sc** 44.96	Titanium 22 **Ti** 47.88	Vanadium 23 **V** 50.94	Chromium 24 **Cr** 52.00	Manganese 25 **Mn** 54.94	Iron 26 **Fe** 55.85	Cobalt 27 **Co** 58.93	Nickel 28 **Ni** 58.69	Copper 29 **Cu** 63.55	Zinc 30 **Zn** 65.39	Gallium 31 **Ga** 69.72	Germanium 32 **Ge** 72.61	Arsenic 33 **As** 74.92	Selenium 34 **Se** 78.96	Bromine 35 **Br** 79.90		Krypton 36 **Kr** 83.80
Rubidium 37 **Rb** 85.47	Strontium 38 **Sr** 87.62	Yttrium 39 **Y** 88.91	Zirconium 40 **Zr** 91.22	Niobium 41 **Nb** 92.91	Molybdenum 42 **Mo** 95.94	Technetium 43 **Tc** (98)	Ruthenium 44 **Ru** 101.1	Rhodium 45 **Rh** 102.9	Palladium 46 **Pd** 106.4	Silver 47 **Ag** 107.9	Cadmium 48 **Cd** 112.4	Indium 49 **In** 114.8	Tin 50 **Sn** 118.7	Antimony 51 **Sb** 121.8	Tellurium 52 **Te** 127.6	Iodine 53 **I** 126.9		Xenon 54 **Xe** 131.3
Cesium 55 **Cs** 132.9	Barium 56 **Ba** 137.3	Lanthanum 57 **La** 138.9	Hafnium 72 **Hf** 178.5	Tantalum 73 **Ta** 180.9	Tungsten 74 **W** 183.8	Rhenium 75 **Re** 186.2	Osmium 76 **Os** 190.2	Iridium 77 **Ir** 192.2	Platinum 78 **Pt** 195.1	Gold 79 **Au** 197.0	Mercury 80 **Hg** 200.6	Thallium 81 **Tl** 204.4	Lead 82 **Pb** 207.2	Bismuth 83 **Bi** 209.0	Polonium 84 **Po** (209)	Astatine 85 **At** (210)		Radon 86 **Rn** (222)
Francium 87 **Fr** (223)	Radium 88 **Ra** (226)	Actinium 89 **Ac** (227)	Rutherfordium 104 **Rf** (267)	Dubnium 105 **Db** (268)	Seaborgium 106 **Sg** (269)	Bohrium 107 **Bh** (270)	Hassium 108 **Hs** (277)	Meitnerium 109 **Mt** (278)	Darmstadtium 110 **Ds** (281)	Roentgenium 111 **Rg** (282)	Copernicium 112 **Cn** (285)	Nihonium 113 **Nh** (286)	Flerovium 114 **Fl** (289)	Moscovium 115 **Mc** (289)	Livermorium 116 **Lv** (293)	Tennessine 117 **Ts** (294)		Oganesson 118 **Og** (294)

Metals

Metalloids

Nonmetals

Cerium 58 **Ce** 140.1	Praseodymium 59 **Pr** 140.9	Neodymium 60 **Nd** 144.2	Promethium 61 **Pm** (145)	Samarium 62 **Sm** 150.4	Europium 63 **Eu** 152.0	Gadolinium 64 **Gd** 157.3	Terbium 65 **Tb** 158.9	Dysprosium 66 **Dy** 162.5	Holmium 67 **Ho** 164.9	Erbium 68 **Er** 167.3	Thulium 69 **Tm** 168.9	Ytterbium 70 **Yb** 173.0	Lutetium 71 **Lu** 175.0
Thorium 90 **Th** 232.0	Protactinium 91 **Pa** 231.0	Uranium 92 **U** 238.0	Neptunium 93 **Np** (237)	Plutonium 94 **Pu** (244)	Americium 95 **Am** (243)	Curium 96 **Cm** (247)	Berkelium 97 **Bk** (247)	Californium 98 **Cf** (251)	Einsteinium 99 **Es** (252)	Fermium 100 **Fm** (257)	Mendelevium 101 **Md** (258)	Nobelium 102 **No** (259)	Lawrencium 103 **Lr** (262)

Figure 1.3

The periodic table of elements, showing the locations of metals, metalloids, and nonmetals.

called *semimetals* or *semiconductors*. As we will see later in this chapter, the metalloid element silicon is the key component in all integrated circuits, known as *chips*, at the heart of all electronic devices.

Your Turn 1.4   The Periodic Table Inside Your Cell Phone

a. Survey the periodic table shown in Figure 1.3. Which elements do you think are found in your cell phone?

b. Most materials that comprise your cell phone may be classified as metals, plastics, or glass. Using the Web as a resource, describe where these materials come from (both the region(s) of the world where they are produced and the raw materials used in their fabrication).

A **molecule** is formed by connecting two or more *nonmetal* atoms. Molecules may either be described as compounds (e.g., CO_2, H_2O, NO_2) or elements. For instance, seven elements exist as two-atom (diatomic) molecules: H_2, N_2, O_2, F_2, Cl_2, Br_2, and I_2.

The elements in the periodic table are organized into vertical columns called **groups**. Groups serve to organize elements according to important properties they have in common and are numbered from left to right. Some groups are given names as well.

For example, the metals in the first two columns, Groups 1 and 2 on the far left side of the periodic table, are referred to as the *alkali metals* and *alkaline earth metals*, respectively. Compounds containing metals from either of these groups will give rise to alkaline (basic) conditions in soil. Additionally, alkaline earth metals are mostly responsible for the hard water found in some vicinities.

The nonmetals in Group 17 are known as *halogens* and include fluorine, chlorine, bromine, and iodine. The final column, Group 18, consists of the *noble gases*—inert elements that undergo few, if any, chemical reactions. You may recognize helium as the noble gas used to make balloons rise, because it is less dense than air. Radon is a noble gas that is radioactive, a characteristic that distinguishes it from the other elements in Group 18.

1.3 | Compounding the Complexity: From Elements to Compounds

> *Learning Objective: Describe the composition of compounds*

Although only 118 elements have been discovered, over 150 million compounds have been isolated, identified, and characterized. Some are very familiar naturally occurring substances such as water, table salt, and sucrose (table sugar). Many known compounds were chemically synthesized by people across our planet. You might be wondering how so many compounds could possibly be formed from so few elements. But consider that millions of words in the English language may be formed from only 26 letters!

Elements have the ability to combine in many different ways. For example, consider iron and oxygen. Anyone who has driven extensively on salty roads during the winter has observed the compound Fe_2O_3, or rust, on the metal sides or undercarriages of cars. A compound of rust will contain two iron atoms and three oxygen atoms. These values never vary, no matter where the rust is found. Every compound exhibits a constant characteristic chemical composition.

However, iron atoms may also combine with oxygen atoms to form a different compound, Fe_3O_4, which is referred to as magnetite. A compound of magnetite will always contain three iron atoms and four oxygen atoms.

Even though both rust and magnetite contain iron and oxygen atoms, the substances exhibit very different properties. As shown in Figure 1.4, not only are the colors of each iron compound different, but they also vary significantly in their magnetic properties. In fact, Fe_3O_4 is the most magnetic naturally occurring mineral on our planet, whereas rust is nonmagnetic. The black stripe across the back of a credit card contains small particles of magnetite that are used to encode your personal identification details, your account number, and the routing number for the banking institution.

(a)

(b)

(c)

Figure 1.4

A comparison of the relative magnetism of various iron-containing solids. Shown are **(a)** pure iron (Fe) filings, **(b)** rust, $Fe_2O_3(s)$, and **(c)** magnetite, $Fe_3O_4(s)$, picking up bits of iron wire.

(a) and (b): GIPhotoStock/Science Source; *(c)*: sciencephotos/Alamy Stock Photo

1.4 | Measuring the Invisible—How Small Are Atoms?

Learning Objective: Convert between decimal and scientific notation, and between different units of measurement

---

Throughout this book, you will see that the world around us may be described by various length scales. Let's now begin our discovery into the submicroscopic depths of your electronic device. You will never look at your cell phone the same way again.

Elements and compounds are made up of *atoms*—the smallest building block that can exist as a stable, independent entity. The word *atom* comes from the Greek word for "uncuttable." Although today it is possible to "split" atoms using high-energy reactions, atoms remain indivisible by ordinary chemical or mechanical means.

Atoms are extremely small. Because they are so tiny, we need colossal numbers of them to see, touch, or weigh. For example, a *single drop of water* contains 5.3×10^{21} atoms. To put this into perspective, this is roughly a trillion times greater than the 8 billion people on Earth—almost enough to give each person a trillion atoms!

In the drop of water example above, we used a particular format, called *scientific notation* ("5.3×10^{21} atoms"). In decimal notation, that number of atoms is 5,300,000,000,000,000,000,000. Instead of using zeros for very large (or small) numbers, it is more convenient to use **scientific notation** (Figure 1.5). This consists of moving a decimal point to an appropriate number of digits, and indicating this shift with either a negative exponent or a positive exponent.

> **Can We "See" Atoms?**
>
> Check out this video to see how scientists can observe both individual and groups of atoms: www.acs.org/cic.
>
>
>
> ©2018 American Chemical Society
>
> *Question to Consider*
> The video described the incredibly small size of atoms by comparing the size of an atom to an apple as the same size difference as an apple to the size of Earth. Measure or look up the size of an apple and the size of the Earth. How many times larger is Earth than a typical apple?

Your Turn 1.5   Scientific Notation

Example Problem:
The Willis Tower in Chicago has 16,100 windows. Express this in proper scientific notation.

Example Solution:
16,100 in proper scientific notation is 1.61×10^4 since the decimal is moved four places to the left.

Your Turn:

a. Express the current U.S. national debt and the world population in scientific notation.

b. Use the Internet to look up the average thickness of office paper in millimeters. Using scientific notation, express this thickness in terms of meters and kilometers.

Although individual atoms are infinitesimally small, we have technology capable of moving them into desired positions and imaging them on a surface. As incredible as this sounds, scientists at Ohio University were able to assemble silver atoms on a surface to create a smiley face (Figure 1.6). **Nanotechnology** is the manipulation of matter with at least one dimension sized between 1–100 nanometers, where 1 nanometer (nm) $= 1 \times 10^{-9}$ m. Whereas individual atoms and small molecules are sized in the sub-nanoscopic range, larger biomolecules such as DNA, hemoglobin, and most viruses

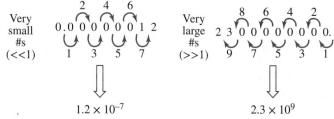

$$1.2 \times 10^{-7} \qquad\qquad 2.3 \times 10^9$$

Figure 1.5

Examples of converting numbers into scientific notation for small and large numbers. The number in front of the exponent must be between 1 and 9.99.

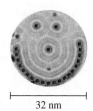

32 nm

Figure 1.6

A nano-sized smiley face formed by the arrangement of individual silver atoms on a surface, as imaged with a scanning tunneling microscope.

©Saw-Wai Hla/Hla Group/Ohio University

Table 1.2	Metric Conversions	
Multiplication Factor	**Prefix**	**Symbol**
10^{12}	tera	T
10^9	giga	G
10^6	mega	M
10^3	kilo	k
10^2	hecto	h
10^1	deca	da
1*	—	—
10^{-1}	deci	d
10^{-2}	centi	c
10^{-3}	milli	m
10^{-6}	micro	μ
10^{-9}	nano	n
10^{-12}	pico	p

*A base unit, such as gram (g), liter (L), second (s).

Note: a micrometer (μm) is also known as a micron.

What Are "Angstroms"?

In addition to the metric unit conversions listed in Table 1.2, this interactive video will introduce another unit that is commonly used in chemistry, the angstrom: www.acs.org/cic.

Chemistry in Context Tutorial

ANGSTROMS

Questions to Consider
a. The width of the helix of DNA is 23 angstroms. How many nanometers is this width?
b. A common object used to compare extremely small objects to is the width of the human hair. How many angstroms wide is a hair that is 75 μm in width?

are nanoscopic in size. Numerous components found in consumer products such as cosmetics, sunscreens, and paints are sized within the nano-regime. The smiley face shown in Figure 1.6 is only 32 nanometers tall and wide. At this size, about 2500 smileys could fit on a cross-section of a human hair!

The conversion of one unit to another, such as nanometer to meter mentioned above, is often referred to as **dimensional analysis**. To convert a quantity into a different sized unit, a conversion factor must be used. A conversion factor is a ratio or fraction that converts a measurement in one unit to another unit. For instance, using the conversion factor where 1×10^{-9} m = 1 nm, the conversion of 32 nm to meters would be:

$$32 \; \cancel{nm} \times \frac{1 \times 10^{-9} \; m}{1 \; \cancel{nm}} = 3.2 \times 10^{-8} \; m$$

Table 1.2 shows some common metric conversion factors and symbols. The next series of activities will introduce another common unit used in chemistry, the Angstrom, and will explore some real-world applications for unit conversions.

@HOME   Dimensional Analysis in the Kitchen

Investigate unit conversions at home using water, a 1-cup measuring cup, and a tablespoon.

1. Measure water with a tablespoon and transfer it to a measuring cup.
2. Repeat Step 1 until the 1 cup measuring cup is full.

Questions to Consider
a. How many tablespoons of water are equal to one cup of water?
b. If there are 8 fluid ounces in 1 cup of water, how many tablespoons of water are in 1 fluid ounce?
c. If there are 4.23 cups in 1 L of water, how many mL is equivalent to 1 cup?
d. If the density of water is 1 g/mL, how many grams would correspond to 1 cup of water?
e. Look up the conversion factor between grams/pounds (lb); how many pounds correspond to 1 cup of water?

Mark Law/Getty Images

Your Turn 1.6   Unit Conversions

Let's explore length scales in the macroscopic world around us and the invisible micro- and sub-microscopic worlds that comprise our cell phones.

a. List some examples of macroscopic objects in your surroundings with dimensions (length, width, height, diameter, etc.) on the order of (i) millimeters, (ii) micrometers (also known as microns), and (iii) meters.
b. Describe the dimensions (length, width, height) of your cell phone or tablet using the three units described in question **a**. Express your answers in both standard decimal and scientific notations.

1.5 | What Makes Atoms Tick? Atomic Structure

Learning Objective: Illustrate the structure of an atom, and compare the relative locations, charges, and masses of the subatomic particles

Atoms, though still indivisible by chemical or mechanical means, contain a nucleus—a minuscule and highly dense center composed of protons and neutrons. Whereas protons are positively charged particles, neutrons are electrically neutral particles. Both species

Table 1.3	Properties of Subatomic Particles		
Particle	**Relative Charge**	**Relative Mass**	**Actual Mass, kg**
proton	+1	1	1.67×10^{-27}
neutron	0	1	1.67×10^{-27}
electron	−1	0*	9.11×10^{-31}

*This value is zero when rounded to the nearest whole number. The electron does indeed have mass, though very small!

Source: McGraw Hill

have almost exactly the same mass and together they account for almost all of the atomic mass. Outside the nucleus are the electrons that define the boundary of the atom. An electron has a mass much smaller than that of a proton or neutron. In addition, electrons carry an electrical charge equal in magnitude to that of a proton, but opposite in sign. Therefore, in any electrically neutral atom, the number of electrons equals the number of protons. The properties of these particles are summarized in Table 1.3. Atoms are held together in part by the attraction of the negative charge of the electrons to the positive charge of the protons in the nucleus.

The number of protons in the nucleus (the **atomic number**) determines the identity of the atom. For example, all hydrogen (H) nuclei contain one proton; hence, hydrogen has an atomic number of 1. Similarly, all helium (He) nuclei contain two protons and have an atomic number of 2. As seen in the periodic table shown in Figure 1.3, the atomic number can be found below the name and above the chemical symbol for each element and increases for each successive element in the periodic table. For example, the nucleus of element 92 (U, uranium) contains 92 protons. Because these atoms are neutral, they must contain the same number of negatively charged electrons as protons. Accordingly, a H atom will contain one proton and one electron, whereas a He atom will contain two protons and two electrons (Figure 1.7).

The **mass number** refers to the number of protons and neutrons in the nucleus of the atom. For instance, the mass number of hydrogen is 1, which indicates that there is one proton and no neutrons. However, helium has a mass number of 4, which means there are two protons and two neutrons in the nucleus (Figure 1.7). The number of neutrons for a given element can vary, unlike the number of protons, which is always the same for a certain element. You will now get to practice counting subatomic particles as you satisfy your sweet tooth!

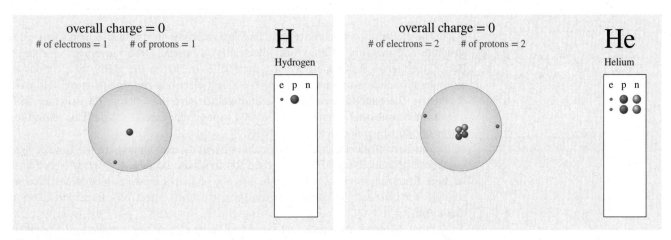

Figure 1.7

Check out this interactive comparison of the atomic structures for hydrogen and helium, showing the location of protons, neutrons, and electrons: www.acs.org/cic. The particle sizes and the relative distance between the nucleus and the electrons are not to scale.

(left and right): The Concord Consortium

@HOME   Candy Atoms!

Make a model of an atom of carbon at home using two popular candies: Nerds™ and M&Ms™.

1. Count out 6 Nerds™. These are going to be your electrons.
2. Count out 6 M&Ms™ of one color. These are going to be your protons.
3. Count out 6 M&Ms™ of a second color. These are going to be your neutrons.
4. Build a model of a carbon atom using the candies you have counted out.

Amy_Michelle/Shutterstock

Questions to Consider

a. What would happen if you took away one of the neutrons? What would happen if you added another neutron to your model? What element(s) would you have?

b. What would happen if you took away one of the protons? What would happen if you added another proton to your model? What element(s) would you have?

c. What would happen if you took away one of the electrons? What would happen if you added another electron? What element(s) would you have?

Your Turn 1.7   Atomic Structure

Example Problem:
Determine the number of protons and electrons in an atom of Ca.

Example Solution:
Finding the element Ca on the periodic table, we see the atomic number is 20. This means there are 20 protons in the nucleus of an atom of Ca. Since the atom of Ca is neutral, there must be an equal number of electrons as protons, so there are also 20 electrons.

Your Turn:
Determine the number of protons and electrons in each of the following atoms:

a. Ga   **b.** Sn   **c.** Pb   **d.** Fe

Determine the number of protons, neutrons, and electrons in each of the following atoms:

a. H (mass number of 2)   **b.** Cr (mass number of 52)
c. Al (mass number of 27)   **d.** As (mass number of 75)

The electrical conductivity of a material depends on its three-dimensional (3D) structure and mobility of electrons. Electricity is basically the movement of charge. Hence, the electrical conductivity of a material is related to the ability of electrons to move from one atom to another. The more easily electrons are able to move, the more conductive the material becomes. Metallic solids have an ordered 3D structure with plentiful electrons that are not tightly bound to individual metal atoms. This allows for extremely effective electrical conductivity.

The flow of electrons via electrical conductivity is analogous to the flow of heat via thermal conductivity. Metals are used for cookware because they effectively transfer heat from the stove to the food in the pot or pan. Likewise, metals are used as conduits for electricity because they transport electrons effectively from one location to another.

What are some materials that you know to be electrically conductive? Copper is conductive. Other metals such as aluminum, silver, and gold are all conductive, too. In fact, among the 118 elements in the periodic table, the metals are the most electrically conductive. It makes sense that when manufacturers create products requiring the conduction of electricity, such as a touchscreen, they will most often use a metal.

1.6 | A Look at the Elements in Their Natural States

Learning Objective: Describe the difference between physical and chemical changes and how elements are extracted from rocks

Atoms of the same element are arranged in different ways to form the bulk, macroscopic elements that we observe in the world around us. These different elemental forms are known as **allotropes**. Some elements are made of diatomic molecules, or molecules that feature two identical atoms such as hydrogen (H_2) or nitrogen (N_2). However, other elements are composed of larger sub-units. For instance, sulfur typically exists as eight-membered rings, S_8, and phosphorus contains an array of four-atom units, P_4. Carbon has three allotropes: graphite, diamond, and a soccer ball array of carbon atoms known as *buckminsterfullerene* (Figure 1.8).

Of the 118 elements listed in the periodic table, 80% are metals. Portable electronic devices contain a variety of metals such as aluminum, copper, nickel, lithium, tin, lead, and traces of others. These metals must be extremely pure, but are not found naturally as pure elements. Wouldn't it be great if we could simply dig into our backyards and find a pure element such as iron, aluminum, or tin? With the exception of some precious metals like gold, the metallic elements do not exist in nature in their pure states. Instead, they must be obtained from compounds.

Let's look at one metal from the periodic table—aluminum. In addition to holding our beverages, aluminum (Al) is used extensively in automobiles because it is extremely lightweight and will not rust like iron does. Some portable electronic devices such as the MacBook Pro and iPads also use aluminum for the case, which makes these gadgets extremely lightweight and highly recyclable.

Even though aluminum is readily found in Earth's crust (Figure 1.9), it does not exist in nature as the pure metal. Many elements, including aluminum, react readily with a common gas in our atmosphere, oxygen (O_2), to form more chemically stable compounds. Consequently, aluminum and many other metallic elements are found within **rocks**, which are heterogeneous mixtures of solid compounds known as **minerals**. Considering the elemental makeup of Earth's crust, it is no surprise that most rocks are complex mixtures of oxygen-containing minerals designated as *oxides*. The combination of oxygen with silicon in minerals results in *silicates*, whereas aluminum and oxygen minerals are known as *aluminates*. As you might imagine, silicon, aluminum, and oxygen atoms might all form some of the compounds found in a rock formation, which is known as an *aluminosilicate* mineral.

To visualize the structure of rocks, let's consider an image of an aluminum-containing rock formation known as *bauxite*, found mostly in Australia, Guinea, China, Indonesia, and

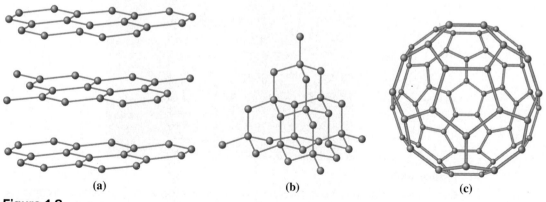

(a)   (b)   (c)

Figure 1.8

Three allotropes of carbon: **(a)** graphite, **(b)** diamond, and **(c)** buckminsterfullerene, C_{60}. For a 3D rendering of these structures, go to www.acs.org/cic.

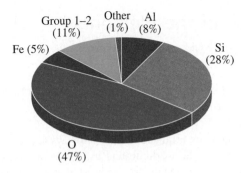

Group 1–2 (11%)
Other (1%)
Al (8%)
Fe (5%)
Si (28%)
O (47%)

Figure 1.9

Elemental composition of the minerals comprising Earth's crust.

Brazil. In the cross-section image of bauxite shown in Figure 1.10, you can see a heterogeneous mixture of solids. However, if you were to pick out one of the individual grains within bauxite, you would get a mineral with a defined composition. Each of these mineral grains is classified as a compound because it is a substance with a defined composition.

Your Turn 1.8   Minerals

In addition to aluminum, other metals such as scandium and yttrium, known as rare earth metals, are also found in cell phones and most electronic devices. What are the natural sources of these metals, and in which parts of the world are these minerals found?

A rock formation that contains a considerable concentration of a desired metal is known as an *ore*. This term is generally used for rocks that can be practically obtained for mining. Even though there are mineral deposits worth millions of dollars, it may not always be economically feasible to extract the metal due to its remote location and/or exorbitant cost of processing. For instance, mineral deposits worth $100 million in a remote part of northern Canada would simply be called rock instead of ore if it would cost more than $100 million to mine and process the deposit! It is no small undertaking to convert ore into pure elements, requiring many purification steps and chemical reactions at high temperatures utilizing expensive equipment.

The process of extracting and purifying metals from ore is known as *metallurgy*. Once the metal is extracted and concentrated from the ore through grinding and leaching, a variety of high-temperature chemical reactions are performed to isolate the metal from its ore.

Iron or copper oxides obtained from ore are reacted with carbon similar in composition to the charcoal you may have used for a summer cookout. During this smelting process, carbon reacts with oxygen in the metal ore at temperatures in excess of 1000 °C

Iron Mining

trubitsyn/Shutterstock

Check out this video for more information about the processes required to open and operate a mine: www.acs.org/cic.

Questions to Consider

a. What sorts of requirements do mine operators have when siting a new mine? What stakeholders need to be consulted when planning a new mine?

b. Why do you think some processing of the raw ore is done at the iron mining site before transporting it to steel mills?

c. Do an internet search to find any mines in your area. What type of material is mined and where is it shipped for processing?

Figure 1.10

Cross-section image of bauxite, illustrating the complex heterogeneous mixture of minerals. For instance, in addition to iron and titanium oxides, bauxite contains grains of gibbsite (composition: $Al(OH)_3$) and boehmite compounds (composition: $AlO(OH)$)—each with a defined composition.

Doug Sherman/Geofile

to isolate the metallic element. For instance, copper may be obtained from a copper ore by this seemingly simple high-temperature reaction:

$$CuO + C \longrightarrow CO + Cu$$

Metals such as tin and lead require lower temperatures; hence, it is no surprise that early civilizations discovered these metals by simply heating minerals over a campfire. Further refining and processing steps are generally required to yield high-purity metals.

When an element is isolated from its ore, a chemical change takes place. In general, a chemical change occurs when the chemical composition of the final product differs from its starting material. For the above reaction, a copper oxide ore is converted to metallic copper, which has distinctly different properties than its starting material. In contrast, a physical change occurs when the phase of the material is altered, without any change in composition. For instance, when ice is melted, there is a conversion from solid to liquid. However, the chemical composition (water, or H_2O) is still the same. Likewise, dissolving sugar in water is also considered a physical change, because you haven't altered the composition of either substance during this process.

Your Turn 1.9   Physical vs. Chemical Change

For each of the following, indicate whether a physical or chemical change has occurred:

a. Burning a match
b. Baking a cake
c. Cracking a piece of glass
d. A piece of apple darkens after being cut
e. Boiling water
f. Rusting of iron

1.7 | Your Cell Phone Started with a Day at the Beach: From Sand to Silicon

Learning Objective: Describe how substance purity is measured and the silicon purity that is required for computer chip manufacturing

---

Your Turn 1.10   Designing Your Own Cell Phone

You have an unlimited budget and are brainstorming ideas about the next design for a cell phone that will revolutionize the world.

a. Describe the properties that are most desirable for your design.
b. Which chemical elements are incorporated in your design and why did you choose them?

As seen in the previous activity, we often focus on properties such as weight and durability and take for granted the processing speeds of our electronic devices. For instance, simply touching the icon for a weather app instantly displays the temperature and weather conditions for your part of the world. Such rapid computational speeds would not be possible without continual improvement of the heart of any electronic device—the microprocessor, known as the *chip*. All microprocessors, whether they control your laptop or desktop computer, your coffee maker, or your cell phone, contain the element silicon (Si).

One of the most intriguing applications of chemistry occurs when ordinary sand is transformed into the ultra-high-purity silicon that is used in every electronic device on the planet. Analogous to aluminum and most other metals, due to the high concentration of oxygen in Earth's crust (Figure 1.9), silicon doesn't exist in nature as a pure element. Instead, this element is found as a compound containing Si and O atoms, SiO_2, which is known as sand or *silica*.

In order to remove the oxygen and produce pure Si, sand is first reacted with carbon to produce silicon with an atomic purity of 95–98%, not yet pure enough to be used

Figure 1.11

A high-purity cylinder (ingot) of silicon and a thin wafer sliced from the ingot.

Johnrandallalves/Getty Images

for electronic applications. This concentration implies that for every 100 atoms of silicon, there are two to five atoms of impurities such as phosphorus, boron, carbon, oxygen, and a variety of metals.

In electronics, silicon must have a purity of at least 99.9999999%, which is known as 9N (9 nines). Some companies today even produce Si with a purity of 99.9999999999%, or 12N! Although we could also indicate the total impurity concentration by a percentage, it would be an extremely small number in this context—only 0.0000001% for 9N silicon, or 0.0000000001% for 12N silicon!

As an alternative to using small numbers with many zeros or negative exponents, it is often preferred to designate such low concentrations as 1 *part per million* (1 ppm) or *part per billion* (ppb). This astounding level of purity implies that only one foreign (non-Si) atom may be present for every million (for ppm) or billion (for ppb) atoms of silicon. For instance, the 9N purity would contain an impurity concentration of 0.001 ppm or 1 ppb. An interesting way to visualize these small concentrations is to think of stacking yellow tennis balls (representing Si atoms) from your front step to the surface of the Moon. If you replace only six of the yellow balls with red ones (representing impurity atoms), that would represent the 9N level of purity. To put the 12N purity in perspective, the maximum number of impurities would correspond to a single red ball within 170 separate stacks of yellow tennis balls stacked from Earth to the Moon!

The various stages of Si processing, from high-purity silicon crystals (Figure 1.11) to the final computer chip, occur over hundreds of steps taking place within specialized rooms known as *clean rooms*. To prevent contamination by dust particles that would render the chip inoperable, clean rooms make extensive use of stainless steel, sloped surfaces to avoid dust accumulation, and perforated floors and special ceiling tiles to promote air circulation. Prior to entering the clean room, personnel must cover their clothing with a white "bunny suit" that has non-lint and anti-static properties (Figure 1.12). To enter the clean room, the worker must also walk over a sticky pad and pass through strong bursts of air (referred to as an air shower) to remove dust particles from shoes and clothing. Clean rooms are rated based on the number of particles measuring 0.5 μm or larger per cubic meter of air. Chip manufacturing clean rooms generally contain less than 350 particles per cubic meter, compared to the ambient air in a typical city environment, which contains 35,000,000 particles per cubic meter.

Although computer chips are now comparable in size to a single grain of rice (Figure 1.13), they still contain billions of individual components known as *transistors* that are used to perform the operations needed by our computers and portable electronic devices. Indeed, the chip that runs a computer or a cell phone is truly an engineering marvel that would not be possible without the conversion of sand into silicon!

Figure 1.12

Technicians work inside a clean room at Sanan Optoelectronics Co., Ltd. in Tianjin, China. This video shows how computer chips are fabricated from silicon: www.acs.org/cic.

©Bradley D. Fahlman

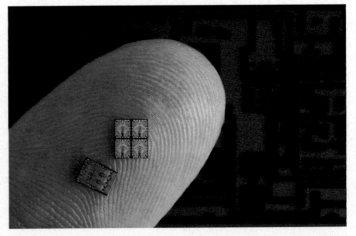

Figure 1.13

Computer processing chips placed onto a single fingertip.

Charle Avice/age fotostock/Alamy Stock Photo

1.8 | More Fun at the Beach: From Sand to Glass

> *Learning Objective: Describe the composition of glass and explain how we can alter its structure to enhance its durability and scratch resistance*

So far, we have talked about the metals and semiconductors used in a portable electronic device. However, those of us who have peered through the cracked screens on our cell phones are very familiar with another component in most portable electronic devices— glass. You might be surprised to learn that sand is used to fabricate not only high-purity silicon, but also the crystal-clear transparent glass that we interact with on mobile devices. You will explore the behavior of light in contact with glass and other materials in this next activity.

@HOME   Light/Matter Interactions

Being careful to protect your eyes, investigate the interactions between light and matter at home using a laser pointer and different materials. Suggested materials include: a glass window, an LCD screen, a plasma TV screen, a concrete sidewalk, an asphalt road, a ceramic plate, and a cotton shirt.

Robin Lund/Alamy Stock Photo

1. Predict how the light from the laser pointer will behave as it comes into contact with each of the materials.

2. Shine the laser pointer at each of the materials to determine how the light behaves.

Questions to Consider

a. For each material, was the light transmitted, reflected, or absorbed during its contact with the material? What evidence do you have to support your claims?

b. How did the results compare to your predictions?

When you consider the front surface of your mobile device, what kind of properties should it have? Qualities that you may look for include transparency, scratch resistance, and shatter resistance. To find a material with these properties, scientists and engineers have taken a page from nature. One of the largest components of Earth's crust is *silica* (silicon dioxide, $SiO_2(s)$), which is found in many different forms. These forms vary by composition and structure, each having different properties.

At the atomic level, silica consists of repeating linkages between silicon and oxygen in a dense, spiderweb-like structure. There are some naturally formed silicon dioxide structures with very well-ordered structures at the atomic level. This ordered structure is called a **crystal**. Pure crystallized silicon dioxide is known as *quartz*, a clear and colorless mineral that is the primary component of sand (Figure 1.14). When small amounts of other elements are present in the crystal, it can give the mineral some color. For example, the yellow color of citrine and the purple color of amethyst are from different forms of iron that are present in trace amounts within the silicon dioxide crystal (Figure 1.15).

In contrast to well-ordered quartz, the structure of glass is disordered on the atomic level with a random array of silicon and oxygen linkages throughout the solid (Figure 1.16). What a tangled web we weave with glass! Disordered materials such as this are called **amorphous** solids. Although relatively brittle compared to crystalline silicon dioxide, glass has the ability to be molten in a fluid-like state and worked into different shapes for various purposes.

Figure 1.14

Light microscope image of sand taken from Big Talbot Island, Florida, illustrating the individual crystals of silica.

Sabrina Pintus/Getty Images

(a)

(b)

Figure 1.15

Photos of **(a)** citrine and **(b)** amethyst—forms of quartz with iron impurities that give it varying colors.

(a): TinaImages/Shutterstock;
(b): Alexander Hoffmann/Shutterstock

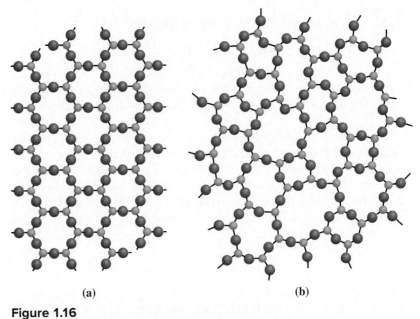

(a)   **(b)**

Figure 1.16

Molecular representations of **(a)** crystalline quartz and **(b)** amorphous glass.

Silica glass is made from heating ordinary sand to its melting point (Figure 1.17a), then cooling the liquid until it hardens to a glass. A variety of additives may be mixed with the silica raw material before melting, which can give the resulting glass a wide range of properties. For instance, the Pyrex glass used for cookware not only contains silicon and oxygen atoms, but also boron (B) and traces of other metals such as sodium (Na), aluminum (Al), and potassium (K). The addition of these elements to glass greatly improves its thermal properties by limiting the extent of expansion at high temperatures, or contraction at low temperatures, to reduce its likelihood of cracking. A technique known as glassblowing (Figure 1.17b) is used to form glass into a desired shape.

As you saw in Your Turn 1.12, when light shines onto a piece of silicon dioxide, whether it is crystalline quartz or amorphous glass, it mostly passes straight through the material. This means that the material is *transparent*. Whenever there are differences in the structure or the composition at the microscopic level, the path of light through the material is altered, potentially making it opaque. Pure crystalline quartz, having the same structure throughout, certainly is transparent. However, if there are imperfections or impurities present, such as in smoky quartz (Figure 1.18), the mineral becomes opaque. Although amorphous glass has a variation in its atomic-level structure distributed randomly throughout the entire material, it will still allow light to pass through the material giving it transparency. Of course, over the past several millennia, glassworkers

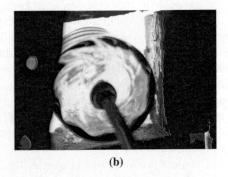

(a)   **(b)**

Figure 1.17

(a) Molten sand at a temperature in excess of 1500 °C being poured from a ceramic crucible.
(b) A glassblower inserting a glass bowl into a high temperature furnace. Check out this link to find out more about glassblowing.

(left); Solobird/Alamy Stock Photo; (right): Glow Images

Figure 1.18

A crystal of smoky quartz.

©Bradley D. Fahlman

Figure 1.19

Stained glass windows from Carthage College chapel

John S. Kirk

have discovered quite a few additives that give glass some color or make the glass opaque. Beautiful examples of this are the stained glass windows commonly found in houses of worship and historical buildings throughout the world (Figure 1.19). Some of this stained glass is colored red by the inclusion of nanoparticles of gold!

If you ever roughhoused in your family's living room when you were young, played softball close to parked cars, or dropped your mobile device, you probably know that glass can be quite fragile. Consequently, much research has gone into improving the strength and scratch resistance of glass. In the 17th century, it was discovered that quickly cooling a drop of molten glass in cold water results in a hardened drop that could withstand a hammer blow. However, a small amount of force to the long tail of these so-called *Prince Rupert's drops* would cause the entire glass piece to shatter explosively into small fragments.

The strength of Prince Rupert's drops comes from the quick hardening of the outer portion of the glass, freezing it into place while the inside of the drop is still cooling. Because cooling down an object tends to shrink its size, but the outer surface of the drop is locked in place, there is a lot of internal stress in the drop as the interior of the glass tries to pull the outer surface inward. Many types of tempered glass have been heat-treated to behave very similarly to these drops. Since heat-strengthened glass tends to shatter into very small pieces when broken, it is often laminated or coated with a thin layer of plastic. For instance, when an automobile windshield is broken, the pieces are quite small and tend to stick together, thus resulting in fewer severe injuries from large pieces of glass (Figure 1.20).

Prince Rupert's Drops

©Bradley D. Fahlman

Check out this video showing the formation of a Prince Rupert's drop (www.acs.org/cic).

Questions to Consider

a. The strength of Prince Rupert's drops have been compared to the strength of steel. The compressive strength of structural steel is about 25,000 psi, meaning it can withstand this much force before it deforms. What is this force in units of pascals (Pa)? (1 Pa = 1.45×10^{-4} psi)

b. When breaking the tail of a Prince Rupert's drop, high-speed cameras have measured cracks traveling along the surface of the glass at speeds of up to 1500 m/s! How fast is this speed in miles per hour?

Figure 1.20

Photo of a broken windshield, showing the retainment of smaller glass fragments by the plastic film coating.

Esa Hiltula/Alamy Stock Photo

Figure 1.21

Illustration of the pressure felt by each foot of a bottom elephant if 100,000 4-ton elephants were stacked on top of each other—certainly, an impossible task!

©Bradley D. Fahlman

In addition to heat treatment, chemical treatments can also strengthen glass. This is precisely the technique that has been used by Corning Corp. to fabricate Gorilla Glass—the tough scratch-resistant glass that is used in a variety of mobile device screens, including cell phones, tablets, and laptop computers. This glass is theoretically able to withstand a pressure of 10 GPa, which is equivalent to the pressure exerted by a stack of 100,000 elephants (Figure 1.21)!

Have fun with thumbtacks and balloons in the next activity to explore how the area of an object influences its pressure acting on a surface.

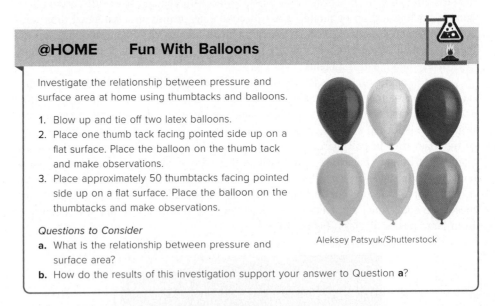

@HOME   Fun With Balloons

Investigate the relationship between pressure and surface area at home using thumbtacks and balloons.

1. Blow up and tie off two latex balloons.
2. Place one thumb tack facing pointed side up on a flat surface. Place the balloon on the thumb tack and make observations.
3. Place approximately 50 thumbtacks facing pointed side up on a flat surface. Place the balloon on the thumbtacks and make observations.

Questions to Consider

a. What is the relationship between pressure and surface area?

b. How do the results of this investigation support your answer to Question **a**?

Aleksey Patsyuk/Shutterstock

The incredible strength of Gorilla Glass is achieved by submerging conventional glass into a bath of molten potassium nitrate (KNO_3). Potassium ions from the bath replace some of the smaller sodium ions close to the surface of the glass. This results in the same types of stresses on the surface of the glass as found in Prince Rupert's drops. It should be noted that Gorilla Glass screens are scratch/shatter-resistant when dropped, but are not scratch/shatter-proof.

Corning and other companies are actively researching the next generation of materials for mobile device screens. These materials include not just amorphous materials like glass, but crystalline materials, too. Sapphire is one of the potential replacements. Sapphire is a natural gemstone that is harder than quartz. In fact, sapphire is the second-hardest material known after diamond. Sapphire is composed of aluminum oxide, which is three times harder than Gorilla Glass.

How to Make Gorilla Glass

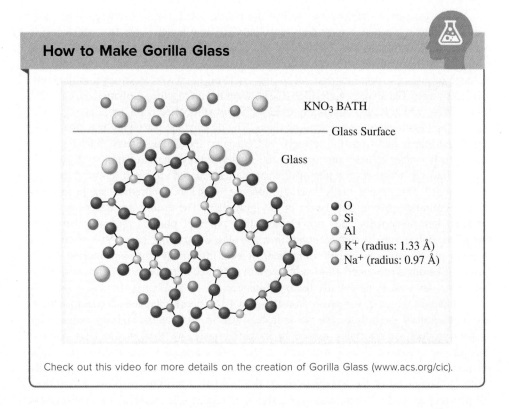

Check out this video for more details on the creation of Gorilla Glass (www.acs.org/cic).

Synthetic sapphire requires heating fine aluminum oxide (Al_2O_3) powder at extremely high temperatures—as high as 1800 °C! The crystal growth process is very slow, taking more than two weeks to grow a single large crystal. Once formed, it is cut into its final size and shape by a diamond saw or a laser. Although sapphire is extremely hard, it is denser than glass. This means that for the same size and thickness of material, sapphire will weigh more. For instance, if you had equivalent sizes of glass and sapphire and the glass had a mass of 100 g, the sapphire would weigh 167 g. The production of synthetic sapphire is also more costly and slower than glass; however, further development of production methods are bringing the cost down. Sapphire is already used for surfaces that see a lot of wear-and-tear such as checkout scanners, airplane windows, and high-end watches. With improvements in production, we may be seeing many more sapphire screens on mobile devices in the near future.

1.9 | From Cradle to Grave: The Life Cycle of a Cell Phone

Learning Objective: Explain how the three pillars of sustainability relate to the fabrication and recycling of portable electronic devices

With the number of active cell phones in the world outpacing the growth of the human population, it is essential that we understand the environmental impacts that their production, use, and disposal have on our planet. The expression *cradle-to-grave* is an approach to analyzing the life cycle of an item, starting with the raw materials from which it came and ending with its ultimate disposal.

Think of items that we take for granted each day such as batteries, plastic water bottles, T-shirts, cleaning supplies, running shoes, and, of course, cell phones—anything that you buy and eventually discard. Where did the item come from? What will happen to the item when you are finished with it? More than ever, individuals, communities, and corporations are recognizing the importance of asking these types of questions. Cradle-to-grave means thinking about *every* step in the process, leading to the product's final disposal.

As a simple illustration, let's follow the plastic packaging that cradled your shiny new cell phone when it was proudly unveiled. The raw material for this packaging is petroleum. Accordingly, the "cradle" of this plastic product most likely was crude oil somewhere on our planet—for example, the oil fields of Alberta, Canada. At the refinery, a range of processes were carried out on the crude oil to convert fractions into the compound styrene. The styrene molecules (C_8H_8) were linked together (polymerized) to form *polystyrene*, which is also commonly used for Styrofoam coffee cups, food take-out containers, and many other commercial products. The polystyrene packaging was then packaged and transported from the refinery in Canada or the United States (burning jet or diesel fuels—other refinery products) to the final assembly plant in China or Taiwan.

However, what was the fate of this packaging material after you removed the new cell phone? This is not really a cradle-to-grave scenario, but rather cradle-to-your-trash—definitely several steps short of any graveyard. The term *grave* describes wherever an item eventually ends up. Unlike other types of plastics that may be easily recycled, polystyrene is not accepted in most plastic recycling bins. As a result, this type of plastic is the principal component of landfills, urban litter, and marine debris where it begins a presumed 1000-year cycle of slow decomposition into carbon dioxide and water, as well as potentially toxic substances.

Cradle-to-a-grave-somewhere-on-the-planet is a poorly planned scenario for plastic packaging. If the polystyrene waste instead was to serve as the starting material for a new product, or creatively reused in its native state, we then would have a more sustainable situation. **Cradle-to-cradle**, a term that emerged in the 1970s, refers to a responsible use of materials in which the end of the life cycle of one item dovetails with the beginning of the life cycle of another so that everything is reused rather than disposed of as waste. When considering the most responsible end-use of a product, one should consider the **three pillars of sustainability** (Figure 1.22):

- Environmental—pollution prevention, natural resource use
- Social—better quality of life for all members of society
- Economic—fair distribution and efficient allocation of resources

The next group activity will allow you to apply the three pillars of sustainability as you debate which energy source is more beneficial to our planet: solar or coal. As you debate this topic, you will see that this issue is not as clear-cut as you once thought!

Let's Debate!   Solar or Coal-Fired Power?

(left): Kevin Pronnecke/imageBROKER/Alamy Stock Photo; (right): Rudmer Zwerver/Image Source/Shutterstock

Form two groups: one group will be in favor of the topic (the proposition) and the other group will be against the topic (the opposition). Both groups will try to persuade a neutral person or judge to agree with them. The topic of the debate is the motion. Here is the motion:

Solar power is more sustainable than coal-based power.

Consider all three pillars of sustainability as well as the cross-cutting areas between the pillars to form your arguments.

Curiosity Conversations

- What is something you learned about the environmental consequences of solar panels that you had not thought about prior to this?
- Describe how your arguments differ from the assertions of popular media.

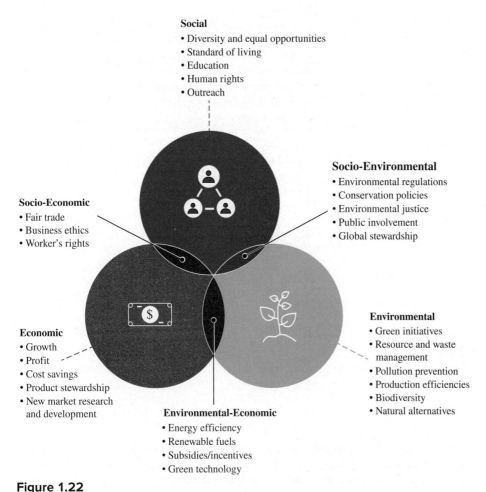

Social
- Diversity and equal opportunities
- Standard of living
- Education
- Human rights
- Outreach

Socio-Environmental
- Environmental regulations
- Conservation policies
- Environmental justice
- Public involvement
- Global stewardship

Socio-Economic
- Fair trade
- Business ethics
- Worker's rights

Economic
- Growth
- Profit
- Cost savings
- Product stewardship
- New market research and development

Environmental
- Green initiatives
- Resource and waste management
- Pollution prevention
- Production efficiencies
- Biodiversity
- Natural alternatives

Environmental-Economic
- Energy efficiency
- Renewable fuels
- Subsidies/incentives
- Green technology

Figure 1.22

The three pillars of sustainability.

©Bradley D. Fahlman

As you would expect, the life cycle of a cell phone—an assemblage of many different types of materials from varying parts of the world—would be much more complex than that of its packaging materials. Among the materials comprising a cell phone, 40% are metals, 40% are plastics, and 20% are ceramics and glass. Properties of metals such as electrical and thermal conductivity, durability, and malleability (ability to be bent into complex shapes) are exploited for the circuit board, battery, and touch-sensitive screen. In contrast, the lightweight, inexpensive, and moldable properties of plastics are well suited for the protective case and chip packaging. Ceramics and glass exhibit brittleness and are electrically insulating. Glass is most often used for the outer screen to protect the underlying display, whereas ceramics are used within the circuit board, speaker, and antenna.

So, how much energy is required to fabricate such a complex design? After all, electronic devices are getting smaller/thinner and more efficient (Figure 1.23), which means less energy will be required to produce them, correct? In fact, it's just the opposite, with their production from raw materials accounting for more than 90% of the energy consumed over their lifetime! This is not the case with low-tech products such as lightbulbs, vacuum cleaners, and ovens that consume much more energy over their lifetimes than was spent for their fabrication. Automobiles used to be in the same "low-tech" category, controlled by analog devices; however, microprocessors now monitor and control every aspect of modern vehicles from the fuel injection system to tailpipe emissions. The increased energy consumption during the production of high-tech devices is primarily because:

- More diverse materials are needed, which requires greater costs for mining and purification, as well as the manufacturing of ceramics and plastics.

What's Inside Your Smartphone?

©2018 American Chemical Society

Check out this video for more details of what's inside your smartphone: www.acs.org/cic.

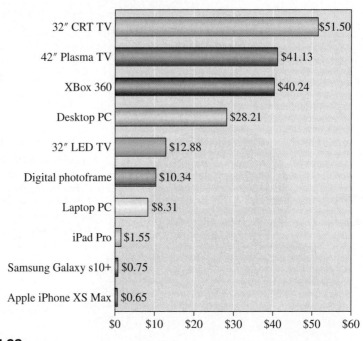

Figure 1.23

Comparison of annual operating costs for various electronic devices. Annual costs are based on an average U.S. residential electricity rate of $0.12/kWh.

- Microprocessors (computer chips) originate from the energy-intensive conversion of sand into ultra-high-purity silicon, and must go through hundreds of complex steps required to fabricate the final integrated circuits.
- Complex devices require many hours of design with teams of people using multiple high-speed computers that run continuously for 24 hours a day, 7 days a week.

Although it is quite easy to determine how much energy an electronic device consumes during its operation, it is very difficult to calculate the energy used in its fabrication. For instance, a new cell phone begins its production many years before it is released, in the hands of engineers who plan out its features and design the complex architecture and computer chips that it will employ. It's hard to estimate how much energy this initiative will consume, because it involves the electricity to power the buildings and laboratories used for research and development. Administrators and members of the sales force also use electricity in their offices and consume fossil fuels during their extensive travel.

Overlooking these pre-manufacturing activities simplifies the situation somewhat, but we still have the problem of globalization. That is, the silicon employed for the computer chips may be purified in Michigan, the circuit board built in California, the lithium in the battery mined and purified in Chile, and the plastics synthesized in China. Some variability in these locations depends on the company's supply chain, which will vary dramatically between electronics companies. The amount of energy required to mine lithium metal in South America would be very different than what is required in Canada. Hence, this makes a general life-cycle analysis very difficult to predict with any level of accuracy without knowing more information about the manufacturing practices of each materials supplier. As an example of how complex the situation is for a single company, Apple has over 180 suppliers worldwide that supply the raw materials and components needed for its product lines.

More companies are becoming transparent about the environmental footprint of their products. This footprint is often reported in units of "kgCO$_2$e" (kg equivalent CO$_2$), which refers to the relative emissions of greenhouse gases per unit of fuel that is consumed. For instance, Apple reports that the 11-inch iPad Pro (4th Gen) is responsible for 100 kgCO$_2$e over its lifetime, down from 184 kgCO$_2$e for the 1st Generation device. For the latest iPad, 82% of emissions are from manufacturing, 5% from transport, 12% from consumer use, and <1% from recycling. In contrast, the iPhone 14 Pro with its smaller energy footprint is reported to release 65 kgCO$_2$e, with 81% generated from production, 15% from consumer use, 3% from transport, and <1% from recycling. However, there is no way to accurately include information about the energy consumption of the supply chain companies. Furthermore, the environmental standards of countries differ greatly, which often results in outsourcing to countries where sustainability is not considered a top priority.

The environmental impacts we have discussed thus far deal only with the direct fabrication, use, and recycling of electronic devices. However, the full life cycle of a device also includes many other energy-intensive activities that are needed to extract, refine, and transport the raw materials from various parts of the world to the central fabrication facility (Figure 1.24). How much energy does it take to extract lithium metal from an ore in Chile? It depends on how difficult the ore is to reach, and what specific techniques the company uses to break apart the ore, extract the metal, and then refine/purify the metal once it is removed. The same may be said about other components of the phone such as the outer screen. Whereas Samsung doesn't expend much energy in attaching the glass to the case in its final assembly plant, how much energy did the glass manufacturer consume to convert sand into a high-strength glass, and then ship large crates of the material to China for final assembly? The answers to these questions are not easily obtainable, and illustrate just how complicated it is to determine the full environmental impact of a high-tech device in our globalized society.

Your Turn 1.11   Smartphone Usage

a. Other than charging, what are some energy requirements of your cell phone?
b. Considering how energy-intensive it is to fabricate cell phones, do you think increasing smartphone usage could cause a decrease in the overall energy consumption in our planet? Explain.

Real Talk   Sustainable Product Design

Check out this interview with former ACS President H. N. Cheng, who describes the importance of understanding life cycle analyses for products: www.acs.org/cic. He also mentions what life is like as an industrial scientist.

Questions to Consider
a. In the video, Dr. Cheng mentions "circular materials design". What are some differences between linear and circular economies?
b. What are some examples of products on the market that have been fabricated using the ideas of circular design?

HollyHarry/Shutterstock

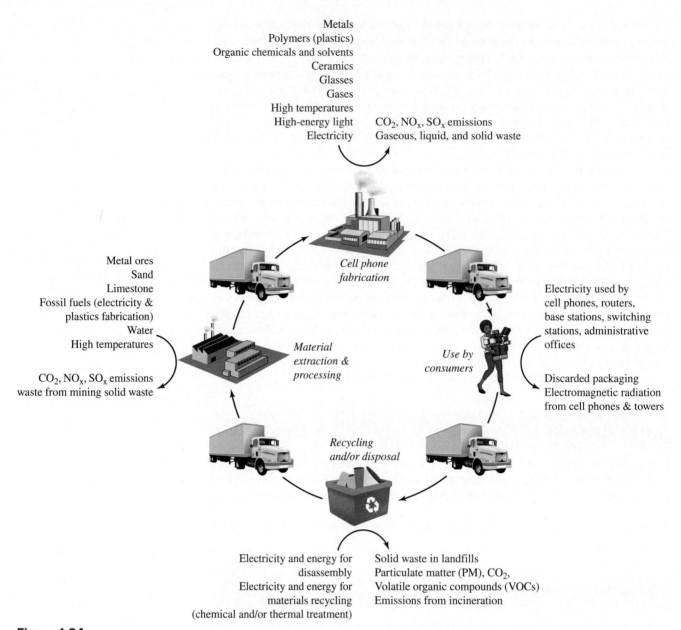

Metals
Polymers (plastics)
Organic chemicals and solvents
Ceramics
Glasses
Gases
High temperatures
High-energy light CO_2, NO_x, SO_x emissions
Electricity   Gaseous, liquid, and solid waste

Cell phone fabrication

Metal ores
Sand
Limestone
Fossil fuels (electricity & plastics fabrication)
Water
High temperatures

CO_2, NO_x, SO_x emissions
waste from mining solid waste

Material extraction & processing

Use by consumers

Electricity used by cell phones, routers, base stations, switching stations, administrative offices

Discarded packaging
Electromagnetic radiation from cell phones & towers

Recycling and/or disposal

Electricity and energy for disassembly
Electricity and energy for materials recycling
(chemical and/or thermal treatment)

Solid waste in landfills
Particulate matter (PM), CO_2,
Volatile organic compounds (VOCs)
Emissions from incineration

Figure 1.24

The life cycle of a cell phone. The years spent for its design and marketing are not included.

1.10 | Howdy Neighbor, May We Borrow a Few Metals? The Importance of Recycling and Protecting Our Supply Chains

Learning Objective: Identify sources and possible alternatives for low-abundant materials used in portable electronics

Although approximately 2 billion new cell phones are purchased each year, over 90% of these phones will either collect dust at home or be sent to landfills after their owners grow tired of them. In 2022, Americans replaced their cell phone approximately every 32 months, with only 20% recycled. However, did you know that each cell phone contains about 300 mg of silver and 30 mg of gold? In fact, the gold and silver used in cell phones sold in 2019 were estimated to be worth more than $2.5 billion! Who would have thought our urban landfills are virtual goldmines? Needless to say, the process of recycling electronics, referred to as *urban mining*, needs to be further developed because the recycling of metals from electronics—while not easy—requires significantly less energy than mining and purifying the metal from its ore. Electronic waste also contains concentrations of precious metals that are 40 to 50 times more abundant than in naturally occurring ore. Some companies are starting to focus on this initiative, such as the Brussels-based company Umicore. Even automakers are developing in-house recycling programs for their electronic devices and batteries in cars they manufacture.

It is estimated that about 50 million tons of electronic waste are discarded annually worldwide, and it contains more than $50 billion worth of metals. As further improvements are made to urban mining practices, and companies continue to expand their electronic trade-in programs for old devices, we can begin to wean ourselves away from traditional mining practices, which would result in less environmental impacts.

Your Turn 1.12   Recycling

An aluminum mining company has claimed that it is less expensive and energy intensive to extract aluminum from ore than it is to recycle aluminum cans. Using the Internet as a resource, consider the costs and energy sources involved in both processes, and decide whether this claim is valid.

Perhaps the most difficult step in electronics recycling is to remove the metals from the device itself. This process consists of boiling the circuit boards in solvents to remove the plastics and then leaching out the metals with strong acids. However, if one is not careful, groundwater could become contaminated with heavy metals and organic waste, possibly contributing to an increase of cancers and other life-threatening illnesses in the surrounding communities. Unfortunately, these recycling practices are often outsourced to countries where environmental regulations are not established and proper safety precautions are not adopted for workers (Figure 1.25).

Figure 1.25

People recycling metal.

Tim Page/Corbis Documentary/Getty Images

Companies are seeking efficiencies in the recycling of electronic devices. For instance, Apple has developed a robot called Daisy to disassemble iPhones for materials recycling. A single Daisy robot is able to disassemble 200 iPhones per hour or about 1.2 million phones per year. For every 100,000 iPhone devices, the following metals are recovered: 1900 kg of aluminum, 1 kg of gold, 7.5 kg of silver, 710 kg of copper, 93 kg of tungsten, 42 kg of tin, 770 kg of cobalt, and 11 kg of rare earth elements.

Other than the precious metals of silver, gold, and platinum, a class of metals that are increasingly important for our society are the rare earth metals (Figure 1.26). These elements are employed for many applications that we rely on every day such as vehicle catalytic converters and fluorescent lighting, as well as memory chips, rechargeable batteries, magnets, and speakers found inside cell phones and portable electronic devices. The military also uses a variety of rare earths for night-vision goggles, advanced weaponry, GPS equipment, batteries, and advanced electronics.

China is the world's leading producer of rare earth metals (Figure 1.27), but is also an increasing consumer for the finished electronic products. Over 90% of the world's supply of rare earth elements are exported from China, which also holds more than 50% of the world's total reserve of these metals.

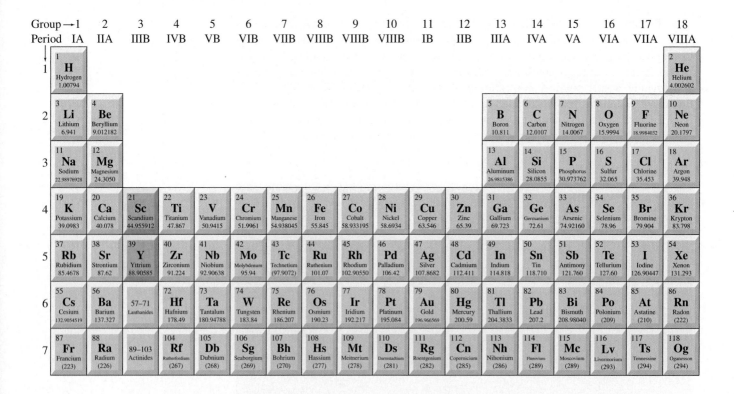

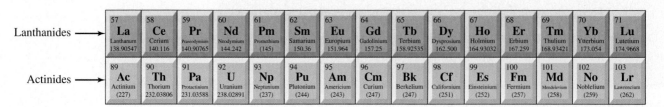

Figure 1.26

Periodic table of the elements. The positions of the rare earth metals are highlighted in blue.

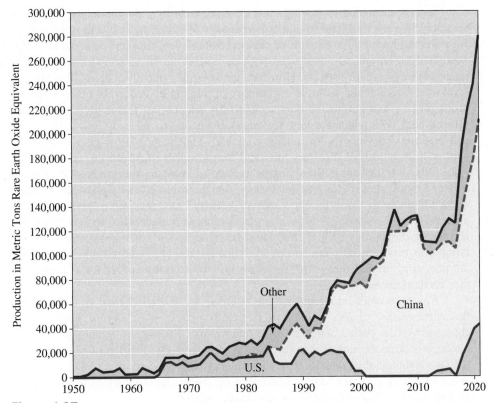

Figure 1.27

Illustration of the dominance of China in the mining of the rare earth metals.

Let's Debate!   Urban Mining of Rare Earths

(left): Alex Floston/Shutterstock; (right): Narat Plaeksakul/Alamy Stock Photo

Form two groups: one group will be in favor of the topic (the proposition) and the other group will be against the topic (the opposition). Both groups will try to persuade a neutral person or judge to agree with them. The topic of the debate is the motion. Here is the motion:

> *We should be able to supply our current/future needs for rare earth elements by increasing urban mining of existing materials. It would be too environmentally harmful to increase mining and refining operations in the U.S for the purpose of supplying our current/future needs for rare-earth elements.*

Both groups should investigate the current and projected future needs for rare earth elements, as well as the current use of urban mining for these elements. In addition, both groups should identify the environmental consequences of mining and refining operations.

Curiosity Conversations

Sometimes science topics can cause strong emotional reactions because they may be counter to your existing view. It is important to recognize when a science topic does this and try to remain curious about the topic. Consider the topic of urban mining for rare-earth elements and its effect on the environment:

- What is something you learned about the environmental consequence of urban mining for rare-earth elements that you had not thought about prior to this?
- What would you like to share to help someone see your point of view?

Indeed, advanced technology can be thought of as a double-edged sword for a society. We are fortunate to enjoy the benefits of faster, lighter, and more powerful portable devices. As a result of this technology, we are more accessible in our businesses, more quickly connected with our social circles, and better able to navigate roadways and city streets. However, we also open ourselves up to a heightened level of risk associated with the availability of our source materials. Once a society progresses beyond antiquated devices and fully adopts new technology, there is no turning back. But what if a key raw material needed to fabricate our cell phones and electronic devices is no longer available? This may arise from a number of causes such as natural disasters, political unrest, energy restrictions, or trade barriers. Whatever the cause, how do we manage to continue the production of electronic devices needed for our businesses and personal lives? And what if these materials are also essential for national security? One answer might be to find alternative raw materials, known as *earth-abundant materials*, that would have similar functionalities. While this could be possible for some substances, for the rare earths this is usually not an option. Although research and development efforts are underway around the world, there are no suitable alternatives for a number of rare earth elements.

It is even more important that we continue to develop alternative technologies that either require smaller amounts of rare earth metals, or none at all. For instance, consumers in the United States are transitioning from fluorescent lighting, which uses a relatively large amount of rare earths, to more energy-efficient light-emitting diodes (LEDs). Although the components responsible for light generation in LEDs, known as *phosphors*, may also be composed of rare earth oxides, these elements are present in a lesser amount than is required for fluorescent lights.

Conclusion

It is hard for many to imagine life without the use of a smartphone or portable electronic device. The latest weather report, our favorite music, and the answers to life's most difficult questions are now only a touch away. Without the role of chemistry, we would not be able to acquire the elements and compounds that comprise our modern electronic devices. Indeed, the chemical transformations of rocks and minerals into pure Si and metals are required for virtually all aspects of our modern lifestyles.

However, there are limited global reserves for some elements used in portable electronics, such as the rare earths. As the world scurries to find more sources for the rare earths—even looking on the ocean floors—we can more easily acquire these and other low-abundant materials that have already been mined. This can be realized by simply developing low-cost (and environmentally friendly) recycling protocols for the used electronic devices sitting in our drawers at home or those discarded in urban landfills.

The next chapter will describe how our manufacturing and end-use practices for electronics affect the very air we breathe. We will move beyond the clean room, where a trace of oxygen will cause problems with computer chip fabrication, to the real world that needs oxygen in order to sustain human life.

LEARNING OUTCOMES

The numbers in parentheses indicate the sections within the chapter where these outcomes were discussed.

Having studied this chapter, you should now be able to:

- classify, compare, and visualize the states of matter. (1.1)
- classify matter according to composition. (1.2)
- describe the composition of compounds. (1.3)
- convert between decimal and scientific notation, and between different units of measurement. (1.4)

- illustrate the structure of an atom, and compare the relative locations, charges, and masses of the subatomic particles. (1.5)
- describe the difference between physical and chemical changes and how elements are extracted from rocks (1.6)

- describe how substance purity is measured and the silicon purity that is required for computer chip manufacturing. (1.7)
- describe the composition of glass and explain how we can alter its structure to enhance its durability and scratch resistance. (1.8)
- explain how the three pillars of sustainability relate to the fabrication and recycling of portable electronic devices. (1.9)
- identify sources and possible alternatives for low-abundant materials used in portable electronics. (1.10)

Questions

The end-of-chapter questions are grouped in three ways:

Emphasizing Essentials

- **Emphasizing Essentials** questions give you the opportunity to practice fundamental skills.

Concentrating on Concepts

- **Concentrating on Concepts** questions are more difficult and often relate to social issues.

Exploring Extensions

- **Exploring Extensions** questions challenge you to go beyond the information presented in the text to solve real-world issues.

Appendix 5 contains the answers to questions with numbers in blue.

Emphasizing Essentials

1. In these diagrams, two different types of atoms are represented by color and size. Characterize each sample as an element, a compound, or a mixture. Explain your reasoning.

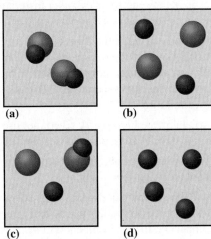

(a)   (b)

(c)   (d)

2. In these diagrams, two different types of atoms are represented by color and size. Characterize each sample as an element, a compound, or a mixture. Explain your reasoning.

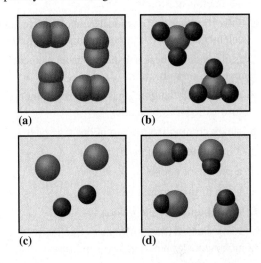

(a)   (b)

(c)   (d)

3. Convert the diameter of the period at the end of this sentence into nanometers.

4. Convert the smallest component of an integrated circuit, approximately 10.0 nanometers, into millimeters.

5. Express each of these numbers in scientific notation.
   a. 1500 m, the distance of a foot race
   b. 0.0000000000958 m, the distance between O and H atoms in a water molecule
   c. 0.0000075 m, the diameter of a red blood cell

6. Express 10.0 m in terms of cm, μm, and nm. Use proper scientific notation in your answers.

7. Consider this portion of the periodic table and the groups shaded on it.

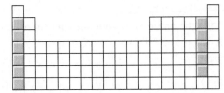

   a. What is the group number for each shaded region?
   b. Name the elements that make up each group.
   c. Give a general characteristic of the elements in each of these groups.

8. Consider the following blank periodic table, which excludes the lanthanide and actinide elements.

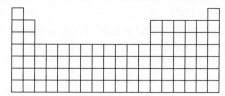

   a. Locate the region of the periodic table in which metals are found.

   b. Common metals include iron, magnesium, aluminum, sodium, potassium, and silver. Give the chemical symbol for each.

   c. Locate the region of the periodic table in which nonmetals are found.

   d. Give the name and chemical symbol for five nonmetals (elements that are not in your shaded region).

9. Classify each of these substances as an element, a compound, or a mixture.

   a. a sample of "laughing gas" (dinitrogen monoxide, also called nitrous oxide)

   b. steam coming from a pan of boiling water

   c. a bar of deodorant soap

   d. a sample of copper

   e. a cup of mayonnaise

   f. the helium filling a balloon

10. Classify each of these substances as either a homogeneous or heterogeneous mixture.

    a. a glass of orange juice

    b. a chocolate chip cookie

    c. a piece of pizza

    d. a sample of vinegar

    e. a sample of beach sand

11. Draw the atomic structure for a lithium atom including electrons, neutrons, and protons.

12. Draw the structures and describe the properties for two allotropes of sulfur. How are these fabricated?

13. Provide the number of protons, neutrons, and electrons for an aluminum atom with a mass number of 27.

14. Provide the number of protons, neutrons, and electrons for an atom of copper with a mass number of 64.

15. Define *electrical conductivity* and *thermal conductivity*, and describe how they are different.

16. Provide the name and group number for each element included in the following compounds.

    a. HfO_2 b. $BeCl_2$ c. $Ti(OH)_4$

    d. FeO e. SiO_2 f. $B(OH)_3$

17. For the following molecules, list the number and type (metal, nonmetal, or metalloid) of atoms that each contain.

    a. CO_2 b. H_2S

    c. NO_2 d. SiO_2

Concentrating on Concepts

18. In the text, we illustrated the 12N purity of silicon in terms of colored tennis balls. Provide illustrations of your own for 9N and 12N purities.

19. From the solids, liquids, or gases that are present in your favorite room or office, list three homogeneous mixtures and three heterogeneous mixtures. Also, provide the names and symbols or chemical formulas of any elements or compounds, respectively.

20. The processor chips in portable and desktop electronics are composed of tiny switches, known as *transistors*. Use the internet to look up the smallest dimensions of the transistors that are used in current processors. Relate these dimensions in terms of nm and km.

21. When early chemists discovered a new element that was unknown before, how did they know where to put the element on the periodic table?

22. Research online how aluminum metal is isolated from its natural ore, as well as the processes involved in its purification. Write a brief outline of what you have learned.

23. The use of sapphire for the screens of portable electronic devices may soon become prevalent. Compare the physical properties, molecular structures, and fabrication techniques for glass and sapphire.

24. Glass is generally thought to be an electrical insulator. However, is it possible to fabricate "conductive glass." Use the internet to research how this material is made and briefly explain what it is made of.

25. Using a molecular perspective, describe the formation of *Prince Rupert's drops* and their violent implosion when the droplet tail is fractured.

26. Describe some components of your cell phone that are in units of cm, mm, µm, and nm.

27. Using the internet as a resource, list three metals that are currently used in cell phones that have a natural abundance in Earth's crust of <50 ppm. Where might these metals come from?

28. Explain why transparency is important for electronic displays. Explore this topic online. Predict and compare the way light interacts with different types of matter, and explain how engineers can alter materials to change or enhance properties such as durability or transparency.

29. Gold is rarely found in the form of nuggets, but rather as smaller flakes in the sediments of streams and rivers. Search the Internet for more information, then explain how gold prospectors separate gold from the sand, dirt, and rock mixture.

30. Can you fabricate high-purity silicon for use in portable electronic devices from plentiful sea sand? Explain.

31. List some waste products generated from the fabrication of high-purity silicon.

32. Critique the accuracy of the following statement: "As cell phones become smaller in size and less expensive, their impact on the environment will decrease."

33. Evaluate the current portable electronics industry in terms of the three pillars of sustainability. For each pillar, provide a letter-grade rating and suggest three possibilities for improvement.

Exploring Extensions

34. What is meant by "Moore's law" and is this still valid? How does Moore's law influence chip design?

35. The crystal structures of many gemstones are based on SiO_2 and Al_2O_3 frameworks. Considering that pure silica and alumina are white solids, describe the origin of the diverse colors exhibited by gemstones.

36. It can be said that "impurities affect the physical properties of most crystalline solids." Explain.

37. "Smart glass" that becomes opaque with a flip of a switch is now being used in businesses and hotel rooms across the world. Investigate this technology on the internet and describe how glass can transition from transparent to opaque with the passage of electrical current.

38. Describe some procedures that have been used to recycle the metals found in cell phones. Use the internet as a resource to learn more about this process.

39. Cell phone companies have advertised "superior toxic substance removal" from their products. Look up online and report which elements have been removed and where were these located within cell phones.

40. Find an example for soil and water pollution that arose from the improper recycling of electronic devices. How could these situations have been prevented?

41. Provide a cradle-to-cradle strategy for the recycling of processor chips found in portable electronics.

42. Investigate online the reactions required to convert SiO_2 sand into high-purity silicon. Draw a flowchart that illustrates this process. What happens to the waste products that are generated in each step? How sustainable is this process?

43. Describe a method for strengthening glass that is used in portable electronic devices other than Gorilla Glass. Use the internet as a resource.

44. Apple's Daisy robot is able to dismantle 200 iPhones per hour, to efficiently recover more materials than traditional recycling processes. Look up online the total number of iPhones sold worldwide last year. What mass (in kg) of Al, Au, Ag, Cu, Co, and rare earth metals could be recovered if 25% of the phones were recycled? What is the combined value of these recycled metals, based on their current prices?

45. Using Internet resources, perform a life-cycle analysis for your cell phone. Try to be as detailed as possible for two scenarios: cradle-to-grave and cradle-to-cradle.

46. Describe some environmental impacts that are involved during the design, research and development, and marketing phases of cell phones before their ultimate production and release to consumers.

47. Using the internet as a resource, compare and contrast the steps, associated costs, and energy use required to extract aluminum from ore vs. can recycling, and rate these practices based on their overall efficiency and sustainability.

48. Consider the image below that shows the increasing global demand for rare earth metals. Calculate the percentage increases in demand for China, Japan/NE Asia, USA, and the rest of the world between 2012 and 2016. Due to rising prices of the rare earths and limited global supplies, more countries are evaluating recycling programs to extract and reuse these elements from existing devices. What devices contain rare earth metals? Based on the number of these devices sold annually, their average lifetimes, and assuming that 100% of available devices are recycled with 100% recovery of the metals, could the U.S. meet its current demand through recycling efforts alone? Explain.

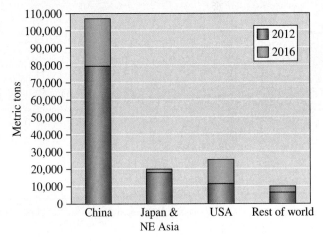

2 The Air We Breathe

Banner credit above: ©elen_studio/Shutterstock; U.S. Department of Defense/Milmotion/ Getty Images

THE COMPONENTS OF AIR

The air we breathe is composed of a variety of substances. Watch the Chapter 2 opening video at www.acs.org/cic and answer these questions.

a. Identify three indoor and three outdoor sources that emit chemicals into the air around you.

b. Briefly describe how each of these chemicals might affect your health.

In this chapter, we will answer the following questions:

1. What is respiration and why is it essential to life?
2. What are the regions of the atmosphere?
3. How do we measure the concentration of pollutants in air?
4. How can we visualize the components present in air?
5. How do we name the substances present in air?
6. What are the health implications of inhaling air pollutants?
7. How do we determine if the air is safe to breathe?
8. How do interpret reported air quality measurements?
9. What are some chemical reactions that create air pollutants?
10. What are combustion reactions and how do these lead to air pollution?
11. How do coal-fired power plants and vehicle use generate air pollutants?
12. How is ground-level ozone produced?
13. What are some harmful components of indoor air?
14. In what ways can we prevent or limit contaminants from polluting our atmosphere?

Introduction

Not unlike our portable electronic devices discussed in Chapter 1, we take the air we breathe for granted and trust that it will always be there. However, what makes up this air that we breathe? Why is it necessary for life? Are there components in the air that can harm us? Together with earth, fire, and water, the ancient Greeks named air as a basic element of nature. Hundreds of years later, chemists experimented to learn more about the composition of air, and found that it is composed of a variety of particles, such as molecules and atoms. Today, we can view air in Earth's atmosphere from outer space. And daily, just like the ancients, we can also peer up through the air to catch a glimpse of the twinkling stars at night.

Our atmosphere completely surrounds us, acting as a thin, invisible veil that separates us from outer space. This chapter describes the atmospheric gases that support life on Earth. Chapter 3 describes the ozone in the stratosphere that protects us from harmful ultraviolet radiation emitted by the Sun. And Chapter 4 describes the greenhouse gases in our atmosphere that protect us from the bitter cold of outer space. Truly, our atmosphere is a resource beyond price.

This chapter also describes how, by our actions, we humans have altered the composition of the atmosphere. Many of these changes have occurred as a result of industrial processing that is needed to fuel our increasing drive for advanced technology (Chapter 1). However, with over 8 billion humans on the planet and counting, it is important to realize that our individual actions—however insignificant they may seem—may have lasting repercussions for our environment. The next activity invites you to think about how our lifestyles, both individually and collectively, can change the air we breathe.

Your Turn 2.1 Footprints in the Air

Hiking boot treads, asphalt pavement, cornfields—each of these is an example of a "ground print" left by humans because each one alters the lay of the land. Similarly, our activities leave "air prints" that alter the composition of our atmosphere.

Identify three indoor and three outdoor sources that emit chemicals into the air around you. For each of these sources, describe whether they (1) hurt the air quality, (2) improve the air quality, or (3) have some effect, but you don't know what it is.

2.1 | Why Do We Breathe?

Learning Objective: Describe and illustrate the process of respiration

Take a breath! Automatically and unconsciously, you do this thousands of times each day. No one has to tell you to breathe—you just do it! Although a doctor or nurse may have encouraged your first breath, nature then took over and you began doing it unconsciously. Even if you were to hold your breath in a moment of fear or suspense, you soon would involuntarily gasp a lungful of that invisible stuff we call air. Indeed, you could survive only minutes without a fresh supply.

Your Turn 2.2   Take a Breath

What total volume of air do you inhale (and exhale) in a typical day? Figure this out. First, determine how much air you exhale in a single "normal" breath. Then, determine how many breaths you take per minute. Finally, calculate how much air you exhale per day. Describe how you made your estimate, provide your data, and list any factors you believe may have affected the accuracy of your answer.

How much did you estimate that you breathe in a day? Typically, an adult breathes more than 11,000 liters (about 3000 gallons) of air per day. The value would be even higher if you had spent the day on a bike trail or hiking in the mountains. Are you surprised by how much air you actually breathe?

@HOME   What's Your Lung Capacity?

A *spirometer* is a special device that can be used to determine the volume of inhaled/exhaled air. However, let's investigate your own lung capacity at home using a simple tape measure and a balloon!

1. Blow up a balloon with a single breath. Keep breathing that one breath until your lungs are emptied. Tie off the balloon.
2. Measure the circumference of the balloon at its widest point using the tape measure. Record your results in centimeters.
3. Find the radius of the balloon. Use the formula for circumference ($C = 2\pi r$).
4. Calculate the volume of a sphere. Use the formula for the volume of a sphere ($v = \frac{4}{3}\pi r^2$).
5. Convert cubic centimeters into milliliters and milliliters into liters using dimensional analysis.

Questions to Consider
a. How does your lung capacity compare to others in your class?
b. What factors do you think can affect lung capacity?

Science Photo Library/Alamy Stock Photo

We breathe in air because it keeps us alive. The air around us contains oxygen, which is essential to our survival. In the process called **respiration**, we take in oxygen to help metabolize the foods we eat. Sugar and oxygen are transformed into carbon dioxide and water, and energy is made available to carry out other essential processes in our bodies. With each breath, we inhale air to obtain oxygen, and we exhale carbon dioxide (and small amounts of water) into the atmosphere.

@HOME   Respiration in the Kitchen

Investigate the process of respiration from your kitchen using sugar, dry yeast, water, sugar, an empty plastic water bottle, and a balloon.

1. Pour approximately two inches of warm water into the empty plastic bottle.
2. Add a spoonful of dry yeast to the water and stir.
3. Place a spoonful of sugar inside the balloon.
4. Carefully wrap the balloon around the lip of the bottle.
5. Lift the balloon up to release the sugar into the dry yeast and water mixture.
6. Make observations!

Questions to Consider
  a. Can you explain what happens to the balloon?
  b. What is inside the balloon and what is it evidence of?

Janette Beckman/McGraw Hill Education

2.2 | Defining the Invisible: What Is Air?

Learning Objective: Categorize the regions of the atmosphere and compare the compositions of inhaled and exhaled air

We are surrounded by a multilayered atmosphere (Figure 2.1), which warms our planet and protects us from harmful radiation. The lowest layer, where we live, is called the *troposphere* and accounts for 75% of the mass of the entire atmosphere. The next layer

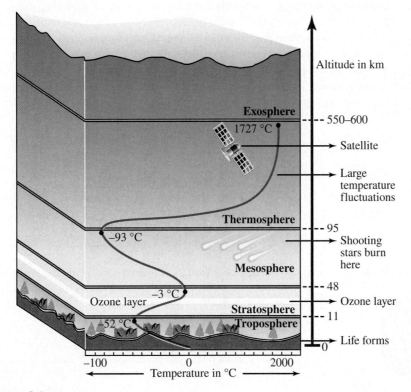

Figure 2.1

Regions of Earth's atmosphere, illustrating respective air temperatures. Note: the temperature (*x*) and altitude (*y*) axes are not to scale.

is the *stratosphere*, followed by the *mesosphere* and *thermosphere*. The last (farthest from Earth) is called the *exosphere*. Temperature fluctuations exist across these levels, and overall pressure decreases the farther away from Earth due to the lower amounts of gases at each higher level.

Air, a collection of gases mixed together in various proportions throughout the atmosphere, is classified as matter. In addition to oxygen needed for respiration, there are several other components of air that we will now examine.

Contrails or Chemtrails?

Check out this video to see why jets emit visible vapor trails: www.acs.org/cic.

©2018 American Chemical Society

Questions to Consider
a. What are some differences between "contrails" and "chemtrails"? What are some conspiracy theories regarding the presence of "chemtrails", and are these ideas backed up by any evidence?
b. From sources found on the Internet, what are the effects of contrails on global warming? Can anything be done to minimize these effects on global climate change?

Your Turn 2.3   What's in a Breath?

Take a breath. What are you breathing in? Exhale. What are you breathing out?

Is there an "ideal" atmosphere in which you might breathe? If so, provide a description.

Although you can't tell by looking, the air you are breathing is not a single pure substance. Rather, it is a mixture—the physical interaction of two or more substances present in variable amounts. In this section, we focus on the pure substances that are in air: nitrogen (N_2), oxygen (O_2), argon (Ar), carbon dioxide (CO_2), and water (H_2O). All are colorless, odorless gases that are undetectable to the nose. Although these gases are generally invisible to the eye, the next activity shows this is not always the case.

Using a pie chart and a bar graph, Figure 2.2 represents the composition of air. Regardless of how we present the data, the air you breathe is primarily nitrogen and oxygen. More specifically, the composition of air by volume is about 78% nitrogen, 21% oxygen, and 1% other gases. **Percent (%)** means "parts per hundred." In this case, the parts are either molecules or atoms.

The composition of the mixture that we call air depends on where you are. Because exhaled air is a slightly different mixture than inhaled air (Table 2.1), we at least temporarily change the air directly in front of us when we breathe.

Notice that nitrogen gas is relatively unreactive, passing in and out of our lungs unchanged. Although nitrogen is essential for life and is part of all living things, its form in the atmosphere is not usable to most organisms. Most plants and animals obtain their nitrogen needs from altered or alternative sources of nitrogen.

In contrast, even though oxygen is less abundant than nitrogen in our atmosphere, it plays a key role on our planet. Oxygen is absorbed into our blood via the lungs when we breathe, and it reacts with the foods we eat to release the energy needed to power chemical processes within our bodies. It is necessary for many other chemical reactions

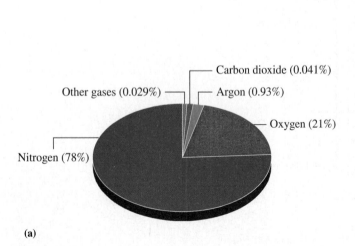

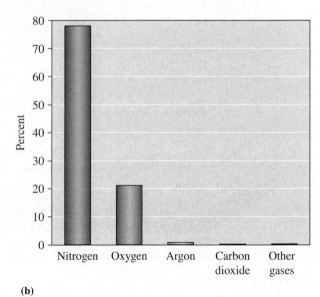

(a)   (b)

Figure 2.2

The composition of dry air by volume represented as a **(a)** pie chart and **(b)** bar graph.

Table 2.1	Typical Composition of Inhaled and Exhaled Air	
Substance	**Inhaled Air (%)***	**Exhaled Air (%)***
Nitrogen (N_2)	78.0	78.0
Oxygen (O_2)	21.0	16.0
Argon (Ar)	0.9	0.9
Carbon dioxide (CO_2)	0.04	4.0
Water (H_2O)	Variable	Variable

*In unit of percent by volume, %(v/v)

as well, including combustion (burning) and oxidation (e.g., rusting). Largely due to its presence in water molecules (H_2O), oxygen is the most abundant element (by mass) in the human body. As we discussed in Chapter 1, its presence in many rocks and minerals also makes it the most abundant element in Earth's crust.

Your Turn 2.4   More Oxygen ...?

We live in an atmosphere of 21% oxygen. A match burns in less than a minute, a fireplace consumes a small pine log in about 20 minutes, and we exhale about 15 times a minute. Life on Earth would be *very* different if the oxygen concentration were twice as high. List at least four ways Earth or our lives might be different with such an increased amount of oxygen.

Nadim Mahmud - Himu/Shutterstock

Given that air is a collection of gases, because of their nature, gas particles can move around and "through" one another. This means that trace amounts of substances can be carried in the air. Often, we can detect these by smell. For example, in areas of France, the scent of lavender may permeate the air outside, whereas in the mountainous areas of the United States, pine may fill your nostrils. Indoors, the aroma of freshly brewed coffee may beckon you to the kitchen, or outside, the smell of the salty sea may invite you to the sands of a beach. Of course, other less desirable substances can be released in the air, too, such as those from fuel combustion, landfills, or flatulence.

@HOME   Your Nose Knows

The air is different in a pine forest, a bakery, an Italian restaurant, and a dairy barn. Blindfolded, you could smell the difference. Our noses alert us to the fact that air contains trace quantities of many substances.

Investigate the movement (diffusion) of gas particles at home using a balloon, a few drops of perfume (or cologne), a string, and a piece of tape.

1. Add a few drops of perfume (or cologne) to the inside of a balloon.
2. Blow up the balloon and tie it off.
3. Tie a string around the bottom of the balloon and use a piece of tape to secure it to a wall.
4. Walk further away from the balloon and make observations.

Questions to Consider
a. How far away from the balloon can you smell the perfume (or cologne)?
b. If the perfume (or cologne) was inside the balloon, why can you smell it outside of the balloon?
c. Name three indoor and three outdoor smells that indicate small quantities of additional chemicals are present in the air.
d. Our noses warn us to avoid certain things. Give three examples of when a smell indicates a hazard or something to be avoided.

Hopefully, the previous activity has you thinking about how changes to the components of air can cause changes in you—in your mood and even in your health. In other words, when the composition of air changes, its properties also change.

2.3 | The Composition of Air

Learning Objective: Describe the concentrations of substances in the air

The percentages we have been using to describe the composition of the atmosphere are based on volume—the amount of space that each gas occupies. If we wanted to, we could closely approximate 100 liters (L) of dry air by combining 78 L of nitrogen, 21 L of oxygen, and nearly 1 L of argon. Because the separate gases would mix completely, the result would correspond to a mixture containing 78% nitrogen, 21% oxygen, and 1% argon. The composition of air can also be represented in terms of the numbers of molecules and atoms present. Equal volumes of gases will contain equal numbers of particles, provided the gases are at the same temperature and pressure. Thus, if you were able to take a sample of air containing 100 particles, 78 would be nitrogen molecules, 21 would be oxygen molecules, and 1 would be an argon atom. In other words, when we say that air is 21% oxygen, we mean that there are 21 molecules of oxygen per 100 of the total molecules and atoms in the air.

@HOME   Air Pollution and Properties of Gases

Check out this video for an experiment to investigate the properties of gases found in the air around you: www.acs.org/cic.

Michael Thomas Mury

Questions to Consider

a. What would result if one used more or less of the baking soda and vinegar reactants?

b. As the narrator asked at the end of the video, what would happen if a lit match or splint were placed in contact with the CO_2 evolved from this reaction? How about for a reaction that generated oxygen gas instead of carbon dioxide?

Although the sum of percentages of nitrogen, oxygen, and argon appear to make up our atmosphere, there are many other components there as well—substances in trace amounts, but still present nonetheless. Some of these make us feel good; for example, breathing in the molecules that make up the scent of fresh bread may trigger a refreshing smile to some faces. However, other components can be harmful to our health, such as the vehicle emissions in an urban environment. These molecules are smaller in concentration—for example, less than 1 molecule in 100. No matter where you live, each lungful of air you inhale contains tiny amounts of substances other than nitrogen and oxygen. Many are present at concentrations less than 1%. Such is the case with carbon dioxide, a gas you both inhale and exhale. In our atmosphere, the concentration of carbon dioxide reached a maximum of 0.0421% in 2022, the highest value recorded since measurements began. This value represents a recent exponential rise that now continues to steadily increase as humans burn fossil fuels.

The value 0.0421% could be described as 0.0421 molecules of carbon dioxide per 100 molecules and atoms in the air, but the idea of a fraction (0.0421) of a molecule is a bit strange. For relatively low concentrations, it is more convenient to use parts per million (ppm). One ppm is a unit of concentration that is 1/10,000 the size of 1%. Here are some useful relationships:

0.0421 % means:

0.0421 parts per hundred

0.421 parts per thousand

4.21 parts per ten thousand

42.1 parts per hundred thousand

421 parts per million

For instance, out of a sample of air containing 1,000,000 molecules and atoms, 421 of them will be carbon dioxide molecules. Therefore, the carbon dioxide concentration is denoted as 421 ppm. Although this is a very small concentration, this doesn't necessarily mean a low impact. Even at low concentrations, carbon dioxide contributes to increasing global temperatures and climate change.

Really One Part per Million?

Check out this video for a demonstration of relative concentrations: www.acs.org/cic.

Some say that a part per million is the same as one second in nearly 12 days. Is this an accurate analogy? How about one step (~2.5 feet) in a 568-mile journey? What about 4 drops (20 drops ≈ 1 mL) of ink in a 55-gallon barrel of water? Check the validity of these analogies, explaining your reasoning. Then, come up with an analogy or two of your own.

Andrey Mihaylov/Shutterstock

One hundred years ago, Earth was home to fewer than 2 billion people. We have now overcome the 8 billion mark, with the majority of people living in urban regions. This growth in population has been accompanied by a massive growth in both the consumption of resources and the production of waste. The waste that we stash in our atmosphere is called air pollution. When large numbers of people do certain activities, like cooking meals over open fires or driving combustion engine vehicles, they tend to pollute the air. For example, Figure 2.3 shows two days of varying pollution levels in Beijing, China. Other large cities such as Los Angeles, Phoenix, Mexico City, Mumbai, and Santiago, Chile, often have dirty air as well. Human activities leave "air prints," both indoors and out.

Certain gases contribute to air pollution at the surface of Earth. One of these gases, carbon monoxide (CO), is odorless; others—ozone (O_3), sulfur dioxide (SO_2), and nitrogen dioxide (NO_2)—have characteristic odors. All can be hazardous to your health, even at concentrations well below 1 ppm.

Figure 2.3

Photographs taken from the same vantage point on different days in Beijing, China.

Kevin Frayer/Getty Images

Your Turn 2.5   Practice with Parts per Million

Example Problem:

Sulfur dioxide is emitted during volcanic eruptions. Different people have different sensitivities to sulfur dioxide, but people who have asthma can have difficulty breathing with as little as 0.2 ppm–0.5 ppm sulfur dioxide in the air. Express this amount as a percentage.

Example Solution:

$$0.2 \text{ ppm:} \quad \frac{1\%}{10{,}000 \text{ ppm}} \times 0.2 \text{ ppm} = 0.00002\%$$

$$0.5 \text{ ppm:} \quad \frac{1\%}{10{,}000 \text{ ppm}} \times 0.5 \text{ ppm} = 0.00005\%$$

Your Turn:

a. In some countries, the limit for the average concentration of carbon monoxide in an 8-hour period is set at 9 ppm. Express this amount as a percentage.

b. Exhaled air typically contains about 78% nitrogen. Express this concentration in parts per million.

2.4 | I Can "See" You! Visualizing Air

Learning Objective: Use macroscopic, molecular, and symbolic representations to describe matter

Chemists typically use three viewpoints to study and understand matter (Figure 2.4). One is the macroscopic view, which consists of viewing matter through the lens of senses, observations, and measurements. Characteristics that can be described in this viewpoint are properties such as color, odor, chemical reactivity, or density. However, we can also describe matter using symbols. As we saw in Chapter 1, these descriptions use letters and numbers within chemical formulas to represent samples of matter (H_2O for water, for example). We can also use symbols within equations to describe various physical relationships of matter (e.g., $d = m/V$ for the relationship of density, mass, and volume, respectively). The third view of matter is the particulate view. In this view, we "see" or imagine what the actual particles, atoms, or molecules look like, and how they might interact (Figure 2.4).

We now apply these concepts to the mixture known as air. Some of its components are elemental substances: nitrogen and oxygen exist as diatomic molecules (N_2 and O_2), while argon and helium exist as single, uncombined atoms (Ar and He). Other components, most notably water vapor (H_2O) and carbon dioxide (CO_2), are compounds. In carbon dioxide, the carbon and oxygen atoms are *not* present as separate entities. Rather, the atoms are chemically combined to form a carbon dioxide molecule, in which the two atoms are held together by a chemical bond (Figure 2.5). More specifically, two oxygen atoms are combined with one carbon atom to form a carbon dioxide molecule.

Figure 2.4

Three viewpoints of water. In the molecular view, water contains two hydrogen atoms (represented in grey) and one oxygen atom (represented in red).

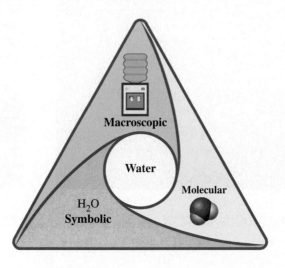

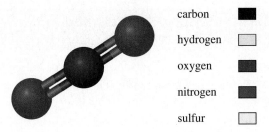

Figure 2.5

Molecular representation of a carbon dioxide (CO_2) molecule, showing a central carbon atom (black) chemically bonded to two oxygen atoms (red). Also shown are the color codes of atoms commonly found in molecular representations.

2.5 | A Chemical Meet & Greet—Naming Molecular Compounds

Learning Objective: Write formulas and names for molecular compounds, including hydrocarbons

If chemical symbols are the alphabet of chemistry, then chemical formulas are the words. The language of chemistry, like any other language, has rules of spelling and syntax. In this section, we help you to "speak chemistry" using chemical formulas and names. As you'll see, each name corresponds uniquely to one chemical formula. However, chemical formulas are *not* unique and may correspond to more than one name.

Right now, we will focus on the chemical names and formulas (known as *nomenclature*) of the compounds relating to the air you breathe. However, rules and practices for naming and symbolizing other categories of chemicals will be shared later in the text. We have already named some of the pure substances found in air, including carbon monoxide, carbon dioxide, sulfur dioxide, ozone, water vapor, and nitrogen dioxide. Although it may not be apparent, this list includes two types of names: systematic and common.

Systematic names for compounds follow a reasonably straightforward set of rules. Here are the rules for **molecular compounds** (compounds composed of two or more nonmetals), such as carbon dioxide (CO_2) and carbon monoxide (CO):

- Name each element in the chemical formula, modifying the name of the second element to end in *-ide*. For example, oxygen becomes *oxide*, and sulfur becomes *sulfide*.
- Use prefixes to indicate the numbers of atoms in the chemical formula (Table 2.2). For example, *di-* means 2, and thus the name carbon *di*oxide means two oxygen atoms for each carbon atom.
- If there is only one atom for the first element in the chemical formula, omit the prefix *mono-*. For example, CO is carbon monoxide, not monocarbon monoxide.

If instead you are writing a chemical formula from a name, remember that the subscript of 1 is not shown in chemical formulas—it is understood. Thus, the chemical formula for carbon dioxide is CO_2 and *not* C_1O_2. Similarly, carbon monoxide is CO, *not* C_1O_1. The next activity gives you a chance to practice naming compounds.

Table 2.2	Prefixes Corresponding to the Number of Atoms for Molecular Compounds		
Number of Atoms	**Prefix**	**Number of Atoms**	**Prefix**
1	*mono-*	6	*hexa-*
2	*di-*	7	*hepta-*
3	*tri-*	8	*octa-*
4	*tetra-*	9	*nona-*
5	*penta-*	10	*deca-*

Your Turn 2.6   Writing Symbols and Naming Oxides

Example Problem:

Write the chemical formula for phosphorus trichloride?

Example Solution:

The "tri-" prefix means "three." Therefore, the formula is PCl_3.

Your Turn:

a. Write chemical formulas for nitrogen monoxide, nitrogen dioxide, dinitrogen monoxide, and dinitrogen tetraoxide.

b. Give the names for SO_2 and SO_3.

Hydrocarbons, molecular compounds comprised exclusively of hydrogen and carbon, also follow naming rules. For these compounds, a prefix indicates the number of carbon atoms that are present in its structure (Table 2.3). Just as *mono-, di-, tri-,* and *tetra-* prefixes are used to count the number of specific atoms, so *meth-, eth-, prop-,* and *but-* are used to count the number of carbon atoms in a hydrocarbon. The suffix tells us something specific about the ratio of carbon atoms to hydrogen atoms in the molecule. For instance, the *-ane* suffix has the general formula C_xH_{2x+2}, but other suffixes are also possible for other C:H ratios.

The prefixes listed in Table 2.3 are very versatile. They can be used not only at the beginning of chemical names, but also within them to represent groups of chains of carbon and hydrogen atoms. Hydrocarbon molecules can contain 50 carbon atoms or more—each with a distinctive chemical formula and name.

Your Turn 2.7   "Mother Eats Peanut Butter"

Many generations of chemistry students have used the mnemonic device "mother eats peanut butter" to remember *meth-, eth-, prop-, but-*. Use this, or another memory aid of your choice, to tell how many carbon atoms are in each of these compounds.

a. Ethanol (a component of adult beverages and a gasoline additive)

b. 1,3-Butadiene (a carcinogen in vehicle exhaust and cigarette smoke)

c. Propane (the major component in liquid petroleum gas, LPG)

Although all compounds have a systematic name, some compounds also have a common name, which is not determined by following a set of rules. For example, consider water (H_2O)—why isn't it called dihydrogen monoxide? Although this would make sense and is actually an accurate name for the compound, water was given its name long before anybody knew anything about hydrogen and oxygen. Chemists, being reasonable folks, did not rename water. Likewise, O_3 goes more often by its common name, ozone (trioxygen is its systematic name), and NH_3 goes by ammonia (its systematic name is nitrogen trihydride). Two of the nitrogen compounds in the previous activity also have common names. NO is also known as nitric oxide, and N_2O is known as nitrous oxide (or sometimes just nitrous). Nitrous oxide is used as a propellant in

Table 2.3		Names of Hydrocarbons Based on the Number of Carbon Atoms			
Chemical Formula	Number of Carbon Atoms	Compound Name	Chemical Formula	Number of Carbon Atoms	Compound Name
CH_4	1	Methane	C_6H_{14}	6	Hexane
C_2H_6	2	Ethane	C_7H_{16}	7	Heptane
C_3H_8	3	Propane	C_8H_{18}	8	Octane
C_4H_{10}	4	Butane	C_9H_{20}	9	Nonane
C_5H_{12}	5	Pentane	$C_{10}H_{22}$	10	Decane

Reddi-Wip whipped cream and as the source of extra oxygen in modified race cars. Unlike systematic names, common names cannot be figured out by simply looking at the chemical formula. You have to know them or look them up.

Your Turn 2.8   Practice With Common Names

Using the Internet, provide the chemical formula and systematic name for the following molecular compounds: **a.** quartz, **b.** laughing gas, **c.** silane, **d.** dry ice, **e.** hydrogen sulfide, and **f.** phosphine.

2.6 | The Dangerous Few: A Look at Air Pollutants

Learning Objective: Identify and explain the health effects of common air pollutants

Why might we want to be concerned about the concentrations of gaseous or solid components in the air? Even at concentrations well below 1 ppm, some of these can be hazardous to your health. For instance, consider the following:

- **Carbon monoxide** (*CO*) has earned the nickname "the silent killer" because it has no color, taste, or smell. When you inhale carbon monoxide, it passes into your bloodstream and then interferes with the ability of your hemoglobin to carry oxygen. If you breathe carbon monoxide, at first you may feel dizzy and nauseous or get a headache—symptoms that could easily be mistaken for another illness. Continued exposure to high concentrations, however, can make you extremely ill or kill you. Both automobile exhaust and charcoal fires are sources of carbon monoxide. Propane-fueled camping stoves (Figure 2.6) can be another.

- **Ozone** (*O₃*) has a sharp odor that you may have detected around electric motors or welding equipment. Even at very low concentrations, ozone can reduce your lung function. The symptoms you experience may include chest pain, coughing, sneezing, or lung congestion. Ozone also mottles the leaves of crops and yellows pine needles (Figure 2.7). Here on Earth's surface, ozone is definitely a harmful pollutant. However, at high altitudes, it plays an essential role in screening harmful ultraviolet (UV) radiation, as you will learn in Chapter 3.

- **Sulfur dioxide** (*SO₂*) has a sharp, unpleasant odor. If you inhale sulfur dioxide, it dissolves in the moist tissue of your lungs to form an acid. The elderly, the young, and individuals with emphysema or asthma are most susceptible to sulfur dioxide poisoning. At present, sulfur dioxide in the air comes primarily from the burning of coal. For example, the 1952 London smog that eventually killed over 10,000 people was in part caused by the SO₂ emissions from coal-fired stoves. The causes of death included acute respiratory distress, heart failure (from preexisting conditions), and asphyxiation. Even those who survived had permanent lung damage, despite their attempts to protect themselves from exposure (Figure 2.8).

- **Nitrogen oxides** (NO$_x$). Nitrogen dioxide (NO₂) has a characteristic brown color and is the primary visible component of urban smog. Like sulfur dioxide, it can combine with the moist tissue in your lungs to produce an acid. In our atmosphere, nitrogen dioxide is produced from nitrogen monoxide (NO; common name: nitric oxide), another pollutant that is a colorless gas. Nitrogen monoxide is formed from the reaction of N₂ and O₂ in the air from anything that is hot, including vehicle engines and power plants. Nitrogen oxides, NO and NO₂ (collectively referred to as NO$_x$), can also form naturally in grain silos and can injure or kill farmers who may inadvertently inhale the gases.

- **Lead** (*Pb*) is a naturally occurring element found in small amounts in Earth's crust. Major sources of lead are ore and metal processing plants, cosmetics, waste incinerators, and plumbing materials, as well as lead-acid battery manufacturing and

Figure 2.6

A propane-fueled camping stove.

Jill Braaten

Figure 2.7

The impact of ozone on pine needles.

©Cathy Middlecamp

Figure 2.8

Masks worn to protect from inhalation **(a)** in 1950s London and **(b)** in modern-day Beijing.

(a): Keystone Press/Alamy Stock Photo;
(b): Hung Chung Chih/Getty Images

(a)   (b)

recycling facilities. Paint formulations once contained lead, but were banned in 1978; hence, renovations of older homes present a serious risk of lead exposure. Although the Environmental Protection Agency (EPA) banned lead-containing compounds from motor vehicle gasoline in the 1980s, the switch to unleaded aviation fuels finally occurred in 2018. When lead is released to the air, it may travel long distances before settling to the ground, where it may accumulate in soil or water reserves. Once inhaled or ingested, lead distributes throughout the body in the blood and accumulates in bones. Depending on the level of exposure, lead may adversely affect the central nervous system, immune system, reproductive and developmental systems, and the cardiovascular system. Infants and young children are especially sensitive to low levels of lead, which may contribute to behavioral problems and learning deficits. Although lead poisoning may be treated, any damage caused by lead exposure cannot be reversed.

- **Particulate matter** (*PM*) is a complex mixture of tiny solid particles and microscopic liquid droplets, and is the least understood of the air pollutants that we have listed. Particulate matter is classified by size rather than composition, and its size is larger relative to the individual molecules we have described thus far (Figure 2.9). The sizes of the particles are inversely correlated with the severity of the health consequence. PM_{10} includes particles

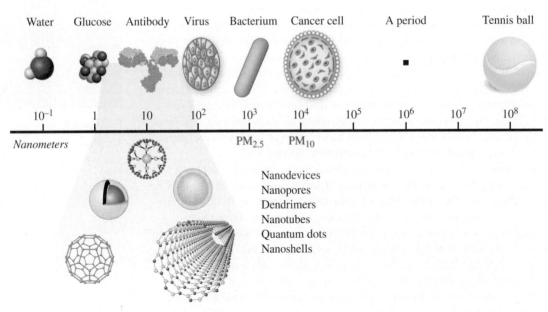

Figure 2.9

Comparison of common length scales of interest to chemistry.

with an average diameter of 10 μm (1.0×10^4 nm), or about one eighth the width of a human hair. $PM_{2.5}$ is a subset of PM_{10} and includes particles with an average diameter of 2.5 μm (2.5×10^3 nm) or less. These tinier and more deadly particles are sometimes called *fine particulates*. Particulate matter originates from many sources, including vehicle engines, coal-burning power plants, wildfires, blowing dust, and even chemistry in the air itself. Sometimes, particulate matter is visible as soot or smoke (Figure 2.10), but the two types described here, PM_{10} and $PM_{2.5}$, are too tiny to see. These particles, when inhaled, go deep into your lungs and cause irritation, possibly even resulting in lung cancer. The smallest particles pass from your lungs into your bloodstream and can cause heart disease.

Figure 2.10

A 2012 wildfire near Denver, Colorado. This fire is releasing particulate matter, some of which is visible as soot.

Helen H. Richardson/The Denver Post/ Getty Images

@HOME   Particulate Matter at Home

Investigate particulate matter at home using tape, string, a paper plate, and petroleum jelly.

1. Take a piece of string and tape it to the back of a paper plate. Leave plenty of string so that you can tie it around something later.
2. Smear petroleum jelly all over the front side of the paper plate into a thin layer.
3. Find a place to hang it outside and tie the string around it. Try to do this when rain is not expected. Leave the plate out for several days to collect particulate matter.
4. Retrieve your plate and lay it on a flat surface. Make observations about your results.

Questions to Consider

a. Did your results surprise you at all? Explain your answer.
b. How would your plate look if you had a microscope to look closer?

Real Talk   EVs: The End of Particulate Emissions?

The release of particulate matter (PM) from vehicles is generally associated with plumes of black smoke resulting from the combustion of diesel fuel in large trucks. However, PM emissions can be released from driving all types of vehicles, including electric vehicles (EVs). Check out this interview with Prof. Barbara Finlayson-Pitts who describes her research related to non-exhaust particulate emissions from vehicles: www.acs.org/cic.

Jim West/Alamy Stock Photo

Questions to Consider

a. What are some sources of non-exhaust particulate matter generated from vehicle use? What factors contribute to a greater release of this pollution?
b. What are some ways that these types of PM emissions can be reduced?

We end this section with a fact that may surprise you. Most of the air pollutants that we just listed can occur naturally! For example, a wildfire (Figure 2.10) produces particulate matter and carbon monoxide, lightning produces ozone and nitrogen oxides, and volcanoes release sulfur dioxide. However, human activity has intensified the emissions of many pollutants. Because they are the same chemicals, whether released from natural or human sources, the pollutants have the same hazards. Their levels or concentrations in the air are what is significant.

Your Turn 2.9   What's in a Polluted Breath?

Take another breath. What components of air are you breathing in?

Due to the production of portable electronics (discussed in Chapter 1) and other consumer products, our air is not as clean as it once was. How do you think the pollutants created from the portable-electronics production process could impact your health?

2.7 | Are You Feeling Lucky? Assessing the Risk of Air Pollutants

Learning Objective: Assess the risk and toxicity of different air pollutants by exposure

Risk is a part of our everyday lives. Although we cannot avoid risk, we still try to minimize it. For example, certain practices are illegal because they carry risks that are judged to be unacceptable. Other activities carry high risks, and we label them as such. For instance, cigarette packages carry a warning label regarding lung cancer. Wine bottles carry warnings about causing birth defects if consumed while pregnant and the dangers of operating machinery under the influence of alcohol. The absence of a warning, however, does not guarantee safety. The risk may be too low to label, it may be obvious or unavoidable, or it may be far outweighed by other benefits.

Warnings are just that. They do not mean that somebody *will* be affected. Rather, they report the likelihood of an adverse outcome. Let's say that the odds of dying from a vehicle accident are one in a million for each 30,000 miles traveled. On average, this means that one person out of every million traveling 30,000 miles would die in an accident. Such a prediction is not simply a guess, but the result of **risk assessment**—the process of evaluating scientific data and making predictions in an organized manner about the probabilities of an outcome.

When is it risky to breathe the air? Fortunately, existing air quality standards can offer guidance. We say *guidance* because standards are set through a complex interaction of scientists, medical experts, government agencies, and politicians. People may not necessarily agree on which standards are reasonable and safe, and standards may change over time as new scientific knowledge is generated or as political decisions change.

In the United States, national air quality standards were first established in 1970, as a result of the Clean Air Act. If pollutant levels are below these standards, presumably the air is healthy to breathe. We say "presumably" because air quality standards usually become stricter over time. If you look worldwide, you will find that air quality regulations vary both in their strictness and in the degree to which they are enforced.

The risks presented by an air pollutant are a function of both **toxicity**, the intrinsic health hazard of a substance, and **exposure**, the amount of the substance encountered. Toxicities are difficult to accurately assess for many reasons, including that it is unethical to run experiments in which people are exposed to harmful substances on purpose. Even if data were available, we would still have to determine the levels of risk that are acceptable for different groups of people. In spite of these complexities, government agencies have succeeded in establishing limits of exposure for the major air pollutants. Table 2.4 shows the National Ambient Air Quality Standards established by the U.S. Environmental Protection Agency (EPA) and those established by the World Health Organization (WHO). Here, **ambient air** refers to the air surrounding us, usually meaning the outside air. As knowledge grows from experimental studies, the standards established by the EPA and WHO are regularly updated.

Exposure is far more straightforward to assess relative to toxicity, because exposure depends on factors that we can measure more easily. These include:

- *Concentration in the air.* The more toxic the pollutant, the lower its concentration must be set. Concentrations are expressed either as parts per million (ppm) or as micrograms per cubic meter ($\mu g/m^3$), as shown in Table 2.4. Earlier, we used the prefix *micro-* with micrometers (μm), meaning a millionth of a meter (10^{-6} m). Similarly, one microgram (μg) is a millionth of a gram (g), or 10^{-6} g.
- *Length of time.* Higher concentrations of a pollutant can be tolerated only briefly. A pollutant may have several standards, each for a different length of time.
- *Rate of breathing.* People breathe at a higher rate during physical activity such as running. If the air quality is poor, reducing activity is one way to reduce exposure.

Suppose you collect an air sample on a city street. An analysis shows that the air contains 5000 $\mu g/m^3$ of carbon monoxide (CO). Is this concentration of CO harmful to

Pollutant	U.S. EPA Standard ppm (μg/m³)	WHO Standard ppm (μg/m³)
carbon monoxide		
1-h average	35 (43,000)	31 (35,000)
8-h average	9 (11,000)	8 (10,000)
24-h average	N/A	0.0035 (4)
nitrogen dioxide		
1-h average	0.100 (188)	0.106 (200)
Annual average	0.053 (100)	0.0053 (10)
ozone		
8-h average	0.070 (140)	0.051 (100)
particulates		
PM₁₀, 24-h average	– (150)	– (45)
PM₂.₅, 24-h average	– (35)	– (15)
PM₂.₅, annual average	– (12)	– (5)
sulfur dioxide		
1-h average	0.075 (210)	N/A
24-h average	N/A	0.015 (40)
lead		
3-mo average	(0.15)	N/A

Table 2.4 Ambient Air Quality Standards

breathe? We can use Table 2.4 to answer this question. Two standards are reported for carbon monoxide, one for a 1-hour exposure and another for an 8-hour exposure. The 1-hour exposure is set at a higher level because a higher concentration can be tolerated for a short time. Because the analyzed CO concentration of 5000 μg/m³ is less than both the 1-h and 8-h exposure limits, the air quality is considered safe to breathe.

Table 2.4 also allows us to assess the relative toxicities of pollutants. For example, we can compare the 8-hour average exposure standards for carbon monoxide and ozone: 9 ppm vs. 0.070 ppm. Doing some quick math, this indicates that ozone is about 130 times more hazardous to breathe than carbon monoxide! Nonetheless, carbon monoxide still can be exceedingly dangerous. As "the silent killer," it may impair your judgment before you recognize the danger.

The Harmful Effects of CO

Check out this video to see why carbon monoxide is so deadly: www.acs.org/cic.

©2018 American Chemical Society

Your Turn 2.10   Estimating Toxicities

a. Which pollutant in Table 2.4 is likely to be the most toxic? Exclude particulate matter. Share a reason for your decision.
b. Examine the particulate matter standards. Earlier, we stated that "fine particles," PM₂.₅, are more deadly than the coarser ones, PM₁₀. Do the values in Table 2.4 support this claim? Why or why not?
c. Is Pb more toxic than particulate matter? Explain your reasoning.

Although the standards for air pollutants are expressed in parts per million, the concentrations of sulfur dioxide and nitrogen dioxide could conveniently be reported in parts per billion (ppb), meaning one part out of one billion.

sulfur dioxide 0.075 ppm = 75 ppb

nitrogen dioxide 0.100 ppm = 100 ppb

As these values reveal, converting from parts per million to parts per billion involves moving the decimal point three places to the right. Even though making this change of units for SO₂ and NO₂ may create a more convenient number, when reporting concentrations across a variety of chemicals in the air, it is beneficial to have a common unit for direct comparison.

Your Turn 2.11   Living Downwind

Sulfur dioxide (SO_2) is released in the air when copper ore is smelted to make copper metal. Let's assume that a woman living downwind of a smelter inhaled 44 µg of SO_2 in an hour.

If she inhaled 625 liters (0.625 m^3) of air per hour, would she exceed the 1-h average for the U.S. National Ambient Air Quality Standards for SO_2? Support your answer with a calculation.

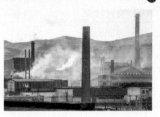

Shutterstock/visualpower

To end this section, we note that our *perception* of a risk also plays an important role. For example, the risks of traveling by car far exceed those of flying. Each day in the United States, more than 100 people die in automobile accidents. Yet some people avoid taking flights because of their fear of a plane crash. Similarly, some people fear living inland near a dormant volcano. Yet, as some extreme hurricanes have demonstrated, living in a coastal area can be a far riskier proposition. Whether perceived as a risk or not, air pollution presents real hazards, to both present and future generations. In the next section, we offer you the tools to assess these hazards.

2.8 | Is It Safe to Leave My House? Air Quality Monitoring and Reporting

> *Learning Objective: Analyze local air quality for air pollutants and make comparisons to national and global trends*

Depending on where you live, you will breathe air of varying quality. Some locations always have healthy air, others have air of moderate quality, and still others have unhealthy air much of the time. As you will see, these differences arise from a number of factors such as population, regional activities, geographical features, prevailing weather patterns, and the activities of people in neighboring regions.

To improve air quality, many nations have enacted legislation (Figure 2.11). For example, we already have cited the U.S. Clean Air Act (1970) that led to the establishment of air

Figure 2.11

Countries that have established legislation concerning air quality standards.

United Nations Environment Programme. Regulating Air Quality: The First Global Assessment of Air Pollution Legislation. Nairobi, Kenya: UNEP, 2021, 47. Figure 15: Countries with legislative instruments containing ambient air quality standards.

quality standards. Like many environmental laws, this one focused on limiting our exposure to hazardous substances. It has been named as a "command and control law" or an "end of the pipe solution," because it tries to limit the spread of hazardous substances or clean them up after the fact.

The Pollution Prevention Act (1990) was a significant piece of legislation that followed the Clean Air Act. It focused on *preventing* the formation of hazardous substances, stating that "pollution should be prevented or reduced at the source whenever feasible." The language shift is significant. Rather than cleaning up pollutants, people should not produce them in the first place! With the Pollution Prevention Act, it became national policy to employ practices that reduce, or ideally eliminate, pollutants at their source.

The decrease in the concentration of air pollutants in the United States has been dramatic (Figure 2.12). Some improvements occurred through a combination of laws and regulations, such as the ones we just mentioned. For instance, the precipitous drop in lead levels is a direct consequence of the Clean Air Act that banned leaded gasoline in 1996. However, aviation fuels still contain a lead-based fuel additive and some countries still use leaded gasoline to improve engine combustion and prevent engine knocking. Likewise, the concentration of SO_2 has dropped significantly since 1990 due to removal of sulfur from diesel fuel. Other decreases in air pollutants stemmed from local decisions. For example, a community may have built a new public transportation system, or an industry may have installed updated equipment. Still others occurred because of the ingenuity of chemists, most notably via a set of practices called "green chemistry," which will be described in the final section of this chapter.

Although air quality may have improved, on average, people in some metropolitan areas breathe air that contains unhealthy levels of pollutants. It is estimated that more than 41% of the U.S. population lives in counties that have unhealthy levels of either ozone or particulate pollution. This corresponds to more than 133 million people living in 215 counties in the U.S.

To help one more quickly assess the hazards, the U.S. EPA developed the color-coded Air Quality Index (AQI) shown in Table 2.5. This index is scaled from 1 to 500, with the value of 100 pegged to the national standard for the pollutant. Green or yellow

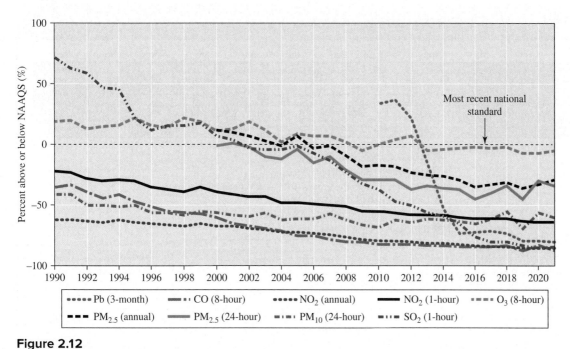

Figure 2.12

U.S. average levels of air pollutants (at selected sites) compared with national ambient air quality standards, 1990–2021. Check out an interactive version of these data: www.acs.org/cic.

Source: EPA

AIR QUALITY INDEX

Source: EPA

Table 2.5	Levels for the Air Quality Index	
When the AQI is in this range:	**... air quality conditions are:**	**... as symbolized by this color:**
0–50	Good	Green
51–100	Moderate	Yellow
101–150	Unhealthy for sensitive groups	Orange
151–200	Unhealthy	Red
201–300	Very unhealthy	Purple
301–500	Hazardous	Maroon

(<100) indicates air of good or moderate quality. Orange (101–150) indicates that the air has become unhealthy for some groups. Red, purple, or maroon (>150) indicates that the air is unhealthy for *everybody* to breathe.

The EPA also provides predictions about the future air quality of subsequent days and recommendations for outdoor activities. Figure 2.13 shows an example of the air quality forecast for ozone and particulates in early June 2023 for Washington, DC. The pollutant of concern was $PM_{2.5}$ that originated from extensive wildfires in Washington and northwest Montana. People were advised to "limit outdoor exertion such as jogging or riding bicycles." With the development of inexpensive and small sensors, it is now even possible to monitor the air quality of your location in real time via small portable devices.

Of course, measurement of air quality is not limited to the U.S. Explore trends in air quality in your vicinity and around the globe in the next few activities.

Your Turn 2.12   Your Local Air Quality (U.S.)

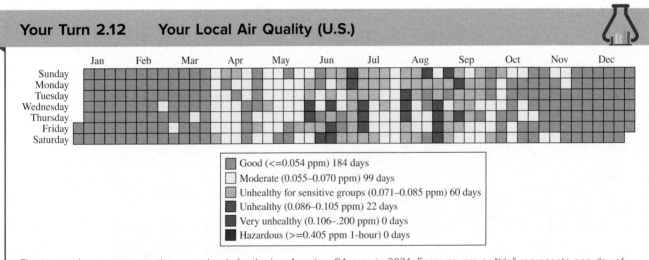

Good (<=0.054 ppm) 184 days
Moderate (0.055–0.070 ppm) 99 days
Unhealthy for sensitive groups (0.071–0.085 ppm) 60 days
Unhealthy (0.086–0.105 ppm) 22 days
Very unhealthy (0.106–.200 ppm) 0 days
Hazardous (>=0.405 ppm 1-hour) 0 days

The image above represents the ozone levels for the Los Angeles, CA area in 2021. Every square or "tile" represents one day of the year and is color-coded based on the AQI for that day. Visit EPA's outdoor air quality tile plot site to see how your local concentrations of air pollutants compare and have changed over time.

Questions to Consider

a. How do the levels of your local CO, NO_2, ozone, PM_{10}, and $PM_{2.5}$ pollutants during the current calendar year compare to major cities in the U.S.?

b. What time of the year are ozone and PM levels highest? What are some reasons for the higher levels during these months?

c. Compare the levels of pollutants in your local area over the past five years. Have there been any improvements in any of the pollutants? If so, what are some reasons why air quality has improved.

Your Turn 2.13 Air Quality Around the World

Of course, air quality concerns are not isolated in the U.S. The interactive map at www.acis.org/cic shows the real-time air quality data around the world for PM$_{2.5}$ pollution.

Questions to Consider

a. What percentage of the world's population is currently experiencing air quality that exceeds the WHO annual PM$_{2.5}$ guideline? How does this compare to the percentage of population that exceeds the EPA guidelines for PM$_{2.5}$ pollution?

b. What age group is most affected by PM$_{2.5}$ pollution globally?

c. What countries are experiencing air quality that currently exceeds the WHO guidelines by over 10 times? What are some reasons why these countries have such high levels of PM$_{2.5}$ pollution?

d. Where are most of the air quality monitors located? (You may need to grab the map and scroll around to view the entire globe.) How might this distribution affect the reported data (and you answers to the questions above)?

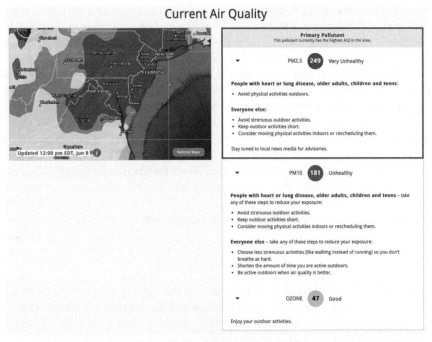

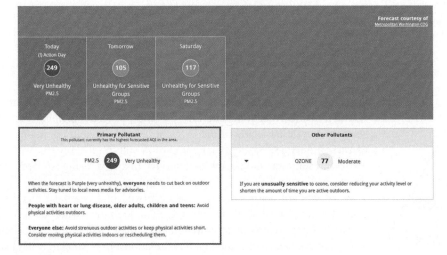

Figure 2.13

Air quality forecast for Washington, DC, June 8–10, 2023, using the colors as defined in Table 2.5.

AirNow/USEPA

2.9 | The Origin of Pollutants: Who's to Blame?

> *Learning Objective: Describe chemical reactions in words and write their balanced chemical equations*

Life on Earth bears the stamp of oxygen. Compounds containing oxygen occur in the atmosphere, in your body, and in the rocks and soils of the planet. Why? The answer is that many different elements combine chemically with oxygen. One such element is carbon. You were already introduced to the compound carbon monoxide, CO, a pollutant listed in Table 2.4. Fortunately, CO is relatively rare in our atmosphere. In contrast, carbon dioxide, CO_2, is far more abundant (0.04%, or 400 ppm). Even so, at this concentration, CO_2 plays an important role as a greenhouse gas. In this section, we explain how both CO_2 and CO are emitted into our atmosphere.

Many animals exhale CO_2 with each breath. Breathing is one natural source of CO_2 in our atmosphere, but it is not a significant cause of the recent increase in atmospheric levels. Carbon dioxide is also produced when humans burn fuels. **Combustion** is the chemical process of burning; that is, the rapid reaction of fuel with oxygen to release energy in the form of heat and light. When carbon-containing compounds burn, the carbon combines with oxygen to produce carbon dioxide (CO_2). When the oxygen supply is limited, carbon monoxide (CO) is likely to form as well.

Combustion is a type of **chemical reaction**, a process whereby substances described as **reactants** are transformed into different substances called **products**. A **chemical equation** is a representation of a chemical reaction using chemical formulas. Chemical equations are the sentences in the language of chemistry. They are made up of chemical symbols (corresponding to letters) that are often combined in the formulas of compounds (the words of chemistry). Like a sentence, a chemical equation conveys information—in this case, about the chemical change taking place. A chemical equation must also obey some of the same constraints that apply to a mathematical equation.

At the most fundamental level, a chemical equation is a qualitative description of this process:

$$Reactants \longrightarrow Products$$

By convention, the reactants are always written on the left and the products on the right. The arrow represents a chemical transformation and can be read as "converted to."

The combustion of carbon (charcoal) to produce carbon dioxide, as shown in Figure 2.14, can be represented in several ways. One is with chemical names:

$$carbon + oxygen \longrightarrow carbon\ dioxide$$

Or, using chemical formulas:

$$C + O_2 \longrightarrow CO_2 \qquad [2.1]$$

This compact symbolic statement conveys a good deal of information. It might sound something like this: "One atom of the element carbon reacts with one molecule of the element oxygen to yield one molecule of carbon dioxide." Using black for carbon and red for oxygen, we can also represent the molecules and atoms involved using spheres, as shown in Figure 2.14.

Atoms are neither created nor destroyed in a chemical reaction. The elements present do not change their identities when converted from reactants to products, although they may change the way their atoms are bonded to one another. This relationship is known as the **law of conservation of matter and mass**. As you will see in this next activity, the mass of the reactants consumed equals the mass of the products formed in a chemical reaction.

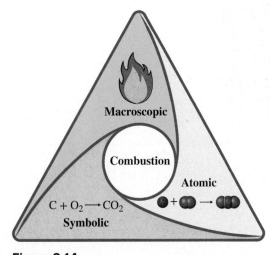

Figure 2.14

Three representations of charcoal burning in air.

©Bradley D. Fahlman

@HOME   Fun With Baking Soda!

Investigate the law of conservation of matter and mass at home using baking soda, vinegar, an empty clear plastic bottle, a latex balloon, and a kitchen scale.

SariSyno/Shutterstock

1. Add 2 inches of vinegar into the plastic bottle.
2. Spoon baking soda into the latex balloon. Carefully take the lip of the balloon and place it over the lip of the bottle. Do not spill the baking soda into the bottle yet.
3. Place the entire setup on the kitchen scale and record the mass.
4. Lift the balloon to drop the baking soda into the vinegar. Record your observations and the mass throughout the reaction.

Questions to Consider
a. What happened to the mass over time throughout the reaction?
b. What changed throughout the reaction and what remained the same?
c. What would happen to the mass if the balloon were not there?

Chemical equations follow this law and are similar to a mathematical expression in that the number and kind of each atom on the left side of the arrow *must* equal those on the right:

$$\text{Left side: 1 C and 2 O} \longrightarrow \text{Right side: 1 C and 2 O}$$

If they do not equal, adjustments may be made only to the amounts of chemicals in the reaction, *never* to the subscripts of the chemicals themselves. This would change the chemical compound to a new substance.

It is possible to pack even more information into a chemical equation by specifying the physical states of the reactants and products. A solid is designated by the italicized symbol *(s)*, a liquid by *(l)*, and a gas by *(g)*. Because carbon is a solid, and oxygen and carbon dioxide are gases at ordinary temperatures and pressures, Equation 2.1 may become:

$$C(s) + O_2(g) \longrightarrow CO_2(g)$$

Equation 2.1 describes the combustion of pure carbon in an ample supply of oxygen, but this is not always the case. If the oxygen supply is limited, carbon monoxide may be one of the products. Let's take the extreme case in which CO is the sole product:

$$C + O_2 \longrightarrow CO \text{ (unbalanced equation)} \qquad \textbf{[2.2a]}$$

This equation is not balanced because there are 2 oxygen atoms on the left, but only 1 on the right. The only way to fix this is to adjust the quantity of elements by adding coefficients. Remember, the subscript indicates the identity of the substance, and that can't change—CO_2 and CO are very different species. In cases like this, the coefficients can be found by trial and error. If a 2 is placed in front of CO, it signifies two molecules of carbon monoxide. This balances the oxygen atoms:

$$C + O_2 \longrightarrow 2 CO \text{ (still not balanced)} \qquad \textbf{[2.2b]}$$

But now the carbon atoms do not balance. Fortunately, this is easily corrected by placing a 2 in front of the carbon on the left side of the equation:

$$2 C + O_2 \longrightarrow 2 CO \text{ (balanced equation)} \qquad \textbf{[2.2c]}$$

Hence, when this reaction occurs, two carbon atoms react with one oxygen molecule to form two molecules of carbon monoxide.

By comparing Equations 2.1 and 2.2c, you can see that more O_2 is required to produce CO_2 from carbon than is needed to produce CO. This matches the conditions we stated for the formation of carbon monoxide; namely, that the supply of oxygen was limited. Consult Table 2.6 for some tips about balancing chemical equations.

Table 2.6	Characteristics of Chemical Equations
Always Conserved	
Identity of atoms in reactants = identity of atoms in products	
Number of atoms of each element in reactants = number of atoms of each element in products	
Mass of all reactants = mass of all products	
May Change	
Number of molecules in reactants may differ from the number in products	
Physical states (*s*, *l*, or *g*) of reactants may differ from those of products	

Your Turn 2.14   Practice with Balancing Equations

For additional practice with balancing equations, check out this interactive simulation: www.acs.org/cic.

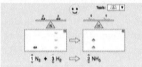

PhET Interactive Simulations, University of Colorado, http://phet.colorado.edu

You may be surprised to learn the origin of the air pollutant nitrogen monoxide. It comes from nitrogen and oxygen found in the air! Both oxygen and nitrogen are safe substances, but can react to create something quite dangerous. These two gases chemically combine in the presence of something very hot, such as an automobile engine or a forest fire:

$$N_2(g) + O_2(g) \xrightarrow{\text{high temperature}} NO(g) \text{ (unbalanced equation)} \quad \textbf{[2.3a]}$$

The equation is not balanced, because 2 oxygen atoms are on the left side but only 1 is on the right. The same is true for nitrogen atoms. Balancing the equation requires placing a 2 in front of NO, which supplies 2 N and 2 O atoms on the right:

$$N_2(g) + O_2(g) \xrightarrow{\text{high temperature}} 2\,NO(g) \text{ (balanced equation)} \quad \textbf{[2.3b]}$$

Hence, we would describe this reaction as one molecule of nitrogen reacting with one molecule of oxygen to form two molecules of nitrogen monoxide.

To further understand the purpose of balancing equations, let's now look at some atomic-level representations of reactions. It is especially useful to visualize the formation of products from reactants, so one can determine if any reactant molecules are left over after the reaction has completed.

Your Turn 2.15   How Does Carbon React with Oxygen?

Examine the two representations of the reactants: solid carbon and oxygen gas. Note that carbon is illustrated as a region of the graphite allotrope, featuring hexagonal rings of carbon atoms. **Hint:** Revisit Figure 1.8.

a. Draw two or more pictures showing the reaction between oxygen molecules and solid carbon as it progresses to form (i) carbon dioxide and (ii) carbon monoxide products.

b. For each of the reactions you have drawn in part **a.**, how many product molecules may be formed from these reactants? Are there any reactant atoms or molecules left over after the reaction has gone to completion? Why or why not?

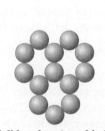

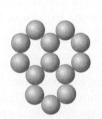

Solid carbon (graphite)   **Gaseous oxygen**

Your Turn 2.16   Modeling Sulfur Dioxide Formation

Consider the two representations of the reactants solid sulfur and oxygen gas shown below. Picture A is correct while picture B is incorrect. Note that sulfur is illustrated as the S_8 allotrope, a common elemental form of sulfur.

a. Describe the features represented in picture **A** that make it correct and the features in picture **B** that make it incorrect.

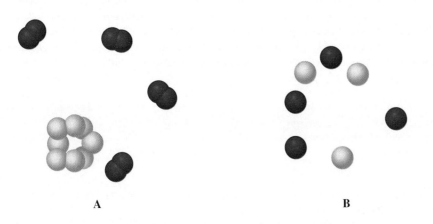

A   B

b. Predict the products (either SO_2 or SO_3) that are formed from the reactants in picture **A**. Draw a picture of the products, including the correct quantities.

c. Would all of the oxygen atoms react completely with all of the sulfur atoms? How did you arrive at your answer?

d. True or False: After the reaction goes to completion, unreacted sulfur atoms are left over. What evidence supports your answer?

As illustrated by the previous activities, the molecular representation of a reaction illustrates many reactant species that collide to form product molecules. The reactant that is fully consumed during the reaction is referred to as the **limiting reagent**, because it limits the number of product molecules that may be formed. As another more common illustration, consider the number of cheese sandwiches that may be made from 12 slices of bread and 12 cheese slices. Because six sandwiches could be made from these ingredients, we would say that bread was the limiting reagent, and cheese would be in excess (i.e., six cheese slices left over, if we didn't double up on cheese!).

@HOME   Limiting Reagents in the Kitchen

Investigate limiting reagents at home using a chocolate bar, large marshmallows, and graham crackers. For the purposes of this activity, a s'more consists of two graham cracker squares, one large marshmallow, and three squares of chocolate.

1. Build a s'more following the given definition of a s'more.
2. Build another s'more and keep repeating until you cannot build anymore.
3. Count the remaining ingredients.

Questions to Consider
a. Which ingredient was your limiting ingredient in making s'mores?
b. Which ingredient(s) did you have in excess?

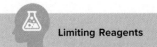

Limiting Reagents

Chemistry in Context
Tutorial

LIMITING REAGENTS

©Bradley D. Fahlman

Check out this video for more details about balancing chemical equations and limiting reagents: www.acs.org/cic.

2.10 | More Oxygen, Please: The Effect of Combustion on Air Quality

> *Learning Objective: Describe the process of complete combustion and its effect on air quality*

In the next two sections, we explore the connection between air quality and the fuels we burn. Hydrocarbons are compounds composed exclusively of hydrogen and carbon. The hydrocarbons we use today are primarily obtained from crude oil. Methane (CH_4), the simplest hydrocarbon, is the primary component of natural gas. Both gasoline and kerosene are mixtures of many long-chain hydrocarbons, as is candle wax.

@HOME   Combustion of Candles

Investigate complete combustion at home using a tea light candle, a lighter, and jar that can cover the candle.

1. Place the tea light candle on a flat surface and use the lighter to light the candle so that a flame is present.
2. Make observations as you watch the candle burn for one minute.
3. Place the jar over the top of the candle so that it is completely covered. Make observations for one minute.

Questions to Consider
a. Why did the two scenarios end with different observations?
b. What is the purpose of the jar in the second part of the investigation?

Given an ample supply of oxygen, hydrocarbon fuels are limiting reagents and therefore burn completely. You may hear this called "complete combustion." In essence, all of the carbon atoms in the hydrocarbon molecule combine with O_2 molecules from the air to form CO_2. Similarly, all the hydrogen atoms combine with O_2 to form H_2O. For example, below is the chemical equation for the complete combustion of methane. This equation is your first peek at why burning carbon-based fuels releases carbon dioxide into the atmosphere.

$$CH_4 + O_2 \longrightarrow CO_2 + H_2O \text{ (unbalanced equation)} \qquad \text{[2.4a]}$$

Note that O appears in *both* products: CO_2 and H_2O. To balance the equation, start with an element that appears in *only one substance* on each side of the arrow. In this case, both H and C qualify. No coefficients need to be changed for carbon, because both sides contain 1 C atom. Balance the H atoms by placing a 2 in front of the H_2O:

$$CH_4 + O_2 \longrightarrow CO_2 + 2 H_2O \text{ (still not balanced)} \qquad \text{[2.4b]}$$

Balance the oxygen atoms last. Four O atoms are on the right side and 2 O atoms are on the left, so we need 2 O_2 molecules to balance the equation:

$$CH_4 + 2 O_2 \longrightarrow CO_2 + 2 H_2O \text{ (balanced equation)} \qquad \text{[2.4c]}$$

In other words, two water molecules and one carbon dioxide molecule are formed by the reaction of one molecule of methane with two molecules of oxygen. A nice feature of chemical equations is that simply counting the number of each type of atom on both sides of the arrow tells you if it is balanced.

Most automobiles run on the complex mixture of hydrocarbons we call gasoline. Octane, C_8H_{18}, is one of the pure substances in this mixture. With sufficient oxygen, octane burns completely to form carbon dioxide and water:

$$2 C_8H_{18} + 25 O_2 \longrightarrow 16 CO_2 + 18 H_2O \qquad \text{[2.5]}$$

Figure 2.15

A winter ice fog in Fairbanks, Alaska.

©Cathy Middlecamp

Both products travel from the engine out the exhaust pipe and into the air. Are these combustion products visible? Usually not. Water, in its gaseous form, and carbon dioxide are both colorless gases. But if you happen to be outside on a winter day, the water vapor condenses to form clouds of tiny ice crystals that you can see. Occasionally, the frozen vapor gets trapped in an inversion layer and forms an ice fog (Figure 2.15).

With less oxygen, the hydrocarbon mixture we call gasoline burns incompletely ("incomplete combustion"). Water is still produced together with both CO_2 and CO. The extreme case occurs when only carbon monoxide is formed, as is shown here for the incomplete combustion of octane:

$$2\ C_8H_{18} + 17\ O_2 \longrightarrow 16\ CO + 18\ H_2O \qquad\qquad [2.6]$$

Compare the coefficient of 17 for O_2 in Equation 2.6 with that of 25 for O_2 in Equation 2.5. Less oxygen is needed for incomplete combustion, because CO contains less oxygen than CO_2.

Your Turn 2.17   Is It Balanced?

Demonstrate that Equations 2.5 and 2.6 are balanced by counting the number of atoms of each element on both sides of the arrow.

What is the actual mixture of products formed when gasoline is burned in your car? This is not a simple question, because the products vary with the fuel, the engine, and its operating conditions. It is safe to say that gasoline burns primarily to form H_2O and CO_2. However, some CO is also produced. The amounts of CO and CO_2 that go out the tailpipe indicate how efficiently the car burns the fuel, which in turn indicates how well the engine is tuned. Some regions of the United States monitor auto emissions with a probe that detects CO. The CO concentrations in the exhaust are compared with established standards—for example, currently 4.2 grams per mile for passenger cars in the state of California. If the vehicle fails the emissions test, it must be serviced to meet at least the minimum emission standards.

Your Turn 2.18   Auto Emissions Report

Evaluate the U.S. auto emissions report shown below. The blue line shows the change in engine speed; the red line shows the change in emissions.

Second-By-Second Emissions Report

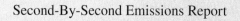

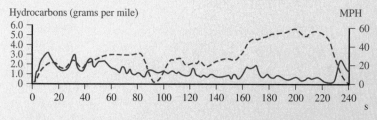

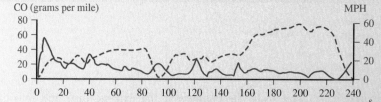

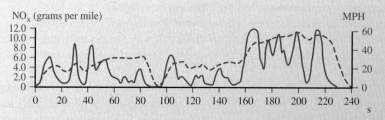

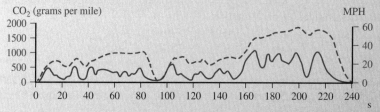

a. The image above reports NO_x emissions in grams per mile. NO_x is a way to collectively represent the oxides of nitrogen. If x = 1 and x = 2, write the corresponding chemical formulas. Also give the chemical names.

b. NO is the primary oxide of nitrogen emitted. What is the source of this compound? *Hint:* Revisit Equation 2.3.

c. Research the emission standards for hydrocarbons, CO, NO_x, and carbon dioxide using the Internet. Would the vehicle shown above pass the current emission standards?

2.11 | Air Pollutants: Direct Sources

Learning Objective: Describe direct sources and formation reactions of air pollutants

In this section, we examine two major sources of air pollutants: coal-fired power plants that generate electricity and motor vehicles. These sources emit a variety of pollutants, which you will now start to investigate.

Sulfur dioxide emissions are linked to the coal that is burned to generate electric power. Although coal consists mostly of carbon, it may contain 1–3% sulfur together with small amounts of minerals. The sulfur burns to form SO_2, and the minerals end up as fine ash particles. If not contained, the SO_2 and ash go right up the smokestack. The millions of tons of coal burned translate into millions of tons of waste in the air. As we will see in Chapter 5, the SO_2 produced by burning coal can dissolve in the water droplets of clouds and fall to the ground as acid rain.

However, the story does not end with SO_2 emissions. Once in the air, sulfur dioxide can react with oxygen to form sulfur trioxide, SO_3:

$$2\ SO_2 + O_2 \longrightarrow 2\ SO_3 \qquad \text{[2.7]}$$

Although normally quite slow, this reaction is faster in the presence of small ash particles. The particles also aid another process. If the humidity is high enough, they help condense water vapor into an aerosol of tiny water droplets. **Aerosols** are liquid and solid particles that remain suspended in the air rather than settling out. Smoke, such as from a campfire or a cigarette, is a familiar aerosol made up of tiny particles of solids and liquids.

The aerosol of concern here is made up of tiny droplets of sulfuric acid, H_2SO_4. It forms because sulfur trioxide reacts readily with water droplets to produce sulfuric acid:

$$H_2O + SO_3 \longrightarrow H_2SO_4 \qquad \text{[2.8]}$$

If inhaled, the droplets of the sulfuric acid aerosol are small enough to become trapped in the lung tissue and cause severe damage.

The good news? Sulfur dioxide emissions in the United States are declining (Figure 2.12). For example, in 1985, approximately 20 million tons of SO_2 was emitted from the burning of coal. Today, the value is closer to 11 million tons. This impressive decrease can be credited to the Clean Air Act of 1970 that mandated many reductions, including those from coal-fired electric power plants. However, combustion of coal is not the only source of sulfur dioxide, as you will discover in this next activity.

Your Turn 2.19 SO_2 from the Mining Industry

Burning coal is not the only source of sulfur dioxide. As you saw in Your Turn 2.10, smelting is another. Silver and copper metal can be produced from their sulfide ores. Write the balanced chemical equations for the following reactions:

a. Silver sulfide (Ag_2S) is heated in air to produce silver and sulfur dioxide.
b. Copper sulfide (CuS) is heated in air to produce copper and sulfur dioxide.
Hint: The ores react with oxygen in air at elevated temperatures.

More stringent regulations were established in the Clean Air Act Amendments and the Pollution Prevention Act of 1990. For example, diesel fuel and gasoline both once contained small amounts of sulfur, but the allowable amounts were drastically lowered in 2006 and in 2017, respectively. In the U.S. and Canada, diesel fuel with <15 ppm sulfur content is currently required for all on-road vehicles. However, the European Union and countries such as Australia and China have adopted more stringent regulations for diesel fuel, with an allowable sulfur content of <10 ppm.

With more than 290 million vehicles, the United States has more vehicles per capita than many other nations. Do these vehicles emit sulfur dioxide? Fortunately, the answer is no since cars have internal combustion engines primarily fueled by gasoline. We already mentioned that the combustion of hydrocarbons in gasoline produces—at best—carbon dioxide and water vapor (Equation 2.5). Since gasoline contains little or no sulfur, its combustion produces little or no sulfur dioxide. Nonetheless, each tailpipe puffs out its share of air pollutants. In addition to carbon dioxide and water, combustion of fuels in automobiles, as well as a variety of industrial processes and power plants, add to the atmospheric concentrations of carbon monoxide, volatile organic compounds, nitrogen oxides, and particulate matter (Figure 2.16). We will now discuss each of these in turn.

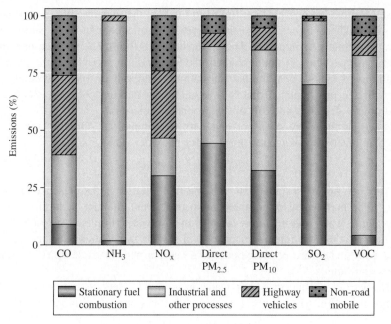

Figure 2.16

Emissions by source. Stationary fuel combustion refers to electrical utilities and industrial boilers. Industrial and other processes refer to industries such as metal smelters, petroleum refineries, cement kilns and dry cleaners. Non-road mobile refers to sources such as recreational and construction equipment, marine vessels, aircraft and locomotives.

Source: EPA

Carbon monoxide pollution comes primarily from automobiles. But think in terms of all the tailpipes out there, not just those attached to cars. Some are attached to heavy trucks, SUVs, and the "three m's": motorcycles, minibikes, and mopeds. Others are on equipment such as farm tractors, bulldozers, motor boats, and lawn mowers. The tailpipes attached to all gasoline and diesel engines emit carbon monoxide.

Cars emit carbon not only in the form of carbon monoxide but also in the form of unburned and partially burned hydrocarbons. This leads us to the topic of VOCs, **volatile organic compounds**. A *volatile* substance readily passes into the vapor phase; that is, it evaporates easily. Gasoline and nail polish remover are both volatile. If you were to spill either of these, the puddle would soon evaporate. One way to think about this: if you can smell it, it is volatile; however, not all VOCs have an odor. An **organic compound** always contains carbon, almost always contains hydrogen, and may contain other elements such as oxygen and nitrogen.

Volatile organic compounds originate from a variety of sources. For example, you can smell naturally occurring VOCs in a spruce or pine forest. VOCs from tailpipes are not so pleasant, because they are vapors of incompletely burned gasoline molecules or fragments of these molecules. The exhaust gas still contains oxygen, as not all of it is consumed in the engine. Section 2.12 will describe the connection between VOCs and ozone formation. However, right now, we want to connect VOCs with the formation of NO_2, which is formed by more complex pathways.

Nitrogen dioxide is brown in color, giving photochemical smog its characteristic brownish tinge. Recall that N_2 and O_2 combine to produce NO, which is a colorless gas (Equation 2.3). But what is the origin of NO_2? Here is a balanced equation that appears to be a likely candidate, which proceeds most rapidly in the presence of VOCs:

$$2 NO + O_2 \longrightarrow 2 NO_2 \qquad \text{[2.9]}$$

Equation 2.10 shows a reaction that predominates in urban settings, where you are likely to find NO. In some cities, this actually lowers the ground-level ozone concentrations along highways congested with vehicles emitting NO:

$$NO + O_3 \longrightarrow NO_2 + O_2 \qquad \text{[2.10]}$$

To further complicate things, on a sunny day, some of the NO_2 converts back into NO. In the presence of VOCs, this results in a net production of ozone.

Based on measurements by the EPA at over 250 sites in the United States, the average CO concentration has decreased almost 60% since 1980. If wildfires are excluded, today's levels are the lowest reported in three decades. The decrease is due to several factors, including improved engine design, computerized sensors that better adjust the fuel–oxygen mixture, and most importantly, the requirement that all cars manufactured since the mid-1970s have catalytic converters (Figure 2.17). Catalytic converters reduce the amount of carbon monoxide in the exhaust stream by catalyzing the combustion of CO to CO_2. They also lower NO_x emissions by catalyzing the conversion of nitrogen oxides back to N_2 and O_2, the two atmospheric gases that formed them. In general, a **catalyst** is a chemical substance that participates in a chemical

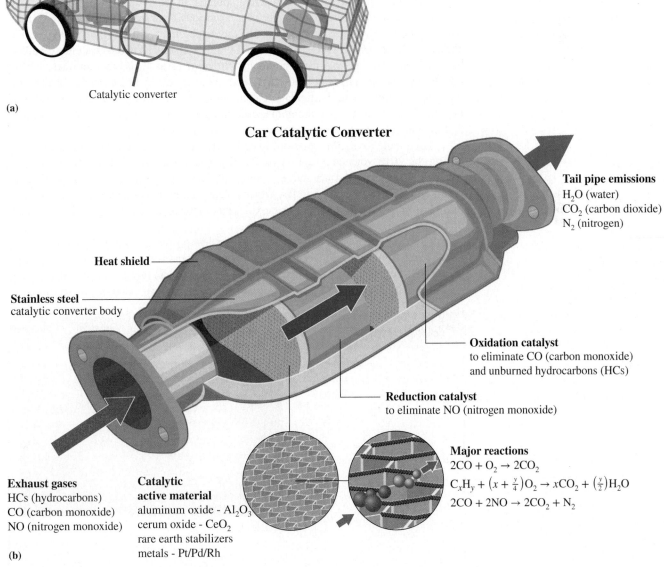

Car Catalytic Converter

(a)

Catalytic converter

Tail pipe emissions
H_2O (water)
CO_2 (carbon dioxide)
N_2 (nitrogen)

Heat shield

Stainless steel
catalytic converter body

Oxidation catalyst
to eliminate CO (carbon monoxide)
and unburned hydrocarbons (HCs)

Reduction catalyst
to eliminate NO (nitrogen monoxide)

Exhaust gases
HCs (hydrocarbons)
CO (carbon monoxide)
NO (nitrogen monoxide)

**Catalytic
active material**
aluminum oxide - Al_2O_3
cerum oxide - CeO_2
rare earth stabilizers
metals - Pt/Pd/Rh

Major reactions
$2CO + O_2 \rightarrow 2CO_2$
$C_xH_y + \left(x + \frac{y}{4}\right)O_2 \rightarrow xCO_2 + \left(\frac{y}{2}\right)H_2O$
$2CO + 2NO \rightarrow 2CO_2 + N_2$

(b)

Figure 2.17

(a) Location of a catalytic converter in a car. **(b)** Cutaway view of a catalytic converter. Metals such as platinum and rhodium serve as catalysts and form a coating on the surface of ceramic beads.

(a) https://oards.com/what-is-inside-of-a-catalytic-converter/; (b) lukaves/123RF

reaction and influences its rate, without itself undergoing permanent change. Catalytic converters typically use expensive metals such as platinum and rhodium as catalysts.

Your Turn 2.20   Save Money with the Way You Drive

Burning less gasoline equates to fewer tailpipe emissions. Which driving practices conserve fuel? Which practices expend it more than necessary? Think about the behavior of motorists on highways, on city streets, and in parking lots. For each of these venues, list at least three ways that drivers could burn less gasoline.

Hint: Consider how you accelerate, coast, idle, brake, and park.

Often present alongside other gaseous pollutants, particulate matter comes in a range of sizes. However, only tiny particles (PM_{10} and $PM_{2.5}$) are regulated as pollutants. Particles of this size can penetrate deeply into your lungs, pass into your bloodstream, and inflame your cardiovascular system. In terms of regulation, particles are the new pollutant on the block. Data collection in the United States for PM_{10} and $PM_{2.5}$ started in 1990 and 1999, respectively.

Particulate matter has many different sources. In the summer, wildfires may raise the concentration of particulate matter to a hazardous level (revisit Figure 2.13). In the winter, wood stoves may produce exactly the same effect. At any time of the year in almost any urban environment, older diesel engines on trucks and buses emit clouds of black smoke. This is also true for tractors in rural areas. Construction sites, mining operations, and the unpaved roads that serve them, also loft tiny particles of dust and dirt into the atmosphere. Particulate matter can even form within the atmosphere itself. For example, the compound ammonia, used in agriculture, is a major player in forming ammonium sulfate and ammonium nitrate in the air, both $PM_{2.5}$.

Even more toxic PM emissions may also develop naturally, such as those observed during the 2018 volcanic eruption on the Big Island of Hawaii. The mix of erupting lava at 2000 °F and seawater sent up plumes of "laze"—a mixture of hydrochloric acid, steam, and volcanic glass particles. Other volcanic emissions denoted as *vog* consisted of a harmful combination of sulfur dioxide, dust, moisture, and other small particulates that could penetrate deep into lung tissue.

Given all these sources, particulate matter has proven a tough pollutant to control. Even so, the EPA reported a decrease of 35% in the annual $PM_{2.5}$ concentrations from 2000 to 2014. However, some of the monitored sites still showed an increase in particle pollution. As this next activity illustrates, what you breathe depends very much on where you live.

Your Turn 2.21   Particles Where You Live

Shown is a map of the continental U.S. and Canada that shows $PM_{2.5}$ data for Oct. 9, 2022.

Source: EPA

a. In terms of air quality, what do the green, yellow, orange, and red colors indicate?

b. Which groups of people are most sensitive to particulate matter?

c. Visit https://www.lung.org/research/sota, a website posted by the American Lung Association. How many days a year does your state have "orange," "red," and "purple" days for ozone and particulate pollution? In your county, calculate the percentages of the various at-risk groups for air-related health concerns.

d. Using the interactive map, select three regions of the U.S., and summarize their composition trends for $PM_{2.5}$ pollution. What are some possible sources for these particulates, and why do their relative concentration profiles vary by quarter?

2.12 | Ozone: A Secondary Pollutant

Learning Objective: Describe the formation of tropospheric ozone

Ozone is a bad actor in the troposphere. Even at very low concentrations, it reduces lung function in healthy people who are exercising outdoors. Ozone also damages crops and the leaves of trees. But ozone does not come out of a tailpipe, and is not directly produced when coal is burned. How is it formed? Before we fill you in on the details, let's look at some ozone trends in the southeastern United States.

Your Turn 2.22   Ozone Around the Clock

Ozone concentrations vary during the day, as shown in Figure 2.18.

a. Near which cities is the air hazardous to one or more groups?
b. At about what time does the ozone level peak?
c. Can moderate levels (shown in yellow) of ozone exist in the absence of sunlight? Assume sunrise occurs around 6 AM and sunset about 8 PM.

This activity raises several related questions. Why is ozone more prevalent in some areas than others? What role does sunlight play in ozone production?

Unlike the pollutants described in Section 2.11, ozone is a **secondary pollutant**. It is produced from chemical reactions involving two directly-emitted pollutants: VOCs and NO_2. Recall that NO, rather than NO_2, comes directly out of a tailpipe (or a smokestack). But, over time and in the presence of VOCs, the NO in the atmosphere is converted to NO_2 (Equation 2.10).

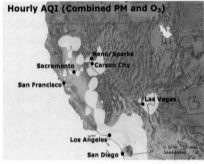

5:00 AM

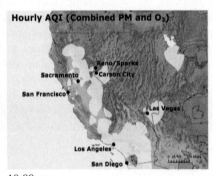

10:00 AM

1:00 PM

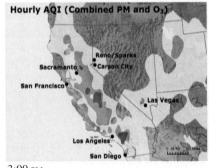

3:00 PM

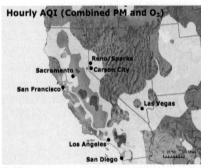

5:00 PM

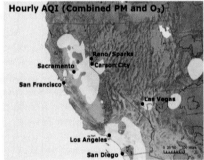

11:40 PM

Figure 2.18

A daily fluctuation in the air quality of California using the colors as defined in Table 2.5.

EPA AirNow

Nitrogen dioxide meets several fates in the atmosphere. The one of most interest to us occurs in the presence of sunlight. The energy provided by sunlight breaks one of the bonds in the NO_2 molecule:

$$NO_2 \xrightarrow{\text{Sunlight}} NO + O \qquad\qquad [2.11]$$

The oxygen atoms produced can react with oxygen molecules to produce ozone:

$$O + O_2 \longrightarrow O_3 \qquad\qquad [2.12]$$

This explains why ozone formation requires sunlight. Sunlight first splits NO_2 to release O atoms. These in turn react with O_2 to form O_3. Thus, once the Sun goes down, the ozone concentrations drop off sharply (Figure 2.18). What happened to the ozone? In just a matter of hours, the ozone molecules react with many things, including animal and plant tissue.

Note that Equation 2.12 contains three different forms of elemental oxygen: O, O_2, and O_3. All three are found in nature, but O_2 is the least reactive and by far the most abundant, constituting about one-fifth of the air we breathe. As you will see in the next chapter, our atmosphere naturally contains tiny amounts of protective ozone in the stratosphere.

Your Turn 2.23   Ozone Summary

Summarize what you have learned about ozone formation by developing your own way to arrange these chemicals sequentially and in relation to one another: O, O_2, O_3, VOCs, NO, and NO_2. Chemicals may appear as many times as you like. You also may wish to include sunlight.

Because sunlight is involved in ozone formation, the concentration of ground-level ozone varies with weather, season, and latitude. High levels of O_3 are much more likely to occur on long, sunny summer days—especially in congested urban areas. Stagnant air also favors the buildup of air pollution. For example, revisit Your Turn 2.11 to explore the air quality data for various cities in the U.S. Ozone was usually the culprit responsible for pollution in cities with sunny climates. In contrast, windy and rainy cities usually have lower levels of ozone.

Your Turn 2.24   Ozone and You

The AirNow website, courtesy of the EPA, provides a wealth of information about ground-level ozone levels in the United States.

a. Let's say that the ozone level is "orange," actually a common occurrence in many U.S. cities during the summer months. Does air of this quality affect you if you have no health concerns but are actively exercising outdoors?
b. How does the air quality in your state compare with others?

Although the Air Quality Index maps often stop at the U.S. border due to the lack of international data, pollution knows no boundaries (Figure 2.19)! Pollutants from the U.S. are easily transported northward into Canada, and vice versa. This is an example of **the tragedy of the commons**. The tragedy arises when a resource is common to all and used by many, but has no one in particular who is responsible for it. As a result, the resource may be destroyed by overuse to the detriment of all who use it. For example, we cannot lay individual claim to the air—it belongs to all of us. If the air we breathe has waste dumped into it, this leads to an unhealthy situation for everyone. Individuals whose activities have little or no effect on the air still suffer the same consequences as those who pollute. The costs are shared by all. In later chapters, we will see other examples of the tragedy of the commons that relate to water, energy, and food.

Pollution Monitoring

©Bradley D. Fahlman

For a hands-on investigation of air pollution, check out this laboratory demo video: www.acs.org/cic.

Questions to Consider
Use the Internet as a resource to answer these questions.

a. What is the purpose of cornstarch and potassium iodide (KI)?
b. Why should the detector strip not be placed in direct sunlight?
c. What are some factors that will affect your results?

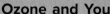

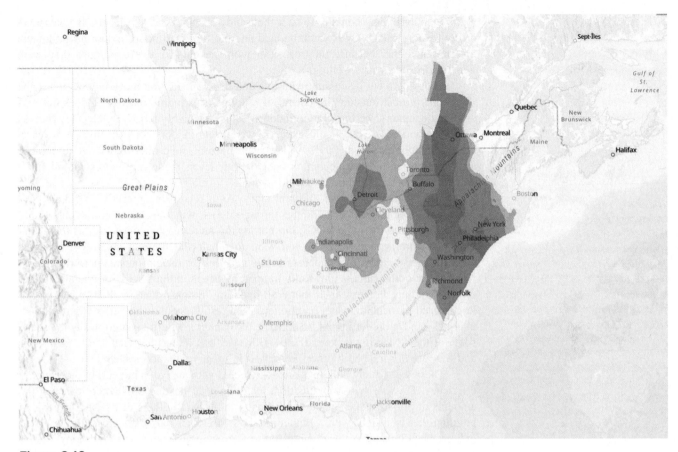

Figure 2.19

Map of PM$_{2.5}$ pollution originating from the Canadian wildfires on June 7, 2023, using the colors as defined in Table 2.5.

Source: EPA AirNow

Ozone pollution, once primarily a local concern, is now a serious international issue. Many cities worldwide have high ozone levels. Couple motor vehicles with a sunny location anywhere on the planet, and you are likely to find unacceptable levels of ozone. Some places, however, are worse than others. London with its cool, foggy days typically has low ozone levels. In contrast, ozone is a serious problem in Mexico City.

It is quite ironic that ozone attacks rubber, thereby damaging the tires of the vehicles that led to its production in the first place! Should you park your car indoors in the garage to minimize possible rubber damage? Should you park yourself indoors if the levels of ozone outside are unhealthy? The next section speaks to the quality of *indoor* air.

2.13 | Are We Really Safe from Polluted Air by Staying Indoors?

Learning Objective: Define and identify sources and health impacts of some indoor pollutants

In *The Wizard of Oz,* Dorothy hugged her dog Toto and exclaimed, "There's no place like home!" She was right, but when it comes to air quality, your home may not always be the best place to be. In fact, the levels of air pollution indoors may far exceed those existing outside. Given that most of us spend over 90% of our time indoors, we should learn about the air that we breathe at home, work, or inside our vehicle.

Indoor air may contain up to a thousand substances at low levels. If you are in a room where somebody is smoking, add another thousand or so. Indoor air pollutants are present either because they came in with the outside air, or because they were generated inside your dwelling.

Let's begin our discussion with this question: Should you move indoors to escape the ozone present outside? In general, if a pollutant is highly reactive, it does not persist long enough to be transported indoors. Thus, for highly reactive molecules such as O_3, NO_2, and SO_2, you should expect lower levels indoors. Indeed, this is the case, because indoor air is typically 10–30% lower in ozone concentration relative to outdoor air. Similarly, sulfur dioxide and nitrogen dioxide levels are lower indoors, although the decrease is not as dramatic as that for ozone.

Carbon monoxide is a different story. As a relatively unreactive pollutant, CO moves freely in and out of buildings through doors, windows, or a ventilation system. The same is true for some VOCs, but not for the more reactive ones such as those that give pine forests their scent. If you want to inhale the fragrant volatile compounds emitted from the bark of Ponderosa pines, it is best to remain outside near the trees.

Some pollutants are physically trapped by filters in the heating or cooling system of the building. For example, many air-handling systems contain filters that remove larger-sized particulate matter and pollen. As a result, people who suffer from seasonal allergies can find relief indoors. Similarly, those near a wildfire can go inside to escape some of the irritating smoke particles. However, gas molecules such as O_3, CO, NO_2, and SO_2 are *not* trapped by the filters used in most ventilation systems.

These days, many buildings are constructed with an eye to increasing their energy efficiency. This is a win–win situation, lowering your heating bills while lowering the amounts of pollutants generated in producing the heat. But there can be a downside. A building that is airtight with a limited intake of fresh air may have unhealthy levels of indoor air pollutants. We need circulation of air to dilute any indoor emissions. Therefore, what appeared initially to be a benefit can turn into a higher risk. In some cases, poor ventilation can cause indoor pollutants to reach hazardous levels, creating a condition known as "sick building syndrome." Clearly, this is an undesirable outcome. Today, architects and builders are finding ways to make buildings more energy efficient, while still maintaining an effective air exchange.

Even with good ventilation, indoor activities can compromise air quality. For example, tobacco smoke is a serious indoor air pollutant, containing thousands of chemical substances. Nicotine is one that you may recognize; others include benzene and formaldehyde. Taken as a whole, tobacco smoke is **carcinogenic**, meaning it's capable of causing cancer.

Combustion of carbon-containing materials also can generate carbon monoxide and nitrogen oxides. For example, carbon monoxide from cigarette smoking in bars can reach 50 ppm, a value well within the unhealthy range. In cigarette smoke, NO_2 levels can exceed 50 ppb. Fortunately, smokers take puffs rather than constantly breathing cigarette smoke.

Your Turn 2.25   Indoor Cigar Party

In 2007, student researchers attended a cigar party and trade show in Times Square, New York. They carried concealed detectors and found levels of 1943 µg of particulate matter per cubic meter of air inside the ballroom.

xpixel/Shutterstock

a. The news article did not report whether the particulate matter was $PM_{2.5}$ or PM_{10}. For either, does the value they measured exceed the air quality standards in the United States?

Hint: Revisit Table 2.4.

b. Assume that the students were measuring $PM_{2.5}$. What are the health implications?

Let's Debate!   Secondhand Smoke From E-Cigarettes

Form two groups: one group will be in favor of the topic (the proposition) and the other group will be against the topic (the opposition). Both groups will try to persuade a neutral person or judge to agree with them. The topic of the debate is the motion. Here is the motion:

Secondhand smoke from e-cigarettes is safe and presents no danger to human health or the environment.

Curiosity Conversations

- Did you find any marketing advertisements that were in opposition to your stance?
- How would your argument(s) change for traditional cigarettes instead of e-cigarettes?

People also burn candles, perhaps to soften the lighting or set a mood. However, candles deplete the oxygen in a room. They also can produce soot, carbon monoxide, and VOCs. Similarly, people may burn incense in their homes. Researchers who studied incense burning in churches in Europe found that the pollutants in smoke from incense and candles may be more toxic than the emissions originating from vehicle engines.

Burning candles or incense can generate fumes more rapidly than they can be removed by your ventilation system or by the breezes that pass through open windows. The next activity gives you the opportunity to further investigate sources of indoor pollutants.

Your Turn 2.26   Indoor Activities

Name 10 activities that add pollutants or VOCs to indoor air. To get you started, two are pictured in Figure 2.20. Remember that some pollutants have no detectable odors.

Figure 2.20

Examples of activities that can cause indoor air pollution.

Source: (left): Image Source/Getty Images; *(right):* pbombaert/Getty Images

Figure 2.21

A home radon test kit.

Keith Eng, 2008

Figure 2.22

A cookstove used for cooking and/or heat also gives off significant air pollution.

Joerg Boethling/Alamy Stock Photo

One naturally occurring air pollutant of concern is radon, a noble gas (Group 18). Radon is a special case of indoor air pollution because it occurs naturally in tiny amounts and is not usually a problem. However, it may reach hazardous levels in basements, mines, and caves. Like all noble gases, radon is colorless, odorless, tasteless, and chemically unreactive. But unlike the others, it is radioactive. Radon is generated in the decay series of uranium, another naturally occurring radioactive element. Because uranium occurs at a concentration of about 4 ppm in all the rocks of our planet, radon is ubiquitous. Depending on how your apartment or residence hall is constructed, the radon produced from uranium-containing rocks may find entry into the basement. Radon is the second-leading cause of lung cancer worldwide behind tobacco smoke. As is the case for other pollutants, the threshold for danger can be estimated but is not precisely known. Radon test kits, such as the one shown in Figure 2.21, are used to measure the radon concentration in living spaces.

The examples of indoor air pollution mentioned in this section already are of concern around the world. However, around 3 billion people on the planet still cook with solid fuels (wood, crop waste, animal dung, or coal) in their homes over open fires and simple stoves (Figure 2.22). Increased concentrations of particulate matter cause a range of health effects, including excess pneumonia, strokes, heart disease, lung cancer, and more. Indoors or out, we need to breathe healthy air; with each breath, we inhale a truly prodigious number of molecules and atoms.

Your Turn 2.27   Deadly Indoor Air Pollution

The interactive figures at www.acs.org/cic show the death rates for indoor pollution between 1990 and 2019.

Questions to Consider

a. In which countries has the death rate from indoor pollution decreased the most since 1990? What are some reasons why the death rates have decreased so dramatically in these countries?

b. What age groups have seen the greatest decrease in deaths due to indoor air pollution since 1990? What are some reasons for this decline?

Armed with knowledge of common air pollutants found both outside and indoors, you will now put on your stethoscope and lab coat to diagnose health-related illnesses due to air pollution.

Your Turn 2.28 **Is There a Doctor in the House?**

You will assume the role of a medical doctor, to use your knowledge of air quality to advise patients. With a partner or group, discuss the following symptoms and provide a probable cause(s) and preventive measures that could have been taken to prevent or decrease the likelihood of these problems.

Evgeniia Siiankovskaia/ Getty Images

a. Patient: John Reyes, 62 years old. You have confirmed the presence of a precancerous growth in his lungs, but he has never smoked or spent much time around others who smoked. His health has been quite good over the years, so he is very surprised by the diagnosis. John worked as a tollbooth operator for the past 30 years, but has now retired. His wife, Laura, is pleased because she can now spend more time with John and doesn't have to worry about washing his dirty work shirts any longer. John stays active by jogging in the morning (between 8 and 9 AM) alongside route I-94, a busy freeway near the city center.

b. Patient: Betty Raymond, 32 years old. She has suffered from asthma in the past, but has been getting worse recently. She often feels shortness of breath and has experienced tightness in her chest. Her eyes frequently burn, which is somewhat alleviated by using over-the-counter eyedrops. About three months ago, she changed her shift at the hospital from working nights to the 7 AM–3 PM shift. After work, she walks 2–3 miles in a park located at the center of the city.

c. Patient: Herbert Moore, 19 years old. This avid hockey player has been complaining of headaches and dizziness lately. Over the past few months, he has been having slower reflexes and often has a hard time staying awake during games. Herbert doesn't smoke and exercises by lifting weights at his home. Smoking was allowed at the former hockey arena, which had a better ventilation system, but there is no smoking allowed inside their new venue due to its small size and weaker ventilation system. The coach is puzzled, because Herbert used to be his star player. The players' locker room is located next to the ice surface, beside the room that houses the Zamboni used to resurface the ice between periods.

d. Patient: Jill McIntyre, 53 years old. Although her health has been fine for many years, she now suffers from high blood pressure and joint/muscle pain. Jill often has headaches and has difficulty remembering details at her job (as a realtor). For the past year, she has been working hard restoring an older home (built in 1968) to its former glory. She eats a healthy diet and regularly exercises at an indoor gym three times a week.

2.14 | Is There a Sustainable Way Forward?

> *Learning Objective: Explain the role of "green chemistry" practices in reducing pollution*

Not unlike the sustainable use of materials discussed in Chapter 1, air pollution provides another context in which we can discuss **sustainability**—making decisions with a concern not only for today's outcomes, but also for the needs of future generations. It makes sense to avoid actions that produce pollutants that can compromise our health and well-being. This is the logic behind the Pollution Prevention Act of 1990, legislation that calls for preventing pollution, rather than cleaning pollution up after it is produced. Likewise, the E.U. has developed a clean air policy that consists of stringent air quality standards as it works toward a zero pollution vision for 2050.

Unfortunately, there is an unequal burden of health concerns related to air pollution. Individuals from marginalized communities are most negatively affected. Residents in these communities tend to live closer to air pollution sources such as major highways, factories, industrial facilities, and ports. They may be more susceptible to air pollution due to social and economic factors such as historic housing policy, access to healthcare, etc.

In low income countries, air pollution is often one of the leading risk factors for death and it has a large impact on the quality of living. People in low-income countries especially suffer from poor indoor air quality due to burning solid fuels for heating and cooking. As countries climb the economic ladder, their indoor air quality improves, but their outdoor air quality gets worse with increased industrialization. Rich countries, on the other hand, have the lowest levels of both indoor and outdoor air pollution. Global sustainability must include environmental justice so that there is no longer a disproportionate environmental burden carried by only some people.

Your Turn 2.29   Air Quality Related Deaths

The interactive figures at www.acs.org show how deaths related to air quality have changed over time by risk factor and location.

Questions to Consider

a. How does air pollution (including both indoor and outdoor sources) rank in terms of risk factors for death? Has this ranking changed over time?

b. How have the trends in deaths from indoor and outdoor pollution changed? What other risk factors have trends that are closest to air pollution.

c. Which countries currently dominate the death rate related to air pollution? Propose some reasons why these countries have the largest death rates. Which countries have improved their death rates due to air pollution in recent years? Why? (**Hint:** Think about changes in economies and government initiatives.)

The Pollution Prevention Act provided the impetus for **green chemistry**, a set of key ideas to guide all in the chemical community, including teachers and students. Green chemistry is "benign by design." It calls for designing chemical products and processes that reduce or eliminate the use or generation of hazardous substances. In the broadest sense, this entails the use of our principles and knowledge to make products and processes better for humans, economics, and the environment.

Begun under the EPA Design for the Environment Program, green chemistry reduces pollution through the design or redesign of chemical processes. The goal is to use less energy, create less waste, use fewer resources, and use renewable resources. Green chemistry is a tool for achieving sustainability, rather than an end in itself.

Innovative "green" chemical methods have already decreased or eliminated toxic substances used or created in chemical manufacturing processes. Some examples include:

- Plastics synthesized from renewable sources instead of typical fossil fuel-derived precursors
- Paints that contain fewer volatile organic compounds
- Cheaper and less wasteful ways to produce pharmaceuticals, pesticides, and consumer products, such as contact lenses and disposable diapers
- Limiting or eliminating the use of organic solvents for dry cleaning and electronics fabrication processes
- Removing arsenic from the touchscreens of portable electronics

Some of the research chemists and chemical engineers who developed these methods have received Presidential Green Chemistry Challenge Awards. Begun in 1995, these presidential-level awards recognize chemists for their innovations on behalf of a less polluted world.

This final video illustrates a real-world application of green chemistry principles to minimize the use of harmful organic solvents for a variety of chemical syntheses.

Real Talk   **Sustainable Development: Solvent-Free Reactions**

Check out this interview with Dr. James Mack, who describes his use of mechanochemistry (pulverization) for reactions instead of using organic solvents: www.acs.org/cic.

Questions to Consider

a. Which of the three pillars of sustainability and principles of green chemistry are addressed by Dr. Mack's research?

b. Using the Internet as a resource, what are some benefits and other applications for mechanochemistry?

James Mack

Conclusion

Nobody wants dirty air. It makes you sick, reduces the quality of your life, and may hasten your death. However, the problem is that many people have become so accustomed to breathing dirty air that they don't notice it. One concept describing this is **shifting baselines**. This refers to the idea that what people expect as "normal" on our planet has changed over time, especially with regard to ecosystems. Burning eyes and breathing disorders have become so common that we have forgotten that they once were not. We have become accustomed to living in *megacities*, urban areas with 10 million people or more, such as Beijing, Tokyo, New York City, Mexico City, and Mumbai. Pollutants such as wood smoke, car exhaust, and industrial emissions often are generated in populated areas, and thus they are concentrated in the troposphere around megacities.

Your knowledge of chemistry can lead you to make better choices to deal with these problems, both as an individual and in your local community. The air we breathe affects both our health and the health of the planet. Our atmosphere contains the essentials for life, including two elements (oxygen and nitrogen) and two compounds (water and carbon dioxide). Our very existence on this planet depends on having a large supply of relatively clean, unpolluted air.

But the air you breathe may be polluted with carbon monoxide, ozone, sulfur dioxide, and the oxides of nitrogen. Emergency room visits correlate with bad air quality. So do shortness of breath, scratchy throats, and stinging eyes. The pollutants that cause us harm are, for the most part, relatively simple chemical substances. They are largely produced as consequences of our dependence on coal for electricity production in power plants, gasoline in internal combustion engines, and the fuels we burn to heat and cook.

Over the past 30 years, government regulations, industry initiatives, and modern technology have reduced pollutant levels. Both catalytic converters on cars and emissions controls on smokestacks have been important players. But it makes more sense not to generate "people fumes" in the first place. Here is where green chemistry plays an important role. By designing new processes that do not produce air pollutants, we do not later have to clean them up.

Indoors or out, the oxygen-laden air we breathe is very close to the surface of Earth. However, Earth's atmosphere extends upward for considerable distance and contains other gases that also are essential for life on this planet. Chapters 3 and 4 will describe two of these: stratospheric ozone and carbon dioxide, respectively. We will see that our human footprints and "air prints" on planet Earth connect in surprising ways to both of these gases.

LEARNING OUTCOMES
The numbers in parentheses indicate the sections within the chapter where these outcomes were discussed.

Having studied this chapter, you should now be able to:

- describe and illustrate the process of respiration (2.1)
- categorize the regions of the atmosphere and compare the compositions of inhaled and exhaled air (2.2)
- describe the concentrations of substances in the air (2.3)
- use macroscopic, molecular, and symbolic representations to describe matter (2.4)
- write formulas and names for molecular compounds, including hydrocarbons (2.5)
- identify and explain the health effects of common air pollutants (2.6)
- assess the risk and toxicity of different air pollutants by exposure (2.7)

- analyze local air quality for air pollutants and make comparisons to national and global trends (2.8)
- describe chemical reactions in words and write their balanced chemical equations (2.9)
- describe the process of complete combustion and its effect on air quality (2.10)
- describe direct sources and formation reactions of air pollutants (2.11)
- describe the formation of tropospheric ozone (2.12)
- define and identify sources and potential health impacts of some indoor pollutants (2.13)
- explain the role of "green chemistry" practices in reducing pollution (2.14)

Questions

Emphasizing Essentials

1. **a.** Calculate the volume of air in liters that you might inhale (and exhale) while you are sleeping for 7.5 hours. Assume that each breath has a volume of about 0.5 L, and that you are breathing 10 times per minute.

   b. From this calculation, you can see that breathing exposes you to a large volume of air. Name five things you can do to improve the quality of the air you and others breathe.

2. Some of the gases found in the troposphere are Rn, CO_2, CO, O_2, Ar, and N_2.

   a. Rank them in order of their abundance in the troposphere.

   b. For which of these gases is it convenient to express concentration in parts per million?

   c. Which of these gases is/are currently regulated as an air pollutant where you live?

   d. Which of these gases is/are found in Group 18 of the periodic table, the noble gases?

3. Identify three sources of particulate matter found in air. Explain the difference between $PM_{2.5}$ and PM_{10} in terms of size and health effects.

4. **a.** The concentration of argon in air is approximately 0.934%. Express this value in ppm.

   b. The air exhaled from the lungs of a smoker has a concentration of 20–50 ppm CO. In contrast, air exhaled by nonsmokers is 0–2 ppm CO. Express each concentration as a percent.

   c. On a very humid day, the water vapor concentration in the air might be 8500 ppm. Express this as a percent.

   d. A sample of air taken from Antarctica was found to contain 8 ppm water vapor. Express this as a percent.

5. Gases found in the atmosphere in small amounts include Xe, N_2O, and CH_4.

   a. What information does each chemical formula convey about the number and types of atoms present?

   b. Write the names of these gases.

6. Air contains trace amounts of substances. Define trace amounts and give two examples of possible trace substances found in air.

7. If you had a sample of 500 particles of air, how many of these particles would be nitrogen, oxygen, and argon?

8. Count the atoms on both sides of the equation to demonstrate that these equations are balanced.

   a. $2\,C_3H_8(g) + 7\,O_2(g) \longrightarrow 6\,CO(g) + 8\,H_2O(l)$

   b. $2\,C_8H_{18}(g) + 25\,O_2(g) \longrightarrow 16\,CO_2(g) + 18\,H_2O(l)$

9. Consider this representation of the reaction between nitrogen and hydrogen to form ammonia (NH_3).

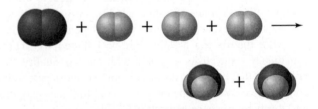

   a. Are the masses of reactants and products the same? Explain.

   b. Are the numbers of molecules of reactants and of products the same? Explain.

   c. Are the total number of atoms in the reactants and the total number of atoms in the products the same? Explain.

10. Write balanced chemical equations to represent these reactions.
    Hint: Nitrogen and oxygen are both diatomic molecules.

    a. Nitrogen reacts with oxygen to form nitrogen monoxide.

    b. Ozone decomposes into oxygen and atomic oxygen (O).

    c. Sulfur reacts with oxygen to form sulfur trioxide.

11. These questions relate to the combustion of hydrocarbons.

    a. LPG (liquid petroleum gas) is mostly propane, C_3H_8. Balance this equation.

    $$C_3H_8(g) + O_2(g) \longrightarrow CO_2(g) + H_2O(g)$$

    b. Cigarette lighters burn butane, C_4H_{10}. Write a balanced equation for the combustion of butane.

    c. With a limited supply of oxygen, both propane and butane can burn incompletely to form carbon monoxide. Write balanced equations for both reactions.

12. Balance the following equations in which ethane, C_2H_6, burns in oxygen.

    a. $C_2H_6(g) + O_2(g) \longrightarrow C(s) + H_2O(g)$

    b. $C_2H_6(g) + O_2(g) \longrightarrow CO(g) + H_2O(g)$

    c. $C_2H_6(g) + O_2(g) \longrightarrow CO_2(g) + H_2O(g)$

    d. Explain why the coefficients for oxygen vary, depending on whether C, CO, or CO_2 is formed.

13. Air is an example of a homogeneous mixture. Provide two more examples of homogeneous mixtures and justify their classification.

14. When a person inhales air, 21% of that air is oxygen. However, only 16% of the air exhaled is oxygen. Provide at least two reasons for why there is a decrease in the amount of oxygen exhaled.

15. What is the troposphere and why is it important for our survival?

16. Dry ice is solid carbon dioxide. Represent carbon dioxide in the following three ways: symbolic, particulate, and macroscopic.

17. Name the following nitrogen-containing compounds: NO_2, N_2O, NO, NCl_3, and N_2O_4.

18. Name the following compounds: CCl_4, SO_3, Cl_2O_6, and P_4S_3.

19. A carbon monoxide detector will go off if the concentration of CO is 400 ppm or greater for a period of 4–15 minutes.

    a. Express 400 ppm as a percent.

    b. Why is carbon monoxide considered to be an air pollutant?

    c. What are the health effects resulting from long-term exposure to carbon monoxide?

20. Sulfur dioxide and nitrogen dioxide are considered to be air pollutants.

    a. Where would you most likely find these pollutants?

    b. Which of these pollutants is more toxic?

    c. Express 0.045 ppm nitrogen dioxide in ppb.

Concentrating on Concepts

21. "Air prints" were mentioned in Your Turn 2.1. Examine these two photographs. The first picture is from a waterfront café on the Greek Island of Hydra, and the other is above a busy street in Tianjin, China. List three ways in which each photo shows the air print of humans.
    Hint: Some may not be visible but rather implied by the photograph.

(both): ©Bradley D. Fahlman

22. The EPA AirNow website states that "air quality directly affects our quality of life." Demonstrate the wisdom of this statement for two air pollutants of your choice.

23. In Your Turn 2.2, you calculated the volume of air exhaled in a day. How does this volume compare with the volume of air in your chemistry classroom? Show your calculations.
    Hint: Think ahead about the most convenient unit to use for measuring or estimating the dimensions of your classroom.

24. According to Table 2.1 the percentage of carbon dioxide in inhaled air is lower than it is in exhaled air. How can you account for this relationship?

25. A headline from the *Anchorage Daily News* in Alaska (January, 17, 2008): "Family in car overcome by carbon monoxide. Fire department saves five after slide into snow bank."

    a. If your car is in the snow bank and the engine is running, CO may accumulate inside your car. Normally, however, CO does not accumulate in the car. Explain.

    b. Why didn't the occupants detect the CO?

26. Consider how life on Earth would change if the concentration of oxygen were cut in half. Give two examples of things that would be affected.

27. Explain why CO is named "the silent killer." Select two other pollutants for which this name would not apply and explain why not.

28. Undiluted cigarette smoke may contain 2–3% CO.

    a. How many parts per million is this? How many parts per billion?

    b. How does this value compare with the National Ambient Air Quality Standards for CO in both a 1-hour and an 8-hour period?

    c. Propose a reason why smokers do not die from carbon monoxide poisoning.

29. In the Northern Hemisphere, the ozone season runs from about May 1 to October 1. Why are ozone levels typically not reported in the winter months?

30. A certain city has an ozone reading of 0.13 ppm for 1 hour, and the permissible limit is 0.12 for that time. You have the choice of reporting that the city has exceeded the ozone limit by 0.01 ppm or saying that it has exceeded the limit by 8%. Compare these two methods of reporting.

31. Here are ozone air quality data for Atlanta, Georgia, from August 1–10, 2015. The primary pollutant was ozone:

    a. In general, which groups of people are the most sensitive to ozone?

    b. The U.S. Environmental Protection Agency determined that air rated above 100 is hazardous for some or all groups. For the data shown, how many days was the air hazardous?

    c. Ozone levels drop off sharply at night. Explain why.

d. During the daytime, the ozone dropped off sharply after August 5. Propose two different reasons that could account for this observation.

32. The graph below is the air quality data for Beijing, China based on the primary pollutant of $PM_{2.5}$.

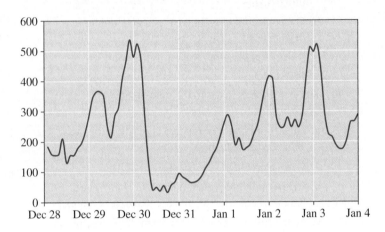

a. In general, which groups of people are the most sensitive to particulate matter?

b. The U.S. Environmental Protection Agency determined that air rated between 51 and 100 is acceptable except for particularly sensitive groups. For the data shown, when was the air acceptable or better?

c. The levels of PM do not necessarily drop off at night the way they do for ozone. Explain.

d. The levels of particulate matter can increase sharply. Propose two different reasons that could account for this observation.

33. Prior to 1990, diesel fuel in the United States could contain as much as 2% sulfur. More recent regulations have changed this, and today most diesel fuel is ultra-low sulfur diesel (ULSD) containing a maximum of 15 ppm sulfur.

    a. Express 15 ppm as a percent. Likewise, express 2% in terms of ppm. How many times lower is the ULSD than the older formulation of diesel fuel?

    b. Write a chemical equation that shows how burning diesel fuel containing sulfur contributes to air pollution.

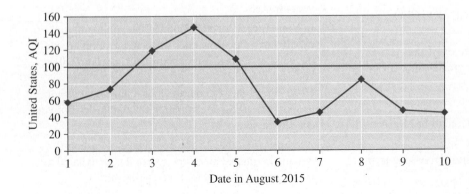

c. Diesel fuel contains the hydrocarbon $C_{12}H_{26}$. Write a chemical equation that shows how burning diesel adds carbon dioxide to the atmosphere.

d. Comment on burning diesel fuel as a sustainable practice, both in terms of how things have improved and in terms of where they still need to go.

34. Consult the U.S. map below of peak ozone data and answer the questions that follow.

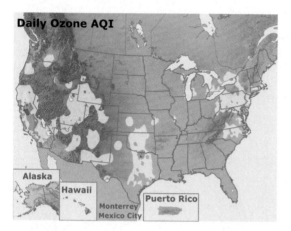

EPA AirNow

a. These data are typical in that most of the ozone pollution is expected in California, Denver, Texas, the Midwest, and the East Coast. Why is ozone pollution so high in these regions of the country?

b. The midwest typically has high ozone levels in the summer, but not on this particular day. Offer a possible explanation.

c. Why are inland areas in California, such as the Sacramento Valley, likely to have worse air quality than the California coast?

35. Look up the air quality data for ozone in two different cities, one hot and dry, and the other cooler and more rainy. Account for any differences you find.

36. At certain times of the year, inhabitants of the beautiful city of Santiago, Chile, breathe some of the worst air on the planet.

a. Driving private cars has been severely restricted in Santiago. How specifically does this improve air quality?

b. Chile has pledged to reduce its reliance on nonrenewable energy sources, such as coal. They aim to obtain 70% of their energy from renewable resources. How will this specifically affect their air quality?

37. One can purchase a carbon monoxide monitor that immediately sounds an alarm if the concentration of CO reaches a threshold. In contrast, most radon detection systems sample the air over a period of time before an alarm sounds. Why the difference?

38. Consumers now can purchase paints that emit only low amounts of VOCs. However, these consumers may not know why it matters to purchase this paint.

a. What would you print on the label of a paint can to make the point that a low-VOC paint is a good idea?

b. We apply paint to many outdoor surfaces, such as buildings, bridges, and fence posts. Comment on the environmental effects of the VOCs that these paints emit.

39. **a.** Explain why running outdoors (as opposed to sitting outdoors) increases your exposure to pollutants.

b. Running indoors at home can decrease your exposure to some pollutants, but may increase your exposure to others. Explain.

40. Select a profession of your choice, possibly the one you intend to pursue. Name at least one way that a person in this profession could have a positive effect on air quality.

Exploring Extensions

41. The images below show an "air inversion" where cold air is trapped below warm atmosphere. Perform the simple experiment described on this website. What effect do air inversions have on local pollution levels? What regions of the U.S. and Europe would inversion layers most likely be formed?

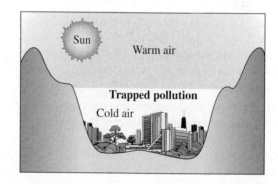

(bottom): Steve Griffin/The Salt Lake Tribune/AP Images

42. Nail polish remover containing acetone was spilled in a room $6\,m \times 5\,m \times 3\,m$. Measurements indicated that 3600 mg of acetone evaporated. Calculate the acetone concentration in micrograms per cubic meter.

43. Mercury, another serious air pollutant, is not described in this chapter. If you were a textbook author, what would you include about mercury emissions? How would you connect mercury emissions to the sustainable use of resources? Write several paragraphs in a style that would match that of this textbook. Be sure to reference your information sources.

44. Evaluate the annual concentration trends for criteria pollutants of ozone, lead, SO_x, NO_x, and PM found in this report. Although concentrations have decreased nationwide since 1990, the EPA provides an updated list of areas that do not meet current air quality standards for criteria pollutants. Check out the EPA Green Book to see where nonattainment regions are located.

    a. Choose one of these regions that is designated as "serious" or "severe" for 8-hr ozone (2015 standard). What concentration of ozone corresponds to the 2015 standard?

    b. How many people are effected by your chosen nonattainment area? What are some factors responsible for these relatively high ozone levels?

    c. Choose an area that is designated as "maintenance area," an area that once violated, but now currently meet the national air quality standards? What steps are involved for a maintenance implementation plan that is required for an area to be reclassified from nonattainment to attainment status?

45. The EPA oversees the Presidential Green Chemistry Challenge Awards. Use the EPA website to find the most recent winners of the award. Pick one winner and summarize in your own words the green chemistry advance that merited the award.

46. Here are two scanning electron micrograph images of particulate matter, courtesy of the National Science Foundation and researchers at Arizona State University. The first is of a soil particle and the second of a rubber particle, and each is about 10 μm in diameter.

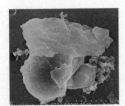 

(both): Hua Xin, Ph.D., Arizona State University/National Science Foundation

    a. Suggest a likely source of the rubber particle. Name two other substances that might contribute PM to the air.

    b. The soil particle is composed mainly of silicon and oxygen. What other elements are commonly present in the rocks and minerals in Earth's crust?

    c. What about these photographs suggests that these particles would inflame your blood vessels?

47. Ultrafine particles have diameters less than 0.1 μm. In terms of their sources and health effects, how do these particles compare with $PM_{2.5}$ and PM_{10}? Use the Internet to locate the most up-to-date information. Be sure to reference your information sources.

48. Consider this graph that shows the effects of carbon monoxide inhalation on humans.

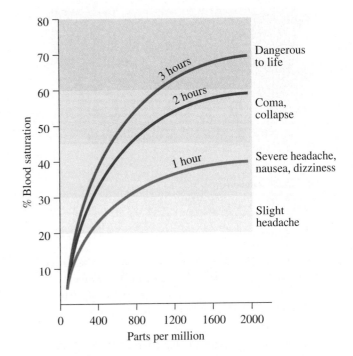

    a. Both the amount of exposure and the duration of exposure have an effect on CO toxicity in humans. Use the graph to explain why.

    b. Use the information in this graph to prepare a statement to include with a home carbon monoxide detection kit about the health hazards of carbon monoxide gas.

49. You may have admired the beauty of hardwood floors. Polyurethane is the finish of choice for floors because it is more durable than varnishes and shellacs. Although polyurethane was once exclusively an oil-based paint, there are now water-based formulations that reduce VOCs by 50–90%. In 2000, the Bayer Corporation was awarded a Presidential Green Chemistry Challenge award for this development. Prepare a summary of this work. Also check stores to see which water-based polyurethanes are available in your area.

50. Homeowners eager to improve their energy bills will use spray-polyurethane foam to add insulation to their attic spaces.

    a. While safe when the reaction is complete, many insulation foams release isocyanate during and immediately after installation. What are the hazards of this volatile organic compound?

b. Professor Richard Wool at the University of Delaware developed bio-based substitutes for high-performance materials, including isocyanate foams. For his innovative work, he won a Presidential Green Chemistry Challenge Award in 2013. Prepare a summary of his accomplishments.

51. A popular social media post has claimed that dangerous levels of benzene, a cancer-causing chemical, are released from the surfaces of car interiors.

a. Perform an extensive Internet search to determine if there is any evidence to support this claim.

b. What vapors contribute to the "new car smell" of vehicles? Are these airborne chemicals harmful to your health?

3 Radiation from the Sun

PROTECTION FROM THE SUN

Watch the opening chapter video at www.acs.org/cic. If you were outdoors on a sunny day, list the ways that you would protect yourself from sunlight. Rank them from what you think is most effective to least effective.

In this chapter, we will answer the following questions:

1. What is sunlight and what are its properties?
2. How do different types of light differ in energy?
3. What are UV-A, UV-B, and UV-C and how do they differ in their properties?
4. What is the link between sun exposure and skin cancer?
5. Where is the ozone layer located within our atmosphere?
6. How is ozone produced in the stratosphere and how does it protect us from the Sun?
7. What is meant by the "ozone hole"?
8. How do chlorofluorocarbons (CFCs) contribute to ozone depletion and why is the ozone hole located above Antarctica?
9. What are some alternatives to using CFCs and what are their impacts on the environment?
10. How do sunscreens and sunblocks work?

Introduction

Although many people love to be outside on a sunny day, it is important for us to understand both the positive and negative implications of the Sun's radiation on Earth. On the positive side, the Sun's radiation provides energy to grow food that you enjoy, forms the important vitamin D in your bloodstream, and maintains Earth's temperature at a livable level. However, the Sun's radiation causes premature weathering of materials and health problems such as sunburns, skin cancer, and damage to your eyes.

Your Turn 3.1   Making Decisions

Over a three-day period, record the types of sun exposure you experience throughout the day. Note the time of day you are exposed to sunlight, the amount of your skin exposed, your location, and the length of time you are exposed to the Sun.

@HOME   The Effect of Sun's Rays

Investigate how the Sun's rays affect our bodies at home using construction paper, sunscreen, and a paintbrush.

1. Dip a paintbrush in sunscreen and paint a picture onto a piece of construction paper.
2. Place the piece of construction paper out in a sunny area for several hours.
3. Make observations about what you see.

Questions to Consider
a. What was different between the areas that did not have sunscreen and the areas that did have sunscreen?
b. What do you think would happen if the SPF of the sunscreen were different?

Ghislain & Marie David de Lossy/
Image Source

In this chapter, we will examine the Sun's radiation by investigating its composition and effects on our health. You may be aware that applying sunscreen protects your skin against sunlight. However, did you know that our atmosphere also provides a natural source of protection against some of these harmful rays? We will expand upon the discussion of the atmosphere from Chapter 2 to pinpoint the components in our atmosphere that provide this natural shielding. We will also describe the chemistry of consumer sunscreen formulations. Throughout this chapter, you will learn how chemistry plays an important role in shielding us from the harmful effects of radiation.

Let's start by answering this question: What is *radiation?*

3.1 | Dissecting the Sun: The Electromagnetic Spectrum

Learning Objective: Describe and calculate the wave-like properties of electromagnetic radiation

Imagine you are relaxing or working in the sun on a clear summer day. The Sun is bright, your skin becomes warm, and your skin may eventually darken—although hopefully the damage does not cause a sunburn! Have you given much thought regarding the composition of the Sun's radiation? What do you think is in a ray of sunlight? Before we discuss these questions, complete the following activity.

Your Turn 3.2   Sun Damage

To get you started thinking about the Sun and the damage it can cause, discuss the following questions with a classmate:

a. What is emitted from the Sun?
b. Why do you think exposure to the Sun can cause damage?
c. What kinds of damage can the Sun cause?
d. What are some of the positive effects of the Sun?

You may have heard the term "radiation" used in several contexts: radiation from the Sun, radiation from nuclear reactions, microwave radiation, or radiation used for medical treatments/diagnoses. In this chapter, we will focus on radiation from the Sun; however, future chapters will address other types of radiation.

One type of **radiation** is energy emitted by a hot object such as the Sun. This energy is then absorbed by other objects such as Earth, plants, or animals. The radiant energy from the Sun takes forms that we can feel (e.g., heat) and see (e.g., visible light). However,

@HOME   Fun With Prisms

Investigate the visible spectrum at home using water, a glass, a sheet of white paper and a flashlight.

1. Fill the glass with water.
2. Set the glass so that it is slightly off of the table or counter but still stable enough not to fall.
3. Place a piece of white paper on the floor so that it is under the glass on the floor.
4. Shine a flashlight at an angle through the glass so that the light ends up shining on the white paper. Record your observations.

Question to Consider
How does this homemade prism work?

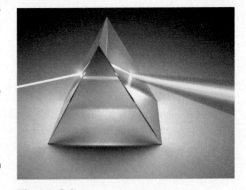

Figure 3.1

Separation of visible light into its components via a diffraction crystal.

KTSDESIGN/Science Photo Library/Getty Images

there are other forms that are actually invisible to the naked eye. One of these is ultra-violet (UV) radiation. As we look at the characteristics of radiation, we will use visible light as an example.

Visible light is continually emitted by the Sun and, after some time, reaches Earth. Have you ever observed a rainbow or the colors generated by a prism or crystal? Prisms and raindrops can separate the visible light reaching Earth into a spectrum of colors (Figure 3.1). As familiar from the acronym "ROY G. BIV," the general spectrum we see contains the colors red, orange, yellow, green, blue, indigo, and violet.

Each color of the visible light spectrum represents a particular range of wave-lengths of radiation. The word *wavelength* correctly suggests that light travels through space like waves travel through a body of water (Figure 3.2). However, while water waves travel exclusively through water, electromagnetic radiation can travel as a wave through air or even through empty space. **Wavelength** (Greek: lambda, λ) simply refers to the distance between successive peaks in a wave and is reported in units of length (Figure 3.3). Sometimes, scientists are more interested in how many waves pass a fixed point in a certain amount of time—the **frequency** (Greek: nu, ν) of the wave. Frequency is also illustrated in Figure 3.3. In this figure, two different waves are represented, with the assumption that both waves are traveling at the same speed across the page. The top wave has a longer wavelength and a lower frequency, whereas the bottom wave has a shorter wavelength and a higher frequency.

Wavelength and frequency are inversely proportional. This means that as either wavelength or frequency increases, the other variable decreases. We can show this relationship by using a simple equation (Equation 3.1), in which c is a constant known as the **speed of light**. The value of c is 3.00×10^8 m/s and represents the velocity that light travels through the air.

$$\text{frequency } (\nu) = \frac{\text{speed of light } (c)}{\text{wavelength } (\lambda)} \qquad \textbf{[3.1]}$$

The velocity or speed of a wave is not limited to visible light. In fact, we can determine the speed of any wave if we know its frequency and wavelength:

$$\text{velocity } (v) = \text{wavelength } (\lambda) \times \text{frequency } (\nu) \qquad \textbf{[3.2]}$$

An Explanation of Color

Bradley D. Fahlman/McGraw Hill

Check out this laboratory experiment for an exploration of the relationship between wavelength of light and the apparent color of materials: www.acs.org/cic.

Figure 3.2

While water waves travel exclusively through water, electromagnetic radiation can travel as a wave through empty space.

Brian Kanof/McGraw Hill

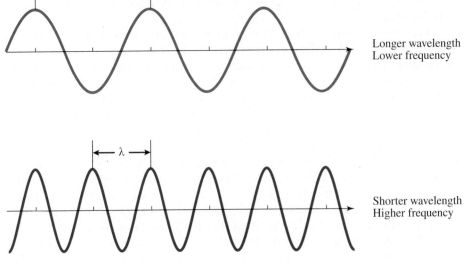

Figure 3.3

Check out Interactive waves for illustrations of wavelength, frequency, and amplitude for a wave.

If we evaluate Equations 3.1 and 3.2, we can identify the units for each of these parameters. For instance, if the speed of light is in units of m/s, and wavelength is in m, then:

$$\text{frequency } (\nu) = \frac{\text{m/s}}{\text{m}} = 1/s \text{ or } s^{-1} \text{ (also known as Hertz, Hz)}$$

The combination of wavelength (m) and frequency (s^{-1}) generates the velocity (or speed) of any wave in units of $m \cdot s^{-1}$ or m/s. However, it may be more convenient to express this velocity in terms of another unit. For instance, to convert the speed of a wave from m/s to km/h, we would need two conversion factors:

i) there are 1000 meters (m) in 1 kilometer (km)
ii) there are 3600 seconds (s) in 1 hour (h); that is, 60 seconds/min × 60 min/h

Accordingly, we could perform the conversion of 3.00×10^8 m/s to km/h as:

$$\left(\frac{3.00 \times 10^8 \text{ m}}{1 \text{ s}}\right) \times \left(\frac{1 \text{ km}}{1000 \text{ m}}\right) \times \left(\frac{3600 \text{ s}}{1 \text{ h}}\right) = 1.08 \times 10^9 \text{ km/h}$$

The human eye can detect only a very small range of wavelengths—namely those between 4×10^{-7} m and 7×10^{-7} m. You will recognize from Chapter 1 that these numbers are in *scientific notation*. You may recall from Section 1.4 that the prefix for 10^{-9} is *nano- (n)*, which means one-billionth. One nanometer is one-billionth the length of a meter. In mathematical relationships, this is:

$$1 \text{ nm} = \frac{1}{1,000,000,000} \text{m} = \frac{1}{1 \times 10^9} \text{m} = 1 \times 10^{-9} \text{ m}$$

Applying this relationship to the visible wavelengths, we can use dimensional analysis to convert from m to nm:

$$7 \times 10^{-7} \text{ m} \times \frac{1 \text{ nm}}{1 \times 10^{-9} \text{ m}} = 700 \text{ nm}$$

Therefore, we can represent the visible range of wavelengths as 400 nm to 700 nm.

Your Turn 3.3   Wavelength and Frequency

Using Equation 3.1, answer the following questions:

a. What is the frequency (in Hz) of green light, with a wavelength of 525 nm?
b. How many waves of green light would pass a fixed point in 1 minute? How about in 1 hour?
c. What is meant by the *amplitude* of a wave? Do changes in amplitude affect the wavelength or frequency of the wave? Explain.

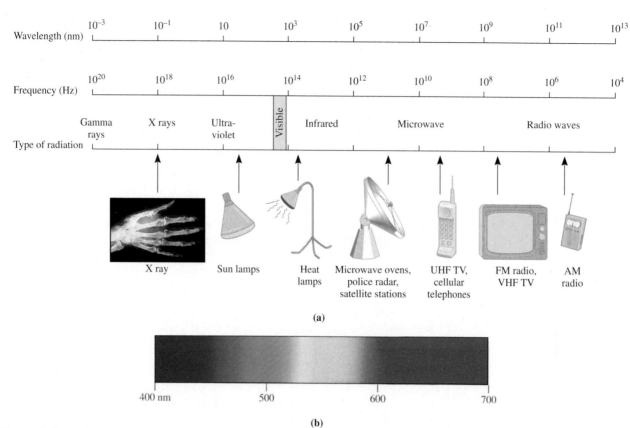

Figure 3.4

Electromagnetic radiation, wavelengths and frequencies. The respective order of colors found in the visible region are shown in **(b).** Note that 10^x is equivalent to 1×10^x.

Xray of hand: McGraw Hill

Now that we have used visible light to define many aspects of radiation, let's examine the entire spectrum of sunlight referred to as the **electromagnetic (EM) spectrum** (Figure 3.4). This term indicates that the waves are composed of both electric and magnetic fields. This spectrum can be thought of as a continuum of waves, from gamma rays that have very small wavelengths—comparable to the diameter of an atomic nucleus (approx. $10^{-14} - 10^{-12}$ m)—to radio waves that have large wavelengths of the same magnitude as the items in our macroscopic world (ca. 1–100 m).

Throughout *Chemistry in Context*, we will examine several regions of the electromagnetic spectrum. Beyond the red side of the visible spectrum, with longer wavelengths, lies the infrared (IR) region. Although you cannot see this radiation, you can

Your Turn 3.4   Analyzing a Rainbow

Example Problem:

Convert light with a wavelength of 300 nm into meters. In which region of the electromagnetic spectrum may this light be found?

Example Solution:

$$(300 \text{ nm}) \times \left(\frac{1 \times 10^{-9} \text{ m}}{1 \text{ nm}} \right) = 3 \times 10^{-7} \text{ m}$$

From evaluating the electromagnetic spectrum shown in Figure 3.4, light of 300 nm is found just below the visible region, which is in the UV range.

Your Turn:

Water droplets in a rainbow act as prisms to separate visible light into its colors.

a. In Figure 3.4, which color has the longest wavelength? The highest frequency?

b. Green light has a wavelength of 500 nm. Express this value in meters.

feel its warming effect. Microwaves (at even longer wavelengths) are also invisible to the naked eye, and are used for radar detection and heating food. The length of a microwave is on the sub-millimeter to centimeter scale—on average, about the diameter of the period that follows this sentence. Finally, at the longest range of wavelengths are radio waves, which are transmitted through the air to facilitate your cell phone conversations.

Beyond the violet side of the visible region are ultraviolet, X-ray, and gamma radiation. The two shortest wavelength ranges in the EM spectrum are exhibited by gamma rays that accompany nuclear radiation, and X-rays used in medical diagnoses. The wavelengths of X-rays are on the Angstrom (Å) scale, or 1×10^{-10} m, which is comparable to the diameter of individual atoms.

Your Turn 3.5   Relative Wavelengths

Example Problem:

Arrange these energies from greatest to smallest frequency: infrared, X-ray, UV and radio.

Example Solution:

From largest to smallest frequency: X-ray > UV > IR > radio

Your Turn:

Consider these four types of energy from the electromagnetic spectrum: infrared, microwave, ultraviolet, visible.

a. Arrange the four energies in order of increasing wavelength.
b. Approximately how many times longer is a wavelength associated with a radio wave than one associated with an X-ray?

Even though the Sun emits many types of radiation, not all of it reaches Earth's surface with equal intensities. In this context, intensity relates to the amount of radiation that impacts a certain area of Earth. Figure 3.5 displays a graph of the relative intensity of solar radiation at Earth's surface as a function of its wavelength.

Your Turn 3.6   Energy from the Sun

Examine the distribution of energy from the Sun shown in Figure 3.5.

a. Label the regions of IR and UV radiation under the curve. Which type of electromagnetic radiation comprises the greatest portion of the energy from the Sun?
b. Which type of radiation is the most intense?

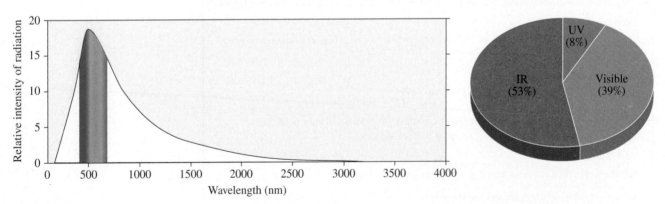

Figure 3.5

Composition of the Sun's radiation reaching the surface of Earth, illustrating the relative position of the visible region of the EM spectrum.

The graph shows that approximately 53% of the total energy emitted by the Sun reaches Earth's surface as infrared radiation, which is a major source of heat for our planet. We will describe the consequences of this phenomenon in Chapter 4. Although only approximately 8% of the sunlight reaching Earth is in the form of UV radiation, we will see in upcoming sections that this is the most harmful type of radiation that strikes our skin.

3.2 | The Personalities of Radiation

Learning Objective: Define photons of electromagnetic radiation and calculate their energies

Let's consider the flame from a propane torch. As the temperature increases, its color shifts from red/orange to yellow and eventually blue (Figure 3.6). If we measure the temperature of the various colored flames, we can see that the hottest is blue (>2000 °C) and the coolest is red (~800 °C). Because temperature is a measure of thermal energy, this observation implies that the *energy* of the flame is related to its wavelength. Hence, in addition to wavelengths and frequencies, each region of the electromagnetic spectrum has a range of energies.

However, because only certain colors are observed in flames of different temperatures, a heated object will emit only certain energies of radiation, rather than a continuum. When energy is distributed in this noncontinuous way, it is said to be **quantized**. As an analogy, consider a staircase as representing various energies of electromagnetic radiation (Figure 3.7). Just as our feline friends are only able to sit on individual steps, and not the space between levels, the Sun emits radiation with only certain energies.

Up to now, we have described radiation in terms of waves. Although this description is well-established and very useful, there is another convenient way to describe radiation. In 1921, Albert Einstein (1879–1955) won the Nobel Prize in Physics for suggesting that this quantized radiation be viewed as a collection of

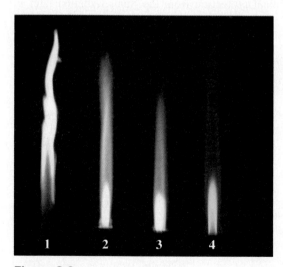

Figure 3.6

Comparison of the colors of Bunsen burner flames based on their temperatures. The oxygen:fuel ratio is increasing for flames 1–4, which results in an increasingly hotter flame.

©Arthur Jan Fijalkowski

Figure 3.7

An analogy for quantization, illustrating that radiation consists of only certain values (steps) of energy.

©Bradley D. Fahlman

Figure 3.8

An illustration of the wave-particle duality of matter—a drop of water (particle-like) suspended above liquid water (wave-like).

Trout55/E+/Getty Images

individual bundles of energy called **photons**. Photons can be considered "particles of light"; however, they do not have the same properties as other particles you may know. One major difference is that photons do not have a mass.

Although scientists have discovered the particulate nature of energy, the wave description is still useful and relevant. Both are valid descriptions for energy. This two-fold nature of radiation is known as **wave-particle duality**, and it may seem to defy common sense. After all, how can light be described as waves and particles at the same time? To add to the confusion, this description is not just limited to electromagnetic radiation, but to all forms of matter. For instance, Figure 3.8 illustrates how water can be visualized as having both wave-like and particle-like properties. As the size of an object gets larger, it becomes more particle-like rather than wave-like. For instance, an apple falling from a tree behaves as a pure particle, being described by its mass, velocity, and momentum, rather than by its wavelength and frequency. In contrast, subatomic species, such as electrons, are best described by their wave-like attributes.

These two views are linked in a relationship that is one of the most important equations in modern science (Equation 3.3). This equation will also assist us as we look at the Sun's radiation throughout this chapter:

$$E = \frac{hc}{\lambda}$$ [3.3]

where: E = the energy of a single photon in units of joules, J
h = Planck's constant, 6.626×10^{-34} J·s
c = the speed of light, 3.00×10^{8} m/s

Using Equation 3.1, we can also relate the energy of radiation to its frequency:

$$E = h\nu$$ [3.4]

The above equations demonstrate that the energy of radiation is inversely proportional to its wavelength, λ, but proportional to its frequency, ν. Therefore, as the wavelength of radiation gets shorter, the energy of its photons increases. Alternatively, high-frequency radiation will have a relatively high energy. These facts will be very important as we look at natural and chemical protection from the Sun's radiation later in this chapter.

Your Turn 3.7   Wavelength, Frequency, and Energy

a. Check out this simulation to relate the wavelength, frequency, and energy of electromagnetic radiation: www.acs.org/cic. Calculate the amount of energy present in a wave of red light and a wave of blue light. Do they have the same amount of energy? Why or why not?

b. In the not-too-distant past, many people used film exposure to take pictures. A "dark" room was a place where this film was developed into the actual images based on how light exposed the film during the taking of the pictures. These dark rooms used red lights during film development because a special dye was added to the film that was not affected by light with a wavelength between 620 and 750 nm. Do you think these rooms could have also used blue lights? Why or why not?

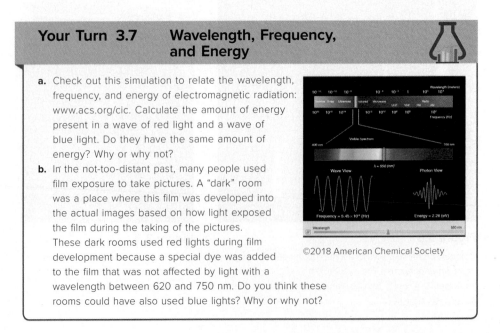

©2018 American Chemical Society

3.3 | The ABCs of Ultraviolet Radiation

> *Learning Objective: Identify the different types of ultraviolet radiation and compare their relative energies*

We know the colors of visible light by their names: red, blue, yellow, and so on. Similarly, we call ultraviolet light by different names. Admittedly, these names are not as colorful: UV-A, UV-B, and UV-C. The UV-A region lies closest to the violet region of visible light and is the lowest in energy; you may know it as "black light." In contrast, UV-C has the highest energy and lies next to the X-ray region of the electromagnetic spectrum. Whereas UV-A is used as a curing agent for gel manicures, UV-C is used to sterilize medical equipment by killing all bacteria on the surface. Table 3.1 shows the characteristics of the different regions of UV light.

Your Turn 3.8　　The ABCs of Solar UV

a. Arrange UV-A, UV-B, and UV-C in order of increasing frequency.
b. Is the order for increasing energy the same as for wavelength? Explain.
c. Should you use a sunscreen that claims to only protect against UV-C? Explain.

With their short wavelengths and high energies, we are fortunate that most UV-B and all UV-C radiation is absorbed by our atmosphere and consequently does not reach Earth's surface. Due to their extreme energies, these regions of the electromagnetic spectrum would cause significant damage to biological tissue if allowed to come into contact with humans, other animals, or plant life. But how are these wavelengths absorbed by the atmosphere? As we noted in Chapter 2, about 21% of the atmosphere consists of oxygen (O_2). Photons with wavelengths of 242 nm or shorter have sufficient energy to break the bond in an O_2 molecule. Because these wavelengths are found in the UV-C region, this radiation is absorbed by reacting with the atmospheric $O_2(g)$.

If oxygen was the only molecule that absorbed UV light from the Sun, Earth's surface and the creatures that live on it would still be subjected to damaging ultraviolet radiation of wavelengths longer than 242 nm (i.e., UV-A, UV-B, and some UV-C). It is at these lower energies that ozone (O_3) plays an important protective role. You may remember from Chapter 2 that ground-level ozone is a harmful component of smog. However, ozone in the upper atmosphere is beneficial and protects humans from ultraviolet radiation. Because the atoms in oxygen and ozone are connected with bonds of different strengths, the O_3 molecule is more easily broken apart than O_2. Accordingly, UV light of lower energy (longer wavelength) is sufficient to separate the atoms in O_3. In particular, photons of wavelength 320 nm or shorter carry enough energy to break the O—O bond in ozone. Because UV-C light is used to break the bonds in oxygen and ozone molecules, effectively none of this harmful radiation reaches Earth's surface. Of course, this is only the case if the ozone layer is completely intact. But what if a hole develops in this protective sheath? We will discuss if and how this is possible and the implications of this worst-case scenario later in this chapter.

Table 3.1	Types of UV Radiation		
Type	**Wavelength**	**Relative Energy**	**Comments**
UV-A	320–400 nm	Lowest energy	Reaches Earth's surface in the greatest quantity and penetrates farthest into the skin.
UV-B	280–320 nm	Medium energy	Most UV-B is absorbed by ozone (O_3) in the stratosphere. UV-B damages the outermost layer of skin.
UV-C	200–280 nm	Highest energy	Although UV-C radiation is very harmful, it is completely absorbed by O_2 and O_3 in the stratosphere.

Your Turn 3.9   Visualizing Ozone Protection

Check out this interactive to explore how the atmosphere protects us from UV radiation: www.acs.org/cic.

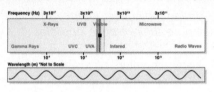

a. Which regions of the EM spectrum have sufficient energy to break apart O—O bonds in ozone?

b. UV-C light (<242 nm) has sufficient energy to break apart O=O bonds in oxygen molecules. In comparison, lower-energy radiation (<320 nm) is capable of breaking apart ozone molecules. How much more energy does a 242-nm photon have compared to a 320-nm photon?

©2018 American Chemical Society

3.4 | The Biological Effects of Ultraviolet Radiation

> *Learning Objective: Explain the biological effects of ultraviolet radiation*

The Sun bombards Earth with countless photons, which are indivisible packages of energy. The atmosphere, the planet's surface, and Earth's living matter all absorb these photons. Radiation in the infrared (IR) region of the spectrum warms Earth and its oceans. The cells in our retinas are tuned to the wavelengths of visible light, which trigger a series of complex chemical reactions that ultimately lead to sight. Compared with animals, green plants capture most of their photons in an even narrower region of the visible spectrum (corresponding to red light). **Photosynthesis** is the process through which green plants (including algae) and some bacteria capture the energy of sunlight to produce glucose ($C_6H_{12}O_6$) and oxygen from carbon dioxide and water.

@HOME   Photosynthesis

Investigate how light affects the process of photosynthesis at home using spinach leaves, two small rocks, light, water, and two cups.

1. Fill both cups with equal amounts of water.
2. Add a fresh spinach leaf to each cup of water. Place a small rock on top of each spinach leaf to keep it submerged in the water.
3. Place one cup of water in a sunny area and the other in a dark area.
4. Return to your cups after 30 minutes and make observations about what you see.

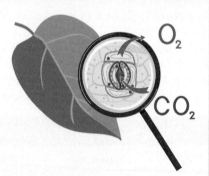

Kazakova Maryia/Shutterstock

Questions to Consider

a. What did you observe in the cup that was placed in a sunny area? What did you observe in the cup that was placed in a dark area? How can you use what you know about photosynthesis to help you explain your results?

b. What would happen if you added a source of carbon dioxide to the water in this experiment? How would your results differ?

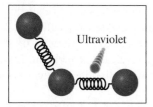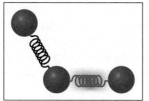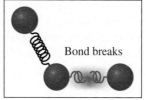

Figure 3.9

Ultraviolet radiation is able to break some, but not all, chemical bonds. Bonds are represented as springs that hold the atoms together but allow the atoms to move relative to each other.

Photons in the UV region of the electromagnetic spectrum have enough energy to remove electrons from neutral molecules, leaving them positively charged. Even shorter UV wavelength photons break bonds, causing molecules to come apart. In living things, such changes disrupt cells and create the potential for genetic defects and cancers. The interaction of UV radiation with chemical bonds is shown schematically in Figure 3.9. It is an irony of nature that this interaction of radiation with matter explains both the damage ultraviolet radiation can cause and the atmospheric mechanism that protects us from it. We will examine both of these effects in more detail in the next two sections. First, we'll see what happens to the skin when it is exposed to UV radiation.

The energy of UV radiation is approximately 10 million times greater than the energy emitted by your favorite radio station. However, whether or not your radio is turned on, you are continuously bombarded by radio waves. Your body cannot detect them, but your radio can. The energy associated with each of the radio photons is very low and not sufficient to produce a local increase in the concentration of the skin pigment, **melanin**, as happens with exposure to ultraviolet radiation. Melanin provides a natural protection against the harmful effects of UV radiation.

Figure 3.10 demonstrates how UV-A and UV-B penetrate into your skin. It is rather counterintuitive that UV-A radiation, which has less energy than UV-B, actually penetrates *deeper* into the skin. This occurs because the energy of the UV-B radiation is aligned well with the energy needed to break chemical bonds; thus, it is rapidly absorbed at the surface of the skin. UV-A, on the other hand, does not have enough energy to break bonds, so it travels farther into the body before its energy is absorbed.

When this radiation hits your skin and is absorbed, it sets off a chain of events. First, the energy from the UV photons is absorbed by skin cells. Lower-energy UV-A light may remove electrons from molecules such as water, creating reactive species

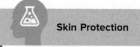

Skin Protection

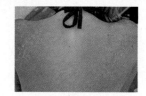

Alonafoto/Shutterstock

Watch this video for more details on how UV radiation causes skin damage and the action of anti-wrinkle creams to offset these harmful effects: www.acs.org/cic.

Questions to Consider

a. Using the Internet as a resource, cite one product that promises to reduce wrinkles and prevent aging. List the ingredients and their properties/benefits for this product.

b. The video mentions the use of snake venom to reduce wrinkles. How does snake venom compare with Botox? Which type of product is more beneficial and safe to use on your face? Are there any other alternatives for Botox and snake venom that can be used to reduce the appearance of wrinkles?

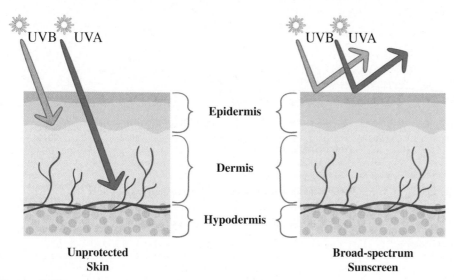

Figure 3.10

UV-A and UV-B radiation penetrate the layers of the skin to different degrees.

M. Malinika/iStock/Getty Images

known as "free radicals." UV-B light, with more energy, can cause some of the chemical bonds to break in nearby molecules. DNA molecules in the skin cells can be damaged, either by the free radicals generated by UV-A absorption or directly by the UV-B radiation. This damage stimulates the production and release of melanin. In all cases, skin darkens as a result. Most of the time, your body repairs the damage or the cell dies.

However, another outcome is possible. Although bonds can break in many different molecules, those broken in the DNA molecules of a skin cell are the most serious. Why? Because this damage may cause the DNA to be mutated in a way that leads to skin cancer. Important points about skin cancer include:

- Most skin cancers are linked to the exposure to sunlight.
- Although skin cancer can appear at any age, it is more common in older people.
- Skin cancer can develop many years after repeated, excessive exposure has stopped.
- The UV-A in sunlight at Earth's surface is most strongly linked with skin cancers. However, UV-B may also play a role.
- Cancer can arise in different types of skin cells. Those in the basal and squamous cells are common, but seldom fatal. In contrast, cancers in the melanocyte cells (melanomas) are more deadly.
- Melanoma is the third-most common cancer diagnosed in young adults between the ages of 15 and 39.

Skin cancer rates are rising in all countries, despite increased awareness of the dangers of exposure to UV radiation. Figure 3.11 presents the trends in skin cancer in the United States. Although everyone is susceptible to skin cancer, it is more common for whites, with the highest rates for white males. For all ethnicities, the rate of new melanoma diagnoses in the U.S. has tripled over the past 35 years.

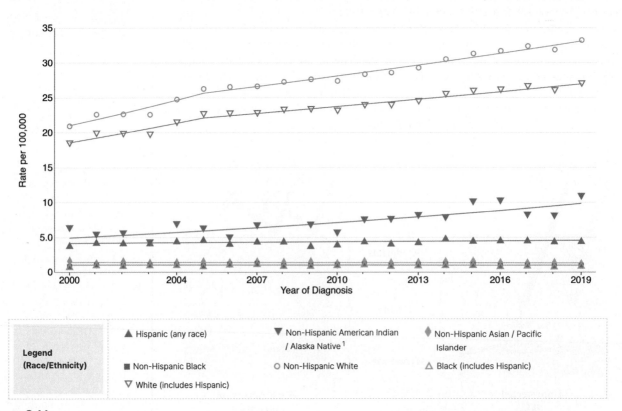

Figure 3.11

The effect of race on the melanoma incidence rates for people (of all ages) in the U.S., 2000–2019. Go to www.acs.org/cic for an interactive version of these data.

Source: National Cancer Institute, SEER Fast Stats, 2021.

Your Turn 3.10 Melanoma

a. Using a reliable source, find data for melanoma cases that have occurred over the past 40 years in your country. Did you find diagnosis or mortality rates? Do they trend in the same direction? Look up at least three other forms of cancer. Are the mortality rates for those cancers rising, dropping, or staying the same?

b. Why do the highest incidences of skin cancer occur for white males?

c. Check out the National Cancer Institute's melanoma risk assessment tool. Based on what you have read thus far, how might you alter your current behavior regarding sun exposure? How might you advise friends and/or relatives?

Currently, your risk of developing skin cancer is related to a complex blend of chemistry, physics, biology, geography, and human psychology. Factors include your altitude and latitude, how well you protect yourself from the Sun when its rays are the most harsh, whether or not you use indoor tanning beds, and how well you respond to public health campaigns for early detection of skin cancer. Our genetic makeup is another important factor that we are unable to change.

Some amount of UV radiation can actually be beneficial. UV-B light stimulates the production of vitamin D, which is important for healthy bones and immune systems. However, most people acquire sufficient levels of vitamin D through the food they eat, so we don't need to be very concerned about getting it from the Sun. Figure 3.12 illustrates the rather tenuous relationship between UV exposure and our health. While we do need some UV radiation, most people can get enough by spending just a few minutes outside daily.

Your Turn 3.11 Exposure

Examine the curve shown in Figure 3.12. Can you think of another example of something that is required for good health, but dangerous to health in high quantities?

How do you know whether the UV levels are at a dangerous level on any given day? A number of countries provide a *UV Index* forecast, which uses computer models to predict the risk of sunburn from overexposure to UV light from the Sun. Important

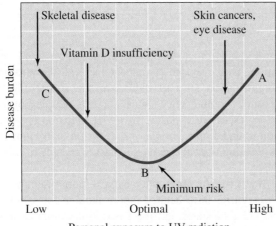

Figure 3.12

The delicate balance of "optimal" exposure to UV radiation and health.

Source: DOI: 10.5694/j.1326-5377.2002.tb04979.x

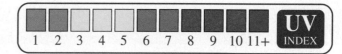

Figure 3.13

The color-coded UV Index scale.

Table 3.2		Suggestions Based on the UV Index Scale
Exposure Category	**Index**	**Tips to Avoid Harmful Exposure to UV**
LOW	0–2	If you burn easily, cover up and use sunscreen.
MODERATE	3–5	Stay in the shade when the Sun's rays are strongest.
HIGH	6–7	Cover up, wear a hat and sunglasses, and use sunscreen. Reduce exposure between 10 AM and 4 PM.
VERY HIGH	8–10	Be especially careful if you are outside on the sand, snow, or water, as these surfaces reflect UV, increasing your exposure. Minimize exposure between 10 AM and 4 PM.
EXTREME	11+	Take full precautions against sunburn. Unprotected skin can burn in minutes. Avoid the Sun between 10 AM and 4 PM.

factors considered by most models include the ozone concentration in the upper atmosphere, elevation, and cloud cover. Daily UV Index values range from 0 (during nighttime) to 15 (extremely high risk) and are color-coded for ease of interpretation (Figure 3.13). These values also are accompanied by suggestions to help protect your eyes and skin from UV damage as shown in Table 3.2.

Your Turn 3.12   UV Index

Many governments provide a daily UV Index forecast for different areas of the country. Use a government website to look up today's UV Index for your location.

a. Compare today's UV Index value to an average value six months ago. Is there much of a difference? Why or why not?

b. Look at a map of your country for the hottest part of the year. What areas correspond to the highest and lowest UV indices? Why?

Although the UV Index focuses on skin damage, this is not the only biological effect of UV radiation. Your eyes can be damaged as well. For example, all people, no matter what the pigmentation level of their skin, are susceptible to retinal damage caused by UV exposure. Another effect is cataracts, a clouding of the lens of the eye that can be caused by excessive exposure to UV-B radiation (Figure 3.14). However, just as proper clothing and sunscreen can cut down on skin damage, we should also protect our eyes by wearing optical-quality sunglasses capable of blocking at least 99% of UV-A and UV-B.

As of 2022, Brazil and Australia have banned indoor tanning for all age groups, while 23 European countries have banned indoor tanning for people younger than 18. Additionally, in the U.S., 44 states and the District of Columbia regulate indoor tanning for minors. The U.S. Centers for Disease Control and Prevention (CDC) also warns that the use of tanning booths, tanning beds, and sun lamps is dangerous, particularly for young people. The reason is straightforward: these booths use primarily

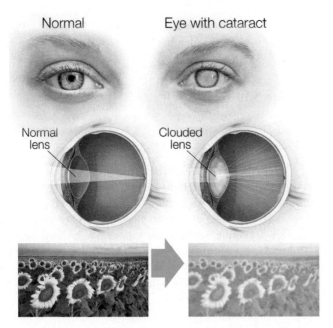

Normal · Eye with cataract

Normal lens · Clouded lens

Figure 3.14

Illustration of cataracts, which can be caused by exposure to UV radiation.

(eyes) sruilk/Shutterstock; (Sunflower) Ronda Kimbrow/iStock/Getty Images

UV-A radiation, the same dangerous wavelengths that reach Earth from the Sun. The CDC points out that the concept of "getting a base tan" is actually a response to skin injury and is not as smart as people may think. Rather, it is wise to protect your skin from the Sun.

Your Turn 3.13   Tanning Beds

What are the laws concerning tanning beds in your state or country? How have these regulations changed over the past decade?

3.5 | The Atmosphere as Natural Protection

Learning Objective: Identify what components of the atmosphere protect us from ultraviolet radiation and where they are found

Without Earth's atmosphere above us, ultraviolet radiation from the Sun would inflict tremendous damage on the planet. In this section, we will look at how the atmosphere protects us from the most harmful radiation from the Sun. In particular, this invisible shield contains an especially active ingredient for protection—the **ozone layer**.

Your Turn 3.14   The Ozone Layer

a. Had you heard of the ozone layer before this course? If so, in which context? What do you know about the ozone layer?

b. Where is the ozone layer located? Which molecules make up the ozone layer?

oxygen molecule, O_2 ozone molecule, O_3

Figure 3.15

Space-filling models of oxygen and ozone molecules.

In the previous activity, you were able to assess your current knowledge about the ozone layer. What exactly is ozone? Let's look more in-depth at this portion of the atmosphere.

If you have ever been near a sparking electric motor or in a severe lightning storm, you may have smelled ozone. Its odor is unmistakable but difficult to describe. Some people compare it to that of chlorine gas; others think the odor reminds them of freshly mown grass. Appropriately enough, the name *ozone* comes from a Greek word meaning "to smell."

Ozone and oxygen molecules differ by only one atom (Figure 3.15). As we will see, this difference in molecular structure translates to significant differences in their chemical properties. One difference is that ozone is far more chemically reactive than O_2. Ozone can be used to kill microorganisms in water and to bleach paper pulp and fabrics. At one time, ozone was even advocated as a deodorizer for air. This use only makes sense, however, if nobody breathes the air during the deodorizing process! In contrast, you safely can (and must) breathe oxygen constantly.

Ozone forms both naturally and as a result of human activities, as you saw in Chapter 2. However, given the high reactivity of ozone, it does not usually persist for long periods of time. If it were not for the fact that ozone is formed anew on our planet, you would not find it except as a curiosity in the chemistry lab.

Ozone can be formed from oxygen, but this process requires energy. A simple chemical equation summarizes the process:

$$\text{Energy} + 3\ O_2 \longrightarrow 2\ O_3 \qquad\qquad [3.5]$$

Equation 3.5 helps explain why ozone is formed from oxygen in the presence of a high-energy electrical discharge, whether from an electric spark or lightning.

Ozone is reasonably rare in the *troposphere*, the region of the atmosphere closest to Earth's surface (Figure 3.16). Here, only somewhere between 20 and 100 ozone molecules typically occur for every billion molecules and atoms that make up the air. This equates to 20–100 ppb, a unit we first saw in Section 1.7. In contrast, the *stratosphere*, farther away from Earth's surface, is mostly where ozone filters some types of ultraviolet light from the Sun. The concentration of ozone in this region is several orders of magnitude greater than in the troposphere, but still very low. As an upper limit, about 12,000 ozone molecules are present for every billion molecules and atoms of gases that make up the atmosphere at this level.

Most of the ozone on our planet, about 90% of the total, is found in the stratosphere. The term *ozone layer* refers to a designated region in the stratosphere with maximum ozone concentration. Figure 3.17 shows the relative ozone concentrations within the troposphere and stratosphere.

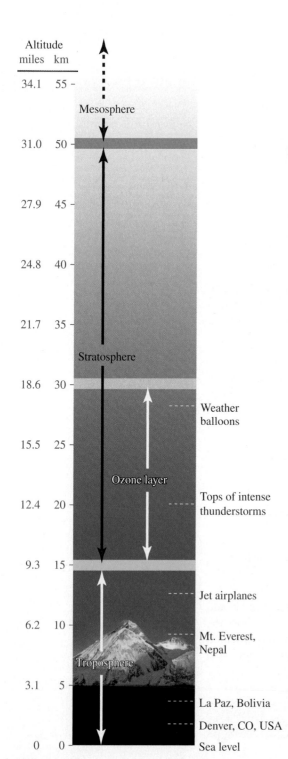

Figure 3.16

The regions of our atmosphere, showing the location of the ozone layer.

Daniel Prudek/123RF

@HOME   Layers of the Atmosphere

Investigate the different densities of the layers of the atmosphere at home using a clear cup or glass, dirt, corn syrup, green liquid dishwashing soap, water, blue and red food coloring, vegetable oil, and honey.

1. Add the dirt to a clear plastic cup or glass. Make sure you pack it down tight.
2. Next, add a layer of honey.
3. Add blue food coloring to some corn syrup, then add a layer of this colored syrup into the cup.
4. Add a layer of liquid dishwashing soap into the cup.
5. Add red food coloring to water and add this to the cup.
6. Finally, add a layer of vegetable oil to the cup.

©Bradley D. Fahlman

Questions to Consider

a. Create a table that matches each layer in your model to the different layers of the atmosphere. *Hint:* The dirt represents the surface of Earth.
b. Which layer of the atmosphere is the densest? How can you tell?

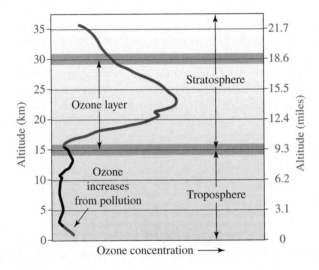

Figure 3.17

The comparative concentrations of ozone within the troposphere and stratosphere.

Your Turn 3.15   Ozone Concentrations

Use Figure 3.17 and values given in the text to answer these questions:

a. What is the approximate altitude of maximum ozone concentration?
b. Express the concentration of ozone found in the stratosphere in units of ppm and ppb.
c. What is the concentration of ozone in the troposphere (in both ppm and ppb) that just meets the EPA limit for an 8-h average?

Because the range of altitudes is so broad, the concept of an "ozone layer" can be a little misleading. No thick, fluffy blanket of ozone exists in the stratosphere. At altitudes of the maximum ozone concentration, the atmosphere is very thin, so the total amount of ozone is surprisingly small.

G. M. B. Dobson (1889–1975), a scientist at Oxford University, invented the first instrument used to quantitatively measure the concentration of ozone in our atmosphere. This technique considers the total amount of ozone in a vertical column of air of known

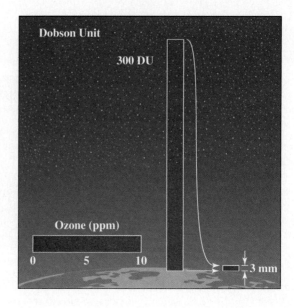

Figure 3.18

An illustration of the definition of the Dobson unit. If all the ozone over a certain area is compressed at 0 °C to form a 3-mm thick slab with a pressure of 1 atmosphere (atm), it is defined as 300 Dobson units (DU). So, if a 5-mm thick slab is produced, the ozone concentration will be 500 DU.

volume (Figure 3.18). The determination can be done from Earth's surface by measuring the amount of UV radiation reaching a detector. The lower the intensity of the radiation, the greater the amount of ozone in the column. Therefore, it is fitting that the unit of such measurements is named for him, referred to as *Dobson units* (DU).

Scientists continue to measure and evaluate ozone levels using ground observations, weather balloons, and high-flying aircraft. However, since the 1970s, measurements of total-column ozone have also been made from the top of the atmosphere. Satellite-mounted detectors record the intensity of the UV radiation scattered by the upper atmosphere. The results are then related to the amount of O_3 present. Since 1926, the daily stratospheric O_3 concentrations have been measured at the Swiss Meteorological Institute. These measurements show that the natural concentration of stratospheric O_3 is not uniform across the globe.

3.6 | How Does Ozone Decompose in UV Light?

Learning Objective: Draw Lewis structures and use them to describe how oxygen and ozone absorb ultraviolet radiation

The process by which ozone protects us from damaging solar radiation involves the interaction of matter and energy from the Sun. To understand these reactions, we must first consider molecular models, which provide information regarding how the atoms are connected within each molecule.

Let's begin with the simplest molecule, hydrogen (H_2). As seen in Chapter 1, each hydrogen atom has one electron. Once two hydrogen atoms become bonded together, the two electrons will be shared by both of the atoms. If we represent each electron by a dot, the two separate hydrogen atoms look something like this:

$$H\cdot \text{ and } \cdot H$$

Bringing the two atoms together yields a molecule that can be represented this way:

$$H:H$$

Because each atom effectively has a share in both electrons, the resulting H_2 molecule has a lower energy than the sum of the energies of two separate H atoms and is therefore more stable. The two electrons that are shared constitute a **covalent bond**. Appropriately,

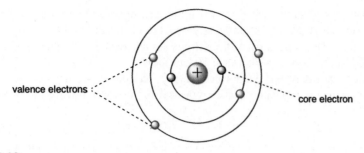

Figure 3.19

The Bohr model of a carbon atom, showing the placement of valence electrons (blue) relative to core electrons (black).

©Bradley D. Fahlman

the name *covalent* implies "shared strength." Recall that the nucleus of an atom is positively charged, whereas electrons are negatively charged. A covalent bond represents the attraction of the nucleus of one atom to the electron or electrons of the other atom.

As shown by the Bohr model of an atom in Figure 3.19, electrons occupy various shells of varying distances from the nucleus. This model of the relative positions of electrons and the nucleus is consistent with energy quantization as discussed in Section 3.2. Only those electrons in the outermost shells, known as **valence electrons**, are involved in covalent bonding.

Lewis structures, also called dot structures, can be drawn for many simple molecules and molecular compounds by following a set of straightforward steps. We first illustrate the procedure with hydrogen fluoride, HF, another simple molecular compound.

Step 1. Note the number of valence electrons contributed by each of the atoms (Figure 3.20).

Hence, for HF:

1 H atom × 1 valence electron per atom = 1 valence electron

1 F atom × 7 valence electrons per atom = 7 valence electrons

Step 2. Add the valence electrons contributed by the individual atoms to obtain the total number of valence electrons available:

1 + 7 = 8 total valence electrons

Step 3. Arrange the valence electrons in pairs. Then, distribute the total number of electrons as dots around each of the atoms:

2 electrons in the case of hydrogen, 8 electrons for most other atoms

$$H\!:\!\ddot{\underset{\cdot\cdot}{F}}\!:$$

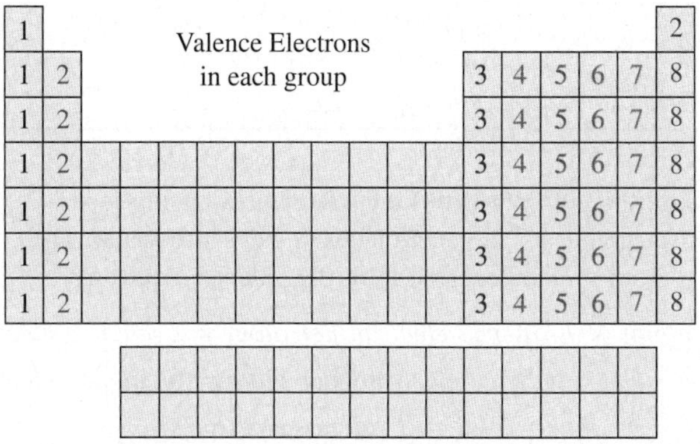

1		Valence Electrons								2
1	2	in each group			3	4	5	6	7	8
1	2				3	4	5	6	7	8
1	2				3	4	5	6	7	8
1	2				3	4	5	6	7	8
1	2				3	4	5	6	7	8
1	2				3	4	5	6	7	8

Figure 3.20

The number of valence electrons of main group elements in the periodic table.

Notice that we surrounded the F atom with 8 dots, organized into 4 pairs. The pair of dots between the H and the F atoms represents the bond that unites the hydrogen and fluorine atoms. The other 3 pairs of electrons are not shared with other atoms; as such, they are called nonbonding electrons, or **lone pairs**.

A **single covalent bond** is formed when 2 electrons (one pair) are shared between two atoms. A line is typically used to represent the two electrons in the bond:

$$H—\overset{\cdot\cdot}{\underset{\cdot\cdot}{F}}:$$

Sometimes, the lone pairs are not drawn on a Lewis structure, simplifying it even more. The result is called a **structural formula**, a representation of how the atoms in a molecule are connected:

$$H—F$$

Recall that the single line in a Lewis structure represents one pair of shared electrons. These two electrons, plus the 6 electrons in the lone pairs, mean that the fluorine atom is associated with 8 valence electrons—whether or not the electrons are explicitly shown. However, the hydrogen atom has no additional electrons other than the single pair it shares with fluorine. It is at maximum capacity with 2 electrons. The fact that electrons in many molecules are arranged so that every atom (except hydrogen) shares 8 electrons is called the **octet rule**.

As another example, consider the Cl_2 molecule—the diatomic form of elemental chlorine. From the periodic table, we can see that chlorine, like fluorine, is in Group 17, which means that its atoms each have 7 valence electrons. Using the scheme given earlier for HF, we first count and add up the valence electrons for Cl_2:

2 Cl atoms × 7 valence electrons per atom = 14 total valence electrons

For the Cl_2 molecule to exist, a bond must connect the two atoms. The remaining 12 electrons constitute 6 lone pairs, distributed in such a way as to give each chlorine atom a share in 8 electrons (2 bonding electrons and 3 lone pairs). Here is the Lewis structure:

$$:\overset{\cdot\cdot}{\underset{\cdot\cdot}{Cl}}—\overset{\cdot\cdot}{\underset{\cdot\cdot}{Cl}}:$$

Your Turn 3.16   Practice with Lewis Structures for Diatomic Molecules

Example Problem:

Draw the Lewis structure of a Ne atom.

Example Solution:

1 Ne atom × 8 valence electrons per atom = 8 total valence electrons
Therefore, the Lewis structure is:

$$:\overset{\cdot\cdot}{\underset{\cdot\cdot}{Ne}}:$$

Your Turn:

Draw the Lewis structure for each molecule.

a. HBr   **b.** Br_2

So far, we have dealt only with Lewis structures having just 1–2 atoms. However, the octet rule also applies to larger molecules. Let's use water, H_2O, as an example. Just as with two-atom molecules, first tally the valence electrons:

2 H atoms × 1 valence electron per atom = 2 valence electrons

1 O atom × 6 valence electrons per atom = 6 valence electrons

Total = 8 valence electrons

In molecules like water that have a single atom bonded to two or more atoms of a different element(s), *the single atom is usually the central one*. Because oxygen is the "single atom" in H_2O, we place it in the center of the Lewis structure. Each of the H atoms bonds to the O atom, using 2 electrons each. The remaining 4 electrons go on the O atom as 2 lone pairs:

$$H \colon \overset{\cdot\cdot}{\underset{\cdot\cdot}{O}} \colon H$$

A quick count confirms that the O atom is surrounded by 8 electrons, so it satisfies the octet rule. Alternatively, we could use lines for the single bonds:

$$H - \overset{\cdot\cdot}{\underset{\cdot\cdot}{O}} - H$$

Chemical formulas show the types and ratio of the atoms present in a compound. In contrast, Lewis structures indicate *how* the atoms are connected, and show the presence of lone pairs of electrons, if present. Note that Lewis structures do *not* directly reveal the shape of a molecule. For example, from the Lewis structure above, it might appear that the water molecule is linear. In fact, the molecule is bent, which has an enormous effect on its properties.

Another molecule to consider is methane, CH_4, the primary component of natural gas. Again, we begin by tallying the valence electrons:

4 H atoms × 1 valence electron per atom = 4 valence electrons

1 C atom × 4 valence electrons per atom = 4 valence electrons

Total = 8 valence electrons

The central carbon atom is surrounded by the 8 electrons, giving carbon an octet of electrons. In the Lewis structure, each H atom uses its single electron to bind with the C atom, for a total of 4 single covalent bonds:

Remember that H can accommodate only one pair of electrons. The next activity gives you the opportunity to practice with other molecules.

In some structures, single covalent bonds do not allow the atoms to follow the octet rule. Consider, for example, the O_2 molecule. Here, we have 12 valence electrons to distribute, six from each of the oxygen atoms. As seen below, there are not enough electrons to satisfy the octet rule for both atoms if only one pair is shared:

However, the octet rule can be satisfied for all atoms if a lone pair becomes shared by two atoms:

A covalent bond consisting of two pairs of shared electrons is called a **double bond**, which is shorter, stronger, and requires more energy to break than a single bond between the same atoms. A double bond is represented by four dots, or by two lines:

The ozone molecule introduces another structural feature. We start again by tallying the valence electrons. Each of the 3 oxygen atoms contributes 6 valence electrons,

---

Lewis Structures

Watch this video for another example of drawing Lewis structures (www.acs.org/cic) and answer the questions below.

Chemistry in Context
Tutorial

LEWIS STRUCTURES

©Bradley D. Fahlman

Draw the Lewis structure for each of these molecules.

a. hydrogen sulfide (H_2S)

b. dichlorodifluoromethane (CCl_2F_2)

c. ammonia (NH_3)

for a total of 18. However, these 18 electrons can be arranged in two ways—each way allows all oxygens to satisfy the octet rule:

$$\ddot{\text{O}}::\ddot{\text{O}}:\ddot{\text{O}}: \qquad :\ddot{\text{O}}:\ddot{\text{O}}::\ddot{\text{O}}$$

a   **b**

Structures **a** and **b** predict that the molecule should contain one single bond and one double bond. In structure **a**, the double bond is shown to the left of the central atom, whereas in **b** it is shown to the right. But experiments reveal that the two bonds in the O_3 molecule are identical, being intermediate between the length and strength of a single and double bond. Structures **a** and **b** are called **resonance forms**, Lewis structures that represent hypothetical extremes of electron arrangements in a molecule. For example, no single resonance form represents the electron arrangement in the ozone molecule. Rather, the actual structure is something like a hybrid of the two resonance forms, **a** and **b**. A double-headed arrow linking the different forms is used to represent the resonance phenomenon:

$$\ddot{\text{O}}=\ddot{\text{O}}-\ddot{\text{O}}: \longleftrightarrow :\ddot{\text{O}}-\ddot{\text{O}}=\ddot{\text{O}}$$

A **triple bond** is a covalent linkage made up of three pairs of shared electrons. For the same atoms, triple bonds are even shorter, stronger, and harder to break than double bonds. Because electrons are negatively charged, and atomic nuclei are positively charged, you may think of electrons as a "glue" that holds the atomic nuclei together. The more electrons shared between two atoms, the more "glue," which results in a much stronger bond between the two atoms.

For an example of a triple bond, consider the nitrogen molecule, N_2. Each nitrogen atom (Group 15) contributes 5 valence electrons, for a total of 10. These 10 electrons can be distributed in accordance with the octet rule if six of them (three pairs) are shared between the two atoms, leaving four to form lone pairs, one lone pair on each nitrogen atom:

$$:\text{N}::\text{N}: \quad \text{or} \quad :\text{N}\equiv\text{N}:$$

Your Turn 3.17   Lewis Structures with Multiple Bonds

Example Problem:

Draw the Lewis structure for carbon dioxide (CO_2).

Example Solution:

Carbon dioxide has a total of 16 valence electrons. Hence, the Lewis structure is:

$$:\ddot{\text{O}}=\text{C}=\ddot{\text{O}}:$$

Your Turn:

Draw the Lewis structure for each compound. Indicate resonance forms, if appropriate.

a. carbon monoxide (CO)   **b.** sulfur dioxide (SO_2)   **c.** sulfur trioxide (SO_3)

It should be noted that like water, the O_3 molecule is not linear as the simple Lewis structure seems to indicate. Remember that Lewis structures tell us only what is connected to what and do not necessarily show the shape of the molecule. The O_3 molecule is actually bent, as in this representation:

$$\ddot{\text{O}}{\Large=}\overset{\overset{\displaystyle ..}{\text{O}}}{}{\diagdown}\ddot{\text{O}}: \longleftrightarrow :\ddot{\text{O}}{\diagup}\overset{\overset{\displaystyle ..}{\text{O}}}{}{=}\ddot{\text{O}}$$

Although you will learn about predicting molecular shapes in the next chapter, at this point you only need to know *how* the bonding in O_2 and O_3 molecules relates to their interaction with sunlight. Let's now look at how this chemistry allows for the oxygen–ozone protective screen.

Earlier, we described how UV radiation reacts with both oxygen and ozone. Here are the corresponding equations, taking into account their Lewis structures and geometries as well as the photon that triggers the reaction:

$$:\ddot{O}=\ddot{O}: \xrightarrow[\lambda < 242 \text{ nm}]{\text{UV photon}} :\ddot{O}: + :\ddot{O}: \qquad [3.6]$$

$$\ddot{\underset{..}{O}}^{\ddot{O}}\underset{..}{\ddot{O}}: \xrightarrow[\lambda < 320 \text{ nm}]{\text{UV photon}} :\ddot{O}=\ddot{O}: + :\ddot{O}: \qquad [3.7]$$

Equations 3.6 and 3.7 describe a key set of chemical reactions that occur in the stratosphere. Note that the energy required to break an O=O double bond is higher (i.e., requires radiation with a shorter wavelength) than that required to break an O—O single bond.

Every day, 300 million (3×10^8) tons of stratospheric O_3 forms, and an equal mass decomposes. New matter is neither created nor destroyed, but merely changes its chemical form. Consequently, the overall concentration of ozone remains constant in this natural cycle. This process, known as the Chapman cycle (Figure 3.21), is an

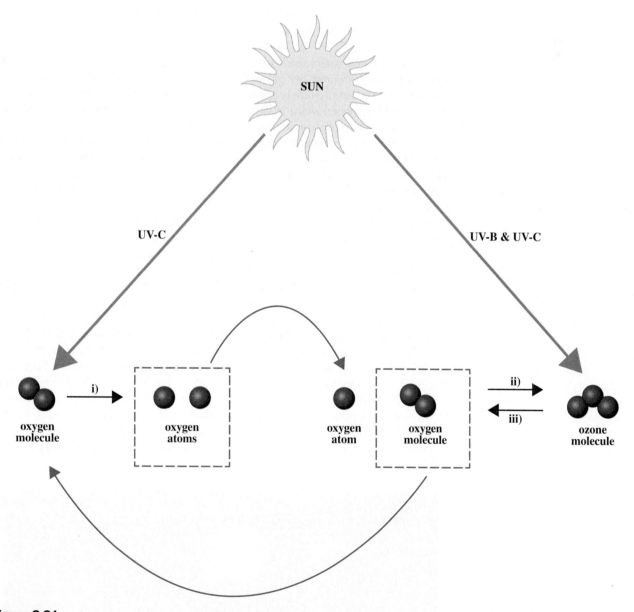

Figure 3.21

The Chapman cycle. Shown are **i)** decomposition of oxygen molecules by UV-C radiation, **ii)** reaction of oxygen atoms with oxygen molecules to form ozone, and **iii)** decomposition of ozone by UV-B and UV-C radiation.

example of a **steady state**, in which stratospheric ozone is formed and decomposed by a natural cycle. In the next section, we will consider what happens when something disturbs the steady state of the cycle, and thus leads to the destruction of our protective stratospheric ozone.

Your Turn 3.18   The Ozone Layer

Based on the reactions and molecules you examined in this section, if left undisturbed, do you think the ozone concentration in the atmosphere would be uniform across the globe at all times? Why or why not?

3.7 | How Safe Is Our Protective Ozone Layer?

Learning Objective: Explain how chlorofluorocarbons and free radical chemistry are related to ozone layer depletion

On average, the total stratospheric ozone concentration increases as one approaches either pole. In previous sections, we showed how UV radiation affects the production and destruction of O_2 and O_3. Ozone production increases with the intensity of the radiation striking the stratosphere, which in turn depends on the angle of Earth with respect to the Sun, and the distance between the Sun and Earth (Figure 3.22). At the equator, the period of highest intensity occurs at the equinox (March and October) when the Sun is directly overhead. Outside the tropics, the Sun is never directly overhead, so the maximum intensity occurs at the summer solstice (June in the Northern Hemisphere, December in the Southern Hemisphere). The angle of Earth with respect to the Sun dominates both ozone production, and the seasons. There is a slight (~7%) increase in solar intensity reaching Earth in early January, when Earth is nearest the Sun, compared with July, when Earth is farthest away. The wind patterns in the stratosphere cause other variations in ozone concentrations—some on a seasonal basis, and others over a longer cycle.

Extraordinary images of Earth are color-coded to clearly show stratospheric ozone concentrations. The dark blue and purple regions indicate where the lowest concentrations of O_3 are observed. A value of 250–270 DU is typical at the equator. As one moves away from the equator, values range between 300 and 350 DU, with seasonal variations. At the highest northern latitudes, values can be as high as 400 DU.

As you probably saw in the previous activity, there is reason for concern regarding the thinning of ozone (the "ozone hole") that occurs seasonally over the South Pole (Figure 3.23). These changes were so pronounced that, when the British monitoring team at Halley Bay in Antarctica first observed it in 1985, they thought their instruments were malfunctioning! The area over which ozone levels are reported to be less than 220 DU is usually considered to be the "hole." From the mid-1990s on, the annual size of the ozone hole has nearly equaled the total area of the North American continent, and in some cases exceeded it.

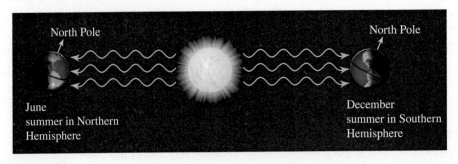

Figure 3.22

Illustration of Earth's tilt and the seasonal variation of the Sun's intensity.

Your Turn 3.19   This Year's Ozone Hole

The image below is the ozone hole over the Antarctic as of Oct. 12, 2022, with the colors representing the concentration of ozone in Dobson units.

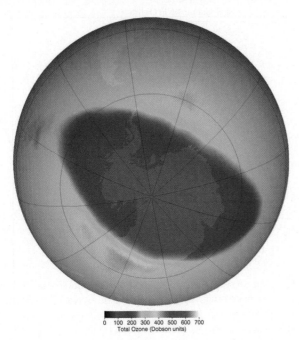

0 100 200 300 400 500 600 700
Total Ozone (Dobson units)

[NASA]

a. Looking at the trends at https://ozonewatch.gsfc.nasa.gov/, plot the average monthly area of the ozone hole over the current year. In which month(s) does the ozone hole appear? How does this compare with recent years?

b. For the current year, what is the lowest reading observed for ozone? Again, how does this compare?

c. Find a video demonstrating the ozone hole above the Antarctic in the month identified in a. above, from 1979 to the present. Describe the trends you observe.

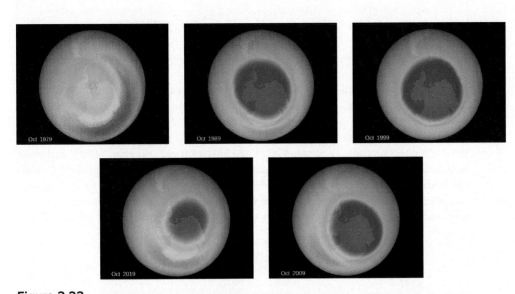

Figure 3.23

Map of the total ozone concentration and size of the ozone hole over Antarctica for a four-decade timespan. Check out www.acs.org/cic for an animation of ozone hole concentrations.

NASA Ozone Watch

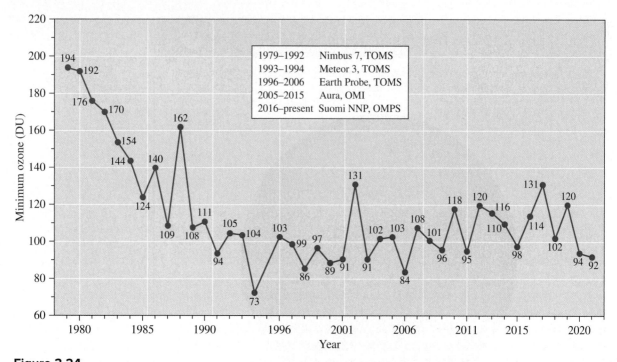

Figure 3.24

The lowest values recorded each Spring (1979–2021) for the stratospheric ozone concentration above Antarctica. *Note:* The high value in 2002 was due to an early breakdown of the vortex that isolates polar from mid-latitude air. No data were acquired in 1995.

Source: NASA Ozone Watch

Check out the dramatic decline in stratospheric ozone levels observed near the South Pole shown in Figure 3.24. In recent years, the minimum has been around 100 DU. Keep in mind that seasonal variation has always occurred in the ozone concentration over the South Pole, with a minimum in late September or early October—the Antarctic spring.

Your Turn 3.20   The Ozone Hole

How would you describe the trend of data present in Figure 3.24? Based on the patterns you observe, what would you predict for the data for years 2022–2027? How might you verify or reject your predictions?

The major natural cause of ozone destruction, wherever it takes place around the globe, is a series of reactions involving water vapor and its breakdown products. The majority of H_2O molecules that evaporate from the oceans and lakes fall back to Earth's surface as rain or snow. But a few molecules reach the stratosphere, where the H_2O concentration is about 5 ppm. At this altitude, photons of UV radiation trigger the dissociation of water molecules into hydrogen (H·) and hydroxyl (·OH) free radicals. You might have heard of "free radicals" in the context of skin creams and other anti-aging cosmetics. A **free radical** is a highly reactive chemical species with one or more unpaired electron(s). An unpaired electron is often indicated with a single dot:

$$H_2O \xrightarrow{\text{photon}} H\cdot + \cdot OH$$

Because of the unpaired electron, free radicals are highly reactive. Thus, the H· and ·OH radicals participate in many reactions, including some that ultimately convert O_3 to O_2. This is the most efficient mechanism for destroying ozone at altitudes above 50 km. As you will see in the next activity, hydroxyl radicals also react with methane.

Your Turn 3.21   Free Radicals

Consider the reaction between the hydroxyl free radical and the molecule methane (CH_4). The three-step process (shown below) occurs in the upper atmosphere. Rewrite these equations using full Lewis structures and identify all of the free radicals present throughout the steps:

$$OH + CH_4 \longrightarrow H_2O + CH_3$$
$$CH_3 + O_2 \longrightarrow OOCH_3$$
$$OOCH_3 + NO \longrightarrow OCH_3 + NO_2$$

Hint: All bonds must be made up of even numbers of electrons.

Water molecules and their breakdown products are not the only agents responsible for natural ozone destruction. Another is the free radical ·NO. Most of the ·NO in the stratosphere is of natural origin. It is formed from dinitrogen monoxide, N_2O, a naturally occurring compound that is produced in the soil and oceans by microorganisms. Natural air currents cause N_2O to gradually drift upward from the troposphere to the stratosphere, where it reacts with oxygen atoms to produce ·NO. Little can or should be done to control this process. It is part of a natural cycle involving nitrogen compounds. However, as we will now see, human activities also play a significant role in ozone destruction.

3.8 | Chemistry to the Rescue or Detriment? Human Roles in the Destruction of the Ozone Layer

Learning Objective: Examine the data to determine the role of human activities in the destruction of stratospheric ozone

Even when the effects of water, nitrogen oxides, and other naturally occurring compounds are included in stratospheric models, the measured ozone concentration is still lower than predicted. Measurements worldwide indicate that the ozone concentration has been decreasing over the past 40 years. There is a good deal of fluctuation in the data, but the trend is clear. The stratospheric ozone concentration at mid-latitudes (60° south to 60° north) has decreased by more than 8% in some cases. These changes cannot be correlated with changes in the intensity of solar radiation, so we must look elsewhere for an explanation. Thus, it is time to turn our attention to the culprit: chemicals that were once quite prevalent in aerosol spray cans and air conditioners.

A major human cause of stratospheric ozone depletion was uncovered through the masterful scientific sleuthing of F. Sherwood Rowland (1927–2012), Mario Molina (1943–2020), and Paul Crutzen (1933–2021), whose work was recognized by the 1995 Nobel Prize in Chemistry. They analyzed vast quantities of atmospheric data and studied hundreds of chemical reactions. As with most scientific investigations, some uncertainties remained. Nonetheless, their evidence pointed to an unlikely group of compounds: the chlorofluorocarbons.

As the name implies, **chlorofluorocarbons (CFCs)** are compounds composed of the elements chlorine, fluorine, and carbon only. Fluorine and chlorine are members of Group 17, the halogens. In their elemental state, most of the halogens are diatomic molecules, but only fluorine and chlorine are gases. Fluorine is extremely reactive and cannot even be contained in a glass receptacle. In contrast, CFC molecules are highly unreactive. To get started with CFCs, let's examine two examples:

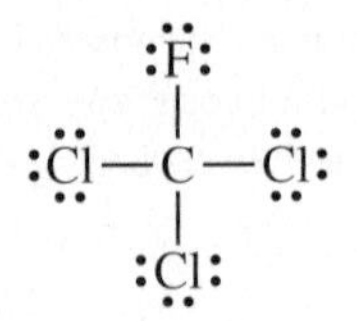

	and	
CCl$_3$F		CCl$_2$F$_2$
trichlorofluoromethane		dichlorodifluoromethane
Freon 11		Freon 12

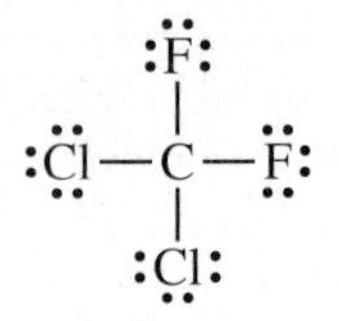

Note how the names of the compounds show the connection of CFCs to methane, CH$_4$. The prefixes *di*- and *tri*- specify the number of chlorine and fluorine atoms that substitute for hydrogen atoms of methane. These two CFCs are known by their trade names, Freon-11 and Freon-12. You may also hear them called CFC-11 and CFC-12, respectively, following a naming scheme developed in the 1930s by chemists at DuPont.

CFCs do not occur in nature; humans synthesized them for a variety of uses. This is an important verification point in the debate over the role of CFCs in stratospheric ozone depletion. As we saw previously, other contributors to the destruction of ozone, such as the ·OH and ·NO free radicals, are formed in the atmosphere from both natural sources and human activities.

The introduction of CCl$_2$F$_2$ as a refrigerant gas in the 1930s was hailed as a great triumph of chemistry and an important advance in consumer safety. It replaced ammonia (NH$_3$) and sulfur dioxide (SO$_2$), two toxic and corrosive refrigerant gases. In many respects, CCl$_2$F$_2$ was (and still is) an ideal substitute. It is nontoxic, odorless, colorless, and does not burn. In fact, the CCl$_2$F$_2$ molecule is so stable that it does not react with much of anything!

Given the desirable nontoxic properties of CFCs, they soon were put to other uses. For example, CCl$_3$F was often blown into mixtures to make foams for cushions and foamed insulation. Other CFCs served as propellants in aerosol spray cans and as nontoxic solvents for oil and grease.

For better or worse, the synthesis of CFCs has had a major effect on our lives. Because these compounds are nontoxic, nonflammable, inexpensive, and widely available, they revolutionized air conditioning, making it readily accessible for homes, office buildings, shops, schools, and automobiles. Beginning in the 1960s and 1970s, CFCs helped spur the growth of cities in hot and humid parts of the world. In effect, a major demographic shift occurred because of CFC-based technology that transformed the economy and business potential of entire regions of the globe.

Ironically, the very property that makes CFCs ideal for so many applications—their chemical inertness—ended up causing harm to our atmosphere. The C—Cl and C—F bonds in the CFCs are so strong that the molecules are virtually indestructible. For example, it has been estimated that an average CCl$_2$F$_2$ molecule can persist in the atmosphere for 120 years before it meets its final decomposing fate. In contrast, it only takes about five years for atmospheric wind currents to bring molecules up to the stratosphere, which is exactly where some of the CFC molecules ended up.

What happens to CFCs in the stratosphere? As altitude increases, the concentrations of oxygen and ozone decrease, but the intensity of UV radiation increases. The high-energy UV-C radiation can break C—Cl bonds. Here is the chemical reaction that releases chlorine atoms from dichlorodifluoromethane:

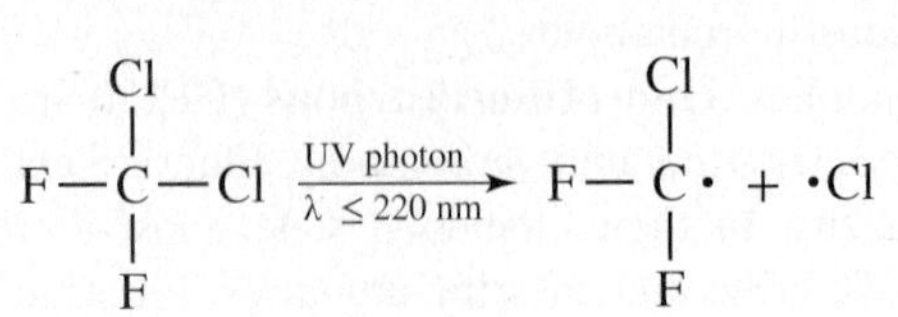

Your Turn 3.22   Visualizing CFC Decomposition

Check out this simulation to discover the effects that EM radiation has on CFC molecules: www.acs.org/cic.

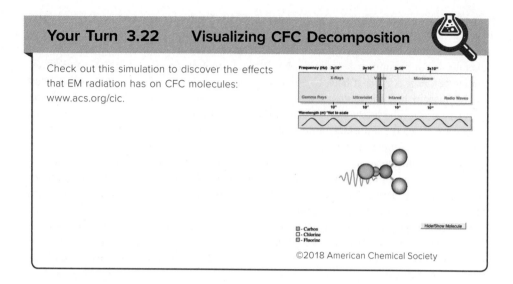

☐ - Carbon
☐ - Chlorine
☐ - Fluorine

©2018 American Chemical Society

The reaction of CFCs with UV light actually produces two free radicals. Through several reactions, the chlorine radical reacts with ozone and forms oxygen gas. Earlier (Equation 3.7), we mentioned that energy was required for the breakdown of ozone into oxygen; in this case, the unstable chlorine radical essentially provides that energy:

$$Cl\cdot + O_3 \longrightarrow ClO\cdot + O_2 \tag{3.8}$$

$$ClO\cdot + O \longrightarrow Cl\cdot + O_2 \tag{3.9}$$

As you can see, the complex interaction of ozone with atomic chlorine provides a pathway for the destruction of ozone. We will explore these pathways in the next activity.

Your Turn 3.23   Ozone vs. Chlorine Concentrations

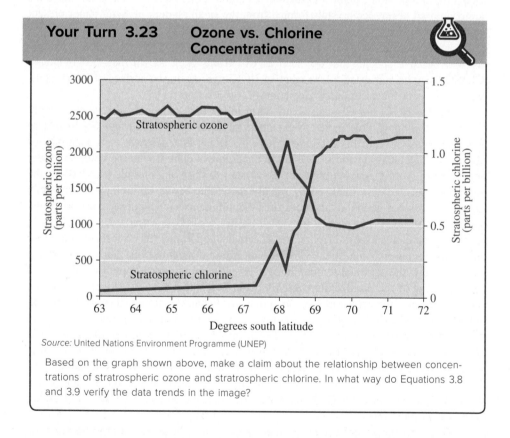

Source: United Nations Environment Programme (UNEP)

Based on the graph shown above, make a claim about the relationship between concentrations of stratospheric ozone and stratrospheric chlorine. In what way do Equations 3.8 and 3.9 verify the data trends in the image?

Notice that Cl· appears both as a reactant in Equation 3.8 and as a product in Equation 3.9. This means that Cl· is both consumed *and* regenerated in the cycle of ozone depletion, with no net change in its concentration. Such behavior is

characteristic of a **catalyst**, a chemical substance that participates in a chemical reaction and influences its speed without undergoing permanent change. Atomic chlorine acts catalytically by being regenerated and recycled to remove more ozone molecules. On average, a single atom can catalyze the destruction of as many as 1×10^5 ozone molecules before it is carried back to the lower atmosphere by winds. The video identified in the margin box illustrates the use of a catalyst to break apart hydrogen peroxide (H_2O_2)—a fun laboratory experiment you can try at home.

Not all of the chlorine implicated in stratospheric ozone destruction comes from CFCs. Natural sources such as seawater and volcanoes produce other chlorinated carbon compounds. However, most chlorine from natural sources is in water-soluble forms. Therefore, any natural chlorine-containing substances are washed out of the atmosphere by rainfall long before they can reach the stratosphere. Of particular significance are the data gathered by NASA and international researchers, which have established that high concentrations of HCl (hydrogen chloride) and HF (hydrogen fluoride) always occur together in the stratosphere. Although some of the HCl might conceivably arise from a variety of natural sources, the only reasonable origin of stratospheric concentrations of HF is CFCs.

Ozone-depleting gases are present throughout the stratosphere. Furthermore, as a result of global wind patterns, CFCs are present in comparable abundance in lower parts of the atmosphere over *both* hemispheres. Why, then, have the greatest losses of stratospheric ozone occurred over Antarctica? Furthermore, given that more ozone-depleting gases are emitted in the Northern Hemisphere, why are their effects felt most strongly in the Southern Hemisphere?

A special set of conditions exists in Antarctica; the lower stratosphere over the South Pole is the coldest spot on Earth. From June to September (Antarctic winter), the winds that circulate around the South Pole form a vortex that prevents warmer air from entering the region. As a result, the temperature may drop to as low as -90 °C. Under these conditions, **polar stratospheric clouds (PSCs)** can form. These thin clouds are composed of tiny ice crystals formed from the small amount of water vapor present in the stratosphere. The chemical reactions that occur on the surface of these ice crystals convert molecules that do not deplete ozone, such as HCl (mentioned previously), to more reactive species that do, such as Cl_2. It has been shown that the compound $ClONO_2$ may be converted to the reactive HOCl species on the surface of ice crystals, which also contributes to ozone depletion.

Neither HOCl nor Cl_2 causes any harm in the dark of winter. But when sunlight returns to the South Pole in late September, the light splits HOCl and Cl_2 to release chlorine atoms. Given this increase in Cl·, a species that destroys vast quantities of ozone, the hole starts to form. Notice the conditions required: extreme cold, a circular wind pattern (vortex), enough time for ice crystals to form and provide a surface for the reactions, and darkness followed by rapidly increasing levels of sunlight. Figure 3.25 shows the seasonal variation, and compares the minimum temperatures above the Arctic and the Antarctic.

As you can see in the figure, the necessary conditions for PSC formation more often are found in Antarctica. Changes in ozone concentrations above Antarctica closely follow the seasonal temperatures. Typically, rapid ozone depletion takes place during spring at the South Pole, which is from September to early November. As the sunlight warms the stratosphere, the polar stratospheric clouds dissipate, thereby releasing Cl· species that attack ozone molecules. However, as the spring season draws to a close, air from lower latitudes flows into the polar region, replenishing the depleted ozone levels. By the end of November, the hole is largely refilled. It turns out that ozone depletion in the Northern Hemisphere is not nearly as severe as it is in the Southern. The difference stems mainly from the fact that the air above the North Pole is not as cold. Even so, polar stratospheric clouds have been repeatedly observed in the Arctic.

What are the global effects of this ozone hole? Decreased stratospheric ozone over the South Pole leads to increased UV-B levels reaching Earth. In turn, skin cancer rates have increased worldwide. Furthermore, studies have shown that increased UV-B radiation poses a serious threat to agricultural production.

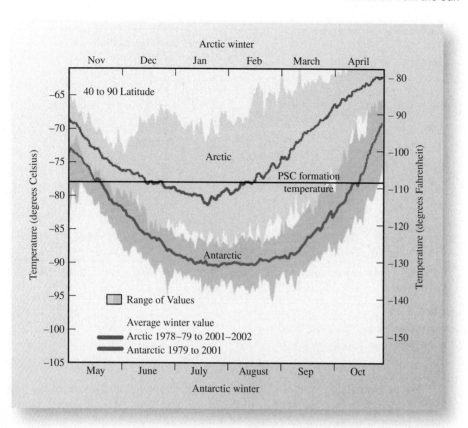

Figure 3.25

Minimum air temperatures in the polar lower stratospheres.

Source: Scientific Assessment of Ozone Depletion: 2002, World Meteorological Organization, UNEP

3.9 | Where Do We Go from Here: Can the Ozone Hole Be Restored?

Learning Objective: Describe efforts used to eliminate the use of chlorofluorocarbons

Once people understood the role of CFCs in ozone destruction, they responded with surprising speed. Individual countries took the first steps. For example, the use of CFCs in spray cans was banned in the United States and Canada in 1978; their use as foaming agents for plastics was discontinued in 1990. The problem of CFC production and subsequent release, however, spanned the globe and required global cooperation to address.

In 1987, 46 countries came together and ratified a treaty (the Montreal Protocol) to eliminate the production of CFCs worldwide. This rare example of worldwide cooperation resulted in a precipitous drop in the global production of CFCs (Figure 3.26).

Your Turn 3.24   Save the Ozone Layer

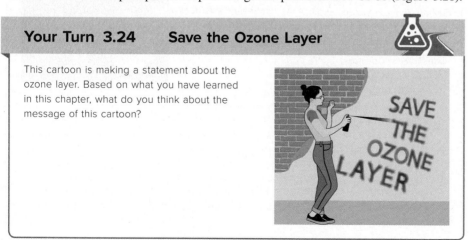

This cartoon is making a statement about the ozone layer. Based on what you have learned in this chapter, what do you think about the message of this cartoon?

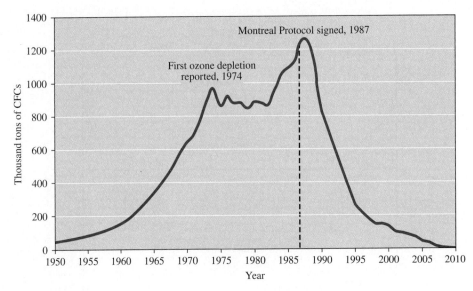

Figure 3.26

Global production of CFCs, 1950–2010.

Source: United Nations Environment Programme (UNEP)

Due to global efforts, the ozone levels have rebounded slightly in recent years and there appears to be a slow repair of the ozone hole. Stratospheric ozone recovery is only possible if the concentrations of reactive halogens in the atmosphere are lowered. Global initiatives to reduce the amounts of reactive halogens in the atmosphere are underway. As of early 2022, the concentration of ozone-depleting substances in the mid-latitude stratosphere had fallen to 50% of levels observed in 1980 (Figure 3.27).

However, the road to stratospheric ozone recovery is slow and complicated. Figure 3.28 illustrates the timeline of stratospheric ozone depletion from its early discovery to the latest scientific assessments and regulations. In finding suitable replacements for CFCs, chemists sought to prepare new compounds that were similar to nontoxic CFCs but without their long-term effects on stratospheric ozone. Any substitute for a CFC should minimize three undesirable properties: toxicity,

---

Your Turn 3.25   CFC Replacements

There have been many possible replacements for CFCs that have been developed. However, there is still no perfect solution. Explore the interactive activity to find out more about how refrigerants work and their properties: www.acs.org/cic.

Ingram Publishing/SuperStock

Questions to Consider:

a. The above activity highlighted some issues regarding the use of the latest refrigerants, HFOs, including the contamination of soil and waters with trifluoroacetate (TFA). Trifluoroacetate and other per- and polyfluoroalkyl (PFAS) substances are referred to as "forever chemicals" due to their durability in nature and difficulty to remove from soil and water samples. What are some potential health effects of PFAS chemicals?

b. What are some other problems generated from the use of hydrofluorocarbons (HFCs) and hydrofluoroolefins (HFOs)? What are some alternatives for next-generation refrigerants beyond HFOs? Describe their properties and some of their advantages and disadvantages.

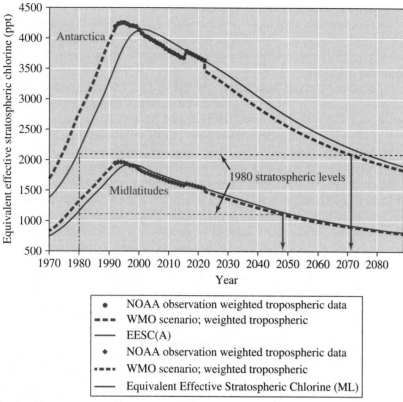

Figure 3.27

Past and projected future changes in reactive halogen concentrations in the atmosphere. The down-pointing arrows mark the estimated dates when concentrations of stratospheric halogen will return to levels present in the 1980s.

Source: The NOAA Ozone Depleting Gas Index, Summer 2022

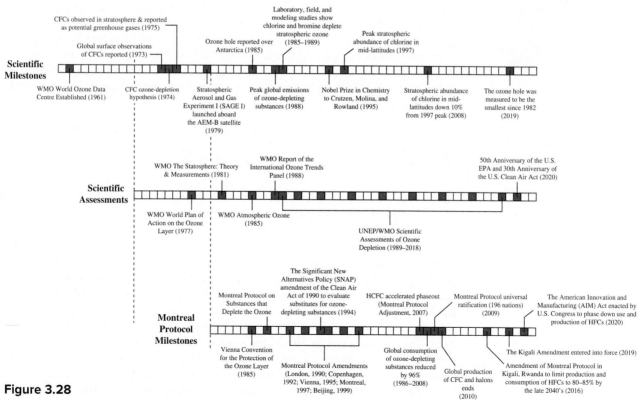

Figure 3.28

Milestones in the history of stratospheric ozone depletion.

https://assets.researchsquare.com/files/rs-948327/v1/c6985167-24a5-4e4a-97e1-ceaedfc3e331.pdf?c=1646934873

flammability, and atmospheric lifetime. At the same time, it should preserve a boiling point that is compatible with those of existing refrigerant gases, typically in the range of −10 to −40 °C. Indeed, obtaining a substitute is a delicate balancing act! Check out the activity below to explore the advantages and disadvantages of some possible replacements for CFCs.

As the above activity highlighted, the use of CFCs as refrigerants and aerosols caused the ozone layer in the stratosphere to be damaged. Replacement of these chemicals with hydrochlorofluorocarbons (HCFCs) and hydrofluorocarbons (HFCs) caused no further damage to stratospheric ozone, but caused the atmosphere below the stratosphere to be heated. Now, with the latest use of hydrofluoroolefins (HFOs), we may have introduced contamination to waters and soils at Earth's surface!

Where do we go from here? Two other refrigerant gases deserve note. One is R-744, better known as carbon dioxide, while the other is ammonia, NH_3. Carbon dioxide was used in refrigeration systems in the 1800s, but suffered from the disadvantage that high pressures are needed to compress it—sometimes more than 100 times atmospheric pressure. It was largely replaced, first by ammonia and then by CFCs. Although today there is renewed interest in using CO_2 and ammonia as refrigerant gases, no revival has yet occurred. Although ammonia is not a greenhouse gas and would not affect climate change, it is quite toxic. Ammonia emissions can also lead to changes in soil and water quality and contribute to the formation of aerosols in the atmosphere.

Another naturally occurring compound that has potential as a CFC replacement is propane, a small hydrocarbon (C_3H_8) that we use for outdoor grilling. Although it is inexpensive and nontoxic, like all hydrocarbons, it is flammable. As a refrigerant gas, propane has properties similar to those of R-22, an HCFC, with much less impact on global warming. However, as discussed in Chapter 2, hydrocarbons such as propane are volatile organic compounds (VOCs) that are involved in photochemical smog production.

As we end this section, an analogy comes to mind, one that Nobel Laureate Mario Molina employed at a 2011 symposium on ozone depletion and global climate change. He commented that some perceive science as a house of cards: If one part is disturbed, the whole house crumbles. He suggested that a better metaphor is a jigsaw puzzle of a kitten, noting that even if pieces are missing, you still can see the kitten. With stratospheric ozone depletion, this was the case. Even with a few key pieces missing, the picture was still discernible. Although with the help of chemistry one can see the picture, clearly there are still missing pieces. Ultimately, the debate among governments and their citizens about how best to protect the stratospheric ozone layer will determine the outcome in the global political arena.

George Doyle/Stockbyte/Getty Images

3.10 | How Do Sunscreens Work?

Learning Objective: Explain and illustrate how chemical compounds may be used to prevent overexposure to sunlight

In addition to our natural source of protection via the ozone layer, there are some simple ways to protect ourselves from UV radiation. The most obvious is to avoid exposure to sunlight when it is at its strongest, especially between 10 AM and 4 PM. We can also wear clothing that minimizes skin exposure; there are several brands of clothing that promise to provide extra UV protection. Be aware that sand, water, and snow all reflect sunlight, and your risk of sunburn increases with gains in altitude. Even cloudy days can be risky, as 80% of the Sun's radiation passes through the cloud cover. If it is not possible to stay out of the sun, then doctors and public health workers recommend you apply a sunblock or sunscreen to scatter or absorb UV radiation.

Mineral-based sunblock formulations contain relatively large particles of the compounds zinc oxide (ZnO) or titanium dioxide (TiO_2). Sunblock products physically block the light from reaching your skin cells, much as tightly woven clothing would. You might have seen examples such as the white opaque sunblock cream used by lifeguards ("lifeguard nose") at a pool or beach. However, the metal oxides may also be present in nanoparticulate form, with dimensions often less than 100 nm in diameter. To put this size in perspective, in the previous chapter we saw that particulate matter pollutants are categorized as PM_{10} or $PM_{2.5}$, depending on the size of the particulates. However, PM is on the micron (or micrometer, μm) scale, which is 1000 times larger than a nanometer.

Your Turn 3.26   Particulate Classification

An individual dust particle is 6 μm in diameter.

a. Would this particle be classified as PM_{10} or $PM_{2.5}$?
b. What is the diameter of this particle in nm?

Because the nanoparticles of ZnO and TiO_2 in sunscreens are so tiny, they do not scatter as much light as the larger particles used in opaque sunblocks. As a result, nanoparticle-based sunscreens are transparent, definitely a cosmetic plus to those who wear them. The nanoparticle products spread more evenly, are cost-effective, and are extremely effective at absorbing, scattering, and reflecting UV radiation. Unlike particulate matter in the air, which can become lodged in the lungs and endanger our health, the nanoparticles used in sunscreens are often suspended in a lotion, so they cannot be inhaled. However, as we will see next, it is important to realize that wearing a sunblock or sunscreen does not mean you are free of risk from damaging UV rays.

Most sunscreens are applied as a lotion or cream; they are also available in aerosol sprays, which are considered less effective. Sunscreens labeled as "broad spectrum" contain chemical compounds that absorb UV-B to some extent together with others that absorb UV-A. The American Academy of Dermatology recommends a sunscreen with a sun protection factor (SPF) of at least 30. The SPF rating refers to the length of time a sunscreen will protect your skin from reddening/burning, as compared to how long it would take to burn without sunscreen protection. However, these ratings assume that a dosage of 2 mg/cm^2 is applied to your skin. This roughly corresponds to 1 teaspoon (or 5 mL) for your face, and a whopping 1 ounce (30 mL—approximately two tablespoons) for a single application to your body. Research has shown that users typically use much less sunscreen per application, resulting in an *actual* SPF that is 20–50% of the value expected from the product label. Furthermore, it is recommended that you re-apply sunscreen every two hours, especially if engaging in activities that may remove sunscreen through contact with water or sweat. This guideline is often overlooked by beachgoers. These next few activities will explore the use of sunscreens for UV protection.

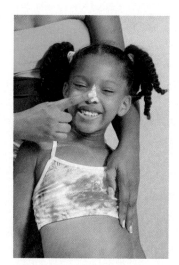

Image Source/Alamy Stock Photo

Your Turn 3.27   Sunscreens

A sunscreen's SPF is a rough indication of how much longer a person can remain in the sun while using the sunscreen, as compared to without any protection. Recently, sunscreens with a SPF of 100+ have been introduced.

a. Estimate how long you think it would take your skin to burn at noon on a midsummer day, assuming you use no sunscreen.
b. If you buy a sunscreen advertising SPF 70, calculate how long you might be able to stay in the sun without burning.
c. What assumptions are being made in predicting a SPF? Why are some dermatologists concerned about the SPF 70 sunscreens?
d. Is a SPF 30 sunscreen twice as effective as a SPF 15 sunscreen?

Earlier in the chapter, we saw the absorption of UV light by ozone and oxygen as a consequence of their reactivity. However, the absorption by physical and chemical sunscreen components is not due to reactions with UV light, but rather is simply due to their structures. As we saw earlier, when atoms are bonded together to form

@HOME   SPF and UV Light

Investigate how sunscreens with different SPFs block UV light at home using UV color-changing beads, a Ziploc bag, a timer, SPF 15 sunscreen, SPF 30 sunscreen, SPF 50 sunscreen, SPF 100 sunscreen, and sunlight.

1. Place UV color-changing beads into five separate clear plastic bags.
2. Using a marker, label one of the bags as the "control". The other bags should be labeled according to the SPF number of the sunscreens to be tested.
3. Measure a teaspoon of SPF 15 sunscreen and apply an even layer onto one side of the appropriately labeled bag.
4. Repeat Step 2 with SPF 30 sunscreen.
5. Repeat Step 2 with SPF 50 sunscreen.
6. Repeat Step 2 with SPF 100 sunscreen.
7. Place all five bags in direct sunlight, including the "control" bag that did not contain any sunscreen. Measure the amount of time it takes for each bead to change color.

Questions to Consider
a. Prepare a table that lists the time it took to change the color of beads for each level of SPF. What are the differences between the control bag and those that had SPF 15, SPF 30, SPF 50, and SPF 100 sunscreen applied to them?
b. What explains the differences between the results?

molecules, their valence electrons are shared between the two atoms to form a covalent bond. When trillions of atoms come together to form a solid, as is the case with the nanoparticle components in sunscreens, the electrons are housed in a very complicated structure of quantized energy levels that allows for light of certain energies to be absorbed. When the light is absorbed, the energy is transferred to electrons in lower-energy levels, which then become excited and move into higher-energy levels. Eventually, electrons lose this excess energy and relax back to their original energy levels, with the release of heat to the surrounding solid (Figure 3.29). The absorption properties of the solid depend on its composition as well as the size of the particulates. For instance, solids of identical composition but with particulate diameters of 20 nm versus 50 nm will absorb radiation of different wavelengths. Based on extensive studies, ZnO appears to have the best broadband protection against both UV-A and UV-B radiation.

While some people are concerned that nanoparticles may present a health risk if they are absorbed into the bloodstream through the skin, studies to date indicate that nanoparticles in sunscreen products do not penetrate beyond the epidermis, the first layer of skin. However, both consumers and government agencies continue to call for further studies to better quantify the risks. As shown in Figure 3.30, a variety of organic-based

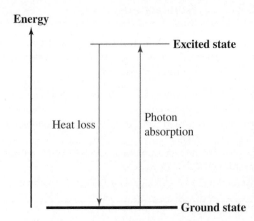

Figure 3.29

Illustration of the absorption of light of different wavelengths by electrons in a lower-energy level (ground state) into a higher-energy excited state. The electrons return to the lowest level through loss of heat to the surrounding solid.

Your Turn 3.28   UV Interactions with Matter

Check out two simulations related to the surface-volume ratio of nanoparticles and their influence on skin penetration of UV radiation: www.acs.org/cic. After exploring these simulations, summarize the influences of nanoparticle size and density on the overall effectiveness of a mineral-based sunscreen.

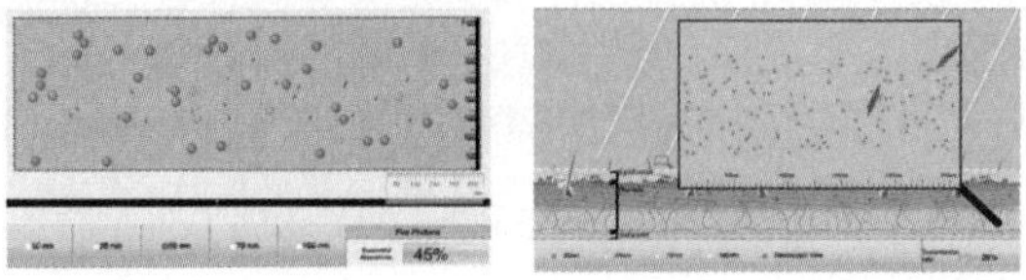

©2018 American Chemical Society

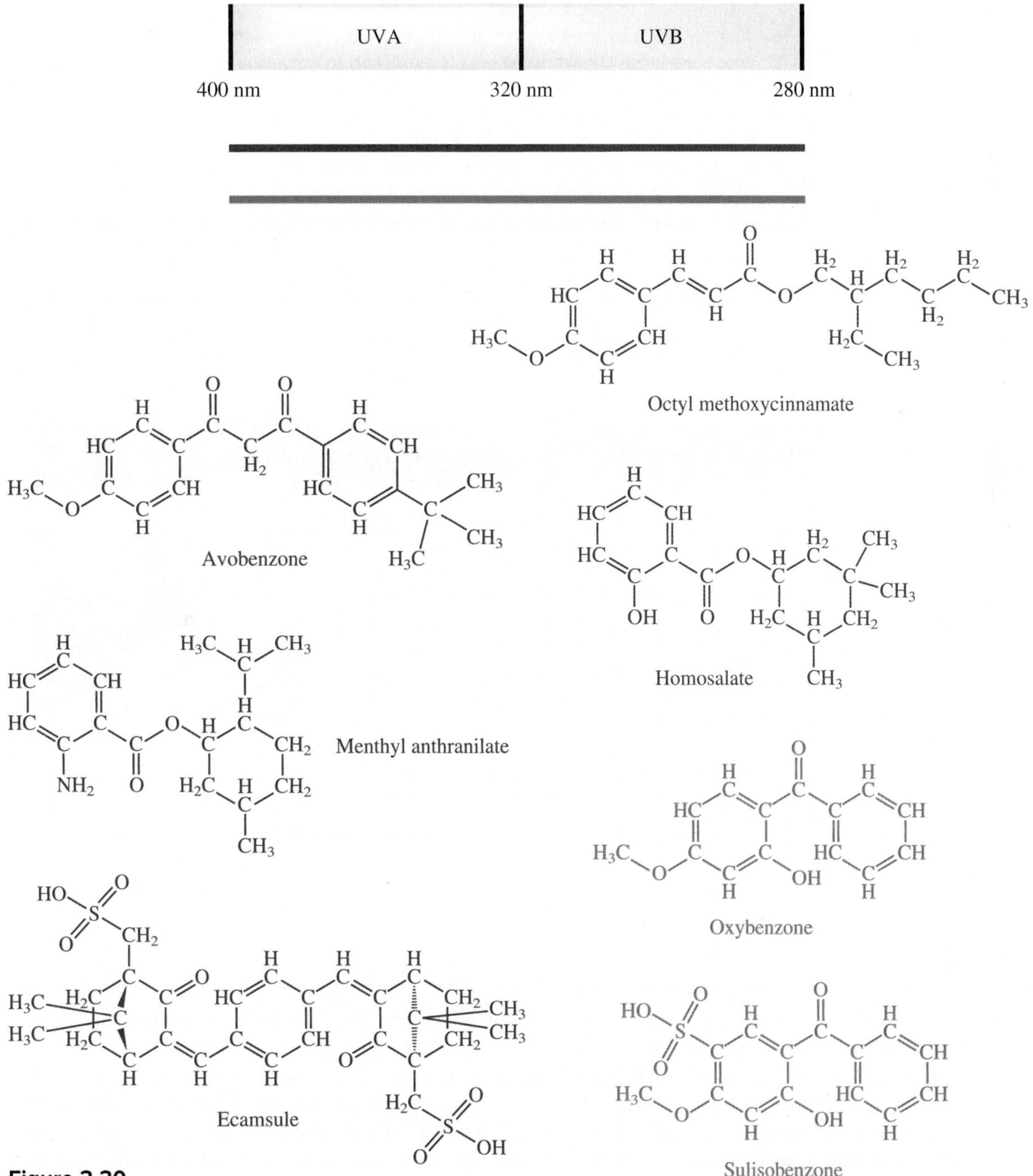

	UVA	UVB
400 nm		320 nm	280 nm

Figure 3.30

Chemical structures and UV-A and UV-B blocking ranges (colored lines) of some organic-based active ingredients in sunscreens.

chemicals may also be incorporated within sunscreen formulations to either enhance or supplant the use of nanoparticles—but they all come with their own set of potential safety risks. Furthermore, in addition to human health concerns, sunscreen ingredients may also pose significant *environmental* impacts, as you will discover in the next activity.

Your Turn 3.29   The Environmental Effects of Sunscreens

The interactive located at www.acs.org/cic describes the effect of sunscreen formulations on aquatic environments.

Questions to Consider

a. Which sunscreen ingredients have been banned in some countries? Which locations have imposed these bans?
b. What are some examples of "UV filter" ingredients in sunscreens?
c. Where are these UV filter components most likely to accumulate in aquatic regions? What are some toxic effects of these ingredients?
d. Are organic-based ingredients such as those listed in Figure 3.30 less harmful to aquatic life relative to inorganic-based nanoparticles?
e. Are sunscreens labeled as "reef-safe" still problematic for use in coastal regions such as Hawaii? Describe some ingredients in these formulations that pose problems for coral.

With so many unknowns regarding the possible health and environmental effects posed by sunscreen use, we close this chapter with a debate regarding whether sunscreens are actually worth using.

Let's Debate!   Are Sunscreens Really Worth It?

Form two groups: one group will be in favor of the topic (the proposition) and the other group will be against the topic (the opposition). Both groups will try to persuade a neutral person or judge to agree with them. The topic of the debate is the motion. Here is the motion:

Sunscreens are not worth using since they cause too much harm to human health and the environment.

Seasontime/Shutterstock

Curiosity Conversation

■ How reliable and thorough is the evidence regarding the potential negative health impacts of using sunscreens?
■ Should certain sunscreen ingredients be banned or regulated?

Conclusion

Most people enjoy being outside, especially when the weather is sunny and warm. We may bask in the sun, play games, go swimming, or climb a mountain, and some jobs require workers to be outside all day.

The Sun's rays are not all the same—in fact, the Sun is sending a range of energies to Earth. We feel some of this energy as heat and see some of it as color. But the invisible ultraviolet energy we receive from the Sun is cause for concern. Ultraviolet radiation contains enough energy to damage human skin. The effects of exposure to too much UV radiation can appear quickly, as in the case with a sunburn. Or the effects can take years to become evident, as with wrinkles and skin cancer. In this chapter, we learned about the damage ultraviolet radiation can cause when it hits our skin.

Luckily for life on Earth, the ozone found in the stratosphere far above us protects us from the majority of harmful rays. However, advancements in technology may not always act in our best interest. In the 20th century, common refrigerant gases began depleting the protective layer of ozone. Fortunately, governments around the world acted relatively quickly to pass the Montreal Protocol, a treaty that limited the production of these harmful chlorofluorocarbons. Over the 35 years since the treaty took effect, the ozone hole has stabilized, and scientists now predict a full recovery by the year 2050. Indeed, chemistry is often a two-edged sword, being used in this case study to defend the very natural source of UV protection that it almost destroyed.

The ozone layer does not absorb all of the UV radiation from the Sun, so people must still take precautions to protect their skin. Chemical- and mineral-based sunscreens and sunblocks, together with protective clothing and headgear, are the most effective means of limiting UV exposure.

LEARNING OUTCOMES

The numbers in parentheses indicate the sections within the chapter where these outcomes were discussed.

Having studied this chapter, you should now be able to:

- describe and calculate the wave-like properties of electromagnetic radiation. (3.1)

- define photons of electromagnetic radiation and calculate their energies. (3.2)

- identify the different types of ultraviolet light and compare their relative energies. (3.3)

- explain the biological effects of ultraviolet radiation. (3.4)

- identify what components of the atmosphere protect us from ultraviolet radiation and where they are found. (3.5)

- draw Lewis structures and use them to describe how oxygen and ozone absorb ultraviolet radiation. (3.6)

- explain how chlorofluorocarbons and free radical chemistry are related to ozone layer depletion. (3.7)

- examine the data to determine the role of human activities in the destruction of stratospheric ozone. (3.8)

- describe efforts used to eliminate the use of chlorofluorocarbons. (3.9)

- explain and illustrate how chemical compounds may be used to prevent overexposure to sunlight. (3.10)

Questions

Emphasizing Essentials

1. How does ozone differ from oxygen in its chemical formula? In its properties?

2. The text states that the odor of ozone can be detected in concentrations as low as 10 ppb. Would you be able to smell ozone in either of these air samples?

    a. 0.118 ppm of ozone, a concentration reached in an urban area

    b. 25 ppm of ozone, a concentration measured in the stratosphere

3. A journalist wrote "Hovering 10 miles above the South Pole is a sprawling patch of stratosphere with disturbingly low levels of radiation-absorbing ozone."

    a. Using the Internet as a resource, how big is this sprawling patch?

    b. Is the figure of 10 miles correct? Express this value in kilometers.

    c. What type of radiation does ozone absorb?

4. It has been suggested that the term *ozone screen* would be a better descriptor than *ozone layer* to describe ozone in the stratosphere. What are the advantages and disadvantages to each term?

5. Describe three differences between air in the troposphere and the stratosphere. In your answer, consider material from both Chapter 2 and Chapter 3.

6. a. What is a Dobson unit?

    b. Does a reading of 320 DU or 275 DU indicate more total column ozone overhead?

7. Using the periodic table as a guide, specify the number of valence electrons for each of these elements.

    a. oxygen (O)

    b. chlorine (Cl)

    c. nitrogen (N)

    d. sulfur (S)

8. Consider this representation of a periodic table:

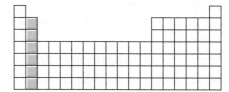

    a. What is the group number of the shaded column?

    b. Which elements make up this group?

c. What is the total number of electrons for a neutral atom of each element in this group?

d. What is the number of valence electrons for a neutral atom of each element of this group?

9. Give the name and symbol for an element with this number of valence electrons.

   a. 4   **b.** 8   **c.** 6

10. Draw the Lewis structure for each of these atoms.

    a. calcium   **b.** chlorine

    c. nitrogen   **d.** helium

11. Assuming that the octet rule applies, draw the Lewis structure for each of these molecules.

    a. CCl_4 (carbon tetrachloride, a substance formerly used as a cleaning agent)

    b. H_2O_2 (hydrogen peroxide, a mild disinfectant; the atoms are bonded in this order: H—O—O—H)

    c. H_2S (hydrogen sulfide, a gas with the unpleasant odor of rotten eggs)

    d. N_2 (nitrogen gas, the major component of the atmosphere)

    e. HCN (hydrogen cyanide, a molecule found in space and a poisonous gas)

    f. N_2O (nitrous oxide, "laughing gas"; the atoms are bonded N—N—O)

    g. CS_2 (carbon disulfide, used to kill rodents; the atoms are bonded S—C—S)

12. Several oxygen species play important chemical roles in the stratosphere, including oxygen atoms, oxygen molecules, ozone molecules, and hydroxyl radicals. Draw Lewis structures for each.

13. Consider these two waves representing different parts of the electromagnetic spectrum.

    Wave 1   Wave 2

    How do they compare in terms of:

    a. wavelength

    b. frequency

    c. speed of travel

14. Use Figure 3.4 to specify the region of the electromagnetic spectrum where radiation of each of the following wavelengths is found.
    Hint: Change each wavelength to meters before making the comparison.

    a. 2.0 cm   **b.** 50 μm

    c. 400 nm   **d.** 150 mm

15. What determines the color of light? Describe the differences between orange and violet light.

16. Arrange the wavelengths in question 14 in order of *increasing* energy. Which wavelength possesses the most energetic photons?

17. Does all light travel at the same speed in a vacuum? Explain your reasoning.

18. Arrange these types of radiation in order of *increasing* energy per photon: gamma rays, infrared radiation, radio waves, visible light.

19. The microwaves in home microwave ovens have a frequency of $2.45 \times 10^9 \ s^{-1}$. Is this radiation more or less energetic than radio waves? Than X-rays?

20. Ultraviolet radiation is categorized as UV-A, UV-B, or UV-C. Arrange these types in order of increasing:

    a. wavelength

    b. potential for biological damage

    c. energy

21. Calculate the wavelength, in nanometers, of the following wave frequencies:

    a. $6.79 \times 10^{14} \ s^{-1}$

    b. $4.44 \times 10^{12} \ Hz$

22. The distance from Earth to the Sun is about 1.50×10^8 km. How long does it take light from the Sun to travel to Earth?

23. Draw Lewis structures for any two different CFCs.

24. CFCs were used in hair sprays, refrigerators, air conditioners, and plastic foams. Which properties of CFCs made them desirable for these uses?

25. **a.** Can a molecule that contains hydrogen be classified as a CFC?

    b. What is the difference between an HCFC and an HFC?

26. **a.** Most CFCs are based on the structure of either methane, CH_4, or ethane, C_2H_6. Use structural formulas to represent these two compounds.

    b. Substituting both chlorine atoms and fluorine atoms for all of the hydrogen atoms on a methane molecule, you obtain CFCs. How many possibilities exist?

    c. Which of the substituted CFC compounds in part b has been the most successful?

    d. Why weren't all of these compounds equally successful?

27. The following free radicals all play a role in catalyzing ozone depletion reactions: $Cl\cdot$, $\cdot NO_2$, $ClO\cdot$, and $\cdot OH$.

    a. Count the number of outer electrons available and then draw a Lewis structure for each free radical.

    b. What characteristic is shared by these species that makes them so reactive?

28. **a.** How were the original measurements of increases in chlorine monoxide and the stratospheric ozone depletion over the Antarctic obtained?

    b. How are these measurements made today?

29. Which graph shows how measured increases in UV-B radiation correlate with percent reduction in the concentration of ozone in the stratosphere over the South Pole? Explain.

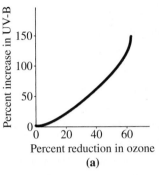
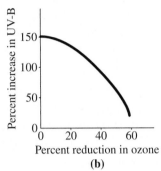

Percent increase in UV-B

Percent reduction in ozone
(a)

Percent increase in UV-B

Percent reduction in ozone
(b)

Concentrating on Concepts

30. The EPA has used the slogan "Ozone: Good Up High, Bad Nearby" in some of its publications for the general public. Explain the message.

31. Nobel Laureate F. Sherwood Rowland referred to the ozone layer as the Achilles heel of our atmosphere. Explain the metaphor.

32. In the abstract of a talk he gave in 2007, Nobel Laureate F. Sherwood Rowland wrote, "Solar UV radiation creates an ozone layer in the atmosphere which in turn completely absorbs the most energetic fraction of this radiation."

    a. What is the most energetic UV fraction?

    b. How does solar UV radiation "create an ozone layer"?

33. "We risk solving one global environmental problem while possibly exacerbating another unless other alternatives can be found." The context of this quote by a U.S. official is phasing out the use of HCFCs.

    a. What compounds were HCFCs being replaced with?

    b. What is the risk of this replacement?

34. It is possible to write three resonance structures for ozone, not just the two shown in the text. Verify that all three structures satisfy the octet rule and offer an explanation of why the triangular structure is not reasonable.

What do you predict the O—O bond lengths will be in ozone? Will they all be the same? Explain your predictions.

36. More ozone is found in the stratosphere than the troposphere. Why is this?

37. Describe why ozone is more reactive than oxygen gas.

38. Consider the Lewis structures for SO_2. How do they compare with the Lewis structures for ozone?

39. Even if you have skin that can produce large amounts of pigment, you cannot get a tan from standing in front of a radio. Why?

40. The morning newspaper reports a UV Index Forecast of 6.5. Given the amount of pigment in your skin, how might this affect how you plan your daily activities?

41. All the reports of the damage caused by UV radiation focus on UV-A and UV-B radiation. Why isn't there more attention paid to the damaging effects of UV-C radiation?

42. What region of UV radiation, and potential health risks, are associated with gel manicures?

43. If all 3.3×10^8 tons of stratospheric ozone that are formed every day are also destroyed every day, how is it possible for stratospheric ozone to offer any protection from UV radiation?

44. How does the chemical inertness of CCl_2F_2 (Freon-12) relate to the usefulness and the problems associated with this compound?

45. Chlorine is a catalyst in chemical reactions involving ozone in the atmosphere. Name two other catalysts found in the atmosphere.

Exploring Extensions

46. DVD players use a laser with a wavelength of 650 nm (a red light) to read the information stored on the discs. Blu-ray players use a 405 nm wavelength laser (blue light). Single-layer DVDs can store 4.7 GB of data compared to 25 GB stored on a Blu-ray disc. Why do Blu-ray discs store so much more data than DVDs?

47. Development of the stratospheric ozone hole has been most dramatic over Antarctica. What set of conditions exist over Antarctica that help to explain why this area is well-suited to studying changes in stratospheric ozone concentration? Are these same conditions not operating in the Arctic? Explain.

48. The free radical $CF_3O\cdot$ is produced during the decomposition of HFC-134a.

    a. Propose a Lewis structure for this free radical.

    b. Offer a possible reason why $CF_3O\cdot$ does not cause ozone depletion.

35. The average length of an O—O single bond is 132 pm. The average length of an O=O double bond is 121 pm.

49. Consider this graph that shows the atmospheric abundance of bromine-containing gases from 1950 to 2100.

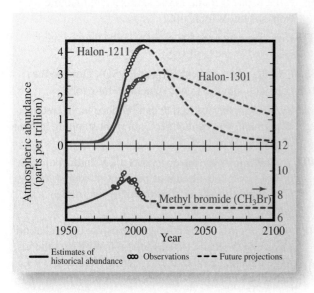

Estimates of historical abundance —— Observations ∞∞ Future projections – – –

a. Halon-1301 is $CBrF_3$ and Halon-1211 is $CClBrF_2$. Why were these compounds once manufactured?

b. Compare the patterns for Halon-1211 and Halon-1301. Why doesn't Halon-1301 drop off as quickly?

c. In 2005, methyl bromide was phased out in the United States except for critical uses. Why is its future use predicted as a straight line, rather than tailing off? Section 2.11 discussed the role of nitrogen monoxide (NO) in forming photochemical smog. What role, if any, does NO play in stratospheric ozone depletion? Are NO sources the same in the stratosphere as in the troposphere?

50. Resonance structures can be used to explain the bonding in charged groups of atoms as well as in neutral molecules, such as ozone. The nitrate ion, NO_3^-, has one additional electron plus the outer electrons contributed by nitrogen and oxygen atoms. That extra electron gives the ion its charge. Draw the resonance structures, verifying that each obeys the octet rule.

51. Some places like China have an issue with smog, a fog created in the reaction of sunlight with nitrogen oxides and carbon compounds released from cars or burning coal. How does smog affect the amount of UV radiation received in these areas? How does it affect the breathing conditions?

52. Ozone depletion has been a concern due to measurements indicating the worldwide ozone concentration has decreased over the past 40 years. Could this issue be solved by releasing ozone into the atmosphere? Why or why not?

53. Although oxygen exists as O_2 and O_3, nitrogen exists only as N_2. Propose an explanation for these facts. *Hint:* Try drawing a Lewis structure for N_3.

54. The chemical formulas for a CFC, such as CFC-11 (CCl_3F), can be figured out from its code number by adding 90 to it to get a three-digit number. For example, with CFC-11 you get $90 + 11 = 101$. The first digit is the # of C atoms, the second is the # of H atoms, and the third is the # of F atoms. Accordingly, CCl_3F has 1 C atom, no H atoms, and 1 F atom. All remaining bonds are assumed to be chlorine.

a. What is the chemical formula for CFC-12?

b. What is the code number for CCl_4?

c. Does this "90" method work for HCFCs? Use HCFC-22 ($CHClF_2$) in explaining your answer.

d. Does this method work for halons? Use Halon-1301 (CF_3Br) in explaining your answer.

55. Many different types of ozone generators ("ozonators") are on the market for sanitizing air, water, and even food. They are often sold with a slogan such as this one from a pool store: "Ozone, the world's most powerful sanitizer!"

a. What claims are made for ozonators intended to purify air?

b. What risks are associated with these devices?

56. The effect a chemical substance has on the ozone layer is measured by a value called its *ozone-depleting potential*, ODP. This is a numerical scale that estimates the lifetime potential stratospheric ozone that could be destroyed by a given mass of the substance. All values are relative to CFC-11, which has an ODP defined as equal to 1.0. Use those facts to answer these questions.

a. Name two factors that affect the ODP value of a compound, and explain the reason for each one.

b. Most CFCs have ODP values ranging from 0.6 to 1.0. What range do you expect for HCFCs? Explain your reasoning.

c. What ODP values do you expect for HFCs? Explain your reasoning.

57. Cooking with an electric stove can have a negative effect on the environment, just as one that uses natural gas can, even though gas is not being burned directly. Why is this? *Hint:* Where does the electricity used to power the stove come from?

58. One mechanism that helps break down ozone in the Antarctic region involves the BrO· free radical. Once formed, it reacts with ClO· to form BrCl and O_2. BrCl, in turn, reacts with sunlight to break into Cl· and Br·, both of which react with O_3 and form O_2.

a. Represent this information with a set of equations.

b. What is the net equation for this cycle?

59. Polar stratospheric clouds (PSCs) play an important role in stratospheric ozone depletion.

   a. Why do PSCs form more often over Antarctica than in the Arctic?

   b. Reactions occur more quickly on the surface of PSCs than in the atmosphere. One such reaction is that of two species that do not deplete ozone, hydrogen chloride and chlorine nitrate ($ClONO_2$), which react to produce a chlorine molecule and nitric acid (HNO_3). Write the chemical equation.

   c. The chlorine molecule produced does not deplete ozone either. However, when sunlight returns to the Antarctic in the springtime, it is converted to a species that does. Show how this happens using a chemical equation.

60. Recent experimental evidence indicates that $ClO\cdot$ initially reacts to form Cl_2O_2.

   a. Predict a reasonable Lewis structure for this molecule. Assume the order of atom linkage is Cl—O—O—Cl.

   b. What effect does this evidence have on understanding the mechanism for the catalytic destruction of ozone by $ClO\cdot$?

61. There are recent studies that have shown that certain chemicals in sunscreen formulations are harming marine ecosystems. Which chemical(s) are responsible and what are some safer alternatives?

Design Elements: Concentrating on Concepts icon, and Simulations/Interactives Icon SAK Design/Shutterstock.

2017 - 2021

°F

Banner credit above: ©RachenPhotographer/Shuttertock; NASA's Scientific Visualization Studio/Kathryn Mersmann

THE GREENHOUSE EFFECT & CLIMATE CHANGE

Watch the chapter opening video. Answer these questions based on your current knowledge.

a. What is the greenhouse effect and how does it impact Earth?
b. Where does the carbon dioxide in the air come from?
c. What is climate change and how is this related to the greenhouse effect?
d. Is there evidence that climate change is occurring now? Explain.

In this chapter, you will explore the following questions:

1. What are the sources of carbon on Earth?
2. How much carbon dioxide is added to the atmosphere every year?
3. What is the importance of the greenhouse effect on Earth?
4. How can you recognize a greenhouse gas?
5. How do greenhouse gases work and what instrumentation is used to measure heat absorption?
6. How do scientists reconstruct past climates and how does the current climate trends differ from the past?
7. How do we know the current climate crisis is anthropogenic, or human-derived?
8. What factors affect the average global temperature?
9. What are the consequences of the modern climate crisis?
10. How can we lessen the impacts of climate change?

Introduction

Since we know that human actions have primarily caused the current climate crisis, we also know that it is our decision what happens next. To make these choices, we must understand the chemistry driving our climate. Before we dig into the science, let's ponder who will be most impacted by the impacts of the changing climate. For this example, we will consider the rise in global sea level.

@HOME   Melting Glaciers and Icebergs

(left): Bernhard Staehli/Shutterstock; (right): Ingram Publishing/SuperStock

Investigate the effects of melted glaciers and icebergs at home using two ice cubes, a large rock with a flat surface on top, two rulers, tape, a marker, water, and two clear glasses.

1. Tape a ruler to the side of each glass so that the bottom of the ruler is touching the table.
2. Place the rock in one glass and add water so that the flat part of the rock is sticking out above the water.
3. Mark the level of the water on the side of the glass using a marker. Then, pour that same amount of water into the other glass. Mark that on the second glass as well. This is your starting height.
4. Add an ice cube to each glass. For the glass with a rock, make sure the ice cube is sitting on the flat part.
5. Let both glasses sit out over time and observe what you see after 30 minutes and one hour.

Questions to Consider
a. Explain which glass represented the glacier and which glass represented the iceberg.
b. What happened to the water level in each glass? How does this relate to climate change and the places that are impacted the most?

Figure 4.1

Lalbagh Fort in Dhaka, Bangladesh and the resulting sea level rise due to 1.5 °C (left) or 3 °C (right) warming. Additional images from around the world can be found at: https://picturing.climatecentral.org/

(Both): Climate Central

An intermediate projection from the Intergovernmental Panel on Climate Change's (IPCC) 2021 report predicts a 2.7 °C (4.9 °F) warming (above preindustrial temperatures) by 2100. Only major reductions in greenhouse gas (GHG) emissions will keep us within a 1.5 °C (2.7 °F) warming. What is the difference between an increase of 1.5 °C and 2.7 °C? Figure 4.1 shows one example.

What does a rise in 3 °C do to sea levels around the world? High-tide would threaten more than 800 million people, 12% of the current global population (600 million of whom live in Asia). This is especially problematic for those who live in small island nations. The livelihoods of the residents and the national economies of these islands are mostly, if not entirely, based on fishing and tourism. At the same time, Pacific islands contribute only 0.03% of total greenhouse gas emissions.

Let's take a closer look at the Maldives, a nation of islands in the Indian Ocean. With a population of 540,500 people, the annual average carbon dioxide emission is 3.7 metric tons per person. While this emission rate is below the global average and far below the rate for most developed nations, the people of Maldives will be greatly impacted by a warmer world because 80% of the 1190 islands (240 km^2 or 92.8 mi^2) are just 1 meter above sea level.

There is a similar trend all over the world. The regions most vulnerable to the impacts of climate change are generally those that contributed the least to the problem and also have the least infrastructure for managing and adapting to climate impacts (Figure 4.2).

In this chapter, we will discuss the chemistry behind climate change. We will present data about the climate system, and help you analyze the data to come up with your own conclusions about what is happening on Earth and what society's next steps should be. This first activity will explore what you know now about this global issue.

Your Turn 4.1   Our Climate

Answer the following questions, which are based on your current knowledge.

a. What have you read or heard about Earth's climate?
b. What causes warming and cooling effects?
c. How has our climate changed in the past 50 years, and what can we expect in the next 50 years?

We will start by examining a very important element, carbon, that plays a critical role in our climate system.

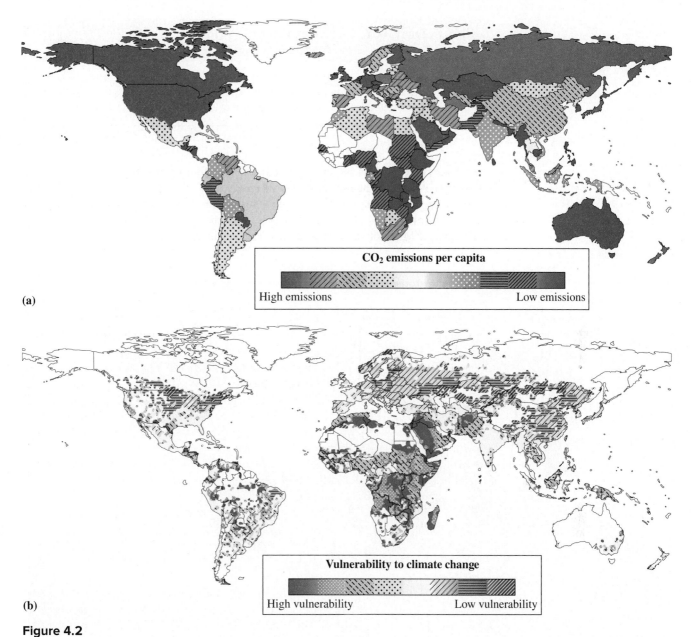

(a)

(b)

Figure 4.2

Distribution of carbon dioxide emissions per capita (a) and vulnerability to climate change (b).

4.1 | Carbon, Carbon Everywhere!

Learning Objectives:

Illustrate, identify, and predict carbon reservoirs and the processes by which carbon moves on Earth

Identify and quantify carbon dioxide emissions from human activities

If you look on the periodic table at element number 6, you will find carbon—one of the most important elements for life on Earth. Chemists consider carbon so important that there is a branch of chemistry devoted to studying carbon and its compounds. **Organic chemistry** is the field that works to understand and produce carbon-based compounds. You may have seen carbon in its elemental form as diamond or as pencil "lead" (actually graphite), but there are millions of compounds that have carbon atoms in them. A few examples include table sugar ($C_{12}H_{22}O_{11}$), methane (CH_4, the primary component of natural gas), acetone (C_3H_6O, a component of many fingernail polish removers), or one compound we will examine extensively, carbon dioxide (CO_2).

As chemists, it is important to understand where we can find carbon atoms on Earth. The global carbon cycle shown in Figure 4.3 represents how carbon-containing substances cycle through nature. Compounds with carbon atoms are found in several places, referred to as pools, or *reservoirs*. You may notice that one reservoir is the atmosphere. Many of the carbon atoms here are in the form of $CO_2(g)$ (~400 ppm),

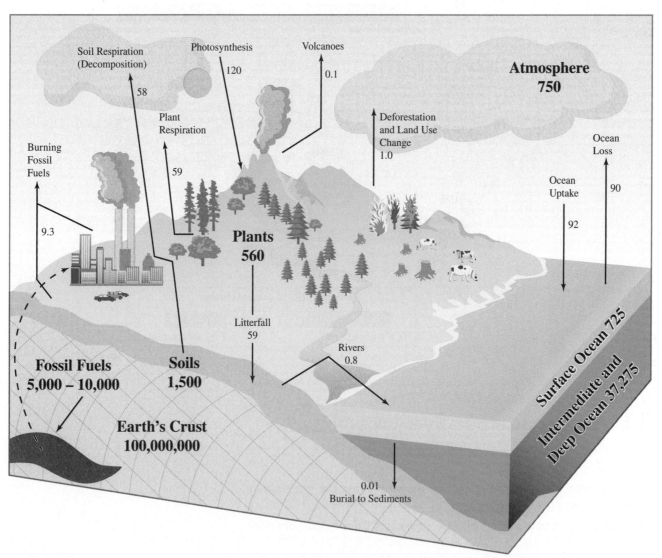

GLOBE® 2017

Data Sources: Adapted from Houghton, R.A. Balancing the Global Carbon Budget. Annu. Rev. Earth Planet. Sci. 007.35:313-347, updated emissions values are from the Global Carbon Project: Carbon Budget 2017. Diagram created by a collaboration between UNH, Charles University and the GLOBE Program.

Figure 4.3

The global carbon cycle. The numbers show the quantity of carbon, expressed in petagrams (Pg), that is stored in various carbon reservoirs (*blue*), or moving through the system in Pg per year (*red*). One Pg of carbon is equal to one gigatonne (Gt) of carbon, which is one billion metric tons, 1×10^{15} grams, about 2.2 trillion pounds, or the weight of 200 million elephants.

Source: The GLOBE program

$CH_4(g)$ (~1.8 ppm), and $CO(g)$ (trace amounts as an air pollutant). Rocks that make up Earth's crust is the largest carbon reservoir. However, the ocean is also an important carbon reservoir. Uptake of carbon dioxide from the atmosphere occurs at the surface of the ocean through photosynthesis, deposition from rainwater, as well as dissolving the gas directly into the ocean surface. Another reservoir for carbon is living creatures—plants and animals—where the carbon atoms are combined with oxygen, hydrogen, nitrogen, and other elements to form carbohydrates, proteins, and lipids.

The carbon cycle demonstrates that carbon atoms are always on the move. Through processes such as combustion, photosynthesis, and sedimentation, carbon atoms move from one reservoir to another (known as carbon flux). It is estimated that an average carbon atom has been recycled from sediment, through the mobile reservoirs of Earth, and back to sediment more than 20 times over Earth's history. Perhaps some of the carbon atoms in your body once belonged to a dinosaur or Julius Caesar! Carbon dioxide gas in the air today may have been released from campfires burning more than a thousand years ago. As you reexamine Figure 4.3, take note that all of the processes are happening simultaneously. However, they may not all occur at the same rate. The amount of time that a carbon atom stays within a reservoir depends on how quickly it can leave the reservoir; for example, the larger the flux out of the atmosphere, the less time a carbon dioxide molecule will remain there.

Your Turn 4.3   Understanding the Carbon Cycle

Using Figure 4.3, answer the following questions.

a. Which processes *add* carbon atoms (in the form of CO_2) to the atmosphere?
b. Which processes *remove* carbon atoms from the atmosphere?
c. What are the two largest reservoirs of carbon atoms?
d. Which parts of the carbon cycle are most influenced by the activities of humans?
e. Why do you think Figure 4.3 is called the "carbon cycle"?

It is important to know how carbon atoms move throughout Earth. For example, the slow transformation of carbon from living organisms into fossil fuels millions of years ago is of great importance to us. Today's rapid transfer of carbon back into the atmosphere by burning fossil fuels will affect future generations, who must deal with the consequences of increased rates of burning these fuels.

Visualizing the Carbon Cycle

Watch this video (www.acs.org/cic) and answer the following questions.

a. Outline the cycle of carbon presented in this video (it highlights the organic carbon cycle). Where are the fossil fuels in this story?
b. Humans are not changing the total amount of carbon on Earth. Why is it a problem to extract carbon from fossil fuels and allow the carbon dioxide produced from combustion to go to the atmosphere?
c. What are some proposed solutions for removing carbon from the atmosphere?

The carbon cycle is a dynamic system, consisting of both natural addition and removal mechanisms. For example, respiration (Equation 4.1) adds carbon dioxide to the atmosphere and photosynthesis (Equation 4.2) removes CO_2; these processes are relatively fast.

$$\text{Respiration: } 6O_2 + C_6H_{12}O_6 \longrightarrow 6CO_2 + 6H_2O \qquad \textbf{[4.1]}$$

$$\text{Photosynthesis: } 6CO_2 + 6H_2O \longrightarrow 6O_2 + C_6H_{12}O_6 \qquad \textbf{[4.2]}$$

**Your Turn 4.4   Visualizing the Annual
                        Carbon Cycle**

Living organisms on Earth "breathe" in CO_2 via photosynthesis and "exhale" CO_2 via respiration. Which process dominates at a particular location will depend on the season. Explore this simulation (www.acs.org/cic) and answer the following questions:

a. Why does the CO_2 concentration in the Northern Hemisphere start to decrease around May/June?

b. Why does the CO_2 concentration in the Northern Hemisphere start to increase around October/November?

c. What is the swirling motion of the carbon dioxide measurements? Is the emission of CO_2 confined to local city centers?

As members of the animal kingdom, we *Homo sapiens* participate in the carbon cycle along with our fellow creatures. As is true for any animal, we inhale and exhale, ingest and excrete, live and die. Human civilization, however, relies on processes that put many more carbon atoms into the atmosphere rather than on processes that remove them (Figure 4.4). Widespread burning of coal, petroleum, and natural gas for electricity production, transportation, and home heating all transfer carbon atoms from the large carbon reservoir located underground into the atmosphere.

Another human influence on CO_2 emissions is manufacturing and construction. Energy-intensive manufacturing includes the production of food (from raw ingredients), paper, pharmaceuticals, textiles, wood products, and steel. Any consumer product that is transformed from raw materials to a final product travels on machinery using energy, most likely electricity. This category also includes the extraction of raw materials from mining, quarrying, and extraction.

While individuals may not have a direct role in decreasing the CO_2 emissions from manufacturing, we do have the power as consumers to choose less-processed items. We also have control over much of the "Buildings" sector, since 62% of these emissions come from residential structures that use combustion sources such as boilers, heaters, generators, etc.

The main source of CO_2 in the "Industry" sector is cement production. Concrete is the most ubiquitous human-made material and has transformed our growing cities. Concrete is a mixture of sand, gravel, a cement binder, and water. Unfortunately, during the production of cement, CO_2 is released as a chemical byproduct. As shown in Equation 4.3, the main component of cement is composed of limestone ($CaCO_3$) that decomposes primarily into lime (CaO) and CO_2 under high heat. The high temperatures (>1000 °C) needed for this reaction require the burning of a fuel, which also releases additional CO_2 to the atmosphere.

$$CaCO_3 + heat \longrightarrow CaO + CO_2 \qquad\qquad [4.3]$$

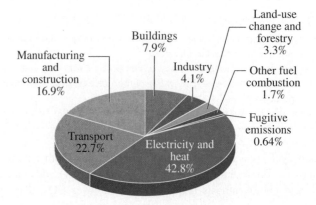

Figure 4.4

Global 2018 carbon dioxide emissions. Percent contribution of direct emissions by sector. Fugitive emissions are the accidental leakage of methane during coal mining.

Source: Data from CAIT Climate Data Explorer via Climate Watch

Your Turn 4.5   CO$_2$ Savings

The combustion of a gallon of gasoline produces 8,887 g of CO$_2$ and there are approximately 1.45 billion cars in the world. Assume every car travels 11,500 miles/year and has a fuel efficiency of 22.0 miles per gallon (mpg). How much CO$_2$ would we save if each car decreased its travel to 10,000 miles/year and was upgraded to a more fuel-efficient hybrid that gets 51.0 mpg? Compare this savings to the annual CO$_2$ emissions from the trucking industry (1.8 Gton CO$_2$/year).

The total quantity of carbon-containing substances released by the human activities of deforestation and burning fossil fuels is about 10.3 Gt per year. About half of this is eventually recycled into the oceans and the biosphere; however, carbon dioxide is not removed as quickly as the rate of addition to the atmosphere. About half of the emitted CO$_2$ stays in the atmosphere, adding 5.2 Gt of carbon per year to the existing base of 750 Gt noted in Figure 4.3. We are concerned primarily with the relatively rapid *increase* in atmospheric carbon dioxide, because *excess* CO$_2$ can affect Earth's climate and the health of Earth's oceans. Therefore, it would be useful to know the mass (Gt) of CO$_2$ added to the atmosphere each year. In other words, what mass of CO$_2$ contains 5.2 Gt of carbon? To answer this question, we will now shift our focus to discuss some fundamental and quantitative aspects of chemistry.

4.2 | Quantifying Carbon Dioxide—First Stop: Mass

> *Learning Objectives:*
>
> *Compare the atomic structures of isotopes and derive relative atomic masses*
>
> *Use Avogadro's number to quantify the mass of atoms*

In order to figure out what mass of CO$_2$ contains 5.2 Gt of carbon, we need the relationship between the mass of C atoms in a sample of CO$_2$ gas and the mass of CO$_2$ molecules. Thus, we must calculate the mass percent of C in CO$_2$, based on the formula of the compound.

First we must ask: How much does an individual atom weigh? The mass of an atom is mainly due to the neutrons and protons in the nucleus. As you learned in Section 1.5, varying elements differ in mass because their atoms differ in composition. Two or more atoms with the same number of protons but different neutrons are known as **isotopes**. Isotopes are designated by their mass numbers; for instance, nitrogen-14 (N-14) refers to a mass number of 14 with 7 protons and 7 neutrons. In contrast, nitrogen-15 (N-15) refers to 7 protons and 8 neutrons.

Instead of using the absolute masses of individual atoms, chemists have found it convenient to employ relative masses—in other words, to relate the masses of all atoms to some convenient standard. The internationally accepted mass standard is carbon-12, the isotope that makes up 98.90% of all carbon atoms. Carbon-12 (C-12) atoms have a mass number of 12 because each atom has a nucleus consisting of 6 protons and 6 neutrons.

The periodic table shows that the atomic mass of carbon is 12.01, not 12.00. This is not an error; it reflects the fact that carbon exists naturally as isotopes. Although C-12 (also written as ^{12}C) predominates, 1.10% of carbon is C-13, the isotope with 6 protons and 7 neutrons. In addition, natural carbon contains a trace of C-14, the isotope with 6 protons and 8 neutrons. The tabulated mass value of 12.01 is called the **relative atomic mass**, a weighted average that takes into consideration the masses and percent of the natural abundances of all naturally occurring isotopes of carbon (Table 4.1). This isotopic distribution and average mass of 12.01

Table 4.1	Isotopes of Carbon		
Isotope	**Mass Number**	**Relative Percent**	**Contribution to Relative Atomic Mass***
C-12	12	98.90%	11.868
C-13	13	1.10%	0.143
C-14	14	~0.001%	0.0001
			Avg. atomic mass of C = 11.868 + 0.143 + 0.0001 = 12.011

*Calculated as the relative percent × mass number. For instance, 98.90% of the C-12 isotope is 0.9890 × 12 = 11.868.

characterize carbon obtained from any natural source—a graphite ("lead") pencil, a tank of gasoline, a loaf of bread, a lump of limestone, or your body. You may notice that no units are used with the atomic masses. These are expressed in *unified atomic mass units* (u), which equals 1.67×10^{-27} kg.

Your Turn 4.6   Isotopes of Nitrogen

Check out the simulation below to learn about isotopes and how their abundances relate to the average atomic mass of an element: www.acs.org/cic.

Example Problems:

a. What is the atomic number, atomic mass, and number of protons, neutrons, and electrons in a neutral atom of B-11?

b. What isotopes do you obtain by adding or subtracting a neutron from B-11?

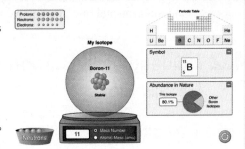

PhET Interactive Simulations, University of Colorado Boulder, http://phet.colorado.edu

Example Solutions:

a. Pull up the simulation and click on the symbol for boron (B). Click on the green "+" icons to expand the symbol and abundance boxes. The atomic number of boron from its isotope symbol is 5. If you click on "atomic mass" at the bottom of the balance, you will get a value of 11.00931 amu for boron. By counting the color-coded subatomic particles, you get 5 protons, 6 neutrons, and 5 electrons for boron. These numbers make sense since the number of positively charged protons and negatively charged electrons must be equal for a neutral atom.

b. The B-10 isotope is obtained by dragging a neutron from the B-11 nucleus to the neutron basket to the left of the balance. The atomic number, and thus number of protons and electrons, does not change, but the isotope now has one fewer neutron. That is, B-10 has 5 protons, 5 neutrons, and 5 electrons with a natural abundance of 19.9%. If you drag an additional neutron to B-11, you get the B-12 isotope, which is unstable.

Your Turn:

Using the above simulation, answer the following questions related to the two naturally occurring isotopes of nitrogen: N-14 and N-15.

a. Use the periodic table to find the atomic number and atomic mass of nitrogen atoms.

b. What is the number of protons, neutrons, and electrons in a neutral atom of N-14?

c. Compare your answers for part **b** with those for a neutral atom of N-15.

d. Given the relative atomic mass of nitrogen, which isotope has the greatest natural abundance?

We now return to the matter at hand—the masses of atoms, particularly the atoms in CO_2. Not surprisingly, it is difficult to weigh a single atom because of its extremely small mass. A typical laboratory balance can detect a minimum mass of 0.1 mg; this corresponds to 5,000,000,000,000,000,000 carbon atoms, or 5×10^{18} carbon atoms!

A unified atomic mass unit (u) is far too small to measure in a conventional chemistry laboratory. Rather, the gram is the chemist's mass unit of choice. Therefore, scientists use exactly 12 g of carbon-12 as the reference for the atomic masses of all the elements. We define relative atomic mass as the mass (in grams) of the same number of atoms that are found in exactly 12 g of carbon-12. This number of atoms is, of course, *very* large; in fact, **Avogadro's number** is 602,200,000,000,000,000,000,000. It is more compactly written in scientific notation as 6.022×10^{23}. This is the incredibly large number of atoms in exactly 12 g of carbon-12, no more than a tablespoon of soot!

Avogadro's number is used to count a collection of atoms, like the term *dozen* counts a collection of eggs. It does not matter if the eggs are large or small, brown or white, organic or not. In all cases, if there are 12 eggs, they are counted as a dozen. However, a dozen ostrich eggs has a greater *mass* than a dozen quail eggs.

Your Turn 4.7   Mass and Half-Dozen

Example Problem:

A single ostrich egg has a mass of 1.35 kg. In comparison, a dozen chicken eggs has a mass of 21 oz. Which variety has a larger mass?

Conrad Stanitski

Example Solution:

To compare masses, their units must be identical. The conversion factor between kg and oz is: 1 kg = 35.274 oz. Hence, to compare the mass of eggs, you can use dimensional analysis:

a single ostrich egg: $1.35 \text{ kg} \times \dfrac{35.27 \text{ oz}}{1 \text{ kg}} = 47.6 \text{ oz}$

Since a dozen chicken eggs have a mass of 21 oz, a single ostrich egg has a greater mass.

Your Turn:

Verify the weight comparison of a half-dozen of each type of ball shown below given that one tennis ball weighs 2.05 oz and one golf ball weighs 45.9 g.

Your Turn 4.8   Marshmallows and Pennies

(left): Anthony Rosenberg/Getty Images; (right): karen roach/Shutterstock

Avogadro's number is so large that the only way to visualize it is through analogies. For example, an Avogadro's number of regular-sized marshmallows, 6.02×10^{23} of them, would cover the surface of the United States to a depth of 650 miles. Or, if you are more impressed by money than marshmallows, assume 6.02×10^{23} pennies were distributed evenly among the approximately 7.8 billion inhabitants of Earth. Every man, woman, and child could spend $1 million every hour, day and night, and half of those pennies would still be left unspent by the time each person passes away. Can these fantastic claims be correct? Check one or both, showing your reasoning. Come up with an analogy of your own.

Knowledge of Avogadro's number and the relative atomic mass of any element permit us to calculate the average mass of an individual atom of that element. Thus, the mass of 6.02×10^{23} oxygen atoms is 16.00 g, the relative atomic mass from the periodic table. To find the average mass of just one oxygen atom, we must divide the mass of the large collection of atoms by the size of the collection. This means dividing the relative atomic mass by Avogadro's number. Fortunately, calculators help make this job quick and easy:

$$\frac{16.00 \text{ g oxygen}}{6.02 \times 10^{23} \text{ oxygen atoms}} = 2.66 \times 10^{-23} \text{ grams per oxygen atom}$$

This very small mass confirms once again why chemists do not generally work with small numbers of atoms. We manipulate trillions at a time. Therefore, practitioners of this art need to measure matter with a sort of chemist's dozen—a very large one, indeed!

Your Turn 4.9   Calculating the Mass of Atoms

Follow these instructions for the three examples (a-c) below:

- Predict whether the value will be large or small.
- Calculate the value.
- Do your calculations match your predictions? Think about whether your predictions were reasonable.

  a. The average mass in grams of an individual atom of carbon.
  b. The mass in grams of 5 trillion carbon atoms.
  c. The mass in grams of 6×10^{15} carbon atoms.

4.3 | Quantifying Carbon Dioxide—Next Stop: Molecules and Moles

Learning Objective: Quantify the mass of an atom in a molecule

Another way to communicate the number of atoms, ions, or molecules is to use the term **mole** (mol), defined as an Avogadro's number of objects. The term is derived from the Latin word to "heap," or "pile up." Thus, 1 mole of carbon atoms is 6.022×10^{23} C atoms and 1 mole of aluminum atoms is 6.022×10^{23} Al atoms. In fact, 1 mol of people would be 6.022×10^{23} people! So, how many atoms will comprise 1 mol of oxygen gas? Before you answer 6.022×10^{23} oxygen atoms, remember that oxygen gas is made up of O_2 *molecules*. Since there are 2 moles of oxygen atoms (O) in every mole of oxygen molecules (O_2), there are 1.204×10^{24} O atoms in one mole of oxygen gas.

As you already know from previous chapters, chemical formulas and equations involving molecular compounds are written in terms of atoms and molecules. For example, consider Equation 4.4 for the complete combustion of carbon:

$$C(s) + O_2(g) \longrightarrow CO_2(g) \qquad \textbf{[4.4]}$$

This equation tells us that one *atom* of carbon combines with one *molecule* of oxygen to yield one *molecule* of carbon dioxide. Thus, it reflects the ratio in which the particles interact. It is equally correct to say that 10 C atoms react with 10 O_2 molecules (20 O atoms) to form 10 CO_2 molecules. Or, putting the reaction on a grander scale, we can say 6.022×10^{23} C atoms react with 6.022×10^{23} O_2 molecules (1.204×10^{24} O atoms) to yield 6.022×10^{23} CO_2 molecules! The last statement is equivalent to saying: "one *mole* of carbon plus one *mole* of oxygen yields one *mole* of carbon dioxide." Thus, the numbers of *atoms and molecules* taking part in a reaction are proportional to the numbers of *moles* of the same substances:

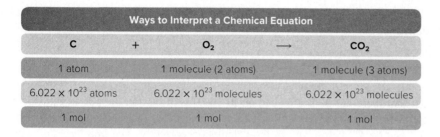

Ways to Interpret a Chemical Equation		
C + O$_2$ $\longrightarrow$		CO$_2$
1 atom	1 molecule (2 atoms)	1 molecule (3 atoms)
6.022×10^{23} atoms	6.022×10^{23} molecules	6.022×10^{23} molecules
1 mol	1 mol	1 mol

The ratio of two oxygen atoms to one carbon atom remains the same regardless of the number of carbon dioxide molecules.

In the laboratory and the factory, the quantity of matter required for a reaction is often measured by mass. The mole is a practical way to relate the number of particles to the more easily measured mass. The **molar mass** is the mass of Avogadro's number, or one mole, of whatever substance is specified. For example, from the periodic table we can see that the mass of one mole of carbon atoms, rounded to the nearest tenth of a gram, is 12.0 g. In comparison, a mole of oxygen atoms has a mass of 16.0 g. But we can also speak of a mole of O$_2$ molecules. Because there are two oxygen atoms in each oxygen molecule, there are two moles of oxygen atoms in each mole of molecular oxygen, O$_2$. Consequently, the molar mass of O$_2$ is 32.0 g, twice the molar mass of O.

The same logic for determining the molar mass of the element O$_2$ applies to compounds, such as carbon dioxide. The formula of carbon dioxide, CO$_2$, reveals that each molecule contains one carbon atom and two oxygen atoms. Scaling up by 6.022×10^{23}, we can say that each mole of CO$_2$ consists of 1 mol of C atoms and 2 mol of O atoms. But remember that we are interested in the molar mass of carbon dioxide, which we obtain by adding the molar mass of carbon atoms to twice the molar mass of oxygen atoms:

$$1 \text{ mol CO}_2 = 1 \text{ mol C} + 2 \text{ mol O}$$

$$= \left(1 \text{ mol C} \times \frac{12.0 \text{ g C}}{1 \text{ mol C}}\right) + \left(2 \text{ mol O} \times \frac{16.0 \text{ g O}}{1 \text{ mol O}}\right)$$

$$= 12.0 \text{ g C} + 32.0 \text{ g O}$$

$$1 \text{ mol CO}_2 = 44.0 \text{ g CO}_2$$

Your Turn 4.10   Molar Mass

Example Problem:
Calculate the molar mass of ammonia, NH$_3$, the primary ingredient of fertilizers.

Example Solution:

$$1 \text{ mol NH}_3 = (1 \text{ mol N} \times 14.0 \text{ g/mol N}) + (3 \text{ mol H} \times 1.0 \text{ g/mol H}) = 17.0 \text{ g}$$

Your Turn:
Calculate the molar mass of each of the following important greenhouse gases that are found in the atmosphere.

a. O$_3$ (ozone)

b. N$_2$O (dinitrogen monoxide or nitrous oxide)

c. CCl$_3$F (Freon-11; trichlorofluoromethane)

Recall that we started out on this mathematical excursion so that we could calculate the mass of CO$_2$ produced from burning 5.2 Gt of carbon. We now have all the necessary pieces assembled. Out of every 44.0 g (1 mole) of CO_2, 12.0 g is C. We can use this ratio to calculate the mass of CO_2 released by burning any known mass of carbon.

For instance, we could calculate the grams of C in 100.0 g CO_2 by setting up the relationship in this manner:

$$100.0 \text{ g } \cancel{CO_2} \times \frac{12.0 \text{ g C}}{44.0 \text{ g } \cancel{CO_2}} = 27.3 \text{ g C}$$

The fact that there are 27.3 g of carbon in 100.0 g of carbon dioxide is equivalent to saying that the mass percent of C in CO_2 is 27.3%. Note that carrying along the units "g CO_2" and "g C" helps you do the calculation correctly. The unit "g CO_2" can be canceled, and you are left with "g C," because one unit is in the numerator and the other is in the denominator. Keeping track of the units, and canceling where appropriate, are useful strategies in solving many problems.

Your Turn 4.11   Mass Ratios and Percents

Example Problem:
Calculate the mass ratio of C atoms in CO molecules.

Example Solution:
The mass ratio is calculated by comparing the molar masses of C and CO:

$$\frac{12.0 \text{ g/mol C}}{28.0 \text{ g/mol CO}} = 0.429$$

Your Turn:

a. Calculate the mass ratio of S atoms in SO_2 molecules.
b. Find the mass percent of S atoms in SO_2 molecules.
c. Calculate the mass ratio and the mass percent of N atoms in N_2O molecules.

Grams of CO_2

This video shows the conversion of 3.3 Gt of C atoms to grams of CO_2: www.acs.org/cic. Using this knowledge and that data shown in this chapter, calculate the grams of CO_2 from the 5.2 gigatonnes of carbon that stays in the atmosphere every year due to human activity.

Chemistry in Context
Tutorial

CO_2 CALCULATIONS

©Bradley D. Fahlman

To find the mass of CO_2 that contains 5.2 Gt of C atoms, we could convert 5.2 Gt to grams, but it is not necessary. As long as we use the same mass unit for C atoms and CO_2 molecules, the same numerical ratio holds. Compared with our last calculation, this problem has one important difference in how we use the ratio. We are solving for the mass of CO_2 molecules, not the mass of C atoms. Look carefully at the units this time:

$$5.2 \text{ } \cancel{\text{Gt C}} \times \frac{44.0 \text{ Gt } CO_2}{12.0 \text{ } \cancel{\text{Gt C}}} = 19 \text{ Gt } CO_2$$

Once again, the units cancel and we are left with Gt of CO_2.

Our burning question of the mass of CO_2 molecules added to the atmosphere each year from the combustion of fossil fuels has finally been answered: 19 Gt! Along the way, we demonstrated the problem-solving power of chemistry and introduced five important ideas: atomic mass, Avogadro's number, mole, molar mass, and mass percent. The video tutorial referenced in the "Grams of CO_2" activity shows another example of calculating the mass of CO_2 derived from carbon.

If you know how to apply these ideas, you have the ability to critically evaluate media reports about releases of carbon or CO_2 (and other substances), and judge their accuracy. One can either take such statements on faith, or check their accuracy by applying mathematics to the relevant chemical concepts. Obviously, there is insufficient time to check every assertion, but we hope that you develop questioning and critical attitudes toward all statements about chemistry and society, including those found in this book.

Your Turn 4.12   Checking Carbon from Cars

A clean-burning automobile engine emits about 5 lb of C atoms in the form of CO_2 molecules for every gallon of gasoline it consumes. The average American car is driven about 12,000 miles per year. Using this information, check the statement that the average American car releases its own weight in carbon into the atmosphere each year. List the assumptions you make to solve this problem. Compare your list and your answer with those of your classmates.

How Should I Report My Data?

In the calculations presented in this chapter, answers are reported with a specific number of *significant figures*. Keeping track of significant figures means that our reported values cannot be more precise than the instrumentation used to collect the data. For instance, a mass of "one gram" that has been determined using a balance with a precision of ±0.01 g should be reported as 1.00 g.

Watch this video (www.acs.org/cic) for more information about determining significant figures (or "sig figs"). Then, answer the questions below.

a. How many significant figures are in: 750 Gt C (the amount of carbon in our atmosphere noted in Figure 4.3)?
b. Add up all of the natural carbon flows into the atmosphere pictured in Figure 4.3 using the sig fig rules.
c. How many molecules of carbon dioxide are in 100.0 g of CO_2?

Stephen Frisch/McGraw Hill

4.4 | Why Does It Matter Where Carbon Atoms End Up?

Learning Objective: Explain Earth's energy balance and the importance of the greenhouse effect

Now that we can quantify the amounts of carbon in different reservoirs, we need to look at what effect carbon-containing material can have in these reservoirs. The energy to heat Earth comes mainly from the Sun; however, this is not the entire story. Based on Earth's distance from the Sun and the amount of solar radiation the Sun emits, the average temperature on Earth should be −18 °C (0 °F) and the oceans should be frozen year round. Thankfully, this is not true; Earth's average surface temperature is currently around 15 °C (59 °F).

Venus (Figure 4.5) is another planet whose temperature is inconsistent with its distance from the Sun. Considered by some to be the brightest and most beautiful body in the night sky, Venus has an average temperature of about 450 °C (840 °F). Based simply on its distance from the Sun, however, the average temperature on Venus should be only 100 °C, the boiling point of water. What do Earth and Venus have in common that would explain these discrepancies? They both have an atmosphere. To see the role that our atmosphere plays, we now examine what happens when solar radiation reaches Earth.

The energy processes that contribute to Earth's energy balance appear in Figure 4.6. Earth receives nearly all of its energy from the Sun, primarily in the form of ultraviolet, visible, and infrared radiation. Some of this incoming radiation is reflected back into space by the dust and aerosol particles suspended in our atmosphere (25%). Other parts of this incoming radiation are reflected by the surface of Earth itself, especially

Figure 4.5
Venus, as photographed by the *Galileo* spacecraft.
Tristan3D/Shutterstock

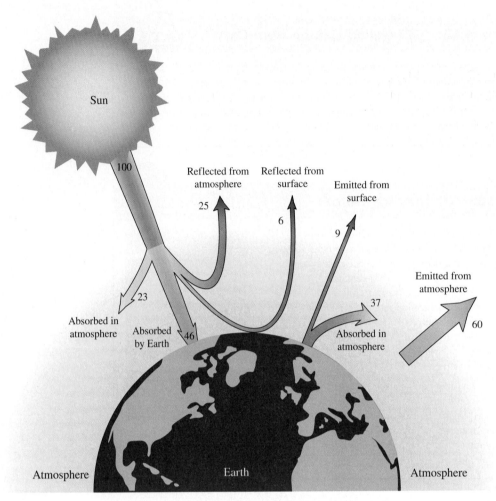

Figure 4.6

Earth's energy balance. Orange represents a mixture of wavelengths of electromagnetic radiation; shorter wavelengths of radiation are shown in blue, longer ones in red. The values are given in percentages of the total incoming solar radiation.

those regions white with snow or sea ice (6%). Thus, 31% of the radiation received from the Sun is reflected (blue arrows). The remaining 69% of the radiation from the Sun is absorbed (orange arrows), either by the atmosphere (23%) or by landmasses and oceans (46%). We can account for all of the Sun's radiation by adding the reflected and absorbed radiation: 31% + 69% = 100%.

In order to maintain Earth's energy balance, all of the radiation that is absorbed from the Sun must eventually go back into space so that incoming energy equals outgoing energy at the top of the atmosphere. Figure 4.6 shows us that:

- 46% of the Sun's radiation is absorbed by Earth's surface.
- Earth's surface has its own energy balance such that it re-emits all of the radiation that it absorbs (37% + 9% = 46%). Since Earth has a lower temperature than the Sun, the energy it emits has longer wavelengths and comes in the form of IR radiation (red arrows).
  - Part of what Earth emits escapes directly into space (9%).
  - The remainder is absorbed by gases in the atmosphere (37%).
  - At any one time, 80% (or 37 ÷ 46 × 100%) of Earth's emitted radiation will be absorbed by the atmosphere.
- The Sun's radiation that is not absorbed by Earth's surface (54%) is either absorbed by the atmosphere (23%), reflected by the atmosphere (25%), or reflected by Earth's surface (6%).
- The 60% of the radiation that is absorbed by the atmosphere, either directly from the Sun (23%) or from Earth's surface (37%), eventually is emitted into space to complete the energy balance.

Figure 4.7

A typical botanical greenhouse.

Jeanie333/Shutterstock

This process of absorption causes collisions between neighboring atmospheric gas molecules, which warms up Earth's atmosphere. As we hope you can see, the gases in Earth's atmosphere hold in heat like a botanical greenhouse (Figure 4.7).

Comparing Energy Balances: Earth and Moon

Watch this video (www.acs.org/cic) and answer the following question: The Moon and the Earth are about the same distance from the Sun, why are their surface temperatures so different?

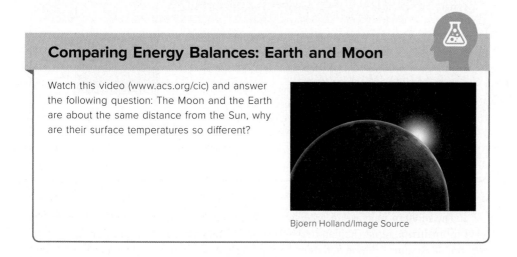

Bjoern Holland/Image Source

What does a greenhouse actually do? If you have ever parked a car with its windows closed on a sunny day, you probably have experienced firsthand how a greenhouse can buildup heat. The car, with its glass windows, operates much the same way as a greenhouse for growing plants. The glass windows transmit visible and UV light from the Sun. This energy is absorbed by the car's interior, particularly by any dark surface. Some of this radiant energy is then re-emitted as longer wavelength IR radiation (heat). Unlike visible light, infrared light is not easily transmitted through the glass windows and so it becomes trapped inside the car. When you re-enter the vehicle, a blast of hot air greets you. The temperature inside a car can exceed 49 °C (120 °F) in the summer in certain climates! Although the physical barrier of the windows is not an exact analogy to Earth's atmosphere, this is the origin of the "greenhouse effect" terminology.

The **greenhouse effect** is a natural process by which atmospheric gases absorb and re-radiate a major portion (about 80%) of the infrared radiation radiated by Earth. Again, Earth's higher-than-expected average annual temperature of 15 °C (59 °F) is a result of the gases in our atmosphere that absorb and re-radiate heat energy towards the surface. The atmosphere of Venus acts in a similar manner but its atmosphere is made up of nearly 96% carbon dioxide, which, as we will see, is a far greater concentration than that in Earth's atmosphere.

Carbon dioxide, which is present in the atmospheres of both Earth and Venus, is a greenhouse gas. **Greenhouse gases** (GHGs) are those gases capable of absorbing and emitting IR radiation, thereby warming the atmosphere. In addition to carbon dioxide, other important GHGs are water vapor, methane, nitrous oxide, ozone, and chlorofluorocarbons (CFCs). The natural presence of these gases (particularly water and carbon dioxide) is essential in keeping our planet at habitable temperatures. Water vapor is the most abundant greenhouse gas in our atmosphere and increases with increasing temperature of Earth's atmosphere. The amount of water vapor is primarily due to natural processes.

The exchange of energy among Earth, atmosphere, and space results in a steady state and a stable average temperature of Earth. However, the increase in concentration of greenhouse gases that is taking place today is changing the energy balance and causing changes in the temperature of the planet. The term **enhanced greenhouse effect** refers to the process in which atmospheric gases absorb and re-emit *more than* 80% of the heat energy radiated by Earth. An increase in the concentration of greenhouse gases will mean that more than 80% of the radiated energy will be returned to Earth's surface, with an accompanying increase in average global temperature. The term **global warming**, which you likely have heard, is frequently used to describe the increase in average global temperatures that results from an enhanced greenhouse effect.

Your Turn 4.13   Greenhouse Simulation

Use this simulation to learn about the impact of greenhouse gases on Earth's surface temperature: www.acs.org/cic. For context, our current average surface temperature is 13.9 °C or 57 °F.

a. Observe what happens when the Sun is initially turned on. Wait for the temperature to stabilize.

b. What happens when you turn on the clouds?

c. What happens when you adjust the amount of greenhouse gases (you can even use the pre-programmed timepoints of Ice Age, 1750, 2020).

PhET/University of Colorado Boulder

The modern increase in greenhouse gases started in the mid-1800s and is attributed to **anthropogenic** (human-derived) sources such as industry, transportation, mining, and agriculture. These activities require carbon-based fuels, which produce carbon dioxide when burned. In the late 19th century, Swedish scientist Svante Arrhenius (1859–1927) considered the problems that increased industrialization might cause by building up CO_2 in the atmosphere. He calculated that doubling the concentration of CO_2 would result in an increase of 5–6 °C in the average temperature of the planet's surface. But, what makes CO_2 a greenhouse gas? The next few sections will answer this key question.

4.5 | The Importance of Molecular Shapes

Learning Objective: Predict the 3D shape of a molecule

Carbon dioxide, water, and methane are greenhouse gases; in contrast, nitrogen and oxygen are not. What is the difference? The answer relates in part to how many atoms a molecule has and its molecular shape. In this section, we will use your knowledge of Lewis structures to predict shapes of molecules. In the subsequent section, we will connect these shapes to molecular vibrations, which can help us explain the difference between greenhouse gases and non-greenhouse gases.

In Section 3.6, you used Lewis structures to predict how electrons are arranged in atoms and molecules; now, we will connect that information to a three-dimensional shape. What is the shape for diatomic molecules such as O_2 and N_2? Here, the shape is unambiguous; the molecule must be linear:

$$:N::N: \quad \text{or} \quad :N\equiv N: \quad \text{or} \quad N\equiv N$$

$$\ddot{O}::\ddot{O} \quad \text{or} \quad \ddot{O}=\ddot{O} \quad \text{or} \quad O=O$$

However, a variety of geometries are possible for larger molecules. The first step in predicting the shape of a molecule is to draw its Lewis structure. Each atom (except hydrogen and helium) is usually associated with four pairs of electrons, known as the *octet rule*. In addition to bonding electrons of the covalent bonds, some molecules may also include nonbonding pairs (also known as *lone pairs*) of electrons.

Opposite charges attract and like charges repel. For instance, pairs of negatively charged electrons repel one another. Thus, when determining shape, the mutually repelling electron pairs around the central atom are placed as far apart as possible to minimize the overall energy of the molecule. We will illustrate the procedure for understanding the experimental shape of a molecule using methane (CH_4), a greenhouse gas.

1. **Determine the number of valence electrons associated with each atom in the molecule**. The carbon atom (Group 14) has four valence electrons; each of the four hydrogen atoms contributes one electron. This gives 8 total valence electrons.

2. **Draw a Lewis structure in which you arrange the valence electrons in pairs to include 8 electrons around the central atom**. This may require single, double, or triple bonds. For the methane molecule, use the eight valence electrons to form four single bonds (four electron pairs) around the central carbon atom. This is the Lewis structure:

$$
\begin{array}{cc}
\text{H} & \text{H} \\
\text{H} \!:\!\ddot{\text{C}}\!:\! \text{H} \quad \text{or} & \text{H}-\text{C}-\text{H} \\
\text{H} & \text{H}
\end{array}
$$

A quick check of the structure reveals this to be a stable structure because carbon has an octet and each hydrogen has two electrons.

3. **Assume that the most stable 3D molecular shape has the bonding electron pairs as far apart as possible**. Although the Lewis structure above implies that the CH_4 molecule is flat, it is not. The four bonding electron pairs around the carbon atom in CH_4 repel one another, and in their most stable arrangement they are as far from one another as possible. As a result, the four hydrogen atoms are also as far from one another as possible. The resulting shape is *tetrahedral,* because the hydrogen atoms correspond to the corners of a *tetrahedron,* a four-cornered geometric shape with four equal triangular sides, sometimes called a triangular pyramid.

Figure 4.8

The legs and the shaft of a music stand approximate the geometry of the bonds in a tetrahedral molecule such as methane.

Kasper Ravlo/Shutterstock

One way to describe the shape of a CH_4 molecule is by analogy to the base of a folding music stand. The four C—H bonds correspond to the three evenly spaced legs and the vertical shaft of the stand (Figure 4.8). The angle between each pair of bonds is 109.5°. The tetrahedral shape of a CH_4 molecule has been experimentally confirmed. Indeed, it is one of the most common atomic arrangements in nature, particularly in carbon-containing molecules.

Your Turn 4.14   Methane—Flat or Tetrahedral?

a. If the methane molecule were flat, as the two-dimensional Lewis structure seems to predict, what would the H—C—H bond angle be?

b. Consider the part of the music stand circled in yellow, shown in Figure 4.8. In the analogy of shape using a music stand, where would the carbon atom be located? Where would each of the hydrogen atoms be situated?

c. Offer a reason why the tetrahedral shape, not the two-dimensional flat shape, is more advantageous for this molecule. For a 3D rendering of this structure, go to www.acs.org/cic.

Hint: Think about the position of electrons around the carbon atom.

Chemists represent molecules in several ways. The simplest, of course, is the chemical formula itself. In the case of methane, this is simply CH_4. Another is the Lewis structure, but again this is only a two-dimensional representation that gives information about the valence electrons. Figure 4.9 shows these two representations, as well as two others that are three-dimensional in appearance. One has a wedge-shaped line that represents a bond coming out of the paper toward the reader. The dashed wedge in the same structural formula represents a bond pointing away from the reader. The two solid lines lie in the plane of the paper. The other, a **space-filling model**, was drawn with a molecular modeling program. Space-filling models enclose the volume occupied by electrons in an atom or a molecule. Seeing and manipulating physical or digital models, in either the classroom or online, can help you visualize the structure of molecules.

However, not all valence electrons reside in bonding pairs. In some molecules, the central atom has nonbonding electron pairs, also called *lone pairs*. A lone pair of electrons occupies more space than a bonding pair of electrons. Consequently, the lone pair repels the bonding pairs more strongly than the bonding pairs repel one another. The shape of a molecule is determined by its arrangement of bonding electron pairs (covalent bonds) and nonbonding electron pairs (lone pairs) around the central atom.

Your Turn 4.15   Shape of Ammonia

a. Write out the Lewis structure for ammonia (NH_3).

b. How many bonded atoms and how many non-bonded electron pairs are around the central atom?

c. Since bonding electron pairs want to be as far apart from each other as possible and lone pairs repel bonding pairs even more strongly, predict the shape of ammonia.

d. Using the music stand analogy from Figure 4.8, where would the nitrogen and hydrogen atoms be located?

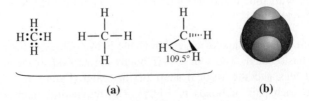

(a)   (b)

Figure 4.9

Representations of molecules of methane, CH_4. Shown are the **(a)** Lewis structures and structural formula, and **(b)** space-filling model. For a 3D rendering of this structure, go to www.acs.org/cic.

Table 4.2 shows some common shapes that can be predicted from knowing the number of bonded atoms and nonbonding electron pairs around a central atom.

Table 4.2	Common Molecular Geometries			
# of Bonded Atoms to the Central Atom	# of Nonbonding Electron Pairs on Central Atom	Geometry	Example	Structural Model
2	0	linear	CO_2	
2	2	bent	H_2O	
3	0	trigonal planar	$COCl_2$	
3	1	trigonal pyramidal	NH_3	
4	0	tetrahedral	CH_4	

Your Turn 4.16   Predicting Molecular Shapes, Part 1

Using Table 4.2 and the strategies just described, predict and sketch the shape for each of these molecules.

a. CCl_4 (carbon tetrachloride, a human-made emission from chemical manufacturing)
b. CCl_2F_2 (Freon-12, dichlorodifluoromethane, a human-made chlorofluorocarbon)
c. H_2S (hydrogen sulfide, emitted from petroleum refining and volcanoes, smells like rotten eggs)

We already looked at the structures of several molecules important for understanding the chemistry of climate change. What about the Lewis structure of the carbon dioxide molecule?

Your Turn 4.17   The Shape of Carbon Dioxide

a. How many valence electrons does carbon dioxide have?
b. What happens if only single bonds form between each C and O?
c. If double bonds are used, how many bonded atoms and how many lone pairs of electrons surround the central atom in a Lewis structure?
d. Predict the shape of carbon dioxide.

We applied the idea of electron-pair repulsion to molecules in which there are four groups of electrons (CH_4, NH_3, and H_2O), and two groups of electrons (CO_2) surrounding the central atom. Electron-pair repulsion also applies reasonably well to molecules that include three, five, or six groups of electrons. In most molecules, the electrons and atoms are still arranged to maximize the separation of the electrons. Use this logic to see the bent shape we associated with the ozone molecule in Section 3.8.

Your Turn 4.18   Shape of Ozone

a. How many valence electrons does ozone (O_3) have?
b. What happens if only single bonds form between each O?
c. How many bonded atoms and how many lone pairs of electrons surround the central atom in the Lewis structure?
d. Explain why the shape of ozone is bent and predict the bond angle.

Your Turn 4.19   Predicting Molecular Shapes, Part 2

Using Table 4.2 and the strategies just described, predict and sketch the shape of molecules of SO_2 (sulfur dioxide) and SO_3 (sulfur trioxide). Verify your answers with the PhET simulation: www.acs.org/cic. It's best to use the "model" mode and build your own molecules.

PhET Interactive Simulations, University of Colorado Boulder, http://phet.colorado.edu

---

Now you see that molecules have different shapes that can be predicted. In the next section, we return to our story of greenhouse gases, using your knowledge of molecular shapes to explore why all gases are not greenhouse gases.

4.6 | How Do Greenhouse Gases Work?

Learning Objective: Describe how greenhouse gases absorb infrared (IR) radiation

How do greenhouse gases absorb and emit heat, keeping our planet at more or less comfortable temperatures? In part, the answer lies in how molecules respond to photons of energy. This topic is complex, but we will provide the basics to understand how the greenhouse gases in our atmosphere absorb and reradiate heat. At the same time, we'll reveal why some gases do *not* act as greenhouse gases.

We begin this topic by revisiting the interaction of UV light with molecules, as discussed in Section 3.6 in relation to the ozone layer. You saw that high-energy photons (UV-C) could break the strong covalent bonds in O_2, and that photons of lower energy (UV-B) could break the weaker bonds in O_3. Put another way, both ozone and oxygen molecules can absorb UV radiation. When this absorption occurs, an oxygen-to-oxygen bond is broken.

IR photons do not contain enough energy to cause chemical bonds to break. Instead, a photon of IR radiation can only add sufficient energy to the molecule to cause vibrations. Depending on the molecular structure, only certain vibrations are possible. For a photon to be absorbed, the energy of the incoming photon must correspond exactly to the vibrational energy of the molecule. This means that different molecules absorb IR radiation at different wavelengths, and thus vibrate at different energies.

The types of vibrations can be described as *asymmetric stretching* (the atoms alternate how much each one stretches from the central carbon), *symmetric stretching* (the central carbon atom is stationary and the oxygen atoms move back and forth (stretch) in opposite directions away from the central atom), and *bending* (the molecule bends from its normal linear shape). We illustrate these movements with a model of the CO_2 molecule below.

Only asymmetric stretching and bending vibrations account for the greenhouse properties of carbon dioxide. To explain why symmetric vibrations are not involved, we must look at what happens to the properties of a molecule during its vibration.

The property of **electronegativity** (EN) is a measure of the ability of an atom to attract bonded electrons. Elements with the highest electronegativity lie in the top-right of the periodic table (toward fluorine, F, which has the highest EN). In contrast, the lowest electronegativity is found for the elements in the bottom-left of the periodic table (toward francium, Fr). In a CO_2 molecule, the average concentration of electrons is greater on the oxygen atoms than on the carbon atom because oxygen has a higher electronegativity.

The greater electronegativity of oxygen versus carbon means that the oxygen atoms carry a partial negative charge (δ^-) relative to the carbon atom, which carries a partial positive charge (δ^+). As the bonds stretch, the positions of the electrons change, thereby changing the charge distribution in the molecule. When a molecule is symmetrical, the δ^+ and δ^- cancel and there is no net separation of partial charges. IR absorption occurs when there is a separation of partial charges during molecular vibrations. Because of the linear shape and symmetry of CO_2, the changes in charge distribution during the symmetric stretch cancel, and no infrared absorption occurs.

Vibrational Modes

Watch this video (www.acs.org/cic) and answer the following questions. The cm^{-1} unit indicated in the video denotes the wavenumber; take the inverse of this number to calculate the wavelength in units of cm. IR wavelengths are typically reported in nm or μm.

a. Note the wavelength where each movement occurs.
b. Rank the vibrations in order of how much energy needs to be absorbed by the molecule. Recall from Section 3.2 that energy and wavelength are inversely related.

Your Turn 4.20   How Will a Molecule Stretch?

For each molecule below, determine its shape, and then determine where in the molecule any asymmetric stretches, corresponding to IR absorption, will occur.

a. NO_2 b. O_3 c. CH_4 d. NH_3

The infrared (heat) energy that molecules absorb can be measured with an instrument called an **infrared spectrometer**. Infrared radiation is passed through a sample of the compound to be studied, in this case gaseous CO_2. A detector measures the amount of radiation, at various wavelengths, which is transmitted through the sample. High transmission means low absorbance, and vice versa. This information is displayed graphically, where the relative intensity of the transmitted radiation is plotted versus wavelength. The result is the infrared spectrum of the compound. Figure 4.10 shows the infrared spectrum of gaseous CO_2.

Your Turn 4.21   Carbon Dioxide Absorption Spectrum

Referring to Figure 4.10, answer the following questions.

a. Do these values correspond with the values you determined in the above "Vibrational Modes" activity?
b. What vibration is missing? Why?

The infrared spectrum shown in Figure 4.10 was acquired using a laboratory sample of CO_2 gas, but the same absorption takes place in the atmosphere. Gaseous molecules of CO_2 that absorb specific wavelengths of infrared energy experience different fates. Some hold that extra energy for a brief time, and then re-emit it in all directions as heat. Others collide with atmospheric molecules like N_2 and O_2, and can transfer some of the absorbed energy to those molecules, also as heat. Thus, CO_2 retains some of the heat emitted by Earth's surface in the atmosphere or re-emits it back toward the surface, keeping our planet comfortably warm. This is what makes CO_2 a greenhouse gas.

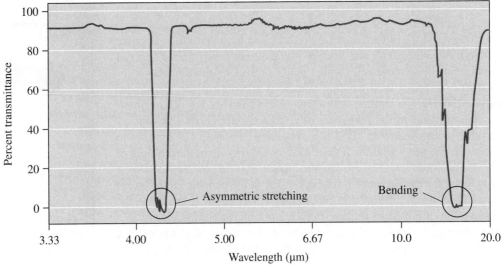

Figure 4.10

Infrared spectrum of carbon dioxide gas.

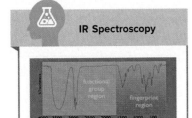

IR Spectroscopy

©Knowbee

Watch this video (www.acs.org/cic) for an introduction to IR spectroscopy (focus your attention on the first 10 minutes) and answer the following questions.

a. Briefly describe how an IR spectrometer works. When do you see an absorption band?

b. Explain the difference between the instrument response for 1-Hz and 2-Hz light on the hypothetical carbonyl sample.

Hence, any molecule that has three or more atoms will be able to absorb photons of IR radiation and behave as a greenhouse gas. Water is by far the most important gas in maintaining Earth's temperature, followed by carbon dioxide. Figure 4.11 shows the IR spectrum of gaseous H_2O, which displays strong absorption bands. However, methane, nitrous oxide, ozone, and chlorofluorocarbons (such as CCl_3F) also strongly absorb IR radiation and help retain planetary heat.

Your Turn 4.22   Diatomic Nitrogen and Oxygen

$$O=O \qquad\qquad N\equiv N$$

Nitrogen (N_2) and oxygen (O_2) are not greenhouse gases.

a. Given what you know about molecular shape, vibrations, and IR absorption, explain the statement above.

b. Why is it important that these two gases are not greenhouse gases?
   Hint: Think about their abundance in the atmosphere.

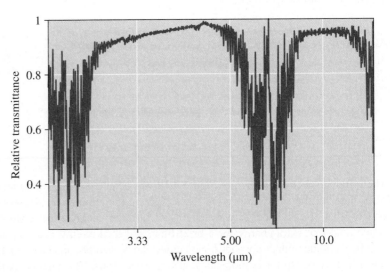

Figure 4.11

Infrared spectrum of water vapor.

Source: NIST Chemistry WebBook

4.7 | How Can We Learn from Our Past?

> *Learning Objectives:*
>
> *Explain the process for collecting historical and current climate data, such as how isotope data allow scientists to determine past trends*
>
> *Analyze historical greenhouse gas and temperature data and describe how current climate trends on Earth differ from the past*

Now that we have examined reservoirs of carbon and determined how the shapes of molecules can influence their ability to be greenhouse gases, we need to look at what has happened to the climate on Earth in the past. However, an important distinction needs to be made between the terms **climate** and **weather**. *Weather* includes the daily high and low temperatures, the drizzles and downpours, the blizzards and heat waves, and the fall breezes and hot summer winds—all of which have relatively short durations. In contrast, *climate* describes regional temperatures, humidity, winds, rain, and snowfall over decades, not days. Although the weather varies on a daily basis, our climate has stayed relatively uniform over the last 10,000 years. The values quoted for the "average global temperature" are but one measure of climate phenomena. The key point is that relatively small changes in average global temperature can have huge effects on many aspects of our climate.

Over the past 4.5 billion years—the approximate age of Earth—both Earth's climate and its atmosphere have varied widely. Earth's climate has been directly affected by periodic changes in the shape of Earth's orbit, the tilt of Earth's axis, and the wobble in the Earth's rotation on that axis. Such changes are thought to be responsible for the ice ages that have occurred regularly during the past million years. Even the Sun itself has changed; its energy output half a billion years ago was 25–30% less than it is today. In addition, changes in atmospheric greenhouse gas concentrations affect Earth's energy balance, and hence its climate. Carbon dioxide was once 20 times more prevalent in the atmosphere than it is today. However, that level was lowered as CO_2 dissolved in the oceans or was incorporated into rocks such as limestone. The biological process of photosynthesis also radically altered the composition of our atmosphere by removing CO_2 and producing oxygen. Certain geological events like volcanic eruptions add millions of tons of CO_2 and other gases to the atmosphere.

Although these natural phenomena will continue to influence Earth's atmosphere and climate in the coming years, we must also assess the role that human activities are playing. With the development of modern industry and transportation, humans have moved huge quantities of carbon from ancient terrestrial and marine sources like coal, oil, and natural gas into the atmosphere in the form of CO_2 over a relatively short period of time. To evaluate the influence humans are having on the atmosphere, and hence on any climate changes, it is important to investigate the fate of this large, rapid influx of carbon dioxide. Indeed, CO_2 concentrations in the atmosphere have increased significantly in the past half-century. The longest record of direct measurements are from the Mauna Loa Observatory in Hawaii, as displayed in Figure 4.12. The red zigzag line shows the average monthly concentrations; the sawtooth shape is due to the seasonal changes in the rates of photosynthesis and respiration (as discussed previously in Your Turn 4.4). The black line is a 12-month moving average that corrects for the average seasonal cycle. Notice the steady increase in average annual values from 315 ppm in 1960 to more than 420 ppm today.

Your Turn 4.23 The Cycles of Mauna Loa

Consult Figure 4.12 and answer the questions below related to carbon dioxide levels at the Mauna Loa observation station.

a. Calculate the rate of increase in CO_2 concentration (in ppm/year) since the beginning of the record (1958).

b. Calculate the rate of increase in CO_2 concentration since 2000 and compare this answer to that from part (a).

c. Estimate the variation in ppm of CO_2 within any given year (you can use the "Last 1 Year" tab to see more specific data points).

d. On average, the CO_2 concentrations are higher each April than each October. Why is this?

Mauna Loa Monthly Averages

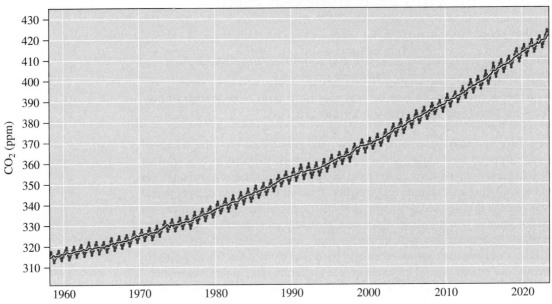

Figure 4.12

Carbon dioxide concentrations from March 1959 to July 2023, as measured at the Mauna Loa Observatory in Hawaii. For an interactive version of these data, see: www.acs.org/cic.

[NOAA]

Your Turn 4.24　　What About the Other Greenhouse Gases?

Visit this website (www.acs.org/cic) and view the data on at least three other greenhouse gases we have discussed. How does the concentration of each of these gases compare to that of carbon dioxide?

How can we obtain data about the composition of our atmosphere before the use of instrumentation that directly measures it? The analysis of ice core samples has revealed an enormous amount of data going back about 800,000 years. Regions on the planet that have permanent snow cover contain preserved histories of the atmosphere, buried in layers of ice. Figure 4.13a shows a dramatic

(a)　　　　　**(b)**　　　　　**(c)**

Figure 4.13

(a) Quelccaya ice cap (Peruvian Andes) showing the annual layers. **(b)** Ice core that can be used to determine changes in concentrations of greenhouse gases over time. **(c)** Microscopic air bubbles in ice.

(*a*): ©Lonnie G. Thompson, Ohio State University; (*b*): Ragnar Th Sigurdsson/ARCTIC IMAGES/Alamy Stock Photo; (*c*): ©W. Berner, 1978, PhD Thesis University of Bern, Switzerland. (D. Lüthi, M. Le Floch, B. Bereiter, T. Blunier, J.-M. Barnola, U. Siegenthaler, D. Raynaud, J. Jouzel, H. Fischer, K. Kawamura, and T.F. Stocker, High-resolution carbon dioxide concentration record 640,000–800,000 years before present, Nature, 453, 379–382, 2008.)

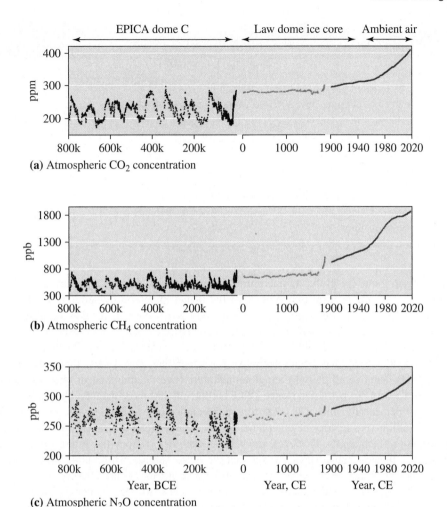

Figure 4.14

Carbon dioxide and other GHG concentrations over the last 800,000 years as measured from Antarctic ice cores and direct ambient atmospheric measurements.

example of annual ice layers from the Peruvian Andes. The oldest ice on the planet is located in Antarctica, and scientists have been drilling and collecting ice core samples there for more than 60 years (Figure 4.13b). Air bubbles trapped in the ice (Figure 4.13c) provide a vertical timeline of concentrations of trace atmospheric gases in the atmosphere; the deeper you drill, the farther back in time you go.

Relatively shallow ice core data show that for the first 800 years of the last millennium, the CO_2 concentration was relatively constant at about 280 ppm. Figure 4.14 combines data from direct measurements of the atmosphere with many different ice core samples from Antarctica. Drilling by a team of Russian, French, and U.S. scientists at various locations in Antarctica yielded ice cores taken from snowfalls of 800 millennia. Beginning in the 1800s, CO_2 and other greenhouse gases (such as methane and nitrous oxide) began accumulating in the atmosphere at an ever-increasing rate, which corresponds to the Industrial Revolution and the accompanying combustion of fossil fuels that powered that transformation.

Most obvious from the graph are the periodic cycles of high and low greenhouse gas concentrations, which occur roughly in 100,000-year intervals. Focusing on carbon dioxide, two important conclusions can be drawn from these data. First, the current atmospheric CO_2 concentration is over 100 ppm *higher* than at any time in the last million years. Also during that time, never has the CO_2 concentration risen as rapidly as it is rising today.

Your Turn 4.25   Temperature and GHGs

Go to www.acs.org/cic for a simulation about the impact of selected greenhouse gases (GHGs) on climate change over the past 800,000 years and answer the following questions.

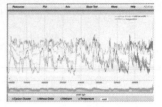

a. What is the relationship between GHG concentrations and temperature?

b. Is this relationship a *correlation* or *causation*? Explain your rationale.

©2018 American Chemical Society

c. Focus in on the most current data. Do you notice anything different about the relationship between temperature and GHG concentrations? What is going on here?

What about the global temperature? Measurements indicate that during the past 40 years or so, the average temperature of the planet increased by about 1 °C (1.8 °F) (Figure 4.15). Given the large heat capacity of Earth's oceans, it takes a substantial amount of heat energy to raise the average annual surface temperature even a little bit. The years 2013–2021 all rank in the top 10 warmest years on record. In fact, each of the last four decades has been successively warmer than any decade that preceded it since 1850. The overall global average temperature (over land and ocean) has increased by 0.08 °C (0.14 °F) per decade since 1880, but the rate has more than doubled to 0.18 °C (0.32 °F) per decade since 1981. Of course, a century or two is relatively an instant in the 4.5-billion-year history of our planet and we cannot read too much into short-term temperature fluctuations. However, the most recent report from the Intergovernmental Panel on Climate Change (IPCC) states that the global surface temperature has increased faster since 1970 than in any other 50-year period over at least the last 2000 years.

Although the trend in temperatures over the last 40 years generally follows an increase in carbon dioxide concentrations, the temperature data from year to year are much less consistent. There are many other factors that affect inter-annual temperatures such as short-term changes in atmospheric circulation patterns like El Niño and La

Global Land and Ocean
January–December Temperature Anomalies

1.00 °C	1.80 °F
0.80 °C	1.44 °F
0.60 °C	1.08 °F
0.40 °C	0.72 °F
0.20 °C	0.36 °F
0.00 °C	0.00 °F
−0.20 °C	−0.36 °F
−0.40 °C	−0.72 °F
−0.60 °C	−1.08 °F

1880   1900   1920   1940   1960   1980   2000   2021

Figure 4.15

Annual global surface temperatures as compared with the 1901–2000 average value. Blue bars indicate cooler than average years and red bars show warmer than average years.

[NOAA]

Niña events. For example, the high average surface temperature in 1998 has been attributed to a very strong El Niño event. This being said, the atmospheric concentrations of carbon dioxide, methane, and nitrous oxide are at the highest concentration in at least the last 800,000 years and IPCC scientists say that human activity, such as the emission of these greenhouse gases, have unequivocally caused the warming observed since the mid-20th century.

Temperature Change Visualization

To help visualize the temperature changes occurring on Earth, watch the "Time series: 1884 to 2021" video at www.acs.org/cic. Discuss the terminology "global warming" and "climate change" to describe what you are seeing.

It is important to realize that an increase in the average global temperature does not mean that across the globe every day is now 0.8 °C warmer than it was in 1970. In fact, many regions on Earth have experienced just a little warming and others have cooled. Yet there are other regions, particularly in the higher latitudes, that have experienced greater-than-average warming. Warming is most drastic in the Arctic, where, not surprisingly, much of the tangible effects of climate change have already been observed.

Your Turn 4.26   Local Temperature Changes

Check out www.acs.org/cic to find data on the temperature changes over the last 100 years in the area where you live. Has the temperature changed much? Now search for a map of the world that shows temperature change. Which areas have changed the most in the last 100 years? Which have changed the least? How does this compare to your area?

When we look back into the past, we see that the global temperature has undergone fairly regular cycles, matching the highs and lows in CO_2 concentration quite remarkably (Figure 4.16). Other data show that periods of high temperature have also been characterized by high atmospheric concentrations of methane and nitrous oxide, two other significant greenhouse gases. The accuracy of these data does not allow an assignment of cause and effect. Evidence other than a simple correlation of the data supports causation, but does not prove it definitively. The scientific method does not allow for definitive proof, rather it allows for the development of an explanation supported by a preponderance of evidence and when consensus is reached, it will become a scientific theory. At any point, if the new evidence is widely accepted by the scientific community, any scientific theory can be disproven. What is clear, however, is that the current concentrations of CO_2 and other greenhouse gases are much higher than any time in the last million years. Notice that the variation from hottest to coldest is 11 °C (20 °F). However, this represents the difference between the moderate climate we have today and ice covering much of northern North America and Eurasia, as was the case during the last glacial maximum 20,000 years ago.

Over the past million years, Earth has experienced 10 major periods of glacier activity and 40 minor ones. Without question, mechanisms other than human-caused greenhouse gas concentrations are involved in the periodic fluctuations of global

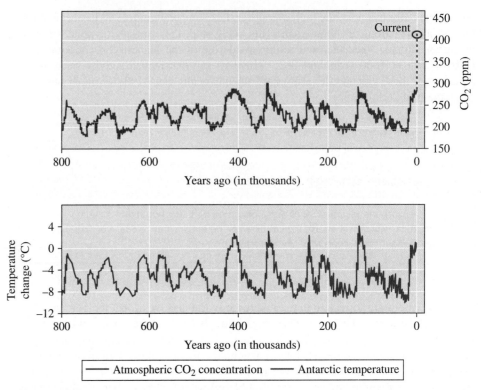

Figure 4.16

Carbon dioxide concentration (orange) and global temperatures (blue) over the last 800,000 years from ice core data. Temperatures are based on isotope signatures of water in the Dome C ice core. Carbon dioxide is measured in air bubbles trapped in ice; data shown are a composite of Dome C and Vostok ice cores. The "current" data point was based on direct measurements in 2019. Based on figure by Jeremy Shakun, data from Lüthi et al., 2008 and Jouzel et al., 2007.

climate. Some of this temperature variation is caused by minor changes in Earth's orbit that affect the distance from Earth to the Sun and the angle at which sunlight strikes the planet. However, this hypothesis cannot fully explain the observed temperature fluctuations. Orbital effects most likely drive a series of feedback cycles across the globe, such as changes in reflectivity, cloud cover, and airborne dust, as well as CO_2 and other greenhouse gas concentrations, like CH_4. The mechanisms that couple these effects together are complicated and not completely understood, but it is likely that the effects from each are *additive*. In other words, the existence of natural climate cycles doesn't preclude the effects on global climate caused by human activities since the Industrial Revolution.

It turns out that isotope data can be used to show that the modern increase in CO_2 is primarily caused by the combustion of fossil fuels. As discussed in Section 4.2, carbon has three isotopes ^{12}C (the dominant isotope), ^{13}C, and ^{14}C (also known as radioactive carbon). Scientists have been monitoring the carbon isotopes in atmospheric CO_2 and there has been a recent downward trend in the $^{13}C{:}^{12}C$ and $^{14}C{:}^{12}C$ ratios (Figure 4.17). Why is this happening?

In all living things, only 1 out of 10^{12} carbon atoms is ^{14}C, which is radioactive. A plant or animal constantly exchanges CO_2 with the environment and this maintains a constant ^{14}C concentration in the organism. However, when the organism dies, the biochemical processes that exchange carbon stop functioning and the ^{14}C is no longer replenished. This means that after the death of the organism, the concentration of ^{14}C decreases with time because it undergoes radioactive decay to form ^{14}N. Coal, oil, and natural gas are remnants of plant life that died hundreds of millions of years ago. Hence, in fossil fuels, and in the carbon dioxide released when fossil fuels burn, the level of ^{14}C is essentially zero. Careful measurements show that the concentration of ^{14}C in atmospheric CO_2 has recently decreased. This

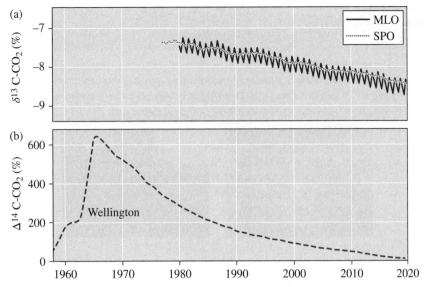

Figure 4.17

Carbon isotope ratios have been decreasing over time. Carbon-13 is measured at both the Mauna Loa Observatory (MLO) and the South Pole Observatory (SPO). The carbon-14 data (shown in red) is from Wellington, New Zealand. Note the sharp increase in carbon-14 prior to 1965 is due to atmospheric nuclear weapons testing.

Source: Climate Change 2021: The Physical Science Basis. Contribution of Working Group I to the Sixth Assessment Report of the Intergovernmental Panel on Climate Change, Figure 5.6 (c).

strongly suggests that the origin of the added CO_2 is the burning of fossil fuels, a human activity.

But what about other old sources of carbon to the atmosphere (like volcanic eruptions)? Now, we turn to ^{13}C. Carbon found in ancient plant life was taken up as CO_2 from the atmosphere through photosynthesis. It turns out that plants prefer to take in ^{12}C over ^{13}C. This, combined with the fact that the lighter ^{12}C will diffuse more easily into the plant, means that plants are enriched in ^{12}C. Thus, when they are burnt, the CO_2 emitted is also enriched in ^{12}C. Careful measurements of carbon dioxide in the atmosphere show a decrease in the $^{13}C:^{12}C$ ratio suggesting that the origin of this carbon dioxide is burning fossil fuels. This decrease in the isotope ratio would not be observed if the extra CO_2 came from an increase in volcanic emissions.

Your Turn 4.27   CO$_2$ Increases from Combustion

How do the data shown below continue to support the idea that the combustion of fossil fuels is the primary source of the increasing CO_2 concentrations in Earth's atmosphere?

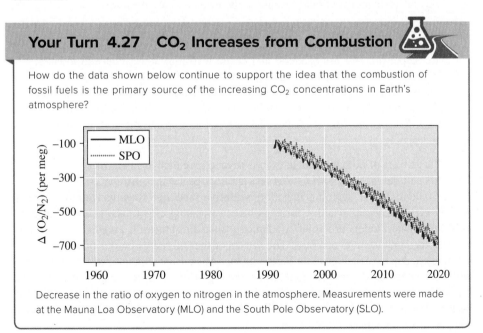

Decrease in the ratio of oxygen to nitrogen in the atmosphere. Measurements were made at the Mauna Loa Observatory (MLO) and the South Pole Observatory (SLO).

4.8 | What Factors Drive Our Climate?

Learning Objectives:

Discuss climate drivers beyond carbon dioxide

Defend the scientific consensus that our modern climate crisis is primarily due to human activity

Understanding the full extent of climate change and all of the factors that impact our climate system is a daunting task. In 1988, the United Nations Environment Programme and the World Meteorological Organization teamed up to establish the UN Intergovernmental Panel on Climate Change (IPCC). After reviewing thousands of published scientific papers, the IPCC provides periodic assessments of the drivers of climate change, their impacts and future risks, as well as mitigation and adaptation strategies to reduce risk. There have been 6 assessment reports; the latest was released in March 2023. With input from thousands of international climate scientists, the evidence clearly shows:

- Human influence on Earth's climate is unequivocal.
- Human activities (primarily the combustion of fossil fuels and deforestation) are responsible for atmospheric and ocean warming, lower concentrations of ice and snow on the planet, and sea level rise.
- Continued emission of greenhouse gases will result in further warming and long-lasting change in Earth's climate system. This will cause an increased likelihood of severe, pervasive, and irreversible impacts for both ecosystems and people.

The drivers of climate change extend beyond greenhouse gases, like CO_2. Climate scientists call the factors (both natural and anthropogenic) that influence the balance of Earth's incoming and outgoing radiation **radiative forcings**. Negative forcings have a cooling effect, whereas positive forcings have a warming effect. The primary forcings used in climate models are solar irradiance ("solar brightness"), greenhouse gas concentrations, land use, and aerosols. The effects of these forcings on our planet's surface temperature due to their impact on Earth's energy balance are summarized in Figure 4.18, with greenhouse gases shown to create a warming effect, and aerosols mostly yielding a cooling effect. Each forcing has scientific uncertainty associated with it (as indicated by the error bars); the larger the error bar, the more uncertain the value. Let's consider some of these factors in more detail.

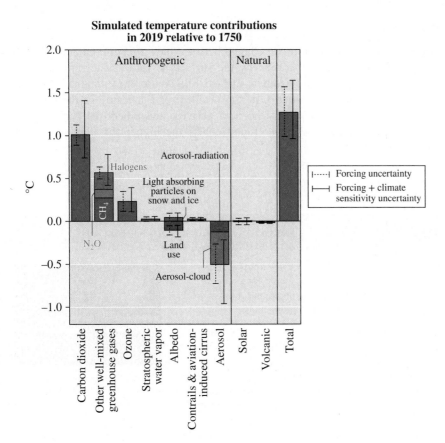

Figure 4.18

The contribution of forcing agents to 2019 temperature change relative to 1750. Solid error bars represent *very likely* (5–95%) ranges for the combined effects of the forcing and climate response uncertainty. Dashed error bars show the contribution of the forcing uncertainty alone.

Source: Climate Change 2021: The Physical Science Basis. Contribution of Working Group I to the Sixth Assessment Report of the Intergovernmental Panel on Climate Change

Greenhouse Gases

Greenhouse gases (GHGs) represent the most dominant anthropogenic forcing (Figure 4.18). In fact, the collective positive forcing from GHGs is more than 2.5 times larger than the magnitude of all other forcings combined. Most active is CO_2, constituting over 55% of the warming from all GHGs. Other notable impacts come from methane (15%), ozone (13%), halogens (e.g., CFCs, HCFCs and HFCs, 10%), and nitrous oxide (6%) (Figure 4.19).

Your Turn 4.28   Climate Impact of the Montreal Protocol

In Section 3.9, you learned that the Montreal Protocol was enacted to protect stratospheric ozone. What is the impact of this treaty on climate change? Use reputable Internet sources to answer this question (e.g., NASA or Science Daily).

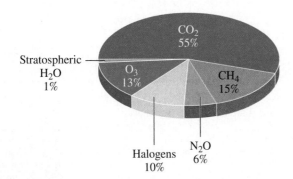

Figure 4.19

The contribution of each greenhouse gas to the overall 2019 temperature change relative to 1750. (Data found in Figure 4.18.)

Land Use

Changes in land use drive climate change because they alter the amount of incoming solar radiation that is absorbed by the surface of Earth. The ratio of electromagnetic radiation *reflected* from a surface relative to the amount of radiation *incident* on it is called the **albedo**. In short, albedo is a measure of the reflectivity of a surface. As you will see in the next activity, the albedo of Earth's surface varies, depending on the type of ground cover; the higher the number, the more reflective the surface.

Your Turn 4.29   Albedo and Land Surfaces

Use the interactive found on www.acs.org/cic to understand the impact of solar energy and albedo on Earth's surface temperature. Note: this interactive does not include atmospheric interactions.

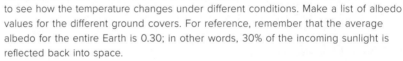

University Corporation for
Atmospheric Research (UCAR)

Questions to Consider

a. How does Earth's temperature change as the Sun becomes brighter or dimmer?

b. How does Earth's temperature change as the albedo (reflectivity) is higher or lower? Use the "Pick a surface…" drop-down menu under the albedo slider to see how the temperature changes under different conditions. Make a list of albedo values for the different ground covers. For reference, remember that the average albedo for the entire Earth is 0.30; in other words, 30% of the incoming sunlight is reflected back into space.

c. Relate albedo to surface color.

d. How would the surface temperature change if there is a loss of snow due to increasing local temperature?

e. How would the surface temperature change if a forest is turned into grassland (deforestation)?

Land use change, mostly from deforestation and an increase in agriculture, has changed the surface albedo (mostly making it more reflective) and altered the local greenhouse gas emissions. Overall, there is a small negative forcing from the changes in land use (revisit Figure 4.18). On the other hand, there is a small positive forcing due to particles deposited on snow and ice, which will be explored in the following activity. High clouds, such as contrails, also contribute to a net warming at the surface.

Your Turn 4.30   Ice Albedo Feedbacks

Open this interactive: www.acs.org/cic. Read and watch the animations to learn about the ice-albedo feedback.

Questions to Consider

a. Summarize the impact of soot covering ice on the Ice-Albedo feedback loop.

b. Reflect on strategies that could be used to limit the ice-albedo feedback loop.

Arterra Picture Library/
Alamy Stock Photo

Your Turn 4.31   Build a Planet

Use this interactive simulation to see how to reestablish Earth's energy balance: www.acs.org/cic. The previous sections have described how a future world could have an enhanced greenhouse effect while the global albedo is decreasing. For example, use the sliders to increase the greenhouse factor to 0.45 (from 0.36) and albedo to 0.25 (from 0.30). What must the surface temperature be in order to reestablish an energy balance?

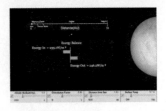

©2018 American Chemical Society

Aerosols

Because they are a complex class of materials, **aerosols**—suspensions of small solid particles in a gas or liquid—have a correspondingly complex effect on climate. Many natural sources of aerosols exist, including dust storms, ocean spray, forest fires, and volcanic eruptions; remember the discussion of aerosols and particulate matter in Section 2.6. However, human activity can also release aerosols into the environment in the form of smoke, soot, fly ash from coal combustion, and gases that eventually condense to form particles.

The effect of aerosols on climate is probably the least understood of the forcings listed in Figure 4.18. Tiny aerosol particles (< 4 μm) are efficient at scattering incoming solar radiation. Other aerosols absorb incoming radiation, and still other particles both scatter and absorb. Both processes decrease the amount of radiation that reaches Earth's surface.

In addition to a direct cooling effect, aerosol particles can serve as nuclei for the condensation of water droplets, and hence promote cloud formation. Clouds reflect incoming solar radiation, although the effects of increased cloud cover are more complex than this. Therefore, in both direct and indirect ways, aerosols counter the warming effects of greenhouse gases.

Your Turn 4.32   Aerosols and Climate

Read the NASA article "Aerosols: Tiny Particles, Big Impact" at www.acs.org/cic and answer the questions below.

a. What is the impact of large volcanic eruptions (like that of Mt. Pinatubo in 1991) on climate?
b. What are the physical characteristics of clouds produced from anthropogenic seeds (pollution-rich clouds)?
c. Reflect on why the radiative forcing (and thus contribution to surface temperature) from aerosols is not as well understood as that from greenhouse gases.

Solar Irradiance

Your Turn 4.33   Sun Skeptics

Some people have stated that changes in the Sun, such as increased solar flares, are causing global climate change. Look at Internet sources, such as Skeptical Science's website, for your research. What are your thoughts?

The amount of solar radiation that reaches our planet varies naturally with changes in Earth's orbit (oscillating slightly over a 100,000-year period), the magnitude of the tilt of Earth's axis and the direction of that tilt (both change over the course of several tens of thousands of years). None of these variations occurs on a timescale short enough to explain the warming since the Industrial Revolution. Additionally, sunspots occur in large numbers about every 11 years when there is increased magnetic activity in the

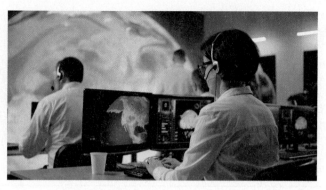

Figure 4.20

Climate scientists use computer simulations to understand future climate change.

Frame Stock Footage/Shutterstock

outer layers of the Sun, thus stirring up a larger amount of charged particles that emit radiation. The solar brightness over the 11-year cycles of sunspot activity varies only by about 0.1%. As you can see from Figure 4.18, this natural variability is the *smallest* of any forcing listed.

Your Turn 4.34   The "Little Ice Age"

Read the article "There Is No Impending 'Mini Ice Age'" by NASA about mini Ice Ages (www.acs.org/cic) and answer the following questions.

a. What is the "Little Ice Age" and comment on its causes.
b. What is the impact of the Grand Solar Minimum if it were to happen today?

The challenge is to understand current climate change well enough to *predict* future changes, and by doing so, to determine the decrease in emissions required to minimize harmful changes. To make predictions, scientists work with models. They design computer models of the oceans and the atmosphere that take into account the ability of each to absorb heat, as well as to circulate and transport matter (Figure 4.20). If that weren't difficult enough, the models must also include astronomical, meteorological, geological, and biological factors—ones that are often not completely understood. Human influences, such as population, industrialization levels, and pollution emissions must also be included. Dr. Michael Schlesinger, who directs climate research at the University of Illinois, remarked: "If you were going to pick a planet to model, this is the *last* planet you would choose."

Your Turn 4.35   Climate Models

Read the article "How do climate models work" from Carbon Brief at www.acs.org/cic and answer the following questions.

a. What is a climate model?
b. What are some example ways climate models have become more complex?

Given the complexity inherent in all the forcings we have just described, you can appreciate that assembling these forcings into a climate model is no easy task. Furthermore, once a model has been built, scientists have difficulty assessing its validity. However, scientists do have one trick up their sleeves. They can test climate models with known data sets as a means to tease apart the contributions of different forcings or different factors that affect Earth's climate. We know the temperature data of the 20th century. In Figure 4.21, the black lines represent the known data. Next, examine

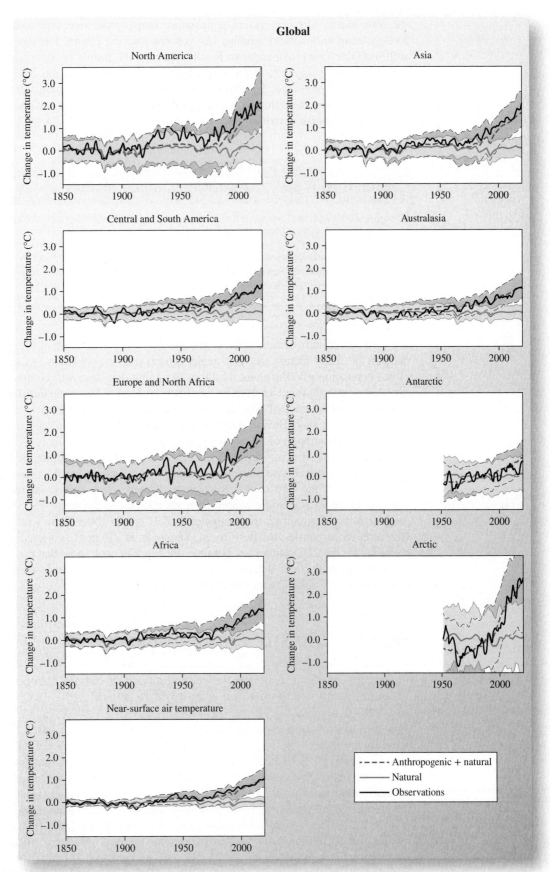

Global

Figure 4.21

Comparison of observed and simulated temperature changes compared to the 1995–2016 average. Black lines are observed values, colored lines indicate simulated data where purple incorporates anthropogenic and natural forcings and pink only simulates natural forcings. The bottom image is the global average trend in the near-surface air temperature.

Source: Technical Summary. In Climate Change 2021: The Physical Science Basis. Contribution of Working Group I to the Sixth Assessment Report of the Intergovernmental Panel on Climate Change

the pink bands. These represent temperature ranges that were predicted by climate models using *only* natural forcings. As you can see, the natural forcings do not map well onto the actual temperatures. Finally, examine the purple bands to see that when anthropogenic forcings are included, the temperature increases of the 20th century can be accurately reproduced. So, although the last 50 years of warming were *influenced* by natural factors, the actual temperatures cannot be accounted for without including the effects of human activities.

Your Turn 4.36   Assessing Climate Models

Figure 4.21 shows that modeled temperatures relying on natural forcings (pink bands) do not match the observed temperatures.

a. Name the forcings included in the models that included only natural forcings.
b. List two additional forcings included in the models that more accurately re-create the temperatures of the 20th century.

The magnitude of future emissions, and hence the magnitude of future warming, depends on many factors. As you might expect, one is population. As of 2023, the global population stood at about 8 billion. Assuming that there will be more people on the planet in the future, we humans are likely to have a larger **carbon footprint**, an estimate of the amount of CO_2 and other greenhouse gas emissions in a given time frame, usually a year. Having more people to feed, clothe, house, and transport will require the consumption of more energy. In turn, this translates to more CO_2 emissions, at least if the rate of using current fuels continues to grow. In addition, scientists who create climate models have to include values for two factors: (1) the rate of economic growth, and (2) the rate of development of "green" (less carbon-intensive) energy sources. Again, as you might expect, both are difficult to predict.

So what, if anything, can computer models tell us about Earth's future climate? Given the uncertainties we have listed, hundreds of different projected temperature scenarios for the 21st century are possible. Most predictions show that the temperature will increase. With some amount of future warming virtually ensured, we now turn our discussion to the consequences of climate change.

4.9 | Impacts of Climate Change

Learning Objective: Examine the consequences of climate change in terms of what is happening now and what might lie ahead

The 2023 IPCC report concluded that emissions of greenhouse gases from human activities are responsible for approximately 1.1 °C of warming since 1850–1900. You may be thinking, "So what?" After all, the temperature changes predicted by most models are only a few degrees. At any single spot on the planet, the temperature fluctuates several times that amount daily. The key point is that relatively small changes in the average global temperature can have huge effects on many aspects of our climate. (Figure 4.22) shows global surface temperatures relative to the 1850–1900 time period for the past 60 million years and projections for the next 300 years. By 2100, there will be a temperature increase of 2.7 °C over preindustrial levels if we follow the intermediate emission scenario (purple line). It has not been that hot on Earth for over 3 million years!

Table 4.3 summarizes the outlook from recent IPCC reports related to global climate change. Regional impacts of climate change on human systems from a 2022 IPCC report are illustrated in Figure 4.23.

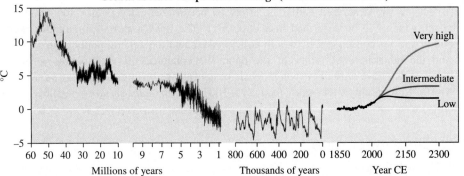

Figure 4.22

Global surface temperature change during the last 60 million years and projections for the next 300 years.

Table 4.3	IPCC Conclusions

The Physical Science

- It is unequivocal that human influence has warmed the atmosphere, ocean and land where widespread and rapid changes occurred.
- Human-induced climate change is already affecting many weather and climate extremes in every region across the globe.
- Warming of 1.5 °C and 2 °C will be exceeded during the 21st century unless deep reductions in greenhouse gas emissions occur in the coming decades.
- Many changes due to past and future greenhouse gas emissions are irreversible for centuries to millennia, especially changes in the ocean, ice sheets and global sea level.

Impacts, Adaptation & Vulnerability

- Human-induced climate change, including more frequent and intense extreme events, has caused widespread adverse impacts and related losses and damages to nature and people, beyond natural climate variability.
- Limiting warming to 1.5 °C would substantially decrease projected losses and damages in human systems and ecosystems, but cannot eliminate them all.
- Approximately 3.3 to 3.6 billion people live in contexts that are highly vulnerable to climate change.

Mitigation of Climate Change

- Global emissions must peak between 2020 and 2025 and global net zero CO_2 emissions must be reached in early 2050s to limit the global warming to 1.5 °C.
- Reducing GHG emissions across the full energy sector requires major transitions, including a substantial reduction in overall fossil fuel use, the deployment of low-emission energy sources, switching to alternative energy carriers, and energy efficiency and conservation.
- International cooperation is a critical enabler for achieving ambitious climate change mitigation goals

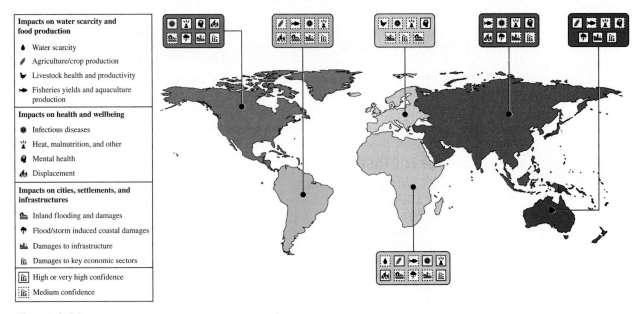

Impacts on water scarcity and food production

- Water scarcity
- Agriculture/crop production
- Livestock health and productivity
- Fisheries yields and aquaculture production

Impacts on health and wellbeing

- Infectious diseases
- Heat, malnutrition, and other
- Mental health
- Displacement

Impacts on cities, settlements, and infrastructures

- Inland flooding and damages
- Flood/storm induced coastal damages
- Damages to infrastructure
- Damages to key economic sectors

- High or very high confidence
- Medium confidence

Figure 4.23

Observed regional impacts of climate change on human systems. The images depicted above each region represent human systems upon which climate change has had an increasing adverse impact. Confidence levels reflect uncertainty in the attribution of the observed impact to climate change.

Source: Data from Summary for Policymakers In: Climate Change 2022: Impacts, Adaptation, and Vulnerability. Contribution of Working Group II to the Sixth Assessment Report of the Intergovernmental Panel on Climate Change, Portions of Figure SPM.2.

1.5° Celsius ...

A global temperature increase of 1.5 °C is a common goal. Watch this animation (www.acs.org/cic) that uses a bucket analogy for the carbon budget. What are some conclusions you can draw?

There are a wide variety of impacts connected to changing climate system; only a select few are described here. The effect on ocean chemistry is also another major impact, which will be described in Section 5.11. The IPCC reports estimate the likelihood of various consequences to help policymakers and the general public better understand the inherent uncertainty of the data. We conclude this section by outlining outcomes that have *high* or *very high* levels of confidence or impacts that are *very likely* (90–100% probability), *extremely likely* (95–100%), or *virtually certain* (99–100% probability) to occur.

Ice Disappearance

IPCC conclusions:

- Human influence is *very likely* the main driver of the global retreat of glaciers since the 1990s and the decrease of Arctic sea ice.
- It is *virtually certain* that the surface ocean has warmed since the 1970s and it is *extremely likely* that human influence is the main driver.

The warming of the upper ocean leads to a decrease in sea ice. For example, the Arctic Ocean will likely become free of September sea ice before 2050. Arctic sea ice is becoming younger and thinner; now about 70% of it is seasonal, meaning it does not last until the next year.

Your Turn 4.37   Arctic Sea Ice

Summarize the trends in Arctic sea ice minima shown on the page "Arctic Sea Ice Minimum Extent" from NASA found at www.acs.org/cic.

Warming also impacts ice sheets (the only two on Earth right now are on Antarctica and Greenland) and mountain glaciers. Explore the following Your Turn activities to figure out how ice sheets are affected by global warming. Loss of ice does not only endanger wildlife habitats and sources of drinking water, but it also decreases the local albedo which will lead to more warming.

Your Turn 4.38   Ice Sheets

a. Describe the trends in mass variation seen in the data from Greenland and Antarctica on the page "Ice Sheets" from NASA.
b. Watch this video and explain how a glacier melts: www.acs.org/cic.

Mountain Glaciers

After watching this video (www.acs.org/cic), explain the importance of mountain glaciers to human communities.

Sea Level Rise

Warmer temperatures result in an increase in sea level because as water warms, it expands. A smaller effect is caused by the influx of fresh water into the ocean from land-based glacier runoff.

IPCC conclusions:

- Thermal expansion explained 50% of sea level rise during 1971–2018, while ice loss contributed 22%, ice sheets 20%, and changes in land storage 8%.
- Global sea level increased by 0.20 (0.15 to 0.25) meters between 1901 and 2018. Human influence is *very likely* the main driver since at least 1971.

Sea level increases are not seen uniformly across the globe. In addition, they are influenced by regional weather patterns. Regardless, even small increases in sea levels can cause erosion in coastal areas and the stronger storm surges associated with hurricanes and cyclones.

Simulations can help city planners anticipate future sea level rises. Launch NOAA's Sea Level Rise Viewer (www.acs.org/cic). Enter a coastal city in the United States or territories in the navigation bar. You may choose to select your home city, your favorite spot to visit, or somewhere you have heard about, as long as it is in a coastal region. Make sure the legend is turned on (in the upper right menu).

a. Make sure the "sea level" rise tab is highlighted on the left. Find the nearest water droplet marker on the map. Click on it and use the water level slider to simulate sea rise height at that location and watch what changes on the map and in the photo.
b. Click on the "local scenarios" tab on the left and select a nearby blue scenario location icon on the map. (You may need to zoom out to find an icon.)
   a. Use "view by scenario" to see sea level height projections for different IPCC scenarios
   b. Use "view by year" to see sea level projections by year.
   c. For each, use the slider to see changes.
c. Click on the "vulnerability" tab to see how resilient the nearby population is to the sea level rise.

More Extreme Weather

IPCC conclusions:

- It is *virtually certain* that hot extremes (including heatwaves) have become more frequent and more intense across most land regions since the 1950s, while cold extremes have become less frequent and less severe. There is *high confidence* that human activity is the main driver.
- It is *very likely* that heavy precipitation events will intensify and become more frequent in most regions.

Warmer temperatures mean more evaporation and warm air can hold more water vapor. Scientists are still defining the relationship between climate change and the jet stream, but the changes to these fast-moving high-altitude air currents are likely causing weather systems to get stuck in place. This exacerbates the regional heat and dryness/rainfall.

Your Turn 4.40   Extreme Weather in the US

One way to measure the intensity of weather is by how costly the impact is for a nation. Let's explore the 323 weather events costing more than $1 billion in the United States from 1980 to 2021. Visit the page "Billion-Dollar Weather and Climate Disasters" on the NOAA website or go to www.acs.org/cic.

a. What are the trends over time (use the "time series" tab to view a chart)?
b. Scroll down to the "county risk assessment" on the "disaster and risk mapping" tab. Cycle through the drop-down menu to assess the distribution of risks (historic and future) as well as socioeconomic vulnerabilities. What are some trends in the data? You can click on any specific county to see data (use the county for the city chosen in the sea level interactive exercise).

Loss of Biodiversity

IPCC conclusions:

- Changes in land biosphere since 1970 are consistent with global warming.
- There is *high confidence* in climate zones shifting poleward and the average growing season lengthening by up to 2 days per decade since the 1950s in the Northern Hemisphere.

A recent study (McElwee, 2021 *Current History*) projects that if warming reaches 2 °C above preindustrial levels, "18% of insects, 16% of plants, and 8% of vertebrate species will lose over half of their geographic range, and localized extinctions are a near certainty." Mass extinction is primarily driven by human development of the environment and only secondarily by climate change.

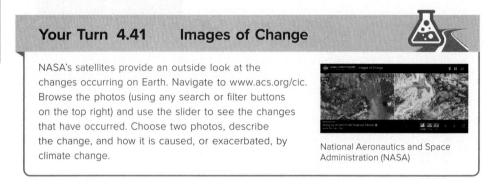

Your Turn 4.41   Images of Change

NASA's satellites provide an outside look at the changes occurring on Earth. Navigate to www.acs.org/cic. Browse the photos (using any search or filter buttons on the top right) and use the slider to see the changes that have occurred. Choose two photos, describe the change, and how it is caused, or exacerbated, by climate change.

National Aeronautics and Space Administration (NASA)

Figure 4.24 summarizes all of the impacts climate change is having on Earth's natural systems, but what about the human toll?

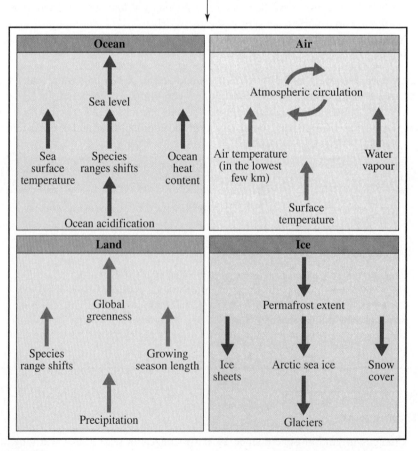

Figure 4.24

Summary of climate change impacts on natural systems.

Harm to Humans

IPCC conclusions:

- There is *very high confidence* that climate change has adversely affected the physical health of people globally and the mental health of regions assessed in the IPCC report.
- There is *very high confidence* that extreme heat events result in human mortality and morbidity and there has been an increase in occurrence of climate-related food-borne and water-borne diseases.
- There is *high confidence* that climate and weather extremes are increasingly driving the displacement in all regions, with small island states disproportionately affected. This contributes to a variety of humanitarian crises (including food insecurity and climate migration).

Your Turn 4.42   Human Health

Climate change can create new health problems or exacerbate existing ones. The chart below, created by the Centers for Disease Control and Prevention, shows the impacts of climate change on human health.

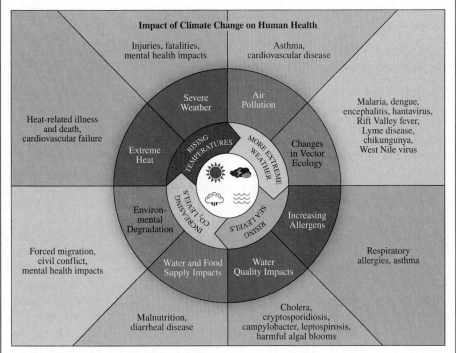

Impacts of climate change on human health.

[Centers for Disease Control and Prevention]

How do you think these impacts pose greater challenges for vulnerable populations or disadvantaged communities? Use reputable websites for your investigation.

In addition to directly affecting human health, climate change has already led to an increase in human migration. The UN High Commissioner for Refugees reported 21.5 million new displacements per year from 2009 to 2019 due to climate-related events. This is more than twice the number of displacements due to conflict and violence.

Preparing for Displacement

To think about how society should prepare for the millions of people who will continually be displaced by climate-related disasters, watch this TED talk by Colette Bichon Battle (www.acs.org/cic) and answer the following questions:

a. Climate migration can also occur in wealthy nations, such as the United States. What is the example given in this talk?

b. Discuss Collete Pichon Battle's use of the term "climate migrant" instead of "climate refugee."

c. Discuss the relationship between the exploitation of natural and human resources. What argument is made by the speaker? What are your thoughts?

Your Turn 4.43   Human Displacement

Climate change is worsening the drivers of human displacement. Read through the interactives and stories found in the "Displaced on the frontlines of the climate emergency" article (www.acs.org/cic).

a. Where are the displacement crises—both in terms of IDP (internally displaced people) and those who migrate?

b. Outline what is happening in one of the highlighted countries (Afghanistan, Bangladesh, Central America, Sahel, Somalia, Yemen, Mozambique).

Read the story of Ayan Muude Adawe, a climate refugee (www.acs.org/cic). Describe her situation.

Now that we have explored some rather concerning possibilities for our future, let's examine what we can do now to try to prevent these worst-case scenarios.

4.10 | Action Plans to Prevent Future Global Catastrophes—Who and How?

Learning Objectives:

Summarize the Paris Agreement and compare major policy initiatives such as "cap and trade" and carbon tax

Propose ways to reach a sustainable planet and reflect on how to reduce your own ecological footprint

Today's scientific data make it clear that Earth's climate is changing. For example, measurements of higher surface and ocean temperatures, retreating glaciers and sea ice, and rising sea levels are unequivocal. In addition, the carbon isotopic ratio found in atmospheric CO_2 leaves little doubt that human activity is responsible for much of the observed warming. However, at issue is what we *can* do and what we *should* do about the changes that are occurring.

Climate mitigation is any action taken to permanently eliminate or reduce the long-term risk and hazards of climate change to human life, property, or the environment. The most obvious strategy for minimizing anthropogenic climate change is to reduce the amount of CO_2 and other greenhouse gases emitted into the atmosphere in the first place. Take a look back at Figure 4.4. It is difficult to imagine curtailing any of these "necessities" to any great extent. Therefore, decreasing our energy consumption will not be easy, at least in the short term. The simplest and least-expensive approach is to improve energy efficiency, which is one of the key ideas in green chemistry. Due to the inefficiencies associated with energy production, saving energy on the consumer end multiplies its effect on the production end three to five times. However, relying on individual consumers worldwide to buy climate-friendly goods and do climate-friendly things will not be sufficient to hold CO_2 emissions below harmful levels.

Regardless of any potential decreases in future emissions, some effects of climate change are unavoidable. As mentioned previously, many of the CO_2 molecules emitted today will remain in the atmosphere for centuries. **Climate adaptation** refers to the ability of a system to adjust to climate change (including climate variability and extremes) to moderate potential damage, to take advantage of opportunities, or to cope with the consequences. Some adaptive methods include developing new crop varieties and shoring up or constructing new coastline defense systems for low-lying countries and islands. The further spread of infectious diseases could be minimized by enhanced public health systems. Many of these strategies are win–win situations that would benefit societies even in the absence of climate change challenges.

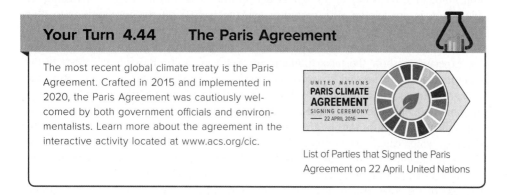

Your Turn 4.44   The Paris Agreement

The most recent global climate treaty is the Paris Agreement. Crafted in 2015 and implemented in 2020, the Paris Agreement was cautiously welcomed by both government officials and environmentalists. Learn more about the agreement in the interactive activity located at www.acs.org/cic.

List of Parties that Signed the Paris Agreement on 22 April. United Nations

The Paris Agreement was derived from a long line of international conferences, reports, and negotiations. Figure 4.25 shows a timeline of events that got us to where we are today.

Compared with the scientific consensus that anthropogenic emissions of greenhouse gases are the primary cause of climate change, there is much less agreement among governments regarding what actions should be taken to limit greenhouse gas emissions. Many countries hesitate to sign global climate treaties. For example, although the Kyoto Protocol went into effect in 2005, the United States never opted to participate. Signing Parties may believe that meeting the reduction requirements set by the treaty

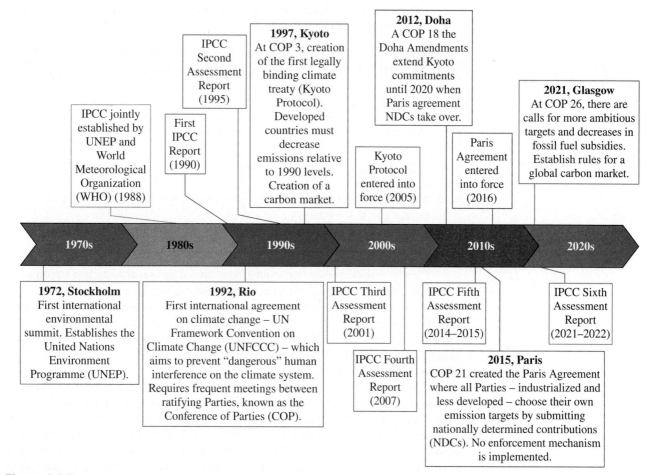

Figure 4.25

Timeline of major policy agreements and climate reports since 1972.

Nicole C. Bouvier-Brown

would cause serious harm to their economy. Historically, fossil fuels have been relatively inexpensive when compared to alternative forms of energy production. In addition, mitigation measures entail significant upfront costs and the cost of mitigation or adaptation is not known with certainty, making it difficult for corporations to plan effectively. Another reason for not ratifying a global treaty is concern about the lack of emissions limitations on developing nations, mainly China and India; those countries are expected to show the most dramatic increases in carbon dioxide emissions in the coming years. This second concern was taken under consideration when drafting the Paris Agreement; thus, all Parties are treated similarly. Many countries pride themselves on their autonomy and do not like a global body making rules for them. Thus, the Paris Agreement is based on Parties choosing their own nationally determined contributions (NDCs), which cannot be enforced. This makes environmentalists nervous because voluntary reduction programs have proved insufficient to reduce emissions. A final problem lies in the fact that the benefits of emissions reductions will not be felt for decades because of the long residence time of CO_2 molecules in the atmosphere. The current economic and political structure may be too short-sighted to see long-term benefits.

Our planet has already warmed 1.1 °C since preindustrial times and the 2021 IPCC Assessment estimates that every 1000 Gton of cumulative CO_2 emissions will increase the global surface temperature by 0.45 °C. To put these numbers in perspective, reducing emissions by only 1 Gton (or 1×10^9 ton or 1 billion tons) would require one of the following changes:

- Capturing and sequestering carbon dioxide at 268 coal-burning power plants or 2515 natural gas power plants.
- Replacing 38 billion incandescent light bulbs with LED lights.
- Increase gasoline fuel efficiency to 38 mpg, assuming every one of the 1.45 billons cars currently gets 30 mpg.
- Eliminating the annual electricity use of 105 cities like Los Angeles.

Your Turn 4.45   Trees as Carbon Sinks

An average medium-sized urban tree, allowed to grow 10 years, absorbs a total of 133 pounds of carbon dioxide. For the purpose of this exercise, let's assume a tree absorbs 13.3 pounds each year. In the United States, the average annual per capita CO_2 emission is 15 tons.

a. How many new trees would be required to absorb the annual CO_2 emissions for an average U.S. citizen?

b. What percentage of annual global emissions from burning fossil fuels could be absorbed by 12 billion trees?

Hint: Refer back to Figure 4.3.

Clearly, emission reduction cannot be achieved by one activity and implementation of any of these ideas will not be accomplished on a purely voluntary basis. Addressing the climate crisis involves a variety of different actions and there is a realization that laws and regulations are needed. Many countries see a path through carbon markets, known as cap and trade (COP 3 established a carbon market and COP 26 established rules, Figure 4.25). Others see more promise with a carbon tax. The two systems are summarized in Figure 4.26.

Cap and trade is a popular idea with those who are comfortable with market-based economics. For example, it has been successful in reducing sulfur dioxide emissions from power plants in the United States. California currently uses a cap-and-trade system to address greenhouse gas emissions. Every year, it determines the total amount of permissible emissions (the "cap") and this cap declines over time. Each entity is assigned allowances that authorize the emission of a certain quantity of greenhouse gases (carbon dioxide equivalents). At the end of a year, each entity must have sufficient allowances to cover its actual emissions. If it has extra allowances, it can trade or sell them to another company that might have exceeded their emissions limit. If a company

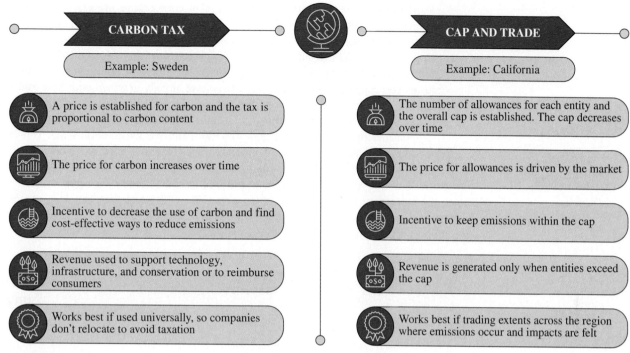

Figure 4.26

Comparing the two prominent policies for decreasing carbon emissions, carbon tax and cap and trade.

[Nicole C. Bouvier-Brown]

has insufficient allowances, it must purchase them. The price for allowances is determined by an auction, but the minimum amount increases over time. In preparation for this program to start in 2013, the auction price of 1 allowance was $10.09, while the May 2021 auction price was $30.85. This program covers 80% of California's greenhouse gas emissions and in 2014, it was linked with the Quebec cap-and-trade system.

The cap-and-trade system has some possible disadvantages, including a potentially volatile market for the emissions permits. Energy providers might experience wide, often unpredictable swings in their energy costs. Those swings would result in large fluctuations in consumer costs. As an alternative to cap and trade, some people advocate a carbon tax program. Instead of limiting emissions and letting the market decide how "best" to comply, a carbon tax simply increases the cost of burning fossil fuels and that cost increases over time. Sweden is one of the first countries to implement a carbon tax. In 1991, when the program began, the cost was SEK 250 ($23.50 USD) per ton of carbon dioxide emitted and in 2021 it was SEK 1200 ($112.75 USD). The idea of a carbon tax is simple: the cost of pollution is paid by those who cause it. Placing an additional cost based on the amount of carbon contained in a certain quality of fuel is intended to make alternative energy sources more competitive in the near term. Of course, levying a tax on carbon fuels or emissions will mean higher prices for consumers as well. To combat this, some programs will not only invest the revenue from the tax, but they will refund some of it back to consumers.

Your Turn 4.46 A Local Response to Climate Change

While many federal governments have been slow to produce binding climate change legislations, individual states/provinces, regions, counties, and cities have taken matters into their own hands.

Prepare a short report of an example of a climate change response that was enacted by your local region. When did this activity occur and have there been any positive outcomes as a result of this response?

While governmental policy and changes in corporations are essential to reverse the trends in greenhouse gas emissions, what can you, as an individual, do? First, let's figure out your own ecological impact. How might you estimate the amount of Earth's natural capital it takes to support the way in which you live? Clearly, this is far more difficult than estimating the gas mileage for your vehicle or counting the number of Calories you consume. Fortunately, scientists have already grappled with how to do the math. They base the calculations on the way in which a person lives coupled with the available renewable resources needed to sustain this lifestyle.

Consider the metaphor of a footprint. You can see the footprints that you leave in sand or snow. You also can see the muddy tracks that your boots leave on the kitchen floor. Similarly, one might argue that your life leaves a footprint on planet Earth. To understand this footprint, you need to think in units of hectares or acres. A hectare is a bit more than twice the area of an acre. The ecological footprint is a means of estimating the amount of biologically productive space (land and water) necessary to support a particular standard of living or lifestyle.

Your Turn 4.47   Footprint Calculations

Investigate some websites that calculate your personal carbon and ecological footprints.

a. For each site, list the name, the sponsor, and the information requested in order to calculate the footprints.

b. Does the information requested differ from site to site? If so, report the differences.

For the average global citizen, the ecological footprint was estimated in 2018 to be about 2.77 hectares (6.8 acres). In other words, on average, it requires biologically productive land and water the size of about 56 basketball courts to provide the resources to feed you, clothe you, transport you, and give you a dwelling with the creature comforts to which you are accustomed. At the same time, we have to consider how much productive land and water is available on our planet. This can be estimated by including regions such as croplands and fishing zones, and omitting regions such as deserts and ice caps. The biocapacity available for each global citizen is about 1.58 hectares. In other words, the productive area available is less than what is needed to sustain our lifestyles. Figure 4.27 shows the ecological footprints and biocapacity values for a few selected countries. Most countries show a biocapacity deficit (a negative difference between how much is consumed and how much ecosystems can renew). A positive difference indicates that the biocapacity can sustain the needs of that country. The trends seen in individual countries depend on consumption as well as available resources.

How much biologically productive land and water is available on our planet? We can estimate this by including regions such as croplands and fishing zones, and omitting regions such as deserts and ice caps. Currently, the value is estimated at about 12 billion hectares (roughly 30 billion acres) of land, water, and sea surface. This turns out to be about a quarter of Earth's surface. Is this enough to sustain everybody on the planet with the lifestyle of people in the United States? The next activity allows you to see for yourself.

Your Turn 4.48   Your Personal Share of the Planet

There is an estimated 12.2 billion hectares (about 30.1 billion acres) of biologically productive land, water, and sea are available on our planet.

a. Find the current estimate for the world population. Cite your source.

b. Use this estimate, together with the estimate for biologically productive land, to calculate the amount of land theoretically available for each person in the world.

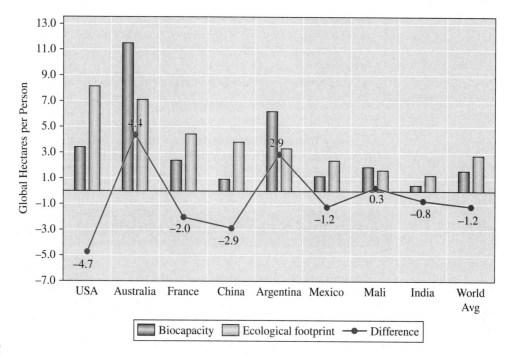

Figure 4.27

A comparison of ecological footprints, in global hectares per person.

Source: York University Ecological Footprint Initiative & Global Footprint Network. National Footprint and Biocapacity Accounts, 2022 edition. Downloaded from https://data.footprintnetwork.org (based on 2018 data).

Why is this important? We have been exceeding Earth's ability to meet our demands since the 1970s. Using the United States as an example, let's do one more calculation to see by how much the average person is exceeding what Earth can support.

Your Turn 4.49   How Many Earths?

In 2018, the United States had an ecological footprint of about 8.12 hectares (about 20 acres) per person.

a. Find an estimate of the current population of the United States. Cite your source.
b. Calculate the amount of biologically productive land that the United States currently requires for this population.
c. What percentage is this amount of the biologically productive space that is available on our planet?

Energy is essential for every human endeavor. Personally, you obtain the energy you need by eating and metabolizing food. As a community or nation, we meet our energy needs in a variety of ways, including burning coal, petroleum, and natural gas. The combustion of these carbon-based fuels produces several waste products, including carbon dioxide. The countries with large populations and those that are highly industrialized tend to burn the largest quantities of fuels and, as a result, emit the most CO_2.

Let's Debate!   Policies for a Sustainable Future

Explore the En-Roads Simulator (www.acs.org/cic) that investigates the summative effects of policies on their greenhouse gas emissions and the resulting global temperature increase in 2100.

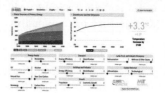

1. Go through the different policies; using the sliders, see what has the largest impact on the global temperature.
2. Propose a way to reach the 1.5 °C aspirational goal. Ideally, use the fewest new policies as possible.

En-ROADS/Climate Interactive

—(continued)

> Debate how different stakeholders would respond to your proposal outlined in question 2 above. Choose at least two of the following stakeholders: industry, government officials, forestry and agriculture, conventional energy, climate justice, clean energy, small island nations, etc.
>
> What are some concerns and where would the priorities lie for each group?

Every fraction of warming makes a difference. There is no single solution that will be suitable for each person to reach a sustainable lifestyle. Given your circumstances, what can you do?

Your Turn 4.50   What Can You Do?

Explore this final activity to learn how your daily decisions can help offset the effects of climate change: www.acs.org/cic

CreateStudio Pro/Bradley D. Fahlman

John Holdren, the senior advisor to President Obama on science and technology issues from 2009 to 2017, summarized our options in dealing with climate change with three words: mitigation, adaptation, and suffering. "Basically, if we do less mitigation and adaptation, we're going to do a lot more suffering," he concluded. But who will be responsible for the mitigation? Who will be forced to adapt? Who will bear the brunt of the suffering? It is likely that significant disagreements will arise regarding answers to these questions. But we can agree that any practical solution must be global in nature, and include a complicated mix of risk perception, societal values, politics, and economics.

Conclusion

(chimpanzees): Fuse/Getty Images; (children): Sajjad Hussain/AFP/Getty Images; (girl): Lussin/Shutterstock; (fisherman): ChameleonsEye/Shutterstock

Every being in the images above—children in Afghanistan, a carefree girl enjoying a fountain, elder Islanders, and even chimpanzees—contributed minimally, if at all, to global climate change. However, they are all extremely vulnerable to the effects of the modern climate crisis and are forced to adapt to the changes that are occurring. In addition, they all lack the political power to alter the trajectory of the countries that are emitting the most carbon dioxide. They are all powerless.

Even Earth itself cannot voice its opinion about policy changes; natural processes simply respond to the system inputs through chemistry. For example, as a greenhouse gas, carbon dioxide plays a role in keeping our planet comfortably warm and able to support life, but there can be too much of a good thing. In comparison with the past, the climate changes we are seeing today are occurring much more rapidly. The geologic evidence indicates past changes occurred over millennia, not decades as they occur today. To date, the most dramatic effects have been observed at the poles. These include quickly receding glaciers, shrinking sea ice, and melting permafrost. So far, the more densely populated lower latitudes have experienced far smaller effects from climate change. Like it or not, we are in the midst of conducting a global experiment, one that will test our ability to sustain both our economic development and our environment.

Throughout this chapter, you have studied carbon and the processes that move it through Earth's reservoirs. You have learned about greenhouse gases and how they are connected to Earth's energy balance. You have discovered the vast array of impacts human activity has caused. You now see what is meant by a changing climate. You have ideas on what should be done collectively and what you can do individually.

Those of us who are lucky enough to not only receive an education, but to be empowered with the truth about the modern climate crisis, must remember those who do not have a voice. What can you do to advocate for climate policy and dispel myths and misinformation about climate change? What mitigation or adaptation strategies should be high priority for developed nations? What can you do to minimize your own ecological footprint?

In the next chapter, you will learn about a chemical that is essential to life on this planet—water. You will understand how our daily actions result in changes to the availability, purity, and properties of water. As we focus on this precious resource, we will continue our discussions related to air quality and climate change, which impact the likelihood of securing suitable water resources for future generations.

LEARNING OUTCOMES

The numbers in parentheses indicate the sections within the chapter where these outcomes were discussed.

Having studied this chapter, you should now be able to:

- illustrate, identify, and predict carbon reservoirs and the processes by which carbon moves on Earth (4.1)
- identify and quantify carbon dioxide emissions from human activities (4.1)
- compare the atomic structures of isotopes and derive relative atomic masses (4.2)
- use Avogadro's number to quantify the mass of atoms (4.2)
- quantify the mass of an atom in a molecule (4.3)
- explain Earth's energy balance and the importance of the greenhouse effect (4.4)
- predict the 3D shape of a molecule (4.5)
- describe how greenhouse gases absorb infrared (IR) radiation (4.6)

- explain the process for collecting historical and current climate data, such as how isotope data allow scientists to determine past trends (4.7)
- analyze historical greenhouse gas and temperature data and describe how current climate trends on Earth differ from the past (4.7)
- discuss climate drivers beyond carbon dioxide (4.8)
- defend the scientific consensus that our modern climate crisis is primarily due to human activity (4.8)
- examine the consequences of climate change in terms of what is happening now and what might lie ahead (4.9)
- summarize the Paris Agreement and compare major policy initiatives such as "cap and trade" and carbon tax (4.10)
- propose ways to reach a sustainable planet and reflect on how to reduce your own ecological footprint (4.10)

Questions

Emphasizing Essentials

1. John Holdren, senior climate advisor to President Obama said: "Global warming is a misnomer, because it implies something that is gradual, something that is uniform, something that is quite possibly benign. What we are experiencing with climate change is none of those things." Use examples to:

   a. explain why climate change is not uniform.

   b. explain why it is not gradual, at least in comparison to how quickly social and environmental systems can adjust.

   c. explain why it probably will not be benign.

2. The surface temperatures of both Venus and Earth are warmer than would be expected on the basis of their respective distances from the Sun. Why is this so?

3. Using the analogy of a greenhouse to understand the energy radiated by Earth, of what are the "windows" of Earth's greenhouse made? In what ways is the analogy not precisely correct?

4. Consider the photosynthetic conversion of CO_2 and H_2O to form glucose, $C_6H_{12}O_6$, and O_2.

   a. Write the balanced equation.

   b. Is the number of each type of atom on either side of the equation the same?

   c. Is the number of molecules on either side of the equation the same? Explain.

5. Describe the difference between climate and weather.

6. **a.** It is estimated that 29 megajoules per square meter (MJ/m^2) of energy comes to the top of our atmosphere from the Sun each day, but only 17 MJ/m^2 reaches the surface. What happens to the rest?

   b. How much energy would leave the top of the atmosphere?

7. Consider Figure 4.16.

   a. How does the present concentration of CO_2 in the atmosphere compare with its concentration 20,000 years ago? With its concentration 120,000 years ago?

   b. How does the present temperature of the atmosphere compare with the 1950–1980 mean temperature? With the temperature 20,000 years ago? How does each of these values compare with the average temperature 120,000 years ago?

   c. Do your answers to parts **a** and **b** indicate causation, correlation, or no relation? Explain.

8. Understanding Earth's energy balance is essential to understanding the issue of global warming. For example, the solar energy striking Earth's surface averages 168 watts per square meter (W/m^2), but the energy leaving Earth's surface averages 390 W/m^2. Why isn't Earth cooling rapidly?

9. Explain each of these observations.

   a. A car parked in a sunny location may become hot enough to endanger the lives of pets or small children left in it.

   b. Clear winter nights tend to be colder than cloudy ones.

   c. A desert shows much wider daily temperature variation than a moist environment.

   d. People wearing dark clothing in the summertime put themselves at a greater risk of heatstroke than those wearing white clothing.

10. Construct a methane molecule (CH_4) from a molecular model kit (or use Styrofoam balls or gumdrops to represent the atoms and toothpicks to represent the bonds). You may also use an online molecular building program such as Molview. Demonstrate that the hydrogen atoms would be farther from one another in a tetrahedral arrangement than if they all were in the same plane (square planar arrangement).

11. Draw the Lewis structure and name the molecular geometry for each molecule.

    a. H_2S

    b. OCl_2 (oxygen is the central atom)

    c. N_2O (nitrogen is the central atom)

12. Draw the Lewis structure and name the molecular geometry for these molecules.

    a. PF_3

    b. HCN (carbon is the central atom)

    c. CF_2Cl_2 (carbon is the central atom)

13. **a.** Draw the Lewis structure for methanol (wood alcohol), H_3COH.

    b. Based on this structure, predict the H—C—H bond angle. Explain your reasoning.

    c. Based on this structure, predict the H—O—C bond angle. Explain your reasoning.

14. **a.** Draw the Lewis structure for ethene (ethylene), H_2CCH_2, a small hydrocarbon with a C=C double bond.

    b. Based on this structure, predict the H—C—H bond angle. Explain your reasoning.

    c. Sketch the molecule showing the predicted bond angles.

15. Three different modes of vibration of a water molecule are shown. Which of these modes of vibration contributes to the greenhouse effect? Explain.

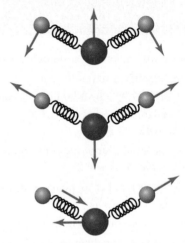

16. If a carbon dioxide molecule interacts with certain photons in the IR region, the vibrational motions of the atoms are increased. For CO_2, the major wavelengths of absorption occur at 4.26 micrometers (μm) and 15.00 micrometers (μm).

    a. What is the energy corresponding to each of these IR photons?

    b. What happens to the energy in the vibrating CO_2 species?

17. Water vapor and carbon dioxide are greenhouse gases, but N_2 and O_2 are not. Explain. Why is this important?

18. Explain how each of these relates to global climate change.

    a. volcanic eruptions

    b. CFCs released from air conditioning units

19. Termites possess enzymes that allow them to break down cellulose into glucose, $C_6H_{12}O_6$, and then metabolize the glucose into CO_2 and CH_4.

    a. Write a balanced equation for the metabolism of glucose into CO_2 and CH_4.

    b. What mass of CO_2, in grams, could one termite produce in one year if it metabolized 1.0 mg glucose in one day?

20. Consider Figure 4.4.

    a. Which sector has the highest CO_2 emission from fossil fuel combustion?

    b. What alternatives exist for each of the major sectors of CO_2 emissions?

21. Estimate how long you watch streaming videos each day, then scale this value up to account for an entire year. If there are 36 g of CO_2 emitted per hour, calculate your emissions from this one activity. (*Note:* this does not account for the 0.2 g CO_2 emitted from each internet search.)

22. Radon has an atomic number of 86.

    a. Give the number of protons, neutrons, and electrons in a neutral atom of the most common isotope, Rn-222.

    b. How do the numbers of protons, neutrons, and electrons in a neutral atom of Rn-220 compare with those of Rn-222?

23. Radon primarily has two naturally occurring isotopes: Ag-107Rn-222 and Rn-220Ag-109. Why isn't the average atomic mass of silver given on the periodic table simply 221?

24. **a.** Calculate the average mass in grams of an individual atom of silver.

    b. Calculate the mass in grams of 10 trillion silver atoms.

    c. Calculate the mass in grams of 5.00×10^{45} silver atoms.

25. Calculate the molar mass of these compounds. Each plays a role in atmospheric chemistry.

    a. H_2O

    b. CCl_2F_2 (Freon-12)

    c. N_2O

26. **a.** Calculate the mass percent of chlorine in CCl_3F (Freon-11).

    b. Calculate the mass percent of chlorine in CCl_2F_2 (Freon-12).

    c. What is the maximum mass of chlorine that could be released in the stratosphere by 100 g of each compound?

    d. How many atoms of chlorine correspond to the masses calculated in part **c**?

27. The total mass of carbon in living systems is estimated to be 7.5×10^{17} g. Given that the total mass of carbon on Earth is estimated to be 7.5×10^{22} g, what is the ratio of carbon atoms in living systems to the total carbon atoms on Earth? Report your answer in percent and in ppm.

28. Using the Internet as a resource, what is the global warming potential (GWP) of a greenhouse gas? Other than atmospheric concentration, what two other properties are included in the calculation of the global warming potential for a substance?

29. In 2009, U.S. Energy Secretary Steven Chu suggested that painting roofs white would be one way to combat global warming. Explain the reasoning behind this course of action.

30. **a.** It is estimated that volcanoes globally release about 2.3×10^7 t (23 million metric tons) of SO_2 per year. Calculate the mass of sulfur atoms in this amount of SO_2 molecules.

    b. If 9.7×10^7 t of SO_2 is released per year by fossil fuel combustion, calculate the mass of sulfur atoms in this amount of SO_2 molecules. (*Note:* this value is down from the 1.5×10^8 t in 1980.)

Concentrating on Concepts

31. Give the number of protons, neutrons, and electrons in each of these neutral atoms.

    a. oxygen-18 ($^{18}_{8}O$)

    b. sulfur-35 ($^{35}_{16}S$)

    c. uranium-239 ($^{239}_{82}U$)

    d. bromine-82 ($^{82}_{35}Br$)

    e. neon-19 ($^{19}_{10}Ne$)

    f. radium-226 ($^{26}_{88}Ra$)

32. Give the symbol showing the atomic number and the mass number for the isotope that has:

    a. 9 protons and 10 neutrons (used in nuclear medicine).

    b. 26 protons and 30 neutrons (the most stable isotope of this element).

    c. 86 protons and 136 neutrons (the radioactive gas found in some homes).

33. For each of the values below, determine the number of significant figures (sig. figs.).

    a. 780.2 g   **c.** 400.0 g/mL

    b. 9.003 L   **d.** 0.0025 μm

34. For each of the following calculations, report the answer to the correct number of significant figures. Remember to include the correct unit for each of the calculations as well.

    a. 5.078 g + 0.04 g   **c.** 7.645 g ÷ 1.87 mL

    b. 3.00 L − 0.1 L   **d.** 56 nm × 98.24 nm

35. A sample of silicon has a mass of 1.25 g and a volume of 0.54 cm^3. Calculate the density of the sample. Do not forget to include the correct number of significant figures for the answer.

36. John Holdren, quoted in the conclusion of the chapter, suggests that we use the term *global climatic disruption* rather than *global warming*. After studying this chapter, do you agree with his suggestion? Explain.

 List three similarities and three differences between climate and weather. Is the term "global warming" a description of climate or weather phenomena? Explain.

37. The Arctic has been referred to the canary in the coal mine for climate impacts that will affect us all.

    a. What does the phrase "canary in the coal mine" mean?

    b. Explain why the Arctic serves as a canary in a coal mine.

    c. The melting of the tundra accelerates changes elsewhere. Give one reason why.

38. Given that direct measurements of Earth's atmospheric temperature over the last several thousands of years are not available, how can scientists estimate past fluctuations in the temperature?

39. A friend tells you about a newspaper story that stated, "The greenhouse effect poses a serious threat to humanity." What is your reaction to that statement? What would you tell your friend?

40. Throughout the chapter, we discussed how anthropogenic influences have changed the concentration of CO_2 in Earth's atmosphere. Are there other reasons why the concentration of CO_2 has changed? Explain.

41. Over the last 20 years, about 120 billion tons of CO_2 has been emitted from the burning of fossil fuels, yet the amount of CO_2 in the atmosphere has risen by only about 80 billion tons. Explain.

42. Carbon dioxide gas and water vapor both absorb IR radiation. Do they also absorb visible radiation? Offer some evidence based on your everyday experiences to help explain your answer.

43. How would the energy required to cause IR-absorbing vibrations in CO_2 change if the carbon and oxygen atoms were connected by single rather than double bonds?

44. The increases in global temperature mentioned throughout this chapter are average values for the entire globe, including land and ocean measurements. Would you expect the temperature changes to be the same over land masses and oceans? Why or why not? What implications does this scenario present?

45. Ethanol, C_2H_5OH, can be produced from sugars and starches in crops such as corn or sugarcane. The ethanol is used as a gasoline additive, and when burned it combines with O_2 to form H_2O and CO_2.

    a. Write a balanced equation for the complete combustion of C_2H_5OH.

    b. How many moles of CO_2 are produced from each mole of C_2H_5OH completely burned?

    c. How many moles of O_2 are required to burn 10 mol of C_2H_5OH?

46. What is meant by the atmospheric lifetime of a greenhouse gas and why is this measurement important?

47. a. Use Figure 4.11 to estimate the wavelengths corresponding to the strongest IR absorbance for water vapor.

    b. Which wavelengths do you predict represent bending vibrations, and which represent stretching? Explain the basis of your predictions. *Hint:* Compare the IR spectrum of H_2O with that of CO_2.

48. Draw a diagram or model that describes how Earth functions as a greenhouse. Explain your representation in words.

49. Compare and contrast stratospheric ozone depletion and climate change in terms of the chemical species involved, the type of radiation involved, and the predicted environmental consequences.

50. Explain the term *radiative forcings* to someone unfamiliar with climate modeling.

51. It is estimated that Earth's ruminants, such as cattle and sheep, produce 73 million metric tons of CH_4 each year. How many metric tons of carbon are present in this mass of CH_4?

52. Nine of the ten warmest years since 1880 have occurred since the year 2000. Does this *prove* that the enhanced greenhouse effect (global warming) is taking place? Explain.

53. A possible replacement for CFCs is HFC-152a, with a lifetime of 1.4 years and a global warming potential (GWP) of 120. Another is HFC-23, with a lifetime of 260 years and a GWP of 12,000. Both of these possible replacements have a significant effect as greenhouse gases and are regulated under the Kyoto Protocol.

    a. Based on the given information, which appears to be the better replacement? Consider only the potential for global warming.

    b. What other considerations are there in choosing a replacement?

54. The emissions of CO_2 from fossil fuel burning can be reported in different ways. For example, the Carbon Dioxide Information Analysis Center (CDIAC) reported in 2014 that China, the United States, and India ranked highest among world nations:

Ranking	Nation	Metric tons of CO_2
#1	China (mainland)	10,291,927
#2	United States	5,254,279
#3	India	2,238,377

    a. Would the rankings change if expressed on a per capita basis? If so, which nation would rank first?

    b. CDIAC reports the per capita rankings on the basis of metric tons of carbon emitted, rather than metric tons of CO_2. Qatar leads the world in per capita emissions at 45.4 metric tons of carbon. Would this value be higher or lower if expressed on the basis of metric tons of CO_2 emitted? Explain.

55. Provide some examples of countries who have adopted a cap-and-trade policy and others who have instituted a carbon tax. What are some positives and negatives that have been reported in the respective countries from the adoption of these policies?

56. When Arrhenius first theorized the role of atmospheric greenhouses, he calculated that doubling the concentration of CO_2 would result in an increase of 5–6 °C in the average global temperature. How far off was he from the current IPCC modeling?

57. The Union of Concerned Scientists publishes a list of the top 15 nations for CO_2 emissions (based on data from International Energy Agency).

a. From what you already know, predict the top 10 nations who have contributed the most CO_2 to our atmosphere since 1750. Check how accurate your predictions were by using the Internet.

b. How would these rankings change if they were listed just for the most recent year?

c. How would these rankings change if they were listed per capita?

Exploring Extensions

58. China's growing economy is fueled largely by its dependence on coal, described as China's "double-edged sword." Coal is both the new economy's "black gold" and the "fragile environment's dark cloud."

a. What are some of the consequences of dependence on high-sulfur coal? *Hint:* Review the impact of high-sulfur fuels discussed in Chapter 2.

b. Sulfur pollution from China may slow global warming, but only temporarily. Explain.

c. What other country is rapidly stepping up its construction of coal-fired power plants and is expected to have a larger population than China by the year 2030?

59. The Quino checkerspot butterfly is an endangered species with a small range in northern Mexico and southern California. Evidence indicates that the range of this species is larger than previously predicted from earlier studies.

a. Describe how this species has been able to adapt to warming climate.

b. What are the projections of Quino's range and likelihood of survival within the next 30 years?

60. Data taken over time reveal an increase in CO_2 in the atmosphere. The large increase in the combustion of hydrocarbons since the Industrial Revolution is often cited as a reason for the increasing levels of CO_2. However, an increase in water vapor has *not* been observed during the same period. Remembering the general equation for the combustion of a hydrocarbon, does the difference in these two trends *disprove* any connection between human activities and global warming? Explain your reasoning.

61. In the energy industry, 1 standard cubic foot (SCF) of natural gas contains 1196 mol of methane (CH_4) at 15.6 °C (60 °F). Refer to Appendix 1 for conversion factors.

a. How many moles of CO_2 could be produced by the complete combustion of 1 SCF of natural gas?

b. How many kilograms of CO_2 could be produced?

c. How many metric tons of CO_2 could be produced?

62. Choose a country that is part of the Paris Agreement. Describe where the chosen country stands with respect to their pledged goals and suggest steps forward for that country if applicable.

63. A solar oven is a low-tech, low-cost device for focusing sunlight to cook food. How might solar ovens help mitigate global warming? Which regions of the world would benefit most from using this technology?

64. In 2013, Quebec, a province in Canada, adopted a cap-and-trade policy for carbon dioxide. Write a short report on the outcomes of this policy, in terms of both the economic result and the effect it has had on greenhouse gas emissions.

65. The consequences of climate change described in this chapter are examples of what are known as external costs. These costs are not reflected in the price of a commodity, such as the price of a gallon of gasoline or a ton of coal, but nevertheless take a toll on the environment. The external costs of burning fossil fuels often are shared by those who emit very little carbon dioxide, such as the people of the island nation of Maldives. Although a rise of sea level of just a few millimeters may not seem like much, the effects could be catastrophic for nations that lie close to sea level. Use the resources of the Internet to investigate how the people of Maldives are preparing for rising sea levels. Also comment on how this is an example of the *tragedy of the commons*.

5 Water Everywhere: A Most Precious Resource

WATER QUALITY AND USE

Watch the chapter-opening video (www.acs.org/cic) and think about the water you drink and use daily.

a. What substances and impurities are found in this water?

b. Where does this water come from, and where does the wastewater eventually go?

c. How do you think the water habits of a community can affect the natural water supply?

In this chapter, you will explore the following questions:

1. What are the unique properties of water?
2. How does the shape of water molecules and intermolecular forces influence their physical properties?
3. Where is water located on Earth, and how readily available are these resources?
4. What are aqueous solutions?
5. How can we quantify the concentration of aqueous solutions?
6. How does water interact with other chemicals?
7. How can we predict whether a substance will be soluble in water?
8. What are the properties of acids and bases?
9. How do acids and bases react with one another?
10. How can we quantify the acidity of water?
11. What are the effects of acids on surface waters?
12. How can we treat water for use and consumption?
13. How can we purify saltwater and polluted waters for safe consumption?

(left): Maryia Bahutskaya/Getty Images; (right): altitudevisual/123RF

Introduction

Calm and rough. Life and death. Thirsty and quenched. Plentiful and scarce. These terms describe the essential resource for life on Earth—water. Water plays a role in nearly everything that takes place on our planet. Scientists look for water when they search for life on other planets. Humans are 60% water, and 71% of Earth is covered with water. The amount of water you drink daily depends on your size, age, health, and level of physical activity. Have you ever imagined a world without water? What if you were not able to take a drink of water? Your body can go weeks without food, but only days without water. If the water content in your body were reduced by 2%, you would get thirsty. With a 5% water loss, you would feel fatigued and have a headache. At a 10–15% loss, your muscles would become spastic, and you would feel delirious; greater than 15% dehydration will kill you.

Although oceans are home to a wealth of plant and animal life, they are not hospitable to the creatures that dwell on land. As Rachel Carson (1907–1964) noted in *Silent Spring,* "By far the greater part of the Earth's surface is covered by its enveloping seas, yet amid this plenty we are in want. By a strange paradox, most of the Earth's abundant water is not usable for agriculture, industry, or human consumption because of its heavy load of sea salts." Those who live on land need fresh water and must obtain it through natural processes such as rain and snowfall or energy-intensive water purification technologies.

Unfortunately, fresh water is not an unlimited resource on our planet. Furthermore, it is not renewable fast enough to meet the needs of our increasing world population. As a result, water has become a strategic resource. Its scarcity brews conflicts and raises questions about who has the right to access and use it.

According to the U.S. Geological Survey (USGS), more than 390 liters (~100 gallons) of water per day are required to support the lifestyle of the average U.S. citizen.

Your Turn 5.1   Keep a Water Diary

Pick a two-day period that represents typical activities for you. Record all of your actions that involve water by time and activity. While writing down your use of water, record the following:

a. The role that water played in your life. For example, are you consuming it? Are you using it in some process? Is it part of your outdoor experience?
b. The source of the water, the quantity involved, and where it went afterward.
c. The degree to which you made the water dirty.

This chapter describes the unique properties of water, how chemicals behave in it, how this affects water quality, and how larger global issues are impacting water in lakes, streams, and oceans.

5.1 | The Unique Properties of Water

Learning Objective: Predict the shape and polarity of molecules and describe their role on physical properties

Water is a compound with unique properties in oceans, lakes, or rivers. Some of these are important in understanding large-scale processes on our planet. For example, water is the only common substance that can exist as a solid, liquid, or vapor at average temperatures on Earth. In its three states, water affects the daily weather of a region and, over a more extended period, the climate. Water is a liquid under **standard temperature and pressure (STP)**; that is, a temperature of 0 °C and pressure of 1 bar (0.987 atm). This is surprising because almost all other compounds with a similar molar mass as water (18 g/mol) are gases under these conditions. For instance, consider these three gaseous components found in air: N_2 (28 g/mol), O_2 (32 g/mol), and CO_2 (44 g/mol).

Water also has an anomalously high boiling point of 100 °C (212 °F); in contrast, liquids with similar molecular structures such as hydrogen sulfide, H_2S, have much lower boiling points. Further, when water freezes, it exhibits another somewhat bizarre property—it expands. However, most liquids contract when they solidify. These and other unusual properties of water are derived from its molecular structure and the interactions among individual molecules.

Recall that water is a covalently bonded molecule with a bent shape (Figure 5.1).

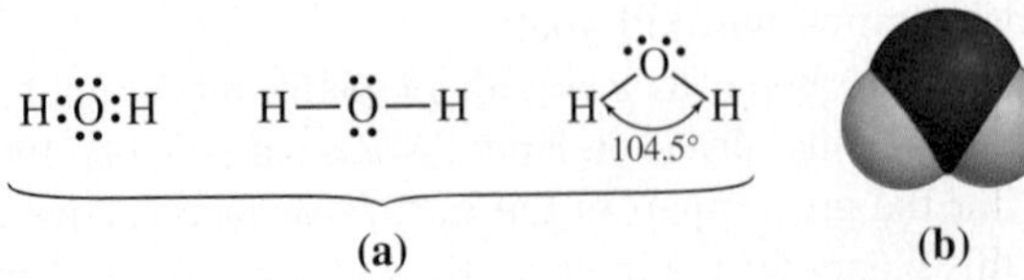

$$H\!:\!\ddot{O}\!:\!H \qquad H-\ddot{O}-H \qquad H\underset{104.5°}{\overset{\ddot{O}}{\diagup\!\diagdown}}H$$

(a)   (b)

Figure 5.1

Representations of water, H_2O. **(a)** Lewis structures and structural formula; **(b)** space-filling model.

Table 5.1	Electronegativity Values for Selected Elements						
Group 1	2	13	14	15	16	17	18
H 2.1							He*
Li 1.0	Be 1.5	B 2.0	C 2.5	N 3.0	O 3.5	F 4.0	Ne*
Na 0.9	Mg 1.2	Al 1.5	Si 1.8	P 2.1	S 2.5	Cl 3.0	Ar*

*Noble gases rarely (if ever) bond to other elements and therefore do not have electronegativity values.

Let's focus on how the electrons are shared in the O—H covalent bond. The electrons are not shared evenly despite drawing a line showing that the two atoms are connected. Experimental evidence indicates that the O atom attracts the shared electron pair more strongly than the H atom. In chemical language, oxygen is said to have a higher electronegativity than hydrogen. **Electronegativity** is a measure of the attraction of an electron toward an atom within a covalent bond. The values have no units and are relative measurements to one another, up to a maximum of 4.0 for fluorine (Table 5.1). The greater the electronegativity, the more an atom attracts the electrons in a covalent bond toward itself.

The greater the difference in electronegativity between two bonded atoms, the more *polar* the bond is. For example, the electronegativity difference between oxygen and hydrogen is 1.4. In contrast, this difference in a S—H bond is only 0.4—much less polar than an O—H bond. The electrons in an O—H bond are pulled closer to the more electronegative oxygen atom. As shown in Figure 5.2, this unequal sharing results in a partial negative charge (δ^-) on the O atom and a partial positive charge (δ^+) on the H atom.

An arrow indicates the direction in which the electron pair is displaced, often referred to as a **bond dipole**. The result is a **polar covalent bond**, a covalent bond in which the electrons are not equally shared but rather are closer to the more electronegative atom. In contrast, a **nonpolar covalent bond** is a covalent bond in which the electrons are shared equally between atoms. As an example of nonpolar covalent bonds, consider molecules existing in our atmosphere, such as N_2 and O_2. Because both atoms in the molecule are identical, there is no electronegativity difference between them, which corresponds to the equal sharing of their electrons.

Your Turn 5.2   Polar Bonds

Example Problem:

Which bond is more polar: H—F or H—Cl? Explain your reasoning.

Example Solution:

H—F is more polar since F is more electronegative than Cl.

Your Turn:

For each pair of bonds, which is more polar? In the bond you select, the electron pair is more strongly attracted to one of the atoms. Which one?

a. N—H or O—H   **b.** N—O or O—S   **c.** H—H or Cl—C

Electronegativity
value
_____
3.5   2.1

δ^-O ⬅════ H δ^+

difference = 1.4

Figure 5.2

Representation of the polar covalent bond between hydrogen and oxygen atoms. The electrons are pulled toward the more electronegative oxygen atom.

We have made the case that bonds can be polar, some more than others. What about molecules as a whole? To help you predict if a *molecule* is polar, we offer two useful generalizations:

- A molecule that contains only nonpolar bonds *must be* nonpolar. For example, the diatomic elements such as O_2, N_2, Cl_2, and H_2 are nonpolar.
- A molecule that contains polar covalent bonds *may or may not be* polar. The polarity depends on the shape of the molecule. For example, the water molecule contains two polar bonds, and the molecule is polar (Figure 5.3). Each H atom carries a partial positive charge (δ^+), and the oxygen atom carries a partial negative charge (δ^-). The water molecule is polar with these two polar bonds and a bent geometry.

You may think of the polar bonds as "tugs-of-war" between the two atoms. If the bond dipoles are not positioned 180° from one another and do not cancel, such as in water, the molecule is *polar*. However, let's now consider CO_2 (Figure 5.4). If you employ the strategy used in Section 4.5 to construct its Lewis structure and determine its shape, you will determine that the molecule is linear instead of bent. Because the "tugs-of-war" between the bond dipoles cancel one another, CO_2 is a nonpolar molecule—even though it contains two polar covalent bonds.

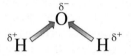

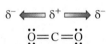

Figure 5.3

H_2O. A polar molecule with polar covalent bonds.

Figure 5.4

Carbon dioxide, CO_2. A nonpolar molecule with polar covalent bonds.

Many of the unique properties of water are a consequence of its polarity. But before we continue the story of water, take a moment to complete this activity.

Your Turn 5.3   Polarity of Molecules

In this activity, you will explore the polarity of molecules using this simulation: www.acs.org/cic.

a. Draw the Lewis structure BeClBr.

b. Are the covalent bonds in BeClBr polar or nonpolar? Use Table 5.1.

c. Use the simulation to build the BeClBr molecule by dragging the atoms into their correct geometrical arrangements. Vary the relative electronegativities of the atoms by using the sliders provided. Similar to Figures 5.3 and 5.4, draw a representation for BeClBr.

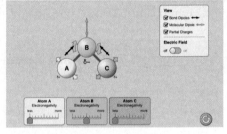

PhET Interactive Simulations, University of Colorado Boulder, http://phet.colorado.edu

d. Is the BeClBr molecule polar or nonpolar? Explain.

e. Complete a–c for the following molecules: COS, SO_2, and I_3^-. Determine the polarity of the covalent bonds and whether the molecules are polar or nonpolar.

5.2 | How Do Water Molecules Interact?

Learning Objective: Predict the intermolecular forces involved among polar molecules and describe their role in physical properties

Consider what happens when two water molecules approach each other. Because opposite charges attract, an H atom (δ^+) on one of the water molecules is attracted to the O atom (δ^-)

on the neighboring water molecule. This is an example of an **intermolecular force**, a force that occurs *between* molecules, as compared to an intramolecular force, a force *within* a single molecule. Compare this to intercollegiate sports being played between schools, whereas intracollegiate (intramural) sports are played within one school.

With more than two water molecules, the story gets more complicated. Examine each H_2O molecule in Figure 5.5 and note the two H atoms and two nonbonding pairs of electrons on the O atom. These allow for multiple intermolecular attractions. This set of attractions among molecules is called "hydrogen bonding." A **hydrogen bond** is an electrostatic attraction between an H atom, which is bonded to a highly electronegative atom (O, N, or F), and a neighboring O, N, or F atom—either in another molecule or in a different part of the same molecule. Do not confuse hydrogen bonds (intermolecular force) with covalent bonds (intramolecular force). Typically, hydrogen bonds are only about one-tenth as strong as the covalent bonds connecting atoms *within* molecules. Also, the atoms involved in hydrogen bonding are farther apart than in covalent bonds. In liquid water, there may be up to four hydrogen bonds per water molecule, as shown in Figure 5.5.

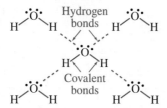

Figure 5.5

The inter- and intramolecular forces within and among water molecules (distances not to scale).

Your Turn 5.4   H-Bonding Representations

a. Explain what the dashed lines between water molecules in Figure 5.5 represent.
b. In the same figure, label the atoms on two adjacent water molecules with δ^+ or δ^-. How do these partial charges help explain the orientation of the molecules?
c. Illustrate hydrogen bonding in four molecules of NH_3.

Your Turn 5.5   Hydrogen Bonding?

Example Problem:

Are PH_3 molecules held together by hydrogen bonding?

Example Solution:

PH_3 molecules contain P—H bonds. There are no hydrogen bonds present since there are no H—O, H—N, or H—F bonds in the molecule.

Your Turn:

Do any of the following molecules have hydrogen bonding? Explain your rationale.

a. HCl   **b.** H_2S **c.** HF   **d.** N_2O **e.** CH_4

Although hydrogen bonds are not as strong as covalent bonds, hydrogen bonds are still quite strong compared with other types of intermolecular forces. The boiling point of water gives us evidence for this assertion. For example, consider hydrogen sulfide, H_2S, a molecule that has the same shape as water but does not contain hydrogen bonds. Since hydrogen sulfide is a polar molecule, the intermolecular forces are denoted as **dipole–dipole forces**, which are generally weaker than hydrogen bonds. As a result, H_2S boils at about $-60\ °C$ and so is a gas at room temperature, whereas water boils at 100 °C. Because of hydrogen bonding, water is a liquid at room temperature and body temperature (about 37 °C). Life's very existence on our planet depends on this property.

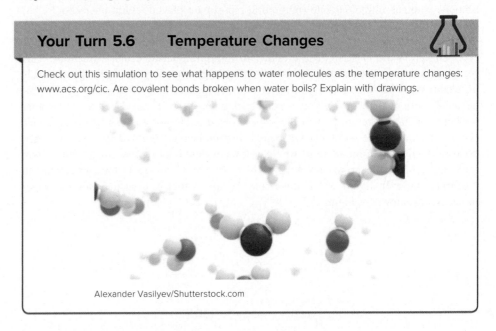

Your Turn 5.6   Temperature Changes

Check out this simulation to see what happens to water molecules as the temperature changes: www.acs.org/cic. Are covalent bonds broken when water boils? Explain with drawings.

Alexander Vasilyev/Shutterstock.com

Hydrogen bonding can also help you understand why ice cubes and icebergs float. Ice is composed of a regular arrangement of water molecules in which every H_2O molecule is hydrogen-bonded to four others (Figure 5.6). Note the empty space in the form of hexagonal channels. When ice melts, the pattern is lost, and individual H_2O molecules can enter the open channels. As a result, the molecules in the liquid state are more closely packed than in the solid state. Thus, a volume of 1 cm^3 of liquid water contains more molecules than 1 cm^3 of ice. Consequently, liquid water has a greater mass per cubic centimeter than ice. This is simply another way of saying that liquid water's **density**, the mass per unit volume, is greater than that of ice.

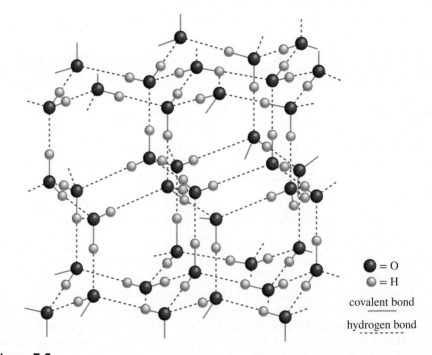

= O
= H
covalent bond
hydrogen bond

Figure 5.6

The hydrogen-bonded lattice structure of the common form of ice. Note the open channels between "layers" of water molecules that cause ice to be less dense than liquid water.

We usually express the mass of water in grams. Expressing its volume is a bit trickier. We use either cubic centimeters (cm^3) or milliliters (mL)—the two units are equivalent. The density of liquid water is 1.00 g/cm^3 at 4 °C, and varies only slightly with temperature. So, for convenience, we sometimes say that 1 cm^3 of water has a mass of 1 g. On the other hand, 1.00 cm^3 of ice has a mass of only 0.92 g, so its density is 0.92 g/cm^3. The bottom line? The ice cubes in your favorite beverage float rather than sink.

Unlike water, most substances are denser as solids (Figure 5.7). The fact that water shows the reverse behavior means that, in winter, ice floats on lakes rather than sinks. This topsy-turvy behavior also means that surface ice, often covered by snow, can act as an insulator and keep the lake water beneath from freezing. Aquatic plants and fish can therefore live in a freshwater lake during cold winters. And when the ice melts in spring, the water formed from the melting ice sinks, which helps mix the nutrients in the freshwater ecosystem. Water's unique behavior has implications for the biological sciences and life itself.

@HOME   Density Comparisons

Figure 5.7

The solid phase of paraffin wax is denser than its liquid phase and thus sinks to the bottom of the container (left). In comparison, the solid phase of water (ice) is less dense than its liquid phase and floats to the top (right).

Richard Megna/Fundamental Photographs

Investigate the densities of a substance at different phases at home using vegetable oil, water, a freezer, and two clear glasses.

1. Freeze vegetable oil and water into solid cubes.
2. Pour some liquid vegetable oil into a glass. Then add the solid cube of vegetable oil. Make observations about what you see.
3. Pour some liquid water into a glass. Then add the solid cube of water. Make observations about what you see.

Questions to Consider

a. What was different between the glass of solid and liquid vegetable oil and the glass of solid and liquid water?
b. Explain your results using the intermolecular forces between the molecules of each substance.

The phenomenon of hydrogen bonding is not restricted to water. It can occur in other molecules that contain covalent O—H, N—H, or F—H bonds. Hydrogen bonds help stabilize the shape of large biological molecules, such as proteins and nucleic acids.

For example, DNA molecules form hydrogen bonds between *different* strands of DNA. In contrast, proteins can form hydrogen bonds with different regions within the *same* molecule. Again, hydrogen bonding plays an essential role in the processes of life.

We end this section by examining one last unusual property of water, its uncommonly high capacity to absorb and release heat. Water absorbs more heat per gram than most other substances, allowing bodies of water on Earth to serve as heat reservoirs. **Specific heat** is the quantity of heat energy that must be absorbed to increase the temperature of 1 gram of a substance by 1 °C. The specific heat of water is 4.18 J/g °C (1.00 cal/g °C). This means that 4.18 joules (J) of energy is needed to raise the temperature of 1 g of liquid water by 1 °C. Conversely, 4.18 J of heat must be removed to cool 1 g of water by 1 °C. Water has one of the highest specific heats of any substance and is said to have a high heat capacity. Because of this, it is an exceptional coolant and can be used to carry away the excess heat in a car radiator, a power plant, or the human body.

@HOME   Water Balloons!

Safety Note: This lab requires a flame, so you will need to wear safety goggles, secure loose clothing, and tie long hair back. You should also use a long wand lighter. If you are in a place where you cannot have smoke or fire, please skip this activity.

Investigate the thermal properties of water at home using two balloons, a long wand lighter, a sink, and some tap water. **You will want to do this investigation over a sink.**

1. Inflate a balloon, tie it off, and add half a cup of water to a second balloon. Inflate this balloon to the same size as the first and tie it off.
2. Hold the long wand lighter to the first balloon. Record what happens.
3. Now, hold the long wand lighter to the balloon's section with the water in it. Record what happens.
4. Finally, hold the long wand lighter to the section of the balloon not filled with water. Record what happens.

Rob Daly/OJO Images/
Getty Images

Questions to Consider
a. Explain how water was able to stop the balloon from bursting.
b. In the text, it says that water can use to remove excess heat from the human body. Why can't you put a flame against your skin?

Because of water's high specific heat, large bodies of water influence regional climate. Water evaporates from seas, rivers, and lakes because the bodies of water absorb heat. By absorbing vast amounts of heat, the oceans and the water droplets in clouds help moderate global temperatures. Because water has a higher capacity to "store" heat than the ground does, when the weather turns cold, the ground cools more quickly. Water retains more heat and can provide more warmth to the areas bordering it for a longer time. Such properties should be familiar to anyone who has ever lived near a large body of water.

Your Turn 5.7   A Barefoot Excursion

Have you ever walked barefoot across a carpeted floor and onto a tile or stone floor? If not, try it and see what you notice. Based on your observation, does carpet or tile have a higher heat capacity? Why?

We have just examined some of the critical properties of water that influence life on our planet. Before we explore its ability to dissolve many different substances, we seek a broader picture of where water comes from, how we use it, and which issues relate to its use.

5.3 | Where, Oh Where, Is All the Water?

> *Learning Objectives:*
>
> *Identify sources of water on Earth and the relative availability of fresh water*
>
> *Describe ways that water may become contaminated and analyze data to evaluate water use, consumption, and level of contamination*

Just as we need clean, unpolluted air to breathe, we also need **potable water**, safe for drinking and cooking. Since tap water is considered a potable water source, we may also use this water for bathing and washing dishes. In contrast, nonpotable water contains particulates from dirt and chemicals that may cause cancer such as GenX, toxic metals such as arsenic, or bacteria that cause cholera. Although not drinkable, nonpotable water still has its uses. For example, water from rivers or lakes may be hauled in trucks (Figure 5.8) and used to wash sidewalks, reduce roadway dust, or irrigate.

Your Turn 5.8   Cholera Contamination

We just mentioned that nonpotable water may contain contaminants such as GenX, arsenic, and bacteria that can cause cholera. In different parts of the world, these are all significant issues.

Questions to Consider
a. What are the sources of each type of contaminant?
b. What are some of the health effects of being exposed to these chemicals in water?

John Wollwerth/Shutterstock

Everyone expects access to water that is safe and devoid of harmful chemicals and microbes. However, the World Health Organization (WHO) estimates that globally over 2 billion people—more than a quarter of our population—live in water-stressed countries (Figure 5.8), with a drinking water source often contaminated with faeces.

Over 3000 infants and children die each day because of contaminated water, sometimes characterized by its appearance (Figure 5.9), but other times not. The water crisis in Flint, Michigan, and current lawsuits related to Camp Lejeune have renewed

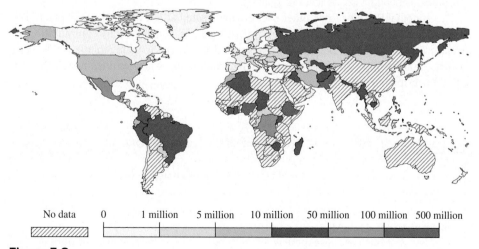

No data   0   1 million   5 million   10 million   50 million   100 million   500 million

Figure 5.8

Number of people without access to safe drinking water, 2000–2020. For an interactive map of these data, see www.acs.org/cic.

Ourworldindata.org

Figure 5.9

A river polluted from mining drainage.

K Steve Cope/Shutterstock

our focus on water contamination, teaching us to be more cautious of the water sources we often take for granted. The water samples taken from homes in the Flint area had lead (Pb) levels seven times higher than the U.S. Environmental Protection Agency (EPA) guidelines. Unfortunately, this contamination may create lifelong problems for children exposed to high levels of lead in drinking water, including low academic performance. Such a preventable situation shows us the importance of regular water monitoring and ensuring that testing is done ethically and transparently.

What makes water safe to drink? "Safe" water doesn't mean it's pure—all water sources will contain dissolved substances. Many of these substances are from natural sources. For example, beneficial minerals found in groundwater contribute calcium and magnesium ions to the water. However, other natural substances can be harmful. The EPA defines a water contaminant as anything physical, chemical, biological, or radioactive that is harmful to human health or degrades the taste or color of the water. The EPA regulates more than 90 substances known to contaminate drinking water. This type of water is known as nonpotable water, meaning that it is not drinkable, but can be used for other purposes, depending on its level of quality.

Nonpotable water finds other uses if first treated at a municipal water plant. This reclaimed water, sometimes called recycled water, is distributed to communities via water trucks or through "purple pipes" (Figure 5.10). It can irrigate athletic fields, flush toilets, or fight fires. To keep water flowing, community water utilities match the type of water available with its best use.

Your Turn 5.9   Matching Form to Function

In some communities, reclaimed or recycled (nonpotable) water is used to wash cars, water gardens, and flush toilets.

a. List three other activities for which nonpotable water could be used.
b. What conditions might prompt a community to use nonpotable water?
c. Does your community use reclaimed or recycled water? Find out for which purposes, if any.
d. Revisit your water diary from the beginning of the chapter. Identify areas of your water usage that could use nonpotable water. How would this affect your water habits?

(a)   (b)

Figure 5.10

(a) Water truck at the University of Alaska, Fairbanks, with a warning that the water is not fit to drink.
(b) Reclaimed water is pumped in purple pipes.

(a): ©Cathy Middlecamp, University of Wisconsin; *(b):* ©Reclaimed water booster pump station piping at City of Surprise, AZ. Courtesy of Malcolm Pirnie, the Water Division of Arcadis/Tim Francis

The most convenient source of fresh water for human activities is **surface water**, the freshwater found in lakes, rivers, and streams (Figure 5.11). Less convenient access is **groundwater**, freshwater found in underground reservoirs, also known as **aquifers**. People worldwide pump groundwater from wells drilled deep into these underground reservoirs. Fresh water is also found in our atmosphere through mists, fogs, and humidity.

In some nations, water is truly a bargain right out of the faucet at home. For example, the average price for 1000 gallons (3800 L) of tap water in the U.S. is about two dollars. So inexpensive, this tap water is supplied free in drinking fountains along streets or public buildings.

How much of the water on our planet is fresh water? Amazingly, only about 3%; the remainder is salt water. As shown in Figure 5.12, about two-thirds of this freshwater is locked up in glaciers, ice caps, and snowfields. Additionally, about 30% is found underground and must be pumped to the surface to use it. Lakes, rivers, and wetlands account for a mere 0.3% of the freshwater. Think of it this way: If the contents of a 2-L bottle represented all the water on our planet, only 60 mL of this would be fresh water. The water easily accessible to us in lakes and rivers would be about four drops!

Figure 5.11

Lakes and reservoirs provide much of our drinking water. This one, Hetch Hetchy, provides water to San Francisco, California.

©Cathy Middlecamp, University of Wisconsin

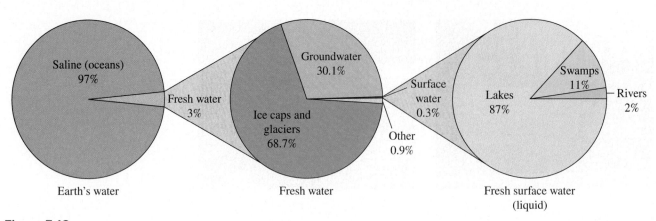

Figure 5.12

The distribution of fresh water on Earth.

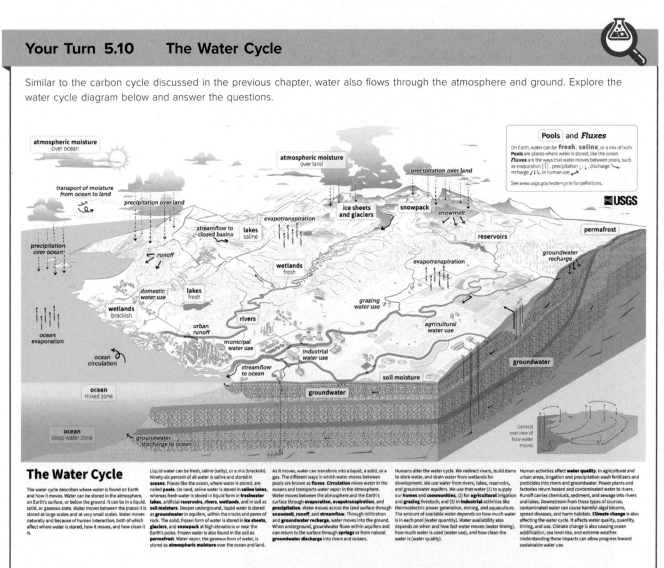

a. In Chapter 4, we discussed the global consequences of climate change. What areas of the water cycle will be most affected by climate change?

b. What would affect water quality the most: fluxes, circulation, water moving between the atmosphere and the surface, water moving across the surface, water moving into the ground, or water returning to the surface?

c. Revisit your water diary. What is one way that you can promote sustainability in your everyday water use?

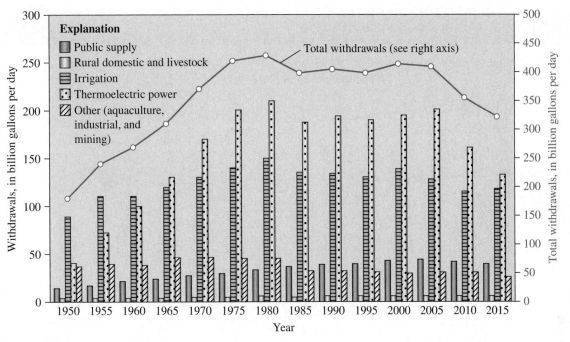

Figure 5.13

Total fresh and saltwater withdrawals in the United States, 2015.

Source: Estimated Use of Water in the United States in 2015, June 2018, USGS.

Your Turn 5.11   A Drop to Drink

We just stated that four drops in 2 liters correspond to the amount of fresh water available for our use. Is this accurate?

Hint: Use the relationships in Figure 5.12, and assume 20 drops per milliliter.

How do we use water? Predictably, the answer depends on where you live. In the United States, the USGS estimates that of the 322 billion gallons of water withdrawn daily, 86% comes from fresh water and 14% from salt water. Figure 5.13 shows four primary activities responsible for this water use, with the largest electricity production. About 132 billion gallons of water daily, or 41% of the total water withdrawn, is used as a coolant in electric power plants—coal, natural gas, and nuclear. The next-largest uses are for crop irrigation and homes, schools, and businesses, accounting for another 37% and 12%, respectively.

In many places, water is being pumped out of the ground faster than the natural water cycle replenishes. For example, much of the bountiful grain harvest from the central U.S. stems from using water from the High Plains Aquifer (Figure 5.14). This vast aquifer trapped water from the last Ice Age and runs from South Dakota to Texas. It is an unsustainable practice to pump water from aquifers faster than they are recharged. Continuous pumping can bring harmful outcomes as well. For example, salt water may intrude into a freshwater aquifer if water is removed from a geologically unstable area near the coast.

Most of the water from this aquifer is pumped out for agricultural reasons. Irrigation is one of the main factors stressing water resources in the Great Plains. In parts of the region, more than 81 trillion gallons of water were withdrawn for irrigation in Texas, Oklahoma, and Kansas from 1950 to 2015. During the same period, water levels in parts of the High Plains aquifer in those states decreased by more than 150 feet (Figure 5.14, dark red areas).

Your Turn 5.12   A Closer Look at Water Use

Check out the interactive map of water use in the U.S. (www.acs.org/cic) and answer the questions below.

Water use in the U.S., 2015

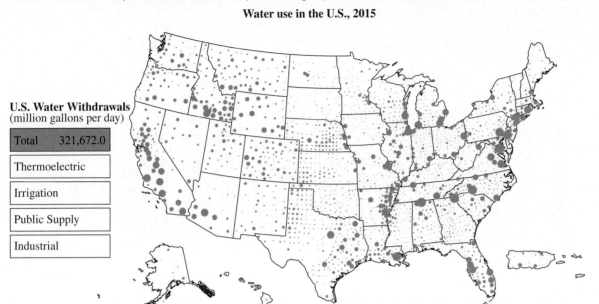

U.S. Water Withdrawals
(million gallons per day)

Total	321,672.0
Thermoelectric	
Irrigation	
Public Supply	
Industrial	

Questions to Consider

a. Which counties in the U.S. had the highest levels of water withdrawal in 2015? What was the population of these counties? Is there a correlation between population and water withdrawals? Suggest a reason(s) for this correlation, or lack thereof.

b. By hovering over the rectangles labeled as "thermoelectric," "irrigation," "public supply," or "industrial uses," you can see the parts of the country that have the highest water withdrawals for these uses. Choose three states and compare the relative water withdrawals for these four applications. Think of some reasons why the relative water uses for these states differ from the national averages shown in Figure 5.13.

c. For the states chosen in **b**. above, provide the breakdown of water use by surface vs. groundwater for various usage categories. You will have to consult other Internet resources to explore these data.

d. Consult this interactive map of worldwide water withdrawals. Which countries reported a higher water use than the U.S.? What are some possible reasons for increased water use in these countries?

Figure 5.14

High Plains aquifer water-level changes, predevelopment (about 1950) to 2015.
[USGS]

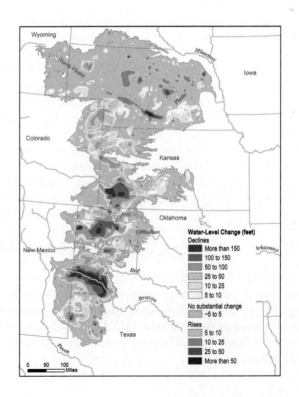

Such water diversion stories present us with examples of the tragedy of the commons (revisit Section 2.12). The water from aquifers and surface water is a resource used in common, yet no one, in particular, is responsible for its use. If water is overdrawn for agriculture or some other purpose, this act can be to the detriment of all who depend on this particular and necessary resource. The availability of water is becoming more paramount as the effects of climate change have led to conditions known as **aridification**, explored in the next activity.

Your Turn 5.13   What Is "Aridification"?

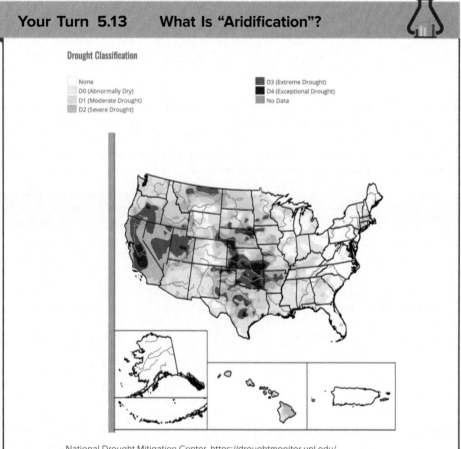

Drought Classification

- None
- D0 (Abnormally Dry)
- D1 (Moderate Drought)
- D2 (Severe Drought)
- D3 (Extreme Drought)
- D4 (Exceptional Drought)
- No Data

National Drought Mitigation Center. https://droughtmonitor.unl.edu/

Aridification is the process of the environment getting so dry due to the effects of climate change, which leads to temperature-driven river flow declines. These effects progressively increase the risk of arid conditions and drought, which is currently being observed in an expanding swath of the western United States. Check out the interactive drought monitor at www.acs.org/cic and answer the questions below.

Questions to Consider

a. Click on a map region, which includes your home state, then click on your state. What levels of drought conditions are included within your state?

b. Click on your local county. What level of drought is currently in your county?

c. Does this level of drought make sense with what you know about the past levels of rainfall within your community? Using the comparison slider, compare how the drought level for your county and state have changed over the past 10 years.

Worldwide, agriculture accounts for about 30% of global water consumption. Crops such as wheat, rice, corn, and soybeans are grown by farmers across the globe, each requiring several thousand liters of water, on average, to produce one kilogram of food. These values, reported in Table 5.2, are examples of **water footprints**—estimates of the volume of freshwater used to create particular goods or provide services. The values in Table 5.2 are global averages; the actual value for a water footprint depends

Table 5.2	Water Footprints for Meats and Grains
Food (1 kg)	**Water footprint (L, global average)**
corn (maize)	1222
wheat	1827
soybeans	2145
rice	2172
chicken	4325
pork	6000
sheep	10,400
beef	15,400

Source: https://www.healabel.com/water-footprints-of-food-list/ 2022

both on the country and on the particular region within the country in which the crop was grown. For example, according to the Water Footprint Network, corn grown in the United States has an average water footprint of 760 L. In comparison, the values in China and India are 1160 L and 2540 L, respectively. Over time, footprint values change if changes in rainfall or agricultural practices occur.

Your Turn 5.14   Beyond Toilets

Flushing a toilet is just one part of your daily water routine. Calculate your water footprint using this Water Footprint Calculator website: www.acs.org/cic.

Questions to Consider
1. What surprises you about your water usage?
2. What actions would you consider taking based on the results of this calculator?

Blablo101/Shutterstock

Your Turn 5.15   Differences in Water Footprints

Based on the data in Table 5.2, how does the water usage for crops compare to meat production? What are some reasons for this?

Water footprints can be estimated for other products as well. For example, consider a 250-mL glass of cow's milk. On average, the volume of water used to produce this is 255 L—almost a thousand times the volume of one glass of milk! This includes the water needed to care for the cow and the water used to grow the food that it eats. It also consists of the water used at a dairy farm to collect the milk and clean the equipment. You can check out the water needed to produce other beverages, foods, and consumer goods in Table 5.3.

Your Turn 5.16   Where Did All the Water Come From?

Choose two items from those listed in Table 5.3 and brainstorm all the areas where water is used in producing those items. Are there ways that the water footprint could be lowered for those items? Describe how this might be done.

Table 5.3	Water Footprints for Various Products
Product	**Water footprint (L, global average)**
1 cup of coffee (250 mL)	140
1 cup of tea (250 mL)	120
1 banana (200 g)	132
1 orange (150 g)	80
1 glass of orange juice (200 mL)	170
1 egg (60 g)	197
1 chocolate bar (100 g)	1700
1 cotton T-shirt (250 g)	2700

Source: Water Footprint Network, 2022

Water footprint values are inexact and, as a result, controversial. Our intent in providing them is not to label items as good or bad. Instead, these values are meant to remind you that water is used to produce goods and to provide you with a more inclusive picture of water use. We hope this section has increased your awareness of how water is used, misused, and sometimes picks up contaminants. What is it about water that allows contamination to happen so easily? Our next section will help us put a "name to a face" as we discover the types of substances that are commonly dissolved in our water.

5.4 | Water As a Solvent

> *Learning Objectives:*
>
> *Describe the composition of aqueous solutions*
>
> *Interpret rules to provide the names of ionic compounds*

Water dissolves a remarkable variety of substances. As we will see, some of them, including salt, sugar, and ethanol, are *very* soluble in water. In comparison, limestone rock, oxygen, and carbon dioxide dissolve only in tiny amounts. To build your understanding of water quality, you need to know *what* dissolves in water, *why* it dissolves, and *how* to specify the concentration of the resulting solution.

Let's begin with some useful chemical terminology. Water is a **solvent**—a substance, often a liquid, capable of dissolving one or more pure substances. The **solute** is the solid, liquid, or gas that dissolves in a solvent. The result is called a **solution**—a homogeneous (of uniform composition) mixture of a solvent and one or more solutes. In this section, we are particularly interested in **aqueous solutions**, in which water is the solvent.

Your Turn 5.17   Solutions

Example Problem:

For salt water, identify the solute and solvent.

Example Solution:

Salt (NaCl) is the solute (minor component of the solution that being dissolved) and water is the solvent (major component of the solution).

Aleksei Lazukov/Shutterstock

Your Turn:

For each of the following solutions, choose the solute and solvent.

a. a carbonated beverage
b. stainless steel
c. sweetened iced tea
d. automotive antifreeze

Because water is such a good solvent, it is practically never 100% pure. Rather, it contains impurities. For example, when water flows over the rocks and minerals of our planet, it dissolves tiny amounts of their constituents depending on their relative **solubility**. Although this usually causes no harm to our drinking water, the ions dissolved in water occasionally are toxic. As mentioned, gases can also dissolve in water. When the water on our planet comes in contact with air, it dissolves tiny amounts of the gases in the air, most notably oxygen and carbon dioxide. Some air pollutants are *very* soluble in water. So when it rains, the water cleans some of the pollutants out of the air, including SO_x and NO_x, resulting in *acid rain*. As we will see later in this chapter, the acidic solutions that form can have serious consequences for the environment.

Elements rarely appear in their pure form in nature but rather as compounds. As opposed to molecular compounds covered in Chapter 2, which are formed from two or more nonmetals, there are also **ionic compounds** formed from a metal and a nonmetal. Don't remember where metals and nonmetals are on a periodic table? Revisit Figure 1.3.

An **ion** is an atom or group of atoms that has acquired a net electric charge due to gaining or losing one or more electrons. A sodium ion, Na^+, is an example of a **cation**, a positively charged ion. In contrast, chloride, Cl^-, is an example of an **anion**, a negatively charged ion. An **ionic bond** is the chemical bond formed when oppositely charged ions attract in the solid state.

Why do certain elements become cations and others become anions? The answer is in the way the electrons are arranged in the outer (valence) shells of the atoms. For example, recall that a neutral sodium atom has 11 electrons and 11 protons. Like all the metals in Group 1, sodium has one valence electron. It is rather loosely attracted to the nucleus because it is in the outermost energy shell and can be lost. When this happens, the Na atom forms Na^+, a cation:

$$Na \longrightarrow Na^+ + e^- \qquad \text{[5.1]}$$

When metals are included in ionic compounds, they all form cations, or positively charged ions, by losing electrons. The Na^+ has a +1 charge because it contains 11 protons and only 10 electrons, the same number of electrons as the noble gas neon (Table 5.4). It is strange that Na has lost something and become positively charged. But remember, it lost an electron, which has a negative charge.

Unlike sodium, chlorine is a nonmetal. A neutral chlorine atom has 17 protons and 17 electrons. Chlorine, like all nonmetals in Group 17, has seven valence electrons. Eight valence (outer) electrons are energetically stable so that chlorine will gain one electron:

$$Cl + e^- \longrightarrow Cl^- \qquad \text{[5.2]}$$

The chloride ion (Cl^-) has 17 protons and 18 electrons, the same number of electrons as the noble gas argon. Since there is one more electron than proton, the net charge is −1 (Table 5.5). When nonmetals are included in ionic compounds, they all form anions, or negatively charged ions, by gaining electrons. Notice that in forming the cation of sodium and anion of chlorine, the resulting ions had the same number of electrons as a noble gas. This is no accident; noble gases are very stable and unreactive, so in forming ions, a neutral atom will either gain or lose electrons to match that of the nearest noble gas.

Table 5.4	Electronic Bookkeeping for Cation Formation	
Sodium Atom	**Sodium Ion**	**Neon Atom**
Na	Na^+	Ne
11 protons	11 protons	10 protons
11 electrons	10 electrons	10 electrons
net charge of 0	net charge of +1	net charge of 0

Table 5.5	Electronic Bookkeeping for Anion Formation	
Chlorine Atom	**Chloride Ion**	**Argon Atom**
Cl	Cl$^-$	Ar
17 protons	17 protons	18 protons
17 electrons	18 electrons	18 electrons
net charge of 0	net charge of −1	net charge of 0

Sodium metal, Na(s), and chlorine gas $Cl_2(g)$ react vigorously when they come into contact. The result is the aggregate of Na$^+$ and Cl$^-$ ions, known as sodium chloride or table salt. In the formation of sodium chloride, the electrons are transferred from one atom to another. However, electrons are not shared as they are in a molecular or covalent compound.

The oppositely charged ions are electrically charged; the evidence lies in the difference between the solid and liquid forms of ionic compounds. Although solid sodium chloride does not conduct electricity, if you melt sodium chloride, the ions are free to move, and the hot liquid conducts electricity. This is evidence that ions are present. This makes sense because, in a solid form, the ions are fixed in place and cannot move and transport charge.

All ionic compounds are hard and brittle; when hit with a hammer, they will shatter rather than flatten. This is because there are strong forces that extend throughout the ionic crystal. There is no such thing as a localized ionic bond like a covalent bond in a molecule. Instead, ionic bonds hold together a large assembly of ions, for example, Na$^+$ and Cl$^-$.

Figure 5.15 shows the most common charges of elements. As you can see, all the elements in Group 1 have a charge of +1, and all elements in Group 2 have a charge of +2. Whereas everything in Group 17 has a −1 charge, almost everything in Group 16 has a −2 charge. Some of the elements in the middle can have more than one charge.

The oppositely charged ions contribute to the name of an ionic compound that may be derived by using the following rules:

- The cation is named first, and the anion is named second.
- Name the cation like it is written, and change the ending of the monatomic anion to -ide.
- All ionic compounds are neutral, so the number of ions will be listed on the formulas to make them neutral. It is often called finding the lowest common denominator to find the number of ions in a formula.

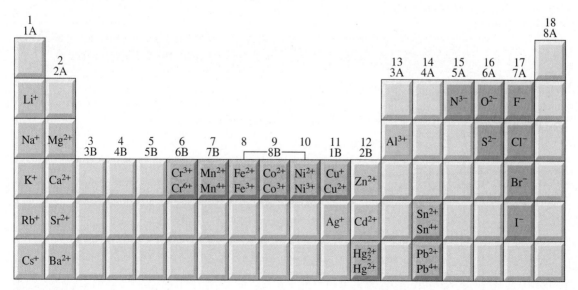

Figure 5.15

Common ions formed from their elements. Ions in **green** (cations) or **blue** (anions) have only one possible charge. Ions in **red** (cations) have more than one possible charge.

The schematic below shows how to apply these rules to name *sodium chloride*, which consists of sodium and chloride ions:

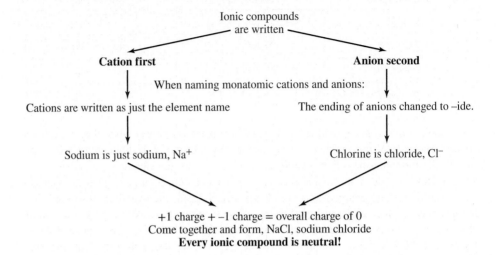

Ionic compounds are written

Cation first   **Anion second**

When naming monatomic cations and anions:

Cations are written as just the element name   The ending of anions changed to –ide.

Sodium is just sodium, Na^+ Chlorine is chloride, Cl^-

+1 charge + –1 charge = overall charge of 0
Come together and form, NaCl, sodium chloride
Every ionic compound is neutral!

When writing a formula for an ionic compound, we need to figure out how many ions the formula has from the rule above that all ionic compounds are neutral. For example, by consulting Figure 5.15, the formula of magnesium fluoride would contain Mg^{2+} and F^- ions (i.e., $2 \times -1 = -2$) to balance the +2 charge of the magnesium ion. Therefore, we add a subscript to indicate that we have more than one ion in a formula. So the formula of magnesium fluoride is MgF_2.

Some compounds are more complex and contain *polyatomic* ions, which are a collection of elements with an overall charge that is either positive or negative (Table 5.6). The rules to name these compounds are the same for monatomic ions discussed above, but there is no need to use suffixes for the anion. For example, let's consider the formula for ammonium nitrate. Consulting Table 5.6, ammonium is a polyatomic cation with the formula NH_4^+, and nitrate is a polyatomic anion with the formula NO_3^-. Since the charges of these ions are equal, we just need to list them in the correct order, cation first and anion second. Hence, the formula is NH_4NO_3. Parentheses are only needed when writing the formula of an ionic compound with more than one polyatomic ion; for instance, the formula of ammonium sulfate composed of two NH_4^+ ions and a single SO_4^{2-} ion would be $(NH_4)_2SO_4$.

Your Turn 5.18   Practice Naming Ionic Compounds

Example Problems:
a. What is the name of K_2O?
b. What is the chemical formula for calcium nitrite?

Example Solutions:
a. Potassium oxide. The compound contains K^+ and O^{2-} ions (there are two K^+ ions to balance the –2 charge of the O^{2-} ion).
b. Using Table 5.6, calcium nitrite contains Ca^{2+} and NO_2^- ions. Since the compound has a +2 charge on the cation and –1 charge on the anion, there must be two anions to balance the overall charge. Hence, the formula is: $Ca(NO_2)_2$.

Your Turn:

Write the chemical formula for the ionic compound formed from each pair of ions.
a. Na^+ and SO_4^{2-}
b. OH^- and Mg^{2+}
c. Al^{3+} and $C_2H_3O_2^-$
d. CO_3^{2-} and K^+

Table 5.6	Common Polyatomic Ions		
Name	**Formula**	**Name**	**Formula**
acetate	$C_2H_3O_2^-$	nitrite	NO_2^-
bicarbonate*	HCO_3^-	phosphate	PO_4^{3-}
carbonate	CO_3^{2-}	sulfate	SO_4^{2-}
hydroxide	OH^-	sulfite	SO_3^{2-}
hypochlorite	ClO^-	ammonium	NH_4^+
nitrate	NO_3^-		

*Also called hydrogen carbonate

Lastly, how do we name a compound if the charge of the metal is not explicitly provided? Revisiting Figure 5.15, there are groups of elements that are stable with more than one charge. For example, iron can have a charge of +2 or +3, and copper can be +1 or +2. For these compounds, we indicate the charge of the metal in the name as a Roman numeral in parentheses. The charge of the metal is determined by context, based on the balancing of positive and negative charges of ions to maintain an overall neutral compound. For example, FeO contains the O^{2-} ion, so iron must be Fe^{2+} to balance the negative charge. We indicate the charge on the iron as iron(II) oxide. For iron(III) oxide that contains both Fe^{3+} and O^{2-} ions, we use the lowest common denominator of their charges ($+3 \times -2 = -6$) to determine the ratio of ions needed in the chemical formula. Hence we need two Fe^{3+} ions (+6 charge) to balance three O^{2-} ions ($3 \times -2 = -6$ charge). Accordingly, the formula for iron(III) oxide is Fe_2O_3. The following activity provide more examples of naming ionic compounds.

Naming Ionic Compounds

Chemistry in Context
Tutorial

NOMENCLATURE

©Bradley D. Fahlman

Watch this video for more examples of naming ionic compounds: www.acs.org/cic

---

Your Turn 5.19 — More Practice With Naming Compounds

Example Problems:

a. Provide the name of aluminum sulfate.
b. What is the name of CuO?

Example Solutions:

a. Two Al^{3+} ions and three SO_4^{2-} ions are required to balance the overall charges of ions. Hence, the formula is $Al_2(SO_4)_3$.
b. Since the anion is O^{2-}, copper must be Cu^{2+} to balance the overall charge of ions. Hence, the name is copper(II) oxide since Cu^{2+} ions are involved.

Your Turn:

Provide the correct names and formulas for the following ionic compounds.
a. Name CaI_2.
b. Write the formula of barium nitride.
c. Name $Mg(OH)_2$.
d. Write the formula of sodium carbonate.
e. Name Co_2S_3.
f. Write the formula of copper(II) hydrogen carbonate.

Now that we can name ionic compounds that are commonly dissolved in water, we will shift our focus to quantify the concentrations of these solutes.

5.5 | How Much Is OK?
Quantifying Water Quality

Learning Objective: Calculate the concentration of solutions in various concentration units

In addition to natural ionic compounds, humans also contribute to the number of substances dissolved in water. When we wash clothes, we add detergent and whatever made our clothing dirty in the first place. When we flush a toilet, we add liquid and solid wastes. Our urban streets add solutes to rainwater during the process of storm run-off, and our agricultural practices add fertilizers and other soluble compounds to water.

What does the property of water as a good solvent mean for our drinking water? To assess water quality, you need to know several things. One is a way to specify *how much* of a substance has dissolved so that you can compare the value with a known standard. In other words, you need to understand the concept of *concentration*, which was first introduced in Section 2.3 regarding the composition of air. Now we examine this concept in terms of substances dissolved in water. As we will see, the units of percent and parts per million we used for air quality are also valid ways of expressing concentrations for aqueous solutions.

Percent (%) means parts per hundred. Percent is used to express the concentration of a wide range of solutions. For example, an aqueous solution containing 0.9 g of sodium chloride (NaCl) in 100 g of the solution is a 0.9% solution by mass. This concentration of sodium chloride is referred to as "normal saline" in medical settings when given intravenously. You may find the antiseptic isopropyl alcohol in your medicine cabinet as a 70% aqueous solution by volume. It contains 70 mL of isopropyl alcohol in every 100 mL of an aqueous solution.

But when the concentration is very low, as is the case for many substances dissolved in drinking water, **parts per million** (ppm) are more commonly used. The water contains substances naturally present in the parts per million range. For example, the EPA acceptable limit for nitrate ions, NO_3^-, found in well water in some agricultural areas is 10 ppm; the limit for fluoride ions, F^-, is 4 ppm.

Parts-per-million is a dimensionless ratio of 1 part of solute per 1 million (10^6) parts of the solution. Although parts per million is a useful concentration unit, measuring 1 million grams of water is not very convenient. We can do things more easily by switching to the unit of a liter. One ppm of any substance in water is equivalent to 1 mg of that substance dissolved in a liter of solution. Here is the math:

$$1 \text{ ppm} = \frac{1 \text{ g solute}}{1 \times 10^6 \text{ g water}} \times \frac{1000 \text{ mg solute}}{1 \text{ g solute}} \times \frac{1000 \text{ g water}}{1 \text{ L water}} = \frac{1 \text{ mg solute}}{1 \text{ L water}} \quad \textbf{[5.3]}$$

Municipal water utilities may use the unit mg/L to report the minerals and other substances dissolved in tap water. For example, Table 5.7 shows a tap water analysis from an aquifer that supplies a Midwestern community in the United States.

Some contaminants are of concern at concentrations much lower than parts per million and are reported as **parts per billion** (ppb). Parts per billion is a dimensionless

Table 5.7		Tap Water Mineral Report	
Cation	**mg/L**	**Anion**	**mg/L**
Calcium ion (Ca^{2+})	97	Sulfate ion (SO_4^{2-})	45
Magnesium ion (Mg^{2+})	51	Chloride ion (Cl^-)	75
Sodium ion (Na^+)	27	Nitrate ion (NO_3^-)	4
		Fluoride ion (F^-)	1

ratio of 1 part of solute per 1 billion (10^9) parts of the solution. In aqueous solutions, 1 ppb = 1 µg/L, whereas 1 ppm = 1 mg/L. Another way to think about ppm and ppb is to assume that 1 ppm corresponds to 1 second in nearly 12 days. Then, 1 ppb corresponds to 1 second in 33 years. Yet another way of looking at these units is that one part per billion corresponds to a few centimeters on the circumference of Earth—very small indeed!

Your Turn 5.20   Mercury Ion Concentrations

Watch this tutorial to explore units of concentration for mercury in water: www.acs.org/cic.

Questions to Consider

a. Five liters of an aqueous solution contains 80 µg of dissolved mercury ion. Express the mercury ion concentration of the solution in ppm and ppb.

b. Would your answer in part **a** comply with the U.S. acceptable limit for drinking water? Explain.

Chemistry in Context
Tutorial

CONCENTRATION UNITS: PPM

©Bradley D. Fahlman

Molarity (M), another useful concentration unit, is defined as a unit of concentration represented by the number of moles of solute present in 1 liter of solution:

$$\text{Molarity (M)} = \frac{\text{moles of solute}}{\text{liter of solution}} \qquad \textbf{[5.4]}$$

The great advantage of molarity is that solutions of the same molarity contain exactly the same number of moles, and hence the same number of molecules (ions or atoms) of solute. The mass of a solute varies depending on its identity. For example, 1 mole of sugar has a different mass than 1 mole of sodium chloride. But if you take the same volume, all 1-M solutions (read as "one molar") contain the same number of solute molecules.

As an example, consider a solution of sodium chloride in water. The molar mass of NaCl is 58.5 g/mol; therefore, 1 mol of NaCl has a mass of 58.5 g. By dissolving 58.5 g of NaCl in some water and adding enough water to make exactly 1.00 L of solution, we would have a 1.00-M sodium chloride aqueous solution, denoted as NaCl(aq).

Your Turn 5.21   Concentrations of Solutions

Check out this simulation to see how the relative amounts of a solute or solvent affect the concentration of the solution: www.acs.org/cic.

Questions to Consider

a. Pull the concentration detector into the liquid, and shake the canister into the liquid to make a solution. How much drink mix can you add? What is the maximum concentration?

b. Is the maximum concentration you can make the same for each? Why or why not?

c. What does saturated mean?

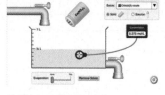

PhET Interactive Simulations, University of Colorado Boulder, http://phet.colorado.edu

---

Figure 5.16 shows the preparation of a 1.00-M solution of sodium chloride. Note the use of a **volumetric flask**, a type of glassware that contains a precise volume of the solution when filled to the mark on its neck. But because concentrations are simply ratios of solute to solvent, there are many other ways to make a 1.00-M NaCl(aq) solution.

Figure 5.16

Preparation of a 1.00-M NaCl aqueous solution. A video illustrating techniques for solution preparation may be found at www.acs.org/cic.

©Bradley D. Fahlman

Another possibility is to use 0.500 mol NaCl (29.2 g) in 0.500 L of solution. This requires using a 500-mL volumetric flask rather than the 1-L flask shown in Figure 5.16.

$$1.00\text{-M NaCl}(aq) = \frac{1.00 \text{ mol NaCl}}{1.00 \text{ L solution}} \quad \text{or} \quad \frac{0.500 \text{ mol NaCl}}{0.500 \text{ L solution}} \qquad [5.5]$$

As you will see in the next activity, the molar mass of the solute must be known to convert from ppm or ppb to molarity.

Your Turn 5.22 — Practice With Concentration Units

Example Problem:

A water sample had a concentration of 150 ppm of dissolved mercury, Hg^{2+}. What is this concentration expressed in molarity, M?

Example Solution:

$$150 \text{ ppm } Hg^{2+} = \frac{150 \text{ mg } Hg^{2+}}{1 \text{ L } H_2O} \times \frac{1 \text{ g } Hg^{2+}}{1000 \text{ mg } Hg^{2+}} \times \frac{1 \text{ mol } Hg^{2+}}{200.6 \text{ g } Hg^{2+}} = \frac{7.5 \times 10^{-4} \text{ mol } Hg^{2+}}{1 \text{ L } H_2O}$$

or 7.5×10^{-4} M Hg^{2+}

Your Turn:

a. Express a concentration of 16 ppb Hg^{2+} in units of molarity.
b. For 1.5-M and 0.15–M NaCl, how many moles of solute are present in 500 mL of each?
c. A solution is prepared by dissolving 0.50 mol NaCl in enough water to form 250 mL of solution. A second solution is prepared by dissolving 0.60 mol NaCl to form 200 mL of solution. Which solution is more concentrated? Explain.
d. A student was asked to prepare 1.0 L of a 2.0-M $CuSO_4$ solution. The student placed 40.0 g of $CuSO_4$ crystals in a volumetric flask and filled it with water to the 1000-mL mark. Was the resulting solution 2.0 M? Explain?

In this section, we made the case that water is an excellent solvent for various substances and that we can express the concentration of these substances numerically. The next section helps you understand how and why substances dissolve in water.

5.6 | A Deeper Look at Solutes

> *Learning Objective: Describe and model the solvation of ionic and molecular compounds in water*

As discussed in Chapter 4, the oceans absorb and emit carbon dioxide. However, it is essential to know that many other gases, as well as liquids and solids, may be dissolved in water. Although some elements react with water such as the Group 1 metals, and others such as chlorine are somewhat soluble in water, the majority of solutes in aqueous solutions are compounds. Refresh your memory regarding the differences in structure and properties between molecular and ionic compounds by completing the following activity.

Your Turn 5.23   Ionic Versus Molecular Compounds

You have seen both ionic and molecular compounds in various contexts, such as air quality, global climate change, and portable electronic devices. Review your current knowledge of the differences in these compounds in the chart below. You may add more rows with new information you learn in this chapter, such as "behavior in water."

	Ionic	Molecular
Structure		
Macroscopic properties		
Bonding model		
Bond strength		
Naming		
Other		

DID YOU KNOW ?

In general, the solubility of a solid solute in a liquid solvent is increased at higher temperatures since more solid may be dissolved. However, for gaseous solutes, the solubility decreases with an increase in temperature. This is due to the greater kinetic energy (energy of motion) of gas molecules at higher temperatures, which makes it easier for them to escape the liquid solvent.

As mentioned, gases can also dissolve in a liquid. Carbonated beverages are an example of this. Carbon dioxide is dissolved in the aqueous solvent to create the fizziness we expect from these beverages. Therefore, soda is a type of aqueous solution with carbon dioxide being the solute and water acting as the solvent. What have you seen happen when sodas warm up? In most cases, more bubbling occurs. When bubbling occurs, carbon dioxide comes out of the solution in a process known as **degasification** (Figure 5.17). The same process can occur in our oceans when the water temperature rises. During degasification, some of the carbon dioxide originally dissolved in the ocean is released into the atmosphere. Despite having much lower CO_2 concentrations than your favorite soft drink, oceans represent an extremely large sink for absorbing CO_2 gas emitted into the atmosphere.

Have you ever been in a cave or seen a rock formation? Some of these structures result from a solid solute slowly dropping out of an aqueous solution, known as **crystallization**. When a solid "falls" out of solution, we call this **precipitation**, and the solid is called a **precipitate** (Figure 5.18). Temperature changes and solute concentrations can affect precipitation; for example, coral reefs form by precipitation reactions under specific conditions. When those conditions change, coral reefs can either grow larger or dissolve. The latter situation is of serious concern and will be detailed later in this chapter.

As we pointed out earlier, about 97% of the water on our planet is found in the saltwater of the oceans. This water source contains a variety of ionic compounds known as salts, however, there are many other varieties present than simple table salt (NaCl). This begs the question: Why are many ionic compounds so soluble in water?

Recall from Section 5.1 that water molecules are polar. When you take salt crystals and dissolve them in water, the polar H_2O molecules are attracted to the ions contained in these crystals. The partial negative charge (δ^-) on the O atom of a water molecule is attracted to the positively charged Na^+ cations of the salt crystal. At the same time, the H atoms, with their partial positive charges (δ^+), are attracted to the negatively charged Cl^- anions. Over time, the salt ions are separated and surrounded by water

Figure 5.17

The explosive escape of carbon dioxide from a soft drink catalyzed by the presence of Mentos® candies. The surface of the candies has microcavities that serve to nucleate carbon dioxide molecules and force the release of carbon dioxide all at once, for a dramatic demonstration.

Holly Hildreth/McGraw Hill

Figure 5.18

A close-up view of the formation of silver chloride precipitate in an aqueous solution of sodium nitrate.

1995 Richard Megna/Fundamental Photographs

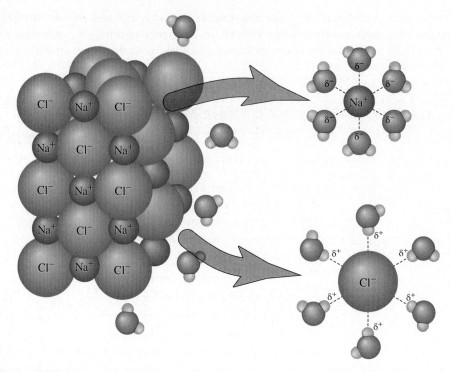

Figure 5.19

Sodium chloride dissolves in water. For a video of this process, see: www.acs.org/cic.

molecules. Equation 5.6 and Figure 5.19 represent the process of forming an aqueous sodium chloride solution.

$$NaCl(s) \xrightarrow{H_2O} Na^+(aq) + Cl^-(aq) \qquad [5.6]$$

This dissolution process is similar for solutions of compounds containing polyatomic ions. For example, the sodium and sulfate ions separate when solid sodium sulfate dissolves in water. Note, however, that the sulfate ion stays together as a unit:

$$Na_2SO_4(s) \xrightarrow{H_2O} 2Na^+(aq) + SO_4^{2-}(aq) \qquad [5.7]$$

Many ionic compounds dissolve in this manner. This explains why almost all naturally occurring water samples contain various amounts of ions. The same is also true for our bodily fluids, as these also contain significant concentrations of ionic solutes called **electrolytes**. When a **strong electrolyte** such as NaCl or another ionic compound is placed into a polar solvent such as water, the compound completely dissociates into positively charged ions and negatively charged ions.

Your Turn 5.24   Electricity and Water Don't Mix

Small electric appliances such as hair dryers and curling irons carry warning labels advising the consumer not to use the appliance near water. Why is water a problem since pure water does not conduct electricity? What is the best course of action if a plugged-in hair dryer accidentally falls into a sink full of water?

If the solubility principles we just described apply to *all* ionic compounds, our planet would be in trouble. When it rained, ionic compounds such as calcium carbonate (limestone) would dissolve and end up in the ocean! Fortunately, many ionic compounds are only *slightly soluble* or have extremely low solubilities, referred to as **weak electrolytes**. These differences arise because of the sizes and charges of the ions, how strongly they attract one another, and how strongly the ions are attracted to water molecules.

Table 5.8	Water Solubility of Ionic Compounds		
Ions	Solubility of Compounds	Solubility Exceptions	Examples
Group 1 metals, NH_4^+	all soluble	none	$NaNO_3$ and KBr. Both are soluble.
nitrates	all soluble	none	$LiNO_3$ and $Mg(NO_3)_2$. Both are soluble.
chlorides	most soluble	silver, mercury(I), lead(II)	$MgCl_2$ is soluble. AgCl is insoluble.
sulfates	most soluble	strontium, barium, lead(II), silver(I)	K_2SO_4 is soluble. $BaSO_4$ is insoluble.
carbonates	mostly insoluble*	Group 1 metals NH_4^+	Na_2CO_3 is soluble. $CaCO_3$ is insoluble.
hydroxides, sulfides	mostly insoluble*	Group 1 metals, NH_4^+	KOH is soluble. $Sr(OH)_2$ is insoluble.

*Insoluble means that the compounds have extremely low solubilities in water (less than 0.01 M). All compounds have at least a very small solubility in water.

When ionic compounds react, they typically do so by **double-displacement**, where the cations and anions are exchanged between reactants. For instance, the reaction between aqueous sodium chloride (NaCl) and aqueous silver nitrate ($AgNO_3$) would yield sodium nitrate ($NaNO_3$) and silver chloride (AgCl). Using Table 5.8 as a guide to solubility, it predicts that the product $NaNO_3$ would be soluble and AgCl would be insoluble to form a solid **precipitate**. As other examples of solubility, calcium nitrate ($Ca(NO_3)_2$), is soluble in water, as are all compounds containing the nitrate ion. Calcium carbonate ($CaCO_3$) is insoluble, as are most carbonates. By similar reasoning, copper(II) hydroxide ($Cu(OH)_2$) is insoluble, but copper(II) sulfate ($CuSO_4$) is soluble. Table 5.9 summarizes some environmental consequences of solubility as they pertain to the dissolution of minerals.

Solubility of Ionic Compounds

Edgieus/Shutterstock

Check out this video for a laboratory investigation that explores the solubility of various ionic compounds in water: www.acs.org/cic.

Predicting Solubility

Watch this video (www.acs.org/cic) that illustrates reactions between aqueous solutions of silver nitrate ($AgNO_3$), copper(II) nitrate ($Cu(NO_3)_2$), sodium sulfide (Na_2S), and sodium chloride (NaCl). Assuming that these reactions occur by double displacement, write the balanced equation for each reaction and predict whether a precipitate will form. Use Table 5.8 as your guide.

©2018 American Chemical Society

   a. Reaction 1: silver nitrate + copper(II) nitrate
   b. Reaction 2: silver nitrate + sodium sulfide
   c. Reaction 3: silver nitrate + sodium chloride
   d. Reaction 4: copper(II) nitrate + sodium sulfide
   e. Reaction 5: copper(II) nitrate + sodium chloride
   f. Reaction 6: sodium sulfide + sodium chloride

Table 5.9	Environmental Consequences of Solubility	
Source	Ions	Solubility and Consequences
salt deposits	sodium and potassium halides*	These salts are soluble. Over time, they dissolve and wash into the sea. Thus, oceans are salty, and seawater cannot be used for drinking without expensive purification.
agricultural fertilizers	nitrates	All nitrates are soluble. The runoff from fertilized fields carries nitrates into surface water and groundwater. Nitrates can be toxic, especially for infants.
metal ores	sulfides and oxides	Most sulfides and oxides are insoluble. Minerals containing iron, copper, and zinc are often sulfides and oxides. If these minerals had been soluble in water, they would have washed out to sea long ago.
mining waste	mercury(I), lead(II)	Most mercury and lead compounds are insoluble. However, they may leach slowly from mining waste piles and contaminate water supplies.

*Halides, such as Cl^- and I^-, are anions of the atoms in Group 17, the halogens.

Your Turn 5.25   Solutes in Your Water Diary

Examine your water diary. Which areas does your water use depend on dissolved ionic and molecular compounds? Do any of your water uses add solutes to the water?

5.7 | *Like Dissolves Like:*
A Pathway for Contamination

Learning Objectives:

Apply the "like dissolves like" principle to predict whether a substance will be soluble in water

Understand how certain chemicals can contaminate water sources

Now that we have examined how ionic compounds dissolve and react in water, let's look at molecular compounds. From the previous discussion, you might have understood that only ionic compounds dissolve in water. But remember that sugar dissolves in water as well. The white granules of "table sugar" that you use to sweeten your coffee or tea are *sucrose*, a polar molecular compound with the chemical formula $C_{12}H_{22}O_{11}$ (Figure 5.20).

Figure 5.20

The structural formula of sucrose. The covalently bound —OH groups are shown in red. For a 3D rendering of this structure, go to www.acs.org/cic.

When sucrose dissolves in water, the sucrose molecules disperse uniformly among the H_2O molecules. However, unlike ionic compounds, the sucrose molecules remain intact and do *not* separate into ions. Evidence for this includes that aqueous sucrose solutions do not conduct electricity (Figure 5.21). Even though the sugar molecules act as **nonelectrolytes**, they still interact with water molecules since they are both polar and are attracted to one another. Furthermore, the sucrose molecule contains eight —OH groups and three additional O atoms that can participate in

(a)   (b)   (c)

Figure 5.21

Conductivity experiments. A conductivity meter such as this apparatus shown here indicates whether electricity is being conducted. The lightbulb will glow only if the electrical circuit is completed, which is possible only if the electrodes are immersed in an electrically conductive solution.

(a–c): GIPhotoStock/Science Source

hydrogen bonding with water. Solubility is always promoted when an attraction exists between the solvent molecules and the solute molecules or ions. This suggests a general solubility rule: *Like dissolves like.* In other words, a *polar* solvent will dissolve a *polar* solute, or a *nonpolar* solute will be soluble in a *nonpolar* solvent.

Let's also consider two other common polar molecular compounds, both highly soluble in water. One is ethylene glycol, the main ingredient in antifreeze, and the other is ethanol, or ethyl alcohol, found in beer and wine. Both molecules contain the polar —OH group and are classified as alcohols (Figure 5.22).

ethanol   ethylene glycol

Figure 5.22

Lewis structures of ethanol and ethylene glycol. The —OH groups are shown in red.

As shown in Figure 5.23, the H in the —OH group of an ethanol molecule can hydrogen bond, just as was the case for water. This is why water and ethanol have a great affinity for each other; the partially negative (δ^-) from water attracts the partially positive (δ^+) in ethanol, and vice versa. As a result, alcohol and water form solutions in all proportions. Again, both molecules are polar and *like dissolves like.*

Ethylene glycol is another example of an alcohol, sometimes called "glycol." Ethylene glycol is added to water, such as the water in a car radiator, to keep it from freezing. In addition to being an antifreeze additive, it is also one of the volatile organic compounds (VOCs) that some water-based paints emit when drying. Examine its structural formula in Figure 5.22 to see that it has two —OH groups available for hydrogen bonding. These intermolecular attractions give appreciable water solubility to ethylene glycol, a necessary property for any antifreeze.

It has often been stated that "oil and water don't mix." The reason for this incompatibility is that water molecules are *polar*, whereas the hydrocarbon molecules in oil are *nonpolar*. When in contact, water molecules attract other water molecules, and hydrocarbon molecules are attracted only toward themselves. Because oil is less dense than water, oil slicks float on top of the water (Figure 5.24).

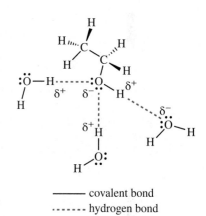

———— covalent bond
- - - - - hydrogen bond

Figure 5.23

Hydrogen bonding between an ethanol molecule and three water molecules.

Figure 5.24

Oil and water are not miscible with each other. For a video of this phenomenon, go to www.acs.org/cic.

Charles D. Winters/Science Source

Your Turn 5.26   More About Hydrocarbons

Hydrocarbon molecules such as pentane (C_5H_{12}) and hexane (C_6H_{14}) contain C—H and C—C bonds. Using the electronegativity values in Table 5.1, predict whether these bonds are polar or nonpolar. Why are these molecular compounds nonpolar?

Hint: Consider their molecular geometries alongside the bond dipoles.

Because water is a poor solvent for grease and oil, we cannot use water to wash these off of our hands or clothing. Instead, we must use soaps and detergents. These compounds are **surfactants** that help polar and nonpolar compounds mix, sometimes called "wetting agents." The surfactant molecules contain polar and nonpolar groups (Figure 5.25). The polar groups allow the surfactant to dissolve in water, whereas the nonpolar ones can dissolve the grease.

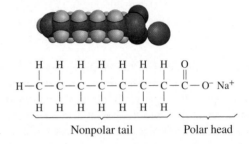

Figure 5.25

The molecules of surfactants contain both polar and nonpolar groups. For a 3D rendering of this structure, go to www.acs.org/cic.

Your Turn 5.27   Surfactants to the Rescue!

On Dec. 8, 2022, more than 14,000 barrels (588,000 gallons) of oil spilled into a northeastern Kansas creek due to a ruptured oil pipeline. With the release of enough oil to fill an Olympic-sized swimming pool, this disaster represented the largest onshore crude pipeline spill in nine years. Consulting a variety of Internet resources, prepare a report that describes the type of surfactants that were used to clean up this oil spill. Draw the structures of these chemical agents, and describe how long this clean up effort will likely take.

DroneBase/AP Images

@HOME   Popping A Balloon With an Orange?

Investigate the effect of a nonpolar substance being added to a nonpolar surface using an orange peel and a balloon.

1. Inflate the balloon and tie it off.
2. Squeeze the orange peel onto the inflated balloon.
3. Make observations about what you see.

Africa Studio/Shutterstock

Questions to Consider
  a. Explain the results with a discussion of solubility and miscibility.
  b. What is another substance that should yield similar results? Why?

Another way to dissolve nonpolar molecules is to use nonpolar solvents. Like dissolves like! Nonpolar solvents (called "organic solvents") are widely used to produce drugs, plastics, paints, cosmetics, and cleaning agents. As another example, dry-cleaning solvents are typically composed of chlorinated hydrocarbons. One example, "perc," is a cousin of ethene. Start with ethene (C_2H_4, also called ethylene), a compound with a C=C double bond, and replace all the H atoms with Cl atoms. The result is tetrachloro-ethylene—also called perchloroethylene or "perc" for short:

ethylene   tetrachloroethylene ("perc")

Perc and other similar chlorinated hydrocarbons are **carcinogens** or suspected carcinogens. They have serious health consequences, whether we are exposed to them in the workplace or as contaminants in our air, soil, or water, as detailed in our next activity.

Your Turn 5.28   What Happened at Camp Lejeune?

From the 1950s to the 1980s, water was being contaminated on Camp Lejeune in North Carolina, one of the busiest and largest Marine Corps bases in the U.S. The water at this military base contained some of the highest levels of water pollutants ever recorded for a large, public water system, with toxin levels 240–3400 times those permitted by safety standards.

Questions to Consider
  a. Do some research into the chemicals that contaminated the water at Camp Lejeune. Have you heard of any of these chemicals before?
  b. Where did these contaminated chemicals come from?
  c. What are some of the effects of these chemicals on human health?
  d. What is the Camp Lejeune Justice Act of 2022 and when was this signed into law?

In an attempt at limiting environmental and health problems, chemists aim to redesign processes so they don't require solvents. But if this is not possible, they try to replace harmful solvents like perc with environmentally friendly ones. One possibility is liquid carbon dioxide. Under high-pressure conditions, CO_2 can condense to form a liquid. Compared with organic solvents, it offers many advantages. It is nontoxic, nonflammable, chemically benign, and non-ozone-depleting and does not contribute to smog formation. Although you may be concerned that it is a greenhouse gas, carbon dioxide used as a solvent is a recovered waste product from industrial processes and is generally recycled.

Adapting liquid CO_2 to dry cleaning posed a challenge, as it is not very good at dissolving oils, waxes, and greases found in soiled fabrics. To make carbon dioxide a better solvent, Joe DeSimone (b. 1964), a chemist and chemical engineer at the University of North Carolina, Chapel Hill, developed a surfactant to use with $CO_2(l)$. For his work, DeSimone received a 1997 Presidential Green Chemistry Challenge Award. His breakthrough paved the way for designing environmentally benign, inexpensive, and easily recyclable replacements for conventional organic and water solvents. DeSimone was instrumental in the beginnings of Hangers Cleaners, a dry-cleaning chain that uses the process he developed.

Real Talk   Carbon Dioxide For Dry-Cleaning?

Watch Dr. Joe DeSimone talk about alternatives to harmful solvents for dry-cleaning applications: www.acs.org/cic.

Questions to Consider

a. How does using liquid carbon dioxide as a solvent compare to organic solvents?

b. Comment on this statement: Using carbon dioxide as a replacement for organic solvents simply replaces one set of environmental problems with another.

Manchan/Photodisc/Getty Images

c. If a local dry-cleaning business switched from "perc" to carbon dioxide, how might this business report improvements to the economy, society, and the environment?

d. In the video, Dr. DeSimone describes some hurdles that he experienced related to bringing a potentially sustainable technology to market. Can you think of any other products or technologies that have been delayed from the market due to governmental regulations, cost, or industry competition?

The tendency of nonpolar compounds to dissolve in other nonpolar substances explains how fish and animals accumulate nonpolar substances such as PCBs (polychlorinated biphenyls) or the pesticide DDT (dichlorodiphenyltrichloroethane) in their fatty tissues. When fish ingest these, the molecules are stored in body fat (nonpolar) rather than in the blood (polar). Once widely used in electrical coolant fluids, PCBs can interfere with the normal growth and development of a variety of animals, including humans, in some cases at concentrations of less than one ppb.

The higher you go in the food chain, the greater concentrations of harmful nonpolar compounds like DDT you will find. This is called **biomagnification**, the increase in the concentration of certain persistent chemicals in successively higher levels of a food chain. Figure 5.26 shows biomagnification and a bioaccumulation process. **Bioaccumulation** is the increase in toxins over time in biological creatures. In 1962, Rachel Carson's publication of *Silent Spring* also linked a decline in the songbird populations to pesticide exposure.

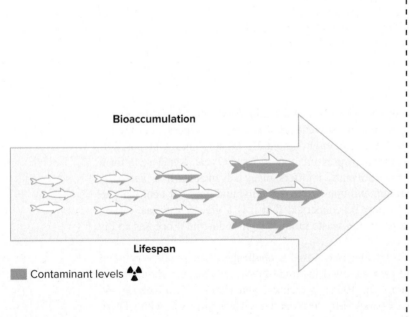

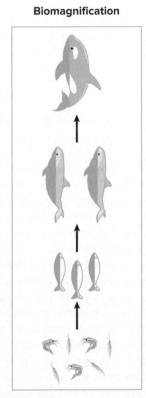

Figure 5.26

Bioaccumulation and biomagnification. Toxins will accumulate in biological life forms over time. Biomagnification is when organisms in the water take up and store a toxin. They are eaten by larger creatures that, in turn, are eaten by still-larger ones. Consequently, creatures highest on the food chain will accumulate the highest concentration of the toxin.

Before we move on to our investigation of water as a solvent for acids and bases, let's consider a class of persistent chemicals that are quickly spreading throughout our ecosystem.

Your Turn 5.29   The PFAS Problem

Check out the interactive map of water contamination with persistent fluorinated alkyl substances (PFAS) in the U.S.: www.acs.org/cic.

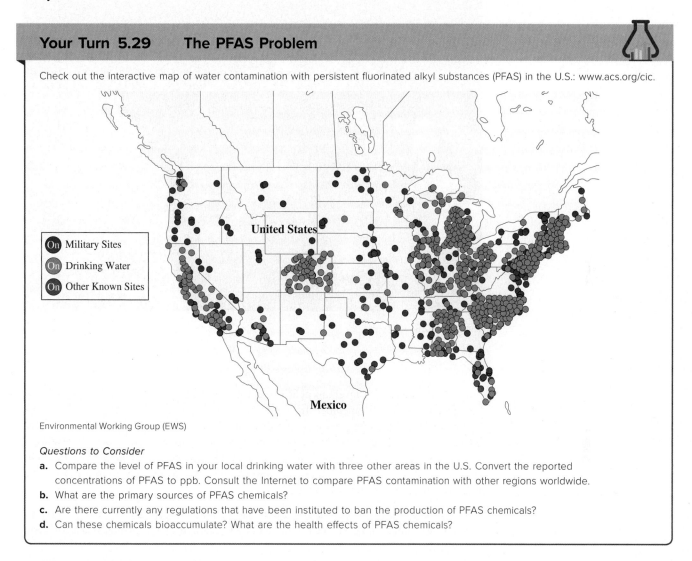

Environmental Working Group (EWS)

Questions to Consider

a. Compare the level of PFAS in your local drinking water with three other areas in the U.S. Convert the reported concentrations of PFAS to ppb. Consult the Internet to compare PFAS contamination with other regions worldwide.

b. What are the primary sources of PFAS chemicals?

c. Are there currently any regulations that have been instituted to ban the production of PFAS chemicals?

d. Can these chemicals bioaccumulate? What are the health effects of PFAS chemicals?

5.8 | The Properties of Acids and Bases

Learning Objective: Identify and classify the species involved in acid–base reactions

Water is the "universal solvent" because it dissolves various solutes. Let's now consider two aqueous solutions that greatly affect Earth and your daily lives: acids and bases. Historically, chemists identified acids by properties such as their sour taste. Although tasting is not a smart way to identify chemicals, you undoubtedly know the sour taste of acetic acid in vinegar. The sour taste of lemons also comes from acids (Figure 5.27). Acids also show a characteristic color change with indicators such as litmus.

Another way to identify an acid is by its chemical properties. For example, under certain conditions, acids can react with and dissolve marble, eggshell, or the shells of marine creatures. These materials all contain the carbonate ion (CO_3^{2-}), as either calcium carbonate or magnesium carbonate. An acid reacts with carbonate to produce carbon dioxide. This gas is the "burp" you experience when carbonate-containing antacid tablets react with acids in your stomach. As we will see in a later section, this chemical reaction also explains the dissolution of the skeletons of carbonate-based sea creatures, such as coral, in acidified oceans (Figure 5.28).

Figure 5.27

Citrus fruits contain both citric acid and ascorbic acid.

Nancy R. Cohen/Stockbyte/Getty Images

Figure 5.28

Example of a coral reef.

Manamana/Shutterstock

At the molecular level, an **acid** is a compound that releases hydrogen ions, H^+, in an aqueous solution. Remember that a hydrogen atom is electrically neutral and consists of one electron and one proton. If the electron is lost, the atom becomes a positively charged ion, H^+. Because only a proton remains, H^+ is often referred to as a **proton**.

Naming an acid depends on the identity of the anion obtained when the acid dissociates. For instance, whether the anion is monatomic or polyatomic. Follow the rules below to properly name an acid:

- If the anion is monoatomic and does not contain oxygen, add a "hydro-" prefix to the name, and a suffix of "-ic acid". An example is HF, which is hydrofluoric acid.
- If the anion is polyatomic and has an "-ate" ending, change the suffix of the corresponding acid to "-ic acid." An example is the sulfate ion, SO_4^{2-}, which as an acid is sulfuric acid, H_2SO_4.
- If the anion is polyatomic and has an ending of "-ite", change the suffix of the corresponding acid to "-ous acid". An example is the nitrite ion, NO_2^-, which as an acid is HNO_2, nitrous acid.

Consider hydrogen chloride (HCl), a compound that is a gas at room temperature. Hydrogen chloride is composed of HCl molecules. These dissolve readily in water to produce a solution that we name hydrochloric acid. As the polar HCl molecules dissolve, they become surrounded by polar water molecules. Once dissolved, these molecules break apart into two ions: $H^+(aq)$ and $Cl^-(aq)$. This equation represents the two steps of the reaction:

$$HCl(g) \xrightarrow{H_2O} HCl(aq) \longrightarrow H^+(aq) + Cl^-(aq) \qquad [5.8]$$

We also could say that HCl *dissociates* into H^+ and Cl^-. No HCl molecules remain in the solution because they dissociate completely in water, a characteristic shared by **strong acids**.

There is a slight complication with the definition of acids as substances that release H^+ (protons) in aqueous solutions. By themselves, they are much too reactive to exist as such. Rather, they attach to something else, such as water molecules. When dissolved in water, each HCl molecule donates a proton (H$^+$) to an H$_2$O molecule, forming H$_3$O$^+$, a **hydronium ion**. As we will see later in this section, an *acid* is a substance that *donates* a proton to a *base*, which is a proton *acceptor*. Here is a representation of the overall reaction in which the acid (HCl) donates a proton to the base (H$_2$O):

$$\underset{\text{acid}}{\text{HCl}(aq)} + \underset{\text{base}}{\text{H}_2\text{O}(l)} \longrightarrow \text{H}_3\text{O}^+(aq) + \text{Cl}^-(aq) \qquad \textbf{[5.9]}$$

The solution represented on the product side in *both* Equations 5.8 and 5.9 is called hydrochloric acid. It has the characteristic properties of an acid because of the presence of H$_3$O$^+$. Chemists often simply write H$^+$ when referring to acids (e.g., in Equation 5.8) but understand this to mean H$_3$O$^+$ (hydronium ion) in aqueous solutions. Figure 5.29 shows the Lewis structure of the hydronium ion.

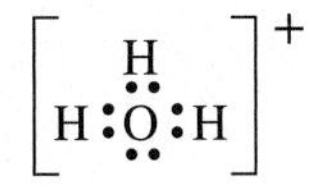

Figure 5.29

Lewis structure for the hydronium ion. Notice that oxygen follows the octet rule in this structure.

Your Turn 5.30   Acidic Solutions

Example Problem:

Write a balanced chemical equation for hydrobromic acid, HBr, in water.

Example Solution:

HBr(aq) + H$_2$O(l) ⟶ Br$^-$(aq) + H$_3$O$^+$(aq)

Your Turn:

For each of the strong acids shown below, write a balanced chemical equation that shows the release of a proton, H$^+$, when dissolved in water. Also, provide an equation that shows the formation of a hydronium ion.

Hint: Remember to include the charges on the ions. The net charge on both sides of the equation should be the same.

a. HI(aq), hydroiodic acid   **b.** HNO$_3$(aq), nitric acid   **c.** H$_2$SO$_4$(aq), sulfuric acid

---

Your Turn 5.31   Are All Acids Harmful?

Although the word *acid* may conjure up all sorts of pictures in your mind, you eat or drink various acids daily. Check the food or beverages' labels and list the acids you find. Speculate on the purpose of each acid.

Hydrogen chloride is one of several gases that dissolve in water to produce an acidic solution. Carbon dioxide, with a current atmospheric concentration of 416 ppm, is another example. Just as solids vary in their water solubility, so do gases. Compared with more polar gases such as SO$_2$ and NO$_2$, carbon dioxide is far less soluble in water. Even so, it dissolves to produce a weakly acidic solution.

Since an acid is defined as a substance that releases hydrogen ions in water, how can carbon dioxide act as an acid? There are no hydrogen atoms in carbon dioxide! The explanation is that when CO$_2$ dissolves in water, it produces carbonic acid, H$_2$CO$_3$(aq). Here are some ways to represent the process:

$$\text{CO}_2(g) \xrightarrow{\text{H}_2\text{O}} \text{CO}_2(aq) \qquad \textbf{[5.10a]}$$

$$\text{CO}_2(aq) + \text{H}_2\text{O}(l) \longrightarrow \text{H}_2\text{CO}_3(aq) \qquad \textbf{[5.10b]}$$

The carbonic acid then dissolves to produce H$^+$ and the hydrogen carbonate ion (HCO$_3^-$), also known as the bicarbonate ion:

$$\text{H}_2\text{CO}_3(aq) \rightleftharpoons \text{H}^+(aq) + \text{HCO}_3^-(aq) \qquad \textbf{[5.10c]}$$

As indicated by a double-arrow symbol (described in more detail later), this reaction occurs only to a limited extent, producing small amounts of H$^+$ and HCO$_3^-$. Accordingly, we say carbonic acid is a **weak acid** that doesn't entirely dissociate in an aqueous solution.

Figure 5.30

Oven-cleaning products may contain NaOH, commonly called lye.

Eric Misko/McGraw Hill

Carbon dioxide is only slightly soluble in water; accordingly, only a tiny amount of the dissolved carbonic acid dissociates to produce H^+. However, these reactions are happening on a large scale across the planet. The carbon dioxide can dissolve in water in the troposphere, resulting in acidic rain or in the planet's oceans, lakes, and streams.

No discussion of acids would be complete without discussing their chemical counterparts, bases. For our purposes, a **base** is a compound that releases hydroxide ions (OH^-) in an aqueous solution. Aqueous solutions of bases have their own characteristic properties attributable to the presence of $OH^-(aq)$ ions. Unlike acids, bases generally taste bitter and do not lend an appealing flavor to foods. Aqueous solutions of bases have a slippery, soapy feel. Common bases include household ammonia (an aqueous solution of NH_3) and NaOH (sometimes called lye). The cautions on oven cleaners (Figure 5.30) warn that lye can cause severe damage to eyes, skin, and clothing. Potassium hydroxide, KOH, is used in the production of dyes, medical products, biodiesel, alkaline batteries, and soft soaps.

Many common bases are compounds containing the hydroxide ion. For example, sodium hydroxide (NaOH), a water-soluble ionic compound, dissolves in water to produce sodium ions (Na^+) and hydroxide ions (OH^-):

$$NaOH(s) \xrightarrow{H_2O} Na^+(aq) + OH^-(aq) \qquad \textbf{[5.11]}$$

Although sodium hydroxide is very soluble in water, most compounds containing the hydroxide ion are not, according to the solubility rules of ionic compounds (Table 5.8). As you might expect, bases that dissociate completely in water, such as NaOH, are called **strong bases**.

Your Turn 5.32   Basic Solutions

Example Problem:

Write a chemical equation for the dissolution of magnesium hydroxide, $Mg(OH)_2$, in water.

Example Solution:

$Mg(OH)_2(s) \xrightarrow{H_2O(l)} Mg^{2+}(aq) + 2\ OH^-(aq)$

Your Turn:

For each of the bases shown below, write a chemical equation that shows the release of a hydroxide ion(s), OH^-, when dissolved in water.

  a. KOH(s), potassium hydroxide.

  b. LiOH(s), lithium hydroxide.

  c. $Ca(OH)_2(s)$, calcium hydroxide.

Hint: Whereas acids with more than one proton (known as *polyprotic acids*) lose one H^+ at a time, bases lose all OH^- groups at once.

Some bases, however, do not contain the hydroxide ion, OH^-, but rather react with water to form it. One example is ammonia (NH_3), a gas with a distinctive sharp odor. Unlike carbon dioxide, ammonia is very soluble in water. It rapidly dissolves in water to form an aqueous solution:

$$NH_3(g) \xrightarrow{H_2O} NH_3(aq) \qquad\qquad \textbf{[5.12a]}$$

On a supermarket shelf, you may see a 5% (by mass) aqueous solution of ammonia called "household ammonia." This cleaning agent has an unpleasant odor; if it gets on your skin, you should wash it off with plenty of water.

As mentioned above, a *base* acts as a *proton acceptor* in aqueous solutions. Hence, when an ammonia molecule reacts with a water molecule, the water molecule acts as the acid by transferring H^+ to the basic NH_3 molecule. An ammonium ion, $NH_4^+(aq)$, and a hydroxide ion, $OH^-(aq)$, are formed. However, this reaction only occurs to a small extent; that is, only a tiny amount of $OH^-(aq)$ is produced:

$$\underset{\text{base}}{NH_3(aq)} + \underset{\text{acid}}{H_2O(l)} \rightleftharpoons NH_4^+(aq) + OH^-(aq) \qquad\qquad \textbf{[5.12b]}$$

The source of the hydroxide ion in household ammonia should now be apparent. When ammonia dissolves in water, it releases small amounts of hydroxide and ammonium ions. Ammonia is an example of a **weak base**—a base that doesn't completely dissociate in an aqueous solution. For the equilibrium reaction shown in Equation 5.12b, the concentration of reactants is much greater than the products. In contrast, the analogous reaction with a strong base such as NaOH would be highly product-favored. Instead of writing the reaction as a reversible equilibrium, reactions involving strong acids and bases are best represented by a traditional arrow:

$$\underset{\text{strong base}}{NaOH(aq)} + \underset{\text{acid}}{H_2O(l)} \xrightarrow[\text{nonreversible}]{100\%} Na^+(aq) + OH^-(aq) + H_2O(l) \qquad \textbf{[5.13]}$$

Note that the $H_2O(l)$ solvent can be removed from both sides of Equation 5.13 to yield Equation 5.11, the simple dissociation of NaOH in an aqueous solution.

Your Turn 5.33   Acids and Bases in Your Water Diary

Revisit your water diary from the beginning of the chapter. Where do acids and bases show up in the diary? Do you need to make adjustments to include acids and bases? Do any of your actions add acidic and basic components to water? Explain.

5.9 | Heartburn? How to Neutralize Acids and Bases

Learning Objective: Describe how acids and bases react with one other

As seen in the previous section, acids and bases react with each other—often very rapidly. This happens not only in laboratory test tubes but also in your home and almost every ecological niche of our planet. For example, an acid–base reaction occurs if you put lemon juice on fish. The acids found in lemons neutralize the ammonia-like compounds that produce the "fishy smell." Similarly, if the ammonia fertilizer on a cornfield comes in contact with the acidic emissions of a power plant nearby, an acid–base reaction occurs.

@HOME   Neutralizing With Baking Soda

Investigate a neutralization reaction at home using baking soda, vinegar, and a bowl.

1. Add baking soda to a bowl.
2. Pour vinegar into the bowl and make observations.
3. Add more baking soda and make observations.
4. Add more vinegar and make observations.

Questions to Consider
a. Identify the acid and base in this reaction.
b. What implications does adding more acid to the reaction have? How can this be applied to current issues of ocean acidification?

Anythings/Shutterstock

Let us first examine the acid–base reaction of hydrochloric acid and sodium hydroxide solutions. When the two are mixed, the products are sodium chloride and water:

$$\underset{\text{acid}}{\text{HCl}(aq)} + \underset{\text{base}}{\text{NaOH}(aq)} \longrightarrow \text{NaCl}(aq) + \text{H}_2\text{O}(l) \qquad \textbf{[5.14]}$$

This is an example of a **neutralization reaction**, a chemical reaction in which a proton from an acid combines with the hydroxide ion from a base to form a water molecule. The formation of water can be represented like this:

$$\text{H}^+(aq) + \text{OH}^-(aq) \longrightarrow \text{H}_2\text{O}(l) \qquad \textbf{[5.15]}$$

What about the sodium and chloride ions? Recall from Equations 5.8 and 5.11 that the HCl(g) and NaOH(s), when dissolved in water, completely dissociate into ions. We can rewrite Equation 5.14 to show all chemical species involved in the reaction, which is often referred to as the *total ionic equation*:

$$\text{H}^+(aq) + \text{Cl}^-(aq) + \text{Na}^+(aq) + \text{OH}^-(aq) \longrightarrow \text{Na}^+(aq) + \text{Cl}^-(aq) + \text{H}_2\text{O}(l) \qquad \textbf{[5.16]}$$

Neither $\text{Na}^+(aq)$ nor $\text{Cl}^-(aq)$ take part in the neutralization reaction and are thus referred to as *spectator ions*; they remain unchanged. Canceling these ions from both sides results in Equation 5.15, which summarizes the chemical changes in an acid–base neutralization reaction:

$$\text{H}^+(aq) + \cancel{\text{Cl}^-(aq)} + \cancel{\text{Na}^+(aq)} + \text{OH}^-(aq) \longrightarrow \cancel{\text{Na}^+(aq)} + \cancel{\text{Cl}^-(aq)} + \text{H}_2\text{O}(l) \qquad \textbf{[5.17]}$$

This form of the neutralization equation that omits the spectator ions is often designated as the *net ionic equation*. Get some practice with total ionic and net ionic equations in the next activity.

Your Turn 5.34   Neutralization Reactions

For each acid–base pair, write a balanced neutralization reaction. Then, rewrite the equation in total ionic and net ionic forms. Identify all spectator ions.

1. $\text{HNO}_3(aq)$ and $\text{KOH}(aq)$
2. $\text{HCl}(aq)$ and $\text{NH}_4\text{OH}(aq)$
3. $\text{HBr}(aq)$ and $\text{Ba(OH)}_2(aq)$

A **neutral solution** is neither acidic nor basic; it has equal concentrations of H^+ and OH^-. The square brackets indicate that the ion concentrations are expressed in molarity, M, and are read as "the hydrogen ion (or proton) concentration." In a neutral solution, $[\text{H}^+] = [\text{OH}^-]$. Pure water is a neutral solution. Some salt solutions are also neutral, such as the one formed by dissolving solid NaCl in water. In contrast, acidic solutions contain a higher concentration of H^+ than $\text{OH}^-([\text{H}^+] > [\text{OH}^-])$, and basic solutions contain a higher concentration of OH^- than $\text{H}^+([\text{H}^+] < [\text{OH}^-])$.

Your Turn 5.35   Acidic or Basic?

These solutions represent either strong acids or strong bases. Classify each as acidic or basic. Then, list all the ions in order of decreasing relative amounts in each solution.

a. KOH(*aq*)   **b.** HNO$_3$(*aq*)   **c.** H$_2$SO$_4$(*aq*)   **d.** Ca(OH)$_2$(*aq*)

It may seem strange that acidic and basic solutions contain hydroxide ions *and* protons. But it is impossible to have H$^+$ without OH$^-$ (or vice versa) when water is involved. A simple, useful, and very important relationship exists between the concentration of protons and hydroxide ions in any aqueous solution:

$$[H^+][OH^-] = 1 \times 10^{-14} \qquad \textbf{[5.18]}$$

That is, when [H$^+$] and [OH$^-$] are multiplied, the product is a constant with a value of 1×10^{-14}, as shown in Equation 5.18. This indicates that the concentrations of H$^+$ and OH$^-$ depend on each other. When [H$^+$] increases, [OH$^-$] decreases, and when [H$^+$] decreases, [OH$^-$] increases. However, both ions are always present in aqueous solutions.

Knowing the concentration of H$^+$, we can use Equation 5.18 to calculate the concentration of OH$^-$ (or vice versa). For example, if rainwater has a H$^+$ concentration of 1×10^{-5} M, we can calculate the OH$^-$ concentration by substituting 1×10^{-5} M for [H$^+$]:

$$(1 \times 10^{-5} \text{ M}) \times [OH^-] = 1 \times 10^{-14}$$

$$[OH^-] = \frac{1 \times 10^{-14}}{1 \times 10^{-5}}$$

$$[OH^-] = 1 \times 10^{-9} \text{ M}$$

Because the hydroxide ion concentration (1×10^{-9} M) is smaller than the hydrogen ion concentration (1×10^{-5} M), the solution is acidic.

In pure water or a neutral solution, the hydrogen and hydroxide ions concentrations are equal: applying Equation 5.18, we can see that [H$^+$][OH$^-$] = $(1 \times 10^{-7}$ M$)$ $(1 \times 10^{-7}$ M$) = 1 \times 10^{-14}$.

More examples of these calculations are provided in the next tutorial video.

Acidic and Basic Solutions

Watch this video for an example of some acid/base calculations: www.acs.org/cic. For parts **a** and **c** below, calculate [OH$^-$]. For **b**, calculate [H$^+$]. Then, classify each solution as acidic, neutral, or basic.

a. [H$^+$] = 1×10^{-4} M
b. [OH$^-$] = 1×10^{-6} M
c. [H$^+$] = 1×10^{-10} M

Chemistry in Context
Tutorial

ACID/BASE CALCULATIONS

©Bradley D. Fahlman

How can we know if the acidity of seawater, rainwater, or another solution is cause for concern? To make a judgment, we need a convenient way of quantifying how acidic or basic a solution is. The pH scale is such a tool because it relates the acidity of a solution to its H$^+$ concentration.

5.10 | Quantifying Acidity/Basicity: The pH Scale

> *Learning Objective: Quantify the relative acidity or basicity of a solution in terms of the pH scale*

The term "pH" may already be familiar to you. For example, test kits for soils and the water in aquariums and swimming pools report acidity in terms of pH. Deodorants and shampoos claim to be pH-balanced (Figure 5.31). And, of course, articles about acid rain refer to pH. The notation pH is always written with a small p and a capital H and stands for "power of hydrogen." In the simplest terms, a pH is a number, usually between 0 and 14, indicating a solution's acidity (or basicity).

Figure 5.31

This shampoo claims to be "pH-balanced" and adjusted to be closer to neutral. Soaps tend to be basic, which can be irritating to the skin.

C.P. Hammond/McGraw Hill

At the midpoint on the scale, pH 7 separates acidic from basic solutions. Solutions with a pH less than 7 are acidic, and those with a pH greater than 7 are basic (alkaline). Solutions of pH 7 (such as pure water) have equal concentrations of H^+ and OH^- and are said to be neutral.

Universal indicator paper is a quick way to determine the pH value of a solution. However, for more accurate results, pH meters are used. The pH values of common substances are displayed in Figure 5.32. You may be surprised that you eat and drink so many acids! Acids naturally occur in foods and contribute distinctive tastes. For example, the tangy taste of McIntosh apples comes from malic acid. Yogurt gets its sour taste from lactic acid, and cola soft drinks contain several acids, including phosphoric acid. Tomatoes are well known for their acidity, but with a pH of about 4.5, they are, in fact, less acidic than many other fruits. The pH may even lie outside the 0 to 14 range for highly acidic or basic solutions.

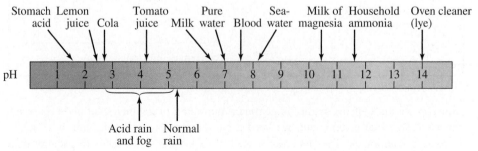

Figure 5.32

Common substances and their pH values.

@HOME   Cabbage Juice Indicator

PART 1

1. Tear red cabbage leaves into smaller pieces and put them into a blender until you have approximately two cups of leaves.
2. Pour hot water over the red cabbage leaves so that they are covered.
3. Blend the red cabbage and water mixture for 30 seconds.
4. Pour off the liquid into an empty plastic bottle. This is your pH indicator.

Pixtal/age fotostock

5. Investigate the pH of common household substances. Some examples may include tap water, vinegar, juice, milk, coffee (liquid), hand soap added to water, cleaning solution, soft drinks, and candies (crushed up and/or dissolved in water).
6. Pour two teaspoons of the liquid you are testing into a clear glass.
7. Add a teaspoon of the cabbage juice pH indicator into the glass and make observations about the color.
8. Repeat Steps 6 and 7 for each liquid you are testing.

Questions to Consider

a. Did any of the pH readings surprise you? Explain.
b. Think about the most acidic and most basic liquids you tested. How could you get each of them to be neutral?

PART 2

Investigate how carbon dioxide affects the acidity of water.

1. Pour an equal volume of room-temperature water into two clear glasses.
2. Place a straw into one of the glasses and blow bubbles into the water for 10 seconds and then take a break for a few seconds.
3. Repeat Step 6 twelve times making sure to take breaks.
4. Add 3 tablespoons of the cabbage juice pH indicator to both cups and make observations.

Questions to Consider

a. What was different between the glass with bubbles and the glass without?
b. How can you apply this to what you know about climate change and ocean acidification?

Equation 5.19 is used to calculate the pH of a solution.

$$[\text{H}^+] \text{ is pH} = -\log [\text{H}^+] \qquad [5.19]$$

As you might have guessed, pH values are related to hydrogen ion concentration. If $[\text{H}^+] = 1 \times 10^{-3}$ M, then the pH is 3. Similarly, if $[\text{H}^+] = 1 \times 10^{-9}$ M, the pH is 9. As pH values rise above 7, $[\text{H}^+]$ decreases and $[\text{OH}^-]$ increases. Remember, Equation 5.18 indicates that the hydrogen ion concentration multiplied by the hydroxide ion concentration is a constant, 1×10^{-14}. When the concentration of H^+ is high, the concentration of OH^- is low. Explore the relationship between the proton concentration and pH in the next activity.

Your Turn 5.36   The pH Scale

Check out the pH simulation at www.acs.org/cic. Click on Macro, and drag the green pH meter into the solution. Measure the pH of a few solutions. Put coffee in the container, and record its initial pH. Then add water to this solution by pulling the knob at the top. Record which way the pH changes. Add more coffee to the solution by pressing the button at the top, and record which way the pH changes.

Repeat these steps with one of the basic solutions.

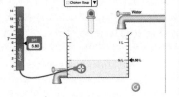

PhET Interactive Simulations, University of Colorado boulder, http://phet.colorado.edu

(continued)

Questions to Consider

a. In which direction did the pH change when adding water to the weak acid coffee? In which way did it change when adding more coffee to the solution?

b. How did the pH change when adding water to the weak base, and vice versa? Go to the My Solutions part of the simulation. Change the pH of the solution by dragging the [OH⁻] or [H_3O^+] tabs on the bar down or up. Record each of these numbers at four different places along the scale. Change the top slider to Quantity (mol), and then click the molecule count at the bottom. Record the molecule count and number of moles for each component at three different places along the scale.

c. How are the [OH⁻] and [H_3O^+] concentrations related to the four places you recorded? Do they always have a common product?

d. How are the molecule count and the number of moles for each component related at each of the three places you recorded around the scale?

As the pH value decreases, the acidity increases. For example, a solution with a pH of 5.0 is 1/10 the acidity of one with a pH of 4.0. This is because a pH of 4 means that the $[H^+] = 0.0001$ M. By contrast, a solution with a pH of 5 has a more dilute $[H^+]$; its $[H^+] = 0.00001$ M. This second solution is less acidic because it has 1/10 the hydrogen ion concentration of a solution of pH 4. Figure 5.33 shows the relationship between pH and hydrogen ion concentration, which will be explored in the following tutorial video.

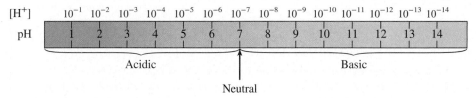

Figure 5.33

The relationship between pH and the concentration of H⁺ in moles per liter (M). As pH increases, [H⁺] decreases and [OH⁻] increases.

Acidic or Basic?

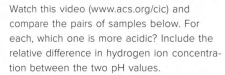

Watch this video (www.acs.org/cic) and compare the pairs of samples below. For each, which one is more acidic? Include the relative difference in hydrogen ion concentration between the two pH values.

a. Rainwater, pH = 5.0; lake water, pH = 4.0.

b. Ocean water, pH = 8.3; tap water, pH = 5.3.

c. Tomato juice, pH = 4.5; milk, pH = 6.5.

Chemistry in Context Tutorial

THE pH SCALE: ACIDIC/BASIC?

©Bradley D. Fahlman

Your Turn 5.37   On the Record

A legislator from the Midwest is on record with an impassioned speech in which he argued that the state's environmental policy should bring the rain's pH down to zero. Assume that you are an aide to this legislator. Draft a tactful memo to your boss to save him from additional public embarrassment.

5.11 | Acid's Effect on Water

Changes in pH can affect the health of our oceans and other water sources. However, before we delve into these details, is the water on our planet acidic, basic, or neutral? Water would be expected to have a pH of 7.0, but Figure 5.32 showed that the pH of water depends on where it is found. "Normal" rain is slightly acidic, whereas seawater is basic. This begs the question: How can seawater be basic when rain is naturally acidic?

According to the National Oceanic and Atmospheric Administration (NOAA), the ocean absorbs an average of 3 billion metric tons of carbon dioxide from the atmosphere each year. Ocean water contains small amounts of three chemical species that arise from dissolved CO_2 and play a role in maintaining the ocean pH at approximately 8.2. These three species—the carbonate ion, the bicarbonate ion, and carbonic acid (Figure 5.34)—also interact with each other. These species also help maintain your blood at a pH of about 7.4. A solution that contains ionic species that help maintain a constant pH is referred to as a **buffer** solution.

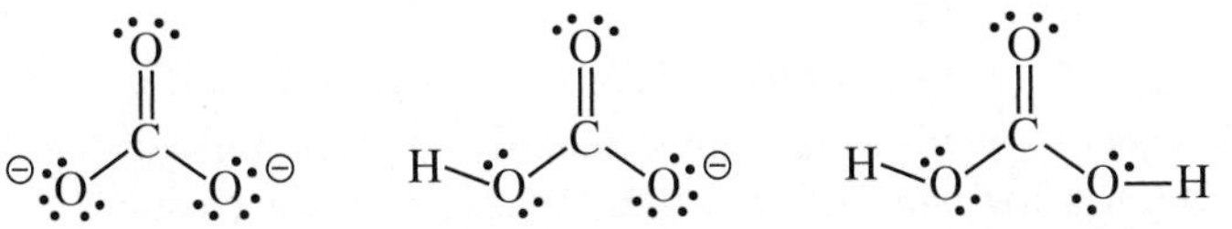

carbonate ion
CO_3^{2-}

bicarbonate ion
HCO_3^-

carbonic acid
H_2CO_3

Figure 5.34

Lewis structures for the carbonate and the bicarbonate ions, as well as carbonic acid.

Changing the amount of carbonic acid in the ocean can affect the concentration of other chemicals, affecting marine life. Mollusks, sea urchins, coral, and other organisms have connections to ocean chemistry because they build their shells out of calcium carbonate, $CaCO_3$. The amount of carbon dioxide released into the atmosphere over the past 200 years has increased (Section 4.9). As a result, more carbon dioxide is dissolving into the oceans and reacting to form carbonic acid. In turn, the pH of seawater has dropped by roughly 0.1 pH unit since the early 1800s. This may sound like a small number; however, remember that each full pH unit represents a 10-fold difference in the concentration of H^+. A decrease of 0.1 pH unit corresponds to a 26% increase in the amount of H^+ in seawater. The lowering of the ocean pH due to increased atmospheric carbon dioxide is called **ocean acidification** (Figure 5.35).

The effect of climate change on ocean acidification were addressed in the 2021 IPCC report, discussed in Section 4.9. Some conclusions drawn from this report include:

- It is *virtually certain* that human-caused carbon dioxide emissions are the main driver of the current global acidification of the surface open ocean.
- There is *high confidence* that since the Industrial Revolution, the oceans have decreased by 0.1 pH unit, increasing their acidity by 26%.

As shown Figure 5.36, the calcium carbonate in the shells of sea creatures then begins to dissolve in response to the decreased concentration of carbonate ions in seawater due to the reaction with H^+ ions:

$$H^+(aq) + CO_3^{2-}(aq) \longrightarrow HCO_3^-(aq) \qquad \text{[5.20]}$$

The interactions of carbonic acid, bicarbonate ions, and carbonate ions are summarized in Figure 5.36. As carbon dioxide dissolves in ocean water, it forms carbonic acid. Carbonic acid dissociates to produce "extra" acidity in the form of H^+. The H^+ ions react with carbonate ions, thereby depleting them and producing more bicarbonate ions. Calcium carbonate then dissolves to replace the carbonate that was consumed.

Figure 5.35

An image of an egg in an acidic solution. The acidity of water can cause the dissolution of an eggshell through a chemical reaction. The eggshells start dissolving in the more acidic solutions due to the interactions between $CO_3^{2-}(aq)$, $HCO_3^-(aq)$, and $H_2CO_3(aq)$.

©Bradley D. Fahlman

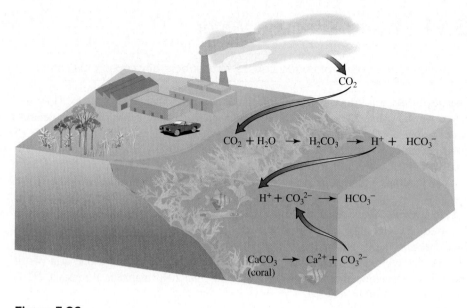

Figure 5.36

Chemistry of CO_2 in the ocean.

Ocean scientists predict that within the next 40 years, the carbonate ion concentration will reach a low enough level that the shells of sea creatures near the ocean surface will begin to dissolve. One study has shown that the Great Barrier Reef off the coast of Australia is already growing at slower and slower rates. However, other factors could be to blame. For example, ocean warming also contributes to the poor health of coral reefs, as you will examine in this next activity.

Coral Reefs

1. Watch this video (www.acs.org/cic) and summarize the chemistry behind ocean acidification. Why is a decrease in ocean pH a concern for coral?

2. Conduct a brief search on one of the following coral reefs and answer the following questions.

   - Red Sea Coral Reef, Israel, Egypt and Djibouti
   - New Caledonia Barrier Reef, South Pacific
   - Meso American Barrier Reef, Atlantic Ocean
   - Florida Reef, Florida
   - Andros Coral Reef, Bahamas
   - Saya Del Malha, Indian Ocean

   a. How is ocean acidification impacting your chosen reef? Discuss the effect of ocean acidification on marine life and how these effects make their way up the food chain. The 2021 IPCC report suggests (with *medium confidence*) that the open ocean pH is as low as at any point in the last 2 million years.

   b. Are there coastal communities whose economy and/or food supply rely heavily upon the coral reef? How might they feel the effects of ocean acidification?

To date, only a small number of researchers have focused on the effects of thinning shells on sea creatures. However, adverse effects on whole ecosystems have been projected. For example, weaker (or missing) coral reefs could fail to protect coastlines from harsh ocean waves. Coral reefs also provide fish species with their habitat, and damage to the reefs would translate into losses of marine life. Finally, weakening the reefs would make them more susceptible to further damage from storms and predators. View how serious the situation is globally in this next interactive activity.

Your Turn 5.38   Life Below Water

The graphic below shows how the coral cover has changed worldwide since 1980. For an interactive version of these data, go to www.acs.org/cic.

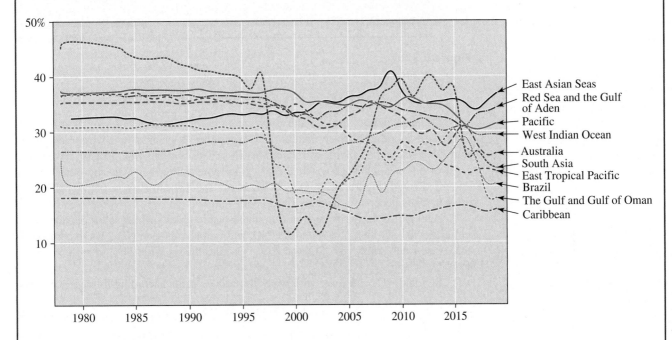

Questions to Consider

a. Which region of the world has experienced the greatest decline in coral growth since 2005? What are some reasons for this significant decline?

b. Have any regions experienced growth in coral cover since 2010? What are some factors that have enabled coral to increase?

Can the ocean heal itself? Although we don't know the answer for sure, nonetheless, we can speculate from what we know of past events. The ocean can compensate when changes in ocean pH have occurred over a very long period. Compensation happens because extensive sediment collections at the bottom of the ocean contain massive amounts of calcium carbonate, mainly from the shells of long-deceased marine creatures. Over long periods, these sediments dissolve to replenish the carbonate lost to reaction with excess H^+. But today's changes in ocean pH have happened rapidly on the geologic time scale. In just 200 years, the pH of the ocean has dropped to a level not seen in the past 400 million years. Because the acidification occurs over a relatively short time and in the water close to the surface, the sediment reserve has not had time to dissolve and counteract the effects of the added acidity.

Even if the amount of carbon dioxide in the atmosphere were to level off immediately, the oceans would take thousands of years to return to the pH measured in preindustrial times. Coral reefs would take even longer to regenerate, and any species lost to extinction would not return.

Your Turn 5.39   A Global Response to Ocean Acidification

In 2008, a group of scientists met in Monaco to raise awareness about ocean acidification. They issued the Monaco Declaration, calling on the world's countries to reverse carbon dioxide emissions trends by 2020. Have more recent gatherings of scientists and negotiators created a worldwide policy to address ocean acidification? Do research of your own and summarize your findings.

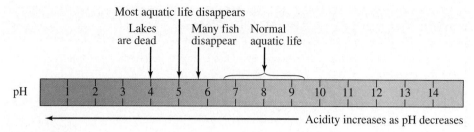

Figure 5.37

Aquatic life and pH.

Ocean creatures are not the only creations bearing the costs of acidification. Organisms in other surface waters also experience a change in environment when acid rain (also called *acidic precipitation*) fills lakes and streams. Healthy lakes have a pH of 6.5 or slightly above. If the pH is lowered below 6.0, fish and other aquatic life are affected (Figure 5.37). Only a few hardy species can survive below pH 5.0. At pH 4.0, lakes become essentially dead ecosystems.

Acid rain is primarily derived from atmospheric gases such as sulfur dioxide (SO_2), sulfur trioxide SO_3, nitrogen monoxide NO, and nitrogen dioxide NO_2. These compounds are affectionately known as "SO_x and NO_x." Chemically, we write SO_x, where x = 2 or 3, and NO_x, where x = 1 or 2.

SO_3 and NO_2 can react with water to produce acids according to these reactions:

$$SO_3(g) + H_2O(l) \longrightarrow H_2SO_4(aq) \qquad \textbf{[5.21]}$$

$$4\ NO_2(g) + 2\ H_2O(l) + O_2(g) \longrightarrow 4\ HNO_3(aq) \qquad \textbf{[5.22]}$$

The two ions that are formed from the dissociation of these acids, sulfate ($SO_4{}^{2-}$) and nitrate ($NO_3{}^-$), can be detected in rainwater. This next activity will explore how SO_x and NO_x emissions, and acid rain deposition, have changed since 1990.

Your Turn 5.40   What Is the Chemistry of Acid Rain?

Go to this website to consider what the EPA has done to combat these types of emissions over the last thirty years: www.acs.org/cic.

Questions to Consider

a. What are the names of the acids H_2SO_4 and HNO_3?

b. Where do you live? What has happened to SO_2 and NO_x emissions where you live since 1990? What factors do you think contributed to this change?

c. What state had the most significant decrease in SO_2 and NO_x emissions since 1990? Which had the biggest increase in SO_2 and NO_x emissions over this time period? What are the biggest contributors to these emissions?

d. Sulfur dioxide gas dissolves in water to produce sulfurous acid. Write the equation that forms sulfurous acid from sulfur dioxide.

Numerous studies have reported the progressive acidification of lakes and rivers in specific geographic regions and concomitant reductions in fish populations. In southern Norway and Sweden, where the problem was first observed, one-fifth of the lakes no longer contain any fish, and half of the rivers have no brown trout. In southeastern Ontario, the average pH of lakes is 5.0, well below the pH of 6.5 required for a healthy lake. In Virginia, more than one-third of the trout streams are episodically acidic or at risk of becoming so.

When acid rain falls on or runs off into a lake, the pH of the lake drops (becomes more acidic) unless the acid is neutralized or used by the surrounding vegetation. In some regions, the surrounding soils may contain bases that can neutralize the acid. The capacity of a lake or other body of water to resist a decrease in pH is called its **acid-neutralizing capacity**, which is explored in this next activity.

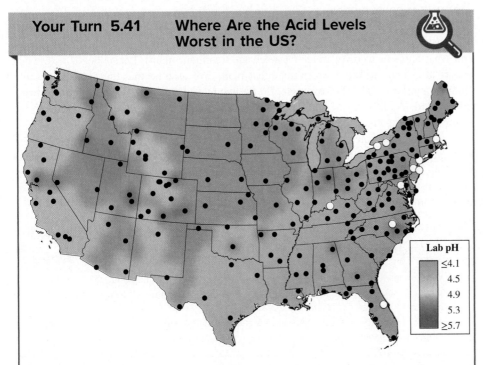

Your Turn 5.41 Where Are the Acid Levels Worst in the US?

Many areas of the midwestern United States have no problem with the acidification of lakes or streams, even though the Midwest is a significant source of acidic precipitation. The surface geology of much of the Midwest is limestone, $CaCO_3$. As a result, lakes in the Midwest have a high acid-neutralizing capacity because limestone slowly reacts with acid rain. Because acid is consumed by the carbonate and bicarbonate ions, the pH of the lake remains more or less constant. Many lakes in New England and northern New York are surrounded by granite, a hard, impervious, and much less-reactive rock. Unless other local processes work, these lakes have minimal acid-neutralizing capacity. Visit this website to see where the EPA monitors acid levels in lakes: www.acs.org/cic.

a. Write a reaction for calcium carbonate, carbon dioxide, and water transforming into Ca^{2+} ions and bicarbonate ions.

b. Where are they monitoring the surface water acid levels? Why are these areas chosen for monitoring?

c. In the image above, you can see where the pH levels around the country are the lowest. Does this fit with what you learned from the EPA website?

d. Which other state would you suggest the EPA test for lake acid levels?

As it turns out, understanding the acidification of lakes and rivers is much more complicated than measuring pH and acid-neutralizing capacities. Annual variations add one level of complexity. Heavy winter snowfalls persist into the spring for some years and then melt suddenly. As a result, the runoff may be more acidic than usual because it contains all the acidic deposits locked away in the winter snow. A surge of acidity may enter the waterways at just the time when fish are spawning or hatching and are more vulnerable. In the Adirondack Mountains of northern New York, about 70% of the sensitive lakes are at risk for episodic acidification, compared with a far smaller percentage chronically affected (19%). In the Appalachians, the number of episodically affected lakes (30%) is seven times those chronically affected.

As you can see, pH differences in water play a huge role in biodiversity, habitats, and the overall environment. This chapter's final sections will discuss strategies to provide clean fresh water for people to use and consume.

5.12 | Treating Our Water

> *Learning Objective: Describe how water can be treated to become safer for use or consumption*

This section explores what happens to clean water (at a local water treatment plant) and what happens to dirty water (at a sewage treatment plant). Let us begin with what occurs at a local drinking water treatment plant. We assume that the plant gets water from an aquifer or lake. For example, if you live in San Antonio, water is pumped from the Edwards Aquifer. Or, if you live in San Francisco, the water comes from a reservoir in the Hetch Hetchy valley, more than 100 miles away.

In a typical water treatment plant (Figure 5.38), the first step is to pass the water through a screen that physically removes large impurities such as weeds, sticks, and beverage bottles. The next step is to add aluminum sulfate ($Al_2(SO_4)_3$) and calcium hydroxide ($Ca(OH)_2$). This next activity will review these chemicals and someothers involved in water treatment.

Your Turn 5.42   Water Treatment Chemicals

a. Write chemical formulas for these ions: sulfate, hydroxide, calcium, and aluminum.

b. What are some compounds that can be formed from these four ions? Write their chemical formulas.

c. The hypochlorite ion (ClO^-) also plays a critical role in water purification. Write chemical formulas for sodium hypochlorite and calcium hypochlorite.

Aluminum sulfate and calcium hydroxide are *flocculating agents*; they react in water to form a sticky floc (gel) of aluminum hydroxide, $Al(OH)_3$ (Equation 5.23).

$$Al_2(SO_4)(aq) + 3\ Ca(OH)_2(s) \longrightarrow 2\ Al(OH)_3(s) + 3\ CaSO_4(aq) \qquad \textbf{[5.23]}$$

This gel collects suspended clay and dirt particles on its surface. As the $Al(OH)_3$ gel slowly settles, it carries particles with it that were suspended in the water. Any remaining particles are removed as the water is filtered through charcoal or gravel and then sand.

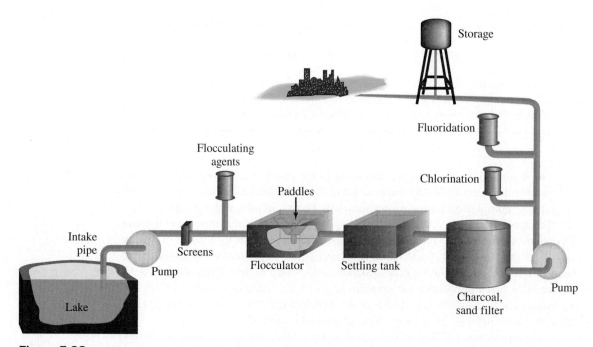

Figure 5.38

A typical municipal water treatment facility. Components are not illustrated to scale.

The crucial step comes next—disinfecting the water to kill disease-causing microbes. In the United States, this is most commonly done with chlorine-containing compounds. Chlorine can only kill the microorganisms with which it comes in contact. Chlorine does not kill bacteria or viruses enclosed inside silt or clay particles, so particles must be removed before the chlorination step. Chlorination is accomplished by adding chlorine gas (Cl_2), sodium hypochlorite (NaClO, found in bleach), or calcium hypochlorite ($Ca(ClO)_2$, used to clean swimming pools). All of these compounds generate the anti-bacterial agent hypochlorous acid, HClO. A very low concentration of HClO, 0.075 to 0.600 ppm, remains to protect the water against further bacterial contamination as it passes through pipes to the user. **Residual chlorine** refers to the chlorine-containing chemicals that remain in the water after the chlorination step. These include hypochlorous acid (HClO), hypochlorite ions (ClO^-), and dissolved elemental chlorine (Cl_2).

Before chlorination, thousands died in epidemics spread via polluted water. In a classic study, John Snow (1813–1858), an English physician, traced a mid-1800s cholera epidemic in London to water contaminated with the excrement of cholera victims. Another example occurred in 2007 in war-torn Iraq. After extremists put chlorine tanks on suicide truck bombs earlier that year, authorities kept tight controls on chlorine. At one point, a shipment of 100,000 tons of chlorine was held up for a week at the Jordanian border amid fears for its safe passage through Iraq. With the water infrastructure disrupted and the quality of water and sanitation poor, levels of fecal bacteria, including pathogenic species, increased dramatically. As a result, thousands of Iraqis contracted cholera.

Despite its benefits, chlorination has its drawbacks. The taste and odor of residual chlorine can be objectionable and are commonly cited as why people drink bottled water or use filters to remove residual chlorine. A more serious drawback is the reaction of residual chlorine with other substances in the water to form by-products in drinking water at concentrations that may be toxic. The most widely publicized are compounds such as $CHCl_3$ (chloroform), $CHBr_3$ (bromoform), $CHBrCl_2$ (bromodichloromethane), and $CHBr_2Cl$ (dibromochloromethane). These **trihalomethanes** (THMs) form from the reaction of chlorine or bromine with organic matter in drinking water. Like HClO, hypobromous acid (HBrO) used to disinfect spa tubs can generate trihalomethanes.

Your Turn 5.43   THMs at a Glance

a. Draw Lewis structures for any two THM molecules.
b. How do THMs differ from CFCs in their chemical composition?
c. How do THMs differ from CFCs in their physical properties?

Many European and a few U.S. cities use ozone to disinfect their water supplies. One advantage is that a lower ozone concentration relative to chlorine is required to kill bacteria. Furthermore, ozone is more effective than chlorine against water-borne viruses. But ozonation also comes with disadvantages. Ozonation is expensive and economical only for large water-treatment plants. Another is that ozone decomposes quickly and hence does not protect water from possible contamination as it is piped through the municipal distribution system. Consequently, a low dose of chlorine must be added to ozonated water as it leaves the treatment plant.

Disinfecting water using ultraviolet (UV) light is growing in popularity. By UV, we mean UV-C, the high-energy UV radiation that can break down DNA in microorganisms, including bacteria. Disinfection with UV-C is fast, leaves no residual by-products, and is economical for small installations, including rural homes with unsafe well water. Like ozone, however, UV-C does not protect the water after it leaves the treatment site. Again, a low dose of chlorine must be added. Depending on local needs, one or more additional purification steps may be taken after disinfection at the water treatment facility. Sometimes the water is sprayed into the air to remove volatile chemicals that create objectionable odors and tastes. If little natural fluoride is present in the water supply, some municipalities add fluoride ions (~1 ppm NaF) to protect against tooth decay. However, this policy is somewhat controversial as you will see in the next activity.

Let's Debate!   Is Fluoridation Safe?

Form two groups: one group will be in favor of the topic (the proposition) and the other group will be against the topic (the opposition). Both groups will try to persuade a neutral person or judge to agree with them. The topic of the debate is the motion. Here is the motion:

Fluoride added to drinking water helps prevent tooth decay and poses no significant health impacts. Hence, fluoridation should be instituted in all communities.

Curiosity Conversation
- What are some possible negative health effects posed by exposure to fluoride and how reliable are these studies?
- Why have some communities opted out of fluoridation programs? What are some socio-economic impacts caused by this decision?

We just described how water is treated before it is ready to drink out of the tap. But once we turn on the tap, we start getting the water dirty again. We add waste to the water each time it leaves our bathrooms in a toilet flush, runs down the drain after a soapy shower, or goes down the sink after we wash the dishes. Using as little water as possible makes sense because if we dirty it, it must be cleaned again before being released into the environment. Remember green chemistry! It is better to prevent waste than to treat or clean up waste after it is formed.

How do we remove waste from water? If the drains in your home are connected to a municipal sewage system, then the wastewater flows to a sewage treatment plant. Once there, it undergoes similar cleaning processes to those for water treatment, except end-stage chlorination, before it is released back to the environment.

Cleaning sewage is more complicated because it contains waste from organic compounds and nitrate ions. To many aquatic organisms, this waste is a source of food! As these organisms feed, they deplete oxygen from surface waters. **Biological oxygen demand** (BOD) measures the amount of dissolved oxygen microorganisms use up as they decompose organic waste in the water. A low BOD is one indicator of good water quality.

Nitrates and phosphates contribute to BOD because these ions are essential nutrients for aquatic life. An overabundance of either can disrupt the normal flow of nutrients and lead to algal blooms (Figure 5.39) that clog waterways and deplete oxygen from the water. In turn, this reduced oxygen can lead to massive fish kills. The problem of reduced oxygen in water is compounded by the fact that the solubility of oxygen in water is so very low in the first place.

Figure 5.39

A pond with algal bloom in Brookmill Park, Great Britain.

DEA/C. BARAGGI/DeAgostini/Getty Images

Some treatment plants use wetland areas to capture nutrients such as nitrates and phosphates before the water is returned to the surface or recharges the groundwater. Plants and soil microorganisms in these wetland areas (marshes and bogs) facilitate nutrient recycling, thus reducing the nutrient load in the water. If the water produced from treated sewage is clean enough, why not just use it as a source of drinking water? Singapore's growing population relies on several potable water sources. One of these, NEWater, is purified wastewater, sufficient to meet up to 40% of Singapore's current needs. The next activity allows you to explore this controversial use of reclaimed water.

Your Turn 5.44　　Toilet to Tap?

With increasing drought conditions in many parts of the world, more communities are considering using reclaimed water as a source of drinking water. Suppose the quality of the water produced from the sewage treatment process matched the water quality in our current drinking water system. Would you accept treated sewage water as drinking water? Comment either way.

Your Turn 5.45　　Water Treatment and Your Diary

Revisit your water diary. How was the water you used over the diary period treated? Research the treatment processes in your community. Are there ways to conserve water that needs to be treated?

With an understanding of how water is treated in sewage and drinking water plants, our final section will examine how scientists and communities use chemistry to enable more people to receive potable water.

5.13 | Water Solutions for Global Challenges

Learning Objective: Demonstrate some strategies used to purify water

As stated by the United Nations [U.N. Sustainable Development Goal 6], "Access to safe water, sanitation and hygiene is the most basic human need for health and well-being." In this final section, we showcase efforts that demonstrate the sustainable use of water. The first relates to the production of fresh water from salt water. The second describes methods for purification of water at its point-of-use.

Fresh Water from Salt Water

The high salt content (3.5%) of seawater makes it unfit for human consumption. While some creatures can live in salt water, humans cannot subsist on drinking it. Today, we can tap the sea as a water source for agriculture and drinking. **Desalination** is any process that removes sodium chloride and other minerals from salty water, thus producing potable water. Currently, approximately 17,000 desalination plants worldwide make more than 95 million liters of water daily. With the demand for freshwater ever increasing, we are now witnessing the construction of many new desalination facilities across the globe—one of the world's largest, in the United Arab Emirates, is shown in Figure 5.40.

One means of desalination is **distillation**, a separation process in which a liquid solution is heated, and the vapors are condensed and collected. Impure water is heated; as the water vaporizes, it leaves behind most of its dissolved impurities. However,

Figure 5.40

A desalination plant at Jebel Ali in the
United Arab Emirates.

Shao Weiwei/Shutterstock

distillation requires energy. Figure 5.41 shows that a Bunsen burner provides this energy
in one case and the Sun in the other. Recall that water has high specific heat and
requires an unusually large amount of energy to convert to vapor. Both properties result
from the extensive hydrogen bonding among water molecules.

Large-scale distillation operations employ new technologies with impressive
names, such as *multistage flash evaporation*. Although these technologies have increased
energy efficiencies over the basic distillation process shown in Figure 5.41a, their energy
requirements are still high and are usually provided by burning fossil fuels. An alternative is to purify water using smaller solar distillation units, as shown in Figure 5.41b.

Other desalination options exist. For example, **osmosis** is the passage of water
through a semipermeable membrane from a less concentrated solution to a more
concentrated one. The water diffuses through the membrane, and the solute does

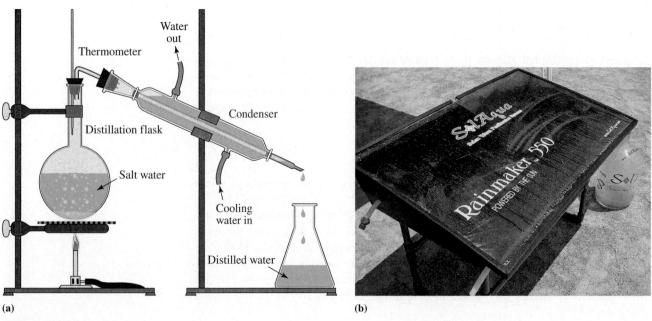

(a)   **(b)**

Figure 5.41

(a) Laboratory distillation apparatus. **(b)** Tabletop solar still.

(b): Courtesy of SolAqua

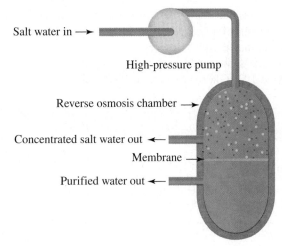

Salt water in →

High-pressure pump

Reverse osmosis chamber →

Concentrated salt water out ←

Membrane →

Purified water out ←

Figure 5.42

Water purification by reverse osmosis.

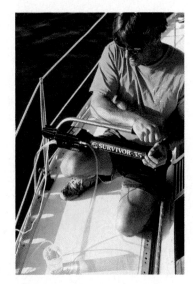

Figure 5.43

A small reverse-osmosis apparatus for converting seawater to potable water.

Courtesy of Katadyn

not. This is why the membrane is called "semipermeable." However, with an input of energy, osmosis can be reversed. **Reverse osmosis** uses pressure to force the movement of water through a semipermeable membrane from a more concentrated solution to a less concentrated one. Pressure is applied to the saltwater side, forcing water through the membrane to leave the salt and other impurities behind to purify water (Figure 5.42).

As expected, producing the required pressure in reverse osmosis systems is also energy intensive. Reverse-osmosis technology can be used to produce high quality drinking water for residential consumption and ultra-pure water used in the microelectronic and pharmaceutical industries. Portable units are also suitable for use on sailboats (Figure 5.43).

Your Turn 5.46   At What Cost?

An Internet blogger proclaimed, "Desalination will make it possible for us to get clean water. This will solve our water shortages." Revisit the green chemistry key ideas to help you refute these claims.

Point-of-Use Straws

With advances in sanitation and management of water-borne diseases over the past century, many high-income countries have access to high-quality drinking water that meets specific standards. However, more than two billion people worldwide are sickened or die yearly due to cholera, typhoid, and other diseases caused by microbes in untreated water. A European company, Vestergaard Frandsen, developed the LifeStraw, which removes virtually all bacteria and protozoan parasites from water. LifeStraws are used in many parts of the globe, including in times of need following a natural disaster.

Aptly named, the personal LifeStraw is a type of pipe filter through which to consume water, as shown in Figure 5.44. This unit can be used to drink from a stream, river, or lake. Lasting about three years, it can purify about 4000 liters of water. The larger LifeStraw Family unit contains a different filter that removes bacteria and improves water quality. The family unit filters up to 18,000 liters of water for a family of five and lasts three years.

However, the personal LifeStraw has limitations. It is not a long-term solution to the lack of potable water. In addition,

Figure 5.44

Children using personal LifeStraws to drink.

©Vestergaard Frandsen

it doesn't remove metals such as arsenic or mercury or the viral microbes responsible for diarrhea. Both types of LifeStraws provide an interim solution in regions where fresh water is contaminated with microbes.

You will explore some other needs and possibilities for water purification in the next activity, followed by a last visit to your water diary.

Your Turn 5.47   The Future of Water

a. Brainstorm two ideas that might help us keep water clean. What are some recent technologies that have been developed for water purification?
b. Identify an essential global water issue. Suggest two factors that make it important. Name two ways in which people currently are addressing this issue.

Your Turn 5.48   Final Analysis of the Water Diary

Revisit your water diary. After studying this chapter, what are your thoughts on your current usage of water? Suggest a different method for tracking your water usage data.

Conclusion

Like the air we breathe, water is essential to our lives. It bathes our cells, transports nutrients through our bodies, provides most of our body mass, and cools us when it evaporates. Water is also central to our way of life. We drink it, cook and clean things in it, use it to irrigate our crops, and manufacture goods with it. However, we add waste to the water as we do these things. Although freshwater purifies through a cycle of evaporation and condensation, humans are dirtying water faster than nature can regenerate clean water.

Furthermore, emissions of acidic oxides—carbon dioxide, sulfur oxides, and nitrogen oxides—are affecting the acidity of the world's oceans, rainfall, lakes, and rivers. In the United States, "acid rain" is not the dire plague once described by environmentalists and journalists. Nor is it a matter to be ignored. It was sufficiently serious that federal legislation, the Clean Air Act Amendments of 1990, were enacted to reduce SO_x and NO_x emissions, precursors to acid deposition. If you have learned anything from this chapter, we hope it has been the recognition that simple strategies cannot solve complex problems. Any failure to acknowledge the intertwined relationships involving the combustion of coal and gasoline, the production of carbon, sulfur, and nitrogen oxides, and the reduced pH of seawater and precipitation, is to deny some fundamental chemistry facts. Knowledge of ecology and biological systems is also needed so that acid deposition can be understood in the context of entire ecosystems, a task requiring experts from several disciplines to collaborate.

Although fresh water is a renewable resource, the demands of population growth, rising affluence, and other global issues amplify shortages of this essential commodity. If our personal, national, and global appetite for fossil fuels grows unchecked, our environment may become warmer and more acidic. Moreover, the problem may be intensified as the supply of petroleum and low-sulfur coals diminish, and we become even more reliant on high-sulfur coal.

In the following chapters, we will discuss the energy produced by fossil fuels and renewable energy sources—nuclear fission, water and wind, biomass, and the Sun itself. All are currently being utilized, and their use will no doubt increase. But we conclude this chapter with the modest suggestion that, for a multitude of reasons, the conservation of energy by industry and collectively by individuals could have profoundly beneficial effects on our environment, including the water we rely on for life itself.

LEARNING OUTCOMES   The numbers in parentheses indicate the sections within the chapter where these outcomes were discussed.

Having studied this chapter, you should now be able to:

- predict the shape and polarity of molecules and describe their role on physical properties (5.1)
- predict the intermolecular forces involved among polar molecules and describe their role in physical properties (5.2)

- identify sources of water on Earth and the relative availability of fresh water (5.3)
- describe ways that water may become contaminated and analyze data to evaluate water use consumption, and level of contamination (5.3)
- describe the composition of aqueous solutions (5.4)

- interpret rules to provide the names of ionic compounds (5.4)
- calculate the concentration of solutions in various concentration units (5.5)
- describe and model the solvation of ionic and molecular compounds in water (5.6)
- apply the "like dissolves like" principle to predict whether a substance will be soluble in water (5.7)
- understand how certain chemicals can contaminate water sources (5.7)

- identify and classify the species involved in acid–base reactions (5.8)
- describe how acids and bases react with one another (5.9)
- quantify the relative acidity or basicity of a solution in terms of the pH scale (5.10)
- explain the causes and effects of ocean acidification (5.11)
- describe how water can be treated to become safer for use or consumption (5.12)
- demonstrate some strategies used to purify water (5.13)

Questions

Emphasizing Essentials

1. In any language, water is the most abundant compound on the surface of Earth.

   a. Explain the term *compound* and also why water is *not* an element.

   b. Draw the Lewis structure for water and explain why its shape is bent.

2. The following are four pairs of atoms. Consult Table 5.1 to answer these questions.

N and C	S and O
N and H	S and F

   a. What is the electronegativity difference between the atoms?

   b. Assume that a single covalent bond forms between each pair of atoms. Which atom attracts the electron pair in the bond more strongly?

   c. Arrange the bonds in order of increasing polarity.

3. Consider a molecule of ammonia, NH_3.

   a. Draw its Lewis structure.

   b. Does the NH_3 molecule contain polar bonds? Explain.

   c. Is the NH_3 molecule polar?

   Hint: Consider its geometry.

   d. Would you predict NH_3 to be soluble in water? Explain.

4. Both methane (CH_4) and water are compounds of hydrogen and another nonmetal.

   a. Give four examples of nonmetals. In general, how do the electronegativity values of nonmetals compare with those of metals?

   b. How do the electronegativity values of carbon, oxygen, and hydrogen compare?

   c. Which bond is more polar, the C—H bond or the O—H bond? Justify your answer.

   d. Methane is a gas at room temperature, but water is a liquid. Explain.

5. This diagram represents two water molecules in a liquid state. What kind of bonding force does the arrow indicate?

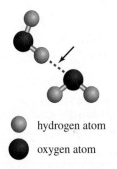

   hydrogen atom

   oxygen atom

6. **a.** Draw the Lewis structure for the water molecule.

   b. Draw Lewis structures for the hydrogen ion and the hydroxide ion.

   c. Write a chemical reaction that relates all three structures from parts **a** and **b**.

7. The density of water at 0 °C is 0.9987 g/cm^3; the density of ice at this same temperature is 0.917 g/cm^3.

   a. Calculate the volume occupied at 0 °C by 100.0 g of liquid water and by 100.0 g of ice.

   b. Calculate the percentage increase in volume when 100.0 g of water freezes at 0 °C.

8. The density of CO_2 at 1 atm is 0.001977 g/cm^3; the density of CO_2 at 49.3 atm is 0.153 g/cm^3.

   a. Calculate the volume occupied by 100.0 g of CO_2 at 1 and 50 atm.

   b. The density increases for gases as the pressure increase. Why do you think this is?

9. **a.** Using dimensional analysis, show how ppb could be denoted as µg/L. How would you express an even smaller unit, parts per trillion (ppt), in terms of mass per liter?

   b. For a solution concentration of 1 ppm, how much water would you need for 1 pound of a solute? How about for a concentration of 1 ppb?

10. Consider these ions: nitrate, sulfate, carbonate, and ammonium.

    a. Give the chemical formula for each.

    b. Write a chemical equation in which the ion (in aqueous form) appears as a product.

11. Write a chemical equation that shows the release of hydroxide ions as each of these bases dissolves in water.

    a. $KOH(s)$, potassium hydroxide

    b. $Ba(OH)_2(s)$, barium hydroxide

12. Calculate the $[OH^-]$ that corresponds to the given $[H^+]$ or vice versa for the species below. Classify the aqueous solutions as acidic, neutral, or basic.

    a. $[H^+] = 1 \times 10^{-8}$ M

    b. $[OH^-] = 1 \times 10^{-2}$ M

    c. $[H^+] = 5 \times 10^{-7}$ M

    d. $[OH^-] = 1 \times 10^{-12}$ M

13. In each pair below, the $[H^+]$ is different. By what factor of 10 is it different?

    a. $pH = 6$ and $pH = 8$

    b. $pH = 5.5$ and $pH = 6.5$

    c. $[H^+] = 1 \times 10^{-8}$ M and $[H^+] = 1 \times 10^{-6}$ M

    d. $[OH^-] = 1 \times 10^{-2}$ M and $[OH^-] = 1 \times 10^{-3}$ M

14. Which of these has the *lowest* concentration of hydrogen ions: 0.1-M HCl, 0.1-M NaOH, 0.1-M H_2SO_4, or pure water? Explain your answer.

15. Based on the generalizations in Table 5.8, which compounds are likely to be water-soluble?

    a. $KC_2H_3O_2$ b. $LiOH$

    c. $Ca(NO_3)_2$ d. Na_2SO_4

16. For a 2.5-M solution of $Mg(NO_3)_2$, what is the concentration of each ion present?

17. Calcium carbonate is a salt. Write its chemical formula. Would you expect calcium carbonate to be soluble or insoluble in water?

18. Classify the following aqueous solutions as acidic, neutral, or basic.

    a. $HI(aq)$

    b. $NaCl(aq)$

    c. $NH_4OH(aq)$

19. Write a chemical equation that shows the release of one hydrogen ion from a molecule of each of these acids.

    a. $HBr_{(aq)}$, hydrobromic acid

    b. $H_2SO_{3(aq)}$, sulfurous acid

    c. $CH_2H_3O_{2(aq)}$, acetic acid

20. Explain how you would prepare these solutions using powdered reagents and any necessary glassware.

    a. Two liters of 1.50-M KOH

    b. One liter of 0.050-M NaBr

    c. 0.10 L of 1.2-M $Mg(OH)_2$

Concentrating on Concepts

21. In some cases, the boiling point of a substance increases with its molar mass.

    a. Does this hold true for hydrocarbons? Explain with examples.

    b. Based on the molar masses of H_2O, N_2, O_2, and CO_2, which would you expect to have the lowest boiling point?

    c. Unlike N_2, O_2, and CO_2, water is a liquid at room temperature. Explain.

22. Consider these liquids:

Liquid	Density, g/mL
dishwashing detergent	1.03
maple syrup	1.37
vegetable oil	0.91

    a. If you pour equal volumes of these three liquids into a 250-mL graduated cylinder, in what order should you add the liquids to create three separate layers? Explain.

    b. Predict what would happen if a volume of water equal to the other liquids was poured into the cylinder in part a, and the contents were mixed vigorously.

23. Let's say the water in a 500-L drum represents the world's total supply. How many liters would be suitable for drinking?

24. Based on your experience, how soluble is each of these substances in water? Use terms such as *very soluble, partially soluble,* or *not soluble.* Cite supporting evidence.

    a. orange juice concentrate

    b. household ammonia

    c. chicken fat

    d. liquid laundry detergent

    e. chicken broth

25. a. Bottled water consumption was reported to be 29 gallons per person in the United States in 2011. The 2010 U.S. census reported the population as 3.1×10^8 people. Given this, estimate the total bottled water consumption.

    b. Convert your answer in part **a** to liters.

26. NaCl is an ionic compound, but $SiCl_4$ is a molecular compound.

    a. Use Table 5.1 to determine the electronegativity difference between chlorine and sodium and between chlorine and silicon.

    b. What correlations can be drawn about the difference in electronegativity between bonded atoms and their tendency to form ionic or covalent bonds?

    c. How can you explain, on the molecular level, the conclusion reached in part **b**?

27. The maximum mercury contaminant level (MCL) in drinking water is 0.002 mg/L.

    a. Does this correspond to 2 ppm or 2 ppb mercury?

    b. Is this mercury in the form of elemental mercury ("quicksilver") or the mercury ion (Hg^{2+})?

28. The acceptable limit for nitrate, often found in well water in agricultural areas, is 10 ppm. If a water sample is found to contain 350 mg/L, does it meet the acceptable limit?

29. A student weighs 5.85 g of NaCl to make a 0.10-M solution. What size volumetric flask do they need?

30. Solutions can be tested for conductivity using this type of apparatus.

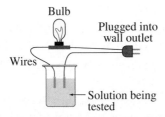

Predict what will happen when each of these dilute solutions is tested for conductivity. Explain your predictions briefly.

    a. $CaCl_2(aq)$ **b.** $C_2H_5OH(aq)$ **c.** $H_2SO_4(aq)$

31. An aqueous solution of KCl conducts electricity, but an aqueous solution of sucrose does not. Explain.

32. **a.** A 5-minute shower requires about 90 L of water. How much water would you save for each minute that you shorten your shower?

    b. Running the water while you brush your teeth can consume another liter. How much water can you save in a week by turning it off?

33. Which gas is dissolved in water to produce each of the following acids?

    a. carbonic acid, H_2CO_3

    b. sulfurous acid, H_2SO_3

34. Write a balanced chemical equation for each of the following acid-base reactions.

    a. Potassium hydroxide is neutralized by nitric acid.

    b. Hydrochloric acid is neutralized by barium hydroxide.

    c. Sulfuric acid is neutralized by ammonium hydroxide.

35. Give names and chemical formulas for five bases of your choice. Name three observable properties generally associated with bases.

36. Give names and chemical formulas for five acids of your choice. Name three observable properties generally associated with acids.

37. Describe the relationship between the strength of an acid ("weak" vs. "strong") and its concentration. For instance, do high concentrations of an acid mean that the acid is considered "strong" (and low concentrations for "weak" acids)?

38. Use the Internet to determine which has the higher water footprint, a 100-gram chocolate bar or a 16-ounce glass of beer. Explain the difference.

39. Explain why water is often called the *universal solvent*.

40. Is there any such thing as "pure" drinking water? Discuss what is implied by this term and how the meaning of this term might change in different parts of the world.

41. Some vitamins are water-soluble, whereas others are fat-soluble. Would you expect either or both to be polar molecules? Explain.

42. At the edge of a favorite fishing hole, a new sign reads, "Caution: Fish from this lake may contain over 1.5 ppb Hg." Explain to a fishing buddy what this concentration unit means and why the caution sign should be heeded.

43. This periodic table contains four elements identified by numbers.

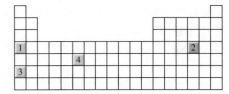

    a. Based on the trends you observe from Table 5.1, which of the four elements would you expect to have the highest electronegativity value? Explain.

    b. Based on trends within the periodic table, rank the other three elements in order of decreasing electronegativity values. Explain your ranking.

44. A diatomic molecule XY that contains a polar bond *must* be a polar molecule. However, a triatomic molecule XY_2 that contains a polar bond *does not necessarily* form a polar molecule. Use some examples of actual molecules to help explain this difference.

45. Imagine you are at the molecular level, watching water vapor condense.

    a. Sketch four water molecules using a space-filling representation similar to this one. Sketch them in the gaseous state and then in the liquid state. How does the collection of molecules change when water vapor condenses to a liquid?

    b. What happens at the molecular level when water changes from a liquid to a solid?

46. Propose an explanation for the fact that NH_3, like H_2O, has an unexpectedly high specific heat.

47. **a.** What type of bond holds together the two hydrogen atoms in the hydrogen molecule, H_2?

    b. Explain why the term *hydrogen bonding* does *not* apply to the bond within H_2.

    c. Explain why the term *hydrogen bonding* does *not* apply to the bond within H_2O, but *does* apply to a sample of water.

48. Consider ethanol, an alcohol with the chemical formula of C_2H_5OH.

    a. Draw the Lewis structure for ethanol.

    b. A cube of solid ethanol sinks rather than floats in liquid ethanol. Explain this behavior.

Exploring Extensions

49. Today, we are creating dirty water faster than nature can clean it for us.

    a. Name five daily activities that dirty the water.

    b. Name two ways in which polluting substances naturally are removed from water.

    c. Name five steps you could take to keep water cleaner in the first place.

50. Life on our planet depends on water. Explain each of the following statements.

    a. Bodies of water act as heat reservoirs, moderating the climate.

    b. Ice protects ecosystems in lakes because it floats rather than sinks.

    c. A water pipe will burst if filled with water when it freezes.

51. The unusually high specific heat of water helps keep our body temperature within a normal range despite age, activity, and environmental factors. Consider some of the ways the body produces and loses heat. How would these differ if water had a low specific heat?

52. If an insoluble salt is added to pure water, will the resulting solution conduct electricity? Explain.

53. As we have discussed in this section, gaseous NO_x and SO_x combine with water molecules to form aerosols of acid rain.

    a. Write an equation that describes how marble limestone, composed primarily of calcium carbonate, $CaCO_3$, may be eroded from contact with acid rain droplets.

    b. What effect does acid rain have on metals such as iron? What can be done to protect metals against the effects of acid rain?

54. a. Assume that coal has the chemical formula of $C_{135}H_{96}O_9NS$. Calculate the coal's fraction and percent (by mass) of sulfur.

    b. A certain power plant burns 1.00×10^6 tons of coal per year. Assuming the sulfur content calculated in part **a**, calculate the tons of sulfur released per year.

    c. Calculate the tons of SO_2 formed from the balanced reaction of this amount of sulfur with oxygen in the atmosphere.

    d. Once released into the atmosphere, the SO_2 is likely to react with oxygen to form SO_3 (revisit Equation 2.7). What can happen next if SO_3 encounters water droplets?

55. Health goals for contaminants in drinking water are expressed as MCLG or maximum contaminant level goals. Legal limits are given as MCL or maximum contaminant levels. How are MCLG and MCL related for a given contaminant?

56. Some areas have a higher than normal amount of THMs (trihalomethanes) in the drinking water. Suppose you are considering moving to such an area. Write a letter to the local water district asking relevant questions about the drinking water.

57. Infants are highly susceptible to elevated nitrate levels because bacteria in their digestive tract convert nitrate ions into nitrite ions, a much more toxic substance.

    a. Give chemical formulas for both the nitrate ion and nitrite ion.

    b. Nitrite ions can interfere with the ability of blood to carry oxygen. Explain the role of oxygen in respiration.

    c. Boiling nitrate-containing water will not remove nitrate ions. Explain.

58. In 2016, testing indicated that the drinking water supply in Flint, Michigan, had concentrations of dissolved lead that far exceeded those established by the Safe Drinking Water Act.

    a. What was the major source of lead in the drinking water?

    b. What are the possible health effects of elevated levels of lead?

    c. Do some research on what is being done to improve water quality in this community.

    d. What other cities or states have recently reported high levels of lead?

59. Explain why desalination techniques, despite proven technological effectiveness, are not used more widely to produce potable drinking water.

60. In 2015, the Great Lakes-St. Lawrence River Basin Sustainable Water Resources Agreement set the stage to coordinate water management and protect water from the use by those outside the region.

    a. List states and provinces involved with this unique transboundary agreement.

    b. What was the impetus behind protecting these waters?

61. Liquid CO_2 has been used successfully for many years to decaffeinate coffee. Explain how and why this works.

62. How can you purify your water when you are hiking? Name two or three possibilities. Compare these methods in terms of cost and effectiveness. Are any of these methods similar to those used to purify municipal water supplies? Explain.

63. Hydrogen bonds vary in strength from about 4 to 40 kJ/mol. Given that the hydrogen bonds between water molecules are at the high end of this range, how does the strength of a hydrogen bond *between* water molecules compare with the strength of an H—O covalent bond *within* a water molecule? Do your values bear out the assertion that hydrogen bonds are about one-tenth as strong as covalent bonds?

64. Levels of naturally occurring mercury in surface water are usually less than 0.5 mg/L.

    a. Name three human activities that add Hg^{2+} ("inorganic mercury") to water.

    b. What is "organic mercury"? This chemical form of mercury tends to accumulate in the fatty tissues of fish. Explain why.

65. We all have the amino acid glycine in our bodies. Here is the structural formula.

$$
\begin{array}{ccc}
& H & O \\
& | & \| \\
H-N-&C-&C-O-H \\
| & | & \\
H & H &
\end{array}
$$

    a. Is glycine a polar or nonpolar molecule? Explain.

    b. Can glycine exhibit hydrogen bonding? Explain.

    c. Is glycine soluble in water? Explain.

66. Hard water may contain Mg^{2+} and Ca^{2+} ions. The process of water softening removes these ions.

    a. How hard is the water in your local area? One way to answer this question is to determine the number of water-softening companies in your area. Use the Internet to find out if your area is targeted for the marketing of water-softening devices.

    b. If you chose to treat your hard water, what are the options?

67. Suppose you are in charge of regulating an industry in your area that manufactures agricultural pesticides. How will you decide if this plant is obeying necessary environmental controls? Which criteria affect the success of this plant?

68. Before the U.S. EPA banned manufacturing in 1979, PCBs were considered valuable chemicals. What properties made them desirable? Besides being persistent in the environment, they bioaccumulate in the fatty tissues of animals. Use the electronegativity concept to show why PCB molecules are nonpolar and thus fat-soluble.

69. The PUR "Purifier of Water" is a point-of-use system.

    a. How does this system work?

    b. Compare it to the personal LifeStraw by listing the benefits offered by each system.

70. In the United States, the EPA has set SMCLs (secondary maximum contaminant levels) for substances in drinking water that are not health threatening. Visit the EPA website to learn more about one of these substances and prepare a summary of your findings.

71. The EPA uses an extensive process to add contaminants to its list of regulated substances. Search the Internet for Unregulated Contaminant Monitoring (UCM) program information.

    a. What is the UCM, and when does it occur?

    b. What is the importance of its Contaminant Candidate List (CCL), and how does it relate to the precautionary principle?

    c. List some general categories of substances included in the CCL. Include one specific substance from the most current list.

72. List a recent theme for World Water Day. Prepare a short presentation of this theme in a format of your choice.

73. "Alkaline water" has been purported to be better for your health than ordinary filtered or tap water. Prepare a report that explains whether this is just marketing hype, or are there truly benefits to drinking this water.

74. The company that produces LifeStraws has a set of FAQs on the Internet. One reads:

    "Does LifeStraw filter heavy metals like arsenic, iron, and fluoride?" What might the "informed chemist" say about the phrasing of this answer: "No, the present version does not filter any of these heavy metals"?

75. A personal LifeStraw is a filter used to remove the worst contaminants in water. Why is this helpful? Where in the world would this help the most?

76. The average American drinks 0.5 gallons of water daily (or 1.9 liters). If the personal LifeStraw filters 4000 liters before it is used up, how long would it last an American?

6 Energy from Combustion

WHAT DOES IT TAKE?

Watch the opening chapter video (www.acs.org/cic) and answer the following questions.

a. Imagine you are going to take a road trip across the United States, from New York City to Los Angeles. The distance of this trip is 4460 km and the vehicle you will use gets 30 miles per gallon of gasoline. How much gasoline would you need for this trip?

b. Now, imagine that the car you are using has the capability of also using an alternative, renewable fuel such as biodiesel or ethanol. However, using ethanol is less efficient than using gasoline. In other words, when using ethanol, you can achieve only 20 miles per gallon. How many gallons of ethanol would you need to cover your entire trip?

c. Currently, the ethanol used for fuel is predominantly made from corn. Assume it takes 26.1 pounds of corn to make 1 gallon of ethanol, and 1 acre of land to produce 7110 pounds of corn. How much land is needed to produce sufficient fuel for your trip?

In this chapter, you will explore the following questions:

1. Why are fossil fuels used to generate energy?
2. What happens when you burn a fuel?
3. What's the difference between energy, heat, and temperature?
4. How can we measure the energy released from combustion?
5. How is energy released during a combustion reaction?
6. How is electricity generated in a fossil fuel power plant?
7. What are the efficiencies of various energy sources?
8. What are the environmental implications of obtaining and using fossil fuels—coal, oil, and natural gas?
9. How can petroleum be made into so many products?
10. What's in gasoline?
11. How can coal be used to make liquid fuel?
12. What are the arguments for and against using ethanol and biodiesel as alternatives to petroleum products?
13. How can we assess the sustainability of fuels?

Introduction

Our modern fuels, the substances we burn or combust, are available in many different forms and are used in a variety of ways. Reflect on your use of fuels in your everyday life in this next activity.

Your Turn 6.1 Combustion in Your Daily Life

What are four examples where combustion provides energy in your daily life?

In this chapter, we will describe fuels and their characteristics, including alternative fuels, to help us meet our current and future energy needs. We begin by describing the properties of fuels and what happens when fuels are burned. We then take you inside a power plant to illustrate how the combustion of fuels is converted to electricity. Because there are many types of fuels, we will also describe how their varying compositions affect not only heat output, but also the gaseous products generated from their combustion. Let's begin our analysis by answering the following questions: What is a fuel, and what happens when it burns?

Don Farrall/Photodisc/Getty Images

6.1 | Fossil Fuels: A Prehistoric Fill-Up at the Gas Station

Learning Objectives:

Identify the characteristics of ideal fuels

Describe the formation of fossil fuels and why they are unsustainable

A **fuel** is any solid, liquid, or gaseous substance that may be combusted (burned) to produce heat or work. Sources of fuel date back to prehistoric times, where solids such as grass and straw were burned for heat. The use of coal as a fuel actually dates back to ancient civilizations, where it was used to isolate copper from ore in northeastern China as early as 1000 BCE. However, the Industrial Revolution in the late 18th century sparked the large-scale use of coal for steam engines and steelmaking (Figure 6.1). The development of drilling technology for oil wells in the mid-19th century gave rise to the petroleum industry and mass consumption of petroleum products for transportation, electricity, heating, and plastics fabrication. More recently, there has been a rise in the use of natural gas because it is seen as a cleaner alternative to coal and oil; technological advances have made it easier to access natural gas deposits. Currently, most of the world's energy needs are provided by burning fossil fuels. With so many choices of fuels available to us, what are some desirable properties of fuels?

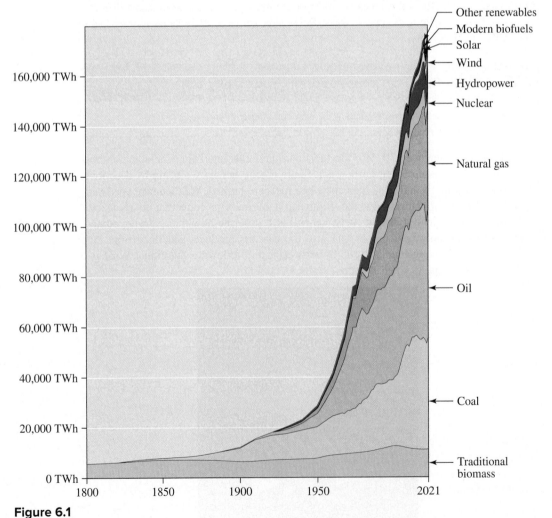

Figure 6.1

Global primary energy consumption by source since 1800 in terrawatt-hours (TWh), or 10^12 watt-hours.

Our World in Data based on Vaclav Smil (2017) and BP Statistical Review of World Energy. For an interactive version, see: www.acs.org/cic.

Your Turn 6.2   The "Best" Fuel?

Consider the various types of fuels that have been used since ancient times.

alexmaster/123RF

a. What properties make a particular fuel desirable or undesirable?

b. What makes a good match between a fuel and its intended use?

c. With the above considerations in mind, propose some reasons why industrialized nations shifted from using wood to coal, and then eventually to petroleum and natural gas as the chief sources of fuel?

d. What fuel properties will be needed to take the place of petroleum in the future?

A fuel is considered valuable if it ignites easily at a low temperature and produces a large quantity of heat during its combustion. In addition, fuels should be inexpensively isolated and have properties that allow for their safe and efficient storage and transport. Lastly, a desirable fuel should leave little residue behind after being burned, and should produce by-products that are not harmful to human health or the environment. Unfortunately, no fuel satisfies all of these conditions.

Humans currently use coal, petroleum products (e.g., gasoline, diesel, propane), and natural gas as our primary sources of fuel. Contrary to a pervasive myth, these so-called **fossil fuels** are not the prehistoric remains of dinosaurs. In fact, most of the fossil fuels we use today were formed from decaying plant life and microorganisms that flourished millions of years before the first dinosaurs appeared.

The process of photosynthesis exhibited by green plants, including primeval plant life, involves the capture of sunlight to produce glucose and oxygen from carbon dioxide and water:

$$6\ CO_2(g) + 6\ H_2O(l) \longrightarrow C_6H_{12}O_6(aq) + 6\ O_2(g) \qquad \text{[6.1]}$$

The solar energy is stored in the chemical bonds of the organic matter (e.g., glucose).

The process of decay or respiration reverses this process, breaking down the organic matter into $CO_2(g)$ and $H_2O(l)$ using $O_2(g)$ and releasing energy.

$$\text{Respiration: } 6\ O_2 + C_6H_{12}O_6 \longrightarrow 6\ CO_2 + 6\ H_2O \qquad \text{[6.2]}$$

Your Turn 6.3   Controlling Decay: Composting

Read this article about composting and differentiate between aerobic and anaerobic composting: www.acs.org/cic. Why might a compost pile be steaming?

Under certain conditions, the carbon-containing compounds that make up the organism only *partially* decompose. In the prehistoric past, vast quantities of plant and animal life became buried beneath layers of sediment in swamps or at the bottom of the oceans. Oxygen failed to reach the decaying material, thus slowing the decomposition process. The temperature and pressure increased as additional layers of mud and rock covered the buried remnants, causing additional chemical reactions to occur. Over time, the plants and photosynthesizing microbes that once captured the Sun's rays were transformed into the substances we call coal, petroleum, and natural gas. In a very real sense, these fossils may be thought of as ancient solar energy (sunshine) stored in solid, liquid, and gaseous states.

Yes, today's photosynthesizers will become tomorrow's fossil fuels. But this will not occur in a time frame useful to humans. It is staggering to realize that the amount of fossil fuels we will have consumed in just a few centuries took nature several millennia to produce. Over geological periods of time—millions and millions of years—changes in temperature and pressure transformed partially decomposed remnants of dead plant and animal life into valuable forms, such as coal or crude oil.

Your Turn 6.4   Where Are the Fossil Fuels?

Consult the interactive trends found in the International Energy Portal link: www.acs.org/cic.

a. Which countries are the world's largest producers of coal, petroleum, and natural gas?
b. Do these countries also have the greatest reserves of these fossil fuels?
c. Propose reasons why the rankings for **a** and **b** are not identical. For instance, Canada is ranked #3 in petroleum resources, but has a production ranking of #4; in contrast, the United States is ranked #10 for world petroleum reserves, but is the #1 producer.
d. Have the trends in production and reserve capacity of particular countries changed over the last 10 years? Propose some socioeconomic factors that may explain these trends.

Considering how long it takes to convert plant life to fossil fuels, the rate at which we are burning coal, petroleum, and natural gas is not sustainable—at least in terms of having enough of it available to meet current and future energy needs. On the other hand, fossil fuels appear to be in adequate supply, because new deposits are always being found and extraction technologies are continually improving to capture more of it. But even if the supply of fossil fuels is infinite (it is not), sustainability involves more than just availability.

Will We Run Out of Fossil Fuels?

Watch this video (www.acs.org/cic) and answer the following questions.

a. What are the "ingredients" required to make fossil fuels?
b. Why are fossil fuels "nonrenewable" even though they are made from decayed biomass?

Using a variety of Internet sources, predict some likely time frames that will exhaust our available supply of nonrenewable coal, petroleum, and natural gas. What factors may worsen or improve these predictions?

Burning fossil fuels for energy fails to meet the criteria of sustainability in two ways. First, the fuels themselves are nonrenewable. Once gone, they cannot be replaced—at least within a useful timescale. Second, the waste products of combustion have adverse effects on our environment, both today and in the future. Chapter 4 described how the concentration of atmospheric $CO_2(g)$—a greenhouse gas—has risen dramatically since the beginning of the Industrial Revolution. These increases will continue to affect our climate. Burning coal also releases pollutants such as soot, carbon monoxide, mercury, and the oxides of sulfur and nitrogen. These emissions affect us right now because they lower the quality of our air, acidify our rain, aggravate existing health conditions, and accelerate global climate change. Before we consider what we can do about this, let's first consider what happens to a fossil fuel when it is burned and why it provides so much of our current energy needs.

@HOME   By-Products of Candle Burning

Investigate the by-products of candle burning at home with an oven-safe clear glass dish, a tea light candle, and a lighter or match.

1. Light the tea candle and make observations as the candle burns.
2. Next, carefully hold the oven-safe clear glass dish over the flame so that the flame just touches it for up to five minutes.
3. Extinguish the flame and make observations.

Ansis Klucis/Shutterstock

Questions to Consider
a. What is the substance that collects on the bottom of the bowl?
b. Where does the substance on the bottom of the bowl go if it is not collected on the bowl? How does this relate to burning coal?

6.2 | Burn, Baby! Burn!
The Process of Combustion

Learning Objectives:

Understand the combustion process and describe it using balanced chemical equations

Write structural formulas for hydrocarbons

The Fire Triangle

Watch this video (www.acs.org/cic) and explain the three components that make up the fire triangle. Include ways to eliminate each component to extinguish a fire.

There are three necessary requirements to generate a fire—a source of heat, a fuel, and an oxidizer. When these components are combined, a chemical reaction takes place that releases a variety of by-products and a significant amount of heat.

@HOME   A DIY Fire Extinguisher

Investigate the fire triangle by making a fire extinguisher at home using baking soda, vinegar, a tea light candle, a match or lighter, a cup that has a lid and a straw, and tape.

1. Tape the area around the straw so that the straw stays in the same position.
2. Add 5 tablespoons of vinegar to the cup.
3. Light the tea candle.
4. Add the baking soda to the cup very quickly and put the lid on. Quickly point the straw at the flame.
5. Make observations about what you see.

Monty Rakusen/Alamy Stock Photo

Questions to Consider
a. What gas is produced by the reaction in the cup?
b. Which part of the fire triangle have you removed by pointing this gas at the flame?

Regardless of the specific source of fuel or oxidizing agent, the general chemical reaction is the same (Equation 6.3). As in all chemical reactions, the species on the left-hand side of the chemical equation are referred to as **reactants** (i.e., fuel and oxidizer molecules for a combustion reaction), whereas those appearing on the right-hand side are known as **products**:

$$\text{Fuel} + \text{Oxidizer} \xrightarrow{heat} \text{Products} \qquad \text{[6.3]}$$

However, the identity of the products will differ, depending on the fuel and oxidizer used for combustion.

The great majority of fuels are **hydrocarbons,** compounds composed only of the elements hydrogen and carbon. Let's review a few basic rules for chemical bonding. The octet rule, introduced in Section 3.6, indicates that carbon atoms in hydrocarbons form chemical bonds that share eight valence electrons. For example, in methane (CH_4), the primary component of natural gas, the central C atom is surrounded by eight electrons that are arranged to form four covalent bonds:

$$
\begin{array}{c}
\text{H} \\
| \\
\text{H} - \text{C} - \text{H} \\
| \\
\text{H}
\end{array}
$$

Your Turn 6.5   Verifying the Octet Rule

Example Problem:

Verify that the structure of octane satisfies the octet rule.

Example Solution:

The Lewis structure of octane is:

$$
\begin{array}{c}
\text{H H} \quad \text{H H} \quad \text{H H} \quad \text{H} \\
\text{H} \diagdown \diagup \diagdown \diagup \diagdown \diagup \diagdown \text{H} \\
\text{C} \quad \text{C} \quad \text{C} \quad \text{C} \quad \text{C} \\
\text{H} \diagup | \diagup \diagdown \diagup \diagdown \diagup \diagdown \text{H} \\
\text{H} \quad \text{H H} \quad \text{H H} \quad \text{H H}
\end{array}
$$

Since each carbon is surrounded by four bonds (8 electrons), the carbons satisfy the octet rule. Also, each hydrogen contains one bond (2 electrons), which also satisfies its bonding requirements.

Your Turn:

Look up the formulas and structures for the following hydrocarbons and verify that each carbon satisfies the octet rule.

a. ethane
b. ethene (also known as ethylene)
c. ethyne (also known as acetylene)
d. benzene

Remember, chemical formulas, such as CH_4, C_2H_4, or C_8H_{18}, indicate the kinds and numbers of atoms present in a molecule, but do not show how the atoms are connected. To get this level of detail, you need a **structural formula.** For example, the structural formula for *n*-butane, C_4H_{10}, a hydrocarbon used to fuel lighters and camp stoves, is shown below. The *n* in the chemical name stands for *normal*, meaning that the carbon atoms are in a straight chain:

$$
\begin{array}{c}
\text{H} \quad \text{H} \quad \text{H} \quad \text{H} \\
| \quad\; | \quad\; | \quad\; | \\
\text{H} - \text{C} - \text{C} - \text{C} - \text{C} - \text{H} \\
| \quad\; | \quad\; | \quad\; | \\
\text{H} \quad \text{H} \quad \text{H} \quad \text{H}
\end{array}
$$

As we discussed in Section 4.5, each of the carbon atoms features a tetrahedral arrangement of C—H and C—C bonds.

A drawback to structural formulas is that they take up a lot of space on the page. To convey the same information more compactly, **condensed structural formulas** are often used. In these forms, the structural formula is understood to contain an appropriate number of bonds to satisfy the octet rule. Here are two condensed structural formulas for *n*-butane, the second more "condensed" than the first:

$$CH_3-CH_2-CH_2-CH_3 \qquad CH_3CH_2CH_2CH_3$$

Although the H atoms in these structures appear to be part of the chain of C atoms, it is understood that they are not, and instead are connected to the prior carbon atom.

If a pure hydrocarbon fuel is burned, the only products that will be generated upon complete combustion are carbon dioxide and water vapor. As an example, consider the combustion of *propane*, C_3H_8:

$$C_3H_8(g) + 5\ O_2(g) \longrightarrow 3\ CO_2(g) + 4\ H_2O(g) \qquad \qquad \textbf{[6.4]}$$

However, in reality, there is likely some carbon monoxide ($CO(g)$) and/or carbon soot generated due to *incomplete* combustion. For combustion of propane, this would occur if the molar ratio of $O_2:C_3H_8$ is less than 5:1, referred to as *oxygen-deficient* conditions.

Propane, coal, and many other fuels also contain sulfur, which results in sulfur oxide emissions (SO_x, where x = 2 or 3). Furthermore, nitrogen oxides (NO_x, where x = 1 or 2) are generated through the high-temperature reaction of $N_2(g)$ and $O_2(g)$ during combustion. Recall the role of NO_x emissions in the formation of harmful ground-level ozone, described in Section 2.12. As we will see throughout this chapter, a diversity of products are possible from the combustion of a fuel.

Your Turn 6.6 Practice with Combustion Reactions

For each of the fuels below, write the balanced combustion reaction.

a. Glucose, sugar ($C_6H_{12}O_6$)
b. Methane, natural gas (CH_4)
c. Butane, fuel in lighters (C_4H_{10})

Hint: Go back to Section 2.9 for information on balancing equations.

Your Turn 6.7 Combustion in Air

The balanced equation presented in Equation 6.4 describes the combustion of propane in pure oxygen. However, this combustion in real life occurs in air, which is predominantly composed of nitrogen. How would you write a balanced equation to illustrate the combustion of propane in air? Assume that the nitrogen is *inert*—that is, it doesn't react with propane or oxygen.

6.3 | What Is "Energy"?

> *Learning Objective: Explain terms relating to types of energy, heat, and temperature with respect to molecular changes*

Creating fire was a critical discovery that promoted the survival of our early Paleolithic ancestors. The energy generated from burning fuels keeps us warm and powers our automobiles, lights, and appliances. In Chapter 3, you explored a form of energy known as UV radiation and its effect on human health. However, what exactly is "energy" and where does it come from?

Energy is a fundamental property of our universe, referring to the ability or capacity of matter to do work or to produce change. For instance, the combustion of fuel in a car drives an engine that turns the wheels. Alternatively, energy is given off as heat and light when fireworks are ignited, or from nuclear reactions occurring in the Sun.

Although there are many different forms of energy, there are two general types. **Kinetic energy** is the energy of motion, which includes the movement of atoms and molecules, as well as our own activities such as walking, climbing, or running. Electromagnetic (EM) radiation results from the movement of electric and magnetic fields; hence, it is best classified as a form of kinetic energy. In contrast, **potential energy** is stored energy, or the energy of position; that is, position with respect to another object. Because different fuels have different compositions and arrangements of atoms, they hold different amounts of potential energy. The magnitude of this energy is dependent on the types of bonds present in the fuel molecules. During combustion reactions, the relative positions of the atoms in the fuel change, which alters its potential energy. Explore how atomic positions affect the potential energy of a molecule in this next interactive activity.

Your Turn 6.8   Too Close for Comfort: The Potential Energy of Diatomic Molecules

This simulation explores the relationship between the potential energy of a diatomic molecule and the distance between its atoms: www.acs.org/cic. Choose a molecule and move the atoms closer together and farther apart using the sliders. What do you notice happens to the potential energy of the molecule as the atomic distance is changed? Explain this general trend (recall that atomic nuclei are positively charged and electrons are negatively charged).

PhET Interactive Simulations, University of Colorado

PhET Interactive Simulations, University of Colorado, http://phet.colorado.edu

---

As a fuel is burned, the energy stored in its atoms and molecules is converted into heat, a form of kinetic energy. The **first law of thermodynamics**, also called the *law of conservation of energy*, states that energy is neither created nor destroyed. It implies that although the forms of energy change, the total amount of energy before and after any transformation remains the same.

@HOME   Shaking Sand

Investigate the first law of thermodynamics at home using sand, two foam cups, tape, and a food thermometer.

1. Add 1 cup of sand to one of the foam cups.
2. Record the temperature of the sand in the foam cup using the thermometer.
3. Turn the other foam cup over and tape it to the lip of the foam cup containing sand.
4. Shake the sealed cup system for 5 to 10 minutes.
5. Remove the tape and second cup from the system.
6. Record the temperature of the sand using the thermometer.

Hutchings Photography/Digital Light Source/McGraw Hill

Questions to Consider
a. What type of energy was shaking the cup?
b. What is the evidence that energy was conserved?

Heat is the kinetic energy that flows from a hotter object to a colder one. When two bodies are in contact, heat always flows from an object at a higher temperature to one with a lower temperature. **Temperature** is a measure of the average kinetic energy of the atoms and/or molecules present in a substance. Everything around us is at some temperature—hot, cold, or lukewarm. An object is "cold" when its atoms and molecules move more slowly, on average, relative to an object that is "warm." Therefore, for the temperature of an object to increase, the kinetic energy of its atoms and molecules must increase. This next activity will explore the history of temperature-measuring devices.

Detecting Temperature

Heat cannot be measured directly. However, changes in heat can usually be detected as changes in temperature. Watch this video (www.acs.org/cic) and describe at least two different types of thermometers, how they are used, and how they work.

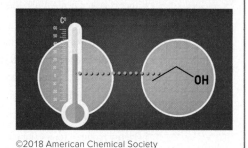

©2018 American Chemical Society

Although the concepts of temperature and heat are related, they are not identical. Your bottle of water and the Pacific Ocean may be at the same temperature, but the ocean contains and can transfer far more heat than the bottle of water. This is because the number of water molecules in the Pacific Ocean is much larger than the number of water molecules in the bottle of water. Indeed, as discussed in Section 5.2, bodies of water can affect the climate of an entire region as a consequence of their ability to absorb and transfer large amounts of heat.

6.4 | How Hot Is "Hot"? Measuring Energy Changes

Learning Objectives:

Describe how to measure the energy of combustion reactions and predict whether a reaction or process will be endothermic or exothermic

Draw and interpret energy diagrams for chemical reactions

The ability of a substance to provide heat energy makes it a good fuel. The **calorie** (cal) was introduced with the metric system in the late 18th century and was defined as the amount of heat necessary to raise the temperature of one gram of water by one degree Celsius. When calorie is capitalized, as in a nutritional Calorie, it means kilo-calorie on the metric scale. The values tabulated on food package labels and in cookbooks are, in fact, kilocalories:

$$1 \text{ kilocalorie (kcal)} = 1000 \text{ calories (cal)} = 1 \text{ Calorie (Cal)}$$

The modern system of units uses the **joule** (J), a unit of energy equal to 0.239 cal. One joule (1 J) is approximately equal to the energy required to raise a 100-g apple to a height of 1 m against the force of gravity. Another way to think of this is that each beat of the human heart requires about 1 J of energy. In terms of calorie equivalents, 1 cal = 4.184 J. Figure 6.2 provides a contextual comparison of various energy magnitudes, in terms of joules.

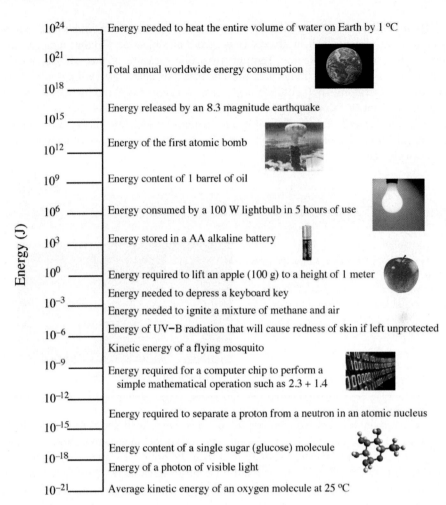

Figure 6.2

A contextual comparison of various energy magnitudes.

(Earth): Studio Photogram/Alamy Stock Photo; *(Bomb):* Library of Congress Prints & Photographs Division [LC-USZ62-36452]; *(Bulb):* Ingram Publishing/SuperStock; *(Battery):* Jeffrey B. Banke/Shutterstock; *(Apple):* lynx/Iconotec/Glow Images; *(Binary):* Mmaxer/Shutterstock

---

Your Turn 6.9   Energy Calculations

Example Problem:

Calculate how many Calories are equivalent to 1.38 kJ.

Example Solution:

Since 1 J = 0.239 cal, and 1 kJ = 1000 J
1.38 kJ × 0.239 cal = 0.330 Cal

Your Turn:

a. A slice of pizza contains 217 kcal (217 Cal). Express this value in kilojoules.
b. Calculate the number of 1-kg books you could lift to a shelf 2 m off the floor with the amount of energy from metabolizing one slice of pizza (at 908 kJ).
c. In part **b.**, a simplifying assumption was made. What was the assumption and is it reasonable? Based on this assumption, is your answer too high or too low? Explain your reasoning.

Since we need food to survive, can we consider food a type of fuel? After all, we don't combust when we eat! The process of combustion is actually an oxidation process—a reaction with oxygen. Carbohydrates (sugars) and fats are categories of biomolecules that provide our bodies with energy. They do so because when they react

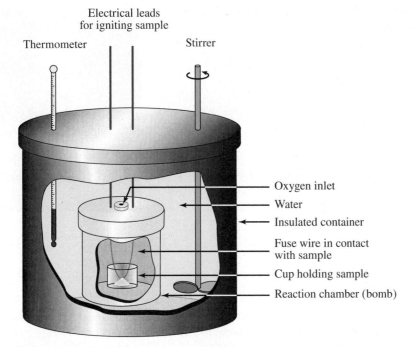

Electrical leads
for igniting sample

Thermometer

Stirrer

Oxygen inlet

Water

Insulated container

Fuse wire in contact
with sample

Cup holding sample

Reaction chamber (bomb)

Figure 6.3

Schematic drawing of a calorimeter. The reaction chamber is often referred to as a "bomb calorimeter" due to large pressures that are created inside a sealed container during combustion.

with oxygen, their products have less potential energy than the reactants. This energy difference is both transformed into usable energy in the body and dissipated as heat throughout the body.

A **calorimeter** is a device used to experimentally measure the quantity of heat energy released in a combustion reaction. Figure 6.3 shows a schematic representation of a heavy-walled stainless steel calorimeter. A sample of known mass (the fuel) and excess oxygen are added into the reaction chamber, which is then sealed and submerged in water. The reaction is initiated with a spark, and the heat evolved by the reaction flows from the reaction to the water and the rest of the apparatus. As a consequence, the temperature of the entire calorimeter system increases. The quantity of heat given off by the reaction can be calculated from this temperature rise by using the known heat-absorbing properties of the calorimeter and the amount of water it contains. The greater the temperature increase (measured in °C), the greater the quantity of energy evolved from the reaction (measured in J).

An experimental measurement using a calorimeter is referred to as the **heat of combustion**, the quantity of heat energy given off when a specified amount of a substance burns in oxygen. Heats of combustion are typically reported in units of kilojoules per mole (kJ/mol), kilojoules per gram (kJ/g), kilocalories per mole (kcal/mol), or kilocalories per gram (kcal/g). For example, the experimentally determined heat of combustion of methane is 802.3 kJ/mol. This means that 802.3 kJ of heat is given off when 1 mole of $CH_4(g)$ is combusted according to this balanced reaction:

$$CH_4(g) + 2\ O_2(g) \longrightarrow CO_2(g) + 2\ H_2O(g) + 802.3\ kJ \qquad [6.5]$$

Notice that the evolved heat can be considered a product in this reaction.

Burning methane is analogous to water flowing from the top of a waterfall. Initially in a state of higher gravitational potential energy, the water drops down to one of lower potential energy. The potential energy of water is converted into kinetic energy, which is then converted into other forms of energy when the water hits the rocks below, such as sound and thermal energy. Similarly, when methane is burned, energy is released when the atoms from the reactants change their interactions and "fall" to a state of

A DIY Calorimeter

Pakhnyushchy/Shutterstock

Watch this video to see how you can measure the energy content of fuels: www.acs.org/cic. After watching the video, compare and contrast the bomb calorimeter and the "can and burner" calorimeter for determining heats of combustion.

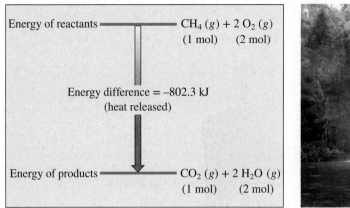

Figure 6.4

Comparison of a waterfall to the energy difference in the combustion of methane—an exothermic reaction.

(Waterfall): Ingram Publishing/SuperStock

lower potential energy as they are transformed into products. Figure 6.4 is a schematic representation of this process. The downward arrow indicates that the energy associated with 1 mole of $CO_2(g)$ and 2 moles of $H_2O(g)$ is less than the energy associated with 1 mole of $CH_4(g)$ and 2 moles of $O_2(g)$.

The combustion of methane is **exothermic**, a term describing any chemical or physical change accompanied by the release of heat. In this reaction, the energy difference is -802.3 kJ. The negative sign attached to the energy change for all exothermic reactions signifies the decrease in potential energy going from reactants to products.

Not surprisingly, the total amount of energy released depends on the amount of fuel burned. To compare fuels, we can use this value to calculate the number of kilojoules released for a gram of fuel, rather than for a mole. The molar mass of CH_4, calculated from the atomic masses of carbon and hydrogen, is 16.0 g/mol. We then can calculate the heat of combustion per gram of methane:

$$\frac{802.3 \text{ kJ}}{1 \text{ mol } CH_4} \times \frac{1 \text{ mol } CH_4}{16.0 \text{ g } CH_4} = 50.1 \text{ kJ/g } CH_4$$

This is a large a high heat of combustion, meaning there is a lot of energy released per gram of fuel. Figure 6.5 compares the energy differences (in kJ/g) of several different fuels. We can make some generalizations based on the chemical formulas of

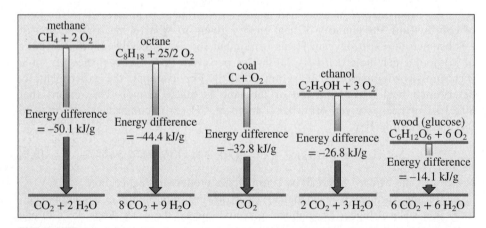

Figure 6.5

Energy differences (in kJ/g) for the combustion of methane (CH_4), *n*-octane (C_8H_{18}), coal (assumed to be pure carbon), ethanol (C_2H_5OH), and wood (presumed to be glucose). Carbon dioxide and water are formed in the gas phase.

the fuels. First, the fuels with the highest heats of combustion are hydrocarbons. Second, as the ratio of hydrogen-to-carbon decreases, the heat of combustion decreases. And third, as the amount of oxygen in the fuel molecule increases, the heat of combustion decreases.

It is no coincidence that the molecular structures of the most viable fuels, other than nuclear fuels, are composed largely of carbon and hydrogen atoms. These atoms are easily oxidized to form stable (low potential energy) carbon dioxide and water molecules, resulting in an overall release of energy from the reacting fuel and oxygen.

Not all chemical reactions are exothermic; some reactions *absorb* energy from the surroundings. We discussed two important examples in earlier chapters. One is the decomposition of O_3 to yield O_2 and O, which requires the input of energy in the form of high-energy (UV-B and UV-C) photons. The other reaction is the combination of N_2 and O_2 to yield two molecules of NO, which requires a high-temperature environment. Both of these reactions are **endothermic**, a term that describes any chemical or physical change that absorbs energy and creates products of a higher potential energy state. The convention for representing an endothermic reaction is to place a positive sign (or no sign at all) in front of the energy value and unit; for example, +29 J means that the reaction has absorbed 29 J of energy.

Photosynthesis is endothermic. This process requires the absorption of 2800 kJ of sunlight per mole of glucose ($C_6H_{12}O_6$) formed, or 15.5 kJ per gram. The complete process involves many steps, but the overall reaction can be described with this equation:

$$2800 \text{ kJ} + 6 \text{ CO}_2(g) + 6 \text{ H}_2\text{O}(l) \xrightarrow{\text{chlorophyll}} \underset{\text{glucose}}{C_6H_{12}O_6(s)} + 6 \text{ O}_2(g) \qquad \textbf{[6.6]}$$

Notice that the absorbed heat can be considered a reactant.

This reaction requires the participation of the green pigment chlorophyll. The chlorophyll molecule absorbs energy from the photons of visible sunlight and uses this energy to drive the photosynthetic process, an energetically uphill reaction. As seen in Section 4.1, photosynthesis plays an essential role in the carbon cycle, because it removes CO_2 from the atmosphere.

A general summary of endothermic and exothermic reactions is as follows (Figure 6.6):

i) If heat added to reactants is larger than the heat evolved by formation of products: *endothermic* (e.g., baking bread, producing sugar by photosynthesis)

ii) If heat added to reactants is less than the heat evolved with products: *exothermic* (e.g., combustion of fuels)

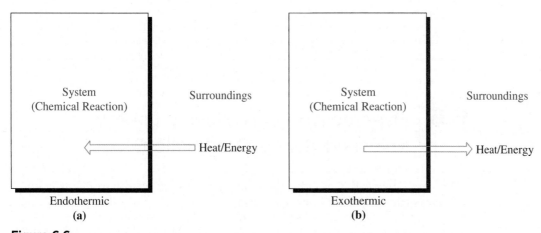

Figure 6.6

An illustration of the net heat flows for an **(a)** endothermic process and **(b)** exothermic process. The term *system* is used to denote the reaction taking place, whereas the surroundings are everything else outside the reactants (glass, countertop, room, etc.).

Endothermic vs. Exothermic Reactions

This video describes the differences between endothermic and exothermic reactions: www.acs.org/cic.

American Chemical Society

@HOME   Making Your Own Hot and Cold Packs

You hurt your ankle while jogging in your favorite 5K race. Luckily, your friend comes to the rescue with a room temperature pack from the first aid closet; one snap and you have a soothing cold pack. But how does this work? Therapeutic hot and cold packs sold in pharmacies or supermarkets consist of isolated compartments containing water and a salt. Once the divider between the two compartments is broken, the salt and water are allowed to mix and the pack gets either hot or cold.

a. Explain which type of reaction (exothermic or endothermic) would be needed to make a hot pack or cold pack?

  1. Obtain a sample of as many of the following salts as possible:

   - calcium chloride ($CaCl_2$—available at hardware or retail stores; salt used as a sidewalk de-icer)
   - water softener salt (mostly $NaCl$—available at hardware or retail stores)
   - magnesium sulfate ($MgSO_4$—also known as Epsom salt, available at pharmacies or retail stores)
   - ammonium chloride (NH_4Cl—available at hardware, retail, or landscaping stores)
   - potassium chloride (KCl—known as "Morton Lite," available at hardware or retail stores)
   - sodium bicarbonate ($NaHCO_3$—known as baking soda, available at grocery or retail stores)

  2. Place 50 mL of water into separate Styrofoam cups (one for each salt), and record the initial temperature of the water using a thermometer. Record the temperature changes that occur when 1 tablespoon of each salt is dissolved in water.
     Note: **Goggles should be worn to protect your eyes from chemical spills or splashes.**

b. For those reactions in which a temperature change is observed, which correspond to endothermic processes, and which correspond to exothermic processes?

c. Which salts would be the most effective choices for hot or cold packs? Are there any other factors that should be considered in making your final decisions?

6.5 | Hyperactive Fuels: How Is Energy Released during Combustion?

Learning Objective: Calculate the amount of energy absorbed or released by breaking or forming chemical bonds

As you have seen in previous chapters, molecular compounds are composed of atoms that are bonded together by covalent bonds. Chemical reactions involve the breaking and forming of these bonds. Energy is required to break bonds, just as energy is required to break chains or tear paper. In contrast, forming chemical bonds releases energy. The overall energy change associated with a chemical reaction depends on the net difference of the energy needed to break bonds, and the energy released when bonds form.

For example, consider the combustion of hydrogen. Hydrogen is desirable as a fuel because, compared with other fuels, it releases a large amount of energy when it burns:

$$2\ H_2(g) + O_2(g) \longrightarrow 2\ H_2O(g) \qquad\qquad \textbf{[6.7]}$$

To calculate the energy released from the combustion of hydrogen, let us assume that all the bonds in the reactant molecules are broken, and then the individual atoms are recombined to form the products. In fact, the reaction does not occur this way, but we are interested in only the relative states of reactants and products, not the mechanistic details.

Bond energy is the amount of energy that must be absorbed to break a specific chemical bond; covalent bond energies are given in Table 6.1. Because energy must be absorbed, breaking bonds is an endothermic process, thus all bond energies in Table 6.1 are positive. The values are expressed in kJ per mol of bonds broken. Note that the atoms appear both across the top and down the left side of the table. The number at the intersection of any row and column is the energy (in kJ) needed to break a mole of the covalent bonds between the two atoms. For example, the energy required to break 1 mole of C—H bonds is 416 kJ. Similarly, the energy to break 1 mole of N≡N triple bonds is 946 kJ, not three times 160 kJ (N—N).

Your Turn 6.10   O₃ Versus O₂

As noted in Section 3.3, ozone absorbs UV radiation of wavelengths less than 320 nm, whereas oxygen requires higher-energy electromagnetic radiation with wavelengths less than 242 nm. Use the bond energies in Table 6.1 plus information about the resonance structures of O₃ from Chapter 3, to explain this difference.

Table 6.1	Covalent Bond Energies (in kJ/mol)							
	H	**C**	**N**	**O**	**S**	**F**	**Cl**	**Br**	**I**
Single Bonds									
H	436								
C	416	356							
N	391	285	160						
O	467	336	201	146					
S	347	272	—	—	226				
F	566	485	272	190	326	158			
Cl	431	327	193	205	255	255	242		
Br	366	285	—	234	213	—	217	193	
I	299	213	—	201	—	—	209	180	151
Multiple Bonds									
C=C	598			C=N	616		C=O*	803	
C≡C	813			C≡C	866		C≡O	1073	
N=N	418			O=O	498				
N≡N	946								

*In CO₂

To determine whether a reaction is endothermic or exothermic, we need to keep track of whether energy is absorbed or released. To do this, we indicate when energy is absorbed with a positive sign. Energy is absorbed when a bond is broken. In contrast, forming a bond releases energy and the sign is negative. For example, when 1 mole of O=O double bonds is broken, the energy change is +498 kJ, and when 1 mole of O=O double bonds is formed, the energy change is −498 kJ.

Now we are ready to apply these concepts and conventions to the burning of hydrogen gas, $H_2(g)$. First, draw the Lewis structures of the species involved so that we can count the bonds that need to be broken and formed:

$$2\,\text{H—H} \;+\; \overset{..}{\underset{..}{\text{O}}}=\overset{..}{\underset{..}{\text{O}}} \;\longrightarrow\; 2 \;\; \underset{\text{H}}{\overset{\text{H}}{\text{O}}} \qquad\qquad \textbf{[6.8]}$$

Remember that chemical equations can be read in terms of moles. Both Equations 6.7 and 6.8 indicate "2 moles of H_2 plus 1 mole of O_2 yields 2 moles of H_2O." To use bond energies, we need to count the number of moles of bonds involved. Here is a summary:

Molecule	Bonds per Molecule	Moles in Reaction	Moles of Bonds	Bond Process	Energy per Bond	Total Energy
H—H	1	2	1 × 2 = 2	breaking	+436 kJ	2 × (+436 kJ) = +872 kJ
O=O	1	1	1 × 1 = 1	breaking	+498 kJ	1 × (+498 kJ) = +498 kJ
H—O—H	2	2	2 × 2 = 4	forming	−467 kJ	4 × (−467 kJ) = −1868 kJ
					Total:	**−498 kJ**

From the last column, we can see that the overall energy change in breaking bonds (872 kJ + 498 kJ = 1370 kJ) and forming new ones (−1868 kJ) results in a net energy change of −498 kJ. Since 498 kJ of heat is released for 2 moles of H_2 (Equation 6.8), the combustion of one mole of H_2 releases 249 kJ of heat. This corresponds to a heat of combustion of −249 kJ/mol for hydrogen gas.

This calculation is diagrammed in Figure 6.7. The energy of the reactants, two H_2 molecules and one O_2 molecule, is set at zero—an arbitrary but convenient value. The green arrows pointing upward signify energy absorbed to break the bonds in the

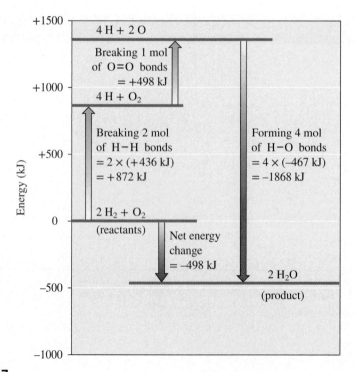

Figure 6.7

The energy changes during the combustion of hydrogen to form water vapor.

reactant molecules and form four H atoms and two O atoms. The red arrow on the right pointing downward represents energy released as these atoms bond to form the product H_2O molecules. The purple arrow corresponds to the net energy change (between products and reactants) of −498 kJ, signifying that the overall combustion reaction is strongly exothermic. The products are lower in energy than the reactants, so the energy change is negative. The net result is the release of energy, mostly in the form of heat.

The energy change we just calculated from bond energies, −498 kJ for burning 2 mol of hydrogen, compares favorably with the experimentally determined value when all of the species are gases. This agreement justifies the use of our rather unrealistic model of analysis: that all the bonds in the reactant molecules are first broken and then all the bonds in the product molecules are formed. You will calculate the heat released per gram of fuel in the next activity.

Heat of Combustion for Acetylene

First, watch this video to review the step-by-step process for calculating the heat of combustion of propane using bond energies: www.acs.org/cic.

Chemistry in Context Tutorial

BOND ENERGIES

©Bradley D. Fahlman

Your Turn 6.11   Heat Energy Per Gram

How much heat is produced per gram of hydrogen combusted? How does this value compare to the other fuels in Figure 6.5?

Not all calculations will result in the same close agreement with experimental data. One possible source of error is that the bond energies listed in Table 6.1 apply only to gases. Hence, calculations are most accurate if all reactants and products are in the gaseous state. Moreover, tabulated bond energies (other than C=O in Table 6.1) are listed as average values, which are based on many different types of molecules. The strength of a bond depends on the overall structure of the molecule in which it is found; it depends on what else the atoms are bonded to. Thus, the strength of an O—H bond is slightly different in water (H_2O), hydrogen peroxide (H_2O_2), or methanol (CH_3OH). Nevertheless, the procedure illustrated here is a useful way of estimating energy changes in a range of reactions. The approach also helps illustrate the relationship between bond strength and chemical energy. Check out the following tutorial for an opportunity to practice calculating the heat of combustion.

Now, try it on your own! Use the bond energies in Table 6.1 to calculate the heat of combustion for ethyne, C_2H_2, also commonly referred to as acetylene. Report your answer both in kJ/mol C_2H_2 and kJ/g C_2H_2.

6.6 | Fossil Fuels and Electricity

Learning Objective: Describe the operation of combustion-based power plants

Beyond using fossil fuels for our direct transportation and heating needs, about 63% of the electricity generated in the world comes from their combustion—primarily coal. But how do electrical power plants "produce" electricity, and what really goes on inside them?

The first step in producing electricity from coal is to burn it. Examine the photographs in Figure 6.8. You can almost feel the heat from the burning coal! In the coal beds of the boilers, the temperature can reach 650 °C (1202 °F). To generate this level of heat, the power plant burns a train car load of coal every few hours. In fact, operating at full capacity, a large power plant can burn up to 10,000 tons of coal per day!

The second step in producing electricity is to use the heat released from combustion to boil water—usually in a closed, high-pressure system (Figure 6.9). The elevated pressure serves two purposes: It raises the boiling point of the water above 100 °C, and it compresses the resulting water vapor. The hot high-pressure steam is then directed toward a steam turbine.

The third and final step generates electricity. As the steam expands and cools, it rushes past the turbine, causing it to spin. The shaft of the turbine is connected to a large coil of wire that rotates within a magnetic field, and the turning of this coil generates an electric current. Meanwhile, the water vapor leaves the turbine and continues to cycle through the system. It passes through a condenser, where a stream of cooling

Figure 6.8

Photos from a small coal-fired electric power plant. Shown are **(a)** piles of coal outside the plant; **(b)** a row of boilers into which the coal is fed; **(c)** behind the blue door in photograph **(b)**; **(d)** a close-up image of coal burning on the boiler bed.

(a–d): ©Cathy Middlecamp

(a)   **(b)**

(c)   **(d)**

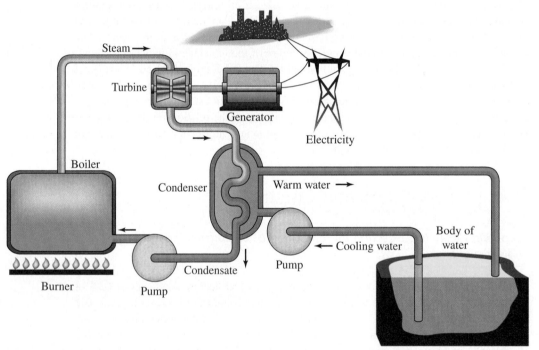

Figure 6.9

Diagram of an electric power plant illustrating the conversion of energy from the combustion of fuels to electricity. Components are not illustrated to scale.

water carries away the remainder of the heat energy originally acquired from the fuel. The condensed water then re-enters the boiler, ready to resume the energy transfer cycle. Nuclear power plants, described in Section 7.3, operate on a similar principle—the use of superheated steam to turn a turbine. The difference is simply the type of fuel used to heat the water into steam.

When fuel molecules combust, their potential energy is converted into heat, which is then absorbed by the water in the boiler. As the water molecules absorb the heat, they move faster and faster in all directions—their *kinetic* energy increases—until they can escape contact with one another and vaporize to steam. As pressurized steam, water

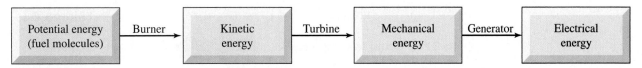

Figure 6.10
Summary of the energy transformations in a fossil fuel–powered electric power plant.

molecules have a tremendous amount of kinetic energy, which is transferred to the turbine that is spun into motion. The kinetic (motion) energy of the water molecules has been transformed into mechanical energy in the turbine. The generator then turns and converts the mechanical energy (a form of kinetic energy) in the turbine into **electrical energy**, another form of kinetic energy. The various energy transformation steps are summarized in Figure 6.10.

Your Turn 6.12   Energy Conversion

Although power plants require several steps to transform potential energy into electrical energy, other devices do this more simply. For example, a battery converts chemical energy into electrical energy in one step. List three other devices that convert energy from one form into another. For each one, name the types of energy and show the path of energy transformations involved.

6.7 | Efficiency of Energy Conversion

Learning Objective: Describe energy conversion and factors that influence its efficiency

According to the first law of thermodynamics, we are assured that the total energy of the universe is conserved. Explore the first law in the next activity, which provides a fun illustration for the conversion between potential and kinetic energies.

Your Turn 6.13   Where Did the Energy Go?

Using the "Intro" simulation (www.acs.org/cic), place the skateboarder at the top of the skate park and follow the energy conversions by clicking on the "pie chart" and "bar graph" icons.

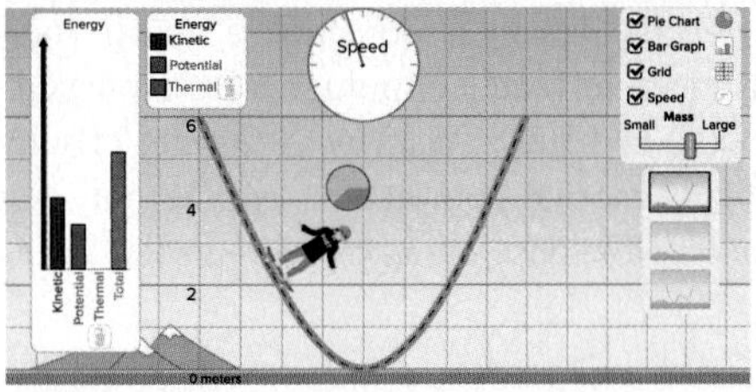

a. When are the gravitational potential and kinetic energies the highest? When are they the lowest? When are both energies the same?
b. What effect does changing the mass of the skateboarder have on the relative amounts of kinetic, potential, and total energies?
c. What happens to the total energy as the skateboarder changes position? Explain this trend.

PhET Interactive Simulations, University of Colorado

PhET Interactive Simulations, University of Colorado, http://phet.colorado.edu

Now, open the "Friction" simulation (www.acs.org/cic) and place the skateboarder at the top of the track. Experiment with varying levels of friction and skateboarder masses.

d. What happens to the skateboarder over time?
e. What effect does friction have on the relative amounts of potential, kinetic, thermal, and total energies over time?
f. Do these results violate the first law of thermodynamics? Explain.

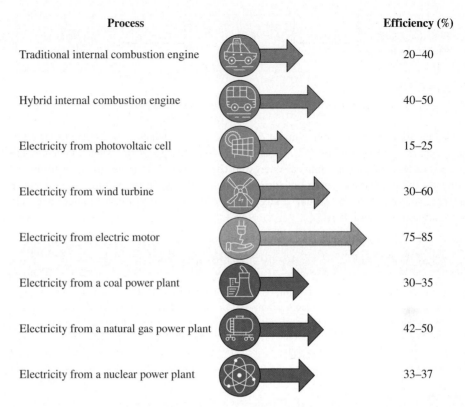

Process		Efficiency (%)
Traditional internal combustion engine		20–40
Hybrid internal combustion engine		40–50
Electricity from photovoltaic cell		15–25
Electricity from wind turbine		30–60
Electricity from electric motor		75–85
Electricity from a coal power plant		30–35
Electricity from a natural gas power plant		42–50
Electricity from a nuclear power plant		33–37

Figure 6.11

The percent of source energy that is used in each process.

Nicole C. Bouvier-Brown

Based on the first law of thermodynamics, no new energy is created during combustion, but none is destroyed either. The problem is that as we burn coal, natural gas, and petroleum, we always convert at least some of the energy stored in the fuels into forms that cannot be easily used or recovered. This is described by the second law of thermodynamics. As energy is transferred or transformed, more and more of it becomes unusable. In other words, inefficiency is inevitable, but it can vary depending on the process (Figure 6.11).

For instance, consider a fossil fuel power plant. Newer boiler systems and advanced turbine technologies have pushed the efficiencies of each step in Figure 6.10 to 90% or better. So, why is the net efficiency of a power plant only 30–50%? Worse yet, an internal combustion engine only has an efficiency of 20–40%! Think of some reasons for these energy losses in the next activity.

Your Turn 6.14   Why So Inefficient?

a. Think of examples as to why the heat energy from fuel combustion in a power plant would not all be converted into electricity.
b. List some of the energy losses that take place when driving a car. Use the Internet to verify and expand your list, if necessary.
c. It is often assumed that 15% of the energy from fuel combustion is used to move the vehicle. Estimate the percent used to move the passengers.

As you will see in the next activity, across the entire energy sector in the United States, 67% (65.4/97.2 quads) of the energy is "rejected" (Figure 6.12). Some of this is due to energy loss that could be recovered with enhanced technology and behavior, but most of it is due to the transformation of energy into useless (or waste) heat.

Your Turn 6.15 "Rejected" Energy

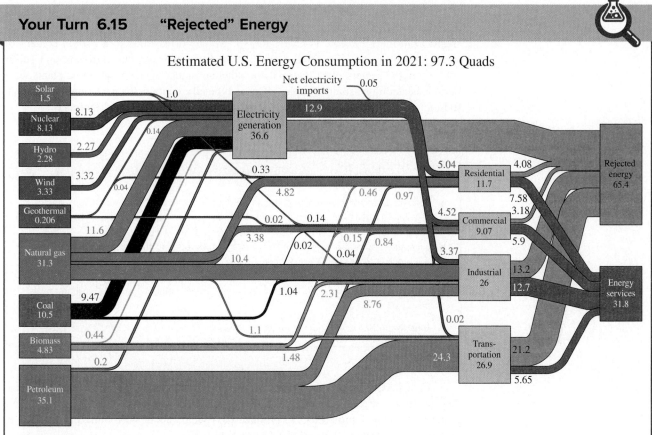

Estimated U.S. Energy Consumption in 2021: 97.3 Quads

Figure 6.12

Estimated 2021 energy consumption in the United States, broken down by source and use. All numerical values are given in Quads, or quadrillion BTU (where 1 BTU is equal to 1055 J).

Lawrence Livermore National Laboratory and the Department of Energy

Using the data shown in the image above, answer the following questions.

a. What percent of electricity generated (from all sources) is unusable (or "rejected")?
b. What percent of the energy used in transportation is unusable (or "rejected")?
c. Looking at the entire energy sector, how much energy is "rejected"? Provide your answer in joules.

6.8 | Power from Ancient Plants: Coal

Learning Objective: Describe the production, consumption, and consequences of coal combustion

About two centuries ago, the Industrial Revolution began the great exploitation of fossil fuels that continues today. In the early 1800s, wood was the major energy source in the world (Figure 6.1). Coal turned out to be an even better energy source than wood because it yielded more heat per gram. By the 1960s, coal became the primary source of generating electricity and today, 38% of global electricity comes from coal—the single largest source of electricity generation. While coal use is declining in many industrialized countries, global coal use continues to rise due, in most part, to developing economies.

Your Turn 6.16 Advantages to Coal

List some advantages to using coal as a fuel.

Although coal is often assumed to be composed of pure carbon, it is a more complex mixture, containing small amounts of many other elements. An approximate chemical formula for coal is $C_{135}H_{96}O_9NS$, which corresponds to a carbon content of about 85% by mass. The smaller amounts of hydrogen, oxygen, nitrogen, and sulfur come from the ancient plant material and other substances present when the plants were buried. In addition, some samples of coal typically contain trace amounts of silicon, sodium, calcium, aluminum, nickel, copper, zinc, arsenic, lead, and mercury.

Your Turn 6.17   Coal Calculations

a. Assuming the composition of coal can be approximated by the formula $C_{135}H_{96}O_9NS$, calculate the mass of carbon (in tons) in 1.5 million tons of coal. This quantity of coal might be burned by a typical power plant in 1 year.

b. Compute the amount of energy (in kJ) released by burning this mass of coal. Assume the process releases 30 kJ/g of coal. Useful conversion factors: 1 ton = 2000 lb and 1 pound = 454 g.

c. What mass of CO_2 would be formed by the complete combustion of 1.5 million tons of this coal?

Hint: In the balanced chemical equation, assume a mole ratio of Coal:CO_2 of 1:135.

Coal occurs in varying grades; however, all are better fuels than wood because they contain a higher percentage of carbon and a lower percentage of oxygen. For example, burning 1 mole of C to produce CO_2 yields about 40% more energy than is obtained from burning 1 mole of CO to produce CO_2.

Your Turn 6.18   Coal Varieties

Using the Internet, provide a comparison of the different grades of coal. Include details about their composition, energy content, and region of origin.

a. How are these different varieties formed?

b. Which variety provides the best energy content?

c. What forms of coal are most abundant?

Although coal is available across the globe and remains an inexpensive and widely used fuel, it has serious environmental costs. First, underground mining is both dangerous and expensive. Mine safety has dramatically improved in most countries. However, during the past century, more than 100,000 workers have been killed by accidents, cave-ins, fires, explosions, and poisonous gases in the United States alone. Many more have been incapacitated by respiratory diseases. Worldwide, the picture is far worse. Use the following Your Turn exercises to explore additional costs of coal extraction and combustion.

Your Turn 6.19   Acidic Mine Drainage

When groundwater infiltrates abandoned mine shafts, or comes in contact with sulfur-rich rock often associated with coal deposits, it becomes acidified. Use the Internet to find out why this happens. What are the environmental implications of acidic mine drainage? You may refer to this video for information: www.acs.org/cic.

Figure 6.13

A photo taken in 1940 at 10:55 AM in Pittsburgh, Pennsylvania, showing the mid-day Sun completely shielded by smoke and soot from the burning of coal in nearby steel mills and power plants.

Walter Stein/AP Images

Mountaintop Mining

When coal deposits lie sufficiently close to the surface, alternative mining techniques are possible. One technique is called *mountaintop mining*. This process calls for scraping away the overlying vegetation, and then blasting off the top several hundred feet of a mountain to reveal the underlying coal seam. Watch this video and explain how this mining technique impacts local communities: www.acs.org/cic.

Another drawback is that coal is a dirty fuel. It is, of course, physically dirty; however, the issue here is the dirty combustion products. Soot from countless coal fires in cities in the 19th and early 20th centuries blackened both buildings and lungs. In fact, during the mid-1940s, the burning of coal in steel mills near Pittsburgh, Pennsylvania, generated enough atmospheric pollution to mask the midday sun (Figure 6.13).

As seen in the previous activity, waste is generated from the combustion of coal. In fact, the process generates 1 ton of coal ash for every 4 tons of raw coal used. In 2016, the world produced almost 2 billon tons of coal ash. Often, ash is left on site (even if temporarily). This presents a hazard to local communities. For example, Figure 6.14 shows the devastation caused by millions of gallons of coal ash sludge that spilled down a valley when the retaining walls of a storage pond failed.

Particles from Coal

Watch this video to see the impacts of coal combustion on local community members: www.acs.org/cic. What are some of the health implications? Use the Internet to explain why these particles would be hazardous.

Figure 6.14

A photo taken in December 2008 showing homes in Kingston, Tennessee, buried by 300 million gallons of coal ash sludge. The cleanup was completed in 2015 and costed $1.1 billion.

Wade Payne/AP Images

In order to prevent environmental harm and enhance sustainability, there has been recent interest in reusing coal ash waste for consumer products. For instance, the addition of ash in concrete is shown to improve its durability, chemical resistance, and shrinkage during hardening. Additionally coal ash can be used as an additive in bricks, ceramic tiles, and plaster, as filler in metal and plastic composites and in paints and adhesives, and as structural fill for road construction.

A final cost may ultimately have the largest impact—burning of coal produces gases that contribute to acid rain (NO_x and SO_x), as well as global warming (CO_2). In fact, coal combustion produces more CO_2 per kilojoule of heat released than either petroleum or natural gas.

Your Turn 6.20   Coal Emissions

In the United States, coal-burning power plants are responsible for two thirds of the sulfur dioxide emissions and one fifth of the nitrogen monoxide emissions.

a. Why does burning coal produce sulfur dioxide? Name another source of SO_2 in the atmosphere.

b. Why does burning coal produce nitrogen monoxide? Name two other sources of NO.

Because of these environmental and health impacts, and given that coal reserves are relatively plentiful in places like the United States, significant research efforts are underway to improve coal technologies. The term "clean coal technology" encompasses a variety of methods that aim to increase the efficiency of coal-fired power plants while decreasing harmful emissions.

Let's Debate!   Clean Coal?

Form two groups: one group will be in favor of the topic (the proposition) and the other group will be against the topic (the opposition). Both groups will try to persuade a neutral person or judge to agree with them. The topic of the debate is the motion. Here is the motion:

Javier Larrea/AGE Fotostock

Clean coal is a viable alternative for power generation and represents our best option for electrical generation in the U.S. and around the world.

Curiosity Conversation

- What is meant by "clean coal" and how can this technology be used to improve the environmental impact of coal based power plants?
- Revisit the three pillars of sustainability to consider the advantages and disadvantages of coal relative to other sources of energy.

What does the future hold for the dirtiest fuel? The answer depends on where you live. Figure 6.15 compares coal consumption in different regions of the globe between 1965 and 2021. Most regions have shown substantial increases in coal use, especially in Asia Pacific countries.

Although many power plants have installed scrubbers to clean combustion exhaust of sulfur dioxide, most have not tackled the more difficult task of carbon capture utilization and storage (CCUS). According to the International Energy Agency (IEA), only one commercial power plant equipped with this technology is in operation. The Boundary Dam project in Saskatchewan, Canada has been operating since 2014 and can capture 90% of the carbon dioxide in the plant's exhaust. While many pilot CCUS projects have occurred, there is no economic incentive to spend the money needed to retrofit existing power plants as long as releasing carbon dioxide into the

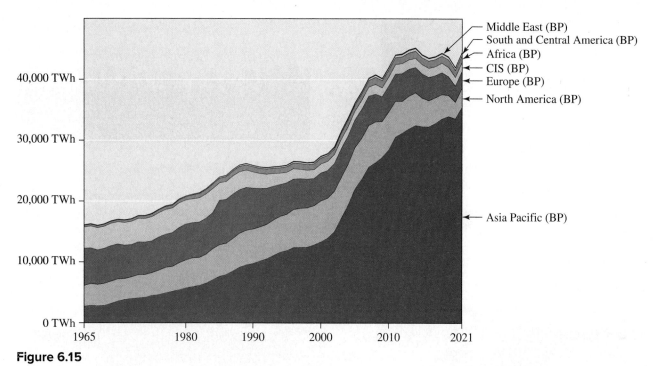

Figure 6.15

Coal consumption in different regions of the world between 1965 and 2021. For an interactive version, see: www.acs.org/cic.

Our World in Data based on BP Statistical Review of World Energy (2022).

air is free. According to the IEA, over 40 CCUS power generation projects are currently being developed. If all 40 projects move forward, combined with the existing plant, there is a capacity to capture about 60 million tons of CO_2 per year in 2030. However, this is well short of the 430 million tons per year needed by 2050 to reach net zero CO_2 emissions from the energy sector in order to have a chance at limiting the global temperature rise to 1.5 °C. In addition, using CCUS does not address the health concerns of the residents living nearby the facilities; these people, who often have little political power, will still be exposed to power plant exhaust. On the other hand, we need to use all mitigation and adaptations strategies to decrease the amount of CO_2 emissions as soon as possible.

6.9 | From Steam Engines to Sports Cars: The Shift from Coal to Oil

Learning Objective: Describe the production, consumption, and consequences of oil combustion

The use of petroleum for waterproofing and construction dates back to ancient civilizations due to natural seepages of asphalt. However, the modern oil era began in 1859 when the first oil well was drilled in Pennsylvania, USA. With the introduction of electricity in 1882, oil was no longer needed for lighting applications and the industry shifted towards powering the newly invented automobile. The oil boom in Texas, U.S. was initiated with the Lucas Spindletop gusher. This type of eruption was never seen before, reaching heights of 150 feet (Figure 6.16) and lasting 9 days. It wasn't until 1938 that the first oil well was drilled in Saudi Arabia tapping into one of the largest oil reservoirs in the world.

Today, oil is mainly used to propel vehicles, heat buildings and produce electricity. For example, gasoline is the most consumed petroleum product in the United States. Petroleum has the distinct advantage of being a liquid, making it easily pumped to the surface

Figure 6.16

Photo of the Lucas Spindletop gusher on January 10, 1901 (left), and the widespread exploration of the oil field a year later (above).

(both): Courtesy of Texas Energy Museum

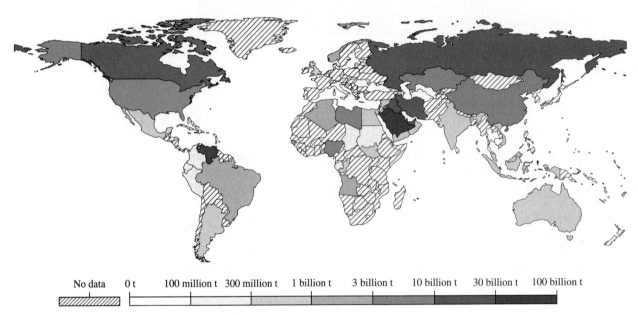

| No data | 0 t | 100 million t | 300 million t | 1 billion t | 3 billion t | 10 billion t | 30 billion t | 100 billion t |

Figure 6.17

Worldwide proven oil reserves as of 2020. For an interactive version, see: www.acs.org/cic

Our World in Data based on BP Statistical Review of World Energy (2022).

and transported via pipelines to refineries. Moreover, petroleum yields about 40–60% more energy per gram than coal. A typical value for petroleum is 48 kJ/g as compared to 30 kJ/g for a high-grade coal. Petrochemical companies also use petroleum as a raw material to produce plastics, solvents, pesticides, and hundreds of other consumer products.

As you saw earlier in Section 6.1, oil is a fossil fuel with a limited total supply. However, there are still plentiful oil reserves located throughout the world (Figure 6.17). In the mid-1950s, the average global oil consumption was about 8 million barrels per day, with a production of 15 million barrels. Today, we consume over 99 million barrels per day and produce over 100 million barrels. A barrel of oil is equivalent to 159 L (42 gal), and 7⅓ barrels have a mass of 1000 kg. The world oil production to consumption ratio is not fixed, but may tip in either direction depending on geopolitical factors and the success of crude oil exploration and recovery efforts (Figure 6.18). For example, consumption dropped more precipitously than production in the early part of 2020 due to quarantine measures to address the COVID-19 pandemic. Thus, some stock was accumulated and consumption outpaced production throughout 2021.

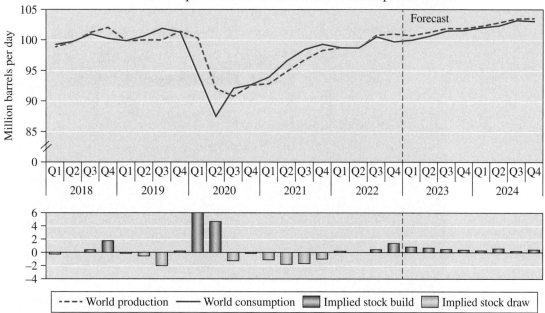

World Liquid Fuels Production and Consumption Balance

Figure 6.18

The world oil production and consumption balance since 2018

Source: US Energy Information Administration (EIA), March 2023

Your Turn 6.21 How Much Is a Barrel?

The petroleum industry uses barrels/day as a standard measure of consumption and production. However, we often speak in terms of the weight of fuels, such as tons, or the volume, such as gallons (gal) or liters (L). Using Appendix 1, convert the current world consumption of crude oil into tons, gallons, and liters.

Contrary to a popular myth, oil is not found in underground pools, but is found within the pores of geologic rock formations such as sandstone, tarry oil sands, and shales (Figure 6.19). As you might expect, a higher degree of porosity within the rock usually means a higher potential to store oil. However, porosity is only part of the story. In order for a rock formation to yield sufficient quantities of oil or gas, the pores must be interconnected, something known as the *permeability* of the reservoir. If the pore spaces are isolated from one another, oil will not be able to flow, regardless of whether there is appreciable porosity and oil content.

A great deal of time and money is spent in identifying the likely locations of oil-rich rock formations. These reservoirs can exist on land or under the ocean, at varying depths from the surface. Using sound waves, scientists can first determine whether underground rocks are likely to contain oil or gas deposits. From there, an exploratory well is drilled, and core samples are taken to the surface where they are examined to determine whether oil is contained within the pores. Due to this time-consuming process, the oil we use today comes from fields that were discovered decades ago.

When drilling is successful, spontaneous flow of oil to the surface may continue for days or years, but will eventually slow or stop due to a loss of pressure. When this happens, well pumps must be used, and deposits of natural gas may also be separated from oil and injected back into the reservoir to increase the pressure and keep

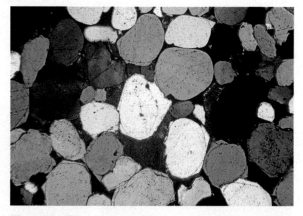

Figure 6.19

Microscopic image of a sandstone reservoir rock, showing the locations of trapped oil deposits within its pores and grain boundaries (cracks). Water and natural gas are often also present within the pores of these rocks.

Doug Sherman/Geolife

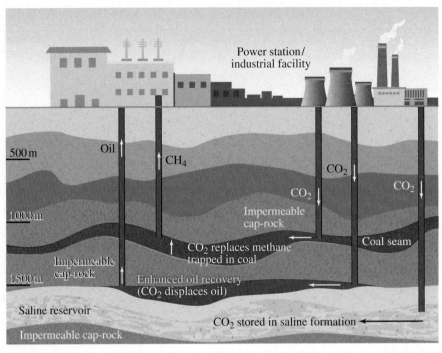

Figure 6.20

Schematic of a CO_2-EOR strategy to use CO_2 that has been sequestered from a fossil fuel-fired power plant for oil recovery.

Oil Drilling

Watch this video (www.acs.org/cic) and outline the process of drilling for oil.

oil flowing. Eventually, these strategies are not sufficient to maintain the flow of oil. Thus, there is an ongoing search to find economical and sustainable strategies to extract more of this trapped oil. Once we tap the easy sources, we must use unconventional oil recovery methods to extract oil found in more difficult locations.

One strategy, carbon dioxide enhanced oil recovery (CO_2-EOR) (Figure 6.20), is currently attracting interest because it provides a sustainable use for a captured greenhouse gas mentioned in the carbon capture utilization and storage (CCUS) process (revisit Section 6.8). The CCUS facility in Saskatchewan, Canada compresses CO_2 from coal combustion; this liquid CO_2 can then be injected into a saline reservoir or used in CO_2-EOR. This process has significantly increased the lifetime and sustainability of the wells.

Another strategy is to extract oil from tar sands. Tar sands, also known as oil sands, are a mixture of sand, clay, water, and bitumen (Figure 6.21). Bitumen is thick, sticky, black oil. This material can then be refined and mixed with lighter oils to produce synthetic crude oil. The largest tar sand deposits are in Alberta (Canada)

Figure 6.21

Images showing the oil extraction process from tar sands. Shown (left-right) is a close-up of the tar sands and open pit mining of tar sands.

(left to right): AdShooter/E+/Getty Images; Blackfox Images/Alamy Stock Photo; dan_prat/E+/Getty Images

and Venezuela. However, there are many concerns about relying on tar sands for oil production; the mining, processing, and transport of material have high environmental and social costs.

While liquid fuel is easier to pump and transport, it is also easier to contaminate water sources. For example, crude oil can spill into oceans from off-shore drill sites, underwater pipes, or from barges. Explore the impacts of oil spills in this next activity.

Tar Sands

Watch these videos ("Oil Sands 101" and "The true cost of oil") at www.acs.org/cic and describe the process of using tar sands for oil, as well as the numerous environmental and social impacts of this practice.

Your Turn 6.22   Oil Spills

Using the Internet, determine the environmental impacts of oil spills and describe how these spills can be cleaned up.

6.10 | Natural Gas: A "Clean" Fossil Fuel?

Learning Objective: Describe the production, consumption, and consequences of natural gas combustion

Natural gas, which is primarily composed of methane (CH_4), was first produced from coal and commercially used by the British around 1785 to light houses and streets. Natural gas was mainly used for lighting until the 1885 invention of the Bunsen burner, which opened up new opportunities for heating and cooking. Today, natural gas accounts for about one-quarter of the global electricity production where the gas is burned in a power plant. Among the fossil fuels, natural gas is relatively clean. When burned, it releases essentially no sulfur dioxide and relatively little particulate matter, carbon monoxide, and nitrogen oxides. No ash residue, which often contains toxic metals, remains after combustion. Although burning natural gas does produce CO_2, a greenhouse gas, the amount is less per unit of energy released than for the other fossil fuels. As such, the demand for natural gas has increased over time. Thus, the production of natural gas is continuing to accelerate around the world (Figure 6.22).

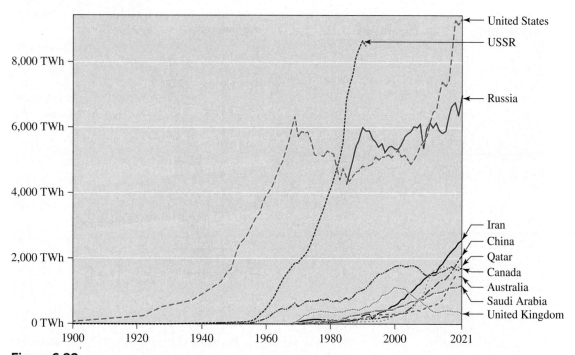

Figure 6.22

Natural gas production in select countries since 1900. For an interactive version, see: www.acs.org/cic

Our World in Data based on BP Statistical Review of World Energy; the SHIFT Project

Your Turn 6.23   Coal Versus Natural Gas

The combustion of 1 gram of natural gas releases 50.1 kJ of heat.

a. Calculate the mass of CO_2 released when natural gas is burned to produce 1500 kJ of heat. Assume that natural gas is pure methane, CH_4.

b. Select one of the grades of coal you researched in Your Turn 6.18. Compare the mass of CO_2 produced when enough of this coal is burned to produce the same 1500 kJ of heat.

Natural gas wells date back to 1821 in New York; however, oil reservoirs are typically accompanied by pockets of natural gas. Conventional vertical drilling methods used to acquire oil are also used to capture natural gas that is relatively easy to access. To access deposits that are more difficult to reach, energy companies are now using horizontal drilling techniques. One such technique uses water to fracture the rock; thus it is called hydraulic fracturing, or **fracking**.

Although fracking has been around since the 1940s, it has boomed in the last few decades as it became more economically viable. Fracking involves drilling down into the gas- or oil-bearing rocks that lie 1–3 miles beneath Earth's surface. This technique is required to obtain natural gas or petroleum from hard rock formations (e.g., shales) that have low permeabilities. However, fracking is also used to increase the rate at which fossil fuels are recovered from porous formations (e.g., sandstones, limestones).

A fracking fluid, consisting of water and a cocktail of substances, is injected under pressure in order to create cracks into which natural gas and petroleum can flow (Figure 6.23). The water also carries fine sand that props open these cracks. This next activity will give you the opportunity to pin down some of the details of fracking.

Figure 6.23

Schematic of fracking, used to extract natural gas from underground reservoirs.

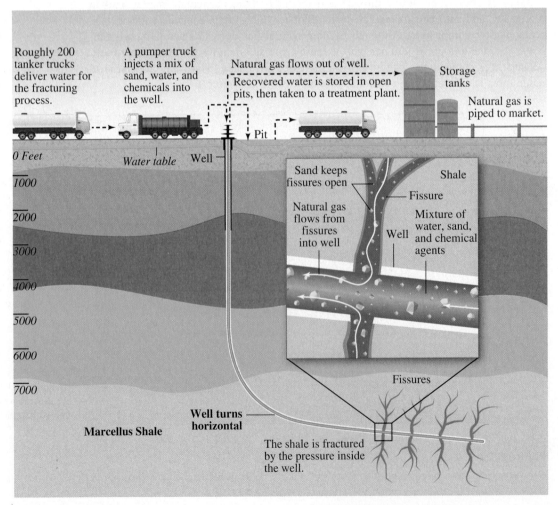

Fracking: Process and Impacts

Watch these videos at www.acs.org/cic: "How does fracking work?" and "Health Concerns Over Fracking Fluids" to gain some background information about fracking. You may also use Internet resources to find the information.

a. During the extraction of natural gas, the shale fracturing is done hydraulically rather than with dynamite. What does the term *hydraulic* mean? Suggest reasons why dynamite wouldn't work.

b. How much water is typically injected into a well? What are the specific ingredients of the "cocktail" that this water contains? How difficult was this information to find?

c. Some of the water returns to the surface as waste water. What are some of the options for handling this waste water?

d. What are the impacts of fracking?

Fracking is highly controversial because it involves high water consumption and results in many environmental and health impacts. Consequently, varying federal and regional regulations have been instituted across the globe, with some regions opting to completely ban this activity, such as Germany, France, Australia, and some U.S. states such as Maryland, Vermont, and New York.

With an understanding of how methane may be acquired, the last activity of this section will show the importance of quantifying and tracking the sources of methane, as well as other greenhouse gases, in the atmosphere.

Real Talk   Emissions Budget

Check out this interview with Dr. Lambert Doezema from Loyola Marymount University, who discusses the importance of calculating an emissions "budget" for methane: www.acs.org/cic.

Questions to Consider
a. Why is it important to track the sources of methane emissions?
b. Using the data presented on www.acs.org/cic, summarize the major sources of methane gas and some reasons for uncertainty in emissions data.

6.11 | What's in a Barrel?
Petroleum Fractionation

Learning Objective: Predict boiling points based on the strength of intermolecular forces and explain how these differences lead to fractionation of crude oil

Petroleum, also known as crude oil, is a mixture of several thousand different compounds, with the great majority being hydrocarbons composed of 5–12 carbon atoms per molecule. Many of the hydrocarbons in petroleum are **alkanes**, hydrocarbons with only single bonds between carbon atoms.

How are products, such as gasoline or diesel oil, produced from petroleum? The process takes place at an oil refinery, the icon of the petroleum industry (Figure 6.24). During the initial step in the refining process, the crude oil is separated into fractions, including the gasoline fraction. Refineries transform crude oil using several processes, including that of **distillation**, a separation process in which a solution is heated to its

Figure 6.24

Photo of a crude oil refinery showing the tall distillation towers. Small amounts of natural gas are flared, as evidenced by the flames.

Keith Wood/Corbis

boiling point and the vapors are collected and condensed. But just what happens to a liquid at the molecular level when its boiling point is reached?

The **volatility** of a liquid refers to how easily it is transformed into its gaseous phase—a process known as **vaporization**. As we saw in Section 5.2, when a liquid is heated, the intermolecular forces (IMFs) that hold together individual molecules are broken. For instance, the high boiling point of water is due to strong hydrogen bonding forces. While hydrogen bonding and dipole–dipole forces are present between polar molecules, the attractive forces between nonpolar molecules, such as hydrocarbons, are known as **London dispersion forces**.

As illustrated in Figure 6.25a, when two nonmetal atoms or nonpolar molecules approach one another, each has a random distribution of electrons, which results in no measurable attraction. However, since electrons are in constant motion, the approach of neighboring atoms/molecules may cause more electrons to be situated on one side of the species than the other. This creates partial negative/positive charges on the neighbors, generating a weak attraction (Figure 6.25b).

London dispersion forces are relatively weak (0.1–2 kJ/mol) compared to the hydrogen bond intermolecular forces discussed in Section 5.2 (5–15 kJ/mol) and the C—H covalent bonds in hydrocarbons (416 kJ/mol, Table 6.1). It should be noted that all molecules, even polar molecules, contain London dispersion attractive forces. However, for polar molecules, the dipole–dipole or hydrogen bonding forces often outweigh London dispersion forces.

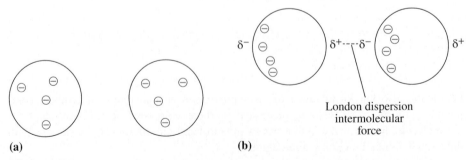

Figure 6.25

Illustration of London dispersion. Check out a video clip of intermolecular forces: www.acs.org/cic. **(a)** Two separate nonmetal atoms or nonpolar molecules with a random distribution of electrons. The positively charged nucleus is not shown. **(b)** Weak London dispersion attraction between neighboring species due to the generation of partial charges, which result from the movement of electrons.

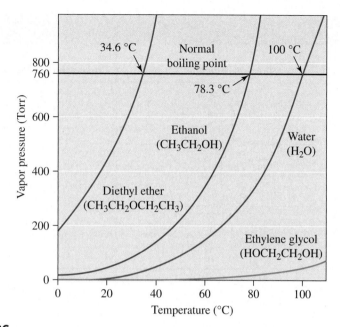

Figure 6.26

The relationship between the vapor pressure of certain liquids and temperature.

At any temperature greater than absolute zero (0 K or −273 °C), all liquids will experience some degree of vaporization. As more molecules are released from the liquid state into the gaseous state, the **vapor pressure** of the liquid increases. The **boiling point** of a liquid is the temperature at which the vapor pressure of the liquid equals the ambient pressure. The **normal boiling point** of a liquid occurs when the vapor pressure of the liquid is equal to 1 atm (101 kPa or 760 Torr). For instance, the normal boiling point of water at sea level, with an ambient atmosphere of 1 atm, is 100 °C.

If we need to increase the vapor pressure of a substance to reach its boiling point, how could that be accomplished? One way is to increase its temperature. As temperature increases, the liquid molecules move faster; thus, more molecules have enough energy to overcome the intermolecular forces holding the liquid molecules together and escape into the gas phase. Therefore, vapor pressure increases for a liquid as temperature increases (Figure 6.26). As you can see, the strength of the intermolecular forces also impacts the boiling point of a liquid. In general, as the strength of IMFs increase, the boiling point of the liquid will also increase.

As the constituent atoms within a molecule increase in size, the electrons are farther from the nucleus, making them easier to redistribute when approaching their neighbors. As a result, molecules that contain larger and heavier atoms will exhibit stronger London dispersion forces than smaller and lighter ones. Furthermore, molecules with similar structures but greater molar masses will also experience stronger London dispersion forces. As we see in Table 6.2, as the alkane chain length increases, the boiling point of the hydrocarbon increases due to stronger London dispersion forces between adjacent hydrocarbon molecules.

Your Turn 6.24   Relative Boiling Points

Rank the following molecules in order of increasing boiling points, based on the relative strengths of their intermolecular forces.

a. Br_2, F_2, I_2, Cl_2
b. CF_4, CBr_4, CI_4, CH_4

Table 6.2	Selected Alkanes (Gases and Liquids)		
Name and Chemical Formula	**Boiling Point (physical state at 25 °C)**	**Structural Formula**	**Condensed Structural Formula**
Methane CH_4	−161 °C (gas)	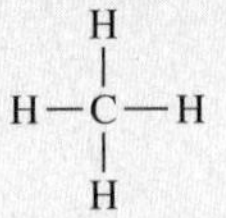	CH_4
Ethane C_2H_6	−89 °C (gas)	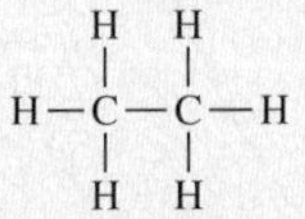	CH_3CH_3
Propane C_3H_8	−42 °C (gas)	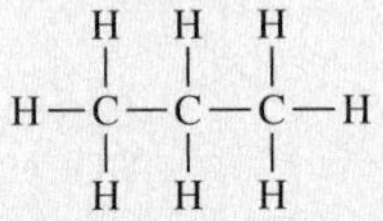	$CH_3CH_2CH_3$
n-Butane C_4H_{10}	−0.5 °C (gas)	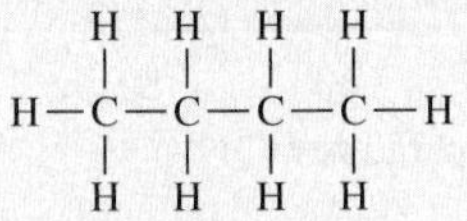	$CH_3CH_2CH_2CH_3$
n-Pentane C_5H_{12}	36 °C (liquid)	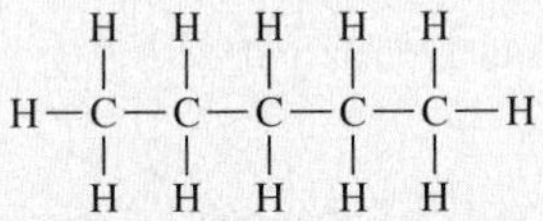	$CH_3CH_2CH_2CH_2CH_3$
n-Hexane C_6H_{14}	69 °C (liquid)	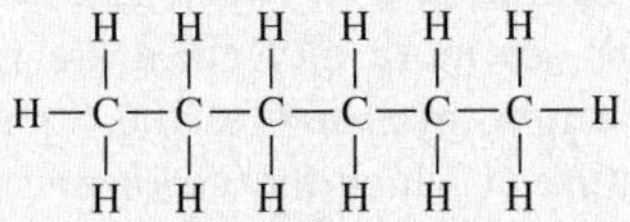	$CH_3CH_2CH_2CH_2CH_2CH_3$
n-Heptane C_7H_{16}	98 °C (liquid)	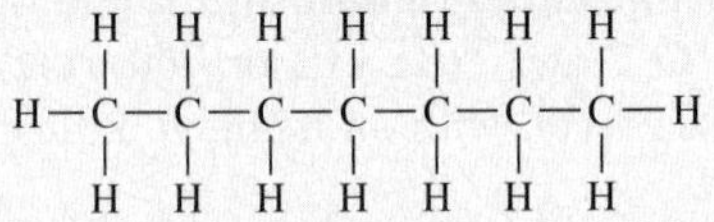	$CH_3CH_2CH_2CH_2CH_2CH_2CH_3$
n-Octane C_8H_{18}	125 °C (liquid)	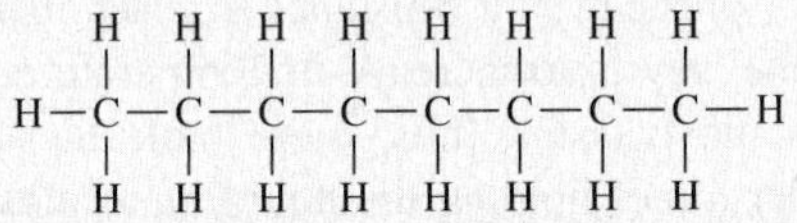	$CH_3CH_2CH_2CH_2CH_2CH_2CH_2CH_3$

Note: n-butane, n-pentane, n-hexane, n-heptane, and n-octane all have other structural forms, known as isomers (see Section 6.12). The n stands for normal, the straight-chain isomer.

To distill crude oil, it must first be pumped into a large vessel (the boiler in Figure 6.27) and heated. The fractionation process then proceeds in the distillation tower where hydrocarbons are separated based on their relative boiling points. Hydrocarbons with lower boiling points (lower molar masses) are vaporized first and rise up the distillation tower. Compounds with higher molar masses and/or more complex chemical structures vaporize at higher temperatures and are collected at lower regions of the distillation tower.

Figure 6.27 illustrates a distillation tower, showing the possible mixtures of compounds that can be obtained. These include solids such as asphalt, liquids such as gasoline, and gases such as methane. The amounts of each depend on the type of crude oil. The "lightest" fraction collected is the *refinery gases*. These compounds have 1–4 carbon atoms per molecule and include methane, ethane, propane, and butane. Refinery gases are flammable and often used as fuel at the refineries. They also can be liquefied and sold for home use. Chemical manufacturers also may use refinery gases as a starting material for making new compounds.

Methane, the main component of natural gas, is collected in the refinery gas fraction. However, as discussed in the previous section, most natural gas is obtained directly from oil and gas wells, rather than from distillation at a refinery. The natural gas piped to your home is practically pure methane, but also includes some ethane (2–6%) and other hydrocarbons of low molar mass. Sulfur-containing compounds such as ethanethiol (CH_3CH_2SH) are added as odorants to assist in the detection of gas leaks.

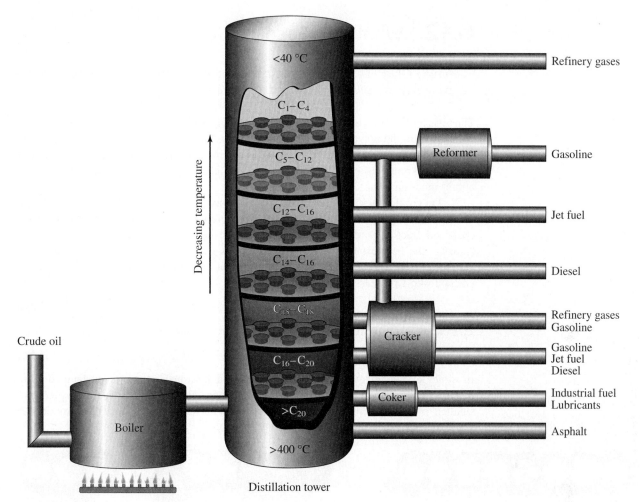

Figure 6.27

Illustration of a distillation tower for crude oil, showing the various fractions and typical applications for the products.

Source: McGraw Hill Education

As seen in Figure 6.28, a wide range of compounds are obtained from a barrel of crude oil. In fact, about 45 gallons of petroleum products are derived from a barrel of crude oil, from which about 37 gallons are burned for heating and transportation. The remaining gallons are used primarily for nonfuel purposes, including the few gallons that serve as "feedstocks" to produce plastics, pharmaceuticals, fabrics, and other carbon-based products.

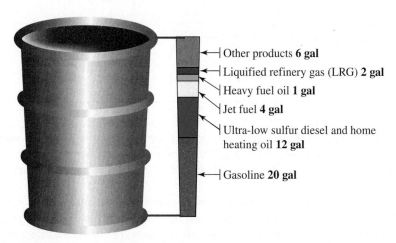

Figure 6.28

Products (in gallons) from the refining of a barrel of crude oil.

Source: U.S. Energy Information Administration, 2018.

6.12 | What's in Gasoline?

> *Learning Objective: Describe the composition of gasoline formulations and explain how gasoline is produced from crude oil*

The distribution of compounds obtained from distilling crude oil does not naturally correspond to the prevailing needs of consumers. For example, the demand for gasoline is considerably greater than that for higher-boiling fractions. Chemists employ several processes, known as *cracking* or *reforming*, to change the natural distribution and to obtain more gasoline of higher quality.

Thermal cracking is a process that breaks large hydrocarbon molecules into smaller ones by heating them to a high temperature. In this procedure, the heaviest crude oil fractions are heated to a temperature between 400 and 450 °C. This heat "cracks" the heaviest tarry crude oil molecules into smaller ones useful for gasoline and diesel fuel. For example, at high temperature, one molecule of $C_{16}H_{34}$ can be cracked into two nearly identical molecules:

$$C_{16}H_{34} \xrightarrow{\text{heat}} C_8H_{18} + C_8H_{16} \qquad \text{[6.9a]}$$

Thermal cracking also can produce different-sized molecules:

$$C_{16}H_{34} \xrightarrow{\text{heat}} C_{11}H_{22} + C_5H_{12} \qquad \text{[6.9b]}$$

In either case, the total number of carbon and hydrogen atoms is unchanged from reactants to products. The larger reactant molecules simply have been fragmented into smaller, more economically important molecules. We can use space-filling models to show the size differences more clearly:

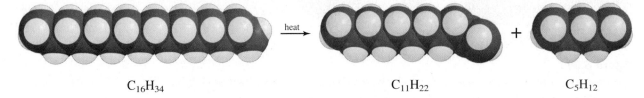

$C_{16}H_{34}$ $\qquad\qquad\qquad\qquad\qquad$ $C_{11}H_{22}$ $\qquad\qquad$ C_5H_{12}

Your Turn 6.25 More Practice with Cracking

a. Draw structural formulas for one pair of products formed when $C_{16}H_{34}$ is thermally cracked (see Equations 6.9a, b).
   Hint: Draw the atoms of the product molecules in a linear chain and include one double bond (only in one product).
b. Write the generic chemical formula for a hydrocarbon with one C=C double bond. Compare this to the general chemical formula of alkanes shown in Section 6.2.
c. Apply your knowledge of intermolecular forces to predict the relative boiling points of $C_{16}H_{34}$, $C_{11}H_{22}$, and C_5H_{12}. Look up the actual values and compare these to your predictions.

Unfortunately, thermal cracking requires a lot of energy to reach the necessary high temperatures. A more selective and inexpensive way to crack larger hydrocarbon molecules is by using catalysts. **Catalytic cracking** can be done at lower temperatures, thus reducing energy use.

Another important chemical process is **catalytic reforming**. Here, the atoms within a molecule are rearranged, usually starting with linear molecules and producing ones with more branches. As we will see, the more highly branched molecules burn more smoothly in automobile engines.

It turns out that molecules with the same molecular formula are not necessarily identical. For example, octane has the formula C_8H_{18}. Careful analysis reveals 18 different

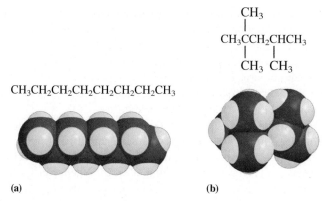

CH₃CH₂CH₂CH₂CH₂CH₂CH₂CH₃

(a)   (b)

Figure 6.29

Structural formula and space-filling model for **(a)** *n*-octane and **(b)** iso-octane. For 3D representations of these structures, go to www.acs.org/cic

compounds with this formula. Molecules with the same molecular formula but with different chemical structures and different properties are called **isomers**. In *n*-octane (*normal* octane), the carbon atoms are all in a continuous chain (Figure 6.29a), whereas in iso-octane, the carbon chain has several branch points (Figure 6.29b). The chemical and physical properties of these two isomers are similar, but they are not identical. Molecules with linear structures will have stronger London dispersion forces than branched structures due to the stronger interaction of C—H groups along the entire chain length.

Your Turn 6.26   IMFs, Boiling Points, and Isomers

Molecular structure influences the boiling points of hydrocarbons because of the London dispersion forces between the molecules in the liquid phase.

a. Explain the two boiling point trends shown below for methane, ethane, propane, and *n*-butane (left), and neopentane vs. *n*-pentane (right):

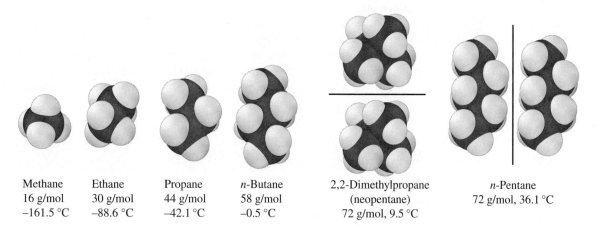

Methane	Ethane	Propane	*n*-Butane	2,2-Dimethylpropane (neopentane)	*n*-Pentane
16 g/mol	30 g/mol	44 g/mol	58 g/mol	72 g/mol, 9.5 °C	72 g/mol, 36.1 °C
−161.5 °C	−88.6 °C	−42.1 °C	−0.5 °C		

b. Predict the boiling point trend between *n*-octane and iso-octane. Explain your answer.

Although the heats of combustion for *n*-octane and iso-octane are nearly identical, the two fuels burn differently in an auto engine. The more compact shape of the latter compound imparts a "smoother" burn. In a well-tuned car engine, gasoline vapor and air are drawn into a cylinder, compressed by a piston, and ignited by a spark. Normal combustion occurs when the spark plug ignites the fuel-air mixture and the flame front travels rapidly across the combustion chamber, consuming the fuel.

Sometimes, however, compression alone is enough to ignite the fuel before the spark occurs. This pre-ignition results in lower engine efficiency and higher fuel consumption because the piston is not in its optimal location when the burned gases expand. "Knocking," a violent and uncontrolled reaction, occurs after the spark ignites the fuel, causing the unburned mixture to burn at supersonic speed with an abnormal rise in pressure. Knocking produces an objectionable metallic sound, loss of power, overheating, and even engine damage when severe.

In the 1920s, knocking was shown to depend on the chemical composition of the gasoline. The "octane rating" was developed to designate a particular gasoline's resistance to knocking. Iso-octane performs exceptionally well in automobile engines and arbitrarily has been assigned an octane rating of 100. When you go to the gasoline pump and fill up with 87 octane, you are buying gasoline that has the same knocking characteristics as a mixture of 87% iso-octane. Higher grades of gasoline also are available: 89 octane (regular plus) and 91 octane (premium) (Figure 6.30). These blends contain a higher percent of compounds with higher octane ratings.

Figure 6.30

Gasoline is available in a variety of octane levels.

Justin Sullivan/Getty Images

Gasoline additives can be used to boost the fuel's octane number. Beginning in the 1920s, the octane-booster tetraethyllead (TEL, $Pb(C_2H_5)_4$) was mixed with gasoline to improve vehicle performance and fuel economy. Lead was preferred over aromatic hydrocarbons or alcohols because of its low production cost. However, there were problems associated with the use of TEL, which you will now discover in this next activity.

Your Turn 6.27   Getting the Lead Out

a. Why was lead banned?
b. Unfortunately, there are still sources of lead around us; use the Internet to identify some key sources.

Due to the health and environmental hazards associated with TEL, the United States began to phase out leaded gasoline in 1973, but it was not fully banned until 1996. Most other high-income countries followed suit. The UN Environment Programme launched a program in 2002 to phase out leaded gasoline around the world; it finally accomplished this goal in July 2021.

However, if lead cannot be used, what else can be done to boost the octane number of gasoline? Although *n*-octane has a poor octane rating, it is possible to catalytically re-form *n*-octane to iso-octane, thus greatly improving its performance. This rearrangement is accomplished by passing *n*-octane over a catalyst consisting of rare and expensive elements such as platinum (Pt), palladium (Pd), rhodium (Rh), or iridium (Ir). The reforming of isomers to improve their octane rating became important as tetraethyllead was being phased out.

Your Turn 6.28   Octane Number

Use the Internet and look up octane numbers for *n*-octane, *n*-heptane, ethanol, methanol, and methyl tertiary-butyl ether (MTBE). Make a table listing your findings and include iso-octane for reference. What patterns do you see in the data?

Elimination of TEL as an octane enhancer necessitated finding substitutes that were inexpensive, easy to produce, and environmentally benign. As you saw in the previous activity, several were evaluated including methanol, ethanol, and MTBE, each with an octane rating greater than 100. Fuels containing these additives are referred to as **oxygenated gasolines**. Because they contain oxygen, these gasoline blends burn more cleanly and produce less carbon monoxide than their nonoxygenated counterparts.

As pointed out in Section 2.11, the volatile organic compounds (VOCs) in conventional gasoline play a role in tropospheric ozone pollution, especially in high-traffic areas. Since 1995, over 100 U.S. metropolitan areas with the highest ground-level ozone levels have adopted the Year-Round Reformulated Gasoline Program mandated by the Clean Air Act Amendments of 1990. This program requires the use of **reformulated gasolines** (RFGs)—oxygenated gasolines that contain a lower percentage of volatile hydrocarbons found in nonoxygenated conventional gasoline. RFGs cannot contain more than 1% benzene (C_6H_6), a cancer-causing hydrocarbon, and must be at least 2% oxygen. Due to their composition, reformulated gasolines evaporate less readily than conventional gasolines and produce fewer carbon monoxide emissions. When RFGs were introduced in the 1990s, MTBE was the oxygenate of choice. However, concerns over its toxicity and its ability to leach from gasoline storage tanks into the groundwater have led many states to ban MTBE and switch to ethanol.

6.13 | New Uses for an Old Fuel

Learning Objective: Illustrate and explain catalyzed chemical reactions

World supplies of coal are predicted to last for hundreds of years, much longer than current estimates of remaining available oil reserves. Unfortunately, the fact that coal is a solid makes it inconvenient for many applications, especially as a fuel for vehicles. Therefore, research and development projects are underway aimed at converting coal into fuels that possess characteristics similar to petroleum products.

Before large supplies of natural gas were discovered, cities were lighted with *water gas*, a mixture of carbon monoxide and hydrogen. Water gas is formed by blowing steam over hot coke, the impure carbon that remains after volatile components have been distilled from coal:

$$\underset{\text{coke}}{C(s)} + H_2O(g) \longrightarrow \underset{\text{water gas}}{CO(g) + H_2(g)} \qquad \textbf{[6.10]}$$

This same reaction is the starting point for the **Fischer–Tropsch process** (Figure 6.31) for producing synthetic gasoline from coal. German chemists Emil Fischer (1852–1919) and Hans Tropsch (1889–1935) developed the process during the early 20th century. At that time, Germany had abundant coal reserves, but little petroleum.

The Fischer–Tropsch process can be described by this general equation:

$$n\, CO(g) + (2n + 1)H_2(g) \xrightarrow{\text{catalyst}} C_nH_{2n+2}(g,\, l) + n\, H_2O(g) \qquad \textbf{[6.11]}$$

Figure 6.31

Pilot plant for coal liquefaction via Fischer–Tropsch. Check out this video to see how Fischer–Tropsch is used to convert biomass into liquid fuel: www.acs.org/cic.

Kim Steele/Photodisc/Getty images

Figure 6.32

Energy-reaction pathway for the same reaction with (**blue** line) and without (**green** line) a catalyst. The green and blue arrows represent the activation energies. The **red** arrow represents the overall energy change for either pathway.

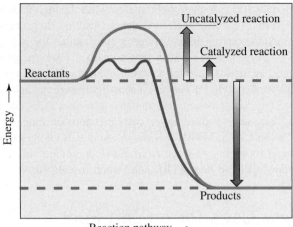

The hydrocarbon products can range from small molecules like methane, to the medium-sized molecules ($n = 5 - 8$) typically found in gasoline. This chemical reaction proceeds when the carbon monoxide and hydrogen are passed over a catalyst containing iron or cobalt.

To better understand the role of the catalyst, consider a typical exothermic reaction, as shown in Figure 6.32. The potential energy of the reactants (left side) is higher than the potential energy of the products (right side) because it is an exothermic reaction. Now, examine the pathways that connect the reactants and products. The green line indicates the energy change during a reaction in the absence of a catalyst. Overall, this reaction gives off energy, but the energy initially *increases* because some bonds need to be broken. The energy necessary to initiate a chemical reaction is called its **activation energy** and is indicated by the green arrow for the uncatalyzed reaction. Although energy must be expended to get the reaction started, energy is given off as the process proceeds to a lower potential energy state. Generally, reactions that occur rapidly have low activation energies; slower reactions have higher activation energies. However, there is no direct relationship between the height of the activation barrier and the net energy change in the reaction. In other words, a highly exothermic reaction can have a large or a small activation energy.

Increasing the temperature often results in increased reaction rates; when molecules have extra energy, a greater fraction of collisions can overcome the required activation energy. Sometimes, however, increasing the temperature isn't a practical solution. The blue line in Figure 6.32 shows how a catalyst can provide an alternative reaction pathway and thus a lower activation energy (represented by the blue arrow), without raising the temperature. However, the net energy change of the chemical reaction (red arrow; energy of products minus energy of reactants) is the same with and without the addition of a catalyst.

In the Fischer–Tropsch process, strong C≡O triple bonds must be broken for the reaction to proceed. Breaking this bond corresponds to an activation energy so large that the reaction simply does not proceed without the help of a metal catalyst. Molecules of CO can form bonds with the metal surface, and when this happens, the C≡O bonds weaken. The hydrogen molecules also attach to the metal surface, completely breaking the H—H single bonds. The rest of the reaction proceeds quickly, producing hydrocarbons with a higher molar mass.

The advantage of a catalyst is that it is not consumed during the reaction and thus it can be used multiple times. In addition, only small amounts of it are needed. Green chemists value catalytic reactions not only because small amounts of catalysts are employed, but also because the reactions often can be carried out at lower temperatures.

Historically, commercialization of Fischer–Tropsch technology has been limited. South Africa has the world's largest implementation of the technology. As a coal-rich and oil-poor nation, it is the only country that synthesizes a majority of its diesel fuel from coal. However, spikes in oil prices, along with plentiful domestic coal supplies and the desire to combust "cleaner" coal products has sparked the use of the Fischer–Tropsch process in other energy-hungry countries such as Qatar, Germany, Finland, India, the U.S., and Malaysia.

Coal, whether solid or converted into liquid fuels, still burns to produce a lot of CO_2. Research by the National Renewable Energy Laboratory indicates that greenhouse gas emissions over the entire fuel cycle for producing coal-based liquid fuels are nearly twice as high as their petroleum-based equivalent. Thus, we need to continue the search for fuels to replace coal.

6.14 | From Brewery to Fuel Tank: Ethanol

> *Learning Objective: Describe the production and use of biofuels (ethanol)*

Some people think that a more sustainable energy future will require the increased use of **biofuels**, a generic term for a renewable fuel derived from a biological source such as trees, grasses, algae, animal waste, or agricultural crops. Biofuels can replace fuels derived from crude oil, such as gasoline and diesel fuel.

Why are biofuels attractive alternatives? The fuel is renewable; plants used to make today's biofuels will regrow within a few years (versus the millions of years needed to form fossil fuels). In addition, when the plant regrows, it will absorb CO_2. In principle, the CO_2 released when this biofuel is used to generate energy is the same amount of CO_2 that will be absorbed when it grows again; thus, it is a **carbon neutral** process. In practice, this is not the case (see Section 6.16). Based on the first law of thermodynamics, whether burned as a fuel or decayed naturally, the plants release the same amount of CO_2 into the biosphere. Biofuels are also attractive because they are a domestic fuel source and they can generate income for farmers.

Wood, the most common biofuel, has been used throughout human history for cooking and heating. Have you ever wondered why wood burns? Wood contains **cellulose**, a naturally occurring compound composed of C, H, and O that provides structural rigidity in plants, shrubs, and trees. Similar to hydrocarbons, cellulose is made up of carbon and hydrogen. Unlike these, however, cellulose also contains oxygen, which lowers its energy content as a fuel. In fact, all of the biofuels we will describe in this section contain some oxygen. As their oxygen content increases, biofuels release proportionately less energy per mass than hydrocarbons during their combustion (revisit Figure 6.5).

Cellulose is a carbohydrate composed of a chain of thousands of glucose molecules linked together. This is why we equated burning wood to burning glucose in Figure 6.5. Thus, the chemical equation for burning wood is the same as that of respiration (Equation 6.2).

Even though widely available in many parts of the world, the supply of wood is insufficient to meet our energy demands. Cutting down trees for fuel also destroys entire ecosystems, not to mention the elimination of vegetation that we depend on to effectively absorb CO_2 from our atmosphere. So, instead of relying on wood, people are eyeing liquid biofuels such as ethanol.

Ethanol (C_2H_5OH) is the same alcohol found in wine, beer, and spirits. It is a clear, colorless, and flammable liquid. Since ancient times, people have known how to ferment grain in order to produce ethanol. Admittedly, their purpose was to brew alcoholic beverages rather than to fuel automobile engines.

Which sugars and grains can be fermented? Almost any will, although some may require a catalyst to nudge the process along. The choice depends on both availability and politics. The largest producer of ethanol as a biofuel is the United States (Figure 6.33) where the feedstock is starch from corn kernels (Figure 6.34). In early human history, corn was not as widely available as it is today. As you'll learn in Chapter 13, the people of the New World bred corn from a wild strain and made a crop that readily grew. Those living elsewhere used other grains such as rice and barley to brew alcoholic beverages.

Production of Ethanol (Million Gallons)

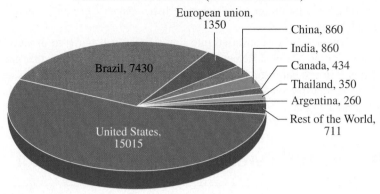

Figure 6.33

Global 2021 ethanol production in millions of gallons.

[Image made by Nicole C. Bouvier-Brown using data from the Renewable Fuel Association (RFA)]

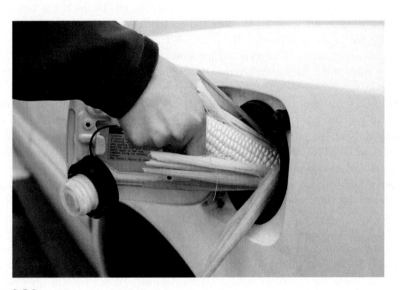

Figure 6.34

An advertisement for ethanol, a renewable fuel that can be produced from many different grains, including corn.

fluxfoto/E+/Getty Images

In the context of oxygenated gasoline, we introduced ethanol in Section 6.12. Below is its Lewis structure:

$$H - \overset{\displaystyle H}{\underset{\displaystyle H}{C}} - \overset{\displaystyle H}{\underset{\displaystyle H}{C}} - \overset{\displaystyle \cdot\cdot}{\underset{\displaystyle \cdot\cdot}{O}} - H$$

ethanol

Ethanol is an example of an **alcohol**, a hydrocarbon substituted with one or more —OH groups (hydroxyl groups) bonded to its carbon atoms. Alcohols are hydrocarbons that contain a **functional group**; that is, a distinctive arrangement of a group of atoms that imparts characteristic properties to the host molecule. To emphasize the hydroxyl group (—OH) present in any alcohol, chemists typically write C_2H_5OH for ethanol rather than C_2H_6O. Just like hydrocarbons, alcohols are flammable and burn to release energy. During complete exothermic combustion, the products are CO_2 and H_2O:

$$C_2H_5OH(l) + 3\ O_2(g) \longrightarrow 2\ CO_2(g) + 3\ H_2O(g) + 1240\ kJ \qquad \textbf{[6.12]}$$

Because alcohols contain one or more —OH groups, their properties differ from those of hydrocarbons. One difference is that humans can safely consume small amounts of ethanol in wine, beer, and other alcoholic drinks. In contrast, hydrocarbons have no appeal as a beverage, and many are *carcinogenic* (cancer-causing). Another difference is their solubility. For example, because it is a polar molecule, ethanol dissolves readily in water, whereas hydrocarbons are insoluble in water.

Several steps are required to obtain ethanol from corn. The first involves making a "soup" of corn kernels and water. The second step uses **enzymes** (a biological catalyst) to breakdown the starch molecules found in these kernels into glucose. Like cellulose, **starch** is a carbohydrate found in many grains, such as corn and wheat. It is a natural polymer of glucose; like cellulose, starch molecules are composed of a chain of thousands of glucose molecules linked together. But unlike cellulose molecules, the links in starch are formed differently such that the enzymes in our bodies can break them. Thus, we can digest the starch in foods such as potatoes and rice; we cannot digest the cellulose found in paper and leafy foods.

The third step, fermentation, converts glucose into ethanol. Yeast cells take over by releasing different enzymes that catalyze this conversion:

$$\underset{\text{glucose}}{C_6H_{12}O_6} \xrightarrow{\text{yeast enzymes}} 2 \underset{\text{ethanol}}{C_2H_5OH} + 2 CO_2 \qquad \textbf{[6.13]}$$

The result is an alcoholic brew with an alcohol content of about 10–20%. Ethanol is finally separated using distillation because ethanol and water differ in their boiling points. Figure 6.35 shows the tall distillation towers at an ethanol plant.

Together, the United States and Brazil account for about 82% of the world's production of ethanol biofuel (Figure 6.33). While corn is the raw material in the United States, Brazil derives almost all of its ethanol by fermenting sugarcane. Why the difference?

To answer this question, we again note that practically any sugar or grain can be fermented to produce ethanol. The substance fermented depends on its availability, economics, and politics. Sugarcane plants are rich in sucrose, also called "table sugar." Just as the glucose produced from the starch of corn kernels can be fermented to produce ethanol, so can sucrose. In Brazil, sugarcane is grown in areas that once were tropical rainforests. In the United States, corn is grown in the Midwest in areas that were once prairies and forests. In both cases, the use of land to produce biofuel remains a controversial topic. Either ecosystems are destroyed to make room for growing biofuels or land once used to grow food is now being used to create fuel.

Ethanol can also be made from cellulose, a compound mentioned earlier that gives support to plants, shrubs, and trees. **Cellulosic ethanol** is ethanol produced from any plant containing cellulose, typically cornstalks, switchgrass, wood chips, and other materials that are inedible by humans. Although you may not recognize the name switchgrass, if you live in the Plains regions of the United States, Canada, or Mexico,

Ethanol Production

Watch this video (www.acs.org/cic) and outline the steps of ethanol production from corn. Comment on the steps that require energy and what the likely sources of that energy would be.

Figure 6.35

The Archer Daniels Midland ethanol plant in Peoria, Illinois.

David R. Frazier Photolibrary, Inc./Alamy Stock Photo

Figure 6.36

Switchgrass, a perennial plant that is native to North America.

Warren Gretz/NREL/U.S. Dept. of Energy

you most likely have seen it. Switchgrass is a native plant; Figure 6.36 shows one of its varieties.

Derived from the inedible parts of plants, cellulosic ethanol has a widespread appeal. For years, chemists have been successful at producing small amounts of cellulosic ethanol in the laboratory. But carrying out the process in batches large enough to obtain millions of gallons of ethanol has turned out to be another matter entirely. Like starch, cellulose does not ferment and therefore must first be broken down into sugars. However, the enzymes that catalyze the breakdown of cellulose are expensive and their rate of reaction is slow, which has greatly limited the use of cellulosic ethanol plants.

Your Turn 6.29   Biofuel from Nonfoods

List three desirable characteristics of sources of cellulosic ethanol, such as wood chips and switchgrass.

Regardless of its source, ethanol doesn't go straight into your gas tank, because most of our automobiles are not engineered to burn it alone. However, auto engines can run on "gasohol," a blend of gasoline with ethanol. Currently, more than 30 countries have mandated that gasoline contain up to 10% ethanol (known as E10) (Figure 6.37). This blend and other "oxygenated" fuels have the added benefit of higher octane ratings and reducing vehicle emissions that produce ground-level ozone.

Whether using E10 or E85, don't confuse its octane rating with gas mileage. The octane rating relates to how smoothly the fuel burns in the engine, rather than to the energy content of the fuel. The octane rating of a gasoline increases as more ethanol is added. But with more ethanol, the gas mileage decreases slightly.

Why? Recall that ethanol releases a lower amount of energy per amount burned than do the hydrocarbons found in gasoline. Using *n*-octane as a representative hydrocarbon, here are the two combustion equations:

$$C_2H_5OH(l) + 3\ O_2(g) \longrightarrow 2\ CO_2(g) + 3\ H_2O(g) + 1240\ kJ \qquad \textbf{[6.14]}$$

$$C_8H_{18}(l) + \frac{25}{2}\ O_2(g) \longrightarrow 8\ CO_2(g) + 9\ H_2O(g) + 5060\ kJ \qquad \textbf{[6.15]}$$

Per gram, the values are 26.8 kJ/g for C_2H_5OH and 44.4 kJ/g for C_8H_{18} (Figure 6.4). Ethanol releases less energy because it contains oxygen. As a fuel, ethanol already is partially oxidized, or "burned."

Figure 6.37

Gasoline is often blended with ethanol to make E10 ("gasohol") that is 90% gasoline and 10% ethanol. E85 is 85% ethanol and 15% gasoline.

(left): Jeffrey Sauger/Bloomberg/Getty Images; *(right):* Ashley Cooper pics/Alamy Stock Photo

Your Turn 6.30 Energy Density of Ethanol

If a car has an 8.0-gallon fuel tank, how much energy would be generated from (a) gasoline versus (b) ethanol? The density of gasoline is 800 g/L and the density of ethanol is 789 g/L. Assume that the combustion of gasoline can be approximated by that of *n*-octane as shown above.

As stated earlier, ethanol is not the only biofuel. The next section is devoted to *biodiesel*.

6.15 | From Deep Fryer to Fuel Tank: Biodiesel

Learning Objective: Describe the production and use of biofuels (biodiesel)

The production of biodiesel has grown in recent years. Its synthesis is so straightforward that you may have carried it out in your chemistry lab. Biodiesel is unique among transportation fuels in that it can be produced economically in small batches by individual consumers, including students.

Although biodiesel is made primarily from vegetable oils, animal fats work equally well. As you may already know, oils and fats are part of your diet and help fuel your body. Although biodiesel could be synthesized from olive oil or butter, both are too expensive (and too tasty) to use as a starting material. Instead, biodiesel is made from soy, rapeseed, or palm oil. It also can be produced from waste cooking oil, such as that used to make french fries. Figure 6.38 shows a carton of oil that is destined for a restaurant fryer, together with the waste cooking oil.

To understand why fats and oils can serve as starting materials for biodiesel, we need to know more about **triglycerides**, a class of compounds that includes both fats and oils. **Fats** such as butter and lard are triglycerides that are solids at room temperature. In contrast, **oils** such as olive oil and soybean oil are triglycerides that are liquids. Triglycerides are the starting material for biodiesel and they occur naturally in both plants and animals.

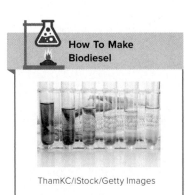

How To Make Biodiesel

ThamKC/iStock/Getty Images

Check out this video to learn how to make your own biodiesel and determine its energy content: www.acs.org/cic. Outline the procedure shown here, including the necessary ingredients. How does this method work? Write out an equation to summarize the chemistry.

(a)

(b)

Figure 6.38

Restaurants purchase oil for frying in large amounts. Shown here are **(a)** a 35-pound box and **(b)** hot oil out of the fryer. Depending on the cook, the cooking oil is changed frequently (shown here) or infrequently. In the latter case, the oil darkens with waste products.

(a–b): ©Cathy Middlecamp

For a 3D illustration of this molecule, go to www.acs.org/cic

Here is the structural formula for glyceryl tristearate, a triglyceride found in animal fat:

$$H_3C-(CH_2)_{16}-\overset{\overset{O}{\|}}{C}-O$$

$$H_3C-(CH_2)_{16}-\overset{\underset{\|}{O}}{C}-O-\overset{H_2}{C}-CH-\overset{H_2}{C}-O-\overset{\underset{\|}{O}}{C}-(CH_2)_{16}-CH_3$$

Actually, any triglyceride would serve our purposes because they all share common structural features. We picked this particular one because you will encounter it later in the context of nutrition (Chapter 11).

Glyceryl tristearate is a complex molecule. Even so, this structure contains three hydrocarbon chains, illustrated above in red. Can you see their resemblance to hydrocarbon fuels? Each of these hydrocarbon chains, if snipped off, could serve as diesel fuel, a mixture of hydrocarbons with 14 to 16 carbon atoms (revisit Figure 6.27). When you eat foods containing triglycerides, each "diesel-like" hydrocarbon chain is metabolized slowly in your body to release energy and produce CO_2 and H_2O. Although diesel fuel burns in an engine much more rapidly and at a higher temperature, the net result is the same. Energy is released, and CO_2 and H_2O are produced.

Although soybean oil and other triglycerides will burn, they should not go straight into your gas tank. Before triglycerides can be utilized as a fuel, they need to be snipped into smaller pieces that are closer in size (for ease of evaporation) to the molecules in diesel fuel. One way to do this is to react them with an alcohol such as methanol (CH_3OH) and a catalytic amount of sodium hydroxide (NaOH). Equation 6.16 lists the chemical reaction, using glyceryl tristearate (an animal fat) as the starting material.

$$C_{57}H_{110}O_6 + 3\ CH_3OH \xrightarrow{\text{NaOH}} 3\ CH_3(CH_2)_{16}\overset{\overset{O}{\|}}{C}OCH_3 + C_3H_8O_3 \qquad [6.16]$$

glyceryl tristearate (a triglyceride)   methyl stearate (a biodiesel molecule)   glycerol

One triglyceride molecule produces three biodiesel molecules that each contain a long chain of carbon atoms. Depending on the fat or oil used as a starting material, other biodiesel molecules are possible.

Your Turn 6.31   Cherries and Bananas

Have you ever been served cherries jubilee or bananas flambé? A chef first douses the fruit in a liquor and then lights it with a match. The ethanol in the liquor burns with a dramatic blue-tinged flame.

Moving Moment/Shutterstock

a. Write the chemical equation for the complete combustion of ethanol—that is, burning it with enough oxygen so that only carbon dioxide and water are produced.

b. Although it wouldn't be tasty, you also could pour cooking oil or biodiesel on fruit and set it afire. Write the balanced chemical equation for the complete combustion of methyl stearate, $C_{19}H_{38}O_2$, the biodiesel product in Equation 6.16.

The previous activity illustrates that ethanol and biodiesel, like fossil fuels, contain carbon and thus produce carbon dioxide when burned. Although the context for the activity was culinary, these chemical reactions also occur in automobile engines.

In general:

- Biodiesel molecules contain a hydrocarbon chain, typically with 16–20 carbon atoms.
- The hydrocarbon chains usually contain one or more C=C bonds, especially if the triglyceride used as a starting material is an oil.
- In addition to the hydrocarbon chain, each biodiesel molecule contains oxygen. The two O atoms form part of the *ester* functional group, which we will describe later in the context of polyesters in Section 9.6.
- Triglycerides (fats and oils) typically produce a mixture of different biodiesel molecules, unlike the triglyceride glyceryl stearate that produced just one product.

Also notice the three molecules of methanol in Equation 6.16. Methanol provides the $-OCH_3$ group to "cap" each carbon chain at the point it was snipped from the larger triglyceride molecule. Other alcohols, including ethanol, can work equally well. No matter which alcohol is used, the net result is three molecules of biodiesel.

Your Turn 6.32   More About Alcohols

In this chapter, you already have encountered two alcohols: ethanol and methanol.

a. Draw the structural formula for each.
b. Here is a new alcohol: $CH_3CH_2CH_2OH$. Suggest a name.
c. There is another alcohol with the same chemical formula as $CH_3CH_2CH_2OH$; that is, an isomer of this alcohol. Draw its structural formula.

It should be evident that many different alcohols can be derived from hydrocarbons. All you need to do is substitute an $-OH$ group for one of the H atoms. Furthermore, alcohols can contain more than one $-OH$ group. A byproduct of biodiesel synthesis is one such multifunctional alcohol, **glycerol**. In Equation 6.16, we gave only its chemical formula, $C_3H_8O_3$. From its structural formula, you can see that glycerol is a "triple" alcohol:

$$\begin{array}{ccccc} & H & H & H & \\ & | & | & | & \\ H- & C & -C & -C & -H \\ & | & | & | & \\ & OH & OH & OH & \end{array}$$

Glycerol is used in many different consumer products. You may know it by the name *glycerin*, a common ingredient in soaps and cosmetics (Figure 6.39). However, every 9 pounds of biodiesel nets 1 pound of glycerol, which has resulted in a glut of glycerol on the market. In 2006, Galen Suppes and coworkers at the University of Missouri earned a Presidential Green Chemistry Challenge Award for a process to convert glycerol into a different alcohol, propylene glycol:

$$\begin{array}{ccccc} H & H & H \\ | & | & | \\ H-C-C-C-H & + & H_2(g) \end{array} \xrightarrow{\text{copper catalyst}} \begin{array}{ccccc} H & H & H \\ | & | & | \\ H-C-C-C-H & + & H_2O(l) \end{array} \quad [6.17]$$
$$\quad\;\; OH\; OH\; OH \qquad\qquad\qquad\qquad\qquad\quad OH\; OH\; H$$

glycerol   propylene glycol

The U.S. Food and Drug Administration (FDA) lists propylene glycol as "generally recognized as safe" and has approved it for use as a food additive. It is used as a preservative in packaged food, as a moisturizer in cosmetics, and as a solvent for some drugs not soluble in water. Propylene glycol produced in this manner is from a renewable resource, not requiring petroleum as a feedstock. Conversion of the glycerol into

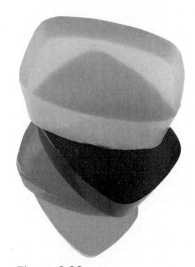

Figure 6.39

Glycerin (glycerol) is one of the ingredients in translucent soaps.

Stockbyte/Getty Images

value-added products lowers the cost of biodiesel production, making it more competitive with petroleum-derived diesel fuel.

Your Turn 6.33   Heat of Combustion for Biodiesel

Revisit the structural formula for a biodiesel molecule.

a. Would you predict that the heat released per gram of biodiesel burned is higher or lower than that of octane? Explain your reasoning.
Hint: Revisit Figure 6.5 to see the values for different fuels.

b. If instead the comparison were made on the basis of 1 mole of biodiesel versus 1 mole of octane, which would release more heat if burned?

c. It is more useful to make comparisons per gram of fuel rather than per mole of fuel. Explain.

Biodiesel is blended with petroleum-based diesel fuel, just like ethanol is blended with gasoline. For example, B20 is 20% biodiesel and 80% diesel fuel (Figure 6.40a). Blends up to 20% are fully compatible with any diesel engine, including those in medium- and heavy-weight trucks. As of 2022, biodiesel blends were available at more than 800 retail locations in the United States, which can easily be located via the Internet (Figure 6.40b).

In these last two sections, we have discussed two biofuels: ethanol and biodiesel. In the process, we have hinted at the complexities and the controversies involved. In the final section of this chapter, we take a critical look not only at biofuels, but also at the larger energy picture. We need to find a sustainable way forward. But are biofuels really as sustainable as proponents claim?

(a)

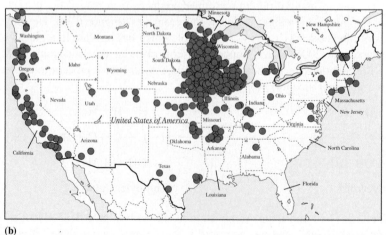

(b)

Figure 6.40

(a) B20 biodiesel is a mixture of 80% petroleum diesel and 20% biodiesel. **(b)** Locations of the biodiesel stations in the U.S. (2022).

(gas station): L.G. Patterson/AP Images

6.16 | Are Biofuels Really Sustainable?

> *Learning Objective: Evaluate the sustainability of biofuels*

Biofuels like ethanol and biodiesel are renewable sources of energy, but are they sustainable? Renewable energy comes from sources that naturally replenish themselves on a human timescale. Additional examples include wind, solar, hydro, and biomass power. Currently, renewable energy sources make up less than 15% of the total global energy consumption, but 28% of electricity use; biofuels are a mere 1% (Figure 6.41). Here, we focus on biofuels because of their connections to gasoline and diesel fuel; as liquids, they can be pumped to fuel vehicles and aircraft. In Chapter 7, we will discuss a variety of alternative energy sources.

Sustainable energy can be maintained for the foreseeable future without compromising or threatening the ability of future generations to meet their needs. Scientists aren't the only ones responsible for a sustainable planet. Recall that in Section 1.9 we mentioned the three pillars of sustainability: the environmental pillar, the social pillar, and the economic pillar. Traditionally, the world of business has been focused on the bottom line–turning a profit. Today, corporations should also be judged on additional factors such as the well-being of their workers and how well they protect the health of the environment, including the quality of the air, water, and land. Taken together, this three-way measure of the success of a business based on its benefits to profit, people, and the planet has become known as the **Triple Bottom Line** (Figure 6.42). The economy must be healthy—that is, the annual reports need to show a profit. But no economy exists in isolation. It connects to a community whose members also need to be healthy. In turn, communities connect to ecosystems that need to be healthy. Hence, Figure 6.42 includes three connecting circles and their intersection is the "Green Zone." This area represents the conditions under which the Triple Bottom Line is met.

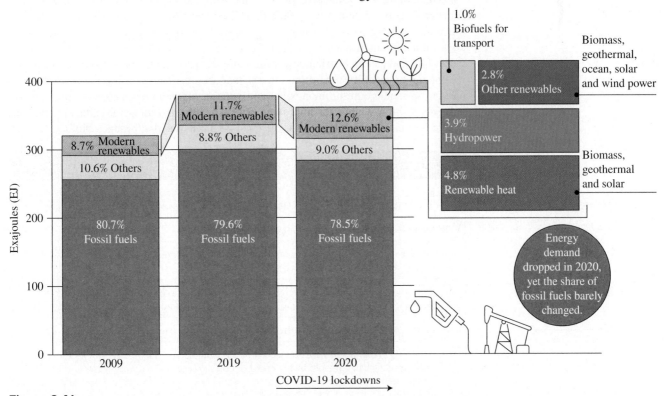

Share of Modern Renewable Energy, 2009, 2019 and 2020

Figure 6.41

Renewable energy share of global energy consumption. "Others" includes traditional biomass (wood, agricultural waste, and animal dung) and nuclear power.

Source: Renewable Energy Policy Network for the 21st Century (2022), Renewables 2022: Global Status Report, Paris, REN21 Secretariat.

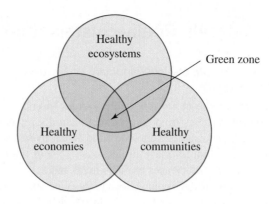

Figure 6.42

A representation of the Triple Bottom Line. The "Green Zone" (where the Triple Bottom Line is achieved) lies at the intersection of these three circles.

Let's Debate!   Concerns about Biofuels

Form two groups: one group will be in favor of the topic (the proposition) and the other group will be against the topic (the opposition). Both groups will try to persuade a neutral person or judge to agree with them. The topic of the debate is the motion. Here is the motion:

Biofuels contribute to a net reduction of greenhouse gas emissions.

Curiosity Conversation

- After searching the Internet and analyzing the ethical principles outlined by the Nuttfield Council on Bioethics, frame your arguments based on healthy ecosystems and communities.
- How do the greenhouse gas emissions differ for first-, second-, and third-generation biofuels?

Are ethanol and biodiesel sustainable? Answers depend on where and how each particular biofuel is produced. Let's begin with the tally for biodiesel because it is somewhat more straightforward than that of ethanol. Recall that biodiesel has several inherent advantages over ethanol. Its synthesis from oils is relatively easy and can be done either in small batches or on a large scale. Biodiesel blends well with existing diesel fuel and can be distributed via the same infrastructure. Like ethanol, and because it contains oxygen, it burns more cleanly than diesel fuel, releasing lower amounts of particulate matter, carbon monoxide, and volatile organic compounds. In the balance, it seems to be a winner for improving public health, assuming that ethical principles (Table 6.3) have been followed for the local communities that produce biofuel. But can

Table 6.3	Ethical Principles to Apply to Current and Future Use of Biofuels
1. Biofuels development should not be at the expense of peoples' essential rights (including access to sufficient food and water, health rights, work rights, and land entitlements).
2. Biofuels should be environmentally sustainable.
3. Biofuels should contribute to a net reduction of total greenhouse gas emissions and not exacerbate global climate change.
4. Biofuels should develop in accordance with trade principles that are fair and recognize the rights of people to just reward (including labor rights and intellectual property rights).
5. Costs and benefits of biofuels should be distributed in an equitable way.
6. If the first five principles are respected and if biofuels can play a crucial role in mitigating dangerous climate change, then depending on additional key considerations, there is a duty to develop such biofuels. These additional key considerations are absolute cost, alternative energy sources, opportunity costs, the existing degree of uncertainty, irreversibility, degree of participation, and the notion of proportionate governance.

Nuffield Council on Bioethics, Biofuels: Ethical Issues, 2011, 84

(a) (b)

Figure 6.43

(a) A plantation of young palm trees in Malaysia; **(b)** a palm oil factory in Malaysia.

(a): kptan/123RF; *(b):* Shariff Che'Lah/123RF

we assume this? Figure 6.43 shows a plantation and factory in Malaysia, one part of the world where palm oil is produced. The next activity offers the opportunity to further explore palm oil production.

Your Turn 6.34 Palm Oil, Biodiesel, and Ethics

In 2008, an Oxfam report noted: "The big losers from the rich countries' biofuel boom are poor people, at risk from spiraling food prices, and a 'scramble to supply' that places their land rights, labor rights, and human rights under threat." Use palm oil as a case study. It is produced in many parts of the world, including Indonesia and Malaysia. Prepare a one-page briefing that identifies the key issues for your classmates.

The sustainability of ethanol production is more difficult to assess. As noted earlier, ethanol originates from several feedstocks, including corn and sugarcane. Understandably, the environmental tally for each one must be assessed separately. No matter which feedstock, scientists and citizens alike are questioning the sustainability of ethanol production.

In terms of energy costs, the good news is that the Sun freely provides the energy for plants to grow. The bad news, however, is that growing corn requires additional energy inputs. Depending on the political climate, tax credits or subsidies might be offered to farmers or ethanol producers to offset costs. Planting, cultivating, and harvesting all require energy. The same is true for watering the crops, producing and applying the fertilizers, manufacturing and maintaining the necessary farm equipment, and distilling the alcohol from the fermented grains. Currently, this energy is supplied by burning fossil fuels at a significant monetary cost and with significant carbon dioxide emissions. The overall energy cost for corn ethanol is difficult to quantify and may outweigh the energy content of the ethanol produced.

Next, we consider the social bottom line. Many rural communities in the Midwest have benefited greatly from booming ethanol demand. Construction of a distillery provides not only jobs to local workers, but also a buyer for locally grown corn. Several communities, hit hard by the depressed viability of family farms, have experienced a resurgence thanks to demand for ethanol. There are certain drawbacks as well. Increased demand for corn leads to increased prices on many other products (especially foods that were displaced by corn). Everyone in that community, and others around the world, must pay these higher prices.

Last, we turn to the environmental bottom line. Growing corn relies on the heavy use of fertilizers, herbicides, and insecticides. The manufacturing and transport of these chemicals requires the burning of fossil fuels, which in turn releases carbon dioxide. Furthermore, these chemicals, once introduced to the field, degrade the soil, air, and water quality. Although corn growers can and do follow responsible practices, they certainly face challenges when called on to produce more corn.

Where to go from here? Clearly, the choices are not easy. We have burned our way through many of the easy and conventional sources of energy. Our future choices are complex and involve trade-offs.

Your Turn 6.35　　Group Activity: Life-Cycle Analysis of Biofuels

Ethanol and biodiesel have both received praise and skepticism. Their supporters tout the fuels as solutions to the environmental problems associated with burning gasoline and diesel, such as global climate change. This is because biofuels are thought to reduce net emissions of carbon dioxide since plant-based feedstocks capture carbon dioxide from the atmosphere via photosynthesis during their growth. However, skeptics believe that the benefits of this "carbon credit" are dwarfed by emissions incurred during the production, distribution, and use of biofuels. One tool, life-cycle analysis (LCA), has been particularly useful in generating information about the environmental benefits and constraints of biofuels.

LCA examines the environmental impact of a product from "the cradle to the grave." Specifically, LCAs quantify and evaluate all raw materials and energy consumed as well as wastes discharged beginning with product design and the sourcing of the raw materials from the earth through manufacturing, distribution, use, maintenance, and disposal.

LCAs can be used in various ways. They are often used to compare two products. For example, questions about the environmental impacts of paper vs. plastic bags have been extensively explored. LCAs can be used by a product's manufacturer to identify processes or materials used that have a high environmental impact and make substitutions to lower the impact, or to educate consumers about a product. In this activity, you will construct the life cycles of two renewable fuels, ethanol and biodiesel. Then, using information from published LCAs, you will evaluate and compare their environmental impacts.

Materials

- Colored beads—20 of each: green, blue, red, orange, and yellow

Activity

1. The steps, inputs, and outputs involved in the production of ethanol and biodiesel are listed below. Arrange these items in flowchart diagrams that you feel best describe the life cycle of each fuel.

   For corn ethanol and biodiesel from soy:

   Ethanol/Biodiesel production at a plant
   Fertilizer
   Pesticides
   Energy
   Refueling station
   Carbon sequestration
   GHG Emissions
   Transportation
   Corn/Soybean cultivation and production
   Water
   Consumer vehicle operation
   Chemicals and materials

   Note: Some of these are processes and some are physical materials. Also, some items may be used once and some will be used multiple times.

2. Your diagrams should illustrate that in the production and use of biofuels, environmental impacts are incurred. However, the amount of resources used or pollutants generated varies greatly between fuels. You will now represent the relative amounts of resources used or pollutants generated by ethanol and biodiesel production and use in your diagrams.

   You have 20 green, blue, red, orange, and yellow beads. Each represents a different resource used or pollutant generated in your life-cycle diagram.

Green = Fertilizer Use (20 beads)
Blue = Water Use (20 beads)
Red = Pesticide Use (20 beads)
Orange = Energy Use (20 beads)
Yellow = GHG Emissions (20 beads)

As a group, divide the beads between your two diagrams based on your predictions of the relative amounts of a pollutant produced or resource used. For example, you have 20 beads representing fertilizer use. Do you know something about corn (ethanol) that would lead you to believe that its cultivation requires more fertilizer than that of soy (biodiesel)? If so, place more green beads on your drawing representing ethanol production and use. Distribute all of your beads in a similar matter.

Tell your instructors when you have finished distributing the beads between your ethanol and biodiesel production diagrams. They will give you some actual values of the relative amounts of GHG emissions, and fertilizer, pesticide, energy, and water use involved in the life cycle of the two fuels. Use these data to redistribute the beads between your diagrams.

Questions to Consider

a. How did your predictions compare to the actual values?
b. Are there certain inputs or outputs that you feel are less harmful for the environment? Justify.
c. What do you conclude about the relative environmental impacts of biodiesel and ethanol? Justify.

Your Turn 6.36   Best Fuel for Cars

Society cannot function without cars, so what is the best fuel for these vehicles? Think about what you have learned throughout this chapter and then watch this video (www.acs.org/cic). Summarize your answer to the question "what is the best fuel for passenger cars?"

Conclusion

Fire! To early humans, fire was a source of security. It warded off animals, brought the ability to cook and preserve food, and minimized the spread of some diseases. Fires were an important social vehicle, a place to gather and share stories. Fire also allowed people to venture into colder regions of the planet.

Today, combustion still remains central to our human community. We use it daily to cook, to heat or cool our dwellings, to produce goods and crops, and to travel the roads, rails, waterways, and skies of our planet. Few chemical reactions have as far-reaching consequences for our health, well-being, and productivity as the ability to burn fuels.

As we have seen in this chapter, the process of combustion converts *energy* into less useful forms. For example, when we burn a mixture of hydrocarbons such as gasoline, we dissipate some of the potential (stored) energy it contains in the form of heat. Although the *forms* of energy change, the total amount of energy before and after any transformation remains the same, according to the first law of thermodynamics.

We have also seen that the process of combustion converts *matter* into less useful and sometimes even undesirable forms. For example, the products of complete combustion—carbon dioxide and water—are not currently usable as fuels. Furthermore, as seen in Chapter 4, CO_2 is a greenhouse gas linked to global climate change. Products of incomplete combustion, such as carbon monoxide and soot, are undesirable because of their effects on human health. The same is true for NO_x air pollutants that are formed in the high temperatures of flames.

Today, fossil fuels power the planet. So do renewable fuels, but to a much smaller extent. What will future generations use for energy sources? Renewable biofuels such as ethanol and biodiesel will compose part of our energy future. Like all fuels, they need to align with our values, including stewardship and sustainability. However, there are other possible ways of satisfying our ever-increasing appetite for energy, such as nuclear and solar energies, which will be discussed in the next chapter.

LEARNING OUTCOMES

The numbers in parentheses indicate the sections within the chapter where these outcomes were discussed.

Having studied this chapter, you should now be able to:

- identify the characteristics of ideal fuels (6.1)
- describe the formation of fossil fuels and why they are unsustainable (6.1)
- understand the combustion process and describe it using balanced chemical equations (6.2)
- write structural formulas for hydrocarbons (6.2)
- explain terms relating to types of energy, heat, and temperature with respect to molecular changes (6.3)
- describe how to measure the energy of combustion reactions and predict whether a reaction or process will be endothermic or exothermic (6.4)
- draw and interpret energy diagrams for chemical reactions (6.4)
- calculate the amount of energy absorbed or released by breaking or forming chemical bonds (6.5)

- describe the operation of combustion-based power plants (6.6)
- describe energy conversion and factors that influence its efficiency (6.7)
- describe the production, consumption, and consequences of coal (6.8), oil (6.9), and natural gas (6.10) combustion
- predict boiling points based on the strength of intermolecular forces and explain how these differences lead to fractionation of crude oil (6.11)
- describe the composition of gasoline formulations and explain how gasoline is produced from crude oil (6.12)
- illustrate and explain catalyzed chemical reactions (6.13)
- describe the production and use of biofuels - ethanol (6.14) and biodiesel (6.15).
- evaluate the sustainability of biofuels (6.16)

Questions

Emphasizing Essentials

1. **a.** List five fuels. Name at least two properties that these fuels share.

   b. Of the fuels you listed, which are fossil fuels or those derived from them?

   c. Of the fuels you listed, which are renewable?

2. The combustion of coal releases several substances into the air.

   a. Of these substances, one is a gas that is produced in large amounts. Give its chemical formula and name.

   b. In contrast, the amount of SO_2 (sulfur dioxide) released is relatively small. Even so, this SO_2 is of concern. Explain why.

   c. Another gas produced in small amounts is NO (nitrogen monoxide). However, coal contains very little nitrogen. What is the origin of the nitrogen in NO?

   d. When coal burns, fine particles of soot may be released. What are the health concerns with $PM_{2.5}$, the smallest of these particles?

3. Revisit Figure 6.9, which depicts the components of an electric power plant.

   a. For a coal-fired power plant, where would the coal appear in the figure?

   b. Water is part of two separate loops. One loop connects the boiler and turbine and is usually under pressure. Explain why.

   c. Another loop brings in (and out) water from a lake or river. Explain why a large body of water is needed.

4. Energy exists in different forms in our natural world. In Figure 6.9, identify where:

   a. Potential (stored) energy of the fuel is converted into heat.

   b. Kinetic energy of water molecules is converted into mechanical energy.

   c. Mechanical energy is converted into electrical energy.

   d. Electrical energy is converted into forms such as heat and light.

5. A coal-burning power plant generates electrical power at a rate of 500 megawatts (MW), or 5.00×10^8 J/s. The plant has an overall net efficiency of 37.5% (0.375) for the conversion of heat to electricity.

   a. Calculate the electrical energy (in joules) generated in one year of operation and the heat energy used for that purpose.

   b. Assuming the power plant burns coal that releases 30 kJ/g, calculate the mass of coal (in grams and metric tons) that is burned in one year of operation. **Hint:** 1 metric ton = 1.3×10^3 kg = 1.3×10^6 g.

6. The energy of sunlight can be converted into the potential energy of glucose and oxygen.

   a. Name the process by which this conversion occurs.

   b. Name three fuels whose energy originates in sunlight.

7. Describe how grades of coal differ. What is the significance of these differences?

8. Intact dinosaur fossils were discovered during oil and gas exploration; some were even found at the top of the oil fields. What does this say about the origin of fossil fuels?

9. Mercury (Hg) is present in trace amounts in coal, ranging from 50 to 200 ppb. Calculate tons of mercury in the coal based on the lower (50 ppb) and higher (200 ppb) concentrations.

10. Name two ways in which all hydrocarbons are alike. Then, name two ways in which they differ.

11. Here are the condensed structural formulas for two alkanes: CH_3CH_3 and $CH_3(CH_2)_2CH_3$.

    a. What are the names for these compounds?

    b. For each one, give the chemical formula and draw a structural formula that shows all bonds and atoms.

    c. Comment on the relative advantages of chemical formulas, condensed structural formulas, and structural formulas in terms of convenience and information provided.

12. Why are hydrocarbons used as a primary source of fuel for energy applications?

13. The structural formulas of straight-chain ("normal") alkanes containing one to eight carbon atoms are given in Table 6.2.

    a. Draw a structural formula for *n*-decane, $C_{10}H_{22}$.

    b. Predict the chemical formula for *n*-nonane (9 C atoms) and for *n*-dodecane (12 C atoms).

14. Here is a ball-and-stick representation for one isomer of butane (C_4H_{10}):

    a. Draw a structural formula for this isomer.

    b. Draw structural formulas for all other isomers. Watch out for duplicate structures.

15. Consider these three hydrocarbons:

Compound, Formula	Melting Point (°C)	Boiling Point (°C)
pentane, C_5H_{12}	−130.5	35.9
triacontane, $C_{30}H_{62}$	65.8	449.7
propane, C_3H_8	−187.7	−42.2

At room temperature (25 °C), categorize each one as a solid, liquid, or gas.

16. During petroleum distillation, kerosene and hydrocarbons with 12–18 carbons used for diesel fuel condense at position C marked on this diagram.

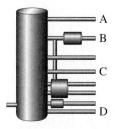

    a. Separating hydrocarbons by distillation depends on differences in a specific physical property. Which one?

    b. How does the number of carbon atoms in the hydrocarbon molecules separated at A, B, and D compare with the number separated at position C? Explain your reasoning.

    c. How do the uses of the hydrocarbons separated at A, B, and D differ from the uses of those separated at position C? Explain your reasoning.

17. The complete combustion of methane is given in Equation 6.5.

    a. By analogy, write the balanced chemical equation for the combustion of ethane, C_2H_6.

    b. Rewrite this equation using Lewis structures.

    c. The heat of combustion for ethane, C_2H_6, is 52 kJ/g. How much heat is produced if 1.0 mol of ethane undergoes complete combustion?

18. a. Write the balanced chemical equation for the complete combustion of *n*-heptane, C_7H_{16}.

    b. The heat of combustion for *n*-heptane is 4817 kJ/mol. How much heat is released if 250 kg of *n*-heptane burns completely to produce CO_2 and H_2O?

19. A single-serving bag of potato chips has 70 Cal (70 kcal). Assuming that all of the energy from eating these chips goes toward keeping your heart beating, how long can these chips sustain a heartbeat of 80 beats per minute?

    Hint: 1 kcal = 4.184 kJ, and each human heartbeat requires approximately 1 J of energy.

20. A 12-oz serving of a soft drink has an energy equivalent of 92 kcal. In kilojoules, what is the energy released when metabolizing this beverage?

21. State whether these processes are endothermic or exothermic.

    a. Charcoal burns in an outdoor grill.

    b. Water evaporates from your skin.

    c. Glucose is synthesized in the leaves of a plant by photosynthesis.

22. Use the bond energies in Table 6.1 to calculate the energy changes associated with each of these reactions. Label each reaction as endothermic or exothermic.

    a. $N_2(g) + 3 H_2(g) \longrightarrow 2 NH_3(g)$

    b. $H_2(g) + Cl_2(g) \longrightarrow 2 HCl_3(g)$

    Hint: Draw Lewis structures of the reactants and products to determine the number and kinds of bonds.

23. Use the bond energies in Table 6.1 to calculate the energy changes associated with each of these reactions. Then, label each reaction as endothermic or exothermic.

    a. $2 H_2(g) + CO(g) \longrightarrow CH_3OH(g)$

    b. $H_2(g) + O_2(g) \longrightarrow H_2O_2(g)$

    c. $2 BrCl(g) \longrightarrow Br_2(g) + Cl_2(g)$

24. Ethanol can be produced by fermentation. Another way to produce ethanol is the reaction of water vapor with ethene (ethylene), a hydrocarbon containing a C=C double bond.

    $$CH_2CH_2(g) + H_2O(g) \longrightarrow CH_3CH_2OH(l)$$

    a. Rewrite this equation using Lewis structures.

    b. Use the bond energies in Table 6.1 to calculate the energy change for this reaction. Is the reaction endothermic or exothermic?

25. Consider two coal-fired power plants that generate 5.0×10^{12} J of electricity daily. Plant A has an overall net efficiency of 38%. Plant B, a proposed replacement, would operate at higher temperatures and therefore have a higher net efficiency of 46%. Assume that coal is pure carbon and it releases 30 kJ of heat per gram combusted.

    a. If 1000 kg of coal costs $36.14, what is the difference in daily fuel costs for the two plants?

    b. How many fewer grams of CO_2 are emitted daily by Plant B, assuming complete combustion?

    c. If a household requires 3.5×10^7 kJ of energy for heat annually, how much more coal would need to be burned at Plant A to heat one house?

26. Here are structural formulas for ethane, ethene (ethylene), and ethanol:

    a. Is ethane an isomer of ethene? Of ethanol? Explain.

    b. Are any other isomers possible for ethene? Explain.

    c. Are any other isomers possible for ethanol? Explain.

27. These three compounds all have the same chemical formula of C_8H_{18}. The hydrogen atoms and C–H bonds have been omitted for simplicity.

    a. For each compound, draw structural formulas that show the missing H atoms. All should have 18 H atoms.

    b. Which (if any) of these structural formulas are identical?

    c. Draw the structural formulas for any two additional isomers of C_8H_{18}.

28. Catalysts speed up cracking reactions in oil refining and allow them to be carried out at lower temperatures. Describe two other examples of catalysts given in previous chapters of this text.

29. Explain why cracking is a necessary part of the refinement of crude oil.

30. Consider this equation representing the process of cracking:

    $$C_{16}H_{34} \longrightarrow C_5H_{12} + C_{11}H_{22}$$

    a. Which bonds are broken and which bonds are formed in this reaction? Use Lewis structures to help answer this question.

    b. Use the information from part a and Table 6.1 to calculate the energy change during this cracking reaction.

31. What is a biofuel? Give three examples.

32. Consider these three alcohols: methanol, ethanol, and *n*-propanol (the straight-chain isomer).

    a. These compounds all contain a common functional group. Name it.

    b. These compounds all are flammable. Give names and chemical formulas for the products.

    c. Predict which one of these compounds has the lowest boiling point. Explain your reasoning.

    d. The chemical structure of one of these compounds is somewhat similar to that of glycerol. Which one and why?

33. Cellulose and starch both can be fermented to produce ethanol.

   a. In terms of chemical structure, how are starch and cellulose similar?

   b. In terms of a food source for humans, how are starch and cellulose different?

34. When glucose, $C_6H_{12}O_6$, is "burned" (metabolized) in your body, the products are carbon dioxide and water.

   a. Write the balanced chemical equation.

   b. The chemical equation for burning wood is essentially the same as that of metabolizing glucose. Explain why.

35. As biofuels, biodiesel and ethanol warrant a close comparison. Use these parameters as the basis:

   a. The source

   b. The chemical reaction that produces the fuel

   c. The combustion products

   d. The solubility in water (reference Chapter 5 for more details)

36. Compare and contrast a molecule of biodiesel with a molecule of ethanol. Use these parameters as the basis of your comparison:

   a. The types of atoms each contains and their approximate relative proportions

   b. The number of atoms each contains

   c. The functional groups each contains

37. Use Figure 6.5 to compare the energy released for the combustion of 1 gallon of ethanol and 1 gallon of gasoline. Assume gasoline is pure octane (C_8H_{18}). Explain the difference.

   Hint: You will need to consider their densities.

38. Explain the terms *conventional* and *unconventional* with respect to crude oil as an energy source. Give an example of each.

Concentrating on Concepts

39. The sustainability of burning coal (and other fossil fuels) to produce electricity involves more than just the availability of coal. Explain.

40. In this chapter, we approximated the chemical formula of coal as $C_{135}H_{96}O_9NS$. However, we also noted that low-grade lignite (soft coal) has a chemical composition more similar to that of wood. Cellulose is a primary component in wood. Given this, predict an approximate chemical formula for lignite.

41. Use Figure 6.1 to compare the sources of energy consumption. Arrange the sources in order of decreasing percentage and comment on the relative rankings.

42. Compare the processes of combustion and photosynthesis in terms of energy released or absorbed, chemicals involved, and the ability to remove CO_2 from the atmosphere.

43. How might you explain the difference between temperature and heat to a friend? Use some practical, everyday examples.

44. Write a response to this statement: "Because of the first law of thermodynamics, there can never be an energy crisis."

45. Bond energies such as those in Table 6.1 are sometimes found by "working backward" from heats of reaction. A reaction is carried out, and the heat absorbed or evolved is measured. From this value and known bond energies, other bond energies can be calculated. For example, the energy change associated with the combustion of formaldehyde (H_2CO) is −465 kJ/mol.

$$H_2CO(g) + O_2(g) \longrightarrow CO_2(g) + H_2O(g)$$

Use this information and the values found in Table 6.1 to calculate the energy of the C=O double bond in formaldehyde. Compare your answer with the C=O bond energy in CO_2 and speculate on why there is a difference.

46. Use the bond energies in Table 6.1 to explain why chlorofluorocarbons (CFCs) are so stable. Also explain why it takes less energy to release Cl atoms from CFCs than it does to release F atoms and connect this to HFCs as replacements for CFCs.

47. Halons are similar to CFCs but also contain bromine. Although halons are excellent materials for fighting fires, they more effectively deplete ozone than CFCs. Here is the Lewis structure for Halon-1211:

   a. Which bond in this compound is broken most easily? How is that related to the ability of this compound to deplete ozone?

   b. In fire extinguishers, C_2HClF_4 is a possible replacement for halons. Draw its Lewis structure and identify the bond broken most easily.

   Hint: The carbon atoms are the central atoms.

48. The energy content of fuels can be expressed in kilojoules per gram (kJ/g), as shown in Figure 6.5. From these values, how do fuels containing oxygen compare to those that do not? Now calculate the energy content for each of these fuels in kilojoules per mole (kJ/mol). What trend do you now observe?

49. A friend tells you that hydrocarbon fuels containing larger molecules liberate more heat per gram than those with smaller ones.

   a. Use these data, together with appropriate calculations, to discuss the merits of this statement.

Hydrocarbon	Heat of Combustion
octane, C_8H_{18}	−5070 kJ/mol
butane, C_4H_{10}	−2658 kJ/mol

b. Based on your answer to part **a**, do you expect the heat of combustion per gram of candle wax, $C_{25}H_{52}$, to be more or less than that of octane? Do you expect the molar heat of combustion of candle wax to be more or less than that of octane? Justify your predictions.

50. The Fischer–Tropsch conversion of hydrogen and carbon monoxide into hydrocarbons and water was given in Equation 6.11:

$$n \, CO + (2n + 1) \, H_2 \longrightarrow C_nH_{2n+2} + n \, H_2O$$

a. Determine the heat evolved by this reaction when $n = 1$.

b. Without doing a calculation, do you think that more or less energy is given off per mole in the formation of larger hydrocarbons ($n > 1$)? Explain your reasoning.

51. Here is a ball-and-stick model of ethanol, C_2H_5OH or C_2H_6O:

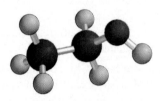

a. Dimethyl ether is an isomer of ethanol. Draw its Lewis structure.

b. People used to refer to "ether" as an anesthetic. What they meant was diethyl ether. Draw its structural formula.

     Hint: Diethyl means there are a total of how many carbons?

c. The ether is a functional group not described in this chapter. Based on your answer to the two previous parts, what common structural feature do all ethers have?

52. Octane ratings of several substances were found in Section 6.12.

a. What evidence can you give that the octane rating is not a measure of the energy content of a gasoline?

b. Octane ratings measure a fuel's ability to minimize or prevent engine knocking. Why is this important?

c. Why do higher octane blends cost more than lower octane ones?

d. A premium gasoline available at most stations has an octane rating of 91. What does this tell you about whether the fuel contains oxygenates?

53. Both *n*-octane and iso-octane have essentially the same heat of combustion. How is this possible given that they have different molecular structures?

54. All of these terms fit under the heading of fuels: renewable fuel, nonrenewable fuel, coal, petroleum, biodiesel, natural gas, and ethanol. Use a diagram to show the relationship among them. Also find a way to show where the terms *fossil fuel* and *biofuel* fit.

55. Use a diagram to show the relationship among these terms that relate to foods we eat: fat, lard, oil, triglyceride, butter, olive oil, and soybean oil. Although biodiesel is not a food, it still connects to these terms. Find a way to represent this connection.

56. On a timescale of a few years, the combustion of ethanol derived from biomass releases a *lower* net amount of CO_2 into the atmosphere than does burning gasoline derived from crude oil. People argue whether this statement is true or not. Outline some points on each side of the argument.

57. Emissions of some pollutants are lower when biodiesel is used rather than petroleum diesel. In the case of biodiesel fuel, suggest a reason for lower emissions of the following:

a. sulfur dioxide, SO_2

b. carbon monoxide, CO

Exploring Extensions

58. Although coal contains only trace amounts of mercury, the amounts released into the environment by the burning of coal have significant consequences. Defend or refute this statement by gathering the appropriate evidence.

59. According to a statement once made by the U.S. EPA, driving a car is "a typical citizen's most polluting daily activity."

a. What pollutants do cars emit?

b. What assumptions does the truth of this statement depend on?

60. An article has suggested that replacing a 75 W incandescent bulb with an 18 W compact fluorescent bulb would save about 75% in the cost of electricity. Electricity is generally priced per kilowatt-hour (kWh). Using the price of electricity where you live, calculate how much money you would save over the life of one compact fluorescent bulb (about 10,000 h). *Note:* Standard incandescent bulbs last about 750 h.

61. Watch this video (www.acs.org/cic) and answer the following question: How is Marisol being an environmental justice advocate?

62. This chapter mentions several nonconventional sources of oil and gas, including drilling deep below seawater, hydraulic fracturing deep below the earth, and extracting oil from shales and tarry oil sands. Pick one, describe it, and provide an analysis using the Triple Bottom Line: economic health, environmental health, and social health.

63. Chemical explosions are *very* exothermic reactions. Describe the relative bond strengths in the reactants and products that would make for a good explosion.

64. The chapter pointed out that the FDA approved propylene glycol for use as a food additive. In which foods is it used and for what purposes?

65. Tetraethyllead (TEL) was first approved for use in gasoline in the United States in 1926. It wasn't banned until 1986. Construct a timeline that includes any four events in the 60 years of its use, including some that led to its ban.

66. Tetraethyllead (TEL) has an octane rating of 270. How does this compare with other gasoline additives? Examine a structural formula for TEL and propose a reason for the value of its octane rating in comparison to other additives.

67. Another type of catalyst used in the combustion of fossil fuels is the catalytic converter that was discussed in Chapter 2. One of the reactions that these catalysts speed up is the conversion of $NO(g)$ into $N_2(g)$ and $O_2(g)$.

   a. Draw a diagram of the energy of this reaction similar to the one shown in Figure 6.32.

   b. Why is this reaction important?

68. Figure 6.7 shows energy differences for the combustion of H_2, an exothermic chemical reaction. The combination of N_2 and O_2 to form NO (nitrogen monoxide) is an example of an endothermic reaction:

$$N_2(g) + O_2(g) \longrightarrow 2\ NO(g)$$
$$:\!\dot{N}\!=\!\ddot{O}\!:$$

The bond energy for N=O is 630 kJ/mol. Sketch an energy diagram for this reaction and calculate the overall energy change.

69. Because the United States has large natural gas reserves, there is significant interest in developing uses for this fuel. List two advantages and two disadvantages of using natural gas to fuel vehicles.

7 Nuclear Energy and Other Alternative Sources

WHERE DOES YOUR ENERGY COME FROM?

Watch the chapter opening video at www.acs.org/cic. Then, visit the U.S. Energy Information Administration's website to find a state-by-state comparison of energy production and consumption.

a. Choose one state or territory and select it on the map. Scroll down to look at the information provided. From which sources does this state or territory get the energy that it consumes? What types of energy sources are used to produce energy in this state? How do these sources differ?

b. Choose a second state and make a comparison with the state you chose in part **a**.

c. Of all the sources of energy you found in parts **a** and **b**, which are derived from fossil fuels? Which are not? This chapter will explore energy sources that are alternatives to fossil fuels.

In this chapter, you will explore the following questions:

1. What are alternative and renewable sources of energy?
2. Why do nuclear reactions release such enormous amounts of energy?
3. How do nuclear power plants produce electricity?
4. What is radioactivity, and what are some applications for radioactive elements in energy production?
5. What are the sources and health effects of nuclear radiation?
6. How long do some substances remain radioactive?
7. What are the risks associated with nuclear power?
8. What are the trends in nuclear power use around the world?
9. How can sunlight be used for electricity generation?
10. How do solar panels work?
11. What are some other types of renewable energy sources?

Introduction

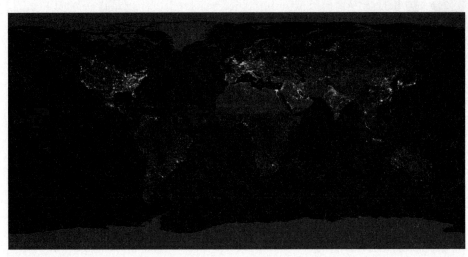

NASA's Earth Observatory/NOAA/DOD

People across the world use energy. As individuals, we use energy to warm and cool our living spaces, cook food, power electronic devices, move ourselves around, and provide light in the darkness, as can be seen from the above NASA Earth Observatory satellite image. Together, the humans on the planet consume more than 500 exajoules of energy each year, which are approximately equivalent to 8.75 billion barrels of oil.

Currently, the majority of energy consumed across the globe comes from the fossil fuels introduced in Chapter 6, including coal, natural gas, and petroleum. The combustion of fossil fuels releases energy stored from sunlight that reached Earth hundreds of millions of years ago.

Your Turn 7.1   Personal Energy

Review the past 24 hours, or even just the past few hours. What kinds of energy did you use? Where did the energy come from?

7.1 | Alternative Energy Sources

> *Learning Objective: Describe the relative use of alternative and renewable energy sources around the world*

The energy we use may come from fossil fuels like natural gas piped into homes and gasoline to fuel cars. Energy is also found in electricity from wall outlets, but where does electricity come from? The answer depends on where you live and the resources available to provide energy.

Your Turn 7.2   Your Electricity Mix

Explore the interactive map (www.acs.org/cic) to see the sources of electricity around the world. Click on "change country" to evaluate the trends in specific countries.

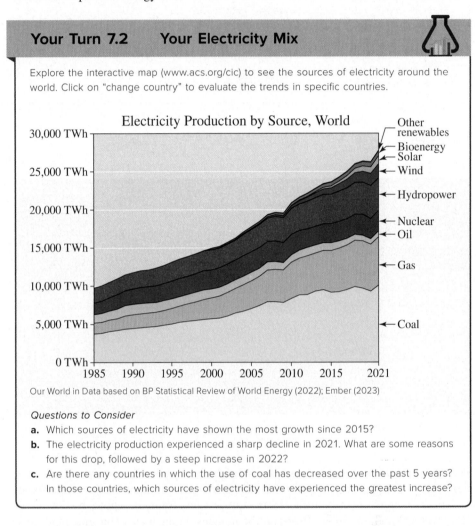

Our World in Data based on BP Statistical Review of World Energy (2022); Ember (2023)

Questions to Consider
a. Which sources of electricity have shown the most growth since 2015?
b. The electricity production experienced a sharp decline in 2021. What are some reasons for this drop, followed by a steep increase in 2022?
c. Are there any countries in which the use of coal has decreased over the past 5 years? In those countries, which sources of electricity have experienced the greatest increase?

Electricity generation in many regions of the world is driven by the availability of natural resources—resources that come from nature. For example, the combustion of natural gas is the dominant source of electricity for those living in the Russian Federation. Electricity in the Pacific Northwest of the United States can be generated by hydroelectric dams. Countries without abundant fossil fuel resources, such as France, may rely heavily on nuclear power. **Alternative sources** of energy are sources of energy that do not rely on the combustion of hydrocarbon-based fuels. Examples of alternative energy sources are: nuclear power, solar panels, wind turbines, and hydroelectric dams. These alternative energy sources are important because they all generate electricity without consuming fossil fuels and thus produce zero carbon dioxide emissions.

Although these energy sources result in zero greenhouse gas (GHG) emissions when they generate electricity, certain processes used to build and fuel the plants do emit GHGs. For instance, there are GHG emissions associated with the construction and maintenance of the facility, as well as (for nuclear power) fuel processing and waste disposal. Massive amounts of methane and other GHGs are produced during the flooding of land for a hydroelectric dam.

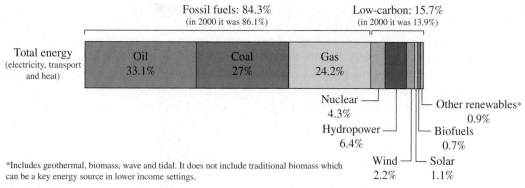

Figure 7.1

Global primary energy consumption by source.

Our World in Data based on BP Statistical Review of World Energy (2020).

Solar, wind, hydroelectric, and geothermal energies are said to be **renewable** energy sources because they are continually and rapidly collected and replenished from natural resources. As shown in Figure 7.1, although renewable sources have increased rapidly in recent years, they still do not contribute as much toward our global energy consumption as fossil fuels. It should be noted that although nuclear energy is a low-carbon energy source, it is not renewable because it relies on the mining of uranium, a finite resource.

Your Turn 7.3   Low-C Energy Use Around the World

The interactive map (www.acs.org/cic) shows the change in low-carbon sources around the world. Use the pull-down menu to select specific regions to see how their trends in low-C energy sources compare to the rest of the world.

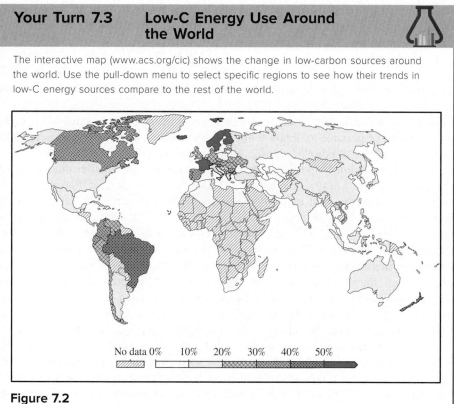

Figure 7.2

Share of primary energy from low-carbon sources, 2021. Low-carbon energy is defined as the sum of nuclear and renewable sources. Renewable sources include hydropower, solar, wind, geothermal, wave, tidal, and bioenergy. Traditional biofuels are not included.

Our World in Data based on BP Statistical Review of World Energy.

Questions to Consider

a. Which countries have experienced the greatest shift toward low-C energy sources over the past 20 years? Using the Internet as a resource, what types of energy sources are used in these countries?

b. Which countries have experienced the lowest increase in low-C energy sources over the past 5 years? What are some reasons for their relatively slow transition away from carbon-based fuels?

As you will see in this chapter, there is no perfect energy source, as all have the potential for negative consequences on the environment or the inhabitants of Earth. As our world shifts toward low-carbon based alternatives to mitigate the effects of climate change, it is important that we carefully examine these sources through a sustainability lens that asks you to consider environmental, social and economic factors. Only then will you appreciate the benefits and potential risks of each type of alternative energy. We now begin our journey with nuclear energy, a controversial option, indeed!

7.2 | From Nuclear Energy to Bombs: The Splitting of Atomic Nuclei

Learning Objectives:

Write balanced nuclear equations and describe nuclear reactions at both the macroscopic and particulate levels

Explain why nuclear reactions release such enormous amounts of energy from a small amount of matter

We begin our discussion of alternative energy sources with nuclear energy. The key to understanding the fundamentals of nuclear reactions is probably the most famous equation in all of the natural sciences, which summarizes the equivalence of energy, E, and matter, or mass, m:

$$E = mc^2 \qquad \text{[7.1]}$$

This equation dates from the early years of the 20th century, and is one of the many contributions of Albert Einstein (1879–1955). The symbol c represents the speed of light, 3.00×10^8 m/s, so c^2 is equal to 9.00×10^{16} m²/s². The large value of c^2 means that it should be possible to obtain a tremendous amount of energy from a very small amount of matter—whether in a power plant or in a weapon.

Nuclear fission is the splitting of a large nucleus into smaller ones with the release of energy. Energy is released because the total mass of the products is slightly less than the total mass of the reactants. It is important to note that both matter and energy are conserved together. When matter "disappears," an equivalent quantity of energy "appears." Alternatively, one can view matter as a very concentrated form of energy; nowhere is it more concentrated than in an atomic nucleus. Remember that an atom is mostly empty space. If a hydrogen nucleus were the size of a baseball, then its electron would be found within a sphere half a mile in diameter. Because almost all the mass of an atom is associated with its nucleus, the nucleus is incredibly dense. Given the energy–mass equivalence of Einstein's equation, the energy content of all nuclei is, relatively speaking, immense.

Although this section is focused primarily on nuclear fission, it should be mentioned that atoms can also be fused as well as split. **Nuclear fusion** takes place when two low-mass isotopes, typically of hydrogen, are combined under extreme pressures and temperatures. These reactions occur in the Sun where hydrogen is converted into helium, providing sunlight and heat that bathes our planet. Although the energy output from nuclear fission is about 1 million times greater than other sources, nuclear fusion results in 3–4 times greater energy than fission! Explore the differences between nuclear fission and fusion in this next activity.

Only the nuclei of certain elements undergo fission, and these only under certain conditions. Three factors determine whether a particular nucleus will split: its size, the numbers of protons and neutrons it contains, and the energy of the neutrons that bombard the nucleus to initiate the fission. For example, relatively light and stable atoms such as oxygen, chlorine, and iron do not split, but extremely heavy nuclei may fission spontaneously. Some heavy nuclei, such as those of uranium and plutonium, can be made to split if hit hard enough with neutrons. Notably, one isotope of uranium fissions

Nuclear Fission vs. Fusion

Check out this video, which illustrates the key differences between nuclear fission and fusion: www.acs.org/cic.

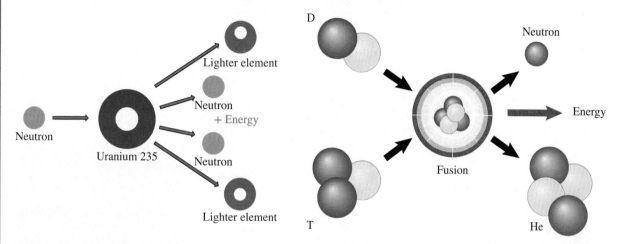

Questions to Consider
a. If nuclear fusion generates much more energy than nuclear fission, why hasn't this been used for nuclear power applications?
b. What isotopes are typically used for nuclear fusion? Why are these isotopes chosen for fusion reactions?
c. In December 2022, scientists at the Lawrence Livermore National Laboratory in the United States announced a major breakthrough related to nuclear fusion. Summarize their findings and its impact on the future of nuclear energy in the U.S. and around the world.
d. Describe the emissions and by-products generated from nuclear fission and fusion reactions.

with neutrons of a more moderate speed, such as those employed in the reactor of a nuclear power plant.

Let's examine uranium more closely. All uranium atoms contain 92 protons. If these atoms are electrically neutral, these protons are accompanied by 92 electrons. From our discussion of isotopes in Section 4.2, recall that you can find the atomic number (or number of protons) on the periodic table. In nature, uranium is found predominantly as two isotopes. The more abundant one (99.3% of all uranium atoms) contains 146 neutrons. The mass number of this isotope of uranium is 238; that is, 92 protons plus 146 neutrons. We represent this isotope as uranium-238, or more simply as U-238. The less abundant isotope (0.7%) contains 143 neutrons and 92 protons; namely, U-235 (Table 7.1).

Your Turn 7.4 Comparing Isotopes

A trace amount of a third isotope, U-234, is also found in nature. How do U-238 and U-234 compare in terms of the number of protons and the number of neutrons? Do their numbers of electrons differ?

Table 7.1	Comparing Isotopes of Uranium			
Name	**Number of Protons (Atomic Number)**	**Mass Number**	**Number of Neutrons**	**Atomic Symbol**
Uranium-235	92	235	235 − 92 = 143	$^{235}_{92}U$
Uranium-238	92	238	238 − 92 = 146	$^{238}_{92}U$

More commonly, we specify an isotope with both its mass number and atomic number. The former is a superscript and the latter a subscript, both written to the left of the chemical symbol. Using this convention, uranium-238 becomes:

$$\text{Mass number = number of protons + number of neutrons} \longrightarrow {}^{238}_{92}\text{U}$$
$$\text{Atomic number = number of protons} \longrightarrow$$

Similarly, U-235 is written as ${}^{235}_{92}\text{U}$. Although ${}^{235}_{92}\text{U}$ and ${}^{238}_{92}\text{U}$ differ by a mere three neutrons, this difference translates to one key difference in nuclear properties. Under the conditions present in a nuclear reactor, ${}^{238}_{92}\text{U}$ does not undergo fission, yet ${}^{235}_{92}\text{U}$ does.

@HOME   Candy Isotopes

Investigate isotopes at home by modeling a fictional element "Chocolatium" using a bowl, a 1-cup measuring cup, a kitchen scale, two paper plates, and M&M candies.

1. Empty a bag of your M&M candies into a bowl. Each individual piece of candy represents an atom; the different colors of candy represent the different isotopes of Chocolatium.
2. Use the measuring cup to scoop out 1-cup of your sample. Pour the contents of the cup onto a paper plate.

Aaron Roeth Photography

3. Separate each of the different isotopes (colors) into groups. Count the number of each isotope and record it.
4. Find the total mass of each group of isotopes and record it.
5. Calculate the mass of one atom for each isotope using the equation:

$$\text{mass of one atom} = \frac{\text{total mass of isotope}}{\text{number of atoms of that isotope}}$$

6. Calculate the percent (%) abundance for each isotope using the equation:

$$\text{\% abundance} = \frac{\text{number of atoms of an isotope}}{\text{total number of atoms in whole sample}} \times 100\%$$

7. Calculate the atomic mass of Chocolatium by multiplying the mass of each isotope by its percent abundance and adding these together.
8. Repeat Steps 2–7 one more time to have a second trial of data.

Questions to Consider
a. How did your calculations vary between trials? Why might this be true? What impact would having a larger sample size have on your results?
b. What do each of the isotopes have in common and how are they different?

Your Turn 7.5   Comparing Atomic Number and Mass Number

Refer back to Your Turn 7.4 and your answers for U-234. What is the atomic number and mass number for U-234? Write the atomic symbol using this information. Are the atomic symbols for U-234 and U-238 the same?

The process of nuclear fission is initiated by neutrons, but also can release neutrons, as seen by this example:

$${}^{1}_{0}\text{n} + {}^{235}_{92}\text{U} \longrightarrow \left[{}^{236}_{92}\text{U}\right] \longrightarrow {}^{141}_{56}\text{Ba} + {}^{92}_{36}\text{Kr} + 3{}^{1}_{0}\text{n} \qquad \textbf{[7.2]}$$

Let's examine the components of Equation 7.2 from left to right. Initially, a neutron hits the nucleus of U-235. This neutron, ${}^{1}_{0}\text{n}$, has a subscript of 0, indicating no positive charges; the superscript is 1 because the mass number of a neutron is 1. The nucleus of $^{235}_{92}U$ captures the neutron, forming a heavier isotope of uranium, $^{236}_{92}U$. This isotope is written in square brackets indicating that it exists only momentarily. Uranium-236 immediately splits into two smaller atoms (Ba-141 and Kr-92) with the release of three more neutrons.

Nuclear equations are similar to, but not the same as, "regular" chemical equations. To balance a nuclear equation, you count the protons and neutrons rather than counting atoms. A nuclear equation is balanced when the sum of the subscripts (and of the superscripts) on the left of the arrow is equal to the sum of those on the right. Coefficients in nuclear equations, such as the 3 preceding the $^{1}_{0}n$ in Equation 7.2, are treated the same way as in chemical equations, multiplying the term that follows it. For example, examine the math to see why nuclear Equation 7.2 is balanced:

	Left	Right
Superscripts:	$1 + 235 = 236$	$141 + 92 + (3 \times 1) = 236$
Subscripts:	$0 + 92 = 92$	$56 + 36 + (3 \times 0) = 92$

When the nucleus of an atom of U-235 is struck with a neutron, many different fission products are formed. The activity that follows acquaints you with two other possibilities.

Writing Nuclear Equations

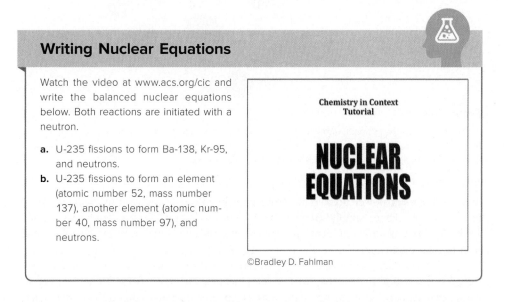

Watch the video at www.acs.org/cic and write the balanced nuclear equations below. Both reactions are initiated with a neutron.

a. U-235 fissions to form Ba-138, Kr-95, and neutrons.

b. U-235 fissions to form an element (atomic number 52, mass number 137), another element (atomic number 40, mass number 97), and neutrons.

Chemistry in Context Tutorial

NUCLEAR EQUATIONS

©Bradley D. Fahlman

Look again at nuclear Equation 7.2. Both sides contain neutrons, which might lead you to think that we should cancel them. Although you might do this in a mathematical expression, don't do it here. The neutrons on both sides of the equation are important! The one on the left initiates the fission reaction, whereas the ones on the right are produced by it. Each neutron that is produced can in turn strike another U-235 nucleus, causing it to split, which will release a few more neutrons. This is an example of a **chain reaction**, a term that generally refers to any reaction in which one of the products becomes a reactant, thus making it possible for the reaction to become self-sustaining. This particular rapidly branching nuclear chain reaction is self-sustaining and spreads in a fraction of a second (Figure 7.3).

A **critical mass** is the amount of fissionable fuel required to sustain a chain reaction. For example, the critical mass of U-235 is about 15 kg (or 33 lb). If this mass of pure U-235 were brought together in one place with a source of neutrons, fission would occur spontaneously and would continue as long as the critical mass persists. Nuclear weapons work on this principle, although the energy released quickly blows the critical mass apart, stopping the fission reaction. But, as you will soon see, the uranium fuel

Visualizing a Nuclear Chain Reaction

Check out this video to see how dominoes can mimic a nuclear chain reaction: www.acs.org/cic.

Larry Washburn/fStop/Getty Images

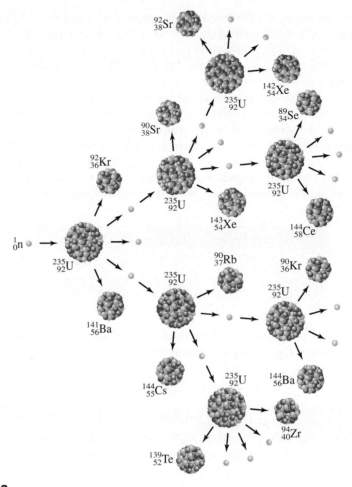

Figure 7.3

A neutron initiates the fission of uranium-235, starting a chain reaction. It should be noted that not every U-235 nucleus that is bombarded by a neutron results in the same products every time. Check out www.acs.org/cic for an animated simulation of a chain reaction.

used in a nuclear power plant is far from pure U-235 and is unable to explode like a nuclear bomb. There simply aren't enough neutrons around (and enough fissionable nuclei for these neutrons to hit) to produce an uncontrolled chain reaction characteristic of a nuclear explosion.

We mentioned earlier that energy is given off during fission because the mass of the products is slightly less than that of the reactants. However, from the nuclear equations we have just written, no mass loss is apparent because the sum of the mass numbers is the same on both sides. In fact, the actual mass does decrease slightly. To understand this, remember that the actual masses of the nuclei are not the mass numbers (the sum of the number of protons and neutrons); rather, they have measured values with many decimal places. A *unified atomic mass unit,* u, is defined as 1/12 the mass of a C-12 atom, or 1.67×10^{-27} kg. This unit is convenient for expressing the mass of individual atoms. It should be noted that a unified mass unit is equivalent to the dalton, Da, another commonly used unit of mass. An atom of uranium-235 has a mass of 235.043924 u or 235.043924 Da. Were you to keep all six decimal places and compare the masses on both sides of the nuclear equation for the fission of U-235, you would find that the mass of the products is less by about 0.1%, or 1/1000. This difference corresponds to the energy that is released.

How much energy would be released if all the nuclei in 1.0 kg (2.2 lb) of pure U-235 were to undergo fission? We can calculate an answer by using an equation closely related to $E = mc^2$; namely, $\Delta E = \Delta mc^2$. Here, the Greek letter delta (Δ) means "the change in," so now with a change in mass we can calculate a change in energy. Because 1/1000 of this mass is lost, the value for Δm, the change in mass, is 1/1000 of

1.00 kg, which is 1.00×10^{-3} kg. Now, we will substitute this value, and $c = 3.00 \times 10^8$ m/s, into Einstein's equation:

$$\Delta E = \Delta mc^2 = (1.00 \times 10^{-3} \text{ kg}) \times (3.00 \times 10^8 \text{ m/s})^2 \qquad [7.3]$$

Completing the calculation gives an energy change in what may appear to be unusual units:

$$\Delta E = 9.00 \times 10^{13} \text{ kg} \cdot \text{m}^2/\text{s}^2$$

However, the unit $\text{kg} \cdot \text{m}^2/\text{s}^2$ is identical to a joule (J). Therefore, the energy released from the fission of an entire kilogram of uranium-235 is a whopping 9.00×10^{13} J, or 9.00×10^{10} kJ!

To put things into perspective, 9.00×10^{13} J is the amount of energy released by the detonation of about 22 kilotons of the explosive trinitrotoluene, better known as TNT. By way of comparison, this is roughly twice the amount of energy released by the atomic bombs dropped on Hiroshima and Nagasaki in 1945! This energy originates from the fission of a single kilogram of U-235, in which a mass of approximately 1 gram (0.1% mass change) was transformed into energy.

Your Turn 7.6   Comparing Nuclear to Coal

Anthracite coal contains an average energy content of 32 kJ/g, which is higher than that of other coal varieties. What mass of anthracite coal would be needed to produce the same amount of energy as the fission of 1 kg of U-235?

As it turns out, one cannot fission a kilogram or two of pure U-235 in one fell swoop. In an atomic weapon, for example, the energy released blasts the fissionable fuel apart in a fraction of a second, thus halting the chain reaction before all the nuclei can undergo fission. Nonetheless, the energy released is enormous—on the order of 10 kilotons of TNT for the atomic bomb dropped on the city of Hiroshima. Figure 7.4 shows an atomic explosion at the U.S. Nevada Test Site. Code named Turk, this test in 1955 had more than four times the explosive power of the bombs at Hiroshima and Nagasaki.

Figure 7.4

The nuclear test "Turk" was exploded on a dry lake bed northwest of Las Vegas, Nevada, on March 7, 1955.

Corbis Historical/Getty Images

Fortunately, the energy of nuclear fission can be harnessed. This is exactly the objective of a nuclear power plant. Here, the energy is slowly and continually released under controlled conditions, as we shall see in the next section.

7.3 | Harnessing a Nuclear Fission Reaction: How Nuclear Power Plants Produce Electricity

> *Learning Objective: Describe how energy is produced in a nuclear power plant*

Section 6.6 described how a conventional power plant burns coal, oil, or some other fuel to produce heat. The heat is then used to boil water, converting it into high-pressure steam that turns the blades of a turbine. The shaft of the spinning turbine is connected to large wire coils that rotate within a magnetic field, thus generating electrical energy. A nuclear power plant operates in much the same way except the energy released from the fission of atomic nuclei, such as U-235, is used to heat the water instead of combustion of a fossil fuel (Figure 7.5). Like any power plant, a nuclear facility is subject to the efficiency constraints imposed by the second law of thermodynamics. While it is impossible to convert heat completely into work in a cyclical process, the theoretical efficiency for converting heat energy into work depends on the maximum and minimum temperatures between which the plant operates. This net efficiency, typically 55–60%, is significantly reduced by other mechanical, thermal, and electrical losses.

The nuclear reactor is the heart of the power station. The reactor, together with one or more steam generators and the primary cooling system, is housed in a special steel vessel within a concrete dome-shaped containment building. The nonnuclear portion contains the turbines that run the electric generator, as well as the secondary

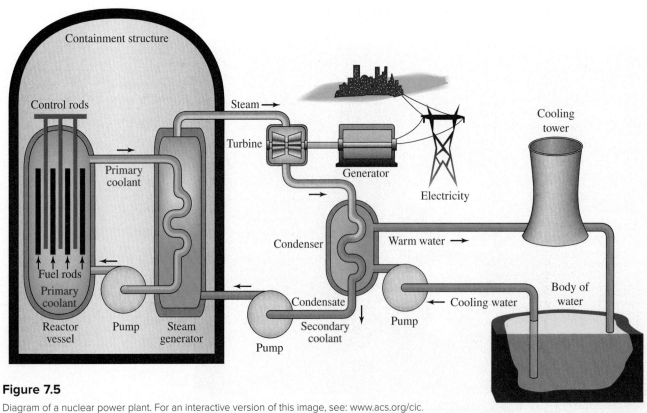

Figure 7.5

Diagram of a nuclear power plant. For an interactive version of this image, see: www.acs.org/cic.

Source: United States Nuclear Regulatory Commission

cooling system. In addition, the nonnuclear portion must be connected to some means of removing excess heat from the coolants. Accordingly, a nuclear power station has one or more cooling towers, or is located near a sizable body of water (or both)—a feature that is shared with fossil fuel electric power plants (revisit Figure 6.9).

The uranium fuel in the reactor core is in the form of uranium(IV) oxide (UO_2), each comparable in height to the width of a dime, as shown in Figure 7.6. About 4–5% of the fuel is composed of U-235 (fissionable uranium) and the rest is U-238. These pellets are placed end-to-end in tubes composed of an alloy of zirconium and other metals, which in turn are grouped into stainless steel–clad bundles (Figure 7.7).

Each rod contains at least 200 pellets. Once started, a fission reaction can sustain itself by a chain reaction. However, neutrons are needed to induce the process (Equation 7.2 and Figure 7.3). One means of generating neutrons is to use a combination of beryllium-9 and a heavier element such as plutonium, which are included in the fuel rods. The heavier element releases helium atoms also called alpha (α) particles, ^4_2He:

$$^{238}_{94}Pu \longrightarrow {}^{234}_{92}U + {}^4_2He \qquad \text{[7.4]}$$
$$\text{alpha particle}$$

These alpha particles in turn strike beryllium atoms, releasing neutrons, carbon-12, and gamma rays, $^0_0\gamma$:

$$^4_2He + {}^9_4Be \longrightarrow {}^{12}_6C + {}^1_0n + {}^0_0\gamma \qquad \text{[7.5]}$$
$$\text{gamma ray}$$

The neutrons produced in this way can initiate the nuclear fission of uranium-235 in the reactor core.

Figure 7.6

Comparative sizes of nuclear fuel pellets and a U.S. dime.

C.P. Hammond/McGraw Hill

Your Turn 7.7   Poo–Bee and Am–Bee

A neutron source constructed with Pu and Be is a PuBe ("poo–bee") source. Similarly, the AmBe ("am–bee") source is constructed from americium and beryllium. Analogous to the PuBe source, write the set of reactions that produce neutrons from an AmBe source. Start with Am-241.

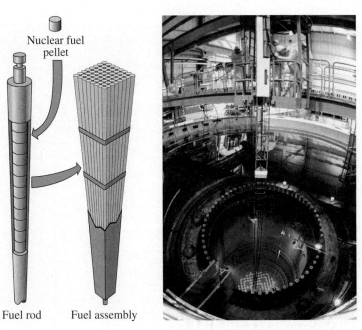

Nuclear fuel pellet

Fuel rod   Fuel assembly

Figure 7.7

Schematic of fuel pellets, fuel rod, and fuel assembly making up the core of a nuclear reactor *(left)*. The fuel assembly is submerged under water in an active reactor core *(right)*.

(nuclear power plant): Toby Talbot/AP Images

Remember—one fission event produces two or three neutrons. The trick is to "sponge up" or remove these extra neutrons, but still leave enough to sustain the fission reaction. A delicate balance must be maintained. With extra neutrons, the reactor runs at too high a temperature; with too few neutrons, the chain reaction halts and the reactor cools down. To achieve the needed balance, one neutron from each fission event should in turn cause another fission reaction.

Metal rods interspersed among the fuel elements serve as the neutron "sponges" because they help remove excess neutrons. These *control rods,* composed primarily of an excellent neutron absorber such as cadmium or boron, can be positioned to absorb fewer or more neutrons. With the rods fully inserted, the fission reaction is not self-sustaining. But as the rods are gradually withdrawn, they are not able to sponge up as many neutrons. As a result, the reactor can "go critical" and become self-sustaining, running at different rates depending on the exact position of the control rods. Over time, fission products that absorb neutrons build up in the fuel pellets, which requires replacement of the reactor fuel bundles.

Your Turn 7.8   Earthquake!

Look ahead to Figure 7.18 to see that earthquakes may occur in the vicinity of nuclear reactors. Reactors near the epicenter should automatically shut down. Should the software be programmed to fully insert the control rods into the reactor core, or should they be pulled out? Explain.

The fuel bundles and control rods are bathed in the primary coolant, a liquid that comes in direct contact with them and removes heat transferred from the reaction. In the Sequoyah nuclear reactor (Figure 7.8) and in many others, the primary coolant is an aqueous solution of boric acid, H_3BO_3. The boron atoms absorb neutrons and thus control the rate of fission and temperature. Like the control rods, the solution serves as a moderator for the reactor, slowing the neutrons, and controlling the rate of fusion. Another major function of the primary coolant is to absorb the heat generated by the nuclear reaction. Because the primary coolant solution is at a pressure more than 150 times normal atmospheric pressure, it does not boil. It is heated far above its normal boiling

Figure 7.8

The Sequoyah nuclear power plant in Tennessee. The two cooling towers (one shown with a cloud of condensed water vapor) are the most prominent features of this plant. The reactors, however, are located in the two small cylindrical containment buildings with white roofs.

Aerial Archives/Alamy Stock Photo

point and circulates in a closed loop from the reaction vessel to the steam generators, and back again. This closed primary coolant loop thus forms the link between the nuclear reactor and the rest of the power plant (Figure 7.5).

The heat from the primary coolant is transferred to what is sometimes referred to as the secondary coolant, the water in the steam generators that does not come in contact with the reactor. At the Sequoyah nuclear plant (Figure 7.8), more than 30,000 gallons of water are converted to vapor each minute. The energy of this hot vapor turns the blades of turbines that are attached to an electric generator. To continue the heat-transfer cycle, the water vapor is then cooled, condensed back to a liquid, and returned to the steam generator (Figure 7.5). In many nuclear facilities, the cooling is done using large cooling towers that are commonly mistaken for the reactors. However, the reactor buildings are not as large.

Your Turn 7.9   Clouds (Not Mushroom-Shaped)

Some days you can see a cloud coming out of the cooling tower of a nuclear power plant, as shown in Figure 7.8. What causes the cloud? Does it contain any nuclear products produced from the fission of U-235? Explain.

Nuclear power plants also use water from lakes, rivers, or the ocean to cool the condenser. For example, at the Seabrook nuclear power plant in New Hampshire, every minute about 400,000 gallons of ocean water flow through a huge tunnel (19 feet in diameter and 3 miles long) bored through rock 100 feet beneath the ocean floor. A similar tunnel from the plant carries the measurably warmer water back to the ocean. Special nozzles distribute the hot water so that the observed temperature increase in the immediate area of the discharge is only about 2 °C. The ocean water is in a separate loop from the fission reaction and its products. The primary coolant (water with boric acid) circulates through the reactor core inside the containment building. However, this boric acid solution is kept isolated in a closed circulating system, which makes the transfer of radioactivity to the secondary coolant water in the steam generator highly unlikely. Similarly, the ocean water does not come in direct contact with the secondary system, so the ocean water is well-protected from radioactive contamination.

Although the electricity generated by nuclear and fossil fuel sources is identical, different sources of energy vary in how much electricity is generated per second. The rate of energy production is referred to as **power**, and a common unit of power is the joule per second (J/s), or **watt** (W). Single nuclear reactors typically have a maximum power capacity between 500 and 1300 megawatts (MW). In comparison, a typical coal-fired power plant generates electricity at a rate of 600 MW.

Your Turn 7.10   The Palo Verde Reactors

One of the most powerful nuclear plants in operation in the United States is the Palo Verde complex in Arizona. At maximum capacity, just one of its three reactors generates 1.2 billion joules of electric energy every second, or 1.4 GW of power. Calculate the total amount of electrical energy produced by the three reactors per day and the mass of U-235 lost each day.

Hint: Start by calculating the quantity of energy generated not per second, but per day. Then, use the equation $\Delta E = \Delta mc^2$ and solve for the change in mass, Δm. Report the mass loss in grams.

The Palo Verde nuclear reactor.

Larry Lee Photography/Corbis

The topics we have been discussing—nuclear fission, uranium, nuclear fuel, nuclear weapons—all rest on an understanding of radioactivity. We now turn to this topic.

7.4 | What Is Radioactivity?

> *Learning Objectives:*
>
> *Explain why some isotopes are radioactive, while others are not*
>
> *Compare the fundamental types of nuclear radiation and their properties*

Our knowledge of radioactive substances is well over 100 years old. In 1896, the French physicist Antoine Henri Becquerel (1852–1908) discovered radioactivity. At the time, his research involved using photographic plates. Prior to use, these plates were sealed in black paper to keep them from being exposed to light. By accident, he left a mineral near one of these sealed plates. When he unwrapped the plate he found that the plate's light-sensitive emulsion had darkened. It was as though the plate had been exposed to light! Becquerel immediately recognized that the mineral emitted powerful rays that penetrated the lightproof paper.

Further investigation by the Polish scientist Marie Curie (1867–1934) (Figure 7.9) revealed that the rays were coming from a constituent of the mineral—the element uranium. In 1899, Curie applied the term **radioactivity** to the spontaneous emission of radiation by certain elements. Subsequent research by Ernest Rutherford (1871–1937) led to the identification of two major types of radiation. Rutherford named them after the first two letters of the Greek alphabet, alpha (α) and beta (β).

Alpha and beta radiation have strikingly different properties. A beta particle (β) is a high-speed electron emitted from the nucleus. It has a negative electric charge (1$-$) and only a tiny mass, about 1/2000 that of a proton or a neutron.

In contrast, an alpha particle (α) is a positively charged particle emitted from the nucleus. It consists of two protons and two neutrons (the nucleus of a He atom) and has a 2+ charge since no electrons accompany the helium nucleus.

Gamma rays frequently accompany alpha or beta radiation. A gamma ray (γ) is emitted from the nucleus and has no charge or mass. It is a high-energy, short-wavelength photon. Just like infrared (IR), visible, and ultraviolet (UV) radiation, gamma rays are part of the electromagnetic spectrum and have more energy than X-rays. Table 7.2 summarizes these three types of nuclear radiation.

The term "radiation" tends to be confusing because people don't always specify whether they mean electromagnetic radiation or nuclear radiation. As seen in Section 3.1, electromagnetic radiation refers to all the different types of light: radio, microwave, infrared, visible, ultraviolet, X-rays, and gamma rays. For example, it is perfectly correct to say "visible radiation" instead of "visible light." **Nuclear radiation**, however, refers to alpha, beta, or gamma radiation emitted from a nucleus. Gamma rays are both electromagnetic radiation and nuclear radiation. When emitted from the nucleus of a

Figure 7.9

Marie Curie won two Nobel Prizes—one in chemistry, the other in physics—for her research on radioactive elements.

Hulton Deutsch/Corbis Historical/Getty Images

Table 7.2		Types of Nuclear Radiation		
Name	**Symbol**	**Composition**	**Charge**	**Change to the Parent Nucleus**
alpha	^4_2He or α	2 protons 2 neutrons	2+	mass number decreases by 4; atomic number decreases by 2
beta	$^0_{-1}e$ or β	1 electron	1$-$	mass number does not change; atomic number increases by 1
gamma	$^0_0\gamma$ or γ	photon	0	no change in either the mass number or the atomic number

radioactive substance, gamma rays are referred to as nuclear radiation. In contrast, when emitted from a galaxy far away, these gamma rays are called electromagnetic radiation.

Your Turn 7.11   Which "Radiation"?

For each sentence, use the context to decipher whether the speaker is referring to nuclear or electromagnetic radiation.

a. "Name a type of radiation that has a shorter wavelength than visible light."
b. "The gamma radiation from cobalt-60 can destroy a tumor."
c. "Watch out for UV rays! If you have lightly pigmented skin, this radiation may cause a sunburn."
d. "Rutherford detected the radiation emitted by uranium."

When either an alpha or beta particle is emitted, a remarkable transformation occurs—the atom that emitted the particle changes its identity. For example, in Equation 7.4, alpha emission resulted in the nucleus of plutonium becoming that of uranium. Similarly, when uranium emits an alpha particle, it becomes the element thorium. This nuclear equation shows the process for uranium-238:

$$^{238}_{92}U \longrightarrow ^{234}_{90}Th + ^{4}_{2}He \qquad [7.6]$$

Notice that the sum of the mass numbers on both sides of the nuclear equation is equal: $238 = 234 + 4$. The same is true for the atomic numbers: $92 = 90 + 2$.

In some cases, the nucleus formed as the result of radioactive decay is still radioactive. For example, thorium-234, formed by the alpha decay of uranium-238, is radioactive. Thorium-234 undergoes subsequent beta decay to form protactinium (Pa):

$$^{234}_{90}Th \longrightarrow ^{234}_{91}Pa + ^{0}_{-1}e \qquad [7.7]$$

In contrast to alpha emission, with beta emission the atomic number increases by 1 and the mass number remains unchanged.

A concept that can help you make sense of this seemingly unusual set of changes is to regard a neutron as a combination of a proton and an electron. Beta emission can be thought of as breaking a neutron apart. Equation 7.8 shows this process, giving us an explanation of how an electron can be emitted from the nucleus:

$$^{1}_{0}n \longrightarrow ^{1}_{1}p + ^{0}_{-1}e \qquad [7.8]$$

During beta emission, the mass number (neutrons plus protons) in the nucleus remains constant because the loss of the neutron is balanced by the formation of a proton. For example, a neutron in thorium "became" a proton in protactinium. Because of this additional proton, the atomic number increases by 1. Again, this model can help you better visualize beta emission, but may not be exactly what is occurring.

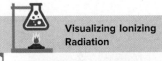

Visualizing Ionizing Radiation

©American Nuclear Society

Check out www.acs.org/cic for a variety of lab experiments and activities related to understanding nuclear reactions and ionizing radiation.

Your Turn 7.12   Alpha and Beta Decay

a. Write a nuclear equation for the beta decay of rubidium-86 (Rb-86), a radioisotope produced by the fission of U-235.
b. Plutonium-239, a toxic isotope that causes lung cancer, is an alpha emitter. Write the nuclear equation.

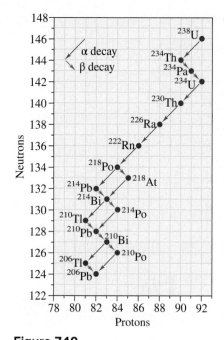

Figure 7.10

The naturally occurring radioactive decay series of uranium-238.

As we noted earlier, a nucleus may decay to produce another radioactive nucleus. In some cases, we can predict this, because *all* isotopes of *all* elements with atomic number 84 (polonium) and higher are radioactive. Thus, all the isotopes of uranium, plutonium, radium, and radon are radioactive because these elements all have atomic numbers greater than 83.

What about the lighter elements? Some of these are naturally radioactive, such as carbon-14, hydrogen-3 (tritium), and potassium-40. Whether an isotope is radioactive (referred to as a **radioisotope**) or stable depends on the ratio of neutrons to protons in its nucleus. With each emission of an alpha or a beta particle, this neutron-to-proton ratio changes. Eventually, a stable ratio is achieved, and the nucleus is no longer radioactive. Most of the atoms that make up our planet are *not* radioactive. They are here today, and you can count on them being here tomorrow—although possibly not located in the same spot you last saw them.

In some cases, radioisotopes may decay many times before producing a stable isotope. For example, the radioactive decay of U-238 and Th-234 (Equations 7.6 and 7.7) are only the first two steps of a 14-step sequence! As shown in Figure 7.10, lead-206 is the end product in this sequence. This sequence is called a **radioactive decay series**; that is, a characteristic pathway of radioactive decay that begins with a radioisotope and progresses through a series of steps to eventually produce a stable isotope. Radon, a radioactive gas, is produced midway in both the U-238 and U-235 decay series. Thus, wherever uranium is present, so is radon.

7.5 | Nuclear Radiation and You

> *Learning Objectives:*
>
> *Identify sources, health impacts, and applications of ionizing radiation*
>
> *Describe sources and concentration levels of background radiation*

Despite what you may have seen in the movies, nuclear radiation is only weakly carcinogenic. Furthermore, when it damages your cells or tissues, your body uses a number of mechanisms to repair a certain level of radiation damage. We live on a planet that naturally contains radioactive substances and, for the most part, we survive. When the repair systems in our bodies are overwhelmed and damage accumulates, then we have reason for concern.

What causes the cell or tissue damage? The answer lies in the alpha particles, beta particles, and gamma rays that radioisotopes emit. All have enough energy to remove an electron(s) from molecules, referred to as ionization. The same is true for X-rays, such as those used to produce medical images, as well as UV-C radiation described in Section 3.3. For this reason, we use the term **ionizing radiation** to refer collectively to X-rays, gamma rays, and UV-C radiation. Cosmic rays from space are also ionizing radiation. In contrast, UV-A, UV-B, visible, infrared, microwave, and radio wave radiation have lower energies and are *nonionizing* types of radiation (Figure 7.11).

Free radicals are generated when ionizing radiation interacts with molecules. As you saw in Section 3.7, free radicals are highly reactive with other molecules, including your DNA if it is nearby. Depending on how the DNA molecule is damaged by the free radical, the cell that contains this DNA may die, may repair itself, or may mutate resulting in cancer. Rapidly dividing cells, including some tumors, are particularly susceptible to damage by ionizing radiation. As a result, nuclear radiation and X-rays can treat certain types of cancer.

Ionizing radiation can treat other diseases as well. For example, people with Graves' disease have an overactive thyroid gland that produces excess hormone which boosts their metabolism, causing many complications. Although the idea of swallowing a radioactive pill may not sound appealing, such a pill provides a cure because it

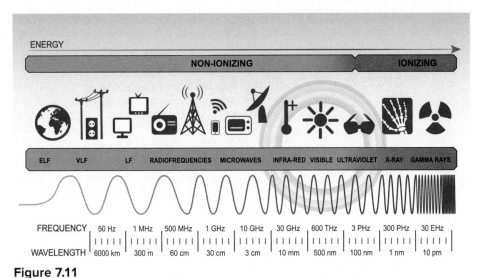

Figure 7.11

Comparison of ionizing and nonionizing regions of the electromagnetic spectrum.

Polina Kudelkina/Shutterstock

contains radioactive I-131 in the form of potassium iodide. Just as dietary iodine is incorporated into the thyroid gland, so is the radioactive iodine. Once in the thyroid gland, the radioactive I-131 destroys the overactive thyroid tissue, in whole or in part. Then, to restore normal metabolic function, most patients take a supplement of a synthetic form of thyroxin, the iodine-containing hormone normally secreted by the thyroid gland.

Our world naturally contains radioactive substances, so your radiation levels can never be reduced to zero. Scientists use the term **background radiation** to describe the level of radiation that, on average, is present at a particular location. It can arise from both natural and human-made sources. As seen in Figure 7.12, the largest natural source of background radiation is radon, a radioactive gas that is formed in the decay series of uranium. Your exposure to radon depends both on the amount of uranium in the rocks and soils where you live and on whether your dwelling allows this radon to accumulate in living spaces. Other terrestrial radiation comes from radioactive elements (primarily uranium, thorium, and radium) in soil, rock, and water sources. In addition, all organic matter from both plants and animals contains radioactive carbon and potassium isotopes. Some of these materials are ingested with food and water, whereas others such as radon are inhaled. As a result, all humans have internal

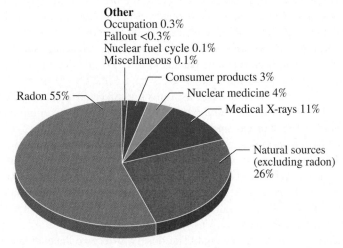

Figure 7.12

Sources of ionizing radiation in the U.S.

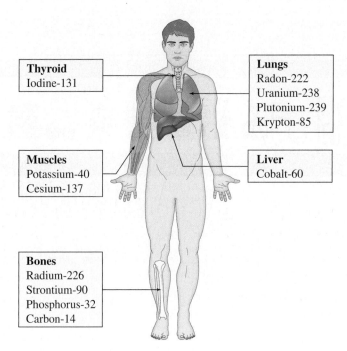

Figure 7.13

All living organisms contain some radioactive isotopes, including these that are located in particular parts of the human body.

radiation, which is mainly from the accumulation of radioactive potassium-40 and carbon-14 isotopes inside their bodies from birth (Figure 7.13).

Decades ago, almost all of our exposure was from natural sources. In recent years, however, medical exposures have entered the picture, such as those from CT scans, X-rays, and diagnostic radioactive tracers. Consequently, about half the exposure to ionizing radiation today is medical. This exposure now greatly exceeds natural background radiation for some patients.

However, what levels of radioactivity do we absorb or emit? The annual dose is given in two units, the **rem** and the **sievert** (Sv), whereby 1 Sv = 100 rem. Both units are a measure of the dose of radiation, taking into account the damage that occurs to human tissue when this dose is absorbed. However, these units represent a relatively high dose of radiation. Because most exposures to radiation are significantly less than a sievert or a rem, smaller units are necessary—microsieverts (mSv) and millirems (mrem).

When you take a pill, it's easy to calculate the drug dose for your body mass. It is simply a function of the concentration of the drug and how much you weigh. However, with ionizing radiation, calculating the dose is far more complicated. One reason is that different types of radiation deliver different doses. For example, the alpha particle packs a big punch, because alpha particles are larger and deposit a greater amount of energy in the tissue when they hit. It takes approximately 20 times as many beta particles to do the same tissue damage as a given number of alpha particles.

Another reason is that the same type of radiation may differ in the energy that it deposits in your tissue. For example, X-rays come in different wavelengths with correspondingly different energies. So, both the sievert and the rem are based on another unit, the **rad**, short for "radiation absorbed dose." The rad is a measure of the energy deposited in tissue and is defined as the absorption of 0.01 joule of radiant energy per kilogram of tissue. Thus, if a 70-kg person were to absorb 0.70 J of energy, a dose of 1 rad would be received. Although this is not much energy, it is localized in a small region where the radiation hits.

All radiation is *not* created equal. The dose you receive depends on the type, as well as whether it is inside or outside of your body. Let's now explore just how much radiation you are exposed to on a yearly basis.

Your Turn 7.13   Your Annual Ionizing Radiation Dose

Explore the simulation at www.acs.org/cic, which demonstrates the various sources of radiation that we are exposed to on a daily basis. Using the U.S. Nuclear Radiation Commission (NRC) radiation calculator, determine your personal annual radiation dose.

Questions to Consider

a. How does your annual dosage compare to the annual dosages of others in your class, or to the average level of radiation exposure in the U.S.?

b. How do geographic factors affect your annual radiation dose?

c. How could you adjust your lifestyle to reduce exposure to ionizing radiation?

7.6 | How Long Do Substances Remain Radioactive?

Learning Objective: Describe and calculate how the amounts of radioisotopes change over time in terms of half-lives

How long does a radioactive sample "last"? The answer depends on the radioisotope. Some radioisotopes decay quickly over a short period of time, whereas others undergo radioactive decay much more slowly. Each radioisotope has its own **half-life** ($t_{1/2}$), the time required for the level of radioactivity to fall to one-half of its initial value. The half-life of a radioisotope is independent of its initial concentration in a sample. For example, plutonium-239, an alpha emitter formed in nuclear reactors fueled with uranium, has a half-life of about 24,110 years. Accordingly, it will take 24,110 years for half of a sample of Pu-239 to decay. After a second half-life (another 24,110 years, or 48,220 years total), the level of radioactivity will be one fourth of the original amount. And in three half-lives (72,330 years total), the level will be one eighth (Figure 7.14). From these times, you can see that it takes a very long time for the sample size of Pu-239 to decrease!

@HOME   Sweet Radioactive Decay!

Investigate radioactive decay and half-life at home using three pieces of licorice, a marker, and a sheet of paper.

1. Make an *x*- and *y*-axis on your graph paper. Label the *x*-axis as "Number of Half-lives" and the *y*-axis as "Amount of Original Sample Remaining".

2. Place one piece of licorice on the sheet of paper at *x* = 1.

3. Break the second piece of licorice in half. Place one of the halves on the sheet of paper at *x* = 2.

OllyDark/Shutterstock

4. Break the remaining piece of licorice from Step 3 in half. Place one of the halves on the sheet of paper at *x* = 3.

5. Keep repeating this until you cannot break the licorice in half anymore. If you need to add another sheet of paper to extend the *x*-axis, you can.

6. Take the third piece of licorice and break it at a random point. Using these two pieces, estimate their placement on the graph.

Questions to Consider

a. If the *y*-axis had numerical values, how could Step 6 be helpful in calculating how much radioactive material is left?

b. Is the licorice ever really gone at the atomic level?

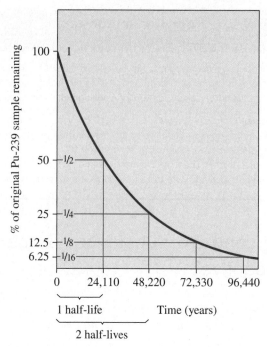

Figure 7.14

Decay of a sample of Pu-239 over time.

Other radioisotopes decay even more slowly. For example, the half-life of U-238 is 4.5 billion years. Coincidentally, this is approximately the age of the oldest rocks on Earth, a determination made by measuring their uranium content. The half-life for each particular isotope is a constant, and is independent of the physical or chemical form in which the element is found. Moreover, the rate of radioactive decay is essentially unaltered by changes in temperature and pressure. Table 7.3 shows that radioisotopes can have half-lives from milliseconds to millennia.

Table 7.3	Half-Life of Selected Radioisotopes	
Radioisotope	**Half-Life ($t_{1/2}$)**	**Found in the Used Fuel Rods of Nuclear Reactors?**
uranium-238	4.5×10^9 years	Yes (present originally in fuel pellet)
potassium-40	1.3×10^9 years	No
uranium-235	7.0×10^8 years	Yes (present originally in fuel pellet)
plutonium-239	24,110 years	Yes (see Equation 7.4)
carbon-14	5715 years	No
cesium-137	30.2 years	Yes (fission product)
strontium-90	29.1 years	Yes (fission product)
thorium-234	24.1 days	Yes (small amount generated in natural decay series of U-238)
iodine-131	8.04 days	Yes (fission product)
radon-222	3.82 days	Yes (small amount generated in natural decay series of U-238)
plutonium-231	8.5 minutes	No (half-life is too short)
polonium-214	0.00016 second	No (half-life is too short)

From Table 7.3, you also can see that Pu-239 and Pu-231 have different half-lives, as do the other isotopes of plutonium. In 1999, Carola Laue, Darleane Hoffman, and a team at Lawrence Berkeley National Laboratory characterized plutonium-231. These researchers had to work fast because the half-life of Pu-231 is a matter of mere minutes! In general, each radioisotope has its own unique half-life, including isotopes of the same element.

We can use the half-life of a radioisotope to determine the percent of a sample that remains at some later point in time. For example, once Pu-231 is generated in a laboratory, what percent of the original sample remains after 25 minutes? To answer this, first recognize that 25 minutes is roughly three half-lives, or 3×8.5 minutes. After one half-life, 50% of the sample has decayed and 50% remains. After two half-lives, 75% of the sample has decayed and 25% remains. And after three half-lives, 87.5% has decayed and 12.5% remains. These values are not exact because 25 minutes is not exactly three half-lives. Nonetheless, these estimations can be useful.

This question also could have been phrased in this way: "After 25 minutes, what percent of a sample of Pu-231 has *decayed*?" This question requires one more step. To find the amount decayed, simply subtract the percent that remains from 100%. If 12.5% remains, then $100\% - 12.5\% = 87.5\%$ has decayed. Table 7.4 summarizes these changes for any radioisotope.

Here Today . . .

. . . and gone tomorrow? People sometimes use the value of 10 half-lives to indicate when a radioisotope will be gone; that is, when only a negligible amount of it will be present. Watch the video at www.acs.org/cic and calculate what percent of the original sample remains after 10 half-lives. Add rows to Table 7.4 so that it shows the mathematics of decay to 10 half-lives.

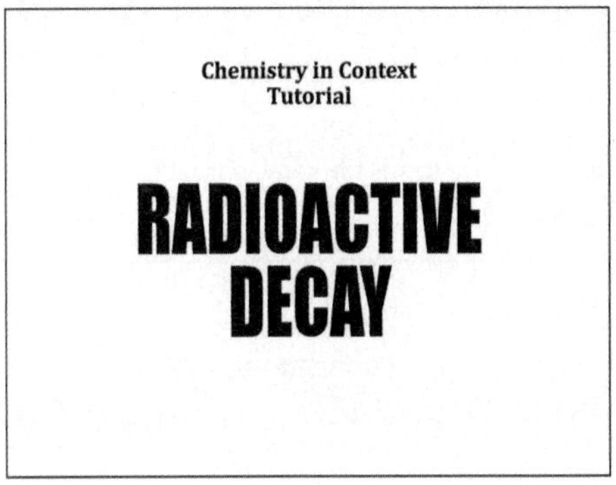

Chemistry in Context Tutorial

RADIOACTIVE DECAY

©Bradley D. Fahlman

Let's do another half-life estimation with a different radioisotope. For example, if you had a sample of U-238, what percent would remain after 25 minutes? To answer this, recognize that minutes, days, or even months would be a mere instant in the span of a 4.5-billion-year half-life. Thus, essentially all of the uranium-238 would remain. The next two activities offer you more practice with half-life calculations.

Table 7.4	Half-Life Calculations	
# of Half-Lives	% Decayed	% Remaining
0	0	100
1	50	50
2	75	25
3	87.5	12.5
4	93.75	6.25
5	97.88	3.12
6	98.44	1.56

Your Turn 7.14   Half-Lives

Example Problems:

a. Carbon-14 has a half-life of 5730 years. If a sample originally contains 70 mg, how much is left after 11,465 years?

b. The half-life of Co-60 is 5.25 years. If 50 g are left after 15.8 years, how many grams were originally present?

c. If 100 g of Au-198 decays to 6.25 g in 10.8 days, what is the half-life of Au-198?

Example Solutions:

a. 11,465/5730 = 2 half-lives. Therefore, the sample would contain 70 mg × 1/2 × 1/2 = 17.5 mg left.

b. 15.8/5.25 = 3 half-lives. Therefore, 50 g × 2 × 2 × 2 = 400 g originally present.

c. 100/6.25 = 16. Therefore, the original concentration has decreased by a factor of 16 in 10.8 days. This corresponds to 4 half-lives: 100 g × 1/2 × 1/2 × 1/2 × 1/2 = 6.25 g. Therefore, one half-life is 10.8 days/4 = 2.7 days.

Your Turn:

Radon-222 is a radioactive gas produced from the decay of radium, a radioisotope naturally present in many rocks.

a. What is the most likely origin for the radium present in rocks?
Hint: See Figure 7.10.

b. Radon activity is usually measured in picocuries (pCi). Suppose that the radioactivity from Rn-222 in your basement were measured at 16 pCi, a high value. If no additional radon entered the basement, how much time would pass before the level dropped to 0.50 pCi?
Hint: In dropping from 16 to 1 pCi, the radioactivity level halves four times: 16 to 8 to 4 to 2 to 1.

c. How much of a 100 g sample is left after 15.2 days?

d. In regard to part **b**, why is it incorrect to assume that no more radon will enter your basement?

Your Turn 7.15   Tritium Calculation

Hydrogen-3 (tritium, H-3) is sometimes formed in the primary coolant water of a nuclear reactor. Tritium is a beta emitter with $t_{1/2}$ = 12.3 years. For a given sample containing tritium, after how many years will about 12% of the original sample remain?

The long half-lives of the isotopes produced from nuclear reactors mean that nuclear waste will emit radiation for thousands of generations to come. Today, most nuclear waste is stored in steel or concrete casks at the site where it is generated, while countries try to develop plans for long-term underground storage facilities. These geological repositories can be designed to require minimal human intervention. In the U.S., Yucca Mountain in Nevada was under consideration since the 1970s as an underground long-term nuclear waste repository. However, after billions of dollars were spent to fund the development of the repository, this project was officially abandoned in 2022 due to scientific and political concerns.

Finland is currently developing the world's first permanent disposal site for high-level nuclear waste near the Olkiluoto Nuclear Power Plant on an island off the west coast of Finland. Using the KBS-3 waste disposal technology developed in Sweden, the waste will be encapsulated in cast iron canisters and placed inside copper alloy capsules. These capsules will then be surrounded by bentonite clay and drilled into crystalline rock about 1400 feet underground. This facility is projected to contain all of Finland's nuclear waste until about the year 2100, and is meant to contain spent fuel rods for 100,000 years. The design also includes multiple barriers designed to prevent water from reaching the waste and carrying it into the water supply.

One final difficulty with reactor waste is that the fission products, if released, may enter and accumulate in your body, with potentially fatal consequences. One culprit is strontium-90, a radioactive fission product that entered the biosphere in the 1950s from the atmospheric testing of nuclear weapons. Strontium ions are chemically similar to calcium ions; both elements are in Group 2 of the periodic table. Like Ca^{2+}, Sr^{2+} accumulates in milk and in bones. Thus, once ingested, radioactive strontium with its half-life of 29 years poses a lifelong threat. Along with I-131, Sr-90 was among the harmful fission products released in the vicinity of the Chornobyl reactor that exploded and caught on fire in Ukraine in 1986.

Your Turn 7.16   Strontium-90

Sr-90 is one of the fission products of U-235 listed in Table 7.3. It forms in a reaction that produces three neutrons and another element. Write the nuclear equation.

Hint: Remember to include the neutron that induces the fission of U-235.

7.7 | What Are the Risks of Nuclear Power?

> *Learning Objective: Describe the short-term and long-term impacts resulting from past nuclear accidents*

All active nuclear plants to date use the process of fission to produce energy, and all produce radioactive fission products. Have these radioactive products posed a danger in the past? In this section, we consider the accidental release of radioisotopes into the environment. Although this is not the sole legacy of nuclear power, it nonetheless is a significant one.

On April 26, 1986, the engineers of the Chornobyl nuclear power plant in Ukraine (Figure 7.15), then part of the Soviet Union, were running a safety test when the reactor overheated. This plant had four reactors, two built in the 1970s and two more in the 1980s. Water from the nearby Pripyat River was used to cool the reactors. Although the surrounding region was not heavily populated, approximately 120,000 people lived within a 30-km radius, including the cities of Chornobyl (pop. 12,500) and Pripyat (pop. 40,000).

Chornobyl stands as the world's worst nuclear power plant accident. So what went wrong in Ukraine? During an electrical power safety test at the Chornobyl Unit 4

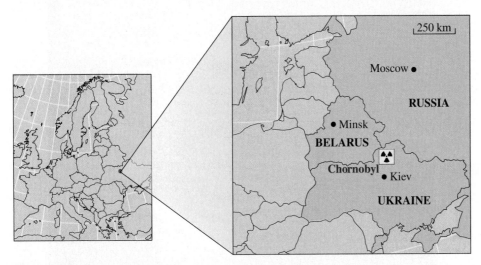

Figure 7.15

Chornobyl, Ukraine, in the former Soviet Union.

Note: The alternate spelling of "Chernobyl" is the transliteration of the Russian pronunciation. Чорнобиль (Chornobyl) is the Ukrainian word.

reactor, operators deliberately interrupted the flow of cooling water to the core. This caused the temperature of the reactor to rise rapidly. In addition, the operators had left an insufficient number of control rods in the reactor and other control rods couldn't be reinserted quickly enough. Furthermore, the steam pressure was too low to provide coolant, due to both operator error and faulty reactor design.

The Chornobyl Disaster: How It Happened

Check out this video (www.acs.org/cic) to see how the Chornobyl disaster happened. An HBO miniseries and podcast entitled "Chernobyl" were released in 2019, which dramatically depicted this disaster and the cleanup efforts that followed.

Igor Kostin/Laski Diffusion/Hulton Archive/Getty Images

Questions to Consider
a. How could this disaster have been avoided?
b. What were some of the short-term and long-term health effects resulting from this disaster? How widespread were these impacts?

A chain of events quickly produced a disaster. An overwhelming power surge produced heat, rupturing the fuel elements and releasing hot reactor fuel particles. These, in turn, exploded on contact with the coolant water, and the reactor core was destroyed in seconds. The heat ignited the graphite used to slow neutrons in the reactor. Graphite is a form of carbon and can burn if the temperature is high enough. When water was sprayed on the burning graphite, the water and graphite reacted to produce hydrogen gas:

$$2 \ H_2O(l) + C(s, \ graphite) \longrightarrow 2 \ H_2(g) + CO_2(g) \qquad \textbf{[7.9]}$$

In turn, the hydrogen exploded upon reaction with oxygen in the air:

$$2 \ H_2(g) + O_2(g) \longrightarrow 2 \ H_2O(g) \qquad \textbf{[7.10]}$$

The explosion blasted off the 4000-ton steel plate covering the reactor (Figure 7.16).

Figure 7.16
An aerial view of the Chornobyl Unit 4 reactor taken shortly after the chemical explosion.
STR/FILE/AP Images

Your Turn 7.17   Hydrogen Explosion

Equation 7.10 represents the combustion of hydrogen; that is, the reaction of hydrogen and oxygen to produce water vapor. Equation 6.8 provided the Lewis structures for this chemical reaction. Using the bond energies for those bonds broken and formed, the energy change for this reaction can be estimated. As shown in Figure 6.7, the value is −498 kJ for burning two moles of H_2.

a. Calculate the energy change per mole and per gram of H_2.
b. Of the fuels listed in Figure 6.5, methane releases the most heat per gram upon combustion. Burning hydrogen releases even more heat. Approximately how many times more?

Fires started in what remained of the building and burned for 10 days. Although a "nuclear" explosion was not possible, the fire and explosions of hydrogen blew vast quantities of radioactive material out of the reactor core and into the atmosphere. People living within 60 km of the power plant were permanently evacuated. The radioactive dust cut a swath across Ukraine, Belarus, and up into Scandinavia, affecting some who had not benefited from the power plant but nonetheless shared in its risks. The human toll was immediate; several people working at the plant were killed outright, and another 31 firefighters died in the cleanup process.

One of the hazardous radioisotopes released was iodine-131, a beta emitter with an accompanying gamma ray:

$$^{131}_{53}I \longrightarrow {}^{131}_{54}Xe + {}^{0}_{-1}e + {}^{0}_{0}\gamma \qquad\qquad \text{[7.11]}$$

If ingested, I-131 can cause thyroid cancer. In the contaminated area near Chornobyl, the incidence of thyroid cancer increased sharply, especially for those younger than age 15. More than 6000 cases of thyroid cancer have been reported among those who were children and adolescents living in Belarus, the Russian Federation, and Ukraine at the time of the accident. Fortunately, with treatment, the survival rate for thyroid cancer is high and most have survived. Apart from the dramatic increase in thyroid cancer incidence among those exposed at a young age, there is no clearly demonstrated increase in the incidence of cancers or leukemia due to radiation in the exposed populations.

Your Turn 7.18   Iodine

When people speak of iodine, they may be referring to an iodine atom, an iodine molecule, or an iodide ion, depending on the context.

a. Draw Lewis structures to show the differences among these chemical forms of iodine.
b. Which one is the most chemically reactive and why?
c. Which chemical form of iodine-131 is implicated in thyroid cancer?

In 2017, construction of a steel arch, large enough to encompass a college football stadium with the Statue of Liberty at midfield, was completed at Chornobyl to encase the ruins of the plant (Figure 7.17). Recognizing that Chornobyl is a global problem, about 30 countries contributed to the cost of the five-year project, estimated at $1.5 billion. To this day, the area around the nuclear reactor has remained unsuitable for human habitation.

Underlying the solemn facts of Chornobyl is the inevitable question: Could it happen again? The closest brush with nuclear disaster in the United States occurred in March 1979, when the Three Mile Island power plant near Harrisburg, Pennsylvania, lost coolant and a partial meltdown occurred. Although some radioactive gases were released during the incident, no fatalities resulted. A 20-year follow-up study concluded in 2002 that the total cancer deaths among the exposed population were not higher than those of

Figure 7.17

View of the containment structure constructed at the site of the Chornobyl disaster.

Pyotry Sivkov/TASS/Getty Images

the general population. Nuclear engineers agree that no commercial nuclear reactors in the United States have the design defects that led to the Chornobyl catastrophe.

The world's most recent nuclear disaster occurred in 2011, when a pair of natural disasters, an earthquake and a tsunami, resulted in the meltdown of three units of the Fukushima Daiichi nuclear power plant in Japan (Figure 7.18).

The tsunami delivered a one–two punch. First, the flood waters knocked out the electrical generators necessary to pump the cooling water at the Fukushima power plant; as a result, the reactor cooling systems failed. The fuel inside reactors 1, 2, and 3 quickly heated, and the heat started a chemical reaction that generated hydrogen gas. Fearing

Figure 7.18

Flooding from the tsunami that followed the 2011 Tohoku earthquake, a 9.0 on the Richter scale.

Kyodo News/AP Images

a hydrogen gas explosion that would rival that of Chornobyl, plant workers vented the hydrogen. At the same time, this action released some of the radioactive fission products, including I-131, to the surrounding countryside. Despite the venting, explosive chemical reactions occurred at four of the six reactors, which released dangerous radioisotopes into the environment.

Over the following two weeks, the government instructed approximately 300,000 people living within 30 km of the power plant to leave. The International Atomic Energy Association recommended in 2013 that most evacuees be allowed to return to their homes, and the Japanese government lifted its evacuation order for the center of Namie in 2017. Nevertheless, as of 2022, fewer than 5% of its original population of 17,613 have returned to their homes due to concerns over radiation levels. In 2016, the World Health Organization estimated that the level of exposure of the Japanese general public was so low that it did not expect to see any radiation-related long-term health effects. However, 80% of the geographical area surrounding Namie is composed of mountains and forest, which is difficult, if not impossible, to decontaminate. Since this accident, the nongovernment environmental organization Greenpeace has taken thousands of radiation readings in vicinities surrounding the Fukushima nuclear plant. A 2018 report concluded that radiation levels in parts of Namie will remain well above international safety recommendations for many decades, presenting higher risks of leukemia and other cancers, especially for children.

Although the post-disaster radiation levels in areas outside of Namie were found to be low, the Fukushima incident caused a significant shift in the Japanese energy policy. After the disaster, Japan's nuclear agency shut down all but two of the 48 nuclear power plants in the country. This led to expansion of renewable sources such as floating wind farms and solar power plants. However, to meet the energy needs of its growing population, Japan reopened four additional nuclear plants alongside many new coal-fired power plants, which now total 91. This dramatic shift from nuclear to fossil fuel sources will have alarming implications for air pollution and will likely affect Japan's ability to meet its pledges of cutting greenhouse gas emissions by 80% from 2013 levels by 2050.

Your Turn 7.19   Zirconium

At the Chornobyl nuclear power plant, hydrogen was generated by a reaction of water with the hot graphite, as described earlier in this section. At Fukushima, however, the hydrogen was generated by a reaction of water with the element zirconium found in the alloy composing the outer casing of the fuel rods.

a. Zirconium is the metal of choice for reactors due to several reasons, including that it does not absorb neutrons. Why is this a desirable property?

b. Zirconium, if heated to a high temperature (such as in an accident at a nuclear plant), has two undesirable properties: (1) it will swell and crack, and (2) it will react with water to produce hydrogen. Explain the dangers that these properties present.

Today, nuclear plants and their past operations continue to be under intense scrutiny. Undoubtedly, nuclear energy will be part of our future; however, at present, it is not clear to what extent future generations will rely on nuclear energy.

7.8 | Is There a Future for Nuclear Power?

> *Learning Objective: Compare the nuclear power capacity of countries around the world*

Should we build more nuclear power plants? The answer depends on both whom you ask and when you ask them. Some longtime opponents of nuclear energy are now in favor of it. Similarly, some who supported it are now questioning its social costs, to both our current generation and those to come.

Eric Misko/McGraw Hill

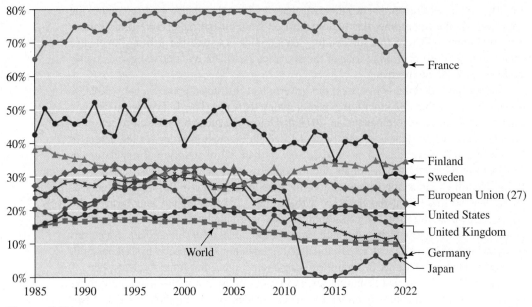

Figure 7.19

Share of primary energy from nuclear power in selected countries, 1957–2021. For an interactive version, see: www.acs.org/cic.

Note: MW is megawatts; MWh is megawatt-hours.

Our World in Data based on BP Statistical Review of World Energy (2022); Our World in Data based on Ember's Yearly Electricity Data (2023); Our World in Data based on Ember's European Electricity Review (2022).

Let's say that you switch on your coffeepot. If you live in the United States, about one fifth of the electricity is coming from a nuclear power plant. If you live in France, Belgium, or Sweden, the percentage is even higher. In either case, you can brew your coffee!

Worldwide, nations differ in the extent to which they employ nuclear energy to generate electricity. For example, in the United States, about 20% of commercial electric power is produced from 96 nuclear reactors, all licensed by the Nuclear Regulatory Commission. As of 2022, these reactors were operating at 59 sites in 29 states. As you can see in Figure 7.19, the share of nuclear energy worldwide has declined to 10% as of 2021, but remains an important source of energy in the U.S. and many European countries.

When you brew your coffee a decade from now, from where will the electricity originate? Nuclear power plants surely will be one source. As of 2022, there are 438 nuclear power reactors currently in operation around the world, with a combined electrical capacity of 394 GW. In addition, there are 60 reactors currently under construction with 103 additional reactors planned and 325 reactors proposed for future construction.

Your Turn 7.20   State-by-State

This map shows the 29 states that have nuclear power plants in the United States.

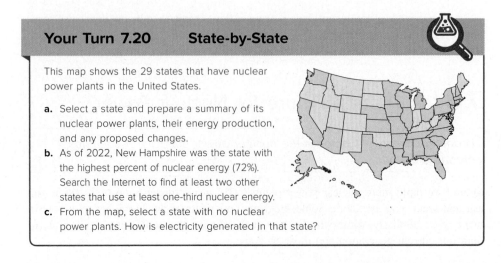

a. Select a state and prepare a summary of its nuclear power plants, their energy production, and any proposed changes.

b. As of 2022, New Hampshire was the state with the highest percent of nuclear energy (72%). Search the Internet to find at least two other states that use at least one-third nuclear energy.

c. From the map, select a state with no nuclear power plants. How is electricity generated in that state?

Figure 7.20

Aerial view of the 550-acre construction site of the Vogtle nuclear power plant.

Construction sites for new nuclear power plants are enormous. They cover hundreds of acres and employ a workforce in the thousands, making them essentially cities themselves. The construction site at Plant Vogtle in Waynesboro, Georgia, (Figure 7.20), even has its own railway! Once finished, this site will be the largest nuclear plant in the United States.

The construction and continued operation of a commercial nuclear power plant is not only a matter of energy supply-and-demand, but also one of public acceptance. People have been lining up on one side or the other of the nuclear fence since nuclear power was first proposed back in the 1970s. Which side are you on?

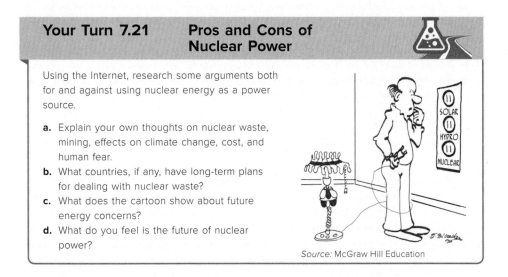

Your Turn 7.21   Pros and Cons of Nuclear Power

Using the Internet, research some arguments both for and against using nuclear energy as a power source.

a. Explain your own thoughts on nuclear waste, mining, effects on climate change, cost, and human fear.
b. What countries, if any, have long-term plans for dealing with nuclear waste?
c. What does the cartoon show about future energy concerns?
d. What do you feel is the future of nuclear power?

Source: McGraw Hill Education

The larger picture of nuclear energy worldwide is one of change. In part, these changes stem from increased energy demand. Major commercial development of nuclear energy is clearly on the agenda of many nations. For example, although India generated only 3.2% of its electricity from 22 reactors in 2018, construction is now underway on seven new plants, with others planned or proposed. As of 2019, China had 46 operational nuclear power reactors, with 11 currently under construction (Figure 7.21).

Annual Change in Nuclear Energy Generation, 2021
Shown is the change in nuclear energy generation relative to the previous year, measured in terawatt-hours.

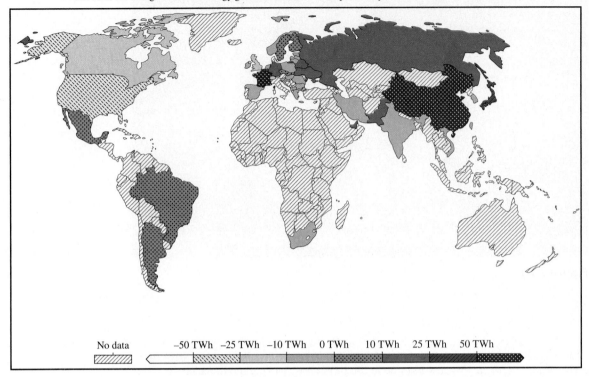

Figure 7.21

Annual change in nuclear energy generation worldwide. For an interactive version, see: www.acs.org/cic.

Our World in Data based on BP Statistical Review of World Energy.

Your Turn 7.22   Worldwide Nuclear Use

a. Where are nuclear reactors located and which countries have shown the greatest expansion of nuclear capacity within the past 10 years? Write a summary of the data in Figure 7.21.

b. Which countries have shown the greatest decrease in nuclear plants over the past 10 years? What are some reasons for this decrease in nuclear capacity?

c. Use the Internet to determine which countries are the top producers of uranium worldwide. How many of these countries have commercial nuclear reactors?

d. Suggest reasons why some countries, more than others, should develop nuclear energy. What are some other applications of nuclear technology beyond electricity generation?

People across the globe share the dream of clean and sustainable sources of energy for the future. Does this dream include nuclear energy? If so, should we build more nuclear power plants to achieve this dream? If you had asked this question in the United States back in the early 1960s, the answer would have been yes. At that time, the United States experienced a dramatic growth in the nuclear power industry, one that lasted until 1979 when the malfunction at Three Mile Island occurred. The fear that accompanied this incident certainly contributed to the end of the growth phase. More important at that time, however, were the economics of nuclear energy. With the retreat of fossil fuel prices and the added costs of nuclear safety and oversight imposed in the 1980s, it simply was not economically feasible for utilities to construct new nuclear plants. However, one new design being developed is for smaller, modular nuclear reactors that can be factory built. These reactors could be deployed with minimal construction on-site, speeding construction and making power plant expansion easier in order to meet increasing energy demands. Furthermore, as development of nuclear fusion continues to improve, the use of nuclear fusion for energy would circumvent problems associated with nuclear waste and the limited availability and safety concerns of radioisotopes.

What are the economic realities today? First, any new reactors will be built with improved designs, especially in light of the earthquake and tsunami that disabled reactors in Japan. Second, these designs will have a higher price tag.

In terms of design, the near future of nuclear power, especially in the United States, is primarily focused on ensuring current nuclear power plants are prepared for extraordinary disasters such as what occurred at Fukushima. The U.S. Nuclear Regulatory Commission (NRC) stated in a report that "a sequence of events like the Fukushima, Japan, incident is unlikely to occur in the U.S.," but an "accident involving core damage and uncontrolled release of radioactivity to the environment, even one without significant health consequences, is inherently unacceptable."

The NRC issued three orders to U.S. nuclear power facilities, in reaction to the events in Japan. The orders included the requirements:

- that all facilities obtain sufficient portable safety equipment, such as devices that could burn off hydrogen generated in an accident, to support all reactors and spent fuel pools at a given site simultaneously. Spent fuel is the radioactive material remaining in the fuel rods that have been used in nuclear reactors. The spent fuel rods are stored under at least 20 feet of water so that people near the pool are not exposed to radiation. This is to ensure that if a disaster affects multiple reactors, there will be protection.
- that certain facilities improve their venting systems for boiling water reactors to ensure protection against a backup of steam and to control the temperature.
- that new equipment be installed in order to monitor water levels in each plant's spent fuel pool. This will ensure that facilities will know water levels throughout the plant.

Another recommendation of the NRC is to use more passive cooling systems in nuclear plants. Passive cooling systems use convection currents rather than mechanical pumps to circulate coolant, making the system less susceptible to malfunctions.

As you can see, there are no easy answers for the issue of nuclear power. Global demand for energy expands daily, as does the mass of radioactive waste from nuclear power plants with which we must cope. The era of climate change has dawned. Yet both real and perceived hazards associated with radioactivity, with mining and enriching uranium, and with nuclear weapons still remain. This presents a classic risk–benefit situation, and the final compromise has yet to be reached. For now, it is clear that nuclear power may not the cure-all for the world's energy woes, but it remains one of the safest and cleanest (Figure 7.22). It is, however, the cause of some environmental and social woes. Even so, it will remain an important source of energy for years to come.

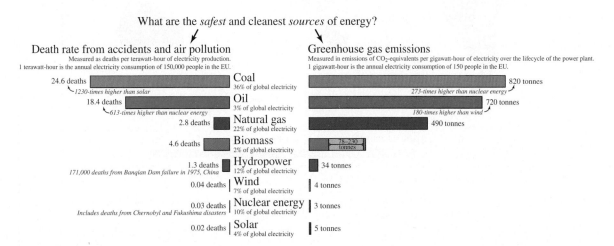

Figure 7.22

Comparison of various sources of energy in terms of safety and greenhouse gas emissions during use.

Our World in Data

7.9 | Solar Power: Electricity from the Sun

Learning Objectives:

Describe global trends in the use of solar energy

Summarize the applications for concentrating solar power arrays and photovoltaic cells

Given our increasing energy needs, it would surely make sense to take advantage of sunlight, a renewable energy source. The Sun's rays hit Earth every hour with enough energy to meet the world's energy demand for an entire year! Currently, however, approximately 1.7% of the electric power generated in the United States comes directly from solar energy. So, why does solar energy currently account for such a small part of our larger energy picture?

Although remarkable amounts of sunshine hit Earth daily, the rays do not strike any one site on the planet for 24 hours a day, 365 days a year. Furthermore, some parts of the planet receive too low an intensity of light to be practical for solar collecting. The disparity arises due to differences in geographic locations and local factors such as cloud cover, aerosols, smog, and haze. For example, examine the map shown in Figure 7.23. The data in the figure are reported in kilowatt-hours (kWh) per square meter per day for a flat-panel solar collector that is stationary. The next activity helps you explore the differences in daily solar energy during a calendar year.

Your Turn 7.23   Where Does the Sun Shine?

The interactive map at www.acs.org/cic shows the use of solar energy throughout the world.

Share of Primary Energy from Solar, 2021

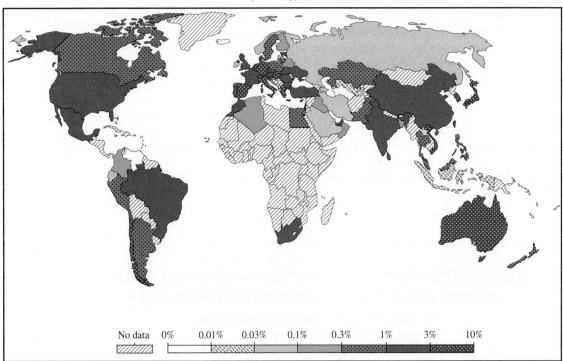

No data   0%   0.01%   0.03%   0.1%   0.3%   1%   3%   10%

Our World in Data based on BP Statistical Review of World Energy (2022).

a. Select a country of your choice and view the trends of solar power over the past 10 years. Which countries have increased their reliance on solar power most significantly? What are some reasons for this expansion?

b. When did the greatest expansion of solar capacity occur worldwide? Is there a reason(s) why solar energy showed such growth beginning in that year?

c. Looking at Figure 7.23, it should come as no surprise that California, Arizona, New Mexico, and Texas lead the United States in average annual solar radiation. Why do some parts of these states have higher values than others?

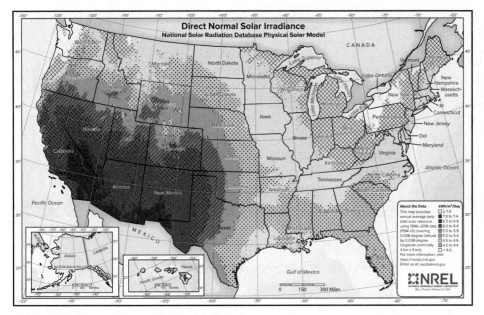

Figure 7.23

The average amount of daily solar energy received by a fixed photovoltaic panel oriented due south.

Note: These energy levels would be higher if the panels tracked the path of the Sun, rather than being stationary.

Source: National Renewable Energy laboratory (NREL) for the U.S. Department of Energy, 2018.

Your Turn 7.24   Could Solar Energy Power a House in Your Neighborhood?

How feasible is solar energy for meeting your everyday energy demands?

a. Use Figure 7.23, or another resource from the Internet, to estimate how many kilowatt-hours (kWh) of solar energy fall on one square meter of land per day in your area.

b. A section of a homeowner's monthly electric bill is shown below. In your area, would this homeowner be able to power their household using only solar energy?

Your energy use

Meter # IN24775778

Schedule 07 (residential rate)

Service Period	Meter Reading
06/23/23	15335
05/21/23	15079
33 days of service	256 kWh

c. Assuming that this homeowner's monthly energy use is typical, how many households could be supported by a nuclear power plant support with a capacity of 750 MW?
Hint: Start by calculating how many megawatt-hours (MWh) of energy the plant would generate in one day and how many kWh of energy the household uses in one day.

d. Does the answer you calculated in part **c.** make sense? If so, explain. If not, describe the assumptions made that might lead to a larger or smaller number than expected.

The challenge, then, is to locate the areas in which the incident average solar energy is high, and to collect this energy in sufficient quantities to produce electricity. There are two primary ways that energy from the Sun can be transformed into electricity to power our lives. Recall that electromagnetic radiation from the Sun is composed of a range of wavelengths with different energies. Thus, both the Sun's heat (longer wavelengths, lower energy) and its light (shorter wavelengths, higher energy) can be used to generate electricity.

(a)   **(b)**

Figure 7.24

(a) An aerial view of the Solar Millennium Andasol project in Spain, which has an approximate capacity of 150 MW of solar-thermal power.
(b) A close-up of a portion of the mirrored array.

(a) Langrock/Solar Millennium/SIPA/Newscom; (b) Boris Roessler/Deutsche Presse-Agentur/Newscom

Concentrating the Sun's energy in order to heat water is a solar-thermal process. When the Sun's heat is used to generate electricity, the process is known as **concentrating solar power** (CSP, Figure 7.24). The mirrored arrays of CSP concentrate sunlight in much the same way that a magnifying glass can focus light to burn a hole in a piece of paper. The heat from this concentrated light is then used to power a steam turbine and generate electricity as coal and nuclear power plants do. As of 2022, there were 114 operational concentrating solar power plants in the world, and another 10 under construction, each generating between 1 and 400 MW of power.

Your Turn 7.25   Solar-Thermal Collectors

Check out "Concentrating Solar Power Projects by Country" (www.acs.org/cic) by the National Renewable Energy Laboratory for a world map of concentrating solar power projects.

All solar collectors focus and concentrate the Sun's rays for the purpose of producing heat. However, they do so in different ways.

a. Searching the Internet, find and describe the designs for three different types of collectors.
b. How is each design matched to its end use? As part of your answer, include the scale of use—that is, for a single home, for a community, or for a business.
c. Name at least one limitation for each type of collector.

Figure 7.25

Photovoltaic (solar) cells are used to improve security, enhance safety, and direct pedestrians and vehicles.

Warren Gretz/DOE/NREL

A second way to tap into the Sun's energy is to use a **photovoltaic (PV) cell**—a device that converts light energy directly into electric energy, often called a *solar cell*. It takes only a few PV cells to produce enough electricity to power your calculator or digital watch. Other common uses for photovoltaic cells include communication satellites, highway signs, security and safety lighting (Figure 7.25), automobile recharging stations, and navigational buoys. Cost savings can be substantial. For example, using solar cells rather than batteries in navigational buoys saves the U.S. Coast Guard several million dollars annually through reduced maintenance and repair.

If more power is required, PV cells can be combined into modules or arrays to make up solar panels, as shown in Figure 7.26. Many people today power their homes and businesses with solar PV systems—especially throughout Europe. Depending on the size of a home, it may require a dozen or more solar panels for power. These panels are usually mounted facing due south. Installing them on a system that rotates to track the Sun's path, and thus maximizing their exposure to sunlight, optimizes their efficiency but has a higher initial cost. For electric utility or industrial applications, hundreds of solar arrays are interconnected to form a large-scale PV system, such as the one shown in a field in Bavaria, Germany (Figure 7.26c).

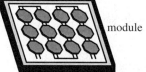
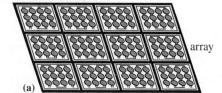

(a)

(b)

(c)

Figure 7.26

(a) Arrangement of photovoltaic cells used to make a module and an array. **(b)** A silicon solar array installed on a roof. **(c)** An aerial view of the Solarpark Gut Erlasee in Bavaria, Germany. At peak capacity, it can generate 12 MW. In comparison, a typical nuclear power plant generates 1000 MW of electricity.

Source: (b): NREL/U.S. Dept. of Energy; (c): dpa picture alliance archive/Alamy Stock Photo

In the next section, we will glance inside a solar panel to see how sunlight is converted into electricity. However, now take a break in the Sun to make some s'mores, as directed in this next activity.

@HOME   S'mores From Sunshine!

Investigate the sheer energy of the Sun at home on a warm day using an old pizza box, scissors, plastic wrap, tape, sticks, rocks, chocolate, graham crackers, and marshmallows.

Arina P Habich/Shutterstock

1. Cut a rectangular flap in the top of the pizza box.
2. Wrap the flap with aluminum foil. Use tape if it needs more securing.
3. In the hole where the flap was, tape plastic wrap so it fully covers the hole.
4. Place a graham cracker inside the pizza box, and then a piece of chocolate on top of it. Finally, add a marshmallow on top of it. Close the lid.
5. Use a stick to prop the flap up so that the flap is at an angle.
6. Leave the pizza box outside in the Sun and observe what happens. Adjust the flap so that the sunlight is being reflected toward the s'more if you are having trouble getting it to melt.

Questions to Consider

a. What are the advantages to making a s'more using solar energy over using a bonfire? What are the disadvantages?

b. How can this setup be used to also explain greenhouse gases?

7.10 | Solar Energy: Electronic "Pinball" Inside a Crystal

> *Learning Objective: Describe how sunlight is converted into electricity within a photovoltaic (solar) cell*

How does a photovoltaic cell generate electricity? The answer lies in the behavior of the electrons in the cell material. When light shines onto a PV cell, it may pass through the cell, be reflected, or be absorbed. If absorbed, the energy may cause an excitation of the electrons in the atoms of the cell. These excited electrons escape from their normal positions in the cell material and become part of an electric current.

Only certain materials behave this way in the presence of light. Photovoltaic cells are made from a class of materials called **semiconductors**, materials that have a limited capacity of conducting an electric current. Most semiconductors are made from a crystalline form of silicon, a metalloid. A crystal of silicon consists of a regular array of silicon atoms, each bonded to four others (to satisfy the octet rule) by means of shared pairs of electrons (Figure 7.27a). These shared electrons are normally fixed in the bonds and are unable to move through the crystal. Consequently, silicon is not a very good electrical conductor under ordinary circumstances. However, if a bonding electron absorbs sufficient energy, it can be excited and released from its bonding position (Figure 7.27b). Once freed, the electron can move throughout the crystal lattice, making the silicon an electrical conductor. The release of a bonding electron leaves behind a positively charged vacancy, referred to as a *hole*.

In reality, pure silicon semiconductors do not allow an electric current to flow unless they are doped. **Doping** is a process of intentionally adding small amounts of other elements, known as *dopants* (sometimes called impurities), to pure silicon. These dopants are chosen for their ability to facilitate the transfer of electrons. For example, a few ppm of gallium (Ga) or arsenic (As) are often introduced into the silicon crystal structure through processing at high temperatures. These two elements, and others from the same groups in the periodic table, are used because their atoms differ from silicon by only a single valence electron. Recall that silicon (Si) is in Group 14 and has four valence electrons. In comparison, gallium (Ga) is in Group 13 and has three valence electrons. Arsenic (As) is in Group 15 and has five valence electrons.

Thus, when an atom of As is introduced in place of a Si atom in the lattice, an extra electron is added. This material with an excess of electrons is referred to as an

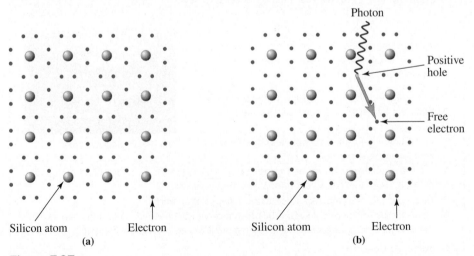

(a) Schematic of bonding in silicon. **(b)** Photon-induced release of a bonding electron in a silicon semiconductor.

Figure 7.27

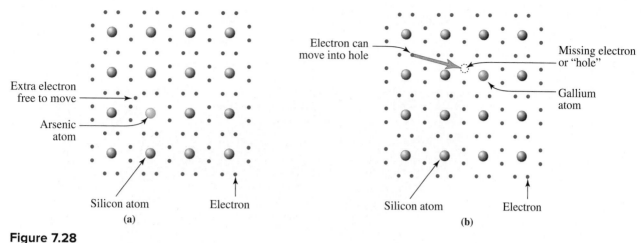

Figure 7.28

(a) An arsenic-doped n-type silicon semiconductor. **(b)** A gallium-doped p-type semiconductor.

n-type semiconductor. In contrast, the replacement of a Si atom with a Ga atom means that the crystal is now one electron "short." This material with a shortage of electrons (or an excess of holes, the lack of an electron at a position in the crystal where one could/did exist) is referred to as a **p-type semiconductor**. Figure 7.28 illustrates doped n- and p-type semiconductors. Both types of doping increase the electrical conductivity of silicon because electrons can now move from an electron-rich to an electron-deficient environment.

Your Turn 7.26   Doping Predictions

Some solar cell designs often use phosphorus and boron to dope silicon crystals.

a. Which will form an n-type semiconductor? Explain your reasoning.
b. Which will form a p-type semiconductor? Explain your reasoning.

To induce a voltage in a PV cell, a layer of n-type and a layer of p-type semiconducting materials are placed in direct contact. To generate an electric current, the light hitting the PV cell must have enough energy to set the electrons in motion from the n-type side to the p-type side through the electric circuit (Figure 7.29). The transfer of electrons generates a current of electricity that can be used to do all the things electricity does, including being stored in batteries for later use. As long as the cell is exposed to light, the current continues to flow, powered only by solar energy.

A photovoltaic cell is typically composed of multiple layers of doped n- and p-type semiconductors in close contact (Figure 7.29). The p–n junctions not only make possible the conduction of electricity, but also ensure that the current flows in a specific direction through the cell. These "sandwiches" of n- and p-type semiconductors are also used in transistors and other miniaturized electronic devices that have revolutionized communications and computing.

Only photons with enough energy can knock electrons free from the doped material. These electrons then become part of the external electric circuit. For a PV cell to convert as much sunlight as possible into electricity, the semiconductors must be constructed in such a way as to make the best use of the photons' energy. If not, the energy of the Sun is not trapped at all or is lost as heat.

The fabrication of photovoltaic cells poses some significant challenges. The first is that although silicon is the second-most abundant element in Earth's crust, it is most frequently found combined with oxygen as silicon dioxide, SiO_2. You know this material by its common name, sand, or more correctly as quartz sand. The good news is

---

Electrons and Holes

Check out this video to see how electrons and holes move through a semiconductor crystal: www.acs.org/cic.

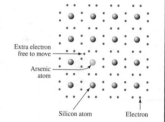

Questions to Consider

a. Glass is also composed of silicon atoms. Could glass also be doped with elements to be useful in PV cells?

b. Using the Internet as a resource, what concentrations (in ppm) of dopants are typically added to silicon for PV applications?

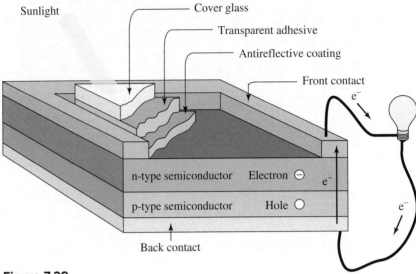

Figure 7.29

Schematic diagram of one layer of a solar cell showing the sandwiching of n-type and p-type semiconductors, and the "electron-hole pairs" generated by photons of sunlight. The movement of electrons and holes in opposite directions results in the flow of electrical current through the external circuit. Check out this video to see how solar cells work: www.acs.org/cic.

that the starting material from which silicon is extracted is cheap and abundant. However, the not-so-good news is that the processes used to extract and purify silicon are expensive. To be useful in solar devices, the silicon must be refined to a purity of at least 99.9999%.

A second challenge is that the direct conversion of sunlight into electricity is not efficient. A photovoltaic cell could, in principle, transform up to 31% of the radiant energy to which it is sensitive into electricity. However, some of the radiant energy is reflected by the cell or absorbed to produce heat instead of an electric current. As of 2022, researchers have developed crystalline silicon PV cells with efficiencies over 25%. This is a significant increase over the first solar cells built in the 1950s, which had efficiencies of around 6%. In Section 6.7, we lamented the 35–50% net efficiency of converting heat into work in a conventional power plant. It might seem that we should be even more distressed at the lower limits that can be achieved by photovoltaics. Remember, however, that the first use of solar cells was to provide electricity in NASA spacecraft. For that application, the intensity of radiation was so high that low efficiency was not a serious limitation and costs were not of paramount concern. However, for commercial use on Earth, costs and efficiency are key limitations. The Sun is an essentially unlimited energy source and converting its energy into electricity, even inefficiently, is free from many of the environmental problems associated with burning fossil fuels or storing waste from nuclear fission. These considerations add impetus to the research and development of solar cells.

One approach to increasing commercial viability is to replace crystalline silicon with the noncrystalline form of the element, known as *amorphous* Si. Photons are more efficiently absorbed by less highly ordered Si atoms, a phenomenon that permits reducing the thickness of the silicon semiconductor to 1/60 or less of its former value. The cost of materials is thus significantly reduced, but the efficiency of conversion to electricity is not as high as crystalline silicon.

Other researchers are developing multilayer solar cells. By alternating thin layers of p-type and n-type semiconductors, each electron has only a short distance to travel to reach the next p–n junction. This lowers the internal resistance within the cell and raises its efficiency. The maximum theoretically predicted efficiencies could improve to 50% for two junctions, to 56% for three junctions, and to 72% for 36 junctions. Furthermore, the use of other semiconductors in contact allows the device to absorb different regions of the electromagnetic spectrum, to cover the entire UV–IR range. In comparison, Si is most sensitive to the absorption of light in the blue region of the electromagnetic spectrum, which results in significant wavelength ranges not being effectively absorbed. As

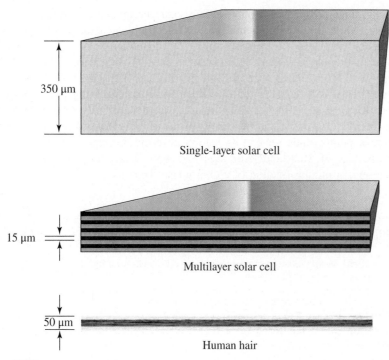

350 μm

Single-layer solar cell

15 μm

Multilayer solar cell

50 μm

Human hair

Figure 7.30

A comparison of the relative thickness of a solar cell layer, in either a single-layer or multilayer cell, to the diameter of an average human hair. *Note:* 1 μm = 10⁻⁶ m.

of 2022, the maximum efficiency actually demonstrated with a multilayer solar cell was 47%. Figure 7.30 gives a sense of the thickness of the layers in multilayer solar cells.

Thin-film solar cells are made from amorphous silicon or non-silicon materials such as cadmium telluride (CdTe) or a combination of copper, indium, gallium, and selenium (referred to as CIGS, $CuIn_xGa_{(1-x)}Se_2$, where x = 0–1). These thin films use layers of semiconductor materials only a few micrometers thick. For comparison, the width of a typical human hair is about 50 μm. Thin-film solar cells can even be incorporated into rooftop shingles and tiles, building facades, or the glazing for skylights because of their flexibility compared with more rigid traditional cells (Figure 7.31). Other solar cells are being made using various materials, such as solar inks employing conventional printing press technologies, solar dyes, nanoparticles known as "quantum dots," and conductive plastics. Concentrated photovoltaics have also been developed that employ lenses or mirrors to improve the light gathering ability of solar panels.

Figure 7.31

Thin-film solar tiles on a roof.

Source: NREL/U.S. Dept. of Energy

Your Turn 7.27   Solar PV Use

How are people today using solar photovoltaics? Search the Internet to answer this question for each group listed below.

a. farmers and ranchers
b. small-business owners
c. homeowners

Long-range prospects for photovoltaic solar energy are encouraging. Its cost is decreasing, while the environmental cost of electricity generated from fossil fuels continues to rise. Despite recent advances in technology, solar energy from either thermal or photovoltaic sources is often more expensive than fossil fuels. However, at energy auctions in places such as South Africa and India, the cost of solar-energy power plants has been cheaper than building a new coal-burning power plant. Nevertheless, there is still the question of land use. At currently attainable levels of operating efficiency, the electricity needs of the United States have been estimated to require a photovoltaic generating station covering an area of 100×100 miles, roughly the size of Maryland! Although photovoltaic power is steadily growing, it still represents a minute fraction of global power supplies.

Your Turn 7.28   Where to Site Solar Power?

Last we heard, Maryland was not volunteering to be converted entirely into a solar farm to power the rest of the United States.

a. Would Maryland be a reasonable geographic location?
   Hint: Revisit Figure 7.23.
b. Which locations in the United States show the most promise for solar energy collection?
c. Location isn't everything. Name two other factors that come into play in dedicating land to solar energy collection.
d. Using the Internet, compare the relative costs and payback of solar power in different regions of the U.S. How does the cost of solar energy in the U.S. compare to other countries around the world?

Because of the diffuse nature of sunlight, photovoltaic technology is well suited to distributed generation, as will be described for fuel cells in the next chapter. More than one third of Earth's population is not hooked into an electric grid. This is due to the costs associated with constructing and maintaining equipment, and supplying the fuel needed to generate the electricity. Because PV installations are relatively maintenance-free, they are particularly attractive for electric generation in remote regions. For example, the highway traffic lights in certain parts of Alaska far from power lines operate on solar energy. A similar but more significant application of photovoltaic cells may be to bring electricity to isolated villages in lower income countries. In recent years, more than 200,000 solar lighting units have been installed in residential units in Colombia, the Dominican Republic, Mexico, Sri Lanka, South Africa, China, and India. Indeed, photovoltaic cells are currently improving the lives of millions of people across our planet (Figure 7.32). What does the future hold for solar power? Check out the interview below for an intriguing possibility in transportation.

Real Talk   Solar-Powered Transportation Systems?

Watch this interview with Dr. Burford Furman who describes an exciting future of solar-powered transportation systems: www.acs.org/cic.

Questions to Consider
a. Using the Internet, describe some examples of solar powered public transportation systems currently in operation.
b. What are some safety issues that may be posed by solar-powered transportation systems?

San José State University

Figure 7.32

Photovoltaics can power water pumps in remote areas of the world where there is no access to electricity.

Source: NREL/U.S. Dept. of Energy

Electricity generated by solar-thermal collectors and PV cells during the day must be stored using batteries for use at night. In 2017, Elon Musk (1971), founder of Tesla and SpaceX, installed a battery the size of a football field in southern Australia. This battery can store energy generated by wind and solar installations during the day and power about 30,000 homes during the night.

Nevertheless, the direct conversion of heat and sunlight into electricity has many advantages. In addition to relieving some of our dependence on fossil fuels, an economy based on solar electricity would reduce the environmental damage of extracting and transporting these fuels. Furthermore, it would help lower the levels of air pollutants such as SO_x and NO_x. It would also help avert the dangers of climate change by decreasing the amount of CO_2 released into the atmosphere. Fossil fuels will certainly remain the preferred form of energy for certain applications. However, for the longer term, we can turn to a number of renewable energy sources, many of which are driven directly by the Sun or are a result of the solar heating of our atmosphere and water. The final section takes a brief look at how we can generate electricity from other sustainable renewable resources.

7.11 | Beyond Solar: Electricity from Other Renewable Sources

Learning Objective: Describe global trends in the use of wind, hydroelectric, and geothermal energy sources to generate electricity

No single source of electricity can meet our global energy needs. We also know that no energy source comes without a cost, such as mining, pollution, greenhouse gases, or setting up distribution networks. Based on the discussions of earlier chapters, it is to our advantage to further develop and add a greater percentage of renewable sources than it is to continue to rely on fossil fuels. We discussed renewable sources such as biofuels and ethanol in Chapter 6. In this section, we turn to wind, water, and the heat given off by the core of our planet as renewable energy sources.

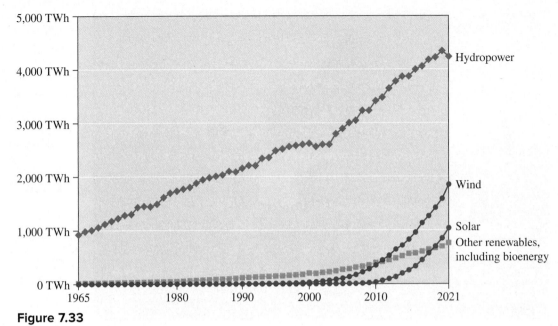

Figure 7.33

Worldwide renewable energy electricity generation by source, 1965–2021. For comparison, in 2021, 2.5 trillion kWh of electricity was generated by the combustion of fossil fuels. For an interactive version, see: www.acs.org/cic.

Our World in Data based on BP Statistical Review of World Energy & Ember.

Wind

Among the possible alternative sources of energy, wind power has seen the largest growth over the last decade. Figure 7.33 shows the amount of electricity generated by alternative energy sources since 1965. While hydroelectric power has generated the most electricity over this time period, wind power has grown exponentially since 2005. In addition to electrical generation for the public, wind power may be used for individual homes and businesses, particularly in areas that are too remote to be on the power grid.

To understand wind power, we must first consider the role of the Sun. The Sun's heat ultimately drives the movement of air on our planet. It does this because Earth's surface is made up of different types of land and water that absorb heat from the Sun at different rates. For example, if air over the land heats up, it causes the air above the warm land to expand. The air then moves from high pressure to low pressure and creates wind. Wind turbines today make use of large blades, sometimes nicknamed by the locals as "pinwheels" that dot the landscape. These spin a shaft that turns a generator to produce electricity. Wind farms are located around the world in order to take advantage of prevailing winds. Such a farm is shown at Pakini Nui (South Point) on the Big Island of Hawaii (Figure 7.34).

Figure 7.34

Pakini Nui Wind Farm, completed in 2007, supplied 20.5 MW of power in 2013.

Radius Images/Corbis

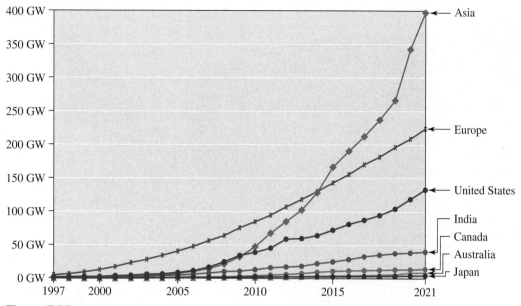

Figure 7.35

Comparative installed wind energy capacity in selected countries worldwide, including both onshore and offshore wind sources. For an interactive version, see: www.acs.org/cic.

Our World in Data based on Statistical Review of World Energy - BP (2022).

The vast majority of electricity generation from wind is from wind farms, which feature hundreds of individual wind turbines. A single typical wind turbine generates 1.5–2.5 MW of electrical power. In comparison, the average nuclear reactor in the U.S. generates about 1000 MW of electricity and coal plants average about 550 MW. However, if turbines are arranged together in a wind farm, they can exceed the output of traditional power plants. For instance, the largest wind farm in the United States, the Alta Wind Energy Center in California, can generate 1548 MW of electricity. As seen by Figure 7.35, Asia and Europe have much greater wind capacity than the U.S. In fact, the largest wind farm in the world, the Gansu wind farm in China, has an upper output of 7900 MW!

In the United States, wind power contributes a significant portion of electricity generation in 41 states (Figure 7.36). The siting of a wind farm requires considerable planning and includes consideration for the maximum wind speed, the fraction of time that sufficient wind exists, land rights, and the impact on adjacent public and private land. There are a few negative environmental impacts caused by wind power. For example, birds and bats are sometimes killed in collisions with turbines and turbine blades. A specific location may have excellent wind conditions with regular sustained winds, but would be a poor site because it is adjacent to a protected wildlife preserve with a large population of migrating birds.

The National Renewable Energy Laboratory (NREL) has surveyed the average wind speed at an elevation of 80 meters, the height of most turbines used by electric utility companies. As shown in the map in Figure 7.37, the U.S. region with the greatest wind speeds is the Great Plains states. At a height of 80 m, there is minimal blockage of wind by most trees, buildings, and other structures. Also, as the height above the ground increases, so does the average wind speed. Wind turbines require wind speeds of between 3.6 and 18 m/s (8 and 40 mph) to generate electricity. Slower speeds are inefficient in generating electricity, and may not even provide enough force to turn the blades. Faster speeds may cause damage to the mechanical and electrical components of the turbine. Wind turbines are turned off or on according to the wind speed to optimize the production of electricity.

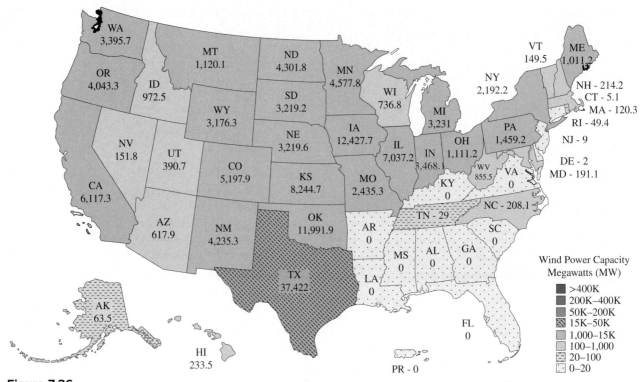

Figure 7.36

Wind-power capacity in the U.S. (2022).

Source: National Renewable Energy Laboratory (NREL) for the U.S. Department of Energy, 2022.

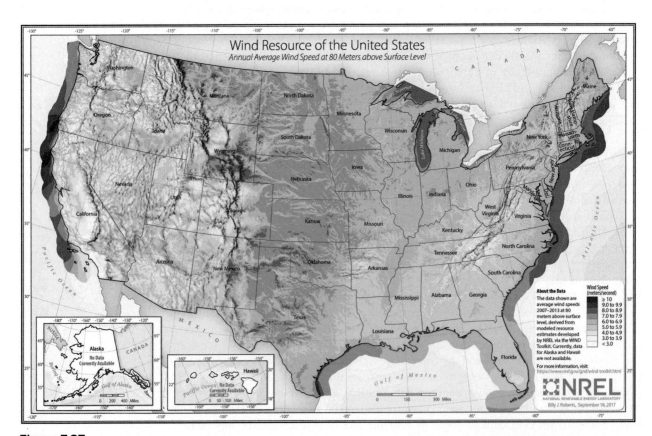

Figure 7.37

Annual average wind speed at 80 m height above ground in the U.S.

Source: National Renewable Energy Laboratory for the U.S. Department of Energy, 2017.

Your Turn 7.29	Regional Variations in Wind & the Future

a. Consider the maps of the United States in Figures 7.35, 7.36, and 7.37. Why do you think there is very little wind power in the southeastern U.S.?

b. Consulting the data found at www.acs.org/cic, calculate the percent increase in wind capacity forecasted to increase in the U.S. between 2020 and 2050.

A wind turbine consists of a tower and blades. The tower elevates the turbine high enough to be exposed to sufficient wind speeds to produce electricity. The tower is mounted in massive concrete blocks of up to 102 m³ (3600 ft³) and reinforced with 30 tons of steel. The body of the turbine, known as the *nacelle*, is mounted at the top of the tower. The nacelle contains the bulk of the mechanical and electrical operating pieces of the turbine, as seen in Figure 7.38. The blades are attached to a shaft that rotates an electromagnet within an electrical generator, similar to the generators found in traditional power plants. In fact, each MW of power generated by a wind turbine is reported to require up to 1 tonne of rare earth magnets! There is also a computer controller within the nacelle, which optimizes the output of the turbine by controlling operations such as rotating the nacelle into the wind and changing the pitch of the blades.

In addition to ground-based wind turbines (also known as onshore turbines), many utility companies are installing turbines that are located off the coast in oceans or large lakes. There are currently no offshore wind farms in the United States, but there are an increasing number in Europe. For instance, more than one quarter of the United Kingdom's wind energy is supplied by offshore wind farms.

Offshore wind offers some advantages compared to onshore wind farms. More than half of the population of the U.S. lives in coastal areas, so the electrical demand in these areas is high and the available land is limited. Offshore winds also typically blow harder and more uniformly than onshore winds, making offshore wind farms more efficient than their onshore cousins.

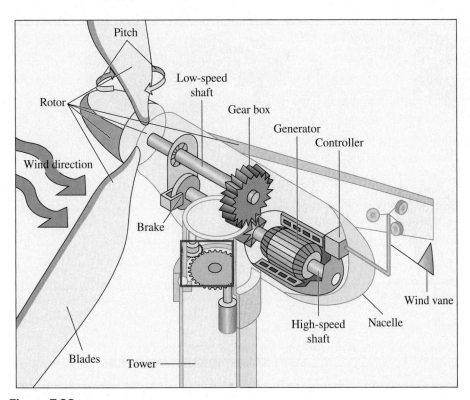

Figure 7.38

Schematic diagram of a wind turbine nacelle. Check out this video for a virtual tour of a wind farm: www.acs.org/cic.

Water

Water power, also known as hydropower, is one of the oldest and largest sources of renewable energy, using the natural flow of moving water (falling or fast-running) to generate electricity or provide power. For centuries, humans have harnessed the movement of water with devices such as water wheels. When water flows over a wheel, this wheel can turn other devices, including stones that can grind grain into flour. Similarly, small- and large-scale hydroelectric dams harness the movement of water. When water falls across turbine blades, the potential energy of water trapped in a reservoir behind the dam is converted into kinetic energy, which in turn is converted to electricity. The movement of ocean water in tides, currents, and waves can be also harvested by a variety of principles to generate electricity. Some involve the turning of turbine blades, while others involve forcing compressed air through a turbine. All involve the kinetic energy of motion to turn a generator to produce electricity.

Worldwide, only a few dams are still being constructed, as most large rivers are already in the service of hydroelectric projects. Figure 7.39 illustrates the use of hydropower globally. The largest areas of growth are located in Canada, Brazil and China. Although hydroelectric power does not release greenhouse gases during production, significant emissions and environmental damage result from the construction of dams and hydroelectric power stations. As an example, two dams have been removed from rivers in the Pacific Northwest because the electricity generated was determined to be less than the cost to the environment. You will debate whether hydroelectric causes more harm than good in the next activity.

Geothermal

Geothermal energy comes from the heat given off at the core of our planet from the slow decay of radioactive particles. It is a renewable energy source because the earth continuously generates heat. Geothermal energy relies on drilling into underground reservoirs containing hot water or steam, and thus drawing heat from Earth. These heated sources of water can then be used to drive generators to produce electricity,

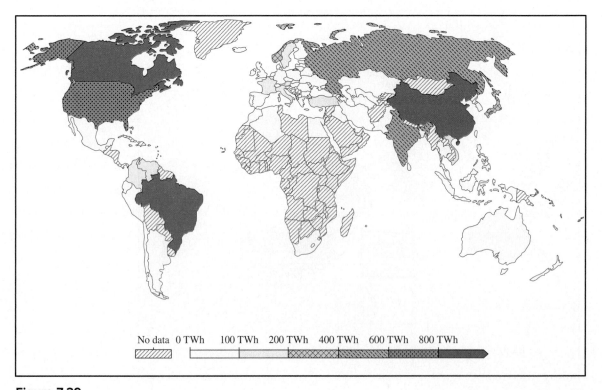

Figure 7.39

Hydroelectric capacity worldwide, 2021. For an interactive version, see: www.acs.org/cic.

Our World in Data based on BP Statistical Review of World Energy.

Let's Debate!   Is Hydroelectric Power Worth It?

Form two groups: one group will be in favor of the topic (the proposition) and the other group will be against the topic (the opposition). Both groups will try to persuade a neutral person or judge to agree with them. The topic of the debate is the motion. Here is the motion:

The environmental cost of hydroelectric power is too high and not worth developing or expanding further relative to other renewable energy sources.

Curiosity Conversations

- Should current hydroelectric dams be decommissioned? Describe the environmental impacts posed by this decision?
- How do the environmental impacts of hydroelectric power, both during construction and during use, compare to other renewables as well as fossil fuels?
- How would the environmental impacts of offshore facilities compare with those of onshore stations?

or the hot water may be used directly to heat a home. Geothermal works well in locations known for volcanic activity that have "hot rock," such as the Big Island of Hawaii, which generates 29% of its energy from geothermal sources. Although the U.S. currently leads the world in total geothermal electric generating capacity, geothermal energy is showing tremendous growth in countries such as Kenya, Turkey, and Indonesia (Figure 7.40). The top five countries ranked in terms of geothermal

Your Turn 7.30   Our Energetic Future

Renewable energy comes from the wind, the oceans, and geothermal sources, not just from the Sun and biofuels (discussed in Chapter 6). Pick one of these renewable energy sources and learn more about the technologies available to harness it.

a. Name the geographic restrictions (if any) to its use.
b. Prepare a list of the reasons to support this technology. Prepare a similar list for the "nay-sayers."
c. Predict how this technology will affect energy production output where you live.

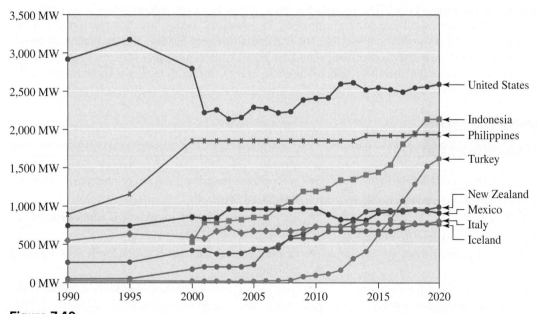

Figure 7.40

Geothermal power output by country, 1990–2020. For an interactive version, see: www.acs.org/cic.

Our World in Data based on BP Statistical Review of World Energy (2021).

electric generating capacity are the U.S. (2.6 GW), Indonesia (2.1 GW), the Philippines (1.9 GW), Turkey (1.6 GW), and New Zealand (984 MW).

Renewable energies are derived from natural sources (wind, water and earth) that are replenished faster than they are consumed. This section offered a renewable energy sampler; no worldview of energy resources would be complete without considering these and other sustainable sources of energy. Generating renewable energy produces lower emissions than burning fossil fuels and will be very useful in our quest to address the climate science crisis. The share that renewable sources occupy on the world's energy scene, as well as their economics, availability, and ease of use, must be improved.

Your Turn 7.31   The World's Energy

Supporters of nuclear energy encourage its use as a source that does not emit greenhouse gases. Bill McKibben, a prominent environmentalist, said that "there is an urgent need to stop subsidizing the fossil fuel industry, dramatically reduce wasted energy, and significantly shift our power supplies from oil, coal, and natural gas to wind, solar, geothermal, and other renewable energy sources." After reading this chapter, how do you think nuclear and renewable energy fit into the world's energy portfolio?

Conclusion

More than 60 years have passed since the first commercial nuclear power plant began producing electricity in the United States. The glittering promise of boundless, unmetered electricity—drawn from the nuclei of uranium atoms—has proved illusory. But the needs of both our nation and the world for safe, abundant, and inexpensive energy are far greater today than they were in 1957. Therefore, scientists and engineers continue their atomic quest, most recently, with exciting developments in nuclear fusion.

However, there are other types of renewables that can be used for electricity, such as photovoltaic cells that are used to tap the energy of the Sun. From photovoltaic roof shingles to vehicles and airplanes with exterior solar panels, the sky is literally the limit for the exciting future of solar power. Advances in research, together with changes in global economies, will continue to make this and other sustainable options such as wind, geothermal, and hydroelectric more fiscally and energetically feasible throughout the world.

Nuclear and renewable sources of energy provide electricity without releasing stored carbon into the atmosphere in the form of carbon dioxide. In particular, wind and solar are growing in popularity, taking over a larger portion of the energy market. However, solar and wind power are both variable; the Sun doesn't shine at a steady rate and the wind doesn't blow constantly. To make the most of these energy sources, they must be paired with technologies that can store the energy for later use. The next chapter will explore the chemistry underlying these energy storage devices we know as batteries.

LEARNING OUTCOMES   The numbers in parentheses indicate the sections within the chapter where these outcomes were discussed.

Having studied this chapter, you should now be able to:

- describe the relative use of alternative and renewable energy sources around the world (7.1)

- write balanced nuclear equations and describe nuclear reactions at both the macroscopic and particulate levels (7.2)

- explain why nuclear reactions release such enormous amounts of energy from a small amount of matter (7.2)

- describe how energy is produced in a nuclear power plant (7.3)

- explain why some isotopes are radioactive, while others are not (7.4)

- compare the fundamental types of nuclear radiation and their properties (7.4)

- identify sources, health impacts, and applications of ionizing radiation (7.5)

- describe sources and concentration levels of background radiation (7.5)

- describe and calculate how the amounts of radioisotopes change over time in terms of half-lives (7.6)

- describe the short-term and long-term impacts resulting from past nuclear accidents (7.7)

- compare the nuclear power capacity of countries around the world (7.8)

- describe global trends in the use of solar energy (7.9)
- summarize the applications for concentrating solar power arrays and photovoltaic cells (7.9)
- describe how sunlight is converted into electricity within a photovoltaic (solar) cell (7.10)

- describe global trends in the use of wind, hydroelectric, and geothermal energy sources to generate electricity (7.11)

Questions

Emphasizing Essentials

1. Name two ways in which one carbon atom can differ from another. Then, name three ways in which *all* carbon atoms differ from *all* uranium atoms.

2. The representations ^{14}N or ^{15}N give more information than simply the chemical symbol N. Explain.

3. **a.** How many protons are in the nucleus of this isotope of plutonium: Pu-239?

   b. The nuclei of all atoms of uranium contain 92 protons. Which elements have nuclei with 93 and 94 protons, respectively?

   c. How many protons do the nuclei of radon-222 contain?

4. Determine the number of protons and neutrons in each of these nuclei.

   a. ^{14}C, a naturally occurring radioisotope of carbon

   b. ^{12}C, a naturally occurring stable isotope of carbon

   c. ^{3}H, tritium, a naturally occurring radioisotope of hydrogen

   d. Tc-99, a radioisotope used in medicine

5. $E = mc^2$ is one of the most famous equations of the 20th century. Explain the meaning of each symbol in this equation.

6. Give an example of a nuclear equation and one of a chemical equation. In what ways are the two equations alike? Different?

7. What is an alpha particle and how is it represented? Answer these same questions for a beta particle and a gamma ray.

8. The nuclear equation below represents a plutonium target being hit by an alpha particle. What do the superscript numbers mean? How about the subscripts? Show that the sum of the subscripts on the left is equal to the sum of the subscripts on the right. Then, do the same for the superscripts.

$$^{239}_{94}\text{Pu} + {}^{4}_{2}\text{He} \longrightarrow \left[{}^{243}_{96}\text{Cm}\right] \longrightarrow {}^{242}_{96}\text{Cm} + {}^{1}_{0}\text{n}$$

9. For the nuclear equation shown in question 8:

   a. Suggest the origin of the $^{4}_{2}$He particle.

   b. $^{1}_{0}$n is a product. What does this symbol represent?

   c. Curium-243 is written in square brackets. What does this notation convey?

   Hint: See Equation 7.1.

10. Californium, element number 98, was first synthesized by bombarding a target with alpha particles. The products were Californium-245 and a neutron. What was the target isotope used in this nuclear synthesis?

11. Explain the significance of neutrons in initiating and sustaining the process of nuclear fission. In your answer, define and use the term *chain reaction*.

12. Nuclear fission occurs through many different pathways. For the fission of U-235 induced by a neutron, write a nuclear equation to form:

    a. Bromine-87, Lanthanum-146, and more neutrons.

    b. a nucleus with 56 protons, a second with a total of 94 neutrons and protons, and a third with two additional neutrons.

13. This schematic diagram represents the reactor core of a nuclear power plant:

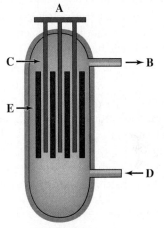

Match each letter in the figure with one of the following terms:

   fuel rods
   cooling water into the core
   cooling water out of the core
   control rod assembly
   control rods

14. Identify the segments of the nuclear power plant diagrammed in Figure 7.5 that contain radioactive materials and those that do not.

15. Explain the difference between the primary coolant and the secondary coolant. The secondary coolant is not housed in the containment dome. Why not?

16. Boron can absorb neutrons and be used in control rods. Write the nuclear equation in which boron-10 absorbs a neutron to produce lithium-7 and an alpha particle.

17. Plutonium-239 decays by alpha emission (with no gamma ray), and iodine-131 decays by beta emission (and emits a gamma ray).

    a. Write the nuclear equation for each.

    b. Plutonium is most hazardous when inhaled in particulate form. Why is this?

c. Iodine-131 can be hazardous if ingested. Where do all isotopes of iodine accumulate in the body?

d. If 25 g of each sample were present, after three half-lives, how much would remain? Which substance would take the longest to decay to this amount? Explain.

Hint: See Table 7.3.

18. Radioactive decay is accompanied by a change in the mass number, a change in the atomic number, a change in both, or a change in neither. For the following types of radioactive decay, which change(s) do you expect?

a. alpha emission

b. beta emission

c. gamma emission

19. Figure 7.10 shows the radioactive decay series for U-238. Analogously, U-235 decays through a series of steps (α, β, α, β, α, α, α, β, α, β, α) to reach a stable isotope of lead. For practice, write nuclear reactions for the first six. Although some steps are accompanied by a gamma ray, you may omit this.

Hint: The result is an isotope of radon.

20. What percent of a radioactive isotope would remain after two half-lives, four half-lives, and six half-lives? What percent would have decayed after each period?

21. Estimate the half-life of radioisotope X from this graph:

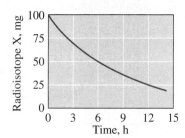

22. Every year, 5.6×10^{21} kJ of energy comes to Earth from the Sun. Why can't this energy be used to meet all of our energy needs?

23. The symbol • represents an electron and the symbol ⬤ represents a silicon atom. The darker purple sphere in the center of the diagram represents either a gallium or an arsenic atom. Does this diagram represent a gallium-doped p-type silicon semiconductor or an arsenic-doped n-type silicon semiconductor? Explain your answer.

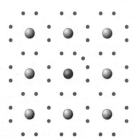

24. Describe the main reasons why solar cells have solar energy conversion efficiencies significantly less than the theoretical value of 31%.

Concentrating on Concepts

25. Alchemists in the Middle Ages dreamed of converting base metals, such as lead, into the precious metals gold and silver.

a. Why could they never succeed?

b. Today, can we convert lead or mercury into gold? Explain using specific nuclear reactions.

c. Can nuclear reactions be used to synthesize consumer quantities of precious metals, such as gold and silver?

26. The isotopes U-235 and U-238 are alike in that they are both radioactive. However, these two isotopes have very different abundances in nature. Recall their natural abundances (found in Section 7.2) and explain the significance of this difference.

27. Consider the uranium fuel pellets used in commercial nuclear power plants.

a. Describe one way in which U-235 and U-238 can be separated.

b. Why is it necessary to enrich the uranium for use in the fuel pellets?

c. Fuel pellets are enriched only to a few percent, rather than to 80–90%. Name three reasons why.

d. Explain why it is not possible to separate U-235 and U-238 by chemical means.

28. **a.** Why must the fuel rods in a reactor be replaced every few years?

b. What happens to the fuel rods after they are taken out of the reactor?

29. At full capacity, each reactor in the Palo Verde power plant uses only a few pounds of uranium to generate 1243 megawatts (MW) of power. To produce the same amount of energy would require about 2 million gallons of oil or about 10,000 tons of coal in a conventional power plant. How is energy produced in the Palo Verde plant, compared with conventional power plants?

30. One important distinction between the Chornobyl reactors and those in the United States is that those in Chornobyl used graphite as a moderator to slow neutrons, whereas U.S. reactors use water. In terms of safety, give two reasons why water is a better choice.

31. If you look at nuclear equations in sources other than this textbook, you may find that the subscripts have been omitted. For example, you may see an equation for a fission reaction written this way:

$$^{235}U + {}^{1}n \longrightarrow [{}^{236}U] \longrightarrow {}^{87}Br + {}^{146}La + 3\ {}^{1}n$$

a. How do you know what the subscripts should be? Why can they be omitted?

b. Why are the superscripts *not* omitted?

32. Coal can contain trace amounts of uranium. Explain why thorium must be found in coal as well.

33. Suppose somebody tells you that a radioisotope is gone after 10 half-lives. Critique this statement, explaining why it could be a reasonable assumption for a small sample, but might not be for a large one.

34. "Bananas are radioactive!" A vice president of nuclear services made this comment in a public lecture in the context of comparing the different sources of radiation to which people are exposed.

    a. Why might he have made such an assertion?

    b. Suggest a better way to have phrased this.

    c. Should you stop eating bananas because they are radioactive? Explain.

35. Fossil fuels have been called the "Sun's ancient investment on Earth." Explain this statement to a friend who is not enrolled in your course.

36. In most locations, the cost of electricity generated by solar-thermal power plants currently is greater than that of electricity produced by burning fossil fuels. Suggest two strategies that might be used to promote the use of environmentally cleaner electricity from photovoltaics.

37. Name two current applications of photovoltaic cells *other* than the production of electricity in remote areas.

38. Describe two other renewable energy sources, besides solar. Describe the pros and cons of each source.

Exploring Extensions

39. Explain the term *decommission*, as in "decommissioning a nuclear power plant." What technical challenges are involved? The resources of the Internet can help you.

40. Einstein's equation, $E = mc^2$, applies to chemical reactions as well as nuclear ones. An important chemical change studied in Chapter 6 was the combustion of methane, which releases 50.1 kJ of energy for each gram of methane burned.

    a. Use Einstein's equation to determine what mass loss corresponds to the release of 50.1 kJ.

    b. To produce the same amount of energy, what is the ratio of the mass of methane burned in a chemical reaction to the mass converted into energy?

    c. Think about the mass conversion and combustion. Why does $E = mc^2$ not apply to a combustion reaction?

41. When 4.00 g of hydrogen nuclei undergoes fusion to form helium in the Sun, the change in mass is 0.0265 g and energy is released.

    a. Which has more mass, the hydrogen or helium nuclei? Explain.

    b. Use Einstein's equation, $E = mc^2$, to calculate the energy equivalent of this mass change.

42. Under conditions like those on the Sun, hydrogen can fuse with helium to form lithium, which in turn can form different isotopes of helium and of hydrogen. The mass of one mole of each isotope is given.

$$\underset{2.01345 \text{ g}}{{}^{2}_{1}\text{H}} + \underset{3.01493 \text{ g}}{{}^{3}_{2}\text{He}} \longrightarrow [{}^{5}_{3}\text{Li}] \longrightarrow \underset{4.00150 \text{ g}}{{}^{4}_{2}\text{He}} + \underset{1.00728 \text{ g}}{{}^{1}_{1}\text{H}}$$

    a. In grams, what is the mass difference between the reactants and the products?

    b. For one mole of reactants, how much energy (in joules) is released?

43. Lise Meitner and Marie Curie were both pioneers in developing an understanding of radioactive substances. You likely have heard of Marie Curie and her work, but may not have heard of Lise Meitner. How are these two women related in time and in their scientific work?

44. Advertisements for Swiss Army watches stress their use of tritium. One ad states that the "hands and numerals are illuminated by self-powered tritium gas, 10 times brighter than ordinary luminous dials." Another advertisement boasts that the "tritium hands and markers glow brightly making checking your time a breeze, even at night." Evaluate these statements and, after doing some Internet research, discuss the chemical form of tritium in these watches, and what its role is.

45. Deciding where to locate a nuclear power plant requires analysis of both risks and benefits associated with the plant. If you were to play the role of a CEO of a major electric utility considering whether to pursue permits for the construction of a nuclear power plant in your area, what risks and benefits would you cite?

46. Provide at least two similarities and two differences between a nuclear-fueled power plant (Figure 7.5) and a coal-fueled power plant (Figure 6.9).

47. At the cutting edge of technology, the line between science and science fiction often blurs. Investigate the "futuristic" idea of putting mirrors in orbit around Earth to focus and concentrate solar energy for use in generating electricity.

48. Building-integrated photovoltaic materials are becoming more popular, due to the relatively unsightly appearance of solar panels on rooflines of homes. Provide some examples of these materials. How do these building materials work and how durable are these materials with respect to extreme weather conditions (snow, sleet, hail, wind, etc.)?

49. Although silicon, used to make solar cells, is one of the most abundant elements in Earth's crust, extracting it from minerals is costly. The increased demand for solar cells has some companies worried about a "silicon shortage." Find out how silicon is purified and how the PV industry is coping with the rising prices.

50. Of the alternative forms of renewable energy presented in Section 7.11, which do you think is most promising? Do some Internet research to find some pros and cons to this energy source. On the basis of what you find, do you think it is a viable option for the future, or is more research necessary to implement it?

51. Figure 7.26c shows an array of photovoltaic cells installed at the Solarpark Gut Erlasee in Bavaria, Germany.

    a. At present, where is the largest photovoltaic power plant located in your country?

    b. Name two other locations of large-scale photovoltaic cell installations.

    c. Name two factors that promote a centralized array rather than individual rooftop solar units.

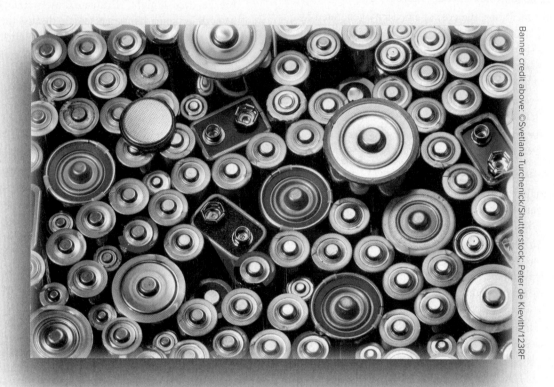

BATTERIES IN YOUR EVERYDAY LIFE

Watch the chapter-opening video (www.acs.org/cic) and consider how your life would change without batteries.

a. List all of your daily activities involving batteries and include the type of battery used for each of these activities.

b. Not all batteries are the same. This chapter will describe chemical reactions that occur inside various types of batteries. Among the batteries you listed in part **a**, which are able to be recharged? For these rechargeable batteries, predict some factors that will influence their usable lifetime (i.e., the number of possible charge/discharge cycles until they can no longer power a device).

In this chapter, you will explore the following questions:

1. How does a battery work?
2. What are the relationships between voltage, current, resistance, and power?
3. How do alkaline batteries work and how do these differ from other types of batteries?
4. What are the differences between rechargeable and non-rechargeable batteries?
5. How do lead–acid batteries work and what are their applications?
6. What type of batteries are used for renewable energy storage?
7. How do lithium-ion batteries work and why are they preferred for electric vehicles?
8. What are the differences between supercapacitors and batteries?
9. What are the advantages of "hybrid" vehicles?
10. How do fuel cells work and what are their benefits for automotive applications?
11. How can we obtain and store hydrogen for fuel cell applications?
12. What is the importance and environmental impact of battery recycling?

Reed Richards/Alamy Stock Photo

Introduction

We rely on a flow of electrons—known as electricity—to heat or cool our living and work spaces, provide light to read by, and power our electronic devices. For some applications, the electricity we use is generated at centralized power plants, such as those powered by fossil fuels (Chapter 6) or fissionable isotopes (Chapter 7). To a lesser extent, we also rely on wind, sun, and geothermal energy, as well as the potential energy of water trapped by dams, as sources to generate electric power.

However, for our mobile lifestyles, we are increasingly dependent on convenient-sized portable sources of electricity, better known as batteries. These long-lasting and reliable power sources fill a special energy niche. They power our cell phones, laptops, and other portable electronic devices, but their recent use in modes of transportation represents a modern renaissance. Many automakers now offer all-electric vehicles across their lineup, including sedans, trucks, and SUVs. Most manufacturers also feature at least one hybrid model, which uses a combination of electrical and fuel sources to power the vehicle.

Benjamin Franklin (1706–1790) first coined the term "battery" in 1749 to describe an apparatus used for his experiments. However, Alessandro Volta (1745–1827) is credited for fabricating the first functional battery, which was referred to as a *voltaic pile*. Explore the history of battery development in this next activity.

Early Battery Designs

Batteries have evolved a great deal since their early designs. Using the Internet as a resource, answer the questions below regarding the following batteries:

  i) a voltaic pile, invented by Alessandro Volta (1745–1827)
 ii) a Daniell cell, invented by John Daniell (1790–1845)
iii) a Planté cell, developed by Gaston Planté (1834–1889)
 iv) a Leclanché cell, developed by Georges Leclanché (1839–1882)
  v) a Gassner cell, invented by Carl Gassner (1855–1942)

  a. What do all of these battery designs have in common?
  b. Are any of these designs still used today? Explain.
  c. Current alkaline batteries, made popular by the Energizer and Duracell corporations, were developed by Lew Urry (1927–2004) at the Eveready Battery Company in 1949. What are some similarities and differences between modern alkaline battery designs relative to the early battery concepts listed above?

Gio_tto/iStock/Getty Images

In this chapter, we will describe the components, operating principles, and safety considerations for various types of batteries. We will also explain the environmental impacts of their production and end-of-use practices. Let's begin by peering through the casing of a battery to discover the chemical reactions that produce electrical energy.

8.1 | How Does a Battery Work?

> *Learning Objectives:*
>
> *Explain how a battery works via redox reactions and flow of electrons between counter electrodes*
>
> *Determine the oxidation states for species involved in a redox reaction*

The ever-increasing use of batteries worldwide is fueled by our steady consumer demand for portable electronics and emerging applications such as electric vehicles and renewable energy storage. The workhorses of batteries are **galvanic cells**—compartments capable of converting the energy released from spontaneous chemical reactions into electrical energy. A collection of several galvanic cells wired together constitutes a **battery**.

All galvanic cells produce useful energy by transferring electrons from one substance to another. This electron transfer process may be divided into two parts. One is **oxidation**, a process in which a chemical species loses electrons. The other is **reduction**, a process in which a chemical species gains electrons. We refer to these two parts as **half-reactions** because each represents half of the overall process occurring in the galvanic cell. Some mnemonics to remember this is OIL RIG, which stands for **O**xidation **I**s **L**oss, and **R**eduction **I**s **G**ain. Another is LEO the lion says GER, which stands for **L**oss of **E**lectrons is **O**xidation, **G**ain of **E**lectrons is **R**eduction.

Half-reactions always occur in pairs within a battery and must include both ions and electrons. Even though electrons cannot be poured from a bottle into a flask, it is still helpful to show them in half-reactions so that you can better understand what is taking place. Note that the electrons show up on either the products or reactants side of the half-reaction, but not on both. If on the products side, then the reactant has lost electrons; this is an oxidation half-reaction. In contrast, if the electrons are on the reactants side, then the reactant gains electrons, which is a reduction half-reaction.

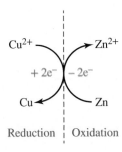

Reduction | Oxidation

Figure 8.1

A scheme for redox reactions, showing the electrons released by an oxidation half-reaction being used for the reduction half-reaction.

As an example, consider this simple Cu-Zn galvanic cell:

$$\text{oxidation half-reaction: } Zn(s) \longrightarrow Zn^{2+}(aq) + 2e^- \qquad \textbf{[8.1]}$$

$$\text{reduction half-reaction: } Cu^{2+}(aq) + 2e^- \longrightarrow Cu(s) \qquad \textbf{[8.2]}$$

In this case, two electrons are lost by zinc in the oxidation half-reaction (Equation 8.1). But where do they go? These electrons were transferred to the Cu^{2+} ions being reduced (Equation 8.2). For the overall equation to balance, the number of electrons lost during oxidation must equal the number of electrons gained through reduction (Figure 8.1).

We can now combine the two half-reactions to obtain the overall balanced equation, known as a **redox reaction** (i.e., **red**uction and **ox**idation):

$$Zn(s) + Cu^{2+}(aq) + \cancel{2e^-} \longrightarrow Zn^{2+}(aq) + Cu(s) + \cancel{2e^-} \qquad \textbf{[8.3]}$$

The electrons that appear on both sides of Equation 8.3 cancel because the copper ions gain the electrons lost by the metallic zinc. So, we can rewrite the overall equation as:

$$Zn(s) + Cu^{2+}(aq) \longrightarrow Zn^{2+}(aq) + Cu(s) \qquad \textbf{[8.4]}$$

This next activity will show an example of a redox reaction involving perhaps the most well known landmark in the United States.

Why is the Statue of Liberty Green?

©2018 American Chemical Society

Ever wonder why the Statue of Liberty is green? Watch this video (www.acs.org/cic) and answer the following questions.

Questions to Consider

a. Would the Statue of Liberty change color if it were composed of aluminum? Explain your reasoning.

b. Using the Internet, what could be done to remove the green patina from the surface of the Statue and return the Statue of Liberty to its original copper color?

@HOME   Redoxing Pennies

Investigate a reduction and oxidation (redox) reaction at home using nonmetal bowls, paper towels, two old pennies (pre-1982 will work best), table salt, and white vinegar.

Isis Hernandez/Shutterstock

1. Record the appearance of the two pennies to be used in this experiment. Sketch any differences in color that appear on the front and back surfaces of the penny.
2. Place a paper towel into each of the two small bowls.
3. Place a penny into each of the bowls.
4. Pour a small amount of vinegar into the first bowl, enough to soak the paper towel.
5. Prepare a salt solution by mixing 1/4 teaspoon of salt and 1 teaspoon of vinegar in a cup. Mix thoroughly by swirling or stirring. Pour this solution into the second bowl, making sure that the paper towel is fully soaked by the solution.
6. Record the appearance of the two pennies every 5 minutes for a total of one hour.

Questions to Consider

a. What is the chemical equation for the reaction that is occurring? What is being oxidized? What is being reduced?

b. What happened in the second bowl? What was the purpose of adding the salt in the acid solution?

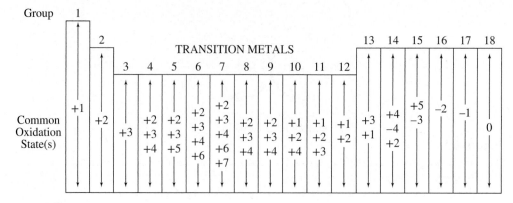

Figure 8.2

The common oxidation states of ions in various groups of the periodic table.

The charge of the metal ion is called its **oxidation state**. The oxidation state is always zero for pure elements in their most stable states, whether Zn, Fe, F_2, Ar, or P_4. In Equation 8.4, the oxidation state of zinc has increased from 0 to +2, and the oxidation state of copper has decreased from +2 to 0. As a general rule of redox reactions, *reduction* always *decreases* the oxidation state; oxidation always *increases* the oxidation state.

Because batteries contain both oxidation and reduction reactions, they must be composed of substances that undergo facile electron loss or gain. For compounds containing ions from Groups 1, 2, or 13 of the periodic table, the oxidation state of the metal is (almost) always +1, +2, or +3, respectively. Consequently, the names of these ionic compounds, such as "sodium chloride," do not need to include the charge of the metal because it is predictable and unchanging. However, as discussed in Section 5.4, transition metals are stable at various oxidation states (Figure 8.2), making them desirable for reduction/oxidation reactions.

Your Turn 8.1   Classifying Half-Reactions

Example Problem:

For the following half-reaction, indicate whether oxidation or reduction has occurred and the change in oxidation state of the metal.

$$Au^+(aq) + e^- \longrightarrow Au(s)$$

Example Solution:

This corresponds to a reduction half-reaction since the oxidation state of the gold decreases from +1 to 0.

Your Turn:

Categorize each as an oxidation half-reaction or a reduction half-reaction. Explain your reasoning, indicating the change in oxidation states of the relevant species.

a. $Al^{3+} + 3e^- \longrightarrow Al$ **d.** $2\,H_2O \longrightarrow 4\,H^+ + O_2 + 4e^-$
b. $Zn \longrightarrow Zn^{2+} + 2e^-$ **e.** $2\,H^+ + 2e^- \longrightarrow H_2$
c. $Mn^{7+} + 3e^- \longrightarrow Mn^{4+}$

Electron transfer in a battery occurs within its **electrodes**—electrical conductors that serve as sites for chemical reactions. At the **anode**, oxidation occurs and is the source of electrons flowing into the circuit. At the **cathode**, reduction takes place. The cathode receives the electrons sent from the anode through the external circuit to complete the reduction process.

The movement of electrons through an external circuit produces **electricity**, the flow of electrons from one region to another driven by a difference in potential energy. The electrochemical reactions provide the energy needed to operate a cell phone, a power tool, and countless other battery-operated devices. The chemical species oxidized and reduced in the cell are physically separated, forcing electrons released during the oxidation process through an electrical path (for example, a wire) to transfer to the reactant being reduced.

8.2 | Ohm, Sweet Ohm!

> *Learning Objective: Use Ohm's law to determine characteristics of an electrical circuit*

You have probably inserted a battery into a flashlight, calculator, or remote control. You may recognize the ones shown in Figure 8.3. One end of each of these batteries is marked with a + sign, while the other end displays a − sign. These markings indicate the electron transfer at work when they are connected.

Figure 8.4 provides an illustration of an electrical circuit, showing the flow of electrons from the negative terminal of a battery through an electrically conductive wire to illuminate a bulb. When the switch is open (as shown), the electrons cannot flow through the circuit from battery to bulb since the circuit has been interrupted. By flipping the switch, the electrical circuit is restored, which allows electrons to flow through the circuit uninterrupted. The cycle of electron flow will continue until all of the electrons are used up from the redox reactions occurring in the battery or until the switch is opened, which again breaks the electrical circuit.

An electrical circuit within a device uses electricity to perform a task, such as powering a lamp or running a handheld vacuum cleaner. Within this circuit, the flow of electrons is referred to as electrical **current**, whereas **voltage** is the difference in electrical energy (also known as electrical potential) between two points. As its name suggests, *resistance* is something that opposes the flow of electrons and is measured in ohms (Ω). In fact, the lightbulb shown in Figure 8.4 is considered a resistor since it slows the flow of electrons through the external circuit. The current, or rate of electron flow, is measured in amperes (amps, A), whereas voltage is measured in volts (V).

An amusing illustration for the connection between voltage, current, and resistance is shown in Figure 8.5. This relationship is known as **Ohm's law**, which is described by a simple equation:

$$V = I \times R \qquad\qquad [8.5]$$

where: V = voltage (measured in volts, V)
I = current (measured in amps, A)
R = resistance (measured in ohms, Ω)

As seen in Equation 8.5, an increase in circuit resistance leads to decreased electrical current. Furthermore, as the voltage increases at a constant circuit resistance, so does the electrical current. Learn more about Ohm's law in this next activity, using water flow as an analogy.

Figure 8.3

Alkaline batteries from size AAA to D, all of which produce 1.5 V.

Eric Misko/McGraw Hill

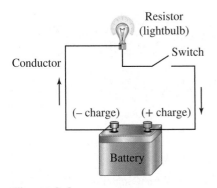

Figure 8.4

A simple schematic for an electrical circuit for a battery-powered lightbulb.

Figure 8.5

An illustration of Ohm's law. Voltage is related to the force of electron flow; current is related to the flow of electrons; resistance works to block the flow of electrons in an electrical circuit.

Electrical Current: A Water-Flow Analogy

Check out this video (www.acs.org/cic) for a fun demo of Ohm's law and answer the following questions.

Questions to Consider

a. Using the Internet, research some types of resistors used in electrical circuits. What are these devices composed of and how do they serve to restrict the flow of electrons?

b. In the video, the host shows an image of various resistors that have various colored rings. Using the Internet, determine what is indicated by these various colored markings.

c. In the video, the host mentions that a sponge would be a denser material than steel wool for the water to flow through. According to this definition, which would be a denser material for water to run through: sand, small pebbles, or large rocks?

d. In Figure 8.4, the lightbulb acts as a resistor in the electrical circuit. What other applications employ resistors?

Your Turn 8.2   Coulombs

The SI unit of electric charge is the coulomb, C, which is equivalent to the charge of 6.242×10^{18} protons. Express amps and ohms in terms of the base SI units of J, C, and s. Perform dimensional analysis for Equation 8.5 to show that volts may be expressed as J/C.

In addition to voltage and current, we are often interested in the power output of an electrical device. **Power** is the rate at which electrical energy is transferred through an electrical circuit. As discussed in Chapters 6 and 7, electrical energy may be *produced* from the combustion of a fuel, conversion of sunlight, or harnessing flows of wind or water. Conversely, electrical energy may also be *consumed* through its conversion into another useful form of energy, such as mechanical work, heat, or light.

The power consumption of a device can be determined if the voltage and current draw through the circuit are known (Equation 8.6):

$$P = V \times I \qquad \textbf{[8.6]}$$

The standard unit of power is volt-amps, known as the watt, W; however, the power consumption in your home is measured in terms of kilowatts (kWs). Substituting Ohm's law into Equation 8.6 will yield the definition of power in terms of resistance and current (Equation 8.7) or voltage (Equation 8.8). The next activity will give you a chance to apply these equations to a few real-world scenarios.

$$P = I^2 \times R \qquad \textbf{[8.7]}$$

$$P = \frac{V^2}{R} \qquad \textbf{[8.8]}$$

Your Turn 8.3   Power & Ohm's Law

Example Problem:

If a toaster is plugged into a 120-V outlet that supplies 8 amps of current, how much power does the toaster consume?

Example Solution:

Using Equation 8.6, $P = (120 \text{ V}) \times (8 \text{ A}) = 960 \text{ W}$

Your Turn:

a. How much power is generated from an electric heating element, with a resistance of 2.5 ohms, if the generator has an output voltage of 120 V?

b. How much current flows through an incandescent 45-W lightbulb that operates at 120 V?

8.3 | Batteries, Batteries Everywhere!

Learning Objectives:

Describe the characteristics and uses of various batteries

Identify the redox reactions that occur within alkaline batteries

Once the electrical circuit is complete, a voltage can be measured across the cell; that is, the difference in electrochemical potential between the two electrodes. The greater the difference in potential between the two electrodes, the higher the voltage and the greater the energy associated with the electron transfer. For example, with a nickel–cadmium (Ni–Cd) cell, the maximum difference in electrochemical potential under the conditions specified is measured as 1.2 V. In contrast, alkaline batteries deliver 1.5 V, and lithium-ion cells are capable of potentials approaching 4 V! To produce the higher voltages needed to power larger devices (e.g., power tools, Figure 8.6), several cells must be connected. Although modern batteries are the product of years of development and optimization, they all originated from Volta's layered design known as a voltaic pile. Have fun building your own in this next activity.

Figure 8.6

This 7.2-V Ryobi portable power drill has two Ni–Cd battery packs and a charging unit.

Jill Braaten/McGraw Hill

@HOME   A DIY Voltaic Pile

Investigate how the first battery was made at home by making your own voltaic pile using vinegar, table salt, two bowls, pennies, nickels, liquid dish soap, aluminum foil, scissors, paper towels, a small plate (any material except metal or paper), and a digital multimeter.

1. Mix 1/4 cup of vinegar and 1 tablespoon of table salt together in a bowl.
2. Place at least 4 pennies and 4 nickels into a different bowl with liquid dish soap and water and wash them to remove dirt. Dry them off with a paper towel.
3. Cut the aluminum foil into a strip that is 8 centimeters long and 2 centimeters wide. Then, fold it into thirds vertically and fold the edges inward so that it is an even thinner strip.
4. Next cut up a paper towel into several 2-centimeter by 2-centimeter squares with rounded edges.
5. Place a dry paper towel on the plate.
6. Add the aluminum foil strip on top of the paper towel in the middle of the plate.
7. Now, place a penny on top of this aluminum foil strip.
8. Take one of the paper towel squares and soak it in the salt-vinegar solution. Remove it and shake it so that it is wet but not dripping. Place it on top of the penny. Then add a nickel on top of the wet paper towel.
9. Place one probe tip of the digital multimeter on the aluminum foil strip and the other probe tip on the nickel at the top of the stack. Record the measurement along with the number of "cells" you have built. One cell is a unit that contains a penny, a soaked paper towel, and a nickel.
10. Repeat steps 7 through 9 at least 3 more times. **Note: Do NOT let the wet paper towels touch.** This will cause a short circuit.

Volodymyr Krasyuk/Rawpixel/Shutterstock

Questions to Consider
a. What trend do you observe in the voltage as you increase the number of cells?
b. What is the purpose of the vinegar and the table salt in the voltaic piles?

The chemical composition of a battery determines its voltage and is *not* related to the size of the battery. You can see from the examples listed in Table 8.1 that different voltages are produced using different chemical reactions. Only a few volts are possible with

Table 8.1	Some Common Galvanic Cells		
Type	**Maximum Voltage (V)**	**Rechargeable?**	**Examples of Uses**
Nickel–cadmium (Ni–Cd)	1.25	Yes	Toys and portable electronic devices, including digital cameras, power tools
Nickel–metal hydride (NiMH)	1.25	Yes	Replacing Ni–Cd for many uses in consumer devices; hybrid vehicles
Alkaline	1.5	No	Flashlights, small appliances, calculators, audio/video remote controls, toys
Lithium (primary)	1.5–3.6	No	LED lighting, smoke alarms, watches, vehicle remotes and key fobs
Lead–acid	2.1	Yes	Automobiles (starting, lighting, and ignition)
Lithium-ion, lithium-polymer	3.6	Yes	Laptop computers, cell phones, portable electronic devices, power tools

a single galvanic cell. But, as noted earlier, higher voltages are possible by connecting cells. For example, to run a 19.2-V power drill, manufacturers sell a "battery pack" containing multiple cells.

All alkaline batteries, from the tiny AAA size to the large D cells, produce the same voltage of 1.5 V. However, larger cells have a greater **capacity** to sustain the flow of electrons longer because they contain more active material. The capacity of a battery is usually given in units of milliamp-hours (mAh); for instance, C, AA, and AAA alkaline batteries have capacities of 3800 mAh, 1100 mAh, and 540 mAh, respectively. Just as a gasoline-powered vehicle can travel farther on a larger tank of gas, an electric device can operate for longer periods of time with a larger-capacity battery pack.

The half-reactions for an alkaline cell are:

oxidation half-reaction (taking place at the anode):

$$Zn(s) + 2\ OH^-(aq) \longrightarrow Zn(OH)_2(s) + 2e^- \qquad \textbf{[8.9]}$$

reduction half-reaction (taking place at the cathode):

$$2\ MnO_2(s) + H_2O(l) + 2e^- \longrightarrow Mn_2O_3(s) + 2\ OH^-(aq) \qquad \textbf{[8.10]}$$

overall cell reaction (sum of the two half-reactions):

$$Zn(s) + 2\ MnO_2(s) + H_2O(l) \longrightarrow Zn(OH)_2(s) + Mn_2O_3(s) \qquad \textbf{[8.11]}$$

For the compound MnO_2 found within alkaline batteries, two O^{2-} species will give a total negative charge of 4−. To maintain an overall zero charge for the entire compound, the manganese must carry a charge of 4+. Hence, this compound is known as *manganese(IV) oxide*. The cell is called "alkaline" because it operates in a basic medium containing OH^- ions.

Compact, long-lasting batteries have even found their way into our bodies. For example, the widespread use of cardiac pacemakers is mainly due to improvements made in the electrochemical cells rather than in the pacemakers themselves. Lithium–iodine cells are so reliable and long-lived that they are often the battery of choice for this application, lasting as long as ten years before needing to be replaced.

8.4 | Power-on-the-Go: Rechargeable Batteries

Learning Objective: Describe the design, operation, advantages, and limitations (including hazards) of rechargeable batteries

Imagine if the battery in your cell phone needed to be discarded and replaced every day after one use. Not only would that cost money and pollute the environment, but it would also be time-consuming for portable electronic devices where the battery is

Figure 8.7

A portable solar-powered charging
device for mobile devices.

trek6500/Shutterstock

seamlessly incorporated into the design. We take for granted the convenience of using a battery that can be recharged on demand in your office, on a plane or train, in your vehicle, or even outdoors by using a portable solar panel (Figure 8.7).

A primary (or disposable) battery, such as the alkaline batteries traditionally used in flashlights, cannot be recharged and must be discarded after the charge is "used up." In contrast, rechargeable batteries are called **secondary batteries** and use electrochemical reactions that can run in both directions. The transfer of electrons occurs during both the forward (discharging) and the reverse (recharging) processes.

For example, let's consider the reversible reactions inside a rechargeable Ni–Cd battery that was once widely used in many portable tools, digital cameras and robot vacuums. As the battery is discharged, atoms of cadmium oxidize to Cd^{2+} at the anode. These ions, in turn, combine with OH^- to form cadmium hydroxide, $Cd(OH)_2$ (Equation 8.12, left-right). Simultaneously, Ni^{3+}, present in the cathode as the hydrated form of nickel(III) oxide hydroxide, $NiO(OH)$, is reduced to Ni^{2+} to form nickel(II) hydroxide, $Ni(OH)_2$ (Equation 8.13, left-right). The movement of electrons when two electrons from the cadmium half-reaction are transferred to the nickel half-reaction is where the power comes from in Ni–Cd batteries.

$$\text{(Anode)} \quad Cd(s) + 2\ OH^-(aq) \underset{\text{charging}}{\overset{\text{discharging}}{\rightleftharpoons}} Cd(OH)_2(s) + 2e^- \qquad \textbf{[8.12]}$$

$$\text{(Cathode)} \quad 2\ NiO(OH)(s) + 2\ H_2O(l) + 2e^- \underset{\text{charging}}{\overset{\text{discharging}}{\rightleftharpoons}} 2\ Ni(OH)_2(s) + 2\ OH^-(aq)$$
$$\textbf{[8.13]}$$

Despite its useful redox chemistry, the use of Ni–Cd batteries has been largely abandoned worldwide due to environmental concerns associated with cadmium pollution resulting from battery disposal. Nickel metal hydride (NiMH) batteries have largely replaced Ni–Cd batteries and are now used in most of the same applications as the more expensive Li-ion batteries described later in this chapter.

The reversible reactions occurring in NiMH batteries are shown in Equations 8.14 and 8.15, which are quite similar to Ni–Cd cells. The anode is composed of an alloy of many possible metals including Ti, V, Ni, Cr, Co, Fe, Zr, and some rare earths. When a NiMH cell is charged, the anode releases hydrogen into the electrolyte, which produces a metal hydride (MH) and hydroxide ions (OH^-). At the cathode, $Ni(OH)_2$ reacts with hydroxide ions to form $NiO(OH)$. The nominal voltage of a NiMH battery is 1.2 V, which is equivalent to Ni–Cd cells.

$$\text{(Anode)} \quad MH(s) + OH^-(aq) \underset{\text{charging}}{\overset{\text{discharging}}{\rightleftharpoons}} M(s) + H_2O(l) + e^- \qquad \textbf{[8.14]}$$

$$\text{(Cathode)} \quad NiO(OH)(s) + H_2O(l) + e^- \underset{\text{charging}}{\overset{\text{discharging}}{\rightleftharpoons}} Ni(OH)_2(s) + OH^-(a, q) \qquad \textbf{[8.15]}$$

Figure 8.8

A photo showing the consequences of a fire in a Li-ion laptop battery. The suspected cause of the fire was failure of the membrane separator, which resulted in a short circuit and thermal runaway.

KYDPL KYODO/AP Images

All batteries, whether primary or secondary, require a **separator** placed between the anode and cathode to ensure that the electrodes do not come into physical contact. As you might expect, if the anode and cathode are allowed to touch in a battery, this would cause a short circuit and failure of the battery—often with dangerous consequences (Figure 8.8). Although early galvanic cells employed a salt bridge separator (Figure 8.9), typical separators in modern batteries are composed of a semipermeable membrane. This membrane effectively separates the electrodes while allowing ions to pass through and retain charge-neutrality during battery operation.

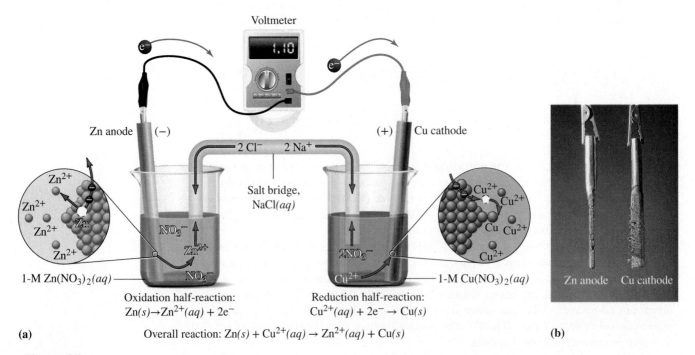

Figure 8.9

(a) An illustration of a two-component Zn–Cu galvanic cell, containing a *salt bridge* separator (composed of aqueous sodium chloride, NaCl), and electrolytes composed of 1-M zinc(II) nitrate ($Zn(NO_3)_2$, anode compartment) and 1-M copper(II) sulfate ($CuSO_4$, cathode compartment). **(b)** A photo showing the loss of metallic Zn by oxidation and gain of Cu metal through reduction. For an animated description of a Zn–Cu galvanic cell, go to www.acs.org/cic.

(b) Stephen Frisch/McGraw Hill

In order to facilitate ionic movement through the separator, an **electrolyte** must be used in all galvanic cells. The electrolyte is composed of an ionic compound (a salt) dissolved in a liquid such as water or an organic solvent such as dimethyl carbonate, $(CH_3O)_2CO$. One problem posed by liquid electrolytes is the possibility of leaking from the battery casing. You may have seen the corrosive mess inside a flashlight or child's toy from an old leaking battery (Figure 8.10).

To avoid issues with electrolyte leakage, many commercial batteries use **dry cells**, in which the electrolyte is immobilized as a paste, with only enough moisture available to allow ions to flow. Unlike wet cells used in early battery designs, a dry cell can operate in any orientation without spilling because it contains no free liquid. For rechargeable NiMH (Figure 8.11a) or primary alkaline (Figure 8.11b) batteries, the electrolyte that fills the pores of the membrane separator is an aqueous paste of potassium hydroxide (KOH), a strong base.

What features make a battery rechargeable? The key is that both the reactants and products are solids. Furthermore, the solid products cling to a stainless-steel grid within the battery rather than dispersing. If a voltage is applied to this grid, these products can be converted back to reactants, thus recharging the battery. Although a rechargeable battery can be discharged and recharged many times, eventually, the accumulation of impurities, a breakdown of the separators, or the generation of unwanted side-reaction by-products can end its useful life.

Figure 8.10

Corroded battery due to electrolyte leakage.

BigNazik/iStock/Getty Images

Your Turn 8.4   Lifetime of Rechargeable Batteries

Rechargeable batteries have a lifespan and no longer function after many charging/discharging cycles. Using the Internet as a resource, investigate the following:

a. What is the "memory effect" in some rechargeable batteries, and what operating conditions cause this phenomenon?
b. Which batteries are most prone to this effect?
c. How can this effect be repaired?
d. Many believe batteries will last longer if stored in a cold environment such as a refrigerator. Do you agree? Do all batteries (primary and secondary) benefit from being stored in a cold environment?

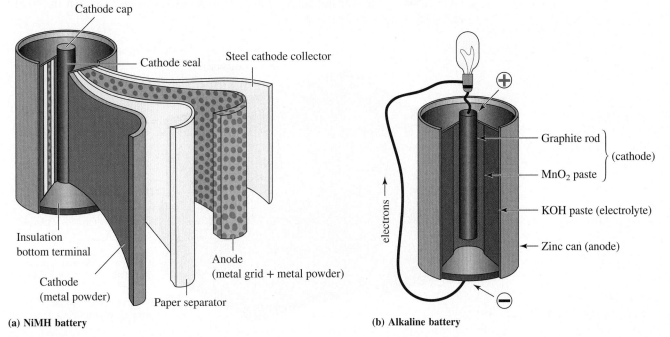

(a) NiMH battery

(b) Alkaline battery

Figure 8.11

Representations of **(a)** a NiMH rechargeable battery, showing how the components are layered to increase the surface area of the electrodes, and **(b)** a primary alkaline battery, in which the zinc container acts as the anode.

Batteries come in many shapes and sizes, each uniquely matched to its use. For example, in a hearing aid, the size and weight of the cell are of paramount importance. In contrast, an automobile battery must last for years and perform at various temperatures. To be successful in the consumer market, batteries must be affordable, last a reasonable time, and be safe to use and recharge. Ultimately, to be successful into the future, batteries must also be designed so their materials can be recycled sustainably.

8.5 | Lead–Acid Batteries

> *Learning Objective: Describe the design, operation, applications, and limitations (including hazards) of lead–acid batteries*

Found under the front hood of most cars, the lead–acid battery is the workhorse of rechargeable batteries, powering automobile starting, lighting and ignition (SLI). Lead–acid batteries are also used in personal transportation devices such as electric scooters and mopeds, golf carts, and bicycles. Although we take lead–acid batteries for granted, these represent one of the key technological developments that ushered in the rapid rise of the automobile, which has forever changed our world. The lead–acid battery comprises six electrochemical cells, each generating 2.0 V for a total of 12 V (Figure 8.12). Here is the overall chemical equation, the sum of the two half-reactions:

$$\underset{\text{lead}}{Pb(s)} + \underset{\text{lead(IV) oxide}}{PbO_2(s)} + \underset{\text{sulfuric acid}}{2\ H_2SO_4(aq)} \underset{\text{charging}}{\overset{\text{discharging}}{\rightleftharpoons}} \underset{\text{lead(II) sulfate}}{2\ PbSO_4(s)} + \underset{\text{water}}{2\ H_2O(l)} \qquad \textbf{[8.16]}$$

The arrows in Equation 8.16 indicate that the battery discharges when the chemical reaction proceeds to the right. For example, using the battery to start the car or the lights and radio discharges the battery. However, when the engine runs, an alternator provides the current needed to reverse the chemical reaction and recharge the battery. Fortunately, the battery can be discharged and recharged many times before it needs to be replaced. A high-quality battery can perform for at least five years, but most need replacing in three to four years.

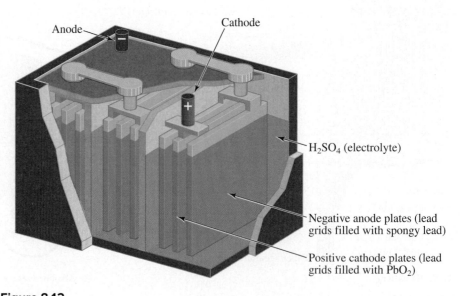

Figure 8.12

Cutaway schematic of a lead–acid battery. This video shows how a lead-acid battery works: www.acs.org/cic

Your Turn 8.5   The Battery in Your Car

Let's take a closer look at the lead–acid storage battery, the one found in most cars (Equation 8.16 and Figure 8.12).

a. Lead occurs in this equation as Pb, PbO_2, and $PbSO_4$, all solids. In which of these is lead in an ionic form? What are the charges (or oxidation states) of the ions? Which one of these symbols represents lead in its metallic form?

b. When lead is converted from its metallic form into an ionic form, are electrons lost or gained? Is this oxidation or reduction?

c. When the battery is discharging, is metallic lead oxidized or reduced?

Because lead–acid batteries have the advantage of being rechargeable and low cost, they may be used together with wind-turbine electric generators. The generator recharges the batteries during favorable windy conditions, whereas the batteries discharge during unfavorable winds. You can also find lead–acid batteries in environments where the emissions from internal combustion engines cannot be tolerated. The forklifts in warehouses, the passenger carts in airports, and the electric wheelchairs in supermarkets are typically powered by lead–acid batteries. Their weight may be an advantage in stabilizing these vehicles.

In an automobile, however, the weight of the lead–acid battery hurts overall fuel economy. Another disadvantage is the nature of the chemical components in the battery. The anode (metallic lead), cathode (lead(IV) oxide), and electrolyte (sulfuric acid solution) pose disposal challenges as toxic or corrosive chemicals. Fortunately, most lead–acid batteries are recycled, which helps to limit the environmental harm resulting from their improper disposal.

8.6 | Renewable Energy Storage

Learning Objectives: Explain how batteries can be used for grid energy storage applications

As we expand our electrical grid energy options beyond fossil fuels, energy storage systems will be essential because of their ability to capture green energy and supply the grid when the Sun isn't shining, or the wind has stopped blowing. Known as *battery energy storage systems* (BESSs), these devices will help meet electricity demands when people need power the most. Experts say that energy storage using batteries is critical in expanding the reach of renewables and accelerating the transition to a carbon-free electrical power grid. Most recently, lithium-ion batteries, the topic of our next section, are the dominant storage technology for large-scale power plants to help electricity grids ensure a reliable energy supply.

The Moss Landing power plant in California (Figure 8.13) was once a natural gas plant, but is now being converted to a lithium-ion energy storage facility. This newly

Figure 8.13

The Moss Landing power plant.

Holly Curry/McGraw Hill

redesigned power plant has enough energy to power about 300,000 California homes for four hours during the evenings, as well as during heat waves and other times when energy demand outstrips supply.

Your Turn 8.6   How Big Is the Battery?

The Moss Landing power plant conversion, the world's largest battery storage system, will be of tremendous use for the expansion of renewables in California. Research this project using a variety of Internet resources and answer the following questions.

Questions to Consider

a. If California homes use, on average, 800 kilowatt hours per month, calculate how much energy in megawatts is needed to power 300,000 California homes for four hours. Assume 30 days a month, and an average hour will be a nighttime hour (1000 kilowatts = 1 megawatt).

b. The battery they are using is 400 megawatts. Do you think they will be able to achieve their goal?

c. When was this project completed and how long did the conversion take? Are there any other similar projects, which convert fossil fuel based power plants to battery storage facilities, proposed in the U.S. or around the world?

Falling prices for battery components and technological progress are due to advances in battery chemistry. Ongoing research and development of battery components have resulted in steady improvements in the capacity of batteries and ability to store and discharge energy over extended periods. In the U.S., state clean energy mandates and tax incentives, paired with lower-cost solar panel installations, have fueled record growth in the adoption of renewables. California is the global leader in balancing green energy supplies through the use of energy storage solutions. However, the rest of the world is following California's lead with plans for megawatt systems in New York, South Florida, London, Lithuania, Australia, Germany, Saudi Arabia, and Chile.

Battery storage systems must have very advanced technologies to be charged by electricity generated from renewable sources such as wind turbines and solar panels (Figure 8.14). Smart battery software uses algorithms to coordinate energy production. Control systems are used to decide when to keep the energy in reserve or when to release it to the electrical grid. Energy is released during peak usage, which keeps costs down and the electricity running.

In California, falling battery prices, and the state's goal of having a carbon-free electrical grid by 2045, have led to many new battery storage solutions. The amount of energy that California will require to meet its goal depends on the future technology mix,

Figure 8.14

The Hornsdale substation utilizes a lithium-ion storage battery.

Lincoln Fowler/Alamy Stock Photo

energy use, and other factors. Experts estimate at least 30 GW (1,000,000 kW = 1 GW) of storage will be needed by 2045.

It has been reported that most fossil fuel power plants in the U.S. and around the world will end their working lives by 2035. The Investment Tax Credit, a 30% tax rebate in the U.S. for installing solar panels, helps push the cost of operating a natural gas plant up because they are not profitable while solar energy is inexpensive. While energy storage projects thrive in California, battery prices must continue to decrease in order to allow for widespread global use. Analysts are optimistic that prices will continue to drop to reach large-scale deployment of BESSs around the globe.

8.7 | Vehicles Powered by Electricity

Learning Objective: Describe the design, operation, advantages, and limitations (including hazards) of lithium-ion batteries

For batteries to be useful in modes of transportation, we need to maximize the distance the vehicle can travel on a single charge and minimize the time it takes to recharge the battery. Current electric vehicles (EVs), such as the Tesla Model S, can travel up to 400 miles between charges (under ideal conditions), which is comparable to the distance one can travel on a gasoline tank before a fill-up. How much time does it take to charge the car battery? And how does the electricity required to charge the battery compare to the cost of filling up a fuel tank? Let's consider these questions in the following activity.

Your Turn 8.7   Tesla Model S Battery Charging

At 400 miles, the Tesla Model S currently features one of the longest ranges for all-electric vehicles per charge. In this activity, you will calculate how long it will take to charge the battery and its associated electricity costs.

a. Compare the power output for a standard 110-V/20-A electrical plug (a "level 1" charger) to that of a 240-V/40-A plug (a "level 2" charger) and that of a "dual charger" (240 V/80 A).

b. Calculate the time required to charge a Tesla Model S using each charging scenario listed in **a**. Assume that the battery's capacity is 100 kWh. Also, assume a charging efficiency of 92%.
   Hint: Use dimensional analysis to ensure that the units properly cancel.

c. Using the Internet, find some locations of "supercharging stations" across the U.S. Using their maximum charging rates, how long would a single charge require?

d. Look up the current cost per kWh of electricity in your region, and calculate the costs per charge for each of the above charging scenarios.

e. Using the current price of gasoline, an average driving distance of 1100 miles per month, and the range of your favorite gasoline-powered vehicle versus the Tesla Model S, determine the cost difference in operating each vehicle per month and year. What factors will influence the maximum driving range of the vehicle? How will these affect the operating costs of the electric car?

As you calculated in the previous activity, it takes a long time to charge the battery in an EV—much longer than filling up a gas tank on a traditional gasoline-powered vehicle. This shouldn't surprise you. It takes an hour or so to charge the battery on your laptop or cell phone, so the much larger battery packs in a vehicle should take significantly longer to charge.

With frequent brown-outs in California and other parts of the U.S. and additional draws on the electrical grid due to climate change, will we have a reliable electrical grid to sustain the basic power needs of a growing population as well as exponential rise in number of EVs on the road? This question is particularly relevant since GM and other major car manufacturers have announced plans to cease production of conventional gasoline-powered vehicles worldwide by 2035. Debate this question in the following activity.

Let's Debate!   Can We Support EVs?

Form two groups: one group will be in favor of the topic (the proposition) and the other group will be against the topic (the opposition). Both groups will try to persuade a neutral person or judge to agree with them. The topic of the debate is the motion. Here is the motion:

The current electrical grid in the U.S. and around the world is not sufficient to sustain the flood of EVs that will hit the consumer market over the next few years. Hence, we need to expand the use of renewables and nuclear power to ensure we have enough capacity to sustain our increasing power needs.

Curiosity Conversations

- Will the capacity for alternative energies be sufficient and reliable enough to support our growing energy needs by 2035?
- The use of renewables will require increased use of batteries for off-peak energy storage. Will we have enough raw materials to meet our demand for new batteries in *both* EVs and energy storage for the electrical grid, while continuing to meet demand in portable electronic devices?
- Will prices go down enough for alternative energies and batteries to make it economically feasible to shift away totally from fossil fuels?
- What are some environmental impacts posed by the expansion of renewables as well as nuclear power plants?

Thankfully, batteries slowly discharge upon use, allowing us to drive many miles, or surf the web on our portable devices over extended periods. The reason for slow charging/discharging in batteries is due to the chemical nature of the battery itself. In previous sections, we described the chemical reactions that occur in batteries. These reactions require time to complete, and their rates are affected by the operating conditions of the battery. For instance, using or storing a battery at elevated temperatures will cause the reactions to proceed faster, which results in faster charging times but shorter battery life. Consequently, using or storing a battery in cold environments will require longer charging times but will generally result in longer battery life.

To understand the rate of these reactions (also known as **kinetics**), let's consider what happens when a pervasive type of rechargeable battery is charged/discharged. Since lithium is the lightest metal in the periodic table, lithium-ion (Li-ion) batteries are most commonly used for portable electronics and vehicle applications. For EV applications, it is paramount that the battery is small and lightweight, maximizing the vehicle's driving range. You can think of this as being equivalent to vehicles of today, composed of plastics and composites, being able to travel farther on a tank of gas than classic cars, which were comprised of heavyweight steel and chrome.

In addition to being lightweight, Li atoms and ions are much smaller than other metals. This means that more ions may be placed within the electrode during charging, which results in longer usage on a single charge. A battery's **energy density** relates to both the number of ions stored in the electrode material and the weight or volume of the battery (Equation 8.17). Figure 8.15 shows a comparison of various types of rechargeable batteries. Battery designs involving Li are the most preferable due to being lightweight (high gravimetric energy density; also known as the *specific energy density*) and smaller (high volumetric energy density).

$$\text{Energy density} = \frac{\text{Voltage} \times \text{\# of movable Li ions in electrodes}}{\text{Total battery weight (gravimetric) or volume (volumetric)}} \quad \textbf{[8.17]}$$

The reactions occurring in a Li-ion battery are much less complicated relative to alkaline or other rechargeable batteries discussed thus far. As shown in Figure 8.16, Li^+ ions simply shuttle back and forth between the two electrodes during charging and discharging. The specific half-reactions for a Li-ion battery are as follows:

$$\text{(Cathode)} \quad LiCoO_2(s) \underset{\text{discharging}}{\overset{\text{charging}}{\rightleftarrows}} Li_{1-x}CoO_2(s) + x\ Li^+ + x\ e^- \quad \textbf{[8.18]}$$

$$\text{(Anode)} \quad x\ Li^+ + x\ e^- + 6\ C(s) \underset{\text{discharging}}{\overset{\text{charging}}{\rightleftarrows}} Li_xC_6(s) \quad \textbf{[8.19]}$$

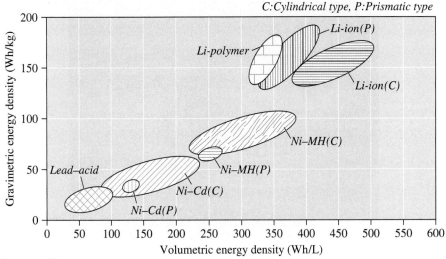

Figure 8.15

Comparison of the energy densities for various rechargeable batteries. Li-polymer batteries have similar electrochemical reactions to Li-ion varieties but differ in their choice of electrolyte and packaging.

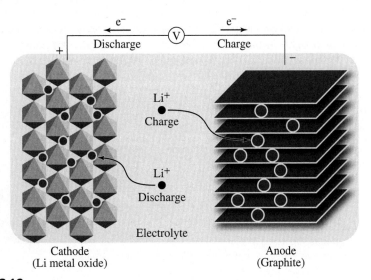

Figure 8.16

Schematic of the Li$^+$ migration during charging and discharging of a Li-ion battery. Check out www.acs.org/cic for an animated version.

The battery can function only if there are available Li$^+$ ions and suitable electrode material to house the ions during its use and recharging cycles. As charging/discharging takes place, the structure and composition of the electrode surfaces will change, which will result in less effective storage of Li$^+$ in subsequent cycles. To improve the lifetime of a Li-ion battery, one should not over-discharge the battery. Allowing the battery to run down to <10% can cause some of the Li$^+$ ions to irreversibly react with the cathode material (Equation 8.20), lowering its overall capacity because fewer ions will be available for intercalation.

$$Li^+ + LiCoO_2 \longrightarrow Li_2O + CoO \qquad [8.20]$$

You may have seen videos or pictures of a Li-ion battery engulfed in flames (Figure 8.17). Despite the widespread media reports of Samsung Note 7 fires in 2016, this is very rare. However, fires and explosions may be caused by many issues, such as separator failure, leaking electrolytes from cracks in the battery pack, or failure of the charging control circuitry. If a Li-ion battery is overcharged, the cathode breaks apart and releases oxygen gas, which causes an immediate exothermic reaction with the Li metal plated on the anode. This then causes a further breakdown of the cathode and products

How Do Li-ion Batteries Work?

©2018 American Chemical Society

Check out this video for tips on how to make your smartphone battery last longer: www.acs.org/cic.

Questions to Consider

a. This video mentions that one should avoid heat to keep your phone battery alive longer. Does this mean that your phone would last longer in cold temperatures? Should you charge your cell phone in the freezer?

b. Have you ever seen a cell phone explode in the heat? How would this happen if the cell phone is left for a while fully discharged?

Figure 8.17

A Zotye M300 electric car catching fire in Hangzhou, China.

Gu feng hz/Imaginechina/ICHPL/AP Images

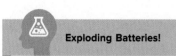

Exploding Batteries!

©2018 American Chemical Society

Check out this video to see why Li-ion batteries could explode: www.acs.org/cic.

Questions to Consider

a. The video describes lithium as a reactive metal. All Group 1 metals are reactive toward oxygen, nitrogen, and water. The electrolyte inside a Li-ion battery is composed of a lithium salt dissolved in an organic-based alkyl carbonate solvent. Could Li-ion batteries use an aqueous lithium salt solution as an electrolyte? Explain your reasoning.

b. Why does a Li-ion battery explode when exposed to the air? Why is it bad for your cell phone battery to produce oxygen gas?

c. What types of solid electrolytes are in development to improve the safety of future Li-ion batteries?

deposited on the anode, which generates sufficient heat to ignite the organic-based electrolyte. There are current efforts to incorporate fire-retardant chemicals within the electrolyte of Li-ion batteries to prevent such a thermal runaway event. The next two activities will shed light on the origin of battery fires and describe some efforts that are underway to improve the efficiency and safety of future batteries.

Real Talk   Next-Generation Batteries

Check out this video with Dr. Ekaterina Pomerant-seva from Drexel University, who describes her work on developing new materials for more power-ful and safer batteries: www.acs.org/cic.

Drexel University

Questions to Consider

a. Would sodium batteries have the same issue as lithium batteries regarding moisture exposure?

b. Dr. Pomerantseva mentions in the video that the smallest distance between the layers they found was 9.6 angstroms when they pre-intercalculated a potassium compound. How many sodium ions could fit in 9.6 angstroms if placed end to end? Sodium ions have a radius of 98 pm and 1 angstrom = 100 pm.

8.8 | Storage Wars: Supercapacitors vs. Batteries

Learning Objective: Describe the structure of supercapacitors and how their properties and applications differ from batteries

Since it takes time for Li^+ ions to migrate into the various layers of anode and cathode compartments, charging and discharging processes within Li-ion batteries take time to complete. The slower the charging rate, the more Li^+ ions can position themselves within the layered electrode structures. The same goes for all other rechargeable batteries—their charging (and discharging) rate is governed by the kinetics of the chemical reactions. However, especially for vehicle applications, we are most interested in rapid charging, which will allow us to get back on the road in a shorter time.

Enter supercapacitors. Like a battery, a supercapacitor consists of two charged electrodes, or plates, immersed in an electrolyte. However, supercapacitors store energy utilizing a static charge instead of storing energy in electrochemical reactions. Think about the last time you walked across a carpeted room while dragging your feet. When you touched a conductive object such as a metal doorknob, the built-up static electricity was released through your finger (Figure 8.18). This buildup of electrical charge is referred to as **capacitance**. As their name implies, supercapacitors (also called *ultracapacitors*) are devices that store a significant amount of charge that may be released as electrical energy to power a device.

As shown in Figure 8.19, the charges in a capacitor or supercapacitor are separated by a certain distance, *d*. Equation 8.21 shows that the device's capacitance, *C*, will increase as the area of the plates, *A*, increases or the distance between the plates decreases. The capacitance units are given in farads, F, with supercapacitors exhibiting values in the 100–12,000 F range. As a frame of reference, a 1-F capacitor can store 1 coulomb of charge at 1 V.

$$C \propto \frac{A}{d} \qquad \textbf{[8.21]}$$

Whereas all rechargeable batteries have a limited lifetime of available cycles, supercapacitors can be charged/discharged for millions of cycles. Imagine virtually unlimited use/recharge cycling of your cell phone without any loss in capacity! Furthermore, charging takes place in seconds, as compared to minutes or hours required to charge batteries. Additionally, supercapacitors can operate under a wider temperature range without premature failure.

With so many desirable characteristics for supercapacitors, why are batteries still preferred for portable devices and vehicles? Supercapacitors suffer from a few critical limitations. As shown in Table 8.2, supercapacitors have very low energy densities as compared to batteries. Whereas a battery stores its potential energy in chemical form, the potential energy in a supercapacitor is stored in an electric field, corresponding to

Figure 8.18

An image showing a shock from touching a doorknob through the buildup of static electricity.

©2006 Richard Megna, Fundamental Photographs, NYC

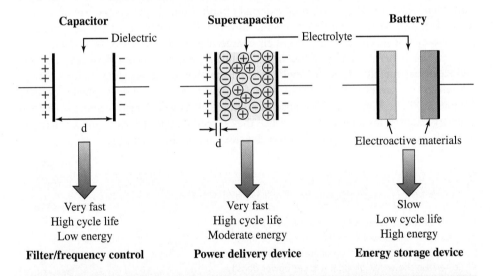

Figure 8.19

Schematics of a capacitor, supercapacitor, and battery.

Table 8.2	Performance Comparison Between Supercapacitors and Li-ion Batteries	
Function	**Supercapacitor**	**Lithium-ion**
Charge time	1–10 seconds	10–60 minutes
Cycle life	1 million	500 and higher
Cell voltage (V)	2.3–2.75	3.6–3.7
Specific energy (Wh/kg)	5	120–240
Specific power (W/kg)	Up to 10,000	1000–3000
Cost per Wh	$10	$0.25–$1.00 (large systems)
Service life (in vehicle)	10–15 years	5–10 years
Charge temperature	−40 to 65 °C (−40 to 149 °F)	0–45 °C (32 to 113 °F)
Discharge temperature	−40 to 65 °C (−40 to 149 °F)	−20 to 60 °C (−4 to 140 °F)

Source: Battery University.

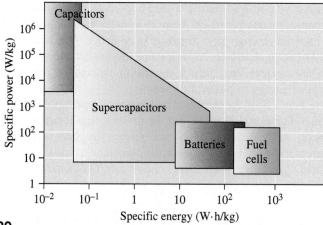

Figure 8.20

Comparison of various energy storage devices. Whereas specific power (power density per unit mass) relates to the acceleration of an electric vehicle, specific energy (energy density per unit mass) relates to its range. Fuel cells will be discussed later in this chapter.

less energy per weight of the material. In contrast, a battery can store a lot of energy in a small amount of mass and perform work for an extended period. For example, a cell phone battery will last most of the day, but to recharge, its needs to be connected to a power source for an hour or more.

Another disadvantage of supercapacitors is **self-discharge**, the loss of charge over time even when not connected to an external circuit. Some rechargeable batteries, such as Ni–Cd, also suffer from this problem; however, Li-ion batteries have an extremely low rate of self-discharge, retaining their charge over long periods of inactivity.

When a battery charges/discharges, it takes time to convert between chemical and electrical energies. From an application perspective, the longer it takes for a battery to deliver its power, the lower its specific power, also known as **power density** (Figure 8.20). In contrast, supercapacitors have high power densities and can charge/discharge almost instantaneously because they directly store electrical energy. Explore a simple analogy for energy and power densities in this next activity.

@HOME   Energy vs. Power Density

Think of two containers holding water as an analogy for energy and power densities. The jug has more water (energy) than the mug does. But it takes longer to release the water, giving it a lower power density. In contrast, the mug has less water (energy) than the jug but can release its contents more quickly. The mug, therefore, has a higher power density.

cloud7days/123RF

Investigate the difference between energy density and power density at home using a jug, a coffee mug, a sink, and water.

1. Fill a jug with water.
2. Fill a coffee mug with water.
3. Have a friend hold the jug of water, and you hold the coffee mug filled with water (or vice versa) over a sink. Pour the contents out of each at the same time and make observations.

Questions to Consider

a. Since a supercapacitor has a high power density, it can output a lot of energy for its mass and can release energy and recharge quickly. Can you think of an example where this would be useful?
b. Which will last longer—Li-ion batteries or supercapacitors?
c. Which stores more energy—Li-ion batteries or supercapacitors?
d. Which has a higher rate of discharge (relieved of its charge more quickly)—Li-ion batteries or supercapacitors?

As you discovered in the previous activity, supercapacitors are most useful in applications where a large amount of power is needed for relatively short periods. For instance, supercapacitors have been used to deliver power for photographic flashes in digital cameras and to provide backup power to computer memory cards. Supercapacitors have also found applications in stabilizing the voltage of power lines, which is especially useful for fluctuating wind or solar sources. Most recently, supercapacitors have been used alongside rechargeable batteries to improve the efficiency of hybrid electric vehicles by allowing faster acceleration from rest while conserving battery capacity. In the next section, we will discuss these *hybrids*—vehicles that use electrical power and gasoline.

Your Turn 8.8   Supercapacitor Myths

Check out this article about supercapacitor myths, which shows the difference between batteries and supercapacitors: www.acs.org/cic.

This section discussed a variety of limitations of supercapacitors relative to batteries for automotive and portable electronics applications. Using various sources, discuss the merit of the following statement: New developments will bring supercapacitors into full competition with rechargeable batteries.

8.9 | Higher MPGs with Fewer Emissions: Gasoline—Electric Hybrid Vehicles

> *Learning Objective: Compare the operation, advantages, and emissions of hybrid electric vehicles with conventional gasoline-powered vehicles*

Due to concerns regarding gasoline prices and the environmental consequences of fossil fuel combustion, the number of electric vehicles on the road has skyrocketed worldwide in recent years. In addition to battery electric vehicles (BEVs) discussed previously, **hybrid electric vehicles (HEVs)**, better known as *hybrids,* are popular among car shoppers. These vehicles are propelled by a conventional gasoline engine in tandem with an electric motor powered by batteries.

In 1999, the Honda Insight, a small two-seater, was the first hybrid sold in the United States. The Toyota Prius became available in Japan in 1997 and three years later in the United States. Today, most manufacturers, including luxury brands such as Porsche (Figure 8.21), produce at least one hybrid car, SUV, or truck model.

Figure 8.21

The 2016 Porsche Panamera S E-Hybrid and cutaway schematic show the drivetrain's position and batteries in the trunk (**red**).

(both): ©2016 Dr. Ing. h.c. F. Porsche AG

In delivering about 50 miles per gallon, the Honda and Toyota hybrids burn about half the gasoline of a conventional car, thus emitting half the amount of carbon dioxide as a conventional vehicle. The electric motor draws power from the batteries to start the car moving and to power it at low speeds. Using a process called *regenerative braking*, the energy of the car's motion (kinetic energy) is transferred to the alternator, which in turn charges the batteries during deceleration and braking. The gasoline engine assists the electric motor during normal driving, with the batteries boosting power when extra acceleration is needed. Although NiMH batteries were once the battery of choice in hybrids, most HEVs now use Li-ion batteries due to their faster charging rates and higher energy density.

The consumption of a gallon of gasoline releases about 19.5 pounds (8.7 kg) of CO_2 into the atmosphere. Therefore, the average gasoline-powered vehicle emits 6 to 9 tons of CO_2 each year. Unlike other vehicle emissions, such as NO_x and CO, pollution-control technologies currently do not reduce CO_2 emissions. Instead, we must reduce the amount of carbon dioxide either by burning less fuel *or* by burning a fuel such as H_2 that does not contain, and therefore does not emit, carbon. Doing the math, each increase of 5 miles per gallon a year (for example, improving from 20 to 25 miles per gallon) can reduce CO_2 emissions by about 18 tons over a vehicle's lifetime. This calculation assumes a vehicle lifetime of 200,000 miles, which results in burning about 2000 fewer gallons of gasoline.

Your Turn 8.9   Yes, Tons of CO_2!

Could an automobile *really* emit 7 tons of carbon dioxide in a year? Do a calculation to verify this. State all the assumptions that you make. If you were to improve your car's gas mileage by 5 miles per gallon, how much CO_2 would you save over a year? Aside from buying a hybrid car, what are some ways people can reduce transportation-related carbon dioxide emissions?

Automobile manufacturers now provide more than one option for fuel-efficient, low-emission vehicles, allowing customers to choose the option that best meets their transportation needs. Chevrolet introduced the first plug-in hybrid electric vehicle (PHEV) to the U.S. market in 2011. The Chevrolet Volt and other plug-in hybrid vehicles use rechargeable batteries for short daily commutes to run an electric motor and switch to a combustion engine to recharge the batteries and travel longer distances (Table 8.3). Furthermore, electrical energy provided by the battery decreases the direct emissions from the tailpipe, and the lower amount of gasoline consumed is a major selling point.

The large lithium-ion battery packs make plug-in hybrid vehicles more expensive than their gasoline-powered and gasoline-electric hybrid peers. However, nearly every major automotive manufacturer currently has research and development teams working on all-electric (EV) and plug-in hybrid electric vehicles (PHEVs). Most of these companies agree that, in the future, plug-in electric cars could easily number in the millions on city streets worldwide, provided that battery technologies continue to improve and suitable economic incentives are developed. China currently has the world's largest market share of electric vehicles, accounting for just under 50% of the world's total. To increase sales in the U.S., a federal tax credit of up to $7500 toward purchasing new plug-in hybrids and all-electric vehicles has been offered since 2011. In addition, other policies and incentives have been developed worldwide to promote charging infrastructure deployment in major cities—a prerequisite for the widespread adoption of this technology. Even so, the numbers are daunting. The 4.5 million electric passenger cars worldwide in 2021 were far outnumbered by over 1 billion gasoline-powered cars already in use.

With such advantages offered by HEVs and PHEVs, has the United States turned into a hybrid nation? The answer most certainly is no. Although the United States has the highest count of hybrid vehicles on its roads, accounting for one-half of all-electric vehicle sales, HEVs and PHEVs represent only ~3% of all cars on the road. The most significant increases in PHEV sales worldwide in recent years were in China, Germany, Sweden, and Norway (Figure 8.22). The following two sections examine another way in which we might power our lifestyles: hydrogen fuel cells.

Rank	Manufacturer/Model	Miles per Gallon (city/highway)
Table 8.3	**Fuel Economy Leaders for the 2022 Model Year**	
	Electric Vehicles (EVs)	
1	Tesla Model 3 RWD	138/126
2	Lucid Air Grand Touring	130/132
3	Chevrolet Bolt EV	131/109
4	Hyundai Kona EV	132/108
5	Tesla Model S	124/115
	Plug-in Hybrid Electric Vehicles (PHEVs)	
1	Toyota Prius Prime	133/79
2	Hyundai Ioniq PHEV	119/72
3	Kia Niro PHEV	105/70
4	Ford Escape PHEV	105/74
5	Toyota RAV4 Prime	94/62
	Hybrid Electric Vehicles (HEVs)	
1	Hyundai Ioniq	58/60
2	Toyota Prius	58/53
3	Hyundai Elantra Hybrid	53/56
4	Honda Insight EX	55/49
5	Toyota Corolla Hybrid	53/52

This list is taken from the top midsize cars, including plug-in hybrids and all-electric vehicles.

Note: For EVs, fuel economy is given in miles per gallon equivalent (MPGe), where 33.7 kWh = 1 gallon of gasoline.

Source: www.cars.com and Fueleconomy.gov

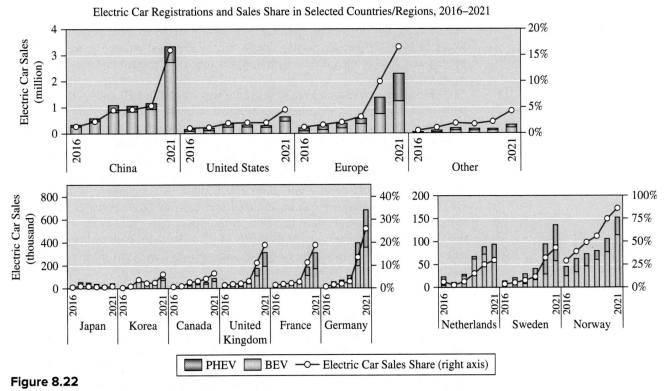

Figure 8.22

Electric car sales and registrations in selected countries, 2016–2021. PHEV is a plug-in hybrid electric vehicle; BEV is a battery electric vehicle.

8.10 | Fuel Cells: The Basics

> *Learning Objective: Describe the design, operation, applications, and advantages of fuel cells*

With fuel cells, we take another step on our journey to find fuels that release high amounts of energy with minimal pollution. In Section 6.4, we compared the energy released, gram for gram, in the combustion of coal, hydrocarbons, and other combustible fuels. As we saw, methane was the winner. Assuming the combustion products to be $CO_2(g)$ and $H_2O(g)$, the heats of combustion of coal and *n*-octane (C_8H_{18}, a major component in gasoline) are 30 and 45 kJ per gram of fuel, respectively. In comparison, the heat of combustion of methane is 50 kJ/g.

However, when paired against methane, hydrogen easily wins the competition:

$$H_2(g) + \tfrac{1}{2} O_2(g) \longrightarrow H_2O(g) + 124.5 \text{ kJ/g} \qquad \text{[8.22]}$$

In other words, hydrogen releases almost three times as much energy as methane per gram when burned! In addition to its superior energy production, using hydrogen raises another tantalizing prospect—the powering of motor vehicles with a fuel that would produce only water vapor as a product. Neither $CO(g)$ nor $CO_2(g)$ would be made, although depending on the engine conditions and temperatures, some $NO_x(g)$ could conceivably form.

Your Turn 8.10   Hydrogen Versus Methane

Is hydrogen *really* that good as fuel? Use the bond energy values from Table 6.1 to find out. Clearly show how you performed your calculation, noting any assumptions you needed to make. Does the value you calculated match that of Equation 8.22?

As with other flammable fuel sources such as methane or gasoline, when hydrogen is directly mixed with oxygen, a mere spark can set off an explosion. With its seven-million cubic feet of hydrogen gas, the *Hindenburg* was to airspace as the *Titanic* was to the high seas. When the airship caught fire in 1937 and plunged many of its passengers and crew to their deaths, hydrogen was indelibly stamped in our consciousness as an explosive fuel.

But suppose someone were to suggest a way to combine H_2 and O_2 to form H_2O without the hazards of combustion. Furthermore, suppose that this person also claimed that the reaction could be carried out without direct contact between the hydrogen and the oxygen. The skeptical chemist in us might dismiss such assertions as sheer nonsense—an impossibility. And yet, the operation of a fuel cell is a case in point. A **fuel cell** is an electrochemical cell that produces electricity by converting the chemical energy of a fuel directly into electricity without its combustion. William Grove (1811–1896), an English physicist, invented fuel cells in 1839. However, these cells remained a mere curiosity until the dawn of the Space Age. Only in the 1980s, when the U.S. *Apollo* spacecraft carried three sets of 32 cells fueled with hydrogen, did fuel cells come into public view. The electricity generated by these cells powered the lights, motors, and computers on board the shuttle.

Unlike conventional batteries, such as those in flashlights, under the hood of a car, or powering your laptop computers, fuel cells operate on an external supply of fuel that is electrochemically oxidized inside the fuel cell. However, they also require an external supply of oxygen gas or another *oxidizing agent* to accept the lost electrons of the fuel. Whereas an **oxidizing agent** is reduced (adds electrons), a **reducing agent** is oxidized (releases electrons) during the redox reaction occurring inside a fuel cell. With the supply of fuel and oxidant (the reactants) continually being replenished by the fuel tank, these so-called *flow batteries* produce electricity. They do not run down or need

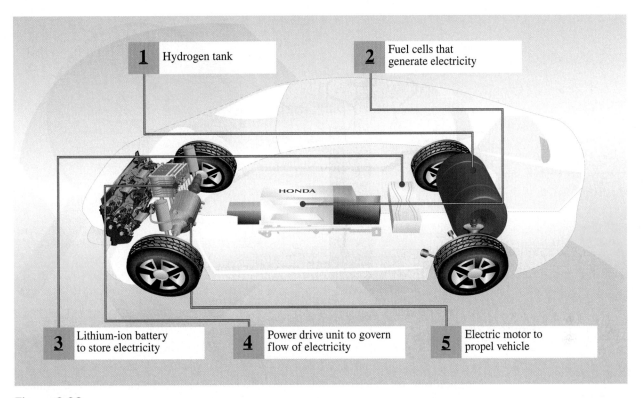

Figure 8.23

The schematic of a Honda FCX powered by fuel cells shows the location of the fuel cell, hydrogen storage tank, and lithium-ion storage battery.
Source: American Honda Motor Company.

to be recharged in the same manner as conventional batteries do. Check out the location of the hydrogen fuel tank in Figure 8.23, a schematic drawing of a fuel cell vehicle (FCV). The hydrogen tanks are pressurized up to 5000 pounds/inch^2 (psi), equivalent to 34 MPa.

What may surprise you about fuel cells is that the chemicals being oxidized and reduced are physically separated; they do not come in direct contact with each other. Oxidation still occurs at the anode, and reduction at the cathode. However, instead of the anode being the source from which electrons are released, the anode is merely an electric conductor that provides a physical location in the cell at which the fuel oxidation occurs. Similarly, the cathode is an electric conductor where oxygen reduction occurs and does not enter the reaction.

The electrolyte separating the anode from the cathode serves the same purpose as in a traditional electrochemical cell: to allow the flow of ions and hence the flow of charge. The earliest commercially available fuel cells used a corrosive acid (phosphoric acid, H_3PO_4) as an electrolyte. As a result, these fuel cells were closed systems that fully contained the liquid, not unlike the closed system of a conventional alkaline battery. Current designs of fuel cells are open systems that require a continuous flow of fuel and oxidant, adding complexity and cost.

Today, fuel cells based on different electrolyte materials have been developed for various applications. One type incorporates a solid polymer electrolyte separating the reactants. We use this type to explain the general operation of fuel cells. The polymer electrolyte membrane, also called a **proton-exchange membrane** (PEM), is permeable to H^+ ions (protons) and is coated on both sides with a platinum-based catalyst (Section 6.13). These electrolytes operate at reasonably low temperatures, typically 70–90 °C, and transfer electrons to rapidly provide electric power. As a result, PEM fuel cells are currently popular with automakers for new fuel cell vehicles and personal consumer applications. A typical design is shown in Figure 8.24.

In fuel cells, hydrogen is used as the fuel in conjunction with oxygen; the oxidation and reduction half-reactions are represented by Equations 8.23 and 8.24,

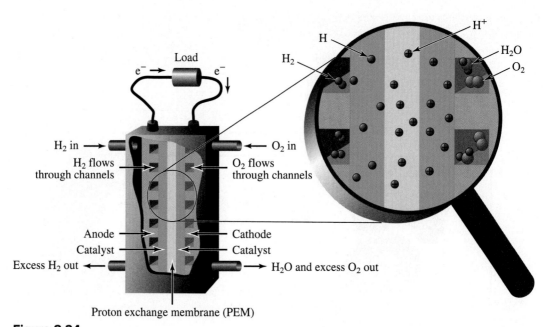

Figure 8.24

Schematic of a PEM fuel cell in which H_2 and O_2 combine to form water without combustion.

respectively. As a molecule of hydrogen (H_2) passes through the membrane, it is oxidized and loses two electrons to form two hydrogen ions:

$$\text{(Anode)} \quad H_2(g) \longrightarrow 2\,H^+(aq) + 2\,e^- \qquad \textbf{[8.23]}$$

The hydrogen ions, H^+, flow through the proton exchange membrane and combine with oxygen (O_2). At the same time, they combine with two electrons to form water:

$$\text{(Cathode)} \quad \tfrac{1}{2}\,O_2(g) + 2\,H^+(aq) + 2\,e^- \longrightarrow H_2O(g) \qquad \textbf{[8.24]}$$

As with galvanic cells, the overall equation is the sum of the two half-reactions:

$$H_2(g) + \tfrac{1}{2}\,O_2(g) + 2\,\cancel{H^+(aq)} + 2\,\cancel{e^-} \longrightarrow 2\,\cancel{H^+(aq)} + H_2O(g) + 2\,\cancel{e^-} \qquad \textbf{[8.25]}$$

The 2 e^- and 2 H^+ appearing on both sides of the arrow can be canceled:

$$H_2(g) + \tfrac{1}{2}\,O_2(g) \longrightarrow H_2O(g) \qquad \textbf{[8.26]}$$

The electrons flowing from the anode to the fuel cell cathode move through an external circuit to do work, which is the whole point of the device. Thus, in a fuel cell, a transfer of electrons occurs from H_2 to O_2. This occurs with no flame, with relatively little heat, and without producing any light. Because of these characteristics, the reaction is not classified as combustion. Suppose only the energy-producing step is considered (admittedly omitting other parts of the energy picture). In that case, hydrogen fuel cells are considered a more environmentally friendly way to produce electricity than coal-fired or nuclear power plants. No carbon-containing greenhouse gases are produced, no air pollutants are emitted, and no spent nuclear fuel needs to be disposed of. Water is the only chemical product if hydrogen is the fuel, an added benefit for the astronauts on the space shuttle, who relied on it as their water source while in space.

The overall reaction (Equation 8.26) releases 249 kJ of energy per mole of water. But instead of liberating most of this energy in the form of heat, the fuel cell converts 45–55% of it into electric energy. This direct production of electricity eliminates the inefficiencies associated with using heat to do work to produce electricity. For instance, internal combustion engines are only 20–40% efficient in deriving energy from fossil fuels. Table 8.4 shows a comparison of traditional fuel combustion with fuel cell technology. The activity that follows will give you a chance to explore an interactive animation of a fuel cell.

Table 8.4		Combustion versus Hydrogen Fuel Cell Technology		
Process	**Fuel**	**Oxidant**	**Products**	**Other Considerations**
Combustion	Hydrocarbons, alcohols, H_2, wood, etc.	O_2 from air	H_2O, CO/CO_2, heat, light, sound	Rapid process, flame present, low efficiency, useful for producing heat
Hydrogen fuel cell	H_2	O_2 from air	H_2O, electricity, heat	Slow process, no flame, quiet, efficient, useful for generating electricity

Your Turn 8.11   Revisiting the PEM Fuel Cell

Spend time exploring the animation of a PEM fuel cell found at www.acs.org/cic.

a. How is a fuel cell different from other batteries described earlier in this chapter?

b. Can a PEM fuel cell be recharged? Explain.

c. Why is the combination of H_2 and O_2 in a fuel cell not classified as combustion? Explain.

Just as batteries, motors, and electric generators come in different sizes and types, so do fuel cells. Although the fuels and principles of operation are essentially the same, different electrolytes give every kind of fuel cell unique characteristics that are appropriate for a given application. Although more attention has been devoted to batteries in recent years, many companies are still experimenting with fuel cell vehicles. Like EVs, fuel cell vehicles (FCVs) are powered by electric motors. But they differ because FCVs create electricity, whereas EVs draw electricity from an external source, storing that energy in an onboard battery.

As a source of electricity, fuel cells have a broad range of applications. Hospitals, airports, banks, police stations, and military installations now use them for standby and backup power applications. Fuel cells are a form of **distributed generation**; that is, they generate on-site electricity right where it is used, avoiding the losses of energy that occur over long electric transmission lines. They serve as an alternative to central electric utility power plants.

Before our societies can fully benefit from hydrogen fuel cell technologies, scientists and engineers need to meet several technological challenges. The first is storing, transporting, and eventually distributing hydrogen to the consumer. A second challenge is to produce enough hydrogen in a sustainable manner to meet the projected demand. The following section examines both of these challenges in more depth.

8.11 | Hydrogen for Fuel Cell Vehicles

Learning Objective: Describe the methods used to generate hydrogen fuel and understand the limitations and challenges of using hydrogen for fuel cells

Imagine refueling your fuel cell vehicle with hydrogen at a "gas station." Currently, refueling stations are few and far between. As of 2022, the U.S. Department of Energy reported that only 107 retail hydrogen fueling stations are in operation in the United States. However, a FCV can travel about 300 miles before refueling, which is certainly competitive with mileage achieved with a conventional gasoline-fueled engine.

Because hydrogen is a gas, it requires a different system for storage and transfer from that used for gasoline. As a gas, hydrogen also takes up a lot of space. For example, at sea level and room temperature, H_2 occupies a volume of about 11 L (almost 4 gallons) per gram. In comparison, 11 L of gasoline has a mass of about 8250 g! To avoid having an enormous fuel tank, your vehicle must store hydrogen in a pressurized gas cylinder. To replenish the hydrogen in this cylinder, you must refuel with an airtight connection through a hose that can withstand high pressures, as shown in Figure 8.25. Although the

Figure 8.25

Refueling a Honda FCX Clarity, a hydrogen-powered vehicle.

Eugene Garcia/The Orange County Register/ZUMAPRESS.com/Newscom refueling process requires a different system than a gasoline pump used for a gasoline-powered vehicle, the process is similar in that there is a nozzle, and you squeeze a trigger to start the hydrogen flow.

Compressing H_2 into metal cylinders results in a heavy and somewhat unwieldy tank. Hence, chemical engineers are investigating other methods for storing and transporting H_2 that could reduce space and preclude the need for high-pressure gas compression. One promising technology is that some compounds can absorb hydrogen molecules like a sponge absorbs water if subjected to high pressure. Then, by either decreasing the hydrogen pressure or increasing the temperature, the H_2 can be re-released on demand (Figure 8.26). For example, metal hydrides can perform in this way. Lithium hydride, LiH, is one example. A chemical formula of LiH may appear strange to you, and for a good reason. The problem is not the lithium ion (Li^+), as this should be an old friend by now. Instead, the hydride ion, H^-, is a chemical species that differs markedly from the hydrogen ion, H^+. The hydride ion, with two electrons instead of one for neutral hydrogen, plays an important role in the reversible storage of hydrogen gas.

Metal-hydride storage systems are ideally suited for PEM fuel cells that require high-purity hydrogen. Because metal hydrides are selective and absorb only hydrogen, not larger gas molecules such as CO, CO_2, or O_2, they act simultaneously as a storage material and a way of filtering out other gases. Using these materials addresses the need for storage technologies that take up less vehicle space needed for people and cargo, while safely allowing more fuel on board for longer-range travel.

Beyond storage, a second challenge is the projected demand for hydrogen as a fuel. Where is all the hydrogen going to come from? On the one hand, things look promising because hydrogen is the most plentiful element in the universe. Over 93% of all atoms are hydrogen atoms! Although hydrogen is not nearly this abundant on Earth, there is still an immense supply of the element. On the other hand, essentially all of the hydrogen on our planet is in some form other than H_2. Hydrogen gas is too reactive to exist for long in this form, so it is primarily found in its oxidized form of H_2O, better known as water. Therefore, to obtain hydrogen for use as a fuel, we must form it from water or other hydrogen-containing compounds, a process that requires energy.

Fossil fuels, including natural gas and coal, are one possible source of hydrogen. In particular, methane, the major component of natural gas, is currently the chief hydrogen source. Currently, around 50% of the world's supply of hydrogen is produced from CH_4 via an endothermic reaction with steam:

$$165 \text{ kJ} + CH_4(g) + 2 \text{ } H_2O(g) \longrightarrow 4 \text{ } H_2(g) + CO_2(g) \qquad \textbf{[8.27]}$$

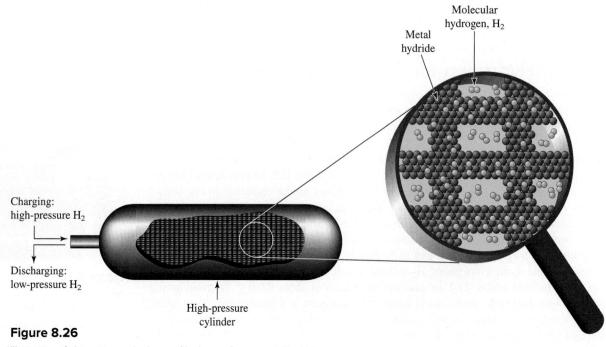

Figure 8.26

Illustration of absorption and release of hydrogen from a metal hydride.

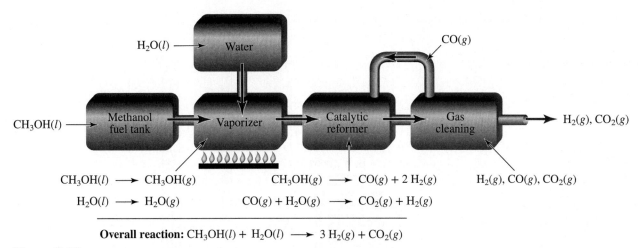

Overall reaction: $CH_3OH(l) + H_2O(l) \longrightarrow 3 H_2(g) + CO_2(g)$

Figure 8.27

This schematic shows how hydrogen gas is obtained from methanol via a reforming process. Red arrows show the direction of fluid flow.

Your Turn 8.12   Back to Bond Energies

a. Use the average bond energy values in Table 6.1 to check the energy required by the reaction in Equation 8.27. Show your work.

b. Did the values you calculated in part **a** match those given in the equation? Explain.

An alternative to fossil fuel combustion is to convert a hydrogen-rich fuel such as methanol or natural gas by a reformer. This device uses heat, pressure, and a catalyst to run a chemical reaction that yields hydrogen as one of the products (Figure 8.27). Because they are liquids under standard conditions, methanol and ethanol could be pumped at conventional gas stations. However, the onboard reformers add cost and maintenance demands to the vehicle.

Still, the reforming process contains a major flaw: greenhouse gases and other air pollutants are produced. Is there another source of hydrogen around? In Jules Verne's 1874 novel, *Mysterious Island,* a shipwrecked engineer speculates about the energy resource that will be used when the world's coal supply has been used up. "Water," the engineer declares, "I believe that water will one day be employed as fuel, that hydrogen and oxygen which constitute it, used singly or together, will furnish an inexhaustible source of heat and light."

Is this simply science fiction, or is it energetically and economically feasible to break water into its elemental components? To assess the credibility of the claim by Verne's engineer, we need to examine the energy requirements of this chemical reaction. In Section 8.10, we noted that the formation of 1 mol of water from hydrogen and oxygen releases 249 kJ of energy. An identical quantity of energy must be absorbed to reverse the reaction to produce hydrogen (Figure 8.28):

$$249 \text{ kJ} + H_2O(g) \longrightarrow H_2(g) + \tfrac{1}{2} O_2(g) \qquad \textbf{[8.28]}$$

The most convenient method of decomposing water into hydrogen and oxygen is by **electrolysis**, the process of passing a direct current of electricity of sufficient voltage

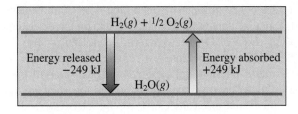

Figure 8.28

Energy differences in the hydrogen–oxygen–water system.

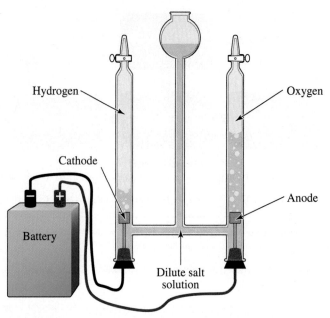

Figure 8.29

Schematic of the experimental setup for the electrolysis of water.

Build Your Own Electrolytic Cell

Stephen Frisch/McGraw Hill

Check out this video to see how to make your own galvanic and electrolytic cells: www.acs.org/cic.

through water to decompose it into H_2 and O_2 (Figure 8.29). This process takes place in an **electrolytic cell**, a type of electrochemical cell in which electrical energy is converted to chemical energy. An electrolytic cell is the opposite of a galvanic cell, where chemical energy is converted to electric energy. When water is electrolyzed in an electrolytic cell, the number of hydrogen molecules generated is twice that of the number of oxygen molecules, as shown in Equation 8.28. This suggests that a water molecule contains twice as many hydrogen atoms as oxygen atoms, testimony to the formula H_2O. Build your own electrolytic cell in this next activity.

Water electrolysis produces no CO or CO_2 and requires about half the energy input per mole of H_2 compared to the methane reaction shown in Equation 8.27. However, the question remains: How will the electricity be generated for large-scale electrolysis? Most electricity in the United States is currently produced by burning fossil fuels in conventional power plants. If we only had to contend with the first law of thermodynamics, the best we could possibly achieve would be to burn an amount of fossil fuel equal in energy content to the hydrogen produced in electrolysis. But we must also deal with the consequences of the second law of thermodynamics, which states that heat spontaneously flows from objects at higher temperatures to those at lower temperatures. Because of the inherent and inescapable inefficiencies associated with transforming heat into work, the maximum possible efficiency of an electric power plant is 63%. Adding the additional energy losses caused by friction, incomplete heat transfer, and transmission over power lines would require at least twice as much energy to produce the hydrogen than we could obtain from its combustion. This is comparable to buying eggs for 10¢ each and selling them for 5¢, which is no way to do business.

Instead of burning fossil fuel to split water, another option is to use a sustainable source of energy, the radiant energy of the Sun. Photons of visible light have enough energy to split water. Unfortunately, water doesn't absorb light at these wavelengths (which is why water is colorless). New materials are being designed to use the power of the Sun to help drive the splitting of water. One type of photoelectrochemical cell, or a galvanic cell, contains a platinum (Pt) cathode and an anode covered with nanoparticles of titanium(IV) oxide (TiO_2) coated with dye molecules. The dye molecules are tuned to absorb light in the most intense part of the solar spectrum. When submerged in an aqueous electrolyte solution and exposed to light, some of the electrons in the dye are promoted to higher-energy states, high enough that they are transferred quickly to the TiO_2. Once there, the electrons can leave the electrode and move through an electric circuit (Figure 8.30).

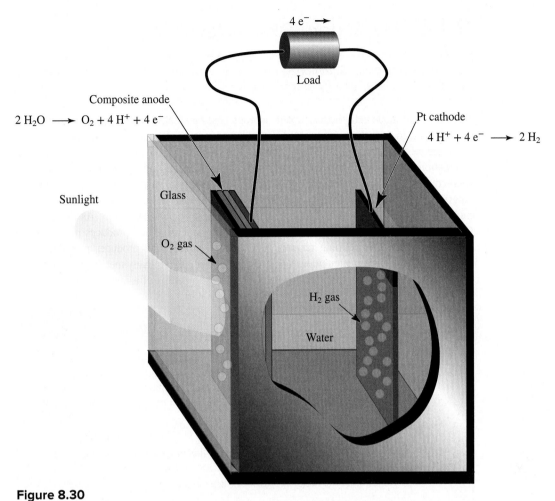

$2 H_2O \longrightarrow O_2 + 4 H^+ + 4 e^-$

$4 H^+ + 4 e^- \longrightarrow 2 H_2$

Figure 8.30

A schematic diagram of a photoelectrochemical cell for splitting water.

As you have learned, the loss of electrons corresponds to oxidation. In this case, the oxygen in water can be oxidized to O_2. After passing through the external circuit, the electrons arrive at the platinum cathode, where they reduce hydrogen ions to H_2. Modern device efficiency is less than 14% but is expected to increase.

Your Turn 8.13   Light That Splits Water

The energy required to promote electrons in the dye to TiO_2 corresponds to a wavelength of 420 nm.
a. Which region of the electromagnetic spectrum does this fall in?
   Hint: Refer to Figure 3.4.
b. It is advantageous to use light energy directly, as opposed to the Sun's heat energy, to split water. Explain.

In a vivid example of *green chemistry*, scientists are looking to biological organisms to produce hydrogen. Certain species of unicellular green algae produce hydrogen gas during photosynthesis (Figure 8.31). The advantage is that sunlight provides energy rather than fossil fuel combustion. With a current efficiency of 10–13%, the efficiency of the process is too low to be commercially viable. However, having new strains of algae that are more efficient in utilizing sunlight could tip the economic balance. A current area of research is to genetically engineer such types of algae—both a promising and a controversial area of inquiry.

Figure 8.31

Some strains of algae can produce hydrogen via photosynthesis.

Pascal Goetgheluck/Science Source

8.12 | My Battery Died—Now What?

> *Learning Objective: Summarize the social, environmental, and economic costs of recycling energy storage devices*

Batteries have made cell phones, tablets, laptops, and handheld calculators so commonplace that we tend to take them for granted. Even lower-income countries have become increasingly reliant on battery-powered electronics. Yet the battery in a cell phone, car, or solar installation costs more than what one pays for it in a store or online. There is also an environmental price tag, an "external cost" that is borne by all. Part of this stems from the "ingredients" found in just about any battery, namely, one or more metals. These metals must be mined from Earth and refined from the ores in which they occur. As discussed in Chapter 1, the process of mining is extremely energy-intensive and produces mine tailings and other waste. The refining process also requires energy and produces pollutants. For example, metal refining often releases sulfur dioxide because so many metals occur naturally as sulfides. The next activity allows you to examine the details of a metal-refining process.

Your Turn 8.14   Metal Refining (Smelting)

The nickel–metal hydride battery used in power tools and the lead–acid battery used in cars require nickel and lead, respectively. These metals are smelted from sulfur-containing ores such as NiS and PbS. Smelting is the process of heating and chemically processing an ore. The smelting of sulfide ores was mentioned in the context of air quality in Section 2.11.

a. Name three attributes that distinguish a metal from a nonmetal.
   Hint: Review Section 1.2.
b. To produce elemental nickel, oxygen gas is reacted with an ore of nickel and sulfur, represented as NiS:

$$NiS(s) + O_2(g) \longrightarrow Ni(s) + SO_2(g)$$

   Is Ni in the ore oxidized or reduced?
c. Write the analogous chemical equation for lead and identify oxidized and reduced species.
d. Why is the release of SO_2 a serious problem?
   Hint: Revisit Chapter 2.

The environmental price tag also includes the disposal of "dead" batteries. Even rechargeable batteries eventually have to be replaced because, at some point, the voltage drops below usable levels, and electrons no longer flow. Although the battery may be dead, the chemicals are still hazardous. Thus, communities either must pay to clean up the landfills where batteries are improperly disposed of, or they must pay to recycle the batteries properly.

Your Turn 8.15   Environmental Justice: E-Waste

Environmental justice means that no population bears an excessive share of negative environmental consequences. But often, those benefiting from the modern lifestyle differ from those who suffer from the adverse environmental and health effects of disposal and recycling. Electronic waste (e-waste) is a problem in wealthy countries that often gets shipped to countries with fewer economic resources.

Questions to Consider
a. Do some research on where the U.S. currently sends batteries for recycling. What type of batteries are shipped and what are the effects on the people and environment in those locations?
b. Are there any Li-ion battery recycling sites in the U.S.? Where are these sites located and what facilities and regulations are in place to prevent environmental harm?
c. What are some environmental issues that have been reported in your local area? Did any of these issues involve the recycling of batteries or electronics, such as heavy metal waste?

One way to reduce battery waste is to think "cradle-to-cradle" (Section 1.9). At the end of a life cycle for one battery, the beginning of a life cycle for another battery should begin. This way, everything is reused rather than added to the waste stream. If each battery served as the starting material for a new product, then the metals these batteries contain would not be lost to the landfill. This also is called "closed-loop recycling." The economics make sense, especially when it is cheaper to extract and reuse a metal (such as from a discarded battery) than to mine new ore and refine it. With a cradle-to-cradle approach, the item to be recycled is sent to a company that pulls out the desired metal and then returns it to another manufacturer. Unfortunately, far too few batteries are recycled in this manner worldwide due to the costs associated with battery recycling.

Keeping toxic materials out of the environment also makes sense. For example, the components of an automobile battery—metallic lead, lead(IV) oxide (PbO_2), and sulfuric acid (H_2SO_4)—are toxic or corrosive. Other metals commonly used in batteries, including cadmium and mercury, are equally, if not more harmful. Disposing of these batteries in landfills contaminates the land, the surface water, and ultimately the ground-water with these metals. In addition, some metals become lost to the manufacturing supply chain as they are too widely dispersed to mine effectively. The next activity explores a possible future scenario if we continue along this path.

Your Turn 8.16   Could Metals Become Extinct?

In 2019, the American Chemical Society published this "Periodic Table of Endangered Elements":

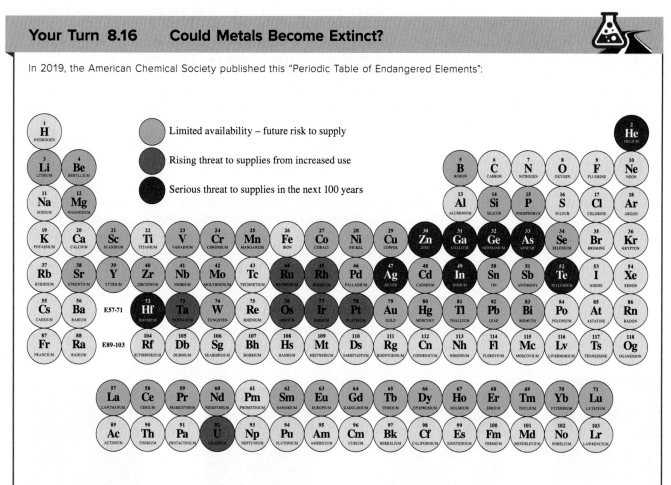

a. Is it really possible for a metal to become "extinct"?

b. Which of the highlighted elements above are currently in demand for energy storage devices (batteries, fuel cells, supercapacitors)?

c. Are there any nonhighlighted elements that are currently being used or tested for future use in energy storage applications?

The lead–acid battery represents one success story. Today, most state laws require retailers that sell lead–acid batteries to also collect them for recycling. Currently, more than 90% of lead–acid batteries are recycled worldwide. The next activity enables you to learn more about battery recycling.

Your Turn 8.17   Battery Recycling

What can you do to keep the metals used in batteries from being lost to a landfill? The answer depends on the battery type. Search the Internet to answer the following:

a. Which types of batteries are more commonly recycled: rechargeables (secondary) or nonrechargeables (primary)?
b. Why is recycling a Ni—Cd battery more critical than recycling an alkaline one?
c. List some reasons why household battery-recycling programs have not been as effective as those for recycling car batteries.

In an article on the future of metals, Thomas Graedel, an industrial ecologist at Yale, pointed out that "metals have limits in the same way that crude oil and clean water do." If metals are to remain available for use in the future, we all need smarter battery designs that allow for efficient recycling of the metals they contain. It makes no sense for cadmium, mercury, nickel, and lead to end up in landfills. Instead, they should end up in new batteries.

There is precedent and good reason for recycling metals. Recall from Section 2.11 that catalytic converters typically contain platinum. The chemical and petroleum industries already have set protocols for recycling platinum catalysts. The item is sent to a company that extracts platinum and returns the metal to a manufacturer for reuse. Although rechargeable batteries can be recycled, only about 10% of all cell phones are recycled.

Lithium is an interesting case in point. Recall from Chapter 1 that lithium is an alkali metal in Group 1 of the periodic table, just like sodium and potassium (Figure 8.32), and that Li, Na, and K all are highly reactive metals with one valence electron. Finally, recall that these metals occur in nature as ions: Li^+, Na^+, and K^+.

Lithium differs from other Group 1 elements in several important ways. Compared to the atoms of sodium and potassium, lithium atoms are smaller and lighter. Having less mass is an advantage when it comes to building portable batteries. Furthermore, being smaller is also an advantage for specific energy considerations. Lithium ions are small enough to fit within certain electrode materials, in contrast to Na^+ and K^+, which may be too large. However, lithium is far less abundant in Earth's crust than sodium and potassium. Lithium deposits tend to be found in remote locations, such as the one shown in Figure 8.33. Lithium is mined primarily from salt lakes (brine lakes) that once were ancient sea beds.

Lithium (stored in oil)

Sodium (removed from oil, being cut)

Potassium (in sealed glass tube)

Rubidium (in sealed glass tube)

Figure 8.32

Selected Group 1 elements.

Stephen Frisch/McGraw Hill

Figure 8.33

One of the world's largest lithium deposits is a salt lake in a desert area of Chile. Lithium is in the form of soluble chloride and carbonate salts, LiCl and Li_2CO_3, which appear as the white residue around the lake's rim.

esteban mazzoncini/iStockphoto/Getty Images

The future availability of lithium is a key point for discussion. As you saw earlier in this chapter, the batteries of choice for EVs and HEVs contain lithium. Using these batteries in millions of cars—each with about 4.5 kg (10 lb) of lithium per battery pack—may severely test our ability to supply lithium to battery manufacturers. At the moment, people are arguing over whether we can do this. On the one hand, our planet's lithium appears to be sufficient to meet our needs. On the other, only some of the lithium deposits are of high enough quality and in accessible enough regions to be extracted economically.

Clearly, it is in everybody's best interest to follow the key ideas of green chemistry as we manufacture batteries now and in the future. This means that we must minimize or avoid the use of toxic metals in batteries *and* use smart battery designs that enable metals in short supply to be efficiently recycled.

Your Turn 8.18   Group Activity: Are Electric Vehicles *Really* Sustainable?

Electric vehicles have the advantage of fewer direct greenhouse gas (GHG) emissions from their use relative to fossil fuel–powered vehicles. However, GHGs are emitted during their production and end-of-life recycling efforts, in addition to emissions from power plants in supplying the electrical grid power needed to recharge the battery in PHEVs and EVs.

As a group, use various sources to answer the following questions and determine whether gasoline-powered or electric vehicles are more sustainable over their lifetime.

a. What is the total global electricity demand for EVs?
b. What is the current power-generation mix in the U.S. (including biomass, renewables, hydroelectric, nuclear, and fossil fuels)? How does this energy distribution differ in other parts of the world? How is this projected to change over the next 10 years?
c. How do the global GHG emissions and other air pollutants compare between EVs and gasoline-powered vehicles? Consider production, use, and end-of-life stages for both types of vehicles.

Conclusion

We look to many different forms of electron transfer to meet our energy needs. Batteries can store chemical energy and convert it into a flow of electrons useful for many applications. Supercapacitors directly store electrical energy and quickly release it to a device on demand. Hybrid vehicles use new battery technologies combined with internal combustion engines to improve fuel efficiency. Fuel cells also represent efficient ways to produce electricity and are becoming more popular for transportation and large-scale electricity production.

We hope that our energy discussions in this book have provided enough background that you have gained a perspective on the complexity of energy issues that we face. We also hope you are in a position to take stock of the situation and look ahead to the future. A few facts seem beyond debate. The world's thirst for energy will not diminish; it most assuredly will continue to grow. Moreover, how we currently generate energy is not sustainable. We have been the beneficiaries of a bountiful resource base from Earth. In turn, we must provide ample sources of energy with effective and low-cost means of energy storage for generations yet to come.

Learning Outcomes

The numbers in parentheses indicate the sections within the chapter where these outcomes were discussed.

Having studied this chapter, you should now be able to:

- explain how a battery works via redox reactions and flow of electrons between counter electrodes (8.1)

- determine the oxidation states for species involved in a redox reaction (8.1)

- use Ohm's law to determine characteristics of an electrical circuit (8.2)

- describe the characteristics and uses of various batteries (8.3)

- identify the redox reactions that occur within alkaline batteries (8.3)

- describe the design, operation, advantages, and limitations (including hazards) of rechargeable batteries (8.4)

- describe the design, operation, applications, and limitations (including hazards) of lead–acid batteries (8.5)

- explain how batteries can be used for grid energy storage applications (8.6)

- summarize the design, operation, advantages, and limitations (including hazards) of lithium-ion batteries (8.7)

- describe the structure of supercapacitors and how their properties and applications differ from batteries (8.8)

- compare the operation, advantages, and emissions of hybrid electric vehicles with conventional gasoline-powered vehicles (8.9)

- describe the design, operation, applications, and advantages of fuel cells (8.10)

- describe the methods used to generate hydrogen fuel and understand the limitations and challenges of using hydrogen for fuel cells (8.11)

- summarize the social, environmental, and economic costs of recycling energy storage devices (8.12)

Questions

Emphasizing Essentials

1. Define the terms *oxidation* and *reduction*. Why must these processes take place together?

2. Which of the following half-reactions represent oxidation and which reduction? Explain your reasoning.
   a. $Fe \longrightarrow Fe^{2+} + 2e^-$
   b. $Ni^{4+} + 2e^- \longrightarrow Ni^{2+}$
   c. $2\,Cl^- \longrightarrow Cl_2 + 2e^-$

3. Which chemical species gets oxidized and which gets reduced in the following overall chemical equation:

$$2\,Zn(s) + O_2(g) \longrightarrow 2\,ZnO(s)$$

4. What is the difference between a galvanic cell and a true battery? Give an example for each.

5. Two common units associated with electricity are the volt and the amp. What does each unit measure?

6. Consider the galvanic cell pictured. A coating of impure silver metal begins to appear on the surface of the silver electrode as the cell discharges.

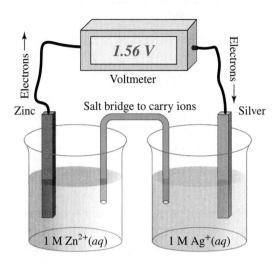

Voltmeter

1.56 V

Electrons →

Electrons ↓

Zinc   Salt bridge to carry ions   Silver

1 M $Zn^{2+}(aq)$ 1 M $Ag^+(aq)$

a. Identify the anode and write the oxidation half-reaction.

b. Identify the cathode and write the reduction half-reaction.

7. In the lithium–iodine cell, Li is oxidized to Li^+ and I_2 is reduced to $2\,I^-$.

a. Write the oxidation half-reaction and the reduction half-reaction that take place in this cell.

b. Write the overall reaction that occurs in the cell.

c. Identify the half-reaction that occurs at the anode and at the cathode.

8. **a.** How does the voltage from a tiny AAA alkaline cell compare with that from a large D alkaline cell? Explain.

b. Can both batteries sustain the flow of electrons for the same amount of time? Explain.

9. Identify the type of galvanic cell commonly used in each of these consumer electronic products. Assume none uses solar cells.

a. laptop computer

b. cell phone

c. digital camera

d. calculator

10. The mercury battery has been used extensively in medicine and industry. Its overall cell reaction can be represented by the following equation.

$$HgO(l) + Zn(s) \longrightarrow ZnO(s) + Hg(l)$$

a. Write the oxidation half-reaction.

b. Write the reduction half-reaction.

c. Why is the mercury battery no longer in common use?

11. **a.** What is the function of the electrolyte in a galvanic cell?

b. What is the electrolyte in an alkaline cell?

c. What is the electrolyte in a lead–acid storage battery?

12. These two *incomplete* half-reactions in a lead–acid storage battery do not show the electrons lost or gained. The reactions are more complicated, but it is still possible to analyze the reactions that take place.

$$Pb(s) + SO_4^{2-}(aq) \longrightarrow PbSO_4(s)$$
$$PbO_2(s) + 4\,H^+(aq) + SO_4^{2-}(aq) \longrightarrow PbSO_4(s) + 2H_2O(l)$$

a. Balance both equations with respect to charge by adding electrons as needed.

b. Which half-reaction represents oxidation and which reduction?

c. One of the electrodes is made of lead; the other is lead(IV) oxide. Which is the anode and which is the cathode?

13. During the conversion of $O_2(g)$ to $H_2O(l)$ in a fuel cell, the following half-reaction takes place.

$$\tfrac{1}{2}\,O_2(g) + 2\,H^+(aq) + 2e^- \longrightarrow H_2O(l)$$

Does this half-reaction represent an example of oxidation or reduction? Explain.

14. How does the reaction between hydrogen and oxygen in a fuel cell differ from the combustion of hydrogen and oxygen?

15. Find the resistance of a circuit that draws 0.10 A with 12 V applied.

16. A transformer is connected to 120 V. Find the current if the resistance is 480 ohms.

17. If a small appliance is rated at a current of 10 A and a voltage of 120 V, calculate the power rating of the appliance.

18. Jill runs her juicer every morning. The juicer uses 100 W of power and the supplied current is 5.0 A. How many volts are needed to run the juicer?

19. This diagram represents the hydrogen fuel cell that was used in some of the earlier space missions.

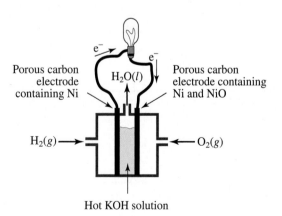

Porous carbon electrode containing Ni

$H_2O(l)$

e^- e^-

Porous carbon electrode containing Ni and NiO

$H_2(g) \longrightarrow$ $\longleftarrow O_2(g)$

Hot KOH solution

The chemistry in the hydrogen–oxygen fuel cell can be represented by these half-reactions:

$$H_2(g) \longrightarrow 2\,H^+(aq) + 2e^-$$
$$\tfrac{1}{2}\,O_2(g) + 2\,H^+(aq) + 2e^- \longrightarrow H_2O(l)$$

Which half-reaction takes place at the anode and which at the cathode? Explain.

20. What is a PEM fuel cell? How does it differ from the fuel cell represented in Question 15?

21. How do PEM fuel cells allow H_2 and O_2 to combine to form water without combustion?

22. In addition to hydrogen, methane has been studied for use in PEM fuel cells. Balance the given oxidation and reduction half-reactions, and write the overall equation for a methane-based fuel cell.

Oxidation half-reaction:

$$\_\_ CH_4(g) + \_\_ OH^-(aq) \longrightarrow \_\_ CO_2(g) + \_\_ H_2O(l) + \_\_ e^-$$

Reduction half-reaction:

$$\_\_ O_2(g) + \_\_ H_2O(l) + \_\_ e^- \longrightarrow \_\_ OH^-(aq)$$

23. Relative to a vehicle with an internal combustion engine, list two advantages offered by hydrogen FCVs.

24. Potassium and lithium both are reactive Group 1 metals. Both form hydrides, highly reactive compounds.

   a. Potassium reacts with H_2 to form potassium hydride, KH. Write the balanced chemical equation.

   b. KH reacts with water to produce H_2 and potassium hydroxide. Write the balanced chemical equation.

   c. Offer a reason why LiH (rather than KH) has been proposed as a means of storing H_2 for use in fuel cells.

25. What challenges keep hydrogen fuel cells from being a primary energy source for vehicles?

Concentrating on Concepts

26. Based on these data, describe the mathematical relationship between voltage and current. What is the resistor value for this model circuit?

27. Explain the concept of energy density of a battery and write out the energy density formula.

28. Describe how a normal AA battery stores and conducts electrons into useful energy.

29. List some differences between a rechargeable battery and one that must be discarded. Use a Ni—Cd battery and an alkaline battery as examples.

30. What is the difference between an electrolytic cell and a fuel cell? Explain, giving examples to support your answer.

31. Provide some differences between a lead–acid storage battery and a fuel cell.

32. Describe the importance of a separator in primary and secondary batteries. What would happen if the anode and cathode were allowed to touch inside a battery? Explain.

33. The company ZPower is promoting its silver–zinc batteries as replacements for lithium-ion batteries in laptops and cell phones.

   a. What advantages do silver–zinc batteries have over current Li-ion batteries?

   b. Write the oxidation and reduction half-reactions for ZPower's proposed battery using this overall cell equation as a guide. Indicate which reactant gets oxidized and which gets reduced.

$$Zn + Ag_2O \longrightarrow ZnO + 2\,Ag$$

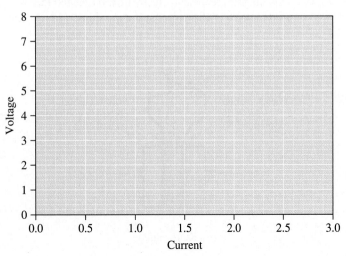

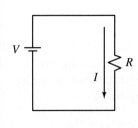

Voltage (V)	Current (mA)
0	0
1.0	0.18
2.0	0.35
3.1	0.56
4.0	0.73
4.9	0.90
6.0	1.1
7.1	1.3
8.0	1.5

34. The battery of a cell phone discharges when the phone is in use. A manufacturer, while testing a new "power boost" system, reported these data:

Time, minutes:seconds	Voltage, V
0:00	6.56
1:00	6.31
2:00	6.24
3:00	6.18
4:00	6.12
5:00	6.07
6:35	6.03
8:35	6.00
11:05	5.90
13:50	5.80
16:00	5.70
16:50	5.60

a. Prepare a graph of these data.

b. The manufacturer's goal was to retain 90% of its initial voltage after 15 minutes of continuous use. Has that goal been achieved? Justify your answer using your graph.

35. Assuming that HEVs are available in your area, draw up a list of at least three questions you would ask the auto dealer before deciding to buy or lease one. Offer reasons for your choices.

36. Describe some advantages and disadvantages of HEVs. If you owned an HEV, how would this affect your lifestyle?

37. You never need to plug in Toyota's gasoline–battery hybrid car to recharge the batteries. Explain.

38. Occasionally, the power goes out. When there is no electricity for an extended period of time, are HEVs, PHEVs, and EVs affected any differently than gasoline-powered vehicles?

39. Hydrogen is considered a relatively environmentally friendly fuel, producing only water when burned in oxygen. Name two positive effects that the widespread use of hydrogen would have on urban air quality.

40. Fuel cells were invented in 1839 but never developed into practical devices for producing electric energy until the U.S. space program in the 1960s. What advantages did fuel cells have over previous power sources?

41. Hydrogen and methane both can react with oxygen in a fuel cell. They also can be burned directly. Which has greater heat content when burned, 1.00 g of H_2 or 1.00 g of CH_4?

Hint: Write the balanced chemical equation for each reaction and use the bond energies in Table 6.1 to help answer this question.

42. Engineers have developed a prototype fuel cell that converts gasoline into hydrogen and carbon monoxide. The carbon monoxide, in contact with a catalyst, then reacts with steam to produce carbon dioxide and more hydrogen.

a. Write a set of reactions that describes this prototype fuel cell, using octane (C_8H_{18}) to represent the hydrocarbons in gasoline.

b. Speculate as to the future economic success of this prototype fuel cell.

43. Consider this representation of two water molecules in the liquid state.

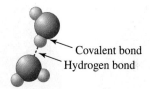

a. What happens when water boils? Does boiling break covalent bonds within molecules, or does it disrupt hydrogen bonds between molecules?

Hint: Revisit Section 5.2.

b. What happens when water undergoes electrolysis? Does this break covalent bonds within molecules, or does it disrupt hydrogen bonds between molecules?

44. Describe some similarities and differences between supercapacitors and traditional batteries.

45. After reading this chapter, which battery source do you think is the most efficient and most effective for your daily activities? Explain.

46. Why isn't the electrolysis of water the best method to produce hydrogen gas?

47. Describe three different ways scientists can extract hydrogen from different compounds found on Earth. Comment on the relative sustainability of each of these methods.

48. Small quantities of hydrogen gas can be prepared in the lab by reacting metallic sodium with water, as shown in this equation:

$$2\ Na(s) + 2\ H_2O(l) \longrightarrow H_2(g) + 2\ NaOH(aq)$$

a. Calculate the grams of sodium needed to produce 1.0 mol of hydrogen gas.

b. Calculate the grams of sodium needed to produce sufficient hydrogen to meet an American's daily energy requirement of 1.1×10^6 kJ.

c. If the price of sodium were \$165/kg, what would be the cost of producing 1.0 mol of hydrogen? Assume the cost of water is negligible.

49. **a.** As a fuel, hydrogen has both advantages and disadvantages. Set up parallel lists for the advantages and disadvantages of using hydrogen as the fuel for transportation and for producing electricity.

   b. Do you advocate the use of hydrogen as a fuel for transportation or for the production of electricity? Explain your position in a short article for your student newspaper.

Exploring Extensions

50. Although Alessandro Volta is credited with the invention of the first electric battery in 1800, some feel this is a reinvention.

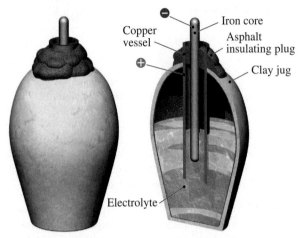

A terracotta pot, rolled copper sheet, and an iron rod, which may date back to the Parthian period (between 250 BCE and 224 CE).
Javier Jaime/123RF

   Using the Internet, summarize some arguments for and against this device being considered a battery. What are some applications that have been proposed for this so-called "Baghdad battery"?

51. Oxidation and reduction also take place during combustion, the process of burning a fuel in oxygen. Because no metal electrodes are present during combustion, the electron transfer is harder to track. In this case, oxidation occurs when a chemical species loses H atoms or gains O atoms. Similarly, reduction occurs when a chemical either gains H atoms or loses O atoms.

   a. Use these new definitions to determine which species is oxidized and which is reduced in the equation below, the combustion of hydrogen. Explain.

   $$H_2(g) + \tfrac{1}{2} O_2(g) \longrightarrow H_2O(g)$$

   b. Determine which species is oxidized and which is reduced in each of the following combustion reactions. Explain.

   $$C + O_2 \longrightarrow CO_2$$
   $$2\,C_8H_{18} + 17\,O_2 \longrightarrow 16\,CO + 18\,H_2O$$

52. "The Earth is a metal-rich rock. I can't see the human race running out of metals when it will be possible to mine in new places or recycle or simply reduce consumption. We probably won't be able to live on the planet due to global warming or other environmental problems before we run into a metal supply problem." These comments were offered by geologist Maurice A. Tivey of Woods Hole Oceanographic Institution in an article in *Chemical & Engineering News* published in June 2009.

   a. Do you agree with the writer's sentiment about not running out of metals? Explain.

   b. Name two challenges connected with increasing the recycling of batteries.

53. What is *the tragedy of the commons*? How does this concept apply to our practice of using metals such as mercury and cadmium in batteries?

54. How can the principles of green chemistry be applied during the development of new technologies for batteries and fuel cells? Give three specific examples.

55. If all of today's technology presently based on fossil fuel combustion were replaced by hydrogen fuel cells, significantly more H_2O would be released into the environment. Is this of concern? Research other consequences that might be anticipated from switching to an economy powered by hydrogen, a so-called hydrogen economy.

56. Consider these three sources of light: a candle, a battery-powered flashlight, and an electric lightbulb. For each source, provide

   a. the origin of the light.

   b. the immediate source of the energy that appears as light.

   c. the original source of the energy that appears as light. *Hint:* Trace this back stepwise as far as possible.

   d. the end products and by-products produced from using each.

   e. the environmental costs associated with each.

   f. the advantages and disadvantages of each light source.

57. Iceland is taking bold steps to cut its ties to fossil fuels. Part of the plan is to demonstrate that the country can produce, store, and distribute hydrogen to power both public and private transportation.

Ragnar Th Sigurdsson/Arctic Images/Alamy Stock Photo

a. Name three factors that motivate Iceland to cut its ties to fossil fuels.

b. What tangible outcomes have resulted to date?

c. Can lessons learned in Iceland be relevant where you live? Explain.

58. What are some differences between prismatic, cylindrical, and pouch cells for Li-ion batteries and which is the preferred design for automotive applications?

RECYCLING PLASTICS

Watch the chapter opening video (www.acs.org/cic), which shows how a soft drink bottle may be recycled into clothing. After watching that video, look at the plastic items around you and answer these questions:

a. Find three different items made from plastic with three different recycling symbols (a number within a triangle).

b. What do these symbols mean?

In this chapter, you will explore the following questions:

1. Where can you find polymers in your everyday life?
2. What is the general structure of polymers?
3. How is gaseous ethylene converted into solid polyethylene?
4. How does the structure of polyethylene govern its properties and applications?
5. What are the structures, properties, and applications of the six most common types of polymers?
6. How are polyesters formed?
7. How is nylon synthesized?
8. What are "the four R's" of recycling?
9. How are polymers recycled?
10. How do plant-based polymers differ from those derived from petroleum?
11. What are some environmental and health implications of using polymers?

The EVOLUTION of CELL PHONES

1980   1985   1995   2002   NOW

Source: https://netherregioniii.files.wordpress.com/2012/06/evolutionofcellphone.jpg

Introduction

Imagine carrying a brick around in your pocket wherever you go. A rough, hard, heavy brick that was your connection to the digital world. It would barely fit in your pocket, if at all. Rubbing your hand or face across the brick would scratch your skin. Dropping it would cause it to shatter, potentially hurting your feet. Without polymers, this would be what your cell phone would look and feel like. Not a convenient tool to carry around to connect with family and friends, but an inefficient, heavy anchor that would likely remain unused and might be easily damaged.

Throughout history, humans have continually improved the tools around them to better conform to their needs and desires by making materials stronger, lighter, and more durable. One class of materials that has transformed our world is known as *polymers*. By definition, a **polymer** is a chemical compound that is in the form of long chains of smaller, repeating units called *monomers*. Some polymers are natural, such as spider silk; others are human-made, or synthetic, such as Kevlar. In either case, polymers are large molecules made from many small starting materials. Human-made, or synthetic, polymers are amazingly tailorable and can be used to create materials with properties ideal for different applications. Need a lightweight but strong material that resists scratches? There's a polymer for that. Need a smooth surface that won't irritate skin or clothing? How about a material that is slightly bendable and doesn't break when dropped? All of these properties can be engineered into a synthetic polymer.

Whereas synthetic polymers were first developed in the early 20th century, natural polymers have been around since life itself began. Cellulose, starch, and other complex carbohydrates are examples of natural polymers. Natural rubber is a polymer obtained from rubber trees. Even the code for life itself, DNA, is a natural polymer.

In this chapter, you will learn about what makes up a polymer, how the structure of the polymer can give rise to a variety of properties, how these amazing materials are created, and how products made of polymers can be recycled into other materials for further use.

9.1 | Polymers Here, There, and Everywhere

Learning Objective: Compare the properties and applications of plastics and polymers

Polymers have revolutionized the world around us, especially in sports. For example, football is often played on artificial turf made from recycled plastics, and hockey players skate on rinks of Teflon or high-density polyethylene when natural ice is unavailable. Polymers are also important in the construction of athletic equipment. In tennis, tennis balls are made from synthetic polymers. In cycling, carbon fibers embedded in plastic resins provide the strength, flexibility, and lightweight construction required in bicycles. Polymers are also useful for protecting athletes. For instance, helmets made of plastic protect the players' heads in a wide variety of sports. Look for some examples of polymers in the following activity.

Your Turn 9.1   Tennis Anyone?

Examine the photo of a tennis player.

a. Choose three applications of polymers in the photo.
b. For the polymers you identified in **a**, describe some characteristics that are important for their intended use.

Maridav/Shutterstock

High Impact Photography/iStockphoto/Getty Images

In the previous paragraph, notice that we referred to both *polymers* and *plastics*. These two terms are related, and are often used interchangeably. Whereas the word *polymer* encompasses both natural and synthetic compounds, the term *plastic* is appropriate only for some synthetic varieties. So, all plastics are polymers, but not all polymers are plastics.

Although polymers are everywhere, you may need to train your eye to recognize them. Not all have the look and feel of a yellow rubber duck! Some are transparent, such as the clear plastic wrap on food, whereas others are opaque, such as the container that holds liquid laundry detergent. Some are rigid, such as nylon automotive parts, while others are more flexible, such as a plastic spatula. Some polymers are drawn into fibers to weave clothing and carpet, whereas others are molded into different shapes.

In principle, synthetic polymers can be made from many different starting materials. In practice, most come from a single raw material: crude oil. As you learned in Section 6.9, oil is no longer as easy to obtain on our planet as it used to be. Today, crude oil is the starting material for many plastics, pharmaceuticals, fabrics, and other carbon-based products. However, plastics can also come from renewable materials, as we will discuss toward the end of this chapter.

Both the origin and fate of polymers are of interest to us. For instance, modern cell phones and electronics contain many plastic components. Of the estimated 7 billion tons of plastic waste that have been generated to date, less than 10% has been recycled! To understand the complexities surrounding the sources and ultimate fate of polymers, you first need to know something about their chemical structures and how they are made—the topic of the next section.

9.2 | Polymers: Long, Long Chains

Learning Objective: Describe the general structure of polymers

Rayon, nylon, and polyurethane. Teflon, Spandex, Styrofoam, and Formica. These common, but seemingly different materials, are all synthetic polymers. What they have in common is most easily evident at the molecular level. Each of these polymers consists of long chains of atoms covalently bonded to one another. A polymer can easily contain thousands of atoms, and have a molar mass of more than 1,000,000 g/mol! As such, polymers are often referred to as macromolecules. It is common to use the unit dalton, Da, to express the molar mass of macromolecules such as polymers. For instance, a molar mass of 10,000 g/mol would be equivalent to 10 kDa.

Monomers (*mono* meaning "one"; *meros* meaning "unit") are the small molecules used to synthesize polymers. Each monomer is analogous to a link in a chain. Polymers (*poly* means "many") can be formed from one monomer, or from a combination of two or more different monomers. The long chain shown in Figure 9.1 may help you to imagine a polymer made from identical monomers; that is, identical links in a chain.

Keep in mind that chemists did not invent polymers. For example, the natural polymers of glucose, cellulose, and starch were described earlier in the context of biofuels (Section 6.14). Other natural polymers include wool, cotton, silk, natural rubber, skin, and hair. Like synthetic polymers, natural ones exhibit a stunning variety of properties. They give strength to an oak tree, delicacy to a spider's web, softness to goose down, and flexibility to a blade of grass (Figure 9.2).

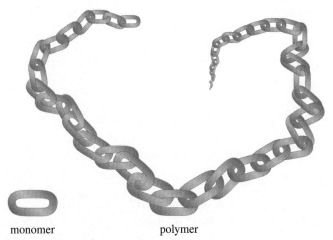

monomer   polymer

Figure 9.1

Representations of a monomer (single link) and a polymer (long chain) made from one type of monomer.

Figure 9.2

Oak logs and grass both contain the natural polymer cellulose. Glucose is the monomer.

(both): ©Cathy Middlecamp, University of Wisconsin

Some early synthetic polymers were developed as substitutes for expensive or rare natural polymers such as silk and rubber. Others were developed to deliver comparable strength at a lower mass. Contrast the density of steel (about 8 g/cm^3) with that of plastics (1–2 g/cm^3). As a result, an automobile body constructed with plastics weighs less than its steel counterpart, thus requiring less fuel to operate. Similarly, plastic packaging reduces weight and helps save fuel during shipping.

@HOME   Plastics from Milk?

Investigate how to make your own biodegradable plastic at home using whole milk, food coloring, a stovetop, a saucepan, white vinegar, a strainer, and paper towel.

Stephanie Ryan

1. Add two cups of whole milk to a saucepan and heat it on a stovetop until it is very hot but not boiling. Add a drop of food coloring to the milk if you want your bioplastic to have a color.
2. Add three tablespoons of white vinegar to the milk and stir.
3. Pour the mixture through a strainer. Be careful because it will be very hot. There will be clumps collecting in the strainer. This is your bioplastic.
4. Rinse your bioplastic with water and after several minutes of cooling, place a paper towel over it to soak up excess liquids.
5. Press the bioplastic into any shape you want it to be. Set it aside on a paper towel to dry for a few days.

Questions to Consider
a. Using the Internet, what is the monomer that is found in your bioplastic?
b. What other types of biodegradable plastics have been discovered? Why are they appealing?

9.3 | Adding Up the Monomers

Learning Objective: Explain the mechanism of addition polymerization reactions

How do monomers combine to make a polymer? In the previous section, we used a chain to represent a polymer but made no mention of how the chain was formed. In this section, we will provide details of how covalent bonds connect the monomers.

Polyethylene is our first example. As the name indicates, polyethylene is a polymer of ethylene, $H_2C=CH_2$ (Equation 9.1). Ethylene is a common name for ethene, the smallest member in the family of hydrocarbons with chemical formula C_xH_{2x}, containing a $C=C$ double bond. In the polymerization reaction, n molecules of the ethylene monomer combine to form polyethylene:

$$n \begin{array}{c} H \\ | \\ C \\ | \\ H \end{array} = \begin{array}{c} H \\ | \\ C \\ | \\ H \end{array} \xrightarrow{R\cdot} \left[\begin{array}{cc} H & H \\ | & | \\ C & C \\ | & | \\ H & H \end{array} \right]_n \qquad [9.1]$$

The coefficient n in front of the ethylene monomer specifies the number of molecules reacting. In turn, this determines the molar mass of the polymer, typically between 10,000 and 100,000 g/mol, but can run into the millions. On the right side, the n appears as a subscript, indicating that each monomer has become part of the long chain. The large square brackets enclose the repeating unit of the polymer.

Polyethylene is the sole product. The monomers add to one another to form a long chain of n units. As a result, we call this **addition polymerization**, a type of polymerization in which the monomers add to the growing chain in such a way that

(a)   (b)

Figure 9.3

(a) Bottles made from polyethylene.
(b) A sign posted on a railway tank car that transports liquefied ethylene. The number 1038 identifies it as ethylene, the red diamond indicates high flammability, and the number 2 indicates moderate reactivity.

(a): ©Cathy Middlecamp, University of Wisconsin

the polymer contains all the atoms of the monomer. Industrial chemists use several synthetic routes to produce polyethylene. The most common uses a metal catalyst.

Notice the R· over the arrow in Equation 9.1. So that you can better appreciate its significance, we will tell you a bit more about ethylene, the monomer. Produced at oil refineries, ethylene is a flammable, colorless gas with a faint gasoline-like odor. These properties are quite unlike the polyethylene product, which is an odorless solid (Figure 9.3a). Although not classified as an air pollutant, ethylene nonetheless is a VOC (volatile organic compound). As you learned in Chapter 2, VOCs in the atmosphere are precursors to photochemical smog. Accordingly, safety precautions are needed when transporting ethylene from refineries to sites at which polyethylene is produced. To conserve space, the ethylene gas is pressurized and refrigerated to liquefy it. In this form, it is transported in tank cars that bear labels like the one shown in Figure 9.3b.

Does liquid ethylene polymerize in the tank car? Fortunately, no. Clearly, the end user of the ethylene would be distressed to receive a tank car full of solid polyethylene! In order to initiate the polymerization reaction, a free radical (R·) is required, as shown over the arrow in Equation 9.1. This free radical represents one of a variety of chemical species (e.g., ·OCH$_3$, ·OH), all with an unpaired electron. Recall that the hydroxyl free radical, ·OH, was described in Section 3.7.

To initiate the process of forming the polymer chain, the R· species attaches to H$_2$C═CH$_2$ (Figure 9.4). To understand what happens next, recall that the double bond in ethylene contains *four* electrons. After an ethylene molecule reacts with R·, only *two* of these electrons remain in a C—C single bond. The other *two* electrons move (shown by the red arrows) to form two new bonds—one to R·, and the other as an unpaired electron at the end of the molecule, thus providing a site at which another monomer can add.

As each ethylene monomer adds, a new C—C bond forms and the chain grows. This process repeats many times. Occasionally, the ends of two polymer chains join and stop the chain growth. Polymerization stops when the supply of monomers is exhausted. The result of all this chemistry is that gaseous ethylene is converted into solid polyethylene.

Figure 9.4

Scheme showing the mechanism for the polymerization of ethylene.

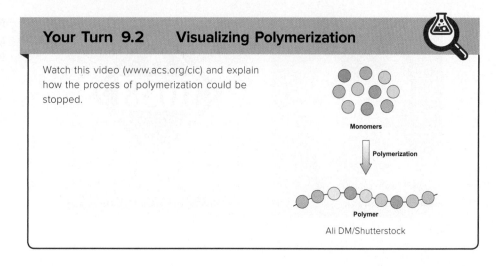

Your Turn 9.2   Visualizing Polymerization

Watch this video (www.acs.org/cic) and explain how the process of polymerization could be stopped.

Monomers

Polymerization

Polymer

Although we placed R· over the arrow in Equation 9.1, we also could have represented the reaction in this way:

$$2\,R\cdot + n \ \begin{matrix} H \\ \diagdown \\ C \end{matrix}\!=\!\begin{matrix} H \\ \diagup \\ C \end{matrix} \longrightarrow R\!-\!\!\left[\begin{matrix} H & H \\ | & | \\ C & C \\ | & | \\ H & H \end{matrix}\right]_n\!\!-\!R \qquad\qquad [9.2]$$

Because the R group that "caps" each end of the molecule is such a small part of the much longer chain, it is common practice to omit R· as a reactant, as we showed in Equation 9.1.

The numerical value of n, and hence the length of the chain, can vary. During the manufacturing process, n will be adjusted in order to create specific properties for the polymer. Moreover, the individual polymer molecules can have varying lengths within a single sample. In every case, however, the molecules contain a chain of carbon atoms. In essence, the molecules in polyethylene resemble those in a hydrocarbon such as octane, except that they are much, much longer.

Your Turn 9.3   Polymerization of Ethylene

In Equation 9.1 for the polymerization of ethylene, assume that $n = 4$.

a. Rewrite Equation 9.1 to indicate this change.
b. Draw the structural formula of the product without using brackets. This will require adding an R group at each end of the chain.
c. In terms of its molecular structure, how does the product differ from octane?
d. Polyethylene is also called polyethene in the United Kingdom, reflecting the fact that ethene (not ethylene) is the systematic name. As you saw in Section 2.5, the prefix *eth-* indicates two carbon atoms; the *-ene* suffix indicates a C=C double bond. What would the common and systematic names be for the three-carbon analogue, $H_2C\!=\!CH\!-\!CH_3$?

9.4 | Got Polyethylene?

Learning Objective: Describe how polyethylene properties depend on their molecular structures and intermolecular forces

Polyethylene is found in many packaging materials, including plastic bags, milk jugs, detergent containers, and "bubble wrap" (Figure 9.5). Yet, as we have seen in the previous section, all polyethylene is made from the monomer ethylene. How can these materials with seemingly different properties all be made from the same substance?

Figure 9.5

Packing material and containers made from polyethylene.

All: Jill Braaten/McGraw Hill

Your Turn 9.4   Polyethylene Hunt

As we will describe in this section, polyethylene containers, bags, and packaging materials are marked either as low-density (LDPE) or as high-density (HDPE). Using the recycling code as your guide, locate several items made from each. Do LDPE and HDPE differ in flexibility? Is one more translucent? Is one more often colored with a pigment than the other? Summarize your findings in a brief report.

LDPE   HDPE

The different properties of polyethylene stem largely from differences in their long molecular chains. Relatively speaking, these molecules are very long indeed. As an analogy, if a polyethylene molecule were as wide as a piece of spaghetti, its length would be almost a half-mile long! In fact, the actual width of a polyethylene molecule is about 0.5 nm—i.e., 0.001% of an individual hair fiber! Now imagine that the polyethylene used to make plastic bags contains molecular chains arranged somewhat like cooked spaghetti on a plate. The strands are not well aligned, although in some regions the molecular chains run parallel to one another. Moreover, the polyethylene chains, like the spaghetti strands, are not covalently bonded to one another.

Recall that in Section 6.11 London dispersion forces were introduced to explain what attractive forces hold hydrocarbons together and give rise to their varying boiling points. These intermolecular forces differ from the covalent bonds that exist *within* molecules arising from shared pairs of electrons. Consequently, chemical changes are governed by the strengths of covalent bonds, whereas physical changes are as a result of intermolecular forces.

London dispersion forces are present in all molecules, but are significant in large molecules, such as polymers. These dispersion forces arise in a polymer because each atom in the long polymeric chain contains electrons. These electrons are attracted to the atoms on neighboring chains; the degree of attraction between strands of polyethylene results from the large number of atoms involved. The attraction is a bit like that between the two halves of Velcro. The larger the surface area of one Velcro strip, the better it will hold to the other. Individual intermolecular forces between atoms are very small, but polymers have significant intermolecular forces because of the large number of these interactions. Discover evidence of the molecular arrangement of polyethylene by doing this next short experiment.

Figure 9.6 represents the necking of polyethylene from a molecular point of view. As the strip narrows, the molecular chains shift, slide, and align parallel to one another in the direction of pull. In some plastics, such stretching (sometimes called "cold drawing") is carried out as part of the manufacturing process to alter the three-dimensional

Figure 9.6

A representation of "necking" at the molecular level.

@HOME   Polyethylene "Necking"

Investigate the molecular arrangement of
polyethylene at home using scissors and a
heavy-duty polyethylene bag.

1. Cut a strip from a heavy-duty polyethylene
   bag using scissors.
2. Grab the two ends of the strip and pull.
3. Make observations.

Questions to Consider

©Bradley D. Fahlman

a. As you should have noticed, when stretching
   polyethylene, a small shoulder forms on the
   wider part of the strip and a narrow neck almost seems to flow from it in a process
   called "necking." Does necking affect the number of monomer units, *n*, in the average
   polymer?
b. Does necking affect the bonding between the monomer units?
c. How does necking differ from stretching a rubber band?

arrangement of the chains in the solid. As the force and stretching continue, the polymer eventually reaches a point at which the strands can no longer realign, and the plastic breaks. Paper, a natural polymer, tears when pulled because the strands (fibers) are rigidly held in place and are not free to slip like the long molecules in polyethylene.

Differences in the physical properties of polymers can also arise as a result of the extent of branching within the polymer chain. This is the case with high-density polyethylene (HDPE) and low-density polyethylene (LDPE), as shown in Figure 9.7. As you may have discovered in Your Turn 9.4, the plastic bags dispensed in the produce aisles of supermarkets are usually LDPE. These bags are stretchy, transparent, and not very strong. Their molecules consist of about 500 monomeric units and the central polymeric chain has numerous branches, like limbs radiating from a central tree trunk (Figure 9.7b).

(a)

HDPE   LDPE

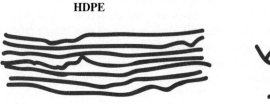

(b)

Figure 9.7

High-density (linear) polyethylene and low-density (branched) polyethylene: **(a)** structural formulas, and **(b)** schematic representations.

Your Turn 9.5   Envisioning a Future Without Single-Use Plastic Bags

Many cities, a few states, and even whole countries have enacted legislation to ban single-use plastic bags. There are supporters and opponents to banning bags. What problems are caused by using single-use plastic bags?

After learning about the problems single-use plastic bags cause, what if you had the ability to time travel to see what the future would be like ten years from now in a community that does not allow single-use plastic bags? Share what this community would be like.

This low-density form was the first type of polyethylene to be manufactured. About 20 years after its discovery, chemists were able to adjust reaction conditions to prevent branching, and thus HDPE was born. In their Nobel Prize–winning research, Karl Ziegler (1898–1973) and Giulio Natta (1903–1979) developed catalysts that enabled them to make linear (unbranched) polyethylene chains of about 10,000 monomer units. With no side branches, these long chains arranged in parallel, unlike the irregular tangle of the polymer chains in LDPE (Figure 9.7b). HDPE exhibits a highly ordered molecular structure and has a slightly higher density, greater rigidity, more strength, and a higher melting point than LDPE.

Your Turn 9.6   HDPE and LDPE

The densities of HDPE and LDPE are 0.96 g/cm³ and 0.93 g/cm³, respectively. Use Figure 9.7 to rationalize the slight difference in densities.

As you might expect, HDPE and LDPE have different uses. High-density polyethylene is used to make many different types of plastic bottles, toys, stiff or "crinkly" plastic bags, and heavy-duty pipes. A newer use of HDPE was spurred by surgery patients with blood-borne diseases, such as HIV/AIDS. Without suitable protection, surgeons would run the risk of being infected. AlliedSignal, Inc. produced a linear polyethylene fiber called Spectra that could be fabricated into liners for surgical gloves. These gloves are reported to have 15 times more resistance to cuts than medium-weight leather work gloves, but are still thin enough to allow a keen sense of touch. A sharp scalpel can be drawn across the glove with no damage to the fabric. Such strength is in marked contrast to the properties of everyday plastic gloves used by health professionals.

It would be a mistake to conclude that polyethylene is restricted to the extremes represented by highly branched or strictly linear forms. By modifying the extent and location of branching in LDPE, its properties can be varied from the soft and wax-like coatings on milk cartons to stretchy plastic food wrap. HDPE is rigid enough to be used for plastic milk bottles. The hot water of a dishwasher will melt neither HDPE nor LDPE, but either may melt if left near a hot frying pan or heating element. Polyethylene has one more property of interest—namely, that it is a good electrical insulator. During World War II, polyethylene was used by the Allied forces to coat electrical cables in aircraft radar installations. Discover other types of polyethylene and their properties in this next activity.

Your Turn 9.7   Other Types of Polyethylene

In addition to LDPE and HDPE, polyethylene is manufactured as "MDPE" and "LLDPE." Use the Internet to find out about these and other types of polyethylene. How do their properties differ from those of LDPE and HDPE?

9.5 | The "Big Six": Theme and Variations

> *Learning Objective: Compare the properties and applications for common types of polymers*

Today, more than 60,000 synthetic polymers are known. Although polymers were developed for many specialized uses, six types account for roughly 75% of those used in both Europe and the United States. We refer to these everyday polymers as the "Big Six," which include polyethylene (low- and high-density), polyvinyl chloride, polystyrene, polypropylene, and polyethylene terephthalate. Table 9.1 provides details about the monomers, properties and applications of these polymers.

Table 9.1	The Big Six		
Polymer Recycle Symbol	**Monomer**	**Properties of Polymer**	**Uses of Polymer**
Polyethylene ⟳ 4 LDPE	Ethylene $H_2C=CH_2$	Translucent if not pigmented. Soft and flexible. Unreactive to acids and bases. Strong and tough.	Bags, films, sheets, bubble wrap, toys, wire insulation.
Polyethylene ⟳ 2 HDPE	Ethylene $H_2C=CH_2$	Similar to LDPE. More rigid, tougher, slightly more dense.	Opaque milk, juice, detergent, and shampoo bottles. Buckets, crates, and fencing.
Polyvinyl chloride ⟳ 3 PVC, or V	Vinyl chloride $H_2C=CHCl$	Variable. Rigid if not softened with a plasticizer. Clear and shiny, but often pigmented. Resistant to most chemicals, including oils, acids, and bases.	Rigid: Plumbing pipe, house siding, charge cards, hotel room keys. Softened: Garden hoses, waterproof boots, shower curtains, IV tubing.
Polystyrene ⟳ 6 PS	Styrene	Variable. "Crystal" form transparent, sparkling, somewhat brittle. "Expandable" form lightweight foam. Both forms rigid and degraded in many organic solvents.	"Crystal" form: Food wrap, CD cases, refrigerator shelves, transparent cups. "Expandable" form: Foam cups, insulated containers, food packaging trays, egg cartons, packaging peanuts.
Polypropylene ⟳ 5 PP	Propylene	Opaque, very tough, good weatherability. High melting point. Resistant to oils.	Bottle caps. Yogurt, cream, and margarine containers. Carpeting, casual furniture, luggage.
Polyethylene terephthalate ⟳ 1 PETE, or PET	Ethylene glycol $HO-CH_2CH_2-OH$ Terephthalic acid	Transparent, strong, shatter-resistant. Impervious to acids and atmospheric gases. Most costly of the six.	Soft-drink bottles, clear food containers, beverage glasses, fleece fabrics, carpet yarns, fiber-fill insulation.

Note: The structures of the first five monomers differ only by the atoms shown in red.

Note: Terephthalate is pronounced "ter-eh-THAL-ate." The "ph" is silent.

All of the above polymers are solids that can be colored with pigments. All are also insoluble in water, although some degrade or soften in the presence of hydrocarbons, fats, and oils. The Big Six are classified as **thermoplastic polymers**, meaning that with heat, they can be melted and reshaped over and over again. However, they exhibit a range of melting points depending on the route by which they were manufactured. Of the Big Six, polyethylene has the lowest melting point, with LDPE and HDPE melting at about 120 °C and 130 °C, respectively. In comparison, polypropylene (PP) melts at 160–170 °C.

In contrast to thermoplastics, some plastics are known as **thermosets**. These solidify or "set" irreversibly with heat, allowing them to be used in higher temperature environments. Examples include rubber-soled footwear and antique Bakelite ovenware.

Depending on the arrangement of their molecules, polymers have varying degrees of strength. At the microscopic level, the molecules in some parts of the polymer may have an orderly repeating pattern, such as one would find in a crystalline solid (Figure 9.8). In these **crystalline regions**, the long polymer molecules are arranged neatly and tightly in a regular pattern. In other parts of the same polymer, you may find **amorphous regions**. Here, the long polymer molecules are found in a random, disordered arrangement and are packed more loosely. Because of their structural regularity, the crystalline regions impart strength and abrasion resistance, such as in HDPE and PP. Although some polymers are highly crystalline, most still include amorphous regions. These regions impart flexibility. For example, the amorphous regions in PP give it the ability to be bent without breaking. The range of properties among polymers means that they are differently suited for specific applications. The next activity provides an opportunity to match polymers with their uses.

Thermoplastics vs. Thermosets

Watch this video (www.acs.org/cic), which shows what happens to thermosets and thermoplastics at high temperature.

Questions to Consider
a. Using the Internet as a resource, are thermosets or thermoplastics easier to recycle? Explain.
b. What are some common types and applications of thermoset plastics?

Your Turn 9.8   Uses of the Big Six

Use Table 9.1 and other information provided about the Big Six to answer these questions.

a. Which polymer would not be suitable for margarine tubs because it softens with oil?
b. Which polymers are transparent? Which one is used in clear soft-drink bottles?
c. Which one is tough and used for bottle caps? Name another application in which toughness is important.
d. Which ones are listed as unreactive to acids and can serve as containers for acidic beverages, such as orange juice?

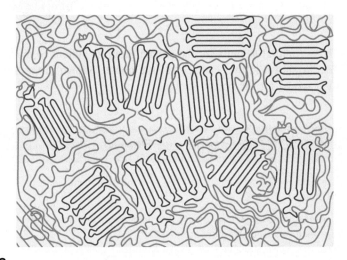

Figure 9.8

A semicrystalline polymer with crystalline regions (**red**) and amorphous regions (**green**).

From Table 9.1, you can see that five monomers are used to make six different polymers. Here, we focus on three monomers structurally related to ethylene: vinyl chloride, propylene, and styrene.

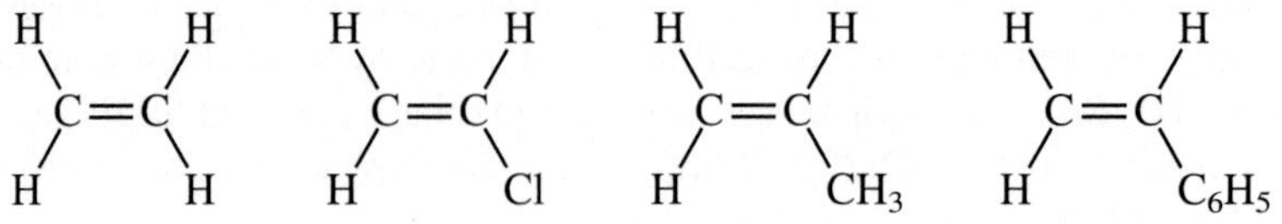

$$
\begin{array}{cccc}
\overset{\displaystyle H}{\underset{\displaystyle H}{}}C=C\overset{\displaystyle H}{\underset{\displaystyle H}{}} & \overset{\displaystyle H}{\underset{\displaystyle H}{}}C=C\overset{\displaystyle H}{\underset{\displaystyle Cl}{}} & \overset{\displaystyle H}{\underset{\displaystyle H}{}}C=C\overset{\displaystyle H}{\underset{\displaystyle CH_3}{}} & \overset{\displaystyle H}{\underset{\displaystyle H}{}}C=C\overset{\displaystyle H}{\underset{\displaystyle C_6H_5}{}}
\end{array}
$$

ethylene   vinyl chloride   propylene   styrene

Your Turn 9.9   Identifying Monomers

Examine the structures of the monomers used in PVC, Saran, and Teflon shown below.

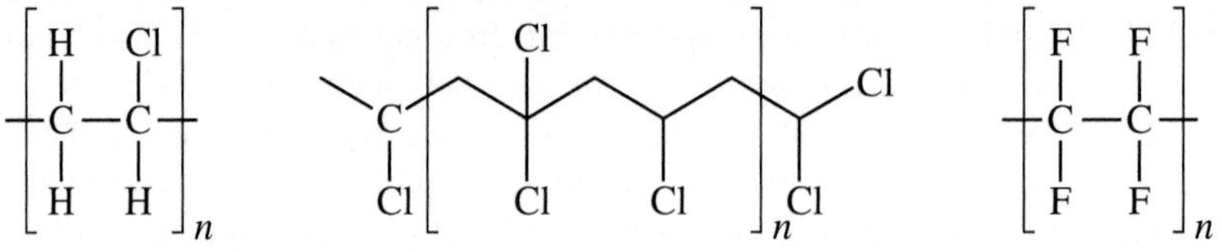

PVC   Saran   Teflon

Compare the repeating units (monomers) of PVC, Saran and Teflon.

a. In what ways are the monomers alike? In what ways do they differ?
b. What are the names of the monomers for PVC, Saran, and Teflon? How do the properties of the polymers differ from those of their monomers?

As seen in the above activity, in vinyl chloride, one of the H atoms of ethylene is replaced by a Cl atom. Similarly, in propylene, one of the H atoms of ethylene is replaced by a *methyl* group ($-CH_3$). In contrast, styrene features a phenyl group, $-C_6H_5$, which replaces one of the H atoms. The *phenyl* group consists of six carbon atoms arranged to form a hexagon:

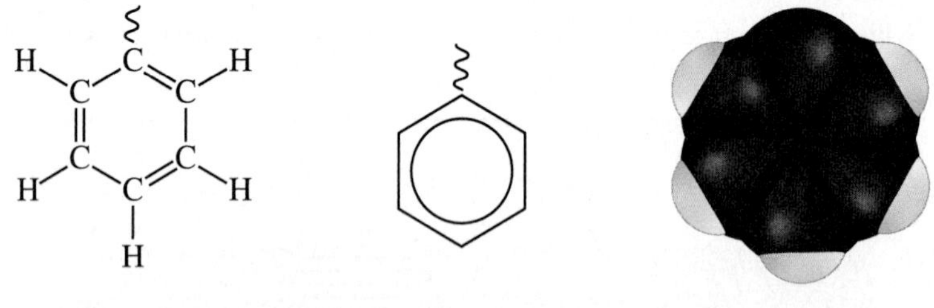

Because the first structural formula for the phenyl group is tedious to draw, the ring is often simplified, as shown in the second structure above. Shown in the third structure is a space-filling model for the phenyl group.

Your Turn 9.10   Benzene and Phenyl

The difference between a phenyl group, $-C_6H_5$, and the compound benzene, C_6H_6, is simply one H atom.

a. Both the phenyl group and benzene have two *resonance* structures. Draw them.
   Hint: Resonance was introduced in Section 3.6.
b. Given these resonance structures, why is the shorthand symbol of a circle within a hexagon a particularly good representation of both benzene and the phenyl group?

As you might suspect, vinyl chloride, propylene, and styrene undergo addition polymerization, just like ethylene. But the results are somewhat different. To see why, let's look at what happens when n molecules of vinyl chloride polymerize to form polyvinyl chloride, PVC.

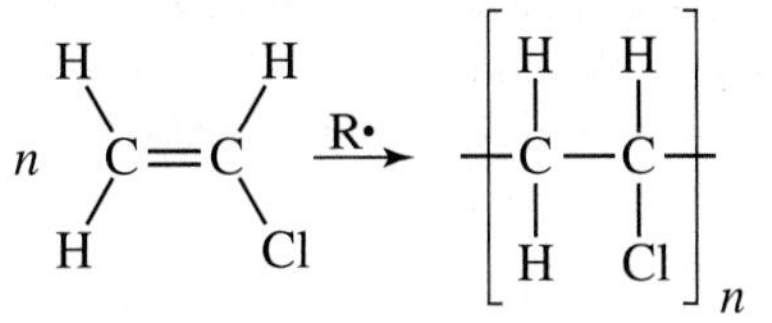

$$n \; \overset{H}{\underset{H}{}}C\!=\!C\overset{H}{\underset{Cl}{}} \; \xrightarrow{R\cdot} \; \left[\; \overset{H}{\underset{H}{}}\overset{|}{\underset{|}{C}} - \overset{H}{\underset{Cl}{}}\overset{|}{\underset{|}{C}} \; \right]_n \qquad \text{[9.3]}$$

In Equation 9.3, the Cl atom could be drawn in any of the four positions that attach to the C atoms in the vinyl chloride monomer. The Cl atom creates an asymmetry in the monomer. Think of the carbon atom bearing two H atoms as the "tail" and the carbon with the Cl atom as the "head."

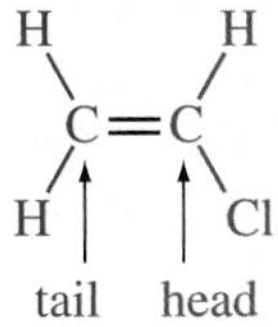

$$\overset{H}{\underset{H \uparrow}{}}C\!=\!C\overset{H}{\underset{\uparrow Cl}{}}$$

tail   head

When vinyl chloride monomers add to form polyvinyl chloride, they orient in one of three ways, as shown in Figure 9.9:

- head-to-tail, with the Cl atoms on every other C atom
- alternating head-to-head/tail-to-tail, with the Cl atoms next to each other
- a random mix of the previous two arrangements

The head-to-tail arrangement is the usual product for polyvinyl chloride.

The arrangement of monomers in the chain is one factor that affects the flexibility of the polymer. Thus, each arrangement of PVC has somewhat different properties, with the most regular one—repeating head-to-tail—being the stiffest because the molecules pack more easily together to form crystalline regions. The stiffer PVC finds use in drain and sewer pipes, credit cards, house siding, furniture, and various automobile parts. The random arrangement is still stiff, but somewhat less so.

PVC can be further softened with **plasticizers**, compounds that are added in small amounts to polymers to make them softer and more pliable. Plasticizers work by fitting in between the large polymer molecules, thus disrupting the regular packing of the

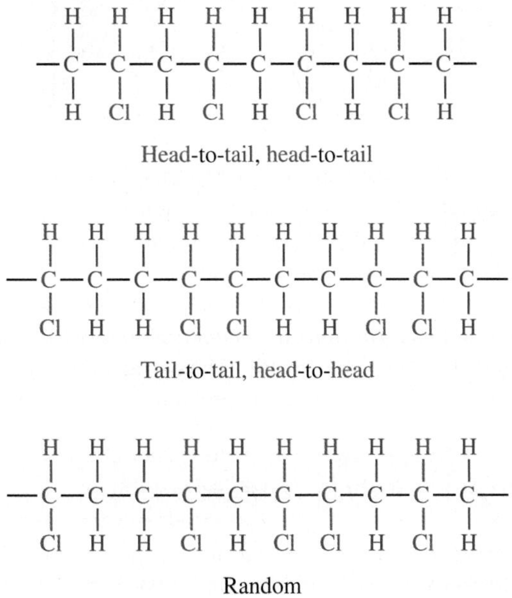

Head-to-tail, head-to-tail

Tail-to-tail, head-to-head

Random

Figure 9.9

Three possible arrangements of the monomers in PVC.

molecules. Flexible PVC that contains plasticizers is familiar in shower curtains, "rubber" boots, garden hoses, clear IV bags for blood transfusions, artificial leather ("patent leather"), and flexible insulation coatings on electrical wires. Flame retardants are also commonly added to polymer mixtures to prevent fires in offices, homes, and small enclosed spaces such as boats and airplanes. Additives to plastics are controversial for several reasons, as we'll see in Section 9.11.

Next, let us consider the formation of polypropylene (PP) from polymerization of propylene:

$$\begin{array}{c} H \\ \diagdown \\ \end{array} C = C \begin{array}{c} H \\ \diagup \\ \end{array}$$
$$H \qquad CH_3$$

Again, several arrangements are possible because of the asymmetry of the mono-mer. A particularly useful form of polypropylene is the repeating head-to-tail, head-to-tail arrangement. This regularity imparts a high degree of crystallinity and makes the polymer strong, tough, and able to withstand higher temperatures. These properties are reflected in its uses. For example, indoor–outdoor carpeting is often made using the strong fibers of polypropylene.

Your Turn 9.11   "The Tough One"

Polypropylene may not be as familiar to you as polyethylene or PET, in part because many polypropylene items don't carry a recycling symbol.

a. As just mentioned, polypropylene can be drawn into fibers such as those used in indoor–outdoor carpeting. Suggest two other uses for polypropylene fiber where toughness is desired.

b. Although HDPE is used in many food containers, polypropylene is used for margarine containers. Toughness is not the issue. So what is the issue?

Finally, let us examine the polymerization of n molecules of styrene to form polystyrene (PS), an inexpensive and widely used plastic. Here is a representation of the addition polymerization scheme:

$$n \quad \begin{array}{c} H \\ \diagdown \end{array} C = C \begin{array}{c} H \\ \diagup \end{array} \quad \xrightarrow{\text{R·}} \quad \left[\begin{array}{cc} H & H \\ | & | \\ -C & -C- \\ | & | \\ H & \end{array} \right]_n \qquad \qquad \text{[9.4]}$$

Polystyrene is a hard plastic with little flexibility. Like the other Big Six, it melts when heated (thermoplastic) and casts well into molds. Transparent cases for DVDs and clear-plastic party glasses and plates are also made from polystyrene. So are hard exteriors of many laptop computers and cell phones.

Most commercial polystyrene has the random arrangement of the monomers shown in Figure 9.10a. In this form, sometimes referred to as general purpose or "crystal" polystyrene, the polymer is hard and brittle. Have you ever squeezed too hard on a clear-plastic party glass causing it to split? It probably was polystyrene (Figure 9.10b).

Your Turn 9.12   Polystyrene Possibilities

Show the arrangement of atoms in a polystyrene chain in the repeating head-to-tail, head-to-tail arrangement. Why do you think this arrangement is favored rather than the head-to-head or tail-to-tail arrangement?

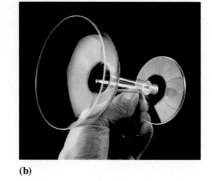

Random

(a)

(b)

Figure 9.10

(a) The random arrangement of polystyrene. **(b)** Partyware made from "crystal" (general purpose) polystyrene.

(b): Jill Braaten/McGraw Hill

Hot beverage cups, egg cartons, and packing peanuts are also made from polystyrene, sometimes called expandable polystyrene (EPS). These items are made from small hard "expandable" polystyrene beads. These beads contain 4–7% of a **blowing agent**; that is, either a gas or a substance capable of producing a gas to manufacture a foamed plastic. For PS, the blowing agent is typically a low-boiling liquid such as pentane, C_5H_{12}. If the beads are placed in a mold and heated with steam or hot air, the pentane vaporizes, which serves to expand the polymer. The expanded particles are fused together into the shape determined by the mold. Because it contains so many bubbles, this plastic foam is not only light but also an excellent thermal insulator.

Chlorofluorocarbons, better known as CFCs, were once on the list of compounds used as blowing agents. Because CFCs destroy stratospheric ozone (Chapter 3), their use was phased out in 1990. Pentane in its vapor form and carbon dioxide were two possible replacements. For example, the Dow Chemical Company developed a process that uses pure carbon dioxide as a blowing agent to produce Styrofoam for packaging material, which eliminated the use of CFC-12. Styrofoam is a brand name of polystyrene foam insulation produced by the Dow Chemical Company. Furthermore, the CO_2 used is a by-product from existing commercial and natural sources, such as cement production and natural gas wells. Thus, it does not contribute additional CO_2, a greenhouse gas, to the atmosphere. Using CO_2 in place of CFCs as a blowing agent illustrates a key principle of green chemistry introduced in Section 2.14: the fabrication and use of substances that are not toxic. Dow received a 1996 Presidential Green Chemistry Challenge Award for developing these alternative reaction conditions.

Russ Widstrand/WidStock/Alamy stock Photo

9.6 | Cross-Linking Monomers

Learning Objectives:

Explain the mechanism of condensation polymerization

Identify and name various functional groups that are present in monomer structures

Monomers make the polymer! As you saw in the previous section, changes in the monomer lead to changes in the properties of the polymer. To understand different monomers, we need to revisit the concept of **functional groups**; that is, distinctive arrangements of groups of atoms that impart characteristic chemical properties to the molecules that contain them (Table 9.2). For example, the *hydroxyl* functional group (—OH) was introduced earlier in the context of ethanol, a biofuel (Section 6.14). This group is present in all compounds classified as alcohols, including one that is of interest in this chapter, ethylene glycol.

Table 9.2	Selected Functional Groups	
Name	**Chemical Formula**	**Structural Formula**
Hydroxyl (in alcohols)	—OH	
Carboxylic acid	—COOH or —C(O)OH	
Ester	—COOC— or —C(O)OC—	
Amine	—NH$_2$	
Amide	—CONH$_2$ or —C(O)NH$_2$	

Let's start with the *carboxylic acid* group:

Although the carboxylic acid group contains an —OH group, it is *not* an alcohol. Rather, think of —COOH as a single unit. As noted in Table 9.2, functional groups containing a carbonyl (C=O) are often represented with parentheses. For instance, —COOH is often denoted as —C(O)OH to indicate that the —OH group is directly bonded to carbon instead of oxygen.

Table 9.3 shows that carboxylic acid groups naturally occur in foods such as vinegar and cheese. If you examine the entries in the table closely, you will see that a molecule can contain more than one carboxylic acid functional group; for example, terephthalic acid and adipic acid. A molecule can also have two different functional groups, such as lactic acid.

Carboxylic acids are closely related to another functional group, the *ester*. An ester can be represented by this structural formula:

Armed with the ability to recognize alcohols, carboxylic acids, and esters, you are now ready to explore *polyesters*, the topic of this section.

The star on the polyester stage is PET, also known as PETE. Both abbreviations stand for polyethylene terephthalate ester. Because PET is semi-rigid, clear, and reasonably gas-tight (Figure 9.11), its most common uses are in beverage containers and in screen protectors for cell phones. Because polyethylene terephthalate contains no polyethylene, it is sometimes written as poly(ethylene terephthalate). This reduces the confusion, at least when the name is written. Spoken, the two names still sound the same. Polyesters can also be drawn into sheets and fibers. For example, Mylar is a trade name for thin plastic sheets of PET such as those used to make shiny festive balloons. When filled with helium, these balloons remain aloft for many hours or days because the

Figure 9.11

Two-liter soft-drink bottles made from PET.

Jill Braaten/McGraw Hill

Table 9.3	Selected Carboxylic Acids	
Name	**Chemical Formula**	**Information**
Ethanoic acid	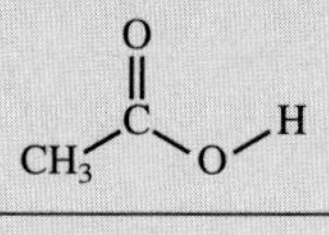 CH_3—C(=O)—O—H	Naturally occurring in vinegar. Also called acetic acid.
Propanoic acid	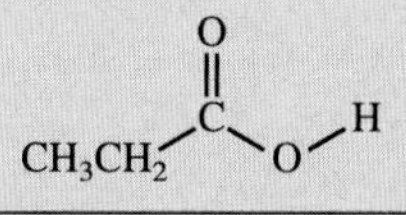 CH_3CH_2—C(=O)—O—H	Naturally occurring in some cheeses, providing a "sharp" taste. Also called propionic acid.
Benzoic acid	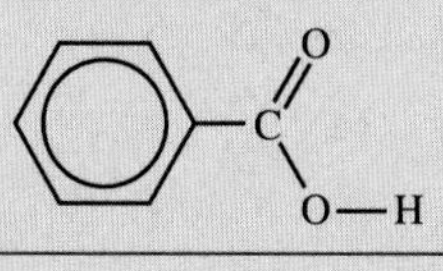 (benzene ring)—C(=O)—O—H	Another naturally occurring carboxylic acid. Used as a food preservative.
Terephthalic acid	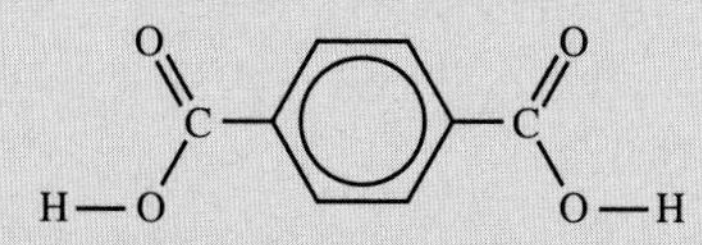 H—O—C(=O)—(benzene ring)—C(=O)—O—H	One of the monomers used to produce PET.
Adipic acid	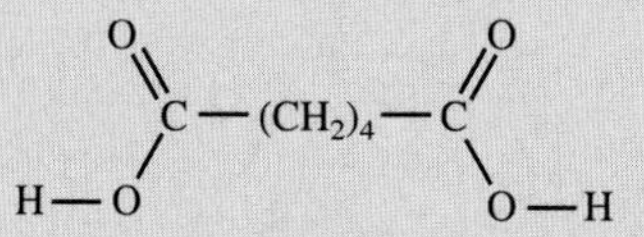 H—O—C(=O)—$(CH_2)_4$—C(=O)—O—H	One of the monomers used to produce a type of nylon.
Lactic acid	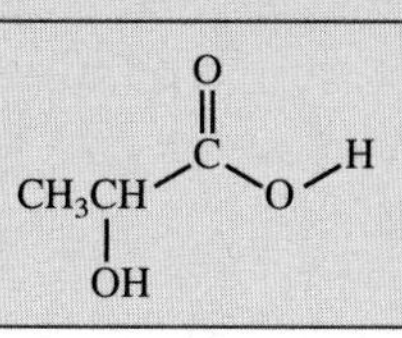 CH_3CH—C(=O)—O—H with OH	The monomer for polylactic acid (PLA), a bio-based polymer.

polyester is so impervious to gases. Eventually, however, the helium atoms escape slowly over time and the balloons deflate.

In contrast to polyethylene, PET is *not* formed by addition polymerization. Rather, it is formed by **condensation polymerization**, a process in which the monomers join by releasing (eliminating) a small molecule, usually water. Thus, condensation polymerization always has a second product in addition to the polymer itself. Many natural polymers are formed by condensation reactions, including cellulose, starch, wool, spider silk, and proteins. Synthetic polymers include Dacron, Kevlar, and different types of nylon.

Also in contrast to polyethylene, PET is *not* formed from a single monomer. Rather, PET is a **copolymer**, a polymer formed by the combination of two or more different monomers. One monomer, ethylene glycol, is an alcohol that contains two hydroxyl groups, one on each carbon atom. The other monomer, terephthalic acid, contains two carboxylic acid groups, one on each side of the benzene ring. In essence, each monomer is ready to react with two functional groups.

To understand how copolymers form, let's work with only one molecule of each monomer. Figure 9.12 illustrates how these monomers can join. Written and circled in red, the —OH in the carboxylic acid group and the H atom in the hydroxyl group react to produce water, HOH. The remaining portions of the alcohol and the carboxylic acid connect to form the ester functional group, highlighted in blue.

Notice that the product has functional groups that are sites for additional chain growth: —COOH on the right end and —OH on the left. The former can react with the —OH of another ethylene glycol molecule; the latter can react with the —COOH of another terephthalic acid molecule. Each time, a molecule of water is released and an ester group is formed. This process occurs multiple times to yield polyethylene terephthalate (Figure 9.12b). The result is a *polyester*, so named because an ester connects the monomers.

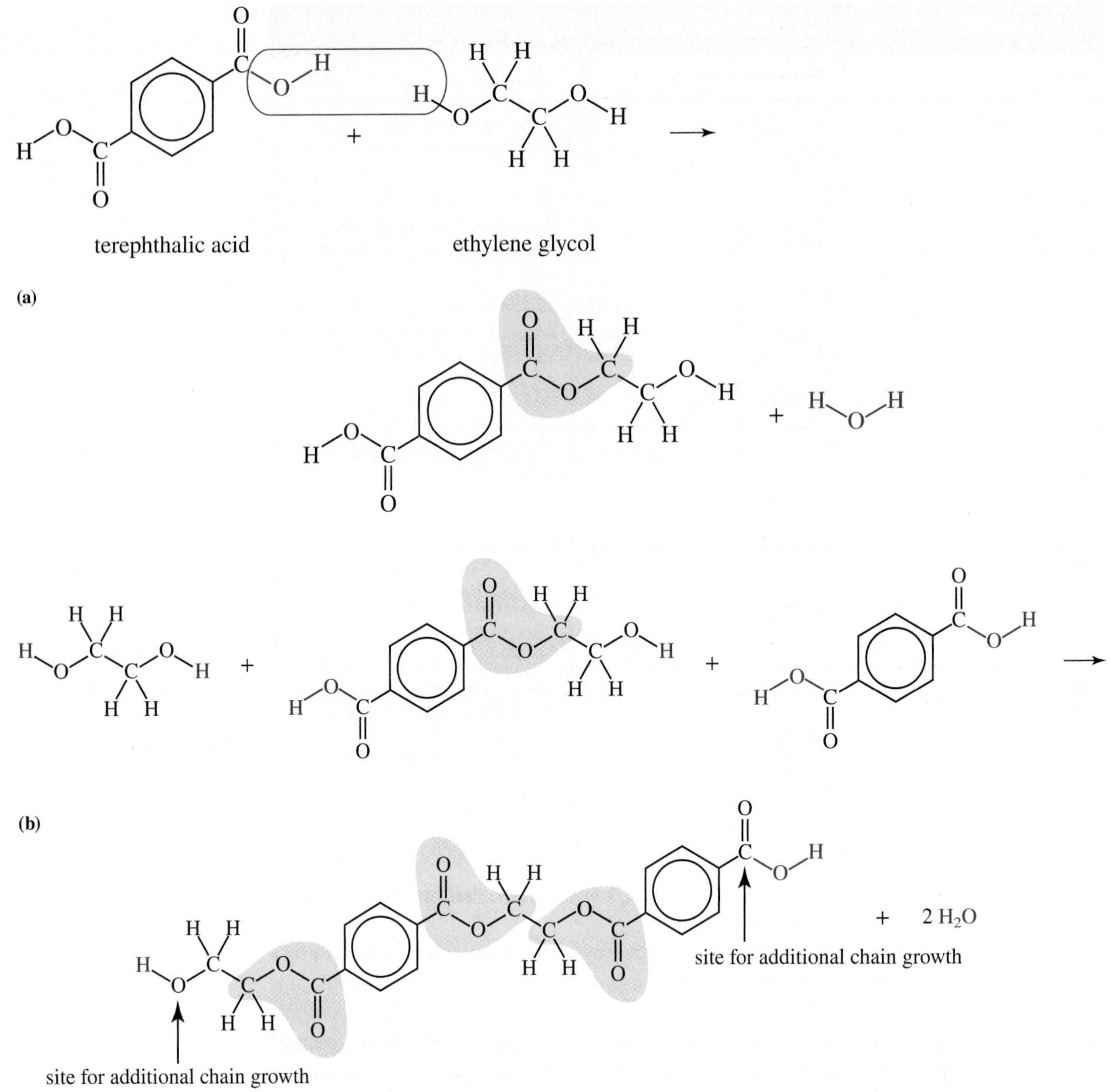

(a)

(b)

Figure 9.12
Illustrations of PET formation from the condensation polymerization of terephthalic acid and ethylene glycol monomers. Whereas **(a)** shows how the reaction of two monomers takes place, **(b)** illustrates how polymer growth takes place as more monomers are introduced. The ester functional groups are highlighted in **blue**, whereas the origins of water in the reactants are colored in **red**.

Your Turn 9.13   Esters and Polyesters

You have seen that terephthalic acid and ethylene glycol can react. Now consider ethanoic acid (acetic acid) and ethanol (ethyl alcohol):

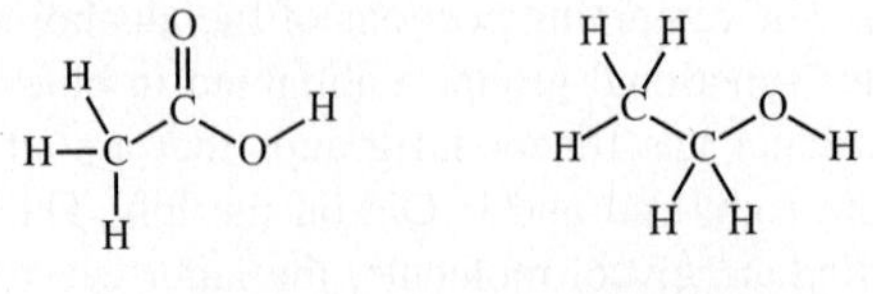

ethanoic acid   ethanol

a. Show how this carboxylic acid and alcohol can react to form an ester.
   Hint: Remember a water molecule is formed as a product.
b. Could ethanoic acid and ethanol react to form a polyester? Explain your reasoning.

PET is not the only polyester in town! By varying the number and type of carbon atoms in the monomers, chemists have synthesized other polyesters with trade names such as Dacron, Polartec, Fortrel, and Polarguard. Polyester spins readily into fibers that are easy to wash and quick to dry. Polyester also blends well with other fibers, such as cotton or wool. The next activity describes polyethylene naphthalate (PEN), a polyester that has better temperature resistance than PET.

Your Turn 9.14   From PET to PEN

In both PET and PEN, the alcohol monomer is ethylene glycol; however, the organic acid monomers differ slightly. Here is the organic acid monomer in PEN, naphthalic acid:

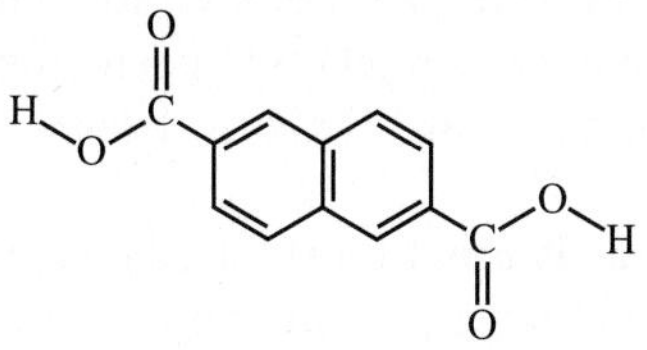

Use structural formulas to show the reaction of two molecules of naphthalic acid with two molecules of ethylene glycol.

In our discussion of polymers so far, all have been based on chains primarily composed of carbons. These may be considered *organic* polymers. Recall that in Section 2.5 we defined organic compounds as those that always contain carbon, usually hydrogen, and sometimes other elements such as oxygen and nitrogen. There are a number of *inorganic* polymers that are based principally on elements other than carbon. The most widely used inorganic polymers are silicones. Rather than carbon, these polymers rely on a backbone of alternating silicon and oxygen atoms:

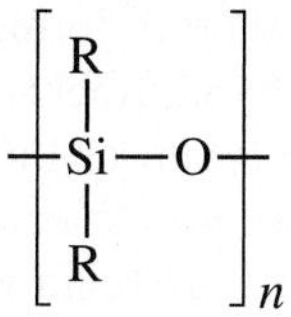

$$\left[\begin{array}{c} R \\ | \\ Si-O \\ | \\ R \end{array}\right]_n$$

Similar to PET, silicones are formed by a condensation reaction. Sometimes water is a by-product of the condensation reaction, but more often an alcohol such as methanol or ethanol is produced instead of water. The side chains designated by R in the structure are typically carbon chains of varying lengths. These side chains greatly affect the properties of the silicone, giving them just as wide a range of properties as their organic polymer cousins. Indeed, due to these tailorable properties, silicones have found applications in a variety of products from lubricating oils and paints, to cooking utensils and caulks.

9.7 | From Proteins to Stockings: Polyamides

Learning Objective: Describe the formation of polyamides such as nylon and proteins

No discussion of condensation polymerization can be complete without examining two specific types of polymers. The first is proteins, which are natural polymers such as those found in our muscles, fingernails, and hair; the second is nylons, which are synthetic substitutes that brilliantly duplicate some of the properties of silk, a naturally occurring protein. According to the Science History Institute, it is estimated that manufacturers worldwide produce around 8 million pounds of nylon annually, accounting for roughly 12% of all synthetic fibers.

Amino acids are the monomers from which our body builds proteins. Each amino acid molecule contains two functional groups: an amine ($-NH_2$) and a carboxylic acid ($-COOH$). Twenty different amino acids occur naturally, each differing in one of the groups bonded to the central carbon atom. This side chain is represented with an R, as shown below in the general structural formula for an amino acid:

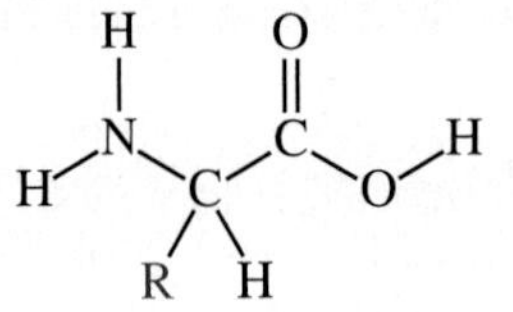

Chemists use R as a placeholder in a molecule. With amino acids, R represents one of 20 side chains. Earlier in this chapter, R· was used to represent a free radical. In some amino acids, R consists of only carbon and hydrogen atoms; in others, R may include oxygen, nitrogen, or even sulfur atoms. Some R groups have acidic properties, while others are basic.

As monomers, amino acids join to form a long chain via condensation polymerization. However, keep in mind three key differences between a condensation polymer such as PET and any given protein:

- PET is a polyester. In contrast, proteins are **polyamides**; that is, condensation polymers that contain the amide functional group ($-CONH_2$).
- PET is built from two monomers, ethylene glycol and terephthalic acid, that are in a 1:1 ratio. In contrast, proteins can contain up to 20 different monomers (amino acids) in any ratio.
- In proteins, each amino acid has two *different* functional groups, $-NH_2$ and $-COOH$; in PET, the two monomers have two *identical* functional groups, either $-OH$ or $-COOH$.

To see how these differences play out, examine this reaction between two amino acids (Figure 9.13). One has the side chain R; the side chain on the other amino acid is labeled as R′. In this reaction, an amide is formed and a molecule of water is eliminated. This amide contains a C—N bond, referred to as a **peptide bond**, the covalent bond that forms when the $-COOH$ group of one amino acid reacts with the $-NH_2$ group of another, thus joining the two amino acids. However, it should be noted that only in the context of proteins is a C—N bond referred to as a peptide bond. In the sophisticated chemical factories of the cells of any organism, this condensation reaction between different amino acids is repeated many times to form the long polymeric chains that we call proteins. Given that 20 different amino acids exist in nature (detailed later in Section 11.6), a great variety of proteins can be synthesized. Some contain hundreds of amino acids, others only a few.

Chemists sometimes attempt to replicate the chemistry of nature. For example, a brilliant chemist working for the DuPont Company, Wallace Carothers (1896–1937)

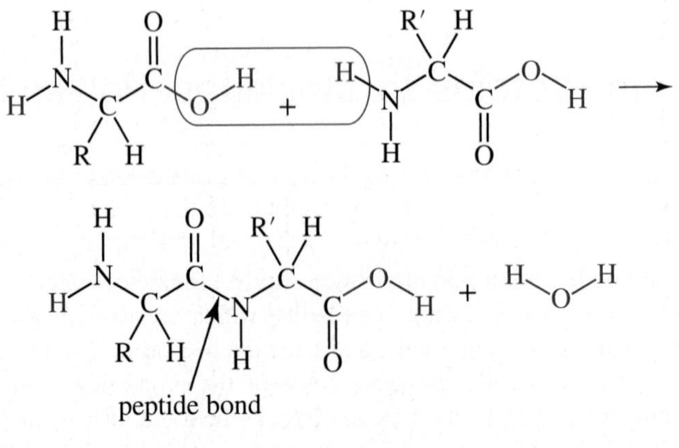

Figure 9.13

Reaction of two amino acids to form a peptide bond.

(Figure 9.14), was studying many polymerization reactions, including the formation of peptide bonds. Instead of using amino acids, Carothers tried combining adipic acid and hexamethylenediamine:

Note that adipic acid has a carboxylic acid at each end of the molecule and hexamethylenediamine has an amine group on each end. As in protein synthesis, the acid and amine groups react to form an amide and release water. But unlike protein synthesis, the resulting polymer, better known as *nylon,* is formed from only two monomers (Figure 9.15):

DuPont executives decided that nylon had promise, especially after company scientists learned to draw it into thin filaments. These filaments were strong, smooth, and very much like the protein spun by silkworms. Therefore, nylon was first introduced to the world as a substitute for silk. Nylon was one of the first **biomimetic materials**—components for use in human applications that are developed using inspiration from nature. The world greeted the release of nylon with bare legs and open pocketbooks! Four million pairs of nylon stockings were sold in New York City on May 15, 1940, the first day they became available (Figure 9.16). But, in spite of consumer passion for "nylons," the civilian supply soon dried up as the polymer was diverted from hosiery to parachutes, ropes, clothing, and hundreds of other wartime uses. By the end of World War II in 1945, nylon had repeatedly demonstrated that it was superior to silk in strength, stability, and resistance to rot. Today, this polymer, with its many modifications, continues to find diverse applications in carpets, sportswear, camping equipment, the kitchen, and the laboratory.

Figure 9.14

Wallace Carothers, the inventor of nylon.

New York Public Library/Science Source

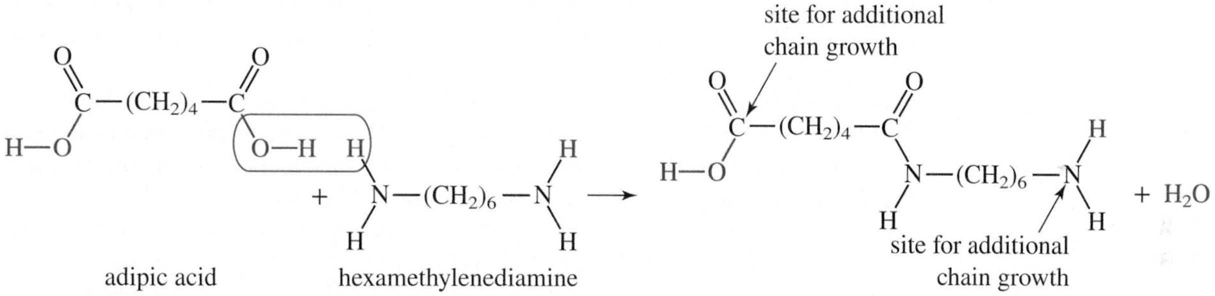

Figure 9.15

Illustration of nylon formation through the condensation polymerization of monomers.

Figure 9.16

Customers eagerly lined up to buy nylon stockings in 1940, when they were first commercially available.

Bettman/Getty Images

Kevlar ends our tale of condensation polymers. The next activity reminds you that it is a polyamide, just like the silk spun by silkworms and spiders.

Your Turn 9.15   Kevlar

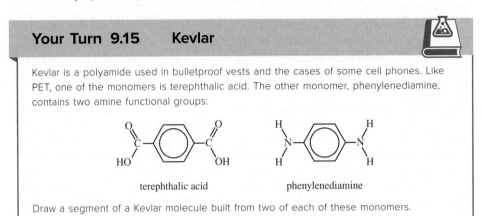

Kevlar is a polyamide used in bulletproof vests and the cases of some cell phones. Like PET, one of the monomers is terephthalic acid. The other monomer, phenylenediamine, contains two amine functional groups:

terephthalic acid   phenylenediamine

Draw a segment of a Kevlar molecule built from two of each of these monomers.

9.8 | Dealing with Our Solid Waste: The Four Rs

Learning Objective: Describe the four Rs of dealing with plastic waste

It is easy to be drawn to the complexity and beauty of spider webs. This natural polymer has many useful properties, including strength, the ability to stretch, and enough stickiness to ensnare prey. Orb spiders, like the one shown in Figure 9.17, are notoriously picky builders and spin new webs each day. This daily web construction could easily exhaust the resources available to the spider. So, how does an orb spider manage to spin so much silk and still survive? Most simply, it recycles! Orb spiders have the ability to ingest old spider silk and recover the raw materials from which they are built. The silk of an orb spider ranks among the toughest biological materials ever studied, an order of magnitude stronger than a similar piece of Kevlar. Although the actual chemical processes are not fully understood, up to two thirds of the existing web goes into making a new one.

Can humans mimic the spider to convert our waste into something useful? After all, we have produced a lot of plastic! In 1950, the value was just under 2 million metric tons worldwide. Over the years, the amount of plastic produced annually has increased steadily, reaching 460 million metric tons worldwide in 2019. It is estimated

Figure 9.17

Golden orb spider and web.

Edwin Remsberg/Alamy Stock Photo

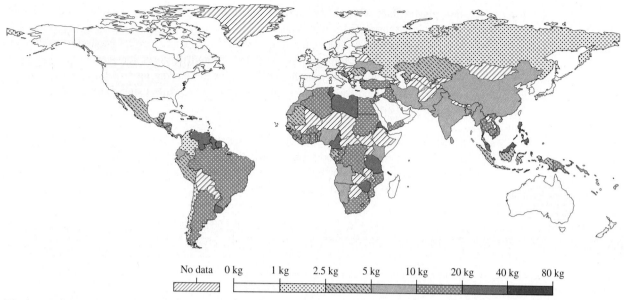

Figure 9.18

Mismanaged plastic waste per capita, 2019. Mismanaged plastic waste is plastic that is either littered or inadequately disposed. A country's total does not include waste that is exported overseas, where it may be mismanaged. For an animated version, see: www.acs.org/cic. Our World in Data

that over 9 billion tons of plastic has been made since the 1950s! Without question, we need sustainable answers to the question of how to deal with plastic waste, since mismanaged plastic waste remains a huge problem worldwide (Figure 9.18).

Most likely, you have taken trash to the curb for pickup. You also may have watched the waste being driven away in a truck, never to be seen again (by you at least). Plastics are part of this waste, and sending plastic to a landfill is far from an ideal solution. Although recycling is a good idea, even better options exist. Here are the Four Rs, ranked in order of their desirability:

- **Reduce** the amount of materials used (e.g., use less plastic in the production of a bottle)
- **Reuse** materials (e.g., repeatedly use your own bags at the grocery store)
- **Recycle** materials (e.g., don't throw beverage bottles away, recycle them)
- **Recover** either the materials or the energy content from materials that cannot be recycled (e.g., burn plastics with high-energy content)

How much plastic do you use and recycle? The next activity asks you to keep a tally.

Your Turn 9.16   Plastic You Toss: Part I

Keep a journal of all the plastic you either throw away or recycle in one week. Include plastic packaging from food and other products that you purchase.

a. Estimate the mass of this plastic. Is it a few grams, a kilogram, or more?
b. Which is greater, the mass of the plastic you throw away or the mass you recycle?

Keep your journal handy because you will be asked to revisit it.

Let's now examine each of the Four Rs as options for dealing with plastics:

Reduce! Source reduction is always the option of choice. This means using less material and generating less waste later on. Source reduction conserves resources, reduces pollution, and minimizes toxic materials in the waste stream. As an example, consider beverage bottles. Through an improved design, a 2-L soda bottle now uses about a third less plastic than when it was introduced in 1970; similarly, a 1-gallon milk jug now weighs less than it did a few decades ago.

Corporations are recognizing that reduced packaging offers economic incentives, such as lower costs for shipping and lower landfill costs for waste. For example, Walmart reported that only slight reductions in the packaging size for a single product category saved $3.5 million on transportation costs, while using 5100 fewer trees and 1300 fewer barrels of oil.

Reuse! Reusing something means not disposing of it after a single use. In the check-out line, supermarket clerks once gave their customers only two choices, "Paper or plastic?" Today, however, they may ask if you brought your own bag. The next activity expands on the idea of reusing bags.

Your Turn 9.17   Paper, Plastic . . . Neither?

The graphic below shows how many times a given grocery bag would have to be reused to have as low an environmental impact as a standard single-use plastic bag.

Grocery bag comparisons of environmental impact
Number of times a given grocery bag type would have to be reused to have as low an environmental impact as a standard single-use plastic bag.

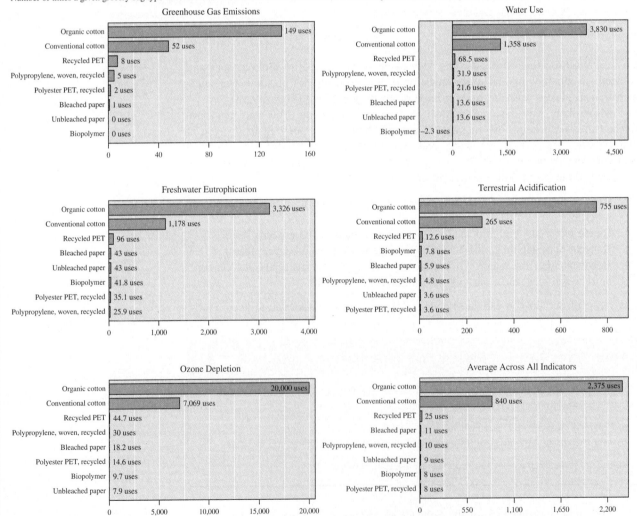

Our World in Data

Grocery stores are not the only place in which people could rethink their use of plastic and paper bags. Provide an example of one other place in which people could rethink their use of plastic and paper bags. Comment on whether you would be willing to change your "bag habits" and reuse your own bag.

As another example, consider how polystyrene foam packing "peanuts" can be reused. Although only a tiny part of the waste stream, these peanuts are a huge nuisance once they escape their intended use. They end up just about everywhere, including waterways, roads, and fields. Because they are only about 5% polystyrene by weight, they have little recycling value. Reusing these peanuts definitely is the option of choice. Actually, the same is true for all polystyrene foam packing materials. If you have worked at a retail store or shipping desk, chances are you have seen some type of "in-house" reuse or recycling.

With many companies aspiring to be "zero waste" in their operations, corporate leaders have realized that sustainability isn't just a way of being cleaner and more efficient, but can also be used to drive innovation. **Sustainable packaging** is the design and use of packaging materials to reduce their environmental impact and improve the sustainability of all practices. Guidance provided in 2022 by the Sustainable Packing Coalition suggested that reusable packaging:

- should have a lower environmental impact
- should displace single-use plastics and eliminate plastic pollution
- should address excessive consumption
- should be motivated primarily by sustainability
- can return to traditional systems like the "milkman model" (i.e., product is delivered in a reusable container that is retrieved as part of the next delivery)
- shouldn't require significant behavior change

As you might expect, polymer chemists and chemical engineers are key players in this endeavor.

Recycle! You probably see recycling containers just about everywhere—in campus buildings, sports centers, airports, and hotels. Reasons for recycling include:

- reduction of waste at landfills and incinerators
- prevention of air, water, and soil pollution during the manufacturing process
- decrease in emissions of greenhouse gases during manufacturing
- conservation of natural resources such as petroleum, timber, water, and minerals

How well are we doing? First the good news. In 2021, the Environmental Protection Agency reported that, on average, each person in the United States recycled 1.2 pounds of their individual waste generation of 5 pounds per day. Other countries have even higher percentages of waste recycling. For instance, Germany (62%), Austria (63%), Belgium (58%), the U.K. (39%), Taiwan (59%), and Singapore (59%).

Items that people deposit in bins or at the curbside include aluminum cans, office paper, cardboard, glass, and plastic containers. In addition, about 0.4 pound per person of waste (such as grass clippings and food scraps) is composted, and another 0.6 pound of waste per person is incinerated to produce energy daily. Given these reductions, the amount of waste sent to the landfill is now averaging 2.3 pounds per person per day.

And now the bad news: roughly 13% of what we discard is plastic, and we recycle plastics at a surprisingly low rate. This may seem at odds with all the milk jugs and plastic bottles you see being recycled in your own community. However, certain plastics are recycled more consistently than others. For example, in the United States,

Your Turn 9.18   Plastic You Toss: Part II

In Your Turn 9.16, you kept a journal of all the plastic you either discarded or recycled in a week. Revisit what you wrote. Do the yearly recycling figures just cited ring true for you? That is, do you throw out far more plastic than you recycle? Briefly report on how your own plastic use stacks up against the national averages. Remember that your journal may not reveal your use of durable and nondurable goods over a longer period of time.

polyethylene milk containers are recycled at a rate of 28% and clear PET soft-drink bottles at 31%. Although these numbers may seem high, nonetheless around 70% of these containers are still being tossed out.

Recover! What about incineration; that is, recovering the energy in plastics by burning them as fuels? Because the Big Six and most other polymers closely resemble hydrocarbon fuels, incineration would seem to be an excellent way to dispose of them, reducing demand on landfills. The chief products of combustion are carbon dioxide, water, and a good deal of energy. In fact, pound for pound, plastics have an energy content 37% higher than U.S. coal and 87% higher than wood. However, in the United States, plastics currently account for only about 13% of the weight of municipal solid waste.

Incineration and landfills create relatively few jobs in comparison to recycling programs. A commitment to increasing recycling can benefit a local economy. Incineration of plastics has drawbacks. The gases produced by combustion may be "out of sight," but they best not be "out of mind." Because their chemical compositions are similar to fossil fuels, burning plastics produces CO_2, a greenhouse gas. Of special concern in incineration are chlorine-containing polymers such as polyvinyl chloride that release hydrogen chloride (HCl) during combustion. Because HCl dissolves in water to form hydrochloric acid, such smokestack exhaust could make a serious contribution to acid rain. Burning chlorine-containing plastics can produce other toxic gases. So, in terms of the overall benefit, including the energy involved, recycling is always preferable to incineration.

Your Turn 9.19   Burning a Plastic

Under conditions of complete combustion, polypropylene burns to produce carbon dioxide and water.

a. Write a balanced chemical equation. Assume an average chain length of 2500 monomers.
b. If the combustion is incomplete, other products form. Name two possibilities.
c. Using the Internet as a resource, list some additives that are commonly added to polymers such as polypropylene. Could these also be released as harmful environmental pollutants during combustion? Explain your answer.

The plastic that we do not reuse, recycle, or recover eventually ends up in a landfill (the "out of sight, out of mind" approach), or as trash in the environment. Both are problematic. Although landfill space is still available, landfills have drawbacks. They take up space in congested areas, they have costs associated with their construction and upkeep, they leak and attract vermin, and they emit methane, a greenhouse gas. In this next activity, you will learn about microplastics and will meet a student who is trying to make a difference in our world by removing plastics from the environment.

The majority of plastics do not biodegrade in the landfill (or anywhere else). Most bacteria and fungi lack the enzymes necessary to break down synthetic polymers. Some microbes, however, possess the enzymes to break down naturally occurring polymers, such as cellulose. For example, the methane produced by livestock is formed when bacteria obtain energy by decomposing cellulose during the animals' digestion. Similarly, methane is generated by natural decomposition of organic materials in landfills due to bacterial activity.

Even natural polymers do not decompose completely in landfills. Modern waste disposal facilities are covered and lined to deter leaching of waste and waste by-products into the surrounding ground. These landfill linings and coverings create anaerobic (oxygen-free) conditions that impede the breakdown of these wastes. As a result, many supposedly biodegradable substances decompose slowly, or not at all. Excavation of old landfills has unearthed old newspapers that are still readable (Figure 9.19). Ideally, landfill liners can last forever. However, over time, liners break down or rupture.

Real Talk   Plastics in the Environment

Check out the video interview with Edgar McGregor from San José State University, who describes his work in removing plastics from the environment: www.acs.org/cic.

As mentioned in the video, plastics represent the most prevalent variety of debris found in oceans and the Great Lakes. The image to the right shows the location of garbage patches in the Pacific Ocean. Although we typically think of plastic contaminants as discarded packaging material or plastic cups, there is another type that is much smaller in dimensions.

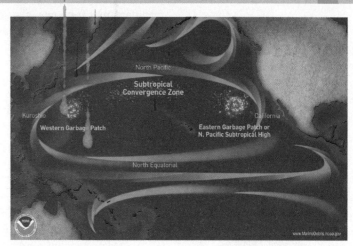

NOAA

Using the Internet as a resource, answer the following questions:

a. What are "microplastics" and where do these materials come from?

b. How widespread are microplastic pollutants in marine environments worldwide?

c. What are some of the potential impacts posed to human health or marine life from exposure to microplastics?

d. What legislation has been initiated to reduce the harmful effects of microplastics? Have these actions resulted in any measurable improvements in the concentration of these marine pollutants, to date?

Figure 9.19

Some buried waste can remain intact for a long time. This newspaper from 1982 was excavated 33 years later.

Bernard Weil//Toronto Star/Getty Images

Your Turn 9.20   Landfill Liners

Landfill liners include natural clay and human-made plastics. For example, thick sheets of high-density polyethylene may be employed. Even the best HDPE liners, however, can crack and degrade.

a. From Table 9.1, which types of chemicals soften HDPE?

b. Name five substances sent to the landfill that could degrade a HDPE liner over time.

Given the problems associated with landfill disposal and incineration of natural and synthetic polymers, *recycling* has an important role to play. However, in contrast to landfilling, recycling polymers requires an input of energy. Furthermore, if the waste plastic is dirty or of low quality, more energy may be needed to recycle it than to manufacture it from new plastic. Nonetheless, recycling is one of several ways to divert plastic from landfills and incinerators. In the next section, we examine the bigger picture of trash.

9.9 | Recycling Plastics: The Bigger Picture

Learning Objective: Describe the process and applicability of recycling plastics

Together with other waste, the plastic that you discard is part of a bigger picture. In the United States, the EPA has been keeping statistics about municipal solid waste—better known as trash—for more than 50 years. **Municipal solid waste** (MSW) includes everything you discard or throw into your trash bin, including food scraps, grass clippings, and old appliances. MSW does not include all sources, such as waste from industry, agriculture, mining, or construction sites. In the United States, municipal solid waste has been averaging about 292 million tons per year. What is the largest single item? Paper, as you can see from Figure 9.20. Materials of biological origin such as paper, wood, food scraps, and yard trimmings make up the majority of the materials classified as municipal solid waste. All of these can be dealt with by one of the Four Rs (reduce, reuse, recycle, recover).

Worldwide, it is estimated that 2 billion tons of MSW is generated each year. How much of this MSW is plastic? Consult Figure 9.20 to see that plastic is roughly 13% of what U.S. citizens discard, the equivalent of 38 million tons in 2022. The U.S. EPA reports data for three types of plastics:

- durable items, such as plastic furniture, bowls, and garden hoses
- nondurable items, such as disposable plastic cups, plates, trash bags, pens, and safety razors
- packaging, such as beverage bottles and food containers

How much of this plastic do we recycle? Discover the answer in the next activity.

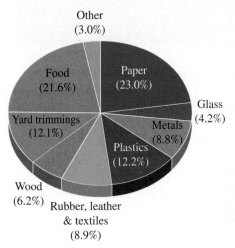

Figure 9.20

What's in your garbage? Composition by weight of municipal solid waste before recycling (292 million tons, 2022).

Source: U.S. Environmental Protection Agency, Advancing Sustainable Materials Management: 2015 Fact Sheet, July 2018

Your Turn 9.21   Plastics Recycling Scorecard

According to the EPA, here is the U.S. plastics recycling scorecard for 2018. Durable goods or products, with a lifetime of three years or more, include items such as luggage, plastic furniture, and garden hoses. Nondurable goods or products, with a lifetime of less than three years, include plastic pens and safety razors. Containers and packaging are products that are assumed to be discarded the same year the products they contain were purchased.

Use of Plastic	Weight Generated (millions of tons)	Weight Recycled (millions of tons)
Durable goods	57.10	10.57
Nondurable goods	50.44	14.19
Containers/packaging	82.22	44.33

a. For each type of plastic, calculate the plastic recycled as a percent of the waste generated.
b. Nondurable and durable goods tend to have low recycling rates. List three examples of each, and suggest reasons why.

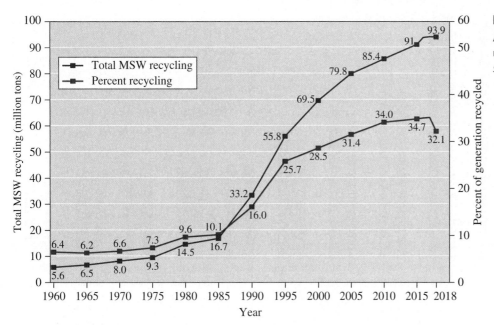

Figure 9.21 shows the amount in millions of tons over the years for *total* MSW recycling in the United States, not just for plastics. The overall rate of recycling and composting of MSW in recent years has reached 34%. In comparison, 48% of the MSW was recycled or composted in the European Union (E.U.) in 2020.

By way of comparison, in 2021 only 5-6% of the plastic in the United States was recycled. Why so little? The devil is in the details! Some items can be easily recycled; others present a logistical nightmare. Furthermore, some types of plastics have a ready market, whereas others do not. Table 9.4 reveals the relatively high recycling rate for plastic bottles in comparison to the overall 5–6% plastic recycling rate. Figure 9.22 further illustrates how plastics recycling compares with other products. As discussed in Chapter 8, lead–acid batteries continue to represent one of the most widely recycled products.

Your Turn 9.22   Pounds or Tons Recycled

Examine the values in Table 9.4 from the American Chemistry Council (ACC).

a. Are these values comparable to those quoted by the EPA in Your Turn 9.21? Assume that the tons quoted were short tons; that is, 2000 pounds per ton.

b. The EPA reported the amounts recycled using one unit (million tons), and the ACC reported in another (million pounds). Several explanations for this difference are possible. Propose one.

Table 9.4	Recycled Plastic Bottles in 2018	
Plastic	**Amount Recycled in 2018 (million pounds)**	**Recycling Rate**
PET	1813	28.9%
HDPE	1007	30.9%
PP	30.6	17%
PVC	0.5	1.7%
LDPE	0.8	1.2%

Source: American Chemistry Council, 2018, National Post-Consumer Plastic Bottle Recycling Report.

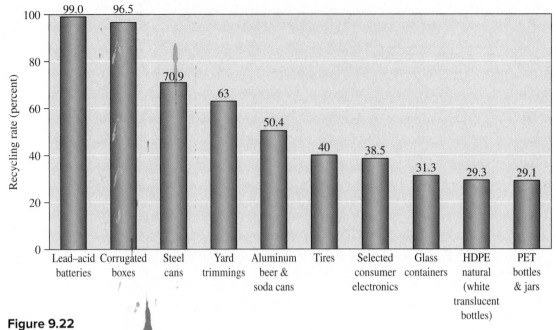

Figure 9.22

Recycling rates of selected products, 2018. The reported rates do not include combustion with energy recovery.

Source: U.S. Environmental Protection Agency, Advancing Sustainable Materials Management: 2018 Fact Sheet, Dec. 2020

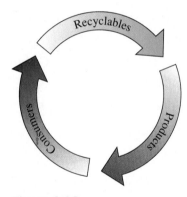

Figure 9.23

Recycling ideally is a never-ending loop.

Source: National Association for PET Container Resources

For recycling to be successful and self-sustaining, a number of factors must be coordinated. These involve not only science and technology, but also economics and sometimes politics—especially at the local level. The best recycling involves a closed loop (Figure 9.23) in which plastics are collected, sorted, and then converted into products that consumers buy, use, and later recycle.

In order to recycle, it is first necessary to collect the plastic. Several options are available: collecting at curbside, at local drop-off centers, and through bottle bill programs involving a deposit and refund. For recycling to be successful, a dependable supply of used plastic must be consistently available at designated locations.

Once collected, the plastic needs to be transported to a facility where it can be sorted (Figure 9.24) and prepared for some marketable commodity, such as the

Figure 9.24

Polymer sorting and processing at a recycling facility.

Jim West/Alamy Stock Photo

manufacturing of outdoor furniture, toys, or even clothing. The symbols that appear on plastic objects (revisit Table 9.1) help facilitate the sorting process. Because of the large volume of material, automated sorting methods have been developed. Once sorted, the polymer is melted, which can then be used directly in the manufacturing of new products. Alternatively, it can be solidified, pelletized, and stored for future use.

If a mixture of various polymers is melted, the product tends to be darkly colored and has different properties depending on the nature of the mixture. This type of reprocessed material is generally good enough to "downcycle," meaning to convert it into lower-grade uses such as parking lot bumpers, disposable plastic flower pots, and inexpensive plastic lumber. Such mixed material is not as valuable as the pure, homogeneous recycled polymer. This underscores the importance of sorting plastics. For similar reasons, manufacturers prefer to use only a single polymer in a product to avoid the need to separate.

Let's Debate!   Is Recycling Worth It?

Check out this video to see how recycling actually works: www.acs.org/cic.

Form two groups: one group will be in favor of the topic (the proposition) and the other group will be against the topic (the opposition). Both groups will try to persuade a neutral person or judge to agree with them. The topic of the debate is the motion. Here is the motion, as stated in a recent Greenpeace report:

teen00000/123RF

"Most plastics simply cannot be recycled. Mechanical and chemical recycling of plastic waste fails because plastic waste is extremely difficult to collect, virtually impossible to sort for recycling, environmentally harmful to reprocess, often made of and contaminated by toxic materials, and not economical to recycle."

Curiosity Conversations

- Does recycling reduce the amount of plastics in landfills and lower the amount of microplastic waste accumulating worldwide?
- Do the benefits of plastics recycling outweigh the economic cost and toxicity claims?
- How should recycling plastics change?
- How could we become less reliant on plastics?
- What can be done to phase out all single-use plastics and shift toward reusing plastics instead of recycling?
- Are there better materials than plastics that we could use?

Given a supply of plastic (ideally clean and sorted), the manufacturers can get to work. The items produced are composed of varying percentages and types of recycled materials. The terminology is confusing. **Recycled-content products** are those made with materials that otherwise would have been in the waste stream. These include items manufactured from discarded plastic, as well as rebuilt items such as plastic toner cartridges that are refilled. Trash bags, laundry detergent bottles, and carpeting are common plastic items that may qualify as recycled-content products. Some playground equipment and park benches also are made from discarded plastic.

Recycled products are now beginning to provide the origin of the recycled material. **Postconsumer content** is material that previously was used individually that otherwise would have been discarded as waste. Recycling this waste—office paper, foam packing, and beverage bottles—is one way to keep it out of the landfill. **Preconsumer content** is waste left over from the manufacturing process itself, such as scraps and clippings. Preconsumer fabrics, such as polyester fabric scraps from the clothing industry, can be recycled rather than discarded.

A product that is designated as *recyclable* simply means that it *can* be recycled. The term may be misleading because a recycling pathway may not exist. Recyclable products do not necessarily contain any recycled materials.

Your Turn 9.23   Recyclable and Recycled

Give three examples of items that you might purchase and recycle. Also give three examples of recycled-content products. Can an item fall into both categories?

Figure 9.25

PET beverage bottles are widely recycled.

Thinkstock/SuperStock

To complete the cycle shown in Figure 9.23, the recycled items are marketed and (ideally) purchased by consumers. Without a product and buyers, recycling programs are doomed to fail. In fact, recycling laws in a number of cities have not been implemented and enforced because one of the links in this polymeric chain of supply—collecting, sorting, processing, manufacturing, and marketing—was missing.

Consult Table 9.4 to see that as consumers, we are moderately adept in dropping PET beverage bottles into recycling bins (Figure 9.25). In fact, more than 1 billion pounds of PET is recycled in the United States! Because PET is more successfully recycled than most plastics, it warrants a closer look. PET soft-drink bottles need special handling before they can be melted and reused. The bottles are usually sorted to remove other types of plastic, such as PVC (from labels, liners, or safety seals) or polypropylene (from bottle caps), which would weaken the final product. The next activity shows how PET can be separated from other polymers by density. This is helpful in the case of PET mixed with PVC because these can look alike.

Your Turn 9.24   Float or Sink?

Here are density values for PET and for three other plastics likely to be found with it in a recycling bin:

Plastic	Density (g/cm^3)
PET	1.38–1.39
HDPE	0.95–0.97
PP	0.90–0.91
PVC	1.30–1.34

When dropped into a liquid, a plastic will float or sink depending on the density of the liquid. Here are the densities for several liquids that do not degrade the four plastics above:

Liquid	Density (g/cm^3)
Methanol	0.79
42% ethanol/water mixture	0.92
38% ethanol/water mixture	0.94
Water	1.00
Saturated solution of $MgCl_2$	1.34
Saturated solution of $ZnCl_2$	2.01

Given a PET sample contaminated with HDPE, PP, and PVC, propose a way to separate the PET from the other three plastics. Assume that all density values were measured at the same temperature.

When you recycle a PET beverage bottle, it may come back to life as part of another beverage bottle. More likely, though, the polyester is downwardly recycled (downcycled) to produce items of lower purity. For example, the PET may be melted and spun into polyester carpeting, clothing (Figure 9.26), "fleece" bed sheets, and the fabric uppers in jogging shoes. Five recycled 2-L bottles can be converted into a shirt or the insulation for a ski jacket; it takes about 450 such bottles to make a 9- × 12-foot polyester carpet.

A reasonable question to ask at this point is what happens to all of these products made from recycled PET. Shaw Industries won a 2003 Presidential Green Chemistry Challenge Award for providing an answer with its development of EcoWorx broadloom

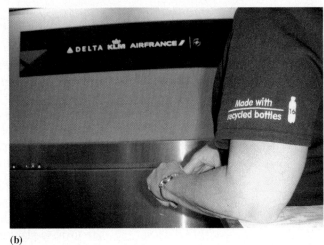

(a)

(b)

Figure 9.26

(a) Activewear made from recycled PET. **(b)** Airline uniform made from recycled PET bottles.

(a): Vanderlei Almeida/AFP/Getty Images; *(b):* ©Cathy Middlecamp, University of Wisconsin

carpet and carpet tiles. By removing the PVC from the backing of these carpets, the product became 100% recyclable into other products. Additional environmental benefits of this carpeting include lower VOC emissions and lower transportation costs, because the carpet tiles are lighter in weight. EcoWorx carpeting contains 44% recycled content and is labeled "We want it back." As more of this carpet is used, recycled, and again recycled, this percent is expected to grow. This is another example of the concept of **cradle-to-cradle** recycling.

Your Turn 9.25   In a Store Near You

Unless people buy products made from recycled plastics, manufacturers will have little financial incentive to produce them. Find five recycled-content plastic items available for sale.

a. Identify the polymer(s) in each, and the % recycled content, if provided.
b. Comment on the consumer appeal of the item, including whether or not you would purchase it.

This section explored the complexities of recycling. But remember that recycling is not the only game in town. There is no single best solution to the problem of plastic waste, or more generally, *all* solid waste. Incineration, reuse, recycling, and source reduction all provide benefits, but all have costs. Therefore, it is likely that the most effective response will be an integrated waste management system that employs multiple strategies. Ultimately, such a system would optimize efficiency, conserve energy and material, and minimize cost and environmental damage.

Your Turn 9.26   More Plastics than Fish?!

The World Economic Forum published a report titled "The New Plastics Economy: Rethinking the Future of Plastics," in which it was asserted that by 2050 there will be more plastics than fish (by weight) in oceans. Download the report from the Internet and describe the assumptions that contribute to this dire forecast. What steps can be taken to prevent this scenario, and what are the biggest threats to realizing a so-called "new plastics economy"?

9.10 | From Plants to Plastics

> *Learning Objective: Explain some differences between plant-based polymers and petroleum-based polymers*

As we mentioned earlier, most synthetic polymers are produced from petroleum, a nonrenewable resource. However some polymers find their origin in renewable materials such as wood, cotton fibers, straw, starch, and sugar. What makes these plant-based polymers different from their petroleum-based cousins?

- They are **compostable**; that is, under the conditions of either a home composter or an industrial composter, they are able to undergo biological decomposition to form a material (compost) that contains no materials that are toxic to plant growth.
- Some polymers can be "unzipped" and converted back into monomers, and remade into virgin polymer.
- Their synthesis generally requires fewer resources, results in less waste, and uses less energy than petroleum-based polymers.
- They do not contain chlorine or fluorine.

Because of these characteristics, such polymers are termed "eco-friendly." However, employ this term with caution. As you will see, the composting of biopolymers is not as straightforward as it may sound. In addition, in order to compare different polymers, the waste and energy costs from its manufacture and use, as well as the end of a product's life, all need to be considered. Again, the best solution is to *reduce* what you consume rather than to switch to any particular type of plastic.

While polylactic acid (PLA) is not the only plastic produced from plants, it serves as an example of eco-friendly plastics. Like the Big Six, PLA is a thermoplastic polymer that softens with heating and can be molded. Because PLA is a polyester that has a similar look and feel to PET, it is used to produce some of the same items as PET, including clear shiny bottles, transparent food packaging, fibers for clothing, and plasticware (Figure 9.27). PLA is also used as a coating on paper cups and plates to make them water-resistant.

Unlike PET with a melting point >250 °C, PLA softens around 60 °C (140 °F). As a result, if you leave an item made from PLA in a car on a sunny day, you may return to find it has melted. So, unless blended with other resins to improve its temperature stability, PLA is limited to lower temperature applications.

(a)

(b)

(c)

Figure 9.27

(a) PLA cups can be colorless, transparent, and water-resistant, just like PET. **(b)** PLA can be pigmented, again like PET. **(c)** Paper cups coated with PLA.

(a): GIPhotoStock X/Alamy Stock Photo; *(b):* Voinakh/Shutterstock; *(c):* Tim Gainey/Alamy Stock Photo

As its name suggests, PLA is built from the monomer lactic acid, which is naturally present in the biosphere. It gives sour milk its taste, and is partly responsible for making your muscles ache after vigorous exercise. This monomer has two functional groups: a carboxylic acid group and a hydroxyl group. Here is its structural formula:

Like PET, polylactic acid is a condensation polymer, and releases a molecule of water each time a covalent bond forms in the polymer chain.

PLA Degradation

Pawarun/Getty Images

Check out this investigation (www.acs.org/cic) to learn more about the degradation of PLA. For full experimental details, consult the Chemistry in Context laboratory manual.

Questions to Consider
a. What are some environmental advantages of using PLA?
b. Why does cutting the plastic make it easier to chemically break down PLA?
c. Instead of microorganisms, what chemical is used to break down PLA?

Your Turn 9.27 The Chemistry of PLA

We don't show the chemical reaction for the formation of PLA from lactic acid because it does not proceed in a single step and is complicated. Even so, you should be able to write a chemical formula for PLA.

a. Circle and label the functional groups in the monomer, lactic acid.
b. When lactic acid polymerizes, explain how you know it is a condensation reaction.
c. What is the repeating unit in PLA?

As an eco-friendly polymer, PLA has its share of controversies. One set arises from its synthesis from corn. Like corn used to produce ethanol (Section 6.14), corn that is used to produce PLA competes with its use as food. Furthermore, the runoff from cornfields may produce nutrient-rich waterways (Section 11.12), and the corn may be genetically engineered to resist pests (Section 13.8). Polylactic acid, marketed by the trade name Ingeo, is primarily manufactured in the U.S. by NatureWorks, a subsidiary of Cargill. It is also manufactured in the Netherlands by TotalEnergies Corbion PLA and by several manufacturers in Japan and China.

PLA can also be produced from carbohydrate-containing plants other than corn. For example, a Dutch company uses sugarcane and a Japanese manufacturer uses tapioca root. In the United States, PLA is largely manufactured from the starch of corn kernels leading to the nickname of "corn plastic."

Unlike petroleum-based polymers, PLA is compostable, but the process goes slowly without the heat supplied by an industrial composter. How slowly? In a backyard compost heap, the process takes up to a year. In contrast, industrial composters do the job in 3–6 months. However, many communities do not have access to such composters, at least not at present. Note that there is no ecological benefit to tossing PLA in a landfill; the actual breakdown of anything in a landfill is slow.

Can PLA be recycled? In theory, yes, but at present no. In fact, having PLA in the recycling stream is of concern to those who recycle PET, because, as one recycler quipped, "The two mix as well as oil and water." Recyclers currently collect and bale PET bottles and then process the plastic, eventually making it into new shirts, containers, fiberfill, or carpeting. PLA, if present in more than small amounts, needs to be separated from the PET recycling stream.

Your Turn 9.28 Detective Work on Your Campus

Meals are served on most college and school campuses, up to thousands each day. Most likely, professionals in your food service department have given serious thought to which cups, plates, forks, spoons, chopsticks, and napkins to use. Be a detective and learn about the sustainable practices on your campus. What happens to plates, cups, and utensils? Are they washed and reused? Are they discarded? If so, are any made from PLA? What are the controversies? Prepare a one-page briefing on a particular item, for example, hot beverage cups.

9.11 | A New "Normal"?

> *Learning Objective: Summarize the environmental and health consequences of using plastics*

Figure 9.28

(a) A 1960s 10-ounce glass Coke bottle; returnable. **(b)** A 1970s 2-L bottle, with plastic cup at the base for additional strength.

(a): jvphoto/Alamy Stock Photo; *(b):* ©Todd Franklin/Neato Cool Creative, LLC

One of the key components of sustainability is the concept of **shifting baselines**; that is, the idea that what people expect as "normal" on our planet has changed over time. Our use of plastics is a good example. Many people are still alive who remember "how it used to be" before the advent of plastics. The baby boomers of the 1950s remember collecting glass bottles such as that shown in Figure 9.28a to reclaim the two-cent deposit.

Think about plastic beverage bottles, for instance. It wasn't until 1970 that the first 2-L (64-oz) bottles appeared on supermarket shelves. The early models were made from PET and fitted with an opaque base cup for added strength (Figure 9.28b). PepsiCo was the first to sell soft drinks in 2-L bottles, and other beverage companies quickly followed suit.

Before the advent of plastic, most beverages were bottled in glass. Even as late as the 1970s, milk was brought to homes in glass bottles, with the empty bottles collected at the time of delivery. Portion sizes were smaller as well. For example, Coca-Cola bottles once held 10 ounces (Figure 9.28a); in contrast, today's aluminum cans hold 12 ounces and plastic bottles are larger still. Rather than dispensing cans or plastic bottles, vending machines of the past dispensed glass bottles, and wooden racks stood nearby to receive the empties. However, glass bottling is not necessarily "greener" or more sustainable than bottling in plastic. The next activity invites you to explore the two options.

Your Turn 9.29   Glass or Plastic?

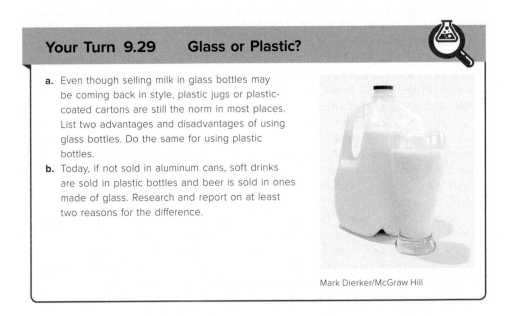

a. Even though selling milk in glass bottles may be coming back in style, plastic jugs or plastic-coated cartons are still the norm in most places. List two advantages and disadvantages of using glass bottles. Do the same for using plastic bottles.

b. Today, if not sold in aluminum cans, soft drinks are sold in plastic bottles and beer is sold in ones made of glass. Research and report on at least two reasons for the difference.

Mark Dierker/McGraw Hill

Plastic debris! Not only has our use of plastic become the norm, but it has also become the norm to find plastic debris everywhere—streets, backyards, streams, beaches, and even wilderness areas. The trouble lies in the durability of plastic. Once a piece of plastic finds its way into the local environment, it does not dissolve, break down in sunlight, or decompose—at least not at any appreciable rate. Rather, it tends to break into smaller and smaller pieces that widely disperse. The very properties that make plastics so useful in the first place mean that the pieces of plastic persist for years and years. Does this sound familiar? See if the next activity helps jog your memory.

Your Turn 9.30   Lessons from Refrigerators Past

Chlorofluorocarbons, better known as CFCs, were once widely used in refrigerators, aerosol sprays, foams, and medical inhalers.

a. Why were CFCs phased out?

b. Some CFCs remain in the atmosphere for 100 years or more. Explain how this property of CFCs is connected to the fact that they have been phased out.

c. Name some properties that polymers such as HDPE, LDPE, PVC, and PS share with CFCs.

d. Unlike CFCs, it is highly unlikely that plastics will be phased out. Offer some reasons why.

e. Some people believe that we cannot sustain our current use of plastics. Give evidence that either supports or contradicts this statement.

There is more to the story than the plastic bottles and wrappers that litter the landscape. Also ubiquitous in nature—including in our bodies—are the invisible substances that leach out of plastics. Leaching is when a chemical substance is detached from its carrier substance by water that is passed over the substance. In plastics, there are substances that are not chemically bonded to plastic called plasticizers that are mixed in to make plastics softer and more pliable. Over time, plasticizers slowly leach out of the plastic and into the biosphere. The next activity introduces you to di-2-ethylhexyl phthalate (DEHP), a controversial plasticizer.

Your Turn 9.31   Meet DEHP

DEHP belongs to a common class of plasticizers called *phthalates* (THAL-ates). Phthalates are esters of phthalic (THAL-ic) acid, an isomer of terephthalic acid, one of the monomers used to synthesize PET.

phthalic acid

a. Explain the meaning of the term *ester*.

b. Below is the structural formula for DEHP. Circle the two ester groups in this molecule.

DEHP

c. Draw a structural formula for the alcohol that reacted with terephthalic acid to form this ester.

As you saw in the previous activity, the DEHP molecule has two long "wavy" side chains attached to a benzene ring. Imagine what happens when DEHP, perhaps as much as 30% by weight, is mixed in with a repeating head-to-tail arrangement of PVC. PVC tends to be stiff because its molecules pack well together and form crystalline regions. However, with the addition of DEHP, the regular packing of the PVC polymer chains is disrupted and the polymer becomes much more flexible.

Why is there a controversy? Debate whether we should continue to use phthalates and bisphenol A (BPA), another additive used in plastics manufacturing, in this last activity.

Let's Debate!   Should We Ban Phthalates?

Chemicals such as DEHP and bisphenol A (BPA) are suspected **endocrine disrupters**, compounds that affect the human hormone system, including hormones for reproduction and sexual development.

Form two groups: one group will be in favor of the topic (the proposition) and the other group will be against the topic (the opposition). Both groups will try to persuade a neutral person or judge to agree with them. The topic of the debate is the motion. Here is the motion:

Despite extensive research activity, there have been no reports to date that directly link phthalates or BPA with increased human mortality. Hence, neither phthalates nor BPA should be banned or regulated until clear evidence of human mortality is discovered.

Curiosity Conversations

- Although the evidence against phthalates and BPA have been mounting for decades, why is it difficult to resolve this controversy?
- From which products do phthalates and BPA originate and what are some possible health effects due to their exposure?
- Have there been studies to investigate the effects of phthalates and BPA on animals or the environment?
- Have phthalates such as DEHP and BPA been banned in any consumer products? Were these bans based on scientific testing on animals or humans, or is this an application of the **precautionary principle**? Explain.

Conclusion

Synthetic polymers are at the very center of modern living, yet their existence depends on a precious resource that we are consuming—crude oil. We have come not only to depend on synthetic polymers, but also in many cases to take them for granted to the point of being wasteful. Once more, we encounter a chemistry topic that has the potential to inspire us to revisit the issue of our lifestyle and its sustainability.

Over time, chemists have created an amazing array of polymers and plastics—new materials that have made our lives more comfortable and more convenient. In many cases, these plastics represent a significant improvement over the natural polymers they replace. Furthermore, products that we use today would be impossible without synthetic polymers: cell phones, automobiles, breathable contact lenses, fleece clothing, kidney dialysis equipment, and artificial hearts. We have become dependent on polymers, and it verges on the impossible to abandon their use.

The chemical industry has responded to consumers. Together with those who work in the corporate world, we must learn to cope with plastic waste, while at the same time save raw materials and energy for tomorrow. To create a new world of plastics and polymers will require the intelligence and efforts of policy planners, legislators, economists, manufacturers, consumers, and chemists. This chapter showed that efforts at reducing, reusing, recycling, and recovering are well under way.

As we've seen in previous chapters, everything is interconnected. In this chapter, we looked at the connections of polymers to their raw materials—petroleum or plants—as well as their connections to waste (or compost) in the environment. The chapter ended with an unexpected connection, that of additives to plastics that leach into the environment and have drug-like properties similar to those of estrogens. Every day, we use many plastic containers and wraps to store and sometimes cook food, so we should keep this in mind as we move into the kitchen laboratory—the topic of the next chapter.

LEARNING OUTCOMES

Having studied this chapter, you should now be able to:

- compare the properties and applications of plastics and polymers (9.1)

- describe the general structure of polymers (9.2)

- explain the mechanism of addition polymerization reactions (9.3)

- describe how polyethylene properties depend on their molecular structures and intermolecular forces (9.4)

- compare the properties and applications for common types of polymers (9.5)

- explain the mechanism of condensation polymerization (9.6)

- identify and name various functional groups that are present in monomer structures (9.6)

- describe the formation of polyamides such as nylon and proteins (9.7)

- describe the four Rs of dealing with plastic waste (9.8)

- describe the process and applicability of recycling plastics (9.9)

- explain some differences between plant-based polymers and petroleum-based polymers (9.10)

- summarize the environmental and health consequences of using plastics (9.11)

Questions

Emphasizing Essentials

1. Give two examples of natural polymers and two of synthetic polymers.

2. Think about your intended profession or career path. How can you contribute in a meaningful way to reducing our solid waste? Suggest three ways.
   Hint: Thinking about the Four Rs may be of help.

3. Equation 9.1 contains an *n* on both sides of the equation. The one on the left is a coefficient; the one on the right is a subscript. Explain.

4. In Equation 9.1, explain the function of the R· over the arrow.

5. Describe how each of these strategies would be expected to affect the properties of polyethylene. Also provide an explanation at the molecular level for each effect.
   a. increasing the length of the polymer chain
   b. aligning the polymer chains with one another
   c. increasing the degree of branching in the polymer chain

6. Figure 9.3a shows two bottles made from polyethylene. How do the two bottles differ at the molecular level?

7. Ethylene (ethene) is a hydrocarbon. Give the names and structural formulas of two other hydrocarbons that, like ethylene, can serve as monomers.

8. Why is a repeating head-to-tail arrangement not possible for ethylene?

9. Determine the approximate number of $H_2C{=}CH_2$ monomeric units, *n*, in one molecule of polyethylene with a molar mass of 40,000 g/mol. How many carbon atoms are in this molecule?

10. A structural formula for styrene is given in Table 9.1.
    a. Redraw it to show all of the atoms present.
    b. Give the chemical formula for styrene.
    c. Calculate the molar mass of a polystyrene molecule consisting of 5000 monomers.

11. Vinyl chloride polymerizes to form PVC in several different arrangements, as shown in Figure 9.9. Which example is shown here?

12. Here are two segments of a larger PVC molecule. Do these two structures represent the same arrangement? Explain your answer by identifying the orientation in each arrangement.
    Hint: See Figure 9.9.

13. Butadiene, H_2C=CH—HC=$CH_2H_2C_5CH$—HC_5CH_2, can be polymerized to make a synthetic rubber. Would this be by addition or condensation polymerization?

14. Which of the "Big Six" most likely would be used for these applications?

    a. clear soft drink bottles

    b. opaque laundry detergent bottles

    c. clear, shiny shower curtains

    d. tough indoor–outdoor carpet

    e. plastic baggies for food

    f. packaging "peanuts"

    g. containers for milk

15. a. Analogous to Equation 9.3, write the polymerization reaction of *n* monomers of propylene to form polypropylene.

    b. Analogous to Figure 9.9, show a random arrangement of the monomers in a segment of polypropylene.

16. Many containers are made from plastic. Check the recycling code on 10 containers of your choice (see Table 9.1). In your sample, which polymer did you most frequently encounter?

17. Name the functional group(s) in each of these monomers:

    a. styrene

    b. ethylene glycol

    c. terephthalic acid

    d. the amino acid in which R=HR=H

    e. hexamethylenediamine

    f. adipic acid

18. Circle and identify all the functional groups in this molecule:

19. Kevlar is a type of nylon called an *aramid*. It contains rings similar to those of benzene. Because of its great mechanical strength, Kevlar is used in radial tires and in bulletproof vests. Your Turn 9.17 gives the structures for the two monomers: terephthalic acid and phenylenediamine. Name the functional groups in both the monomers and in the polymer.

20. Table 9.3 gives structural formulas for ethanoic acid and propanoic acid. From these two names, you should be able to determine the naming pattern.

    a. How would a carboxylic acid containing five carbon atoms be named?

    b. Methanoic acid is the smallest carboxylic acid. Also known as formic acid, it is one of the components in the sting of an ant bite. Draw the structural formula for methanoic acid.

    c. Butanoic acid, like propanoic acid, has a sharp smell. Draw the structural formula for butanoic acid.

21. Silk is an example of a natural polymer. Name three properties that make silk desirable. Which synthetic polymer has a chemical structure modeled after silk?

22. The Dow Chemical Company won a Presidential Green Chemistry Challenge Award for developing a process that uses CO_2 as the blowing agent to produce Styrofoam packaging material.

    a. What is a blowing agent?

    b. What compound does CO_2 likely replace in the process, and why is this substitution environmentally beneficial?

23. Suggestions for reducing your waste include (1) buying in bulk and/or economy sizes and (2) avoiding individually packaged servings. Let's say you followed this practice for these cases. Which plastic would you use less of? Would you use more of something else?

    a. For use in your refrigerator, buying a half-gallon plastic jug of milk rather than 2 quarts.

    b. For guests at a reception, purchasing 2-liter bottles of lemonade rather than individual bottles.

    c. Buying more concentrated laundry detergent in a smaller plastic bottle.

24. Recycled products now are beginning to provide the origin of the recycled material.

    a. Give examples of postconsumer content and of preconsumer content.

    b. Do recyclable products contain recycled materials?

Concentrating on Concepts

25. Draw a diagram to show the relationships among these terms: *natural, synthetic, polymer, nylon, protein.* Add other terms as needed.

26. Currently, many 2-liter beverage bottles are made of PET with polypropylene caps. Why is polypropylene a good choice for a bottle cap? What difficulty does using polypropylene present in the recycling of PET bottles?

27. Glucose from corn is the source of some new bio-based polymer materials. Glucose also is the monomer in cellulose. Earlier in this text, you encountered glucose in the chemical reaction of photosynthesis. What is photosynthesis and from what compounds is glucose produced?

28. The properties of a polymer depend, in part, on which chemical elements it contains. Name three additional things that influence the properties of a particular polymer.

29. Many monomers contain a C=C double bond. Select such a monomer and draw its structural formula together with the corresponding polymer. Describe the similarities and differences between the monomer and the polymer.

30. What structural features must a monomer possess to undergo addition polymerization? Explain, giving an example. Do the same for condensation polymerization.

31. This equation represents the polymerization of vinyl chloride. At the molecular level as the reaction takes place, how does the Cl—C—H bond angle change?

$$n \quad \overset{H}{\underset{H}{}} C = C \overset{H}{\underset{Cl}{}} \xrightarrow{R\cdot} \left[\begin{matrix} H & H \\ | & | \\ C & C \\ | & | \\ H & Cl \end{matrix} \right]_n$$

32. Polyacrylonitrile is a polymer made from the monomer acrylonitrile, CH_2CHCN.

   a. Draw the Lewis structure for this monomer.
   Hint: The N atom is attached via a triple bond.

   b. Polyacrylonitrile is used in making Acrilan fibers used widely in rugs and upholstery fabric. If ignited, this fiber can release a poisonous gas. In the case of a fire, what danger might rugs and upholstery made of this polymer present?

33. Roy Plunkett, a DuPont chemist, discovered Teflon while experimenting with gaseous tetrafluoroethylene. Here is the monomer:

$$\overset{F}{\underset{F}{}} C = C \overset{F}{\underset{F}{}}$$

   a. Analogous to Equation 9.1, write the chemical reaction for the polymerization of *n* molecules of tetrafluoroethylene to form Teflon.

   b. Why is a repeating head-to-tail arrangement not possible for this polymer?

   c. Teflon is a solid and CFC-12 (CCl_2F_2) is a gas. Nonetheless, they both contain C—F bonds. What other characteristics do Teflon and CFC-12 have in common?

34. Equation 9.1 shows the polymerization of ethylene. From the bond energies of Table 6.1, is this reaction endothermic or exothermic?

35. Would your answer from Question 34 differ if tetrafluoroethylene were used as the monomer? See Question 33 for the monomer.

36. Do you expect the heat of combustion of polyethylene, as reported in kilojoules per gram (kJ/g), to be more similar to that of hydrogen, coal, or octane, C_8H_{18}? Explain your prediction.

37. Recycling is not the same as waste prevention. Explain.

38. Here is a recycling symbol that is more colorful than the standard ones used on many plastic containers.

PLA

   a. What is PLA?

   b. Why is corn depicted in the center of the symbol?

   c. This symbol is printed in green ink, presumably to convey that this polymer is "green." Give two reasons why PLA is considered an eco-friendly polymer.

   d. For each of your reasons in the previous part, provide information counter to your argument.

39. Consider the polymerization of 1000 ethylene molecules to form a large segment of polyethylene.

$$1000 \ CH_2{=}CH_2 \xrightarrow{R\cdot} (CH_2CH_2)_{1000}$$

   a. Calculate the energy change for this reaction.
   Hint: Remember that polystyrene foam is made with blowing agents.

   b. To carry out this reaction, must heat be supplied or removed from the polymerization vessel? Explain.

40. Here is the structural formula for Dacron, a condensation polyester:

$$\left[O{-}CH_2{-}CH_2{-}O \overset{\displaystyle O}{\underset{\displaystyle}{\parallel}} C \bigcirc C \overset{\displaystyle O}{\parallel} \right]_n$$

Dacron is formed from two monomers, one with two hydroxyl groups (—OH) and the other with two carboxylic acids (—COOH). Draw a structural formula for each monomer.

41. When you try to stretch a piece of plastic bag, the length of the piece of plastic being pulled increases dramatically and the thickness decreases. Does the same thing happen when you pull on a piece of paper? Why or why not? Explain on a molecular level.

42. Consider Spectra, AlliedSignal Inc.'s HDPE fiber, used as liners for surgical gloves. Interestingly, Spectra is linear HDPE, which is usually associated with being rigid and not very flexible.

   a. Suggest a reason why LDPE cannot be used in this application.

   b. Name two other possible uses of a fabric made of Spectra.

43. The Four Rs are reduce, recycle, reuse, and recover.

   a. Give an example of each, naming the plastic involved.

   b. A possible fifth R is "rethink." For example, plastic waste can be rethought in terms of benefits to public health. Give an example of a connection between waste reduction and public health.

44. All the Big Six polymers are insoluble in water, but some dissolve or at least soften in hydrocarbons (see Table 9.1). Use your knowledge of molecular structure and solubility to explain this behavior.

45. Polystyrene foam packing peanuts are degraded when immersed in acetone (a solvent in some nail-polish removers). If the acetone is allowed to evaporate, a solid remains. What is this solid? Explain what happened. *Hint:* Remember that polystyrene foam is made with blowing agents.

46. Today, some packing peanuts are made from plant-based materials rather than polystyrene foam.

   a. Starch is one of the options. What is starch and what is its source?

   b. Name two advantages and two disadvantages of starch packing peanuts.

   c. Name an option for disposing of starch packing peanuts.

47. Explain the concept of shifting baselines. Then, give two examples each in regard to:

   a. plastic items used in packaging.

   b. plastic item contamination in waterways.

48. DEHP is a plasticizer that is an example of a phthalate, an ester of phthalic acid.

   a. What is a plasticizer?

   b. Why are plasticizers such as DEHP added to PVC?

   c. DEHP has been banned for some uses. Name two and explain why.

Exploring Extensions

49. Find the structural formula for the acetone molecule mentioned in Question 45. What functional group does it contain? Why does polystyrene dissolve in acetone? Would polystyrene also dissolve in water or ethanol? Explain.

50. Cotton, rubber, silk, and wool are natural polymers. Consult other sources to identify the monomer in each of these polymers. Which are addition polymers and which are condensation polymers?

51. A Teflon ear bone, fallopian tube, or heart valve? A Gore-Tex implant for the face or to repair a hernia? Some polymers are biocompatible and are now used to replace or repair body parts.

   a. List four properties desirable for polymers used *within* the human body.

   b. Other polymers are used *outside* your body but in close contact with it, such as those in contact lenses. What are contact lenses made of? What properties are desirable?

52. PVC, also known as "vinyl," is a controversial plastic. Comment on the controversies, either from the standpoint of a consumer or from a worker in the vinyl industry.

53. Learn the story of the discovery of Kevlar. This polymer was originally sought for use in radial tires but found other applications as well. Write a short report citing your sources.

54. Isoprene polymerizes to form polyisoprene, a natural rubber. Here is the structural formula of isoprene, with its carbons numbered:

$$CH_2 = \underset{1}{\overset{2}{C}} \diagdown \overset{CH}{\underset{3}{\diagup}} \overset{}{\diagdown} \underset{4}{CH_2}$$
$$\mid$$
$$CH_3$$

When isoprene monomers add, polyisoprene has a C=C double bond between carbon atoms 2 and 3. How does this double bond form? *Hint:* Each C=C bond contains four electrons. Each new C—C bond that forms to link two monomers needs only two electrons, one from each of the monomers that joined to form it.

55. Synthetic rubber is usually formed through addition polymerization. An important exception is silicone rubber, which is made by the condensation polymerization of dimethylsilanediol. Here is a representation of the reaction:

$$n\ HO - \underset{\underset{CH_3}{|}}{\overset{\overset{CH_3}{|}}{Si}} - OH \longrightarrow \left[O - \underset{\underset{CH_3}{|}}{\overset{\overset{CH_3}{|}}{Si}} - O \right]_n + n\ H_2O$$

   a. Predict two properties for this polymer. Explain the basis for your predictions.

   b. Silly Putty is a popular form of silicone rubber. Name two of its properties.

   c. Name two other household uses for silicone rubber.

56. Given the number of personal computers in use today, there is good reason to keep keyboards, monitors, and "mice" out of the landfill.

    a. Which polymers do your computer and its accessories contain?

    b. What are the options for recycling the plastics in computers?

57. Although PVC is traditionally considered to be an unrecyclable plastic, describe how scientists are working on the development of recycling pathways for this material.

58. Some regions in the United States have bottle bills that require a deposit on some or all containers. Some grocers, beverage companies, and bottle associations stand strongly against bottle bills. In contrast, some consumer groups and environmental groups argue strongly for them. Draft a one-page position statement that speaks either for or against bottle bills.

59. Cargill won a 2007 Presidential Green Chemistry Challenge Award for using soybeans instead of petroleum to produce polyols. What is a polyol? How are polyols used to produce "soybean plastics"?

YOUR FAVORITE MEAL PREP

Watch the chapter opening video (www.acs.org/cic), in which cooking and baking are described as similar to what one would do in the chemistry laboratory. Consider making your favorite meal. Identify which steps in its preparation involve a physical change and which steps involve a chemical change.

In this chapter, you will explore the following questions:

1. How does your tongue detect flavors?
2. What are some of the health benefits of chocolate?
3. How can we ensure that a food product is prepared consistently?
4. How can we convert ingredient quantities listed in a recipe to other units?
5. Why do we need to adjust recipes when cooking at higher altitudes?
6. What is *sous vide* cooking?
7. How does a microwave oven cook our food?
8. What are some ways we can preserve our food?
9. How can we tell if our food is ready to consume?
10. How are the three states of matter exploited for food preparation?
11. What are the applications of fermentation for food and drink?
12. How is alcohol metabolized in our body?
13. How are flavors extracted from coffee and tea?

Introduction

Accompanying this paragraph are words of wisdom from that fictional, though masterful, chocolatier and candy maker, Mr. Willy Wonka. From its wonderful smell, smooth texture, soothing taste, and lingering finish, chocolate represents a remarkable culinary and scientific creation. Ironically, it all starts from raw cacao beans that have a repulsive taste and appeal. However, through a laborious set of chemical and physical processes, the beans become the wonderful chocolate many of us crave and enjoy.

> *"Invention, my dear friends, is 93% perspiration, 6% electricity,*
> *4% evaporation, and 2% butterscotch ripple."*
>
> —Willy Wonka

We eat and drink for a variety of reasons. Fundamentally, foods and beverages are necessary for our survival, providing the nutrients and energy we need. However, many of us also eat food because it tastes good, makes us feel good, and is part of our customs and traditions. Regardless of whether we are herbivores, carnivores, or omnivores, our food types, the ways in which we prepare food, and our consumption practices are often tied to our principles, lifestyles, families, cultures, and histories. These traditions involve stories, rituals, choices, and even religious observations. This chapter surveys the chemistry of food and drink.

Health Disclaimer: This chapter is meant to discuss the science of different ways we prepare food and beverages for consumption. Although we do touch on general ideas about food and our health, we do not discuss any details of benefits or detriment to one's health in these preparations. Although the next chapter will provide information about food, nutrition, and health, you may also want to conduct your own research with trained medical and nutritional health specialists about any health impacts of the foods and beverages we discuss.

2014 Wake Forest University. Photo by Ken Bennett

@HOME   What's in a Mouthful?

Traditionally, regions of the tongue have been identified that are consistent with a particular type of taste. Gather a sample of lemon juice, sugar, and salt. With a toothpick, apply samples of each to the tip, sides, and back of the tongue. Which part of your mouth responds to each food? Draw a map of your tongue that shows which regions responded to each type of food.

Your Turn 10.1   The Truth Behind the Tongue Map

Tongue map—true or false? In the previous activity, you were asked to map different types of taste to regions of your tongue. Were different regions of your tongue able to detect sweet or sour tastes? Do a bit of research on the Internet to find out the origin and current thoughts regarding the so-called "tongue map."

10.1 | What's in a Mouthful? The Science of Taste

Learning Objective: Describe how your tongue detects flavors

The bottom-line answer to the question posed in this section title is "a whole bunch of chemical interactions!" Our tongue is a remarkable organ and an essential nerve center that also helps move food around to aid in our chewing. Our tongue also detects taste—or the more technical term, *gustatory*—sensations. Most scientists, chefs, and other experts agree that there are six basic tastes our tongue detects: sweet, salty, bitter, sour, savory (umami), and fatty (oleogustus). From those six basic tastes, countless flavor combinations can be made and experienced. The mouth also helps determine a food's chewiness, oiliness, texture, and viscosity. A common misconception is that the tongue contains specific regions that detect each flavor, often illustrated by a "tongue map" (Figure 10.1a). However, research has shown that diverse tastes

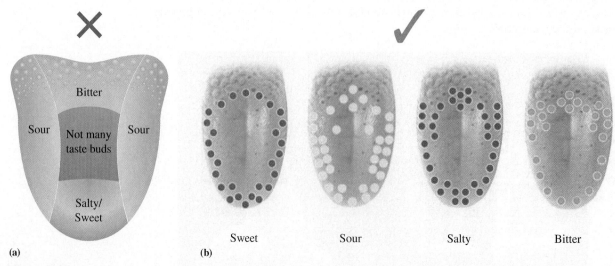

Figure 10.1

(a) A traditional diagram of taste buds on the tongue. (b) An illustration of taste buds, indicating that the entire tongue can sense diverse tastes more or less equally.

are sensed by all parts of the tongue equally, with the sides being more sensitive than the middle overall (Figure 10.1b). The video to the right delves more deeply into our sense of taste and how our taste buds may be tricked by the presence of substances such as toothpaste.

The molecules in food and drinks interact with the macromolecular sensors on the surface of our tongue. Like other places in our bodies, our tongue has nerve cells that have both chemical receptors and ion channels. We perceive taste because a particular molecule binds to these receptors, which triggers a particular signal to the brain. Alternatively, brain signals may also be generated by a change in the number of ions passing through an ion channel. For example, as you saw in Section 5.6, strong electrolytes such as sodium chloride completely dissociate into ions when dissolved in water (the solvent of our saliva):

$$NaCl(s) \xrightarrow{H_2O(l)} Na^+(aq) + Cl^-(aq) \qquad \textbf{[10.1]}$$

The sodium-ion channels of our tongue detect the increased concentration of sodium ions from these food sources, and send signals to our brain that trigger the detection of saltiness. Similarly, proton-ion channels detect acidity:

$$\underset{\substack{\text{vinegar} \\ \text{(acetic acid)}}}{CH_3COOH} \rightleftharpoons \underset{\substack{\text{acetate} \\ \text{ion}}}{CH_3COO^-} + \underset{\text{proton}}{H^+} \qquad \textbf{[10.2]}$$

These experiences do not stop at the tongue. The chemicals from the substances we consume can also interact in many other locations in our body that affect our mood and health. There have been significant research efforts to understand receptors and chemical reactions occurring on the tongue. However, many more studies are needed to understand all of the processes that help us enjoy those first contacts with food and drinks on our tongue.

What Affects Our Sense of Taste?

Check out this video (www.acs.org/cic) to see how toothpaste affects our sense of taste.

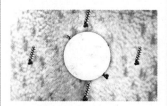

2018 American Chemical Society

Questions to Consider
a. What other substances have you noticed will affect your sense of taste?
b. Using the Internet, describe some reasons why some people with the COVID-19 virus lost their sense of taste and smell. Why didn't this affect everyone with this virus?

@HOME   Taste Testing!

Assemble an assortment of:

- Squares of semi-sweet and milk chocolate
- Fruit squares (watermelon, apple, oranges, grapes, etc.)
- Ice cubes
- Garlic powder
- Limburger or other smelly cheese
- Powdered ginger
- Cinnamon graham crackers
- Garlic crackers
- Tonic water
- Lemon juice
- Table salt
- Sweet solution: ½ teaspoon table sugar dissolved in 2 tablespoons water
- Salty solution: ⅛ teaspoon table salt dissolved in 2 tablespoons water
- Sour solution: ½ teaspoon vinegar dissolved in 2 tablespoons water
- Bitter solution: coffee, brewed or made from instant

Devise a taste-testing procedure to answer the following questions:

a. Can the same food/drink taste different to different people?
b. How does the sense of smell affect food/drink flavors?
c. How do food/drink mixtures (different textures and/or flavors) interfere with the ability to identify tastes?
d. How would salt change the flavor of the foods/drinks? Would the presence of differing amounts of salt change these effects on flavor?

10.2 | Chocolate!

> *Learning Objective: Describe the health benefits of various chocolate ingredients*

The smell of food is where our experience starts. Those wonderful aromatic compounds enter our nostrils and bind with odor receptors in the nasal cavity. Messages are then sent to our brain centers, setting up what we expect to taste in the food. For many, smells may elicit powerful memories—grandma's fresh bread, a summer barbecue, or hot spiced apple cider around Halloween. The aroma of chocolate may be no different—its beckoning in hot chocolate after a winter's snowball fight, or its welcoming in a dessert cake as it is pulled from the oven for the end of a family meal. In fact, more than 600 chemicals come together to give chocolate its aroma and taste. It is no wonder that such a product can produce such a pleasing response.

@HOME   How Does Aroma Affect Taste?

You will need samples of apple and potato chopped into small pieces. Close your eyes and have someone mix up the samples. Pinch your nose closed and taste both of the samples, taking a drink of water between the two different foods. Can you tell the difference between the apple and the potato? Explain what you experienced. Try this with other foods—for example, flavor extracts or jelly beans.

(top): www.nick-moore.com/Moment Open/Getty Images; *(bottom):* Andreas Altenburger/Shutterstock

Let's return to chocolate. The Aztecs called it Theobroma—"food of the gods." Food researchers have studied this food and have discovered a number of healthy attributes for our bodies. Its smell originates soothing feelings, and once chocolate touches the tongue, the cascade of reactions to our taste centers in the brain are initiated. No matter how you eat chocolate, its distinctive taste is likely to put one in a pleasing or calm mood, even if just for a moment.

There are chemical reasons for the relaxing feeling we have after eating chocolate. One is that chocolate contains *theobromine* (Figure 10.2), which may help regulate blood pressure and stimulate brain activity. Chocolate also contains *tryptophan*, a chemical precursor for the body's synthesis of *serotonin*, a neurotransmitter responsible for making us feel happy. Chocolate has also been found to contain *anandamide*, another neurotransmitter that targets the same brain receptors as *tetrahydrocannabinol* (THC), the active component of marijuana.

Mood and health are certainly connected, and chocolate also contains health-beneficial chemicals. Among them are *flavonols* (Figure 10.2), or phytochemicals, that are also present in teas, red wine, and fruits and berries. These compounds have antioxidant properties and support a healthy heart. Furthermore, recent research has indicated that these compounds may even have a role in cancer prevention. Chocolate also contains magnesium, phosphorus, and potassium, essential elements for the body.

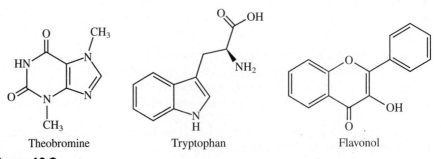

Theobromine   Tryptophan   Flavonol

Figure 10.2

Chemical structures for theobromine, tryptophan, and flavonol—key chemicals found in chocolate.

Dark chocolate tends to be healthier than milk or white chocolate, because it has less saturated fat. Although one could argue whether or not chocolate is a "perfect" food, it certainly is among those that provide the total package—great taste, good feelings, and good health. You can get your chocolate fix for the day by first making a delicious cup of hot chocolate using the recipe below. The activities that follow will allow you to explore the full flavors of chocolate varieties.

Cocoa Nib Hot Chocolate
(Yield: four 6-oz servings)

Ingredients
1 ounce cocoa nibs
16 ounces whole milk
6 ounces 65% to 70% bittersweet chocolate, finely chopped
3 ounces sugar
2 ounces water
¼ teaspoon kosher salt

Procedure
- Coarsely chop the cocoa nibs using a food processor or a spice grinder.
- Add the milk to the chopped cocoa nibs in a 1-quart microwave-safe bowl. Microwave until the milk reaches 160 °F (about 3 minutes on high). Let the solution steep at room temperature for 30 minutes.
- While the nibs are steeping, combine the bittersweet chocolate, sugar, water, and salt in a 1-liter French press. Set aside.
- Once the nibs have steeped, microwave the mixture until it simmers or reaches 185 °F (about 2 minutes on high). Using a fine-mesh strainer, strain the nibs into the French press carafe. Let it sit for 1 minute, and then stir.
- Pump the French pressplunger 10 to 15 times to thoroughly mix the hot chocolate. It should be frothy.

Source: Recipe credited to Alton Brown, 2011.

The Different Types of Chocolate

Which is "better"—milk or dark chocolate? Check out this video: www.acs.org/cic.

Questions to Consider
a. This video compares milk chocolate to dark chocolate. Do you have a preference for one over the other. What do you like or dislike about chocolate?
b. How does dark chocolate compare to milk chocolate for health benefits, texture, flavor and cost to produce?
c. Research the sustainability of chocolate production. Is chocolate production sustainable? Why or why not?

©2018 American Chemical Society

It has been suggested that letting chocolate melt and spread across your tongue is the best way to enjoy its full flavors. This warms the chocolate and allows the various components, including flavorful oils, to "bloom" and release their full flavor. Melting it in this way also allows the chocolate to spread across your tongue and contact more of those vital flavor receptors. Try it yourself in the next activity!

So, what's in a mouthful? Nothing should be taken for granted. What we consume directly affects our longevity, energy, and mood. In order to eat and experience these wonderful sensations, though, we must first prepare the food—the topic of our next few sections.

@HOME
A Chocolate Taste Test!

Find three different bars of plain chocolate. They could be different brands, or dark, milk, and white chocolate. Put a small sample of each in your mouth, one at a time. Try tasting it in two different ways:

1. By putting it in your mouth and chewing it.
2. Now, repeat using the same three types of chocolate, but this time place an ample amount of each in your mouth, let it melt, and think about the flavors.

Questions to Consider
What is the texture of each sample? How does the sample taste? What do you hypothesize to be different about each type of chocolate?

10.3 | The Kitchen Laboratory

> *Learning Objective: Explain the importance of recipes to produce the desired taste, texture, and consistency of prepared food*

We transform the food we grow, cultivate, and raise. Oftentimes, we do this to make it safe to eat, such as by reaching a temperature that will kill any unwanted organisms that might make us sick. At other times, we transform our food to make it taste better. As a result, the kitchen, in so many ways, becomes a place to experiment, explore, and refine the possibilities for enhancing the experiences and even the nourishment we gain from food and drink.

Maybe you know someone who seems to make perfect bread or biscuits every time they try, but who never lifts a measuring cup or spoon. They cook by feel and by sight. They likely follow a practiced and carefully acquired regiment that has amazing consistency because they practiced. Although there are certainly wonderful moments and opportunities to be spontaneous in the kitchen, good cooking does require—like good scientific research—a careful set of protocols and replication methods.

Just as a different pathway can be used to synthesize many chemicals in the laboratory, there are many ways to produce a food product. Examine Table 10.1, compiled from a randomly selected set of five recipes from an Internet search for chocolate chip cookie recipes.

You might at first be tempted to state that we have just contradicted ourselves. That is, there is much variation in these recipes for chocolate chip cookies. And you would be correct until you examine the details of the properties of the cookies made from each recipe. Each variation leads to a different version of the product. Some are soft and chewy, whereas others are crisp and crunchy. So, yes, they all produce chocolate chip cookies, but not all cookies are the same!

monkeybusinessimages/iStock/ Getty Images

The "Best" Chocolate Chip Recipe

Explore the complexities of cookie baking with this video (www.acs.org/cic) and answer the questions that follow.

Questions to Consider
Maybe you have an image of the ideal chocolate chip cookie. What change in ingredient(s) do you predict causes a crunchy versus a chewy cookie? Can you confirm your predictions by making changes to these ingredients? What can you conclude after conducting your experiments?

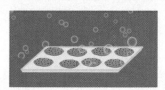

©2018 American Chemical Society

Table 10.1	Selected Ingredient Measurements Per Dozen Chocolate Chip Cookies				
Recipe	**Flour (cups)**	**Total Sugar (cups)**	**Butter (cups)**	**Baking Soda (teaspoons)**	**Brief Description of Resulting Cookies**
A	0.45	0.30	0.20	0.20	"Perfect blend of textures"
B	0.65	0.45	0.27	0.20	"Wonderfully combined textures"
C	0.50	0.27	0.20	0.15	"Crisp bottom, soft top"
D	0.60	0.40	0.20	0.15	"Crisp and crunchy"
E	0.45	0.30	0.20	0.10	"Soft and chewy"
Average	0.53	0.34	0.21	0.16	
Standard deviation	0.09	0.08	0.03	0.04	

10.4 | The Science of Recipes

> *Learning Objective: Convert between measured quantities used in cooking*

Amounts matter. More specifically, amounts in proportion matter. That is, if you change the amounts of one ingredient, the amount of the other ingredient(s) needs to change in proportion. One dozen chocolate chip cookies may be nice, but two dozen may be even better. The cookies that are produced in either amount, though, still need to taste the way you want. Otherwise, why make more?

Your Turn 10.2   Cookie Recipes

You need to make 100 cookies for your school bake sale. One cookie recipe makes 25 cookies and uses the following ingredients:

Ingredients for Chocolate Chip Cookies

½ cup butter
1 cup chocolate chips
½ cup brown sugar
½ cup white sugar
1 egg
½ teaspoon vanilla
1¼ cups flour
¾ teaspoon baking soda
¼ teaspoon salt

How much of each ingredient do you need to make the requested 100 cookies?

First introduced in Section 2.9, the term limiting reagent is often used in chemistry to describe the reactant that is totally consumed during a chemical reaction and thereby limits the amount of product that may be formed. Check out the use of limiting reagents in the kitchen in this next activity.

Your Turn 10.3   "Limiting Reagents" in the Kitchen

The "perfect quesadilla" can be made with two large flour tortillas (200 g per tortilla) and 1 cup of cheese (50 g).

a. You open the refrigerator and find 350 g of cheese that is about to expire. If you use all of the cheese, how many quesadillas can you make?
b. If you have a total of eight tortillas, which ingredient will be completely consumed, and which will be left over?

Recipes are carefully designed chemical reactions that produce the product(s) we want. We can change amounts but only in proportion to the other ingredients. Too much flour in the gravy—no good. Too much salt in the cake—no good. Too much sugar in the dressing—no good. Too much butter—well, maybe that's good (though perhaps not for your health, as we will detail in Chapter 11)! Once again, we change the food we harvest, and not haphazardly. We carefully follow steps so that the product is something we can make, eat, and enjoy again and again!

@HOME   Sugar Cookies!

Investigate ratios in recipes using a pre-packaged sugar cookie mix, eggs, granola, chocolate chips, sticks of butter, five bowls, a teaspoon, a tablespoon, a measuring cup, a baking pan, a large mixing bowl, sticky notes, and an oven. You will be using a ratio of 1 cup sugar cookie mix, 1 egg, 2 tablespoons granola, 1 tsp chocolate chips, and 1/2 stick of butter.

1. Separate your ingredients into separate bowls. Place the measuring cup in the sugar cookie bowl, the tablespoon in the granola bowl, and the teaspoon in the chocolate chip bowl. In front of each bowl add the number of the ingredient you need. 1 for the sugar cookie mix, 1 for the egg, 2 for the granola, 1 for the chocolate chips, and 1/2 for the butter.
2. Go down the line and add each ingredient as it says on the sticky note to the mixing bowl. When you get to the end of the line, you can start again IF you have enough ingredients down the line to complete a full pass-through.
3. When you run out of ingredients, mix the ingredients in the mixing bowl. Pour them into the baking pan. Then, follow the baking instructions on the back of the sugar cookie mix.

Questions to Consider
a. What was the limiting factor in your recipe? What ingredients did you have in excess?
b. How did modifying the recipe to have more butter and eggs likely affect it?

If you review a recipe from a different country, you might find that their system of measuring ingredients is different from the one you use. For example, a country from Europe might use milliliters (mL) to measure the volume of a liquid, whereas in the United States the recipe might use ounces (oz) or teaspoons (tsp). Again, emphasizing the importance of protocols, as in good scientific experimentation, you need to use the correct amounts of ingredients for the recipe. Therefore, it is important to know how to convert between the various systems of measurements.

Your Turn 10.4   Apple Pie Using Metric Units

(left): Mitch Hrdlicka/Photodisc/Getty Images; *(right):* Jiratthitikaln Maurice/Shutterstock

Below is an apple pie recipe using the metric system to describe the amounts of ingredients to use. Conduct an Internet search to translate between the metric system and the English system—used primarily in the United States. Convert the units below into units of cups, teaspoons, and tablespoons.

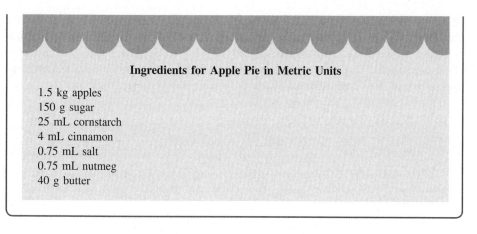

Ingredients for Apple Pie in Metric Units

1.5 kg apples
150 g sugar
25 mL cornstarch
4 mL cinnamon
0.75 mL salt
0.75 mL nutmeg
40 g butter

In a stir-fry or soup recipe, one can usually get away with a little extra dash of this or that. In baking, though, precision and accuracy are critical for creating the particular tastes and textures referred to in many of the examples in the preceding text. Although there is an ongoing debate regarding which is better to use—volumes or masses for ingredients—many professionals use masses for their recipes (Table 10.2).

Think for a minute about flour. Flours are fine powders, and their particles can pack differently based on how you measure the flour from the bag or jar. Scooping the measuring cup into the bag versus using a spoon to transfer the flour to the measuring cup actually can produce different amounts of flour (by mass) that ends up in the recipe. Also consider brown sugar. Many recipes call for the cook to pack the brown sugar into a measuring cup. If we pack the sugar differently, do you think

Table 10.2	Conversions Between Volumes and Masses for Common Baking Ingredients	
Ingredient	**Volume (cups)**	**Mass (grams)**
Butter (salted or unsalted)	1 cup	226 grams
Flour (all purpose or plain)	1 cup	130 grams
Flour (cake)	1 cup	120 grams
Flour (whole wheat)	1 cup	130 grams
Potato Flour	1 tablespoon ½ cup	12 grams 80 grams
Cornstarch (corn flour)	1 tablespoon	10 grams
Ground almonds (almond meal or flour)	1 cup	90 grams
Cornmeal	1 cup	120 grams
Sugar (granulated white sugar)	1 cup	200 grams
Sugar (brown) (lightly packed)	1 cup	210 grams
Confectioners' sugar (powdered or icing)	1 cup	120 grams
Chocolate chips	1 cup	170 grams
Cocoa powder (regular unsweetened or Dutch processed) (can vary by brand)	1 tablespoon 1 cup	6 grams 100 grams
Graham cracker bread crumbs	1 cup	100 grams
Old fashioned rolled oats	1 cup	95 grams

that different amounts (again by mass) would be put into the recipe? At the end of the day, maybe these small variations between using volumes and masses may not be completely noticeable in the everyday kitchen. However, in many restaurants, where it is critical to serve a repeatable and reliable product with almost no variation, these discrepancies in measurements may create unwanted and noticeable differences in their food.

Your Turn 10.5   Density Calculations

Based on the metric conversions you found in Your Turn 10.4, determine the density (in units of g/mL) for a few of the cooking ingredients from Table 10.2. Which flour has the highest density?

10.5 | Kitchen Instrumentation: Flames, Pans, and Water

> *Learning Objectives:*
>
> *Explain why you need to adjust recipes when cooking at higher altitudes*
>
> *Describe how heat energy is used to physically and chemically change food during cooking*

After adhering to the careful measurements of recipe ingredients, the next step is to assemble and transform the ingredients. Let's start with a process that seems rather simple in the kitchen: boiling pasta. Here is one set of instructions:

Step 1: Make sure there is at least enough water to cover the pasta. Too little water will all get absorbed by the pasta before it has finished cooking. Too much is not a problem; it will just take longer for the water to reach a boil. Add some salt and a swirl of oil, if desired.

Step 2: Add uncooked pasta to boiling water.

Step 3: Return water to boil and set a timer to a cook time for *al dente*. This is an Italian phrase for "to the tooth," which means the pasta has a good, firm, but cooked texture.

Step 4: Remove pasta from hot water and drain. Serve with your favorite *gravy* (the true Italian phrase for pasta sauce!).

Your Turn 10.6   Varying Steps in Cooking Instructions

You just read a set of instructions that will give a particular product of cooked pasta. Now, imagine changing aspects of those steps. What would happen if you do not add salt? What happens if you cook your pasta longer than the *al dente* cook time? What if you are cooking pasta at higher-than-sea-level altitudes—should the cooking time remain the same?

Using hot water for cooking may seem very simplistic; however, this technique actually employs a remarkable number of scientific principles. First and foremost, boiling water cooks our food because of the laws of thermodynamics. Heat is transferred from objects at a higher temperature to objects at a lower temperature. Our pasta or vegetables cook when placed into boiling water because heat is transferred from the hot water to the foods initially held at room or refrigerator temperatures.

Water is a good solvent to use for cooking because of its high **latent heat**. In other words, its ability to absorb significant amounts of heat before undergoing a phase

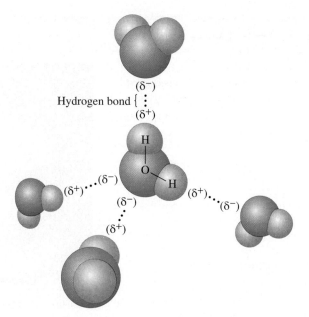

Figure 10.3

An illustration of intermolecular forces, known as hydrogen bonds, among water molecules.

change from liquid to gas. This is partly accounted for because of water's chemical composition (H_2O) and the ways in which the water molecules interact with one another via hydrogen bonding (Figure 10.3 and Section 5.2). This is why water can absorb much heat before changing state; breaking the relatively strong hydrogen bonds requires a significant amount of energy. At its boiling point, the water molecules themselves have more energy than at room temperature. Accordingly, some of this energy is transferred to the food to cook it.

Why do different altitudes require different cooking times when using boiling water? To answer this, we need to return to the concept of boiling. Boiling water is not just about achieving a particular temperature. Water boils at a certain temperature *and* pressure. Recall from Section 6.11 that we define the boiling point as the temperature at which the vapor pressure above the liquid has at least matched the ambient pressure acting on the liquid. Additionally, when a substance is undergoing a phase change, its temperature does not change until the phase change is complete (Figure 10.4). This is because the energy being transferred during a phase change is used to either break or form intermolecular forces and not used to change the kinetic energy of the molecules themselves. Hence, when we cook something with boiling water, we can no longer increase the temperature of the water molecules until they have all changed to steam.

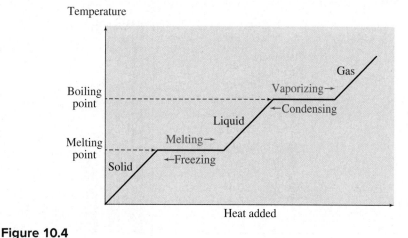

Figure 10.4

Heating curve for all substances that change between solid, liquid, and gaseous states. Check out the simulation (www.acs.org/cic) for an interactive heating curve.

@HOME   From Ice to Steam . . .

Investigate the heating curve for water by observing it in action at home using ice cubes, a saucepan, a food thermometer, a smartphone and a stovetop.

1. Fill a saucepan 1 inch from the top with ice cubes.
2. Clip a food thermometer on the edge of the saucepan and into the ice so that it is not touching the bottom of the saucepan. Set a smartphone or video camera so that it is recording the thermometer readings.
3. Turn on the stovetop burner and observe the thermometer readings.
4. Continue watching as the ice cubes go through phase changes.
5. Don't forget to turn off the stovetop.

Questions to Consider

a. Plot a heating curve, similar to that in Figure 10.4. What shape did your heating curve take?
b. What is happening to the molecules of water at the plateaus in your graph?

Charles D. Winters/Timeframe Photography/McGraw Hill

Since higher altitudes have lower atmospheric pressures, it takes less energy to have the vapor pressure of the liquid reach and exceed the external air pressure. As a result, water will reach its boiling point at a lower temperature in higher altitudes (Figure 10.5), which necessitates longer cooking times.

Your Turn 10.7   Pasta Cooking Times

Predict the difference in boiling time for spaghetti noodles in Denver, Colorado (the "Mile High City") versus the city of Los Angeles (approximately at sea level). Will it take a shorter amount of time or a longer amount of time?

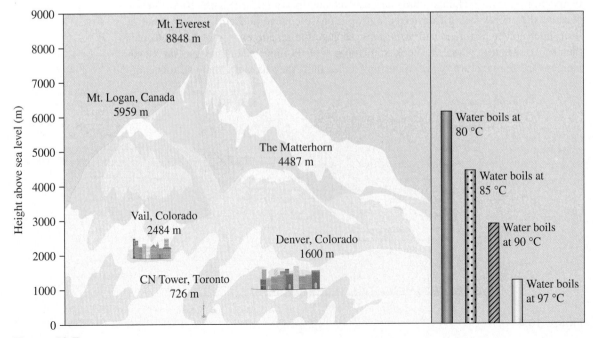

Figure 10.5

The boiling point of water with changes in altitude. Check out www.acs.org/cic to compare the boiling point of water at various locations around the world.

Table 10.3	Altitude Adjustment of Pressure Cookers			
Altitude (ft.)	**15 psi Pressure Cooker (psi)**	**14.5 psi Pressure Cooker (psi)**	**13 psi Pressure Cooker (psi)**	**11.5 psi Pressure Cooker (psi)**
10,000	10.4	9.9	8.4	6.9
9,000	10.8	10.3	8.8	7.3
8,000	11.2	10.7	9.2	7.7
7,000	10.8	10.3	8.8	7.3
6,000	12.1	11.6	10.1	8.6
5,000	12.5	12	10.5	9.0
4,000	13	12.5	11	9.5
3,000	13.5	13	11.5	10
2,000	14	13.5	12	10.5
1,000	14.5	14	12.5	11
0	15	14.5	13	11.5

Pressure cookers also take advantage of the relationship between boiling temperature and pressure. Because a pressure cooker is sealed, once the water is turned to steam, it is trapped in the pot. The pressure above the water also increases so that the boiling point of water increases, raising the temperature of the water (and food). However, one must also take into account a change in altitude when cooking with pressure cookers. The higher the altitude, the lower the actual pressure in the cooker (Table 10.3). In previous chapters, we discussed pressure in units of atmospheres (atm). Table 10.3 uses pressure units of pounds per square inch (psi), which is also commonly used in the United States when referring to pressures of other objects such as tire pressures and inflatable sports equipment. One psi is equivalent to 0.068 atm.

Regardless of the temperature used in cooking, we use heat to transform our foods physically (e.g., melting ice) and chemically. For instance, let's consider the breakfast staple food of poultry eggs, which may be boiled, fried, poached, or baked. Each of these preparations requires heat transfer to physically and chemically change the raw egg to the version we wish to eat. When cooking on a natural gas stove, you might ignite a burner around medium-high heat. Then, you would select a small pan and possibly add some oil or butter. Once the butter or oil sizzles slightly, you would crack the egg into the pan. The translucent portion of the egg first begins to turn white, and the yolk remains a thick yellow liquid. After a few minutes, you might gently take a spatula under the egg and turn it over (the bolder cook may lift the pan by its handle and flip the egg), taking care not to "break" the yolk. You let it go a minute or two more to complete the cooking of the egg white, but not the yolk because you want the "over easy" product. If done properly, you will have accomplished a slightly brown and crisp edge to your egg, which is then slid onto a plate and enjoyed with a piece of toast.

Now, let's take a closer look at the physical and chemical changes that have occurred during the routine practice of cooking an egg. This symphony that produced such a wonderful food was conducted under the laws of thermodynamics. In chemical or physical processes, energy is neither created nor destroyed, but can be transferred and/or transformed. Additionally, within these laws, we also know that when energy is in the form of heat, it can only be transferred from a hotter object to a less hot (or cold) object. When you turn on the stove, a flame emerges or a coil warms. The energy of the stovetop is transferred to the pan. The energy of the pan is then transferred to the oil or butter, causing it to melt, thus creating a physical change. The final transfer of thermal energy to the raw egg results in a chemical change (Figure 10.6). These alterations to the molecules result in changes to the egg's properties, and the clear white liquid is transformed into an opaque white solid. Because the composition of

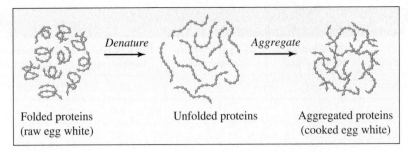

Figure 10.6

Schematic of the denaturation process of egg whites.

the egg yolk is different from the white, it does not respond to the same amount of heat in an identical manner. The amount of heat transferred during the same time duration is not quite sufficient to transform it from a golden yellow gel into a light yellow solid. However, leaving the egg in the heated pan will eventually accomplish this product if it is desired.

The edge of the white has a further opportunity to turn crispy brown. The transferred heat has initiated yet another chemical reaction called the **Maillard reaction**. This reaction occurs at high temperatures and features a chemical reaction between the functional groups present in sugars (the primary component of carbohydrates) and amino acids (the building blocks of proteins) within foods (Figure 10.7). The products of this reaction are the browned crust that forms on cooked eggs, meats, bread, cakes, and cookies, which adds a delightful texture and slightly roasted sweetness to the flavor.

The Maillard Reaction

©2018 American Chemical Society

Want more details about the Maillard reaction? Check out this video: www.acs.org/cic.

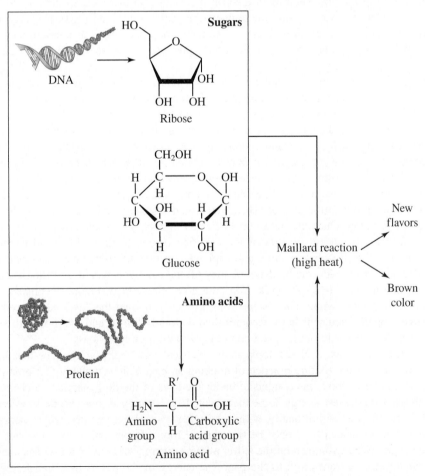

Figure 10.7

Schematic describing the browning and flavoring of foods via the Maillard reaction. More details regarding the structure and function of DNA and proteins will be provided in later chapters.

10.6 | Cooking in a Vacuum: Not Just for Astronauts!

Learning Objective: Describe the technique and benefits of sous vide *cooking*

Another means of cooking food is called *sous vide*. This method started as a French technique and literally means "under vacuum." This method carefully controls the temperature at which food is cooked, which is usually the same temperature at which it is to be eaten. *Sous vide* still uses water as the medium for cooking; however, unlike boiling or steaming, the food never actually contacts the water. *Sous vide* works by using a temperature-controlled water bath appliance, much like what might be used in a chemistry laboratory. In this appliance, circulated water is brought to a certain temperature, say 49 °C (120 °F) for a salmon fillet (Figure 10.8). The food is seasoned as desired and placed in a plastic bag that is vacuum sealed. Upon placing the bag into the water bath, temperature control is maintained, and the food begins to cook. *Sous vide* is usually a longer process for cooking; for example, several meat recipes require 48–72 hours of cooking. Despite the long cook time, the food will not overcook as long as the temperature of the water bath is maintained. However, because the entire piece of meat is at the same temperature, char or browning is not present when the food comes out of the bag. Many recipes suggest a quick turn on a hot grill or skillet to accomplish the Maillard reaction if desired.

So, why cook without air? The answer is really of a practical nature. Unpackaged foods get messy in a water bath, and bags with air bubbles float in the water, resulting in uneven heating of the food. There are scientific reasons for doing so as well. Consistent temperature control is critical to most recipes; however, air is a poor conductor of heat and can create uneven heating. Additionally, water, the important reason food stays juicy, will evaporate if there is available space. If you remove the air from the package and seal the food, the packaging does not allow the water in the food to escape. Hence, some benefits of *sous vide* cooking are a better consistency of temperature and texture of the food product. You will get a chance to practice *sous vide* with the following recipe. The activities that follow will delve into some possible environmental and health implications of this technique.

Figure 10.8

Sous vide cooking.

Roulier/Turiot/Image Professionals GmbH/Alamy Stock Photo

Fragrant *Sous Vide* Salmon
(Yield: 4 servings)

Part I: Ingredients for Fish Spice Mix
50 g hazelnuts
44 g sesame seeds
12 g coriander seeds
10 g poppy seeds
4 g dried, ground ginger
4 g table salt
2.5 g dried chamomile blossoms

Procedure for Fish Spice Mix

- Roast the hazelnuts at 350 °F until the skins turn dark brown. Start checking for color after 10 minutes. Once the nuts have cooled, remove the skins by rubbing them with a cloth. Chop the skinned nuts.
- In a frying pan, stir the sesame seeds over medium heat until they begin to pop. This should take about 3 minutes. Grind the seeds using a mortar and pestle or a coffee grinder.
- In a frying pan, heat the coriander seeds over medium heat until they are fragrant and golden brown. This should take about 3 minutes. Grind the seeds using a mortar and pestle or a coffee grinder.
- Combine all ingredients and grind to a coarse powder. Work in batches if necessary.

Part II: Ingredients for Fragrant Salmon
1 kg water
50 g table salt
40 g sugar
600 g salmon fillets (about four fillets)
30 g of either olive oil or melted butter per fillet
80 g unsalted butter
15 g fish spice mix

Procedure for Fragrant Salmon

- Make a brine by dissolving the salt and sugar in the water.
- Put the brine in a zip-top bag and add the salmon fillets to the bag. Refrigerate for 3 to 5 hours.
- Heat a large water bath to 115 °F. The water bath must be able to maintain this constant temperature.
- Remove the salmon fillets from the brine solution and place each fillet in its own zip-top bag with oil or melted butter. Remove as much air as possible from the bags and seal them.
- Submerge the fillets in the zip-top bag in the water bath. Let the salmon reach a core temperature of 113 °F. It should take about 25 minutes for fillets that are 2.5-cm (1-in) thick.
- Right before the fillets are cooked, in a frying pan, melt the unsalted butter over medium-low heat.
- Add the fish spice to the melted butter and increase the heat until the butter bubbles lightly.
- To the butter–spice mixture, add the fillets and baste for about 30 seconds per side.
- Serve immediately.

Source: Recipe from N. Myhrvold and M. Bilet, *Modernist Cuisine at Home* (The Cooking Lab, 2012).

> **Your Turn 10.8   The Sustainability of Cooking Techniques**
>
> Think about the different ways of cooking that we have talked about in this chapter. Make a list of the pros and cons of each method. Based on your list, which method is most environmentally sustainable? Explain.

> **Your Turn 10.9   Dangers from Chemical Leaching?**
>
> *Modernist Cuisine at Home,* one of the defining volumes for *sous vide* cooking, claims that bags made expressly for cooking are safe.
>
> **a.** What type(s) of polymers are used in such specialized *sous vide* bags?
> **b.** Using the Internet as a resource, are these specialized cooking bags immune to the leaching of unwanted chemicals into our food? Consider what happens to the polymer(s) at the temperatures commonly used for *sous vide* cooking.
> **c.** Would the threat of chemical leaching into food be greater if other inexpensive bags (e.g., Ziploc) were used instead of the specialized *sous vide* variety? Explain.

10.7 | Microwave Cooking: Fast and Easy

> *Learning Objective: Explain how a microwave oven cooks our food*

As our work schedules get busier and workdays get longer, many of us do not always have sufficient time to prepare a complex meal from scratch. In our increasingly busy society, microwave oven cooking has risen to become the most popular form of cooking because of its speed and simplicity. However, food quality is not nearly as desirable as that cooked over a stove or in an oven, where one has more control over the chemical reactions and the resulting development of flavors.

> **Your Turn 10.10   The Electromagnetic Spectrum Revisited**
>
> In Chapter 3, we discussed the various regions of the electromagnetic (EM) spectrum.
>
> **a.** Describe the relative energies and wavelengths of the UV, IR, and microwave regions, and diagram how each of these energies would affect a water molecule (i.e., bond breaking, bond stretching/vibration, or molecular rotation).
> **b.** A company claims to have a new type of cooking apparatus using radio waves to cook food. Its claim is that the food is cooked more uniformly and that food quality is better than using microwaves. Do you believe their claims? Explain your answer.

Due to its lower energy and longer wavelength than UV, visible, or IR regions of the EM spectrum, microwave radiation is not sufficient to cause rupturing of individual chemical bonds. Instead, microwave radiation is absorbed by the water, fat, and sugar molecules in food, which causes these molecules to rotate (Figure 10.9). Because the molecules rotate some 2.5 million times per second, they can easily bump into and rub against one another, resulting in the production of heat due to frictional forces. Microwaves penetrate the food item to a depth of approximately 1–1.5 inches. Accordingly, in thicker pieces of food, microwaves don't effectively reach the center. Unlike a conventional oven in which food is heated by hot air, food cooked in a microwave oven normally does not become brown and crispy because the air inside the oven is at room temperature.

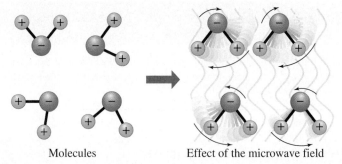

Molecules   Effect of the microwave field

Figure 10.9

Illustration of the rotation of polar molecules in response to an applied external microwave field. Check out this animated simulation of microwave radiation: www.acs.org/cic.

Question to Consider
Sketch and compare the differences in heat transfer occurring as food is cooked in a microwave oven versus a conventional oven. Some claim that microwave ovens cook food from the inside out. Is this accurate?

Metals do not absorb microwave radiation and, instead, reflect these wavelengths. Therefore, a metal such as aluminum is used on the sides of the oven to prevent the microwaves from escaping and irradiating objects outside the oven, such as yourself! If you look closely at the front window of your microwave oven, you will notice a metal mesh. This is designed to be transparent to visible light but opaque to microwave radiation. That is, the holes in this mesh are smaller than the wavelengths of the microwaves but large enough so that visible light may still pass through.

If metal is used inside the microwave oven, why is it so dangerous to place a metal object inside the oven? As you discovered in Chapter 1, metals are great conductors of electricity. Hence, when microwaves irradiate a metal, electrons on its surface move rapidly to the side, which prevents the microwaves from being absorbed by the food item. The oscillation of the microwaves can produce a concentrated electric field at the corners or an edge of a metallic object, which ionizes the surrounding air. This results in an arc (visible sparks) between the metal object and the oven walls, which can cause failure of the microwave source, known as a *magnetron*, and can often damage the walls of the oven.

Contrary to popular belief, it can actually be safe to place small amounts of a metal, such as aluminum foil, into a microwave oven. However, it should never be used to completely cover a food item, because the microwaves would not be absorbed by the food and would result in the dangerous reflection/arcing situation described above. Small pieces of *nonwrinkled* aluminum foil may be used to cover certain areas of foods, such as poultry drumsticks, to prevent overcooking.

Wrinkled foil should not be used as it can cause increased reflection of microwaves and possible sparking. Additionally, the foil should be placed no closer than one inch from the oven walls. If the oven has metal shelves or a metal turntable, foods should not be placed within foil containers or metal pans, and the foil used for food shielding should not touch the metal shelves/turntable to prevent arcing.

Let's Debate!   Sustainability of Cooking Methods

Form two groups: one group will be in favor of the topic (the proposition) and the other group will be against the topic (the opposition). Both groups will try to persuade a neutral person or judge to agree with them. The topic of the debate is the motion. Here is the motion:

It is more sustainable to boil water using a microwave oven relative to using a standard electrical or gas cooktop stove.

Curiosity Conversations

■ How does the electricity consumption and greenhouse gas (GHG) emissions compare among these cooking methods?

■ Considering the average time spent per year on cooking, would there be a significant benefit for sustainability in primarily using one of these cooking methods over the others?

10.8 | Food Preservation

> *Learning Objective: Describe the physical and chemical processes involved in food preservation*

Just like us, the harmful microbes that may be present in food need water to survive. Accordingly, we can preserve food items by simply removing water. Drying and curing have long been a part of our human practices for helping to sustain our food supplies—especially during harsh environmental conditions when food items could not grow or readily reproduce. A slow drying process simply removes the water present in fruits, vegetables, and meats, thus preserving them (Figure 10.10). You will get practice with drying techniques in the next activity, to prepare fresh basil.

@HOME   Fresh Basil Preparation

Investigate different drying techniques at home using fresh basil, paper towels, a microwave-safe plate, a baking sheet, an oven, a food processor, and a microwave.

Oven Method

1. Preheat the oven to its lowest setting possible and leave the oven door partly open.
2. Place the basil leaves in a single layer on a baking sheet and into the oven.
3. Check after 1 hour and every 15 minutes thereafter. The leaves should be crunchy and dark olive green, but not burnt.
4. Grind the dried basil leaves into a fine powder with a food processor.

Microwave Method

1. Place a paper towel on a microwave-safe plate.
2. Place the basil leaves in a single layer on top of the paper towel. Then add a paper towel on top of them.
3. Place them in the microwave for 30 seconds and remove. Observe their color.
4. Repeat Step 3 two more times.
5. Because microwaves differ, you may need to keep microwaving, but do 10 second intervals after this. The leaves should be crunchy and dark olive green, but not burnt.
6. Grind the dried basil leaves into a fine powder with a food processor.

DJTaylor/Shutterstock

Questions to Consider
a. Which method produces the better flavor and why?
b. Air drying is an option for drying basil, but why could this be a dangerous method?

Figure 10.10

Dried fruit and cured meats.

(left): Pakhnyushchy/Shutterstock; *(right):* Pixtal/age fotostock

As an application of osmosis that was first introduced in Section 5.13, sodium chloride is used to remove water from a food item in a process known as *salting*. Salting that includes nitrites (NO_2^-) and nitrates (NO_3^-) also kills bacteria due to their antioxidative properties, and adds flavor to meat. The nitrites and nitrates break down further in the meat into nitric oxide (NO), which binds to iron in hemoglobin and prevents further oxidation of the meat.

Smoking is one of the oldest food preservation techniques. Smoke, usually formed from some type of wood burning, interacts with the meat over many hours. The drying action of the smoke tends to preserve the meat; however, some of the chemicals present in wood smoke (e.g., alcohols, formaldehyde) also serve as natural preservatives by killing microbes and slowing fat oxidation. The specific protein that is key in the smoking process is *collagen*, found mostly in the connective tissue of the animal. Over suitable times and temperatures, the collagen converts into gelatin (Figure 10.11). Gelatin is more water-soluble than collagen and gives way to a desirable tender meat product. Additionally, the much-preferred *smoke ring*, the pink discoloration just under the meat crust, forms because nitrogen dioxide (NO_2) from the smoke interacts with meat compounds and forms a more acidic environment that changes the meat to a pink color.

Interestingly, smoking is not just for meats. Potatoes, corn, eggplant, zucchini, and tomatoes are among the other favorites for smoking, and create a variety of tantalizing flavors that do not resemble their raw counterparts. Nuts, too, can be smoked and eaten as-is, or even transformed into unique varieties of nut-based, nondairy cheeses.

Another way to cause chemical changes to our food is through the use of acidic marinades. Upon extended contact with fish fillets, citrus juices will impart a wonderful flavor and also serve to denature proteins, much like the heat of a skillet or grill. There are a variety of styles for *ceviche* derived from different cultures, but all involve some type of acid as a primary ingredient. The higher-acidic environment can also kill many harmful microbes, which can be used to preserve fish for months at a time. As you will see in the next video, sushi was first developed as a way to preserve fish through exposure to lactic acid, which prevents undesirable microbes from growing.

Pickling is a wet-curing process that uses both salts (and/or nitrates) and acids. Unlike preparing ceviche or sushi, pickling usually involves much longer periods of time exposure—weeks or months—to alter the food product. For example, consider sauerkraut or kimchi, two food products derived from different cultural traditions. At

Sushi Preparation

Check out this video (www.acs.org/cic) to discover the interesting history of sushi preparation.

©2018 American Chemical Society

Question to Consider
Using the internet as a research tool, determine what acids are commonly used in the preservation of foods.

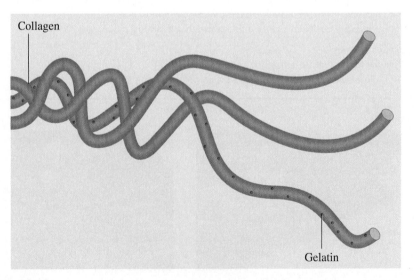

Figure 10.11
Illustration of the thermal unfolding of the triple-helix structure of collagen to form gelatin strands.

the simplest level, both of these dishes involve exposing raw cabbage to vinegar in an *anaerobic* (oxygen-free) environment over time in the presence of bacteria in order to change the vegetable into a new food product. This process is known as *fermentation* and will be described in more detail in Section 10.11.

@HOME   Quick Pickling!

Investigate the process of quick pickling at home using water, vinegar, table salt, a cucumber, garlic (optional), fresh herbs (optional), dried herbs, a jar with a lid, and a refrigerator.

Chatham172/Shutterstock

1. Thinly slice your cucumber and place the slices into the jar.
2. Add equal parts water and vinegar to the jar so that the cucumber slices are covered. For every cup of water and cup of vinegar you add to the jar, add one tablespoon of salt.
3. If you are choosing to add garlic, fresh herbs, or dried herbs, add them to the jar.
4. Add the lid to the jar and shake to mix.
5. Place the jar in the refrigerator for several days and enjoy!

Note: These pickles have not been canned and need to be refrigerated.

Questions to Consider
a. What is the purpose of the salt?
b. How does quick pickling differ from fermentation pickling? Explain the two processes in terms of pH and reactants and products.

Your Turn 10.11   Identifying the Preparation of Your Food

Keep a journal of the foods you eat over three days. From the foods you consumed, can you identify any of the food preparation techniques you just learned about (drying and curing, smoking, acid marination, or pickling)?

10.9 | How Can I Tell When My Food Is Ready?

Learning Objective: Explain how quantitative and qualitative measurements are used to determine the readiness of food

Because of the replication and refinement of recipes over time, most that involve cooking food provide some type of temperature for heat transfer (e.g., "set oven at 425 °F") and time for maintaining the transfer. These protocols provide a reproducible way for the food to reach the appropriate levels of flavor and texture.

Various types of thermometers are used to help detect the optimal temperature for heat transfer to produce the desired food product. In addition to checking the internal temperature, reliable thermometers are necessary when frying foods in oils. For example, in order to fry chicken, several recipes state that the oil should be 149–163 °C (350–375 °F), and then maintained between 177–191 °C (300–325 °F) while the chicken cooks. At lower temperatures, one risks producing unhealthy, undercooked, and soggy chicken.

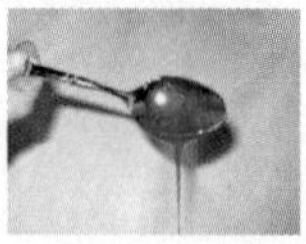

Thread stage

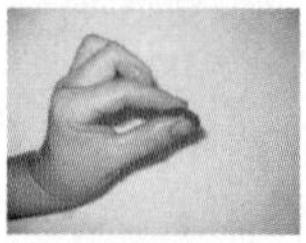

Soft ball stage

Firm ball stage

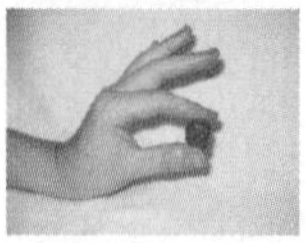

Hard ball stage

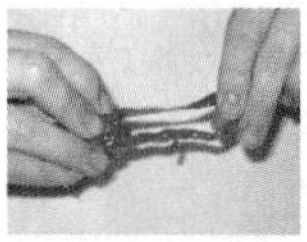

Soft crack stage

Hard crack stage

Brown liquid stage

Figure 10.12

The seven stages of candy making.

©Elizabeth LaBau, www.sugarhero.com

Your Turn 10.12   Cooking Temperatures

Using the Internet or your favorite culinary book, list the recommended temperatures required to cook the meats below. Which meat preparations are below the minimum temperatures suggested by the U.S. Department of Health & Human Services?

a. Beef and lamb steaks, rare
b. Beef and lamb steaks, medium-rare
c. Beef and lamb steaks, medium
d. Beef and lamb steaks, medium-well
e. Beef and lamb steaks, well
f. Chicken
g. Turkey
h. Ham

A good thermometer is also necessary for the creation of hard candy (Figure 10.12). The type of hard candy is generally determined by the stage of sugar cooking (Table 10.4). Hence, a good candy thermometer is necessary to help decipher among these critical stages and the desired candy outcome.

In addition to the **quantitative** information provided by thermometers, changes in color, texture and firmness can also indicate the relative cooking completion in a **qualitative** manner. Recognize once again that all of these property changes that we have described in reaching cooking outcomes result because the chemicals in the food are reacting and being transformed into new chemicals with new properties.

In Table 10.4, the first column could be interpreted as texture assessments. Now, one does not verify these different stages by touching the sugar at the temperature listed. Instead, these are determined by taking a small sample of the sugar at the measured temperature and dropping it into cold water. Once cool enough, the pinch test can help one determine the stage. Give it a try. Find the proper equipment, including a good pot, a candy thermometer, sugar, recipes, cold water, some "pure imagination," and go make some candy. The recipe below should get you started with a candy classic—creamy caramels.

Creamy Caramels
(Yield: 64 candies)

Ingredients
½ cup finely chopped pecans
2 cups sugar
2 cups heavy whipping cream
¾ cup light corn syrup
½ cup margarine or butter

Procedure
• Spread pecans in a buttered square baking pan.
• In a saucepan, stirring constantly over medium heat, bring sugar, cream, corn syrup, and butter to a boil.
• Using a candy thermometer, cook to 245 °F. If you don't have a candy thermometer, you know it's done when you drop a small amount of the mixture in very cold water and it forms a firm ball.
• Spread caramel in pan with pecans and allow to cool.
• Cut into squares.

Source: Recipe from *Betty Crocker's Cookbook,* General Mills, Inc. (New York: Prentice Hall, 1991).

Table 10.4	Stages and Temperatures* Involved in Candy Making	
Stage	**Temperature Range (°C)**	**Sugar Concentration (%)**
Thread (e.g., syrup)	110–112	80
Soft ball (e.g., fudge)	112–116	85
Firm ball (e.g., soft caramel candy)	118–120	87
Hard ball (e.g., nougat)	121–130	90
Soft crack (e.g., saltwater taffy)	132–143	95
Hard crack (e.g., toffee)	146–154	99
Clear liquid	160	100
Brown liquid (e.g., liquid caramel)	170	100
Burnt sugar	177	100

*All temperatures correspond to 1 atm, the atmospheric pressure at sea level.

Similarly, the color of cooked vegetables can provide an indication of their readiness to consume. During cooking, the dark green color of vegetables such as broccoli, asparagus, or spinach is converted into a bright hue. This transformation is due to the heat-induced expansion and rupture of gas pockets that cloud the green chloroplasts of plant cells, thus revealing their true bright color. However, the resulting bright green color of the vegetables will convert back to a dull green or brown shade upon further cooking. If boiling vegetables, this color fading is due to the bright chlorophyll draining out into the surrounding hot water. If sautéing vegetables in a skillet, the heat causes the chemical decomposition of chlorophyll into products with a grayish-yellow color. To keep the intense green color from fading, one can add some baking soda to the cooking water. This increases the basicity of the solution to help to slow the decomposition reaction of chlorophyll. Alternatively, cooking vegetables in a copper pot will help maintain their bright green color. The free copper ions of the cookware help prevent the chlorophyll molecules from decomposing. Of course, one can also simply remove the vegetables at their greenest peak and plunge them into an ice bath to stop the cooking process.

Density—the ratio of mass to volume in a substance or mixture—can also reveal vital information about the suitability for consumption of a cooking product. If you have seen the original "Willy Wonka" movie (1971), you may recall that Mr. Wonka used this idea to detect the good from bad golden eggs laid by his geese. He checked the mass of eggs of the same volume; bad eggs float in tap water, good eggs float in some concentrations of salt water. Why is this? Egg shells are actually a bit porous, and over time bacteria can creep into the egg. As the bacteria grow inside, they produce hydrogen sulfide (H_2S), a gas that produces that foul "rotten egg" smell. Sure, if you crack open a bad egg you will certainly detect the malodor instantly. However, within the intact egg, the gas builds up and makes the egg less dense than a good egg, and thus you have the reason a bad egg floats.

Vintners (wine makers) also take advantage of changes in density caused by changes in sugar content during fermentation. A **hydrometer** is an instrument used to measure the changes in density in liquids (Figure 10.13). As the yeast consumes the sugar, the density of the wine mixture decreases, and the hydrometer sinks to a different level. The vintner can then decide when to stop the fermentation process based on the desired sugar level.

Refractometry represents another method that may be used to determine the sugar content of a solution. All solutions absorb light and cause light to bend (i.e., to be *refracted*); a refractometer measures the extent of this refraction or how much the light has bent. Because solutions represent a different medium than air for the passage

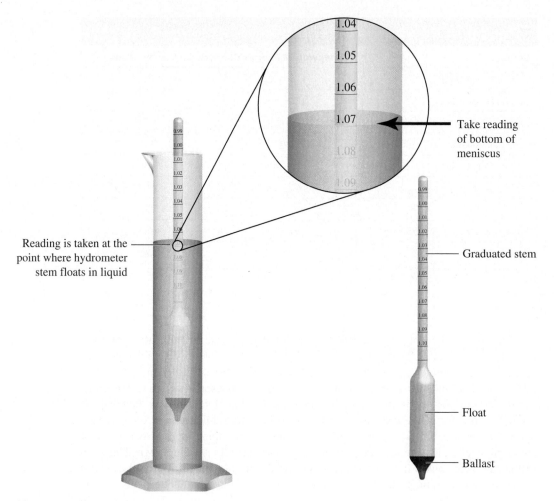

Take reading of bottom of meniscus

Reading is taken at the point where hydrometer stem floats in liquid

Graduated stem

Float

Ballast

Figure 10.13

A simple hydrometer. The difference in the specific gravity readings of a liquid before (original gravity) and after the fermentation process (final gravity) is used to calculate its alcohol by volume (ABV).

of light beams, light travels at different speeds through solutions compared to air. The higher the density of a solution, the more the light will bend while passing through it. For all solutions, there is a linear relationship between its refractive index and **specific gravity**—the ratio of the density of solution to the density of the pure solvent, usually water (Figure 10.14).

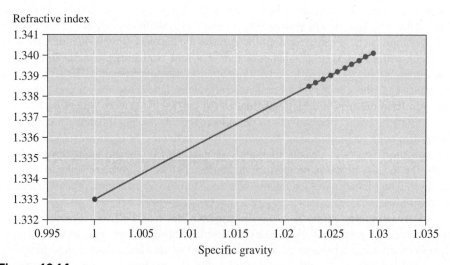

Figure 10.14

Linear relationship between the refractive index and specific gravity of a seawater solution (at 20 °C).

The **Brix scale** (Figure 10.15) is used to quantitatively express the sugar content of an aqueous solution, in which one degree Brix (°Bx) is equal to 1 g of sucrose per 100 g of solution (i.e., 1%, also referred to as a Brix%). Because the temperature of a solution will affect its density, the refractive index or specific gravity measurements should be determined at a constant temperature, typically 20 °C (68 °F).

Your Turn 10.13   The Brix Scale

For the refractometer reading shown in Figure 10.15, what is the concentration of sugar in the aqueous solution? Express the concentration as both % and as molarity (M).

Hint: The density of the solution is 1.06 g/mL.

Your Turn 10.14   Fickle Soft Drink Cans

Maybe you have been to a party or outdoor barbecue where they placed a variety of regular and diet soft drinks in a cooler or tubs with ice water. Are all the cans floating or at the bottom? Or do some float and some sink? Think about what might be causing this difference, in terms of the compositions of the various types of soft drink. Conduct some research to determine why this occurs.

10.10 | Exploiting the Three States of Matter in Our Kitchen

Learning Objectives:

Use Henry's law to determine the concentration of a dissolved gas in solution

Exploit the "like dissolves like" principle in food preparation techniques

The combination of chemicals in our foods, and the diversity of preparation techniques, involve a variety of physical states. The wonderful aromas described earlier come from aromatic chemicals, which frequently form gases at slightly elevated temperatures. These chemicals then entice us—for example, the chemicals that provide the aroma of chicken soup—or repulse us—such as the hydrogen disulfide in those rotten eggs! Water vapor (steam) is used to cook food because of its increased ability to transfer heat to food. However, liquids and solids also show up in countless ways during food preparation. What can be more unique, though, are the ways in which these states can interact to form a variety of food products.

Carbonated beverages are rather common to us. Soft drinks, beer, and champagne all take advantage of dissolved carbon dioxide. If you view a sealed plastic soda bottle, you probably do not notice any bubbles. However, the minute you open the bottle (break the seal), bubbles scatter up through the liquid and escape the surface. Sometimes this can cause a rather intense eruption in the liquid that is caused by the change in pressure. In sealed carbonated beverage containers, the pressure above the liquid is high, which aids in keeping the carbon dioxide gas dissolved in the liquid. This is actually an application of the "like dissolves like" principle introduced in Section 5.7. Since CO_2 is nonpolar and water is polar, there is an incompatibility between the solute and solvent, which acts to drive carbon dioxide from the liquid. As a result, the minute the seal is broken, the pressure above the liquid is decreased and the gaseous carbon dioxide now has sufficient energy under the reduced pressure to escape from the liquid. Check out a fun demonstration of this concept in the following video.

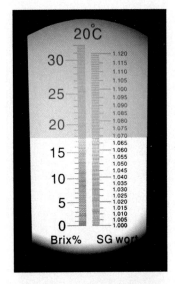

Figure 10.15

A refractometer and Brix scale for determining the sugar content of aqueous solutions. A few drops of liquid are placed onto a prism and light is allowed to pass through. The values of the specific gravity (SG) and °Bx (or %) sugar content of the solution are simply read from the scale where the blue and white regions intersect.

(both): Mark Dierker/McGraw Hill

Henry's law describes the relationship between the pressure of a gas and the concentration of gas dissolved in the liquid (Equation 10.3). This equation can be used for a wide range of applications such as determining the concentration of dissolved oxygen in our bloodstream based on the partial pressure of oxygen gas in our lung alveoli.

$$C = kP \qquad\qquad \textbf{[10.3]}$$

Where: C = concentration of gas in the liquid (mol/L, M)
k = Henry's law constant (mol/L·atm; varies for each gas)
P = pressure of the gas above the solution (atm)

Your Turn 10.15   Flat Cola

Using Henry's law, determine how many grams of carbon dioxide are dissolved in a sealed 500-mL bottle of Coca-Cola with a CO_2 pressure of 1.25 atm. The Henry's law constant for carbon dioxide is 0.031 mol/L·atm.

The trapping of a gas in a liquid or solid may result in a frothy mixture known as a *foam*. Maybe you are familiar with foams already—whipped cream, shaving cream, or foam soap. In the culinary world, these can be used to create subtle nuances of flavor and texture to a dish. How does something like *apple pie with vanilla and cinnamon cream foam* sound? Interesting? Delicious? Foam foods can be made by starting with a thick flavored liquid, thickened with starch, gelatin, eggs, or agar. Then, as seen in the recipe below, a device such as a siphon is used to inject a gas such as nitrous oxide (N_2O) into the liquid under high pressure. *Et voilà*, when the nozzle is triggered, a foam is produced. Tiny gas bubbles get injected into, and trapped in, the thick liquid, creating an edible foam. Although you may have heard that nitrous oxide can cause a "high" when inhaled, when eaten it is a safe, nonbuzz ingredient!

jantenthousand/Shutterstock
©www.MolecularRecipes.com

Strawberry Juice with Apple Foam
(Yield: four 6-oz servings)

Ingredients
24 ounces strawberry juice (buy, or make your own with juicer!)
12 ounces fresh green apple juice (again, use a juicer, if desired)

Procedure
• Add 6 ounces of strawberry juice to four chilled glasses.
• Add apple juice to a whipping canister—do not overfill it.
• Use one cartridge of nitrous oxide to charge the canister. Shake continuously for 5 to 10 seconds.
• Top the strawberry juice with apple foam from the canister.

Source: Recipe from N. Myhrvold and M. Bilet, *Modernist Cuisine at Home* (The Cooking Lab, 2012), pp. 18 and 161.

Foams can also be formed with a little elbow grease by whisking air into the liquid. This is certainly a more laborious process, but can also create wonderful results that you may be familiar with—whipped cream and meringues are just two. Give it a try yourself with this simple recipe.

Homemade Whipped Cream
(Yield: two cups)

Ingredients
1 cup (240 mL) cold heavy cream or heavy whipping cream
2 tablespoons confectioners' sugar or granulated sugar
1/2 teaspoon vanilla extract

Procedure
- Using a hand or stand mixer fitted with a whisk attachment, whip the heavy cream, sugar, and vanilla extract on medium-high speed for 3–4 minutes, or until medium peaks form.
- If you accidentally over-whip the cream, with a curdled and heavy appearance, add in a bit more cold heavy cream and fold it in gently by hand with a spatula until it smooths out.
- Use immediately or cover tightly and chill in the refrigerator for up to 24 hours. You may also freeze whipped cream, but it may lose its creamy texture.

Source: Recipe from https://sallysbakingaddiction.com/homemade-whipped-cream/

Foams consist of two components, a dispersed phase and continuous phase (Figure 10.16). The sensation our mouths experience when tasting a foam is related to the particle sizes of the dispersed phase and the **viscosity** of the continuous phase. The human tongue can distinguish particle sizes larger than about 30 μm (microns) in size. Hence, bubbles smaller than 30 microns will be perceived as a creamy foam; the presence of larger particles will result in a bubbly sensation. In addition, the more viscous the continuous phase of the foam, the smoother its texture will feel on our palettes.

If you have ever made pasta and added a drizzle or two of oil to the water, you likely noticed that the oil floats on the surface. Once again, with "like dissolves like," oil and water do not mix: oil is nonpolar, water is polar. With some assistance, though, the two can be held together. This is the basis for mayonnaise, which is a type of mixture called an **emulsion**. Whereas foams are composed of a gas dispersed in a liquid, emulsions contain a liquid dispersed in another liquid. A considerable amount of shaking of the oil and water mixture can produce a short-lasting emulsion. However, if let to set, the two substances separate once again.

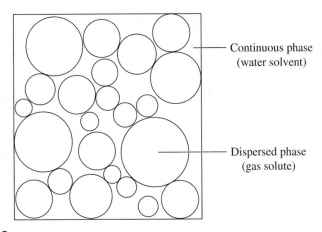

Figure 10.16

Illustration of a foam. Check out this website for more details about the properties and preparation of foams for cocktails: www.acs.org/cic.

©Bradley D. Fahlman

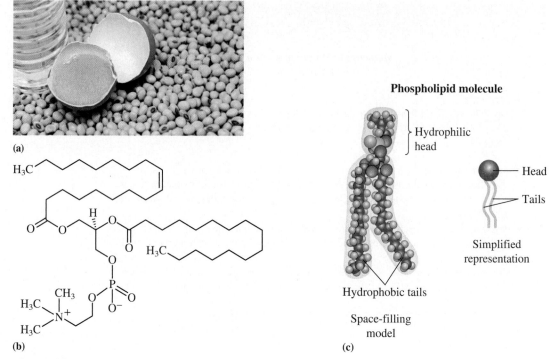

(a)

(b)

Phospholipid molecule

Hydrophilic head

Hydrophobic tails

Space-filling model

(c)

Head

Tails

Simplified representation

Figure 10.17

Three perspectives of lecithin. Shown are **(a)** the macroscopic view of egg yolks, **(b)** structural view of lecithin, and **(c)** space-filling model and symbolic view of a phospholipid molecule that comprises lecithin.

(a): PeoGep/Shutterstock

A way to create a more permanent emulsion, like mayonnaise, is to add an *emulsifying agent* such as egg yolks. Among other chemicals, the egg yolks contain *lecithin*. Lecithin contains phospholipid molecules that are **amphiphilic**, which means they are attracted to both nonpolar and polar molecules (Figure 10.17). In forming mayonnaise, the polar portion of the lecithin attracts water molecules, whereas the nonpolar portion of the lecithin attracts the oil. These two incompatible chemicals are then held together by this molecular hinge.

Garlic & Red Pepper Aioli
(Yield: 1¼ cups)

Ingredients
6 large garlic cloves, peeled
¾ teaspoon kosher salt
2 egg yolks
1 teaspoon freshly squeezed lemon juice
½ cup roasted red peppers (jarred or homemade), roughly chopped
¾ cup virgin olive oil

Procedure
• In a food processor, finely chop the garlic.
• Add everything but the oil to the food processor, and pulse until the ingredients are well combined.
• Thicken the mixture by gradually adding the oil while pulsing the food processor. Keep pulsing until the mixture is uniform.

Source: Emeril Lagasse.com

Another application of intermolecular forces in the kitchen is the process of **spherification** (Figure 10.18). Much like a latex balloon holds a gas, a gel-like, edible membrane can be used to contain small "bites" of flavorful liquid. One version of spherification uses sodium alginate, which is a water-soluble carbohydrate compound. When in solution, the long carbohydrate chains float around unconnected; however, if calcium ions are added, the chains become interconnected. The calcium ions act like fasteners, snapping the chains together and forming a larger molecular network in the form of an impermeable gel around the remaining water solution. Adding a flavored sodium alginate solution dropwise to an aqueous solution of calcium ions produces little solid spheres with the flavored liquid trapped inside. These spheres then "pop" and release their liquid when you bite into them. Try this out yourself in the next recipe.

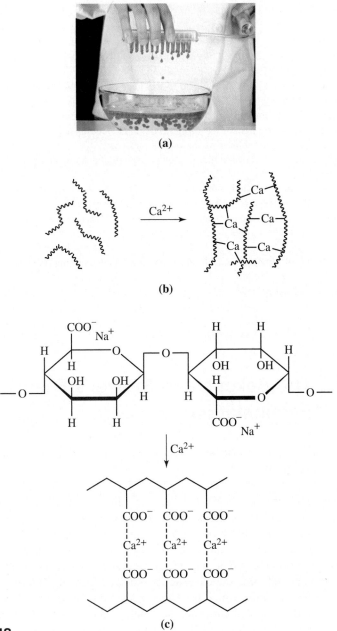

(a)

(b)

(c)

Figure 10.18

Three views of the process of spherification. Shown are **(a)** the macroscopic view of the product; **(b)** a symbolic view, showing the aggregation of individual alginate strands caused by the presence of calcium (Ca^{2+}) ions; and **(c)** the molecular view, showing the chemical equation of alginate molecules bound together by replacing sodium (Na^+) with calcium (Ca^{2+}) ions. Check out this video for a demonstration of spherification: www.acs.org/cic.

(a): ©www.MolecularRecipes.com (www.facebook.com/MolecularGastronomy)

Popping Boba
(Yield: approx. 1 cup)

Whereas traditional boba in bubble tea is made from tapioca starch, popping boba is made from fruit juice. These fruit spheres may be used as a topping for frozen yogurt or served in fruity bubble tea.

Ingredients
1 cup fruit juice
1/4 teaspoon sodium alginate
1 teaspoon calcium chloride

Procedure
- Place one cup of juice into a blender. If you don't have access to a blender, place the juice into a bowl and whisk aggressively.
- Add 1/4 teaspoon of sodium alginate to the juice and run the blender on high for 2 minutes. Alternatively, you could aggressively whisk for several minutes.
- Run the juice solution through a mesh strainer to remove lumps.
- Refrigerate for a few hours or overnight so the air bubbles pop.
- Skim any additional bubbles off the top the next day.
- Create a salt bath in a bowl by pouring one teaspoon of calcium chloride into 2 cups of water.
- Pour plain water into another bowl.
- Use your food syringe to suck up some of the juice solution.
- Carefully drip little bubbles/spheres of juice into the calcium chloride solution.
- Swish the bubbles around with a spoon, then rinse them in the bowl of plain water.
- Drain the juice bubbles from the water and enjoy your popping boba!

Source: Ypsilanti District Library

10.11 | The Baker's and Brewer's Friend: Fermentation

Learning Objective: Describe the process of fermentation used for food and drink preparation

Fermentation is a natural process for the anaerobic metabolism of sugar in the presence of microorganisms such as yeast or bacteria. We already provided examples of fermentation used for food preservation in Section 10.8, for pickling and preparation of foods such as sauerkraut, kimchi, and miso. This process involves the addition of lactic acid producing bacteria, which convert the vegetable sugars into acids and flavored compounds. Yogurt is also prepared through the addition of harmless bacteria to milk, which produces lactic acid. Once the pH is around 4.4, the fermented milk is cooled to yield the final product. Oftentimes, a final heat treatment is performed to kill the viable microorganisms in order to extend the shelf life of the yogurt. A variety of flavors may also be added after fermentation is complete.

In ethanol fermentation, microorganisms take simple sugars like glucose or fructose and convert them into molecules that provide the organism with energy. As seen in Figure 10.19, the by-products of this process are ethanol (C_2H_5OH) and carbon dioxide (CO_2).

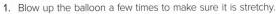

Glucose + Yeast $\xrightarrow{CO_2(g)}$ Pyruvate $\xrightarrow{CO_2(g)}$ Acetaldehyde $\xrightarrow{NADH + H^+ \quad NAD^+}$ Ethanol

Figure 10.19

The chemical reactions involved during fermentation. Glucose combines with yeast to form pyruvate. An enzyme known as *nicotinamide adenine dinucleotide (NAD)* assists in the transformation of acetaldehyde into ethanol.

Yeast that is used in bread making provides the leavening agent, carbon dioxide. Within bread dough, yeast ferments and produces the ethanol and carbon dioxide by-products. The carbon dioxide forms bubbles in the dough, which are trapped in a semi-solid matrix formed by the long-chain molecules—usually proteins or polysaccharides. Holes are left in the bread as it sets, while the ethanol evaporates during the baking process. The process results in bread with a wonderful, airy texture. Discover the properties of yeast yourself in this next activity and the video that follows.

@HOME   Fun with Yeast!

Investigate yeast respiration at home using a balloon, an empty plastic bottle, a packet of fast-acting dry yeast, sugar, and warm water.

1. Blow up the balloon a few times to make sure it is stretchy.
2. Pour a few inches of warm water into the clear plastic bottle.
3. Add the fast-acting dry yeast and swirl.
4. Add a scoop of sugar. Quickly place the balloon over the lip of the bottle.

Wassika Srijudanu/Shutterstock

5. Gently swirl the bottle and set it on a flat surface. Observe the bottle and balloon over time. This can take awhile. A suggestion is to set up a smartphone and take a time-lapse video of the bottle and balloon while you wait!

Questions to Consider

a. What was the purpose of adding the balloon over the lip of the bottle? What was inside the balloon at the end of the investigation? What was the purpose of the sugar?
b. Why did the water need to be warm? What would have happened if the water had been cold?

The Ultimate Donut Battle

Check out this video (www.acs.org/cic) for an application of fermentation—the ultimate donut battle.

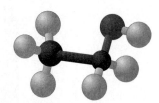

©2018 American Chemical Society

In the production of beer, wine, and spirits, the same process is used, except the ethanol is not allowed to escape. Because the ethanol is contained, a variety of products may be formed, depending on the starting mixture. The primary agents used in alcoholic beverage production are fruits (wines and ciders), grains (beer, whiskey, and vodka), rice (sake), or honey, sugarcane, and molasses (mead and rums). Alcohol itself has a rather undesirable, bitter taste. However, the alcohol derived from fermentation processes results in flavors that are unique and distinctive.

Alcohol is both fat- and water-soluble. The dual solubility property of ethanol is due to the very short hydrocarbon component and the —OH group. This allows ethanol to travel just about anywhere in the human body, and it does. About 10–20% of the consumed amount is readily absorbed by the stomach, with the remaining amount taken in by the small intestines. Its high solubility allows for passage into and through cell membranes, flooding nearly all of our organs. The effects can include relaxation, lifted inhibitions, possible euphoria, false confidence, silliness, slurred speech, slower response rate and motor skills, loss of orientation, an impaired sense of equilibrium, and loss of consciousness. What a remarkable array of effects from a small molecule composed of merely nine atoms, with a molar mass of only 46 g/mol (Figure 10.20).

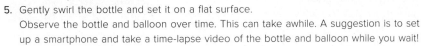

Figure 10.20

A ball-and-stick molecular model for ethanol. Check out www.acs.org/cic for a 3D rendering of ethanol.

10.12 | From Moonshine to Sophisticated Liqueurs: Distillation

> *Learning Objectives:*
>
> *Describe how distillation is used to prepare alcoholic beverages*
>
> *Explain how alcoholic beverages are metabolized in our body*

As seen in Section 6.11 for petroleum refining, distillation is a separation and purification technique that is used in industrial chemistry laboratories throughout the world. However, this process is also used in the production of many alcoholic beverages. Beer, wine, and ciders, though fermented, are not distilled, but spirits such as brandies and vodka are. How does this process work?

The products of fermentation form a liquid mixture. The liquid components of these mixtures, including the drinkable alcohol, have different boiling points. And so, if continuous heat is added to the mixture, the substance with the lowest boiling point will vaporize first. As you can see in Figure 10.21, the distillation apparatus is sealed in such a way that the evaporated liquid is captured and moves through water-cooled tubing, known as the **condenser**. This process of purification of the fermented mixture creates refinements and nuances in the spirit. Depending on what primary agents (e.g., fruits or grains) were used in the fermentation process, the stages of distillation can produce different refinements, purities, and flavors.

How Is Whiskey Made?

How is whiskey made? Check out the video at www.acs.org/cic.

©2018 American Chemical Society

Questions to Consider

a. Describe the distillation of ethanol and methanol. In your description, consider how intermolecular forces affect the distillation of the two types of alcohol.

b. Ethanol is described as being really good at dissolving flavor molecules. Think again about intermolecular forces to discuss what it is about its structure that makes it good at dissolving flavor molecules.

c. Research the sustainability of whiskey production. Based on your research, how would you describe the balance between economic growth, environmental care and social well-being in whiskey production?

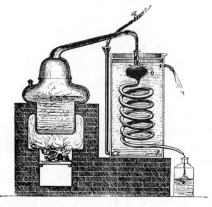

(a)

(b)

Figure 10.21

(a) Schematic of a distillation apparatus used for whiskey production. **(b)** Photo of the production of brandy from the distillation of wine.

(a): Morphart Creation/Shutterstock; *(b)* Carl Court/Getty Images

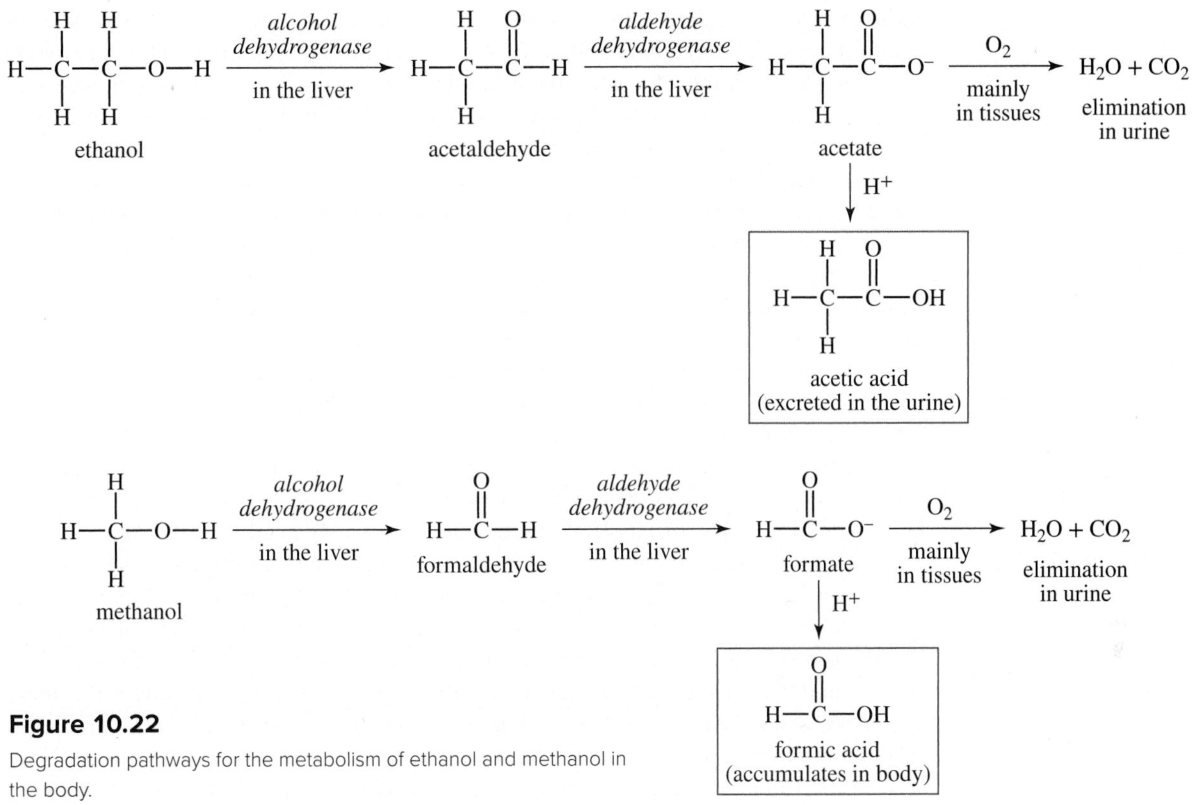

Figure 10.22

Degradation pathways for the metabolism of ethanol and methanol in the body.

Most states require a license for distillation, because, without proper equipment, one can inadvertently produce **methanol** (CH_3OH) as a by-product, which may lead to adverse health effects or death. As little as 10 mL of methanol can break down in the body to make formic acid (Figure 10.22), which reacts with the optic nerve causing damage and resulting in blindness. As little as 30 mL of pure methanol—the volume of a standard shot glass—can be fatal. Commercial distillers carefully control the fermentation and distillation processes to reduce methanol production, which is confirmed by stringent quality-control testing in federally accredited laboratories.

One of the results of distillation is to increase the concentration of alcohol in the spirit. In the United States, the proof level of the spirit is displayed on the label to identify how much alcohol the beverage contains. As shown in Table 10.5, the U.S. proof number is two times the alcohol percent, by volume. A beverage only begins to be called a *spirit* at 20% alcohol by volume or greater, and it can contain no added sugars or flavorings. Spirits with added flavorings and/or sugars are called *liqueurs*.

When alcohol is consumed, it enters the bloodstream and causes the pituitary gland in the brain to block the creation of the hormone *vasopressin*. As a result, the

Table 10.5	Alcohol Content in Typical Alcoholic Beverages	
Beverage	**Typical Proof**	**Percent Alcohol by Volume (% ABV)**
Beer	<20	<10
Wine	25	12.5
Fortified wine	40–60	20–30
Vodka and brandy	80	40
Whiskey	100	50
Tsikoudia (a grape-distilled spirit from Eastern Europe)	120	60
Absinthe	140	70
Neutral grain alcohol	180–200	90–100

kidneys send water directly to the bladder instead of reabsorbing it in the body. According to studies, drinking about 250 mL (8.5 oz) of an alcoholic beverage causes the body to expel 800–1000 mL of water—three to four times as much liquid is lost as ingested! This so-called diuretic effect decreases as the alcohol in the bloodstream decreases; however, this process is the leading cause of hangover symptoms because the body loses so much water. Electrolytes such as sodium, potassium, and magnesium are lost through urination, but are necessary for proper cell, nerve, and muscle functions. As the electrolyte concentrations diminish, headache, nausea, and fatigue set in.

Hangover symptoms are also a result of ethanol metabolites that form in the bloodstream as ethanol is being processed by the body. Through the action of the enzyme *alcohol dehydrogenase,* ethanol is converted into acetaldehyde, which is highly unstable and readily forms free radicals in the body. Long-term exposure to this toxic chemical can damage organs—especially the liver and kidneys.

Perhaps the greatest contributor to hangovers is the presence of methanol, another fermentation product found in alcoholic beverages. As discussed above, methanol is highly toxic; however, its concentration in beverages is quite low (approx. 18 g/L or 0.7% by volume). The metabolism of methanol in our body occurs via the same enzyme as ethanol, resulting in the formation of formaldehyde, which gives rise to more pronounced headaches, nausea, dizziness, and fatigue. As shown in Figure 10.22, formaldehyde is then converted into formic acid, which accumulates in the body and is the main reason for toxicity and alcohol-related deaths. In contrast, acetic acid does not accumulate in our bodies and is readily excreted in the urine. Once the body metabolizes the alcohol, the body sends a message to replenish its water deficiency, usually in the form of a dry mouth. Additionally, the water-deprived organs in the body steal water from the brain, which pulls on the membranes that connect the brain to the skull. This results in a headache.

Since the *alcohol dehydrogenase* enzyme prefers to metabolize ethanol, the conversion of methanol only occurs after all the ethanol has been metabolized, typically by the next morning. This explains why some choose to consume more alcohol the next morning to combat the effects of a hangover. The introduction of more ethanol will allow the methanol to be passed from the body in the urine without being converted into formaldehyde or formic acid. A drug called A*ntabuse* (also known as disulfiram) is one of three drugs approved by the FDA to treat alcohol dependence by blocking the enzyme *aldehyde dehydrogenase* that breaks down acetaldehyde or formaldehyde. Symptoms such as severe vomiting and headaches result from a high residual concentration of acetaldehyde, which makes one wary of their next drink.

10.13 | Extraction: Coffees and Teas

> *Learning Objective: Describe how extraction is used to acquire flavors from coffees and teas*

Coffees and teas have an important role in many cultures and traditions throughout the world, dating back to the earliest civilizations. In brewing coffees and teas, we are interested in extracting compounds from a solid: leaves or beans. Similar to how different chocolates can be made, so too can a variety of teas and coffees. The variations are endless, from the different starting materials used, to different drying and roasting techniques, to the inclusion of different additives for flavoring.

To create drinkable coffee, the starting point is a far cry from that dark liquid in your mug. Raw, green coffee beans virtually have no flavor. Pouring hot water over them will simply produce an astringent and bitter tasting brew. However, roasting the beans initiates a host of reactions that generates a large number of products, all of which add to the varied nuance of coffee flavors. Some tea leaves are also roasted, which results in the various flavors comprising green and black teas.

Brewing coffee or tea is itself another application of a laboratory technique: extraction. The beans and leaves of coffees and teas are solid mixtures. Thus, to assist with the extraction from coffee beans, we grind them to increase their surface area.

The Chemistry of Coffee

Check out this video to discover the complex chemistry of coffee: www.acs.org/cic.

©2018 American Chemical Society

Smaller tea leaves and powders also have more surface area. Then, we expose these materials to a hot extraction medium (water). Fortunately, the desirable components of the coffee or tea tend to be the first to percolate out since they have lower molar masses. The undesired components take longer to extract, and so we don't keep the hot water in contact with the solids for too long as over-extraction leads to significant bitterness.

Many of us are drawn to coffees and teas because they contain caffeine (Figure 10.23). Appropriate doses of caffeine affect the brain and increase mental alertness. And again, like chocolate, teas and coffees contain antioxidants and flavonoids, the chemicals we described earlier that have been shown to improve heart health and interfere with cancer growth.

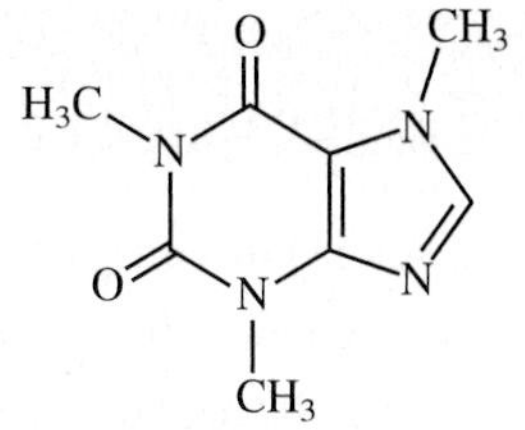

Figure 10.23

The caffeine molecule. For a 3D rendering of this molecule, go to www.acs.org/cic.

@HOME   Caffeine Extraction

This experiment is another application of "like dissolves like." As a polar compound, caffeine is soluble in both water and alcohols. However, if salt water is prepared, the caffeine will prefer to dissolve into the alcohol layer, a technique known as "salting out."

Required Materials
- Coffee (10 g)
- Large thermos or coffee mug (500 mL)
- Liquid measuring cup/glass (100 mL)
- Sea salt (non-iodized; 80 g)
- Rubbing alcohol (90%; 200 mL)
- Turkey baster

Procedure
1. Add 10 g of ground coffee to the thermos/mug, followed by 300 mL of hot water (just below boiling).
2. Fasten the cap and mix the coffee solution by shaking.
3. Add 80 g of sea salt and mix the solution by shaking.
4. Leave the solution to cool down to room temperature.
5. Filter the solution through a coffee filter into another large container to separate the coffee grounds from the solution.
6. Add 200 mL of rubbing alcohol to the resulting room-temperature solution and mix the solution for 10 minutes.
7. Leave the solution for 15 minutes for both phases to separate into layers.
8. After 15 minutes, the alcohol phase will contain caffeine and the aqueous phase will contain only coffee and water.
9. Use a turkey baster to remove the top alcohol layer that contains the caffeine.
10. Pour the solution into a cup or dish and heat over the stove to evaporate the water. The solid that remains will be caffeine.

Questions to Consider
a. Describe the appearance of caffeine (color, texture, etc.).
b. How could you have altered the experiment to quantify the amount of caffeine that is contained per unit mass of coffee?
c. What are some factors that would affect the extraction efficiency of caffeine and flavors from coffee?

Conclusion

The food and drinks we prepare and ingest provide us with nutrients, taste, and even satisfaction and pleasure. We transform food by changing its molecular compositions and structures to yield new substances with different properties, such as flavors. Or, we merely change it physically to a different state or separate it from other components in a mixture, so as to extract or purify it. Most of the time, we cook food by transferring heat in some way. This, however, occurs through a variety of methods (e.g., steak cooked on the grill or *sous vide*) and leads to an equal variety of outcomes (char and cool center, or equal temperature throughout). The recipes, techniques, and utensils we use can often mimic those found in a research lab—for example, glassware, water baths, and digital balances, just to name a few. At the end of the day, good cooking relies on good chemistry. So happy cooking and bon appetit!

LEARNING OUTCOMES

Having studied this chapter, you should now be able to:

- describe how your tongue detects flavors (10.1)
- describe the health benefits of various chocolate ingredients (10.2)
- explain the importance of recipes to produce the desired taste, texture, and consistency of prepared food (10.3)
- convert between measured quantities used in cooking (10.4)
- explain why you need to adjust recipes when cooking at higher altitudes (10.5)
- describe how heat energy is used to physically and chemically change food during cooking (10.5)
- describe the technique and benefits of *sous vide* cooking (10.6)
- explain how a microwave oven cooks our food (10.7)

- describe the physical and chemical processes involved in food preservation (10.8)
- explain how quantitative and qualitative measurements are used to determine the readiness of food (10.9)
- use Henry's law to determine the concentration of a dissolved gas in solution (10.10)
- exploit the "like dissolves like" principle in food preparation techniques (10.10)
- describe the process of fermentation used for food and drink preparation (10.11)
- describe how distillation is used to prepare alcoholic beverages (10.12)
- explain how alcoholic beverages are metabolized in our body (10.12)
- describe how extraction is used to acquire flavors from coffees and teas (10.13)

Questions

Emphasizing Essentials

1. The amount of salt typically used in cooking is 5 g/dm³. Convert this into the following units: g/mL, mg/L, mg/m³.

2. The addition of 1 mole of any solute to 1 kg of water raises the boiling point of water by 0.52 °C. Calculate the boiling point of water that contains the concentration of salt provided in Question 1. How much salt would need to be added to 1 L of water in order for its boiling point to reach 101 °C?

3. What is the density of pure water (in mg/L), and how is this affected by dissolving solutes such as salt or sugar?

4. It has been suggested that some people can taste salt to levels of 0.5 g/dm³, but this varies significantly between testers. How could you experimentally determine your own threshold of salt tasting?

5. Cooking involves a series of chemical reactions. The rate of these reactions roughly doubles for every 10 °C rise in temperature. If the pressure of a cooker is increased from normal atmospheric pressure (1 atm) to 2 atm, how many times faster will food be cooked?

6. Convert the following cooking units into their appropriate SI units (provided in parentheses):

   a. 2 cups of water (L)

   b. 2 teaspoons of salt, NaCl (kg)

   c. 3 hours at 300 °F (seconds and °C)

7. Flavor-causing molecules must be volatile in order to reach the nose and be detected. What structural features of a molecule affect its volatility?

8. **a.** Using the Internet as a resource, describe the process of browning of fruit, such as apples and bananas, once they have been cut or bruised. How is this similar to the browning of skin through exposure to UV radiation?

   b. Is this the same process as that resulting from carmelization or the Maillard reaction? Explain.

   c. How does the addition of Vitamin C work to slow down the browning of fruit? What other methods can be used to prevent this browning reaction?

9. Which of the following would affect the melting speed of an ice cube? Explain.

   a. Crushing the ice

   b. Water temperature

   c. Melting in air vs. water

   d. The shape of the ice cube

10. Explain why cakes may "fall."

Concentrating on Concepts

11. Using the Internet, determine if there are any differences between using iodized salt vs. non-iodized salt for cooking or baking.

12. Provide a molecular description of why the color and texture of green vegetables are altered during overcooking.

13. Can sea salt be labeled as "pure" but still contain trace elements such as magnesium and calcium? Explain.

14. Hikers often complain that it is difficult to make a decent cup of tea at high altitudes. Suggest why this problem may be encountered.

15. Is it possible to properly cook a boiled egg at the top of Mt. Everest? Explain.

16. Using balanced chemical equations, provide an explanation of why baking powder is an ingredient in many baked goods.

17. Bread rises due to fermentation. Research approximately how much alcohol remains behind in bread. What factors affect the alcohol content of bread?

18. Suggest which of the following molecules will respond to microwave radiation in the same manner as water: carbon tetrachloride (CCl_4), ammonia (NH_3), and carbon dioxide (CO_2). Explain your choices.

19. What is the name of the substance responsible for the hot flavor of chili peppers, and what is its molecular structure? Use the Internet to find your answers. Also, when chilies are placed in your mouth, why is their burning sensation not greatly affected by drinking water? What drinks would be more effective in reducing this "burning"?

20. Describe why silver cookware tarnishes, especially in the presence of sulfur-containing foods such as eggs.

Exploring Extensions

21. Pick a couple of your favorite flavors—perhaps those such as banana, peppermint, or buttered popcorn. Research the identity of the key chemicals in these flavors and draw their structures. Do they have anything in common with other flavor molecules you looked up or saw in this chapter? What are those features? How are these molecules different from one another?

22. There is a fun movement happening in the culinary world called "flavor tripping" (Mr. Wonka would have loved it!). This technique involves eating something called a "miracle berry" and consuming other foods that you may not normally eat in their raw or isolated forms. For example, moments after eating a miracle berry, an ordinary lemon when eaten tastes like sweet lemonade. Given what you have learned about taste in this chapter, and after conducting some of your own research, can you provide a chemical explanation for why this berry changes the way things taste? What other foods might be interesting to try in this way?

23. The food we eat does not affect just our bodies. How food is grown or raised, and then produced and processed, can also impact the environment. Current modern agricultural practices are actually a leading contributor to climate change. Over the next couple of days, keep a food journal of what and how much you eat. After you complete your journal:

    a. Select three foods you think ranked highest in promoting your health (state the criteria on which you base your ranking).

    b. Select the three foods you think ranked highest in promoting the health of the environment (again, state the criteria on which you base your ranking).

24. What are "food miles"? Select three of your favorite foods and conduct a food mile analysis. What are the ways we can reduce food miles? Do you think it is important to do so? Why or why not?

25. Select your favorite baked good; for example, chocolate chip cookies, blueberry muffins, rye bread, or something else. Conduct an Internet search to find different recipes for this same item.

    a. What do these recipes have in common?

    b. How are these recipes different?

    c. Can you predict how any variations among the recipes will lead to different outcomes in the baked good?

26. One section in this chapter described the Maillard reaction, a key reaction that brings distinctive flavors to browned foods. Food that is boiled in water or cooked in a microwave does not go through this reaction. Why not?

27. The various types of cooking oils have a range of properties such as taste, color, and their physical state at room temperature. Three commonly used oils—canola, olive, and coconut—provide one set of examples of these property differences. Conduct some research to discover the chemical nature of these oils and how this accounts for their varying properties in cooking and taste.

28. In the coffee industry, equipment ranging from simple French presses to elaborate machines costing thousands of U.S. dollars is used to "brew" coffee. Conduct some research about different types of coffee-making equipment and describe the scientific principles they rely on to make a unique cup of "joe."

29. Teflon-coated cookware is widely used in kitchens throughout the world.

    a. What is the molecular structure of Teflon and how does this coating prevent food from sticking to the pan?

    b. If Teflon acts as a nonstick coating, how does Teflon itself stick to the pan?

30. If fresh pineapple, kiwi, or papaya fruit is added to Jell-O, will it prevent the Jell-O from setting? Explain.

31. Using your knowledge of intermolecular forces and solubility, describe the chemical and physical processes involved in making ice cream.

32. Describe the origin of "asparagus pee"—the odd odor in one's urine after eating asparagus.

33. What types of products in the food industry may be formed from lactic acid bacteria?

34. Provide a list of some coffee constituents that are extracted by hot water. Rank them in order of their elution and explain your rationale. Which of these components would correspond to the enhanced bitterness that occurs after long extraction times?

FOOD CHOICES

Watch the chapter opening video (www.acs.org/cic), which describes the variety of food choices we face daily. So let's reflect on your diet. What did you eat yesterday?

a. Make a list, starting with your first cup of coffee (or however you chose to start your day), through your last evening snack.

b. From your list, select three food items you believe ranked highest in promoting your health. Name the criteria on which you based your ranking.

c. Select from your list the three food items you believe ranked highest in promoting the health of our planet's land, air, and water. Again, name your criteria.

In this chapter, you will explore the following questions:

1. What practical information can we glean from food labels?
2. What are triglycerides and how are they formed?
3. Why are *trans* fats bad for your diet?
4. How are our blood sugar levels controlled?
5. How do artificial sweeteners compare with natural sugars?
6. How are amino acids combined to form proteins?
7. Why are vitamins and minerals important for your health?
8. How do our food choices and levels of exercise affect our weight?
9. How safe is our food supply?
10. How much water and land is required to produce a food item?
11. What are some benefits of eating locally?
12. What are the environmental implications of agricultural practices?
13. How do we feed our hungry and ever-expanding world population?

Valentyn Volkov/Shutterstock

Introduction

Imagine never eating another hamburger. Don't even think about picking up a piece of fried chicken. And your eggs definitely will come *without* bacon from now on. Would this be your worst nightmare? To some people, being a vegetarian is akin to being deprived of those foods they most crave—well, except for coffee and chocolate. All too often, though, we set up our food selection in terms of all or nothing. No ice cream because it has too many Calories. No red meat because it is unhealthy. No soft drinks because they are loaded with sugar. And no diet soft drinks either because they contain artificial sweeteners. No, no, NO!

Could your choices possibly be more nuanced? Unless dictated by an allergy, a specific health concern, or a deeply held belief, what you eat need not be an all-or-nothing proposition. For instance, if you were to eat no meat one day a week, you could improve the odds that, as you age, you would have a higher quality of life. Why exactly is this? As a vegetarian, you will likely get more of the whole grains, fruits, and vegetables your body needs. At the same time, you are likely to get less saturated fat. Do the math—one day a week *does* matter. One day in seven is just short of a 15% change in your food intake. Make that two days a week, and you have changed your diet by more than 25%.

In this chapter, we will make the case that both what you dine on and what you skip affect your health and our planet's health. If you are one of the people on the planet fortunate enough to have adequate food to eat, some very good reasons exist to eat simply and to vary your diet. In fact, *what* you eat and *how much* you eat may be two of the most important decisions you make over the course of your life.

11.1 | You Are What You Eat

Learning Objectives:

How to use food labels to compare the nutritional value of processed foods

Compare the composition of the human body with common foods

Whether you sit down to a gourmet meal or gobble junk food on the run, you eat because you need the energy and nutrients that food provides. We need food as an energy source to power muscles, send nerve impulses, and transport molecules and ions in our bodies. In addition, food serves as the raw material for bodies, including new bone, blood cells, enzymes, and hair. Food also supplies nutrients essential for **metabolism**—the complex set of chemical processes that are essential in maintaining life.

The foods you eat also provide an important source of water, which plays an essential role in our everyday health. This compound serves as both a reactant and a product in metabolic reactions, as a coolant and thermal regulator, and as a solvent for the countless substances that are essential for life. In fact, our human bodies are approximately 60% water.

Eating properly means more than filling your stomach. It is possible to eat, even to the point of being overweight, and still be malnourished. **Malnutrition** is caused by a diet lacking in proper nutrients, even though the energy content of the food may be adequate. Contrast malnutrition with **undernourishment**, a condition in which a person's daily caloric intake is insufficient to meet metabolic needs, which remains a problem in many parts of the world (Figure 11.1).

Although it may seem paradoxical, obesity can be considered a form of malnutrition. In fact, a variety of studies have shown a higher incidence of nutrient deficiencies

Your Turn 11.1   Trends in Worldwide Obesity

Consult the World Obesity Atlas and obesity trends for the U.S. located at www.acs.org/cic and answer the questions below.

Questions to Consider

a. What five countries are predicted to have the highest levels of obesity for men and women by 2030? Calculate the percent increase in obesity for men and women that is predicted to occur between 2020 to 2030.

b. In the U.S., which states have the highest levels of obesity? What trends in obesity are found for various ethnicities, age groups, and education levels?

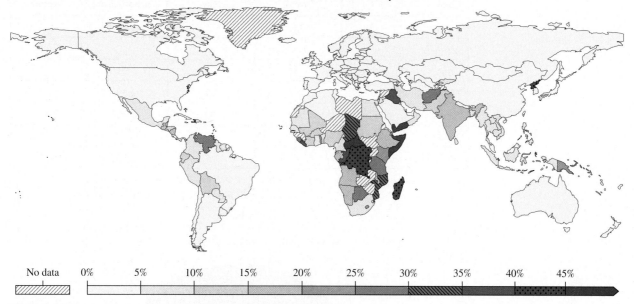

Share of the Population that is Undernourished, 2019

Share of individuals that have a daily food intake that is insufficient to provide the amount of dietary energy required to maintain a normal, active, and healthy life.

No data   0%   5%   10%   15%   20%   25%   30%   35%   40%   45%

Figure 11.1

Worldwide undernourishment, 2019.

Our World in Data

among people with excessive body weight relative to those with a normal body weight. In 2022, the World Obesity Foundation estimated that 20% of men and 17% of women are obese, defined as having a body mass index (BMI) greater than 30 kg/m². However, the obesity level is much higher in the U.S., with 47% of both men and women being classified as obese. This epidemic of obesity is caused by several factors, including eating unhealthy or too much food and lacking physical activity.

Think about the foods you ate yesterday. Did they come to you with minimal or no processing, such as an apple, a baked potato, or a juicy pork chop? Or did you obtain these same foods as applesauce, a bag of frozen french fries, or sliced smoked bacon? The latter are **processed** foods that have been altered from their natural state by techniques such as canning, cooking, freezing, and adding chemicals such as thickeners or preservatives. The typical diet in many countries contains numerous processed foods.

Processed foods in the United States must list nutritional information on their labels. These labels, such as the one shown in Figure 11.2, include the **macronutrients**—fats, carbohydrates, and proteins that provide essentially all of the energy and most of the raw material for body repair and synthesis. Sodium and potassium ions are present in much lower concentrations, but these ions are essential for the proper electrolyte balance in the body. Several other minerals, and an alphabet soup of vitamins, are listed in terms of the percent of recommended daily requirements supplied by a single serving of the product. All of these substances, whether naturally occurring or added during processing, are chemicals. In fact, all food is inescapably and intrinsically chemical, even food that claims to be organic or "natural."

Table 11.1 indicates the mass percentages (grams of component per 100 g of food item) of water, fat, carbohydrate, and protein in several familiar foods. For this particular selection of foods, the variation in composition is considerable. But in every case, these four components account for almost all of the mass present. Water ranges from a high of 89% in 2% milk to a low of 1% in peanut butter. Peanut butter is comparable to steak and fish in terms of protein; among these foods, it leads in fat content. Chocolate chip cookies are the highest in carbohydrates because of their high sugar and refined flour content.

Nutrition Facts	
8 servings per container	
Serving size	**2/3 cup (55g)**
Amount per serving	
Calories	**230**
	% Daily Value*
Total Fat 8g	**10%**
Saturated Fat 1g	5%
Trans Fat 0g	
Cholesterol 0mg	**0%**
Sodium 160mg	**7%**
Total Carbohydrate 37g	**13%**
Dietary Fiber 4g	14%
Total Sugars 12g	
Includes 10g Added Sugars	20%
Protein 3g	
Vitamin D 2mcg	10%
Calcium 260mg	20%
Iron 8mg	45%
Potassium 235mg	6%

* The % Daily Value (DV) tells you how much a nutrient in a serving of food contributes to a daily diet. 2,000 calories a day is used for general nutrition advice.

Figure 11.2

Nutrition information reported on a food label.

Table 11.1	Percent Water, Fat, Carbohydrate, and Protein in Selected Foods			
Food	**Water**	**Fat**	**Carbohydrate**	**Protein**
White bread	37	4	48	8
2% milk	89	2	5	3
Chocolate chip cookies	3	23	69	4
Peanut butter	1	50	19	25
Sirloin steak	57	15	0	28
Tuna fish (canned)	63	2	0	30
Black beans (cooked)	66	<1	23	9

Source: U.S. Department of Agriculture, Agricultural Research Service, Home and Garden Bulletin, 72, 2002.

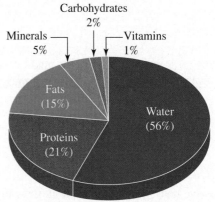

Figure 11.3

Composition of the human body, in percent by mass.

Compare this table with similar data for the human body (Figure 11.3). You are what you eat, but only to a certain extent. You are more like steak than chocolate chip cookies. You also contain more water and fat than bread and contain more protein than milk. From these data, you can determine that a 150-pound (68-kg) person consists of about 90 pounds (41 kg) of water and about 30 pounds (14 kg) of fat. The remaining 30 pounds is almost all protein, carbohydrate, calcium, and phosphorus in the bones. The other minerals and vitamins weigh less than 1 pound (0.5 kg), indicating that a little bit of each goes a long way, a point that will be discussed in Section 11.7.

In the following sections, we take a look at each food component in turn—fats, carbohydrates, proteins, minerals, and vitamins. As you will see, each one is unique in several regards.

11.2 | From Buttery Popcorn to Cheesecake: Lipids

> *Learning Objective: Describe the relationship between lipids and triglycerides, and outline the reaction scheme for forming triglycerides from fatty acids*

From your experiences with ice cream, butter, and cheese, you probably know that fats can help impart a desirable flavor and texture to food. More generally, fats are greasy, slippery, and low-melting solids that are not soluble in water. When melted, fats float on the top of broth or soup because their densities are less than the underlying aqueous mixture. Some of the most delectable foods, such as cakes, frosting, sour cream, and most pastries, are loaded with fats (and Calories, as we will see later).

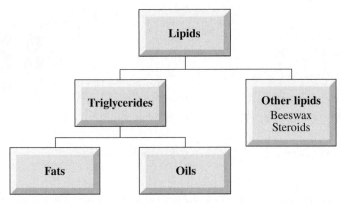

Figure 11.4

Types of lipids.

You may also know about oils, such as those obtained from corn and soybeans. Perhaps you have noticed that peanut oil often forms a layer on top of your peanut butter. Maybe you like eating bread dipped in olive oil. Or maybe in the past, you have prepared a loaf of nut bread using canola oil as the shortening. Many oils are of plant origin and share many of the properties of animal-based fats. However, unlike fats, they are liquids at room temperature.

As we pointed out when describing biodiesel in Section 6.15, the molecules that make up fats and oils share a common structural feature. They are both **triglycerides,** that is, molecules that contain three *ester* functional groups:

$$\begin{array}{c} O \\ \parallel \\ R \diagdown C \diagup O \diagdown R' \end{array}$$

where R and R′ are abbreviations for any hydrocarbon group attached to the rest of the molecule (e.g. —CH_3, —CH_2CH_3).

As we will see shortly, triglycerides are formed from a chemical reaction between three fatty acids and the alcohol *glycerol*. **Fats** are triglycerides that are solids at room temperature, whereas **oils** are triglycerides that are liquids at room temperature. In turn, all triglycerides are **lipids**, a class of compounds that includes not only triglycerides but also related compounds such as cholesterol and other steroids. Figure 11.4 shows the lipid family tree.

We introduced several new terms in the previous paragraph, and we will now describe each one by one. First up is *fatty acids*; examine Figure 11.5 to see an example known as stearic acid. Like all fatty acids, the stearic acid molecule has two

$CH_3(CH_2)_{16}COOH$

condensed structural formula

$CH_3CH_2CH_2CH_2CH_2CH_2CH_2CH_2CH_2CH_2CH_2CH_2CH_2CH_2CH_2CH_2CH_2$—$\underset{OH}{\overset{O}{\underset{\parallel}{C}}}$

semicondensed structural formula

line-angle drawing

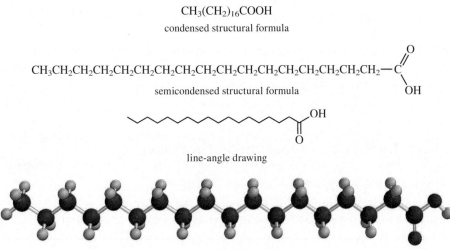

ball-and-stick model

Figure 11.5

Molecular representations of stearic acid, $C_{17}H_{35}COOH$, an example of a fatty acid. Check out this 3D illustration of stearic acid: www.acs.org/cic.

CH$_2$(OH)CH(OH)CH$_2$OH

condensed structural formula

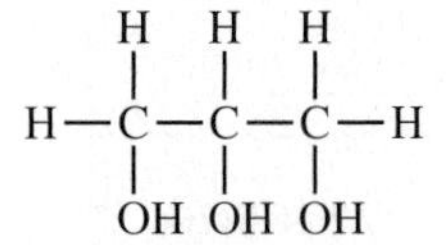

$$\begin{array}{ccccc} \text{H} & & \text{H} & & \text{H} \\ | & & | & & | \\ \text{H}-\text{C}-&\!\!\!\text{C}-&\!\!\!\text{C}-\text{H} \\ | & & | & & | \\ \text{OH} & & \text{OH} & & \text{OH} \end{array}$$

structural formula

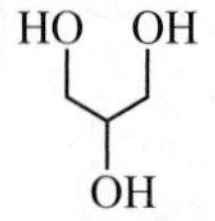

line-angle drawing

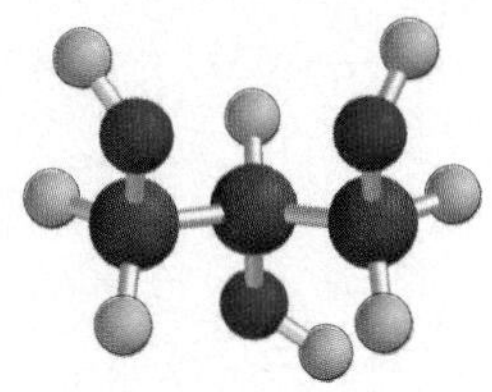

ball-and-stick model

Figure 11.6

Molecular representations of glycerol, an alcohol. Check out this 3D illustration of glycerol: www.acs.org/cic.

important characteristics. One is a long hydrocarbon chain with an even number of carbon atoms, typically 12 to 24. This hydrocarbon chain gives fats and oils their characteristic greasiness. The other is the carboxylic acid group, —COOH, at the end of the hydrocarbon chain. The carboxylic acid group accounts for the "acid" in the names of these compounds.

Chemists routinely use a **line-angle drawing** to represent the structure of a molecule. One carbon atom is assumed to occupy each vertex position. Any line extending from the backbone signifies another carbon atom unless the symbol for another element is given. Hydrogen atoms are not indicated in the line-angle drawing but are implied as required by the octet rule.

Next, we turn to the term **glycerol**, a sticky, syrupy liquid that is sometimes added to soaps and hand lotions. The molecular representations of a glycerol molecule (Figure 11.6) show that it is an alcohol with three —OH groups.

Of interest to us here is that each —OH group of the glycerol molecule can react with a fatty acid to form an ester. For example, three stearic acid molecules can combine with a glycerol molecule to form glyceryl tristearate, a triglyceride. The three ester functional groups in glyceryl tristearate are highlighted in red in Equation 11.1.

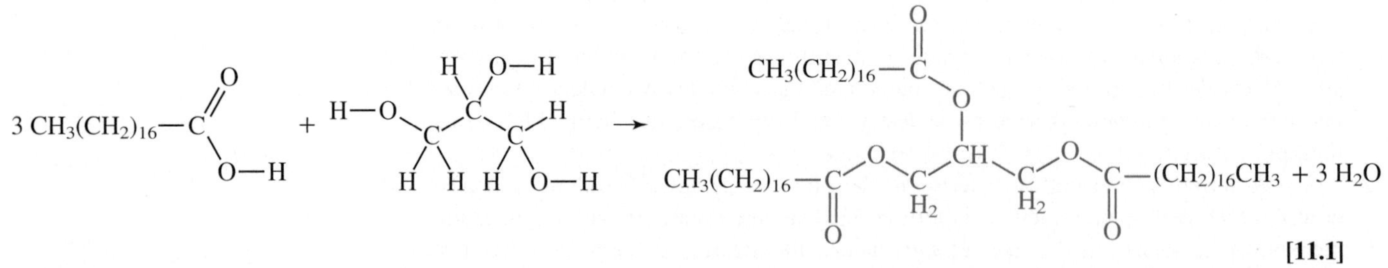

3 stearic acid molecules   glycerol   glyceryl tristearate

$$\text{[11.1]}$$

The process shown in Equation 11.1 is the basis for forming most animal fats and vegetable oils. In most cases, enzymes catalyze this process. Almost all of the fatty acids in our bodies are transported and stored as triglycerides.

Your Turn 11.2   Triglyceride Formation

Like stearic acid, palmitic acid is a component of animal fats.

a. Construct a line-angle drawing for palmitic acid, CH$_3$(CH$_2$)$_{14}$COOH.
b. Name the functional group responsible for the acidic properties of fatty acids.

$$CH_3(CH_2)_7CH=CH(CH_2)_7COOH$$

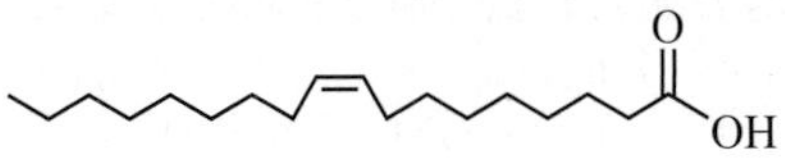

(a)　　oleic acid, a **monounsaturated** fatty acid

$$CH_3(CH_2)_4CH=CHCH_2CH=CH(CH_2)_7COOH$$

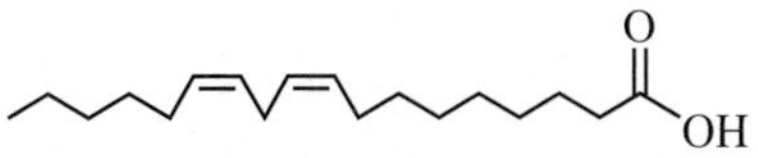

(b)　　linoleic acid, a **polyunsaturated** fatty acid

$$CH_3CH_2CH=CHCH_2CH=CHCH_2CH=CH(CH_2)_7COOH$$

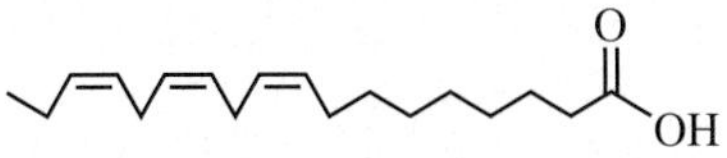

(c)　　linolenic acid, a **polyunsaturated** fatty acid

Figure 11.7

Examples of unsaturated fats. Check out the 3D illustrations of oleic acid, linoleic acid, and linolenic acid at www.acs.org/cic.

The properties of a particular fat or oil depend on the nature of the fatty acids incorporated into the triglyceride. Of key importance is whether the fatty acid molecule contains one or more C=C double bonds. A fatty acid is **saturated** if the hydrocarbon chain contains only single bonds between the carbon atoms. In a saturated hydrocarbon chain, the C atoms contain the maximum number of H atoms that can be accommodated and are therefore saturated in hydrogen. This is the case with stearic acid. In contrast, fatty acids are **unsaturated** if they contain one or more C=C double bonds.

Unsaturated fatty acids are either monounsaturated or polyunsaturated. For example, oleic acid (Figure 11.7a), with only one double bond between carbon atoms per molecule, is classified as **monounsaturated**. In contrast, linoleic acid (two C=C double bonds per molecule, Figure 11.7b) and linolenic acid (three C=C double bonds per molecule, Figure 11.7c) are both examples of **polyunsaturated** fatty acids, which contain more than one double bond between carbon atoms. In Figure 11.7, each of the unsaturated fatty acids contains 18 carbon atoms but differs in the number and placement of the C=C double bonds. This next activity will allow you to explore different unsaturated fatty acids.

Your Turn 11.3　　Unsaturated Fatty Acids

a. What structural feature identifies oleic, linoleic, and linolenic acids as unsaturated fatty acids?

b. Lauric acid, $CH_3(CH_2)_{10}COOH$, is a component of palm oil. Draw the line-angle representation for lauric acid, and classify it as either saturated or unsaturated.

The three fatty acids that form a triglyceride molecule can be identical, two can be the same, or all three can be different. Moreover, these fatty acids can be saturated or unsaturated. All these factors contribute to the variety of fats and oils we find in

Table 11.2	Comparing Fatty Acids		
Name	Number of C Atoms per Molecule	Number of C=C Double Bonds per Molecule	Melting Point (°C)
Saturated Fatty Acids			
Capric acid	10	0	32
Lauric acid	12	0	44
Myristic acid	14	0	54
Palmitic acid	16	0	63
Stearic acid	18	0	70
Unsaturated Fatty Acids			
Oleic acid	18	1	16
Linoleic acid	18	2	−5
Linolenic acid	18	3	−11

animals and plants. Solid or semisolids such as animal fats, lard, and cheddar cheese tend to be high in saturated fats. In contrast, olive, safflower, and other plant oils mostly contain unsaturated fats.

Table 11.2 indicates some trends within a given family of fatty acids. For example, in saturated fatty acids, the melting points increase as the number of carbon atoms per molecule (and the molar mass) increases. On the other hand, in a series of fatty acids with similar numbers of carbon atoms, increasing the number of C=C double bonds decreases the melting point. Thus, when the melting points of the 18-carbon fatty acids are compared, saturated stearic acid (no C=C double bonds) is found to melt at 70 °C, oleic acid (one C=C double bond per molecule) melts at 16 °C, and linoleic acid (two C=C double bonds per molecule) melts at −5 °C. These trends carry over to the triglycerides containing the fatty acids and explain why fats rich in saturated fatty acids are solids at room and body temperatures, whereas ones with a high degree of unsaturation are liquids.

Your Turn 11.4   Melting Points

Draw the molecular structures for some saturated fatty acids and those with C=C double bonds within their framework. Provide a rationale for decreasing melting points of the fatty acids with increasing degrees of unsaturation.

Hint: Recall the concept of intermolecular forces described in Chapter 6.

Your Turn 11.5   Hydrocarbons, Triglycerides, and Biodiesel

a. How does a molecule of octane compare with one of a fat or oil? List two differences.
b. How does a biodiesel molecule compare with one of a fat or oil?

Hint: Revisit Section 6.15.

As you might guess, fats and oils not only differ in their physical properties, but they also differ in how they affect your health. We turn to this topic in the next section.

11.3 | Fats and Oils: Not Necessarily a Bad Thing!

Learning Objective: Distinguish saturated from unsaturated fats, and compare their respective chemical and physical properties, uses, and health effects

People tend to be preoccupied with dietary fat because fats contain more Calories than any other nutrient. But fats are far more than just fuel. Fats enhance our enjoyment of food, improve our "mouth feel," and intensify certain flavors. Almost every dessert tastes better with a bit of whipped cream! Fats are also essential for life. They provide insulation that retains body heat, which helps to cushion internal organs. Moreover, triglycerides and other lipids, including cholesterol, are the primary components of cell membranes and nerve sheaths. In fact, our brains are rich in lipids, which are important for normal brain function.

Fortunately, our bodies can synthesize almost all fatty acids from the foods we eat. The exceptions are linoleic and linolenic acids (Figure 11.7b and 11.7c). These two fatty acids must be present in our diet because our bodies cannot produce them. Generally, this does not pose a problem because many foods, including plant oils, fish, and leafy vegetables, contain linoleic and linolenic acids.

Figure 11.8 reveals some surprising differences in the composition of fats and oils we consume. For example, flaxseed oil is particularly rich in alpha-linolenic acid (α-linolenic acid, or ALA), a polyunsaturated fatty acid that is being studied for its health benefits. Palm kernel and coconut oil contain much more saturated fat than corn and canola. Ironically, the coconut oil used in some nondairy creamers contains about 87% saturated fat, far more than the percentage found in the cream it replaces. In fact, coconut oil contains more saturated fat than pure butter fat. Concern over the high degree of saturation in coconut and palm oil accounts for the statement sometimes printed on food labels: "Contains no tropical oils."

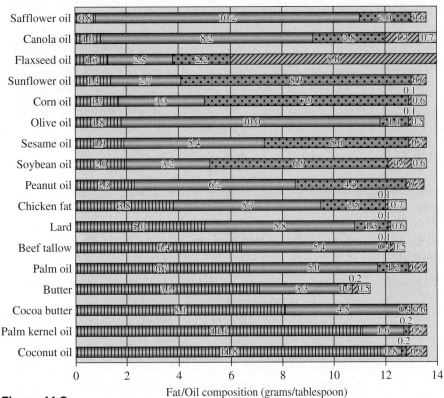
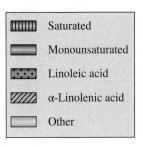

Figure 11.8

Comparative compositions of various saturated and unsaturated oils and fats. Check out the simulation at www.acs.org/cic to investigate the content of an oil.

Your Turn 11.6   The Chemistry of Cooking Oil

a. Consider this label from a popular brand of cooking oil. Is the major component likely to be safflower oil, canola oil, or soybean oil? Explain.

b. This cooking oil brand has one unusual ingredient: vitamin E. Do you think this is a part of the oil itself or an added component? We will provide more details about vitamins in Section 11.7.

Nutrition Facts
Serving Size 1 Tbsp (15 mL)
Servings Per Container about 63

Amount Per Serving		
Calories 120	Cal. from fat 120	
		% Daily Value*
Total Fat 14g		21%
Saturated Fat 1g		6%
Trans Fat 0g		
Polyunsaturated 11g		
Monounsaturated 2g		
Cholesterol 0g		0%
Sodium 0g		0%
Total Carbohydrate 0g		
Protein 0g		
Vitamin E 20%		

Not a significant source of dietary fiber, sugars, vitamin A, vitamin C, calcium, and iron
*Percent Daily Values are based on a 2,000 Calorie diet.

Despite certain health benefits, the higher degree of unsaturation in oils comes with a drawback. You may have noticed the slight rancid odor that oils acquire over time. The reason for this is that C=C double bonds are more susceptible to reaction with the oxygen in the air than are C—C single bonds. As you will learn in this next activity, the "off-flavor" that you may detect in an oil is most likely a result of such reactions with oxygen.

The Chemistry of Olive Oil

Why is olive oil awesome? Check out the video at www.acs.org/cic.

Questions to Consider

a. What are some differences between extra virgin and regular olive oil in terms of processing and health benefits?

b. What are the relative percents of monounsaturated, polyunsaturated, and saturated fats in olive oil?

©2018 American Chemical Society

To prevent oxidation, oils are sometimes treated to increase their saturation, which improves the shelf life of the food containing the oil. One way to more fully saturate an oil or fat is by **hydrogenation**, a process in which hydrogen gas, in the presence of a metal catalyst, adds to a C=C double bond and converts it to a C—C single bond (Equation 11.2a).

$$
\underset{\underset{}{\overset{}{}}}{-\text{C}=\text{C}-} + \text{H}_2 \xrightarrow{\text{metal catalyst}} -\text{C}-\text{C}- \qquad [\textbf{11.2a}]
$$

When oils are hydrogenated, some or all of their C=C double bonds are converted to C—C single bonds, increasing the degree of saturation and raising the melting point. As a result, the oil becomes more margarine-like, semisolid, and spreadable. Converting all C=C double bonds to C—C single bonds would create an undesirable hard-to-spread solid. By carefully selecting the temperature and pressure, the extent of hydrogenation can be controlled to yield products with the desired melting point, softness, and spreadability. Equation 11.2b shows this reaction with linoleic acid, one of the fatty acids in the triglycerides of peanut oil.

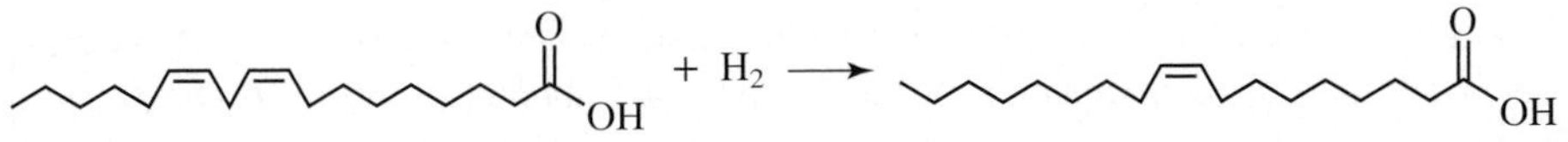

[11.2b]

Note that only one of the double bonds in linoleic acid was hydrogenated. The resulting customized fats and oils are used in margarines, cookies, and candy bars.

Although shelf life and spreadability are important considerations, they are not the only ones. It turns out that some fats and oils are healthier for your heart than others. To understand why we need to look more closely at the geometry of the hydrogen atoms attached to the C=C double bonds of triglycerides. In most natural unsaturated fatty acids, the hydrogen atoms attached to the carbon atoms are on the *same* side of the C=C double bond. We call this bonding arrangement *cis*:

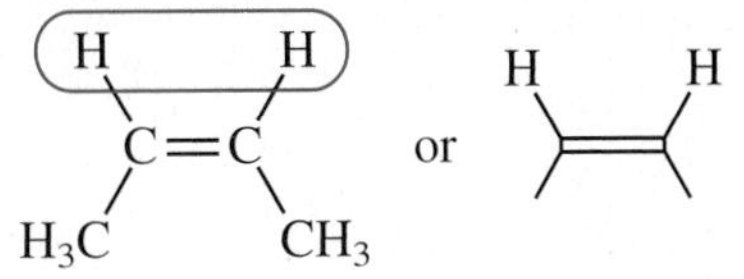

cis-2-butene

Alternatively, the hydrogen atoms can be on the opposite side of the C=C double bond. We call this bonding arrangement *trans*:

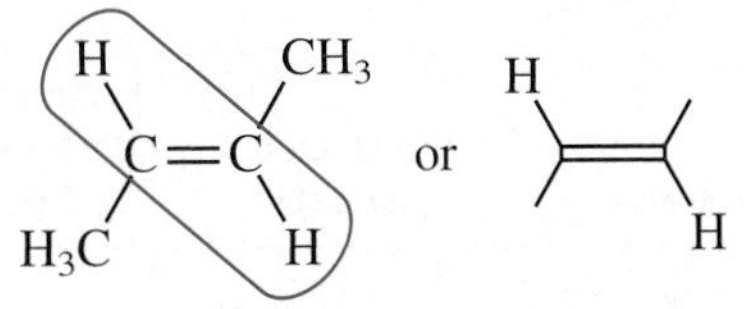

trans-2-butene

For example, oleic acid and elaidic acid are monounsaturated fatty acids that both have the same chemical formula, $C_{18}H_{34}O_2$, so they are *isomers* of one another. However, their properties, uses, and health effects are different. Oleic acid, a *cis* fatty acid, is a major component of the triglycerides in olive oil. In contrast, elaidic acid, a *trans* fatty acid, is found in some soft margarines made via hydrogenation. Compare the structures shown in Figure 11.9.

Trans fats are triglycerides that are composed of one or more *trans* fatty acids. Scientific studies show that *trans* fats raise triglycerides and "bad" cholesterol levels in the blood. Low-density lipoprotein (LDL) cholesterol is referred to as "bad" because of its tendency to build up on artery walls, causing heart attacks and strokes. This finding was somewhat of a surprise because partially hydrogenated fats still contain some C=C double bonds, and unsaturation is definitely a plus in a healthy diet. However, *trans* fats are similar in properties to saturated fats. With their long "straight"

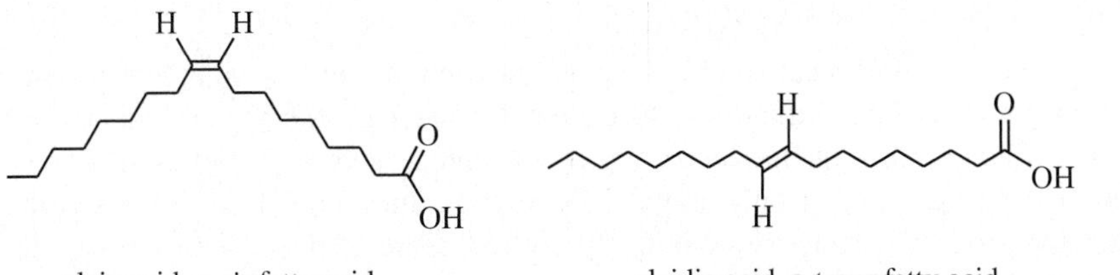

oleic acid, a *cis* fatty acid   elaidic acid, a *trans* fatty acid

Figure 11.9

Molecular structures of oleic acid (left) and elaidic acid (right). Check out a simulation of hydrogenation and 3D renderings of oleic acid and elaidic acid at www.acs.org/cic.

hydrocarbon chains, saturated fat molecules tend to pack well together, one reason they are solids at room temperature. With their *cis* geometries, the molecules of naturally occurring unsaturated edible oils have "bends" that do not pack as well, which is one reason they are liquids at room temperature.

Revisit Equation 11.2b to see that hydrogenation is not an all-or-nothing proposition. When partial hydrogenation occurs, some C=C bonds in the oil or fat remain. We might expect these molecules to be *cis* because C=C bonds naturally occur this way in fats and oils. However, the process of hydrogenation converts some of these *cis* C=C bonds to the *trans* configuration (Equation 11.2c). The fats containing these *trans* fatty acids more closely resemble the shape of saturated fats and therefore behave similarly within the body. What types of fats and oils are in your margarine? Check out the next activity.

$$+ H_2 \longrightarrow \qquad [11.2c]$$

Your Turn 11.7   Margarines and Fat Content

The following table lists the fat content for butter and three margarines, each corresponding to a serving size of 1 tbsp (14 g).

	Butter	Land O' Lakes (stick)	I Can't Believe It's Not Butter (tub)	Benecol Spread (tub)
Total fat	11 g	11 g	9 g	8 g
Saturated fat	7 g	2 g	2 g	1 g
Trans fat	0 g	2.5 g	0 g	0 g
Polyunsaturated fats	1 g	3.5 g	3.5 g	2 g
Monounsaturated fats	3 g	2.5 g	2 g	4.5 g

a. Which margarine has the highest percentage of saturated fat? How does it compare with butter?

b. In butter, what percent of the total fat is polyunsaturated?

c. Conduct a mini-survey. List the fat content for three different butters or margarines. Among those listed in the table above and those in your survey, which would be the most "healthy" for your diet? Explain your rationale.

d. The rise in obesity has long been blamed on fat consumption in the diet. However, the consumption of Calories from fat has decreased since the 1980s, while the rate of obesity continues to grow. Investigate reports about the current medical recommendations for avoiding obesity to find other factors that are likely responsible.

A variety of clinical studies over the past few decades have revealed that a participant's risk of heart disease is strongly influenced by the type of dietary fat consumed. Eating *trans* fat has been shown to increase the risk substantially, whereas saturated fat increased it only slightly. In contrast, unsaturated fat decreased the risk. Therefore, total fat intake alone is likely not the cause of heart disease.

In March 2003, Denmark became the first country to strictly regulate foods containing *trans* fats. Canada followed suit in 2004. Starting in January 2006, the U.S. Food and Drug Administration (FDA) required that food labels include values for *trans* fat. In 2018, *trans* fats were deemed unsafe by the FDA and were banned in food

@HOME   Fatty Foods

Investigate the amount of fats in foods at home using paper bags, scissors, a kitchen scale, foods around your home such as peanut butter, an apple, warmed cheese, a hot dog, a potato chip, and a warmed cookie.

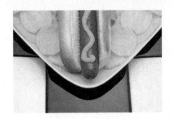

Steve Cukrov/Shutterstock

1. Cut a paper bag into small strips and label them with the foods that you are going to place on them.
2. Find the mass of each small strip before it has food placed on it. Then, place the food on it and find the mass again. Let the food sit on the paper bag strip for several hours.
3. Remove the piece of food from the paper bag strip and find the mass of the paper bag strip again.
4. Find the difference in the mass for the paper bag strip for the amount of fat it absorbed from the piece of food.
5. Repeat Steps 2–4 for each piece of food you are investigating.
6. Calculate the percent fat each food had by taking the mass of fat and divide it by the total mass of the food and multiply it by 100.

Questions to Consider
a. Which food had the most fat? Did this surprise you at all?
b. Why did the cheese and the sausage need to be warmed for this investigation?

---

products without specific regulatory authorization. Also in 2018, the World Health Organization proposed an aggressive plan to eliminate *trans* fats from the global food supply by 2023. The so-called REPLACE action plan consists of the following aspects:

- **Re**view dietary sources of industrially produced *trans* fat and the landscape for required policy change.
- **P**romote the replacement of industrially produced *trans* fat with healthier fats and oils.
- **L**egislate or enact regulatory actions to eliminate industrially produced *trans* fat.
- **A**ssess and monitor *trans* fat content in the food supply and changes in *trans* fat consumption in the population.
- **C**reate awareness of the negative health impact of *trans* fat among policy-makers, producers, suppliers, and the public.
- **E**nforce compliance with policies and regulations.

If successful, the worldwide ban of *trans* fats is estimated to save more than 10 million lives by reducing cardiovascular disease. However, as of 2022, some foods on the market may still contain *trans* fats if they were produced before the ban went into effect. Fried foods may also contain high levels of *trans* fats since the high cooking temperatures used in oil fryers can increase the concentration of *trans* fats; these concentrations may also increase each time the oil is reused for frying.

Although most medical professionals now recommend that consumers avoid products with *trans* fats, this may not be easy because of how foods are labeled. In the United States, a label may show "zero grams *trans* fat" as long as this food contains less than 0.5 grams of *trans* fat per serving. Thus, with multiple servings of what is not truly 0 grams, the amount of *trans* fat can add up.

Manufacturers had to respond to product bans by looking for substitutes for *trans* fats. However, using tropical oils such as palm or coconut oil would not be acceptable to consumers because of their high percentage of saturated fats. As a result, some manufacturers have added other oils to their products that are polyunsaturated, such as sunflower oil or flaxseed oil.

Food chemists have also been busy discovering alternatives to hydrogenation that produce semisolids but do not produce *trans* fats. **Interesterification** is any process in

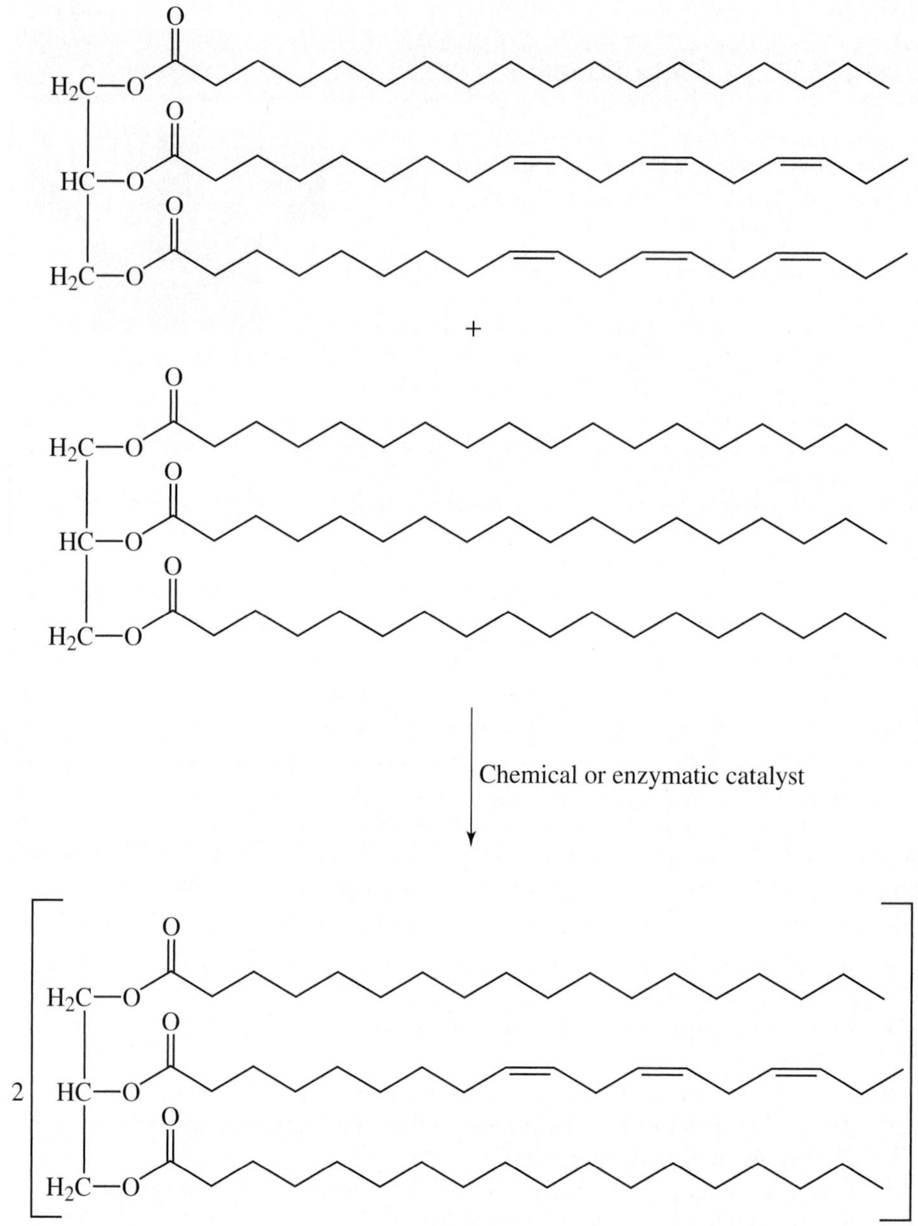

Figure 11.10

An example of interesterification. A triglyceride with two linolenic acid side chains (top) and a fully saturated triglyceride (middle) are combined to produce two molecules with an intermediate degree of chain saturation (bottom).

which the fatty acids on two or more triglycerides are scrambled to produce a mixture of different triglycerides (Figure 11.10). If you perform this process with a low-melting triglyceride (an oil) and a high-melting triglyceride (a fat), the result is a mixture of triglycerides with an intermediate melting point, a semisolid fat.

One way of carrying out an interesterification reaction uses a base as a catalyst. However, the use of basic solutions raises concerns for worker safety, results in significant loss of oils, requires large amounts of water, and produces waste that is high in biological oxygen demand, BOD (Section 5.12). Despite being inexpensive and easy to scale up, base-catalyzed interesterifications also suffer from a lack of specificity with little/no control over the position of the fatty acids in the final product.

In order to better control the structure and properties of the final product, enzymes can also be used to catalyze this reaction. Using enzymes came with a high price tag until Novozymes, and the Archer Daniels Midland Company teamed up to refine the enzymatic reaction. For their efforts, they shared a Presidential Green Chemistry

Challenge Award in 2005. Not only is their method more cost-effective, but it also has environmental benefits. These include a large reduction in water usage and the BOD of the aqueous waste streams. Furthermore, there is less decomposition of oils during the process, and the need for a base catalyst is eliminated.

Your Turn 11.8   Next-Generation Interesterification

a. Revisit the key ideas of green chemistry listed on the first page of this textbook. Which of the key ideas are met by the use of enzymes for interesterification?

b. How do you think food companies should label their products to differentiate if the *trans* fats were made with chemical or enzymatic interesterification?

With this, we end our discussion of fats and oils. The next section tells the sweet tale of sugars and their not-as-sweet relatives, starches.

11.4 | Carbohydrates: The Sweet and Starchy

Learning Objectives:

Illustrate the structure and metabolism of carbohydrates

Describe how insulin and glucagon regulate blood glucose levels

Carbohydrates are compounds that contain carbon, hydrogen, and oxygen, with H and O atoms found in the same 2:1 ratio as in H_2O. This composition gives rise to the name *carbohydrate,* which implies "carbon plus water." However, the H and O atoms in carbohydrates are not in the form of H_2O molecules, but rather as a part of a larger molecule.

Sugars are the sweet-tasting members of the carbohydrate family. Examples that you may recognize include glucose and fructose, both naturally occurring in fruits, vegetables, and honey. In addition, glucose and fructose are components of high-fructose corn syrup (Figure 11.11).

As seen in Figure 11.12, glucose and fructose have the same chemical formula, $C_6H_{12}O_6$, but different structures. These *isomers* are relatively easy to tell apart: glucose has a six-membered ring, and fructose has a five-membered ring. In contrast, the alpha (α) and beta (β) isomers of glucose are hard to distinguish without a model (Figure 11.13).

Both glucose and fructose are examples of a **monosaccharide**, a single sugar. However, sucrose (table sugar) is a **disaccharide**, a "double sugar" formed by joining two monosaccharide units. In forming a sucrose molecule, an α-glucose and β-fructose unit are connected by a C—O—C linkage created when an H atom and an —OH group are split off to form a water molecule. Does this sound familiar? It is a condensation reaction analogous to those in Section 9.6 that formed polyesters. Figure 11.14 shows the chemical reaction that forms the sucrose molecule (a disaccharide) and releases a water molecule.

Monosaccharides can also combine to form much bigger molecules. **Polysaccharides** are condensation polymers made up of thousands of monosaccharide units. As the

Figure 11.11

High-fructose corn syrup is a mixture of fructose, glucose, and water. In pure form, sugars are white crystalline solids.

Mark Dierker/McGraw Hill

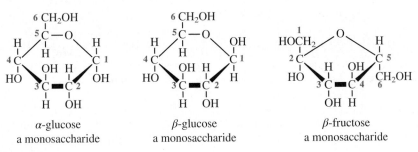

α-glucose
a monosaccharide

β-glucose
a monosaccharide

β-fructose
a monosaccharide

Figure 11.12

Molecular structures of glucose and fructose. To view 3D renderings of these structures, go to www.acs.org/cic.

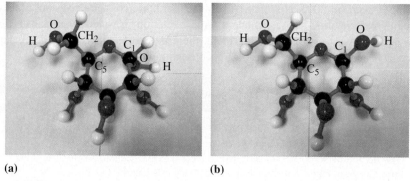

Figure 11.13

Molecular models illustrating the structural differences between **(a)** α- and **(b)** β-glucose. Notice that in α-glucose, the —OH group on carbon 1 is on the opposite side of the ring from the —CH₂OH group attached to carbon 5. In β-glucose, the —OH and —CH₂OH groups are on the same side of the ring.
(both): ©Bradley D. Fahlman

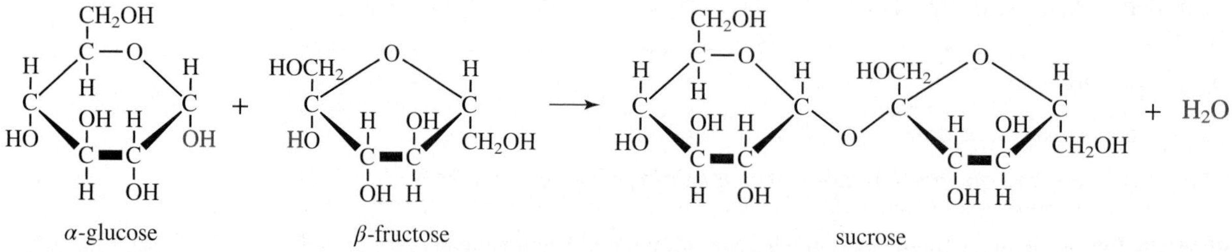

α-glucose   β-fructose → sucrose + H₂O

Figure 11.14

Formation of sucrose, a disaccharide. Check out the 3D rendering of sucrose at www.acs.org/cic.

name implies, these macromolecules contain "many sugar units." Analogous to the formation of sucrose (Figure 11.14), the formation of a polysaccharide releases a water molecule each time a monosaccharide is incorporated into the polymer chain. Familiar examples of polysaccharides include starch and cellulose (Figure 11.15), which lack the sweetness of simple sugars.

Starch, a polymer of glucose first introduced in Section 6.14, is found in nearly all types of grains, including potatoes and rice. Although pleasing to our taste buds, starch lacks a sweet taste and takes a bit longer to digest than sugars. Whether sweet or starchy, carbohydrates have the job of providing energy to the cells in our bodies. However, the starch in corn kernels is also currently fermented to produce millions of gallons of ethanol yearly. But corn starch is not unique. The sugar or starch of almost any plant can be fermented to produce ethanol, including that found in alcoholic beverages.

Our bodies can digest starch by breaking it down into glucose; in contrast, we cannot digest cellulose. Consequently, we depend on starchy foods such as potatoes or pasta

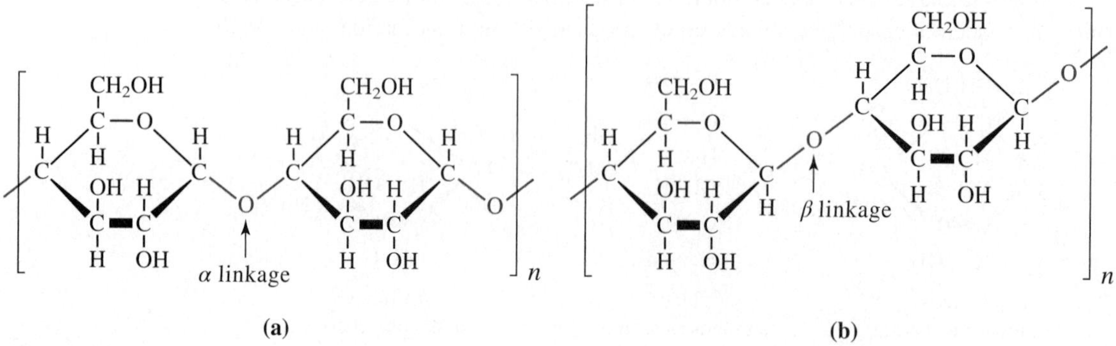

Figure 11.15

The bonding between glucose units in **(a)** starch and **(b)** cellulose. Check out the 3D representations of starch and cellulose at www.acs.org/cic.

rather than devouring paper or toothpicks. The difference in digestibility stems from a subtle difference in how the glucose monomers are connected. In Figure 11.15, compare the α-linkage between the glucose units in starch with the β-linkage between glucose units in cellulose. The enzymes of many mammals—including humans—cannot catalyze the breaking of β-linkages in cellulose. Consequently, we can't dine on grass or trees.

In contrast, cows, goats, and sheep manage to break down cellulose with a little help. Their digestive tracts contain bacteria that decompose cellulose into glucose. The animals' own metabolic systems then take over. Termites also contain cellulose-hungry bacteria, which is why these insects can damage wooden structures.

When we have excess glucose in our bodies, it is polymerized to glycogen with the help of insulin and stored in our muscles and liver. When our glucose levels slip below normal, the glycogen is converted back into glucose. Glycogen has a molecular structure similar to that of starch, except its chains of glucose units are longer and more branched. Glycogen is vitally important because it stores energy for use in our bodies. It accumulates in muscles, especially in the liver, where it is a quick source of internal energy.

Are "carbs" the culprit in rising obesity rates? Diet books about carbohydrate intake have spanned the spectrum from banning all carbs to taking a "good carb versus bad carb" approach. The latter claims that "bad" carbohydrates cause a quick rise in blood sugar, followed by a spike in blood insulin levels. As shown in Figure 11.16, **insulin** is a hormone secreted by the pancreas that allows the cells in your body to absorb and store sugar that is in the blood.

The sugar that is not immediately broken down for energy is converted into fat and stored in your cells. Conversely, glucagon is another hormone secreted by the pancreas that essentially has the opposite effect of insulin: It promotes the use of stored glucose in cells. The release of glucagon decreases after a "glucose spike," such as results from eating "bad carbs," but it is known to increase after consuming proteins. Therefore, some "low-carb" diets promote eating more protein as a way to use up stored Calories. The long-term health effects of this approach are yet to be seen. Clearly, both nutrition and dieting involve a complex bit of chemistry!

As you may know, a healthful diet derives more carbohydrates from polysaccharides than simple (and sweet) sugars. In the next section, we delve into topics related to sweetness.

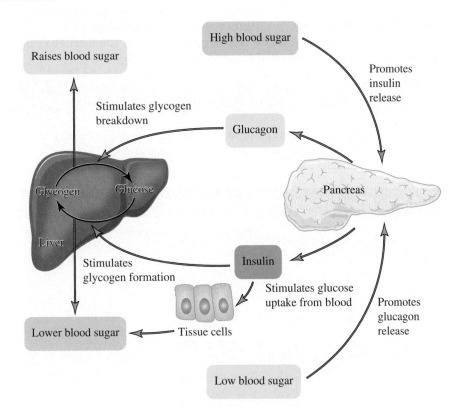

Figure 11.16

The role of insulin and glucagon in controlling the blood sugar levels in our body.

11.5 | How Sweet It Is: Sugars and Sugar Substitutes

> *Learning Objective: Compare and contrast the chemical properties of natural sugars and artificial sweeteners*

Got a sweet tooth? We seem to be born with a preference for sweets, and for most of us, this lasts throughout our lives. Sweeteners come as syrups, small crystals in packages, and cubes or tablets that you can drop into a cup of coffee. Some of these sweeteners are natural, whereas others are artificial (synthetic). Some are very sweet, while others are less so. Some have been available since antiquity, while others are relatively new on the market.

How sweet is sweet? Table 11.3 indicates the relative sweetness of some common natural sweeteners, all compared with sucrose, which is assigned a value of 100. According to this table, one would have to use less fructose to equal the sweetness of a teaspoon of sucrose (table sugar). In contrast, lactose (milk sugar) would require more than 6 teaspoons. As a frame of reference, honey is primarily composed of fructose and glucose; both cane sugar and beet sugar are primarily sucrose.

Your Turn 11.9   The Sweet Story

a. Propose a reason why these sugars differ in their degree of sweetness despite similarities in their chemical formulas.

b. Explain why these sugars are almost identical in their Calories per gram.

As you can see from the previous activity, the sugars we consume are remarkably similar in their chemical composition and Calories, but differ in their sweetness. Our taste buds detect a more intense sweetness from fructose than they do from glucose or sucrose. So, does it matter which sugars you eat? Actually, there is a better question to ask: How *much* sugar are you eating? The quick answer may be "too much." A healthful diet derives more carbohydrates from polysaccharides than simple sugars. If you overindulge in sugar, you increase your risk both of becoming obese and for the diseases that accompany obesity, such as diabetes and high blood pressure.

Let's begin by getting a handle on how much sugar you consume. The next activity allows you to explore one possible source of sugar intake.

Your Turn 11.10   Your Favorite Cola or UnCola

Soft drinks have been referred to as "liquid candy." Is this a fair characterization? Make an argument, *pro* or *con*. In either case, cite the grams of sugar involved.

Table 11.3		Approximate Relative Sweetness Values of Common Sugars		
Sugar	**Also Known As**	**Sweetness**	**Calories/gram**	**Chemical Formula**
Lactose	Milk sugar	16	3.94	$C_{12}H_{22}O_{11}$
Maltose	Malt sugar	32.5	3.86	$C_{12}H_{22}O_{11}$
Glucose	Blood, grape, or corn sugar	74.3	3.87	$C_6H_{12}O_6$
Honey	N/A	97	3.84	N/A
Sucrose	Table, powdered, or granulated sugar	100	4.01	$C_{12}H_{22}O_{11}$
Fructose	Fruit sugar	173	3.87	$C_6H_{12}O_6$

Nutrition Facts

Serving Size 2/3 cup (55g)
Servings Per Container About 8

Amount Per Serving

Calories 230	Calories from Fat 72

	% Daily Value*
Total Fat 8g	12%
Saturated Fat 1g	5%
Trans Fat 0g	
Cholesterol 0mg	0%
Sodium 160mg	7%
Total Carbohydrate 37g	12%
Dietary Fiber 4g	16%
Sugars 12g	
Protein 3g	

Vitamin A	10%
Vitamin C	8%
Calcium	20%
Iron	45%

* Percent Daily Values are based on a 2,000 calorie diet.
Your daily value may be higher or lower depending on
your calorie needs.

	Calories:	2,000	2,500
Total Fat	Less than	65g	80g
Sat Fat	Less than	20g	25g
Cholesterol	Less than	300mg	300mg
Sodium	Less than	2,400mg	2,400mg
Total Carbohydrate		300g	375g
Dietary Fiber		25g	30g

(a)

Nutrition Facts

8 servings per container
Serving size   2/3 cup (55g)

Amount per serving

Calories	230

	% Daily Value*
Total Fat 8g	10%
Saturated Fat 1g	5%
Trans Fat 0g	
Cholesterol 0mg	0%
Sodium 160mg	7%
Total Carbohydrate 37g	13%
Dietary Fiber 4g	14%
Total Sugars 12g	
Includes 10g Added Sugars	20%
Protein 3g	

Vitamin D 2mcg	10%
Calcium 260mg	20%
Iron 8mg	45%
Potassium 235mg	6%

* The % Daily Value (DV) tells you how much a nutrient in
a serving of food contributes to a daily diet. 2,000 calories
a day is used for general nutrition advice.

(b)

Figure 11.17

(a) Food label format before 2020 and **(b)** the required new label format for food products.

Source: U.S. Food and Drug Administration

One reason that you consume sugars is that foods naturally contain them. All the sugars listed in Table 11.3 occur naturally. For example, fructose is found in many fruits, and lactose is found in milk. Another reason you consume sugar is that you add it either during cooking, at the table, or because food companies add it during processing. For instance, manufacturers of processed foods add sugar to peanut butter, spaghetti sauce, and bread. These and many other products have small amounts of sugar added to improve the taste, texture, or shelf life. According to the 2020–2025 Dietary Guidelines for Americans, 13% of the total Calories consumed by people in the United States are from added sugars. This translates to about 17–22 teaspoons of added sugar daily. At about 4 grams of sugar per teaspoon and 4 Calories per gram, this corresponds to about 270–350 Calories daily from added sugar!

To more clearly identify the added sugar content of food (as well as *trans* fats and other minerals), the U.S. FDA required new labels since 2020 (Figure 11.17b), which currently appear on food products.

Your Turn 11.11   Sugar Consumption in the U.S.

Using the Internet, review the latest sugar consumption data from the U.S. National Health and Nutrition Examination Survey.

a. Describe the trends in sugar consumption with sex, age, race and ethnicity, and income. Are any other factors responsible for differing levels of sugar consumption among these groups?

b. How do the trends in sugar consumption between females, males, and children compare to the dietary sugar suggestions provided by the American Medical Association, American Heart Association, and the World Health Organization?

c. How does the sugar consumption (and country-specific dietary sugar recommendations, if present) in other countries compare to the U.S.? Can you think of some reasons for these discrepancies?

d. Using a bag of sugar, measure how much sugar you consumed at breakfast, lunch, and dinner. Compare this with others in your class to see if there are any trends.

High-fructose corn syrup (*HFCS*) is used to sweeten many beverages and foods. Depending on where you live, HFCS goes by different names. In Europe, it is called *isoglucose*; in Canada, it goes by *glucose-fructose*. Corn syrup primarily contains glucose. But if you treat corn syrup with enzymes, you can convert the glucose into fructose, which is sweeter. Depending on the end use, several different "blends" of HFCS exist. For example, a typical blend used in soft drinks is about 55% fructose, with the remainder being glucose. Despite its prevalence in food and drink products in the U.S., the use of HFCS remains controversial. Some countries in the E.U., as well as Australia and Canada, have imposed restrictions on their use.

Let's Debate!   High-Fructose Corn Syrup

Form two groups: one group will be in favor of the topic (the proposition) and the other group will be against the topic (the opposition). Both groups will try to persuade a neutral person or judge to agree with them. The topic of the debate is the motion. Here is the motion:

Despite negative media reports, there is no evidence that the ingestion of high-fructose corn syrup will negatively affect your health.

Curiosity Conversations
- Why is HFCS added to foods?
- Is HFCS metabolized differently than sucrose?
- Are there any reports or scientific evidence that provides a link between HFCS and obesity, relative to other sweeteners?
- Have any companies phased out the use of HFCS in their food products?

We end this section by turning to artificial (synthetic) sweeteners, also called sugar substitutes. As you can see from Table 11.4, these are far sweeter than natural sugars. The Calorie content per gram of aspartame is about the same as sucrose. However, because it is about 200 times sweeter than sucrose, 1/200 of a teaspoon of aspartame is used as the equivalent of a teaspoon of sucrose. Saccharin, aspartame, sucralose, neotame, and acesulfame potassium are the five artificial sweeteners approved in the United States. Other countries have a slightly different list.

The Chemistry of Artificial Sweeteners

Why do things taste sweet? Check out the video at www.acs.org/cic.

Questions to Consider
a. Using the Internet and this article (www.acs.org/cic) as resources, what are some health risks posed by artificial sweeteners?
b. How can a substance that tastes sweet have zero Calories? Describe the metabolism of artificial sweeteners relative to natural sugars.

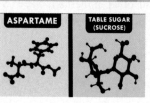

©2018 American Chemical Society

Table 11.4	Approximate Relative Sweetness Values*			
SYNTHETIC SWEETENERS				
Acesulfame Potassium	**Aspartame**	**Saccharin**	**Sucralose**	**Neotame**
200	200	300	600	7,000–13,000

As a reference for artificial sweeteners, sucrose has a sweetness value of 1.

Source: International Food Information Council.

So, are artificial sweeteners the way to go? These compounds certainly do offer the advantage of fewer Calories. For example, the sugar in a 12-ounce can of soft drink accounts for about 140 Calories. In terms of a 2000-Calorie diet, this translates to about 7%. By contrast, the same beverage sweetened with a pure synthetic sweetener would have 0 Calories. Although people have been concerned about the health effects of using artificial sweeteners, most studies indicate that the sweeteners currently on the market are safe for most people. However, a small percentage of the population must avoid aspartame, which we will describe in the next section.

11.6 | Proteins: First among Equals

Learning Objectives:

Illustrate the importance of amino acid side chains and how amino acids combine to form proteins

Explain why the consumption of aspartame is harmful for certain individuals

The word *protein* is derived from *protos,* Greek for "first." However, the name is misleading. Life depends on the interaction of thousands of chemicals, and to assign primary importance to any single compound or class of compounds is too simplistic. Nevertheless, proteins are an essential part of every living cell. They are major components in hair, skin, and muscle. They also transport oxygen, nutrients, and minerals through the bloodstream. Many of the hormones that act as chemical messengers are proteins, as are most enzymes that catalyze life's chemistry.

A **protein** is a polyamide or polypeptide, a polymer built from amino acid monomers. The great majority of proteins are made from various combinations of 20 different naturally occurring amino acids. Molecules of amino acids share a common structure. As shown below, four chemical species are attached to a carbon atom: a carboxylic acid group (green), an amine group (yellow), a hydrogen atom, and a side chain designated R (red):

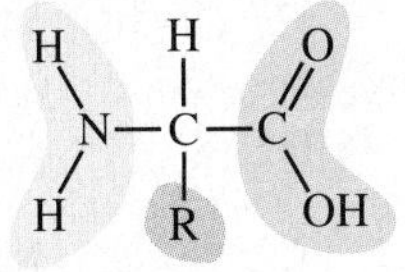

Variations in the R side chain differentiate individual amino acids. Figure 11.18 shows that the structure of the R group controls the chemical properties of an amino acid. Some amino acids feature polar side groups, which are *hydrophilic* ("water-loving") and can form hydrogen bonds with water. In contrast, others contain nonpolar groups and are thereby repelled by water—referred to as *hydrophobic.* Still, other amino acids contain positively or negatively charged groups that lead to pH-dependent properties. There are approximately 300 different types of amino acids in nature; however, only 20 of these serve as building blocks of proteins, which are known as *proteinogenic* amino acids.

Polar side chains can lead to more types of interactions: ionic or hydrogen bonds. Side chains that contain acidic (carboxylic acid, —COOH) or basic (amine, —NH$_2$) groups often become charged ions (—COO$^-$ and —NH$_3^+$, respectively) and attract their opposites to form ionic bonds. If you examine Figure 11.18 closely, you should see potentially favorable interactions between amino acids such as lysine and aspartate through the side-chain functional groups: an amine and a carboxylic acid. Polar side chains often contain hydroxyl groups (—OH; e.g., serine) or amides (—CONR$_2$; e.g., glutamine), which can form hydrogen bonds.

Nonpolar side chains tend to be hydrophobic, meaning they specifically associate with each other to avoid interaction with water. Hydrophobic groups also form vital energetic interactions, called London dispersion forces, discussed in Section 6.11. If you examine the options in Figure 11.18, you will see that nonpolar amino acids are as numerous and varied as polar amino acids. A side chain with a planar phenyl group (phenylalanine) will create a different protein shape relative to an amino acid with a longer carbon chain (such as isoleucine).

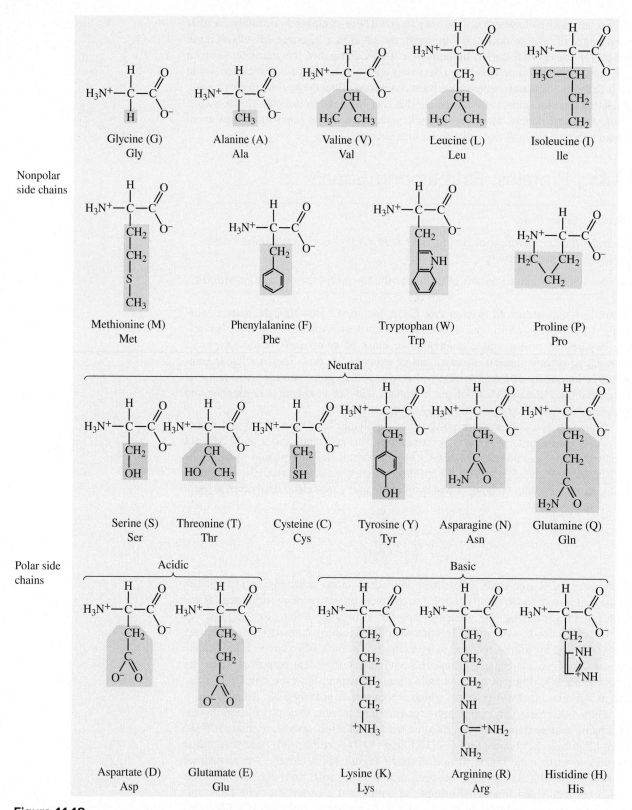

Nonpolar side chains

Glycine (G)
Gly

Alanine (A)
Ala

Valine (V)
Val

Leucine (L)
Leu

Isoleucine (I)
Ile

Methionine (M)
Met

Phenylalanine (F)
Phe

Tryptophan (W)
Trp

Proline (P)
Pro

Neutral

Serine (S)
Ser

Threonine (T)
Thr

Cysteine (C)
Cys

Tyrosine (Y)
Tyr

Asparagine (N)
Asn

Glutamine (Q)
Gln

Polar side chains

Acidic

Basic

Aspartate (D)
Asp

Glutamate (E)
Glu

Lysine (K)
Lys

Arginine (R)
Arg

Histidine (H)
His

Figure 11.18

A comparison of the R group side chains (highlighted) of the 20 naturally occurring amino acids.

Two amino acids can combine by a condensation reaction between the amine group on one amino acid and a carboxylic acid group on the other. For example, glycine can react with alanine, as shown in Equation 11.3a:

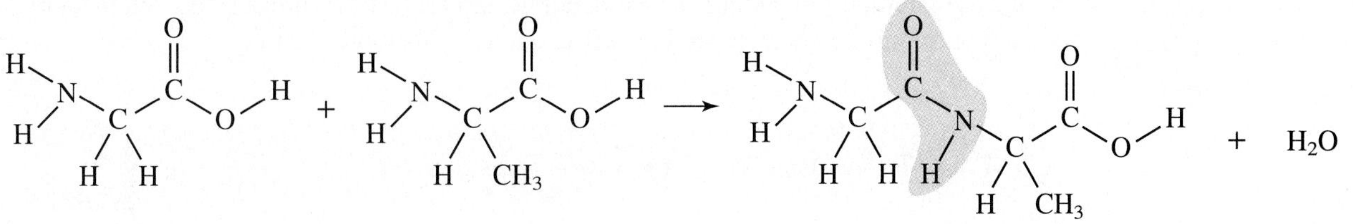

glycine   alanine   dipeptide   water   **[11.3a]**

Note that the acidic —COOH group of the glycine molecule reacts with the —NH$_2$ group of the alanine molecule. In the process, the two amino acids link through the C—N peptide bond shown in the blue shaded area. In addition, a H$_2$O molecule is produced. Once incorporated into the peptide chain, the amino acids are called **amino acid residues**.

Equation 11.3a labels the product as a **dipeptide**, a compound formed from two amino acids. Glycine and alanine can form two different dipeptides. Equation 11.3b shows the other possibility:

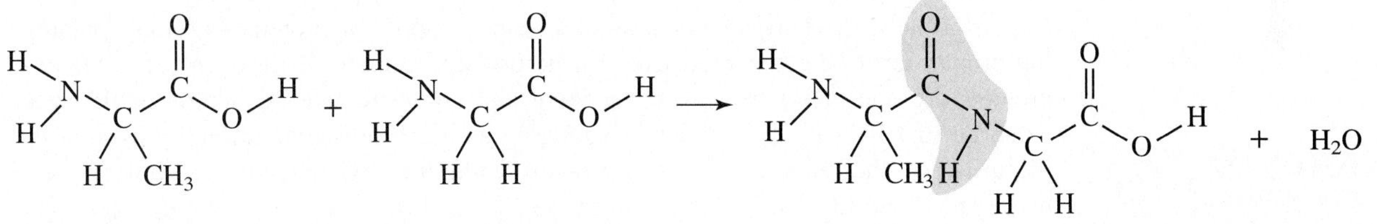

alanine   glycine   dipeptide   water   **[11.3b]**

This time, alanine provided the —COOH group and glycine the —NH$_2$ group.

Look closely, and you will see that the two dipeptides in Equations 11.3a and b are different. In the first dipeptide, the unreacted amine group is on the glycine residue, and the unreacted acid group is on the alanine residue. In the second dipeptide, the —NH$_2$ is on the alanine residue, and the —COOH is on the glycine residue.

The point of all this is that the order of amino acid residues in a peptide makes a difference. A particular protein structure depends not only on which amino acids are present but also on their sequence in the protein chain. Assembling the correct amino acid sequence to make a particular protein is like putting letters in a word: If they are in a different order, a completely new meaning results. Thus, a tripeptide consisting of three different amino acids is like a three-letter word containing the letters *a, e,* and *t*. There are six possible combinations of these letters. Three of them—*ate, eat,* and *tea*—form recognizable English words; the other three—*aet, eta,* and *tae*—do not. Similarly, some sequences of amino acids may be biological nonsense.

Your Turn 11.12   Proteins from Amino Acids

Check out this link to visualize how amino acids combine to form proteins: www.acs.org/cic.

Questions to Consider

a. The protein chain in the simulation had a globular shape. Why is this shape observed instead of a linear chain of amino acid fragments?

b. Some amino acid residues contain a yellow-colored atom. What atom and amino acid residue does this correspond to?

Primary structure of ribonuclease

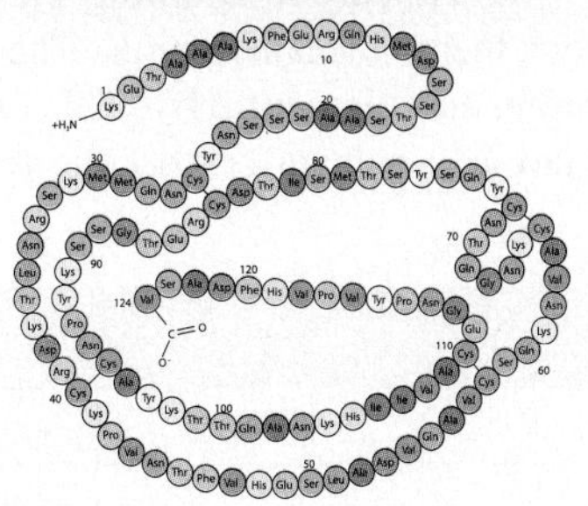

Emre Terim/Shutterstock

Still restricting ourselves to three-letter words and only the letters *a, e,* and *t,* but allowing the duplication of letters, we can make perfectly good words such as *tee* and *tat,* and lots of meaningless combinations such as *aaa* and *tte.* There are, in fact, a total of 27 possibilities, including the six identified earlier. Just as words can use letters more than once, most proteins contain specific amino acids more than once.

Your Turn 11.13   Making Tripeptides

The equations in this section show that glycine (Gly) and alanine (Ala) can combine to form two dipeptides: GlyAla and AlaGly. If these two amino acids can be used more than once, two other dipeptides are possible: GlyGly and AlaAla. Thus, four different dipeptides can be made from two different amino acids. Eight different tripeptides can be made from supplies of two different amino acids, assuming that each amino acid can be used once, twice, three times, or not at all. Use Gly and Ala symbols to write down representations of the amino acid sequence in all eight tripeptides.

Hint: Start with GlyGlyGly.

Normally, the body does not store a reserve supply of protein, so foods containing protein must be eaten regularly. As the principal source of nitrogen for the body, proteins are constantly being broken down and reconstructed. A healthy adult on a balanced diet is in nitrogen balance, excreting as much nitrogen (primarily as urea in the urine) as they ingest. However, growing children, pregnant females, and persons recovering from long-term debilitating illnesses or burns have a positive nitrogen balance. This means they consume more nitrogen than they excrete because they use the element to synthesize additional protein. A negative nitrogen balance exists when more protein is being decomposed than is being made. This occurs in starvation, when the body's energy needs are unmet from the diet, and muscle is metabolized to maintain physiological functions. In effect, the body feeds on itself.

Another cause of a negative nitrogen balance may be a diet that does not include sufficient quantities of **essential amino acids**. These components are those required for protein synthesis but are not synthesized by the human body. Of the 20 natural amino acids that make up our proteins, we can synthesize only 11 in our bodies from simpler molecules. Consequently, we must obtain the other nine amino acids from the foods we eat. If your diet lacks any of the nine essential amino acids identified in Table 11.5, the result can be severe malnutrition.

Good nutrition thus requires protein in sufficient quantity and suitable quality. Beef, fish, and poultry contain all the essential amino acids in approximately the same proportions found in the human body. Therefore, all of these are "complete" proteins. However, most people of the world depend on grains and other vegetable crops rather than on meat or fish. If such a diet is not sufficiently diversified, some essential amino acids may be lacking. For example, Mexican and Latin American diets tend to be rich in corn and corn products, a protein source that is *incomplete* because corn is low in tryptophan, an essential amino acid. A person may eat enough corn to meet the total protein requirement but still be malnourished because of insufficient tryptophan.

Fortunately for millions of vegetarians, a reliance on vegetable protein does not doom them to malnutrition. The trick is to apply a principle that nutritionists call **protein complementarity**. This consists of combining foods that complement essential amino acid content so that the total diet provides a complete supply of amino acids

How Proteins Build Muscle

©2018 American Chemical Society

Check out this video to see how protein builds muscle: www.acs.org/cic.

Questions to Consider

a. The video mentions a couple of recommendations of protein needed per body weight to add muscle. Calculate the grams of protein you would need to consume to build muscle.

b. Using the Internet as a resource, how do steroids allow an individual to build muscle faster relative to consuming more protein?

Table 11.5	The Essential Amino Acids	
Histidine	Lysine	Threonine
Isoleucine	Methionine	Tryptophan
Leucine	Phenylalanine	Valine

used for protein synthesis. Although some may worry that vegetarians must strictly adhere to this principle, most are likely to do it automatically. Say, for example, you eat a peanut butter sandwich. Bread is deficient in lysine and isoleucine, but peanut butter supplies these amino acids. On the other hand, peanut butter is low in methionine, but the bread provides it. The traditional diets in many countries also tend to meet protein requirements. For example, in Latin America, beans are used to complement corn tortillas, and soy foods are eaten with rice in parts of Southeast Asia and Japan. People in the Middle East combine bulgur wheat with chickpeas or eat hummus—a paste made from sesame seeds and chickpeas—with pita bread. In India, lentils and yogurt are eaten with unleavened bread. Thus, it is likely that if you follow a balanced vegetarian diet, you will ingest sufficient quantities of essential amino acids, assuming that you are eating an adequate number of Calories.

We end this section by revisiting sweetness, the topic of the previous section. It may surprise you to learn that *aspartame*, a sugar substitute, is a dipeptide. Aspartame is composed primarily of the amino acids aspartic acid and phenylalanine (Figure 11.19). It is one of the most highly studied food additives, and for most consumers, aspartame is a safe alternative to sugar. One group of people, however, definitely should not use aspartame. The warning on packets of artificial sweeteners and products containing aspartame is explicit: "Phenylketonurics: Contains Phenylalanine."

This is a case where one person's treat is another person's poison. Phenylalanine is an essential amino acid converted in the body into tyrosine, a different amino acid. Individuals with phenylketonuria, a genetically transmitted disease, lack the enzyme (phenylalanine hydroxylase) that catalyzes this transformation. Consequently, the conversion of dietary phenylalanine into tyrosine is blocked, and the phenylalanine concentration in blood and tissues rises (Figure 11.20). To compensate for the elevated

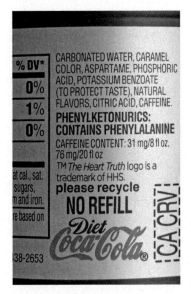

Figure 11.19

The structural formula of aspartame. The blue shading illustrates the peptide bond linking the amino acids aspartic acid and phenylalanine. Shown on the right is a typical warning label found on a soft drink bottle.

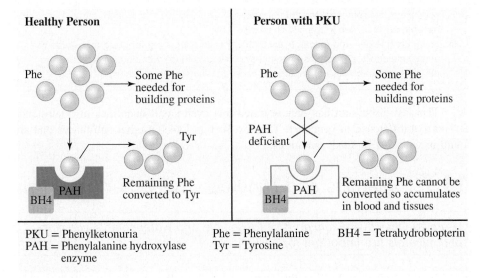

Figure 11.20

A schematic illustration of phenylketonuria, PKU. The BH4 cofactor is an iron-containing compound that acts as a helper molecule to the PAH enzyme for converting phenylalanine into tyrosine.

Carboxylic acid group

Ketone group

Figure 11.21

The molecular structure of phenylpyruvic acid. Because the structure contains both a ketone and carboxylic acid group, it is referred to as a *keto acid*.

Note: A ketone functional group contains a central carbonyl (C=O) connected to two R groups.

phenylalanine, the body converts it into phenylpyruvic acid, excreting large quantities of this acid in the urine. Phenylpyruvic acid is termed a *keto acid* because of its molecular structure (Figure 11.21); hence, the disease is known as phenyl*keto*nuria or PKU. People with the disease are called phenylketonurics.

Excess phenylpyruvic acid causes severe intellectual disabilities. Therefore, the urine of newborn babies is tested for this compound using special test paper placed in the diaper. Infants diagnosed with PKU must be put on a diet severely limited in phenylalanine. This means avoiding excess phenylalanine from milk, meats, and other sources rich in protein. Because phenylalanine is an essential amino acid, a minimum amount of it must still be available, even in phenylketonurics. Supplemental tyrosine may also be needed to compensate for the absence of the normal conversion of phenylalanine into tyrosine. A phenylalanine-restricted diet is recommended for phenylketonurics, at least through adolescence. Adult phenylketonurics also must limit their phenylalanine intake and hence curtail their use of aspartame. However, is aspartame *really* safe for all individuals without phenylketonuria? Explore this for yourself in the following activity.

Is Aspartame Safe?

Is aspartame *really* safe? Check out the video at www.acs.org/cic.

Questions to Consider

a. The video mentions that fruit juice can contain 5 times more methanol than a 12-oz diet soda. What concentration of methanol is present in fruit juice? How does this concentration compare to the methanol found in alcoholic beverages, discussed in the last chapter?

©2018 American Chemical Society

b. Using the Internet as a resource, what other health effects are reported for aspartame and its metabolites? Are these health risks limited to those with phenylketonuria? Are the possible health risks of aspartame backed up by scientific studies?

This last discussion has demonstrated that even small quantities of a substance can make a difference in your diet. The next section looks at other substances that are found in your diet in small quantities.

11.7 | Vitamins and Minerals: The Other Essentials

Learning Objective: Examine how fat-soluble and water-soluble vitamins and minerals are important for body health

Vitamins and minerals are **micronutrients**, substances that are needed only in minuscule amounts but still are essential to life. Nearly everyone in the United States knows that vitamins and minerals are important, but a thriving multimillion-dollar supplement

vitamin A, a lipid-soluble vitamin   vitamin C, a water-soluble vitamin

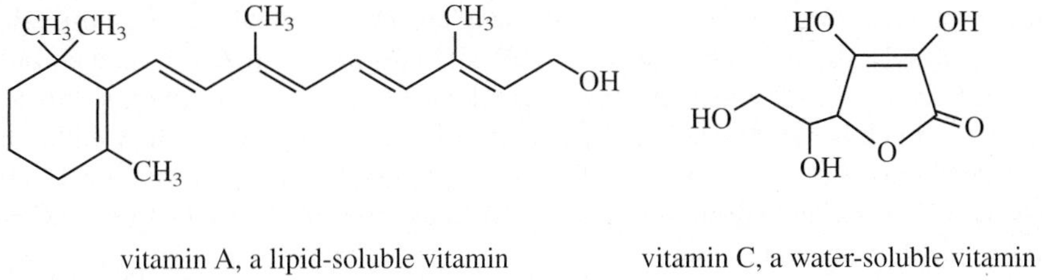

Figure 11.22

Examples of lipid-soluble and water-soluble vitamins. Check out the 3D renderings of vitamin A and vitamin C at www.acs.org/cic.

industry reminds any who forget. Unfortunately, many processed foods that are high in sugars and fats lack essential micronutrients.

Only relatively recently have we understood the role of vitamins and minerals in our diet. Over the ages, humans learned that they became ill if certain foods were lacking. For instance, studies of vitamin-deficient illnesses, such as scurvy, were carried out in the 18th century. However, systematic studies only began in the early 20th century with the discovery of "Vitamine B_1" (thiamine).

Vitamins are organic compounds with a wide range of physiological functions. Although only small amounts are needed in our diet, vitamins are essential for good health, proper metabolic functioning, and disease prevention. In general, vitamins are not energy sources for the body; however, some help break down macronutrients, such as proteins and carbohydrates. They can be classified based on being water- or fat-soluble. For example, examine the structural formula of vitamin A shown in Figure 11.22 to see that it contains almost exclusively C and H atoms. As a result, vitamin A is a nonpolar compound, lipid/fat-soluble, and similar to hydrocarbons derived from petroleum. In contrast, water-soluble vitamins such as vitamin C (Figure 11.22) often contain several —OH groups that can hydrogen bond with water molecules.

> **DID YOU KNOW ?**
>
> The particular designation of "B_1" was the label on the test tube in which the sample was first collected. In 1912, Polish biochemist Casimir Funk (1884–1967) first coined the name *vitamine* (short for *vital amine*), to describe a class of compounds he was studying that were vital for life and contained the amine (—NH_2) group. The final "e" in its spelling disappeared in 1920 after it was suspected that not all vitamins contained amines.

Your Turn 11.14   Classifying Vitamins

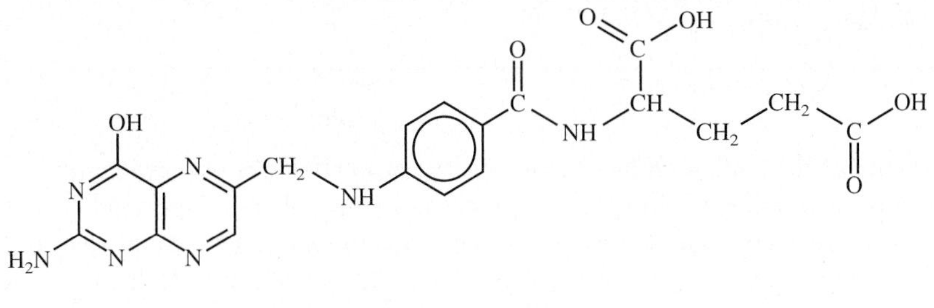

Folic acid (above) helps prevent certain types of anemia and aids in nucleic acid synthesis. This vitamin is particularly important for pregnant females. Do you expect it to be soluble in fat tissue (lipids) or in the bloodstream and cell tissue (water)? Explain your reasoning.

The solubility of vitamins has significant health implications. Because of their fat solubility, vitamins A, D, E, and K are stored in cells rich in lipids, which are available on biological demand. If swallowed in excess, fat-soluble vitamins can build up to a toxic level. For example, high doses of vitamin A can result in troublesome symptoms such as fatigue and headache and more serious ones such as blurred vision and liver damage. Although the toxic level of vitamin D is unknown, similar illnesses can result if too much vitamin D is ingested, including heart and kidney damage. In 2022, a case of vitamin D overdose resulted from a British man taking supplements with several times the daily recommended dose of the vitamin. The man was ill for more

Vitamin Supplements

©2018 American Chemical Society

Do vitamin supplements really work? Check out this video to see for yourself: www.acs.org/cic.

Questions to Consider

a. Using the Internet as a resource, what are the most common types of dietary supplements used by consumers?

b. Does the U.S. or other countries around the world have any regulations concerning the use and distribution of vitamin supplements?

than two months after ceasing to take the supplements, eventually being hospitalized for eight days. High levels of these fat-soluble vitamins are not reached via diet; rather, they result from excessive use of vitamin supplements. The one exception seems to be vitamin D, which is synthesized in the skin by using the energy of sunlight rather than ingested. Research on vitamin D has led more physicians to check vitamin D blood levels as part of an annual physical exam and to use the results to determine whether taking a supplement is necessary.

In contrast, water-soluble vitamins are excreted in the urine rather than stored in the body. As a result, you need to eat foods containing these vitamins frequently. Unfortunately, even water-soluble vitamins can accumulate at toxic levels when taken in large doses, although such cases are rare. A balanced diet provides the necessary vitamins and minerals for most people, making vitamin supplements unnecessary.

We would be remiss not to mention vitamin E, which actually consists of several closely related fat-soluble vitamins rather than a single compound. Vitamin E is only synthesized by plants and in varying amounts. Vegetable oils and nuts are good sources of it. Nonetheless, it is so widely distributed in foods that it is difficult to create a diet deficient in vitamin E. Since the 1990s, this vitamin has been in the news as part of the antioxidant system that protects the body from chemically active and damaging free radicals. Although taking vitamin E supplements was recommended at one time, this is no longer the case. Skin preparations are another matter, though. Many products contain vitamin E and claim it prevents or helps heal skin damage. Investigate these claims for yourself in the next activity.

Your Turn 11.15   Vitamin E and Your Skin

Check the advertisements, and you will see that many hand lotions and beauty creams contain vitamin E.

a. Identify three skin products that contain vitamin E.
b. How is vitamin E thought to help protect your skin?
c. Although it might seem logical that vitamin E would be good for the skin, it is difficult to find evidence. Investigate this topic to see for yourself. Use the resources on the Internet to assist you.

Many of the water-soluble vitamins serve as **coenzymes**, molecules that work in conjunction with enzymes to enhance their activity. Members of the vitamin B family are particularly adept in acting as coenzymes. Niacin plays an essential role in energy transfer during glucose and fat metabolism. The synthesis of niacin in the body requires the essential amino acid tryptophan. Thus, a diet deficient in tryptophan may lead to niacin deficiency. Such a deficiency causes pellagra, a serious condition that is characterized by "the 4Ds" of diarrhea, dermatitis, dementia, and death. This disease is still common today in parts of the world.

Some vitamins were discovered when observers correlated diseases with the lack of specific foods. For example, vitamin C (ascorbic acid) must be supplied in the diet, typically via citrus fruits and green vegetables. An insufficient supply of the vitamin leads to scurvy, a disease in which collagen, an important structural protein, is broken down.

The link between citrus fruits and scurvy was discovered more than 200 years ago when it was found that feeding British sailors limes or lime juice on long sea voyages prevented the disease. Thanks to Nobel Laureate Linus Pauling (1901–1994), who in 1970 authored *Vitamin C and the Common Cold*, vitamin C continues to be in the public eye.

Minerals are ions or ionic compounds that, like vitamins, have a wide range of physiological functions. You may be familiar with minerals such as sodium and

Svetlana Foote/Shutterstock

Your Turn 11.16   Megadoses of Vitamin C

Decades ago, Linus Pauling claimed that large doses of vitamin C were therapeutic in preventing the common cold.

a. What range of vitamin C daily constitutes a "megadose"?

b. Find evidence to either support or refute the claim of preventing the common cold. Cite your sources.

c. Using the Internet, find actual cases where the consumption of large amounts of high vitamin C-containing foods such as carrots or oranges caused skin pigmentation to turn yellow/orange. Do you think the observed skin discoloration is due to water-soluble or fat-soluble components in these foods? Explain.

d. Interview three people of different age groups, including a nurse or physician, if you can. Ask if they take vitamin C, and if so, why.

©Bradley D. Fahlman

Is there iron in your cereal? Check out the lab experiment at www.acs.org/cic.

Questions to Consider

a. How would you modify this experiment to quantify the amount of iron per mass in your cereal?

b. What other foods could be used for this experiment, which would yield an isolable amount of free iron?

calcium, but actually, the list is much longer. Depending on how much of them you need, minerals are classified as either macro, micro, or trace:

- **Macrominerals:** Ca, P, Cl, K, S, Na, and Mg. As ions, these elements are necessary for life but not nearly as abundant in our bodies as O, C, H, and N. You need to ingest macrominerals daily, typically in the range of 1–2 g.
- **Microminerals:** Fe, Cu, and Zn. The body requires lesser amounts of these. You may recognize iron as a component of hemoglobin, a protein in the blood that carries oxygen.
- **Trace minerals:** I, F, Se, V, Cr, Mn, Co, Ni, Mo, B, Si, and Sn. These usually are measured in microgram (μg) quantities. Although the total amount of trace elements in the body is tiny, their small quantity is required for good health.

The periodic table in Figure 11.23 displays the essential dietary minerals. The metals exist in the body as cations; for example, Ca^{2+} (calcium ion), Mg^{2+} (magnesium ion), K^+ (potassium ion), and Na^+ (sodium ion). The nonmetals are typically present as anions. For example, chlorine is found as Cl^- (chloride ion), and phosphorus appears as PO_4^{3-} (phosphate ion).

As illustrated in Figure 11.24, the physiological functions and sources of minerals are widely diverse. Calcium is the most abundant mineral in the body. Along with phosphorus and smaller amounts of fluorine, it is a major constituent of bones and teeth. Blood clotting, muscle contraction, and transmission of nerve impulses all require the calcium ion, Ca^{2+}. Sodium is also essential for life, but not in the excessive amounts supplied by the diets of many people today. The salt we eat does not necessarily come from the salt shaker. Rather, you perhaps unknowingly add salt to your diet from sauces, snack foods, fast foods, and even canned soup! Labels are required to list the "sodium" content, meaning the number of milligrams of Na^+ per serving. For example, different brands of tomato soup may have between 700 and 1260 mg of Na^+ per serving. Compare this with the recommended daily value of no more than 2400 mg (2.4 grams) of Na^+. The major concern with excess dietary sodium is its correlation with high blood pressure for some individuals. Their doctors may advise them to limit their sodium intake.

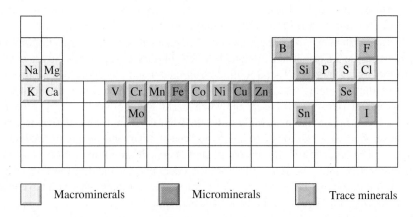

Figure 11.23

Periodic table indicating the dietary minerals necessary for human health.

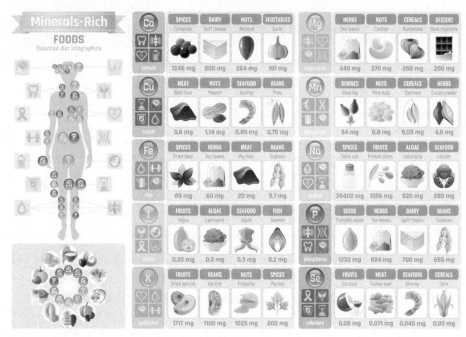

Figure 11.24

Common sources and health benefits of various minerals.

Olga Chernyak/123RF

Your Turn 11.17   Sodium in Your Diet

Compare the sodium content for foods in the same category, such as different brands of pretzels, bread, frozen pizza, salad dressing, or even tomato soup. Have your findings surprised you or influenced your future choices?

Hint: 1 g = 1000 mg.

Oranges, bananas, tomatoes, and potatoes help supply the recommended daily requirement of 2 grams of potassium (in the form of K^+), another essential minerals. You may have heard both K^+ and Na^+ referred to as the "electrolytes" of sports drinks. Sodium and potassium ions are close chemical cousins, as they are both in Group 1 of the periodic table. They have similar chemical properties and physiological functions. Within cells, the concentration of K^+ is considerably greater than that of Na^+. The reverse situation holds true in the lymph and blood serum outside the cells, in which the concentration of K^+ is low, and that of Na^+ is high. The relative concentrations of K^+ and Na^+ are especially important for the rhythmic beating of the heart. Individuals who take diuretics to control high blood pressure may also take potassium supplements to replace K^+ excreted in the urine. However, such supplements should be taken only under a physician's advice because they can dramatically alter the potassium–sodium balance in the body and lead to cardiac complications.

In most instances, microminerals and trace elements have very specific biological functions and are incorporated in relatively few biomolecules. Iodine is an example. Most of the body's iodine is found in the thyroid gland and incorporated into thyroxine, a hormone that regulates metabolism. Excess thyroxine production is associated with hyperthyroidism (Graves' disease), in which basal metabolism is accelerated to an unhealthy level, rather like a racing engine. In contrast, a thyroxine deficiency, sometimes caused by a lack of dietary iodine, slows metabolism and results in tiredness and listlessness. The tendency of the thyroid gland to concentrate iodine makes possible the use of radioactive I-131 in treating thyroid disorders and in imaging the thyroid gland for diagnostic purposes. The next activity allows you to combine what you have learned about radioactive I-131 and iodine as a trace mineral.

Your Turn 11.18   Radioactive Iodine

Hyperthyroidism (Graves' disease) is also referred to as having an overactive thyroid gland.

a. I-131 is used to treat hyperthyroidism. Explain how ingestion of this radioisotope can lead to a reduction in the function of the thyroid gland.

b. I-131 treatment has both risks and benefits for a patient. List two of each.

c. Patients treated with I-131 temporarily carry a source of radioactivity in their bodies. After 10 half-lives, a radioisotope can be said to be "gone." How much time is this for a patient treated with I-131?

Hint: See Table 7.3.

11.8 | Food for Energy

Learning Objective: Compare the energy content of foods and explain how food choices and level of exercise can affect your weight

The energy needed to keep our bodies warm and to run our complex chemical, mechanical, and electrical systems comes from the food we eat: fats, carbohydrates, and proteins. As noted in several earlier chapters, this energy initially arrives on Earth in the form of sunlight and then is absorbed by green plants. In the process of photosynthesis, CO_2 and H_2O are combined to form $C_6H_{12}O_6$ (Equation 11.4). Hence the Sun's energy is stored in chemical bonds of the monosaccharide we know as glucose ($C_6H_{12}O_6$):

$$\text{energy (from sunshine)} + 6\ CO_2 + 6\ H_2O \xrightarrow{\text{chlorophyll}} C_6H_{12}O_6 + 6\ O_2 \qquad \textbf{[11.4]}$$

During respiration, the outcome of photosynthesis is reversed (Equation 11.5). Glucose is converted into simpler substances (ultimately, in most cases, to CO_2 and H_2O), and energy is released:

$$C_6H_{12}O_6 + 6\ O_2 \longrightarrow 6\ CO_2 + 6\ H_2O + \text{energy} \qquad \textbf{[11.5]}$$

The energy balance between Equations 11.4 and 11.5 is schematically illustrated in Figure 11.25.

In addition to having a supply of sufficient energy, our bodies must have some way of regulating the rate at which the energy is released. Without such control, your body temperature would wildly fluctuate. The automobile provides an analogy. Dropping a lighted match into the fuel tank would burn all the gasoline at once and likely the entire car as well! Under normal operating conditions, just enough fuel is delivered

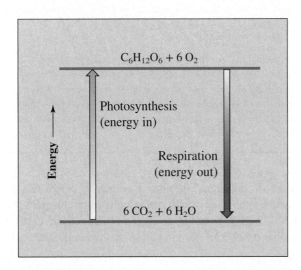

Figure 11.25

Glucose energy balance (photosynthesis and respiration).

Table 11.6	Average Energy Content of Macronutrients
Macronutrient	**Energy Content (Cal/g)**
Fats	9
Carbohydrates	4
Proteins	4

to the ignition system to supply the automobile with the energy it needs without raising the temperature of the car and its occupants beyond reason. By releasing a little energy at a time, the efficiency of the process is enhanced. So it is with our bodies. The conversion of foods into carbon dioxide and water occurs over many small steps, each one involving enzymes, enzyme regulators, and hormones. As a result, energy is released gradually, as needed, and body temperature is maintained within normal limits.

The energy in Calories associated with the metabolism of a gram of fats, carbohydrates, and proteins is given in Table 11.6. On a Calorie-per-gram basis, fats provide about 2.5 times as much energy as proteins and carbohydrates. This observation makes it easy to understand the popularity of low-fat diets for losing weight. Although proteins, like carbohydrates, yield about 4 Cal/g if metabolized, proteins are not used in the body primarily as an energy source. Rather, proteins are used for building skin, muscles, tendons, ligaments, blood, and enzymes.

Given how many tasty foods contain fat, it is easy to get an unhealthy percentage of our daily Calories from fats, as illustrated in the following activity. The activity that follows will give you an opportunity to explore dietary advice.

Your Turn 11.19   "Low-Fat" Cheese

A popular brand of low-fat shredded cheddar cheese advertises that it provides 1.5 g of fat per serving. Of this 1.5 g of fat, 1.0 g is saturated fat. In addition, a serving of this cheese is 28 grams (or 1/4 cup) and has 50 Calories, with 15 of these coming from fat. Is this a "low-fat" cheese? Support your answer with some numbers. Note that the dietary recommendations provided by the U.S. Department of Agriculture (USDA) and the U.S. Department of Health and Human Services (HHS) are that no more than 20–35% of Calories in adults should come from fat.

Your Turn 11.20   Dietary Advice

Which foods should you eat less of, and which ones more? One day the experts say one thing; the next day, they seem to say the opposite. With so much information available, you may be confused or feel overwhelmed.

a. Using the Internet, search for the dietary advice illustrated by "the Basic Four" (1958), "the Food Pyramid" (1991), "MyPyramid" (2005), and "MyPlate" (2015). What do all of these have in common? What are the key differences among these suggestions?

b. Cite research studies that support the above dietary suggestions. What are the merits and weaknesses of MyPlate versus those of MyPyramid? Are there additional components that should be included in these plans?

The reason for the difference in energy content between fats and carbohydrates is evident from their chemical compositions. Compare the chemical formula of a fatty acid, lauric acid ($C_{12}H_{24}O_2$) with that of sucrose (table sugar, $C_{12}H_{22}O_{11}$). Both compounds have the same number of carbon atoms per molecule and very nearly the same number of hydrogen atoms. When molecules such as these "burn" as fuel in your body, the C and H atoms that they contain combine with oxygen to form CO_2 and H_2O, respectively. But more oxygen is required to burn a gram of lauric acid, $C_{12}H_{24}O_2$, than one gram of sucrose, $C_{12}H_{22}O_{11}$. Examine these two reactions:

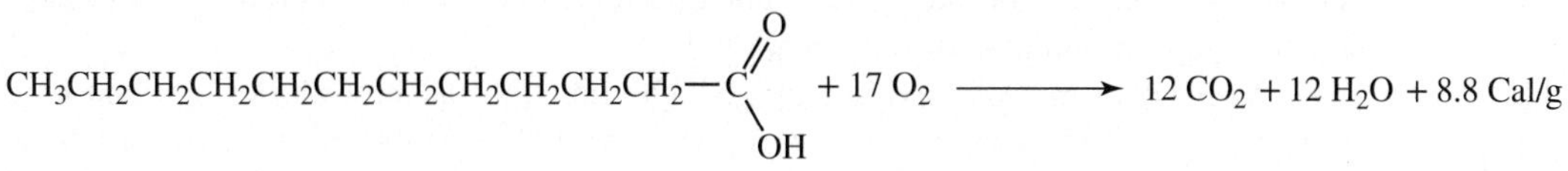

$$CH_3CH_2CH_2CH_2CH_2CH_2CH_2CH_2CH_2CH_2CH_2 \overset{O}{\underset{OH}{-C}} + 17\,O_2 \longrightarrow 12\,CO_2 + 12\,H_2O + 8.8\,Cal/g$$

lauric acid

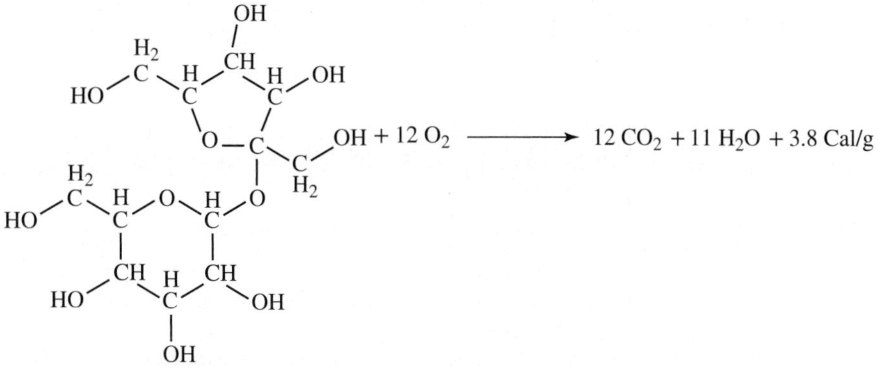

$$ \cdots\cdots + 12\,O_2 \longrightarrow 12\,CO_2 + 11\,H_2O + 3.8\,Cal/g $$

sucrose

In the language of chemistry, the sugar is already more "oxygenated" or more "oxidized" than the fatty acid. Weaker C—H bonds (416 kJ/mol) have already been replaced by stronger O—H bonds (467 kJ/mol) in the sucrose. The result is that even though fewer O=O double bonds (498 kJ/mol) must be broken for sucrose to combine with O_2, less energy is released overall than is the case for the combustion of lauric acid.

So, how many Calories does a person need? The answer is, "It depends." The number of Calories your diet should supply each day is a function of your level of activity, the state of your health, your sex, age, body size, and a few other factors. Table 11.7 summarizes the daily food energy intake that has been recommended for people in the United States. The estimated Calorie requirements are presented by sex and age groups

Table 11.7	Estimated Calorie Requirements (United States)		
	ACTIVITY LEVEL		
Age (years)	**Sedentary***	**Moderately Active†**	**Active‡**
Females			
14–18	1800	2000	2400
19–30	1800–2000	2000–2200	2400
31–50	1800	2000	2200
51+	1600	1800	2000–2200
Males			
14–18	2000–2400	2400–2800	2800–3200
19–30	2400–2600	2600–2800	3000
31–50	2200–2400	2400–2600	2800–3000
51+	2000–2200	2200–2400	2400–2800

*Sedentary *means a lifestyle that includes only the light physical activity associated with typical day-to-day life.*

†Moderately active *means a lifestyle that includes physical activity equivalent to walking about 1.5–3 miles per day at 3–4 miles per hour and the light physical activity associated with typical day-to-day life.*

‡Active *means a lifestyle that includes physical activity equivalent to walking more than 3 miles per day at 3–4 miles per hour and the light physical activity associated with typical day-to-day life.*

Source: Dietary Guidelines for Americans, 2020–2025, USDA

at three different activity levels. Growing children (not included in the table) need a larger energy intake to fuel their high activity level and provide the raw material for building muscle and bone. Children are particularly susceptible to undernourishment and malnutrition. Indeed, mortality rates among infants and young children are disproportionately high in famine-stricken countries.

Where does all this food energy go? The first call on the Calories you consume is to keep your heart beating, your lungs pumping air, your brain active, all major organs working, and your body temperature at about 37 °C. These requirements define the **basal metabolic rate** (BMR), the minimum amount of energy required daily to support basic body functions. This corresponds to approximately 1 Calorie per kilogram (2.2 pounds) of body mass per hour, although it varies with gender, size, and age, as you will see in this next activity.

Your Turn 11.21   Calories by Sex and Age

Consider the information in Table 11.7 and the *Dietary Guidelines for Americans*, which can be found on the Internet.

a. Do individuals of the same age require the same number of Calories for the same activity level? Explain.

b. As an active person grows older, how does the estimated Calorie requirement change?

c. Do other countries have similar dietary guidelines? If so, how do these differ from those established in the U.S.?

Consider a 20-year-old female weighing 55 kg (121 pounds). If her body has a minimum requirement of 1 Cal/(kg·h), her daily basal metabolic rate will be 1 Cal/(kg·h) × 55 kg × 24 h/day, or about 1300 Cal/day. According to Table 11.7, the recommended daily energy intake for a female of this age and weight is a maximum of 2200 Cal if she is moderately active. This means that 59% of the energy derived from this food goes to keep her body systems going.

Where do the rest of the Calories go? The first law of thermodynamics decrees that the energy must go somewhere. If she "burns off" the extra Calories through exercise, none will be stored as added fat and glycogen. But, if the excess energy is not expended, it will accumulate in chemical form.

How hard and how long we have to work (or play) to burn dietary Calories is reported in Table 11.8. In Table 11.9, exercise is related to readily recognizable units such as hamburgers, potato chips, and cookies. Of course, by combining the

---

Table 11.8		Energy Expenditure for Common Physical Activities*	
Moderate Physical Activity	**Cal/h**	**Vigorous Physical Activity**	**Cal/h**
Hiking	370	Running (10 mph)	1050
Light gardening/yard work	245	Heavy yard work (chopping wood)	440
Dancing (ballroom, fast)	315	Swimming (freestyle laps)	510
Golf (walking, carrying clubs)	245	Aerobics	480
Bicycling (< 10 mph)	279	Bicycling (12–14 mph)	559
Walking (3.5 mph)	196	Jogging (5 mph)	490
Weight lifting (light workout)	140	Weight lifting (vigorous workout)	350
Stretching	105	Basketball (competitive game)	490

Values taken from CalorieLab include resting metabolic rate and activity expenditure for a 154-pound (70-kg) person. Calories burned per hour are higher for persons heavier than 154 pounds and lower for persons who weigh less.

Table 11.9	How Much Must I Exercise if I Eat This Food?*		
Food	Calories	Walking at 3.5 mph (min)	Jogging at 5 mph (min)
Apple	125	27	13
Beer, 8 ounces	100	21	10
Chocolate chip cookie	85	11	5
Hamburger	350	75	35
Ice cream, 4 ounces	175	38	18
Pizza, cheese, 1 slice	180	39	18
Potato chips, 1 ounce	108	23	11

Values include resting metabolic rate and activity expenditure for a 154-pound (70-kg) person.

information in this section with the information in earlier parts of this chapter about the types of nutrients in food, it should be clear that a healthful diet cannot be achieved simply by consuming the correct number of Calories. A 2000-Calorie diet of only potato chips and cookies would leave a person malnourished. Proper nutrition is not simply a matter of how much food but also what kind of food a person consumes. To close out this section, explore how food choices and exercise can affect your weight in the next two activities.

Your Turn 11.22   Basketball and Calories

A 154-pound person consumes a meal consisting of two hamburgers, 3 oz of potato chips, 8 oz of ice cream, and an 8-oz beer. Calculate the number of Calories in the meal and the number of minutes the person would have to vigorously play basketball to "work off" the meal.

Your Turn 11.23   Eating and Exercise

Check out this simulation to explore the relationship between diet, exercise, and weight.

Source: PhET Interactive Simulations, University of Colorado Boulder.

1. Input your own age, height and weight on the left side of the simulation underneath the body. The program will calculate your body fat.
2. Explore different combinations of meals and exercise by using the arrows to scroll through the options. Drag all the foods you would consume per day onto the plate and drag your daily exercise into the notebook. Click the play button and observe how it affects your weight. You can pause the simulation to see what happens after a set duration.
3. Reset all your choices and try selecting a lot of food and little/no exercise. Observe how it affects your weight.
4. Reset all your choices and try selecting a lot of exercise and little/no food. Observe how it affects your weight.

Questions to Consider
a. What combinations of food choices and exercises resulted in decreasing your weight?
b. What can you do to increase the number of Calories burned?
c. What combinations of food choices and exercises resulted in a starvation level?

11.9 | Food Safety: What Else Is in Our Food?

> *Learning Objective: Evaluate methods used for maintaining a safe food supply*

It is estimated that 400,000–500,000 deaths occur yearly in the U.S. due to unhealthy eating and inactivity. The dietary suggestions outlined in the previous section are designed to prevent diet-related diseases such as hypertension (high blood pressure), heart disease, liver disease, cancer, stroke, diabetes, and obesity. Although much less frequent, there are more than 3000 deaths and 150,000 cases of food poisoning in the U.S. annually due to the presence of **foodborne illnesses**—the presence of bacteria, viruses, parasites, and chemical toxins that can go undetected in our foods. This begs the question: How safe is our food supply?

To ensure that food is safe to consume, the processes of food production, processing, transportation, and handling/storage must all be carried out in such a way as to prevent contamination with harmful substances. No food producer or restaurateur would purposely wish to harm their consumers; however, sometimes shortcuts in food preparation may be taken to increase profits or save time. Consequently, many countries have devised some food legislation, and international food laws have been developed to protect the quality of foods that are shipped between countries. The first food legislation was instituted in the UK with the Baking Laws of 1155. This law banned the use of sand in bread that was used to make it heavier!

For food laws to be successful, there have to be some means of policing to check that the food industry is in compliance with the legislation. There are three common policing methods used to determine whether levels of contaminants exceed the limits established in the law(s). Depending on the offense, court action could be taken.

- **Food surveillance:** Random samples of food and produce are taken from farms or grocery stores and analyzed for residues of chemicals and/or microbes. Staple foods such as bread, milk, and potatoes are analyzed most frequently because any contamination would affect large populations of consumers.
- **Total diet surveys:** Samples of consumers' meals are analyzed for contaminants. Typically, a group of consumers is recruited and asked to prepare duplicate meals for a prescribed period of time. They eat one meal, and the other is sent to a laboratory for analysis.
- **Enforcement sampling:** Samples are taken when/where there is concern that a food safety problem may exist and this might lead to a food recall. It should be noted that food recalls are mostly due to the voluntary reporting of a food safety issue by a producer without the need for external policing. For instance, in October 2015, Hormel Foods found metal shavings in some cans of Skippy peanut butter due to malfunctioning equipment in their packaging plant. The company announced a product recall after a pallet of the contaminated product was inadvertently released before their quality assurance laboratory discovered the contaminants.

Legislation and monitoring are significantly more complex in our globalized society. Notwithstanding the benefits of eating locally described later, we are accustomed to consuming bananas from Costa Rica, coffee from Colombia, wine and cheese from France, and olive oil from Italy. But when we import food from another country, should we assume their national surveillance protocols properly checked the food item? Not necessarily, thus requiring rigorous testing facilities at major ports of entry. The Food Safety Modernization Act of 2010 has also added requirements to increase the number of routine inspections of food facilities that export to the United States under the jurisdiction of the U.S. FDA. This is designed to help identify potential safety issues before products arrive in the United States and determine facilities' compliance with the FDA's safety standards.

DID YOU KNOW ?

Another source of diet-related illness is food allergies, which only affect certain people and cannot be prevented by legislation and food inspection protocols.

Real Talk   Monitoring Food Safety

Check out the interview with ShaRhonda Dennis, a chemist who works with the FDA: www.acs.org/cic.

metamorworks/Shutterstock

Questions to Consider
Using the Internet, find a recent example of a food recall.

a. How was this food safety issue discovered, and how long did it take for it to be detected?
b. Who was to blame for the food recall? Were there any legal actions taken against this party?
c. Do you have concerns about the overall safety of our foods? What other methods could be used to detect food safety issues?

Your Turn 11.24   Food Additives

Although legislation and monitoring practices are quite effective in preventing widespread cases of foodborne illnesses, more than 1000 food additives are currently unregulated by the FDA.

a. Using the Internet as a resource, list five of these unregulated food additives. What types of processed foods contain these ingredients, and what is the purpose of them?
b. Describe the potential health impacts of these additives. Evaluate the reliability of the claimed health implications based on the presence, or lack, of scientific evidence.

Armed with your knowledge of the ingredients contained in food items and how they are regulated, let's now focus on the bigger picture of producing food on our planet.

11.10 | The Real Costs of Food Production

Learning Objective: Determine the water and land use associated with the production of various food items

Throughout history, food has played a pivotal role in human health and well-being. Millions have starved from a lack of food, while others have died from eating too much of it. Wars have been fought over food, and countries have been destabilized by the lack of it. In previous chapters, we discussed how the air we breathe and the water we drink are connected to regional and global issues. We now do the same for food. For instance, producing food connects to several water-related issues, including:

- Depletion of aquifers and drying up of rivers due to irrigation
- Contamination of groundwater by insecticides and herbicides
- Increase of nutrient pollution due to fertilizer runoff

Every day, you consume water. However, our water consumption involves more than what we obtain directly from bottles, fountains, or our favorite cuisine. Water is also indirectly consumed during the *production* of food. The United Nations Educational, Scientific, and Cultural Organization (UNESCO) estimates that it requires about 15,000 liters of water to bring one kilogram of beef protein to the table—enough to make only nine quarter-pounder hamburgers! It's not that cattle drink this much water. Rather, this value reflects how much water is used to produce the grain that feeds the cattle.

Your Turn 11.25   Your Water Footprint

Check out this link (www.acs.org/cic) for an interactive water footprint calculator for various food items. Determine the water footprint and location of water use for a variety of foods.

Questions to Consider

a. Determine the water footprint for the foods consumed in your typical daily diet. What parts of the world are most impacted by water withdrawals related to these foods?

b. Did the water footprint for any food item surprise you? Explain.

Blablo101/Shutterstock

Let's now examine the connections between food and land use.

We use land for both raising crops and grazing animals. In turn, these activities connect to many issues, including:

- Loss of forest ecosystems to create land for agriculture
- Erosion of topsoil from overplanting and overgrazing
- Loss of biodiversity from the repeated planting of single crops

As you can see from the bar graph of Figure 11.26a, foods differ in the amount of land required to produce them. One reason for the differences is that grain may be needed to feed the animals, as shown in Figure 11.26b. However, as estimates, these values do not necessarily apply to all meat and dairy products. For example, grass-fed animals consume little or no grain in some regions.

Estimates, such as those shown in Figure 11.26, are based on a set of assumptions. As you will see shortly, the values may be higher or lower depending on which assumptions are used. For example, this particular set of values is somewhat high because the food (in kg) brought to the table includes only the edible parts of the

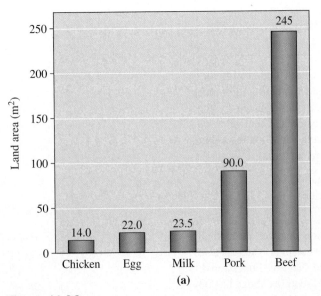

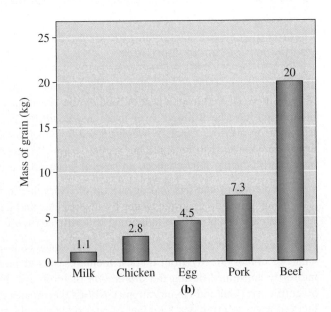

Figure 11.26

Land and grain estimates for food production. **(a)** Square meters of land required to bring 1 kg of each food to the table. **(b)** Kilograms of grain required to bring 1 kg of each food to the table.

Source: Feeding the World: A Challenge for the Twenty-First Century, *Smil, V., MIT Press: 2001.*

animal, not the whole animal. No matter which assumptions are used, the trend is clear: Grain-fed livestock require more land and grain than other varieties of animals. The next activity helps you to be more discerning as you evaluate the land use required for livestock.

Your Turn 11.26   Checking the Assumptions

Below are some factors that affect an estimate of the amount of land or feed grain needed to bring one kilogram of beef to the table. Examine each factor. Does it raise, lower, or have no effect on the estimate?

a. The yield of grain (corn or soy) per acre.
b. The portion of the animal's lifespan included in the estimate.
c. The breed of livestock (e.g., Angus, Hereford, Jersey, Texas Longhorn).

In general, meat consumption and production are increasing worldwide (Figures 11.27 and 11.28). Fueled largely by the rising affluence of people in lower-income countries, this trend is expected to continue through 2030 and beyond. However, can Earth continue to feed its people a few decades from now if we continue on this path of increased meat consumption? To answer this question, let's do some math. From Figure 11.27, we can obtain the approximate values for the projected annual meat consumption (in kg/person/year) in 2030. We then can pair each type of meat with its corresponding grain requirement from Figure 11.26b:

$$10 \text{ kg beef/person} \times 20 \text{ kg grain/kg beef} = 200 \text{ kg grain/person}$$
$$15 \text{ kg pork/person} \times 7.3 \text{ kg grain/kg pork} = 110 \text{ kg grain/person}$$
$$17 \text{ kg chicken/person} \times 2.8 \text{ kg grain/kg chicken} = 48 \text{ kg grain/person}$$

Doing the math, this gives us 358 kg of grain per person in 2030. If we assume a world population of 8 billion people in 2030 (a low estimate), the world grain production would need to be about 2.9 trillion kilograms to produce this much meat. By way of comparison, in 2017, the world grain production *for all uses* was approximately 2.3 trillion kilograms (2.3 billion metric tons). Even if a lower set of values

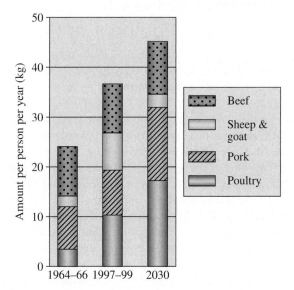

Figure 11.27

Worldwide average meat consumption.

Source: Agriculture, FAO, 2012

Figure 11.28

Worldwide average meat production, 2010–2012.

Source: Meat Atlas, Heinrich Boll Foundation, 2014

for estimated grain requirements were used than those supplied in Figure 11.26b, there would still be insufficient grain available. Although the shortfall could possibly be made up by increasing crop yields, this is unlikely, as we will see in the final sections of this chapter.

11.11 | From Field to Fork I: The Carbon Footprint of Foods

> *Learning Objective: Describe some benefits of eating locally and the effects of food production on global climate change*

We will now describe the connections among food production, energy use, and global climate change. We launch our discussion with the topic of eating locally. **Food miles** are the approximate distance a food item travels from where it was grown to where it was consumed. The food could travel a few yards from your garden, or several thousand miles from another country. Most of our foods must be transported, resulting in varying energy use and atmospheric emissions (Table 11.10). Would it help if all food were produced locally? Not necessarily. For example, growing tomatoes locally in a greenhouse may be less energy-efficient than importing them from a warmer climate. Transportation, as well as other energy costs, must be balanced against food production costs.

People desire to eat locally for many reasons. Picking a tomato from the vine or visiting a local farmer's market truly can be a pleasure. Table 11.11 lists five other reasons people choose to eat locally. In the activity that follows, you will have the opportunity to examine the validity of these claims.

Table 11.10	Energy Use and Emissions for Different Modes of Transportation			
	Rail	**Water**	**Road**	**Air**
Primary energy consumption (kJ/tonne-km)	100	160	2300	7300
Specific total emissions (TgCO₂e)				
Carbon dioxide	37	39	1500	167
Methane	0.1	0.3	1.2	0.0
Dinitrogen monoxide	0.3	0.5	13	1.5
HFCs	0.1	2.9	42	0.0

Note: 1 metric ton (tonne) = 1000 kg or 2205 lb. 1 Tg (teragram) = 1 million metric tons

Sources: Transportation Energy Data Book, September 2018, Oak Ridge National Laboratory; U.S. Transportation Sector Greenhouse Gas Emissions, 1990–2016, Environmental Protection Agency

Table 11.11	Reasons to "Eat Local"

1. Eating local means more for the local economy. A dollar spent locally generates twice as much income for the local economy. Money leaves the community with every transaction when businesses are not owned locally.

2. Locally grown produce is fresher and tastes better. Produce at your local farmer's market is often picked within 24 hours of your purchase. This freshness affects not only the taste of your food but also its nutritional value. Have you ever tried a tomato that was picked within 24 hours?

3. Locally grown fruits and vegetables have longer to ripen. Because the produce is handled less, locally grown fruit does not have to stand up to the rigors of shipping. You will get peaches so ripe that they fall apart as you eat them and melons that were allowed to ripen on the vine until the last possible minute.

4. Buying local food keeps us in touch with the seasons. By eating with the seasons, our foods are at their peak taste and less expensive.

5. Supporting local producers supports responsible land development. When you buy locally, you give those with local open space—farms, and pastures—an economic reason to resist further development.

Steve Bower/Shutterstock

Source: Adapted from "10 Reasons to Eat local Food," Jennifer Maizer, EatlocalChallenge.com

Your Turn 11.27   Why Eat Locally?

Arina P Habich/Shutterstock

Critically examine each statement in Table 11.11.

a. For each statement, provide both a supporting example and an example that opposes eating locally.

b. Suggest two other entries to add to the list.

c. Suggest an item on this list that should be removed or revised. Explain.

Figure 11.29

Carbon footprint of food before it reaches a grocery chain in Great Britain.

Source: Berners-Lee, M. *How Bad Are Bananas? The Carbon Footprint of Everything*, Greystone Books: Vancouver, BC, 2011.

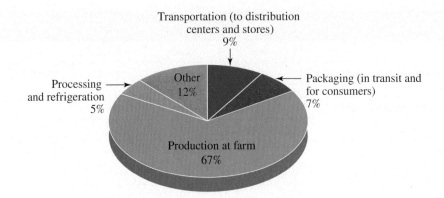

If your primary reason for eating locally is to reduce fossil fuel consumption, you may be interested in a relevant study from Carnegie Mellon University. The researchers reported that transportation represents only 11% of life-cycle greenhouse gas emissions, while final delivery from producer to retail contributes only 4%. So, if the majority of emissions and energy costs with food are not transportation-related, where do they originate?

To answer this question, examine Figure 11.29, which shows carbon footprint data for food reaching a grocery chain in Great Britain. From Section 4.10, recall that carbon footprints estimate carbon dioxide emissions in a given time frame, typically over one year. However, determining a "carbon footprint" for a particular food requires a set of assumptions; consequently, you will see different values for the same food product.

As seen in Figure 11.29, most of the carbon footprint (67%) comes from food production on the farm. For example, operating farm machinery produces carbon dioxide, and using fertilizer stimulates soil microbes to produce nitrous oxide. Farm animals also produce methane. In addition to the production of food, the figure points to other areas in which food contributes to the carbon footprint:

- **Transportation (9%):** Relatively low, unless air freight
- **Packaging (7%):** Primarily from the disposal of the packaging
- **Processing and refrigeration (5%)**

Many small factors, including the leakage of refrigerant gases (all greenhouse gases), also contribute. Although not included in Figure 11.29, food waste and packaging also increase the carbon footprint, which in some countries is up to 25% of all purchased food.

Let's Debate!   Organic Foods

Organic foods are defined as being grown without synthetic fertilizers or pesticides and are not processed using solvents, irradiation, or synthetic food additives.

Form two groups: one group will be in favor of the topic (the proposition) and the other group will be against the topic (the opposition). Both groups will try to persuade a neutral person or judge to agree with them. The topic of the debate is the motion. Here is the motion:

The cost, quality, nutritional value, energy consumption, and overall environmental impacts of organic food production is worse than conventional farming.

Curiosity Conversations

- What are the advantages and disadvantages of organic farming relative to conventional food production?
- How do the crop yields compare between conventional and organic farming?
- How does land and water use differ between organic and conventional farming?
- Why are organic foods generally more expensive than foods produced by conventional farming?
- Considering the three pillars of sustainability discussed in Section 1.9, which type of food is "better" for us and the planet?

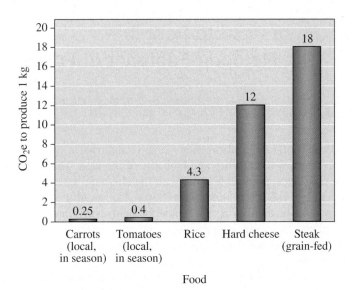

Figure 11.30

Carbon dioxide emissions (CO_2e) from food production.

Source: Berners-Lee, M. *How Bad Are Bananas? The Carbon Footprint of Everything,* Greystone Books: Vancouver, BC, 2011.

Some foods on your dinner table have higher carbon footprints than others. Looking at the graph in Figure 11.30, you probably notice that steak and cheese are the highest of the items included. Why might this be the case? Revisit Figure 11.26 to see that the production of beef, unless grass-fed, requires grain production. In turn, both of these are associated with greenhouse gas emissions. You also may have noticed the low values for tomatoes and carrots. However, these values would increase if these items had been transported instead of grown in your backyard. Similarly, the carbon footprint value would be higher if the tomatoes were grown in a greenhouse.

Before we bring this section to a close, we return to the theme mentioned at the start of this chapter: eating vegetarian. According to the Carnegie Mellon food-miles study mentioned earlier, replacing a single meat-based meal with a vegetarian option saved the equivalent of driving about 1200 fewer miles annually. In contrast, an all-local diet *every day of the week* saved the equivalent of driving 1000 fewer miles annually. You can connect these numbers to carbon dioxide emissions in the next activity.

> **DID YOU KNOW ?**
>
> The term *carbon dioxide equivalent*, CO_2e, includes all greenhouse gases, not just CO_2. For example, in its ability to warm the atmosphere, 1 kg of methane is equivalent to 21 kg of carbon dioxide.

Your Turn 11.28   Local Produce and CO₂ Emissions

In Your Turn 4.12, we noted that a clean-burning automobile engine emits about 5 pounds of carbon (~2300 grams) in the form of CO_2 for every gallon of gasoline burned.

a. How many pounds of CO_2 is this?

b. If you saved the equivalent of 1000 fewer miles each year by eating locally, approximately how many pounds of CO_2 is this? List any assumptions you make.

Your Turn 11.29   Lowering the Carbon Footprint of Your Foods

In this section, we have described the concept of food miles and the benefits of eating locally.

a. What actions make the most sense to lower our food's carbon footprint?

b. What are some assumptions used in determining the carbon footprint of food items?

c. List five general suggestions that are likely to help you lower the carbon footprint of the foods you consume.

In this section, we have examined ways to decrease the carbon footprint of our food. However, in the next section, you will discover that agricultural practices also affect where nitrogen compounds are found and how they move through the air, water, and land of our planet.

11.12 | From Field to Fork II: The Nitrogen Footprint of Foods

Learning Objective: Examine the chemical pathways of the nitrogen cycle, including the production of ammonia

In previous decades, agricultural practices largely employed a tillage rotation, which was thought to improve the fertility and moisture of the land. This consisted of growing a crop on land once every two years, with the alternate years used for soil recovery by maintaining the land in a plowed (or "fallowed") state. The so-called *crop-fallow* rotations resulted in higher yields for fields sown into plowed land versus *crop-crop* rotations. However, frequent plowing exposes the soil to greater loss by wind (Figure 11.31) and water erosion, as well as contributing to increased salinity (salt content) in some soils. To decrease land erosion and increase farm profits without acquiring more land, agricultural practices have now largely shifted toward continuous cropping (minimum/zero tillage), in which the frequency of cropping is extended to *crop-crop-fallow* or even *crop-crop-crop* rotations. However, when such planting frequency is increased, higher rates of nitrogen-based fertilizer are required to maintain high crop yields. Furthermore, problems are often associated with the buildup of fungal or bacterial diseases and herbicide-resistant weeds. To mitigate the effects of diseases and weeds, most continuouscroppers rotate the type of crop from one year to the next, such as corn-soybean-corn, and so on. The overall yield is much higher for crops grown with some degree of rotation relative to those in which the same type of crop is continually planted on the same land (Figure 11.32).

A **fertilizer** is any natural or synthetic material applied to soils or plants to supply plant nutrients and promote plant growth. The elemental composition of fertilizers typically consists of nitrogen, phosphorus, potassium, sulfur, carbon, and hydrogen. Although many of these elements are readily available in soils for uptake by plants, usable forms of nitrogen are scarce; hence, we need to supply this in the form of fertilizers.

You might wonder how nitrogen levels can be so low in soils when N_2 makes up so much of our atmosphere. Although abundant, the nitrogen molecule is *not* in a chemical form that most plants can use. As you may recall, N_2 is far less reactive than O_2.

Figure 11.31

Photo of the Dust Bowl of the 1930s— a vivid reminder of the effects of unsustainable agricultural practices.

Photo by George E. Marsh. National Oceanic and Atmospheric Administration, Dept. of Commerce

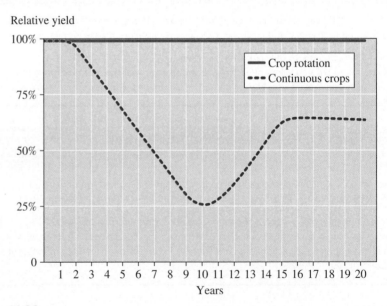

Figure 11.32

A general comparison between the typical yield of continuous crops and those with crop rotation. It may take about a decade to develop sufficient quantities of soil microbes that can reverse the effects of soil-borne diseases.

Table 11.12	Some Reactive Forms of Nitrogen*
Name	**Chemical Formula**
Nitrogen monoxide (nitric oxide)	NO
Nitrogen dioxide	NO_2
Dinitrogen monoxide (nitrous oxide)	N_2O
Nitrate ion	NO_3^-
Nitrite ion	NO_2^-
Nitric acid	HNO_3
Ammonia	NH_3
Ammonium ion	NH_4^+

*These forms of nitrogen are all naturally occurring.

Your Turn 11.30   Unreactive Nitrogen

How does the bond energy of the triple bond in N_2 compare with other bond energies, such as the O—H bond in water the or O=O bond in O_2?

Hint: Refer to Section 6.5.

To grow, plants need access to a form of nitrogen that reacts more easily than N_2, such as, the ammonium ion (NH_4^+), ammonia (NH_3), or the nitrate ion (NO_3^-). These and other forms are called **reactive nitrogen**, the compounds of nitrogen that cycle through the biosphere and interconvert with each other. Some reactive forms of nitrogen are listed in Table 11.12. As you might imagine, the air pollutants NO and NO_2 that were discussed in Chapter 2 are also included in the list. These forms of nitrogen all occur naturally and are present on our planet in relatively small amounts. Other forms of reactive nitrogen also exist, such as the amine functional group (—NH_2).

Although we categorized N_2 as generally unreactive, one reaction involving the nitrogen molecule is of utmost importance: biological nitrogen fixation. Plants such as alfalfa, beans, and peas "fix" (remove) N_2 from the atmosphere (Figure 11.33). To be more accurate, it is not the plants themselves but rather the bacteria living on or near the roots of these plants that fix the nitrogen.

As part of their metabolism, **nitrogen-fixing bacteria** remove nitrogen gas from the air and convert it into ammonia. When the ammonia dissolves in water, it produces the water-soluble ammonium ion, NH_4^+ (revisit Section 5.8). This polyatomic ion is one of two forms of reactive nitrogen that most plants can absorb. Equation 11.6 shows the pathway for ammonium ion formation; the icon ( ) represents the bacteria responsible for the interconversion of nitrogen-containing species.

$$N_2 \xrightarrow[\text{nitrogen fixation}]{} NH_3 \xrightarrow{H_2O} NH_4^+ \qquad \text{[11.6]}$$

The other form of reactive nitrogen that plants can absorb is the nitrate ion, NO_3^-. Again, compounds containing the nitrate ion are always soluble in water. **Nitrification** is the process of converting ammonia in the soil into the nitrate ion. Bacteria are involved in this two-step process:

$$NH_4^+ \xrightarrow[\text{bacteria in the soil}]{} NO_2^- \xrightarrow[\text{bacteria in the soil}]{} NO_3^- \qquad \text{[11.7]}$$

Figure 11.33

Nodules on the root of a soya plant that contain nitrogen-fixing bacteria.

Nigel Cattlin/Alamy Stock Photo

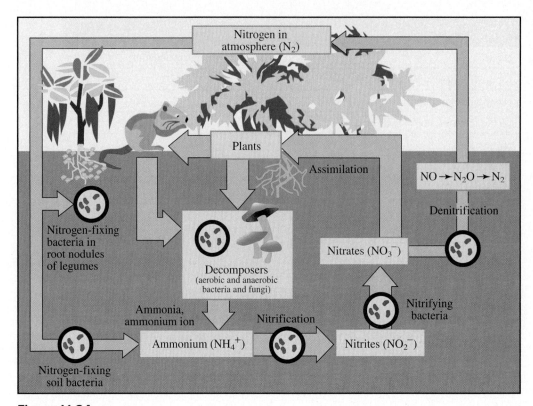

Figure 11.34

The nitrogen cycle (simplified), a set of chemical pathways whereby nitrogen moves through the biosphere.

Finally, to come full circle, bacteria help with **denitrification**, the process of converting nitrate into nitrogen gas. In doing so, these bacteria harness the energy released when the stable N_2 molecule forms. Recall from Section 6.5 the large amount of energy released in forming the triple bond found in the nitrogen molecule, a very stable molecule indeed!

Depending on the soil conditions, the pathway may occur in steps that include NO and N_2O. Thus, these reactive forms of nitrogen can also be converted into N_2 and released from the soil:

$$NO_3^- \xrightarrow[\text{the soil}]{\text{bacteria in}} NO \xrightarrow[\text{the soil}]{\text{bacteria in}} N_2O \xrightarrow[\text{the soil}]{\text{bacteria in}} N_2 \qquad \text{[11.8]}$$

All of these pathways are part of the **nitrogen cycle**, a set of chemical pathways whereby nitrogen moves through the biosphere. Figure 11.34 assembles the separate pathways shown above into a simplified version of the nitrogen cycle. In this cycle, all species are forms of reactive nitrogen except for N_2.

Your Turn 11.31   The Nitrogen Cycle

a. Ammonia (NH_3) is applied to soil as anhydrous ammonia, that is, ammonia without water. Ammonia is very soluble in water. Explain why.

b. When ammonia is mixed with water, the resulting solution is basic. Explain why, including the ammonium ion (NH_4^+) as part of your answer.
   Hint: Revisit Section 5.8.

c. According to Figure 11.34, which chemical species of nitrogen is used (assimilated) by plants?

d. Microbes in the soil interconvert the chemical species of nitrogen. Before plants can assimilate the ammonium ion, another chemical species is formed. What is it?

Remember that reactive forms of nitrogen are needed for plant growth. Because bacteria cannot supply these in the amounts needed for optimal plant growth, farmers use fertilizers. A few centuries ago, fertilizers were obtained by mining deposits of saltpeter (ammonium nitrate) from the deserts of Chile or by collecting guano, a nitrogen-rich deposit from bird and bat droppings in Peru. Neither source, however, was sufficient to meet the demand of the growing world population. An additional drain on the supply of nitrates was their use in producing gunpowder and other explosives such as TNT. By the early 1900s, the search was on for a way to synthesize reactive nitrogen compounds from the abundant N_2 in the air.

How are fertilizers obtained in the large quantities needed for present-day agriculture? The answer lies in a second important reaction of N_2, one that literally captures it out of the air to synthesize ammonia:

$$N_2(g) + 3\ H_2(g) \xrightarrow[\text{450 °C, 200 atm pressure}]{\text{Ru or Fe catalyst}} 2\ NH_3(g) \qquad \textbf{[11.9]}$$

Equation 11.9 is known as the **Haber-Bosch process**. This procedure allows for the economical production of ammonia, enabling the large-scale production of fertilizers and nitrogen-based explosives. However, the use of fossil fuels as the feedstock for hydrogen gas results in significant carbon dioxide emissions. Hence, there is increasing interest in using carbon-capture and renewable production techniques to more sustainably meet the growing demand for ammonia (Figure 11.35). Fertilizer represents the most common application for ammonia, which can be directly applied to the soil or can be applied as ammonium nitrate or ammonium phosphate. However, ammonia is useful in many other applications such as refrigeration, textiles, explosives, pharmaceuticals, and hydrogen storage for fuel cells.

The reactive forms of nitrogen in the nitrogen cycle continuously change chemical forms. Thus, the ammonia that starts out as a fertilizer may end up as NO, in turn increasing the acidity of the atmosphere. Alternatively, the NO may end up as N_2O, a greenhouse gas that is currently rising in atmospheric concentration. Or, the ammonium ion, instead of being tightly bound to the soil, may end up being converted and leached out as the nitrite or nitrate ion, in turn contaminating a water supply. For example, algal blooms in the Gulf of Mexico are caused by the runoff of fertilizers

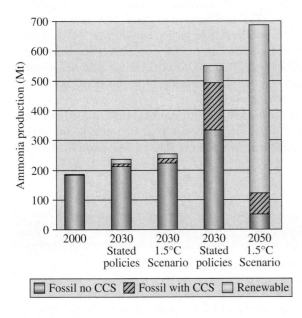
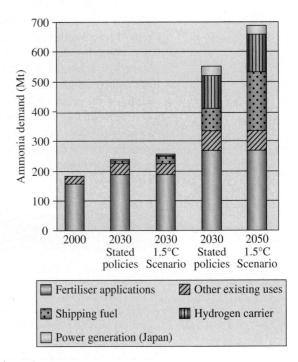

Figure 11.35

Current and projected ammonia production by source and demand by sector.

Note: CCS refers to carbon-capture technology; the 1.5 scenario refers to the Paris Agreement goal of keeping the global temperature rise within 1.5°C.

Figure 11.36

Brown, nutrient-rich water from the Mississippi River meets the Gulf of Mexico, creating a "dead zone."
The zone is approximately 6000–7000 square miles, varying in size seasonally.

©Nancy Rabalais, Louisiana Universities Marine Consortium

from farms in the watershed of the Mississippi River in states from as far away as Montana, Minnesota, and Pennsylvania (Figure 11.36). These blooms affect the livelihood of all those who use the common waters of the Mississippi, including the fishing industry in the Gulf. Although unintended consequences can be minimized by using less fertilizer and timing its application more carefully, we have not yet successfully avoided such harm.

Now we can fully understand the effects of NO emissions on our environment. First, the oxides of nitrogen form ground-level ozone in the presence of sunlight, contributing to photochemical smog, as we saw in Chapter 2. Second, NO_x emissions are a form of reactive nitrogen, just like the fertilizers used for food production. NO is formed from unreactive N_2 in the air when fuels are burned. The more fuels are burned, the more N_2 is changed into a reactive form. Both NO_x and fertilizer use are affecting the balance within the nitrogen cycle on our planet. Finally, NO_x emissions increase the acidity of the precipitation falling from the sky.

Now that we have addressed the carbon and nitrogen footprints of food products, we are still faced with the problem of ensuring enough food is available and safe to consume for an ever-growing population. We address this critical issue in the final section of the chapter.

Your Turn 11.32   Putting It All Together

Using your knowledge of the carbon cycle (Section 4.1) and the nitrogen cycle (Figure 11.34), perform a life-cycle analysis (LCA) on one of your favorite foods. Ensure that you deal with the energies and environmental emissions involved during the fabrication, transportation, and end-of-use of the food packaging, in addition to the lifetime of the food item itself. Compare your results to the LCAs of similar food products that are available on the Internet.

11.13 | Food Security: Feeding a Hungry World

> *Learning Objectives:*
>
> *Explain the concept of food security*
>
> *Assess global trends associated with food production including population, land use, and food consumption*

Our ancient ancestors were hunter-gatherers, spending most of each day searching for their next meal. About 10,000 years ago, humans learned to grow crops and domesticate animals, thus launching the agricultural revolution. At that time, the population of the entire Earth was estimated to be 4 million people, or roughly that of Los Angeles today.

Over the next 8000 years, the global population grew to 170 million, or about half the size of the current U.S. population. By 1000 CE, the population had risen to 310 million, and 200 years ago, the population finally topped 1 billion people. Today, the world population is over 8 billion, and within the next 30 years will likely rise to 9 billion (Figure 11.37). Clearly, we have many mouths to feed.

The concept of **food security**, according to the World Health Organization, deals with three primary issues:

- **Food availability:** having sufficient quantities of food available consistently.
- **Food access:** having sufficient resources to obtain appropriate foods for a nutritious diet.
- **Food use:** appropriate use of food based on knowledge of basic nutrition and care and adequate water and sanitation.

Thomas Malthus (1766–1834) and, more recently, entomologist Paul Ehrlich (b. 1932) predicted that Earth's human population would outstrip food production. During the first 100 years after Malthus published his essay, the population of Earth grew by 60% to 1.6 billion. During the next 100 years (the 20th century), the population exploded by an additional 4.4 billion people. The Food and Agriculture Organization (FAO) of the United Nations estimated that we have managed to increase the world food supply to currently meet the needs of over 89% of the people on Earth. As mentioned in Section 11.1, *undernourishment* refers to the number of people unable to consume enough food to conduct an active and healthy lifestyle (Figure 11.38). Although <2.5% of the population is undernourished in higher income countries in regions such as North America and Europe, lower income countries in regions such as Africa and Asia have seen increasing undernourishment rates since 2014, currently approaching 20% in some regions.

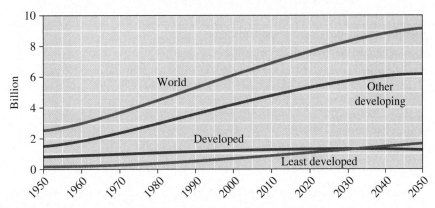

Figure 11.37

World population growth, 1950–2050 (projected).

Source: Food and Agriculture Organization of the United Nations.

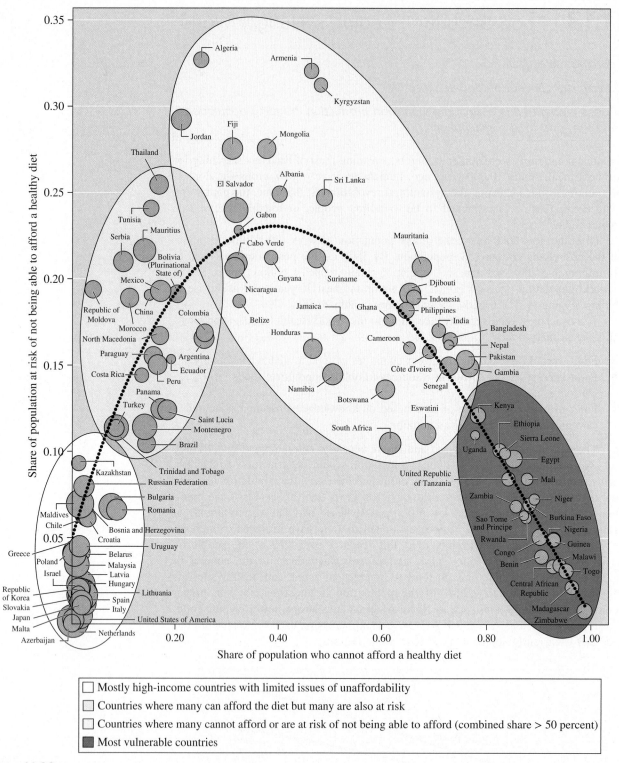

Figure 11.38

Economic access to a healthy diet for selected countries, 2016–2019.

Food and Agriculture Organization of the United Nations. Reproduced with permission.

Given that the population continues to grow, so must food production. This is particularly important considering that global food consumption per person is projected to steadily increase into the foreseeable future (Figure 11.39).

Two methods have largely been responsible for food increases of the past: planting crops on more land and increasing crop yields. However, can this increase in our food supply continue? Neither method has much room for growth. According to FAO

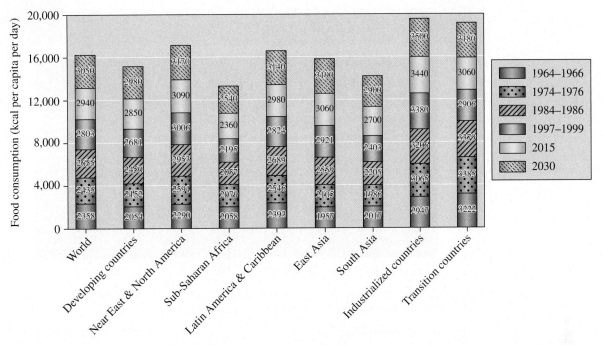

Figure 11.39

Global and regional per-capita food consumption from 1964 to 2030 (projected).

Source: Food and Agriculture Organization of the United Nations, 2015

estimates, almost all of the world's biologically productive land is now in use, and arable land is likely to be increased by only 5% from current levels as of 2050 (Figure 11.40). Furthermore, in many areas, cropland is actually shrinking due to factors such as desertification, soil nutrient depletion, erosion, and urban development. Enter the realm of "vertical farming," which you will explore in this next activity.

The 1940s saw the beginning of the *Green Revolution*. In the following decades, agricultural productivity per acre of corn, rice, and wheat more than doubled. Many factors were responsible, including the use of fertilizers and pesticides, irrigation, mechanization, double-cropping, and, most importantly, the advent of high-yielding crop varieties. Billions of people across the world benefited.

The Green Revolution also resulted in economic, environmental, and social costs despite its successes. As discussed in previous sections, producing crops requires water and energy, which results in greenhouse gas emissions. Furthermore, modern agricultural practices require the use of supplemental nitrogen fertilizers such as ammonia, urea, or nitrates, all of which are forms of "reactive nitrogen."

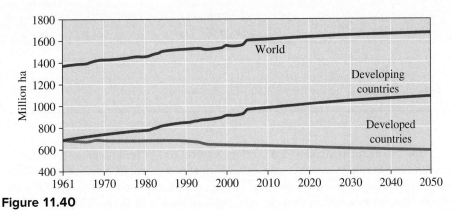

Figure 11.40

Trends in arable land use from 1961 to 2050 (projected).

Source: Food and Agriculture Organization of the United Nations, 2015

What Is "Vertical Farming"?

©2018 American Chemical Society

Watch the video (www.acs.org/cic) on vertical farming and list the benefits and disadvantages of vertical farming relative to traditional agriculture operations.

Questions to Consider
a. Where are vertical farming practices currently in use worldwide?
b. What are some factors that will affect the widespread adoption of this technique?

The use of synthetic pesticides, though integral in raising crop yields, represents yet another environmental cost of the Green Revolution. Like nitrogen fertilizers, pesticides impact our health and the health of our planet. The use of insecticides saves millions of pounds of crops from being devoured by pests in the field, but at the same time, these chemicals can kill beneficial insects along with the intended target. Many persistent pesticides also end up in the environment as a chemical cocktail, the consequences of which we are still deciphering. More environmentally friendly pesticides that target only specific organisms or elicit a plant's own defenses have been developed.

Methyl bromide is an extremely effective pesticide used in sterilizing soils and as a fumigant. It kills a wide variety of insects and is particularly useful in preparing the soil for crops such as strawberries and tomatoes. However, given its toxicity and ability to deplete stratospheric ozone, the use of methyl bromide was phased out in 2005 under the Montreal Protocol, except for some specialized uses.

DID YOU KNOW ?

As of 2022, allowable specialized uses ("critical exceptions") of methyl bromide include its use in California for growing strawberries and preventing certain insect infestations in dry cured pork products, for which no suitable alternatives exist.

Isabelle Rozenbaum/PhotoAlto Agency/Getty Images

Your Turn 11.33   Methyl Bromide and the Ozone Hole

Draw the structural formula for methyl bromide. Why would you expect it to deplete the ozone layer?

Hint: Revisit Chapter 3.

The search for pesticides that are less harmful to our health and to the health of the planet presents a challenge to green chemists. However, the EPA has since identified many alternatives for a variety of uses. Harpin, a naturally occurring protein, has also been designated as a replacement for some uses of methyl bromide. EDEN Bioscience Corporation won a Presidential Green Chemistry Challenge Award in 2001 for its discovery that applying harpin to the stems and leaves of plants triggered a plant's natural defense mechanisms against diseases caused by bacteria, fungi, and nematodes. Some advantages of harpin include the following:

- It does not directly kill, so the pest is unlikely to acquire resistance to it.
- It is produced from the fermentation of a genetically engineered laboratory strain of *E. coli* that is benign.
- It is produced from renewable materials and yields biodegradable waste products.
- It is classified with the lowest hazard potential by the U.S. EPA.
- It can be applied to fields using smaller concentrations than many other pesticides.
- It is rapidly decomposed by sunlight and microorganisms.

Although harpin has many advantages, it is not a perfect solution. Some plants respond to it better than others. Furthermore, harpin needs to be reapplied at several-week intervals.

Your Turn 11.34   Food (In)Security?

Even with sustainable agricultural practices, many other factors could influence the availability of food for Earth's growing population. List five factors that are unpredictable but could cause the price of food to rise, causing widespread food insecurity.

We end this final section by returning to the earlier question of whether eating less meat translates to more available food. Indeed, eating less meat has benefits in terms of energy, land, and water use. Eating the saturated fats of beef in moderation (together with eating more fruits, vegetables, and whole grains) is also part of a recommended diet. But those who study these issues have asserted that cutting meat consumption is only a small contribution to improving global food security. Even so, the small contributions of many people add up to larger benefits. What actions should each of us take? This final activity offers personal advice in simple terms.

Your Turn 11.35   Group Activity: Food Choices Revisited

In the opening Reflect activity of the chapter, you created a list of your food choices. Share your responses with other class members, and review your ranking of foods you believed were the highest in promoting your health, as well as the health of the planet.

a. Based on the topics discussed in this chapter, how would you re-rank your food choices?
b. What changes (if any) are you planning to make to your diets based on your group members' health and the planet's health?
c. Use the Internet to find out how much food is lost (during production or processing) or wasted (discarded by retail markets and consumers) each year. List some practices that could reduce food waste.

Conclusion

Even though our individual tastes vary, our biological needs are much the same. We need carbohydrates and fats as our energy sources; fats for cell membranes, metabolic control, and lubrication; proteins to build muscle and create the enzymes that catalyze the intricate chemistry of life; and vitamins and minerals to help make that chemistry happen.

What and how much we eat affects not only our own health but also the health of the planet. In this chapter, we saw some of the human health and environmental consequences of our food choices. Some foods, especially beef, disproportionately use water, grain, fuel, and land to produce. Some crops, including corn, can stress not only the land of the farmer but also the ecosystems many miles downstream.

Meeting the dietary needs of all on our planet is one of the greatest challenges of our time. A little chemistry knowledge allows us to raise and answer good questions. But chemical knowledge alone cannot lead us to a more peaceful, prosperous, and healthier world. Individual and community choices, which are determined by wisdom from economic, social, religious, and political communities, will also help lead the way forward. In the next chapter, we will build on your knowledge of food chemistry and nutrition as we delve into chemistry's role in treating illnesses.

Learning Outcomes

The numbers in parentheses indicate the sections within the chapter where these outcomes were discussed.

Having studied this chapter, you should now be able to:

- how to use food labels to compare the nutritional value of processed foods (11.1)
- compare the composition of the human body with common foods (11.1)
- describe the relationship between lipids and triglycerides, and outline the reaction scheme for forming triglycerides from fatty acids (11.2)
- distinguish saturated from unsaturated fats, and compare their respective chemical and physical properties, uses, and health effects (11.3)
- illustrate the structure and metabolism of carbohydrates (11.4)
- describe how insulin and glucagon regulate blood glucose levels (11.4)
- compare and contrast the chemical properties of natural sugars and artificial sweeteners (11.5)
- illustrate the importance of amino acid side chains and how amino acids combine to form proteins (11.6)

- explain why the consumption of aspartame is harmful for certain individuals (11.6)
- examine how fat-soluble and water-soluble vitamins and minerals are important for body health (11.7)
- compare the energy content of foods and explain how food choices and level of exercise can affect your weight (11.8)
- evaluate methods used for maintaining a safe food supply (11.9)
- determine the water and land use associated with the production of various food items (11.10)
- describe some benefits of eating locally and the effects of food production on global climate change (11.11)
- examine the chemical pathways of the nitrogen cycle, including the production of ammonia (11.12)
- explain the concept of food security (11.13)
- assess global trends associated with food production including population, land use, and food consumption (11.13)

Questions

Emphasizing Essentials

1. Eating properly involves more than filling your stomach. Explain the difference between malnutrition and undernourishment.

2. It has been claimed that you will eat about 700 times your adult body weight during your lifetime. Is this statement in the ballpark? Do a calculation to find out. State your assumptions clearly.

3. Macronutrients provide a source of energy and raw materials for your body.

   a. Name the three different types of macronutrients.

   b. How do macronutrients differ in energy content?

4. Although water is not considered a macronutrient, it clearly is essential to maintaining health. Name three roles that water plays in our bodies.
   Hint: Refer back to Chapter 6.

5. Consider this pie chart:

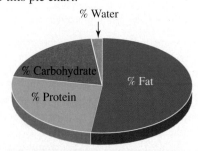

Based on the relative percentages of protein, carbohydrate, water, and fat, is this pie chart more likely to represent steak, peanut butter, or a chocolate chip cookie? Explain your choice.

6. Of the foods listed in Table 11.1,

   a. identify the best sources of carbohydrates and arrange them in decreasing order.

b. identify the best sources of protein and arrange them in decreasing order.

c. identify two foods that should be avoided if you are controlling your fat intake.

7. At a local restaurant, an 18-ounce steak is the manager's special. Use Table 11.1 to calculate the ounces of protein, fat, and water in a portion of this size.

8. Although fats and fatty acids are related, they differ in terms of the size of their molecules, their functional groups, and their role in your diet. Elaborate on each of these differences.

9. Here is the condensed structural formula for lactic acid: $CH_3CH(OH)COOH$.

a. Draw a structural formula for lactic acid that shows all the bonds and atoms.

b. Would lactic acid be saturated or unsaturated if considered a fatty acid?

c. Is lactic acid a fatty acid? Explain.

10. Both unsaturated fats and saturated fats are triglycerides. Explain how they differ in terms of their observable properties, chemical structures, and role in your diet.

11. Using Figure 11.7, identify the fat or oil that contains the highest number of grams per tablespoon of

a. polyunsaturated fat.

b. total unsaturated fat.

c. monounsaturated fat.

d. saturated fat.

12. Compare and contrast a *trans* fat to unsaturated and saturated fats in terms of

a. chemical structure.

b. physical properties.

c. effects on your health.

13. Name foods in which you would be likely to find these carbohydrates:

a. lactose

b. sucrose

c. fructose

d. starch

14. Explain each term and give an example.

a. monosaccharide

b. disaccharide

c. polysaccharide

15. Starch and cellulose are both polysaccharides. How are these two compounds similar in terms of their chemical structures? How are they different in terms of our ability to digest them?

16. Fructose, $C_6H_{12}O_6$, is an example of a carbohydrate.

a. Rewrite the chemical formula of fructose to show that a *carbohydrate* can be thought of as "carbon plus water."

b. Draw a structural formula for any isomer of fructose.

c. Do you expect different isomers of fructose to have the same sweetness? Explain.

17. Fructose and glucose both have the chemical formula of $C_6H_{12}O_6$. How do their structural formulas differ?

18. How many grams of sucralose are required to match the sweetness of 1 g of sucrose?
Hint: See Table 11.3.

19. Chemical names, especially for organic compounds, can give information about the structure of the molecules that these compounds contain. What does the term *amino acid* suggest about its molecular structure?

20. Proteins are polymers, sometimes referred to as polyamides. Similarly, nylons also are polymers and polyamides.

a. What is the amide functional group?

b. Compare and contrast proteins and nylon in terms of

i. the functional groups present on the monomer(s).

ii. the variety of different proteins as compared to that of different nylons.

21. Analogous to Equation 11.3a, show how glycine and phenylalanine react to form a dipeptide.

22. Some amino acids are called "essential amino acids." Explain why.

23. Why can people with phenylketonuria drink beverages sweetened with sucralose but should avoid those sweetened with aspartame?

24. Explain the nutritional significance of the elements shaded on this periodic table.

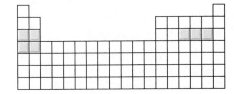

25. In recent decades, which two methods have primarily been used to increase food production?

26. One theme in this chapter is that what you eat affects not only your health but also the health of the planet. Provide two examples that illustrate this theme.

27. Suggest at least two connections between food production and water quality. Do the same for water use.

28. Generally, it requires more water and land to produce beef, chicken, and pork than it does to produce grains such as corn and soybeans. Give two reasons why.

29. Plants require nitrogen in a "reactive" form.

a. Explain the meaning of the term *reactive nitrogen*.

b. Give an example of a form of nitrogen that is nonreactive.

c. Describe one natural and one unnatural source of reactive nitrogen.

Concentrating on Concepts

30. The old adage "You are what you eat" is everywhere, including in this chapter. List your food consumption over the past 24 hours and compare it to Figure 11.3. Does your diet contradict or confirm the adage?

31. Explain why it is impossible to go on a highly advertised "all organic, chemical-free" diet to a friend.

32. Indicate whether each statement is true, true only in some circumstances, or false. Explain your reasoning.

    a. Plant oils are lower in saturated fat than animal fats.

    b. Given time and exposure to the air, fats and oils become rancid.

    c. Fats are not necessary for our diet because our bodies can manufacture them from other substances we eat.

33. The label of a popular brand of soft margarine lists "partially hydrogenated soybean oil" as an ingredient. Explain the term *partially hydrogenated*. Why must the label report partially hydrogenated soybean oil rather than simply soybean oil?

34. A well-known medical clinic used the phrase "double trouble" to refer to *trans* fatty acids. Explain the logic behind this.

35. Explain why the process of interesterification is a useful alternative to hydrogenation.

36. Although interesterification is a useful alternative to hydrogenation, the process had several drawbacks until a green chemistry solution was developed.

    a. What were the drawbacks?

    b. What solution was developed?

    c. Which of the green chemistry key ideas did this solution address?
    Hint: Revisit Chapter 2.

37. Some people prefer to use nondairy creamer rather than cream or milk. Some but not all nondairy creamers use coconut oil derivatives to replace the butterfat in the cream. Is a person trying to reduce dietary saturated fats wise to use nondairy creamers such as these? Explain.

38. Low-Calorie and zero-Calorie substitutes have been developed for fat ("fake fats" such as Olean) and sugar (sucralose and aspartame). Why aren't comparable zero-Calorie substitutes for protein being developed? Even so, some protein substitutes exist. Which groups of people might choose them?

39. Your friend wants to cut food costs and has learned that peanut butter is a good protein source. What additional information should your friend consider before eating peanut butter as a major dietary protein source?
    Hint: See Table 11.1.

40. The table below provides the composition of a fast-food meal. Do calculations to determine whether the meal meets the guideline that 8–10% of total Calories should come from saturated fats.

	Cheeseburger	French Fries	Shake
Calories	330	540	360
Calories from fat	130	230	80
Total fat (g)	14	26	9
Saturated fat (g)	6	4.5	6
Cholesterol (mg)	45	0	40
Sodium (mg)	830	350	250
Carbohydrates (g)	38	68	60
Sugars (g)	7	0	54
Proteins (g)	15	8	11

41. American diets depend heavily on bread and other wheat products. A slice of whole wheat bread (36 g) contains approximately 1.5 g of fat (with 0 g saturated fat), 17 g of carbohydrate (with about 1 g of sugar), and 3 g of protein.

    a. Calculate the total Calorie content in a slice of this bread.

    b. Calculate the percent of Calories from fat.

    c. Do you consider bread a highly nutritious food? Explain your reasoning.

42. Describe three ways in which agriculture is connected to the use of fossil fuels. Which aspects will eating locally alter?

43. Ethanol is an example of a biofuel.

    a. From which macronutrient does it originate: fats, carbohydrates, or proteins?

    b. Name two foods now used to produce ethanol for vehicles.

    c. By what process is the ethanol produced from these foods?

    d. Describe one of the current controversies in producing ethanol.

44. Biodiesel is another example of a biofuel. Answer the same questions for biodiesel that were asked in Question 43 about ethanol.

Exploring Extensions

45. Considering your diet over the past 24 hours, use the ideas in this chapter to describe two ways to improve your diet for your health and two ways to improve your diet for the planet's heath.

46. Nature holds some surprises! An avocado is a tropical fruit; in fact, it is a fruit with a high fat content. However, this fat differs from the type found in coconut or palm oil. What is the primary type of fat found in avocados? How does this fat compare to tropical oils (coconut and palm) in terms of its effects on your health?

47. Estimate your average yearly intake in grams of sugar from soft drinks and fruit juices, listing the assumptions you made in arriving at this estimate. By what amount (in grams) would this estimate increase if you included the sugar you added to beverages such as coffee and tea?

48. Here is information about the sugar content of different foods:

Food Product	Sugar	Calories	Serving Size
Altoids, peppermint	2 g	10	3 pieces (2 g)
Ginger snaps	9 g	120	4 cookies (28 g)
Critic's Choice Tomato Ketchup	3 g	15	1 tbsp (13 g)
Del Monte Pineapple Cup	13 g	50	Individual cup (113 g)
Dr Pepper soft drink	40 g	150	1.5 cups
French Vanilla Coffee-Mate	5 g	40	1 tbsp (15 mL)
Hostess Twinkies	14 g	150	1.5 ounces
LifeSavers, Wint O Green	15 g	60	4 mints (16 g)
Tropicana HomeStyle Orange Juice	22 g	110	8 ounces (1 cup)
Snickers bar	29 g	200	2.1 ounces
Sunkist orange soft drink	52 g	190	1.5 cups
Wheatables crackers	4 g	130	13 crackers (29 g)

a. Examine this list. Which item has the highest ratio of grams of sugar to the number of Calories (g sugar/Cal) in one serving?

b. The sugar content of some of these foods may surprise you. If so, which ones?

c. Do you predict that the type(s) of sugar found in Dr Pepper would be the same as those found in Sunkist orange soft drink? In the orange juice or in the pineapple cup? Explain.

d. The complete label for Wint O Green LifeSavers shows 16 g of total carbohydrates per serving, 15 g of which is sugars. What might account for the other 1 g of carbohydrates?

49. A yellow packet of Splenda sugar substitute contains the compound sucralose. Use the resources of the Internet to answer these questions.

a. How many Calories does a packet of Splenda contain?

b. Splenda's slogan is "Made from sugar, so it tastes like sugar." Does it appear to you to be a helpful statement or misleading advertising? Explain your point of view.

50. Consider this structural formula for one of the forms of vitamin K.

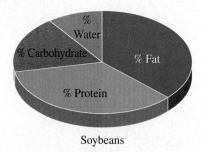

a. Do you expect it to be water-soluble or fat-soluble? Explain.

b. What role does vitamin K play in your body?

c. People rarely, if ever, experience vitamin K deficiencies. Propose a reason why.

51. Compare these two pie charts for the percentage of macronutrients in soybeans and wheat.

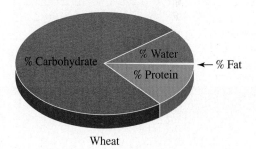

Soybeans

Wheat

a. Explain why the World Health Organization has helped develop soy- rather than wheat-based food products for distribution in parts of the world where protein deficiency is a problem.

b. Suggest cultural reasons why soy might be preferable to wheat in some areas of the world.

52. Some people view climate change and food security as "the twin grand challenges" we face today. Name two ways in which these challenges connect to each other.

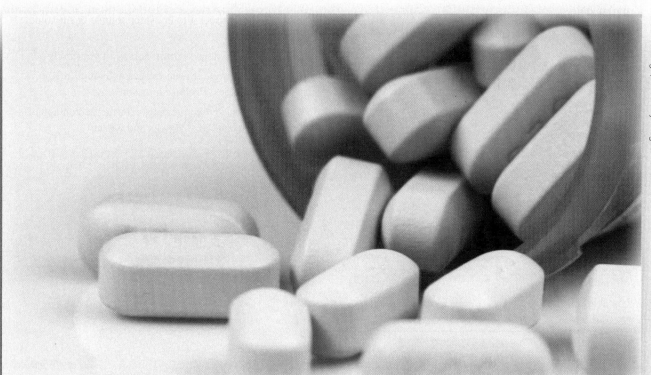

DELVING INTO MEDICINAL CHEMISTRY

Watch the chapter opening video (www.acs.org/cic) and reflect upon your own interactions with doctors. How often do you consider the chemistry behind your health, their diagnostic tools, or their advice? List five common medical treatment paths. For each, decide whether it cured a problem, treated a symptom, or prevented a future problem.

In this chapter, you will explore the following questions:

1. What roles do chemical equilibria have in our bodies?
2. How do our bodies maintain a constant pH level?
3. How can we represent the structures of complex organic molecules?
4. What is the importance of functional groups on the properties of an organic molecule?
5. How can molecules with the same chemical formula behave so differently in our bodies?
6. What are the biological roles of proteins?
7. How do proteins bind to target molecules?
8. What are the biological functions of steroids?
9. How were aspirin and penicillin discovered and developed?
10. How are new drugs developed?

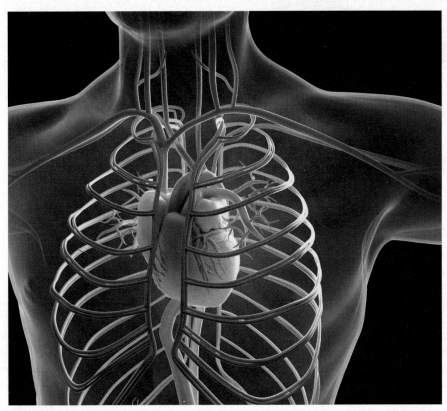

Andrzej Wojcicki/Science Source

Introduction

Grasp a pencil. This momentary (and possibly pointless) act required a conscious choice, followed by a series of stepwise, though rapid, events. You had to scan for a pencil, move your hand, confirm by feel and sight that you were applying pressure, then tighten your hand around it. Each of these individual steps required the rapid firing of neurons; some signals traveled as far as your brain and back to your fingertips. All the while, the heme in your blood cells pulled oxygen from your lungs to transport it toward tissues in need. Your digestive system processed your most recent meal and, perhaps, signaled for a switch in whether your organs store or use sugars. Neurons fired electrochemical codes to create new links that recorded this information and maybe associated it with the music you heard or the feel of that pencil in your hand.

Most of these actions simply happened in an automated way; however, on the molecular level, these processes are very complex.

The human body is like a factory full of many interconnected assembly lines keeping the whole system operational. Chemists, often within the specialty of biochemistry, study the mechanisms by which each line carries out its specific role. Although each line can be considered in isolation, the connections among them are what drive the complexity that defines you. What happens when one or many assembly lines malfunction—when the proper functioning of a bodily mechanism goes haywire? From your perspective, you see malfunctions as a set of symptoms or a disease.

Biologists, biochemists, and chemists study the functioning and malfunctioning of the human body to understand illness and design treatments. From your perspective, you miss this cause-and-effect. When you go to your local urgent care center with severe abdominal pain, the physician is trying to determine the specific pathology of your ailment quickly. The doctor will aim to narrow down a list of possible diagnoses to cure the ailment or, at a minimum, alleviate the symptoms. During the springtime, many people develop seasonal allergy symptoms of watery eyes, runny nose, and sneezing. Again, people flock to their primary care physician to get prescriptions for nasal steroid sprays and other anti-allergy medicines to reduce these disruptive symptoms. Most of us don't even notice how our bodies work until something goes wrong. While we are content to apply modern solutions to return us to comfort and a functioning physiology, we do not see the chemistry and biochemistry behind the symptoms and treatment.

This chapter will delve into the complex inner workings of our bodies and explore the medicinal drugs that can be used to keep our bodies functioning properly.

12.1 | A Life Spent Fighting Against Equilibrium

Learning Objectives:

Describe and exemplify a system that is in equilibrium

Relate reaction quotients to equilibrium constants and write equilibrium expressions

Admittedly, this textbook has generally implied that reactions proceed in one direction—from reactants to products. While this statement can be true, most reactions are *reversible*, proceeding toward either products or reactants until a balance is reached. In this state, which has reached **equilibrium**, the system will contain a constant concentration of products balanced by a constant concentration of reactants. Check out the video below for a simple analogy of equilibrium.

Visualizing Equilibrium

To see equilibrium in action, check out the video at www.acs.org/cic.

©2018 American Chemical Society

Questions to Consider

a. Describe what happens to the water levels in the containers from the start of the video to the end.

b. Does the size of the containers influence the transfer? What do you think would happen if a small (5-mL) sized container and a large (1000-mL) sized container were used?

c. Describe how the containers and the tube that connects them are like a chemical reaction. What do the containers represent? What does the tube connecting the containers represent?

@HOME Snap-Bead Equilibria

Investigate an equilibrium reaction at home using craft snap beads, a small box, a blindfold, a timer, and graph paper. This activity needs four people.

1. Pour a bag of 300 craft snap beads into a small box. Even though the beads might be different sizes and colors, they are all going to be the same reactant: "snapium," or Sp.
2. One person is the "forward-reaction" person. Their job is to connect two snapium atoms and form pairs. The other person is blindfolded and they are the "reverse-reaction" person. Their job is to disconnect any pairs of snapium atoms in the box. The third person is the "shaker." Their job is to gently shake the box while the "forward-reaction" and "reverse-reaction" people have their turns at a constant pace. The fourth person is the "counter." Their job is to have everyone pause at 1-minute intervals and record how many paired "disnapium" molecules there are and how many snapium atoms there are.
3. Continue this process 15 times.
4. Graph your results on graph paper.

Questions to Consider
a. What was the purpose of the shaking? What trends did you see over time?
b. What is the chemical equation for this simulated chemical equilibrium reaction?

Stephanie Ryan

Reactions appear to halt when the system reaches equilibrium, but do not be fooled—there is still much going on! At the molecular level, the reactants will continue to form products, and the products will reform into reactants. These processes create a stable ratio of product concentrations to reactant concentrations, which varies for each specific reaction. This value is so important that we provide it with a special term: the **equilibrium constant**, K_{eq}, which is unitless.

To understand the general expression of the equilibrium constant, let's start with a simple conversion of A to B (Equation 12.1). In this system, K_{eq} is simply the ratio of the concentration of the product, B, over the concentration of the reactant, A (Equation 12.2). By convention, a reactant or product enclosed in brackets indicates its concentration in units of moles/L or M. For instance, $[Na^+]$ indicates the concentration of sodium ions.

$$A \rightleftharpoons B \qquad \qquad \textbf{[12.1]}$$

$$K_{eq} = \frac{[B]}{[A]} \qquad \qquad \textbf{[12.2]}$$

Once you look up or calculate the equilibrium constant, you can immediately determine whether the reactant or product is favored. By "favored," we mean which has a greater concentration after equilibrium has been established. If $K_{eq} > 1$, then the concentration of the product, [B], must be favored. In contrast, if $K_{eq} < 1$, then the concentration of the reactant, [A], must be favored (Figure 12.1). However, what if $K_{eq} = 1$? This simply indicates that both reactant and product concentrations are equivalent, and neither is favored.

Your Turn 12.1 Finding Equilibrium

As seen in the last chapter, glucose and fructose are monosaccharides that are isomers of each other. In other words, they have the same numbers and types of atoms, but in different structural arrangements. Unlike some isomers, glucose and fructose can be reversibly interconverted to each other.

a. Write the balanced chemical reaction for the conversion of glucose into fructose.
b. Write the formula for the equilibrium constant, K_{eq}.
c. In a sample at equilibrium [glucose] = 6.02 mM and [fructose] = 4.45 mM. Calculate the equilibrium constant.

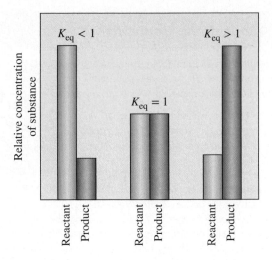

Figure 12.1

An illustration of the magnitude of the equilibrium constant relative to the concentrations of products (**purple**) and reactants (**green**).

We can apply this expression to any reversible reaction. For instance, take the arbitrary conversion of two reactants into two products (Equation 12.3). The concentrations of each are taken into account, as well as the number of moles of each required in the balanced chemical reaction. Specifically, the lowercase letters in this equation represent the number of moles of each in the balanced chemical reaction.

$$a\,A + b\,B \rightleftharpoons c\,C + d\,D \qquad K_{eq} = \frac{[C]^c[D]^d}{[A]^a[B]^b} \qquad \text{[12.3]}$$

By comparing the equilibrium constant of a reaction to the current concentrations of reactants and products, we can make predictive statements about the relative changes in concentrations that will occur *en route* toward establishing equilibrium. The ratio of [products]/[reactants] at a given point in time is known as the **reaction quotient**, Q, and it may be greater than, less than, or equal to the equilibrium constant, K_{eq}.

Before a reaction starts, the reactants are present, but there are no products. The product concentrations are equal to zero because no product has been made. If this is inserted into the equilibrium equation [12.3], it would be 0 M/[reactants] = 0. The reaction quotient (Q) as zero would be less than the equilibrium constant, and we could write this as: $Q < K_{eq}$. When the reaction begins, and the reactants are converted to products, there is a decrease in the concentration of reactants as the concentration of products must increase at the expense of the reactants. However, once products are formed, they can potentially react to regenerate the reactants. We can then compare the equilibrium conditions with our reaction quotient to let us know whether we are approaching equilibrium or are at equilibrium. If the system exhibits conditions where $Q > K_{eq}$, the system would prefer to regenerate some of the reactants by consuming products. However, if $Q = K_{eq}$, then the system is at equilibrium, and there will be no net increase in the concentrations of products or reactants. These different possibilities are summarized in Figure 12.2.

Once a system is at equilibrium, it may be altered by introducing an external stress. The stress may cause the reactants to produce more products, or the stress may cause the products to produce more reactants. These effects are a consequence of **Le Châtelier's principle**, which states that if an equilibrium is disturbed by changing its conditions, the position of the equilibrium will shift in the direction to counteract the change.

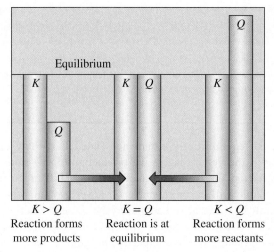

Figure 12.2

The relationship between the reaction quotient, Q, and equilibrium constant, K.

Le Châtelier's Principle

This video (www.acs.org/cic) demonstrates the effects of adding/removing a reactant or product and what happens when a system's temperature or pressure is changed.

Questions to Consider

a. Provide some real-world examples of Le Châtelier's principle.

b. Does the value of the equilibrium constant, K, change when the concentration of reactants/products is changed? Explain.

c. How are the pressure and volume of a gaseous equilibrium related?

d. Does adding a catalyst change the position of an equilibrium? Explain.

©Bradley D. Fahlman

To study biological systems, we need to expand our concept of reactions. Instead of just chemical changes, we must also consider the possibility of moving between two positions or states. In a biological system, the act of a chemical moving from free to bound states or from one side of a membrane to another can be far more important than chemical changes. For the purposes of the equilibrium constant, all other calculation aspects are the same. An illustrated example of the binding/release of two molecules is shown in Figure 12.3. In biochemistry, we discuss the equilibrium constant for the dissociation reaction so frequently that we give it a special constant, K_d.

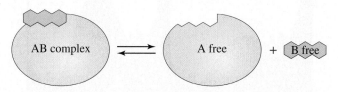

Figure 12.3

The reversible dissociation of molecules A and B shown from the complex (bound state) to their free states.

Your Turn 12.2   Finding Equilibrium in Binding

Epinephrine (or adrenaline) is a small organic molecule that binds to your cells at a specific site called the β-adrenergic receptor to give you that familiar heightened energy level. The release of epinephrine from the receptor has a K value equal to 5×10^{-6}. For a solution of epinephrine with the β-adrenergic receptor, does the epinephrine prefer to stay bound or free?

On its own, the equilibrium constant does not tell us how quickly we can go between states, only if the system favors the production of one state to another. Imagine pushing a stone from one valley to another with a hill or barrier between them (Figure 12.4). The stone will be more stable on lower ground, whether the barrier is low or high. The study of the relative energies of the two end states (A and B) is known as **thermodynamics**. As you may recall from Section 6.4, we used the principles of thermodynamics to determine whether a reaction is endothermic (requiring the addition of heat) or exothermic (releasing heat).

However, thermodynamics only provides one piece of the puzzle regarding a reaction or process. The study of the speed, or *rate*, of a reaction or conversion between chemical states is known as **kinetics**. For instance, graphite is more thermodynamically stable than diamond; however, do not fear—your diamond ring will not be converted into the material found in pencil lead anytime soon. The conversion between diamond to graphite occurs extremely slow in the absence of high pressures/temperatures, requiring millions of years. Thus, on our human timescale, it is indeed appropriate to state "a diamond will last forever!" Revisiting the analogy of a reaction presented in Figure 12.4, the higher the barrier in between states A and B, the more difficult it will be to push the stone between the valleys. The height of this barrier between two states is known as its **activation energy**, which is inversely related to the rate of the reaction. That is, the *higher* the activation energy of a reaction or process, the *slower* it will proceed. As discussed in Section 6.13, a catalyst is typically used to lower the activation energy and speed up a reaction.

Your Turn 12.3   Equilibrium in Your Blood

Your blood cells are rich in an iron-carrying molecule called hemoglobin. One primary responsibility of hemoglobin is picking up oxygen in our lungs and depositing it in our tissues by transferring it to a second molecule called myoglobin. Do you expect myoglobin or hemoglobin to have a higher equilibrium constant for this exchange to be favorable? Explain your reasoning.

To survive, all organisms concentrate the specific chemicals necessary for function (such as nutrients) and pump out or block any material that would be harmful. In short, life requires adjustments to establish or maintain equilibrium—the topic of our next section.

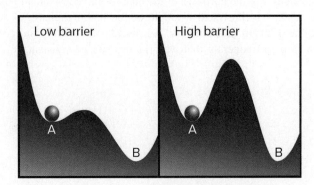

Figure 12.4

Visualizing barriers to a reaction progress. The reaction on the left proceeds faster since it has a lower energy barrier (activation energy).

12.2 | Keeping Our Bodies in Equilibrium

Learning Objective: Calculate the pH of buffers and describe how they are used in solutions to maintain a constant pH

You likely recognize the symptoms of too much acid in your stomach: a gurgling, often unpleasant sensation, followed by potentially harmful acid reflux, which indicates that the chemical environment in your stomach is out of balance. Although long-term and frequent acid reflux may indicate a physiological imbalance worthy of consulting your physician, the most immediate cause is likely something acidic you ate. Lucky for you, these symptoms do not extend to other areas of your body. Your tissues, blood and even the interior compartments of your cells have a defined and well-regulated pH (concentration of hydrogen ions). These areas are resilient to adding acidic or basic chemicals because they employ buffers or chemicals that help maintain a constant pH.

As you may recall from Section 5.8, acids and bases are specific chemicals that alter the concentrations of protons (H^+; also known as the hydronium ion, H_3O^+) and hydroxide (OH^-) ions in a solution. As shown in Equation 12.4, the base created when a weak acid such as formic acid (HCOOH) loses its proton is called the **conjugate base**. Likewise, the acid formed when a weak base such as ammonia (NH_3) adds a proton is called the **conjugate acid** (Equation 12.5). In conjugate acid-base pairs, the conjugate acid will have one more proton than the original weak base; a conjugate base will have one less proton than the original weak acid.

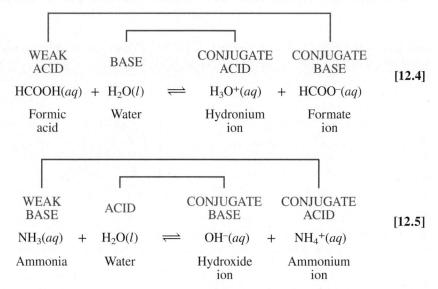

The two equation diagrams:

[12.4]

WEAK ACID · BASE · CONJUGATE ACID · CONJUGATE BASE

$HCOOH(aq) + H_2O(l) \rightleftharpoons H_3O^+(aq) + HCOO^-(aq)$

Formic acid · Water · Hydronium ion · Formate ion

[12.5]

WEAK BASE · ACID · CONJUGATE BASE · CONJUGATE ACID

$NH_3(aq) + H_2O(l) \rightleftharpoons OH^-(aq) + NH_4^+(aq)$

Ammonia · Water · Hydroxide ion · Ammonium ion

Equilibrium constants for reactions of weak acids and water have a special symbol, K_a (Equation 12.6). For instance, the equilibrium expression for Equation 12.4 would be:

$$K_a = \frac{[H_3O^+][HCOO^-]}{[HCOOH]}$$

[12.6]

The value of K_a for the dissociation of a *weak acid* is extremely small (e.g., 1.8×10^{-4} for HCOOH). This tells us that the reaction mixture will contain mostly unreacted acid. For formic acid, there will be very few hydronium (H_3O^+) and formate ions ($HCOO^-$) produced during the reaction. In comparison, the equilibrium constant of a *strong acid*, such as HCl, is extremely high (1×10^6), meaning that the reaction mixture will be composed primarily of hydronium (H_3O^+) and chloride (Cl^-) ions; hence, there will be none of the original hydrochloric acid (HCl) remaining.

Equilibrium constants for weak bases in water have a special symbol, K_b (Equation 12.7). The following base equilibrium constant expression for ammonia shown in Equation 12.5:

$$K_b = \frac{[NH_4^+][OH^-]}{[NH_3]}$$

[12.7]

These types of acid or base dissociations would not be favored if water was not present. Still, even though the interaction with water is a necessary part of the reaction, H_2O is not included in the equilibrium equations for K_a or K_b. Since pure liquids and solids are present in large excess and therefore have constant concentrations, they can be excluded from acid/base equilibrium constant expressions.

Practice devising the dissociation reactions for weak acids/bases and their K_a/K_b expressions in the next activity. The activity that follows will give you a chance to work with a simulation to see the effect of acids or bases on the pH of a solution.

Your Turn 12.4   Acids and Their Conjugate Bases

1. Write the balanced acid dissociation reaction(s) for each weak acid or base below, and then put a box around the conjugate species. Provide the expression for K_a or K_b for each of the acids or bases.

a. CH_3COOH (acetic acid, a weak acid)
   Hint: Only the hydrogen on the right of the chemical formula can dissociate as an ion.

b. H_2CO_3 (carbonic acid, a weak acid)
   Hint: There will be multiple conjugate bases for acids with multiple hydrogens that can dissociate; provide the reaction and equilibrium constant expression for dissociation of the first proton only.

c. CH_3NH_2 (methylamine, a weak base)

Your Turn 12.5   Acid Strength & pH

Check out this simulation (www.acs.org/cic) to see how the relative concentrations of strong or weak acids and bases affect the pH of the solution.

1. Begin with the Introduction tab. Very methodically, go through each section of the tool. Select water, then view the solution through the magnifying glass. The equation underneath provides a key. Under "Views", try out each view and be mindful of what it shows you. The tools allow you to measure pH with either a pH meter, pH paper, or a conductivity tester. You must drag the tool, so it is inserted

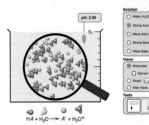

PhET Interactive Simulations, University of Colorado

Source: University of Colorado Boulder

into the solution, then select "play" to see what happens. After you select water, try another solution (strong acid, weak acid, strong base, or weak base), and compare the solutions in terms of how they appear in the solution, the equation that represents them, and how they measure with the tools.

2. Next, select the "My Solution" tab. Notice all of the features are similar to what you had in the introduction, but this time you may select acid or base followed by concentration. You may start with acid, try out the weakest concentration, and select the strength of the acid (weak or strong). Once again, notice how the solution is represented in the magnifying glass with the equation and try out the different tools. Increase the concentration and note how the features change. Do the same for bases.

Questions to Consider

a. Introduction tab—How does the pH of the solution change as you switch from a) water to strong acid, b) strong acid to weak acid, c) water to strong base, and d) strong base to strong base? Does the conductivity change too? How?

b. My Solution tab—Select the conductivity tool, insert it into a strong acid solution, and change the concentration from weak to strong. How does the conductivity change as the solution concentration increases? Examine the magnifying glass; what is happening to the solution that might account for the conductivity change?

A **buffer** is a system that responds only gradually or slightly to an added acid or base. The term "buffer" is quite apt here because its behavior corresponds well to the familiar verb. A good wool coat buffers you against cold gusts of wind, preventing your body from changing its warm temperature. Analogously, buffers—composed of acids and bases—prevent our bodies from large shifts in pH.

Buffer solutions are composed of a weak acid and its conjugate base, or a weak base and its conjugate acid. These solutions are able to maintain a relatively constant pH—even after the addition of a strong acid/base—because they contain species that preferentially react with the protons or hydroxide ions that are added. For instance, the conjugate base will react with added protons from an acid, or the conjugate acid will react with added OH^- ions from a base. In contrast, if a strong acid or base is added to pure water, there will be a significant increase in the concentration of H^+/OH^- ions, leading to a large pH change.

Just as pH = $-\log [H_3O^+]$ (revisit Section 5.10), which can be used to compare the relative acidity or basicity of a solution, one can determine the pK_a of an acid (Equation 12.8) or pK_b of a base (Equation 12.9) to compare their relative strengths.

$$pK_a = -\log [H_3O^+] \qquad \textbf{[12.8]}$$

$$pK_b = -\log [OH^-] \qquad \textbf{[12.9]}$$

As the K_a of an acid increases, its acid strength will increase (and pH will decrease) since more H_3O^+ ions are generated in solution. For instance, an acid with a K_a of 1×10^{-5} will be much stronger than an acid with a K_a value of 1×10^{-8}. Using Equation 12.8, the relative pK_a values for these acids are 5 and 8, respectively; hence, the smaller the pK_a of an acid, the greater its acidity. Likewise, stronger bases will have increasingly smaller pK_b values.

A useful formula that can be used to calculate the pH of a buffer is known as the Henderson–Hasselbalch equation (Equation 12.10). You will see this equation in action in the next activity, which explores the quantitative effect of buffers on the pH of a solution.

$$pH = pK_a + \log \frac{[\text{base}]}{[\text{acid}]} \qquad \textbf{[12.10]}$$

pH Changes of Buffers

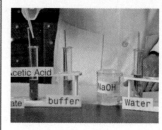

©McGraw Hill Education

Check out this video to see the effect of buffers on pH changes: www.acs.org/cic.

Acids and Buffers

This tutorial video (www.acs.org/cic) shows how to determine the pH of buffer solutions.

Questions to Consider

1. Calculate the pH of a buffer solution made from equal amounts of 0.3 M hydrofluoric acid and 0.7 M sodium fluoride. The K_a of HF is 7.1×10^{-4}.

2. Calculate the pH of the above buffer after 0.08 moles of NaOH is added to 1 L of solution. Assume there is no appreciable change in total volume.

Chemistry in Context Tutorial

ACIDS & BUFFERS

©Bradley D. Fahlman

Since 0.08 moles of base is added, it will react with the acid component of the buffer. Hence, the acid concentration will drop by 0.08 moles and the base concentration will increase by 0.08 moles. Since the solution volume is 1 L, the [acid] will be 0.22 M and the [base] will be 0.78 M after addition of the strong base.

A buffer is most useful when the pH of the solution equals its pK_a value. This is achieved by mixing equal moles of the acid and conjugate base. Table 12.1 lists the K_a and pK_a values for a variety of weak acids. A fair amount of acid can be added to a buffer solution before the conjugate base in the system is depleted, and the pH begins to decrease (becoming more acidic) in proportion to the acid added. We can also add a base to react with the conjugate acid in the opposing direction. Once the conjugate acid is depleted, the pH increases (becoming more basic). In fact, a buffer will resist the addition of acids or bases up to ±1 pH unit around the value of its pK_a (Figure 12.5).

Table 12.1	Dissociation of Some Weak Acids with K_a and pK_a Values		
Acid (name)	**Conjugate Base (name)**	**pK_a**	**K_a**
H_3PO_4 (phosphoric acid)	$H_2PO_4^-$ (dihydrogen phosphate ion)	2.1	7.24×10^{-3}
HCOOH (formic acid)	$HCOO^-$ (formate ion)	3.8	1.78×10^{-4}
CH_3COOH (acetic acid)	CH_3COO^- (acetate ion)	4.8	1.74×10^{-5}
H_2CO_3 (carbonic acid)	HCO_3^- (bicarbonate ion)	6.3	5.1×10^{-7}
$H_2PO_4^-$ (dihydrogen phosphate ion)	HPO_4^{2-} (hydrogen phosphate ion)	6.9	1.39×10^{-7}
NH_4^+ (ammonium)	NH_3 (ammonia)	9.3	5.62×10^{-10}
HCO_3^- (bicarbonate ion)	CO_3^{2-} (carbonate ion)	10.3	5.62×10^{-11}
HPO_4^{2-} (hydrogen phosphate ion)	PO_4^{3-} (phosphate ion)	12.4	3.98×10^{-4}

Figure 12.5

Useful range of buffering against acid or base additions around the pK_a.

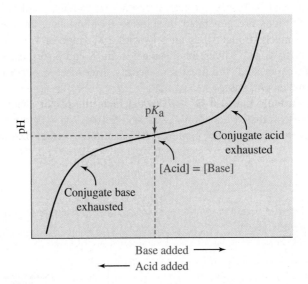

Biological systems require the buffering traits of a weak acid and base mixtures to absorb any local changes in hydrogen ion concentrations. The pH of your tissues will be resilient to most of your activities, even if the reactions occurring within your cells produce or use up hydrogen ions. Importantly, not all of your systems use the same buffer or the same pH. Your tissues will have preferential pH levels. Even within your cells, you will find compartments buffered at different pH levels. These systems are all ideally set at pH values that optimize the reactions and biological machinery that operates within them.

Your Turn 12.6   A Buffer Choice

A practicing scientist must judge a potential buffer for its utility in mimicking different cellular environments. Consider the list of target pH values and match them to the best possible chemical mixture from Table 12.1 for creating that pH.

a. pH 2.5   **b.** pH 7.4   **c.** pH 6.8   **d.** pH 5.4

Thus far, we have described biology in terms of the chemical balance required for a healthy system. Still, to completely understand the chemical transformations and the development of medicines, we need to gather some tools to understand the chemical framework of biology, built upon the versatile carbon atom.

12.3 | Carbon: The Essential Building Block of Life

> *Learning Objective: Illustrate the structures of hydrocarbon isomers using a variety of structural representations*

Carbon is the basis of all life forms on our planet. This element is so ubiquitous in nature that a subdiscipline of chemistry, **organic chemistry**, is devoted to studying carbon compounds. In practice, most organic chemists investigate compounds—of biological origin or human design—in which carbon is combined with a relatively small number of other elements such as hydrogen, nitrogen, oxygen, phosphorus, and sulfur. The name *organic* is historical and suggests a biological origin, but this is not necessarily true.

Your Turn 12.7   Compound Categories

Using the elements listed in the above paragraph, list three organic and three inorganic compounds. For each, describe the context in which you learned about the compound.

To specify an organic compound from the myriad of possibilities, you must be able to name it correctly. The International Union of Pure and Applied Chemistry (IUPAC) has established and periodically updates a formal set of nomenclature rules so that each known compound can be uniquely and distinctly named. However, many of these compounds have been known for a long time by common names such as alcohol, sugar, and morphine. When a headache strikes, even chemists do not call out for *2-acetoxybenzoic acid*; they ask for aspirin. We test our blood glucose levels, not our blood *(2R,3S,4R,5R)-2,3,4,5,6-pentahydroxyhexanal* levels! Chemical names are important and convey useful information about the molecules, but as you can see from our examples, common names are quite convenient.

We remind you that a few basic rules for bonding in organic molecules can help you find order in the chaos of millions of organic compounds. We introduced one of these in Section 3.6, the *octet rule*. Each carbon atom has a share in eight electrons, an octet when bonded. Eight electrons can be arranged to form four bonds, with a pair of shared electrons in each covalent bond. Carbon almost always forms four bonds, including four single bonds or some combination of single, double, and triple bonds. Some possibilities are illustrated in Figure 12.6:

$$-\overset{|}{\underset{|}{C}}- \qquad \overset{\backslash}{\underset{/}{C}}=\overset{/}{\underset{\backslash}{C}} \qquad -C\equiv C-$$

$$O=\overset{/}{\underset{\backslash}{C}} \qquad N\equiv C-$$

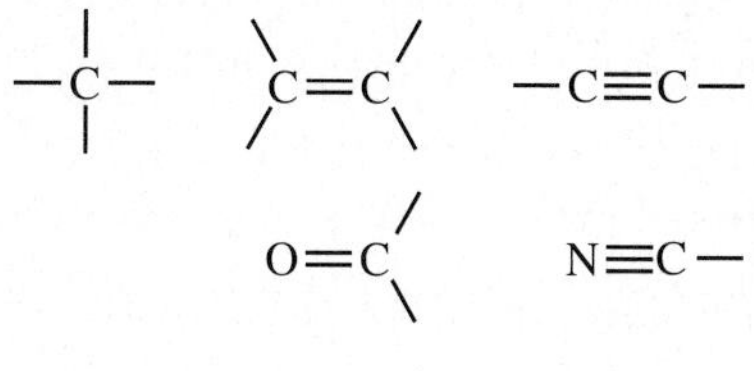

Figure 12.6

Some common bonding arrangements for carbon.

Your Turn 12.8   Checking on Carbon

a. Examine the carbon atoms illustrated in Figure 12.6. Does each one follow the octet rule?

b. From the contexts discussed in previous chapters, can you recall a molecule in which carbon does not form four bonds?

Other elements exhibit different bonding behaviors in organic compounds. A hydrogen atom is always attached to another atom by a single covalent bond. An oxygen atom typically forms two single bonds (to two different atoms) or one double bond (to a single atom). A nitrogen atom commonly forms three single bonds (to three different atoms) but also can form either a triple bond (to one other atom) or a single and double bond (to two different atoms). Both oxygen and nitrogen satisfy the octet rule because they have at least one pair of nonbonding electrons in addition to these chemical bonds.

Your Turn 12.9   The Octet Rule

An extremely valuable resource for chemists is the freeware molecular visualization program MolView. The Help menu provides detailed instructions on how to draw structural formulas and view 3D representations of molecules. Use MolView to build structures for each of the molecules below. For each structure, confirm that all of the carbon, nitrogen, and oxygen atoms satisfy the octet rule. Are hydrogen atoms considered exceptions to the octet rule? Explain.

a. NH_3, ammonia

b. N_2, nitrogen

c. N_2H_2, diazene

d. H_2O, water

e. H_2CO (the C is the central atom), formaldehyde

f. CO_2, carbon dioxide

g. CH_4, methane

h. C_2H_2, acetylene

i. C_2H_4, ethylene

The same number and kinds of atoms can be arranged differently, helping to explain why there are so many different organic compounds. Isomers are molecules with the same chemical formula (same number and kinds of atoms) but with different structures and properties. For instance, in Section 6.12, you encountered two of the isomers of C_8H_{18}: *n*-octane (straight chain) and iso-octane (branched).

In this chapter, we revisit the concept of isomers, this time with butane, C_4H_{10}. Analogous to C_8H_{18}, we can draw both a straight chain and a branched isomer, as represented by the structural formulas below.

n-butane   isobutane

Convince yourself that although these compounds have the same chemical formula; the atoms are connected differently. Both the linear isomer and the more complex isobutane can be represented with a condensed structural formula. Here we show the options for isobutane:

$$CH_3-\underset{\underset{CH_3}{|}}{CH}-CH_3 \quad \text{or} \quad CH_3CH(CH_3)CH_3 \quad \text{or} \quad CH_3CH(CH_3)_2$$

These condensed structural formulas are trickier to interpret. The parentheses around the —CH_3 groups indicate that they are attached to the C atom to their left. Although a —CH_3 group at the left end of a chain could be drawn as H_3C—, for ease, we often reverse the order to CH_3— with the understanding that the bond is to the C atom, not to the H atoms. Note that with three —CH_3 groups attached to the central C atom, a "branch" has been introduced into the molecule.

Figure 12.7 shows three representations of *n*-butane and isobutane. The first column shows the simple structural formula, and the second is a ball-and-stick model. The third column shows a space-filling model that presents a more realistic view of the molecular shape.

Only two isomers of C_4H_{10} exist. However, as the number of atoms in a hydrocarbon increase, so does the number of possible isomers. For instance, in addition to *n*-octane and iso-octane, there are 16 other isomers for C_8H_{18}. And for $C_{10}H_{22}$, there are 75 isomers!

Structural formula	Ball-and-stick formula	Space-filling model

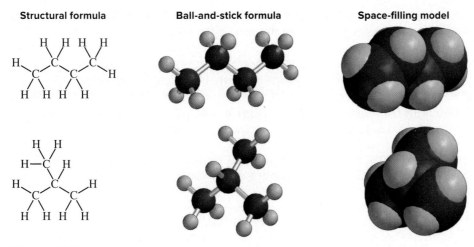

Figure 12.7

Three representations of the isomers of *n*-butane (top) and isobutane (bottom).
Check out 3D representations of n-butane and isobutane at www.acs.org/cic.

Your Turn 12.10   Switching Representations

Expand each of these condensed structural formulas into a complete structural formula.

a. $CH_3CH(CH_3)CH_2CH_3$ **b.** $CH_3CH(CH_2CH_3)CH_3$ **c.** $CH_3CH_2C(CH_3)_3$

As seen in Section 11.2, line-angle drawings are often used to represent the structural formulas for large molecules. In these illustrations, the hydrogen atoms are omitted for clarity in order to give focus to the carbon backbone of the molecule. Hence, it is important to remember that each carbon atom has four bonds, sharing a total of eight electrons. Table 12.2 shows various ways to represent *n*-butane, isobutane, and two

Table 12.2	Molecular Representations		
Compound	**Chemical Formula**	**Structural Formula**	**Line-Angle Drawing**
n-Butane	C_4H_{10}		
Isobutane	C_4H_{10}		
n-Hexane	C_6H_{14}		
Cyclohexane	C_6H_{12}		

other simple molecules. Recall that in a line-angle drawing, one carbon atom is assumed to occupy each vertex position and the ends of the line.

Your Turn 12.11   Practice with Line-Angle Drawings

Revisit the compounds in Your Turn 12.10. For all three, draw line-angle representations.

Your Turn 12.12   Practice with Isomers

a. Are *n*-butane and isobutane isomers? Explain.
b. Are *n*-hexane and cyclohexane isomers? Explain.
c. Three isomers have the formula C_5H_{12}. For each, draw a structural formula, a condensed structural formula, and a line-angle drawing.

Many molecules, including thyroxine, dopamine, aspirin, and glucose, have carbon atoms arranged in a ring. For example, examine the structure of cyclohexane, C_6H_{12}, in Table 12.2. The ring in cyclohexane has six carbons. Rings most commonly contain five or six carbon atoms because this number seems to confer some stability—not too floppy and not too constricted. In epinephrine (Figure 12.8), however, the six-membered rings are based on benzene, C_6H_6, (Figure 12.9) rather than on cyclohexane.

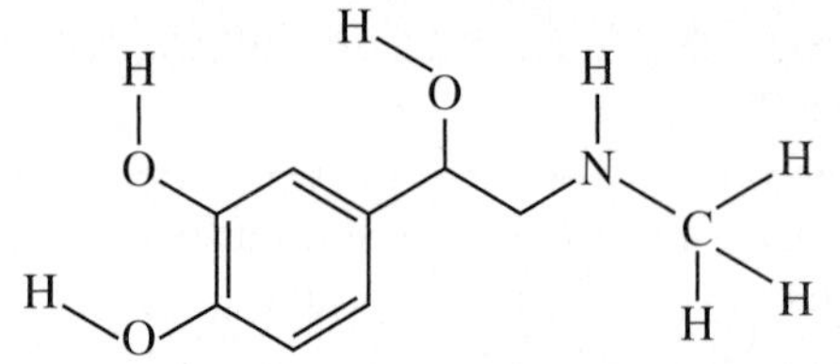

Figure 12.8

The structure of epinephrine, more commonly known as adrenaline, is a vital chemical signal between neurons and a hormone that regulates metabolism. Check out the 3D representation of epinephrine at www.acs.org/cic.

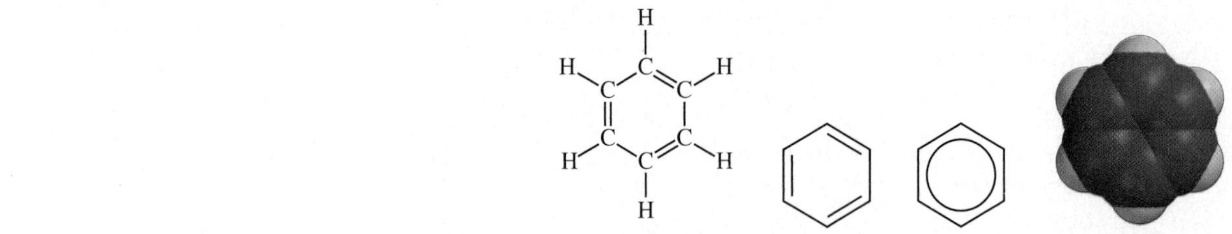

Figure 12.9

Representations of benzene, C_6H_6. Check out the 3D representation of benzene at www.acs.org/cic.

Figure 12.9b shows a line-angle drawing for benzene, illustrating alternating single and double bonds. However, this is only one possibility; two equivalent structures for benzene are known as resonance forms. The structure of benzene is a hybrid of the two resonance structures, which is represented by the dashed structure in Figure 12.10 or the circle shown in Figure 12.9c. Because all of the C—C bond lengths are experimentally determined to be equivalent, each C—C bond in benzene may be thought of as a hybrid between a C—C single bond and a C=C double bond. A typical C—C single bond is 0.154 nm in length, whereas a C=C double bond is 0.134 nm. In benzene, the C—C bond lengths are all 0.139 nm. Lastly, the hexagonal structure of benzene is found in the —C_6H_5 phenyl functional group that is part of many molecules, including polymers such as polystyrene (Section 9.5).

Figure 12.10

Resonance structures of benzene (hydrogen atoms are omitted for clarity).

12.4 | Functional Groups

Learning Objective: Recognize various functional groups commonly found in organic molecules and describe their effect on physical properties such as solubility

Central to the study of biological interactions is functional groups. Functional groups are distinctive arrangements of groups of atoms that impart characteristic physical and chemical properties to the molecules that contain them. These groups are so important that we often show them in structural formulas and represent the remainder of the molecule with an R. The R is generally assumed to include at least one carbon or a hydrogen atom. You already encountered some functional groups in Section 9.6. For instance, the generic formula for an alcohol is ROH, as in methanol (CH_3OH, an alcohol derived from the degradation of wood) and ethanol (CH_3CH_2OH, an alcohol derived from the fermentation of grains and sugar). The presence of the —OH group attached to a carbon makes the compound an alcohol. In addition, the —OH group is *covalently* bonded to the rest of the molecule. This is different from the hydroxide ion, OH^-, which is electrostatically attracted to a cation.

Similarly, a carboxylic acid group, commonly written as ![carboxylic acid structure], —COOH, —C(O)OH, or —CO_2H confers acidic properties to the molecule. In aqueous solutions, a H^+ ion (a proton) is transferred from the —COOH group to a H_2O molecule to form a hydronium ion, H_3O^+. This leaves behind an anion that is commonly written as —COO^-.

We represent carboxylic acids with the general formula RCOOH. In acetic acid (CH_3COOH, the acidic component in vinegar), the R is —CH_3, known as a methyl group. The carboxylic acid is what gives vinegar its characteristic sharp-sour smell and taste. Table 12.3 lists eight functional groups commonly found in drugs and other organic compounds; for example, Figure 12.11 displays the thyroxin molecule, which contains a variety of functional groups. Get some practice with functional groups in this next tutorial video activity.

Functional groups in a biomolecule are often involved in interactions with other molecules. As seen in Chapter 5, functional groups distinguish parts of a molecule as polar or nonpolar. For example, polar functional groups may experience dipole–dipole interactions or hydrogen bonds with other substances. In contrast, nonpolar functional

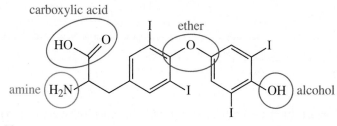

Figure 12.11

The structural formula of thyroxin, a vital hormone precursor molecule, with selected functional groups (an amine, an ether, an alcohol, and an acid) circled. Check out the 3D representation of thyroxin: www.acs.org/cic.

Table 12.3 Some Important Organic Functional Groups

Functional Group	Generic Formula	SPECIFIC EXAMPLES		
		Name	**Structural Formula**	**Condensed Structural Formula**
Alcohol	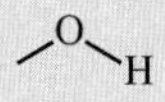	Ethanol (ethyl alcohol)	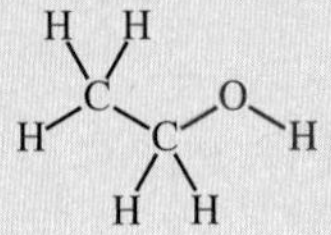	CH_3CH_2OH
Ether	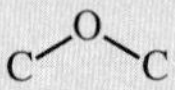	Dimethyl ether	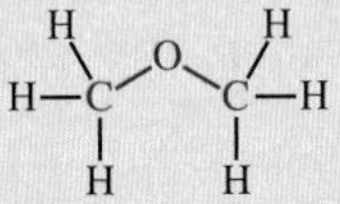	$CH_3{-}O{-}CH_3$ or CH_3OCH_3
Aldehyde	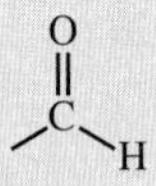	Propanal	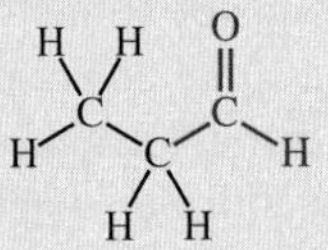	$CH_3CH_2{-}\overset{O}{\overset{\parallel}{C}}{-}H$ or CH_3CH_2CHO
Ketone	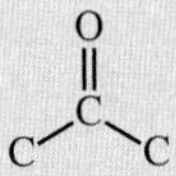	2-Propanone (dimethyl ketone, acetone)	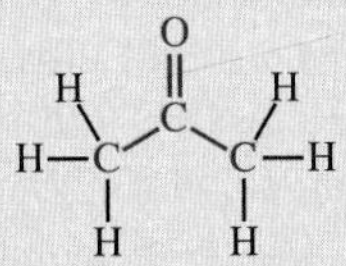	$CH_3{-}\overset{O}{\overset{\parallel}{C}}{-}CH_3$ or CH_3COCH_3
Carboxylic acid	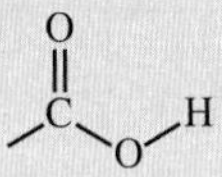	Ethanoic acid (acetic acid)	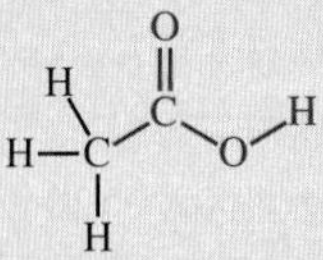	$CH_3{-}\overset{O}{\overset{\parallel}{C}}{-}OH$ or CH_3CO_2H or CH_3COOH
Ester	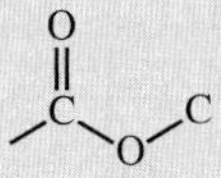	Methyl ethanoate (methyl acetate)	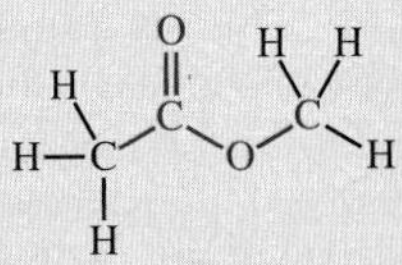	$CH_3{-}\overset{O}{\overset{\parallel}{C}}{-}OCH_3$ or CH_3COOCH_3
Amine	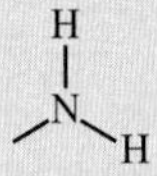	Ethylamine	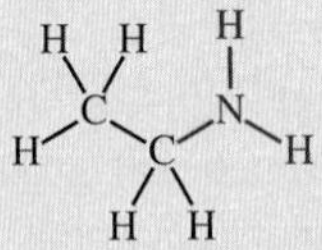	$CH_3CH_2NH_2$
Amide	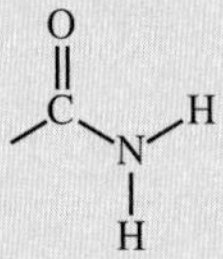	Propanamide	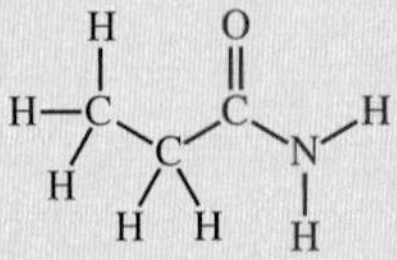	$CH_3CH_2{-}\overset{O}{\overset{\parallel}{C}}{-}NH_2$ or $CH_3CH_2CONH_2$

groups will be attracted to the nonpolar groups in other substances through London dispersion intermolecular forces. The consequences of having an assortment of functional groups can affect physical properties, such as how the molecule folds or how soluble the substance is in water or another solvent. It can also affect chemical properties, such as how substrates bind to enzymes and react.

As mentioned, functional groups can play a role in the solubility of a compound, an important consideration in the targeting, regulation, uptake, and rate of reaction for any chemical in the body. The general solubility rule "like dissolves like," which was introduced in Section 5.7, applies to reactions within in our body as well as those occurring in the lab. Functional groups containing oxygen and nitrogen atoms (e.g., —OH, —COOH, and —NH₂) usually increase the polarity of a molecule. This enhances a molecule's solubility in other polar substances, such as water.

By contrast, hydrocarbons and other organic molecules that do not contain such functional groups are typically nonpolar and will not dissolve in polar solvents. For example, *n*-octane (C_8H_{18}) is nonpolar and insoluble in water. However, it does dissolve in nonpolar solvents such as hexane (C_6H_{14}) and dichloromethane (CH_2Cl_2). Biomolecules with significant nonpolar character tend to accumulate in cell membranes and fatty tissues that are largely hydrocarbon and nonpolar.

Your Turn 12.13   Functional Groups in Dopamine

Using the MolView visualization program, draw the structure of dopamine ($C_8H_{11}NO_2$) in both structural and line-angle representations. How does this structure differ from that of epinephrine (Figure 12.8)? Draw a box around the ring(s) and a circle around the various functional groups. What are the names of these functional groups? Do you think this molecule would be soluble in fatty tissue? Explain.

12.5 | Give These Molecules a Hand!

Learning Objective: Define chirality and its importance in biological activity

Biochemistry is further complicated because many of its interactions involve a common but subtle phenomenon called *optical isomerism* or *chirality*. **Chiral** molecules, or **optical isomers**, have the same chemical formula but differ in their three-dimensional molecular structure and their interaction with *polarized light*. Chirality most frequently arises when four different atoms, or groups of atoms, are attached to a carbon. A chiral compound can exist in two different molecular forms that are non-superimposable mirror images of each other. One optical isomer will rotate polarized light in a clockwise manner, which is called the dextro (D) isomer. The other isomer is called the levo (L) isomer, which rotates polarized light in a counterclockwise manner.

Non-superimposible mirror images should be familiar to you. You carry two of them around with you all the time—your hands. If you hold them palms up, you can recognize them as mirror images. For example, the thumb is on the left side of the left hand and on the right side of the right hand. Your left-hand looks like the reflection of your right hand in a mirror. But your two hands are not identical. Figure 12.12 illustrates this relationship for both hands and molecules. In fact, the term chiral is derived from the Greek word for "hand."

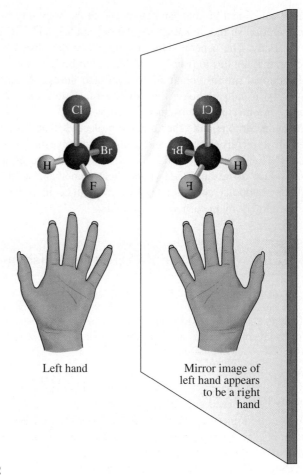

Figure 12.12

The mirror image of a molecular model and a hand. The molecule CHBrClF is chiral, and its shape is tetrahedral.

The molecule in Figure 12.12 could be drawn as a wedge-dash drawing also to convey its 3D character:

$$\begin{array}{c} \text{Cl} \\ | \\ \text{H}-\text{C}\text{····}\text{Br} \\ \text{F} \end{array}$$

The Cl and the H atoms are in the plane of the page, whereas the dashed line to the Br indicates that the Br atom extends behind the page, away from the viewer. The solid wedge toward the F indicates that the F atom is in front of the page, oriented toward the viewer.

As shown in Figure 12.13, the four atoms bonded to a central carbon are in a tetrahedral arrangement. The positions of these four atoms correspond to the corners

Figure 12.13

A tetrahedron has four triangular faces.

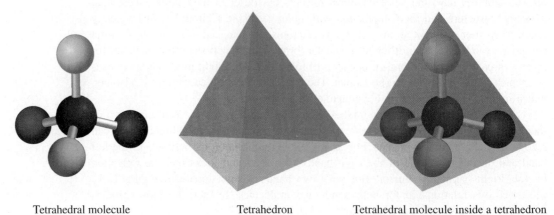

Tetrahedral molecule   Tetrahedron   Tetrahedral molecule inside a tetrahedron

of a three-dimensional tetrahedron with equal triangular faces. When these atoms are different from one another, the molecule is chiral since it cannot be superimposed onto its mirror image.

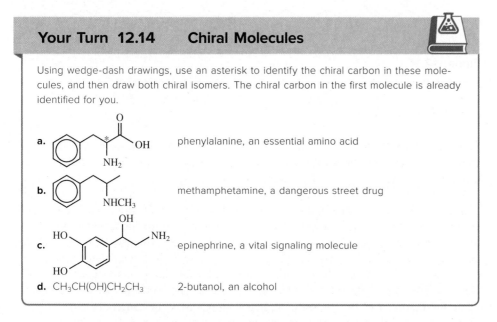

Your Turn 12.14 Chiral Molecules

Using wedge-dash drawings, use an asterisk to identify the chiral carbon in these molecules, and then draw both chiral isomers. The chiral carbon in the first molecule is already identified for you.

a. phenylalanine, an essential amino acid

b. methamphetamine, a dangerous street drug

c. epinephrine, a vital signaling molecule

d. $CH_3CH(OH)CH_2CH_3$ 2-butanol, an alcohol

Many biologically important molecules, including sugars and amino acids, exhibit chirality. This is significant because, although most chemical and physical properties of a pair of optical isomers are very nearly identical, their biological behavior can differ markedly. Generally, the explanation for this difference is related to the necessity of a good molecular fit between a molecule and its receptor site.

You can illustrate the relationship between chirality and biological activity by taking things into your own hands. Just as your right hand fits only into a right-handed glove, the shape of a molecule will govern its binding to a particular site. For chiral molecules, only one of its optical isomers will typically fit into a particular receptor site (Figure 12.14).

The extreme specificity imparted by chiral molecules complicates the study of biochemistry and the medicinal chemist's task of synthesizing drugs, but dramatically improves the sensitivity and nuance of biological interactions. In order to interact with a receptor site, a molecule must include the appropriate functional groups in the correct three-dimensional configuration to give the molecule its desired biological activity.

In many chemical reactions, the "right" and "left" optical isomers are produced simultaneously. When equal amounts of each optical isomer are produced, this is referred to as a **racemic mixture**. Interestingly, when new molecules are synthesized in biological systems, they all have inherent chirality. If a chemist wants to create

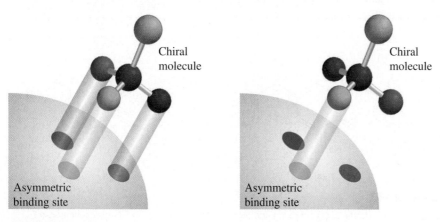

Figure 12.14

A chiral molecule binding to an asymmetrical site (left); its mirror image cannot bind.

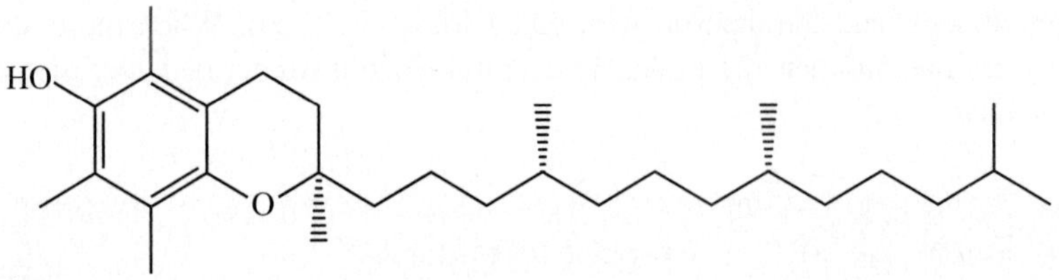

Figure 12.15

The α-form of tocopherol (better known as vitamin E). The only difference between the alpha (α) form and the gamma (γ) form is a methyl group (shown in red). Check out the 3D representation of vitamin E at www.acs.org/cic.

synthetic mimics of biologically active molecules, the task can prove to be extremely challenging. Since the physical properties of optical isomers are often identical, it is difficult to isolate or separate the racemic mixture of isomers.

It is not uncommon for only one optical isomer to be biologically active in our bodies. For example, some antibiotics, hormones, and certain drugs have only one biologically active isomer. However, racemic mixtures can provide an interesting and diverse set of chemical properties. For instance, Vitamin E exhibits potent antioxidant and anticancer properties, and also plays a vital role in fetal development. Vitamin E has several chiral centers, which results in a complex mixture of compounds. Biologically, the D-isomer of the α-form of vitamin E is particularly noted in fetal development, while the D-isomer of the γ-form is most potent against some cancers (Figure 12.15). Your body carefully selects between optical and structural isomers to absorb and concentrate specific forms into appropriate tissues. The isomer that is not selected is likely flushed from your system as waste and does you no harm.

However, a racemic mixture can sometimes be damaging if the optical isomers are both active in your system. Naproxen (Figure 12.16a), a common pain reliever, is one example of many in which chiral purity is not only preferred, it is required. One form of naproxen relieves pain; the other causes liver damage.

Another example involves a treatment for Parkinson's disease. L-DOPA (Figure 12.16b) is a synthetic precursor of the neurotransmitter dopamine. The initial use of racemic DOPA for treating this disease brought adverse effects such as anorexia, nausea, and vomiting. However, using the single optical isomer greatly reduced these side effects.

Your Turn 12.15   Examining Naproxen and L-DOPA

Carefully examine the structural formulas for naproxen and L-DOPA given in Figure 12.16.

a. For each drug, which is the chiral carbon atom?
b. Identify all of the functional groups present in both drugs.
c. Draw the structural formulas for the biologically inactive isomers for both molecules.

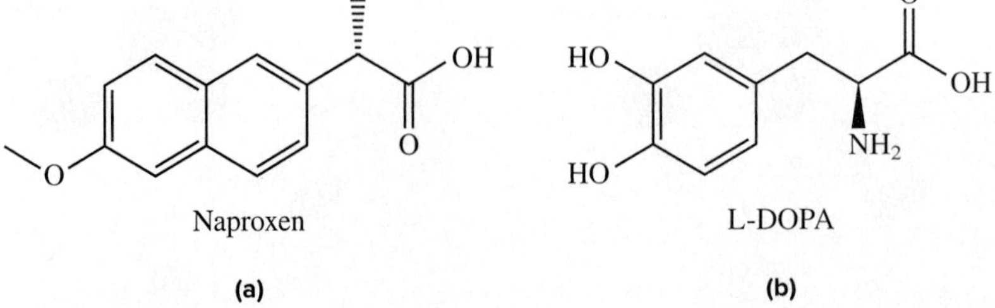

Naproxen

(a)

L-DOPA

(b)

Figure 12.16

Biologically active forms of **(a)** naproxen and **(b)** L-DOPA. Check out the 3D representations of naproxen and L-DOPA at www.acs.org/cic.

12.6 | Life via Protein Function

> Learning Objective: Describe how proteins are used to process glucose
> for energy, route chemical signals between cells, and transport oxygen
> through the blood

Small molecules in our bodies do not act in isolation. They elicit your physiological
response through their interactions with the comparatively massive macromolecules in
your systems. You may already be familiar with some of these large molecules. For
instance, recall grabbing that pencil at the beginning of this chapter. In an automated
sort of way, your body underwent numerous actions—only some of which were a direct
result of your conscious action. At some point, every action required a change in the
behavior of a macromolecule.

Life depends on four major classes of macromolecules: lipids, carbohydrates, nucleic
acids, and proteins (Figure 12.17). As described in Section 11.2, lipids (such as fats)

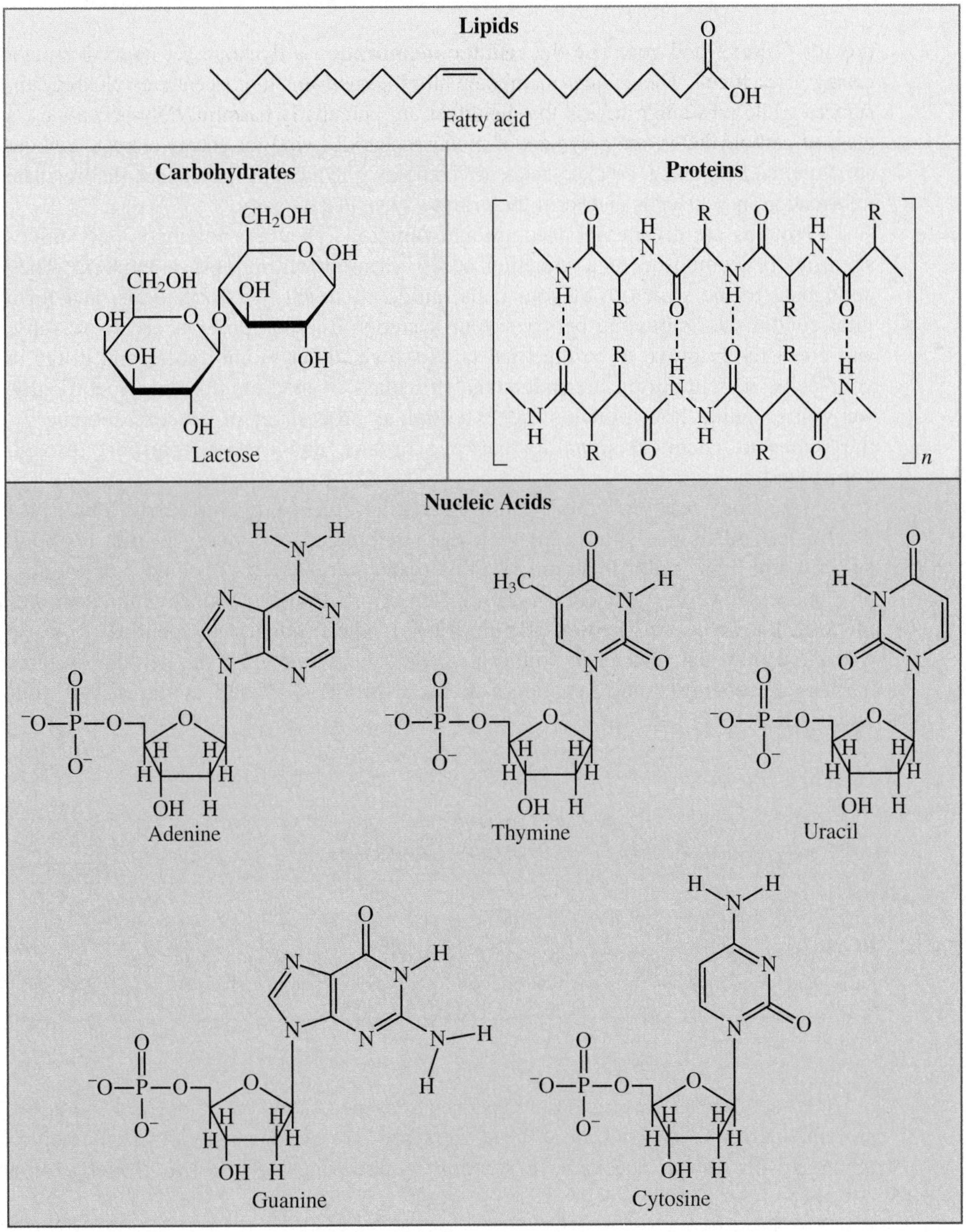

Figure 12.17

Macromolecule classes:
lipids, carbohydrates,
proteins, and nucleic acids.

(a)

(b)

Figure 12.18

Two contrasting examples of sugar use. **(a)** The combustion of simple sugars observed by the burning of wood, which produces heat and light. This starkly contrasts with **(b)**, the use of sugars in respiration to create muscle contraction.

(a): ZoonarAlkimson/age fotostock; *(b):* Izf/Shutterstock

provide Calories and comprise the **cellular membrane**—a dynamic yet protective outer casing of each cell. The cellular membrane allows organisms to concentrate valuable substances while preventing access to others that are potentially harmful. Polysaccharides, a class of carbohydrates, are polymers of monosaccharides vital for energy storage, cellular structure, and signaling. Nucleic acids are complex polymers of nucleotides that regulate information in your cells and form the primary basis for heredity.

Proteins are diverse in structure and function. They are polymers, polyamides specifically, composed of a selection of 20 different amino acid monomers. They contribute to the structure of your cells, guide chemical transformations, and form vital conduits for signaling between your systems. Because proteins are so versatile and consistently prove to be the final targets for smaller biomolecules and drugs in your cells, it is important to understand this class of macromolecules. Specifically, we will examine how proteins are essential to processing of glucose for energy, the routes of chemical signaling between organs, and oxygen transport through your blood.

Your body uses the simple monosaccharide glucose as a universal fuel across your cells and tissues. You have seen the combustion of glucose in this textbook and in vivid living color in a bonfire—this is the combustion of cellulose, a polymer of glucose. However, your body uses a different process for complete combustion of glucose, known as **respiration** (Figure 12.18). The products are identical to those of combustion, but instead of simply providing heat and light, the cellular systems harness the energy stored in glucose to keep the cell—as well as the rest of your body—alive.

Your Turn 12.16   Glucose Reaction

The chemical formula for glucose is $C_6H_{12}O_6$.

a. Write the balanced reaction for the complete chemical combustion of glucose.
b. Determine how much oxygen is required to completely combust 10 grams of glucose.

To coordinate the chemical transformations necessary for complete respiration, proteins known as *enzymes* are utilized. **Enzymes** are biological catalysts common to all organisms, which accelerate a reaction by lowering the *activation energy* of a

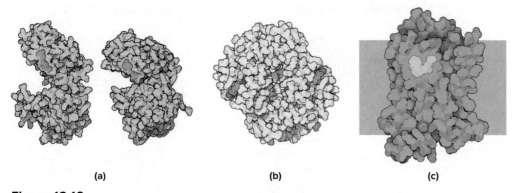

Figure 12.19

Proteins, a vital class of macromolecules, in 3D. **(a)** The enzyme phosphoglycerate kinase performs the seventh step of many necessary to thoroughly convert glucose into useful cellular energy. **(b)** Hemoglobin holds oxygen on the iron in the center of the organic molecule heme (**red**). **(c)** A hormone receptor on a cellular membrane (**gray**) bound to its hormone (**orange**). For more details about hemoglobin, check out www.acs.org/cic.

reaction (revisit Section 12.1). Enzymes contain an **active site** or catalytic region to accelerate and control reactions. In the active site, the enzyme selectively binds only specific reactant(s) to promote the desired reaction (Figure 12.19). As we will describe in the next section, the specificity of the catalyzed reaction is enabled by the specific shape and functional groups of both the reactant and the active site.

To break down glucose fuel in the body, your system requires something more complex than a lit match. Cellular respiration in more complex organisms requires three stages involving the carefully regulated chemical coordination of at least 25 enzymes in three distinct cellular compartments (Figure 12.20).

Stage 1 of cellular respiration is the most universal. **Glycolysis** converts intracellular glucose into higher-energy three-carbon sugars, releasing useful chemical energy. The term *glycolysis* is derived from the Greek term for the process of breaking (*lysis*)

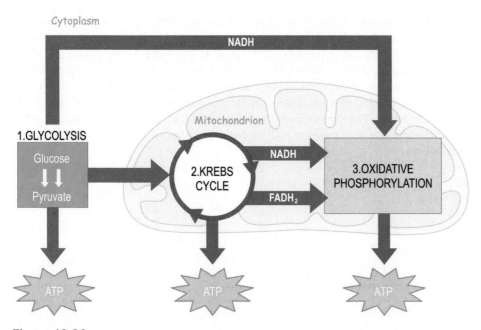

Figure 12.20

The three stages of anaerobic cellular respiration.

Designua/Shutterstock

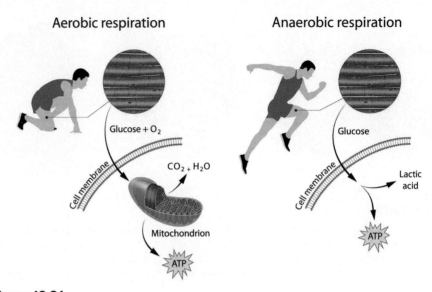

Figure 12.21

Comparison of aerobic and anaerobic cellular respiration.

Aldona Griskeviciene/Shutterstock

something sweet (*glycol*). As shown in Figure 12.21, glycolysis without further respiration produces *lactic acid*—believed to be the chemical source responsible for the familiar and painful ache after rapid exercise. When our muscles require energy immediately and do not have time to wait for slower aerobic processes, we can depend on glycolysis alone. Glycolysis that operates in the absence of oxygen is referred to as *anaerobic respiration*—required for many lower-order microorganisms (e.g., yeasts, bacteria, internal parasites), used in applications such as alcohol fermentation (Section 10.11). This first stage of respiration can be effectively reversed to convert smaller sugars, as well as the products of fatty acid or protein oxidation, back into glucose for storage in the form of *glycogen*, a polysaccharide.

The remaining stages of respiration shown in Figure 12.20 require oxygen and produce carbon dioxide, while gathering far more energy per molecule of glucose. Specifically, through a complex series of reactions involving high-energy electron carriers NADH and $FADH_2$, each glucose molecule can be converted into about 32 molecules of adenosine triphosphate (ATP, Figure 12.22). ATP is the chemical fuel of choice for almost all transformations and changes in the cell. The molecule ATP provides a conveniently sized universal currency for the regular work of the cell. Other fuels, such as fats and proteins, follow similar pathways to feed into the same processes as carbohydrates. The overall process of manipulating energetic molecules, either breaking them down to produce ATP or building them up to store energy, is called **metabolism**.

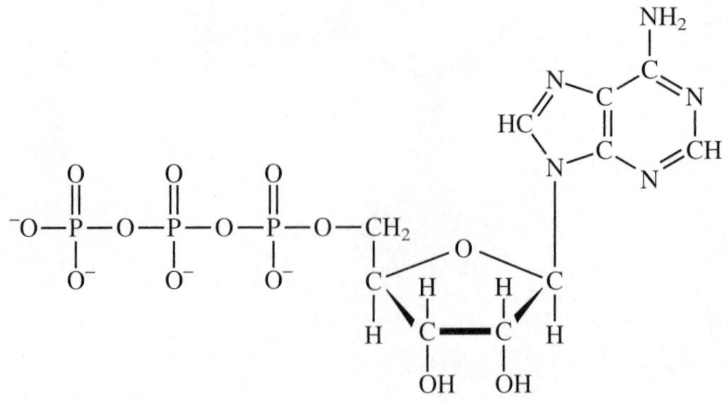

Figure 12.22

Adenosine triphosphate, ATP. Check out the 3D representation of ATP at www.acs.org/cic.

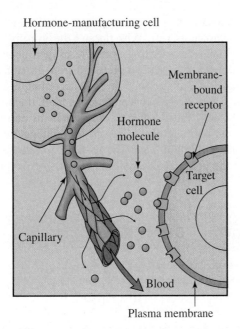

Hormone-manufacturing cell

Membrane-bound receptor

Hormone molecule

Target cell

Capillary

Blood

Plasma membrane

Figure 12.23

An illustration of chemical communication in the body. Hormone molecules travel through the bloodstream from the cell where they are positioned to the target cell containing the correct receptor.

Cells do not spontaneously use glucose; this cellular function and many others are coordinated across your body and triggered. We normally think of internal communication as consisting of electrical impulses traveling along a network of nerves. This is true for the system that triggers movement, breathing, heartbeats, and reflex actions. However, most of the body's messages are conveyed not by electrical impulses, but through chemical processes.

The chemical signals produced by your endocrine glands to regulate systems in your body are called **hormones** (Figure 12.23). These messengers encompass a wide range of functions and a similarly wide range of chemical compositions and structures. One example is *thyroxine*, a hormone secreted by the thyroid gland that is essential for regulating metabolism. The ability of the body to carry a stable quantity of glucose through the blood depends on another hormone known as *insulin*. This hormone, a small protein built from 51 polymerized amino acids, is secreted by the pancreas. Persons who suffer from diabetes are often required to take daily injections of insulin. Another well-known hormone is *adrenaline* (epinephrine), a small molecule that prepares the body for "fight or flight" in the face of danger.

Most hormones remain outside the cell, so they require receptors on the cells they affect. A **receptor** is a specific class of biomolecule generally embedded in the cellular membrane. To respond to extracellular hormones, the receptor must span the membrane so it can sense and respond to the presence of the hormone while also promoting activity within the cell. On the outside of the cell, a membrane receptor is specifically engineered to respond perfectly and selectively to one type of hormone. Through the cellular membrane, the receptor changes slightly to convey the message "epinephrine is here," like a doorbell for the cell. This results in a cascade of events tuned exactly to the purpose of the hormone. The receptor changes in shape, allowing the hormone to adhere but not pass, while still permitting information to be transferred through the membrane. Cells in different organs will have different receptors to tailor their response to specific hormones according to the big-picture role of the organ.

As previously mentioned, the final stages of complete respiration of glucose require oxygen. Since glucose is a universal fuel, the oxygen you inhale must be spread throughout your tissues. Your blood cells carry oxygen molecules from your lungs to your tissues, but they cannot do it without the biomolecule **hemoglobin**. This protein is specifically designed to transport oxygen by binding to an iron-containing organic molecule called *heme*.

How Do Hormones Work?

Kateryna Kon/Shutterstock

Check out this video (www.acs.org/cic) for a summary of how hormones work to regulate a variety of functions in your body.

Typically, the exposure of iron to oxygen gas will result in the formation of iron(III) oxide, Fe_2O_3, known as rust. Why doesn't the iron in your blood oxidize to form rust? The hemoglobin carefully transports the oxygen bound to the biomolecule while preventing the direct reaction of oxygen and iron. In addition, hemoglobin has tuned its binding strength (K_d, revisit Section 12.1) according to the amount of oxygen available. In the lungs, where oxygen concentration is high, the binding strength is very high, so hemoglobin picks up four molecules of oxygen—its maximum capacity. In the tissues where the oxygen concentration is lower, the binding strength falls, so hemoglobin releases all of the oxygen at once. The message to release oxygen is reinforced by changes in pH and the presence of small molecules that are produced by tissues in need of oxygen.

Your Turn 12.17   Extra Protein Structures

The structures of proteins, such as the ones shown in Figure 12.19, have been gathered in a repository online, the Protein Data Bank. Explore the Molecule of the Month archive on the site to find a protein not discussed here. Summarize your discovery. What role does the protein play in biology? What features of the protein are most interesting?

12.7 | Life Driven by Noncovalent Interactions

Learning Objective: Sketch how proteins interact and bind with target molecules

Each scenario described in the previous section required controlled and specific interactions between small molecules and proteins. How can this class of polyamides provide the sensitivity necessary for all of these complex functions? Proteins have the same polyamide backbone that you may find familiar from nylons (Section 9.7), but they also have a diverse variety of amino acids that extend from this backbone structure (Figure 12.24).

Your Turn 12.18   Polyamides: Some Context

List three examples of polyamides that you have explored previously in earlier chapters, and the context of each.

Recall that proteins are polymers composed of amino acid monomers. A diverse range of proteins may be generated from different combinations of the 20 amino acids. Each amino acid has a carboxylic acid (—COOH), an amine (—NH₂), and a side-chain group. The side chain plays no part in the process of condensation polymerization, but interacts with other side chains, as well as the molecules around the protein. Reexamine Figure 11.18 to explore the variety of functional groups and preferred intermolecular forces.

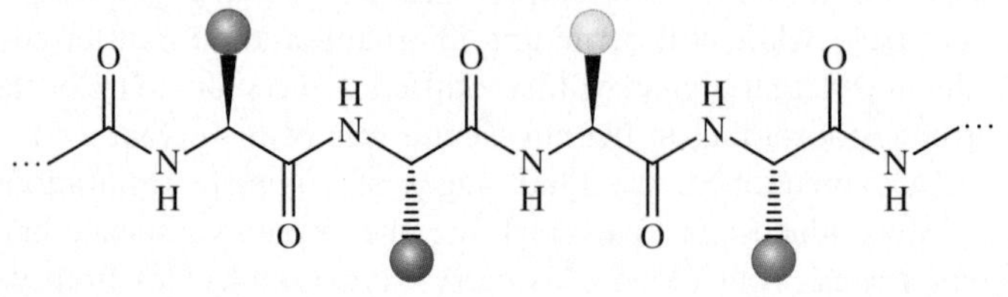

Figure 12.24

A symbolic view of protein diversity. Various amino acid side chains (colored spheres) extend from the polyamide backbone of the protein segment.

Your Turn 12.19 Revisiting Dipeptides

Choose two amino acids of your choice and draw their combination to form a dipeptide. Is your dipeptide polar, nonpolar, or mixed?

Hint: Remember that amines and carboxylic acids react to form the amide functional group.

The Dangers of Synthetic Drugs

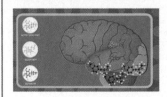

©2018 American Chemical Society

Check out this video to see why synthetic drugs are so dangerous: www.acs.org/cic.

Questions to Consider

a. The video mentioned "Kratom," but didn't describe its origin or effects. Using the Internet as a resource, describe how this drug differs from the other synthetic drugs mentioned in the video.

b. There are medications available that are designed to treat and reverse the effects of an overdose. How do these medications work?

The side groups of amino acids present in large proteins interact with relevant small-molecule targets through a variety of intermolecular forces. The three-dimensional structures of proteins have a limited range of motion, but their overall architecture is relatively well-defined by the amino acid side chains. The result is that small molecules must have complimentary chemical makeup to bind with high affinity and specificity to a receptor protein or within an enzyme active site. For example, the small molecule might be chiral and polar and able to experience dipole-dipole or hydrogen bonding intermolecular forces with the active site of the enzyme. The polarity and chirality of a molecule often influences how the enzyme binds to it over closely related molecules, even those with subtle differences such as optical isomers.

Although the structures of proteins have relatively well-defined shapes, they do have some room for motion. On most proteins, this literal wiggle-room changes the overall shape rather imperceptibly, but the implications for binding to smaller molecules can be quite significant. Three models describe the breadth of possible binding modes, given this limited variable of protein movement. In the *lock-and-key* model (Figure 12.25a), the exact shapes of both molecules are set before the association occurs. Selectivity occurs because the small molecule has the correct functional groups positioned perfectly to interact with the well-defined positions of amino acids in the protein target.

The *induced fit* and *conformational selection* models (Figures 12.25b and 12.25c, respectively) include the concept of motion in either the protein or its interacting partner. Induced fit suggests that the molecules shift into the correct binding position only upon binding. Conformational selection suggests that one of the interacting partners

(a) Lock and key

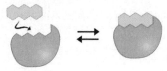

(b) Induced fit

(c) Conformational selection

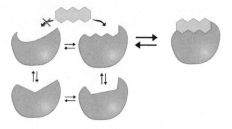

Figure 12.25

A comparison of three binding models in biological interactions.

switches between multiple three-dimensional shapes before the binding event. The binding partner picks out the specific shape necessary to associate correctly. Specificity in these models comes from the fact that only the correct molecule can reach that perfect binding interaction.

Your Turn 12.20   Exploring Interactions

Reexamine the structure of epinephrine in Figure 12.8. Which category of amino acids shown in Figure 11.18 (nonpolar, acidic, basic, or neutral polar) should be associated with each functional group to bind with and respond to adrenaline?

12.8 | Steroids: Essential Regulators for Life (and Performance Manipulators!)

Learning Objective: Recognize the structural features common to steroids and give examples of their biological function

As a key family of compounds, **steroids** are a class of naturally occurring or synthetic fat-soluble organic compounds that share a similar structure. Biologically, they serve as the best illustration of the relationship between form and function. The naturally occurring members of this ubiquitous group of substances include structural cell components, metabolic regulators, and the hormones responsible for secondary sexual characteristics and reproduction. Secondary sexual characteristics are physical characteristics that first appear during puberty and do not have a direct reproductive function. Table 12.4 lists some of the diverse functions of steroids. These compounds have also been used for artificial manipulation of our biology as synthetic drugs for birth control, abortion, bodybuilding, and sports performance enhancement.

Table 12.4	Steroid Functions
Function	**Example Molecules**
Regulation of secondary sexual characteristics	Estradiol (an estrogen), testosterone (an androgen)
Regulation of the female reproductive cycle	Progesterone
Regulation of metabolism	Cortisol
Digestion of fat	Cholic acid
Component of cellular membranes	Cholesterol
Stimulation of muscle and bone growth	Gestrinone, trenbolone

Despite their tremendous range of physiological functions, all steroids are built on the same molecular skeleton. Although this may seem like a risky recipe for loss of specificity, these compounds actually provide a marvelous example of the chemical economy in living systems. The common characteristic of steroids is a molecular framework consisting of 17 carbon atoms arranged in four rings, illustrated below:

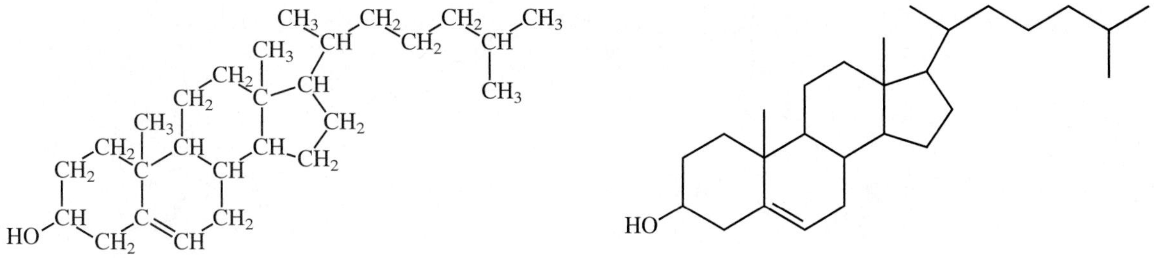

Figure 12.26

Structural representations of cholesterol. Check out the 3D representation of cholesterol at www.acs.org/cic.

Recall that in the above line-angle representation, carbon atoms are assumed to occupy the vertices of the rings but are not explicitly drawn. Also missing are the hydrogen atoms that are present to ensure each carbon atom fulfills the octet rule. The three six-membered carbon rings of the steroid skeleton are designated A, B, and C, and the five-membered ring is designated D. Although the steroid framework appears flat as drawn, it is actually three-dimensional in shape. The dozens of natural and synthetic steroids are all variations of this backbone structure. Some differ only slightly in structural detail but have radically different physiological functions. Extra carbon atoms or functional groups at critical positions on the rings are responsible for these variations.

The steroid *cholesterol* is a major component of cell membranes and is shown in Figure 12.26. The structure on the left includes all the atoms in the molecule; the one on the right gives the skeletal representation using a line-angle drawing.

A careful examination of Figure 12.27 illustrates how subtle molecular differences can result in profoundly altered physiological properties. The difference between a molecule of estradiol and one of testosterone lies only in one of the rings. Each of these forms a family of molecules with even more-subtle chemical changes.

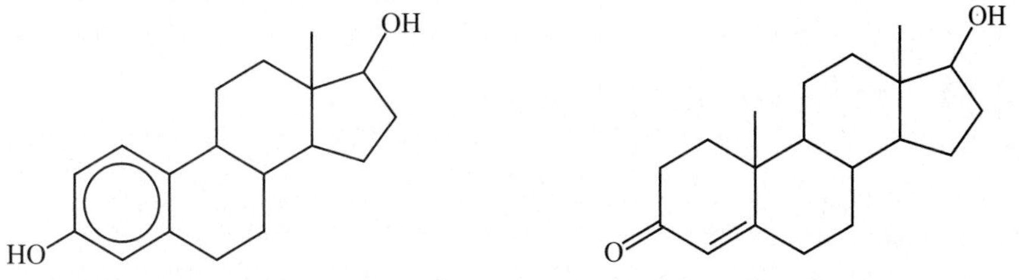

estradiol   testosterone

Figure 12.27

Estradiol and testosterone structures. Check out the 3D renderings for estradiol and testosterone at www.acs.org/cic.

Your Turn 12.21   Estradiol and Testosterone

How are the hormones estradiol and testosterone different in properties and molecular structure?

Your Turn 12.22   Structural Similarities of Steroids

Using the Internet, evaluate the pairs of steroids listed below. Write their chemical formulas and identify some structural similarities in each pair.

a. estradiol and progesterone
b. corticosterone and cortisone
c. cholic acid and cholesterol

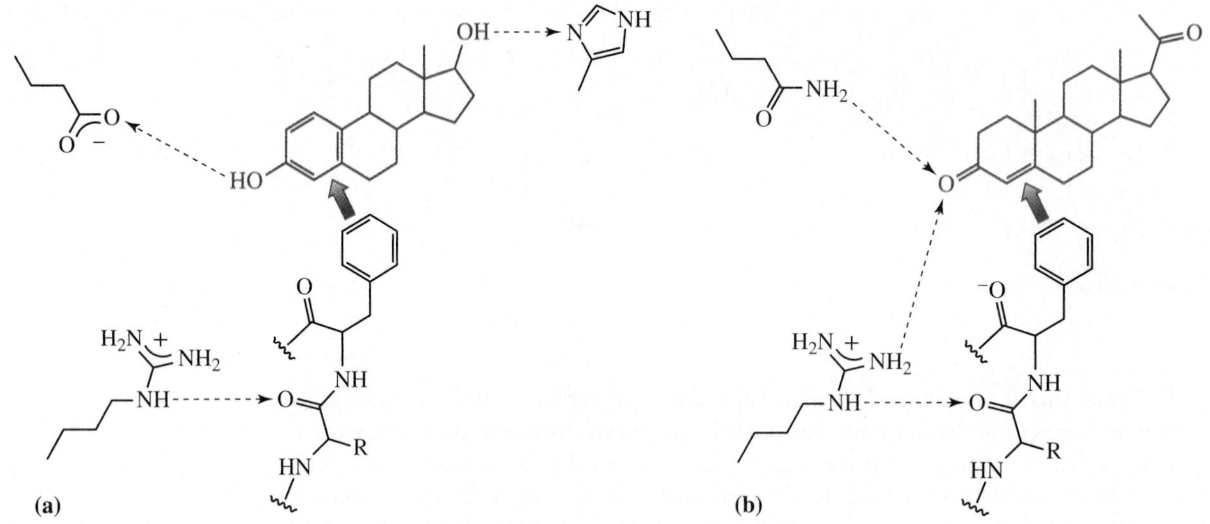

Figure 12.28

Selective receptor binding pockets for estradiol (an estrogen) versus progesterone (an androgen).

Source: (a–b) *National Academy of Sciences*, 1998, 95, 11, Figure A3.

To complete our study of steroids, let's examine how they interact with the proteins that bind them. Figure 12.28 illustrates the difference in binding between an estrogen and an androgen and its target receptor protein. Although there are some similarities between the steroids, their slight differences allow each to more perfectly fit the active site of different receptors. Chemically speaking, differing amino acid residues in the active site result in differences in intermolecular forces, which lead to preferential binding of one molecule relative to another for each protein receptor.

12.9 | Modern Drug Discovery

Learning Objective: Describe the history and development of aspirin and penicillin

Pharmaceuticals are therapeutic substances intended to prevent, moderate, or cure illnesses. **Medicinal chemistry** is the science that involves the discovery or design of new therapeutic chemicals and their development into useful medicines. We are very fortunate that modern medicine has developed drugs that can readily halt bothersome symptoms and cure pathologies. Therapeutic drugs have radically improved the longevity and quality of our lives.

Modern pharmacology has its origins in folklore. The use of herbs, roots, berries, and barks for relief from illnesses can be traced to antiquity, as illustrated in documents recorded by ancient Chinese, Indian, and Near East civilizations. More recently, chemists have designed, synthesized, and characterized numerous prescription and over-the-counter drugs. Today, drugs help patients regulate their blood sugar, blood pressure, cholesterol, and allergies. More recently, when SARS-CoV-2—the virus that causes COVID-19—was first identified, researchers used genetic sequencing and shared coronavirus data to determine the viral sequence of the virus. This allowed scientists to construct mRNA vaccines in record time, which allow our own cells to recognize the proteins on the virus and attack them. The HIV/AIDS epidemic has also brought about new treatment options, including preventive drugs. Other important discoveries have led to drugs that can relieve pain, treat cancer, and even manage mental disorders that once were thought untreatable. How is it that modern chemists are able to expand our medical options so quickly? Learn more about the exciting world of vaccine discovery in the next activity.

Real Talk   Vaccine Discovery

Check out this interview with Dr. Toyin Asojo, who describes her experience designing vaccines for tropical diseases: www.acs.org/cic.

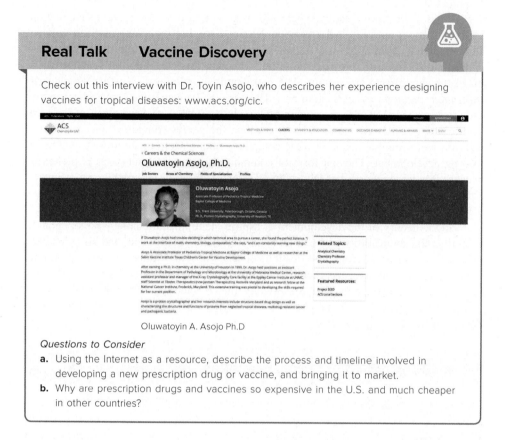

Oluwatoyin A. Asojo Ph.D

Questions to Consider

a. Using the Internet as a resource, describe the process and timeline involved in developing a new prescription drug or vaccine, and bringing it to market.

b. Why are prescription drugs and vaccines so expensive in the U.S. and much cheaper in other countries?

The evolution of willow bark tea to a painkiller, and further structural modifications to this medication, is a fantastic case study in historical and modern drug design. In the 4th century BCE, Hippocrates prepared a tea from boiling willow bark (Figure 12.29) in water. The concoction, common to many different cultures, was said to be effective against fevers. Over the centuries, the folk remedy ultimately led to the synthesis of a true wonder drug, *aspirin*, a drug with the continued potential to impart pain relief to millions of people.

From the 4th century BCE to the 18th century CE, the possible pharmaceutical benefits of willow bark were left unexplored. Edmund Stone (1702–1768), an English clergyman, set the stage for modern explorations with a report to the Royal Society in 1763 on the success of powdered willow bark as a treatment. Chemists were subsequently able to isolate small amounts of yellow, needle-shaped crystals of a substance from the willow bark extract. Because the tree species was *Salix alba*, this new substance was named *salicin*. Experiments showed that salicin could be chemically separated into two compounds; however, only one of these compounds, *salicyl alcohol*, was shown to reduce fevers and inflammation. Once within the body, metabolism converts an alcohol in this active compound into a carboxylic acid and generates the true active ingredient in willow bark tea, *salicylic acid* (Figure 12.30).

Figure 12.29

The white willow tree, *Salix alba*, the original source of aspirin.

Ilya Golubev/Alamy Stock Photo

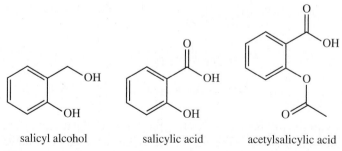

salicyl alcohol   salicylic acid   acetylsalicylic acid

Figure 12.30

Chemical variations to a wonder drug: salicyl alcohol, salicylic acid, and acetylsalicylic acid, a compound more commonly known as aspirin. Check out the 3D renderings for salicyl alcohol, salicylic acid, and acetylsalicylic acid at www.acs.org/cic.

After its discovery, salicylic acid was used to treat pain, fever, and inflammation. Unfortunately, it not only had a very unpleasant taste, but its acidity also led to acute stomach irritation in some individuals. The acidic functional group of salicylic acid created a difficult drug design conundrum. Although it is necessary for function in the body, the side effects are serious. The most intuitive solution is a rather simple *neutralization reaction* (revisit Section 5.9). The acid can be neutralized with a base, either sodium hydroxide (NaOH) or calcium hydroxide (Ca(OH)$_2$), to form a salt of the acid. The resulting salts have fewer side effects than the parent compound. Generating the salt of a promising but acidic or basic compound continues to be an important strategy in drug development. The salt form of common organic acids and bases is preferable because it is less reactive, has less of an odor, and is more water-soluble. An estimated half of all drug molecules used in medicine are administered as salts that improves their water solubility and stability, increasing their shelf life.

In the history of the development of aspirin, neutralization alone was not enough. To further the applicability of the drug, Felix Hoffmann (1868–1946) and his colleagues explored a series of covalent modifications of salicylic acid. Rather than simply neutralizing the carboxylic acid, the chemists modified the structure of salicylic acid by adding an additional carboxylic acid. Specifically, they reacted the carboxylic acid with the alcohol on salicylic acid. Remember that carboxylic acids and alcohols react to form a novel functional group, the ester.

Your Turn 12.23   Ester Formation

Draw structural formulas for the esters that form when these alcohol and acid pairs react.

a. CH_3CH_2OH + (acetic acid structure) $\xrightarrow{H^+}$

b. (propanoic acid structure) + CH_3OH $\xrightarrow{H^+}$

While Hoffmann and his colleagues generated a variety of new esters with different carbon lengths and functional groups, the best hit was one of the simplest. The reaction of salicylic acid with acetic acid produced *acetylsalicylic acid* (Equation 12.11). Although the carboxylic acid is untouched, the ester modification (indicated in yellow) still changes the compound's traits enough to reduce nausea and other adverse side effects greatly.

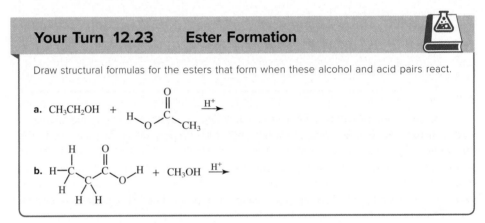

salicylic acid   acetic acid   acetylsalicylic acid   water   [12.11]

Because aspirin retains the —COOH group of the original salicylic acid, it still has some undesirable acidic properties of the parent compound. However, the ester group makes the compound more palatable and less irritating to the stomach lining. Once aspirin reaches the bloodstream, Equation 12.11 is reversed. The ester splits into acetic acid and active salicylic acid; the latter compound exerts its antipyretic (fever-reducing) and analgesic (pain-reducing) properties.

Common Drugs, Common Structural Features

Approximately 40 alternatives to aspirin have been produced—with ibuprofen and acetaminophen being the most familiar. These compounds have similar molecular structures that often share useful physiological properties. Check out this video that shows how the compounds in various pain medications can be distinguished from one another: www.acs.org/cic.

a. Look up the structures for ibuprofen and acetaminophen, and identify two shared and two different structural features with aspirin.
b. Look up the symptoms that each medication is advertised to treat. Do you find any similarities or differences?
c. Look up the top side effects of each medication. Do you find any similarities or differences?

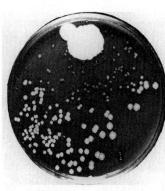
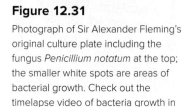

Figure 12.31
Photograph of Sir Alexander Fleming's original culture plate including the fungus *Penicillium notatum* at the top; the smaller white spots are areas of bacterial growth. Check out the timelapse video of bacteria growth in Petri dishes at www.acs.org/cic.
Biophoto Associates/Science Source

Penicillin is another example of a miracle drug whose origin lies in natural sources. Molds had been used for treating infections for 2500 years, although their effects were unpredictable and sometimes toxic. The penicillin story includes an accidental discovery in 1928 by the British bacteriologist Alexander Fleming (1881–1955). Fleming's curiosity was aroused by the chance observation that in a container of bacterial colonies, the area contaminated by the mold *Penicillium notatum*, was largely free of bacteria (Figure 12.31). Spores from the mold, part of an experiment in a nearby lab, drifted into Fleming's laboratory and accidentally contaminated some Petri dishes containing *Staphylococcus* (bacteria) growing on a nutrient medium. He correctly concluded that the mold produced a substance that inhibited bacterial growth, and he named this biologically active material *penicillin*.

@HOME   What Kind of Bacteria?

Investigate different types of bacteria on surfaces around you at home using disposable muffin liners, gelatin, zipper plastic bags, beef bouillon granules, a muffin pan, a measuring spoon, water, sugar, a saucepan, a stovetop, a refrigerator, and cotton swabs.

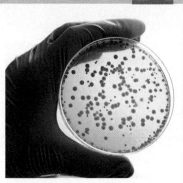
Sompraaong0042/Shutterstock

1. Add 4 packets of gelatin, 4 cups of cold water, 8 teaspoons of sugar, and 4 teaspoons of beef bouillon granules to the saucepan and stir. Turn on the stovetop and bring it to a boil. Stir continuously.
2. Line the muffin tin with disposable muffin liners in each slot.
3. Turn off the stovetop and allow the liquid in the saucepan to cool slightly. Pour the liquid into each muffin liner, so they are half full. Place the muffin tin into the refrigerator to cool until they become solid. Do not touch them. You need to use these within 3 days of making them.
4. Label zipper plastic bags with the samples you plan to test in your home. For example, if you plan to swab your front door knob, label it the front door knob. Then, you will take a cotton swab and gently rub it all over the front door knob. Then, gently rub that same end of the cotton swab onto the solid gel you made in the muffin liner. Then gently lift the liner out of the tray, place it into the zipper plastic bag, and seal. You will place this in a warm area in your home for 3–5 days and make observations later. Repeat for other areas in your home.
5. For disposal, do not open the bags and throw in the trash.

Questions to Consider
a. Did any of the areas of your home surprise you with the number of bacteria or fungi present?
b. What do you think would happen if you added a drop of hand sanitizer to the sample or an antibiotic like penicillin?

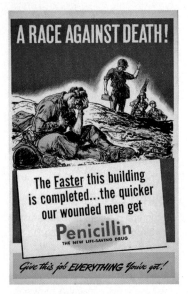

Figure 12.32

A sign posted at the entrance of a new facility for penicillin production during World War II.

Library of Congress Prints and Photographs Division, LC-USZC4-1986

Fleming's experience allowed him to interpret the chance phenomenon, recognizing that some unknown substance produced by the *Penicillium* was a potential antibacterial agent. "The story of penicillin," Fleming wrote, "has a certain romance in it and helps to illustrate the amount of chance, of fortune, of fate, or destiny, call it what you will, in anybody's career." Nevertheless, the discovery would likely not have happened without Fleming's powers of observation and insight.

The process of taking penicillin from the Petri dish to the pharmacy was not much different from what is done today. The first step was a systematic effort to isolate the active agent produced by *Penicillium notatum*. Once identified, the substance had to be purified and concentrated by analytical and chemical techniques. Finally, the efficacy of penicillin in treating humans had to be demonstrated. World War II gave increased impetus to this research and to the development of new methods for preparing large quantities of penicillin. Because the scientists were successful in doing so, thousands of lives were saved during the war (Figure 12.32), and millions since then.

Treatments of infections once considered incurable—pneumonia, scarlet fever, tetanus, gangrene, and syphilis—were revolutionized by Fleming's discovery. The discovery of penicillin may have been serendipitous, but developing the next several generations of antibiotics in this class involved systematic and careful research. Small structural changes to this drug have resulted in more than a dozen different penicillins currently in clinical use, including penicillin G (the original discovered by Fleming and the form that causes an allergic reaction in about 20% of the population), ampicillin, oxacillin, cloxacillin, penicillin O, and amoxicillin (the pink, bubble-gum-flavored concoction you might have been given as a child). Amoxicillin is still available in capsule form and widely prescribed since it is effective against a broad spectrum of bacteria and is usually well tolerated.

Your Turn 12.24   Drugs by Chance

Modern drug discovery methods involve systematic studies of compounds with only small variations in structure and computer modeling, among other techniques. Sometimes side effects of a drug may open the door for its usefulness in treating other illnesses. However, like the discovery of penicillin, there are many examples where a new drug was discovered by "chance." Find an example of a drug discovered by unusual circumstances and describe its discovery.

DID YOU KNOW ?

Overprescribing medication is the main reason for the emergence of antimicrobial resistance. Another might be through the release of medications to the environment.

The effectiveness of penicillin has, unfortunately, led to extreme overuse. As a result, cunning bacteria have developed mechanisms for rendering penicillin (along with other antibiotics) useless. We are now witnessing strains of resistant bacteria, or "superbugs," a phenomenon Fleming predicted back in 1945. Bacteria develop resistance to penicillin by secreting an enzyme that attacks the penicillin molecule before it can act. Some of the newer antibiotics differ in their effectiveness at killing certain bacteria and their susceptibility to the enzymes the organisms produce. Closely related to the penicillins are the cephalosporins (cephalexin or Keflex) that are particularly effective against some resistant strains of bacteria. Careful research on structural modifications has led to other important medicines like cyclosporine, which prevents tissue rejection. Its development made possible the revolutionary success of organ transplant surgery.

Although these examples of aspirin and penicillin have been steeped in history, natural sources and medicinal lore continue to be explored with the aim of discovery. Potentially, each new organic molecule may have a distinct or interesting medicinal purpose—the topic of our final section.

12.10 | New Drugs, New Methods

Learning Objectives:

Describe how medicinal chemists develop new drugs

Explain the structure vs. activity relationship of opioid pain medications

Modern drug discovery is a challenging field because the targets of the drugs can change. Some targets are organisms themselves—like bacteria and viruses—that evolve resistance to our interference methods. Some targets, like cancer, require extreme precision because the biology of cancerous tissue shares many similarities with noncancerous tissue. This section explores the methods and challenges associated with developing new pharmaceuticals.

Drugs can be broadly classified into two groups: those that produce a physiological response in the body and those that inhibit the growth of substances that cause infections. Aspirin, synthetic hormones, and psychologically active drugs (like alcohol, caffeine or heroin) produce a physiological response in the body. These compounds typically initiate or block a chemical action that generates a cellular response, such as a nerve impulse or protein synthesis. The second group of drugs, those that inhibit the growth of substances that cause infections, consists of antibiotics. Antibiotics are drugs that prevent the reproduction of foreign invaders. They do so by inhibiting an essential chemical process in the infecting organism. Thus, they are particularly effective against bacteria.

The basic work of finding and optimizing a new potential pharmaceutical begins with identifying a chemical that has the potential to work against a target disease, either a specific protein in a physiological pathway or a whole organism. The first step typically involves computer modeling to examine the 3D structure of target proteins and to predict the binding affinity of a variety of potential drug compounds for the target protein. Experimental drug testing is then carried out to examine how well a drug performs, meaning how well the drug inhibits the target protein.

Another method of optimizing potential pharmaceuticals consists of modifying cells that display known traits of specific problematic diseases. Scientists are able to generate entire organisms, from simple worms to zebrafish and mice, in both healthy and diseased states to learn how they compare and contrast. Through these efforts, scientists are able to study cancer and neurodegenerative disorders, such as Huntington's and Alzheimer's. Against these model diseases, researchers can test potential compounds that may work to destroy the disease and they can also examine side effects of the medications.

The number and type of potential compounds that are examined depend greatly on our biochemical knowledge of the structure and function of the target system. Pharmaceutical companies and national research institutes house large libraries of chemicals that can be used for processes in drug development. The chemical library database contains information on chemical compounds such as structure, purity and quantity.

Real Talk   Real-World Drug Design

Check out these videos that describe real-world modern approaches to drug design: www.acs.org/cic.

Questions to Consider

a. In the video: "How Does Mark Noe Discover New Medicines?," Dr. Noe states that creating a medicine is like solving a complex jigsaw puzzle? What makes the process puzzling?

b. How do medicinal chemists use protein crystallography in the discovery of medicines?

Luis Alvarez/DigitalVision/Getty Images

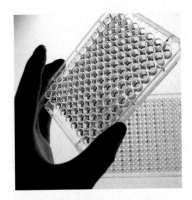
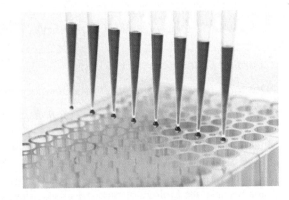

Figure 12.33

High-throughput screening of new drug candidates.

(left): Tek Image/Science Source; *(right)*: dra_Schwartz/E+/Getty Images

An important strategy in designing a drug is to determine its **pharmacophore**, the three-dimensional arrangement of atoms or groups of atoms responsible for the biological activity of a drug molecule. Medicinal chemists then synthesize a molecule containing the specific active fragment within a much simpler, nonactive framework to meet the requirements of the receptor site. With each round of chemical design, chemists can perform further activity tests against model proteins, cells, and organisms (Figure 12.33). The process of systematically changing the structure of a drug molecule and assessing its change in activity is known as a **structure–activity relationship (SAR) study**. The chemist aims for a compound that best fits the target site while minimizing any interactions with other systems. However, it is important for the chemist to evaluate how the human body responds to the chemical compound, such as how the body metabolizes or breaks down the drug.

Your Turn 12.25   What Should a Drug Be Like?

Make a list of the properties you think a drug should have. Then, compare your list with others. Note similarities and differences.

a. Describe the motivation for the items on your list.
b. Are any items missing from your list that you should now include?
c. Are any items present on your list that you should now delete?

An outstanding example of a SAR study is provided by opiate drugs such as morphine. Morphine, a complex molecule, is difficult to synthesize. However, the pharmacophore responsible for opiate activity has been identified and is highlighted in Figure 12.34. The flat benzene ring (the hexagon with a circle inside it) of the

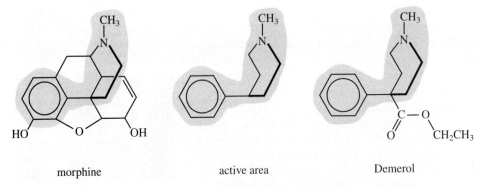

Figure 12.34

Molecular structures of morphine and meperidine (Demerol). The highlighted "active areas," or pharmacophores, are the portions of the molecule that interact with the receptor. The darker lines indicate that these bonds are in front of the others or are coming out of the plane of the page.

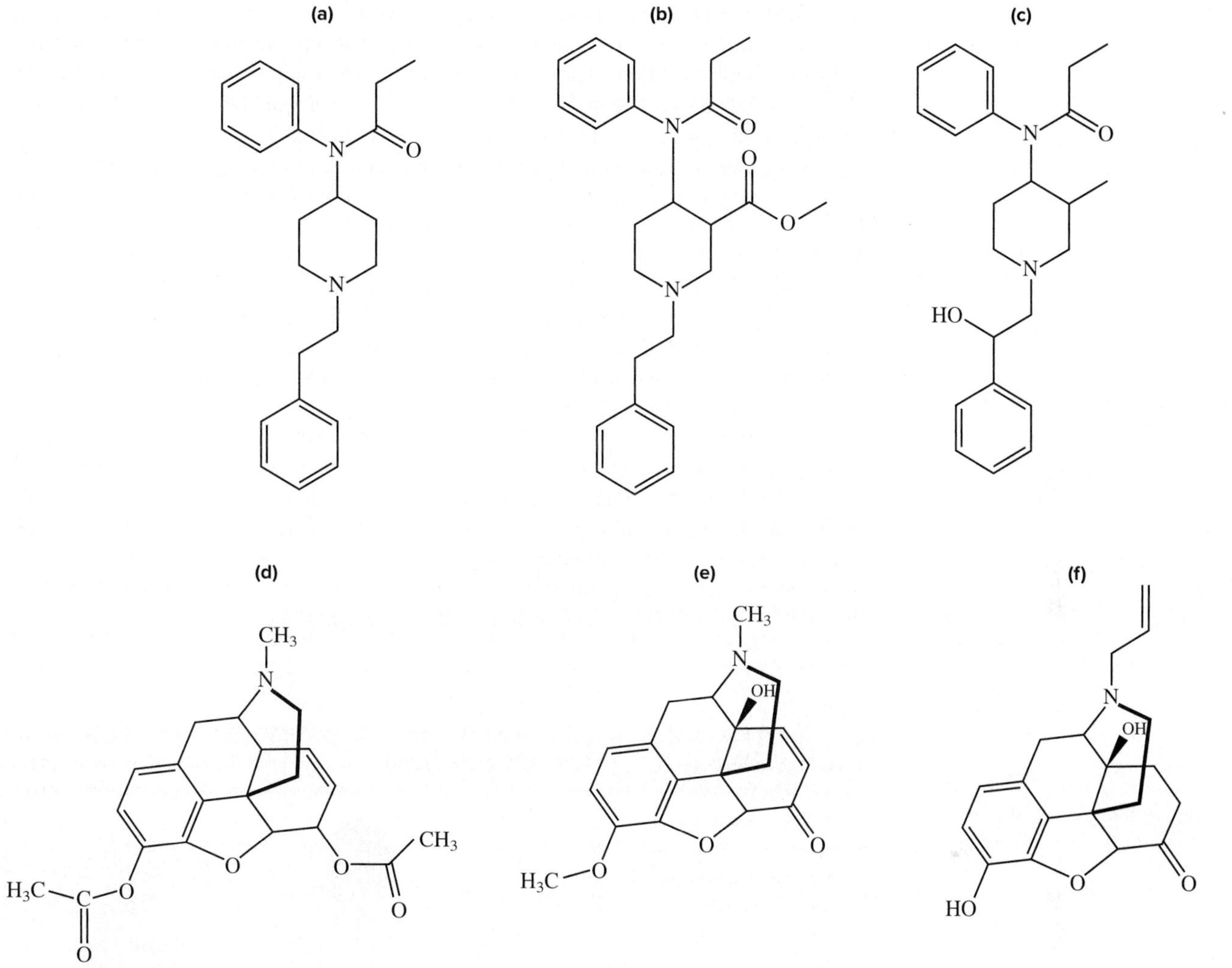

Figure 12.35

Molecular structures of common opioid drugs. Shown are a) fentanyl, b) carfentanyl, c) ohmefentanyl, d) heroin, e) oxycodone, f) naloxone (Narcan).

pharmacophore fits into a corresponding flat area of the receptor, while the nitrogen atom of the pharmacophore binds the drug molecule to the site. Incorporating this particular portion into other less-complex molecules, such as meperidine (more commonly known by its brand name, Demerol), creates opiate activity. Demerol is much less addictive than morphine but also less potent.

Another powerful synthetic opioid that is up to 100 times more potent than morphine is *fentanyl*. It is both a prescribed drug, as well as one that has been made and used illegally, contributing to recent increases in synthetic opiod overdose deaths. According to the Drug Enforcement Administration (DEA), fentanyl is one of the country's deadliest drug threats. As shown in Figure 12.35, the structures of fentanyl and its derivatives are quite different than other opioids such as morphine, heroin, and oxycodone. This structural simplicity of fentanyl may be responsible for its high potency; the organic backbone may more easily fit into active pocket of the opioid receptor sites located in the brain.

As a testament to how small structural changes result in drastic differences in physiological function, consider some derivatives of fentanyl. Fentanyl is 50 times stronger than heroin, and 100 times stronger than morphine. Carfentanyl is a derivative that is 20 times stronger than fentanyl (i.e., 1000 times stronger than heroin). As shown in Figure 12.35, the only difference between fentanyl and carfentanyl is a —$C(O)OCH_3$ ester functional group. However, if you think carfentanyl is bad, consider ohmefentanyl—perhaps best pronounced "oh my" fentanyl due to its potency. With the simple addition of a hydroxyl (—OH) group and a methyl (—CH_3) instead of —H in fentanyl, carfentanyl is 126 times stronger than fentanyl (i.e., 6300 times more potent than heroin)! An extremely scary derivative, indeed.

How can we rationalize such a dramatic increase in strength from such a small change in structure? Recall our discussion of intermolecular forces. Whereas fentanyl primarily binds to the receptor sites via London dispersion forces from the organic backbone, it can also hydrogen bond through the —N groups, and bind via dipole-dipole interactions through the carbonyl, C═O, group. The addition of an ester group, —$C(O)OCH_3$, adds the possibility of more dipole–dipole interactions that are possible in carfentanyl. For ohmefentanyl, the addition of a hydroxyl, —OH, group allows for stronger hydrogen bonding, which likely allows for much tighter bonding to receptor sites and overall potency.

To reverse the effects of an overdose, naloxone (Narcan) may be administered, which results in immediate, though often short-term, relief. As observed in Figure 12.35, naloxone exhibits the same structural backbone as other opioids such as heroin, morphine, and oxycodone. However, why does naloxone preferentially bind to the opioid receptor sites in the brain, effectively replacing the bound opioid drug? Most likely, this is due to the presence of more hydroxyl (—OH) groups in naloxone, which leads to stronger hydrogen bonding interactions with the receptor sites.

Let's Debate! The Opioid Crisis

Check out the interview with Dr. Chris Gill, who describes the dangers of opioid drugs and his work on developing drug testing methods: www.acs.org/cic.

Are we able to treat pain using virtual reality (VR) instead of using addictive opioid-based drugs? Check out this interview with Dr. Derek Fitzgerald who discusses this possibility: www.acs.org/cic.

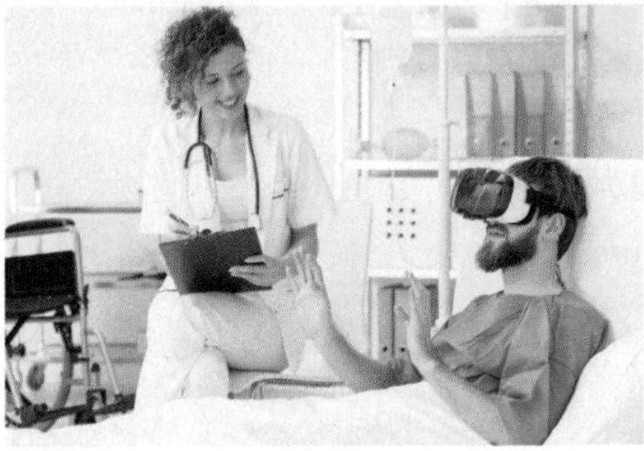

KatarzynaBialasiewicz/iStock/Getty Images

Form two groups: one group will be in favor of the topic (the proposition) and the other group will be against the topic (the opposition). Both groups will try to persuade a neutral person or judge to agree with them. The topic of the debate is the motion. Here is the motion:

> *The opioid epidemic is the extreme overuse of opioid medications, both from medicinal prescriptions and from illegal sources. It has led to a substantial increase in overdose death rates. The majority of heroin users started their addictions by taking prescription pain medications, such as oxycontin, which were often medically prescribed and found at home. To resolve this problem, opioid-based pain medications should be banned and no longer medically prescribed. Other pain management alternatives, which are just as effective without the dangerous side effects, should be used to remediate pain.*

Curiosity Conversations

- Would banning opioids help with existing cases of opioid addiction?
- How would banning opioids in the U.S. affect the illegal drug trade of heroin, fentanyl, and other opioids worldwide?

- How might banning prescription opioid pain medications impact people who rely on them for chronic pain management? What are some negative consequences of banning opioid medications?
- What other alternatives (medications and other treatments) could be used for pain remediation? How effective are these treatments relative to opioid-based medications?
- As Dr. Gill mentions in his interview, some countries such as Canada supply free test kits, Narcan (to treat an overdose), and equipment (pipes, instructions, etc.) to drug users in order to help maintain a safe supply and prevent deaths. Is this helping the opioid pandemic, or serving to enable prolonged drug use? What are some benefits and disadvantages of this approach?

There is no question that theoretical modeling, which employs sophisticated computer programs, has revolutionized the world of drug design. Researchers are now able to model potential drugs and receptor sites in a fraction of the time once required by isolation and synthesis in the laboratory. These 3D representations allow medicinal chemists to "see" how drugs interact with a receptor site from all angles. Computers can then be used to search for compounds that have structures similar to that of an active drug and can modify structures to visualize how the new compounds will function. Discover the 3D structures of various common prescription drugs in this next activity.

Your Turn 12.26   3D Drugs

See for yourself the shapes of drug molecules by visiting the MolView website (www.molview.org). You can rotate three-dimensional representations of best-selling drugs such as Lyrica, Lipitor, and Xarelto on-screen.

a. Select several drugs and examine their three-dimensional structure. How do these computer representations differ from the structural formulas of drugs shown in this chapter?

b. What advantages do computer representations have over two-dimensional drawings? What are their limitations compared with "real" molecules? Are there any disadvantages?

Outside of classic drug development, medicinal chemistry has significantly improved medical imaging. Although chemotherapy is vital in treating cancer, the best and most successful course of treatment is still surgical. We require high-resolution methods to image the detailed inner workings of our systems and detect the exact position of problem areas, ideally before we go under a knife.

Doctors regularly use radioactive iodide to image and later treat thyroid cancer. This process works due to a few key dependable chemical processes. First, the thyroid must collect and concentrate iodide for use in the synthesis of the hormone thyroxin. We can depend on our thyroid to respond to the ingestion of iodide by pulling it from the blood into the gland. Fortunately, we cannot depend on the thyroid to differentiate between isotopes of iodide due to their very similar molar masses. Second, the radioactive iodide dependably and predictably decays on a known timescale measured by its half-life. Upon decay, the release of high-energy particles will be detectable through the tissues and skin. These traits, in combination, result in fantastic images of problematic thyroid glands.

Your Turn 12.27   Isotopes and Ionizing Radiation

Review the types of radiation you have seen in previous chapters or on the Internet: alpha, beta, gamma, or positron emission. Which would be best for detection through tissues and skin? Which is the worst? Explain your choices.

Our era of medicine is unique in the use of nuclear isotopes for both treatment and imaging, and the difference between detection and damage is rather fine. The idea itself is not novel. Pierre and Marie Curie hypothesized that the ionizing rays of radiation would be medically beneficial; sadly, both Marie and their daughter Irene died from diseases likely caused by their unprotected handling of these risky materials. The same radiation that allowed us to detect the thyroid, and the boundaries of a tumor within it, can instead ionize compounds within the tissues. These ionization events create reactive free radicals within biological systems and may irreversibly alter the cells. As a natural process, damaged cells should undergo *apoptosis* or programmed cell death. However, too much damage can disrupt the apoptosis pathways and lead instead to uncontrolled cancerous growth. Under controlled circumstances, selective and targeted use of ionizing radiation is useful to promote the death of cancerous cells.

Your Turn 12.28   Treatment After Terror

Imagine a train carrying waste from a nearby nuclear plant that recently crashed in your city or town. You know from Chapter 7 that the area is likely contaminated with radioactive I-131.

a. What simple salt can you take immediately to prevent damage to your thyroid? (**Hint:** Revisit Chapter 7)
b. What symptoms in the population should medical professionals be watchful for?

Unfortunately, most tissues will not conveniently and automatically concentrate their own imaging agents the way that iodide collects in the thyroid. Instead, traditional medicinal chemistry tools must be applied. **Radiopharmaceuticals** are organic molecules carrying radioactive isotopes that are optimized to create contrast between tissue areas. Through the development of selective and targeted pharmaceuticals, we can bring the sensitivity of thyroid imaging and treatment to a broader range of medical concerns.

Your Turn 12.29   Group Activity: The "Big Question" Revisited

Form a group and consider the introductory video again, then answer the questions below based on your current knowledge.

a. Considering the principles of equilibrium, evaluate the potential consequences of low iodide concentrations.
b. Describe the role of the thyroid and the consequences of its malfunction for Megan.
c. Based on your understanding of drug design, what variables might you consider for Megan's hormone replacement therapy?

Conclusion

Medicinal chemistry as a field has become rather complicated. Gone are the days of "magic bullets" against foreign invaders. Our medicinal targets now require a nuanced understanding of you—your body, your physiology, and every detailed tangled function of your cells, proteins, and biomolecules. Although our understanding of tissue biochemistry, molecular function, and genetic diversity has seen dramatic and sudden improvements in recent history, new treatment options will appear only through integrated, interdisciplinary teamwork. Eric Lander, a principal leader of the Human Genome Project and former presidential advisor on science and technology, gave the following advice regarding the recent avalanche of data on human systems: "It's just a parts list. And if you have a parts list for a 747, that doesn't mean you can fly the plane. It doesn't mean you can build the plane. It doesn't even mean you know what the parts do. It just means you have a list of all the parts." These comments pertained to the complexity of understanding the human genetic code, the topic of our next chapter. However, this advice is nonetheless relevant across fields that are working to bring about a bright new medical future. We have a wonderful, promising, though possibly overwhelming, parts list, and only through interdisciplinary teamwork across fields can we hope to influence and change the future of health and medicine positively.

LEARNING OUTCOMES

The numbers in parentheses indicate the sections within the chapter where these outcomes were discussed.

Having studied this chapter, you should now be able to:

- describe and exemplify a system that is in equilibrium (12.1)

- relate reaction quotients to equilibrium constants and write equilibrium expressions (12.1)

- calculate the pH of buffers and describe how they are used in solutions to maintain a constant pH (12.2)

- illustrate the structures of hydrocarbon isomers using a variety of structural representations (12.3)

- recognize various functional groups commonly found in organic molecules and describe their effect on physical properties such as solubility (12.4)

- define chirality and its importance in biological activity (12.5)

- describe how proteins are used to process glucose for energy, route chemical signals between cells, and transport oxygen through the blood (12.6)

- sketch how proteins interact and bind with target molecules (12.7)

- recognize the features common to steroids and give examples of their biological function (12.8)

- describe the history and development of aspirin and penicillin (12.9)

- describe how medicinal chemists develop new drugs (12.10)

- explain the structure vs. activity relationship of opioid pain medications (12.10)

Questions

Emphasizing Essentials

1. The field of chemistry has many subdisciplines.

   a. What do biochemists study?

   b. What do organic chemists study?

2. As seen in Section 11.5, high-fructose corn syrup (HFCS) is produced from the partial conversion of glucose into fructose. The equilibrium constant of this reaction is

   $$K = 0.74 \frac{[\text{fructose}]}{[\text{glucose}]}$$

   a. Is glucose or fructose favored in this reaction?

   b. In a system at equilibrium if [glucose] = 0.22 mM, what must the [fructose] be?

3. Assume that you have a simple reaction of A $\rightleftharpoons$ B at equilibrium. Use your understanding of the reaction

quotient, Q, and equilibrium constant, K, to explain the result of each of the following.

   a. The amount of A decreases by half.

   b. The amounts of both A and B are doubled.

4. Write the equilibrium constant for the dissociation shown in Figure 12.3.

5. Nitrous acid (HNO_2) has a K_a value of 4.0×10^{-4}, which corresponds to a pK_a of 3.39.

   a. What salt would you use to prepare a good buffer solution with nitrous acid?

   b. If NaOH were added to the buffer solution, what would it react with?

   c. Compare how the pH would change if 0.05 mole of NaOH was added to the buffer solution compared to a container with just water.

6. Which is the best acid to use to make a buffer solution with a pH = 4.70. Acetic acid (pK_a = 4.74); chlorous acid (pK_a = 1.95) or formic acid (pK_a = 3.74)?

7. Use the Henderson–Hasselbalch equation and Table 12.1 to calculate the pH of these solutions:

   a. 0.05-M formic acid and 0.1-M sodium formate.

   b. 0.2-M ammonium chloride and 0.1-M aqueous ammonia.

   c. 0.1-M acetic acid and 0.1-M sodium acetate.

8. Write the structural formula and line-angle drawings for each different isomer of C_6H_{14}.
   Hint: Watch out for duplicate structures.

9. Consider the isomers of C_4H_{10}. How many different isomers could be formed by replacing a single hydrogen atom with an —OH group? Draw the structural formula for each.

10. For each compound, identify the functional group present.

   a. CH_3 —O— CH_3

   b. CH_3CH_2 —C(=O)—O—H

   c. CH_3CH_2 —C(=O)— CH_3

   d. CH_3CH_2 —C(=O)— NH_2

   e. CH_3CH_2 —C(=O)—OCH_3

11. Draw the simplest compound containing each of these functional groups. In some cases, only one carbon atom is required; in other cases, two.

   a. an alcohol

   b. an ether

   c. an ester

   d. a carboxylic acid

   e. an aldehyde

   f. a ketone

12. For each of these, identify the functional group. Then, draw an isomer that contains a different functional group.

   a. CH_3CH_2 —OH

   b. CH_3CH_2 —C(=O)—H

   c. CH_3CH_2 —C(=O)—OCH_3

13. Histamine is a vital, often annoying, part of your body's immune response. It causes runny noses, red eyes, and other symptoms. Here is its structural formula:

   a. Give the chemical formula for this compound.

   b. Circle the amine functional groups in histamine.

   c. Which part (or parts) of the molecule make the compound water-soluble?

14. Estradiol is relatively insoluble in water but readily soluble in most organic solvents. Explain this solubility behavior based on its structural formula.
   Hint: See Figure 12.27.

15. Usually, carbon forms four covalent bonds, nitrogen three, oxygen two, and hydrogen only one bond. Use this information to draw structural formulas for:

   a. A compound that contains one carbon atom, one nitrogen atom, and as many hydrogen atoms as needed.

   b. A compound that contains one carbon atom, one oxygen atom, and as many hydrogen atoms as needed.

16. Which of these molecules has chiral forms?

   a. CH_3 —C(NH$_2$)(OH)— CH_3

   b. H —C(OH)(CH$_3$)— CO_2H

   c. CH_3 —C(NH$_2$)(C≡N)— CO_2H

   d. CH_3 —C(OH)(CH$_3$)— CO_2H

17. Which of these molecules has chiral forms?

   a. CH_3 —C(NH$_2$)(OH)— CH_2CH_3

   b. H —C(OH)(H)— C_2H_5

   c. CH_3 —C(NH$_2$)(CH$_2$OH)— CO_2H

   d. CH_3 —C(OH)(CH$_2$SH)— CO_2H

18. Define and relate the two terms: hormone and receptor. Can one function without the other?

19. Refer to Figure 11.18. Select two examples of amino acids with side chains that are categorized as nonpolar. Describe what characteristics make them nonpolar.

20. Refer to Figure 11.18. Select two examples of amino acids with side chains that are categorized as acidic. Describe what characteristics make them acidic.

21. Molecules as diverse as cholesterol, sex hormones, and cortisone contain common structural elements. Use a line-angle drawing to show the structure they share.

22. Ester formation reactions were vital in the discovery of aspirin. Draw structural formulas for the esters formed when acetic acid reacts with these alcohols:

    a. *n*-propanol, $CH_3CH_2CH_2OH$

    b. isopropanol, $(CH_3)_2CHOH$

    c. *tert*-butanol, $(CH_3)_3COH$

23. Identify the functional groups in morphine and meperidine. Can these molecules be assigned to a particular compound class (i.e., alcohol, ketone, or amine)? Explain.
    Hint: See Table 12.3 for structural formulas.

24. What is meant by the term *pharmacophore*?

25. Sulfanilamide is the simplest sulfa drug, a type of antibiotic. It appears to act against bacteria by replacing para-aminobenzoic acid with sulfanilamide, an essential nutrient for bacteria. Use these structural formulas to explain why this substitution is likely to occur:

sulfanilamide

para-aminobenzoic acid

Concentrating on Concepts

26. Explain why an equilibrium constant cannot tell how quickly a reaction will turn into a product.

27. Use the information in Table 12.1 to redraw Figure 12.5 for these acids and bases:

    a. acetic acid

    b. ammonia

    c. carbonic acid
    Hint: Carbonic acid can be deprotonated twice!

28. Draw structural formulas for each of these molecules, and determine the number and type of bonds (single, double, or triple) for each carbon atom.

    a. H_3CCN (acetonitrile, used to make a type of plastic)

    b. $H_2NC(O)NH_2$ (urea, an important fertilizer)

    c. H_6H_5COOH (benzoic acid, a food preservative)

29. In Your Turn 12.12, you were asked to draw structural formulas for the three isomers of C_5H_{12}. One student submitted this set with a note saying that six isomers had been found. Help this student see why some of the answers are incorrect.
    Note: The hydrogen atoms have been omitted for clarity.

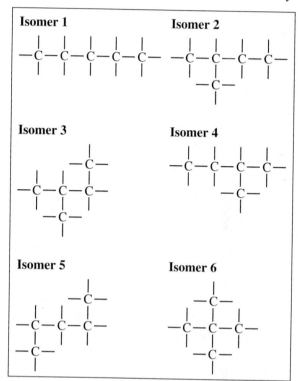

30. Styrene, $C_6H_5CH_5CH_2$, the monomer for polystyrene described in Section 9.5, contains the phenyl group, $-C_6H_5$. Draw structural formulas to show that this molecule, like benzene, has resonance structures.

31. Antihistamines are widely used drugs for treating symptoms of allergies caused by reactions to histamine compounds. This class of drug competes with histamine, occupying receptor sites on cells normally occupied by histamine. Here is the structure for a particular antihistamine:

    a. Give the chemical formula for this compound.

    b. What similarities do you see between this structure and that of histamine (shown in Question 12) that would allow the antihistamine to bind to the same receptor as histamine?

32. Explain why the type of functional groups on a substrate can be important for specific binding to enzymes.

33. Explain why specific positions of functional groups on a substrate can be important for specific binding to enzymes.
Hint: Revisit Figure 12.14.

34. Pair the three categories of macronutrients from Chapter 11 with the following roles in biology:

    a. Cellular membrane

    b. Energy storage as glycogen

    c. Enzymes in metabolism

35. Redraw Figure 12.4, showing the relative barrier for a process with and without an enzyme.

36. Figure 12.23 represents chemical communication within the body. Write a paragraph explaining what this figure means to you in helping to explain chemical communication.

37. Describe the lock-and-key analogy for the interaction between drugs and receptor sites. Use the analogy in a discussion as if you were explaining this to a friend.

38. The lock-and-key analogy for binding clearly references a common everyday object. Create a similar visual analogy for the induced-fit model of binding.

39. As stated in the chapter: "Drugs can be broadly classed into two groups: those that produce a physiological response in the body and those that inhibit the growth of substances that cause infections." Into which class does each of the following drugs fall?

    a. aspirin

    b. estrogen

    c. (Keflex) antibiotic

    d. penicillin

    e. morphine

40. The text states that some racemic mixtures contain chemicals that are effective against disease, others are ineffective but harmless, and still others are potentially harmful. What methods in the text might help determine which of these three categories a recently discovered substance fits?

41. Consider the structure of morphine in Figure 12.34. Codeine, another strong analgesic with narcotic action, has a very similar structure in which the —OH group attached to the benzene ring is replaced by an —OCH₃ group.

    a. Draw the structural formula for codeine and label its functional groups.

    b. The analgesic action of codeine is only about 20% as effective as morphine. However, codeine is less addictive than morphine. Is this enough evidence to conclude that the replacement of —OH groups with —OCH₃ groups in this class of drugs will always change the properties in this way? Explain.

42. Dopamine is found naturally in the brain. The drug L-DOPA is found to be effective against the tremors and muscular rigidity associated with Parkinson's disease. Identify the chiral carbon in L-DOPA, and comment on why L-DOPA is effective, whereas D-DOPA is not.
Hint: Revisit Figure 12.16 for the structure of L-DOPA.

Exploring Extensions

43. Return to Figure 11.18 to examine the side chains of the amino acids lysine and aspartic acid. Propose how a pH buffer would affect the behavior of these side chains.

44. Before the cyclic structure of benzene was determined (see Figure 12.9), there was a great deal of controversy about how the atoms in this compound were arranged.

    a. Count valence electrons for C and H in C_6H_6. Then, draw the structural formula for a possible linear isomer.

    b. Give the condensed structural formula for your answer in part a.

    c. Compare your structure with those drawn by classmates. Are they all the same? Why or why not?

45. Thalidomide was first marketed in Europe in the late 1950s. It was used as a sleeping pill and to treat morning sickness during pregnancy. At that time, it was not known to cause any adverse effects. By the late 1960s, however, the drug was banned. Use the Internet to gather information to write a short paper that describes the optical isomers of thalidomide, why the drug was banned, and why the FDA did not approve thalidomide for use in the United States until recently. For what purpose has the FDA approved the use of thalidomide?

46. Use the lock-and-key model discussed in Section 12.7 to explain why individuals who suffer from lactose intolerance can digest sugars such as sucrose and maltose but not lactose. Use the resources on the Internet to find the structure of lactose.

47. This chapter presented example proteins in important biological roles, including the metabolic enzyme phosphoglycerate kinase, in Section 12.6. Explore the Molecule of the Month archive of the Protein Data Bank to find another enzyme involved in metabolism. Summarize your discovery. What role does the protein play in biology? What features of the protein are most interesting?

48. Merck and Codexis dramatically improved their synthetic route to sitagliptin, a treatment for type 2 diabetes, through the use of enzymes. The work won a 2010 Presidential Green Chemistry Challenge Award. How does this process differ from the earlier one for manufacturing sitagliptin? Do the changes match your understanding of enzymes? Write a brief report on your research, citing your sources.

49. The steroids in our body require carrier molecules to transport them through our blood to different organs. Examine the structures of steroid molecules in Section 12.8 to propose why steroids require assistance to travel in the blood.

50. Danco Laboratories, the U.S. company that produces RU-486 (mifeprex, the "abortion pill"), makes claims about the safety of this steroidal drug, including a comparison to aspirin. Do some research on RU-486 and write a short report on the drug. Include its structure, mode of action, and safety record.

51. One avenue for successful drug discovery is to use the initial drug as a prototype for developing other similar compounds called analogs. Cyclosporine, a major anti-rejection drug used in organ transplant surgery, is considered an example of a drug discovered in this way. Research the discovery of this drug to verify this statement. Write a brief report describing your findings, citing your sources.

52. Dorothy Crowfoot Hodgkin first determined the structure of a naturally occurring penicillin compound. What in her background prepared her to make this discovery? Write a short report on the results of your findings, citing your sources.

53. The antibiotic ciprofloxacin hydrochloride (Cipro) treats bacterial infections in many different body parts. This drug made headlines in 2001 for use in patients who had been exposed to the inhaled form of anthrax. Use the Internet or another source to obtain the structure of Cipro. Draw its structure and identify the functional groups.

54. The doctor told a friend who suffers from heart disease to take one aspirin tablet daily. To save money, your friend often buys a large 300-tablet bottle of aspirin. You, on the other hand, rarely take aspirin but cannot pass up a good bargain. You also buy a large bottle.

   a. Why is the "giant economy-size" bottle of aspirin not as good a deal for you as it is for your friend?

   b. What chemical evidence supports your opinion?

55. Is it dangerous to consume drugs past their expiration date? Explain what happens to pharmaceuticals over time and how expiration dates are chosen.

56. Explain the differences in FDA regulations for pharmaceutical drugs relative to dietary supplements? Are you in agreement with these regulations?

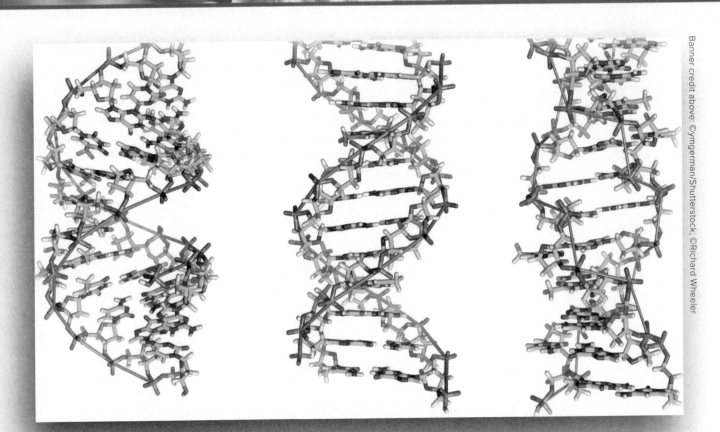

Banner credit above: ©ymgerman/Shutterstock; ©Richard Wheeler

GENETIC TRAITS

Watch the chapter-opening video at www.acs.org/cic, and list five physical traits that you believe are genetically inherited. Survey at least ten people in your family or class to determine which traits are found in these individuals. What similarities do you find? What differences do you find? For those people related to you, are all of the traits the same? Are all of the traits different for those not related to you?

In this chapter, you will explore the following questions:

1. What are genomes?
2. What is the chemical composition of DNA?
3. How does the cell replicate genetic information?
4. How are proteins synthesized from amino acids?
5. How can small changes in protein structure lead to disease?
6. How do genetic engineering techniques such as CRISPR work?
7. What are some applications of genetic engineering?
8. Why is the production and use of GMOs controversial?

A DNA sculpture becomes an intersection of art, science, and play at the Lawrence Hall of Science in Berkeley, California.

Susan Pease/Alamy Stock Photo

> *No branch of science has created more acute or more subtle and interesting ethical dilemmas than genetics. . . . It is genetics that makes us recall, not simply our responsibilities to the world and to one another, but our responsibilities for how people will be in the future. For the first time we can begin to determine not simply who will live and who will die, but what all those in the future will be like.*
>
> *—Justine Burley and John Harris, Eds.,* Companion to Genethics, *2002.*

Introduction

Do you know someone who has diabetes? It is one of the oldest-known diseases, with historical evidence of its occurrence dating back to 1500 BCE. People with diabetes have an imbalance in a hormone called **insulin** that is produced in the pancreas. As a result, their bodies have difficulty regulating blood sugar. So much so that people with diabetes excrete sugar in their urine; one way to test for this thousands of years ago was to pour urine on the ground to see if it attracted ants. Some people are born with diabetes (called type 1), whereas others develop it at an older age (type 2). In either case, diabetes can be a deadly disease if not treated and controlled.

Only recently have effective treatments for diabetes been developed. Well into the 1900s, the only treatment doctors could offer children diagnosed with type 1 diabetes was a starvation diet, which usually did not help extend the children's life expectancy beyond their teenage years. In the early 1920s, Canadian researchers confirmed the role of the insulin hormone in regulating blood sugar. Within a few months, the first children in a diabetes ward at Toronto General Hospital were injected with insulin extracted from the pancreases of calf fetuses. The results were impressive when the bovine insulin was sufficiently purified and given to afflicted children in appropriate doses. Comatose children on the brink of death regained consciousness; with careful monitoring and ongoing treatment, the children could live long lives. The first human to be injected with insulin was Leonard Thompson, a 14-year-old boy who suffered from late stages of type I diabetes in 1922. After insulin treatments, Thomson lived another 13 years before dying from pneumonia at 27.

Soon, the pharmaceutical company Eli Lilly and Company purchased the patent from the University of Toronto and took over insulin production on a mass scale. The proximity of the Indianapolis-based company to the slaughterhouses in Chicago allowed access to an abundance of cow and pig pancreases, and soon insulin was on sale in the United States. Great care had to be taken in purifying the insulin from animal sources because impurities could lead to severe allergic reactions in the patients. Although similar to human insulin, the hormone derived from cow and pig pancreases is not identical, and some diabetes patients suffered from inflammation around the injection site.

Millions worldwide rely on daily insulin doses to control their diabetes and allow them to be active and healthy. However, most insulin today is no longer derived from animal sources. Instead, bacteria have been *genetically engineered* to produce human insulin. This chapter is about the chemistry of life by discussing the fundamental building blocks of genetics—the system of molecules that control the traits of living organisms.

13.1 | A Route to Synthetic Insulin

> *Learning Objective: Understand that the genome contains a code of instructions for life*

For over 50 years, animal-derived insulin was the standard treatment for diabetes patients. Then, in 1978, scientists at the biotechnology company Genentech and the City of Hope National Medical Center in California developed a process to synthesize mass quantities of human insulin with utmost purity. In a process that remains the standard today, the genetic code for human insulin is inserted into the existing code of a bacterial cell, a single-celled organism. The bacterial cell then produces insulin. The resulting synthetic insulin is chemically identical to the insulin produced in the human pancreas. Genentech's synthetic insulin, marketed under the trade name Humulin, was approved for sale in 1982. Today, millions of people worldwide rely on it to control their diabetes.

Your Turn 13.1   Insulin via Bacteria?

Since insulin was introduced to the public, it has undergone major transformations in its production. Before 1982, it was produced from the organs of cows and pigs. However, it took 8000 pounds of animal pancreas glands to produce one pound of insulin. The costs have decreased and the amounts made have increased since the advent of using bacteria to create insulin.

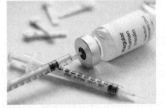

Sherry Yates Young/Shutterstock

Questions to Consider

a. Look at the green chemistry principles in this textbook on the first page. Which key ideas are met by the use of using bacteria to produce insulin?

b. Why were there issues using animal insulin in humans? Why aren't those same issues observed when using bacterial insulin?

How can we get the bacterium to produce human insulin? We do this by "teaching" it to make some new chemicals. Inside each cell is the complete set of instructions—a guidebook on how to grow and reproduce. The guidebook passes from one generation to the next, often completely unchanged. This guidebook, termed the **genome**, is the primary route for inheriting the biological information required to build and maintain an organism.

The genome is divided into short sections of instructions to produce specific outcomes known as chemicals, or events, in the cell. These specific pieces are the basic units of heredity, **genes**—short pieces of the genome that code for the production of proteins. A change within a gene changes an inheritable trait. For a corn plant with more genes in its genome than a human, a change in the gene for color may switch the corn kernels from light yellow to white. But small changes within a gene are not enough to make a cell produce a new chemical, such as insulin. We need a more dramatic change.

We need to insert a whole new set of instructions (i.e., a gene) into the bacterium's genome. By inserting the genetic code for human insulin into the bacterium, we create bacteria that can produce the essential insulin hormone. The bacterium of choice is a strain of *Escherichia coli*. All of us have *E. coli* strains in our intestines. However, the particular strain chosen for the synthesis of insulin is designed to be unable to survive in living organisms, so there is no risk that a person could be a host to a colony of insulin-producing bacteria. Also, although some *E. coli* strains, such as O104 or O157, cause severe illness and death, hundreds of other varieties are harmless to humans. Several of these forms are now used to produce antibiotics, vaccines, cancer-targeting agents, and even biofuels such as butanol.

Bacteria and other living organisms are more complex than you might think. What exactly are we modifying when we genetically alter something? We turn to this topic in the next section.

13.2 | DNA: A Chemical That Codes Life

> *Learning Objective: Describe the chemical composition and function of deoxyribonucleic acid (DNA)*

With each passing second, a single cell is a host to millions of chemical reactions. One very special and necessary chemical lies at the heart of this dazzling chemical complexity. As cells grow and multiply, this chemical must replicate itself without error. It must remain largely unharmed and unchanged by its environment. This one chemical must organize and securely store a lot of information. This information is context-sensitive, as some reactions are ongoing, whereas others start and stop depending on specific signals. In short, we need a highly advanced database in chemical form.

The chemical we have just described is **deoxyribonucleic acid** (DNA)—the biological polymer that carries genetic information in all species. DNA is the template of life, containing all of the biochemical information to make an entire corn plant, for example. DNA can replicate easily, transfer information, and respond to feedback within the cell.

Like other plants and animals, you have a special template of life written on a tightly coiled thread of DNA. Unraveled, the DNA in *each* of your cells is about 2 meters (roughly 2 yards) long. If all the DNA in all your cells were placed end to end, the resulting ribbon would stretch from here to the Sun and back more than 600 times! How is there so much DNA in your cells? A human, on average, contains 30 trillion human cells and 39 trillion bacterial cells. But as you will soon discover, this astronomical figure is far from the most astounding feature of this amazing molecule.

Any strand of DNA—long or short—consists of three fundamental chemical units: phosphate groups, deoxyribose sugars, and nitrogen-containing bases. All are illustrated in Figure 13.1.

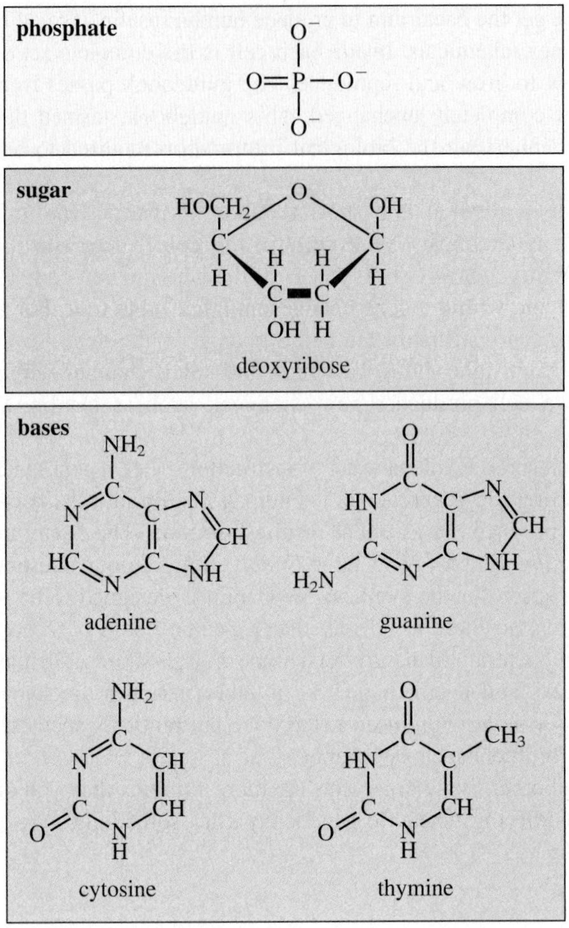

phosphate

$$O=\overset{\displaystyle O^-}{\underset{\displaystyle O^-}{\overset{|}{\underset{|}{P}}}}-O^-$$

sugar

deoxyribose

bases

adenine

guanine

cytosine

thymine

Figure 13.1

The components of deoxyribonucleic acid (DNA).

DNA contains four nitrogen-containing bases, each with different sizes and shapes. The larger bases, adenine (A) and guanine (G), have a six-membered ring and a five-membered ring of atoms fused together. The smaller bases, cytosine (C) and thymine (T), have only a six-membered ring. Notice that all of these compounds have nitrogen atoms embedded in their rings, leading to the name "nitrogen-containing bases." These bases also contain oxygen atoms that can participate in hydrogen bonding.

Your Turn 13.2   Slight Differences

The size or number of rings is not the only difference between the nitrogen-containing bases. Let's examine each of the four bases more closely.

a. Draw a Lewis structure for each of the four bases. Be sure to show the lone pairs of electrons on the nitrogen and oxygen atoms.

b. Identify the H atoms that could form hydrogen bonds with water or other nucleic acids.

c. Now identify the other (nonhydrogen) atoms that could participate in hydrogen bonding.

Hint: Revisit Section 5.2 for information on hydrogen bonds.

DNA molecules are built from sugar molecules. Unlike the nitrogen-containing bases in DNA, only one sugar is present, deoxyribose. Deoxyribose is a monosaccharide, a "single sugar," with the chemical formula $C_5H_{10}O_4$ (Figure 13.1). The next activity allows you to learn more about this sugar.

Your Turn 13.3   Chemical Cousins Ribose and Deoxyribose

Ribose is a close molecular cousin to deoxyribose. It is the monosaccharide found in ribo-nucleic acid (RNA). RNA acts as a messenger, carrying instructions from DNA to control the synthesis of proteins. Compare the structural formula of deoxyribose (Figure 13.1) with that of ribose:

a. Give the chemical formula for each sugar.
b. How do the structural formulas of these two sugars differ?
c. Carbohydrates typically have the chemical formula $C_nH_{2n}O_n$, as illustrated in Section 11.4. Do both ribose and deoxyribose fit this pattern?
d. Which atoms would form hydrogen bonds with water or other nucleic acids in these two sugar molecules?

In addition to a nitrogen-containing base and a sugar, DNA molecules also contain a phosphate group. However, depending on the pH, phosphate may be in the form of PO_4^{3-}, HPO_4^{2-}, or $H_2PO_4^-$ (Figure 13.2). If all three oxygen atoms in the phosphate ion are paired with H^+, the chemical form is H_3PO_4, or *phosphoric acid*. These hydrogen ions impart acidity to the nucleic acid.

Your Turn 13.4   Speciation of Phosphate Ions

By interpreting the data represented in Figure 13.2, describe the fractional (or percent) composition of the phosphate ions present in aqueous solutions at these pH values:

a. pH = 1 (acidic)
b. pH = 10 (basic)
c. pH = 5.5 (acidic)
d. pH = 13.5 (basic)
e. pH = 7 (neutral solution)

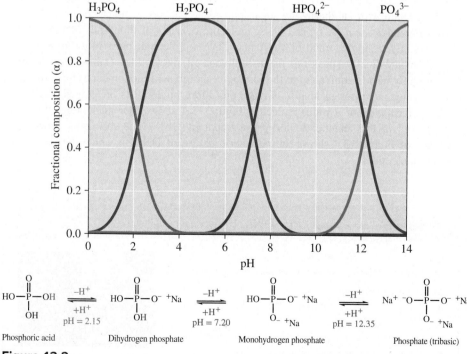

Figure 13.2

The composition of phosphate species present in aqueous solutions at varying pH ranges.

Figure 13.3

A nucleotide monomer built from a phosphate group, deoxyribose, and the base adenine.

All three of the fundamental chemical units—the nitrogen-containing base, the sugar, and the phosphate group—have a valuable role in DNA structure. Joined together, these three pieces make up one **nucleotide** monomer, which, in turn, polymerizes to form DNA. For example, Figure 13.3 shows the nucleotide named *adenine phosphate*. You can see that the sugar is bonded both to the phosphate group and to the base adenine. Similar nucleotides can be formed using the other three nitrogen-containing bases of DNA: guanine, cytosine, and thymine.

Your Turn 13.5   Another Nucleotide

Analogous to Figure 13.3, draw the structural formula for the nucleotide containing cytosine.

Note in Figure 13.3 that one —OH on the deoxyribose ring remains available to react. It does so with the phosphate group of another nucleotide via a condensation reaction (revisit Section 9.6). If this repeatedly happens between nucleotides, the result is a long chain with an alternating sugar–phosphate backbone, better known as DNA— *deoxyribonucleic acid*. A typical DNA molecule consists of thousands of nucleotides. Consequently, a single strand of DNA may have a molecular mass in the millions.

Mechanistically speaking, the assemblage of nucleotides (monomers) to form DNA (a polymer) is equivalent to the joining of amino acid monomers to form proteins, The polymer increases in length as more and more nucleotides are joined, each time with the formation of a water molecule. Figure 13.4 shows four nucleotides that have been linked in this manner to form a segment of DNA. The schematic drawing in the inset of Figure 13.4 shows the polymeric nature of DNA in which the monomers are the nucleotides.

@HOME   Extracting DNA From Strawberries!

Check out this video (www.acs.org/cic) to extract and characterize DNA from strawberries and try it yourself!

Questions to Consider

a. Why was it necessary to mash the strawberry prior to DNA extraction?

b. What is the purpose of the liquid detergent, salt, and rubbing alcohol used in the extraction experiment?

Tiplyashina Evgeniya/Shutterstock

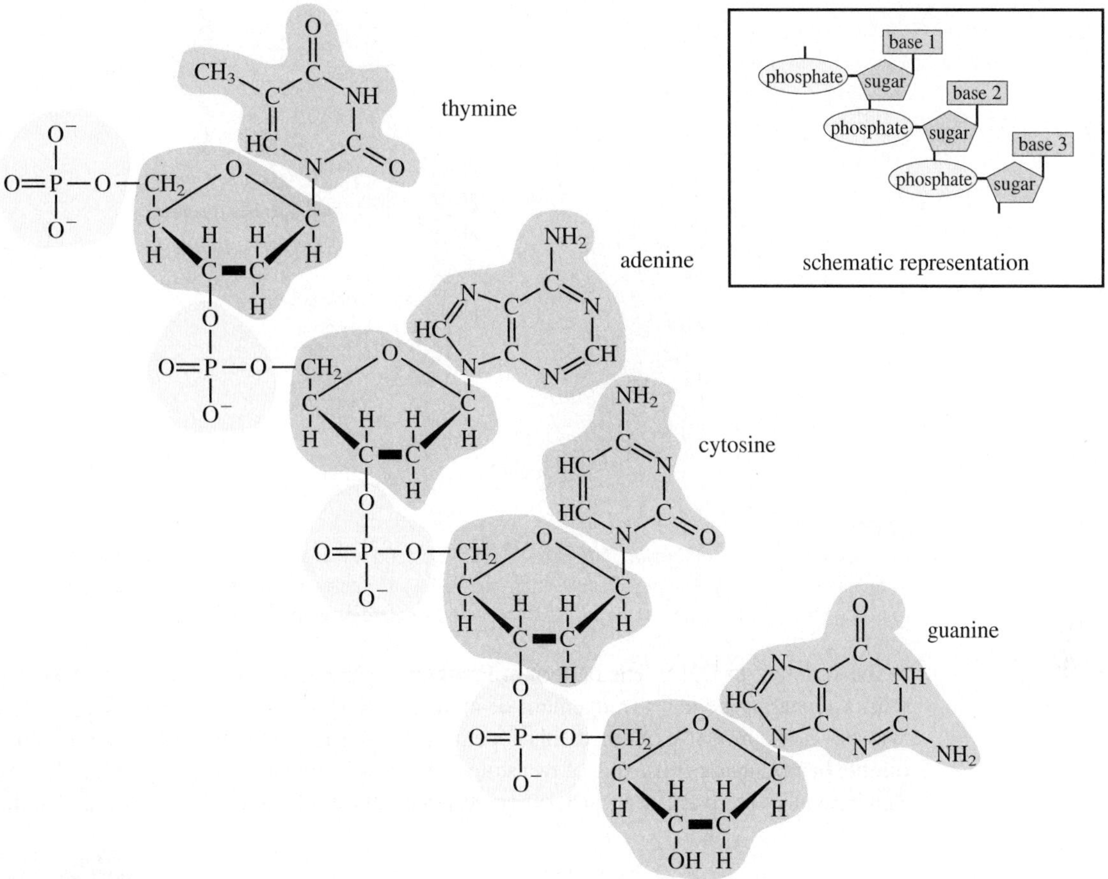

Figure 13.4

A segment of DNA represented chemically and schematically (inset). The phosphate group connects one deoxyribose to an adjacent one. Each of the four bases—thymine (T), adenine (A), cytosine (C), and guanine (G)—is attached to a deoxyribose sugar.

13.3 | The Double Helix Structure of DNA

Learning Objectives:

Interpret experimental evidence for the double-helix structure of DNA

Understand the structural basis of DNA replication

DNA is a gorgeous molecule. The opening photo of this chapter displays a sculptor's rendition of DNA with two silvery strands curving in a gentle spiral, both elegant and simple. Hidden within its structural simplicity is a powerful chemical code for information. The structure—how the nucleotides are covalently bonded and the strands pack together—contributes to the function of DNA. Understanding how DNA performs its many functions first required solving the puzzle of the DNA structure.

To see the shape and submicroscopic details of DNA, scientists turned to the technique of X-ray diffraction. This technique has revolutionized our understanding of molecular structures and chemistry by helping us visualize chemical shapes. **X-ray diffraction** is an analytical technique where a crystal is hit by a beam of X-rays to generate a pattern that reveals the positions of the atoms in the crystal. The X-ray photons interact with the electrons of the atoms in the crystal and are diffracted or scattered (Figure 13.5). The crucial point is that X-rays are only scattered at certain angles related to the distance between atoms, which can be used to determine the structures of various crystalline materials. The X-ray diffraction pattern of a DNA fiber was obtained in late 1952 by the British crystallographer Rosalind Franklin (1920–1958) (Figure 13.6).

James Watson (b. 1928) and Francis Crick (1916–2004) combined Franklin's X-ray diffraction data with earlier chemical and biological analyses to create a model

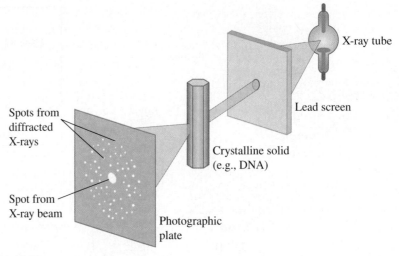

Figure 13.5

An illustration of X-ray diffraction. A crystalline solid is placed within the X-ray beam and rotated. Information regarding the molecular structure of the solid is obtained from the symmetry of spots resulting from the diffraction of the incident X-ray beam by atoms in the solid.

of the structure of DNA. The pattern in Franklin's diffraction photograph was consistent with a repeating helical arrangement of atoms, similar to a loosely coiled spring. The Watson–Crick model explained this repetition by twisting the strands of DNA into a **double helix**, a spiral consisting of two strands that coil around a central axis (Figure 13.6b). The base pairs are parallel to each other, perpendicular to the axis of the DNA molecule,

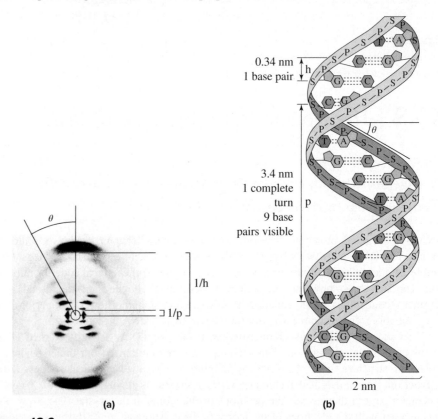

(a)   **(b)**

Figure 13.6

(a) Rosalind Franklin's X-ray diffraction photo for a hydrated DNA fiber. The cross in the center indicates a helical structure, and the darkened arcs at the top and bottom are due to the stack of base pairs. The value h is 0.34 nm, the size of one base pair. One complete turn of 9 base pairs of DNA is the value p and measures 3.4 nm. **(b)** A model of DNA with P = phosphate group; S = the sugar, deoxyribose; and the bases A = adenine; T = thymine; C = cytosine; G = guanine. The sugar and phosphate groups alternate on the backbone, and the four bases attach to this backbone.

(a): Omikron/Science Source

and separated by 0.34 nm, the same distance calculated from the diffraction pattern. In addition, Franklin's results also suggested a second repetition pattern separated by 3.4 nm. Watson and Crick took this as the length of a complete helical turn consisting of 10 base pairs.

@HOME   A Candy Model of DNA

Investigate the structure of a DNA molecule at home using licorice strands, four different colored gum drops, and toothpicks. If you do not have any gumdrops, you can use gummy bears.

Brent Hofacker/Shutterstock

1. Assign each color of gum drop a base pair.
2. Make a base pair pattern with at least five base pairs on the left side with your gum drops. Then, find its complementary base pair gum drop and place it on the right side.
3. Take a toothpick and place it through though one pair. Repeat for each of the other pairs. Make sure that the pointed ends of the toothpicks have some space next to the gum drops to be used in the next step.
4. Place the pointed ends of the toothpicks into a licorice strand on the left and into another licorice strand on the right.
5. Hold one end and twist the other end 180 degrees. Now you have a double helix!

Questions to Consider

a. Revisit the photograph that opens the chapter. It shows a sculpture of DNA that people can climb on. As a piece of art, it represents some parts of the DNA molecule better than others. List three disadvantages of that DNA representation. What chemical details are omitted? What information is lost? Compare it to your candy DNA representation.
b. Now list three advantages of the piece of art. What information is highlighted? What is gained? Now, compare it to your candy DNA representation.
c. Find another artistic rendition of DNA, possibly on your campus or on the Internet, and repeat parts a and b. Cite your sources.

X-Ray Diffraction

Check out this video to see an X-ray diffractometer in action: www.acs.org/cic.

©Bradley D. Fahlman

Questions to Consider

a. Why is X-ray radiation most appropriate to study atomic structure relative to using other regions of the electromagnetic spectrum?
b. Using the Internet as a resource, describe how synchrotron radiation may be used to determine the structure of solids. What are the benefits of using synchrotron radiation relative to the source used in laboratory-scale X-ray diffraction instruments?

The letters in our English words have directionality. For example, the words *ward* and *draw* have the same letters, but the meaning is different because their order is reversed. The same is true of DNA, with the bases within its backbone structure being analogous to letters and the structure of the DNA backbone defining the directionality. For example, the base string *T-A-C* does not have the same meaning as *C-A-T*. Look carefully at the alternating phosphate and deoxyribose groups in the DNA backbone (Figure 13.4) to see how the deoxyribose ring connects directly to the phosphate below it, and the one above it links through another carbon. The different types of chemical bonds make one direction different from the other. When the two strands of the DNA double helix come together, one strand must run in the opposite direction from the other.

Your Turn 13.6   DNA vs. Protein Structure

Looking back at Section 11.6, we see that proteins can also be thought of as words with different letters.

Questions to Consider

a. How many "letters" are in the protein code? How many "letters" are in the DNA code?
b. Were you surprised that only four base pairs make up all forms of life?
c. DNA is in a double helix structure, whereas proteins can be in very different structures. Contemplate this difference based on the number of parts that can make up each structure.

Early chemical analyses of DNA showed that the nitrogen-containing bases come in pairs. No matter the species, the percentage of A almost exactly equals that of T

Table 13.1	Percent of Base Compositions of DNA for Various Species				
Specific Name	**Common Name**	**Adenine**	**Thymine**	**Guanine**	**Cytosine**
Homo sapiens	human	31.0	31.5	19.1	18.4
Drosophila melanogaster	fruit fly	27.3	27.6	22.5	22.5
Zea mays	corn	25.6	25.3	24.5	24.6
Neurospora crassa	mold	23.0	23.3	27.1	26.6
Escherichia coli	bacterium	24.6	24.3	25.5	25.6
Bacillus subtilis	bacterium	28.4	29.0	21.0	21.6

Source: I. Edward Alcamo, DNA Technology: The Awesome Skill 2e, McGraw Hill Education, 2000.

(Table 13.1). Similarly, the percentage of G is nearly identical to that of C. The structural model of DNA validated these rules. Adenine (A) and thymine (T) bases fit together perfectly, like jigsaw puzzle pieces. A closer look shows these two bases linked by two hydrogen bonds (Figure 13.7). Similarly, cytosine (C) and guanine (G) are linked by three hydrogen bonds. This base-pairing phenomenon is the molecular basis underlying the structure and much of the function of DNA. To summarize: A pairs with T, and G pairs with C.

Your Turn 13.7   Complementary Base Sequences

Adenine and thymine are said to be *complementary bases*. So are cytosine and guanine. In both cases, the bases form hydrogen bonds when they pair. Using one-letter codes, write out the complementary base sequences for each of these codes.

a. ATACCTGC   **b.** GATCCTA

The structure of DNA, featuring the puzzle-piece pairing of its nucleotides, inspired another vital discovery. One side of the DNA strand contains all the information required to generate its partner strand! Thus, a single strand of DNA can guide the

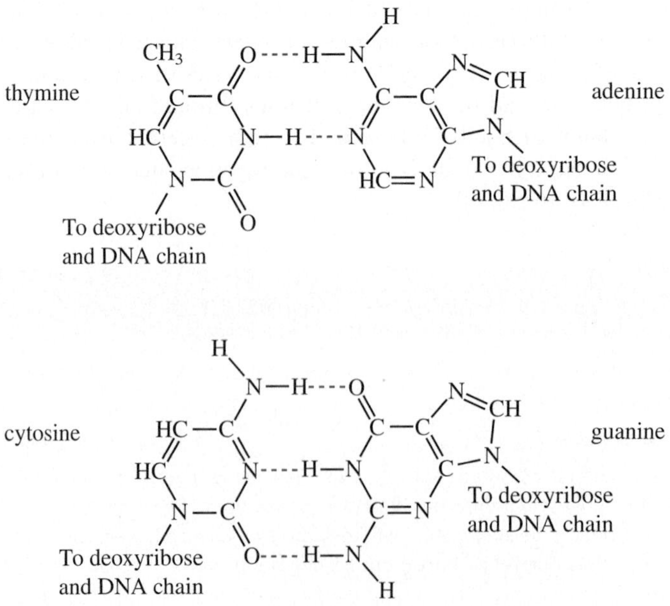

Figure 13.7

Base-pairing of adenine with thymine and cytosine with guanine in DNA. Chemical bonds are solid black lines, and hydrogen bonds are dashed **red** lines.

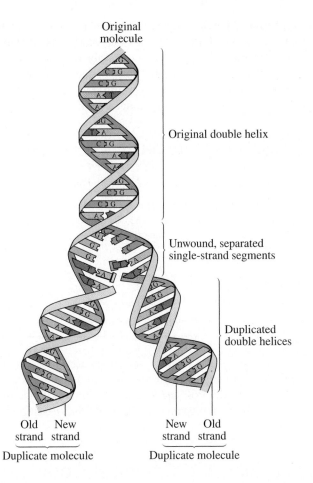

Original
molecule

Original double helix

Unwound, separated
single-strand segments

Duplicated
double helices

Old   New
strand strand

New   Old
strand strand

Duplicate molecule   Duplicate molecule

Figure 13.8

Diagram of DNA replication. The original DNA double helix (top portion of figure) partially unwinds, and the two complementary portions separate (middle). Each strand serves as a template for synthesizing a complementary strand (bottom). The result is two complete and identical DNA molecules.

Visualizing DNA

Check out this video for an illustration of DNA coiling and replication: www.acs.org/cic.

Keith Chambers/Science Source

Questions to Consider

a. How does DNA replicate if it is tightly packed in nucleosomes before the cells split?

b. The video mentions that the speed at which the DNA is replicated is as fast as a jet engine. Estimate how much faster we would need to watch that clip if we looked at real DNA replicating.

generation of its complement. **Replication** is the process of cell reproduction in which the cell must copy and transmit its genetic information to its progeny. The process is well understood and is diagrammed in Figure 13.8.

Before a cell divides, the double helix rapidly, but only partially, unwinds. This results in a region of separated strands of DNA, as pictured in the middle portion of Figure 13.8. Free nucleotides in the cell are selectively hydrogen-bonded to these two single strands that serve as templates for a new DNA molecule: A to T, T to A, C to G, and G to C. Held in these positions, the nucleotides are held together by the action of an enzyme, a biological catalyst. The DNA strands continue to separate, and complementary strands are made until two complete DNA strands exist. By this mechanism, each strand of the original DNA generates a complementary copy of itself. The original strand and its newly synthesized complement coil form a new molecule identical to the first—and visa versa with the other strand. Thus, where there was originally one double helix, now there are two identical copies.

Your Turn 13.8   DNA Sequence Repair

DNA must be copied with utmost perfection. After replication, enzymes scan over strands of DNA to identify and correct errors in base pairing.

ATGCCATGAA
TACGGTATTT

a. Find the error in base pairing for this set of DNA strands and circle it.

b. Do you expect the mismatched strands to be more or less stable than a correct pair? Explain your reasoning.

In most organisms, the newly copied DNA does not remain extended as a double helix but becomes coiled even further. This not only saves space but also organizes and protects genetic information. The coiling is carefully regulated so small portions of DNA can be accessed when specific stored information is needed. This complete set of genetic information is packaged into **chromosomes**, rod-shaped, compact coils of DNA, and specialized proteins packed in the nucleus of cells.

Your Turn 13.9   Is Your DNA Doin' the Twist?

The distance between base pairs in the double helix structure of DNA is 0.34 nm.

a. Calculate the length in centimeters (cm) of human chromosome 11 when extended in a double helix. Chromosome 11 consists of 135,000,000 base pairs.

b. Chromosomes can be visualized best immediately before cell division. In this compact state, the longest axis of chromosome 11 is approximately 4 μm. By what factor has the DNA been further condensed?

   Hint: 1 μm is 1×10^{-6} m.

c. Suggest a reason why this level of compaction is necessary.

   Hint: A typical human cell is only about 50 μm in diameter.

Every time a cell splits to reproduce, the complete set of chromosomes must be uncoiled and replicated perfectly so that each new cell contains an identical set. Some cell types, such as those in the skin and a cancerous tumor, divide more rapidly than others. These cells are more susceptible to collecting and passing on DNA damaged by ionizing radiation, free radicals, or chemical agents.

13.4 | Cracking the Chemical Code

Learning Objective: Explain the structure and function of codons in protein synthesis

It can be overwhelming to think about the complex chemistry happening in our cells every minute, chiefly because DNA molecules organize a *lot* of information! Even in plants, this is mind-boggling; billions of base pairs are repeated in every corn cell to provide the blueprint to produce one corn plant. These base pairs are ordered into specific sequences and grouped, sometimes into genes, to code for the production of proteins. Other information is also present in the DNA, too, but our understanding of how it is used is still in its infancy.

Although the information is carried in DNA, it is expressed in other (smaller) molecules. The best understood are proteins. Proteins are found throughout our bodies in skin, muscle, hair, blood, and the thousands of enzymes that regulate the chemistry of life. By directing the synthesis of proteins in the ribosomes of a cell, DNA can dictate many of the characteristics of an organism.

Proteins are large molecules formed by the linking of amino acids. Recall that the 20 amino acids that commonly occur in proteins can be represented by this general structural formula, which is reproduced from Section 11.6:

$$
\begin{array}{ccc}
H & H & O \\
\diagdown & | & \diagup\!\diagup \\
N\!\!-\!\!C\!\!-\!\!C & & \\
\diagup & | & \diagdown \\
H & R & OH
\end{array}
$$

The amine group is —NH_2, the carboxylic acid group is —COOH, and R represents a different side chain for each of the 20 amino acids. In a condensation reaction, the —COOH group of one amino acid reacts with the —NH_2 group of another. A peptide bond is formed in the process, and a molecule of H_2O is formed. When many amino acids are connected, the result is a protein, a polymer built from amino acid monomers. We can also describe a protein as a long chain of **amino acid residues**, which implies that the amino acids have been incorporated into the peptide chain. Polypeptides are a long chain of amino acid residues that form part or all of a protein molecule.

The information in a sequence of DNA nucleotides translates via a code into a sequence of specific amino acids in a protein. The code cannot be a simple one-to-one correlation between bases and amino acids because each has differing numbers. DNA has only four bases, so it may only encode four amino acids. However, 20 amino acids appear in our proteins. Therefore, the DNA code must consist of at least 20 distinct code "words," with each word representing a different amino acid. Furthermore, the "words" must be selected from a pool of only four letters—A, T, C, and G—or, more accurately, the bases corresponding to those letters.

Some simple statistics can help us determine the minimum length of these code words. To find out how many words of a given length can be made from an alphabet of known size, raise the available number of letters to the power n, corresponding to the number of letters per word:

$$\text{words} = (\text{letters})^n \qquad \text{[13.1]}$$

For example, using four letters to make two-letter words generates 4^2, or 16, different words. Thus, DNA bases taken in pairs (akin to two letters per word) could code for only 16 amino acids, which is insufficient to provide a unique representation for each of the 20 amino acids. So, we repeat the calculation and assume that the code is based on three sequential bases or, if you prefer, three-letter words. Now, the number of different triplet-base combinations is 4^3, or $4 \times 4 \times 4 = 64$. This system provides more than enough capacity to do the job.

Your Turn 13.10   Quadruplet–Base Code

Suppose the DNA code used four sequential base pairs instead of a triplet-base code. How many different four-base sequences would result?

The three-letter groupings of nucleotides are the basis of the information transfer from DNA to proteins. Each grouping, or **codon**, is a sequence of three adjacent nucleotides that either guides the insertion of a specific amino acid or signals the start or end of protein synthesis. If you were to use the letters A, T, C, and G in a game of Scrabble, you could generate 64 different three-letter combinations. For example, a few, CAT, TAG, and ACT are English words. However, most combinations, such as AGC, TCT, and GGG, are meaningless—at least in English. Nature does far better than that; 61 of the 64 possible triplet codons specify amino acids. Thus, the codon sequence CAC in a DNA molecule signals that a molecule of the amino acid histidine should be incorporated into the protein, TTC codes for phenylalanine, and CCG stands for proline. The three-base sequences that do not correspond to amino acids are signals to stop the synthesis of the protein chain. Figure 13.9 illustrates a simplified example of a nine-base nucleic acid segment and how it codes for three amino acids.

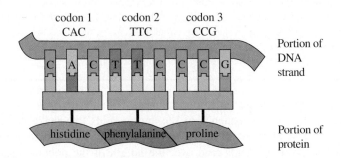

Figure 13.9

A nine-base nucleic acid sequence showing three codons. Check out this simulation of protein coding from DNA: www.acs.org/cic.

Your Turn 13.11   Duplicate Codons

Suggest some advantages of a genetic code in which several codons represent the same amino acid.

Cellular Protein Factories

Sanan Media

Check out this video to learn about genomes: www.acs.org/cic.

Questions to Consider

a. How much of your genome codes for proteins? How much doesn't have any function as far as we can tell?

b. Many different kinds of DNA mutations can affect a person. Research what kind of DNA mutation sickle cell anemia is. What is the difference between a missense and a nonsense mutation? Which is more disruptive?

c. Were you surprised to realize that you share 41% of the DNA of a banana?

Our calculation above showed that three-letter groupings are the minimum necessary to cover the 20 amino acids but did not show us how all 64 codons are used. The code has redundancy. Many amino acids have more than one codon. For example, leucine, serine, and arginine have six codons each. Also, three different codons tell protein synthesis to "stop." However, two amino acids (tryptophan and methionine), and the signal to start protein synthesis, are represented by only a single codon.

The genetic code is identical in all living things. With only a handful of exceptions, the instructions to make people, bacteria, and trees are written in the same molecular language of those 64 codons. The genetic code has a "Rosetta Stone" to translate any genetic sequence from any organism. The significance of this statement will become apparent later in this chapter when we discuss an example of genetically modified corn. We end this section with the following video, which describes how proteins are made from DNA codons.

13.5 | Proteins: Form to Function

Learning Objectives:

Discuss the primary, secondary, and tertiary structure of proteins

Exemplify how small changes in a protein sequence may cause disease

Proteins are polymers. Admittedly, they don't much resemble the clear polymer PET that may hold your favorite soft drink. Similarly, they don't seem to have much in common with the tough polypropylene used to make carpets. Nonetheless, proteins are big molecules built from little ones.

More specifically, proteins are polyamides. Like nylon (Section 9.7), they are built from the chemical reaction of carboxylic acids and amines. Unlike nylon, a protein is built from 20 different amino acid monomers (revisit Figure 11.18). Proteins are complex three-dimensional molecules with very specific forms rather than the long strands found in nylon. The final shape may look messy, but that exact shape is necessary for the function performed in the organism.

Your Turn 13.12   How Is Hamburger Like Nylon?

Take a moment to refresh your knowledge about two polyamides: the nylon from a sports jersey and the protein in hamburger.

a. What functional groups do nylon and meat proteins have in common?

b. Nylon is usually synthesized from two types of monomers. In contrast, proteins are synthesized from one type of monomer. What functional group(s) do the monomers for each contain?

c. Are nylon and protein addition polymers or condensation polymers?

We begin our discussion of protein shapes with the **primary structure** of a protein; that is, the unique sequence of the amino acids that make up each protein (Figure 13.10). The primary structure is the first and most basic identifier of a protein—the list of amino acids read over the length of the polymer. Knowing that a short protein contains three valines (Val), two glutamic acids (Glu), and one histidine (His) may tell you the size and a few other details, but it is not sufficient to

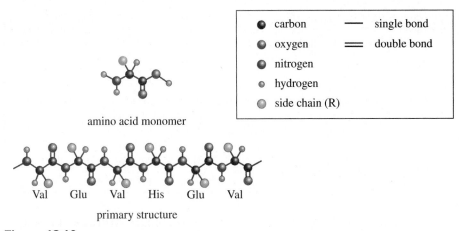

carbon   —— single bond

oxygen   === double bond

nitrogen

hydrogen

side chain (R)

amino acid monomer

Val   Glu   Val   His   Glu   Val

primary structure

Figure 13.10

Representation of the primary structure of a protein.

specify a protein and explore its shape and function. The order and sequence of the amino acids matter. For example, ...-Val-Glu-Val-His-Glu-Val-... is a different protein from ...-Val-Val-Val-His-Glu-Glu-...

Recall that each amino acid has a side-chain group. These side chains interact with one another, or with surrounding molecules. Side chains can attract and "lock" together, and in doing so, they hold a protein in a particular overall shape. The order and identity of the amino acids define how and where those side-chain links can form. Each amino acid plays a role; changing just one can change the shape and, as a result, the function of a protein.

We will group the side chains into *polar* (either charged or neutral) or *nonpolar* (revisit Section 11.6). Like oil and water, nonpolar and polar side chains tend to separate. In the typical aqueous environment of a protein, the polar side chains will be attracted to water molecules via hydrogen bonding. Uncharged, yet polar, side chains often contain hydroxyl groups or amides that will be attracted to one another by forming hydrogen bonds. In contrast, nonpolar side chains will aggregate inside the protein, avoiding unfavorable interactions with water in favor of attractive London dispersion forces between the side chains.

Your Turn 13.13   Polarity of Side Chains

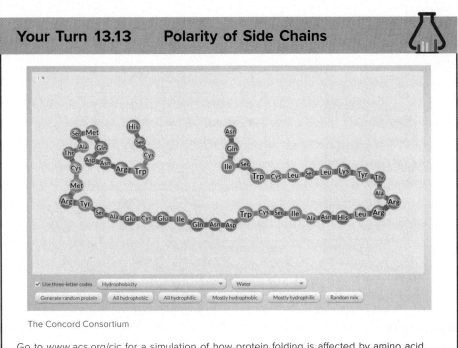

The Concord Consortium

Go to www.acs.org/cic for a simulation of how protein folding is affected by amino acid side chains.

—(continued)

Hydrophobic means water-repelling, so the amino acids that do best are the nonpolar residues that don't like water. Hydrophilic means water-loving, so the amino acids that do best are the polar residues that like water. Click play at the bottom.

Questions to Consider

a. What color of the amino acids are polar and nonpolar?

b. Will it work better in a water or oil solution? The blue background indicates a water solution.

c. Change the environment of the protein to oil. What changes happen to the protein?

d. Change the type of protein to all hydrophobic. What does it do in the oil? In the water? Is this what you expected? Why?

e. Change the type of protein to all hydrophilic. What does it do in the water? In the oil? Is this what you expected? Why?

We continue our discussion by examining the **secondary structure** of a protein, that is, the folding pattern within a segment of the protein chain. Many, but not all, protein chains form regular, repeating structures from the particular bond angles and attractions between neighboring amino acids. The two most common are the α-helix, a spiraling strand, and the β-pleated sheet, extended strands stretching alongside each other with a slight zigzag. The shape of both structures are governed by the presence of hydrogen bonding.

Both forms of secondary structure depend on the tendency of the protein backbone to form intramolecular hydrogen bonds. Figure 13.11 shows the hydrogen bonds between the backbone O and the N—H of the amide group as dotted lines. The number and spacing of the hydrogen bonds can pull together and align a protein strand, stabilizing the secondary structure. The choice of secondary structure, or even the complete lack of it, can be loosely predicted from the primary structure. Some side chains tend to pack well into β-pleated sheets, others tend toward the α-helix, and others even predispose the chain to disorder.

> ### Your Turn 13.14   Intra- Versus Intermolecular Hydrogen Bonding
>
> Using the Internet as a resource, illustrate three examples of polymers exhibiting intermolecular hydrogen bonding and three examples of polymers with intramolecular hydrogen bonding.

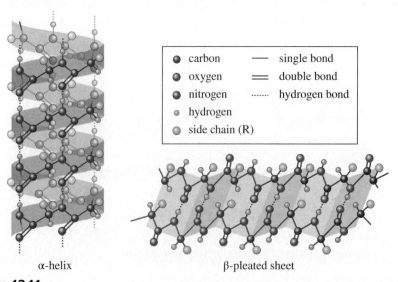

● carbon	— single bond
● oxygen	═ double bond
● nitrogen	······ hydrogen bond
○ hydrogen	
○ side chain (R)	

α-helix   β-pleated sheet

Figure 13.11

Representations of secondary structures of a protein. The two major types of secondary structures are the *α*-helix and the *β*-pleated sheet. Check out this video to explore how the combination of amino acids affects the structure and properties of a protein: www.acs.org/cic.

Table 13.2	Forces Involved in Levels of Protein Structure		
Bonds	**Primary Structure**	**Secondary Structure**	**Tertiary Structure**
Peptide bonds	X		
Hydrogen bonds		X	X
Disulfide bonds			X
Ionic bonds (salt bridges)			X
London dispersion forces			X

Proteins are large, three-dimensional molecules, but both primary and secondary structures are relatively flat. We need a more "global" description of their shape, designated as its **tertiary structure**. A tertiary structure represents the overall molecular shape of the protein that is defined by the interactions between amino acids far apart in sequence but close in space (Table 13.2).

For instance, one unique amino acid, cysteine, contains a thiol group (—SH) in its side chain. In proteins, thiol groups perform an important and highly specialized function; two thiol groups can react to form disulfide (S—S) bonds between the adjacent sulfur atoms. These strong bonds covalently link two different regions of a protein together. Other polar side chains can lead to ionic attractions. Side chains that contain acidic or basic groups (carboxylic acids or amines, respectively) often become charged ions and attract their opposites in the tertiary structure.

From only 20 amino acids, proteins make a variety of shapes and serve a large array of functions. Their three-dimensional shapes dictate specific functionalities. For instance, as described in Section 12.7, the shape of enzymes create an **active site** in an enzyme that binds only specific reactants and accelerates the desired reaction (Figure 13.12). Enzymes are the most commonly discussed type of protein, but several other examples exist. Some proteins bind DNA either to protect it or to send a signal. When these proteins fold, they display positively charged side chains to attract and bind the negatively charged DNA. Another type of protein funnels material through the membrane of the cell. This is accomplished by forming channels that shuttle a specific chemical across the outer layers of the cell while keeping the cell impermeable to undesirable chemicals. Again, form follows function.

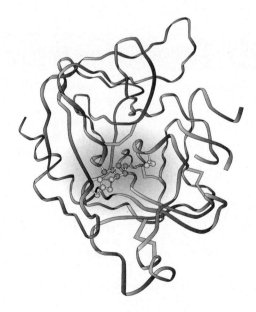

Figure 13.12

The tertiary structure of the enzyme chymotrypsin showing its active site. The "ribbon" portion (gray) represents the amino acid chain; the central colored portion is the active site at which enzymatic chemistry occurs.

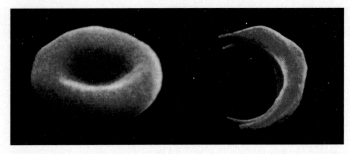

Figure 13.13

An image from a scanning electron microscope showing both normal red blood cells (left) and those distorted into a sickle or crescent shape (right).

Mary Martin/Science Source

A subtle change in the primary structure of a protein can have a profound effect on its properties. Notice the word *can* in the previous statement. Sometimes, an amino acid change leaves the shape and function of the protein unchanged. For instance, a nonpolar leucine can be switched to a nonpolar valine by removing a —CH₃ group. The protein may be a little less stable but, on the whole, the same. Change the wrong glutamic acid (in which the side chain is often negatively charged) to a nonpolar valine, however, and the disease *sickle-cell anemia* occurs. Sickle-cell disease affects a sizeable population, more than 70,000 individuals in the United States alone, and lowers life expectancy from almost 79 years to an average of 55 years.

As discussed in Section 12.6, hemoglobin is the blood protein that transports oxygen. The single alteration of a particular glutamic acid to valine in the primary structure of hemoglobin creates a variant called hemoglobin S and a condition called sickle-cell disease. This substitution causes hemoglobin to convert into an abnormal shape at low oxygen concentrations, forcing red blood cells to distort into a rigid sickle or crescent shapes (Figure 13.13). Because these cells lose their normal flexibility, they cannot pass through the tiny openings of the capillaries (the smallest vessels in the blood circulatory system) in the spleen and other organs. Some of the sickle cells are destroyed, and anemia results. Others clog organs so badly that the blood supply to these organs is reduced.

Your Turn 13.15   Function Follows Form

In sickle-cell anemia, a glutamic acid residue in the sequence of hemoglobin is replaced with a valine residue.

valine   glutamic acid

a. Describe the structural difference between these two amino acids: valine and glutamic acid.
b. Predict the solubility for each of these amino acids in water.
c. Explain how these differences could give rise to the deformed cells typified by sickle-cell anemia.

13.6 | The Process of Genetic Engineering

Learning Objective: Explain how genetic engineering is used to alter the DNA makeup of an organism

We started this chapter by describing how bacterial cells generate the vast amounts of human insulin required to maintain the health of the 537 million people worldwide who struggle with diabetes. For this challenge and many others solved by genetic engineering, we needed to modify the DNA in the cell. If we change the genes, we change the proteins synthesized by these genes. Ultimately, we change the chemistry of the cell.

Throughout history, humans have manipulated genes. This may surprise you, as you might think our ability to modify genes has come about only recently. But consider, for example, how we have cultivated plants. We tended to grow plants with specific traits, such as better taste or appearance. The others we rejected. To produce these strains with new and unique traits, we crossbred different strains. The process took many years, but eventually, we "domesticated" plants and created the crops that feed us today. Our crops are so far removed from their wild forebears that we would hardly recognize them.

Corn is an excellent example being indigenous to the Americas. The region's native people manipulated the genes of the teosinte plant (one that bore seeds on the end of its stalks rather than on the body of the plant) to get the growth pattern seen in today's corncob (Figure 13.14). Domesticating the teosinte plant led to a food that was both more nutritious and more abundant.

Domesticating a plant is a process of *genetic modification*. Even without understanding the chemistry, we selected plants with certain sequences of DNA and rejected ones with other sequences. Similarly, disease resistance that was not immediately required was lost.

A plant may have a bacteria that is killing it, and a gene that may not have mattered before is the key to survival. The stress on the plant may also take other forms, such as a three-year drought or a more aggressive weed. But the process is the same. Nature generally selects for more self-sufficient plants; humans tend to select plants that look and taste better. In either case, the result changes the genome of the plant. Selective breeding, either in the wild or in human agriculture, is a long, slow, and somewhat random route toward modifying the gene pool. Either natural or artificial selection, typically used to describe food products, is designated as genetic

Figure 13.14

Corn's early, much smaller ancestor, teosinte, is shown here next to a modern larger ear of corn.

©& Courtesy of Hugh Iltis/The Doebley Lab

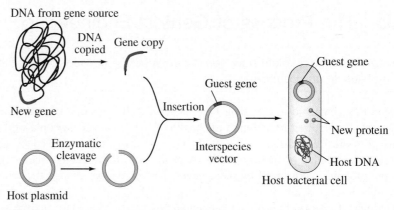

Figure 13.15

A general representation of the process of genetic engineering.

modification. In contrast, **genetic engineering**, is the direct manipulation of the DNA in an organism.

The easiest organisms to manipulate are small single-celled bacteria. These bacteria contain **plasmids**, or rings of DNA, in addition to their chromosomes. Scientists can now routinely remove, change, and replace the plasmids to create new chemistry in the bacteria. They use special enzymes to cut open the plasmid at specific sites. Scientists then copy the DNA containing a promising gene from another organism, and that DNA is inserted into the plasmid ring from the bacterium. The result is a new interspecies DNA plasmid, or **vector**—a modified plasmid used to carry DNA back into the bacterial host (Figure 13.15). Once inside the cell, the chemistry of the bacterium takes over. The bacterium grows rapidly and produces new cells. Soon, the scientist has millions of copies of the "guest" gene and its protein product.

Inserting foreign DNA becomes trickier as one climbs the evolutionary ladder from bacteria to plants. Higher organisms are better at protecting themselves against foreign invaders. For example, many organisms have chemical mechanisms designed to detect and destroy foreign DNA. Even so, scientists have found ways to get around such defenses. One is to hijack a special soil bacterium that can infect plants. This bacterium creates a bridge into the plant cell and, in the process, transfers its DNA into the genome of the plant. The bacterial genes induce points of large abnormal growth (Figure 13.16). You may have seen these growths (tumors) on various plants and trees, including apple trees, rosebushes, and some vegetable plants.

Figure 13.16

A crown gall tumor (*Agrobacterium radiobacter*) on a chrysanthemum plant.

Nigel Cattlin/Science Source

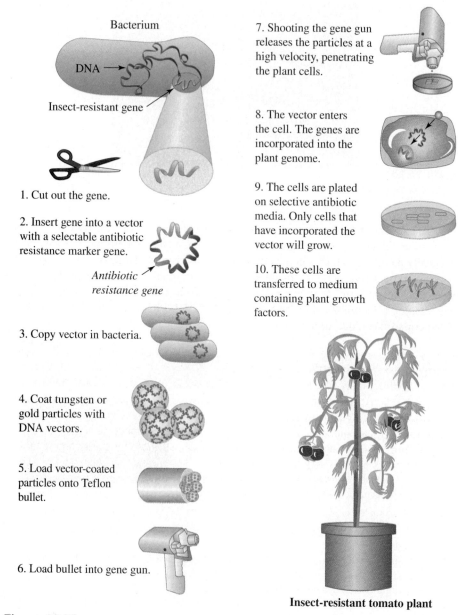

Bacterium

DNA →

Insect-resistant gene

1. Cut out the gene.

2. Insert gene into a vector with a selectable antibiotic resistance marker gene.

Antibiotic resistance gene

3. Copy vector in bacteria.

4. Coat tungsten or gold particles with DNA vectors.

5. Load vector-coated particles onto Teflon bullet.

6. Load bullet into gene gun.

7. Shooting the gene gun releases the particles at a high velocity, penetrating the plant cells.

8. The vector enters the cell. The genes are incorporated into the plant genome.

9. The cells are plated on selective antibiotic media. Only cells that have incorporated the vector will grow.

10. These cells are transferred to medium containing plant growth factors.

Insect-resistant tomato plant

Figure 13.17

An illustration of how an insect-resistant tomato plant is produced through genetic engineering.

Scientists are now adept at modifying the bacterial genome by inserting a different gene of interest. Suppose the gene of interest is the specific sequence from the soil bacterium, *Bacillus thuringiensis* (or *Bt*), that produces a *Bt* toxin for insect resistance. In that case, the plant acquires the instructions to make the toxin itself (Figure 13.17). Not only is there *Bt* corn, but also *Bt* cotton, *Bt* potatoes, *Bt* tomatoes, and *Bt* rice, all producing toxins against crop-destroying pests!

The *Bt* toxin is but one example of genetically engineered possibilities. Not all examples of genetically engineered crops are **transgenic**, in which a transfer of genes occurs across species. Genetic engineering achieves the same aims as selective breeding but with far more speed and control. Transgenic rice plants have been developed for use in sub-Saharan Africa, where the yellow mottle virus destroys much of the rice crop yearly (Figure 13.18). Instead of securing the resistance trait by selective breeding, one can find and copy just the resistance gene from the wild plant, then splice it into the weak but otherwise perfect crop. Technically, the genetic engineering process is identical, but the result is a plant with only rice genes. Scientists are finding new potential genes by exploring the reason for viral resistance found in some rice varieties.

Figure 13.18

Virus-resistant transgenic rice.

Alexis DUCLOS/Gamma-Rapho/Getty Images

The advent of facile gene-editing techniques such as CRISPR/Cas9 have allowed scientists to selectively alter the genome of any organism. For instance, mice can be genetically modified to host a human disease so scientists can study disease progression and develop treatment options. Imagine being able to modify mosquitos so they can no longer transmit malaria, or revive extinct species from closely related relatives. All of these exciting possibilities are within the realm of CRISPR! You will learn more about this emerging technique in the next two activities.

What Is "CRISPR"?

There are many techniques that scientists may use to edit genomes. A recent method known as CRISPR, an acronym for "clustered regularly interspaced short palindromic repeats," has been touted as being much better than older gene splicing and editing techniques.

Watch this video (www.acs.org/cic) and use the Internet as a resource to answer the questions below.

©2018 American Chemical Society

Questions to Consider

a. Describe how CRISPR/Cas9 works and why this technique is more efficient than older methods.

b. What applications have recently been reported for CRISPR/Cas9? Can this technology be used to create human "designer babies"? If so, describe some ethical concerns regarding this controversial application.

c. Describe how climate change may be solved using CRISPR. How can plants be improved to help mitigate climate change?

d. "Borgs" are pieces of protein-wrapped DNA or RNA that float from bacteria to bacteria infecting them. What advantage might these foreign invaders have for us?

Real Talk   The CRISPR Revolution

Check out this interview with Dr. Kevin Yehl, who describes how CRISPR can be used for future medical developments: www.acs.org/cic.

Questions to Consider

a. Dr. Yehl gives an analogy of "genetic scissors" to describe CRISPR/Cas9. What is meant by this statement?

b. In Section 12.9, we discussed antibiotic resistance as a problem for modern medicine. Using the information from the video and the Internet as resources, describe how CRISPR is promising to address this issue.

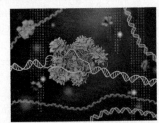

Meletios Verras/Shutterstock

c. Dr. Yehl discusses how easy it is to run a CRISPR experiment. Using the Internet as a resource, describe the techniques used in the lab to perform a CRISPR experiment and how one determines whether the experiment has been successful.

Perhaps you have a trait in mind but have no source for the gene. Traditional selective breeding and whole-gene transfer won't work in such a case. Neither of these can function if a trait does not already exist. Although, in theory, you could wait for a trait to evolve through natural, random mutations, most likely, you would run low on patience. To speed up the evolution process, scientists use chemicals or ionizing radiation to produce random mutations in a batch of seeds. Upon planting, some seeds do not grow, others grow but with no apparent changes, and others show unique traits. Once a useful trait appears, scientists can use either selective breeding to refine a new crop with the positive trait or genetic engineering to isolate the gene responsible for the trait and transfer it into a different plant.

The sheer power of genetic engineering and the idea of transgenic organisms can frighten us. In the next section, we describe the benefits of genetic engineering to the chemical industry, both current and projected. Later, we delve into other uses and potential risks of genetic engineering.

13.7 | Better Chemistry Through Genetic Engineering

Learning Objective: Describe some applications of genetic engineering

Pest and herbicide resistance are not the only reasons for genetic modification. Scientists have also modified corn, soybeans, and wheat genes to make them more resistant to disease, more tolerant of stresses such as salt, heat, and drought, and more nutritious. Although farmers have benefited, many other people have as well. Scientists have designed plants to absorb toxic metals from contaminated soil. Some are developing crops, such as soybeans, that produce high biofuel yields per acre. Others are engineering bacteria to detect and remediate radioactive contamination. Of interest to us in this section is that genetic engineering can incorporate the key ideas of green chemistry into large-scale chemical production.

A traditional synthetic route to a desired chemical may require toxic chemicals, large amounts of solvents, and high temperatures. Although such processes yield many useful chemicals, they also produce a staggering amount of waste—up to 100 times the weight of the desired compound! One way to reduce the ecological footprint is to use enzymes. These "biological machines" perform reactions as you would conventionally within flasks or beakers, but faster and safer with fewer toxic reagents, lower temperatures, and less waste—all hallmarks of green chemistry. Another benefit is that enzymes can be used over and over again.

Growing bacteria with natural or artificial genes is now a step in manufacturing small-molecule drugs such as insulin (Figure 13.19). As discussed earlier, this application of genetic engineering is not only one of the most rapidly growing uses of the technology but also the oldest, dating back to the development of synthetic insulin in the late 1970s. The process is straightforward when the gene, or something similar, already exists. However, special tricks are sometimes required, as was the case in synthesizing the drug atorvastatin (Figure 13.20), in which enzymes for specific reaction steps could not be found, so they had to be evolved. Scientists mimic natural selection by creating an environment where the bacteria must evolve a new trait to survive—a process known as **directed evolution**. Typically, scientists start with a random variety of DNA sequences transferred into a population of bacteria. A new enzyme emerges over multiple generations from growth under certain conditions, such as providing only certain chemicals as food.

Engineered organisms can even produce plastics. Most synthetic polymers are fabricated in large chemical plants using a process that consumes large amounts of chemical reagents and energy. Furthermore, these chemical reagents are often derived from petroleum. Scientists at Metabolix have engineered organisms to improve the sustainability of traditional polymer syntheses. These organisms produce monomers from renewable materials such as corn, sugarcane, and vegetable oil and catalyze

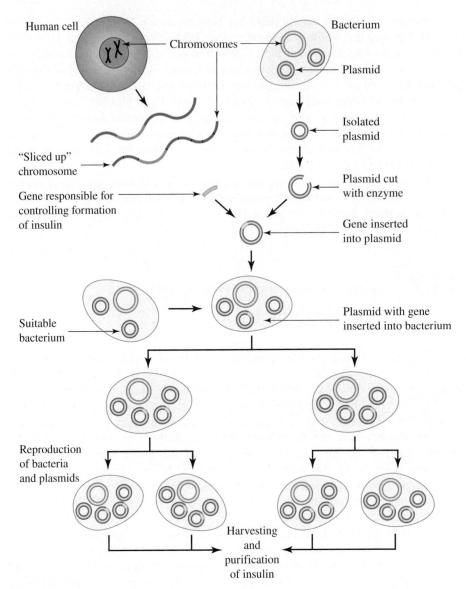

Figure 13.19

An illustration of how human insulin is produced through the use of bacteria.

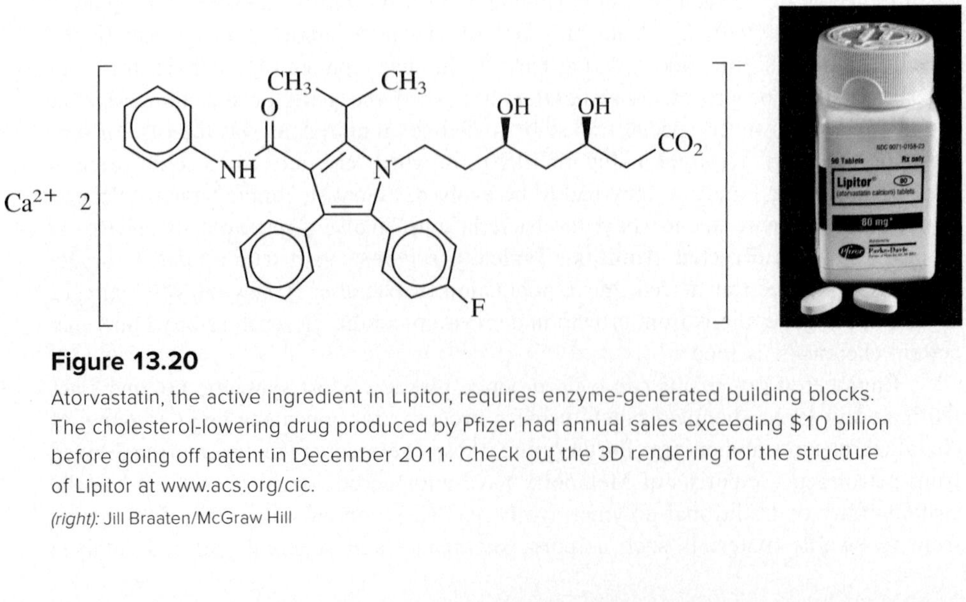

Figure 13.20

Atorvastatin, the active ingredient in Lipitor, requires enzyme-generated building blocks. The cholesterol-lowering drug produced by Pfizer had annual sales exceeding $10 billion before going off patent in December 2011. Check out the 3D rendering for the structure of Lipitor at www.acs.org/cic.

(right): Jill Braaten/McGraw Hill

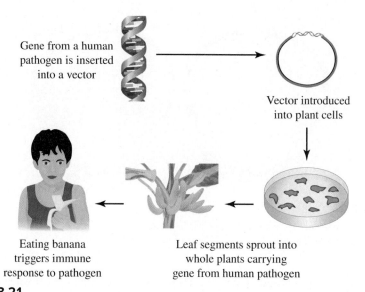

Figure 13.21

The fabrication of transgenic plants with edible vaccines using a recombinant DNA approach.

the polymerization reaction. One example of a resulting polymer is polyhydroxybuty-rate (PHB). This bioplastic is used in items such as plastic utensils and coatings for cups, much like polypropylene (PP). However, unlike PP, PHB is biodegradable. This process uses low-toxicity materials, is extremely efficient, and lowers green-house gas emissions.

Imagine using our farms to create polymers, drugs, and vaccines. This is no longer a utopian dream because our transgenic plants can now produce both. For exam-ple, vaccines against infectious diseases of the intestinal tract have been made in pota-toes and bananas (Figure 13.21). Anticancer antibodies have been expressed and introduced into wheat. Peptide drugs against HIV/AIDS have been produced from tobacco fields. Also, most vaccines require refrigeration or other special handling, together with trained professionals to administer them. Vaccines produced within edible products may be difficult to dose correctly but would be easy to administer and trans-fer. In addition to yielding crops for food, these fields could provide the hope of low-cost, readily available vaccines. Thus, fields of transgenic plants can go hand-in-hand with a good public health policy.

Another case study illustrating the utility of genetic engineering is related to papaya production in Hawaii. In the 1990s, an outbreak of the papaya ringspot virus took a severe toll on fruit production (Figure 13.22). Researchers at the University of Hawaii developed a type of papaya that incorporated DNA from the ringspot virus, making papayas resistant to the disease. Seeds for the genetically engineered papaya were then distributed to Hawaiian farmers for free, and now more than 80% of the papaya crop in Hawaii consists of this virus-tolerant variety.

Genetically engineered crops are also being used in the fight against malnutrition. Researchers at the Swiss Federal Institute of Technology and the University of Freiberg developed a strain of rice that provides a significant amount of vitamin A (or beta-carotene), a vitamin critical in the development and function of eyesight. An estimated 250,000 – 500,000 children worldwide become blind, and 667,000 worldwide die yearly from vita-min A deficiency. The regions of the world with the highest risk for this deficiency are Southeast Asia, Africa, and South America. This strain of rice, called "golden rice" because of the yellow color of the beta-carotene in the grain (Figure 13.23), provides the daily recommended allowance of vitamin A in less than half a cup of rice. The rice developers provide the rice royalty-free to farmers making less than $10,000 annually, representing 99% of all farmers in the regions most affected by vitamin A deficiency.

Compared with traditional methods, a genetically engineered enzyme, microorganism, or crop can yield high-purity products while reducing waste and by-products. Genetically engineered routes increase the yield with fewer steps and often eliminate labor, energy, and

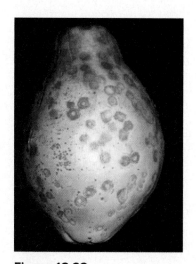

Figure 13.22

Photo of the papaya ringspot virus.

Source: Courtesy S. Ferreira

Figure 13.23

A comparison of genetically modified golden rice with traditional white rice.
Ingram Publishing/Alamy Stock Photo

resource-intensive purification processes. They avoid using toxic and corrosive chemicals, which is better for the environment and increases overall worker safety.

This all being said, the promise of developing technology and the desperate needs of our society can lead to rash choices. In light of this, the risks of genetically engineered cures may be large, perhaps unacceptably so. Let's now delve into the potential risks of genetic engineering.

13.8 | The Great GMO Debate

Learning Objective: Analyze controversial issues associated with genetically modified foods

Although genetic modification has many advantages, it is not a technology without its detractors. Indeed, the topic of GMOs has sparked protests, labeling campaigns, and environmental vandalism (Figure 13.24). The arguments against GMOs and those that

Figure 13.24

Protesters burning transgenic soy seeds in Curitiba, Brazil, in 2006.
Orlando Kissner/AFP/Getty Images

support their use are often heated. This section explores these arguments, starting with the question: Are GMOs regulated?

GMO Regulations

Are GMOs federally regulated in the U.S. and around the world? Check out the video (www.acs.org/cic) and use the Internet as a resource to answer the questions below.

©2018 American Chemical Society

Questions to Consider

a. Which countries are the leading producers of GM crops?

b. CRISPR modifies plants to improve their quality, disease, and herbicide resistance. But why does this technique easily pass through the regulatory processes, when traditional GMOs can not?

c. What do you think about the regulatory agencies in the U.S. not having the authority to regulate GMOs? Do you think this is likely to change in the future?

Some arguments against GMOs include:

Genetic engineering is not natural.
Many people feel that inserting DNA from a different species into a plant, called *transgenetic manipulation*, is not a natural process. People argue that the transfer of genes in this way is beyond the role of humans.

- **Counterpoint.** Introducing foreign DNA into the genome of a plant is a natural process. Recent research has shown that bacteria similar to those genetic engineers use to modify corn with the *Bt* gene inserted their own bacterial DNA into sweet potatoes in South America at least 8000 years ago. What used to be a normal plant root became swollen with starch and was integrated into the diet of early Native Americans. Several varieties of sweet potato now exist throughout the world, and it is the world's seventh most important staple crop, being the primary source of Calories for millions of people. The exchange of DNA across species is not a rare event and most typically occurs in single-celled organisms such as bacteria, but the transfer of genes has also occurred between millet and rice.

GMOs cause health problems such as allergies and cancer.
There is some concern that genetic engineering would introduce dangerous compounds such as allergens or toxins into the food chain. Some of these compounds may be carcinogenic and lead to tumors in those who eat GMOs.

- **Counterpoint.** Although a new gene can express a protein that is an allergen or a toxin, research scientists work with food regulators and rigorously test their creations because they would be legally liable for any toxicity. For example, in 1996, a company trying to insert a beneficial Brazil nut gene into soybeans halted the project when it was discovered that the resulting soybeans also carried a nut allergen. Although most testing data is treated as proprietary information by biotech companies and not released to the public, thousands of published safety reports on GMOs have shown no evidence that the foods are dangerous.

Let's Debate!   "Superweeds"?

Monsanto came out with a new way to control pests in 1996. They marketed genetically engineered soybeans and corn resistant to the weed killer glyphosate.

Read the following paper: https://www.nature.com/articles/497024a

Form two groups: one group will be in favor of the topic (the proposition) and the other group will be against the topic (the opposition). Both groups will try to persuade a neutral person or judge to agree with them. The topic of the debate is the motion. Here is the motion:

Genetically modified organisms have led to increased use of herbicides and the growth of "super weeds."

Curiosity Conversations

- Has the increased use of glyphosate herbicides led to a decrease in the use of other more toxic herbicides?
- What are the environmental implications of GM crops relative to those grown traditionally?
- Would getting rid of GM crops stop the spread of herbicide-resistant weeds?

Farmers can't replant genetically modified seeds.

There is a belief that some seed companies use what is called a "terminator gene" to prevent the seeds from GMO crops from being able to be replanted.

- **Counterpoint.** To date, there have been no crops developed that incorporate this technology. Seed companies often do, however, require agreements that prevent farmers from replanting seeds gathered from a GMO crop, ensuring future purchases of new seeds every year. Before the advent of genetic engineering, most of the corn grown in the United States and the European Union was commercially developed hybrids that, when replanted, led to a mixture of inferior corn variants with lower yields. Beginning in the 1930s, seed companies began developing high-yield hybrid corn, and by 1965, 95% of the corn crops planted in the U.S. were hybrids, requiring the planting of the newly purchased seed to keep yields high. Although this trend started with corn hybrids, farmers growing other crops also began repurchasing new seeds rather than replanting saved seeds. Today, even those farmers not using GMOs prefer to purchase new seeds yearly to ensure a higher harvest yield.

As you can see, many arguments have been back and forth over the use and safety of genetically modified organisms. The first commercially approved GMO, a transgenic tomato more resistant to rotting after being picked, went to market in 1994. Many other crops have been approved and commercialized in the decades since then. Interesting developments include a type of apple that doesn't brown when cut and a potato that produces significantly less acrylamide (a possible carcinogen) when fried.

You may be surprised that you almost certainly have eaten a genetically modified food item. The U.S. Department of Agriculture has estimated that >90% of the corn planted in the United States was genetically modified. The vast majority (94% in the U.S.) of soybeans and cotton were also genetically modified. Considering the variety of products that include high-fructose corn syrup or soy-based vegetable oil, you likely have already consumed a genetically engineered product.

Your Turn 13.16   GMO Labeling

Use the Internet to find out the current status of mandatory GMO food labeling in the U.S. and around the world. List some pros and cons of the proposed legislation. Do you think these food labels would be useful for consumers? Justify your answer based on this section's arguments for/against GMOs.

Your Turn 13.17   Always Consider the Source!

The peer-review process is designed to ensure scientific validity for a study, which is useful to consider the pros and cons of the arguments presented in this section. However, there are instances that required the retraction of a scientific paper based on inconclusive evidence.

a. Using the Internet, find two instances of academic dishonesty, which resulted in retracting the scientific article from a peer-reviewed journal. What were the issues surrounding the study?

b. Find additional scientific studies from peer-reviewed journals that refute the common arguments against GMOs. Find opposing views on GMOs from the Internet, magazines, or newspapers. Do these opponents cite the results of any scientific studies? If so, where have these studies been published (e.g., online, newspapers, etc.), and have the results been allowed to be reviewed by other experts in the field (i.e., the "peer-review" process of publishing scientific data)?

c. Evaluate the report published in 2016 by the National Academies of Science (NAS) about genetically engineered crops. Do you agree with all of their findings? If not, explain your viewpoint, citing sources that contradict the conclusions of the NAS.

Conclusion

More than 30 years after the development of Humulin, and more than 20 years after the commercialization of the first genetically modified crop, there is still controversy surrounding genetic engineering. There are many parallels between the cases of climate change and genetically modified organisms. In both, most scientists in the field are convinced of the soundness of the science. An incomplete understanding of the science has led to misinformation and fanatical opposition to using the knowledge gained from scientific research.

A common accusation of those against GMOs is that the science supporting the use of genetically modified organisms is funded by the agricultural industry. While seed companies do fund research in this area, the majority of research takes place in academic labs supported by nonprofit and public funding agencies. For instance, the University of Hawaii and Cornell University developed ringspot-resistant papaya. Grants from the Rockefeller Foundation and the Bill and Melinda Gates Foundation partly supported research on golden rice.

Regardless of the funding source of research, scientists have a fundamental duty to approach their work unbiasedly. Although a researcher will often expect a certain result from their experiments, a poorly designed and biased experiment will lead to false conclusions, which will waste time and money and may even result in the loss of lives. The checks and balances of peer-reviewed work and scientific consensus ensure that scientific research—especially on controversial topics—is based on facts and logical reasoning.

Learning Outcomes

The numbers in parentheses indicate the sections within the chapter where these outcomes were discussed.

Having studied this chapter, you should now be able to:

- understand that the genome contains a code of instructions for life (13.1)
- describe the chemical composition and function of deoxyribonucleic acid (DNA) (13.2)
- interpret experimental evidence for the double-helix structure of DNA (13.3)
- understand the structural basis of DNA replication (13.3)
- explain the structure and function of codons in protein synthesis (13.4)

- discuss the primary, secondary, and tertiary structure of proteins (13.5)
- exemplify how small changes in a protein sequence may cause disease (13.5)
- explain how genetic engineering is used to alter the DNA makeup of an organism (13.6)
- describe some applications of genetic engineering (13.7)
- analyze controversial issues associated with genetically modified foods (13.8)

Questions

Emphasizing Essentials

1. The theme of this chapter is that DNA guides the chemistry of every living organism on the planet. Name three traits you possess that your DNA dictates.

2. List two industries that have changed and propose one that you expect to change with the rise of genetic engineering.

3. What are some differences between a protein and DNA?

4. Consider the structural formulas in Figure 13.1.

   a. What functional group(s) are found in the adenine molecule?

   b. What functional group(s) are found in the deoxyribose molecule?

   c. From what you learned in Section 11.4, why is deoxyribose a sugar and adenine is not?

5. a. What three units must be present in a nucleotide?

   b. What type of bonding holds these three units together?

6. Consider the acid–base equilibria of the phosphate ion in Figure 13.2. Describe the fractional composition of the phosphate ions inside the mitochondria where the pH is ~8.

7. DNA contains four bases: adenine, cytosine, guanine, and thymine. Name two similarities among the four bases. Highlight one feature unique to each.

8. Circle and name the functional groups in this nucleotide. Also, label the sugar, the base, and the phosphate group.

9. Compare the DNA segment in Figure 13.4 with the nucleotide shown in Question 8.

   a. Circle the two functional groups that react to form a polymer similar to DNA.

   b. The sugar in this nucleotide is ribose rather than deoxyribose. Suggest an appropriate name for a polymer of this nucleotide.

10. a. What does each letter in DNA stand for?

    b. Examine Figure 13.4. Which aspects of the DNA molecule does the name DNA highlight?

    c. Which aspects of the DNA molecule are not part of the name?

11. Here is the structural formula for the base thymine, as attached to the DNA chain:

    a. Label the H atoms that can hydrogen bond with a water molecule.

    b. Use electronegativity differences to explain why only these H atoms can form hydrogen bonds.

    c. Which atoms would form hydrogen bonds with water or other nucleic acids?

12. Explain why the base sequence ATG differs from the base sequence GTA.

13. Given a short sequence of DNA: TATCTAG

    a. Write a DNA code that complements the sequence given.

    b. Draw connecting lines between the sequences to represent the number of hydrogen bonds between each base pair.

14. Amino acids are the monomers used to build proteins.

    a. Draw the *general* structural formula for an amino acid.

    b. Name the functional groups in the structural formula you just drew.

15. Define a *codon* and its role in the genetic code.

16. Polar amino acids can be classified as acidic, basic, or neutral.

    a. Draw an example of a possible amino acid for each of the three types of polar amino acids.

    b. For each example, determine if the side chain can make hydrogen bonds, ionic bonds, or both.

    c. Describe each category in more detail. What functional groups would you expect in the side chains?

17. Describe what is meant by the primary, secondary, and tertiary structure of a protein.

18. Explain how an error in the primary structure of a protein caused by a DNA mutation in hemoglobin causes sickle-cell anemia.

19. Explain one similarity and one difference between selective breeding and the genetic engineering of plants.

20. Describe three benefits of using designed enzymes in organic synthesis.

Concentrating on Concepts

21. Diagram the steps to produce insulin from a cow or pig versus synthetic insulin from bacteria.

22. Figure 13.4 represents a segment of DNA.

    a. What part of each nucleotide becomes the backbone of the polymer?

    b. What part hangs off the backbone?

    c. How is the backbone represented in the sculpture at the beginning of the chapter and Figure 13.6?

23. Use Figure 13.7 to explain why adenine-thymine base pairs are less stable than cytosine-guanine base pairs.

24. a. What are X-rays?

    b. How is X-ray diffraction used to determine a crystal structure?

    c. The first X-ray diffraction patterns were of simple salts, such as sodium chloride. The X-ray diffraction studies of nucleic acids and proteins did not come until much later. Suggest two reasons why.

25. Explain why enzymes, or biological catalysts, can use one strand of DNA as a template to create a perfect complement strand.

26. Ionizing radiation (or the free radicals it generates in the cell) can break covalent bonds to disrupt one or more strands of DNA. A cell can more easily repair damage to a single strand in a DNA double helix than repair a double-strand break. Why is this?

27. Many compounds damage DNA and very effectively kill bacterial cells. Why are these compounds not typically used as antibacterial medicines?

28. Errors in DNA duplication can permanently alter the base sequence of a strand. But not all of these errors result in the incorporation of an incorrect amino acid in a protein for which the DNA codes. Explain why a single base change may not change the amino acid.

29. Almost all organisms use the same four bases and the same codons. Explain how the discussion of genetically engineering bacteria to produce insulin demonstrated this fact.

30. Insulin production through the genetic engineering of *E. coli* has received less public concern than genetic engineering in plants. Propose two reasons insulin production has not received the same public concern as genetically engineered crops.

31. Consider the idea of mixing genes as an improvement on nature.

    a. Describe what is meant by the term *transgenic organisms*.

    b. Consider using selective breeding, genetic engineering, and ionizing radiation to alter the genetic makeup of plants. List an advantage and disadvantage of each.

Exploring Extensions

32. Several companies were involved in the early production of insulin. Find one recent press release each from Lilly and Genentech. In one to two paragraphs, summarize the press releases, and then compare this recent work to the company's role in the development of insulin.

33. Of the major players in the discovery of the structure of DNA, Rosalind Franklin was not included in the 1962 Nobel Prize given for solving the structure of DNA. What was her background and experience that enabled her to make significant contributions? Explore the reasons why she did not receive adequate credit and recognition for her work. Summarize your findings, citing your sources.

34. The genetic traits leading to sickle-cell disease are more common in people of African, African-American, or Mediterranean heritage. Using the Internet, explain a proposed reason that the sickle-cell trait has persisted rather than being discarded through evolution.

35. Finding the structure of proteins can be challenging. Explore the program Foldit, a game and protein structure simulator created and supported by the University of Washington.

    a. In one paragraph, describe how the game combines human ingenuity and computer simulations.

    b. Explore one of the results from the program and write a two-paragraph report.

36. List two advantages and two disadvantages of issuing patents for genetically modified plants and seeds.

37. To clone or to breed? Grieving owners may find the opportunity to clone (genetically replicate) a beloved dog too tantalizing to resist. Even so, many professional dog breeders and their organizations advise against cloning. Phil Buckley, a spokesperson for the Kennel Club in England, argues, "Canine cloning runs contrary to the Kennel Club's objective to promote in every way the general improvement of dogs." Explain why cloning cannot improve dog breeds.

38. Transgenic plants have not been widely accepted in all countries.

    a. For the European Union, find two examples of transgenic plants that have been banned. Compare these to the two that were allowed. Discuss the differences between those allowed and those rejected.

    b. Create a timeline of five events that were key to the rapid increase or subsequent leveling (or both) of the adoption of transgenic crops in the United States. Briefly discuss each of your choices.

39. One reason why science fiction is successful is that it starts with a known scientific principle and extends, elaborates, and sometimes embroiders it. *Jurassic Park* began with the known scientific principle of copying and manipulating DNA and expanded on it to focus on the production of prehistoric creatures. Now it is your turn. Choose any scientific principle from this text. Then write a one- or two-page outline for a story based on that principle. Be sure to identify the chemical concepts and any pseudoscience you employ.

40. Recently developed techniques have dramatically changed our ability to alter DNA in human cells. Use the Internet to gather information on this rapidly advancing medical tool. Write a one- to two-page report on gene therapy, including specific examples of how diseases are being treated and how patients are faring.

41. Find a transgenic organism not discussed in the text. Describe the motivation for engineering this organism, its gene source, and a general description of the genetic modification.

42. You are the head of a government facing another year of a long drought and a severe risk of famine. Another nation has offered you genetically modified rice to feed your people. List two advantages and two disadvantages of accepting the aid. Decide whether you would accept it and explain why.

Design Elements: Concentrating on Concepts icon, and Simulations/Interactives Icon SAK Design/Shutterstock.

14 Who Killed Dr. Thompson? A Forensic Mystery

FORENSIC EVIDENCE COLLECTION

As mentioned in the chapter opening video (www.acs.org/cic), a number of techniques are required to analyze evidence in a forensics investigation. However, the most crucial component of the investigation is related to the crime scene itself. With emotions running high, those among the first to respond to a crime must take steps to ensure that the crime scene is not contaminated. Furthermore, forensic investigators must collect and record all relevant evidence in an appropriate manner. For each of the following personnel, list three important operating protocols that you think would help protect the integrity of the crime scene:

a. Initial responder (police officer, firefighter, or emergency medical technician [EMT])
b. Detectives
c. Forensic scientists

In this chapter, the chemical principles discussed in earlier chapters will be used to solve a fictitious murder mystery. The questions we will be answering are "Who killed Dr. Thompson?" and "Why?"

Some topics from previous chapters that will be woven into the storyline include:

- The periodic table and mixtures
- Air quality
- The electromagnetic spectrum
- Isotopes
- Polarity and intermolecular forces
- Combustion reactions
- Solar power
- Electric vehicles
- The composition of plastics
- Enzymes
- Drug design
- DNA extraction

14.1 | *Friday, Aug. 1—7:08 PM: A Relaxing Evening Interrupted*

After another long and productive work week, Professor David Thompson relaxed with a glass of wine, contemplating the future with joyful anticipation. In just two weeks, he would deliver a landmark presentation at the American Chemical Society meeting in Boston—marking the culmination of his 25-year research career.

For the last 20 years, his research has focused on the development of "smart" cancer treatment drugs. His most promising molecule, called Zeta-12, features a targeting agent that attaches itself to individual cancer cells, and then releases the drug to kill the cells with high specificity and efficiency. He has received continual funding from the National Institutes of Health (NIH), which helped fund the basic studies and extensive clinical trials for the drug. His presentation at the upcoming meeting would detail the successful completion of Phase III clinical testing, leading to FDA approval, and a possible trillion-dollar market. Next week, he plans to partner with a major pharmaceutical company that would supply Zeta-12 to cancer treatment centers across the globe.

Your Turn 14.1   Cancer Treatment Drugs

Using the Internet, find examples of "smart" cancer treatment drugs that are currently in development.

a. What are the possible benefits of these drugs, relative to those currently used for chemotherapy?
b. In drug discovery, what are the differences between Phase I, Phase II, and Phase III clinical testing?

Thompson's phone rang, interrupting his train of thought. *I'll just let it ring, since it's probably Julie complaining about her alimony*, he thought. It had been six months since their heated divorce—an all-too-common ending for marriages with a workaholic spouse. Convinced that even his ex-wife couldn't rob him of his joy, he walked across the room to his desk and answered the call anyway, before it went to voicemail.

"Dr. Thompson?" the baritone voice asked, the tone urgent.

"Yes," Dr. Thompson replied, concerned.

"There's been an accident at your laboratory, and we need you to tell us—"

"What kind of an accident?" Dr. Thompson interrupted.

"A raging fire. The hazmat group is on site right now, and we need to know what chemicals you have so we can safely extinguish the fire."

"Are you kidding? We have solvent stills in the lab that contain sodium metal."

"Which solvents?"

Upset by the seemingly irrelevant question, Dr. Thompson stammered. "Toluene, . . . THF, ether, . . . acetonitrile, hexanes, and, um, . . . dichloromethane."

"Was there anyone working in your lab tonight?"

"No."

"Thanks, Dr. Thompson. We'll do what we can to save your lab."

Before slamming down the phone, Dr. Thompson shouted "I'm on my way!"

14.2 | Solvent Stills: An Effective but Dangerous Way to Purify Solvents

> Learning Objective: Differentiate between extinguishing media used for various types of fires

In Section 6.11, we discussed the process of distillation for the separation of various fuel products from crude oil. A **solvent still** (Figure 14.1) employs a similar process to remove moisture and oxygen from organic solvents. The resultant "dry" and

Figure 14.1

Schematic of a solvent still. Check out the video to see a solvent still in action at www.acs.org/cic.

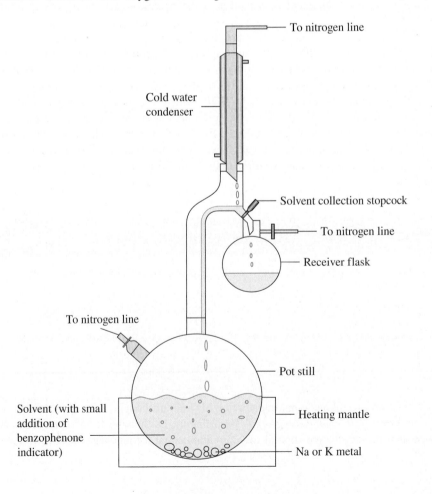

To nitrogen line

Cold water condenser

Solvent collection stopcock

To nitrogen line

Receiver flask

To nitrogen line

Pot still

Solvent (with small addition of benzophenone indicator)

Heating mantle

Na or K metal

oxygen-free solvents are used for many chemical reactions, in which traces of water or oxygen would result in contamination and unsuccessful syntheses. As the organic solvent is heated to boiling, its vapor rises up the glassware and is condensed back to its liquid state using a water-filled condenser. The repeated process of heating a solvent to boiling and condensing its vapor is known as **reflux**. The moisture/air-free solvent is collected in the receiver flask, which may be removed for use via syringe. A nitrogen bubbler system is used to flush the solvent still with an inert gas to prevent moisture and oxygen from dissolving in the solvent, and prevent a buildup of pressure that would result in a dangerous explosion.

An alkali metal such as sodium or potassium serves as a drying agent, reacting with both water and oxygen dissolved in the solvent, thereby removing these impurities. Nonchlorinated organic solvents, such as benzene, toluene, pentane, and hexanes, are purified by adding benzophenone (an O_2/H_2O indicator) and an alkali metal, and then heating the mixture to reflux in a pure nitrogen environment (Figure 14.1). Recall from Section 8.11 that alkali metals such as sodium are readily oxidized and therefore easily lose electrons. Since these electrons serve to reduce another species present in the redox reaction, alkali metals are considered strong **reducing agents**.

As shown in Figure 14.2, each sodium atom donates an electron to a benzophenone molecule, resulting in a sodium ion (Na^+) and a radical anion, evidenced by a characteristic blue-purple color. Within a few hours of adding sodium to the colorless organic solvent, the solution becomes a royal blue-purple color, which indicates that a radical anion has been formed. The radical anion then reacts with trace amounts of water and oxygen present in the solvent to form a number of reaction products (Figure 14.2). Hence, a persistent blue-purple color indicates that the solvent is free (<10 ppm) from both moisture and oxygen.

Some safer alternatives for solvent purification consist of stainless-steel columns packed with silica (SiO_2) or alumina (Al_2O_3) powders with high surface areas (>200 m²/g) and small particle sizes (~5 nm), molecular sieves (aluminosilicate minerals with an average pore size of 4 Å), and a copper catalyst powder—used to react with/remove oxygen.

Although Dr. Thompson often heard of explosions and fires from solvent stills, he meticulously maintained his stills and provided rigorous training for his students to ensure this would not happen on his watch. As he rushed to put on his shoes and

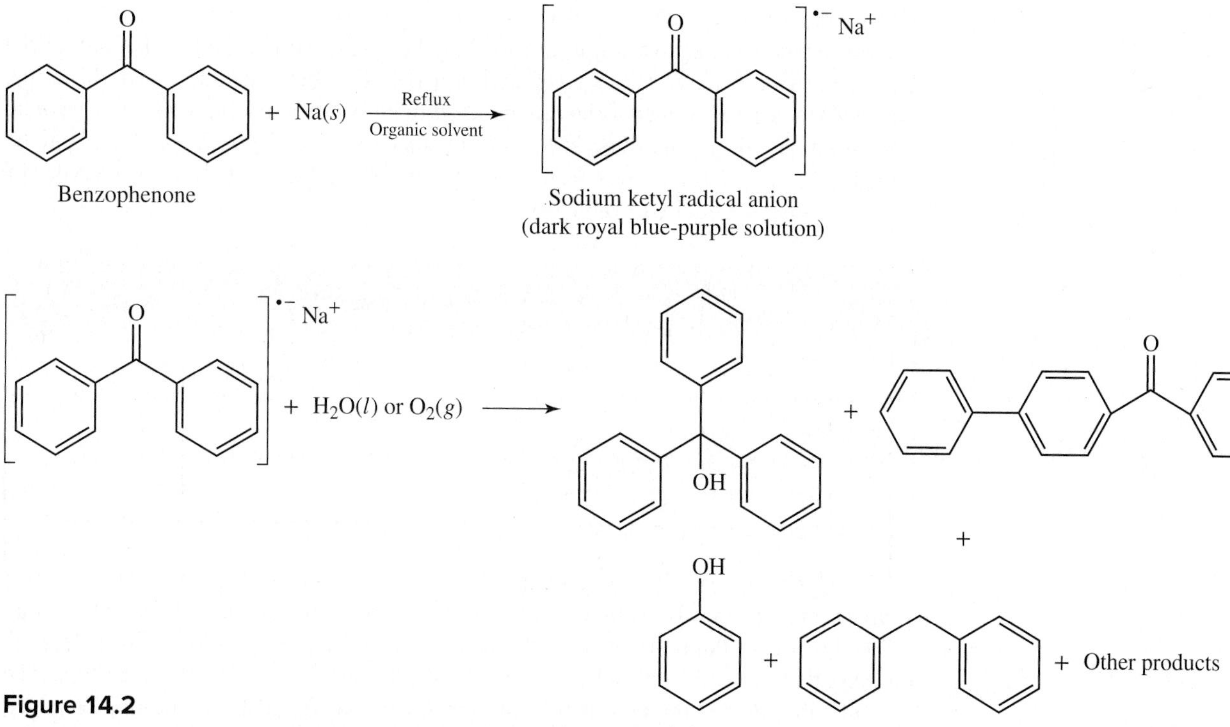

Figure 14.2

The reactions of sodium and benzophenone with oxygen and water within a solvent still.

Table 14.1	The Five Classes of Fires and Required Fire Extinguishers			
Class of Fire	**Type of Fire**	**Type of Extinguisher**	**Extinguisher identification**	**Symbol**
A	Ordinary combustibles: wood, paper, rubber, fabrics, and many plastics	Water, dry powder*, halotron**	A (triangle)	
B	Flammable liquids and gases: gasoline, oils, paint, lacquer, and tar	Carbon dioxide, dry powder*, halotron**	B (square)	
C	Fires involving live electrical equipment	Carbon dioxide, dry powder*, halotron**	C (circle)	
D	Combustible metals or combustible metal alloys	Dry powder*	D (star)	
K	Fires in cooking appliances that involve combustible cooking media: vegetable or animal oils and fats	Wet chemical***	K (hexagon)	

Notes:

*Dry powder is graphite or sodium chloride powder pressurized by nitrogen gas.

**Halotron is a mixture of tetrafluoromethane (CF_4) and argon gases.

***Wet chemical is an aqueous salt solution such as potassium carbonate (K_2CO_3) with added detergents to improve the wettability of grease/oils, which are nonpolar.

Source: County Fire Protection

fumbled with his car keys, he recalled a stainless-steel solvent purification system he saw months earlier at a conference. *Why didn't I upgrade to this safer alternative?* he thought. Speeding through the neighborhood and taking the on-ramp to the freeway, his mind raced to inventory the items that could be lost in the fire. How could he ever replace the lab notebooks that detailed his life's work? What if the notebooks were needed for the patent prosecution process? As he screeched to a halt at the side of the building, he was relieved to see so many fire trucks and personnel already on the scene. His feelings of panic waned as he remembered that his lab notebooks were in a separate lab. *Surely, the firefighters would be able to stop the blaze before it reached them*, he thought. The hazmat team was well aware of the five classes of fires, and the types of fire extinguishers required (Table 14.1).

Your Turn 14.2   Tricky Firefighting

a. From what you know so far about what is present in Dr. Thompson's laboratory, what should be used by the hazmat crew to extinguish the fire? What would happen if they simply used water?

b. Speculate why it would not be wise to use sodium metal and benzophenone to purify chlorinated solvents, such as dichloromethane.

Dr. Thompson's lab contained wooden cabinets and desks; however, the fuel sources were primarily organic solvents (Class B) and combustible sodium metal (Class D). Accordingly, the fire crew entered the burning lab, with one team focused on containing the fire around the solvent stills using dry-powder extinguishers. The second crew sprayed the rest of the lab with water to extinguish the burning of paper, wood, and plastic materials.

Your Turn 14.3 The Fire Triangle Revisited

Explain how dry-powder, wet-chemical, and dry-chemical fire extinguishers work, by referring back to the fire triangle discussed in Chapter 6.

Under the sounds of breaking glass, crackling flames, and pounding metal, Dr. Thompson stood helpless, watching the firefighters scramble to take control of the fire. He scurried back and forth along the perimeter of the scene, asking crew members to divulge details about his lab, to no avail. At long last, firefighters began exiting the building and congregated outside the building, a sign that the fire was finally extinguished. Thompson approached the fire commander, who was speaking with witnesses.

"Can I have a look at the damage to my lab?"

"Not yet. You won't be allowed in for at least a few days until we finish our investigation. We'll call you if we have any questions. Please go home and try to get some sleep."

"Will you let me know if the firefighters were able to save my lab notebooks?"

"Sure. Where were they located?"

"In 305B, the workroom connected to my synthetic lab."

"Okay, will do. We'll be in touch."

Dr. Thompson headed to his car, taking one last look at the biochemistry building, unsure when he would be able to evaluate the extent of the damages.

DID YOU KNOW ?

Many labs have now transitioned to secure electronic laboratory notebooks stored in a cloud. Although e-notebooks are now commonplace in industry, they are not yet so prevalent in academic research laboratories.

14.3 | *Friday, Aug. 1—10:13 PM: The Aftermath*

Luckily, no one was hurt or injured by the devastating fire. Two graduate students from another research lab had reported the fire after hearing a loud explosion. Their office was located along the same hallway as Dr. Thompson's lab. One of the students thought he heard the door of Dr. Thompson's laboratory shut about 30 minutes before the explosion, which was typically very quiet during the evening hours. Neither Dr. Thompson nor his students typically worked at this time in the evening, and the janitors passed through much later, so it was strange that someone would be in his lab.

City fire department investigator Jim Williams was called to the scene to determine whether the fire was natural, accidental, or deliberate in nature. As was his ordinary protocol for fire investigations, Investigator Williams suited up into disposable coveralls that covered his hair, gloves, and shoes. The smell of smoke still saturated the stairwell, which became more pronounced as he climbed the flights of stairs toward the third-floor laboratory. Although he had investigated many lab fires across the state, what he saw as he stepped into Dr. Thompson's laboratory was still surprising (Figure 14.3).

Wow! It looks like a bomb went off in here! Williams thought as he stood in the doorway. He decided to use a grid search pattern because of the extent of the disarray, and took careful note of the doorways, windows, and other possible entry/exit ways. Caution tape was fastened across the main doorway to prevent unauthorized people from entering the lab until his investigation was completed.

The laboratory space featured two adjoining rooms (Figure 14.4). The first was a large (15 m × 15 m) synthetic lab with numerous organic and inorganic chemicals, solvents, and associated storage areas. A workroom was connected through an open doorway, which contained desks, filing cabinets, and a bookshelf that held

Figure 14.3

The extensive fire damage to a student desk in Dr. Thompson's laboratory.

Vladimir Mulder/Shutterstock

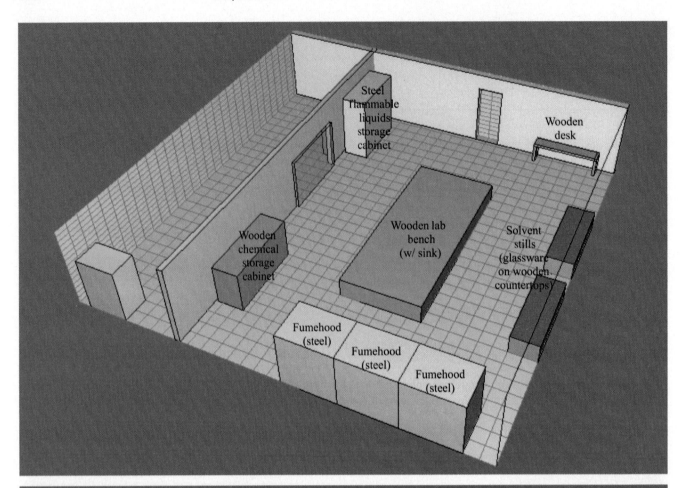

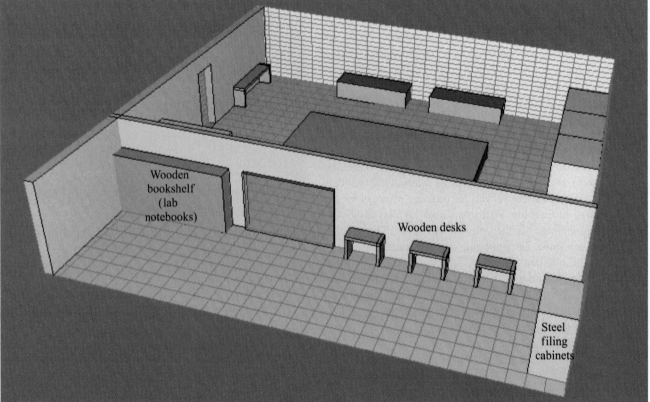

Figure 14.4

Schematics of Dr. Thompson's laboratory. Shown is the main synthetic laboratory (top), which housed the solvent stills believed to be the primary source of the fire. The adjoining room (bottom) housed student desks, filing cabinets, and a bookshelf with laboratory notebooks that documented the many years of research by Dr. Thompson and his employees.

Dr. Thompson's treasured lab notebooks. Sadly, both labs were totally decimated by the impacts of fire, soot, and smoke. Although the light switch in the lab was in the on position, darkness filled the room; the electricity was turned off by the firefighters. Upon inspection of the room with a flashlight, Williams noticed that the lightbulbs had become deformed and cracked due to the extreme heat.

The extent of damage to the workroom was somewhat suspicious to Williams. Even though both labs were connected by an open door, he was skeptical that the fire could spread so efficiently to the bookshelf that was tucked around the corner in the adjoining room. In comparison, there was much less fire and smoke damage to the desks and filing cabinet on the opposite side of the workroom. Perhaps most concerning was the presence of ruptured solvent containers near the stills, and empty acetone canisters outside of the flammable cabinet, near the doorway that connected the two rooms. Lastly, as reported by witnesses, Williams was also suspicious about the possibility of someone present in Dr. Thompson's laboratory just minutes before the fire was reported. He drew detailed sketches and took numerous photographs of both labs, compiling notes about fire damage to the walls and floors. Williams soon realized this case might be out of his expertise: so many chemicals, unfamiliar equipment, and complicated methods. He decided to call the state arson forensic expert to help with the investigation.

14.4 | Saturday, Aug. 2—8:05 AM: Accidental or Deliberate?

> *Learning Objective: Determine the factors that influence the flammability of a liquid*

A photographer and an arson investigator arrived on the scene, to follow up on Williams's questions and determine whether the fire was started deliberately. Williams discussed his findings with the forensic arson investigator, Dr. Keisha White, who asked additional questions about the conditions of the lab, Dr. Thompson's whereabouts at the time of the fire, and whether there were any witnesses or if anyone was seen entering or leaving the building at the time of the fire.

Dr. White was the lead forensic arson investigator for the state and was skilled at determining probable causes of fires at both residences and businesses. Fitted with protective coverings, her forensic team processed the scene by taking photographs, compiling detailed notes about the scene conditions, and collecting evidence.

As expected, the most extensive fire damage was observed near the flammable solvent stills. Two groups of solvent stills were housed in the laboratory. On the left were toluene, hexanes (*n*-hexane + other isomers), and tetrahydrofuran (THF); on the right were diethyl ether, dichloromethane, and acetonitrile. Because these solvents were thought to be the origin of the fire, the arson investigation team needed to consider the relative flammabilities of the solvents present in Dr. Thompson's laboratory.

Your Turn 14.4   Isomers Revisited

One of the solvents is listed as "hexanes." Draw the structural formulas and line-angle diagrams for five possible isomers of hexane. **Hint:** Review Section 6.12.

The flammability of a liquid is related to the concentration of its vapor in air. As you may recall from Chapter 6, the vapor pressure of a liquid is directly dependent on its temperature. As its temperature is increased, the intermolecular forces among the liquid molecules are broken, which results in higher concentrations of gaseous molecules. The lowest concentration of solvent vapor that can ignite in air is known as the **lower flammability limit** (LFL). The **flash point** of a solvent occurs when the vapor pressure of a solvent equals its LFL, and combustion is therefore possible in the presence of an ignition source. The flash point of a fuel will allow combustion, but will not sustain it.

CH₃

Toluene

Diethyl ether

O

Tetrahydrofuran
(THF)

n-Hexane
(+ other isomers)

Dichloromethane
(also known as methylene chloride)

H₃C—C≡N
Acetonitrile

Check out the 3D renderings for toluene, diethyl ether, THF, n-hexane, dichloromethane, and acetonitrile at www.acs.org/cic.

Table 14.2		Relative Fire Hazards of Liquids	
Class	Flash Point (°C)	Boiling Point (°C)	Examples
IA	<22.8	<37.8	Ethylene oxide, methyl chloride, pentane
IB	<22.8	≥37.8	Acetone, benzene, ethanol, gasoline, iso-propanol
IC	≥22.8	<37.8	Butanol, diethyl glycol, styrene, turpentine
II	≥37.8	<60	Camphor oil, diesel fuel, pine tar, Stoddard solvent
IIIA	≥60	<93	Creosote oil, formaldehyde, formic acid, fuel oil #1
IIIB	≥93	>93	Castor oil, coconut oil, fish oil, olive oil

In contrast, the **autoignition temperature** corresponds to the minimum temperature at which the vapor spontaneously ignites, even in the absence of an ignition source.

In order to properly assess the dangers associated with the storage and use of flammable and combustible liquids, a classification system has been devised by the National Fire Protection Association (NFPA). This scheme ranks the relative fire hazards of liquids based on their flash and boiling points (Table 14.2).

Your Turn 14.5   Flash Points

a. At the flash point, is a chemical or physical change taking place? How about at the boiling point?
b. What are some differences and similarities between "flammable" and "combustible" liquids?
c. The NFPA classification scheme presented above is based on an atmospheric pressure of 1 atm (sea level). Would the flash points of the liquids be higher or lower at higher altitudes? What would this mean with respect to storing or handling these liquids at a higher altitude location, such as Denver, Colorado?

Your Turn 14.6   Fire Hazard Ranking

Using the NFPA classification scheme, provide the Class ranking for each of the solvents present in Dr. Thompson's laboratory.

Table 14.3 lists some thermal parameters for solvents found in Dr. Thompson's laboratory. The boiling point of a solvent occurs when the vapor pressure of the solvent equals the external atmospheric pressure. However, the flash point often occurs at temperatures much less than its boiling point—as long as the LFL is exceeded, combustion is possible. In assessing the overall flammability of a solvent, the flash point is more diagnostic than its boiling point. For instance, toluene has a relatively high boiling point (low volatility), but a relatively low flash point (high flammability). In contrast, dichloromethane has a lower boiling point (more volatile), but a much higher flash point (less flammable). This difference is related to the relative reactivity of a particular solvent with oxygen during combustion. The heat of combustion for toluene is 3900 kJ/mol, whereas the heat of combustion for dichloromethane is only 605 kJ/mol. The more heat that is generated by the reaction of a fuel and an oxidizer, the easier it is to exceed the **activation energy** of the reactants, resulting in ignition of the fuel/oxidizer mixture. Therefore, dichloromethane is less flammable than toluene because its combustion is less exothermic—i.e., less energetically preferred.

Table 14.3	Thermal Properties for the Solvents Found in Dr. Thompson's Lab		
Solvent	Boiling Point (°C)	Flash Point (°C)	Autoignition Temp. (°C)
Diethyl ether	35	−45	160
Dichloromethane	40	100	600
Acetone	56	−20	465
Tetrahydrofuran	66	−14	321
Hexanes	69	−26	223
Ethanol	78	17	365
Acetonitrile	82	2	523
Toluene	111	6	530

Your Turn 14.7   Spontaneous Combustion?

a. Consider the solvents listed in Table 14.3. If an ignition source (e.g., a spark from a match) is present, which solvent vapors will catch fire at room temperature (20–25 °C)?

b. Based on the autoignition temperatures of the solvents found in Dr. Thompson's laboratory, could any of these solvents combust without the presence of an outside source of ignition?

Your Turn 14.8   Combustion Reactions and Air Quality

a. Write balanced equations for the combustion of three solvents of your choosing found in Dr. Thompson's lab.

b. What other products, such as CO and particulate matter, will likely be generated from the combustion of solvents and common building materials that may be present in the lab? Which of these products will likely exceed the air quality standards discussed in Chapter 2?

c. Atmospheric monitoring of Dr. Thompson's lab discovered a mercury concentration of 3000 ng/m^3—10 times higher than the EPA safe standard of 300 ng/m^3. List some possible sources of mercury resulting from the lab fire. It should be noted that Dr. Thompson's laboratory did not contain any mercury-based thermometers, thermostats, manometers, or bottles of elemental mercury.

Broken glassware from the solvent stills was scattered throughout the lab. The wall immediately behind the solvent stills, and the flooring in front and along the sides of the stills, were severely damaged. In addition, the wooden desks in the lab were heavily burned, charred, and covered with a thick layer of soot. Due to the presence of water-reactive metals, the lab did not have an automatic sprinkler system, which would have limited the extent of damage to wooden desks, tables, shelves, and cabinets.

The arson investigation team is trained to pay particular attention to the following suspicious items that may be present in the aftermath of a fire:

- Suspicious burn patterns on floors and/or walls, such as fire trails or multiple points of origin
- Degree of charring of materials
- Flaking of plaster or concrete
- Distortion of materials softened or weakened by heat
- Soot and smoke damage

Figure 14.5

Simulated wall burn patterns from solvent still fires in Dr. Thompson's laboratory. Shown are **(a)** early, and **(b)** later, stages of the fire. Simulations were performed using PyroSim software (Thunderhead Engineering).

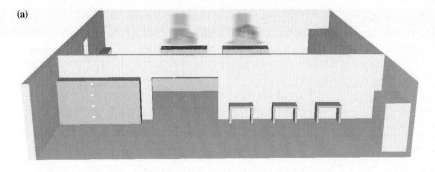

(a)

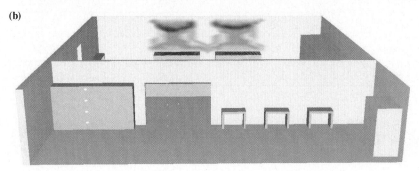

(b)

- Empty containers of fuel at the scene, especially those found without a cap fastened
- Presence of an accelerant, such as open containers or pools of unburned ignitable liquids
- Evidence of tampering with a sprinkler system, if present

For the combustion of a fuel located near a wall, the resultant burn pattern will adopt a characteristic shape at early stages of a fire (Figure 14.5a). As the flames become more intense and travel upward to intersect the ceiling, the wall burn pattern will become more columnar, with the appearance of circular burn patterns on the ceiling (Figure 14.5b). However, a more complex burn pattern may be present if an accelerant has been poured or sprayed onto the wall by an arsonist. To be complete, Dr. White and her team also looked for what *may not be* present, which would indicate an accidental or natural cause for the fire.

Diagnostic burn patterns can also be found on floors of an arson scene that match the pool and flow from pouring a liquid accelerant, known as a *fire trail*. In addition, the floor will often display an intermixed array of light, medium, and heavy burn patterns (e.g., Figure 14.6a), depending on the type of accelerant used.

However, some solvents will result in little or no scorching of the floor (Figure 14.6b and c). Why does a solvent such as gasoline cause so much damage to the underlying floor, whereas the combustion of acetone or ethanol is hardly noticeable? During

Figure 14.6

Burn patterns in commercial vinyl flooring from **(a)** gasoline, **(b)** acetone, and **(c)** ethanol.

©Bradley D. Fahlman

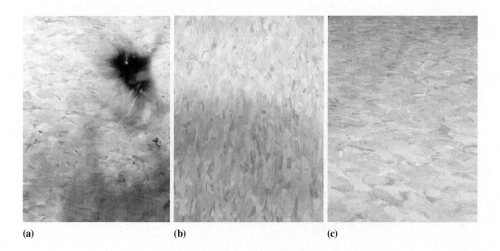

(a)   (b)   (c)

the combustion of an organic solvent, oxygen must combine with the gaseous solvent molecules in an appropriate concentration. As the oxygen content of the fuel increases (e.g., ethanol/acetone contains more oxygen than gasoline), the fuel burns more completely, leaving behind little or no residue.

Gasoline releases a tremendous amount of heat and burns more slowly than oxygenated fuels such as acetone or ethanol. Whereas the combustion of gasoline releases 47 kJ of heat per gram of fuel, ethanol and acetone release less heat, 29.7 kJ/g and 30.8 kJ/g, respectively. In order for damage to occur to underlying flooring as a result of fuel combustion, sufficient heat must be transferred from the burning fuel to the floor material (vinyl tile, etc.). Accordingly, it takes more acetone or ethanol to generate enough heat to cause fire damage to surrounding materials, relative to gasoline. This is especially the case with volatile solvents, because the liquid fuel tends to protect the flooring via evaporation rather than causing damage from combustion.

As noted the previous evening by first responders, Dr. White confirmed the presence of ruptured canisters of toluene and hexanes near the solvent stills, and empty cans of acetone located near the doorway that connected the workroom to the main lab. Not knowing whether it was Dr. Thompson's policy to store solvent near the stills instead of inside the flammable solvents cabinet, she placed a call to his home.

"Hello, Dr. Thompson?"

"Yes . . ."

"This is Dr. Keisha White, an arson investigator with the State Forensic Laboratory working the investigation into the fire at your lab."

"Good evening . . . Um, is there a problem?"

"No, just a routine investigation. I do have some questions for you, if you have a minute."

"Sure."

"Do you keep your solvent canisters inside the flammable solvents cabinet?"

"Yes, of course."

"Is there a possibility that you left some solvent containers near the stills? Maybe when you last refilled the flasks, for instance?"

"No, not a chance. I'm cautious when it comes to safety and I always ensure that solvents are put back in the cabinet after I use them."

"Who else has access to your lab?"

Dr. Thompson considered that question for a moment. "Just me."

Dr. White was puzzled by his response, "You don't have any students currently working in your lab?"

"No. My last student just graduated, and I'm still waiting for the international visas to be processed for two new postdocs."

"Are there any former employees that may still have access?"

"No. Our department policy is that all students have to turn in their keys before they can officially graduate. I did have a postdoc working with me a few months ago, but he no longer works at the University."

"Does he still have a key to your lab?"

"No. I'm sure he turned it in to the department office before he left."

"Okay. Is your lab always locked?"

"Yes. My lab is secured at all times."

"Is it possible that someone made a copy of your lab key before turning it back in? I can tell you that there was no sign of forced entry, or tampering with the lock."

"Hmmm. The keys have 'do not copy' on them, but I guess it's possible."

"We found some ruptured solvent canisters near the stills in your lab. We also found some empty cans of acetone near the door connecting your workroom. Does this seem suspicious to you?"

"Yes. I know for a fact that these bottles were always stored in the flammable cabinet."

"Thanks, that's all I wanted to know. We'll continue our investigation and will let you know more details as soon as possible."

"Thanks. Please call me anytime if you have questions."

DID YOU KNOW ?

It should be noted that burn patterns on a floor or multiple points of origin may also be a result from molten substances that have fallen from the ceiling, rather than the presence of a liquid accelerant. For example, heat from a fire may melt and ignite the asphalt from a collapsing ceiling or the polystyrene light diffusers in ceiling light fixtures. There are documented cases of fires erroneously classified as arson due to floor burn patterns, a problem derived from drawing a conclusion based on only one piece of evidence.

Suspicions regarding the role of arson continued to rise when a member of Dr. White's investigative team discovered some matchsticks—two used, and one unused—in the workroom near the bookshelf. A photograph was taken to mark the location of the matchsticks and Dr. White was called over to view the evidence herself. The matches were carefully placed into plastic evidence bags and labeled for later analyses in the crime lab.

In addition, the investigative team had to determine whether acetone was poured on the floor or lab notebooks to ignite the fire in the workroom. If solvent cans were usually stored inside a storage cabinet, their placement on the floor of the lab indicates that the solvent may have been used to start a fire in the workroom. The investigators postulated that the arsonist might have poured acetone on the lab notebooks in the workroom, perhaps creating a trail toward the solvent stills in the adjacent lab. As the fire reached the solvent canisters in front of the solvent stills, it would heat the canisters to the point of rupture. The ruptured cans could then continue to spray fuel into the fire for a long time, with additional fuel being added explosively as the solvent stills caught fire. With suspicions of arson heightened, the canisters of solvent found in the lab and workroom were packed to test for the presence of fingerprints at the crime lab.

When an accelerant is used to start a fire, trace residues are often found at the source of the fire, which are still measurable within 24–72 hours after the fire has been extinguished. This is especially the case when an accelerant such as gasoline is poured onto a combustible material, such as newspaper or books. Because air may not completely reach the interior of the material, a small amount of fuel remains trapped during combustion. To determine the presence of trace solvent, pieces of vinyl flooring were cut out from locations in front of the solvent stills, between the lab and workroom, and in front of the bookshelf in the workroom. In addition, the charred remains of the lab notebooks were collected and sealed for transport to the arson lab for analysis. In order to prevent evaporation and the loss of volatile liquids, the samples were placed into new, airtight paint containers for transport to the crime lab. Final photographs were taken of both labs, and the rooms were sealed off until further analyses could be completed at the crime lab.

Your Turn 14.9   Evidence Collection

Using the Internet, provide details about how samples suspected to contain residues of fire accelerants are best collected and transported to a forensics lab for testing.

14.5 | Fire Modeling

> *Learning Objective: Use the autoignition temperature of substances to investigate the source and probable temperature of a fire*

Using detailed dimensions and drawings of Dr. Thompson's laboratory, Dr. White performed a computer simulation of the fire, assuming that the solvent stills were the primary sources of fuel. Simulations, such as those using the computer program PyroSim, are powerful techniques to provide information about the potential origin point(s) of a fire. In addition, the simulated burn and smoke patterns can be compared to those actually observed in the lab—an important tool to either implicate or rule out arson as a probable cause of a fire.

The simulated ambient temperature of the lab resulting from a fire originating at the solvent stills is shown in Figure 14.7. Whereas the wall behind the solvent stills likely reached temperatures as high as 800–1000 °C, the rest of the lab exhibited temperatures in the range of 400–650 °C. From the autoignition temperatures of common materials (Table 14.4), it is no surprise that the experimental lab was mostly consumed by the fire.

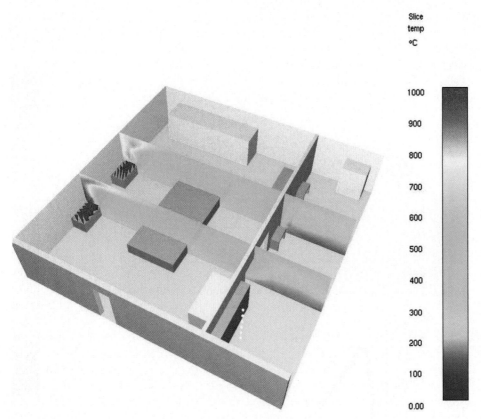

Figure 14.7

A simulated cross-sectional temperature profile of the synthetic lab, for late stages of a fire originating from both solvent stills. The white dots in front of the bookshelf in the workroom show the location of temperature sensors used in the modeling calculations. Calculations performed using PyroSim software (Thunderhead Engineering).

The calculated temperature profile for the workroom is shown in Figure 14.8, which shows temperatures in the range of 160–230 °C for most of the room, with higher temperatures (290–290 °C) near the ceiling of the connecting doorway. For modeling purposes, temperature sensors were added in front of the bookshelf in the workroom

Table 14.4	Autoignition Temperatures for Common Materials
Material	**Autoignition Temp. (°C)**
Paper	218–246
Leather	200–212
Cotton	267
Nylon	289–377
Polycarbonate (PC)	478
Polyethylene (PE)	226
Polyethylene terephthalate (PET)	460
Polypropylene (PP)	201
Polystyrene (PS)	226
Poly(vinylchloride) (PVC)	455
Natural rubber	191–331
Wood	300–482

Figure 14.8

(a) A simulated cross-sectional temperature profile of the workroom, for late stages of a fire originating from both solvent stills. The white dots in front of the bookshelf in the workroom show the location of temperature sensors used in the modeling calculations. (b) The calculated temperature profile from sensors in front of the bookshelf, at a height halfway from the ceiling. Calculations performed using PyroSim software (Thunderhead Engineering).

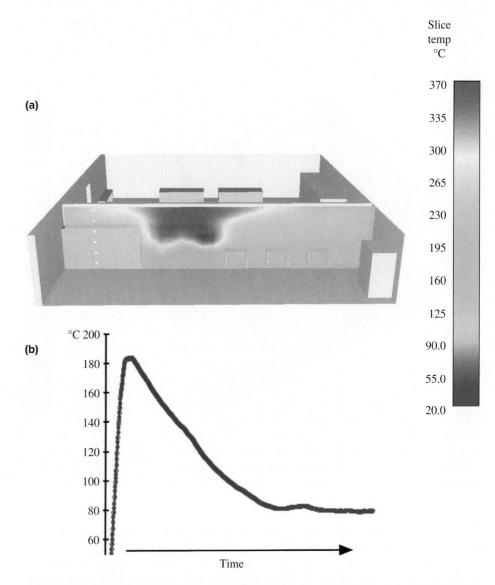

to determine the extent of heat likely experienced by Dr. Thompson's lab notebooks. Calculations showed that the highest temperature of these sensors (Figure 14.8) was only approximately 185 °C.

Your Turn 14.10   Workroom Fire Damage

The key question posed by Williams's initial investigation of the lab concerned whether sufficient heat could be produced by the solvent stills to ignite the lab notebooks in the adjacent workroom. Using the simulated temperature profiles for the workroom shown in Figure 14.8, and the simulated temperature range of the sensors in front of the bookshelf, do you think that sufficient heat could be produced *in the absence of an accelerant* to ignite and burn the lab notebooks? Note that the covers of the lab notebooks were made of imitation leather—cotton fibers that are coated with the polymer polyvinyl chloride (PVC).

DID YOU KNOW ?

The term **flashover** is used when the majority of the exposed surfaces in a room are heated to their autoignition temperatures, and emit flammable gases, which provide additional fuel to the fire.

On the basis of computer models, the temperatures found in the workroom were probably not high enough to ignite the vinyl-bound lab notebooks. However, the main experimental laboratory likely experienced temperatures above the autoignition point of wood, paper, and most plastics found in the lab. This would add to the heat evolved from the main laboratory, and would likely result in *flashover* conditions and significantly higher temperatures in the workroom than predicted by the calculations. Even so, the presence of empty and damaged solvent canisters and an unburned matchstick near the bookshelf strongly indicate arson as a probable cause of the lab fire.

14.6 | *Behind-the-Scenes at the Crime Lab*

Learning Objectives:

Determine how fingerprint analysis is carried out in a forensic lab

Explain some laboratory techniques used to determine if arson is the likely cause of a fire

Perhaps the most straightforward and diagnostic analysis to identify possible suspects in a crime is fingerprint analysis. This information is extremely useful because no two sets of fingerprints are considered to be the same, including those of identical twins.

Latent fingerprints must be made visible before they can be analyzed. One common method of fingerprinting consists of distributing a powder onto a surface, which adheres to the residue deposited from the finger's touch. To avoid smudging the print, sometimes a magnetic powder is used in which the powder is poured onto a surface and then spread evenly using a magnet, instead of dusting the powder with a brush. To allow the best visibility, powders of contrasting or fluorescent colors may be used, depending on the color of the surface to be treated. Once the fingerprint is visible, photographs are taken. The powdered impression is then removed via fingerprint lifting tape and packaged for analysis and identification at the crime lab.

As part of their ongoing investigation, police officials collected fingerprints from Dr. Thompson and were able to obtain fingerprints from some of his former graduate students who still resided in the county. It was also university policy to fingerprint all janitorial staff prior to their employment since they have direct contact with students under the age of 21. Most of the postdoctoral researchers who had worked with Dr. Thompson had since left the state or country.

Fingerprint analysis was carried out on lifted prints from the doorknobs, cabinets, and fume hood sashes by the State Forensic Laboratory at Dr. Thompson's request. However, in a fire scenario, the presence of fingerprints is complicated by the presence of soot. In these areas, the prints were visualized using alternate light sources (ALSs) to illuminate the prints at varying wavelengths. The residue that makes up a latent print will fluoresce and, using special filters and goggles, the prints can be seen and photographed without being touched (Figure 14.9). The full and partial fingerprints taken from various surfaces in the lab were mostly from Dr. Thompson and janitorial crew members. The other unidentified latent prints were entered into the Integrated Automated Fingerprint Identification System, or IAFIS, a forensic database of known and crime scene prints. No match was found in the database for the unidentified latent prints taken from the laboratory.

Next, the forensics investigators analyzed the empty solvent canisters found at the crime scene. For this analysis, the solvent containers were placed into an airtight tank known as a fuming chamber (Figure 14.10). A few drops of liquid cyanoacrylate (**I**), the main ingredient of superglue, was then added to an open container and heated to a temperature of 49–65 °C to vaporize the liquid. The gaseous cyanoacrylate reacts with the amino acids, fatty acids, and proteins in the exposed latent fingerprint. Upon further contact with residual moisture in the fuming chamber, a white, sticky film is generated that adheres to the ridges of the fingerprint.

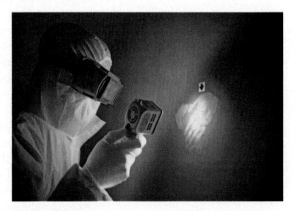

Figure 14.9

A forensic scientist using an alternative light source to reveal a handprint.

Monty Rakusen/Cultura Creative (RF)/Alamy Stock Photo

Figure 14.10

A portable superglue fuming chamber used for fingerprint analysis. Check out the 3D rendering of the cyanoacrylate molecule at www.acs.org/cic.

©2016 Sirchie. All Rights Reserved.

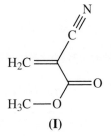

(I)

A number of fingerprints were found on the solvent canisters resulting from the fuming technique. Not surprisingly, most of the prints matched those from Dr. Thompson and from his former graduate students. However, once again, there were thumb and index-finger prints that were not identified. Dr. Thompson would be contacted in the following days to obtain the names of those who might have had access to his laboratory, and whose fingerprints were not yet on file with the State Police.

With fingerprinting of the lab completed, Dr. White and her team of forensic scientists turned their attention to detecting the presence of accelerants in the flooring samples taken from Dr. Thompson's laboratory. A small hole was punched into the top of the airtight container and the hole covered with a silicone septum. Upon heating the container to 60 °C for 30 minutes, any volatile residue present in the debris was evolved and trapped in the airspace of the container, referred to as its *headspace*. A few microliters of the vapor were removed by a syringe and analyzed using a technique known as **gas chromatography** (GC).

Gas chromatography is a simple and rapid analytical technique commonly used to identify separate components in mixtures of liquids or gases. It consists of a liquid or gas sample being injected onto the beginning of a column. The components of the sample are moved through the column by a flow of an inert gas, known as the *mobile phase*. The column contains a material—known as the *stationary phase*—that binds the components of the mixture to a varying degree, based on similar polarities. For instance, if a nonpolar stationary phase is used, it will bind the nonpolar components of the sample more strongly, which allows the more polar substances to pass through the column more quickly (Figure 14.11). As such, a chromatographic separation employs the principle of "like retains like"; this is analogous to the "like dissolves like" principle discussed in Section 5.7.

As the various components reach the detector, a peak in the detector signal is recorded by a plotter or viewed on the screen of a connected computer. The time it takes for a component to flow through a column and reach the detector is known as the *retention time* of the substance. The longer a retention time for a component, the more closely its polarity matches that of the column material.

Your Turn 14.11   Gas Chromatography Analysis

The chromatogram shown below is from a mixture of caffeine, toluene, *n*-heneicosane, pyridine, and *n*-octyl acetate injected onto a column containing a nonpolar stationary phase. Using the Internet as a resource, draw the molecular structures for each compound, and rank them in order of increasing polarity. Identify the five peaks shown in the chromatogram below.

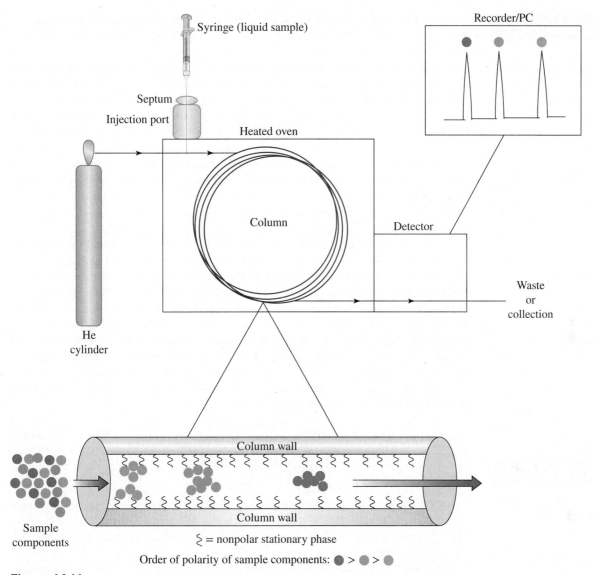

Figure 14.11

Schematic of the components of a gas chromatograph, showing separation of three sample components based on their relative polarities. The most nonpolar component is retained the most strongly by the nonpolar stationary phase. Check out this video to see a gas chromatograph in action: www.acs.org/cic.

The forensic scientists performed GC analysis on the floor samples taken from Dr. Thompson's laboratory. In the samples taken near the solvent stills, they found trace residues of benzene, toluene, dichloromethane, and a variety of compounds likely resulting from the burning of vinyl flooring and wood cabinets. No residues of an accelerant such as acetone were obtained from GC analysis of fragments of unburned lab notebooks, or floor samples taken from either lab. Although this may indicate that arson is not the cause of the fire, the lack of accelerant residues may be due to the fire extinguishing procedure and high volatility of the suspected accelerants. Empty containers of acetone were found in the doorway between the two rooms; however, acetone is miscible with water. When firefighters sprayed water onto the burning bookshelf and workroom, any unconsumed acetone would have been washed away, leaving no residue behind. Near the solvent stills where firefighters used dry chemical to extinguish the fire, the remaining acetone could have evaporated overnight, due to its extremely high volatility, before the forensics team arrived to collect samples. This is why fire investigations are so complicated, not only at the scene but also in the laboratory analysis.

The last pieces of evidence taken from the lab were the matchsticks found in the workroom. Dr. White had recalled reading about an arson case in the United Kingdom where **stable isotope analysis** was successfully used to match the wood from a matchstick found at the crime scene to the wood contained in matchsticks found at the perpetrator's apartment.

Depending on the growing conditions of trees such as temperature, humidity, and nutrient supply, there will be observable differences in the ^{13}C isotopic fraction among different samples of wood. In addition, the ^{18}O concentrations of wood will vary depending on the relative uptake of oxygen atoms within trees from atmospheric CO_2 and O_2, as well as the water taken up through the root system. The ^2H isotopic concentration is also diagnostic regarding the climate and geographic location where a tree was grown. Not surprisingly, there is significant variation in the concentration of each of these isotopes between trees grown in different plantations, as well as variability among trees grown in the same location. Hence, the ^2H:^{18}O:^{13}C ratio is extremely diagnostic regarding the geographic origin of the wood, which may even be used for partially burned matches.

Sections of wood 2–3 mm in length were cut from the matchsticks found at Dr. Thompson's laboratory. The wood samples were frozen, ground into small pieces, and placed into a desiccator containing diphosphorus pentoxide (P_2O_5) to remove traces of residual water from the samples. The most common analysis to determine isotopic ratios is **mass spectrometry**. As shown in Figure 14.12, vapor from a sample is ionized by bombarding it with electrons. The ions are then separated according to their mass-to-charge ratio by accelerating them in a vacuum (pressures of $10^{-6} - 10^{-8}$ Torr) and passing them through a magnetic field. Ions of the same mass:charge ratio will experience the same degree of deflection and reach the detector at the same time.

Variations in the natural abundance of stable isotopes are expressed using the delta (δ) notation as shown in Equations 14.1 and 14.2:

$$\text{Ratio (R)} = \frac{\text{Abundance of the heavy isotope}}{\text{Abundance of the light isotope}} \qquad \textbf{[14.1]}$$

$$\delta = \left(\frac{R_{sample} - R_{standard}}{R_{standard}} \right) \qquad \textbf{[14.2]}$$

The δ values are typically multiplied by 1000 to yield units per thousand (‰), often designated as permille. Hence, a negative ‰ value indicates a lower isotopic concentration (depleted in a particular isotope) relative to the standard. Triplet analyses of each wood piece resulted in an average ^2H:^{18}O:^{13}C ratio of $-112 \pm 1.3‰$: $-5.35 \pm 0.8‰$: $-27.3 \pm 2.1‰$. These results were entered into the main case file to provide a matching profile, should matchsticks be found on the person or premises of a future suspect.

The match heads were also analyzed using **scanning electron microscopy** (SEM). Instead of using light to observe sample features via a traditional microscope,

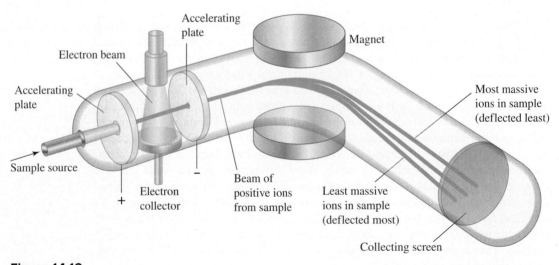

Figure 14.12

The operating principle of a mass spectrometer. Check out this video for a video illustrating the operation of a mass spectrometer: www.acs.org/cic.

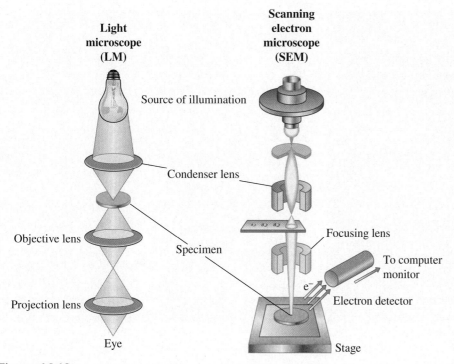

Figure 14.13

Comparison between a light microscope and scanning electron microscope. Check out this video to learn more about electron microscopy: www.acs.org/cic.

an electron microscope uses a high-energy beam of electrons to glean information about the surface topography and composition of a sample (Figure 14.13). When an electron beam contacts the surface of a sample, energy is transferred to the sample atoms, which releases electrons (known as *secondary electrons*). The observed contrast in an SEM image is due to the emission of secondary electrons from different regions of the surface, which provides details about its topography. A *backscattered electron* image may also be obtained, which is related to scattering of the incident electron beam due to interactions with the atomic nuclei present within the sample. Furthermore, X-rays are also emitted from the atoms of the sample; their observed wavelengths are characteristic of the type and concentration of atoms that are present on the surface.

The SEM image shown in Figure 14.14a is the head of a safety match, the type Dr. Thompson normally used in his research laboratory. Safety matches require you to strike the match on the surface of the matchbox. The tips of safety matches contain potassium chlorate ($KClO_3$) and glass (SiO_2). The striking surface also contains glass and an allotrope of phosphorus called red phosphorus. Red phosphorus is reasonably stable at room temperature, but the heat and friction generated by striking the match head against the striker surface converts the red phosphorus into the much more reactive white phosphorus. White phosphorus readily reacts with oxygen producing a large amount light and heat energy, igniting the rest of the match. The glass material in both the match head and striking surface provide rough surfaces that easily generate the friction required to ignite the phosphorus. Potassium chlorate is added to matches as a source of oxygen. Chlorates decompose under heat to form a salt and oxygen gas. The increased amount of oxygen speeds the combustion of the phosphorus and the wood or cardboard base of the match.

"Strike anywhere" matches, in contrast to safety matches, contain phosphorus in the match head itself. Thus, the match can be struck on any rough surface, generating friction and heat to ignite the phosphorus of the match. "Strike anywhere" matches (shown in Figure 14.14c) are less commonly found than the safety match variety as they are at higher risk of accidentally igniting.

DID YOU KNOW ?

Whereas common light microscopes can provide images of surface features with dimensions of >1 μm, a scanning electron microscope typically has a resolution of <10 nm.

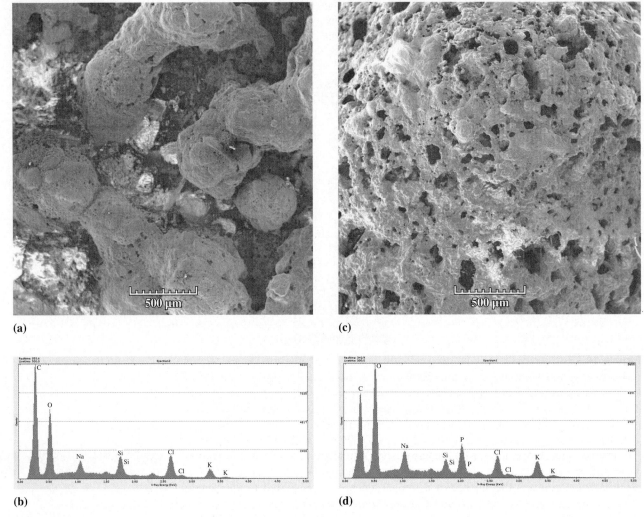

(a)   **(c)**

(b)   **(d)**

Figure 14.14

SEM images and X-ray elemental analyses of struck match heads from a safety match (**a** and **b**, respectively) and from a match found near the bookshelf (**c** and **d**, respectively).

(a, c): ©John S. Kirk VIII

Figures 14.14b and 14.14d compare the X-ray spectrum of a match normally stocked in the Thompson lab to that of a match found near the bookshelf after the fire. X-ray analysis of the matches found in Dr. Thompson's lab revealed phosphorus as a significant component of the match head. This indicated that the matches were of the "Strike anywhere" variety, and were brought in from outside the lab, possibly being used as the ignition source for the fire.

14.7 | *Wednesday, Aug. 13—1:03 PM: Access to the Lab Restored*

Almost two weeks had passed since the lab fire, so Dr. White called Dr. Thompson to update him on the status of the investigation.

"Hello, Dr. Thompson?"

"Yes."

"This is Dr. White, from the State Forensic Laboratory investigating your lab fire."

"Oh, good! Have you completed your investigation?"

"Yes, we have. We have now cleared your lab and you can access it to begin the cleanup."

"Great! So . . . was it really arson?"

"I'm afraid I can't say, it's still an open investigation. I can tell you we didn't find any residue of accelerants in either lab or on your notebooks. However, as we discussed a week or so ago, the empty acetone containers were suspicious. We also found some matchsticks, which should not have been there . . . correct?"

"Correct. My last postdoc smoked. But it's a smoke-free campus. My goodness, I hope he never smoked in the lab!"

"I really would like to know your thoughts on other people who might have access to your lab. Since we last spoke, were you able to think of anyone? We found some fingerprints that didn't match you or your former graduate students. Did anyone else work with you who might have gained entry? Again, the door was locked and there was no sign of tampering or forced entry, so it would have to be someone with a key. Is there *anyone* you can think of who might wish you harm?"

"Professionally, no." Dr. Thompson barked a short laugh. "My ex-wife, Julie Morrison, told me a couple of months ago that she would destroy my life if she got the chance."

"Really? Would she have a key to your lab?"

"Actually, now that you mention it, she might. Julie had an affair with a former postdoc from my lab, who has since left the country. However, I don't remember if he turned in his keys when he left the campus . . ."

"Okay. We will definitely get her in for fingerprinting. Does anyone else have access now or previously?"

"Yes. A couple of former postdocs: Dr. John Littleton and Dr. Avery Smith."

"Did you have a good relationship with them?"

"Yeah . . . it was fine . . . I can't think of them having a reason to do this."

"Okay, thanks. We'll speak with them as well to see if they can come down for fingerprinting. Of course, even if their prints match those that were unidentified from your lab, that doesn't put them in your lab during the night of the fire."

"Yeah, that's true. However, if Julie's prints are there, she is definitely guilty!"

"Well, let's not get too ahead of ourselves. We will let you know. Thanks again for your time, and please let me know if you think of anyone else, or notice anything else out of the ordinary when you get back to your lab."

"Okay, will do. Thanks a lot. I'm heading to my lab now."

As he hung up the phone, he noticed a voicemail left by his ex-wife earlier that morning.

"Hi Dave. I heard what happened to your lab. Have fun cleaning that up! Karma sucks, doesn't it?"

Infuriated by the voicemail, Dr. Thompson almost turned the car around to confront Julie. *How could she do something like this? I'm going to the cops tomorrow to have them listen to that voicemail. She's going to jail for this one!*

Dr. Thompson didn't recall the drive to his lab taking this long. As he maneuvered down side streets to get there faster, he thought about how long it would take to get his lab back up and running. Undoubtedly, he would have to hire a few students to help him set up again. He pulled into the faculty parking lot on campus, choosing a spot nearest the sidewalk to shorten his walk to the building. Now, finally in the building, he bounded up the stairs two at a time. *One more floor. There, made it.*

As Dr. Thompson entered the lab, his knees buckled. He became nauseated and fell to the floor. Power had been restored to the lab, but the lighting fixtures were still inoperable. Investigators had left five large portable work lights at different parts of both rooms, which provided sufficient light to observe the full extent of the damage. There was no question that his entire lab would have to be renovated. Floors, ceilings, and furnishings were all severely burned. Soot covered the walls and surfaces of both rooms, completely masking the light blue color he proudly chose last year for a refreshing paint job. Whether it was an accident or deliberate didn't matter right now. Dr. Thompson was present at the University of California, Irvine, on July 23, 2001, and had witnessed the infamous accidental lab fire in Bill Evans's laboratory. He couldn't believe that the same fate had now befallen his own facility.

Dr. Thompson sat expressionless on the soot-covered student desk in the workroom, poking through a pile of lab notebooks from his past years of research. Although

DID YOU KNOW ?

The accidental explosion of a benzene still in Dr. Evan's laboratory at the University of California, Irvine resulted in an injury to a Ph.D. student and a fire that caused approximately $3.5 million in damage, including repair/refurbishment costs.

some pages were still partially legible, most of the books would not be useful anymore. As he reminisced on the excitement of the past accomplishments of his research group, the hours passed. During the early years of his work, he had never dreamed of his research getting to the point of commercialization.

He still had his data, manuscripts, and analyses saved on his laptop, but the original discoveries recorded in his lab notebooks were mostly gone forever, reduced to a pile of soot. Throughout the day and early evening, a constant line of students and colleagues came into the lab to express their sympathies. Filled with disbelief about the damage, many offered to help with the cleanup, but Dr. Thompson, though greatly appreciative, seldom looked up from his beloved notebooks.

14.8 | Wednesday, Aug. 13—10:57 PM: What Now?

With a heavy heart, Dr. Thompson put aside the remains of his notebooks and stood up. He shuffled through the lab one last time, closing the door behind him. He walked slowly to his car, thinking about the presentation he would deliver in a few days. All he wanted to do was go home and sit on the patio to tweak his PowerPoint slides. The car started immediately as it always had, but then stalled. Repeatedly, he turned the ignition, without any success. Even holding the accelerator to the floor as the ignition was turned did nothing to revive his car. *Great! Now what?*

He reached for his phone to call roadside assistance, but the battery on his cell phone was dead. *Damn it! Why can't this phone make it through an entire day? Can anything else go wrong today?* Dr. Thompson often admitted his disgust with the current state of battery technology to his coworkers and friends. Although he thought of upgrading his cell phone to a new model with a supposedly longer battery life, he was skeptical of manufacturer claims and felt his phone worked just fine for his meager needs. Dr. Thompson was close friends with a researcher at Argonne National Laboratory, and joked that he could solve the battery dilemma in two years if given the resources.

Dr. Thompson got out of his car and began walking toward the biochemistry building to call roadside assistance. He always carried his black leather laptop case, which contained important data, manuscripts, and presentations. It was quiet on campus, as no one was around this time of night. As he took a step off the sidewalk to cross the street, a car sped toward him without warning. Unable to dodge the speeding vehicle, Dr. Thompson was hit dead-on, flipping him over the roof, and killing him instantly.

Under the cloak of darkness from the moonless night, the driver stopped the car, careful not to screech to a halt and leave identifying skid marks. With considerable effort, the murderer pulled Dr. Thompson's body off the road to delay its discovery until morning.

The driver quickly surveyed the damage to the car, retrieving a large piece of plastic from the front bumper, and making sure to collect Dr. Thompson's briefcase en route back to his vehicle. The car then sped away into the night.

14.9 | Thursday, Aug. 14—5:42 AM: A Gruesome Discovery

Sheila Jackson, a junior undergraduate student, was in the midst of training for a full marathon, and her early-morning run often took her through campus. Today she was scheduled for 15 miles, and was ready for the challenge. At mile three, Sheila passed by the faculty parking lot, and noticed something that grabbed her attention. She slowed to a stop, and discovered a body that lay motionless on the side of the road. She knew immediately this was serious, as she saw blood coming from his head and a blood trail leading from his body to the center of the roadway. Sheila called 9-1-1 from her smartwatch, and waited on the other side of the street for police to arrive.

Within minutes, campus police arrived on the scene, with the city police showing up a few minutes later. The police questioned Sheila about her discovery, and then drove her home. The emergency medical technicians (EMTs) called a local medical doctor to obtain an official pronouncement of death. At 6:58 AM, Dr. Thompson was pronounced dead at the scene by Marlene Jacobs, MD. His internal body temperature was 83 °F, indicating his death had occurred many hours prior.

Investigators closed all roads around the hit-and-run accident, and began their survey of the crime scene (Figure 14.15). The streets were dry and the ambient temperature was 58 °F. A few small fragments of clear plastic were found on the roadside near the victim's body, which might have originated from the perpetrator's vehicle. The lack of any visible skid marks on the street near the location of the body indicated that this could be an intentional murder. If someone had accidentally hit a person crossing an intersection, they would have braked hard or swerved to avoid a collision. Either of these preventive actions would have left skid marks on the road. However, another possibility is that the perpetrator had been distracted by texting or talking and didn't notice a person crossing in front of his/her vehicle.

There was a visible pattern of blood leading from the middle of the road, where the victim's body probably first fell, to the side of the road where the body was found. Perhaps the victim had mustered just enough strength to crawl off the roadway by himself. However, it might also be possible that the perpetrator had dragged Dr. Thompson's lifeless body off the roadway after the collision. The astute police investigator examined

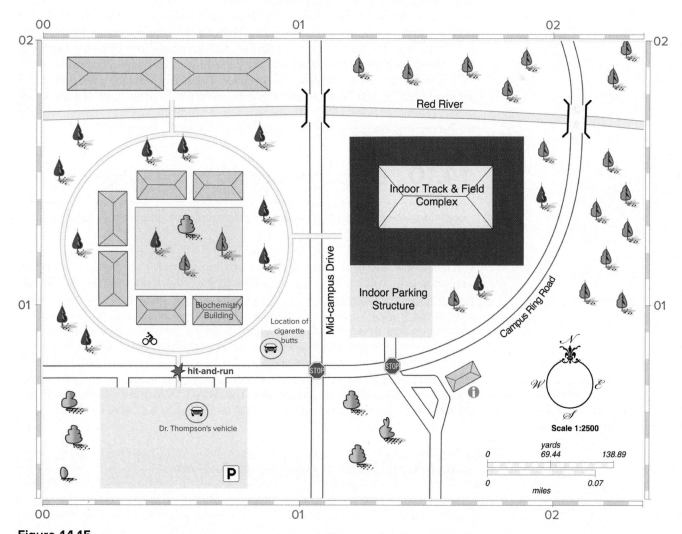

Figure 14.15

Map of the area surrounding the hit-and-run crime scene.

Dr. Thompson's wrists, shoes, ankles, and neck with a forensic light source (UV light with a wavelength of 365 nm). A variety of fingerprint and palm prints were noticeable on Dr. Thompson's wrists. These prints could have been deposited by the perpetrator as he/she dragged the body off the roadway.

The police searched the surrounding area for additional pieces of evidence. Other bits of glass or plastic from the vehicle might have fallen off after the collision. The sides and middle of the roadway were carefully searched in a 200-yard radius from the scene of the crime. Although no additional fragments from a vehicle were discovered, a collection of five cigarette butts and used matchsticks were found in a temporary parking/pickup area about 100 yards from the crime scene. Were these from the perpetrator, who might have patiently waited for Dr. Thompson to cross the street? The butts and matchsticks were collected with plastic forceps and placed in a labeled plastic bag for analyses at the crime lab. Additionally, a black cell phone with a cracked screen and no remaining battery power was found on the side of the road, about 5 meters from where Dr. Thompson's body was found. This was also carefully packaged for transport to the forensics crime lab.

In order to re-create the events of this tragic evening, investigators had to determine whether Dr. Thompson was walking toward, or returning from, his car that evening. Car keys were found in the right front pocket of his trousers, and were successfully used to unlock Dr. Thompson's vehicle, a 2019 white Ford Focus that was parked a short distance from the hit-and-run scene. Investigators dusted the door handles, steering wheel, and gear selector for fingerprints using black powder and forensic lifting tape. After a thorough examination of the vehicle for signs of suspicious activity, the lead investigator attempted unsuccessfully to start Dr. Thompson's car. The battery and starter were deemed to be functional, so there must have been another reason for the vehicle failure. A tow truck was called to deliver Dr. Thompson's car to the crime lab for a more thorough investigation. The investigator surmised that Dr. Thompson might have experienced difficulties in starting the car, and was returning to his office to call for help when he was struck down by a vehicle. But if that were the case, why didn't he simply use his cell phone to call roadside assistance from his car? The black cell phone, believed to belong to Dr. Thompson, was too badly damaged to surmise whether the battery was charged at the time immediately before the accident.

14.10 | *Behind-the-Scenes at the Crime Lab*

Learning Objective: Explain how paint samples are analyzed in forensic investigations

Dr. White was informed of the fatal accident and contacted the police to offer her services to assist with the investigation. Although she primarily focused on cases involving arson and gun-related homicides, it was quite possible that the hit-and-run accident was related to the fire in Dr. Thompson's laboratory. Her interest in this vehicular homicide peaked when she heard of matchsticks being found a short distance away.

The isotopic ratios of the wood from the matchsticks found near the crime scene were determined using the same procedure as before (revisit Section 14.6). In order to match samples from two different locations, extreme attention was given to details regarding sample preparation and analysis. The $^2H:^{18}O:^{13}C$ ratio of samples taken near the hit-and-run accident were $-114 \pm 2.8‰ : -5.51 \pm 1.3‰ : -26.9 \pm 2.1‰$.

Your Turn 14.12   Matchstick Analysis

Based on the isotopic ratios of the wood from the matchsticks found in Dr. Thompson's laboratory, and those found at the hit-and-run crime scene, do you think there is a match? If so, does this prove that the same person was involved in both crimes? Explain your reasoning.

The fingerprints found on Dr. Thompson's lifeless wrists were also analyzed at the crime lab. Magnetic Jet Black powder was applied to the surfaces of his wrists using a magnetic brush. The fingerprint (Figure 14.16) was carefully transferred onto lifting tape for entry into the system for fingerprint matching. Fortunately for investigators, the area with the highest density of latent prints was in the hairless region just below the victim's palm, giving rise to high-quality impressions. In addition, the perpetrator likely exerted significant pressure in depositing their fingerprints onto the victim's wrists while moving the body. From other published reports of this type of forensic analyses, the overall quality of prints is found to be strongly dependent on the pressure used while depositing the marks.

After four hours of waiting, the fingerprint database did not provide a match for those found on Dr. Thompson's wrists. However, it was noted that the prints found in this crime matched some of those found on the solvent bottles in Dr. Thompson's laboratory. Furthermore, the partial and full prints lifted from various portions of Dr. Thompson's wrists and arms were from only one person, so that person was somehow involved.

An autopsy was performed on Dr. Thompson. Not surprisingly, the cause of death was determined to be multiple blunt force traumas to the head and torso. The injuries were consistent with being struck by a vehicle. Toxicology results did not show any observable concentrations of alcohol in his system. His stomach was nearly empty, indicating a long period (>6 h) since he had eaten his last meal, with the presence of starch grains and caffeine. His medical records were obtained to determine whether suicide was a factor. Perhaps Dr. Thompson was depressed about the lab fire, and/or had been diagnosed with a terminal illness. After viewing his medical records, it was determined that Dr. Thompson did not suffer from a terminal illness, and was in good physical condition.

Figure 14.16

Image of a fingerprint developed from the application of a magnetic black powder.

©Matej Trapecar

Your Turn 14.13   Dr. Thompson's Last Meal

Based on the contents of Dr. Thompson's stomach, as revealed by the autopsy, what are some possible foods he might have consumed at his last meal? If he drank a glass of red wine at his last meal, approximately how long would it have taken for traces of ethanol to be detected in his blood? What factors influence the rate of alcohol metabolism in the body?

Your Turn 14.14   Healthy or Not?

During his annual health evaluation performed one month earlier, a blood test for Dr. Thompson revealed the following levels:

Total cholesterol: 229 mg/dL
HDL cholesterol: 42 mg/dL
LDL cholesterol: 187 mg/dL
Triglycerides: 158 mg/dL

Using the Internet and information provided in Chapter 11 as resources, determine whether Dr. Thompson was in good health (relative risk of heart disease).

Investigators also examined Dr. Thompson's car to determine the cause of its mechanical issue and potential relevance to the murder case. No blockage in the tailpipe was noted, but the gasoline contained large amounts of water. Although placing sugar in a gas tank is widely thought to incapacitate a vehicle, this has been shown to be a myth.

DID YOU KNOW?

Episode 15 in the 2004 season of "MythBusters" evaluated the popular myth of adding sugar to a gas tank to render a car inoperable.

In fact, water is much more dangerous to engines than sugar because of the concepts of "like dissolves like" (Section 5.7) and density (Section 5.2). Figure 14.17 illustrates the molecular structures of gasoline, sugar, and water. Whereas the components of gasoline are nonpolar, both sugar and water molecules are quite polar. Therefore, sugar will not dissolve and water will not be miscible in gasoline. So, why is one of these components "better" than the other, as far as causing damage to internal combustion engines? The answer is related to the relative densities of these substances. The density of gasoline is in the range of 0.71–0.77 kg/L, as compared to 1.587 kg/L and 1.0 kg/L for sucrose and water, respectively. Therefore, gasoline will float on top of water, and sugar will drop to the bottom of the gas tank. If a few cups of water are poured into the gas tank, the fuel pump will fill the fuel lines with water instead of gasoline, which would result in major engine damage.

DID YOU KNOW?

The addition of sugar into a gas tank would probably enact the same response as dumping in sand—a nasty mixture to clean up, with the solids dropping to the bottom of the tank instead of being transported into the fuel lines.

The picture was beginning to make more sense now to investigators. The murderer must have poured water into Dr. Thompson's gas tank, and then waited for him to return to his vehicle. Once he was unable to start his car, Dr. Thompson returned to his office to call for assistance, most likely due to a dead battery in his cell phone. He would have to cross the street, which was always desolate at that time of the night. But how did the perpetrator know Dr. Thompson's cell phone would be nonfunctional? Perhaps it was an educated guess, based on the known inability of his phone to last through an entire day. However, it is conceivable that there would have been a backup plan if Dr. Thompson would have been able to use his cell phone to call for assistance that night.

Traces of dark blue paint were found on Dr. Thompson's clothing, which likely originated from the perpetrator's vehicle. The location of the paint was carefully documented, and the victim's clothing was then scraped over clean, white craft paper. The debris collected from scraping was examined using a variety of methods, starting with a light microscope to determine the presence of fibers, flakes, and other microparticulates.

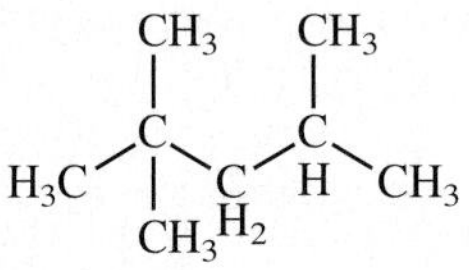

iso-octane (nonpolar)

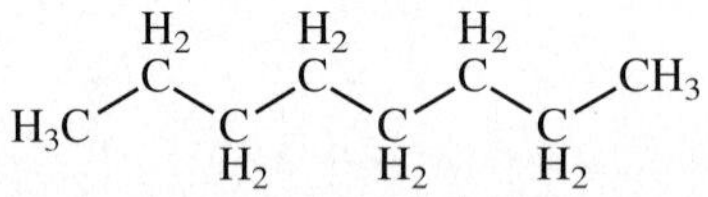

n-octane (nonpolar)

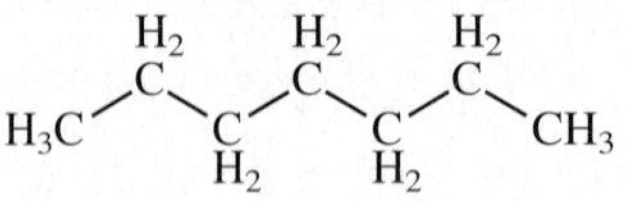

n-heptane (nonpolar)

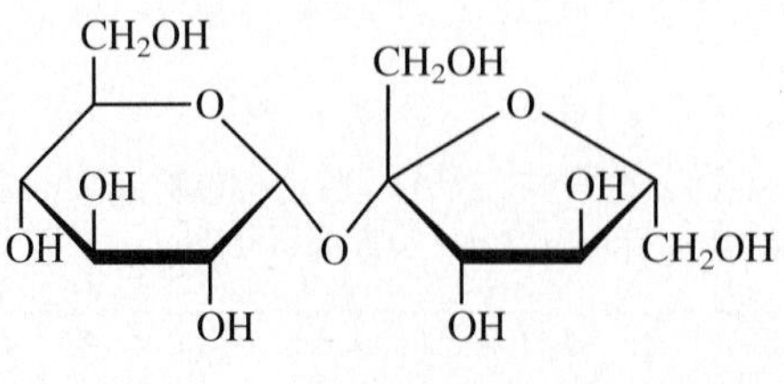

sucrose (polar)

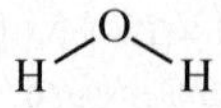

water (polar)

Figure 14.17

Molecular structures of some major components of gasoline (iso-octane, *n*-octane, and *n*-heptane), table sugar (sucrose), and water.

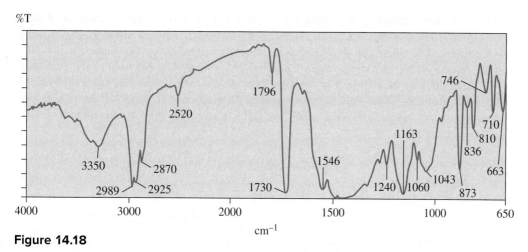

%T

3350 2989 2870 2925 2520 1796 1730 1546 1240 1060 1163 1043 873 746 836 810 710 663

4000   3000   2000   1500   1000   650

cm⁻¹

Figure 14.18

Infrared spectrum of the paint fragment recovered from the victim's clothing.

Infrared (IR) spectroscopy, a technique first introduced in Section 4.6, was used to determine the functional groups present in the paint formulation (Figure 14.18), which was then entered into the Paint Data Query (PDQ) database. The paint matched that of a 2021 metallic-blue Tesla Model S. The use of an electric car for the murder was clearly an insidious choice. Whereas Dr. Thompson could have easily heard the approach of a gasoline-powered vehicle, the much quieter operation of an electric car would have provided no warning to the victim, and no opportunity to jump to safety.

Your Turn 14.15   Peak Matching

Using the Internet as a resource, assign functional groups that correspond to five major peaks that appear in the IR spectrum in Figure 14.18.

Your Turn 14.16   IR Spectrum of Polycarbonate

The image below shows the IR spectrum for the polycarbonate fragments found at the crime scene that had likely come from the headlight lens of the perpetrator's vehicle. Assign functional groups to the regions labeled 1–6.

Transmittance (%)

95.7 94 92 90 88 86 84

2972 ① 1780.58 ② 1500.96 ③ 1189.47 ④ 1004.29 ⑤ 554.28 ⑥

4000   3000   2000   1500   1000   400

Wavenumber (cm⁻¹)

Investigators were pleased that a match was made to a vehicle with relatively low production numbers. In 2021, 24,980 Tesla Model S vehicles were sold worldwide, with 12,553 sold in the U.S. The State Police searched for registered owners of a 2021 Tesla S within the state, which identified 231 households. However, only 35 owners took delivery of a Tesla S in metallic blue. As Dr. White examined the list of owners sent from the State Police, her eyes immediately honed in on Julie Thompson—the victim's ex-wife. The police called her and requested she visit headquarters to answer some questions.

14.11 | Friday, Aug. 22—9:03 AM: The Questioning of Julie Morrison

Julie Morrison sat expressionless in Interrogation Room B.

"Ms. Morrison, thanks for coming in. We have a few questions regarding your whereabouts the night of August 13."

Without looking up from her focused gaze on the floor, she uttered "I was at home."

"So, were you alone?"

"Yes."

"Okay . . . How about the night of August 1?"

"I don't recall."

"That's the night your ex-husband's lab burned down. You don't remember where you were?"

"No. I think I was shopping at the time."

"Ms. Morrison, you should realize that you are a suspect in both crimes. If I were you, I would take these questions more seriously."

"I am. If I can't recall, I can't recall . . ."

"Do you have keys to your ex-husband's lab?"

"No. Why would I?"

"Do you own a metallic-blue Tesla Model S?"

"Yeah."

"We noticed you drove another vehicle today. May I ask where your Tesla is?"

"It's in the shop. I hit a deer just outside of town last week."

"Did you file a claim with your insurance company?"

"Of course. They told me they should have it repaired sometime next week."

Establishing that Julie had in fact damaged the front end of her metallic-blue Tesla S, police continued to question her about her whereabouts during the evenings of the hit-and-run and lab fire incidents, as well as her relationship with Dr. Thompson.

Her bitterness and anger boiled over at various times during the interview—"I hate him right now, but I wouldn't do something stupid like burn down his lab or try to kill him!"

The investigative team contacted Julie's insurance company, which confirmed her story of an accident involving wildlife. The tow truck driver also corroborated her account, stating that he was called to a wooded area about five miles from town, and a deer was clearly the cause of her accident. Julie's fingerprints were taken by the lead detective. Her fingerprints were not a match for those found on the acetone bottles in the lab. In fact, her fingerprints did not match any of the unmatched prints found anywhere in the lab. Furthermore, her prints did not match those found on the wrists of Dr. Thompson's body at the scene of the hit-and-run incident. Although Julie said she never smoked, a swab of her saliva was taken for DNA matching on the cigarette butts found at the crime scene. Julie was then sent home and told not to leave the area until the police investigation was completed.

14.12 | Monday, Aug. 25—8:31 AM: The Questioning of Dr. Littleton

With Julie released as a primary suspect, the attention of the police shifted to other possible suspects. The phone records for Dr. Thompson's cell phone indicated that he called a number from a neighboring town just hours before he was murdered. The number was tracked to Dr. Littleton in the biochemistry department at a nearby college. Police investigators placed calls to his office and home to ask him to come in for questioning. The next day, Dr. Littleton arrived at police headquarters.

"Thanks for coming in so quickly, Dr. Littleton. I am Detective Bentley and this is my colleague Detective Stevens."

"My pleasure. Anything to help out."

"Please describe your relationship with Dr. Thompson."

"I was a postdoc in his lab for a total of three years, and then moved on to start my independent career a few months ago."

"Did you have a good working relationship with him?"

"Yes. No problems."

"Was there a reason why he called you the night of his murder?"

"I'm not sure. I've been trying to understand that myself. He said that his lab was destroyed, but I already heard about that from a colleague in his department."

"From the accounts of others in Dr. Thompson's department, it seems you two didn't part on the best of terms. Is that correct?"

"I'm really not sure what that means. I landed a great faculty position, so he must have written me a good letter of recommendation."

"Did you know Dr. Thompson was scheduled to speak at the American Chemical Society meeting in Boston a few weeks ago?"

"No, I wasn't aware of it. He hasn't attended meetings for a while . . ."

"Where were you on the evenings of August 1 and August 13?"

"I was out of town at a conference between August 9 and the 14th, and I was at an event hosted by the Dean of our college on August 1."

"Where was your conference?

"In Atlanta."

"What day and time was your presentation?"

"I gave my talk in the afternoon . . . around 2:00 on August 13."

"So . . . that's about two hours away from your former employer?"

"More like three-plus with traffic . . . It is Atlanta, you know!"

"Is there a way to confirm your attendance at either of these events?"

"Well . . . I have confirmation of my conference registration, and you could ask my colleagues if they remember seeing me at the fundraiser on the 1st."

"Okay, thanks. Please send me a copy of your conference registration, and we'll contact a few of your colleagues."

Dr. Littleton, figuring the interrogation was over, sat up as if to stand to his feet, and remarked "Sounds good, will do."

Detective Stevens also sat up in his chair, "By the way . . . do you know of anyone who owns a Tesla Model S?"

"No, I wish! Those are nice cars! Why a Tesla?"

"We have reason to believe the vehicle that struck Dr. Thompson was a Tesla."

"Oh . . . A shame to damage such a nice car!"

"Well . . . let's not forget about the person it hit!"

"Yeah, you're right . . . "

The questions continued for another two hours, with attempts to unveil more details about Dr. Littleton's relationship with Dr. Thompson. Based on all accounts from acquaintances, Dr. Littleton was extremely angry with his former advisor, and wanted appropriate credit for his part of the work. As was his policy, Dr. Thompson had all students sign a nondisclosure agreement prior to beginning their research with him. Despite his leading role in the research project, Dr. Littleton was not named as a co-inventor on any of his former advisor's patents, which were poised to bring in tremendous financial gain to Dr. Thompson and the University. One faculty member in the Biochemistry department reported that Dr. Littleton often lamented about his unfair treatment by his former advisor.

Apparently, Dr. Littleton had asked Dr. Thompson numerous times to include him on patents but was always met with opposition. This created increasing conflict between the two scientists, which ended less than 5 months ago with Dr. Littleton leaving the university to start his own research group at a nearby college. Dr. Littleton denied these claims, and said they had worked out their differences before he left. Without any evidence to link him to either crime, the police let him go.

Dr. White strongly believed that Dr. Littleton wasn't telling investigators the whole truth. The search for registered Teslas in the state of Georgia revealed two rental agencies in the Atlanta area. Could Dr. Littleton have rented a Tesla and driven back to kill his former advisor?

Your Turn 14.17   Battery Range

The distance from the scene of the hit-and-run accident to the convention center in Atlanta, Georgia, is 140 miles each way. Based on the reported battery range of the Tesla Model S (assume a 100-kWh battery), would the driver have to stop for recharging between these locations, or could the car make a round trip on one full charge? What factors would affect the battery range of this vehicle?

14.13 | Tuesday, Aug. 26—2:05 PM: Road Trip to Atlanta

Police detectives Bentley and Stevens drove to Atlanta to follow up on the suspicions of Dr. White. Exotic Rentals was visited first, and a warrant was secured to search the rental records. No mention of Dr. Littleton was found in the log book, and the lone Tesla Model S was parked in front of the parking lot, with no visible front-end damage. The Tesla was being charged with a solar-powered charger.

Your Turn 14.18   Solar Charging Station

Using the Internet, research the cost and configuration of three different solar charger kits. How long would it take for these units to fully charge a Tesla Model S battery (assume 100 kWh)? Compare the costs/kW for the different solar charging kits.

The agents visited the second rental agency, Dream Cars, which was located about 25 minutes away. As Agent Stevens flipped through the pages of the log book, he asked the shop manager,

"Is your Tesla currently being rented? I don't see it out front."

"No. The last guy who rented it hit a deer. That thing did a lot of damage. It'll be a while before we can get it back on the road."

As the investigator inspected the log book one last time, the name *Johnathon Littleton* jumped off the page. They had their suspect. The agents called back to headquarters to issue an arrest warrant for Dr. Littleton, and a search warrant for his home and office.

14.14 | Back in the Crime Lab

Learning Objective: Explain how DNA is collected and analyzed by forensic scientists

Dr. Littleton was taken into custody on the evening of August 27 and remained in jail overnight. Fingerprints were taken, as well as a cheek swab to retrieve a sample of his DNA. DNA analysis was performed on the cigarette butts found near the hit-and-run scene and was compared with that of Dr. Littleton.

As discussed in Chapter 13, DNA is located in the nucleus of cells throughout the body. However, DNA must first be extracted from other cellular material, as well as debris such as clothing or cigarette butts. Commonly, a mixture with equal parts water, phenol (C_6H_6OH), and chloroform ($CHCl_3$) is used to extract DNA from its host matrix. However, as you will see in this next activity, DNA can also be extracted using common household chemicals.

@HOME   DNA Extraction

Check out this simple experiment to extract DNA from various food items: www.acs.org/cic. In addition, an online virtual lab that illustrates the extraction of DNA may be found at www.acs.org/cic.

©Bradley D. Fahlman

Questions to Consider

a. Repeat the experiment with other DNA sources. Which source gave you the most DNA and how can you compare them?

b. Experiment with different soaps and detergents. What is the role of the surfactant and rubbing alcohol used in this procedure?

c. Which works best for DNA extraction, cold or hot water? Why is one water temperature better than the other?

d. What are some impurities that could impact the quality of DNA that was extracted?

Your Turn 14.19   Solvent Extraction

Look up the densities of water, phenol, and chloroform.

a. Once the DNA-extraction mixture is shaken vigorously, only two layers form, one having twice as much volume as the other. What are the compositions of each of the layers?

b. Based on your knowledge of DNA structure presented in Chapter 13, in which layer will the DNA be found?

The DNA that has been extracted into the aqueous phase is then amplified using a technique known as **polymerase chain reaction** (PCR). As illustrated in Figure 14.19, the double-stranded DNA is denatured into individual strands by heating to high temperatures (about 94 °C). Once the single strands are obtained, the solution is cooled to 72 °C, and an enzyme known as *Taq polymerase* is added. This catalyst serves to add complementary nucleotides to the single strands of DNA, giving rise to two identical DNA molecules. This process is repeated many times, to yield billions of copies of the original DNA extracted from a sample. The PCR technique has made it possible to

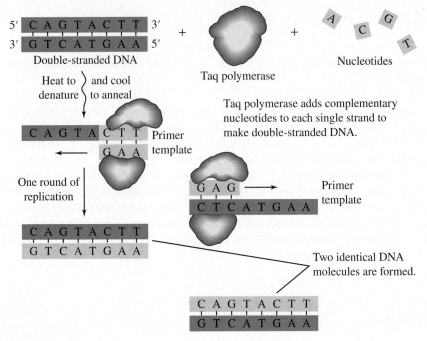

Figure 14.19

The polymerase chain reaction. Check out this interactive website to learn more about the PCR technique: www.acs.org/cic.

produce significant quantities of DNA from extremely small samples, which is essential to properly identify suspects.

Once the DNA fragments have been amplified, they are separated and detected using a technique known as **electrophoresis**. As illustrated in Figure 14.20, electrophoresis involves the DNA molecules being transported through a gel under the influence of an applied electric field. Because the phosphate backbone of DNA is negatively charged, the DNA is pulled through the gel toward the positive electrode. The larger the DNA molecule, the slower it will migrate through the small passages in the gel.

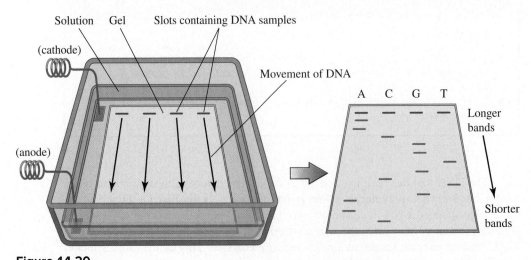

Figure 14.20

Schematic of gel electrophoresis.

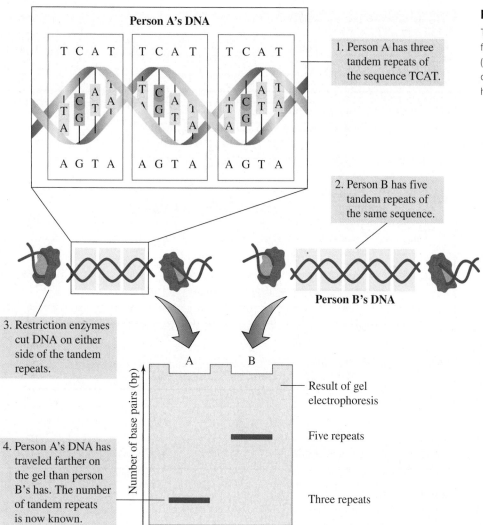

1. Person A has three tandem repeats of the sequence TCAT.

2. Person B has five tandem repeats of the same sequence.

Person B's DNA

3. Restriction enzymes cut DNA on either side of the tandem repeats.

Result of gel electrophoresis

Five repeats

4. Person A's DNA has traveled farther on the gel than person B's has. The number of tandem repeats is now known.

Three repeats

Figure 14.21

The steps involved in DNA fingerprinting. This website (www.acs.org/cic) provides more details regarding how this technique has improved over the past 30 years.

The repeat patterns of everyone's DNA are different, which gives rise to differences in the resultant electropherogram, the output from the electrophoresis (Figure 14.21).

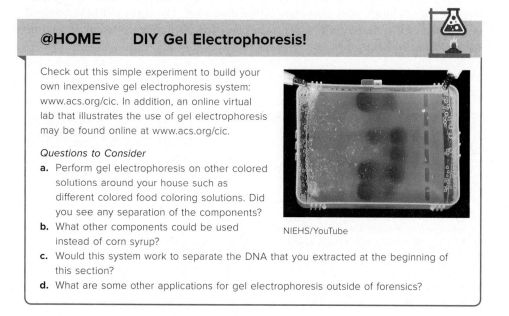

@HOME   DIY Gel Electrophoresis!

Check out this simple experiment to build your own inexpensive gel electrophoresis system: www.acs.org/cic. In addition, an online virtual lab that illustrates the use of gel electrophoresis may be found online at www.acs.org/cic.

NIEHS/YouTube

Questions to Consider

a. Perform gel electrophoresis on other colored solutions around your house such as different colored food coloring solutions. Did you see any separation of the components?

b. What other components could be used instead of corn syrup?

c. Would this system work to separate the DNA that you extracted at the beginning of this section?

d. What are some other applications for gel electrophoresis outside of forensics?

gopixa/Shutterstock

DNA Fingerprinting

Check out a video to learn more about how DNA fingerprinting is performed: www.acs.org/cic.

Questions to Consider

a. The video mentions possibilities for contamination. What are some steps that must be taken during DNA extraction, purification, and DNA fingerprinting to prevent contamination by impurities?

b. What are some of the moral issues associated with DNA fingerprinting?

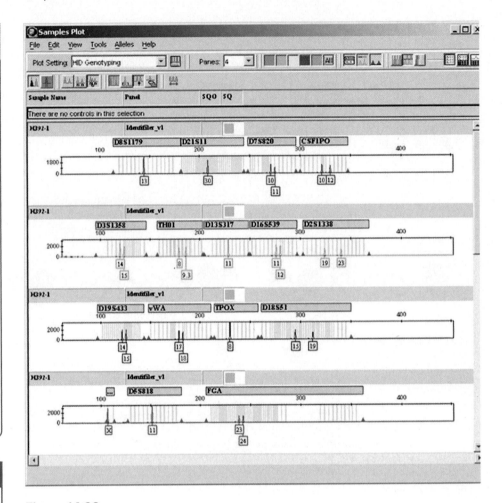

Figure 14.22

A DNA profile for an individual using an automated DNA analyzer.

Thermo Fisher Scientific Inc.

For modern DNA profiling, instead of reading the output from gel electrophoresis, faster and more automated capillary electrophoresis is used, whereby DNA passes through a small tube called a capillary rather than a gel. The pattern of peaks created is a person's DNA fingerprint, which may be matched to evidence found at a crime scene (Figure 14.22).

Let's Debate! Guilty?

Dr. Littleton's DNA matched that taken from the cigarette butts found near the location where Dr. Thompson was murdered. In addition, his fingerprints matched those found on the empty acetone containers in Dr. Thompson's lab, as well as those lifted from Dr. Thompson's body.

Does all this evidence unequivocally prove that he is the murderer? Choose two groups to debate the innocence of Dr. Littleton; one group will serve as the prosecution, while the other group will serve as defendant's counsel. Make sure that a third group (or your instructor) is chosen who will serve as the jury, to consider both arguments and render a decision concerning his guilt beyond a reasonable doubt.

14.15 | *Charge: Murder-1!*

With evidence mounting against Dr. Littleton, the police searched his home and office for additional pieces of evidence. He was an avid smoker, and investigators found wooden matchsticks at both his home and office. Stable isotope analysis at the crime lab identified these as the same brand found at Dr. Thompson's laboratory, as well as at the hit-and-run crime scene. Another round of questioning ensued.

"Look. . . Dr. Littleton, your DNA was on cigarette butts found near Dr. Thompson's body," the investigator asserted.

"I collaborate with Dr. Jones on that campus, so of course you would find some of my cigarette butts! You can't prove I threw them there the night Dr. Thompson was killed!"

"Were you waiting for him to cross the street that night?"

"No. I told you I was in Atlanta that day. I gave a presentation late that afternoon."

"You had plenty of time to drive back and kill him. We confirmed the mileage on the rental car. By the way, why didn't you tell us you rented a Tesla the first time we questioned you?"

"It's really none of your business what type of vehicle I rented. I drove around the area during breaks, and put on that mileage in the Atlanta area. Have you ever driven around Atlanta? It's so spread out, you can easily rack up miles on a car! Also, didn't they tell you I hit a deer?"

"Yeah, that's what they told us. Of course, there's no one to confirm your story. Where is Dr. Thompson's laptop? It wasn't found by his body, and he always had it with him. You know you'd be stupid to try to publish papers based on his work, right?"

"I DON'T KNOW! I am in the same field, so of course I could publish drug-design work. He stole all my ideas, anyway, so that work is mine!"

"You know, we interviewed many people who were at the fundraiser the night of August 1. . . Would you be surprised to learn that NO ONE recalled seeing you that entire night?"

"There were more than 100 people at the event! I still have my invitation . . ."

"We have your prints on solvent bottles from Dr. Thompson's lab that we collected the day after the fire."

"Of course you do—I worked there for many years! You're detaining me because of that? You have nothing on me!"

The lead investigator stood to closely face Dr. Littleton. With perspiration now dripping down his forehead, he yelled "We also found your prints on Dr. Thompson's wrists. How do you explain that?"

"I . . . it can't be. I want to speak to my attorney . . ."

Without a credible alibi for the night of the arson, and having the motive and means to murder his former advisor, Dr. Littleton was charged with first-degree murder and arson.

Real Talk   Real-World CSI

Check out this interview with Dr. Max Houck, who will describe how forensics is carried out in the real world: www.acs.org/cic.

Questions to Consider

a. What is "forensics anthropology" and what role does this play in a crime scene investigation?

b. Using the Internet as a resource, what are some applications and case precedents for forensic entomology?

Images By Kenny/Shutterstock

c. What is the "Forensic Science Code of Ethics"? Using the Internet as a resource, provide a case study where some of these ethical guidelines were compromised.

Conclusion

This capstone chapter of *Chemistry in Context* incorporated concepts from many of the previous chapters. In this story, a number of scientific techniques were used to gather, examine, and analyze the evidence collected at the crime scenes. As you can see, chemistry plays a key role in forensic investigations, making it utterly impossible to commit the "perfect crime."

LEARNING OUTCOMES

The numbers in parentheses indicate the sections within the chapter where these outcomes were discussed.

Having studied this chapter, you should now be able to:

- differentiate between extinguishing media used for various types of fires (14.2)

- determine the factors that influence the flammability of a liquid (14.4)

- use the autoignition temperature of substances to investigate the source and probable temperature of a fire (14.5)

- determine how fingerprint analysis is carried out in a forensic lab (14.6)

- explain some laboratory techniques used to determine if arson is the likely cause of a fire (14.6)

- explain how paint samples are analyzed in forensic investigations (14.10)

- describe how DNA is collected and analyzed by forensic scientists (14.14)

Questions

Emphasizing Essentials

1. What kind of fire extinguisher would be best to put out fires in each of the following situations?

   a. Frying oil caught on fire at a fast-food restaurant

   b. Fire from an overloaded extension cord

   c. A carpet ignited by a flying ember from a fireplace burning wood

2. Dr. Thompson used traditional solvent stills that contain a reactive Group 1 metal. What are some other safer alternatives that can be used in the laboratory to dry solvents?

3. In the text, the heats of combustion for toluene and dichloromethane are given in kJ/mol. How do these heats of combustion compare to those of ethanol and acetone given later in the text?

4. Cyanoacrylate was used to reveal fingerprints on objects taken from Dr. Thompson's lab. What functional groups are present in this molecule?

5. List some applications for light microscopy and electron microscopy in forensic investigations.

6. List three possible sources of contamination of a crime scene by the first responders.

7. Forensic science encompasses more than criminal investigations. List three sections in earlier chapters of *Chemistry in Context* that could be considered forensic investigations. Explain your choices.

8. Using the Internet or a forensic science textbook as a resource, construct a flowchart that illustrates the process of responding to a violent crime, securing a crime scene, collecting and analyzing evidence, and identifying potential suspects.

9. Although rare, crime scenes may be staged by a perpetrator to misguide investigators. List five possible indicators of a staged crime scene.

10. In earlier chapters, we described the "four R's" of sustainability: reduce, reuse, recycle, and recover. In comparison, forensic investigations rely on the "three R's" of recognize, recover, and record. Elaborate on the importance of these three components to the overall success of a forensic investigation.

11. Using the Internet or textbook as a resource, describe what is meant by the "chain of custody" for evidence collected from a crime scene.

12. In surveying a crime scene, how do forensic scientists use hypotheses to decide which evidence is useful?

Concentrating on Concepts

13. What solvents in Table 14.2 would ignite with a match inside a cold room maintained at −20 °C?

14. X-ray spectrometers often measure the energy of X-ray photons in units of kiloelectron volts (keV). This is the amount of energy gained by an electron after being accelerated by 1000 V of electricity. One keV is equivalent to 1.6×10^{-16} J. What is the wavelength of an X-ray from a phosphorus atom with an energy of 2.0 keV?

15. Infrared spectra are typically measured as a function of wavenumbers instead of wavelengths. Wavenumbers are simply the inverse wavelength ($1/\lambda$). A typical

infrared spectrum will range from 4000 cm^{-1} to 400 cm^{-1}. What is this range in nm?

16. It has been said that "all evidence found at a crime scene has been transferred from one location, object, or person to another." Explain this concept, which is referred to as the *exchange principle*, and provide three examples of transfer evidence commonly found at a crime scene.

17. Polymerase chain reaction (PCR) is typically carried out in modern equipment called a *thermocycler*. Thermocyclers heat and cool a mixture of DNA, enzymes, single nucleotides, and other reagents to optimize the conditions for enzyme function. For each cycle in temperature, the number of DNA strands doubles. So, for n cycles, the number of DNA strands is 2^n. How many cycles would be required to create a million copies from just one DNA strand?

18. Is it possible to obtain latent fingerprints from fabric surfaces? Explain.

19. Inspect the IR spectrum below and list the functional groups that are likely present in the sample. Can you indicate the concentrations of functional groups based on their relative peak heights? Explain.

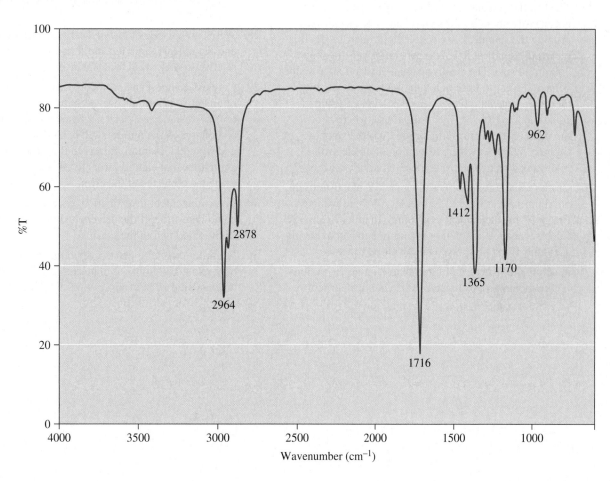

20. Could you discriminate between the three compounds below using IR spectroscopy? What other techniques could be used to determine the identity of each compound?

21. A sample containing heptane, 2,3-dimethyl octane (an isomer of decane), octane, pentane, and methyl cyclohexane is injected onto a GC column with a nonpolar stationary phase. What will be the order of elution for these compounds?
Hint: Consider their relative boiling points.

22. At airports, a swab is taken from the hands and bags of random travelers and then inserted into an instrument to detect the presence of explosives.

a. How does this work?

b. What is the detection limit of this technique?

c. You apply a skin lotion that contains glycerin or nitrates just before entering the security check. You are selected for random screening and your hands are swabbed for explosives detection. Could this result in a false-positive reading (and subsequent long delays!)? Explain.

23. A handgun is found at a crime scene, but the serial number has been severely scratched to prevent it being traced back to the owner. Is there a way to chemically treat the surface to reveal the original serial number? Explain.

24. What other biometrics might be collected from a crime scene besides fingerprints?

Exploring Extensions

25. The smallest resolution that a microscope can achieve is generally about half the wavelength of the particle probing the sample. For a visible light microscope, this means the smallest features that can be see are about 300 nm in size. The electrons in an electron microscope behave similar to light in a light microscope, having their own wavelength that can be calculated using the equation: $\lambda = (6.63 \times 10^{-34} \text{ J·s})/(m \times v)$, where m is the mass of the electron (in kg) and v is the velocity (in m/s). What is the best resolution of an electron microscope using electrons traveling at 6.0×10^7 m/s?

26. Unlike what is commonly shown on TV, bloodstains do not illuminate as bright spots under UV light without first applying a chemical compound. In order to observe fluorescent spots, a solution of luminol and an oxidizing agent must first be sprayed on the blood spots.

   a. Draw the structure of luminol and name the various functional groups present in its chemical structure.

   b. What component(s) of blood catalyze the fluorescence of luminol?

   c. What is the purpose of the oxidizing agent in the luminol solution?

27. A forensic scientist sprayed a luminol solution on a pair of pants found at a suspect's home. Under UV light, the dark stains on the denim appeared as bright blue fluorescent spots.

   a. Are these spots definitely composed of blood? Explain.

   b. Can DNA testing still be performed on bloodstains after they have been sprayed with a luminol solution?

28. Watch an episode of *Crime Scene Investigation* (*CSI*) and list at least three inaccuracies of their investigation relative to real-world forensic investigations.

29. Careless crime scene protocols will affect the outcome of an entire criminal case. Perhaps one of the most famous instances of such negligence was found for responders to the murders of Nicole Brown Simpson and Ronald Goldman. Research this case and list three anomalies that caused jurors to question the validity of evidence collected at the crime scene.

30. In a car accident investigation, describe a technique that could determine if the drivers had their headlights illuminated before the crash.

31. In earlier chapters, we described the investigation of data related to global climate change. Can climatology be classified as forensic science? Explain.

Measure for Measure

Metric Prefixes, Conversion Factors, and Constants

Metric Prefixes

Prefix	Symbol	Value	Scientific Notation
tera	T	1,000,000,000,000	10^{12}
giga	G	1,000,000,000	10^{9}
mega	M	1,000,000	10^{6}
kilo	k	1000	10^{3}
hecto	h	100	10^{2}
deka/deca	da	10	10^{1}
deci	d	1/10 or 0.1	10^{-1}
centi	c	1/100 or 0.01	10^{-2}
milli	m	1/1000 or 0.001	10^{-3}
micro	μ	$1/10^{6}$ or 0.000001	10^{-6}
nano	n	$1/10^{9}$ or 0.000000001	10^{-9}
pico	p	$1/10^{12}$ or 0.000000000001	10^{-12}

Conversion Factors

Length

1 centimeter (cm) = 0.394 inch (in.)

1 meter (m) = 39.4 inches (in.) = 3.28 feet (ft)
    = 1.08 yards (yd)

1 kilometer (km) = 0.621 mile (mi)

1 inch (in.) = 2.54 centimeters (cm) = 0.0833 foot (ft)

1 foot (ft) = 30.5 centimeters (cm) = 0.305 meter (m)
    = 12 inches (in.)

1 yard (yd) = 91.44 centimeters (cm) = 0.9144 meter (m)
    = 3 feet (ft) = 36 inches (in.)

1 mile (mi) = 1.61 kilometers (km)

Volume

1 cubic centimeter (cm^{3}) = 1 milliliter (mL)

1 liter (L) = 1000 milliliters (mL)
    = 1000 cubic centimeters (cm^{3})
    = 1.057 quarts (qt)

1 quart (qt) = 0.946 liter (L)

1 gallon (gal) = 4 quarts (qt) = 3.78 liters (L)

Mass

1 gram (g) = 0.0352 ounce (oz) = 0.00220 pound (lb)

1 kilogram (kg) = 1000 grams (g) = 2.20 pounds (lb)

1 pound (lb) = 454 grams (g) = 0.454 kilogram (kg)

1 tonne or metric ton (t) = 1000 kilograms (kg)
    = 2200 pounds (lb)
    = 1 long ton (t) = 1.10 tons (T)

1 ton (T) = 909 kilograms (kg)
    = 2000 pounds (lb)
    = 1 short ton (T)
    = 0.909 tonne (t)

Time

1 year = 365.24 days

1 day = 24 hours (h)

1 hour (h) = 60 minutes (min)

1 minute (min) = 60 seconds (s)

Energy

1 joule (J) = 0.239 calorie (cal)

1 calorie (cal) = 4.184 joules (J)

1 exajoule (EJ) = 10^{18} joules (J)

1 kilocalorie (kcal) = 1 dietary Calorie (Cal)
    = 4184 joules (J)
    = 4.184 kilojoules (kJ)

1 kilowatt-hour (kWh) = 3,600,000 joules (J)
    = 3.60×10^{6} J

Constants

Speed of light (c) = 3.00×10^{8} meters per second (m/s)

Planck's constant (h) = 6.63×10^{-34} joule-seconds (J · s)

Avogadro's number = 6.02×10^{23} objects per mole
    (objects/mol)

Unified atomic mass unit (u) = 1 dalton (Da)
    = 1.67×10^{-27} kg

Appendix 2

The Power of Exponents

Scientific (or exponential) notation provides a compact and convenient way of writing very large and very small numbers. The idea is to use positive and negative powers of 10. Positive exponents are used to represent large numbers. The exponent, which is written as a superscript, indicates how many times 10 is multiplied by itself. For example,

$$10^1 = 10$$
$$10^2 = 10 \times 10 = 100$$
$$10^3 = 10 \times 10 \times 10 = 1000$$

Note that the positive exponent is equal to the number of zeros between the 1 and the decimal point. Thus, 10^6 corresponds to 1 followed by six zeros, or 1,000,000. This same rule applies to 10^0, which equals 1. One billion, 1,000,000,000, can be written as 10^9.

When 10 is raised to a negative exponent, the number being represented is always less than 1. This is because a negative exponent implies a reciprocal, that is, 1 over 10 raised to the corresponding positive exponent. For example,

$$10^{-1} = 1/10^1 = 1/10 = 0.1$$
$$10^{-2} = 1/10^2 = 1/100 = 0.01$$
$$10^{-3} = 1/10^3 = 1/1000 = 0.001$$

It follows that the larger the negative exponent, the smaller the number. The negative exponent is always one more than the number of zeros between the decimal point and the 1. Thus, 1×10^{-4} is equal to 0.0001. Conversely, 0.000001 in scientific notation is 1×10^{-6}.

Of course, most of the quantities and constants used in chemistry are not simple whole-number powers of 10. For example, Avogadro's number is 6.02×10^{23}, or 6.02 multiplied by a number equal to 1 followed by 23 zeros. Written out, this corresponds to $6.02 \times 100,000,000,000,000,000,000,000$, or 602,000,000,000,000,000,000,000. Switching to very small numbers, a wavelength at which carbon dioxide absorbs infrared radiation is 4.257×10^{-6} m. This number is the same as 4.257×0.000001, or 0.000004257 m.

Your Turn A2.1   Scientific Notation

Express these numbers in scientific notation.

a. 10,000	**b.** 430	**c.** 9876.54
d. 0.000000004	**e.** 0.007	**f.** 0.05339

Answers

a. 1×10^4	**b.** 4.3×10^2	**c.** 9.87654×10^3
d. 4×10^{-9}	**e.** 7×10^{-3}	**f.** 5.339×10^{-2}

Your Turn A2.2   Decimal Notation

Express these numbers in conventional decimal notation.

a. 1×10^6	**b.** 3.123×10^6	**c.** 2.5×10^4
d. 1×10^{-5}	**e.** 6.023×10^{-7}	**f.** 1.723×10^{-16}

Answers

a. 1,000,000	**b.** 3,123,000	**c.** 25,000
d. 0.00001	**e.** 0.0000006023	**f.** 0.0000000000000001723

Clearing the Logjam

You may have encountered logarithms in mathematics courses but wondered if you would ever use them. In fact, logarithms (or "logs" for short) are extremely useful in many areas of science. The essential idea is that they make it much easier to deal with very large *ranges* of numbers, for example, moving by powers of 10 from 0.0001 to 1,000,000.

It is likely that you have seen logarithmic scales without necessarily knowing it. The Richter scale for expressing magnitudes of earthquakes is one example. On this scale, an earthquake of magnitude 6 is 10 times more powerful than one of magnitude 5. An earthquake of magnitude 8 would be 100 times more powerful than one of magnitude 6. Another example is the decibel (dB) scale. Each increase of 10 units represents a 10-fold increase in sound level. Therefore, a normal conversation between two people 1 m apart (60 dB) is 10 times louder than quiet music (50 dB) at the same distance. Loud music (70 dB) and extremely loud music (80 dB) are 10 times and 100 times louder than a normal conversation, respectively.

In Section 5.10, the concept of pH is introduced as a quantitative way to describe the acidity of a substance. A pH value is simply a special case of a logarithmic relationship. It is defined as the negative of the logarithm of the H^+ concentration, expressed in units of molarity (M). Square brackets are used to indicate molar concentrations. The mathematical relationship is given by the equation $pH = -\log [H^+]$. The negative sign indicates an inverse relationship; as the H^+ concentration decreases, the pH increases. Let us apply the equation by using it to calculate the pH of a beverage with a hydrogen ion concentration, or $[H^+]$, of 0.000546 M. We first set up the mathematical equation and substitute the hydrogen ion concentration into it.

$$pH = -\log [H^+] = -\log (5.46 \times 10^{-4} \text{ M})$$

Next, we take the negative logarithm of the H^+ concentration by entering it into a calculator and pressing the log button, then the "plus/minus" button to change the sign. It may display 3.262807357 if you have not preset the number of digits, but common sense prompts you to round the displayed value. That is, the number of *significant figures* (revisit Section 4.3) in the concentration will define the number of *decimal places* in the pH value, and vice versa. For instance, the pH of a solution with $[H^+] = 5.46 \times 10^{-4}$ M (3 sig figs) should be reported as 3.263 (3 figures after the decimal place). Apply the same procedure to calculate the pH of milk with a hydrogen ion concentration of 2.2×10^{-7} M (pH = 6.66).

If we can convert hydrogen ion concentration into pH, how do we go in the reverse direction, that is, how to convert pH into a hydrogen ion concentration? Your calculator can do this for you if it has a button labeled "10^x." Alternatively, it may use two buttons: first "Inv" and then "log." To demonstrate the procedure, suppose you wish to find the hydrogen ion concentration of human blood with a pH of 7.40. Proceed as follows: Enter 7.40, use the "plus/minus" button to change the sign to negative, and then hit 10^x (or follow whatever steps are appropriate for your calculator). The display should give the hydrogen ion concentration as 4.0×10^{-8} M, using the appropriate number of significant figures. Now apply the same procedure to calculate the H^+ concentration of an acid rain sample with a pH of 3.60.

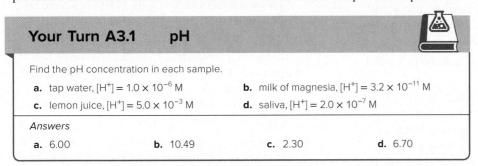

Your Turn A3.1   pH

Find the pH concentration in each sample.

a. tap water, $[H^+] = 1.0 \times 10^{-6}$ M
b. milk of magnesia, $[H^+] = 3.2 \times 10^{-11}$ M
c. lemon juice, $[H^+] = 5.0 \times 10^{-3}$ M
d. saliva, $[H^+] = 2.0 \times 10^{-7}$ M

Answers

a. 6.00
b. 10.49
c. 2.30
d. 6.70

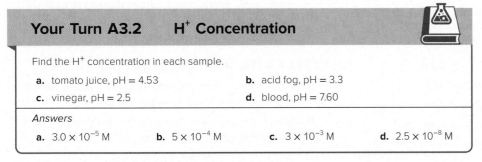

Your Turn A3.2 H^+ Concentration

Find the H^+ concentration in each sample.

a. tomato juice, pH = 4.53
b. acid fog, pH = 3.3
c. vinegar, pH = 2.5
d. blood, pH = 7.60

Answers

a. 3.0×10^{-5} M
b. 5×10^{-4} M
c. 3×10^{-3} M
d. 2.5×10^{-8} M

Appendix 4

Answers to Your Turn Questions

Chapter 1

REFLECT (What's in Your Cell Phone?)

   a. Answers will vary; some possibilities are lightweight, thin, fast, small dimensions, etc.

   b. Answers may vary; a possibility is Si and O_2 to form SiO_2, the composition of glass.

   c. Depending on the source, it is estimated that the average lifespan of a smartphone is 18–24 mo.

1.1 Most of the listed materials would not elicit a response from a touchscreen. However, objects that conduct electricity and have a large diameter, similar to that of a finger, such as a battery will elicit a response.

@HOME (Soup or Salad?)

Part I: No, sugar and baking soda are examples of solid phases since these substances are composed of solid crystals that may be viewed under a microscope. As the crystal size of a solid decreases, the solid may flow similar to a liquid and will share some similar properties with liquids, such as taking the shape of a container. However, the individual particles exhibit some friction against other particles and the container. The flow of the solid and taking the shape of the container requires some external force, such as shaking by your hand, in order to overcome this interparticle resistance. Liquids pour readily and take the shape of the container without any coaxing.

Part II: **a.** Heat was removed from the water as it was placed into the freezer.

      b. Upon freezing, the individual parts of the water seem locked in place. This suggests the molecules have stopped moving. When the ice melts, the water molecules must be able to move around since it no longer holds the shape of the ice. The molecules move away from each other to pool in the bottom of the bowl.

1.2 **a.** Solids: Yes; Liquids: Yes; Gases: No

    b. Solids: Yes; Liquids: No; Gases: No

    c. Solids: No; Liquids: Yes; Gases: Yes

    d. Solids: No; Liquids: No; Gases: Yes

1.3 It can be concluded that these boxes are not containers but zoomed-in areas. The boxes do not follow the properties shown in the table. For example, the solid appears to take the shape of the container.

This representation shows how the atoms are positioned with respect to each other:

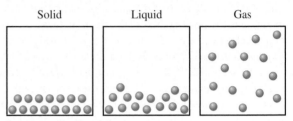

REFLECT (Classification of Matter)

   a. Heterogeneous mixture

   b. Element

   c. Homogeneous mixture (although many wines have sediments suspended in them, which would make them heterogeneous mixtures)

   d. Element

   e. Homogeneous mixture

@HOME (Classifying Matter Around Your Home)

   1. The soft drink is a homogenous mixture of carbon dioxide gas and soda.

   2. The addition of ice makes this a heterogeneous mixture. The microwave melts the ice adding water to the mixture making it a homogenous mixture. Heating the mixture causes the bubbles to come out of solution but it is still a homogeneous mixture of different liquid ingredients.

   3. The salt is a compound. The salt water would be classified as a homogeneous mixture. When we add more salt it becomes a saturated solution with a heterogenous layer. When it is heated it becomes a homogenous mixture again.

   4. The sugar is a compound. The sugar water would be classified as a homogeneous mixture. When we add more sugar it becomes a saturated solution with a heterogenous layer. When it is heated it becomes a homogeneous mixture again.

REFLECT (The Race to Invent the Periodic Table)

   a. Eka-silicon is now referred to as germanium, which is placed right below Si in the periodic table.

   b. Some examples are: Ar (39.95) vs. K (39.10), Co (58.93) vs. Ni (58.69), Th (232) vs. Pa (231).

1.4 **a.** Elements with measurable abundances that are found in a cell phone include H, Li, Be, C, N, O,

S, Mg, Al, Ti, V, Mn, Cd, Fe, Co, Si, Cu, Zn, As, Nb, Mo, Ag, Sn, Sb, Ba, Ta, W, Au, Pb, and Ni.

b. Answers will vary depending on the materials chosen and use of reputable websites.

REFLECT (Can We "See" Atoms?)

The diameter of an average apple is approximately 8 cm. In comparison, the diameter of Earth is approximately 7926 mi, which is 12,756 km or 1.2756×10^9 cm. Hence, Earth is 1.6×10^8 times larger than an apple.

1.5 a. U.S. National debt = 3.05×10^{13} dollars (as of June 2022)

World population = 7.9×10^9 people

b. Average office paper thickness = 0.10 mm

0.10 mm = 1.0×10^{-4} m

0.10 mm = 1.0×10^{-7} km

@HOME (Dimensional Analysis in the Kitchen)

a. 16 Tbsp

b. $1 \text{ fluid oz} \times \dfrac{1 \text{ cup}}{8 \text{ fluid oz}} \times \dfrac{16 \text{ Tbsp}}{1 \text{ cup}} = 2 \text{ Tbsp}$

c. $1 \text{ cup} \times \dfrac{1 \text{ L}}{4.23 \text{ cups}} \times \dfrac{1000 \text{ mL}}{1 \text{ L}} = 236 \text{ mL}$

d. $236 \text{ mL} \times \dfrac{1 \text{ g}}{1 \text{ mL}} = 236 \text{ g}$

e. There are about 453.6 g in one pound, so

$236 \text{ g} \times \dfrac{1 \text{ pound}}{453.6 \text{ g}} = 0.520 \text{ pounds}$

Interactive (What Are "Angstroms"?)

a. 2.3 nm

b. 750,000 angstroms

1.6 a. Some common macroscopic objects with dimensions on the order of . . .

(i) millimeters: the length of an ant, the width of a notebook

(ii) micrometers: the width of a hair fiber, the length of a skin cell

(iii) meters: the length of a car, the width of a football field

b. A cell phone that is 138 mm × 67 mm × 7 mm is 138000 μm or 0.138 m long, 67000 μm or 0.067 m wide, and 7000 μm or 0.007 m thick.

@HOME (Candy Atoms!)

a. You would still have the same element, carbon, but it would have different mass numbers.

b. You would have a different element. 5 protons would be boron with an extra electron and the correct number of neutrons. Since the atom would have an extra electron, it would have an overall charge of −1. Adding a proton would give a mass number of 13; however, the element would be nitrogen with one fewer electron that

would give an overall charge of +1. Atoms with a negative or positive charge are called *ions*.

c. You would still have the same element, carbon, but it would have a net charge other than zero. By removing an electron, there would be more protons than electrons and the charge would be +1. Adding an electron results in a −1 charge.

1.7 a. 31 protons and 31 electrons

b. 50 protons and 50 electrons

c. 82 protons and 82 electrons

d. 26 protons and 26 electrons

a. 1 proton, 1 electron, 1 neutron

b. 24 protons, 24 electrons, 28 neutrons

c. 13 protons, 13 electrons, 14 neutrons

d. 33 protons, 33 electrons, 42 neutrons

1.8 Scandium and yttrium are found in naturally occurring minerals with other elements. Originally, these rare earth metals were found in minerals in Scandinavian countries, but now they are found in minerals all over the world. The mineral known to contain the most scandium is pretulite ($ScPO_4$), which is found in Austria. The mineral with the most yttrium is limoriite ($Y_2(SiO_4)(CO_3)$), which is found in Japan.

REFLECT (Iron Mining)

a. In addition to ready access to water resources and transportation infrastructure, mine operators must ensure that the new mine will not pose negative consequences to the environment or the health of humans or animal populations. Local and regional authorities must be consulted to ensure that there are sufficient resources available and that the mine will benefit the surrounding community.

b. Processing of ore on-site is often carried out due to lower cost relative to transporting to an off-site location.

c. Answers will vary.

1.9 a. chemical   **b.** chemical

c. physical   **d.** chemical

e. physical   **f.** chemical

1.10 a. Desirable properties for a cell phone might include a battery with a longer life or a screen that is indestructible.

b. Incorporating elements like lithium could help make a phone with a longer battery life. Using aluminum and oxygen in the form of sapphire glass could help make a more durable screen.

@HOME (Light/Matter Interactions)

a. glass window: transmitted; LCD screen: reflected; plasma TV screen: reflected; concrete sidewalk: reflected; an asphalt road: absorbed; a ceramic plate: reflected; a cotton shirt: absorbed

b. Answers will vary.

REFLECT (Prince Rupert's Drops)

a. $25000 \text{ psi} \times \dfrac{1 \text{ Pa}}{1.45 \times 10^{-4} \text{ psi}} = 1.7 \times 10^{8} \text{ Pa}$

b. $\dfrac{1500 \text{ m}}{s} \times \dfrac{1 \text{ mi}}{1609 \text{ m}} \times \dfrac{60 \text{ s}}{1 \text{ min}} \times \dfrac{60 \text{ min}}{1 \text{ hr}}$

= 3400 miles per hour

@HOME (Fun With Balloons)

a. Pressure is a measure of a force spread over an area. Therefore, pressure and area are inversely proportional.

b. Just as a high heel exerts more pressure on the floor than a flat-soled shoe, the single thumb tack had more pressure acting on the balloon because there was less surface area. This caused it to pop. The group of thumbtacks had a large surface area and therefore less pressure acting on the balloon and it did not pop the balloon.

1.11 a. Other than charging, your smartphone uses energy every time you search the Internet, watch a video, or use social media.

b. Answers will vary. One should also describe the implications of increased Internet traffic (data centers, large supercomputing facilities, servers, etc.), as well as the energy consumption of smartphones relative to desktop or laptop computers.

Real Talk (Sustainable Product Design)

a. Circular materials design pertains to recovery of spent materials that are then collected, sorted, reprocessed, and used for another application. In contrast, linear materials are discarded after their end of use has been reached.

b. Some examples of circular materials are plastics, metals, recycled cardboard, and natural fibers.

1.12 The Aluminum Association has published several statistics regarding the cost and energy that are saved by recycling aluminum as opposed to aluminum mining from ore. For example, recycling aluminum saves more than 90% of the energy that would be needed to create a comparable amount of the metal from raw materials.

Chapter 2

REFLECT (The Components of Air)

a. Possible answers are indoor: paint, perfumes, deodorants, cooking, incense; outdoor: flowers, decaying leaves, plastics from a hot dish in the summertime, cooking (e.g., barbecue).

b. Most of these chemicals are harmless to human health; however, some people may have allergies toward certain chemicals that are quite serious in some instances.

2.1 Answers will vary, depending on the answers to the previous activity. Basically, each of those indicated above will not have an appreciable effect on air quality, because they are emitted in such small concentrations. However, if pollutants are selected that were mentioned in the introductory video such as ozone, then it should be indicated that they will have a negative effect on air quality.

2.2 $\dfrac{0.5 \text{ L}}{\text{breath}} \times \dfrac{12 \text{ breaths}}{\text{min}} \times \dfrac{60 \text{ min}}{\text{h}} \times \dfrac{24 \text{ h}}{\text{day}} = \dfrac{7,000 \text{ L}}{\text{day}}$.

This calculation is based on an estimate that the average person in a resting state inhales 0.5 L in each breath, and takes about 12 breaths in one minute. The average person in a resting state inhales 0.5 L in each breath. Activities such as exercise, anxiety, or sleep may cause the quantity to change.

@HOME (What's Your Lung Capacity?)

a. Answers will vary.

b. Some possible answers include: weight/height, age, and activity levels.

@HOME (Respiration in the Kitchen)

a. A gas is released from the reaction, which inflates the balloon.

b. carbon dioxide from the reaction of yeast with sugar.

2.3 We inhale a mixture of nitrogen, oxygen, argon, carbon dioxide, water, and other gases in trace amounts. We exhale a mixture of the same chemicals although the relative amounts of each chemical change; most notably the oxygen quantity decreases and the carbon dioxide quantity increases.

An "ideal" atmosphere would consist entirely of N_2 and O_2 gases. However, the addition of other gases such as Ar, H_2O, and CO_2 would not negatively affect our health and could therefore also be present in an ideal breath.

REFLECT (Contrails or Chemtrails?)

a. Contrails result from condensation of aircraft exhaust gases due to the cold temperature at high altitudes. Contrails are primarily composed of water particles and ice crystals. In contrast, chemtrails refer to the belief that these clouds are due to distributing chemical or biological contaminants into the atmosphere. The alleged objectives posed by conspiracy theories range from population control to behavior alterations to climate manipulation. There is no evidence for the distribution of chemtrails.

b. It is estimated that contrails may account for up to 57% of the entire climate impact of aviation, which includes the emissions of carbon dioxide from combustion of aviation fuels. Contrails absorb outgoing heat thereby adding to the greenhouse effect discussed in Chapter 4. It has been proposed that lowering the altitude of flights by a few thousand feet could reduce the formation of contrails or effect their lifetime in the atmosphere.

2.4 With an increased amount of oxygen in the atmosphere, corrosion would occur more quickly and combustion reactions would occur more readily and burn more efficiently. Although this may not seem detrimental in all aspects (think better gas mileage!), the global ramifications of this change would be enormous; things that previously only got hot or smoked, like burnt toast, would now readily catch on fire.

@HOME (Your Nose Knows)

  a. Answers will vary.

  b. The balloon has small pores, which allows the perfume/cologne to escape.

  c. Possible answers include: indoor: air fresheners, candles, paint, bleach, ammonia; outdoor: gasoline, dairy farms, flowers, fresh-cut grass.

  d. Possible answers include: methane (added sulfur-containing compounds), gasoline, smoke.

@HOME (Air Pollution and Properties of Gases)

  a. If more baking soda or vinegar were used, then there would be some of the excess reactant left over.

  b. The lit match would be extinguished if brought into contact with carbon dioxide; if brought in contact with oxygen, the lit match would burst into a flame due to a more intense reaction.

REFLECT (Really One Part per Million?)

- one second in nearly 12 days

$$12 \text{ days} \times \frac{24 \text{ hours}}{1 \text{ day}} \times \frac{60 \text{ minutes}}{1 \text{ hour}} \times \frac{60 \text{ seconds}}{1 \text{ minute}}$$
$$= 1,036,800 \text{ seconds}$$

- one step in a 568-mile journey

$$\frac{1 \text{ step}}{2.5 \text{ ft}} \times \frac{5280 \text{ ft}}{1 \text{ mile}} \times 568 \text{ miles} = 1,199,616 \text{ steps}$$

- four drops of ink in a 55-gallon barrel of water

$$55 \text{ gal} \times \frac{3785 \text{ mL}}{1 \text{ gal}} \times \frac{20 \text{ drops}}{1 \text{ mL}}$$
$$= 4,163,500 \text{ drops of water}$$

So all are pretty fair estimations.

2.5 **a.** 9 ppm is equal to 0.0009%.

  b. 78% nitrogen is equal to 780,000 ppm.

2.6 **a.** NO, NO_2, N_2O, N_2O_4

  b. SO_2 is sulfur dioxide and SO_3 is sulfur trioxide.

2.7 **a.** 2 carbon atoms

  b. 4 carbon atoms

  c. 3 carbon atoms

2.8 **a.** SiO_2, silicon dioxide

  b. N_2O, dinitrogen monoxide

  c. SiH_4, silicon tetrahydride

  d. CO_2, carbon dioxide

  e. H_2S, dihydrogen monosulfide

  f. PH_3, phosphorus trihydride

@HOME (Particulate Matter at Home)

  a. Answers will vary.

  b. A variety of small particulates would be more noticeable.

Real Talk (EVs: The End of Particulate Emissions?)

  a. Particulate matter can be generated from a variety of vehicle-related sources such as tires, roadways, and brakes due to frictional forces acting on surfaces. Hotter temperatures and higher driving speeds cause more PM to be released from tires and asphalt.

  b. The switch from traditional to electric vehicles, which feature regenerative braking (i.e., using electric motors instead of friction against brake pads), will reduce the PM emissions from brake dust while generating energy to replenish the car's battery. On the other hand, the battery packs add weight to vehicles which does influence PM emissions from tires. A greater use of public transportation would reduce PM emissions from vehicles.

2.9 For ideas, return to this chapter's opening video. When we breathe, we are taking in both the expected substances that make up air (nitrogen, oxygen, argon, carbon dioxide, and water) as well as any other gases and fine particulate pollutants that may be present. These pollutant emissions could be NO_x, SO_x, as well as metallic or carbon-based particulates.

2.10 **a.** O_3. This is a hard call, as no common exposure period exists on which to base the comparison. Clearly, CO is not the most toxic, as all its standards are higher. It is not NO_2, because SO_2 has a lower 1-h average standard. Between SO_2 and O_3, ozone has the stricter standard because of the lower 8-h average in comparison to the 3-h average standard for sulfur dioxide.

  b. Claim is supported because the levels of exposure for $PM_{2.5}$ are lower than for PM_{10}.

  c. Each pollutant has a different effect when ingested, but the EPA's limit for exposure to lead is lower than PM. So smaller concentrations of lead have a more detrimental effect than PM.

2.11 She would not exceed the 1-h limit of 210 $\mu g/m^3$. That is, 44 μg per hour/0.625 m^3 of air per hour only equals an exposure of 70 $\mu g/m^3$.

2.12 **a.** Answers will vary.

  b. Ozone is highest during the summer months when sunlight and heat are more plentiful; particulate matter is present during most months; the midwest shows more PM during winter months due to more use of wood burning stoves.

  c. Answers will vary.

2.13 **a.** As of November 23, 2022, 85.3% of the world's population has air quality that exceeds the WHO guideline for $PM_{2.5}$. The percentage of population exceeding the EPA guidelines would be less because the EPA standard for $PM_{2.5}$ is higher than what is established by the WHO.

b. The 20-39 age group is most affected.

c. Answers will vary. As of November 23, 2022, large portions of India and China, and smaller parts of midwestern U.S. and Europe (depending on the time of day) have exceeded these guidelines. Possible reasons may include the use of wood-burning stoves during cold months and use of diesel vehicles in large metropolitan areas.

d. Europe, U.S., and China appear to have the greatest density of air quality monitors. This means that we can only see data from these select locations. We might be missing additional large PM sources in locations without monitors. We might be overlooking vulnerable populations in Africa or South America, for example, simply due to the locations of air quality monitors.

@HOME (Fun With Baking Soda!)

a. The mass of the whole system stayed the same throughout the reaction.

b. The mass of the vinegar and the mass of the baking soda decreased. The mass of the carbon dioxide, as evidenced by the balloon inflating, increased. The mass of water increased.

c. The mass of the system would decrease because the system would no longer be closed. The gas would escape.

2.15 a. Answers will vary; for example:

i)

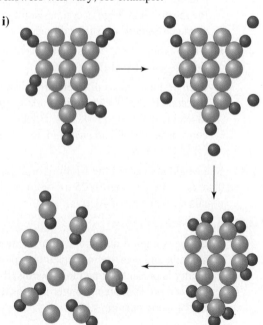

ii)

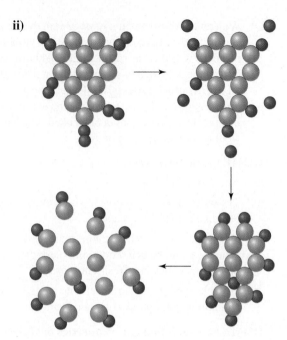

b. $13\ C(s) + 5\ O_2(g) \longrightarrow 5\ CO_2(g) + 8\ C(s)$; $13\ C(s) + 5\ O_2(g) \longrightarrow 10\ CO(g) + 3\ C(s)$; there are 8 C atoms left over when forming CO_2 and 3 C atoms left over when forming CO.

2.16 a. Picture A shows molecular forms of the elements; Picture B shows the elements as individual atoms

b. $S_8(s) + 4\ O_2(g) \longrightarrow 4\ SO_2(g) + 4\ S(s)$; $S_8(s) + 4\ O_2(g) \longrightarrow 2\ SO_3(g) + SO_2(g) + 5\ S(s)$

c. Yes. There are no O_2 molecules left over after reaction in **b.**

d. True. There are 4 S atoms left over when forming SO_2; there are 5 S atoms left over when forming SO_3/SO_2 mixtures, as shown in **b.** When balancing the reactions, there are more S atoms left after forming the product(s).

@HOME (Limiting Reagents in the Kitchen)
Answers will vary depending on the starting amounts of ingredients.

@HOME (Combustion of Candles)

a. In the first scenario the candle had excess oxygen for combustion and in the second scenario, it was limited.

b. The jar closes the system.

2.17 Equation 2.5 contains 16 C, 36 H, and 50 O on each side. Equation 2.6 contains 16 C, 36 H, and 34 O on each side.

2.18 a. For x = 1, nitrogen monoxide (NO) is the emission of interest. For x = 2, nitrogen dioxide (NO_2) is the emission of interest.

b. Nitrogen monoxide is produced when nitrogen and oxygen atoms from air react in the high pressure and temperature conditions of an engine.

c. No, the vehicle would not pass inspection because the emissions data falls above the acceptable range.

2.19 a. $Ag_2S(s) + O_2(g) \longrightarrow 2\ Ag(s) + SO_2(g)$

b. $CuS(s) + O_2(g) \longrightarrow Cu(s) + SO_2(g)$

2.20 Driving practices that conserve fuel are driving slower (maximizing the fuel efficiency of your vehicle), driving on the highway, rolling the windows down instead of using air conditioning, and turning the car off while parked. Conversely, driving practices that expend more fuel than necessary include idling stationary for long periods of time, speeding, running the air conditioning, and driving in congested city traffic.

2.21 a. The colors indicate the relative safety of the air in those regions. Green is good, yellow is moderate, orange is unhealthy for sensitive groups, and red is unhealthy.

b. Some groups at the highest risk from particle pollution are children, the elderly, people who work outdoors, and those with chronic diseases such as asthma, emphysema, heart disease, and diabetes.

c–d. Answers will vary based on location.

2.22 a. The air is hazardous to one or more groups in red and orange areas. Cities closest to these areas are Los Angeles and Sacramento.

b. The ozone level peaks around 5:00 PM.

c. Yes. Once the Sun sets, the ozone levels drop. Sunlight and heat are necessary for ozone to result from VOCs and NO_x. Some ozone may linger, but peak levels are experienced when the Sun is shining.

2.23 An example of this might be: The process of ozone formation begins with inefficient combustion in a car engine. Two products of this inefficient combustion are NO and VOCs. Over time, NO is converted to NO_2 through reaction with the produced VOCs. NO_2 reacts with sunlight to form NO and free O atoms, which can then react with O_2 in the atmosphere to make O_3.

2.24 a. No, at this level the ozone concentration will pose no risk to a healthy person who wishes to exercise outside.

b. Answers will vary depending on location.

REFLECT (Pollution Monitoring)

a. When potassium iodide (KI) reacts with ozone, it produces iodine: $2\ KI + O_3 + H_2O \longrightarrow 2\ KOH + O_2 + I_2$ Once formed, iodine (I_2) reacts with the starch in cornstarch, which results in a blue/purple color.

b. If the detector strip were placed in direct sunlight, the blue/purple color would be erased due to reaction with UV light.

c. Some factors include the wind strength, relative humidity, and sunlight.

2.25 a. A concentration of 1943 μg of particulate matter per cubic meter of air exceeds the National Ambient Air Quality Standards for both PM_{10} and $PM_{2.5}$.

b. Breathing fine particles at this level is hazardous for everybody. The primary danger is to the cardiovascular system, as the particles, when inhaled, pass into the bloodstream and cause or further aggravate heart disease.

2.26 Indoor activities that generate pollutants include burning incense, painting or varnishing (except if using low-VOC paint), cigarette or cigar smoking, frying foods (especially when something burns), using some cleaning products (e.g., ammonia, spray oven cleaner, VOC-containing products such as PineSol), using aerosol hair sprays and some hair-coloring products, using some furniture polishes, or using spray insecticides.

2.27 a. African countries have shown the greatest decrease in death rates from indoor pollution since 1990. A possibility for these sharp decreases is due to using more modern equipment for cooking relative to using fire indoors that would release toxins and particulate pollution.

b. The <5 year and 5–14 age groups have shown the greatest decrease in death rates. Small children are typically most sensitive toward air pollution such as ozone and particulate matter.

2.28 a. PM exposure; John should avoid jogging alongside a busy freeway.

b. Ozone exposure; avoid exercising along busy roads at times of day when ozone levels are the highest.

c. CO poisoning; the arena should provide more suitable ventilation for operation of the Zamboni.

d. Pb exposure; Pb-based paints were the norm until the 1970s. If working on older homes, Jill should wear a dust mask or respirator and gloves.

2.29 a. Air pollution represents the third-highest risk factor for deaths. The total number of deaths due to air pollution has steadily increased; smoking overtook air pollution in 1998 as the 2nd highest global risk factor.

b. In 2007, outdoor air pollution overtook indoor air pollution in number of deaths. Modernization in some countries may be a reason for declining deaths due to indoor air pollution, whereas more widespread use of modes of transportation may explain the increase in deaths due to outdoor air pollution. Smoking and high blood pressure currently pose similar risks as air pollution.

c. Papua New Guinea, China, India, Nepal, Haiti, and mid-African nations currently have the highest death rates due to air pollution. Some countries that have seen the greatest declines in deaths include the U.S., Canada, China, Australia, and Russia. Government regulations and more efficient vehicle technology, in addition to better access to medical services, may explain the large declines in death rates found in these countries.

Real Talk (Sustainable Development: Solvent-Free Reactions)

  a. Dr. Mack's research involves the economic and environmental pillars since grinding powders to fabricate materials generates less waste and can be less costly than using organic solvents. From the list located online (https://www.acs.org/content/acs/en/greenchemistry/principles/12-principles-of-green-chemistry.html), some of the 12 principles of green chemistry addressed by Dr. Mack's research include: i) it is better to prevent waste than to treat or clean up waste after it has been created; ii) wherever practicable, synthetic methods should be designed to use and generate substances that possess little or no toxicity to human health and the environment; iii) the use of auxiliary substances (e.g., solvents, separation agents, etc.) should be made unnecessary wherever possible and, innocuous when used.

  b. Mechanochemistry has been used to make a variety of materials such as magnets, catalysts, metal alloys, and nanomaterials. In addition, mechanochemistry has been used as an alternative to organic solvents.

Chapter 3

REFLECT (Protection from the Sun)

  Possible answers are protective clothing, sunscreens, sunblocks, umbrella, shade from a tree, building, etc. The most effective will be those that don't allow any sun rays to contact skin (clothing); less effective will be sunscreens (effectiveness depends on how much the user puts on and if he/she covers all exposed skin areas).

  3.1 Answers will vary.

@HOME (The Effect of Sun's Rays)

  a. The areas that had sunscreen will be the same color as when the experiment started. The areas that did not have sunscreen will be lighter and bleached by the sun.

  b. The higher SPF sunscreens will likely keep the color of the paper the same longer than lower SPF sunscreens. For short periods of time, the results will likely appear the same.

  3.2 a. The Sun emits energy in the form of ultraviolet, infrared, and visible light radiation.

    b. Being out in the Sun too long causes painful burns and blisters on the surface of the skin. It can only be assumed that damage is also occurring under the skin that we cannot see. The UV light has shorter wavelengths than visible or IR rays. Our skin and eyes are sensitive to these photons and can become damaged by this form of light energy.

    c. Lasting effects of sun damage are evidenced by wrinkled, leathery skin, and the presence of solar lentigines (sunspots) on the skin. Also, cataracts and other eye problems can develop from prolonged UV exposure.

  d. The Sun is necessary for the skin's production of vitamin D, a nutrient that is necessary for your body to be able to absorb calcium.

@HOME (Fun With Prisms)

  White light is made from all of the colors in the visible spectrum. Each of the colors refracts at different angles, which separates the colors from one another.

  3.3 a. $525 \text{ nm} \times \dfrac{1 \times 10^{-9} \text{ m}}{1 \text{ nm}} = 5.25 \times 10^{-7} \text{ m}$;

$$\dfrac{3.00 \times 10^8 \text{ m} \cdot \text{s}^{-1}}{5.25 \times 10^{-7} \text{ m}} = 5.71 \times 10^{14} \text{ s}^{-1}$$
$$= 5.71 \times 10^{14} \text{ Hz}$$

    b. $5.71 \times 10^{14} \text{ s}^{-1} \times \dfrac{60 \text{ s}}{\text{min}} = 3.43 \times 10^{16}$ waves/min;

$$\dfrac{3.43 \times 10^{16} \text{ waves}}{\text{min}} \times \dfrac{60 \text{ min}}{\text{h}}$$
$$= 2.06 \times 10^{18} \text{ waves/h}$$

    c. Amplitude is the height of the wave. It has nothing to do with the wavelength, frequency, or speed of the wave; it has to do with the intensity of the wave.

  3.4 a. Wavelength and frequency are inversely related. Red light has the longest wavelength at 700 nm (and therefore the lowest frequency), meaning that violet light has the highest frequency with a wavelength of 400 nm.

    b. 500 nm is equal to 500×10^{-9} m, which in proper scientific notation is 5×10^{-7} m.

  3.5 a. In order of increasing wavelength: ultraviolet radiation < visible radiation < infrared radiation < microwave radiation.

    b. A radio wave is on the order of 10^1 m, while an X-ray is on the order of 10^{-10} m. This is a difference of 12 orders of magnitude, which means X-rays are 10^{12} times as energetic as radio waves!

  3.6 a. In the electromagnetic spectrum, IR radiation lies to the right of the visible, whereas UV radiation is located to the left of the visible region. The greatest portion of energy from the Sun reaches Earth as infrared radiation.

    b. The most intense radiation emitted by the Sun is visible light with a wavelength around 500 nm (green visible light).

  3.7 a. Using the simulation, red light emits around 700 nm and blue light emits around 475 nm. Any value between 620 and 750 nm for red light and between 450 and 495 nm for blue light is correct.

    Using Equation 3.3:

$$E_{\text{red light}} = \dfrac{(6.626 \times 10^{-34} \text{ J} \cdot \text{s})(3.00 \times 10^8 \text{ m/s})}{700 \times 10^{-9} \text{ m}}$$
$$= 2.84 \times 10^{-19} \text{ J}$$

$$E_{\text{blue light}} = \frac{(6.626 \times 10^{-34} \text{ J} \cdot \text{s})(3.00 \times 10^8 \text{ m/s})}{475 \times 10^{-9} \text{ m}}$$
$$= 4.18 \times 10^{-19} \text{ J}$$

Although both of these values are incredibly small, compared to one another, blue light emits 150% more energy than red light.

b. No, blue light could not have been used. Blue light is too energetic and would overexpose the images being developed. The low-energy red light allows for more control of developing the images.

3.8 a. In order of increasing frequency
UV-A < UV-B < UV-C

b. No, the numerator in Equation 3.3 is a constant, making energy inversely proportional to wavelength.

c. No, UV-C radiation is absorbed by O_2 and O_3 in the stratosphere prior to reaching Earth's surface. Sunscreen needs to protect against UV-A and UV-B radiation that reaches Earth's surface.

3.9 a. Gamma rays, X-rays, and both UV-B and UV-C have sufficient energy to break apart ozone molecules.

b. 320 nm/242 nm = 1.32

A 242-nm photon has roughly 130% more energy than a 320-nm photon!

OR

$$E_{\text{242-nm photon}} = \frac{(6.626 \times 10^{-34} \text{ J} \cdot \text{s})(3.00 \times 10^8 \text{ m/s})}{242 \times 10^{-9} \text{ m}}$$
$$= 8.21 \times 10^{-19} \text{ J}$$

$$E_{\text{320-nm photon}} = \frac{(6.626 \times 10^{-34} \text{ J} \cdot \text{s})(3.00 \times 10^8 \text{ m/s})}{320 \times 10^{-9} \text{ m}}$$
$$= 6.21 \times 10^{-19} \text{ J}$$

$$\frac{8.21 \times 10^{-19} \text{ J}}{6.21 \times 10^{-19} \text{ J}} = 1.32$$

Using these calculations, a 242-nm photon has roughly 130% more energy than a 320-nm photon!

@HOME (Photosynthesis)

a. The spinach leaf had bubbles all over it when it was placed in a sunny area and no bubbles on it when it was placed in a dark area. Photosynthesis produces oxygen gas, so the spinach leaf was producing oxygen gas when it was in a sunny area and that was what was inside the bubbles.

b. If baking soda or another source of carbon dioxide was added to the water the spinach leaf would produce even more oxygen gas in the sunlight because it would have even more reactants to use.

REFLECT (Skin Protection)

a. Answers will vary. Many of the ingredients are added to improve its spreadability on skin and

impart a pleasant aroma. This website provides a list of active ingredients commonly found in these formulations.

b. Both snake venom and Botox are used to prevent wrinkles by preventing muscle movement in the areas where the product was applied (venom) or injected (Botox). Both substances are extremely lethal toxins that appear to be safe if applied in very small doses, although more studies are needed to assess the extent of side effects, especially over long term use. Some alternatives include Daxxify, Xeomin, and Jeaveau, which are all neuromodulators.

3.10 a. According to the American Academy of Dermatology, the incidence rate has more than doubled since 1975, while the mortality rate has remained essentially constant. According to the Centers for Disease Control (CDC), the incidence rates of other cancers such as breast and prostate cancer are also higher than they were 40 years ago; incidence rates of colorectal cancer have declined dramatically over the past 15 years. Additionally, mortality rates of all three cancers have dropped in the past 40 years and five-year survival rates have increased.

b. Lighter skin has less of the melanin THAT provides natural protection against the harmful effects of ultraviolet radiation. In addition, males are typically less likely to apply sunscreen than females and men commonly spend more time in the sun than females. There also may be a link to estrogen levels and natural protection against melanomas. More details may be found here.

c. You should think about protecting yourself not only for short-term UV exposure such as for a day at the beach, but also for UV exposure that builds up over time. For example, according to the Skin Cancer Foundation, UV exposure we receive while driving in the car or shining through office windows can lead to significant skin damage over time. To avoid this low but constant exposure, it's advisable to have UV protection on car or office windows, or to avoid direct sun exposure, even when indoors.

3.11 A well-known example of something that is required for good health but dangerous in high quantities is water. Water is necessary to regulate body temperature and to maintain a number of important processes within your body. However, drinking too much water too quickly can lead to serious complications and death.

3.12 a. Unless you live near the equator, the UV index is higher in summer than it is in winter. This is due to the fact that the Sun is at a steeper angle and there are more hours of daylight.

b. During the summer months, the UV index is higher closer to the equator in states like Hawaii and gets lower as you get farther away from the equator. This is because the Sun's rays are most intense closest to the equator.

3.13 U.S. data may be found online at: https://www. aimatmelanoma.org/legislation-policy-advocacy/ indoor-tanning/

International data may be found online at: https://www.mdpi.com/2227-9067/9/6/768/htm

The regulations have been increased worldwide, with many countries currently regulating tanning for minors. However, South Africa currently has no regulations and those established in Asia and India are not available.

3.14 **a.** Perhaps in the discussion of climate change one has heard of ozone depletion or the hole in the ozone layer. In the 1990s, the ozone hole was a topic covered heavily in the media, whereas in recent times the phrases "global warming" and "climate change" are more of the focus.

b. The ozone layer is located within the stratosphere and is made up of both O_2 and O_3 molecules.

@HOME (Layers of the Atmosphere)

a.

Model Layer	Atmosphere Layer
vegetable oil	exosphere
water	thermosphere
liquid dishwashing soap	mesosphere
corn syrup	stratosphere
honey	troposphere
dirt	earth

b. The troposphere because it is at the bottom.

3.15 **a.** The approximate altitude of maximum ozone concentration is 23 km (14 miles) above sea level.

b. The highest concentration of ozone in the stratosphere is 12,000 ozone molecules per billion molecules and atoms of all types, which is 12,000 ppb.

c. In ambient air, ozone levels can be between 20 and 100 ppb, or more. The EPA suggests limiting your ozone exposure to 75 ppb in an 8-h period.

3.16 **a.** 1 H atom × 1 valence electron per atom = 1 valence electron

1 Br atom × 7 valence electrons per atom = 7 valence electrons

Total = 8 valence electrons

Here is the Lewis structure:

H—Br:

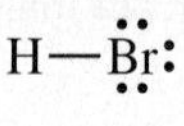

b. 2 Br atom × 7 valence electrons per atom = 14 valence electrons

Here is the Lewis structure:

:Br—Br:

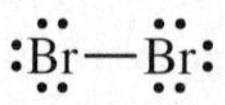

REFLECT (Lewis Structures)

a. 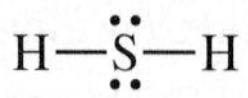 H—S—H

b. 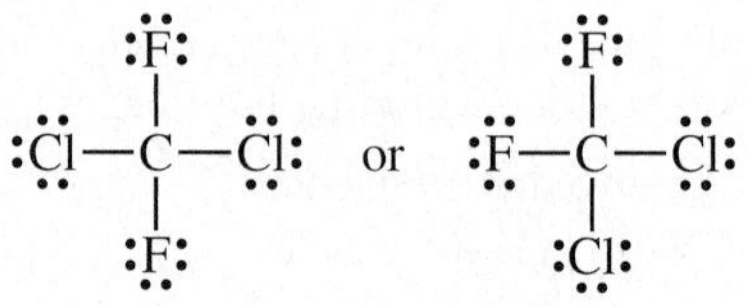 :Cl—C—Cl: or   :F—C—Cl:
        | |
        F   Cl

c. H—N—H
         |
         H

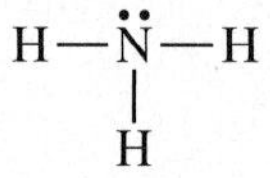

3.17 **a.** 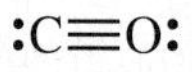 :C≡O:

b. O=S—O: or   :O—S=O

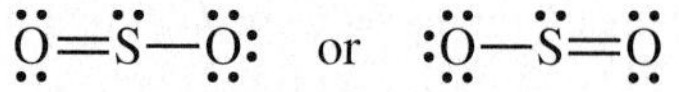

c.
:O
   \
    S=O:
:O/

⬍

:O   :O
  \\   \
   S—O: ⟷   S=O:
:O/   :O/

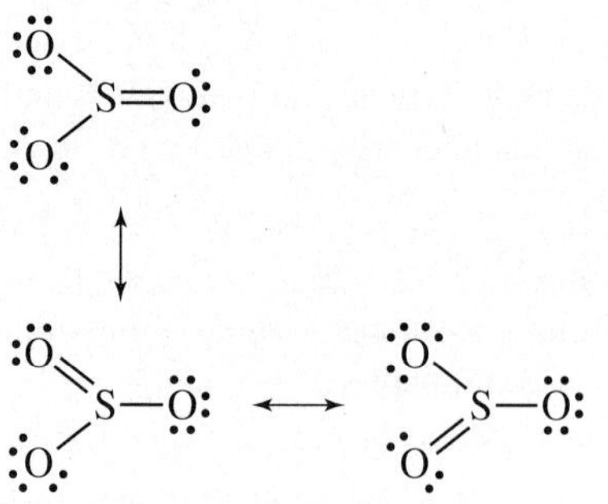

3.18 Ozone concentration will vary based on the relative amount of sunlight a region receives. Because UV radiation is necessary for ozone production, it would be expected that higher concentrations of ozone would be found near the equator or in regions experiencing summer. Conversely, lower concentrations of ozone would be found at the poles or in regions experiencing winter.

3.19 **a.** The average area of the ozone hole over Antarctica in 2018 was 22.9 million km^2. The ozone hole begins to appear in August, grows to its maximum in September, and slowly disappears throughout the month of December. This ranks 13th-largest out of 40 years of NASA satellite observations.

b. The mean lowest reading observed for ozone was 116.5 DU. This is the lowest recorded value since 2011.

c. Regardless of the month selected (August–December), the ozone hole grows in size and the level of ozone decreases dramatically from 1979 to the present.

3.20 Since the late 2000s, ozone levels have (on average) rebounded since reaching all-time lows in the mid- to late 1990s. Although these values are much lower than those observed in the late 1980s, it seems efforts to curb ozone destruction are making a difference. Given the recent trends, it could be predicted that ozone levels should continue to rise over the next three years. This prediction can be monitored using the Ozone Hole Watch website used in Your Turn 3.19.

3.21 Radical compounds are starred.

$$\overset{*}{\cdot}\ddot{O}\!-\!H \;+\; H\!-\!\overset{H}{\underset{H}{C}}\!-\!H \;\longrightarrow\; H\!-\!\ddot{O}\!-\!H \;+\; \overset{*}{\cdot}\overset{H}{\underset{H}{C}}\!-\!H$$

$$\overset{*}{\cdot}\overset{H}{\underset{H}{C}}\!-\!H \;+\; \ddot{O}\!=\!\ddot{O} \;\longrightarrow\; \cdot\ddot{O}\!-\!\ddot{O}\!-\!\overset{H}{\underset{H}{C}}\!-\!H$$

$$\cdot\ddot{O}\!-\!\overset{*}{\ddot{O}}\!-\!\overset{H}{\underset{H}{C}}\!-\!H \;+\; \ddot{N}\!=\!\overset{*}{\ddot{O}} \;\longrightarrow\; \overset{*}{\ddot{O}}\!-\!\overset{H}{\underset{H}{C}}\!-\!H \;+\; \ddot{O}\!-\!N\!=\!\ddot{O}$$

3.23 From the figure, stratospheric ozone and stratospheric chlorine levels are inversely related. As chlorine levels in the stratosphere rise, ozone levels deplete by an equal amount. Conversely, when stratospheric chlorine is removed from the stratosphere, ozone levels increase proportionately. As shown in Equations 3.8 and 3.9, chlorine radicals react with ozone to create more chlorine radicals capable of destroying more ozone.

3.24 The message of this cartoon is satirical in nature because a youth uses something known to be harmful to the ozone layer—an aerosol can of paint—to spread a conservation message about the ozone layer. However, on a much deeper level, this cartoon highlights the importance that each of the more than 8 billion people inhabiting Earth should consider regarding the global implications of their actions.

3.25 **a.** TFA is a widespread contaminant in water samples throughout the world. In some European countries, a limit of 60 ppb has been set as a target level in drinking water. However, there are currently no data regarding the long-term exposure effects of TFA and other PFASs, for humans or aquatic life. These effects will only be discovered through worldwide research efforts that are currently underway. By the time health effects are identified, remediation efforts may be either unfeasible or exceedingly expensive since these chemicals continue to be added to our ecosystem in an unregulated manner.

b. HFCs are greenhouse gases (GHGs) with a global warming potential thousands of times greater than carbon dioxide and thus greatly contribute to climate change (see Section 4.6 for details

regarding GHGs). Besides the TFA issue, HFOs are slightly flammable and are greenhouse gases that contribute to climate change, though to a much lesser extent than HFCs. However, recent studies have indicated that HFOs may decompose into HFCs in the atmosphere, thus worsening their effect on global warming. Ammonia and carbon dioxide have been proposed to replace HFCs and HFOs. These are gases; ammonia is toxic whereas carbon dioxide is odorless and nontoxic. Although CO_2 is a greenhouse gas, it is much less potent for global warming than CFCs or HFCs. It should be noted that ammonia is not a greenhouse gas since it is removed from the atmosphere during precipitation.

3.26 **a.** To be classified as $PM_{2.5}$, the particle diameter must be no larger than 2.5 μm in diameter, so a dust particle that is 6 μm in diameter would be classified as PM_{10}.

b. Because 10^3 nm = 1 μm: $6 \text{ μm} \times \dfrac{1 \times 10^3 \text{ nm}}{1 \text{ μm}}$

$= 6 \times 10^3$ nm or 6,000 nm.

3.27 **a.** The UV index on a sunny summer day would probably fall in the high or very high category, depending on your location. Unprotected skin could burn in as few as 10–20 minutes on a day like this and even fewer if you have sensitive skin.

b. SPF refers to the ratio of time a person can stay in the sun without burning both with and without sunscreen, so it can be calculated by multiplying the SPF by the number of minutes it would take you to get a sunburn. For example, if you typically develop a sunburn after 20 minutes of unprotected sun exposure, applying an SPF 70 sunscreen should give you 20 × 70 or 1400 minutes (23 hours) of protection. But this is assuming you don't sweat or lose the layer of protection in other ways.

c. Although an SPF 70 sunscreen should in theory provide all-day protection from UV radiation, many guidelines are often overlooked. For example, SPF ratings are based on applying a much larger "dose" than most people apply. It is estimated that the average person is getting only 20–50% of the protection advertised on the label by not applying enough of the sunscreen. Additionally, it is recommended that sunscreen be reapplied every two hours to ensure the best protection.

d. The difference between SPF 15 and SPF 30 sunscreen is the amount of time one could theoretically remain in the sun without getting a sunburn. An SPF 30 sunscreen should provide twice the amount of time of protection than an SPF 15 sunscreen, and this is achieved by blocking more of the harmful UV rays emitted by the Sun.

@HOME (SPF and UV Light)

a. The bags with higher SPF levels had a lighter color and took longer to change colors.

b. The sunscreen with higher SPF levels blocked UV light, which affected the amount of light interacting with the beads.

3.28 The bigger the particle, the less protection because there are more holes between them for the UV radiation to travel. Just as sand grains are able to pack more tightly than marbles, smaller particles are more densely packed. The more dense the particles are packed, the less space between them, and the more protection they provide.

3.29 a. Concern about potential environmental impacts has led to the ban of some active ingredients in sunscreen products, including a legislated ban on the sale and distribution of sun protection products containing oxybenzone or octinoxate in the state of Hawaii without a prescription (Hawaii SB 2571, Act 104). Similar bans exist in the U.S. Virgin Islands, Palau, Bonaire, Aruba, and some parts of Mexico.

b. UV filters are the active ingredients in sunscreens that reduce the level of UV radiation reaching the skin by absorbing, reflecting, and/or scattering the sun's UV rays, thereby reducing their ability to reach the skin. There are currently 17 UV filters that are currently found in sunscreen formulations. Some examples of organic-based UV filters are shown in Fig. 3.30; two common inorganic UV filters are ZnO and TiO_2.

c. UV filters may adsorb onto organic matter such as suspended particulates, including phytoplankton, and other aquatic plants can also accumulate such chemicals into their tissues. UV filters may also adsorb onto, or accumulate within, larger algae and vascular plants, which may be consumed by a variety of species such as invertebrates, turtles, aquatic birds, and some mammals. Studies have shown acute toxicity for some UV filters; however, there are no standardized testing methods available and many factors such as solubility and concentrations vary widely among sunscreen formulations.

d. Studies are ongoing; some reports have found that organic-based UV filters are more harmful to aquatic species.

e. Yes, even sunscreens marketed as "reef safe" may pose problems to coral. A 2019 analysis of 97 of the most popular sunscreens (those with at least 150 reviews on Amazon) found that 52 (54 percent) were labeled as "reef safe," and that 48 percent of those labeled as "reef safe" did not meet the criteria identified by NOAA or the legislative bans in Hawaii and Florida, with many containing organic UV filters such as avobenzone and oxybenzone.

Chapter 4

REFLECT (The Greenhouse Effect & Climate Change)

Answers will vary. One should determine that CO_2 is from combustion of hydrocarbon fuels (gasoline, coal, etc.). It would also be interesting to discuss how "climate change" differs from "global warming".

@HOME (Melting Glaciers and Icebergs)

a. The glass with the ice cube floating freely represents an iceberg; the glass with an ice cube on a rock represents a glacier.

b. The water level remains the same for the iceberg glass; the water level increases for the glacier. This represents an increasing water level and flooding due to melting ice as a result of climate change.

4.1 a. Answers will vary.

b. Some possible answers include: Sun intensity and gases in the atmosphere cause warming; cloud cover and wind currents cause cooling.

c. Answers will vary; some students will remark that global climate has become warmer, while others may remark that parts of the world are not warming but cooling.

4.2 In a superficial sense, carbon is important to life in that "all living things have carbon." In addition to being the fourth most abundant element on the planet, its unique properties allow carbon to bond in many ways and with many other elements. Carbon is the fundamental component for the macromolecules that make up proteins, nucleic acids (RNA and DNA), carbohydrates, and lipids in living organisms, In addition, carbon plays an important role in the structure of the food needed to sustain our lives.

4.3 a. CO_2 is added to the atmosphere through respiration (from plants and animals on land and in the ocean), soil decomposition, volcanic eruptions, deforestation and other land use changes, and combustion of fossil fuels. There is a net increase of CO_2 added to the atmosphere due to human activities because natural processes such as respiration are balanced by photosynthesis.

b. CO_2 is removed from the atmosphere through photosynthesis (land and ocean plants), deposition from rainwater into oceans, and directly dissolving into the surface ocean water.

c. The largest reservoirs of carbon atoms are rocks containing carbon sediment - inorganic carbon (like carbonate) crustal rocks and organic carbon found in fossil fuels.

d. Human activity has increased the concentration of CO_2 in the atmosphere due to combustion of fossil fuels and deforestation (and other land use changes).

e. As discussed in Your Turn 4.2, the unique properties of carbon allow it to move to and from different reservoirs based on how it is bonded to other atoms and its location on Earth. For example, a carbon atom exhaled as a CO_2 molecule may be converted into starch ($C_6H_{12}O_6$) and stored in a plant through photosynthesis. The plant may be eaten and used as fuel by another organism, or it may decompose and become part of the soil. Ultimately, carbon found in the atmosphere will travel through the various reservoirs and one day end up in the atmosphere again.

REFLECT (Visualizing the Carbon Cycle)

a. Carbon dioxide is taken up by a plant. Eventually the plant dies (an animal might ingest the plant, but it will die too). As the organic material decays, some carbon is emitted to the atmosphere, but some stays in the soil and can lithify into new rock material. To complete the cycle, this carbon is emitted back into the atmosphere through volcanic eruptions. Occasionally, the carbon does not decay fast enough and gets stuck in the ground to form fossil fuels.

b. Humans are changing where the carbon is. We are moving the carbon from reservoirs that take millions of years to form ("slow carbon") and putting it into the atmosphere where it can only be taken up through the "fast carbon" cycles, such as photosynthesis. Thus, the sinks cannot keep up with the increased source to the atmosphere.

c. Increased plant growth (having more plants, decreasing deforestation, restoring grasslands), use carbon in the "fast carbon cycle," like plants, for combustion and sequester the carbon emissions, direct capture of CO_2 from the atmosphere.

4.4 a. In the Northern Hemisphere spring and summer, the rate of photosynthesis is greater than the rate of respiration because plants are growing rapidly, thus the concentration of CO_2 starts to decrease in the atmosphere in May every year.

   b. In the Northern Hemisphere fall and winter, the rate of respiration is greater than the rate of photosynthesis due to surviving plants saving energy and decay processes by bacteria, plants, and animals. The concentration of CO_2 starts to increase in the atmosphere every October.

   c. The motion is due to atmospheric circulation (wind and weather patterns). While the Northern Hemisphere does have greater CO_2 emissions because it contains the majority of Earth's biomass as well as most of the industrial emissions from combustion, it is important to note that CO_2 travels easily. It is well-mixed in the atmosphere and can have a global impact.

4.5 First, calculate how much CO_2 would be emitted using the less-efficient data:

$$\frac{8887 \text{ g } CO_2}{\text{gal}} \times \frac{\text{gal}}{22.0 \text{ mi}} \times \frac{11500 \text{ mi}}{\text{year}} \times 1.45 \times 10^9 \text{ cars}$$

$$\times \frac{1 \text{ Gton}}{1 \times 10^{15} \text{ g}} = \frac{6.74 \text{ Gton } CO_2}{\text{year}}$$

Next, calculate the emissions using the other set of data:

$$\frac{8887 \text{ g } CO_2}{\text{gal}} \times \frac{\text{gal}}{51.0 \text{ mi}} \times \frac{10000 \text{ mi}}{\text{year}} \times 1.45 \times 10^9 \text{ cars}$$

$$\times \frac{1 \text{ Gton}}{1 \times 10^{15} \text{ g}} = \frac{2.53 \text{ Gton } CO_2}{\text{year}}$$

Subtracting the two values gives a savings of 4.21 Gton CO_2 per year.

This savings is 2.3 times more than the contribution of the trucking industry.

4.6 a. The atomic number of nitrogen is 7 and the atomic mass of nitrogen is 14.01 u.

   b. A neutral atom of N-14 has 7 protons, 7 neutrons, and 7 electrons.

   c. A neutral atom of N-15 has 7 protons, 8 neutrons, and 7 electrons.

   d. An atomic mass of 14.01 means N-14 has the greatest natural abundance (> 99%).

4.7 To compare masses, they must be in the same unit. Looking up a conversion factor, there are 28.35 g in one ounce. We also know that a half-dozen means we are comparing 6 balls tennis balls with 6 golf balls.

For this example, we will convert ounces to grams:

$$\frac{2.05 \text{ oz}}{1 \text{ ball}} \times \frac{28.35 \text{ g}}{1 \text{ oz}} \times 6 \text{ balls} = 349 \text{ g}$$

Thus, 6 tennis balls weigh 349 g. If each golf ball is 45.9 g, then six of them would give a total of 275 g.

As shown in the figure, the tennis balls weigh more. Like atoms of different elements, the masses of tennis and golf balls differ even when the number of units is the same.

4.8 To verify these claims, let's start with marshmallows. The surface area of the United States is roughly 3.8×10^6 miles^2 and the dimensions of a single marshmallow are roughly 1 inch long $\times$ 1 inch wide (we will ignore height for now).

First, calculate how many marshmallows are in each square mile of the United States:

$$\frac{6.02 \times 10^{23} \text{ marshmallows}}{3.08 \times 10^6 \text{ mi}^2}$$

$$= \frac{1.95 \times 10^{17} \text{ marshmallows}}{\text{mi}^2}$$

Next, calculate the number of marshmallows that can fit in a single layer in a single square mile:

$$1 \text{ mi}^2 \times \frac{27{,}878{,}400 \text{ ft}^2}{1 \text{ mi}^2} \times \frac{144 \text{ in}^2}{1 \text{ ft}^2} = 4.01 \times 10^9 \text{ in}^2.$$

A single marshmallow occupies a surface area of 1 in^2, so there are 4.01×10^9 marshmallows/mi^2.

Finally, using the two values calculated above, you can determine how many marshmallows thick each layer would be and convert that to miles:

1.95×10^{17} marshmallows/mi^2/4.01×10^9 marshmallows/mi^2 = 4.9×10^7 layers, in 1-inch-tall marshmallows, which equates to 4.9×10^7 inches $\times$ 1 mile/63,360 inches. Converting to miles gives about 760 miles per marshmallow layer—that's still a pretty tall tower of marshmallows.

For the second analogy, determine how many pennies the entire population could spend in a lifetime (assuming each person started spending on their first day of life and lives to be 100 years old).

1 million dollars = 100,000,000 pennies.

$$\frac{1 \times 10^8 \text{ pennies}}{1 \text{ h}} \times \frac{24 \text{ h}}{1 \text{ day}} \times \frac{365 \text{ days}}{1 \text{ year}} \times \frac{100 \text{ years}}{1 \text{ lifetime}}$$

$\times\, 7 \times 10^9$ lifetime = 6.13×10^{23} pennies spent.

However, because it is an unreasonable assumption that every person on Earth would spend for 100 full years, it is not unreasonable that half of the pennies would be left over.

Ideas for additional analogies that may help demonstrate this concept are people lined up around the globe, or stacks of objects to reach the moon.

4.9 a. $12 \text{ g}/6.022 \times 10^{23}$ C atoms = 2.0×10^{-23} g/C atom

b. The mass of 5×10^{12} carbon atoms $\times\, 2.0 \times 10^{-23}$ g/C atom = 9.95×10^{-11} g = 1×10^{-10} g.

c. The mass of 6×10^{15} C atoms $\times\, 2.0 \times 10^{-23}$ g/C atom = 1×10^{-7} g.

Answers will vary. Whereas one may predict these numbers to be small, others may have thought that a large number of atoms, such as 5 trillion, might actually be enough to measure. In reality, you would need almost 5 trillion billion (10^{20}) atoms in order to weigh out 0.01 g of carbon atoms!

4.10 a. 1 mol O_3 = 3 mol O

$\qquad$ = 3 mol O $\times$ 16.0 g O/1 mol O

$\qquad$ = 48.0 g O_3

b. 1 mol N_2O = (2 mol N $\times$ 14.0 g/mol N) + (1 mol O $\times$ 16.0 g/mol O) = 44.0 g N_2O

c. 1 mol CCl_3F = (1 mol C $\times$ 12.0 g/mol C) + (3 mol Cl $\times$ 35.45 g/mol Cl) + (1 mol F $\times$ 19.1 g/mol F) = 137.4 g CCl_3F

4.11 a. The mass ratio is found by comparing the molar mass of S with the molar mass of SO_2:

$$\frac{32.1 \text{ g/mol S}}{64.1 \text{ g/mol SO}_2} = 0.501$$

b. To find the mass percent of S in SO_2, multiply the mass ratio by 100:

0.501 S $\times$ 100 = 50.1% S in SO_2

c. 1.0 mol N_2O = 44.0 g N_2O (2 mol N/1 mol N_2O)

(28.0 g N/44.0 g N_2O) = 0.636 N/N_2O

0.636 N $\times$ 100 = 63.6% N in N_2O

REFLECT (Grams of CO$_2$)

$$19 \text{ Gt CO}_2 \times \frac{1 \times 10^9 \text{ tonne}}{1 \text{ Gt}} \times \frac{1 \times 10^3 \text{ kg}}{1 \text{ tonne}}$$

$$\times\, \frac{1 \times 10^3 \text{ g}}{1 \text{ kg}} = 1.9 \times 10^{16} \text{ g CO}_2$$

4.12 According to www.autos.com, the weight of cars and trucks can vary from 1.5 tons (compact cars) to 6 tons (full-size pickups and SUVs). Gas mileage varies proportionally with the size of the vehicle. Assuming a car with an average weight of 2 tons (2000 pounds) and an average gas mileage of 25 miles per gallon, the carbon emissions can be calculated as follows:

12,000 miles/25 miles per gallon = 480 gallons of gas used per year.

480 gallons of gas used per year $\times$ 5 pounds of C atoms per gallon = 2400 pounds of C atoms released per year.

Using these data, the statement should be amended to say that the average American car releases *more* than its weight in carbon into the atmosphere each year. However, this estimation is based on the assumption that the car engine in question is clean-burning. In reality, many vehicles on the road do not burn fuel as efficiently and they emit much higher quantities of CO_2 than accounted for by this assumption.

REFLECT (How Should I Report My Data?)

a. Two significant figures; trailing zeros do not count unless they are after a decimal point.

b. The natural carbon flows are: soil respiration (58 Gt), plant respiration (59 Gt), volcanoes (0.1 Gt), and ocean loss (90 Gt). Disregarding sig fig rules and adding these values into a calculator gives 207.1 Gt. But the values do not all have the same precision. The least precise value is 90 Gt - it only has one significant figure in the tens place. Thus, the answer would be 210 Gt because we would have to round to the tens place. To eliminate confusion about the number of sig figs in your answer, you may want to report it as 2.1×10^2 Gt.

c. To figure out the number of molecules, we first need the number of moles in 100.0 g of CO_2 because we know there are Avogadro's number of molecules in 1 mole. Grams are converted to moles through molar mass. We need to use a molar mass that matches or exceeds the precision of the measured value (so that the measurement always dictates the precision of the answer). Because the initial measurement (100.0 g) has 4 sig figs, we need to use a periodic table with at least 4 sig figs (this is found at the back of the textbook).

$12.01 + 2(16.00) = 44.01$ grams in one mole of CO_2

The "2" multiplied by the atomic mass of oxygen does not affect the sig figs because this is an exact number. There are exactly 2 moles of oxygen, so the "2" has an infinite number of significant figures.

$$100.0 \text{ g } CO_2 \left(\frac{1 \text{ mol } CO_2}{44.01 \text{ g } CO_2} \right)$$

$$\left(\frac{6.022 \times 10^{23} \text{ molecules } CO_2}{1 \text{ mole } CO_2} \right)$$

$$= 1.368 \times 10^{24} \text{ molecules } CO_2$$

Because the 100.0 g value has 4 significant figures and each of the conversion factors (grams to moles and moles to molecules) are similarly precise, the answer will have 4 significant figures.

REFLECT (Comparing Energy Balances: Earth and Moon)

Because Earth and the Moon are similar distances from the Sun, the amount of incoming radiation is the same. The reason why the Moon is so cold, while Earth is warm enough to support life (with liquid water), is due to Earth's atmosphere. The atmosphere can absorb energy from Earth's surface (gases that are good at absorbing thermal IR are called greenhouse gases) and radiate heat in all directions. Some of that heat warms the surface of the planet.

4.13 a. When the Sun is initially turned on, the temperature starts to rise; it eventually stabilizes around 2 °C.

b. As the clouds are turned on, temperature decreases since more sunlight is reflected back into space (and thus less would be absorbed by the surface).

c. During the Ice Age, the concentration of greenhouse gases decreased (7 °C), in 1750, the temperature stabilizes to around 13 °C, and there is an increase in greenhouse gas concentrations in 2020 giving a surface temperature around 15 °C. Just using the slider to change the amount of greenhouse gases shows how sensitive the surface temperature is to these alterations.

4.14 a. 90° (at right angles). The two H atoms across from each other would be at 180°.

b. The C atom would occupy the central space where the three leg "bonds" and vertical shaft "bond" meet. The H atoms would be attached to the end of each of the four legs (bonds).

c. The tetrahedral shape is the most stable arrangement because it allows the C—H bonds (made up of electrons) to maximize their distance from one another with bond angles of 109.5°.

4.15

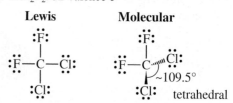

(a)   (b)

For a 3D rendering of this structure, go to 3D model of ammonia.

Nitrogen follows the normal pattern of being surrounded by eight electrons; in this case, there are three bonding pairs and one lone pair of electrons. This enhanced repulsion forces the bonding pairs closer to one another, creating a H—N—H angle slightly less than the predicted 109.5° associated with a regular tetrahedron. The experimental value of 107.3° is close to the tetrahedral angle, again indicating that our model is reasonably reliable. The hydrogen atoms of a NH_3 molecule form a triangle with the nitrogen atom above them at the top of the pyramid. Thus, a molecule of ammonia is said to have a tripod-like shape known as *trigonal pyramidal*.

Going back to the analogy of the folding music stand (Figure 4.8), you could expect to find hydrogen atoms at the tip of each leg of the music stand. This places the nitrogen atom at the intersection of the legs with the shaft, with the nonbonding electron pair corresponding to the shaft of the stand.

4.16 a. CCl_4: 32 valence e^-

Lewis   **Molecular**

~109.5°   tetrahedral

b. CCl_2F_2: 32 valence e^-

Lewis   **Molecular**

~109.5°   tetrahedral

c. H_2S: 8 valence e^-

Lewis   **Molecular**

H—S—H   bent
                                                   <100°

4.17 a. With a total of 16 valence electrons, the C atom contributes four electrons, and six come from each of the two oxygen atoms.

b. If only single bonds were involved, each atom would not be surrounded by eight electrons.

c. If the central carbon atom forms a double bond with each of the two oxygen atoms, thus sharing four electrons with each oxygen atom.

A Lewis structure reveals 2 oxygen atoms bonded to the central carbon and no lone pair electrons on the central carbon atom.

$$\ddot{\text{O}}::\text{C}::\ddot{\text{O}} \qquad \ddot{\text{O}}=\text{C}=\ddot{\text{O}} \qquad \ddot{\text{O}}=\text{C}=\ddot{\text{O}}$$
$$180°$$

(a)   (b)

Representations of a carbon dioxide molecule, CO_2. Shown are the (a) Lewis structures and structural formula, and (b) space-filling model. For a 3D rendering of this structure, go to 3D model of CO_2.

d. Again, groups of electrons repel one another, and the most stable configuration provides the farthest separation of the negative charges. In this case, the groups of electrons are the double bonds, and these are farthest apart with an $O=C=O$ bond angle of 180°. The model predicts that all three atoms in a CO_2 molecule will be in a straight line, and that the molecule will be *linear*. Only electrons around the central atom affect the bond angle. So in CO_2, the electrons around the oxygen atoms have no influence on its overall shape.

4.18 a. Ozone (O_3) has 18 valence electrons.

    b. If only single bonds are used, each atom would not be surrounded by eight electrons, thus we know to incorporate a double bond.

    c. The Lewis structure shows a central atom surrounded by two bonded atoms (one by a single bond and one by a double bond) and one lone pair of electrons. The ozone molecule is best represented by two equivalent resonance structures:

$$\ddot{\text{O}}::\ddot{\text{O}}:\ddot{\text{O}}: \qquad \ddot{\text{O}}=\ddot{\text{O}}-\ddot{\text{O}}: \qquad$$
$$117°$$

(a)   (b)

Representations of an ozone molecule, O_3. Shown are the (a) Lewis structures and structural formula for one resonance form, and (b) space-filling model. For a 3D rendering of this structure, go to 3D model of ozone.

    d. The central O atom has a total of three groups of electrons: the pair that makes up the single bond, the two pairs that constitute the double bond, and the lone pair. These three groups of electrons repel one another, and the minimum energy of the molecule corresponds to their farthest separation. This occurs when the electron groups are all in the same plane, and at an angle of about 120° from one another. Thus, we might predict that the

O_3 molecule should be bent and that the angle made by the three atoms should be approximately 120°. In reality, experiments show the bond angle to be 117°, just slightly smaller than the prediction because the nonbonding electron pair on the central oxygen atom occupies a greater volume than the bonding pairs of electrons, causing a greater repulsion force responsible for the slightly smaller bond angle.

4.19 SO_2: 18 valence e⁻

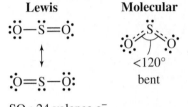

       Lewis   **Molecular**

                                        < 120°
                                        bent

SO_3: 24 valence e⁻

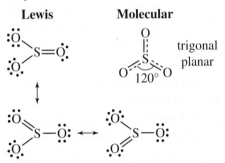

       Lewis   **Molecular**

                                    trigonal
                                    planar
                        120°

REFLECT (Vibrational Modes)

    a. Unit conversion will be demonstrated using the 2338 cm⁻¹ asymmetric stretch.

$$\text{wavelength} = \frac{\text{cm}}{2338} \times \frac{1 \text{ m}}{100 \text{ cm}} \times \frac{1 \times 10^6 \ \mu\text{m}}{1 \text{ m}}$$

$$= 4.277 \ \mu\text{m}$$

Carbon Dioxide Vibration Wavelengths

	Wavenumber (cm⁻¹)	Wavelength (μm)
Asymmetric Stretch	2338	4.277
Symmetric Stretch	1294	7.728
Bending	581	17.2

    b. Because the shortest wavelength requires the most energy, the vibrations are listed in order of decreasing amount of energy required: asymmetric stretch requires the most energy, then symmetric stretch, and finally bending. If you have ever examined a spring, you probably noticed that more energy is required to stretch it than to bend it. Similarly, more energy is required to stretch a CO_2 molecule than to bend it. This means that more energetic photons, those with shorter wavelengths, are needed to cause stretching vibrations, than to cause bending vibrations.

4.20 a. NO$_2$: 17 valence e$^-$

Lewis   Molecular

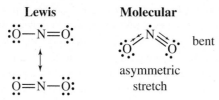

bent

asymmetric stretch

b. O$_3$: 18 valence e$^-$

Lewis   Molecular

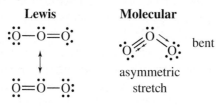

bent

asymmetric stretch

c. CH$_4$: 8 valence e$^-$

Lewis   Molecular

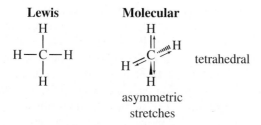

tetrahedral

asymmetric stretches

d. NH$_3$: 8 valence e$^-$

Lewis   Molecular

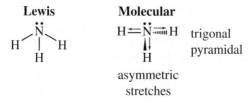

trigonal pyramidal

asymmetric stretches

4.21 a. Yes, these wavelengths correspond to the 4.277, 7.728, 17.2 μm calculated earlier in this section.

b. The symmetric stretch that requires 7.7 μm of energy is missing from the figure. Recall that IR absorption only occurs when molecular vibrations create centers of opposite charge within the molecule. There are no partial charges created during the symmetric stretch of carbon dioxide.

REFLECT (IR Spectroscopy)

a. Light shines on a sample. When the frequency (or wavelength) of that light matches that of a vibrational mode, the light is absorbed and does not make it to the detector. The detector measures the light that is not absorbed by the sample, but rather transmitted through the sample.

b. The 1-Hz example does not correspond to one of the carbonyl vibrational modes, so it is transmitted through and the sample and the graph shows 100% transmission at 1 Hz. The 2-Hz example corresponds to the hypothetical carbonyl stretching mode. Since this light is absorbed by the molecule, the readout indicates an absorption band and a transmittance near 0%.

4.22 a. Both N$_2$ and O$_2$ are linear. In addition, they are both diatomic gases made of two identical atoms. Although molecules consisting of two identical atoms do vibrate, the overall electric charge distribution does not change during these vibrations. Earlier, we discussed this lack of change in the overall electric charge distribution as the reason why the symmetric stretching vibration was not responsible for the greenhouse gas behavior of CO$_2$.

b. Thank goodness N$_2$ and O$_2$ are unable to absorb and reemit heat because they make up 78% and 21% of all gas molecules in the atmosphere, respectively. The higher the concentration of greenhouse gases in the atmosphere, the warmer our planet's surface will be. Carbon dioxide is only 0.0410% of the gas molecules, and we know its presence makes a difference to Earth's radiative balance.

4.23 a. The atmospheric CO$_2$ concentration in May 1958 was 317.51 ppm; in May 2023 it was 423 ppm. The increase is 423 ppm − 317.51/65 years = 1.6 ppm/year

b. The atmospheric CO$_2$ concentration in May 2000 was 371.75 ppm. The increase is 423 ppm − 371.75/23 years = 2.2 ppm/year. The rate of increase has accelerated.

c. Within a given year, the atmospheric CO$_2$ concentration varies by 6–8 ppm.

d. Photosynthesis removes CO$_2$ from the atmosphere. Spring begins in the northern latitudes in April; October is the start of spring in the southern latitudes. Both the landmasses (and number of green plants) are greater in the northern hemisphere, so the seasons in the northern hemisphere control the fluctuations.

4.24 The concentration of greenhouse gases such as N$_2$O, CH$_4$, and SF$_6$ have increased while the concentration of CFCs such as CFC-11, CFC-12, and CFC-13 are slowly decreasing due to the Montreal Protocol. However, CFC replacements such as HCFCs and HFCs are still being added to the atmosphere, which also have large greenhouse warming potentials. All of these trace gas concentrations are much smaller than that of carbon dioxide; however, their global warming potentials (GWPs) are larger on a per-molecule basis.

4.25 a. While carbon dioxide most closely correlates with temperature (increasing temperature, increasing concentrations), all three of the gases generally show a similar relationship.

b. Although there is a *correlation*, we cannot prove *causation*. We do not have enough evidence from these data to show direct causes.

c. The very edge of the dataset shows that each GHG concentration has increased dramatically while the temperature has yet to respond. A likely explanation for the different correlation patterns is the difference in climactic drivers. Prior to industrialization, it is thought that changes in Earth's orbit were primarily responsible for altering the amount of solar energy reaching the planet. This would have changed the temperature which then drove geological and/or biological mechanisms that change the atmospheric composition. In other words, carbon dioxide did not initiate the warming, but it did amplify it. On the other hand, the current (post-industrialization) situation is reversed. Humanity is adding extra GHGs into the atmosphere, and these gases are driving the temperature changes.

REFLECT (Temperature Change Visualization)

Terminology like "global warming" is used because, on average, the temperature of the entire globe is increasing. But the visualization clearly shows that the warming is not uniform across the globe, and some locations actually experience a cooling from time to time. Thus a term such as "climate change" indicates that the climate is changing, but not necessarily warming in all areas. Typically, these terms are used interchangeably.

4.26 Depending on location, a significant temperature change may not be reported for over the last 100 years and regional temperature changes can vary quite significantly. For the continental United States, the average regional temperature extremes from January to May have changed anywhere from a slight decrease to 6 °C in the last 5 years (www.ncdc.noaa.gov). Globally, North America, Africa, and Asia have experienced similar trends in increasing temperature, and the polar regions have experienced the most noticeable increase in temperature over the last 100 years.

4.27 A combustion reaction requires oxygen. Thus, we expect that as combustion of fossil fuels has increased, not only will carbon dioxide concentrations increase (a product of the reaction), but the amount of reactants will decrease.

These changes are very small compared to the total amount of oxygen present in the atmosphere (21%). Thus, we do not have to worry about running out of oxygen due to combustion processes.

Real Talk (Tracking the Causes & Effects of Climate Change)

a. They collect data that will be useful for policymakers. Specifically, NOAA deals with oceans, fisheries, weather, and atmospheric composition (including health and climate impacts). All of these sectors inform our understanding of the causes and effects of climate change.

b. NOAA scientists collect data, so they must design and build instruments that can withstand the harsh conditions of their sample site (including withstanding vibrations, temperature changes, etc. on ships or in aircraft). It takes weeks to months to prepare for a field project. Then the scientists collect data for weeks to months in the field. Once the project is complete, data is analyzed and provided to the scientific community as well as the public. Dr. Gilman works to understand VOC emission sources and the impact of changes in policy. Specifically, she has studied impacts of wildfires, energy production, and landscape changes. These all play vital roles in the climate crisis.

4.28 A 2007 paper (Velders et al., *PNAS*) noted that the climate protection from the Montreal Protocol at the time of the study was larger than the promises made under the first commitment period of the Kyoto Protocol.

More recently, a 2021 study (Young et al., *Nature*) showed that without the Montreal Protocol, there would be an extra 0.8 °C from the extra carbon in the atmosphere due to the inability of plants to take up this resource since they would have suffered UV damage. In addition, the unabated CFCs would have contributed 1.7 °C to the overall greenhouse effect. In total, the Montreal Protocol could save us an additional warming of +2.5 °C by 2100.

4.29 a. Earth's temperature increases as the Sun becomes brighter; this makes sense because more energy is reaching Earth's surface. As the Sun dims, the temperature decreases.

b. As albedo (reflectivity) increases, Earth's temperature decreases. The more reflective surfaces will decrease the amount of energy absorbed by Earth's surface. The highest albedo comes from bright, white surfaces like snow and clouds. The lowest albedo comes from dark surfaces like asphalt, the ocean, and forests (or even the Moon!).

Albedo Values for Different Ground Covers

Surface Type	Albedo
Forest	15%
Grassland	20%
Desert Sand	37%
Ocean	10%
Fresh Ice and Snow	85%
Old, Melting Snow and Ice	55%
Sea Ice	45%
Asphalt	7%
Clouds	65%
Moon	13%

c. The whiter the surface, the higher the albedo; the darker the color, the lower the albedo.

d. When a snow-covered area melts, the albedo decreases and more sunlight is absorbed, creating a positive feedback loop and additional warming. This effect will lead to greater increases in average temperature observed in the Arctic, where the amount of sea ice and permanent snow cover is decreasing. Similarly, when glaciers retreat and expose darker rock, the albedo decreases, causing further warming.

e. Grasslands or crops reflect more incoming light than does the dark green foliage of the rainforests, causing an increase in the albedo that results in a cooling effect.

4.30 **a.** We have already learned that melting snow or ice reveals a darker surface underneath, thus lowering the surface albedo and resulting in an increase of surface temperature. When dark-colored soot is deposited on the surface of snow or ice, it will also decrease the albedo of the surface; more sunlight will be absorbed. As ice sheets melt, it concentrates these dark particles and makes the melting process faster. Once the warming begins, more melting can occur, thus creating a feedback loop leading to additional warming.

b. Ideas may vary, but here are some suggestions: The ice-albedo feedback loop is accelerated with surface warming, so any strategy that decreases the amount of greenhouse gases in the atmosphere, will decrease the speed at which ice is melting (specific ideas will be presented in Section 4.9). To decrease the amount of soot particles, we could decrease the amount of ship traffic in the polar regions. We could also address the type of fuel used in cargo ships (gas has fewer particulate emissions, electricity that is not generated near ice/snow). We could physically filter the particles out of the exhaust so that they are not spread discriminately over the surface.

4.31 Using the slider by "surface temperature," a temperature of 28 °C must be achieved in order to match the 256 W/m^2 of incoming radiation. This is about 13 °C hotter than today's average surface temperature.

4.32 **a.** Mt. Pinatubo injected 20 million tons of SO_2 as high as 60 km into the atmosphere. In addition to providing spectacular sunsets for several months, the sulfur dioxide transformed into sulfate aerosols in the atmosphere, which caused temperatures around the world to drop slightly (around 0.6 °C) for almost two years. Thus volcanic eruptions have a slight cooling effect (negative forcing). The most reliable climate models were able to reproduce the cooling effect caused by the eruption.

b. Gas emissions from human activity tend to create particles that are more water soluble. Thus, when they seed clouds, the clouds are made up of more smaller water droplets. Small droplets make the cloud look brighter (due to the increase in reflecting surface area), increasing albedo. Smaller droplets often mean that there is less local precipitation since droplets must grow large enough to be rained out of a cloud.

c. Aerosols have a variety of impacts (reflecting and/ or absorbing sunlight), which depends on their chemical composition and location in the atmosphere. Particles are not easy to describe because most of them are a combination of many chemical components. They also do not stay in the atmosphere long (about 1 week) and are not evenly distributed throughout the globe. In addition, the cloud-formation process is difficult to measure. In contrast, greenhouse gas impacts are purely based on their physical relationship with infrared light. Since these gases exist in the atmosphere for years (or thousands of years) at a time, scientists can measure their concentrations.

4.33 Based on the things we've learned throughout this chapter, the claim that the Sun is responsible for global warming is not substantiated. There is no evidence of Earth receiving increased amounts of radiation from the Sun, and data have actually shown that solar irradiance and global average temperature are trending in opposite directions.

4.34 **a.** The "Little Ice Age" occurred from 1650 to 1715 and coincided with the last time our Sun experienced a Grand Solar Minimum where it has fewer sunspots and gives off less energy. The "Little Ice Age" period resulted in a combination of lower solar activity and cooling due to volcanic activity.

b. If the Grand Solar Minimum happened to day, it would provide cooling up to 0.3 °C. For comparison, the total amount of warming provided by all greenhouse gases in Figure 4.26 is 1.83 °C, or six times that of the cooling provided by this solar minimum.

4.35 **a.** Hundreds of scientists over many years write computer code that simulates the climate system. Using supercomputers, the mathematical equations (written as thousands or millions of lines of code) are similar to weather forecast models, but focus on longer timescales. Models take into account physical interactions like radiative properties or dynamic flow, chemical mechanisms, and biological processes.

b. There has been an increased understanding of physical, chemical, and biological processes on Earth. For example, modern models include a lot more atmosphere-land interactions. Newer models have a smaller spatial scale, trying to look at local and regional interactions. Models also have smaller time steps, recalculating data on a 30-minute time scale. In summary, models have increased their level of detail.

4.36 a. Aerosols from volcanic eruptions and solar irradiance. The baseline temperature includes natural greenhouse gas emissions.

b. Increases in greenhouse gas concentration due to human activity and changes in Earth's albedo due to human activity.

REFLECT (1.5° Celsius . . .)

Answers will vary. Examples are:
- Prior to 1945, significant carbon emissions were only coming from the US and EU.
- The "bucket" (or budget to reach the temperature goal) was half full in 1995, but it will be 100% full by about 2030.
- By 2014, the US, EU, China, and India made up 50% of the carbon budget. The rest of the world contributed another 30%.
- We only have around 8% left in the budget to stay within the temperature goal.

4.37 Arctic sea ice extent has declined significantly in all months since satellite measurements began in 1979, with Septembers showing the largest declines annually. The last 15 Septembers show the lowest values. On average, the sea ice area is decreasing by 13% per decade (relative to 1981–2010 average). The smallest Arctic ice extent was seen in 2012 (3.39 million km^2).

Melting of sea ice does not impact sea level, but it does influence the salinity, and therefore the dynamics of the regional ocean. For more information about the dynamics of the disappearing sea ice, watch this video: https://www.youtube.com/watch?v=hlVXOC6a3ME

4.38 a. There is a steady decrease in land-based ice from both Greenland and Antarctica. Between 2002 and 2021, Greenland has lost 274 billion metric tons of ice per year while Antarctica has lost 152 billion metric tons. All of this water runs to the oceans and raises the sea level.

b. The glacier is heated at the surface, but not all of the melting occurs at the surface. Meltwater from the surface trickles down the glacial ice to the bedrock and out to sea. The plume of meltwater allows warm ocean water to melt the glacier from below.

REFLECT (Mountain Glaciers)

Glaciers provide fresh water to over 1 billion people on Earth. With the acceleration of glacial melt, the water from snow and ice can lead to landslides and flooding.

4.39 City chosen: New York City, New York

a. In Manhattan, there is a marker for Fulton Street. Using the sea level slider, it begins to flood at 1.2 m (4 ft). The 2021 IPCC report's intermediate emissions scenario would result in a global mean sea level rise exceeding 1 m (above 1995–2014 average level) by around 2160.

b. The local scenario: there will be 1.2 m before 2080 in the High Scenario. At about 2100, using the Intermediate scenario, there are many low-lying areas (shown in green) in nearby New Jersey.

c. The East side of Manhattan and just inland from the New Jersey coast have high vulnerabilities.

4.40 a. There has been an increase in the number of $1 billion events (and thus also an increase in cost). The total cost of the last 5 years is more than 1/3 the cost of all disasters in the last 42 years (1980–2021). There is an increased number of severe storms and many tropical cyclones (hurricanes).

b. So many of the events and vulnerability depend on heat (which is highest in the south and west of the U.S.). Many risks correlate with socioeconomic vulnerabilities.

REFLECT (Importance of Biodiversity)

Biodiversity sustains jobs/economies (60 million people are in the fishery/aquaculture sector; 75% of our crops depend on pollinators). More biodiversity increases resilience (e.g., mangroves protect millions of people from flooding). Nature is protecting us from the true cost of greenhouse gas emissions by acting as carbon sinks. Healthy ecosystems make up 50–90% of the livelihoods of rural poor—they tend to depend more directly on nature.

4.41 Answers will vary, depending on photos chosen.

4.42 Information stated here was gathered from the US Health and Human Services website (https://www.hhs.gov/climate-change-health-equity-environmental-justice/climate-change-health-equity/index.html).

Vulnerable populations or disadvantaged communities tend to be the most exposed, most sensitive, and have the less resources (and political power) to prepare for, and respond to, health hazards. Examples include: children have a higher risk of heat stroke, older adults are vulnerable to events that require evacuation, locations with multiple pollutants tend to have a disproportionate number of people of color, low-income communities tend to have limited access to healthcare, vulnerable people have limited ability to relocate or rebuild after a disaster.

In summary, disadvantaged communities tend to bear the brunt of climate-induced health risks.

4.43 a. The IDP crisis is currently centered in countries such as Columbia, Syria, Democratic Republic of the Congo, Yemen, Afghanistan, Somalia, and Nigeria. The refugee crisis is currently centered among countries such as Venezuela, Syria, Afghanistan, South Sudan, Sudan, Myanmar and the Democratic Republic of the Congo.

b. Answers will vary, depending on location.

Ayan Muude Adawe was a farmer and fled Ethiopia due to drought. Ayan ended up in a coast town in

Northeast Somalia where large storms used to be rare, but have become an almost yearly occurrence. Cyclone Gati hit in November 2020 and was the strongest hurricane on record for the country. The resulting floods destroyed Ayan's shelter and washed away most belongings. The disaster even killed one of Ayan's children. Now Ayan lives in a makeshift settlement camp near the seashore because it is close to work—washing clothes for families that live in a nearby town. Too much rain alternates with too little, without enough time to recover between disasters.

REFLECT (Preparing for Displacement)

a. The speaker, Collete Pichon Battle, discusses her own experience of human displacement from the New Orleans area after Hurricane Katrina in 2005.

b. The speaker says that the term "climate refugee" is often misused, especially when referring to people displaced within their own country. "Refugee" implies that the people don't belong—they are "the other."

c. The speaker says that climate change is a symptom of the social, political, and economic systems that put profit above all else. Profit is more important than the preservation of nature or human welfare. These systems have incentivized consumption and extraction. Repair of Earth (and the climate system) must go hand-in-hand with that of human communities. We must work to make both resilient together.

4.45 a. $15 \text{ tons } CO_2 \times \dfrac{2000 \text{ lb } CO_2}{1 \text{ ton } CO_2} \times \dfrac{1 \text{ tree}}{13.3 \text{ lb } CO_2}$

$= 2256 \text{ or } 2.3 \times 10^3 \text{ trees.}$

b. Figure 4.3 estimates global CO_2 emissions from fossil fuel combustion at 9.3 Pg or 9.3 Gt (18.6 trillion pounds) and assuming each tree absorbs 13.3 lb CO_2 every year:

$1.2 \times 10^{10} \text{ trees } \times (13.3 \text{ lb } CO_2/1 \text{ tree})$
$= 1.6 \times 10^{11} \text{ pounds } CO_2 \text{ absorbed}$

$(1.6 \times 10^{11} \text{ pounds } CO_2/1.86 \times 10^{13} \text{ pounds}) \times 100$
$= 0.9\% \text{ of global emissions absorbed.}$

4.47 a. Using an Internet search to find "carbon footprint calculator" will lead one to many options. Commonly requested information might include the number of people in your household, geographic location, size of your home, types of vehicles and miles driven per year, and types of appliances used in your home. Other questions relate to efforts made in your daily life to reduce your footprint, such as recycling, composting, turning off and unplugging appliances when not in use, carpooling, and growing your own food.

b. The information requested by each site does vary slightly, but all sites require information related to individual usage of the largest sources of carbon

emissions such as motor vehicle and home energy use.

4.48 a. According to the U.S. Census Bureau, the world population is estimated at roughly 8 billion people.

b. Using these figures, there is less than 4 acres of land available for each person in the world.

4.49 a. According to the U.S. Census Bureau, the U.S. population is estimated at roughly 333 million people.

b. 3.33×10^8 people $\times 20$ acres/per person $= 6.7 \times 10^9$ acres

c. 6.7×10^9 acres needed for U.S. population/ 3.01×10^{10} acres available $\times 100 = 22\%$ of global biologically productive space needed just for the U.S. population.

Let's Debate! (Policies for a Sustainable Future)

1. Setting a high price for carbon seems to make the largest impact (decreasing the 2100 temperature by 1 degree Celsius).

2. One proposal would be: set the carbon price high, have high growth in technological advances, highly reduce methane and other gases from land and industry, and highly incentivize electrification of buildings and industry. Expand each category for more details and specificity.

Chapter 5

REFLECT (Water Quality and Use)

Answers will vary. Some sources of water are municipal water treatment plants or wells. The water-use habits of a community such as overuse or pollution from the use of pesticides or herbicides can affect the availability of clean water for others.

5.1 a. Answers will vary. However, one may discuss showering/bathing as a cleaning process. Also one might discuss drinking water to stay hydrated and using water to wash clothing/pets/cars.

b. Water sources will vary depending on location, well, municipal water, etc. One could take it even further and say where the municipal water comes from.

c. The degree to which they got the water dirty will also vary, given the task.

5.2 a. O—H; The electron pair is more attracted to the O atom.

b. N—O; The electron pair is more attracted to the O atom.

c. Cl—C; The electron pair is more attracted to the Cl atom.

5.3 a. $:\!\ddot{C}l\!-\!Be\!-\!\ddot{B}r\!:$

b. The covalent bonds in BeClBr are polar since the atoms have different electronegativities.

c.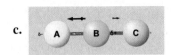

d. Similar to CO_2, the BeClBr molecule is linear, but the bond dipoles don't cancel since Cl is more electronegative than Br.

e. $\ddot{O}=C=\ddot{S}$

C=O polar covalent bond O=S polar covalent bond; linear shape, so nonpolar molecule

polar covalent bonds; bent, so polar molecule

I—I nonpolar covalent bonds; nonpolar molecule

5.4 a. The dashed lines represent hydrogen bonds that are intermolecular forces between the hydrogen on one water molecule and the oxygen of another water molecule.

b. The partial charges show how the partial positive of the hydrogen lines up with the partial negative of the oxygen in the water molecules.

c.

5.5 a. HCl has no hydrogen bonding because it does not contain an HF, HO, or HN bond.

b. H_2S has no hydrogen bonding because it does not contain an HF, HO, or HN bond.

c. HF has hydrogen bonding because it contains an HF bond.

d. N_2O has no hydrogen bonding because it does not contain an HF, HO, or HN bond.

e. CH_4 has no hydrogen bonding because it does not contain an HF, HO, or HN bond.

5.6 No, covalent bonds are not broken when water boils. Rather, the hydrogen bonds between the molecules are broken. The drawing shows that the water molecules are still composed of H_2O before and after boiling.

@HOME (Density Comparisons)

a. The solid vegetable oil sank to the bottom of the glass of the liquid vegetable oil. The solid water floated on top of the glass of liquid water.

b. The intermolecular forces between the vegetable oil molecules are weaker than hydrogen bonding, so when it forms a solid, it will contract, producing a solid with a higher density and sink. The intermolecular forces of water are much stronger. Hence, water will expand when it forms a solid and has a lower density than liquid water. So it will float on liquid water.

@HOME (Water Balloons!)

a. The balloon is made of rubber, which will heat up, soften, and burst quickly. The heat capacity of water is high, so it takes a lot of heat to produce a relatively small temperature rise in water. The water conducts the heat away from the hot spot.

b. Skin is a very sensitive organ, but if it is thicker, like on the bottom of your feet, you can take hot temperatures, as hot coal walkers can.

5.7 The carpet feels warm, and the stone or tile floor feels cool to your bare feet. The stone floor has a lower heat capacity than the carpet. It is denser than the carpet and a better conductor of heat. It takes less energy to make a 1 °C temp change for a substance with a lower heat capacity, so you notice a more drastic temperature change on the tile because heat is being more rapidly removed from your feet.

5.8 a. GenX comes from manufacturing plants, arsenic comes from sedimentary deposits in volcanic areas, and bacteria that cause cholera come from inadequate water treatment, poor sanitation, and inadequate hygiene.

b. These chemicals make people sick and can cause premature deaths.

5.9 a. Nonpotable water could be used to wash clothing and dishes and for decoration purposes in fountains.

b. Answers may vary, but an answer could be: A community might use nonpotable water if fresh drinking water is scarce.

c. Answers will vary.

d. Answers will vary.

5.10 a. As the global temperature becomes warmer due to climate change, there will be more water flowing into oceans from melting of ice sheets, glaciers, and snowpack. This will cause an increase in water levels in oceans, lakes, and rivers. With increasing storm activity, precipitation and runoff into groundwater will also be affected. Basically, every aspect of the water cycle will be affected by climate change.

b. Water quality is affected by irrigation and precipitation, runoff, and returning contaminated water. So the main ways how water quality is affected are water moving between the atmosphere and the surface, water moving across the surface, and water moving into the ground.

c. Answers will vary.

5.11 Answers will vary. Sample math is shown.

$$2000 \text{ mL} = \text{Earth's water}$$

$$2000 \text{ mL} \times 0.03 = 60 \text{ mL} = \text{Fresh water}$$

$$60 \text{ mL} \times 0.003 = 0.18 \text{ mL} = \text{Surface water}$$

$$0.18 \text{ mL} \times 89\% = 0.1602 \text{ mL} = \text{Lakes + Rivers}$$

$$0.1602 \text{ mL} \times \frac{1 \text{ drop}}{0.05 \text{ mL}} = 3.204 \text{ drops of water}$$

5.12 **a.** Answers will vary; some counties that had high water withdrawals were Lancaster, PA, St. Charles Parish, LA, and Calvert County, MD. These counties have relatively sparse populations relative to others; for instance, Los Angeles county had a relatively small withdrawal rate. Hence, there is no direct correlation between population and water withdrawals. One possible reason is that irrigation and thermoelectric power represent two primary uses of water; power plants and land used for agriculture are typically situated well outside of major cities where large populations reside.

b. Answers will vary. Irrigation withdrawals will be higher in arid parts of the country (western states), whereas industrial uses will be higher in places with more industries, such as coastal Louisiana where petrochemical industries are present.

c. Answers will vary.

d. China and India had greater water withdrawals than the U.S. These countries have much larger populations, so more water is needed for thermoelectric power and agriculture, in addition to industries and other applications.

5.13 Answers will vary.

5.14 Answers will vary.

5.15 Crops use far less water than meat. This is because the animals used for meat need water to stay hydrated, and the food they eat includes crop water usage as well.

5.16 Answers will vary. Suggestions for reducing water footprints could be to use local ingredients, less processing of foods, better irrigation practices, etc.

5.17 **a.** CO_2 is the solute, and water is the solvent.

b. Iron is the solvent, and carbon and chromium are the solutes.

c. Sugar and tea particles are the solutes, and water is the solvent.

d. Ethylene glycol is the solute, and water is the solvent.

5.18 **a.** Na_2SO_4

b. $Mg(OH)_2$

c. $Al(C_2H_3O_2)_3$

d. K_2CO_3

5.19 **a.** Calcium iodide

b. Ba_3N_2

c. Magnesium hydroxide

d. Na_2CO_3

e. Cobalt(III) sulfide

f. $Cu(HCO_3)_2$

5.20 **a.** $5 \text{ L} \times \dfrac{1000 \text{ mL}}{1 \text{ L}} \times \dfrac{1 \text{ g}}{1 \text{ mL}} = 5000 \text{ g H}_2\text{O}$

$$80 \text{ μg mercury ion} \times \frac{1 \text{ g}}{1{,}000{,}000 \text{ μg}} = 0.00008 \text{ g}$$

$$\frac{0.00008 \text{ g mercury ion}}{5000 \text{ g H}_2\text{O}} = 0.000000016 \times 10^6$$

$$= 0.016 \text{ ppm} \times 10^9 = 16 \text{ ppb}$$

b. The EPA cutoff for Mercury in water is 2 ppb (0.002 ppm). This would not be an acceptable limit.

5.21 **a.** You can add a lot of drink mix. The maximum concentration is 5.960 M.

b. The maximum concentration is not the same for each. This is because different compounds have different solubility limits in water.

c. Saturated means that a substance has reached the maximum solubility limit.

5.22 **a.** $16 \text{ ppb} = \dfrac{16 \text{ μg}}{1 \text{ L}} \times \dfrac{1 \text{ g}}{1{,}000{,}000 \text{ μg}} = \dfrac{16 \text{ g}}{1{,}000{,}000 \text{ L}}$

$$= 0.000016 \, \frac{\text{g}}{\text{L}}$$

$$\frac{0.000016 \text{ g Hg}^{2+}}{1 \text{ L}} \times \frac{1 \text{ mol}}{200.59 \text{ g Hg}^{2+}} = 7.98 \times 10^{-8} \text{ M}$$

b. $\dfrac{1.5 \text{ mol NaCl}}{1 \text{ L}} \times 0.5 \text{ L} = 0.75 \text{ mol NaCl}$

$$\frac{0.15 \text{ mol NaCl}}{1 \text{ L}} \times 0.5 \text{ L} = 0.075 \text{ mol NaCl}$$

c. 1st Solution:

$$\frac{0.05 \text{ mol NaCl}}{0.250 \text{ L}} = 2 \text{ M}$$

2nd Solution:

$$\frac{0.60 \text{ mol NaCl}}{0.200 \text{ L}} = 3 \text{ M}$$

The second solution is more concentrated.

d. $40 \text{ g CuSO}_4 \times \dfrac{1 \text{ mol}}{159.609 \text{ g}} = 0.251 \text{ mol CuSO}_4$

$$\frac{0.251 \text{ mol}}{1 \text{ L}} = 0.251 \text{ M}$$

The student made a solution that was 0.251 M and not 2.0 M. The student would have needed to add 2 mol or 319.28 g of copper(II) sulfate to 1000 mL to get this molarity.

$$2 \text{ M} = \frac{x \text{ mol CuSO}_4}{1 \text{ L}}$$

$$x = 2 \text{ mol CuSO}_4 \times \frac{159.609 \text{ g}}{1 \text{ mol CuSO}_4} = 319 \text{ g CuSO}_4$$

5.23

	Ionic	**Molecular**
Structure	An extended solid held together by attractive forces between anions and cations	Each molecule held together by covalent bonds; discrete molecules held together by various types of intermolecular forces (e.g., hydrogen bonding, etc.)
Macroscopic properties	Form crystals; higher melting point; mostly soluble in water; high boiling point	Can be solids, liquids, or gases; low melting point; solubility in water depends on the polarity of the molecule; low boiling point
Bonding model	Diagrams will vary; transferred/localized electrons	Diagrams will vary; shared electrons between atoms in the bond
Bond strength	The charges pull the cations and anions together, resulting in a generally stronger bond than molecular compounds	Not as strong as ionic for the most part. Triple bonds are stronger than double bonds, which are stronger than single bonds
Naming	Name the cation and then the anion. Monoatomic anions have an -ide suffix; only use Roman numerals for metals with multiple possible charges	Elements are put in order of increasing group numbers; the number of each atom type gets a prefix (mono, di, tri, etc.); the last element has -*ide* suffix
Other	Answers will vary	Answers will vary

5.24 Water contains dissolved ions that conduct electricity. The best course of action is to unplug the hair dryer from the wall without touching the hair dryer or the water. Then, remove the hair dryer from the water.

5.25 Answers will vary.

REFLECT (Predicting Solubility)

   a. $AgNO_3(aq) + Cu(NO_3)_2(aq) \longrightarrow$
                  no reaction (reform reactants)

   b. $2\,AgNO_3(aq) + Na_2S(aq) \longrightarrow Ag_2S(s) + 2\,NaNO_3(aq)$

   c. $AgNO_3(aq) + NaCl(aq) \longrightarrow AgCl(s) + NaNO_3(aq)$

   d. $Cu(NO_3)_2(aq) + Na_2S(aq) \longrightarrow CuS(s) + 2\,NaNO_3(aq)$

   e. $Cu(NO_3)_2(aq) + 2\,NaCl(aq) \longrightarrow$
            $CuCl_2(aq) + 2\,NaNO_3(aq)$ (no ppt)

   f. $Na_2S(aq) + NaCl(aq) \longrightarrow$
                  no reaction (reform reactants)

5.26 Using the electronegativity values, one would expect the C—H bond and the C—C bond to be nonpolar. These molecules are nonpolar because the electronegativity differences between the atoms are very small.

@HOME (Popping A Balloon With an Orange?)

   a. Adding oils from the orange peel caused the balloon to pop quickly. Since "like dissolves like," the oil permeated the balloon and caused air to be released.

   b. Any nonpolar liquid, such as olive, vegetable, or automotive engine oils, should have the same effect.

5.28   **a.** The main contaminants were tetrachloroethylene, or "perc," trichloroethylene (TCE), and benzene.

   b. They came from the Marines dumping oil and industrial wastewater in storm drains. They also came from a dry-cleaning shop near the base that dumped their wastewater down the drains. It also came from underground storage tanks of gasoline that had leaked and spilled. (For a cleaner alternative to dry-cleaning, see the video below about using CO_2.)

   c. The effects that are listed as compensation from the Camp Lejeune Justice Act of 2022 are cancer of the esophagus, lung, breast, bladder, or kidney; leukemia; multiple myeloma; myelodysplastic syndromes; renal toxicity; hepatic steatosis; female infertility; miscarriage; scleroderma; and neurobehavioral effects and non-Hodgkin's lymphoma.

   d. The Camp Lejeune Justice Act of 2022 was signed into law on Aug. 10, 2022 by President Joe Biden. This will allow veterans and their families to seek legal compensation from the government for the harm caused by exposure to polluted water at this military base.

Real Talk (Carbon Dioxide For Dry-Cleaning?)

   a. Liquid carbon dioxide is a nonpolar solvent with greater solvent power than some organic solvents.

   b. Answers will vary. Although all applications will have some environmental impact, the environmental impact of carbon dioxide as a solvent is less than that of organic solvents. It isn't toxic or flammable.

   c. The social aspect will increase with fewer toxic chemicals being used around their employees. The environmental aspect will be increased with fewer toxic chemicals affecting the environment. The economic aspect will decrease as the carbon dioxide method costs more than the perc method.

   d. Answers will vary. For instance, electric vehicles were in development since the 1990s but were not widely available to consumers until recently, largely due to battery improvements, as well as government regulations, consumer demands, and pressure from petrochemical industries. Another possibility is building-integrated solar panel shingles instead of solar panels.

5.29 **a.** Tests of tap water, military bases and industrial sites have found PFAS contamination in more than 2854 locations in 50 states. Drinking water for up to 200 million Americans may be contaminated with PFAS. There are not a lot of data for worldwide PFAS levels, although it is estimated to be problematic worldwide. One world map can be found online (https://scholar.harvard.edu/ccwagner/quantifying-global-pfos-releases).

b. PFAS compounds are a class of non-stick, waterproof, stain-resistant compounds used in consumer products and industry. Best known are PFOA, formerly used to make Teflon, and PFOS, formerly in Scotchgard.

c. There are currently no regulations that have banned PFAS chemicals. However, this is an active area that will likely lead to many such regulations in the short-term (https://www.washingtonpost.com/health/2022/11/05/plastic-pfas-chemical-health-dangers/).

d. It is currently not known if PFAS chemicals can bioaccumulate. Studies are currently evaluating the health effects of these chemicals. Very low exposure to some PFAS chemicals has been linked to cancer, thyroid disease, weakened childhood immunity and many other health problems.

5.30 **a.** $HI(aq) + H_2O(l) \longrightarrow I^-(aq) + H_3O^+(aq)$

b. $HNO_3(aq) + H_2O(l) \longrightarrow NO_3^-(aq) + H_3O^+(aq)$

c. $H_2SO_4(aq) + H_2O(l) \longrightarrow HSO_4^-(aq) + H_3O^+(aq)$
$HSO_4^-(aq) + H_2O(l) \longrightarrow SO_4^{2-}(aq) + H_3O^+(aq)$

5.31 Answers will vary. Examples could be lemon juice or vinegar as acids. They could be added for taste.

5.32 **a.** $KOH(s) \xrightarrow{H_2O(l)} K^+(aq) + OH^-(aq)$

b. $LiOH(s) \xrightarrow{H_2O(l)} Li^+(aq) + OH^-(aq)$

c. $Ca(OH)_2(s) \xrightarrow{H_2O(l)} Ca^{2+}(aq) + 2\,OH^-(aq)$

5.33 Answers will vary depending on their experiences.

@HOME (Neutralizing With Baking Soda)

a. Vinegar (acetic acid) is the acid; baking soda (sodium bicarbonate) is the base.

b. As more acid is added, it becomes harder to neutralize the solution—more baking soda is needed to neutralize the solution. As the oceans become more acidic, they will dissolve more carbonate-containing formations such as shells of mollusks and coral.

5.34 Neutralization reactions:

a. $HNO_3(aq) + KOH(aq) \longrightarrow KNO_3(aq) + H_2O(l)$

b. $HCl(aq) + NH_4OH(aq) \longrightarrow NH_4Cl(aq) + H_2O(l)$

c. $2\,HBr(aq) + Ba(OH)_2(aq) \longrightarrow BaBr_2(aq) + 2\,H_2O(l)$

Spectator ions: a) NO_3^- and K^+; b) Cl^- and NH^{4+}; c) Br^- and Ba^{2+}

Total ionic equations:

a. $H^+(aq) + NO_3^-(aq) + K^+(aq) + OH^-(aq) \longrightarrow$
$K^+(aq) + NO_3^-(aq) + H_2O(l)$

b. $H^+(aq) + Cl^-(aq) + NH_4^+(aq) + OH^-(aq) \longrightarrow$
$NH_4^+(aq) + Cl^-(aq) + H_2O(l)$

c. $2\,H^+(aq) + 2\,Br^-(aq) + Ba^{2+}(aq) + 2\,OH^-(aq) \longrightarrow$
$Ba^{2+}(aq) + 2\,Br^-(aq) + 2\,H_2O(l)$

Net ionic equations:

a. $H^+(aq) + \cancel{NO_3^-(aq)} + \cancel{K^+(aq)} + OH^-(aq) \longrightarrow$
$\cancel{K^+(aq)} + \cancel{NO_3^-(aq)} + H_2O(l)$
$H^+(aq) + OH^-(aq) \longrightarrow H_2O(l)$

b. $H^+(aq) + \cancel{Cl^-(aq)} + \cancel{NH_4^+(aq)} + OH^-(aq) \longrightarrow$
$\cancel{NH_4^+(aq)} + \cancel{Cl^-(aq)} + H_2O(l)$
$H^+(aq) + OH^-(aq) \longrightarrow H_2O(l)$

c. $2\,H^+(aq) + \cancel{2\,Br^-(aq)} + \cancel{Ba^{2+}(aq)}$
$+ 2\,OH^-(aq) \longrightarrow \cancel{Ba^{2+}(aq)} + \cancel{2\,Br^-(aq)} + 2\,H_2O(l)$
$2\,H^+(aq) + 2\,OH^-(aq) \longrightarrow 2\,H_2O(l)$

5.35 **a.** Basic, $[OH^-] > [K^+] > [H^+]$

b. Acidic, $[H^+] > [NO_3^-] > [OH^-]$

c. Acidic, $[H^+] > [HSO_4^-] > [SO_4^{2-}] > [OH^-]$

d. Basic, $[OH^-] > [Ca^{2+}] > [H^+]$

REFLECT (Acidic and Basic Solutions)

a. $[H^+] = 1 \times 10^{-4}$ M

$[OH^-] = \dfrac{1 \times 10^{-14}}{1 \times 10^{-4}} = 1 \times 10^{-10}$ M

Acidic

b. $[OH^-] = 1 \times 10^{-6}$

$[H^+] = \dfrac{1 \times 10^{-14}}{1 \times 10^{-6}} = 1 \times 10^{-8}$ M

Basic

c. $[H^+] = 1 \times 10^{-10}$ M

$[OH^-] = \dfrac{1 \times 10^{-14}}{1 \times 10^{-10}} = 1 \times 10^{-4}$ M

Basic

@HOME (Cabbage Juice Indicator)

a. The pH is lower for the glass with bubbles (carbon dioxide).

b. As carbon dioxide is dissolved in water, the pH is lower, which affects aquatic life.

5.36 **a.** The pH increased when adding water and decreased when adding more coffee.

b. The pH decreased when adding water to the base and increased when adding more base.

c. They are related by this equation: $[H^+] \times [OH^-] = 1 \times 10^{-14}$. They do have a common product, it is 1×10^{-14}.

d. They are related by Avogadro's number, 6.0×10^{23}.

REFLECT (Acidic or Basic?)

a. The lake water with a pH of 4.0 is more acidic. There are 10 times more hydrogen ions in lake water than in rainwater.

b. The tap water, with a pH of 5.3, is more acidic. There are 1000 times more hydrogen ions in tap water than the ocean water.

c. The tomato juice with a pH of 4.5 is more acidic. There are 100 times more hydrogen ions in tomato juice than the milk.

5.37 Answers will vary but should include the fact that a pH of 0 would be extremely acidic and bad for the environment. The legislator likely means a pH of 7 to be completely neutral.

REFLECT (Coral Reefs)

a. CO_2 is slightly soluble in water and dissolves to form carbonic acid. Thus, there is an increase in H^+ ions in the water. Some coral is made up of $CaCO_3$. With excess H^+ ions, the carbonate (CO_3^{2-}) ions can react with the H^+ instead of the Ca^{2+} ions. Consequently, it is difficult for marine organisms to maintain the integrity of their shells and skeletons.

b. The answer given here is for the Great Barrier Reef, which might be the most famous coral reef on Earth. As one of the seven wonders of the natural world, the Great Barrier Reef contains the world's largest collection of coral reefs, with 400 types of coral, 1500 species of fish, and 4000 types of mollusk (UNESCO). A recent study (Guo et al., 2020 Geophysical Research Letters) found that ocean acidification has significantly impacted the skeletal growth of a key reef-building organism. The declining capacity of coral to build skeletons, in turn, reduces their capacity to create protective habitat for the Reef's marine life, which can have far-reaching consequences on the marine food chain and local economy. The social, economic, and iconic value of the Great Barrier Reef was estimated to be worth $56 billion by Deloitte Access Economics in 2017. It supports 64,000 total jobs, of which 39,000 directly use this resource, making the Reef one of the largest "employers" in Australia. The Great Barrier Reef Marine Park attracts millions of visitors annually; the tourist industry is worth $29 billion. In summary, the destruction of the reefs not only puts entire marine ecosystems at risk but can have a significant economic impact on the coastal communities.

5.38 a. The East Tropical Pacific and Australia regions appear to have experienced the greatest decline in coral growth since 2005. Most regions have shown significant decreases in more recent years, since 2010, for example. Some possible reasons include warmer oceans since they are less deep in these regions than others and are more significantly affected by global warming.

b. Although a few countries have experienced a growth in coral cover since 2005, only the Caribbean has shown a small increase. Copious amounts of sunlight and clear water help these regions grow their coral.

5.39 Answers will vary. For example, in June 2017, scientists met at the United Nations headquarters in New York City for "The Ocean Conference."

5.40 a. sulfuric and nitric acids

b, c. Answers will vary.

d. $SO_2(g) + H_2O(l) \rightarrow H_2SO_3(aq)$

5.41 a. $CaCO_3(s) + CO_2(g) + H_2O(l) \longrightarrow Ca^{2+}(aq) + 2\ HCO_3^-(aq)$

b. They are monitoring them in the Northeast and the Appalachian mountains, because they have a low acid-neutralizing ability due to the granite in the soil.

c. Yes, this does fit with the EPA data because the sites are in the Northeast and the Appalachian mountains, two regions where the map shows low pH levels.

d. Answers may vary. Southeastern states, Florida, Georgia, South Carolina, or Alabama are all possibilities.

5.42 a. Sulfate: SO_4^{2-}; Hydroxide: OH^-; Calcium: Ca^{2+}; Aluminum: Al^{3+}

b. $CaSO_4$, $Ca(OH)_2$, $Al_2(SO_4)_3$, $Al(OH)_3$, among other compounds that include additional species.

c. Sodium hypochlorite: $NaClO$; Calcium hypochlorite: $Ca(ClO)_2$

5.43 a. Answers will vary, but two examples are:

$$H\!:\!\overset{\displaystyle :\!\ddot{C}l\!:}{\underset{\displaystyle :\!\ddot{C}l\!:}{C}}\!:\!\ddot{C}l\!: \qquad H\!:\!\overset{\displaystyle :\!\ddot{B}r\!:}{\underset{\displaystyle :\!\ddot{B}r\!:}{C}}\!:\!\ddot{B}r\!:$$

$$CHCl_3 \qquad\qquad CHBr_3$$

Chloroform   Bromoform

b. THMs have three halogens and one hydrogen. CFCs contain only chlorine and fluorine halogens and no hydrogens.

c. THMs have higher boiling points than CFCs.

5.45 Answers will vary depending on their water diary. Conservation efforts can include not letting the tap run while brushing teeth, collecting rainwater, and running water appliances on "eco-friendly."

5.46 Answers will vary. One answer may include: Desalination is an energy-intensive process that would result in fossil fuel burning. The burning of fossil fuels would then add to greenhouse gases. This would violate the "Design for Energy Efficiency" and "Less Hazardous Chemical Syntheses" key ideas.

5.47 a. Answers will vary. For instance, communities can pick up garbage from streams from streams, or farmers can plant plants near riverbanks to prevent fertilizer runoff.

b. Answers will vary. One answer may include water shortages in Third World countries. It is important because the water is dirty, and there isn't an abundance of it. Diseases are spread, and much time is spent trying to find clean water. Two ways to address this are to collect rainwater and to dig wells.

5.48 Answers will vary. A suggestion for a different method of tracking water usage could be to view the meter to see actual usage. One can use timers to estimate how long they took a shower, etc.

Chapter 6

REFLECT (What Does It Take?)

a. $4460 \text{ km} \times \dfrac{0.6214 \text{ mi}}{1 \text{ km}} \times \dfrac{1 \text{ gal}}{30 \text{ mi}} = 92 \text{ gal}$

b. 139 gal

c. $\dfrac{26.1 \text{ lb corn}}{1 \text{ gal}} \times 139 \text{ gal} \times \dfrac{1 \text{ acre}}{7110 \text{ lb corn}} = 0.51 \text{ acre}$

Note that this does not take into account the energy (likely from fossil fuels) that is needed to grow, harvest, and process the corn.

6.1 Some example answers include:
- Coal or natural gas is used in power plants to generate electricity.
- Gasoline or diesel oil (or electricity from a power plant) is used to run cars.
- Natural gas or electricity is used to cook our food.
- Natural gas or heating oil is used to warm homes.
- Charcoal, propane, natural gas, or wood is used to barbeque food.
- Wood or gas is used in a fireplace to create heat.

In each of these cases, *using* fuels means *burning* them.

6.2 a. Desirable fuels ignite easily at low temperatures and produce a large amount of heat when they combust. Desirable fuels are also inexpensive to produce and can be stored and transported safely.

b. A good match between a fuel and its intended use would be one that does not require excessive amounts of fuel to get the job done and leaves behind the least amount of residue.

c. Solid fuels like wood and coal have several disadvantages. In addition to both producing harmful gases, leaving solid residues, and being nonrenewable resources, coal dust has been proven to cause significant health problems. Liquid and gaseous fuels like petroleum and natural gas burn more smoothly than solid fuels, although both still produce gases that are known to be harmful to the environment.

d. A better alternative to petroleum would burn cleanly in addition to being inexpensive to produce, could be stored and transported safely, and would produce a large amount of heat when it combusts.

6.3 Aerobic composting requires oxygen (provided by mixing the pile). The decomposition reaction follows Equation 6.2 with carbon dioxide as the notable product. This process is best for slow decomposition processes like yard waste.

Anaerobic composting does not require oxygen. It produces methane (a very potent greenhouse gas), acids, and ammonia in addition to carbon dioxide. Thus, special equipment is needed to capture these gases. This process is ideal for food waste.

There are two reasons you might see steam rising from a compost pile. A new compost pile might simply have escaping water vapor. For a pile that has actually started to "compost," the bacteria that eat the raw material in the heap produce heat as a by-product of their metabolism. This heat can raise the temperature of the pile to as high as 70 °C (160 °F)!

6.4 a. China produces the most coal. The United States produces the most petroleum and the most natural gas.

b. The U.S. has the largest reserve of coal. Venezuela and Saudi Arabia have the largest reserves of petroleum. Russia and Iran have the largest reserves of natural gas.

c. Natural resources are often purchased from countries with large reserves by countries with large demand. While the United States ranks only 10th in petroleum reserves, being the #1 producer reduces domestic dependence on foreign oil and prevents countries with large petroleum reserves from driving up oil prices.

d. The trends have not changed dramatically over the last 10 years, although the United States became the largest producer of petroleum and increased its reserves of natural gas in that time. China has remained the largest producer of coal for the past 30 years, but has almost tripled its production in the last 10 years. A driving force for the change in U.S. production and reserve capacity of fossil fuels is most likely linked to attempts to rebuild the struggling economy.

REFLECT (Will We Run Out of Fossil Fuels?)

a. Different types of fossil fuels are created depending on the type of organic matter, temperature, time, and pressure conditions. Coal comes from ancient plant material while oil and natural gas are a result of zooplankton and algae.

b. While fossil fuels are made from biomass, they took millions of years to form the products we use today. We are quickly depleting these resources on the timescale of 100s of years. The supply will not be replenished in time for humans to use it.

According to the U.S. Energy Information Administration, given the current U.S. production and U.S. reserve estimates, there is enough coal to last about 325 years, enough natural gas to last about 90 years, and enough petroleum to last about 53 years (as of 2016). However, these values do not take into account global production or reserve values, changes in fossil fuel consumption, geopolitics, or the research being done to improve current energy technology. Reserves that were formally unreachable have become accessible due to technological advances. We should not be asking how long will fossil fuels last, rather when will we be able to stop using them.

@HOME (By-Products of Candle Burning)

a. Soot/carbon

b. The soot (particulate matter) would be released into the atmosphere, which is what happens when coal is burned. A scrubber is often used in coal-fired power plants to collect this PM so it doesn't enter the atmosphere.

REFLECT (The Fire Triangle)

The Fire Triangle consists of fuel, an oxidizer, and heat (or ignition source)

The fuel is any combustible material (in the video it was sawdust). To extinguish a fire, one could take away the fuel, like making a "fire break".

The oxidizer is usually oxygen ($O_2(g)$) in our atmosphere. Covering the fire or closing off the air supply will extinguish a fire (closing a hood in the laboratory, placing a fire blanket on an object, etc.). Other sources may be used such as ozone ($O_3(g)$), hydrogen peroxide ($H_2O_2(l)$), nitrous oxide ($N_2O(g)$),

or many other oxygen-containing compounds. Although much less common, it is also possible to have a combustion reaction with a halogen serving as the oxidizing agent, such as fluorine ($F_2(g)$), chlorine ($Cl_2(g)$), bromine ($Br_2(l)$), or even halogen-containing compounds.

Heat is needed to raise the fuel's temperature until it reaches its ignition point. Once a fire is generated, the heat or ignition source is no longer needed. The fire will continue to burn until either the oxygen or fuel source is removed. The video showed that blowing out a match or using a conductive material to decrease the heat can extinguish fire.

@HOME (A DIY Fire Extinguisher)

a. Carbon dioxide

b. An oxidizer

6.5 a. C_2H_6

$$H-\overset{\overset{\displaystyle H}{|}}{\underset{\underset{\displaystyle H}{|}}{C}}-\overset{\overset{\displaystyle H}{|}}{\underset{\underset{\displaystyle H}{|}}{C}}-H$$

c. C_2H_2

$$H-C\equiv C-H$$

b. C_2H_4

$$\underset{H}{\overset{H}{\diagdown}}C=C\underset{H}{\overset{H}{\diagup}}$$

d. C_6H_6

Whether it is all single bonds (like in methane or ethane), double bonds (ethene and benzene), or triple bonds (ethyne), there are a total of four bonds (8 electrons) associated with each carbon atom.

6.6 a. $C_6H_{12}O_6 + 6\,O_2 \longrightarrow 6\,CO_2 + 6\,H_2O$

b. $CH_4 + 2\,O_2 \longrightarrow CO_2 + 2\,H_2O$

c. $2\,C_4H_{10} + 13\,O_2 \longrightarrow 8\,CO_2 + 10\,H_2O$

6.7 $C_3H_8 + 5\,O_2 + N_2 \longrightarrow 3\,CO_2 + 4\,H_2O + N_2$

Notice N_2 is, on both the reactants and products side, unchanged. Since nitrogen does not react, we typically omit it from the balanced chemical equation.

6.8 Answers will vary. The general trend is that potential energy will increase dramatically as the atoms are moved closer to each other, and vice versa as the atoms are moved farther apart.

@HOME (Shaking Sand)

a. Shaking the cup was adding kinetic energy to the system.

b. The kinetic energy that was being added to the system was converted into thermal energy and the evidence for this was the temperature increase in the sand. Some of the kinetic energy was also lost as sound energy.

REFLECT (Detecting Temperature)

Bulb thermometers (or liquid thermometers) have been made of mercury and red wine, but now most of them are dyed ethanol. Depending on the temperature of the thermometer, the molecules within the liquid move faster or slower (hot = faster motion, cold = slower motion). This molecular motion causes the liquid within the glass tube to expand (if hot) or contract (if cold). These types of thermometers are universally used for a wide range of applications.

Digital thermometers use a thermistor–a resistor that conducts electricity differently, depending on the temperature–made of a metal oxide or semiconductor materials.

IR thermometers are a kind of digital thermometer which are sensitive to the infrared radiation coming off of an object (using specific semiconductors). These are even used for temperature measurements of Earth from satellites.

6.9 a. $217 \text{ kcal} \times \dfrac{1000 \text{ cal}}{1 \text{ kcal}} \times \dfrac{4.184 \text{ J}}{1 \text{ cal}} \times \dfrac{1 \text{ kJ}}{1000 \text{ J}} = 908 \text{ kJ}$

b. 1 J is equivalent to lifting 100 g a distance of 1 meter (acceleration due to gravity = 10 m/s^2). Lifting 1 kg a distance of 2 m requires 20 times more energy, or 20 J. To burn 908 kJ (or 9.08×10^5 J), you would have to lift 45,400 textbooks!

c. The previous problem assumed that all of the Calories in the pizza could be metabolized into energy available to do work. That is not a reasonable assumption. In reality, the number of books this piece of pizza would allow you to lift would be much lower, as not all of the energy released from metabolizing food is transformed into usable energy.

REFLECT (A DIY Calorimeter)

The bomb calorimeter is a sealed container filled with oxygen. Heat from the combustion reaction transfers to the surrounding water. In contrast, the "can and burner" calorimeter is open to the surroundings and oxygen for the reaction comes from the ambient air. Heat produced from the reaction can escape into the air, thus the resulting heats of combustion are lower using this method. The "can and burner" setup is cheaper and easier to use. Both calorimeters rely on the change in temperature of the water to determine the amount of heat produced from the reaction.

@HOME (Making Your Own Hot and Cold Packs)

a. A hot pack is the result of an exothermic reaction, and a cold pack is the result of an endothermic reaction.

b. When calcium chloride or magnesium sulfate is added to water it is exothermic, whereas adding sodium chloride, ammonium chloride, potassium chloride, or sodium bicarbonate to water is endothermic.

c. Calcium chloride would be an effective salt for a hot pack, whereas ammonium chloride would be effective for a cold pack. One should consider those salts with the largest temperature changes to be the most effective.

6.10 The bond energy for O_3 is intermediate between the O—O single bond (146 kJ/mol) and O=O double bond (498 kJ/mol) values; that is, it is less than the bond energy for the O=O double bond in O_2. Energy is inversely proportional to wavelength, so the higher bond energy of O_2 requires radiation of shorter wavelength to break its bonds.

6.11 $249 \dfrac{\text{kJ}}{\text{mol}}\left(\dfrac{1 \text{ mol}}{2.016 \text{ g}}\right) = 124 \dfrac{\text{kJ}}{\text{g}}$

This is more than twice the energy density of the most potent fuel listed in Figure 6.5 $\left(\text{methane at } 50.1\tfrac{\text{kJ}}{\text{g}}\right)$

REFLECT (Heat of Combustion for Acetylene)

$$2 \, C_2H_2 + 5 \, O_2 \longrightarrow 4 \, CO_2 + 2 \, H_2O$$

$$2\,H{-}C{\equiv}C{-}H \; + \; 5\,\overset{..}{\underset{..}{O}}{=}\overset{..}{\underset{..}{O}} \longrightarrow 4\,\overset{..}{\underset{..}{O}}{=}C{=}\overset{..}{\underset{..}{O}} \; + \; 2\,{}_{H}\overset{\overset{..}{O}}{\diagdown}{}_{H}$$

Bonds broken:	Bonds formed:
2 C≡C $2 \times (+813$ kJ/mol$)$	8 C=O $8 \times (-803$ kJ/mol$)$
4 H—C $4 \times (+416$ kJ/mol$)$	4 H—O $4 \times (-467$ kJ/mol$)$
5 O=O $5 \times (+498$ kJ/mol$)$	
+5780 kJ/mol	−8292 kJ/mol

Heat absorbed/released = 5780 kJ − 8292 kJ
$$= -2512 \text{ kJ}/2 \text{ mol } C_2H_2$$

Heat of combustion = −1256 kJ/mol C_2H_2, or

$$-\dfrac{1256 \text{ kJ}}{\text{mol}} \times \dfrac{1 \text{ mol}}{26 \text{ g}} = -48.3 \text{ kJ/g } C_2H_2$$

6.12 Examples include engines that convert chemical energy into mechanical energy (potential energy stored in bonds is released upon combustion to power parts of a machine), lightbulbs that convert electrical energy into heat and light, or a wood stove that converts chemical energy into heat.

6.13 a. The potential energy is the greatest at the ends of the U-shape and is the lowest at the bottom. The trends for kinetic energy are opposite—highest at the bottom and lowest at the top.

b. As the mass of the skateboarder is increased, the total energy increases, as does the magnitude of the potential and kinetic energies at various stages of the skate park.

c. The total energy remains constant; energy is converted between potential/kinetic forms, but the law of conservation of energy states that the total energy of the system must remain constant (assuming there are no frictional losses).

d. With friction added, eventually the skateboarder will lose energy and stop.

e. The total energy again remains constant, but both kinetic and potential energies are converted into thermal energy (heat), which eventually equals total energy when the skateboarder movement is stopped.

f. No. The total energy remains constant, but is shifting between different forms.

6.14 a. Answer will vary, but one should highlight the "loss" of useful energy and the production of "waste" heat at each transformation or transmission of energy. Consider, for example, the high-temperature steam that initially spins the turbines. As the steam transfers energy to the turbines, the kinetic energy of the steam decreases, it cools, and its pressure drops. It isn't long before the steam does not have enough energy to spin the turbines anymore. Yet the production of this "unused" steam still requires a significant amount of energy—energy that is not converted into electricity.

b. The internal combustion engine uses the gaseous combustion products (CO_2 and H_2O) to push a series of pistons, thus converting the potential energy of the gasoline or diesel fuel into mechanical energy. Eventually, this mechanical energy is transformed into the kinetic energy of the vehicle's motion. Examples of some energy losses that occur while driving include the heat and gas emissions produced in the engine during fuel combustion.

c. If it only takes 15% of the energy from fuel combustion to move an average-sized car (~2000 pounds), it would take an additional 2–4% of energy to move the passengers (assuming a passenger load of 300–500 pounds).

6.15 a. $\dfrac{23.7 \text{ Quads}}{36.6 \text{ Quads}} \times 100\% = 64.8\%$

b. $\dfrac{21.2 \text{ Quads}}{26.9 \text{ Quads}} \times 100\% = 78.8\%$

c. $65.4 \text{ Quads} \left(\dfrac{1 \times 10^{15} \text{ BTU}}{1 \text{ Quad}}\right)\left(\dfrac{1055 \text{ J}}{1 \text{ BTU}}\right)$
$= 6.90 \times 10^{19} \text{ J}$

6.16 Compared to wood, coal has a higher energy density. Coal is widely available, and since it comes as a solid, it is relatively easy to transport and store (as compared to liquids or gases). Coal is cheap, if you do not consider the significant health and environmental costs.

6.17 a. Calculate the approximate molar mass of coal.

$$135 \text{ mol C} \times \dfrac{12.0 \text{ g C}}{1 \text{ mol C}} = 1620 \text{ g C}$$

$$96 \text{ mol H} \times \dfrac{1.0 \text{ g H}}{1 \text{ mol H}} = 96 \text{ g H}$$

$$9 \text{ mol O} \times \dfrac{16.0 \text{ g O}}{1 \text{ mol O}} = 144 \text{ g O}$$

$$1 \text{ mol N} \times \dfrac{14.0 \text{ g N}}{1 \text{ mol N}} = 14.0 \text{ g N}$$

$$1 \text{ mol S} \times \dfrac{32.1 \text{ g S}}{1 \text{ mol S}} = 32.1 \text{ g S}$$

The sum of these elemental contributions for $C_{135}H_{96}O_9NS$ is 1906 g/mol. Therefore, every

1906 g of coal contains 1620 g of carbon. Similarly, 1906 tons of coal contain 1620 tons of carbon.

Mass of carbon = 1.5×10^6 tons $C_{135}H_{96}O_9NS$
$\times \dfrac{1620 \text{ tons C}}{1906 \text{ tons } C_{135}H_{96}O_9NS} = 1.3 \times 10^6 \text{ tons C}$

b. $1.5 \times 10^6 \text{ tons} \times \dfrac{2000 \text{ lb}}{1 \text{ ton}} \times \dfrac{454 \text{ g}}{1 \text{ lb}} \times \dfrac{30 \text{ kJ}}{\text{g}}$
$= 4.1 \times 10^{13} \text{ kJ}$

c. 4.7 million tons

6.18 a. The higher grades of coal, bituminous and anthracite, have been exposed to higher pressures and temperatures for longer periods of time. During that process, they lost more oxygen and moisture to become harder and more dense, and exhibit a higher crystallinity.

b. Anthracite and bituminous have higher energy content, whereas lignite and peat have the lowest energy content, with the amount of energy released only slightly greater than that of wood. Anthracite and bituminous grades have higher percentages of carbon and low sulfur content. That composition makes these grades more desirable for combustion since fewer harmful SO_x emissions will be released.

c. Lignite coal is most abundant worldwide, whereas anthracite reserves are almost all depleted in the U.S. and other countries. Bituminous and sub-bituminous varieties are also widely available globally.

6.19 According to the US Geological Survey (USGS), water (from runoff, snowmelt, etc.) that comes in contact with pyrite (an iron sulfide) in the exposed rock forms sulfuric acid and dissolved iron. Some of the iron can form red, orange, or yellow sediments at the bottom of streams. Heavy metals (such as copper, lead, and mercury) can then be leached from rocks that come into contact with this acidic water. This mine drainage can damage water sources for many communities.

The video shows one particular stream in Kentucky that has a pH of 2.9! This is so acidic, that the stream is essentially dead. Because the aquatic insects are gone, everything else up the food chain is also missing from this ecosystem. In addition, fish and amphibians are especially sensitive to the low pH particularly during their reproductive cycles.

REFLECT (Mountaintop Mining)

Mountaintop mining destroys the ecosystem on the mountain. It creates massive quantities of rubble (referred to as "overburden") that are often pushed over the edge of the mountain into nearby river valleys. Many minerals are then dissolved in these waterways and the large quantities are not well tolerated by the organisms that live there. Mining waste is stored on site and as the pressure builds, there can be seepage into the drinking water supply. Neighborhood wells are also contaminated.

As if these environmental issues were not serious enough, the airborne coal-dust particulates generated by this practice have been implicated in a number of health issues such as lung cancer and birth defects.

REFLECT (Particles from Coal)

When coal is burned, microscopic particulates known as fly ash are generated in the exhaust gas. They are released into the air if filtration and capture is not enforced through regulation. Coal combustion also results in larger/heavier particles known as bottom ash, which are more easily collected at the power plant. Although coal is mostly carbon, its ash has very little carbon remaining. The composition of coal ash is similar to that found in Earth's crust, featuring oxides of silicon, aluminum, iron, and calcium, as well as smaller concentrations of magnesium, potassium, sodium, titanium, and sulfur. Coal contains minor amounts (50–200 ppb) of mercury, lead, and cadmium; however, these toxic elements become concentrated in the ash and may leach into the environment.

6.20 **a.** Coal contains small amounts of sulfur that combine with oxygen during combustion to produce SO_2. Volcanoes are a natural source of SO_2 in the atmosphere.

b. Although coal does contain small amounts of nitrogen that combine with oxygen during combustion, the major portion of NO is formed from the reaction of N_2 and O_2 from the air at the high temperatures produced during the combustion process. Other sources of NO include engine exhaust, lightning, and grain silos.

6.21 Conversion factors:

7.33 barrels = 1 metric ton

1 barrel = 42 gallons; 1 gallon = 3.8 L

According to Figure 6.18, worldwide consumption of petroleum is currently 99 million barrels/day which is equivalent to 4.2 billion gallons/day (99×10^6 barrels (42 gal/barrel), 16 billion liters/day (4.2 billion gal (3.8 L/gal)), or 14 million metric tons/day (99 million barrels (1 metric ton/7.33 barrels)).

REFLECT (Oil Drilling)

A diamond-tipped drill bit powers through the soil and rock and is surrounded by drilling fluid. This fluid flushes out the crushed rock material back to the surface, which also cools the drill bit. This also maintains the pressure in the drilling hole so that it does not collapse under the weight of the surrounding rock and does not allow any seepage from the surroundings. When an oil well first penetrates a reservoir, the pressure is released, which pushes oil from the pores of rock up to the surface through the well.

In case dramatic pressure changes are experienced, a blow-out preventer is used to control the pressure inside the well, to prevent an uncontrolled gusher such as the previously discussed Lucas Spindletop.

Drilling can also occur on a slant (not always vertical) which allows the machinery at a single platform to reach many different deposits.

Once the borehole is deemed viable, a cement casing is added.

REFLECT (Tar Sands)

It's important to note that bitumen is so sticky, it cannot flow through conventional drilling apparati. Thus, the material is typically gathered through open-pit mining or underground. Enormous trucks (400-ton carrying capacity) are used to carry the tar sands where they are subsequently crushed and mixed with hot water to separate the bitumen from the sand. Underground methods (*in situ*) rely on pumping steam underground in horizontal wells in order to liquify the bitumen which is then pumped to the surface. Either way, bitumen must be processed and refined before it is usable as a petroleum product.

There is a larger amount of greenhouse gas emissions from this process than from conventional oil drilling and refining due to the extra steps in extracting the material. The process of acquiring bitumen is very water-intensive, and most of that water comes from local freshwater sources. Wastewater (tailings from the mining) is traditionally stored onsite, which has the potential to contaminate local water supplies. If local rivers are contaminated, so are the fish that live there. The open-pit mining process is extremely destructive to the land surface. Much of the land where the tar sands are found or the products are transported are inhabited. Contaminated water and food supplies often affect those who are impoverished or do not have a lot of political power.

6.22 Oil spilled into the water can kill marine organisms by physically coating waterfowl and aquatic species that can impede their natural buoyancy or waterproofing (e.g., preventing birds from flying). Exposure to large amounts of toxic components in the oil, like benzene, can cause health problems for the organisms, like heart damage, stunted growth, immune system abnormalities, issues with reproduction and death. Animals can also inhale the oil, which affects their lungs. Even if the effect is not lethal, increasing toxicity of marine organisms can put humans at risk if they are consuming fish and shellfish affected by the spill. Contaminated food supplies affect the entire food chain over time. In addition, many coastal communities that are largely impacted by oil spills rely on tourism; thus, the oil spill can hurt local economies.

6.23 **a.** $1500 \text{ kJ} \times \dfrac{1 \text{ g CH}_4}{50.1 \text{ kJ}} \times \dfrac{1 \text{ mol CH}_4}{16 \text{ g CH}_4}$

$\times \dfrac{1 \text{ mol CO}_2}{1 \text{ mol CH}_4} \times \dfrac{44 \text{ g CO}_2}{1 \text{ mol CO}_2} = 82 \text{ g CO}_2$

b. For bituminous coal from Maryland, which releases 30.7 kJ/g on average, 150 g of CO_2 is released when 1500 kJ of heat is produced.

REFLECT (Fracking: Process and Impacts)

a. "Hydraulic" refers to work done by the pressure created by forcing water, oil, or another liquid through a comparatively narrow pipe or orifice. In fracking, water is pumped down a well where the pressure builds and fractures the shale. Dynamite would not work for this purpose because natural gas is flammable and the use of dynamite would result in dangerous explosions due to combustion reactions with the gas that is trying to be extracted.

b. According to the U.S. Geological Survey, on average, it takes 5 million gallons of water per well, but it can be as much as 9.7 million gallons. In contrast, an average vertical well takes 671,000 gallons of water per well. It is very difficult to find out what specific chemicals are used in the fracking fluid (they are deemed trade secrets by the industry), but it usually involves acids for clearing debris and dissolving minerals, chemicals to reduce friction, and disinfectants to prevent bacterial growth. Fracking fluid also uses small particulates such as sand to help keep the fissures open.

c. According to the EPA, the wastewater from fracking can be disposed of in deep injection wells, treated and disposed of in surface water bodies, or recycled for use in future hydraulic fracturing operations.

d. One major impact is the effect on local water supplies. First, the process uses vast amounts of local freshwater. Then, it contaminates that water with fracking chemicals, radioactive materials, salts, heavy metals, and hydrocarbons that can leak into local groundwater supplies. Another impact is the chemical emissions into the air from normal operations. For example, some of the fracking chemicals can vaporize and become airborne. Also, since methane, the major component of natural gas, is so light, it will easily leak out of the fracking infrastructure. Methane is a much more potent greenhouse gas (on a per-molecule basis) than carbon dioxide, so we should limit its emissions. Additionally, there is concern about the pressure changes underground caused by the fracking process; it has been associated with earthquakes and other seismic activity. Most importantly, the fracking chemicals found in the air or water are likely causing health issues such as respiratory ailments, skin lesions, nausea, and headaches.

Real Talk (Emissions Budget)

a. Methane is a greenhouse gas, so understanding its sources and tracking its concentration over time is important to: 1) craft policy strategies that aim to lower greenhouse gas emissions globally and 2) to assess the effectiveness of those changes.

b. Methane emissions into the atmosphere come from agriculture, waste, natural sources (like wetlands), and fossil fuels. About 60% of these emissions are anthropogenic. Methane emissions can be estimated in a number of ways such as measuring atmospheric concentrations via "top-down" methods like airborne or satellite sensors or "bottom-up" methods like measuring leak rates of equipment and scaling those values given the number of facilities using a certain type of equipment. There are many estimates that go into either methodology, and each "best guess" introduces measurement uncertainty. In addition, the emission rates and total concentration attributed to each category constantly shifts. In fact, accidents generally result in the largest emission of methane. Satellite-based detectors struggle to provide accurate measurements in many environments such as offshore areas, mountains, snowy/ice-covered or cloudy regions, and at high latitudes. Hence, detection technologies with higher resolution and coverage are needed to improve the accuracy of these data. Ground-based detectors have a limited range of direct measurement and are reliant on indirect estimates to provide regional or global data. These data should be measured and validated regularly. To decrease uncertainty, both "top-down" and "bottom-up" methods are used.

Although every measurement comes with uncertainty, this completely normal measurement "error" is relatively small, and it does not invalidate the reported methane budgets and source data.

6.24 **a.** Boiling point is predicted to be higher for molecules with stronger intermolecular forces and for molecules with a greater number of electrons (for those with the same type of IMFs). Therefore, in order of increasing boiling point, $F_2 < Cl_2 < Br_2 < I_2$.

b. Following the same rules as in part **a**, $CH_4 < CF_4 < CBr_4 < CI_4$.

6.25 **a.** One possible pair of products is C_8H_{18} and C_8H_{16}. Their structural formulas are:

b. C_nH_{2n} (where n is an integer). The generic formula for alkanes is C_nH_{2n+2}.

c. Based on the rules used in Your Turn 6.24, relative boiling points should be (and are) C_5H_{12} (36.1 °C) < $C_{11}H_{22}$ (192.7 °C) < $C_{16}H_{34}$ (287 °C).

6.26 a. For all of the straight-chained molecules (methane, ethane, propane, n-butane, and n-pentane), the boiling point increases with molecular mass. As the molar mass of the molecule increases, the strength of its IMF (in this case London dispersion) also increases. With a stronger IMF holding the molecules in the liquid phase, it will take more energy (thus a higher boiling point) to overcome the IMF holding the molecules together and changing them into the gas molecules.

Comparing the other two isomers (2,2-Dimethylpropane and n-pentane) shows that the boiling point increases with larger surface area. The straight-chained molecules have stronger London dispersion intermolecular forces holding them close to their neighboring molecules. With a stronger IMF, the boiling point increases. For 3D representations of these structures, see neopentane and n-pentane

b. n-octane and iso-octane are isomers. Because n-octane is a straight-chained hydrocarbon, it will have stronger London dispersion forces, and thus a higher boiling point. In fact, the boiling point of n-octane is 125 °C, compared with 99 °C for iso-octane.

6.27 a. Lead was banned to reduce the chance of "lead poisoning", especially in children. According to the World Health Organization, at high levels, lead can cause attacks in the central nervous systems of children, causing coma, convulsions, and even death. Exposure to lead during development can cause developmental delays.

b. According to the U.S. Department of Labor Occupational Safety and Health Administration, lead exposure is most common to: Industrial workers involved in the production, use, maintenance, recycling, and disposal of lead materials and products such as plumbing fixtures, solder, rechargeable batteries, lead bullets, leaded glass, brass or bronze objects, and radiators. Construction workers involved in the removal, renovation, or demolition of structures painted with lead pigments; or the installation, maintenance, and demolition of lead pipes and fittings, lead linings in tanks and radiation protection, leaded glass, soldering, or other work involving lead metal or lead alloys. Hobbyists involved in the use of objects listed above, such as firing ranges, radiator repair, or lead–acid battery recycling. Children living in places with high levels of lead in water, food, or old deteriorating paint.

6.28

Octane Ratings of Several Compounds

Compound	Octane Number
n-octane	−10
n-heptane	0
iso-octane	100
ethanol (CH_3CH_2OH)	113
methanol (CH_3OH)	114
methyl tertiary-butyl ether (MTBE, $CH_3OC(CH_3)_3$)	116

Like n-octane, n-heptane is a straight-chain hydrocarbon, but with one fewer —CH_2 group. It also has a high tendency to knock and is assigned an octane rating of 0. On the other hand, MTBE, ethanol, and methanol all have oxygen; they have much higher octane numbers. In general, branched and oxygenated hydrocarbons have higher octane numbers.

REFLECT (Ethanol Production)

The corn is ground to expose the starch and then mixed with water to form mash. The mash is heated and mixed with enzymes to convert the starch to sugars. More heat is applied to kill any bacteria. The mash is then cooled and sent to a fermentation vessel. More enzymes are added to break the sugar down to the simplest sugars and then yeast consumes the sugars (this takes 40–80 hours). The sugar becomes ethanol and carbon dioxide. Once fermentation is complete, the alcoholic mixture ("beer") is 17–18% alcohol by volume. The mixture is distilled using a series of columns that allows separation of water and ethanol due to their different boiling points. Lastly, ethanol is condensed back into a liquid. A molecular sieve is used to extract any remaining water to create ethanol that is 99% pure.

Putting aside the vast amount of energy that is needed to grow and harvest corn, this process is quite energy-intensive. Grinding, heating (especially to high temperatures), cooling, and mixing require energy. Distillation also involves heating and cooling the mixture. In addition, a molecular sieve would require energy to pump the liquid through so that filtration could occur. The vast majority of this energy comes from fossil fuel sources.

6.29 The three most desirable characteristics of using switchgrass and wood chips are that they are abundant, relatively renewable, and, unlike corn and sugarcane, are not feedstocks.

6.30 a. $8.0 \text{ gal} \left(\frac{3.7854 \text{ L}}{1 \text{ gal}}\right)\left(\frac{800 \text{ g}}{\text{L}}\right)\left(\frac{44.4 \text{ kJ}}{\text{g}}\right)$

$= 1.1 \times 10^6 \text{ kJ}$

b. Using the same conversion scheme, but with the different density (789 g/L) and energy value (26.8 kJ/g), the answer is $6.4 \times 10^5 \text{ kJ}$

Ethanol provides 60% less energy than gasoline.

In the Laboratory (How To Make Biodiesel)

Heat the vegetable oil to about 45 °C and add methyl alcohol (methanol) with a small amount of potassium hydroxide. Stir for 25 minutes. The larger hydrocarbons are broken into smaller chains that mimic what is found in diesel fuel (14–20 carbons). The mixture is poured into a separatory funnel and the top layer is the fuel. The chemistry can be summarized as:

$$C_{57}H_{110}O_6 + 3\ CH_3OH \xrightarrow{NaOH} 3\ CH_3(CH_2)_{16}\overset{\overset{\displaystyle O}{\|}}{C}OCH_3 + C_3H_8O_3$$

glyceryl tristearate (a triglyceride) | methyl stearate (a biodiesel molecule) | glycerol

6.31 **a.** $C_2H_5OH + 3\ O_2 \longrightarrow 2\ CO_2 + 3\ H_2O$

   b. $2\ C_{19}H_{38}O_2 + 55\ O_2 \longrightarrow 38\ CO_2 + 38\ H_2O$
   (*Hint:* Balance the C and H atoms first.)

6.32 **a.** Structural formulas for methanol and ethanol

$$H-\underset{\underset{\displaystyle H}{|}}{\overset{\overset{\displaystyle H}{|}}{C}}-O-H \qquad H-\underset{\underset{\displaystyle H}{|}}{\overset{\overset{\displaystyle H}{|}}{C}}-\underset{\underset{\displaystyle H}{|}}{\overset{\overset{\displaystyle H}{|}}{C}}-O-H$$

   methanol | ethanol

   b. Propanol, because (like propane) it has three carbon atoms. More properly, this compound is *n*-propanol or 1-propanol.

$$CH_3 - \underset{\underset{\displaystyle OH}{|}}{CH} - CH_3$$

   c. This is isopropanol (2-propanol), better known as "rubbing alcohol."

6.33 **a.** Because of the oxygen present, you could expect there would be less heat released. And because of fewer carbons, you would also expect less heat to be released. So overall, biodiesel releases less energy.

   b. You would have to know the molar mass of the biodiesel to find its energy content per mole.

   c. Depending on how you are measuring the two substances to be compared, you could argue either value is useful. Particle per particle, the per mole measurement is more meaningful, but if you are measuring your fuels by mass, then per gram is more useful.

6.34 Key points include:

   • Palm oil is already the most widely consumed vegetable oil in the world, and is used in a variety of food products as well as soaps, detergents, and shampoos.
   • The use of palm oil as a biofuel increases the demand for palm oil and creates problems for workers in countries that do not have labor laws in place. Often, these workers are forced to work longer hours for little pay to meet the demand.

   • Additionally, this demand drives up the price of products that use this oil so that the cost of living is driven up and those already in tough financial situations are hurt even further.

6.35 Students will draw their own life-cycle analysis diagrams. Their diagrams illustrate that in the production and use of biofuels, environmental impacts are incurred. The proper bead distribution is as follows: ethanol (75 beads total): green (17 beads), blue (10 beads), red (17 beads), orange (15 beads), yellow (16 beads); biodiesel (25 beads total): green (3 beads), blue (10 beads), red (3 beads), orange (5 beads), yellow (4 beads).

6.36 Answers will vary, but here are some main takeaways:

   • hydrocarbons from crude oil are ideal because they are energy dense and easily transportable as a liquid fuel. They cannot be the "best fuel" because of their many negative impacts on our environment (water and air pollution, human health, climate change)

   • other liquid fuels would be ideal since they could use the existing infrastructure, but their production needs to be carbon neutral. Neither the hydrogen gas or the biofuels have the same energy density as traditional gasoline or diesel fuels.

   • electricity requires a lot of new infrastructure and the energy used to generate the energy needs to come from sustainable sources.

   • the best fuel must be cost-effective so that it can be implemented in everyday life.

Chapter 7

REFLECT (Where Does Your Energy Come From?)

Answers will vary depending on the chosen states. Sources such as natural gas or coal are derived from fossil fuels, whereas photovoltaic, wind, or hydroelectric are not.

7.1 Examples of types of energy one may use in a typical day include mechanical energy in the form of driving a car, walking to class, or exercising; chemical energy in the form of using battery-powered electronics; and electrical energy in the form of using lights and other appliances that use electricity.

7.2 **a.** Solar, wind, natural gas and other renewables have shown the most growth since 2015.

   b. The electricity production declined in 2021 due to the effects of the COVID-19 pandemic. For instance, less electricity was needed for companies as most employees worked from home. In 2022, the electricity rebounded as companies started to ramp up production again.

   c. There are many possible answers. One possibility is Europe, which showed a sharp decline in coal, but strong increases in wind and solar over the past 5 years.

7.3 a. Some countries that have experienced large shifts include Belarus, Estonia, Australia, Israel, and Poland. These countries have established aggressive targets to reduce their carbon footprints within the next 20 years and have greatly expanded their solar and wind capacities to help meet these goals.

b. Bangladesh, Taiwan, and South American and African countries have been slow to move away from fossil fuels due to copious coal and oil resources and economies that are not able to support renewables that are more expensive to develop and institute.

REFLECT (Nuclear Fission vs. Fusion)

a. Nuclear fusion requires extremely high temperatures and pressures; hence, it typically takes more energy to initiate the reaction than is generated by nuclear fusion. Even if fusion occurs with more energy produced than required, it is also difficult to maintain the high temperatures required to sustain the fusion reactions over time. Hundreds of Tokamak reactors have already been built around the world, but are in need of optimization before widespread nuclear power is possible via this route.

b. Hydrogen isotopes (deuterium and tritium) are typically used for fusion reactions since they are the smallest with the lowest charge and thus easiest to combine under extreme conditions.

c. For the first time, scientists recorded a greater energy output than was required to initiate the fusion reaction. They used 192 lasers that irradiated a small frozen target of hydrogen encased in diamond. In 100 trillionths of a second, 2.05 megajoules (MJ) of energy, the equivalent of a pound of TNT, bombarded the hydrogen pellet. This resulted in neutron particles and 3 MJ of energy, a factor of 1.5 in energy gain. Although decades of further research and optimization are needed before commercializing this technology for power plant use, this would provide the possibility of generating power with zero emission of greenhouse gases. Hydrogen is the most abundant element in our universe, which would provide an unlimited supply of energy. Further, the deuterium from a glass of water could power a home for a year—an intriguing possibility, indeed!

d. Both nuclear fission and fusion reactions do not generate any greenhouse gas emissions during their use. Nuclear fission generates a variety of radioactive by-products; however, the only by-product produced from nuclear fusion is helium, which is not a greenhouse gas and will be contained within the reactor to sustain fusion reactions.

7.4 U-234 has 92 protons and 142 neutrons, while U-238 has 92 protons and 146 neutrons. Since isotopes differ only in the number of neutrons, both U-234 and U-238 have 92 electrons.

7.5 Atomic number: 92; mass number: 234. $^{234}_{92}U$. This symbol differs from U-238 because they both have different numbers of neutrons.

REFLECT (Writing Nuclear Equations)

a. $^{1}_{0}n + {}^{235}_{92}U \longrightarrow {}^{138}_{56}Ba + {}^{95}_{36}Kr + 3{}^{1}_{0}n$

b. $^{1}_{0}n + {}^{235}_{92}U \longrightarrow {}^{137}_{52}Te + {}^{97}_{40}Zr + 2{}^{1}_{0}n$

7.6 For anthracite coal:

$$9.0 \times 10^{10} \text{ kJ} \times \frac{1.0 \text{ g anthracite coal}}{30.5 \text{ kJ}} \times \frac{1 \text{ kg}}{1000 \text{ g}}$$
$$= 3.0 \times 10^{6} \text{ kg anthracite coal}$$

The equivalent masses for the other grades are calculated in the same way: bituminous coal, 2.9×10^{6} kg; sub-bituminous coal, 3.8×10^{6} kg; lignite (brown coal), 5.6×10^{6} kg; peat, 6.9×10^{6} kg.

7.7 $^{241}_{95}Am \longrightarrow {}^{237}_{93}Np + {}^{4}_{2}He$

$^{9}_{4}Be + {}^{4}_{2}He \longrightarrow {}^{12}_{6}C + {}^{1}_{0}n + {}^{0}_{1}\gamma$

7.8 In an emergency such as an earthquake, the fuel rods should be inserted in preparation for shutdown to slow the nuclear fission reaction in the reactor core.

7.9 The cloud is composed of tiny droplets of condensed water vapor. Some people call this "steam"; steam, however, is invisible until it condenses. The cloud does not contain any nuclear fission products.

7.10 $1.2 \times 10^{9} \text{ J/s} \times \frac{60 \text{ s}}{1 \text{ min}} \times \frac{60 \text{ min}}{1 \text{ h}} \times \frac{24 \text{ h}}{1 \text{ day}} \times 3 \text{ reactors}$
$= 3.1 \times 10^{14}$ J of electric energy generated in a single day when the Palo Verde complex is operating at maximum capacity.

Rearranging $\Delta E = \Delta mc^{2}$ to solve for the mass of U-235 lost gives:

$\Delta m = \Delta E/c^{2} = 3.1 \times 10^{14} \text{ J}/(3.00 \times 10^{8} \text{ m/s})^{2} = 0.0034$ kg or 3.4 g of U-235 lost each day.

7.11 a. electromagnetic radiation

b. nuclear radiation

c. electromagnetic radiation

d. nuclear radiation

7.12 a. $^{86}_{37}Rb \longrightarrow {}^{86}_{38}Sr + {}^{0}_{-1}e$ **b.** $^{239}_{94}Pu \longrightarrow {}^{235}_{92}U + {}^{4}_{2}He$

7.13 a. Answers will vary; the average level of radiation exposure in the U.S. is 0.62 rem.

b. The concentration of cosmic radiation is greater at higher elevations.

c. Skip airport scanners and X-ray machines, unless absolutely necessary. Perform radon testing to determine if your basement has unsafe levels of radon gas.

@HOME (Sweet Radioactive Decay!)

a. The y-axis would relate to the amount of radioactive material remaining.

b. No

REFLECT (Here Today . . .)

After 10 half-lives, 0.0975% of the original sample remains.

# of half-lives	% decayed	% remaining
7	99.22	0.78
8	99.61	0.39
9	99.805	0.195
10	99.9025	0.0975

7.14 **a.** Radium is a product in the natural decay series of uranium-238, the isotope of uranium with highest natural abundance. Therefore, if uranium is present in the rocks and soils, radium will be present as well.

b. If 0.5 pCi remains, that means 0.5 pCi/16 pCi = 3.12% remain, so five half-lives (5 × 3.8 days = 19 days) are required for the level of radioactivity to drop.

c. Radon has a half-life of 3.8 yrs. Hence, 15.2/3.8 = 4 half lives. Therefore, 100 × 1/2 × 1/2 × 1/2 × 1/2 = 6.25 g

d. Radon will continue to enter your basement because the uranium in the soils and rocks underneath your home continuously produce it.

7.15 Using Table 7.4 to verify, after 36.9 years (3 half-lives), 12.5% of the original tritium will remain.

7.16 $_{0}^{1}n + _{92}^{235}U \longrightarrow _{54}^{143}Xe + _{38}^{90}Sr + 3_{0}^{1}n$

REFLECT (The Chornobyl Disaster: How It Happened)

a. Operators were conducting a "safety test," which caused the meltdown. Unqualified and tired operators at 1:00 am were primarily responsible for this disaster. An automatic safety interlock would have prevented this accident since it would have prevented the surge in power due to all of the control rods being withdrawn. All new reactor designs now feature automated controls that would insert control rods and flood the reactor with water at the first sign of a power surge.

b. Many people around the Chornobyl site were exposed to high radiation levels; this has been implicated in a large number of DNA mutations and cancers. In addition, plant and animal life were impacted by high radiation levels. Long-term effects are still under debate, but have likely resulted in high levels of thyroid cancers and contamination of the food supply. Radiation levels appear to be finally recouping to pre-accident levels in recent years.

7.17 **a.** −498 kJ/2 mol H_2 = −249 kJ/mol

−249 kJ/mol (1 mol H_2/2 g H_2) = −125 kJ/g

b. From Figure 6.5, methane releases −50.1 kJ/g during combustion. (−125 kJ/g)/(−50.1 kJ/g) = 2.5; therefore, hydrogen releases about two and half times more energy!

7.18 **a.** :Ï·   :Ï:Ï: [:Ï:]⁻

atom   molecule   ion

b. The iodine atom is the most reactive because it has one unpaired electron, making it a free radical.

c. The element iodine (including its radioisotope, I-131) is taken up by the thyroid gland in the chemical form of I⁻, the iodide ion.

7.19 **a.** Low neutron absorption is an important property for any structural material used in a nuclear reactor because the neutrons that are generated need to interact with the nuclear fuel in order to sustain the chain reaction taking place in the reactor's core.

b. In the event of an accident at a nuclear power plant, it is important that the fuel rods can be cooled until the reaction stops. In the event that the cooling mechanism fails, the fuel rods will get too hot (thousands of °C) and the zirconium metal becomes capable of oxidizing residual water in the core. This reaction "rusts" the protective zirconium casing of the fuel rods and produces highly flammable hydrogen gas. The hydrogen gas is capable of producing an explosion if it is not vented quickly enough. If hot enough, the zirconium alloy, the nuclear fuel, and any machinery remaining in the core can melt and form a dangerous radioactive substance called corium that is capable of eating through the concrete walls of a containment building.

7.20 **a.** Answers will vary. The Nuclear Energy Institute provides helpful information about nuclear power usage in the United States and provides fact sheets for each state summarizing their current energy usage.

b. South Carolina (63.9%), New Hampshire (62.9%), Illinois (53.9%), Connecticut (48.7%), and Tennessee (47.7%) use the highest percentages of nuclear energy.

c. Answers will vary. Currently, 21 states do not have nuclear power plants. Depending on the state, energy sources could be coal, natural gas, hydroelectric, oil, or another renewable energy source.

7.21 **a.** The most important aspect in forming an opinion on any of the listed topics is to make an informed decision using reputable resources. Nuclear power has demonstrated its viability as an alternative energy source; however, it is still a developing technology that will continue to be improved upon.

b. Answers will vary. A good resource is the Nuclear Waste Management Organization's website.

c. The cartoon depicts a future in which there are renewable energy alternatives available to power household appliances.

d. One school of thought is that nuclear power is a viable alternative to fossil fuels and is a solution to the global energy crisis, whereas others feel that although nuclear power may be a solution to dwindling fossil fuel reserves, the radioactive waste generated from this process has the potential to put public safety at risk if it is not handled correctly.

7.22 **a.** Currently, nuclear reactors are located on four continents, with the largest number of operational reactors in Europe, North America, and Asia. Over the past 10 years, some European countries (France, Germany, Belgium, Denmark), China, and Pakistan have shown the largest expansion of nuclear capacity. There is less public fear regarding nuclear energy in these countries and expansion of population and establishment of more rigorous environmental goals dictates the use of other sources of energy beyond fossil fuels.

b. Africa, Canada, and some European countries such as Switzerland and the Netherlands have shown significant decrease in their nuclear capacity. Due to the controversial nature of nuclear energy, these countries have opted to expand solar and wind power instead of nuclear.

c. According to the World Nuclear Association, Australia has the largest uranium resources in the world (30%), followed by Kazakhstan (14%), Canada (8%), and Russia (8%). However, as of 2017, Kazakhstan produces the largest share of uranium (39%), followed by Canada (22%) and Australia (10%). Despite its large uranium deposits, Australia has never had a nuclear power station! Canada has 19 nuclear reactors that produce about 15% of the nation's energy needs. Russia currently has 35 nuclear reactors in operation, supplying about 18% of the country's energy needs. The most nuclear reactors are found in the U.S. (99 currently operational), supplying 20% of total electricity use. France has 58 nuclear reactors, which supply almost 72% of the country's electricity needs.

d. According to the International Atomic Energy Agency, developing countries would benefit the most from increasing their nuclear energy capacity. Nuclear technology can be used in the fight against cancer, cardiovascular and other noncommunicable diseases, as well as to help combat child malnutrition.

7.23 **a.** Some countries that have greatly expanded their solar capabilities include Morocco, Poland, Croatia, Argentina, Vietnam, and Turkey. The price of solar has decreased dramatically in recent years, so this is now more affordable; some of these countries have also provided financial incentives to invest in solar and have aggressive targets for the replacement of fossil fuels with renewables.

b. The use of solar worldwide started increasing dramatically in 2009–2010. At that time, the cost of solar was finally at a point that was relatively affordable for industrialized nations. Governments also started subsidies to entice their population to adopt this technology.

c. Solar radiance depends on local factors such as cloud cover, aerosols, smog, and haze.

7.24 **a.** Answers will vary. Google's Project Sunroof provides online solar irradiance estimates for many locations.

b. Using 256 kWh in 33 days means, on average, this homeowner would need to produce 7.76 kWh of solar energy per day in order to power their home. Consulting the data illustrated in Figure 7.23, only homes located in southwest U.S., within the darkest red regions on the map, would be able to be powered exclusively using solar energy.

c. 7.50×10^8 W/7.76×10^3 W = 9.7×10^4 houses that could be powered by this nuclear plant.

d. Some things to keep in mind would be that nuclear power plants do not continually run at full capacity and that continually supporting 10,000 homes would mean that the nuclear reactor would continually be depleted. Therefore, the amount of homes that could be supported by this nuclear power plant would be much lower.

7.25 **a.** From The NEED project, examples of solar thermal collectors include parabolic troughs, solar power towers, and dish/engine systems.

b. Parabolic troughs use long reflecting troughs that focus the sunlight onto a pipe. The fluid circulating inside the pipe collects the energy and produces steam by transferring the heat to a heat exchanger. Depending on size, this type of solar collector could be used to power large communities. A solar power tower uses a large field of rotating mirrors to track the Sun and focus the sunlight onto a thermal receiver on the top of a tall tower. The fluid in the receiver collects the heat and uses it to generate electricity or store it for later use. This type of collector can also be large enough to provide power to large communities of people. Dish systems concentrate sunlight and use an engine located at the focal point to create electricity. Due to the small size of these units, they are best used to power a single home or small business.

c. All three types of collectors require a continuous supply of strong sunlight, such as that found in desert regions, to function at their full potential.

@HOME (S'mores From Sunshine!)

a. Using solar will not result in particulate matter, which is generated in a bonfire. However, one would need to do this during the day when sunlight is present; a bonfire may be used at any time of day—typically in the evening. Also, it is much faster to melt a marshmallow using a bonfire.

b. The heat generated from reflection of the box is similar to trapped heat due to greenhouse gases in our atmosphere.

7.26 a. Doping with phosphorus forms an n-type semiconductor. Phosphorus is in Group 15 and has more electrons per atom than a silicon atom does.

    b. Doping with boron forms a p-type semiconductor. Boron is in Group 13 and has one fewer electron per atom than a silicon atom.

REFLECT (Electrons and Holes)

a. No. Glass is an insulator and is amorphous (no long-range order of its structure) rather than crystalline. If dopants were added to the lattice, the structure wouldn't be conductive enough to allow for transport of electrons through the solid.

b. 1–2 ppm of dopant atoms are typically added.

7.27 a. Uses include pumping water for livestock and lighting in areas without electricity. Even on farms and ranches with electricity, solar PVs can reduce electric utility bills.

    b. Uses include providing electricity to power lighting, heating, and other electrical needs in a building or in a process used by the business. For example, PVs have been used to generate enough electricity to power the electrical needs of a small microbrewery.

    c. Uses include supplying electricity to part or all of a home. At present, some homes that have a PV system can be used for electrical back-up locations in places where storms occur that may disrupt the power.

7.28 a. The climate of Maryland is not continually hot and dry, and turning Maryland into a solar farm would not be an efficient use of that land.

    b. The desert regions of the southwestern United States would be the most effective place to construct a solar farm.

    c. Large solar farms can damage the local desert ecosystem if not managed carefully. It would be important to protect other vital resources in the process of building and operating a solar farm.

    d. Answers will vary.

Real Talk (Solar-Powered Transportation Systems?)

a. Some examples include: solar-powered Byron Bay train in Australia, and the use of solar-powered buses in parts of Europe and China.

b. Answers will vary. One possibility might include the sudden loss of power that would create the chance for a crash with road vehicles if on the same grid.

7.29 a. The largest challenge to wind power in the southeastern U.S. is the Appalachian mountain range, which blocks the wind and creates uneven territory.

b. The total wind capacity of the U.S. in 2050 is predicted to be 404.25 GW. In comparison, the total wind capacity was 113.43 GW in 2020. This corresponds to a 256% increase.

7.30 Answers will vary. The National Energy Education Development Project's Energy Infobook provides an excellent summary of the many types of renewable energy sources available, which includes not only how each type of energy works, but also the advantages and limitations of each type.

Chapter 8

REFLECT (Batteries in Your Everyday Life)

a. Answers will vary. Some possibilities include flashlights (alkaline), cell phones (Li-ion), and car starters (lead–acid).

b. Li-ion and lead–acid are rechargeable. Some factors include operating temperature, battery age, and extent of discharge.

REFLECT (Early Battery Designs)

a. All batteries contain electrodes and an electrolyte.

b. Planté cells (Pb–acid), Leclanché cells (Zn–C dry cells), and Gassner cells (alkaline batteries) have been refined but are still in use today.

c. The electrodes and overall design are largely the same; the biggest difference is the electrolyte, which has been modified to increase storage density and shelf life.

REFLECT (Why is the Statue of Liberty Green?)

a. Yes, the Statue of Liberty would change color since aluminum also readily oxidizes to form a layer of aluminum oxide. It would corrode and turn white to white-grey in color.

b. The copper oxide can be removed with a mixture of salt and a weak acid—like vinegar. But the copper oxide actually makes the statue stronger. If it were to be cleaned, some copper would be removed, weakening the statue.

@HOME (Redoxing Pennies)

a. For the first bowl, the copper has been oxidized through contact with oxygen. The presence of acetic acid helps to promote the oxidation. While copper is oxidized, oxygen is reduced.

$$2\,Cu(s) + O_2(g) \longrightarrow 2\,CuO(s)$$

b. In the second bowl, the penny became shiny. The presence of an acid and salt (NaCl) serve to dissolve copper oxide on the surface, which exposes shiny copper. More explicitly, the chloride ions (Cl^-) replace the surface oxide (O^{2-}), resulting in $CuCl_2$ that is soluble in water. Hence, some of the surface copper ions are removed through reaction with sodium chloride, which exposes fresh metallic copper.

8.1 a. Reduction. The oxidation state of the aluminum ion has been reduced from +3 to 0 and therefore has gained three electrons.

b. Oxidation. The oxidation state of zinc metal has increased from 0 to +2 and therefore has lost two electrons.

c. Reduction. The oxidation state of the manganese ion has been reduced from +7 to +4 and therefore has gained three electrons.

d. Oxidation. Two water molecules were split into four hydrogen ions, one oxygen molecule, and four electrons. For each molecule of water, the oxidation state of the oxygen ion has increased from −2 to 0 in O_2.

e. Reduction. The oxidation state of each hydrogen ion (proton) has been reduced from +1 to 0 and therefore has gained one electron. However, since there are two protons, a total of two electrons are needed for the reduction to form H_2.

REFLECT (Electrical Current: A Water-Flow Analogy)

a. Answers will vary. A good resource for this information is located at: https://electronicguidebook.com/does-a-resistor-reduce-voltage-or-current/.

b. Answers will vary. A good resource for this information is located at: https://www.electronics-tutorials.ws/resistor/res_2.html.

c. Sand would be the densest material because it would be the most difficult to get the water to flow through.

d. Answers will vary, as there are many possibilities. In addition to incandescent lighting, devices such as electric toasters, kettles, and coffee makers contain heating elements, which heat up when electrons flow through them. Another application is volume controls of televisions or radios. Resistors are also used to limit the current flowing to LEDs and transistors, since too high a current would damage these sensitive devices.

8.2 $1\ \text{amp} = \dfrac{C}{s}$ and $1\ \Omega = \dfrac{J \cdot s}{C^2}$

Substituting these values into Equation 8.5 gives

$V = \left(\dfrac{\cancel{C}}{\cancel{s}}\right)\left(\dfrac{J \cdot \cancel{s}}{\cancel{C^2}}\right)$, which simplifies to $V = \dfrac{J}{C}$.

8.3 a. Using Equation 8.8:

$P = \dfrac{(120\ \text{V})^2}{2.5\ \Omega} = 5800\ \text{watts},$

or 5.8 kW (2 significant figures)

b. Rearranging Equation 8.6:

$I = \dfrac{45\ \text{W}}{120\ \text{V}} = 0.38\ \text{A}$

@HOME (A DIY Voltaic Pile)

a. The voltage increases as the number of cells increase.

b. The vinegar and table salt serve as the electrolyte, which is used to allow ions to migrate in the voltaic pile.

8.4 a. The memory effect occurs when rechargeable batteries are not fully charged and discharged between charging cycles. The battery "remembers" the shortened lifetime, reducing capacity. This term is more accurately referred to as "voltage depression."

b. Nickel–cadmium (Ni–Cd) batteries are most prone to suffering from voltage depression.

c. This effect can often be repaired by fully charging and discharging a battery.

d. Batteries should be stored in a cool, dry environment, not in hot or humid conditions. The ideal temperature to store batteries is 15 °C (∼60 °F). So, storing batteries in the refrigerator might be a good strategy depending on the climate where you live. However, a pack of silica gel should be placed with the batteries to absorb moisture.

8.5 a. Lead exists as Pb^{4+} in PbO_2 and as Pb^{2+} in $PbSO_4$. The symbol Pb represents lead in its metallic form.

b. Electrons are lost from Pb to form cations. This is oxidation.

c. Metallic lead is oxidized during the discharge process, as two electrons are lost to form the Pb^{2+} ion.

8.6 a. 800 kilowatt hours/month × 1 month/30 days × 1 day/24 hours × 4 hours = 4.44 kilowatt hours × 300,000 homes = 1,333,333.33 kilowatt hours = 1,333 megawatts.

b. 1,333 megawatts is 3x the size of the battery they have. But they did mention that the power will be used in the evenings. This is when the house has cooled down and should use less air conditioning. So hopefully, the assumption that the average hour is a nighttime hour is correct.

c. The Moss Landing project was commissioned in 2020, but suffered a number of outages due to the COVID-19 epidemic and other interruptions. As of June 2022, the system is fully operational, offering 568 MW of power storage capacity. Regarding other possible conversions, answers will vary. Some possibilities may include the Hayden Generating Station in Colorado, and Reuter Power Plant in Berlin, Germany. With the impending phaseout of coal-fired power plants around the world, there are many companies evaluating technology to convert these plants into energy storage facilities. One such proposal is found online here.

8.7 a. Using $W = V \times I$; Level 1: (110 V)(20 A) = 2200 W

Level 2: (240 V)(40 A) = 9600 W

Dual: (240 V)(80 A) = 19,200 W

b. Level 1: $100 \text{ kWh} \times \dfrac{1000 \text{ W}}{1 \text{ kW}} \times \dfrac{1}{(0.92)(2200 \text{ W})}$
= 49 h

Level 2: $100 \text{ kWh} \times \dfrac{1000 \text{ W}}{1 \text{ kW}} \times \dfrac{1}{(0.92)(9600 \text{ W})}$
= 11 h

Dual: $100 \text{ kWh} \times \dfrac{1000 \text{ W}}{1 \text{ kW}} \times \dfrac{1}{(0.92)(19{,}200 \text{ W})}$
= 5.7 h

c. According to Tesla, there are currently more than 40,000 supercharging stations across the globe. These charging stations deliver as much as 250 kW while charging. This relates to a charge time of 100 kWh/250 kW = 0.40 h or 24 min. However, the charge rate is 24 minutes to 80% of battery capacity and over 60 minutes to reach 100% of battery capacity.

d. Using an example rate of 17¢ per kWh,

Level 1: 2.2 kW × 49 h × $0.17/kWh = $18.33

Level 2: 9.6 kW × 11 h × $0.17/kWh = $17.95

Dual: 19.2 kW × 5.7 h × $0.17/kWh = $18.60

Supercharger (250 kW): 250 kW × 0.40 h × $0.17/kWh = $17

e. Starting with the Tesla Model S, which is said to get 400 miles per charge:

1100 miles per month/400 miles per charge = 2.75 charges per month

Level 1: $18.33 per charge × 2.75 charges per month = $50.41 per month

Level 2: $17.95 per charge × 2.75 charges per month = $49.36 per month

Dual: $18.60 per charge × 2.75 charges per month = $51.15 per month

Supercharger: $17 per charge × 2.75 charges per month = $46.75 per month

Using an average car with a 13-gallon gas tank that gets 25.1 miles per gallon and using the national average of $3.94 per gallon (as of September, 2023):

13 gallons per tank × 25.1 miles per gallon = 326.3 miles per tank

1100 miles per month/326.3 miles per tank = 3.37 tanks per month

13 gallons/tank × $3.94/gallon = $51.22 per tank

$51.22 per tank × 3.37 tanks per month = $172.61 per month

A Tesla Model S is the more economical choice. It also requires less maintenance because there are no oil changes. However, as mentioned above, Tesla's charging time is longer than estimated because the charger is designed to reduce charging power as the battery reaches capacity. Another factor to consider is the cost of the car itself. A Tesla Model S has a price tag of $75,000 before customizations. There are other less expensive electric cars—the Chevy Bolt starts at $26,500, but they are still more costly than their nonelectric counterparts.

REFLECT (How Do Li-ion Batteries Work?)

a. Cell phones don't work as well in cold weather. When the electrolyte fluid becomes harder in cold weather, the resistance of the battery increases, this means higher power consumption and a voltage drop, rapidly draining the battery. So do not charge your cell phone in the freezer. Best to leave your cell phone at room temperature.

b. Answers will vary.

REFLECT (Exploding Batteries!)

a. Lithium is very reactive with water, so if the electrolyte was a salt dissolved in water, it would react with the Li ions and cause a major fire hazard. Accordingly, Li-ion batteries must use nonaqueous electrolytes.

b. A battery explodes when exposed to the air because air is made from oxygen and nitrogen, which are two molecules that lithium reacts with violently. Battery packs must be hermetically sealed to prevent exposure to air. If a battery produces oxygen as a byproduct of heating, it will also explode because of the lithium in the battery.

c. Glass is a better electrolyte because it has fluid-like properties and does not contain any chemicals that lithium reacts with violently. In addition, a variety of polymers and ceramics have been proposed as solid electrolytes. For a list of some potential options, check out this site.

Real Talk (Next-Generation Batteries)

a. Yes, sodium-ion batteries would have the same issues as lithium-ion batteries regarding moisture exposure. They are both Group 1 metals and react violently with water. In fact, sodium is much more reactive than lithium.

b. 98 pm × 2 = 196 pm as the diameter of a sodium ion. 196 pm × 1 angstrom/100 pm = 1.96 angstroms = diameter of a sodium ion. 9.6 angstroms wide/1.96 angstroms = 4.9 atoms thick. So roughly around 5 atoms could fit in this distance across!

@HOME (Energy vs. Power Density)

a. An example would be a camera flash where power is needed quickly and needs to be recharged quickly so it can work again.

b. Li-ion batteries will last longer because they release energy slowly and consistently. Supercapacitors release energy quickly to be used as a burst of power.

c. Li-ion batteries store more energy than supercapacitors. Supercapacitors store more power than batteries.

d. Supercapacitors will be relieved of their charge much quicker than a Li-ion battery due to a high self-discharge rate.

8.8 A 2019 report by the IDTechEx Company details the performance achievements and objectives of the 80 current manufacturers of supercapacitors over the next 20 years. This report highlights the current efforts underway to improve supercapacitor technology, given the incredible need for energy alternatives.

8.9 The first example below shows how one might perform the calculation by using the following assumptions and example data: 1 gallon of gasoline burned = 19.5 pounds of CO_2 released, an average 25.1 mpg, and an average monthly mileage of 1100 miles.

1100 miles × 12 months = 13,200 miles per year
13,200 miles per year/25.1 miles per gallon = 525.9 gallons of gas used per year.

525.9 gallons per year consumed × 19.5 pounds of CO_2 released per gallon = 10,300 pounds (~5 tons of) CO_2 released.

The next example uses a back-calculation to determine the mpg of a car emitting 7 tons of CO_2 in an average year. Again, using the value that one gallon of gasoline burned = 19.5 pounds of CO_2 emitted and using an average yearly driving distance of 13,200 miles:

7 tons CO_2 = 14,000 pounds CO_2

14,000 pounds CO_2/19.5 pounds CO_2 per gallon = 717.95 gallons of gas used in a year

13,200 miles per year/717.95 gallons per year = 18 mpg. A car that gets 18 mpg (such as a small pickup truck) will emit 7 tons of carbon dioxide in a year. A larger vehicle, such as a standard pickup truck, gets only 15 mpg and will emit almost 8 tons of CO_2 in a year!

Improving your car's gas mileage by 5 miles per gallon would give an assumption of 30.1 mpg. Let's leave everything else the same.

13,200 miles per year/30.1 mpg = 438.5 gallons per year consumed.

438.5 gallons per year consumed × 19.5 pounds of CO_2 released per gallon = 8550 pounds of CO_2 released ~4 tons of CO_2 released so that you would save about a ton a year!

Other things to save gas are walking, riding a bike, or taking public transportation when you can.

8.10 In Equation 8.22, 1 mol of H—H bonds and $^1/_2$ mol of O=O bonds are broken.

= 1 mol (436 kJ/mol) + $^1/_2$ mol (498 kJ/mol)
= 436 kJ + 249 kJ
= 685 kJ

In Equation 8.22, 2 mol of O—H bonds are formed.
= 2 mol (467 kJ/mol)
= 934 kJ

For the overall reaction, (+685 kJ) + (−934 kJ) = −249 kJ/mol or −124.5 kJ/g.

The negative sign indicates the reaction is exothermic, meaning 124.5 kJ of energy is released per gram of H_2 combusted. In contrast, methane produces 50 kJ/g.

8.11   **a.** A battery makes electricity from stored energy, whereas a fuel cell produces electricity in a fuel tank. A battery has a finite amount of energy it can hold, while a fuel cell can continually make more energy.

   b. A PEM fuel cell does not need to be recharged because it relies on an external fuel source, such as a hydrogen fuel tank. You can exchange the tank for a full one when the tank runs out.

   c. The chemistry that occurs within a fuel cell is not combustion because the fuel is not burned.

8.12   **a.** In Equation 8.27, 4 mol of C—H bonds and 4 mol of O—H bonds are broken.
   = 4 mol (416 kJ/mol) + 4 mol (467 kJ/mol)
   = 1664 kJ + 1868 kJ
   = 3532 kJ

   In Equation 8.27, 4 mol of H—H bonds and 2 mol of C=O bonds are formed.
   = 4 mol (436 kJ/mol) + 2 mol (803 kJ/mol)
   = 1744 kJ + 1606 kJ
   = 3350 kJ

   For the overall reaction, (+3532 kJ) + (−3350 kJ) = +182 kJ.

   b. Although there is general agreement (+182 kJ vs. +165 kJ), remember that Table 6.1 gives *average* bond energies, not specific energies associated with the bonds in these compounds (except CO_2). Also, states are not considered for the data in Table 6.1.

8.13   **a.** Energy with a wavelength of 420 nm falls in the visible region (violet).

   b. Heating water to thermally decompose it into H_2 and O_2 has not yet proven to be effective; temperatures higher than 3000 °C are required.

8.14   **a.** Metals are elements that are good conductors of electricity and heat. They have a shiny appearance. In contrast, nonmetals that are poor conductors of electricity and heat are not shiny. Additionally, metals have low electronegativity values, whereas nonmetals do not. This helps explain why metals lose electrons to form cations, whereas nonmetals gain electrons to form anions.

   b. NiS contains the Ni^{2+} ion, which is reduced in the reaction; the ion gains two electrons to form Ni metal.

   c. $PbS(s) + O_2(g) \longrightarrow Pb(s) + SO_2(g)$. In this reaction, Pb^{+2} is reduced because it gains two electrons to form Pb metal. Additionally, sulfur is oxidized because it loses six electrons to go from S^{2-} to S^{4+}. Oxygen is also reduced because each oxygen atom gains two electrons to form O^{2-}.

d. Sulfur dioxide released from smelting-containing ores contributes to air pollution. The EPA monitors the level of SO_2 as part of ambient air quality standards due to the serious nature of this air pollutant as a respiratory irritant.

8.15 a. The U.S. ships up to 60% of lead–acid batteries to Mexico and Asia to recycle. Lead is a dangerous neurotoxin that leads to high blood pressure, abdominal pain, and kidney disease in adults, and serious developmental delays and behavior problems in children because it interferes with brain development.

b. There are five main companies that recycle lithium-ion batteries, two in Canada, one in Germany, one in Finland, and one in Australia. However, China and India are becoming players in Li-ion battery recycling.

c. Answers may vary—for instance, in North Carolina, the main environmental issues are PFAS, climate change, and mining. So heavy metals might be in there, just not with recycling, but with finding the raw materials.

8.16 a. No, metals cannot become extinct, but they can become completely unavailable for commercial applications.

b. Some examples currently used in energy-storage applications include silver, lithium, cadmium, lead, mercury, platinum, nickel, and tin.

c. A 2011 *JACS* publication demonstrated the first metal-free electrocatalyst for fuel cell applications that utilized carbon nanotubes.

8.17 a. Some statistics show that almost 90% of all rechargeable lead–acid batteries are recycled in the United States, whereas less than 2% of consumer disposable batteries are recycled.

b. If a Ni–Cd battery is not disposed of properly and ends up in a landfill, the metal cylinder will corrode, and the cadmium will pollute the water supply. This represents a serious environmental concern.

c. Household batteries are often referred to as "disposable batteries." This name implies that the battery should be disposed of after use. Most people who throw away batteries are unaware of the toxic metals housed within the battery and the implications this has for the environment. On the other hand, car batteries are often replaced by auto mechanics who know how to recycle the battery correctly.

8.18 a. The global electricity demand for EVs was 60 TWh in 2021 and is projected to grow to as much as 2300 TWh in 2040—still just 8% of the projected global electricity consumption.

b. U.S. (as of 2021): 60.8% fossil fuels, 18.9% nuclear, 20.1% renewables. Other countries will vary, usually in the amount of renewables in the mix; for instance, Germany has 39.7% renewables in its electricity mix.

c. A good article to compare the GHG emissions from EVs based on their production and use cycles: Ellingsen, L. A-W.; Singh, B.; Stromman, A. H. *Environmental Research Letters* **2016**, *11*, 054010. (DOI: 10.1088/1748-9326/11/5/054010)

Chapter 9

REFLECT (Recycling Plastics)

a. Answers will vary. For instance: milk jugs, water bottles, or shampoo bottles.

b. Milk jugs: 2 (high-density polyethylene), water bottle: 1 (polyethylene terephthalate), shampoo bottle: 3 (polyvinyl chloride).

9.1 Answers will vary. Answers may include the tennis racquet, tennis court flooring, and clothing. Polymers are flexible, which is good for a tennis racquet frame/strings because they affect how the ball comes off the racquet after being hit. Stretchiness is a good feature for tennis clothing, as the athletes are running and moving around and need flexibility. The tennis court flooring can absorb impact, which is good for athletes' joints!

@HOME (Plastics from Milk?)

a. Protein molecules known as casein act as the monomer.

b. There are many possibilities; some common examples include poly(lactic acid), PLA, and polyhydroxyalkanoates, PHAs. Biodegradable plastics naturally degrade, so they will not persist in the environment and landfills for hundreds of years like traditional polymers.

9.2 Polymerization reactions can be stopped if the monomers are completely used up or through a termination reaction.

9.3 a.

b.

c. Octane is C_8H_{18}. Although the product molecule similarly has eight carbon atoms, it has two fewer hydrogen atoms and two R groups at the ends of the molecule, so its chemical formula is $C_8H_{16}R_2$.

d. Common name: propylene; systematic name: propene

9.4 Answers will vary. An example of LDPE is plastic wrap. An example of HDPE is plastic bags. LDPE is more flexible than HDPE. LDPE is more translucent than HDPE. HDPE is more often colored with a pigment than LDPE.

@HOME (Polyethylene "Necking")

a. Necking does not change the number of monomer units.

b. Necking does not affect the bonding between the monomer units with the polymer chain; it's the intermolecular forces that are agitated when necking, not the intramolecular forces (bonds).

c. Necking is not reversible; in contrast, when a rubber band is stretched, it returns to its original form.

9.5 Answers will vary. Pros may include the energy needed to recycle plastic vs. paper and durability. Cons may include how long plastic takes to biodegrade in the environment, the number of bags used from one shop, the fact that most people won't recycle or reuse their single-use bags, etc.

9.6 LDPE does not line up in rows because of its branching. HDPE is able to stack rows, adding to its density.

9.7 Answers may vary. One answer may include: MDPE is medium-density polyethylene. It is more resistant to stress-cracking than HDPE. LLDPE is linear low-density polyethylene. It is more resistant to punctures than LDPE.

REFLECT (Thermoplastics vs. Thermosets)

a. Thermoplastics are easier to recycle since they are easily melted and reformed. The weaker intermolecular interactions between polymer chains break down at much lower temperatures than thermosets; however, the individual polymer chains do not decompose and are able to be remolded into new shapes.

b. Common examples of thermoset plastics and polymers include epoxy, silicone, polyurethane and phenolic. Applications include construction equipment, electrical components, insulators, circuit breakers, and a variety of motor and automotive components that experience high heat such as brakes.

9.8 a. Polystyrene, because it is degraded in many organic nonpolar solvents.

b. LDPE if not pigmented, polystyrene if in crystal form, polyethylene terephthalate. Soft-drink bottles are made from polyethylene terephthalate.

c. Bottle caps are made from polypropylene. Answers may vary, but may include that toughness is also important in luggage material.

d. Polyethylene terephthalate, polyvinyl chloride, HDPE, and LDPE. However, LDPE is typically used for flexible materials. The most likely material would be HDPE.

9.9 a. The monomers all have a carbon backbone; they differ by the side groups or atoms that are bonded to the carbon atoms.

b. PVC stands for polyvinyl chloride; the monomer is vinyl chloride. The monomer for Saran is vinylidene chloride. The monomer for Teflon is tetrafluoroethylene. The rigid form of PVC is used in construction for pipes. It is a white, brittle solid that is also used in making plastic bottles and plastic cards and food-covering sheets. In comparison, the monomer vinyl chloride is a colorless gas that burns easily. Saran is a thin plastic fiber used for sealing food. It is found in plastic wrap, cling wrap, and Saran wrap. In comparison, the vinylidene chloride monomer is a clear colorless liquid that is flammable. Teflon is used as a non-stick coating for pans and other cookware. Its monomer, tetrafluoroethylene, is a colorless odorless gas.

9.10 a.

Phenyl group   Benzene

$-C_6H_5$ C_6H_6

b. The two possible resonance structures indicate that the electrons uniformly distribute themselves around the ring. All the C—C bonds are of equal strength and length. The circle inside indicates the uniformity of these six bonds in the ring.

9.11 a. Answers will vary. An example is that polypropylene fibers are used in rope and as an additive to concrete to reduce cracking in case of earthquakes.

b. Polypropylene is resistant to oils.

9.12

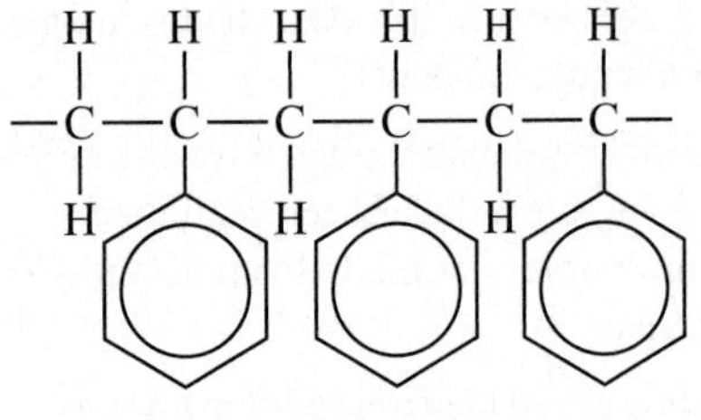

Head-to-tail

Head-to-tail is favored because the phenyl groups are large and would have strain if they were immediately next to one another.

9.13 a.

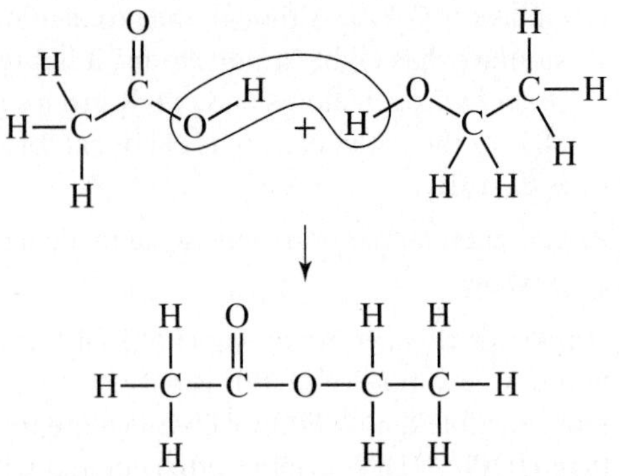

b. No, the reaction ends after the carboxylic acid and alcohol react. In the final product, there are no remaining carboxylic acid or alcohol groups to continue reacting to make another ester group.

9.14

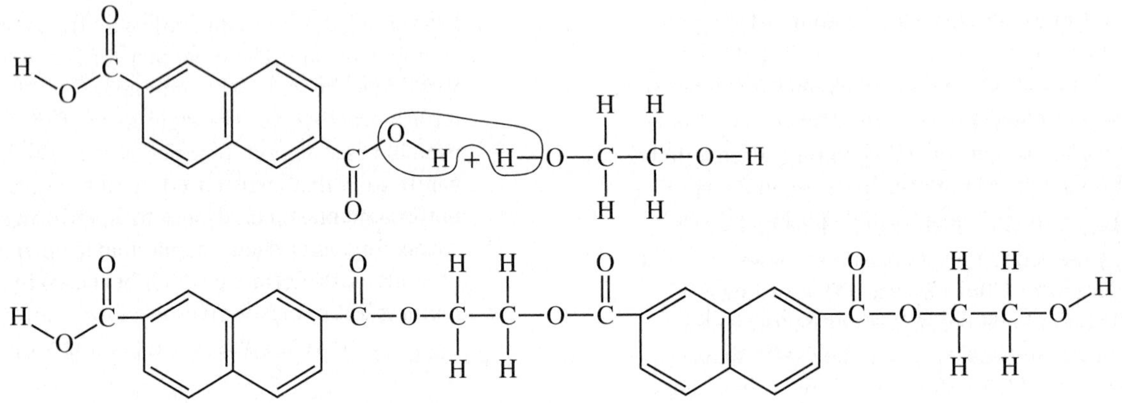

9.15

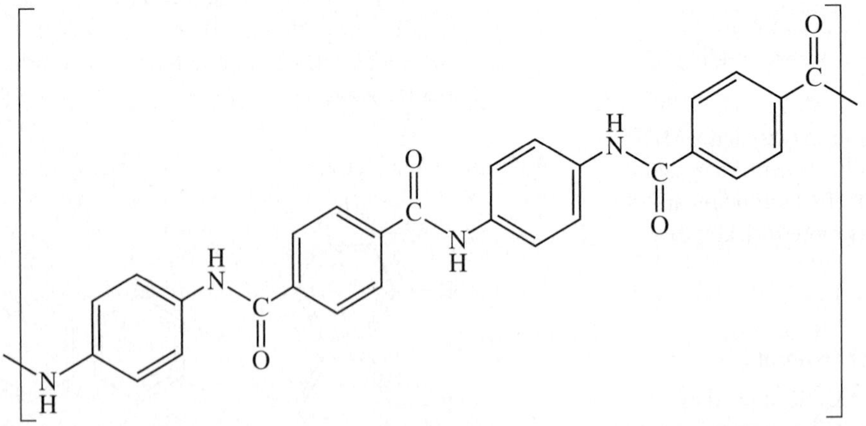

9.16 a. Answers will vary. Some less obvious sources of plastic include Styrofoam or reusable grocery bags composed of polyesters.

b. Answers will vary (but hopefully you are recycling more than you have thrown away!).

9.17 Answers will vary. Many different types of stores offer programs where someone could bring their own reusable bag.

9.18 Answers will vary depending on their plastic-use journals.

9.19 a. $2500\ C_2H_4 + 7500\ O_2 \longrightarrow 5000\ CO_2 + 5000\ H_2O$

b. Incomplete combustion produces CO and particulate matter (soot)—both air pollutants.

c. Answers will vary. Plasticizers are often added to impart flexibility to a polymer. However, biodegradable plasticizers can be added so that the plastic will biodegrade. Fillers can be added to reduce the total cost of the material. Pigments can be added to polymers to achieve desired colors. All of these diverse additives could be released as harmful environmental pollutants during combustion.

Real Talk (Plastics in the Environment)

a. Although a micron is 1/1000 of a mm, the term "microplastics" refers to plastic particles that are smaller than 5 mm across. A variety of plastic materials may decompose into microplastics; plastic food containers and baby bottles have been shown to shed microplastics in hot water. This link describes the main sources of microplastics, which are primarily from synthetic textiles, tires, and city dust.

b. Microplastics have been found virtually everywhere on Earth—in deep oceans, in the ice of both Arctic and Antarctica, in shellfish, and even in the air and rain.

c. Research is still needed to determine the health effects of microplastics. It is suspected that microplastics may have similar effects as particulate matter pollution described in Chapter 2. In addition, many additives in polymers are toxic and interfere with endocrine (hormonal) systems; these effects may be worsened by the presence of small particulates if they are not rapidly cleared from our bodies.

d. Global regulations have only recently been proposed and many more are in development. Hence, it is too soon to determine if these regulations have resulted in any measurable improvements in marine environments. In March 2022, the United Nations Environmental Assembly (UNEA) hosted its 5th meeting with representatives from 175 countries to discuss a legally binding international "Plastic Treaty" to address plastic pollution by 2024. In 2019, the European Chemical Agency (ECHA) proposed extensive restriction on microplastics in consumer products such as cleansers, cosmetics, and fertilizers. Both the United States and Canada have passed federal laws banning plastic microbeads, a type of microplastic. In 2015, the U.S. passed the Microbead-Free Waters Act that prohibits the manufacture, packaging, and distribution of cosmetic products that contain plastic microbeads. In 2018, Canada amended the Canadian Environmental Protection Act of 1999 to include microbeads in toiletries regulations. With the ban of single-use plastics including plastic bags, stir sticks, straws, beverage six-pack holders, cutlery/plates, and food packaging, Canada is aggressively moving toward a zero plastic waste policy by 2030.

9.20 a. Nonpolar chemicals such as oils can soften HDPE.

b. Examples: cooking oil, shoe polish, alcohols, vinegar, and lighter fluid.

9.21 a. Durable goods: 18.51%; Nondurable goods: 28.13%; Containers/packaging: 53.91%

b. Answers will vary. Durable goods may include computers, TV sets, office chairs, carry-on luggage, lawn furniture. Nondurable goods may include disposable silverware, mechanical pencils, Post-it notes. People may not recycle durable items because they are large and difficult to dispose of. People also may not know that they are recyclable. People may not recycle nondurable goods because they seem flimsy or not worth the effort because they are small. Some products have a coating or are painted, which hides the recyclable parts of the product.

9.22 a. $1753 + 1112 + 1.4 + 36.6 + 2.6 = 2906$ million pounds recycled, or 1.45 million tons. This is significantly lower than the values shown in Your Turn 9.21. However, the values reported by the EPA were not broken down into plastic bottles, which were listed in Table 9.4.

b. Answers will vary. The metric ton could have been used for international comparisons, as it is based on the metric system. ACC is a U.S.-based company, and the U.S. tends to use pounds rather than metric units such as tons.

9.23 Answers will vary. Purchase and recycle items might include soda bottles, shampoo bottles, milk jugs. Recycled-content products might include notebooks, egg crates, construction paper. An item could fall into both categories.

9.24 The four plastics could be placed into a saturated solution of $MgCl_2$. PET would sink, but all of the PVC, HDPE, and PP would float on the surface.

9.25 a. Answers will vary depending on the five items they found.

b. Answers will vary depending on the five items they found and their own personal preferences and usage.

9.26 Answers will vary. The assumption is that the use of plastics will double again in the next 20 years and that recycling rates will remain low. Another assumption is that wildlife populations will remain the same over time and not increase. The steps that can be taken to prevent this scenario are described as (1) create an effective after-use plastics economy, (2) drastically reduce the leakage of plastics into natural systems and other negative externalities, and (3) decouple plastics from fossil feedstocks. One threat to realizing this "new plastics economy" is the fact that current plastics are not "bio-benign."

REFLECT (PLA Degradation)

a. PLA is compostable.

b. The chemical reaction only occurs at the material's surface. Cutting the plastic increases its surface area.

c. Sodium hydroxide breaks the ester groups linking together the lactic acid monomers.

9.27 a.

carboxylic acid

hydroxyl
group/alcohol

b. A condensation reaction because the OH from the carboxylic group and the H from the hydroxyl group form a water molecule, which is a product.

c.

9.28 Answers will vary depending on your campus.

9.29 a. Two advantages of plastic jugs: They are lighter and cost less to transport. They are recyclable. Two disadvantages of plastic jugs: Cannot be recycled again into milk jugs because of bacteria from milk. Most milk jugs are not recycled. Two advantages of glass bottles: They can be sterilized to be reused. There is a lot of sand in this world that is available to make more glass. Two disadvantages of glass bottles: It takes a lot of energy to create glass. Glass bottles are heavy and would add to transportation costs.

b. Beer will go flat faster in a plastic bottle than a glass bottle because plastic is more porous than glass. Chemicals can also leach from the plastic to the beer. Soft drinks are sold in plastic bottles because they are often found in vending machines and they would break if they were glass. The cost of transporting them would also outweigh the money made from selling them.

9.30 a. CFCs are linked to the reduction of stratospheric ozone in the atmosphere, creating the "ozone hole."

b. CFCs break down in the atmosphere to release a chlorine free radical. The free radical acts as a catalyst to break down ozone into oxygen. The reaction propagates and results in another chlorine free radical until a termination step happens. If CFCs stay in the atmosphere for 100 years or more, the initiation of this process could continue for that long.

c. Until a termination step is executed, a reaction can happen for a long time. In the case of CFCs, this involves breaking down ozone. In the case of polymers, this involves adding monomers together to create long polymer chains.

d. Answers will vary. One answer could be that plastics are so versatile and are critical for use in existing consumer products. Many industries have evolved and adapted to using plastic. Phasing out plastic would create a huge financial and economic burden for many industries.

e. Answers will vary. One answer could be that we cannot sustain our current use of plastics because we will run out of space for disposal. Because plastics do not decompose quickly, they will end up stacking up in landfills or floating in the oceans. Fossil fuels are the source of many plastics, and these are not unlimited commodities.

9.31 a. An ester is created when an alcohol and a carboxylic acid react. They are characterized by a C—O—C bond and a C=O group on one of the carbons.

b.

c.

Chapter 10

@HOME (What's in a Mouthful?)

Answers may vary. One possibility is the traditional diagram with sour on the sides of the tongue, bitter on the back, and salty/sweet on the front. However, one may have very sensitive palates, and others may have a very dulled sense of taste.

10.1 Research published in *Nature* shows that taste buds across the entire tongue can sense all of the tastes, instead of tastes limited to regions of the tongue. There is also an additional taste called umami, which is a savory taste. This will serve as a good discussion with "foodies" about the sensitivity of taste buds and mixing of flavors to achieve delicious results in meals.

REFLECT (What Affects Our Sense of Taste?)

a. Answers will vary.

b. Scientists are still trying to understand this effect. Some possible reasons are discussed here.

@HOME (Taste Testing!)

Answers will vary based on their various food combinations.

@HOME (How Does Aroma Affect Taste?)

Answers may vary. Some may be able to correctly guess the identity of each sample without smelling it, but a majority will be incorrect in their guesses. One may discuss experience with texture and a difficulty discerning between the tastes they are sensing. For the jelly bean example, one may be able to tell that the samples are overwhelmingly sweet, but without a sense of smell they will report difficulty in discerning flavors.

@HOME (A Chocolate Taste Test!)

White, milk, and dark chocolate each contain different amounts of cocoa beans, thus having different amounts of cocoa butter (fat) and cocoa. This gives them different consistencies/textures and flavors. If the chocolate is allowed to melt in your mouth, you will notice that each melts differently. The higher the percentage of cocoa beans, the lower the amount of sugar and the chocolate will taste less sweet.

REFLECT (The Different Types of Chocolate)

a. Answers will vary depending on preferences.

b. One possible website that lists the benefits of either dark or milk chocolates: https://toakchocolate.com/blogs/news/dark-chocolate-vs-milk-chocolate-which-is-better

c. One possible website that describes the sustainability of chocolate production: https://news.mongabay.com/2022/08/delectable-but-destructive-tracing-chocolates-environmental-life-cycle/

REFLECT (The "Best" Chocolate Chip Recipe)

Answers will vary as each person has a different favorite part of a chocolate chip cookie. Crunchy cookies have more flour, sugar, and baking soda (and overall a greater mass of ingredients) than chewy cookies. To test the effects of each ingredient or combination of ingredients, one could make several batches of cookies that vary in the individual and combinations of ingredients. The only thing that can be confirmed after the experiment is the effects of each ingredient on the outcome, not which is the best—that is subjective!

10.2 The entire recipe needs to be multiplied by four. This yields 2 cups butter, 4 cups chocolate chips, 2 cups brown sugar, 2 cups white sugar, 4 eggs, 2 teaspoons vanilla, 5 cups flour, 3 teaspoons baking soda, and 1 teaspoon salt.

10.3 a. If you use all of the cheese and have an unlimited supply of tortilla shells, you could make seven "perfect quesadillas."

b. If you have only eight tortillas, you can make four "perfect quesadillas" and will use 200 g of cheese. The tortillas will be consumed first, and there will be 150 g of cheese left over.

@HOME (Sugar Cookies!)

a. It depends on the starting number of eggs, sticks of butter, etc.

b. More butter and eggs will likely affect the consistency and cooking time of the cookies. It likely won't

solidify properly, which will require a longer cooking time to work.

10.4 Answers may vary. The number of grams per cup/teaspoon/tablespoon is dependent on the ingredient, because each ingredient has a different density. 1.5 kilograms of apples is roughly 12.5 cups. 150 grams of sugar is roughly ¾ cup. 25 mL of cornstarch is roughly 16 grams and this converts to 2 tablespoons. 4 mL of cinnamon is roughly 1¼ teaspoons. 0.75 mL of salt is between ¼ and ⅛ teaspoon. 0.75 mL of nutmeg is between ¼ and ⅛ teaspoon. 40 grams of butter is roughly 3 tablespoons. This illustrates how using metric units is more accurate and leads to replicable results!

10.5 The average densities for flours are as follows: all purpose flour = 0.55 g/mL, cake flour = 0.51 g/mL, whole wheat flour = 0.55 g/mL, potato flour = 0.68 g/mL, corn flour = 0.68 g/mL, and almond flour = 0.38 g/mL. Potato flour and corn flour have the highest densities. (These calculations used 236 mL = 1 cup, and 16 tbsp = 1 cup.)

10.6 If you do not add salt, it will only change the flavor of the pasta, despite what you may have been told (salt does not make water boil fast; if anything, it boils at a higher temperature, but only if you add a significant amount of salt to the water). If you cook the pasta longer than the *al dente* cook time, it will become mushy. The cooking time will be longer at higher altitudes because the boiling point of water is lower at high altitudes.

@HOME (From Ice to Steam . . .)

a. The heating curve should look like that in Figure 10.4.

b. The molecules of water are undergoing a phase change during the plateaus.

10.7 The spaghetti noodles in Denver will take longer to cook than in Los Angeles. This is because the boiling point of water is lower in Denver; thus, the noodles need a longer time to cook.

10.8 *Sous vide* cooking allows for consistent temperature and texture throughout the meat. However, it can take longer to cook this way and would use more electricity. Boiling food is a relatively fast process, but it can only be used with foods that aren't messy in water. Boiling points also vary by altitude, so there can be variability between locations. Pressure cookers reduce cooking time and result in less electricity being used. However, the pressure inside a pressure cooker varies with altitude and can result in variability of outcomes. Cooking with heat/flame allows for the Maillard reaction to occur, which adds to the taste and texture of food. However, some of the by-products of the Maillard reaction or from charred food are carcinogenic. Using coals or flames can also produce by-products that are pollutants. Answers may vary as to which method is the most environmentally stable depending on the focus of their responses.

10.9 a. *Modernist Cuisine at Home* notes that the vast majority of *sous vide* bags are made with high-density polyethylene (HDPE), low-density polyethylene (LDPE), or polypropylene.

b. According to the Pacific Northwest Pollution Prevention Resource Center, it has not been confirmed that harmful chemicals leech into food from the bags. However, the Center notes that oily and acidic foods could have the potential to increase leeching. Research studies do show the migration of chemicals, but not the exact formulas to find toxicity, nor do the studies include the *sous vide* specifically. The temperatures of *sous vide* are not high enough to break down the polymer.

c. Inexpensive bags and films in the kitchen can have more harmful leeching than *sous vide* bags because they can contain plasticizers and other polymer additives, which aren't healthy to ingest.

10.10 a. Microwave regions have the longest wavelength and lowest energy of the three. UV has the shortest wavelength and highest energy of the three. IR falls in between them. Microwaves cause the molecules to rotate. IR waves cause molecules to vibrate and stretch. UV can have enough energy to break bonds.

b. Their claims are not believable. Radio waves have a longer wavelength and even less energy than microwaves. Microwaves already have inconsistent heating because they cannot penetrate food all the way. Radio waves would be able to penetrate even less. This would result in even more inconsistent heating than with microwaves. Additionally, it would take longer to cook, and thinner foods would need to be used.

REFLECT (Is Microwave Cooking Dangerous?)

A microwave oven uses waves to penetrate the food and rotate the molecules. A conventional oven heats the air around the food and the heated air cooks the food through conduction. Microwaves do not cook food from the inside out, but the molecules at a certain depth inside the food are excited by the waves, which makes it seem like they are cooking from inside out.

@HOME (Fresh Basil Preparation)

a. Answers will vary.

b. Drying in air may lead to spoilage due to oxidation, which may cause sickness.

REFLECT (Sushi Preparation)

Common acids used as preservatives include benzoic acid, acetic acid, sorbic acid, and propionic acid. All are weak organic acids that contain the carboxylic acid functional group.

@HOME (Quick Pickling!)

a. The salt prevents bacteria growth in the pickling solution.

b. A video that describes the differences between quick and fermentation pickling: https://www.youtube.com/watch?v=CgAiUwHR-fQ

10.11 Answers will vary depending on the foods selected. One may have had cured meats, smoked fish, pickled gherkins, or dried fruits. This could be a good point of discussion about the shelf-life of the different foods and even if different cultures or regions have higher prevalence of different methods.

10.12 a. 125 °F; lower than 145 °F on foodsafety.gov

b. 135 °F; lower than 145 °F on foodsafety.gov

c. 145 °F; equal to 145 °F on foodsafety.gov

d. 150 °F; above 145 °F on foodsafety.gov

e. 160 °F; above 145 °F on foodsafety.gov

f. 165 °F; equal to 165 °F on foodsafety.gov

g. 165 °F; equal to 165 °F on foodsafety.gov

h. 140 °F; lower than 145 °F on foodsafety.gov

10.13 17.8 °Bx is equal to 17.8% (w/w). This converts to 17.8 grams of sucrose per 100 g of solution. Recall from Section 5.5 that molarity is moles of solute per liter of solution. To find the moles of solute (the sucrose), 17.8 grams of sucrose is:

$$17.8 \text{ g} \times \frac{1 \text{ mol}}{342.3 \text{ g}} = 0.052 \text{ moles sucrose}$$

To determine the volume of the solution in liters, use the mass and density of the solution:

$$100 \text{ g of solution} \times \frac{1 \text{ mL}}{1.06 \text{ g}} \times \frac{1 \text{ L}}{1000 \text{ mL}} = 0.094 \text{ L}$$

So, $\dfrac{0.052 \text{ moles sucrose}}{0.094 \text{ L solution}} = 0.55$ M sucrose solution.

10.14 Some of the cans are floating and some are at the bottom. Regular soft drinks contain sugar and diet soft drinks contain sugar substitutes that are much sweeter than sugar. They require less sweetener because of this. Regular soft drinks are therefore more dense than diet and will sink in water. The diet soft drink cans will float in water. One may suggest that the amount of carbonation or density of other ingredients in the soft drink could affect the outcome. Or the amount of air between the can and liquid could affect whether the can sinks or floats if two cans of regular soft drinks differ.

REFLECT (The Mentos Fountain)

a. As the temperature is increased, more CO_2 would escape the liquid, resulting in a more dramatic Mentos fountain.

b. No. As seen in Section 5.7, "like dissolves like." Hence, if a polar gas were dissolved in a polar liquid, it would not be as desirable for the gas to escape the liquid. In contrast, for the Mentos fountain demo, a nonpolar gas (CO_2) is dissolved in a polar liquid (water).

10.15 The concentration of CO_2 in the bottle is
$C = 1.25$ atm $\times$ (0.031 mol/L $\cdot$ atm) = 0.039 M.
This means there are $\dfrac{0.039 \text{ mol}}{1 \text{ L}} \times 0.500$ L = 0.019 moles of CO_2. Converting to grams, 0.019 mol $\times \dfrac{44 \text{ g}}{1 \text{ mol}} = 0.84$ g of carbon dioxide are dissolved in the Coca-Cola.

@HOME (Fun with Yeast!)

a. The balloon was used to trap the carbon dioxide produced in the reaction. The sugar was consumed by the yeast to produce carbon dioxide via fermentation.

b. Warm water allows the fermentation process to occur more quickly. Carbon dioxide would still be produced in cold water, but it would be much slower.

REFLECT (How is Whiskey Made?)

a. Both ethanol and methanol contain London dispersion intermolecular forces. These forces are stronger for ethanol since it has a larger molar mass. Hence, the boiling point for ethanol will be higher than methanol. If both alcohols are present in the same solution, the methanol will be removed first followed by ethanol.

b. Flavor molecules are typically polar, which should dissolve in a polar solvent such as water or ethanol. This is an application of the "like dissolves like" principle.

c. Answers will vary. This website provides a summary of some sustainability efforts related to whiskey production.

@HOME (Caffeine Extraction)

a. Caffeine is a white granular powder when extracted from coffee.

b. You could weigh the extracted caffeine and compare to the weight of original coffee.

c. For example, temperature and extraction time.

Chapter 11

REFLECT (Food Choices)

Answers will vary. Although fruits and vegetables are healthy choices for our diet, these foods may not be grown in an environmentally sustainable manner.

11.1 a. For both men and women, the highest obesity rates are found in American Samoa, Cook Islands, Nauru, Palau, and Tuvalu. However, these countries have small populations, which may contribute to a higher % obesity rate. The highest obesity rate for women in more populated nations is found in Egypt (52%), Turkey (50%), South Africa (50%), the U.S. (47%), and Algeria (46%). For men, the highest levels of obesity are in the U.S. (47%), Saudi Arabia (41%), Canada (39%), the U.K. (37%), and Argentina (35%). Between 2020 and 2030, it is predicted that obesity among women will increase by 3% and obesity among men will increase by 2% globally.

b. In the U.S., Kentucky and West Virginia have the highest levels of obesity, followed by many mid-western states. The levels of obesity for African American adults was generally higher than either Hispanic or non-Hispanic Caucasian populations. Obesity is generally lower for those with higher levels of education and those who are young adults. The level of obesity appears to increase with age.

11.2 a.

b. carboxylic acid group, —COOH

11.3 a. Oleic, linoleic, and linolenic acids all have at least one C=C double bond.

b. It is a saturated fatty acid.

11.4 Answers will vary. Sample saturated and unsaturated fatty acids are shown.

Lauric acid, saturated fatty acid

$$CH_3CH_2CH_2CH_2CH_2CH_2CH_2CH_2CH_2CH_2CH_2-C\overset{\displaystyle O}{\underset{\displaystyle OH}{\parallel}}$$

Oleic acid, monounsaturated fatty acid

$$CH_3CH_2CH_2CH_2CH_2CH_2CH_2CH_2CH=CHCH_2CH_2CH_2CH_2CH_2CH_2CH_2-C\overset{\displaystyle O}{\underset{\displaystyle OH}{\parallel}}$$

Double bonds change the structure so that they are less able to line up and have intermolecular interactions. Because there are fewer intermolecular forces between the molecules, they are more easily separated, and the melting points are lower.

11.5 **a.** A molecule of octane has fewer carbon atoms than a fat or oil. Octane contains neither a carboxylic acid functional group nor any C=C double bonds. In contrast, fats and oils may contain these groups.

b. A biodiesel molecule has a similar number of carbon atoms to fats and oils. Biodiesel contains a carboxyl group (C=O), like fats and oils, but no C=C bonds.

11.6 **a.** It is likely soybean oil because it contains mostly polyunsaturated fats and nearly equal amounts of saturated and monounsaturated fats. This matches the composition of soybean oil, having large amounts of linoleic acid (a polyunsaturated fat).

b. Answers will vary depending on background knowledge, but the correct answer is that vitamin E is part of the oil itself.

REFLECT (The Chemistry of Olive Oil)

a. Extra virgin olive oil has gone through minimal processing to extract the oil from olives. In general, this oil has been cold-pressed, is unfiltered, and has not been (or minimally) heated during processing. The process of cleaning and heating has been shown to strip away some of the oil's flavors and some antioxidants that are naturally present. As a result, extra virgin olive oil is richer in vitamins, polyphenol-based antioxidants, and other natural ingredients, but generally has a shorter shelf life.

b. Olive oil is composed of about 73% monounsaturated, 11% polyunsaturated, and 14% saturated fat.

11.7 **a.** I Can't Believe It's Not Butter has the highest percentage of saturated fat (2 g/9 g = 22%). It is still lower than butter (7 g/11 g = 64%).

b. 9% (1 g/11 g) of the total fat in butter is polyunsaturated.

c. Answers will vary. Unsaturated fats are healthier than trans fats and saturated fats. Butter contains 36% unsaturated fats, Land O' Lakes 55%, I Can't Believe It's Not Butter 61%, and Benecol 81%.

d. Answers will vary. One sample answer is the introduction of more sugar into the diet to cover the taste that is missing from removing fats from the diet, which leads to obesity and health issues.

@HOME (Fatty Foods)

a. Answer will vary depending on the size of food items used. A hot dog will likely contain the most fat per unit mass among these items.

b. The fat is saturated in these items, which requires heating to be removed from the food item.

11.8 **a.** The design reduces or eliminates the use or generation of hazardous substances. It uses fewer resources and uses renewable resources.

b. Either chemical or enzymatic interesterifications will produce similar products, so there wouldn't be sufficient difference in structure or properties to justify the need to indicate the processing technique on a label.

11.9 **a.** Answers will vary. They may contain descriptions of the hydrogen bonding between the sugar and the receptors on the taste buds. Different molecules have different properties, including the degree of sweetness.

b. They are all broken down the same way in the body and thus provide the same amount of energy per gram. Energy per gram is dependent on the chemical bonds being broken and formed. Each of these compounds can be represented by the same chemical formula, releasing the same amount of energy.

11.10 Answers will vary. One answer could include: Yes, soft drinks have as much, if not more, added sugar than a candy bar. A soft drink can contain up to 40 grams of sugar per can. A candy bar such as Snickers has 47 grams of sugar. They are comparable. One could also compare a diet soft drink with a candy bar and say that it is not a fair characterization because it contains artificial sweeteners, not sugar.

11.11 **a.** The summary of data shows that non-Hispanic Asian individuals consumed fewer Calories from sugar than Hispanic and non-Hispanic White and Black individuals. The number of Calories from sugars in the total diet declined as age and income increased in adults. For children, the number of Calories from sugars in the total diet increased with age.

b. Answers will vary, but generally, if 5–15% of the Calories should come from added fats and sugars. The values across all groups for added sugars alone were very high.

c. Answers will vary. In the United Kingdom, it is recommended that no more than 5% of the Calorie intake be from added sugars each day. The World Health Organization (WHO) recommends less than 10% of the Calorie intake be from added sugars each day. Discrepancies may be due to local food ingredient regulations or lifestyles.

d. Answers will vary.

REFLECT (The Chemistry of Artificial Sweeteners)

a. More scientific research is needed to determine whether artificial sweeteners pose any negative health impacts. Some health risks that have been purported for these sweeteners include: obesity, heart disease, diabetes, stroke, and high blood pressure.

b. A sweet taste corresponds to binding of the molecule to receptor sites on our tongues. Hence, this doesn't mean that the molecules are metabolized. In fact, these sweeteners are not metabolized by our bodies, so do not result in any Calories. A nice summary of this may be found online at https://www.scientificamerican.com/article/how-can-an-artificial-swe/.

11.12 a. Intermolecular forces among the amino acid residues are responsible for this overall shape. Serine, asparagine, and glutamine contain —OH or —NH$_2$ groups can hydrogen bond to others in another part of the chain; other nonpolar amino acid residues are attracted to one another by London dispersion forces.

b. A yellow-colored atom is sulfur. The yellow atom is found within the group —CH$_2$—CH$_2$—S—CH$_3$, which is methionine.

11.13 Gly-Gly-Gly, Gly-Gly-Ala, Gly-Ala-Gly, Ala-Gly-Gly, Gly-Ala-Ala, Ala-Gly-Ala, Ala-Ala-Gly, Ala-Ala-Ala

REFLECT (How Proteins Build Muscle)

a. Answers will vary. The recommended range of protein consumption was from 0.8 g to 1.8 g of protein per kg of body weight.

b. The use of steroids increases the formation of new muscle fibers and growth of muscle cells, known as hypertrophy.

REFLECT (Is Aspartame Safe?)

a. The concentration of methanol in fruit juice ranges from 1–640 ppm. In contrast, alcoholic beverages can contain up to 7200 ppm.

b. A recent review is found here. The consumption of aspartame has been linked to a variety of health risks such as skin problems, neurodegeneration, and even autism, which are not limited to those with phenylketonuria.

11.14 Folic acid is water-soluble because it is a polar molecule (—NH and —OH functional groups in the molecule make it polar).

REFLECT (Vitamin Supplements)

a. The most common types of supplements include vitamins, minerals, botanicals and herbs, and amino acids.

b. The U.S. currently regulates vitamin supplements as food items, not drugs. However, in contrast to prescription and over-the-counter drugs, dietary supplements do not normally need approval from the U.S. Food and Drug Administration (FDA) prior to being marketed. Around the world, this report provides a comprehensive comparison of supplement regulations around the world.

11.15 a. Answers will vary. They may include wrinkle creams, body balms, and lotions.

b. It is thought that the antioxidant vitamin E protects the skin from free radicals, which damage the skin.

c. Answers will vary. An example claim could be that taking antioxidants has no real preventive or therapeutic value unless the deficiency is your problem.

11.16 a. A megadose is an amount far above the average daily recommended amount (usually considered to be more than twice the recommended daily amount). For adult men, the recommended daily dose is 90 mg, and for adult women, it is 75 mg.

b. Answers will vary. According to an NIH study, long-term daily supplements in large doses do not appear to prevent colds. It could reduce the duration of colds. Taking too much vitamin C can decrease the amount absorbed in the body.

c. The observed skin discoloration is due to the fat-soluble component called carotene. It collects in the body rather than being excreted daily.

d. Answers will vary.

@HOME (Iron in Cereal)

a. The iron collected from a known mass of cereal could be weighed and compared with the original mass of cereal.

b. Vegetables high in iron content such as lentils or crushed multivitamins should yield a measurable amount of iron.

11.17 Answers will vary depending on the items chosen.

11.18 a. I-131 is absorbed in the bloodstream and collects in the thyroid gland. Once there, it destroys thyroid cells, thus leading to a reduction in the function of the thyroid gland.

b. Answers will vary. Risks are that the patient has to take medication for the rest of their life because the cells in the thyroid gland are permanently damaged. Salivary glands may be permanently damaged from the treatment. The benefits are decreased recurrence of hyperthyroidism and decreased mortality.

c. The half-life of I-131 is 8.0197 days. If it takes 10 half-lives to be "gone" from the body, that would be 80.197 days.

11.19 The FDA states that "low-fat" should have 3 grams or less total fat per serving. For it to be low in saturated fat, it has to be 1 gram or less and 15% or less of Calories from saturated fat. This would meet the guidelines for total fat but not saturated fat (10 Calories out of 50 Calories is greater than 15%).

11.20 a. The Basic Four were vegetables and fruits, milk, meat, and cereals/breads. The Food Pyramid indicated that every day we should have 6–11 servings of bread/cereal/rice/pasta, 2–4 servings of fruit, 3–5 servings of vegetables, 2–3 servings of milk/yogurt/cheese, 2–3 servings of meat/poultry/fish/dry beans/eggs/nuts, and should use fats/oils/sweets sparingly. MyPyramid doesn't offer serving suggestions but rather an abstract view of amounts: the most being grains, then equal amounts of vegetables and milk, then fruit is smaller in size, then protein even smaller, and a very small sliver for oils and sweets. It also includes steps to indicate the importance of exercise and watching what you eat. MyPlate divides a plate into largely vegetables and grains and smaller amounts of fruits and proteins. There is also a glass to indicate dairy.

b. Answers will vary. Some claim that MyPlate has flaws because it is missing fats and oils. Others claim that it is unrealistic to have a meal that fits those guidelines. Another claim is that it doesn't highlight healthier choices over others (red meat vs. chicken/fish) and is missing daily servings. A benefit could include that it is easier to interpret with its pie chart format.

11.21 a. No, males require more Calories than females for all ages for the same activity level. This is because males have a greater mass compared to females. The more mass you have, the more energy you need.

b. The estimated Calorie requirement decreases with age.

c. Answers will vary depending on the countries chosen. For the most part, answers will be similar but may be in different units; for example, some countries report food-energy content in joules.

11.22 1524 Calories: $\left[\dfrac{350 \text{ Cal}}{1 \text{ burger}} \times 2 \text{ burgers}\right]$

$+ \left[\dfrac{108 \text{ Cal}}{1 \text{ oz}} \times 3 \text{ oz}\right] +$

$\left[\dfrac{175 \text{ Cal}}{4 \text{ oz}} \times 8 \text{ oz}\right] + \left[\dfrac{100 \text{ Cal}}{8 \text{ oz}} \times 12 \text{ oz}\right],$

or around 3 hours $\dfrac{1524 \text{ Cal}}{490 \text{ Cal/h}}$.

11.23 Answers will vary.

Real Talk (Monitoring Food Safety)

a. Answers will vary depending on the recent event selected.

b. Answers will vary depending on the recent event selected.

c. Answers will vary depending on how concerned one is about food safety. Analytical techniques can be used to identify foodborne pathogens. Regulations can be tightened to decrease instances of food safety issues.

11.24 a. Answers will vary depending on the chemicals chosen. Some examples from this list are classified as GRAS (generally recognized as safe).

b. Answers will vary depending on the chemicals chosen.

11.25 Answers will vary.

11.26 a. If the yield in grain per acre increases, there will be less land necessary to bring one kilogram of beef to the table. If the yield in grain per acre decreases, more land is necessary to bring one kilogram of beef to the table.

b. Early in life, beef cattle may graze before heading to the feedlot. Estimates for land use are higher if more of a cow's life span is included. *Note:* Depending on the land quality and the practices of the farmer or rancher, a cow may require from a few acres to more than 30 acres of grazing land.

c. The livestock breeds were likely combined to find an average for this data. There will be no measurable effect on the estimate.

11.27 a. Answers will vary. Some possible supporting/counter examples may include: 1. Supporting your local farmers at a farmers' market vs. supporting farmers in impoverished countries who need money to support their families; 2. Fresh fruit have no preservatives and taste more natural; food items without preservatives have a much shorter shelf life; 3. Vine-ripened fruit is tasty and juicy; allowing fruit to ripen after harvest may better ensure that it is properly ripened in the grocery store; 4. Fruit that is consumed during its local season will be of highest quality; northern regions will need to wait months to consume favorite foods such as peaches, grapes, etc. and some parts of the world will never be able to

grow certain produce (e.g., bananas in Canada);
5. supporting local farmers will cause them to
be more thoughtful about their environmental
footprint so their crops remain productive; as local
farmers make more income, they too will look to
expand their production, which will devote more
land and water use to agriculture. Other parts of
the world may have a more responsible use of land
and water than local vicinities.

b. Answers will vary.

c. Answers will vary.

11.28 a. $5 \text{ lb C} \times \dfrac{453.591 \text{ g}}{1 \text{ lb}} \times \dfrac{1 \text{ mol C}}{12.01 \text{ g/mol C}}$

$= 188.84 \text{ mol C}$

$188.84 \text{ mol CO}_2 \times \dfrac{44.01 \text{ g CO}_2}{1 \text{ mol CO}_2} \times \dfrac{1 \text{ lb}}{453.591 \text{ g}}$

$= 18.32 \text{ lb CO}_2 \ 20 \text{ lb to 1 sig fig}$

b. $1000 \text{ mi} \times \dfrac{1 \text{ gal}}{30 \text{ mi}} = 33.33 \text{ gal}$

$\dfrac{33.33 \text{ gal}}{x \text{ lb CO}_2} = \dfrac{1 \text{ gal}}{18.32 \text{ lb CO}_2}$

$x = 610.67 \text{ lb CO}_2$ in one year saved 600 lb CO_2
to 1 sig fig

Assumptions are how many miles per gallon are
used. A sample answer is to use 30 miles per
gallon.

11.29 a. Answers will vary but may include: Reduce the
amount of transportation necessary, reduce the
packaging of materials, and make the remaining
packaging recyclable.

b. Answers will vary but may include: Carbon
footprint calculations assume that all consumed
resources can be tracked/quantified. It assumes
that all acres are equivalent.

c. Answers will vary but may include eating less
meat and/or eating food grown locally.

11.30 The bond energy for the triple bond in N_2 is very high
(946 kJ, nearly double) compared to the O=O bond
in O_2 (498 kJ) and the O—H bond in water (467 kJ).
O=O and O—H bonds have similar bond energies to
one another.

11.31 a. Ammonia is soluble in water because it is very
polar and can form hydrogen bonds.

b. When ammonia is mixed with water, the
ammonium ion (NH_4^+) and a hydroxide ion (OH^-)
are formed ($NH_3 + H_2O \longrightarrow NH_4OH$). The
NH_4OH is soluble in water, so the [OH^-] increases
in the solution, the solution is basic.

c. Nitrates (NO_3^-)

d. Nitrites (NO_2^-)

11.32 Answers will vary depending on the favorite food
chosen.

REFLECT (What Is "Vertical Farming"?)

Benefits: Because vertical farming takes place in a
controlled environment, no insects or garden pests
attack growing plants. Additionally, fewer crops will
be lost to inclement weather, such as drought, hail,
and frost. Also, the carbon footprint related to food
transportation will be reduced because vertical farms
are located nearby city centers. Perhaps the greatest
advantage is water usage. Conventional farming uses
approximately 70% of the world's drinkable water,
which becomes polluted by fertilizers and pesticides.
In contrast, vertical farming uses minimal water
resources; even water from transpiration is reused
for irrigation. Disadvantages: Fewer jobs are needed
relative to conventional farming. Pollination costs
are higher for vertical farming because there are no
insects to pollinate the crops. Also, vertical farming
depends on the electrical grid; if a farm loses power
for a day, it will result in a huge loss in production. If
fossil fuels are used to power vertical farms, the net
environmental effect may be negative. Given costs,
this technology may not be feasible for lower-income
countries—transportation costs will still be needed
to ship produce to these countries.

a. The Association for Vertical Farming's website
has a clickable map showing vertical farming
projects' locations.

b. Cost and efficiency of lighting systems, significant
energy resources, and land use is still required in
metro areas, where land is not readily available.

11.33

$$\begin{array}{c} \text{Br} \\ | \\ \text{H}-\text{C}-\text{H} \\ | \\ \text{H} \end{array}$$

When the molecule is in the stratosphere, UV light
causes methyl bromide to break apart into $\cdot CH_3$ and
a bromine radical. The bromine radical reacts with
ozone to produce oxygen, similar to how chlorine
reacts, thus depleting the ozone layer.

11.34 Many possible answers exist, such as acts of
terrorism, financial collapse, climate change, rising
oil prices, herbicide-resistant weeds, etc.

11.35 a. Answers will vary depending on their initial
rankings.

b. Answers will vary depending on the changes they
are planning to make, if any.

c. Answers will vary, but according to a 2012 report
by the NRDC, 40% of food in the U.S. is unused.
Some practices to reduce food waste may include
cutting irregularly shaped products into more
desirable products (ugly carrots vs. baby carrots),
selling at farmers markets, donating food to food
banks, reevaluating sell-by/best-by dates to be
more accurate, having a bargain-shelf for nearly
expired food, etc.

Chapter 12

REFLECT (Delving into Medicinal Chemistry)

Answers will vary. Some possibilities include medication (treated a symptom), surgery (cured a problem), pacemaker implant (prevented a future problem).

REFLECT (Visualizing Equilibrium)

a. The water level in one container drops and the level increases in the other container until their levels are equivalent.

b. Yes, the size of container will influence how long it takes to reach equilibrium (the same water level in both containers). If a small container would be connected to a large container, it will take longer to reach equilibrium, since the water level would take longer to increase in the larger container. You would also need to pour water more slowly to prevent overflowing the first container.

c. The two containers are analogous to the reactants and products; the tube is related to the pathway between reactants/products. A larger-diameter tube implies that equilibrium is reached faster (a faster reaction occurs); a smaller-diameter tube refers to a longer time to reach equilibrium since the water flow would be impeded (the conversion between reactants/products occurs more slowly).

@HOME (Snap-Bead Equilibria)

a. Shaking simulates the random motion of the reactants and products in the container.

b. $2A \longleftrightarrow A_2$

12.1 a. The balanced reaction must follow the language of the sentence closely and show glucose as the reactant and fructose as the product:

$$\text{glucose} \longrightarrow \text{fructose}$$

The two compounds can be written in chemical formulas, but (as this is a structural rearrangement only) the difference between reactant and product is unclear.

$$C_6H_{12}O_6 \longrightarrow C_6H_{12}O_6$$

b. The equilibrium constant must follow the format of concentration of product over concentration of reactant. Again, as the chemical formulas do not show us the important structural differences between glucose and fructose, we should use chemical names.

$$K_{eq} = \frac{[\text{fructose}]}{[\text{glucose}]}$$

c. Here, we are given the concentrations of our product and reactant already at equilibrium. We input these values into the formula we have generated above to calculate our equilibrium constant.

$$K_{eq} = \frac{[\text{fructose}]}{[\text{glucose}]} = \frac{4.45 \text{ mM}}{6.02 \text{ mM}} = 0.739$$

REFLECT (Le Châtelier's Principle)

a. There are many examples; for instance, the Haber process of making ammonia ($N_2(g) + 3 H_2(g) \longrightarrow 2 NH_3(g)$) that may be tweaked by changing the pressure, temperature, and concentrations, or by adding a catalyst. As another example, Le Châtelier's principle operates to control the equilibrium responsible for a pH-buffered blood system involving H_2O, CO_2, and carbonic acid (H_2CO_3).

b. No. The value of the equilibrium constant remains the same, but the position of the equilibrium (toward reactants or products) changes.

c. If the volume increases, the pressure decreases, and vice versa. As the pressure increases, the equilibrium shifts to the side (reactants or products) with fewer gaseous molecules.

d. No. It speeds up both forward and reverse reactions equally.

12.2 We are discussing the release of epinephrine from the receptor, similar to what we have seen in Figure 12.3. The reactant side is the bound complex, whereas the free receptor and epinephrine are the two products. The K_{eq} constant (or given its special name in biochemistry the K_d) value is very low (and much lower than 1). Referring back to Figure 12.1, this means that the reactants are heavily favored, so the complex must be favored over the free state. This means that the epinephrine prefers to stay bound.

12.3 From the context given, we must decide what is necessary for the relative K_{eq} values of the *release of oxygen* from hemoglobin versus myoglobin. For this transfer to happen, hemoglobin must prefer to release oxygen at the same conditions that myoglobin prefers to stay in complex with oxygen. This means that hemoglobin must have a higher K_{eq} than myoglobin.

12.4 a. $CH_3COOH + H_2O \rightleftharpoons H_3O^+ + \boxed{CH_3COO^-}$;

$$K_a = \frac{[H_3O^+][CH_3COO^-]}{[CH_3COOH]}$$

b. $H_2CO_3 + H_2O \rightleftharpoons H_3O^+ + \boxed{HCO_3^-}$;

$$K_a = \frac{[H_3O^+][HCO_3^-]}{[H_2CO_3]}$$

c. $CH_3NH_2 + H_2O \rightleftharpoons OH^- + \boxed{CH_3NH_3^+}$;

$$K_b = \frac{[OH^-][CH_3NH_3^+]}{[CH_3NH_2]}$$

12.5 a. The pH of water is 7.00, the pH of a strong acid is 2.00, the pH of a weak acid is 4.50, the pH of a strong base is 12.00, and the pH of a weak base is 9.50. Conductivity–water does not conduct. Strong acid and strong base conduct very well (strong), and weak acid and weak base conduct weakly.

b. Conductivity increases as strong acid or strong base concentration increases. It does the same, though not as strongly for weak acids and bases. There are more dissociated ions in the strong acid and strong base solutions.

REFLECT (Acids and Buffers)

a. $\text{pH} = -\log(7.1 \times 10^{-4}) + \log\left(\frac{0.70}{0.30}\right) = 3.52$

b. $\text{pH} = -\log(7.1 \times 10^{-4}) + \log\left(\frac{0.78}{0.22}\right) = 3.70$

12.6 **a.** Phosphoric acid and its conjugate base, dihydrogen phosphate, is an excellent mixture for stabilizing a solution at pH 2.5 because it is within one pH unit of its pK_a at 2.1.

b. The dihydrogen phosphate ion, with its conjugate base, the hydrogen phosphate ion, is an excellent buffer for stabilizing a solution at pH 7.4 because it is within one pH unit of its pK_a at 6.9.

c. Carbonic acid or dihydrogen phosphate ion would be a good buffer for stabilizing a solution at pH 6.8 because they are within one unit of their pK_a at 6.3 and 6.9, respectively.

d. Acetic acid or carbonic acid would be a good buffer for stabilizing a solution at pH 5.4 because they are within one unit of their pK_a at 4.8 and 6.3, respectively.

12.7 Answers will vary but may include: Octane, pentane, methane, ethene, and ethanoic acid are organic compounds. Inorganic compounds include sodium chloride, copper(II) nitrate, hydrogen bromide, calcium carbonate, and nitric acid. Contexts will vary depending on experiences.

12.8 **a.** Each carbon makes four bonds and contains no lone pairs in all of the common arrangements presented. In some cases, it has four single bonds (involving eight electrons); in others, it has a double bond and two single bonds (involving eight electrons); and in some cases, it has a single bond and a triple bond (involving eight electrons). They all follow the octet rule.

b. Carbon monoxide does not form four bonds (but still follows the octet rule).

12.9 All structures satisfy the octet rule. The lone pairs of nonbonding electrons are not shown for clarity.

a.
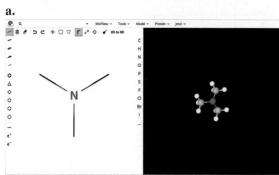

b.
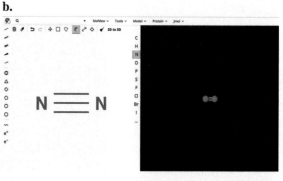

c.
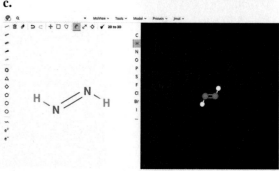

d.
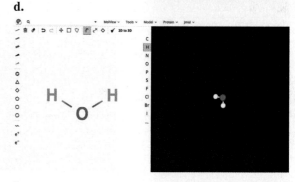

e.
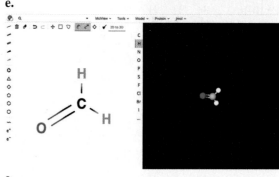

f.
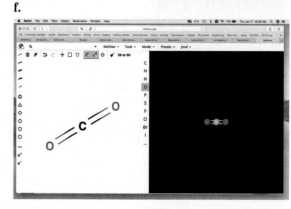

g.

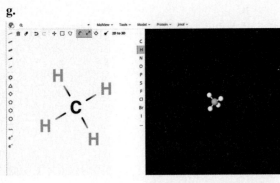

h.

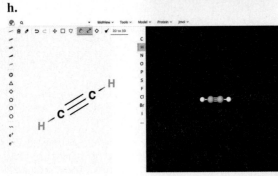

i.

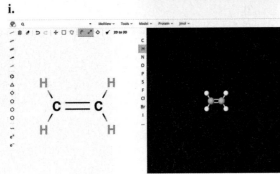

12.10 a.

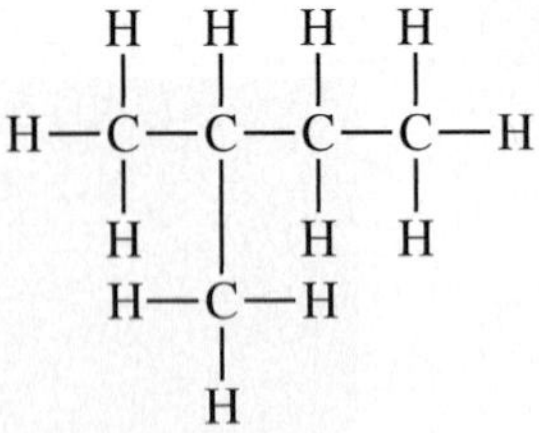

H—C—C—C—C—H with H's and H—C—H, H branches

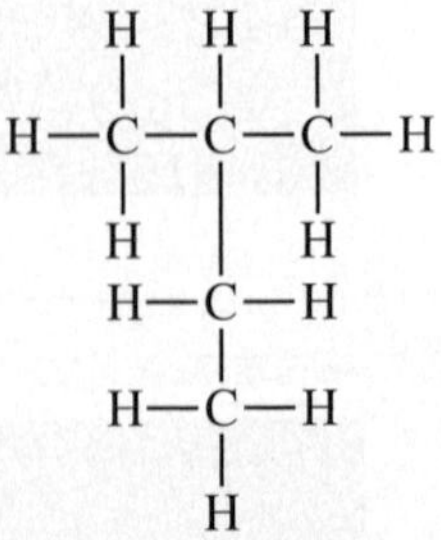

b.

H—C—C—C—H with branches H—C—H, H—C—H, H

c.

H—C—H top, H—C—C—C—C—H, H—C—H bottom H

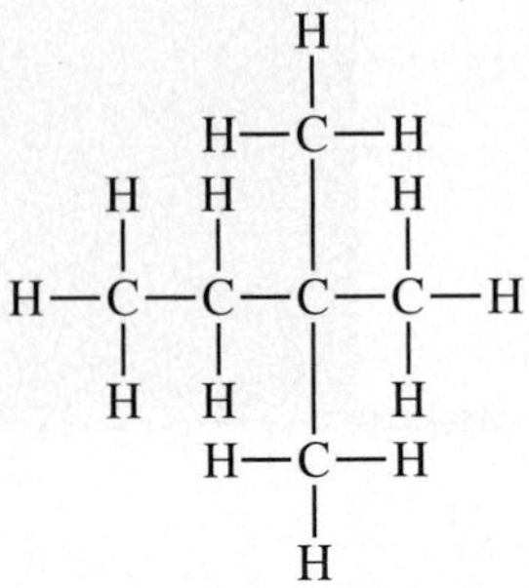

12.11 a.

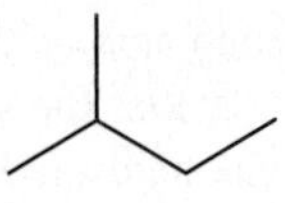

b.

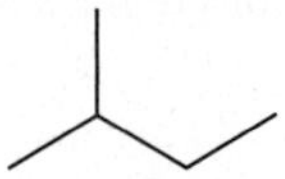

c.

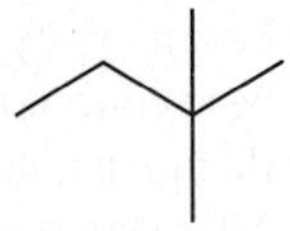

12.12 a. Yes, *n*-butane and isobutane are structural isomers that have the same chemical formula C_4H_{10}.

b. No, *n*-hexane and cyclohexane are *not* isomers. They do not have the same chemical formula. The ring structure eliminates two of the hydrogens of hexane.

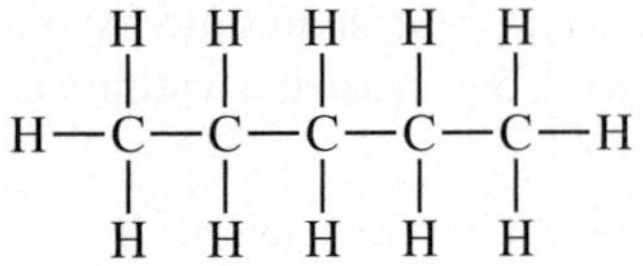

H—C—C—C—C—C—H (with H's)

$CH_3CH_2CH_2CH_2CH_3$

c.

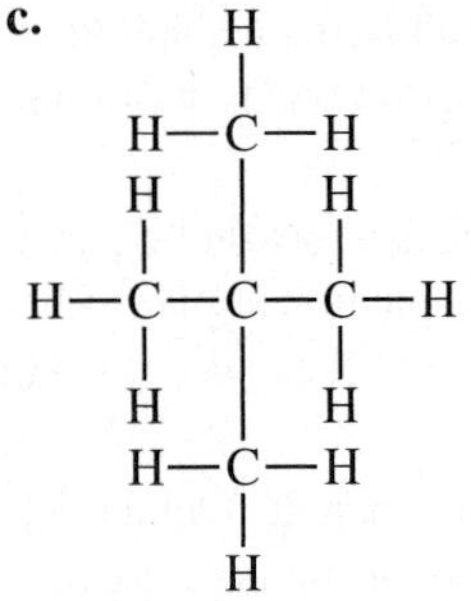

H—C—H; H—C—C—C—H; H—C—H; H

$CH_3C(CH_3)_3$

H—C—C—C—C—H with H—C—H branch

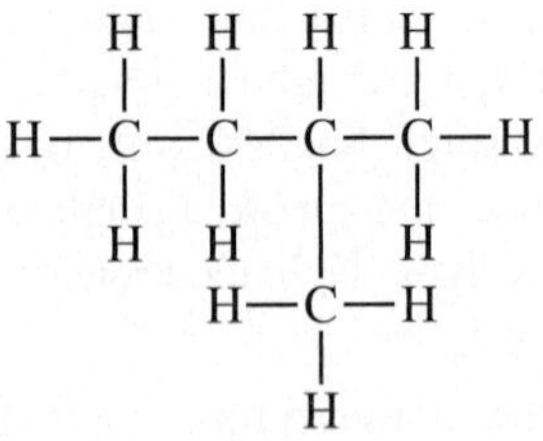

$CH_3CH(CH_3)CH_2CH_3$

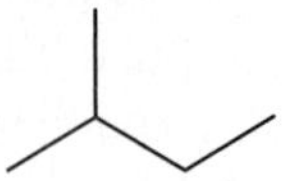

REFLECT (Functional Groups)

a. ketone

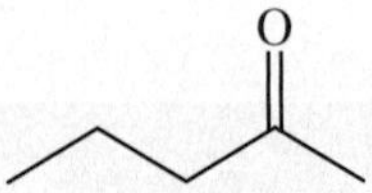

b. alcohol OH

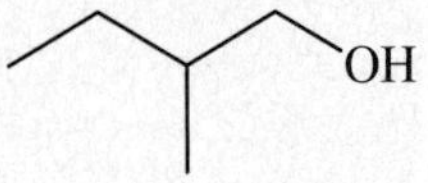

c. amine NH_2

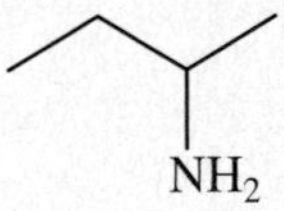

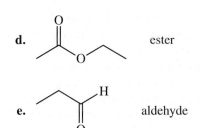

d. ester

e. aldehyde

12.13 Epinephrine differs from dopamine in that there is an additional —OH on the carbon chain and only an NH (not NH_2) followed by another carbon. Their chemical formulas only differ by one carbon, two hydrogens, and one oxygen. Since this molecule has many —OH and —NH groups, it would be quite polar and would therefore NOT be soluble in fatty tissue since this is nonpolar.

12.14 a.

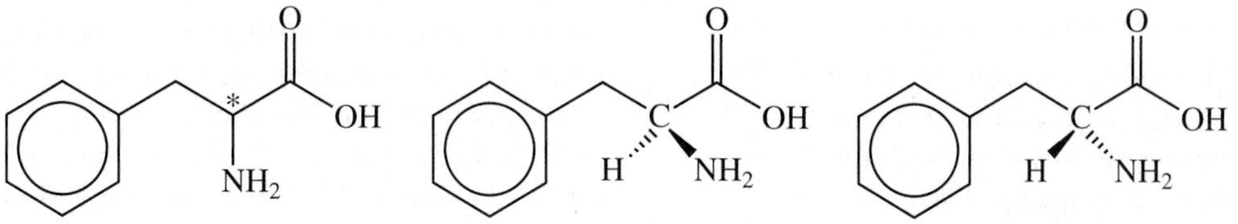

b.

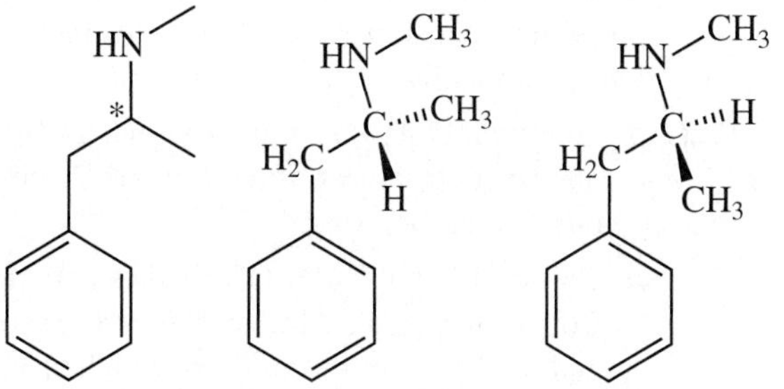

c.

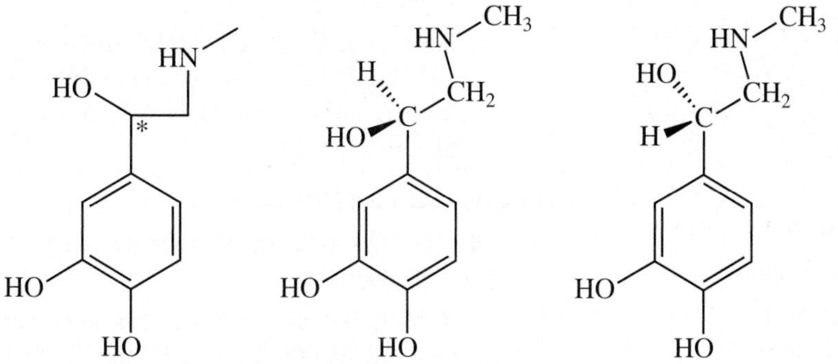

d.

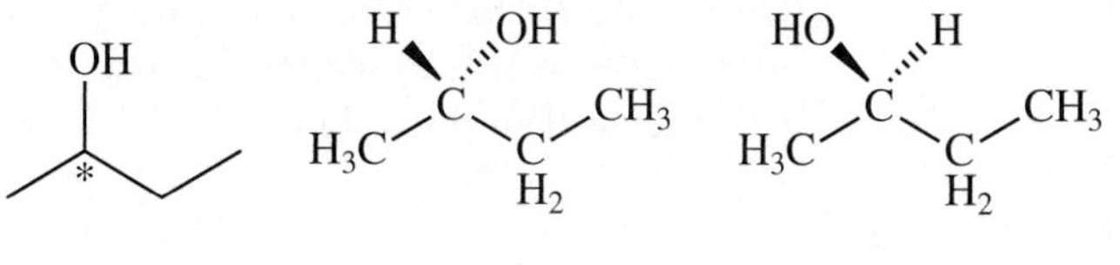

12.15 a.

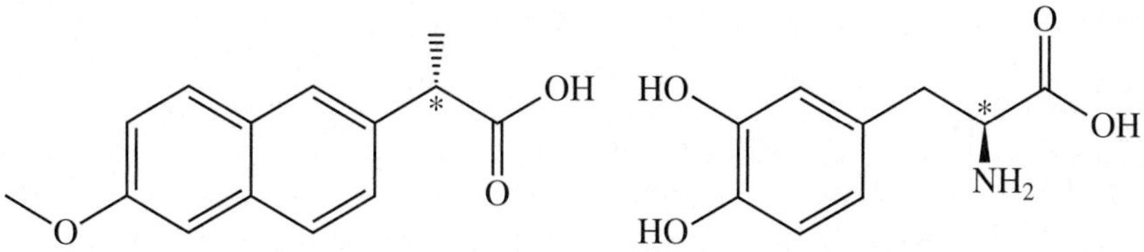

Naproxen   L-DOPA

b. Both drugs contain a carboxylic acid functional group. Naproxen also contains an ether functional group. L-DOPA also contains two alcohol (hydroxyl) functional groups and an amine functional group.

c.

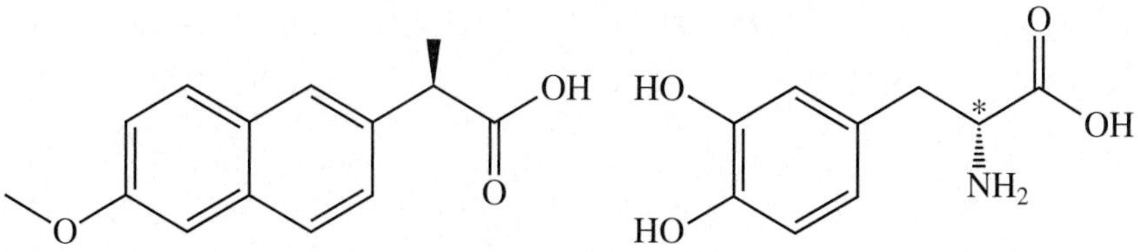

12.16 **a.** $C_6H_{12}O_6 + 6\,O_2 \longrightarrow 6\,CO_2 + 6\,H_2O$

b. 10 grams of glucose = 0.055 mole:

$10\ \text{g} \times \left(\dfrac{1\ \text{mol}}{180.2\ \text{g}}\right)$. To find the number

of moles necessary for oxygen:

$0.055\ \text{mol glucose} \times \dfrac{6\ \text{mol O}_2}{1\ \text{mol}} = 0.33\ \text{mol O}_2.$

$0.33\ \text{mol O}_2 \times \dfrac{32\ \text{g O}_2}{1\ \text{mol O}_2} = 11\ \text{grams of O}_2.$

12.17 Answers will vary depending on their selection within the Molecule of the Month archives.

12.18 Previously, we have seen discussions of polyamides as polymers in Chapter 9—as artificial polymers like Kevlar in bulletproof vests, nylon, and natural polymers like silk spun by spiders. Finally, we have also seen proteins discussed as a vital macronutrient in Chapter 11 and more-specific food examples in the white of an egg or the collagen melted during the smoking of meats in Chapter 10.

12.19 Answers will vary depending on the two amino acids chosen. A sample answer is a glycine and alanine. The dipeptide is nonpolar because glycine and alanine both have nonpolar side chains.

REFLECT (The Dangers of Synthetic Drugs)

a. A comprehensive website describing Kratom is: https://nida.nih.gov/research-topics/kratom

b. These drugs work by preferentially binding to receptor sites, which kicks off the synthetic drug thereby reversing the effects of an overdose.

12.20 A key concept in this chapter is that molecules interact at multiple points. A strong and selective interaction should take advantage of all of epinephrine's functional groups: three hydroxyls and an amine. We could expect that the negatively charged side chains of acidic amino acids would interact strongly with the amine group because it can act as a base and become positively charged. Polar (neutral as well as charged) amino acids should be able to hydrogen bond with the two hydroxyl groups.

12.21 Answers will vary. The structures may differ in only one carbon and four hydrogens, but the structures are very different regarding functional groups. Estradiol contains a benzene ring with one alcohol group attached, whereas testosterone contains cyclohexene with a ketone and an additional methyl. Because of these structural differences, they have different properties that change which proteins they interact with specifically.

12.22 **a.** Estradiol: $C_{18}H_{24}O_2$; progesterone: $C_{21}H_{30}O_2$. Both have ring structures: three six-carbon rings and one five-carbon ring.

b. Corticosterone: $C_{21}H_{30}O_4$; cortisone: $C_{21}H_{28}O_5$. Both have ring structures: three six-carbon rings and one five-carbon ring. They both also contain ketone and alcohol groups.

c. Cholic acid: $C_{24}H_{40}O_5$; cholesterol: $C_{27}H_{46}O$. Both have ring structures—three six-carbon rings and one five-carbon ring. They both contain alcohol functional groups.

Real Talk (Vaccine Discovery)

a. The FDA describes the various steps involved in drug development.

b. As of 2021, the median launch price of a new drug in the U.S. is $180,000 for a year's supply. This cost is then transferred to the consumer through high drug prices. Other countries directly regulate drug prices, which keeps their prices lower.

12.23

REFLECT (Common Drugs, Common Structural Features)

The structures of ibuprofen and acetaminophen are shown below.

Ibuprofen   Acetaminophen

a. All three molecules have a benzene ring and a C=O or carbonyl functional group. All three have two (and only two) groups attached to the benzene ring. As differences, only aspirin has an ester, and only acetaminophen has an amide. In addition, only acetaminophen has an alcohol group, and only ibuprofen has a nonpolar carbon chain.

b. Aspirin: pain, fever, inflammation. Ibuprofen: inflammation and fever. Acetaminophen: pain and fever. They all treat fever.

c. Aspirin: rash, nausea, stomachache. Ibuprofen: nausea, bloating, headache, rash, ear ringing. Acetaminophen: nausea, dark urine, jaundice. Nausea is a common side effect.

12.24 The most popular and common example of a chance drug is Viagra: It was researched to treat high blood pressure initially. Botox is an injection used for cosmetic purposes but can also treat excessive sweating and migraines. A reference website that contains many examples is ncbi.nlm.nih.gov/pmc/articles/PMC3181823.

12.25 Answers will vary depending on the chosen properties. Answers may include targeted treatment, no side effects, no accumulation in the body, no unwanted chemicals when it is metabolized, etc. Their motivations, similarities, and difference on how they might change will vary.

12.26 a. These models allow the viewer to see the orientation of the atoms in three dimensions rather than as dashes and wedges. The elements are also color coded. For some, this may dramatically change how they consider the position of functional groups relative to each other.

b. It helps explain the shape of the molecule, where points of the molecule have space for a reaction to occur, and where it is too crowded for a reaction to occur. It also allows for the view to rotate the molecules in different directions to see

how they might interact with other molecules. Disadvantages may include that you can't physically hold the molecules and that specific technology is necessary to view them. It may also lead the viewer to believe that a molecule's shape is very static even when it is not. Both 2D and 3D models have limitations as compared to real molecules. They do not have all of the information in a molecule, but rather the information chosen to be portrayed in the model. The scale may also be a limitation, but that is also a benefit because we cannot see atoms with the naked eye.

12.27 Gamma radiation is the best to use for detection through tissues and skin because the rays can pass through the layers of tissue necessary to generate images of structures deep inside the body. Beta emissions penetrate the skin similarly but will not go as deeply into tissues. Positron emissions are similar in size to beta emissions. Beta and positron-emitting isotopes are often used due to a concurrent gamma emission. Alpha radiation is the worst to use for detection through tissues and skin because it doesn't have much penetrating power. A simple sheet of paper can stop it.

12.28 a. A salt containing iodate or iodide.

b. Thyroid cancer, lower heart rates, goiters, etc. One may list any of the symptoms of hypothyroidism because the thyroid would be damaged.

12.29 a. It slows metabolism and induces fatigue. Without enough iodide, the thyroid gland can't produce enough thyroxine.

b. If Megan doesn't keep up with her hormone therapy, she risks being tired.

c. It's important to ensure hormone therapy levels are closely monitored to best mimic the naturally occurring gland.

Chapter 13

REFLECT (Genetic Traits)

Answers will vary based on family traits of eye color, hair color, tongue rolling, ear lobe connectivity, or others.

13.1 a. One key idea: it is better to prevent waste; animal slaughter would have caused a lot of waste. Another key idea: It is better to minimize the amount of materials used; 8000 pounds of animal pancreas are required to produce a single pound of insulin.

b. The issue with using animal insulin is that it needed to be highly purified because animal and human insulin might be incompatible. It caused inflammation around the injection site and other more serious complications. Bacterial insulin does not cause these problems because it is chemically identical to human insulin.

13.2 a.

Thymine

Cytosine

Guanine

Adenine

b. Any H atom attached to an N atom could form a hydrogen bond.

c. Any O and N atom could participate in hydrogen bonding.

13.3 a. Deoxyribose formula: $C_5H_{10}O_4$. Ribose formula: $C_5H_{10}O_5$.

b. Ribose has three carbons in the ring with an alcohol functional group. In deoxyribose, the middle carbon of those three has only an H atom instead.

c. Ribose fits this pattern, but deoxyribose does not.

d. The oxygen and hydrogen atoms that are part of the alcohol functional groups can participate in hydrogen bonding.

13.4 a. Very close to 0.9 (90%) of H_3PO_4 and 0.1 (10%) of $H_2PO_4^-$.

b. About 0.99 (99%) HPO_4^{2-} and 0.01 (1%) PO_4^{3-}.

c. About 0.99 (99%) $H_2PO_4^-$ and 0.01 (1%) of HPO_4^{2-}.

d. Very close to 0.1 (10%) HPO_4^{2-} and 0.9 (90%) of PO_4^{3-}.

e. Very close to 0.5 (50%) of HPO_4^{2-} and 0.5 (50%) of $H_2PO_4^-$.

13.5

@HOME (Extracting DNA From Strawberries!)

a. The strawberries were mashed prior to DNA extraction to increase the surface area & improve extraction efficiency. Also, mashing helps break open cells in the strawberries, where the DNA is located.

b. The detergent breaks apart the fats/lipids and proteins that make up the cell membranes, which allows DNA to be extracted. Salt is used to break apart the bonds between the coiled DNA strands and the protective membrane of the cells. Rubbing alcohol is used to precipitate DNA from the extraction mixture.

REFLECT (X-Ray Diffraction)

a. The wavelength of X-rays is about the same size as individual atoms. Hence, X-rays are diffracted as a result of collision with planes of sample atoms.

b. Synchrotron radiation sources produce ultrahigh-energy X-rays, which are used to determine the structures of complex molecules such as proteins and other macromolecules. The benefits of using synchrotron radiation is its high intensity, tunable wavelength and high penetrating ability, which allows for more structural detail to be gleaned from a sample relative to the much weaker cathode tubes used in lab-scale instruments.

@HOME (A Candy Model of DNA)

a. That representation does not show the phosphate or sugar backbone. It also oversimplifies how the strands are a double helix, making them more circular. This representation also does not show the phosphate or sugar backbone at the molecular level.

b. The artistic representation highlights the base pairs that consistently complement one another, how they fit together, and how the pairs appear to rotate as the helix turns. The color coding of the base pairs is a beneficial visual aid. The candy model also uses color coding but it does not show how they fit together.

c. Answers will vary.

13.6 a. Proteins have 20 possible amino acids, and DNA has four possible base pairs.

b. Answers may vary.

c. 20 amino acids have many more degrees of freedom than four base pairs, so it is not strange that proteins can have many more configurations than DNA.

13.7 a. TATGGACG

b. CTAGGAT

REFLECT (Visualizing DNA)

a. DNA replicates with the help of special enzymes (helicases) to break it apart and others (polymerases) to make copies of the DNA.

b. Jet engines cycle at a rate of about 3000 revolutions per minute or 60 revolutions per second. In the video, probably 2–3 base pairs were replicated every second. So it would need to speed up by at least 20–25 times to replicate what real DNA would do.

13.8 a. The eighth base (G) is where the error occurs; the complement shows a T when it should be a C.

b. The mismatched pair will be less stable than a correct pair because there would not be as much hydrogen bonding. Additionally, the sizes of T and C are different, which would cause a disorganized orientation of base pairs.

13.9 a. $\dfrac{0.34 \text{ nm}}{1 \text{ base pair}} \times \dfrac{1 \text{ m}}{1 \times 10^9 \text{ nm}} \times \dfrac{1 \times 10^2 \text{ cm}}{1 \text{ m}}$

$\times \dfrac{1.35 \times 10^8 \text{ base pairs}}{\text{Chromosome 11}} = \dfrac{4.6 \text{ cm}}{\text{Chromosome 11}}$

b. $4 \text{ } \mu m = 0.0004 \text{ cm}$

$4.6 \text{ cm} = 46000 \text{ } \mu m$

$\dfrac{46000 \text{ } \mu m}{4 \text{ } \mu m} = 11,500$

c. Humans have 23 pairs of chromosomes, for a total of 46 chromosomes. The compaction needs to occur; otherwise, they wouldn't fit.

13.10 $4 \times 4 \times 4 \times 4 = 4^4 = 256$ different four-base sequences.

13.11 This allows for possible mistakes in the genetic code to still code for the same amino acid.

REFLECT (Cellular Protein Factories)

a. 2% codes for proteins, and 30% regulates those proteins. The rest is unknown what it does.

b. Sickle cell anemia is a substitution point mutation. A missense mutation happens when a substitution changes the amino acid. A nonsense mutation happens when a substitution causes a premature stop codon. Both are disruptive, but a premature stop codon would cause more nonfunctional proteins.

c. Answers will vary.

13.12 a. Both contain $C{=}O$ and $N{-}H$ groups.

b. Typically, nylon is synthesized by the reaction of a monomer containing two carboxylic acids with a monomer containing two amine groups. For example, refer back to Figure 9.15. In contrast, proteins are synthesized from a monomer (an amino acid) that contains one carboxylic acid and one amine group.

c. They are both condensation polymers.

13.13 a. The polar amino acids are green, and the nonpolar amino acids are yellow.

b. Most of the amino acids are yellow or nonpolar. It would work ok in either environment—it looks about equal in polar and nonpolar residues.

c. When changing the protein to be in oil, little appears to change, but the nonpolar resides should try to work their way to the outside, and the hydrophilic residues should try to work their way to the middle.

d. The hydrophobic protein will completely unfold in the oil and bunch up in the water. Yes, this is what I expected because the oil and hydrophobic residues are attracted to each other. They are trying to get as compact as together with the water because they are repelling it.

e. The hydrophilic protein will completely unfold in the water and bunch up in the oil. I expected this because the water and hydrophilic residues attract each other. They are trying to get as compact as possible in the oil because they are repelling it.

13.14 Answers will vary. For example, one may draw hydrogen bonding between polyamide molecules or Kevlar to illustrate intermolecular hydrogen bonding (hydrogen bonds that occur between two separate molecules). One may also draw proteins that form α-helixes or β-pleated sheets to illustrate intramolecular hydrogen bonds (hydrogen bonds that occur within a molecule).

13.15 a. Valine and glutamic acid differ in their side chains. Although they contain three carbons, glutamic acid also has a carboxylic acid group at the end of that carbon chain.

b. Glutamic acid has a polar side chain, and valine has a nonpolar side chain. Glutamic acid would be predicted to have a higher solubility in water.

c. Valine is nonpolar and, thus, hydrophobic. This change to the surface of the molecule reduces its solubility and causes the sickle shape to result.

REFLECT (What Is "CRISPR"?)

a. Good references that describe how CRISPR/Cas9 works and its applications are NIH, the National Cancer Institute, and Origene Technologies.

b. Answers will vary. Some new applications include cancer treatment, allergy-free foods, biofuel production, and many others.

c. Plants can have their uptake of CO_2 increased to trap more CO_2. They can be designed to withstand pressures by becoming more resilient to temperature changes and water shortages. They can be produced to enhance the yield of plants, improve the nutritional value of food products, extend the shelf-life of fruits and vegetables, and enhance their natural defense mechanisms.

d. There have been borgs discovered that carry genes directly involved in methane oxidation. These are of immense importance to climate researchers because there are bacteria that can reduce methane, a potent greenhouse gas.

Real Talk (The CRISPR Revolution)

  a. The analogy of genetic scissors refers to the fact that the technique CRISPR with the Cas9 enzyme can cut a sequence of DNA at any base pair and extract this piece of DNA. This is helpful if you are trying to remove a protein from a cell. A transformative technology means that CRISPR will change how medicine is done.

  b. Answers will vary.

  c. A useful reference for CRISPR is located here. There are a couple of different ways to determine if your CRISPR worked. You can run gel electrophoresis to see if the DNA you meant to cut is gone. You can also sequence the DNA of the new cell to see if the gene has been cut. You will need to run PCR to get enough DNA to do this. You can also check the RNA by running a qRT-PCR. You can check the protein product by running a western blot or mass spectrometry.

REFLECT (GMO Regulations)

  a. The U.S. and Canada are the leading producers of GM foods, but most parts of the world are also producers, as seen on this website.

  b. CRISPR-modified plants are modified by scientists designing genetic code or introducing DNA from nature-generated genetic variations within the plant. They are not modified by inserting DNA from another organism. This eradicates the fear that foreign DNA is present.

  c. Answers will vary.

13.16 Answers will vary. New GMO labeling rules went into effect in the U.S. on Jan. 1, 2022. The new labels will feature "bioengineered" instead of "GMO," or a QR code that provides information about the food product. This latter option is not the best since consumers without cell phones would not be able to read the information. It is not known if and how enforcement would take place, as well as how the USDA defines GMOs. Around the world, the EU has more extensive rules regarding GM food labeling. Canada, a leading producer of GM crops, currently has no requirements for food labeling.

13.17 **a.** A good resource to track retracted papers is retractionwatch.com.

  b. Many websites claim GMOs are unhealthy or dangerous. For instance, the Institute for Responsible Technology lists the dangers of GMOs but doesn't include references to scientific studies. The Center for Nutrition Studies lists the possible dangers of GMOs with references to scientific literature.

  c. Answers will vary, related to whether one agrees with the report, which states that research findings thus far do not support claims that GM-based foods are less safe than foods derived from non-GM crops. Also, more research and regulations are needed to assess whether GM foods will threaten food security or other unintended results caused by genetic engineering.

Chapter 14

REFLECT (Forensic Evidence Collection)

  a. Survey the situation and secure the area; ensure the scene is safe by extinguishing the fire or apprehending the criminal.

  b. Carefully document the crime scene and interview witnesses.

  c. Carefully collect evidence using appropriate techniques; proper attire must be worn to prevent contamination of the crime scene.

14.1 **a.** The benefits of "smart" cancer-treatment drugs are that they have fewer side effects because they target the cancer cells without also targeting healthy cells.

  b. Phase I studies have healthy volunteers to determine the side effects of the drug. This phase gives information on metabolism and excretion of the drug. Phase II studies are conducted only if Phase I doesn't show high death rates or serious side effects. This phase is focused on determining whether the drug is effective in people who have a certain condition. Phase III studies begin only if sufficient evidence is shown that the drug works in Phase II. This phase is focused on studying effects of varied dosages, of being taken with other medications, and in different populations.

14.2 **a.** The hazmat crew should use either carbon dioxide or dry-chemical extinguishers (e.g., ammonium phosphate, sodium bicarbonate, potassium chloride), because the solvents are considered to be Class B and flammable. However, the sodium metal is considered to be Class D and also requires a dry-powder extinguisher (graphite or sodium chloride powders pressurized with nitrogen gas). Firefighters should not simply use water because it can spread the flammable liquid and fire or produce a violent exothermic reaction (explosion!) when it comes into contact with sodium.

  b. Sodium is a strong reducing agent. Each sodium atom donates an electron to a benzophenone molecule, resulting in a sodium ion and a ketyl radical anion. Using sodium metal to purify a chlorinated solvent such as dichloromethane will result in the formation of products that are shock-sensitive explosives!

14.3 A good resource for this question is the Fire Equipment Manufacturer's Association located here. Dry-powder extinguishers are used to either remove the heat from the fire or separate the fuel from the oxygen. Dry chemical extinguishers stop the chemical reaction and can act as a barrier between the fuel and the oxygen. Wet-chemical extinguishers remove heat from the fire and act as a barrier between the fuel and the oxygen. When barriers are created between the fuel and the oxygen, this can prevent the fire from being reignited.

14.4

H_3C—$\overset{H_2}{\underset{}{C}}$—$\overset{}{\underset{H_2}{C}}$—$\overset{H_2}{\underset{}{C}}$—$\overset{}{\underset{H_2}{C}}$—$CH_3$

H_3C—$\overset{}{\underset{H_2}{C}}$—$\overset{H_2}{\underset{}{C}}$—$\overset{H}{\underset{CH_3}{C}}$—$CH_3$

H_3C—$\overset{}{\underset{H_2}{C}}$—$\overset{CH_3}{\underset{}{CH}}$—$\overset{}{\underset{H_2}{C}}$—$CH_3$

H_3C—$\overset{H}{\underset{CH_3}{C}}$—$\overset{CH_3}{\underset{}{CH}}$—$CH_3$

H_3C—$\overset{CH_3}{\underset{CH_3}{C}}$—$\overset{H_2}{\underset{}{C}}$—$CH_3$

14.5 **a.** At the flash point, a chemical change occurs. At the boiling point, a physical change takes place.

b. Both combustible and flammable liquids can burn. However, it isn't the liquid itself that is burning but the vapor and air. They differ in the temperature at which they burn. Flammable liquids can ignite at normal temperatures, whereas combustible liquids ignite at temperatures above normal. The flash point is lower for flammable liquids than it is for combustible liquids.

c. It has been shown experimentally that flash points decrease with altitude, thus making substances more flammable at normal temperatures. This would mean that flammable liquids would need to be handled with more care and stored at lower temperatures when used at high altitudes.

14.6 Toluene: Has a flash point of 6 °C and boiling point of 111 °C; Class IB

Hexane: Has a flash point of −26 °C and boiling point of 69 °C; Class IB

Tetrahydrofuran: Has a flash point of −14 °C and boiling point of 66 °C; Class IB

Diethyl ether: Has a flash point of −45 °C and boiling point of 35 °C; Class IA

Dichloromethane: Has a flash point of 100 °C and boiling point of 40 °C; Class IIIA

Acetonitrile: Has a flash point of 2 °C and boiling point of 82 °C; Class IB

Ethanol: Has a flash point of 17 °C and boiling point of 78 °C; Class IB

Acetone: Has a flash point of −20 °C and boiling point of 56 °C; Class IB

14.7 **a.** Diethyl ether, acetone, tetrahydrofuran, hexane, ethanol, acetonitrile, and toluene will all catch fire at room temperature with the aid of a match. They have flash points that are less than room temperature.

b. None of the solvents found in Dr. Thompson's laboratory could combust without the presence of an outside source of ignition. The autoignition temperatures are all well above room temperature.

14.8 **a.** Some examples include:

Toluene

$C_7H_8 + 9\ O_2 \longrightarrow 7\ CO_2 + 4\ H_2O$

Hexane

$2\ C_6H_{14} + 19\ O_2 \longrightarrow 12\ CO_2 + 14\ H_2O$

Tetrahydrofuran

$2\ C_4H_8O + 11\ O_2 \longrightarrow 8\ CO_2 + 8\ H_2O$

Diethyl ether

$(C_2H_5)_2O + 4\ O_2 \longrightarrow 4\ CO_2 + 5\ H_2O$

Dichloromethane

$2\ CH_2Cl_2 + 3\ O_2 \longrightarrow 2\ CO_2 + 2\ H_2O + 2\ Cl_2$

Acetonitrite

$4\ C_2H_3N + 15\ O_2 \longrightarrow 8\ CO_2 + 6\ H_2O + 4\ NO_2$

Ethanol

$C_2H_6O + 3\ O_2 \longrightarrow 2\ CO_2 + 3\ H_2O$

Acetone

$C_3H_6O + 4\ O_2 \longrightarrow 3\ CO_2 + 3\ H_2O$

b. Particulate matter such as soot is likely to exceed air quality standards. Nitrogen oxides will also form and likely exceed air quality standards.

c. Compact fluorescent lightbulbs (CFLs) contain mercury vapor if cracked during a fire. The mercury contamination lasts many hours after the fire has been extinguished.

14.9 One good online resource for this discussion is https://www.interfire.org. If the material is easy to cut away, like a piece of wet carpet, it is cut away and stored. If the material is difficult to remove from the scene of the crime, an absorbent can be applied to the surface and the absorbent collected for analysis. Metal paint cans that are unlined are best for storage because they can be sealed. Plastic bags can be punctured and glass jars can be shattered. The samples are stored in cool places because of unknown flash points.

14.10 The highest temperatures of the sensors near the laboratory notebooks were about 185 °C. The autoignition temperatures of paper, cotton, and PVC are all higher than this value. Hence, it is not likely that these items burned without the use of an accelerant.

14.11

Caffeine

Toluene

Pyridine

n-heneicosane

n-octyl acetate

Least polar to most polar: *n*-heneicosane, toluene, *n*-octyl acetate, pyridine, caffeine.

The least polar molecules take longer to make it through the column. Thus, the first peak (2.5 min) is caffeine, the second peak (2.75 min) is pyridine, the third peak (8.5 min) is *n*-octyl acetate, the fourth peak (16.75 min) is toluene, and the final peak (19 min) is *n*-heneicosane.

14.12 The isotopic ratios for wood from matchsticks found in Dr. Thompson's laboratory and those found at the hit-and-run crime scene were similar and within the uncertainty ranges. However, this does not prove that the same person was involved but rather that the same type of matchstick was used by a person at both scenes. If the same person was at both scenes, it does add evidence that could place that person at the crime scene, but it is not enough to prove guilt.

14.13 His stomach was nearly empty, indicating a long period since he had eaten his last meal. The contents that were there included starch grains and caffeine. Answers will vary, but a potential meal could have been a piece of toast and coffee earlier that day. According to the company Intoximeters, the ethanol would have reached his bloodstream in approximately 30 minutes. The proportions of fat and water in his body, the alcohol concentration in the wine, whether he drank the wine in one gulp or in small sips, and whether he had eaten prior to drinking could have an effect on the metabolism of alcohol in the body.

14.14 Dr. Thompson's total cholesterol was borderline high. His HDL cholesterol was near the risk-factor range and not in the desirable range. His LDL cholesterol was high. His triglycerides were borderline high. Overall, he was at risk for heart disease.

14.15 Peak at 3350 is from OH stretching (hydroxyl functional group).

Peaks between 2870 and 2989 are from C—H bond stretching (alkane functional groups).

Peak at 1730 is a C=O bond (carbonyl functional group).

Peak at 1163 is due to N—H bending (amine functional group).

Broad band at around 1500 is due to aromatic C=C (phenyl functional group).

14.16 **1-** 2972: C—H

2- 1780.58: C=O

3- 1500.96: C=C aromatic

4- 1189.47: C—O

5- 1004.29: C—O

6- 554.28: C—H bending or C=C ring torsion

14.17 Tesla states the range as 405 miles for a 100-kWh battery. However, the speed of the car (faster speed = fewer miles), the outside temperature (higher temperature = more miles), and whether the air conditioning or heat is on (on = fewer miles) affect the battery range. It is unlikely that a driver could maintain exactly the speed and temperature conditions when driving on a highway for 140 miles each way to yield the ideal battery range. Despite the above factors leading to a lower range than ideal, the driver should have easily been able to make this trip without the need to recharge.

14.18 Answers will vary. One possibility is: a 5-kW solar charging station. If the Tesla battery is 100 kWh, it would take 20 hours to fully charge the battery.

@HOME (DNA Extraction)

a. Answers will vary depending on the sources chosen.

b. The surfactant is used to break apart the cell membranes needed to extract DNA. The rubbing alcohol is used to precipitate the DNA.

c. Cold water works best to keep the DNA intact during extraction.

d. Some impurities include other sources of proteins, carbohydrates, and lipids.

14.19 Density of water = 1 g/cm^3; density of phenol = 1.07 g/cm^3; density of chloroform = 1.49 g/cm^3.

a. When shaken, the water rises to the top and the phenol and chloroform become a mixture in another layer. Water is polar, phenol and chloroform are nonpolar and mix. The densities of the nonpolar substances are greater than the density of water, therefore that layer sinks.

b. The DNA will be found in the water layer because it is polar.

@HOME (DIY Gel Electrophoresis!)

a. Answers will vary.

b. For instance, glycerol, maple syrup, honey.

c. Yes, this should work.

d. Outside of forensics, gel electrophoresis could be used to identify types of illnesses, or separate complex mixtures of synthetic polymers.

REFLECT (DNA Fingerprinting)

a. Check out this website for a good source that outlines the steps that should be taken during collection and analysis of DNA.

b. Human privacy concerns and interpretation errors are often cited as potentially problematic for DNA profiling.

Real Talk (Real-World CSI)

a. Forensics anthropology is the analysis of human skeletal remains to identify human remains and interpret trauma.

b. A good summary of forensic entomology may be found online at: https://www.ncbi.nlm.nih.gov/pmc/articles/PMC3296382/

c. Forensic scientists must be impartial, accurate, thorough, and truthful.

Appendix 5

Answers to Selected End-of-Chapter Questions Indicated in Blue in the Text

Chapter 1

1. **a.** Compound (two molecules of one compound made up of two different elements).

   b. Mixture (two atoms of one element plus two atoms of another).

   c. Mixture (three different substances, two elements and one compound).

   d. Element (four atoms of the same element).

3. Exact answer will vary depending on viewing size of text. An approximate measurement for the period could be 0.25 mm. Converting this to nanometers:

   $$0.25 \text{ mm} \times \frac{10^{-3} \text{ m}}{1 \text{ mm}} \times \frac{1 \text{ nm}}{10^{-9} \text{ m}}$$
   $$= 2.5 \times 10^5 \text{ nm or } 250{,}000 \text{ nm}$$

6. 1×10^2 cm, 1×10^6 μm, 1×10^9 nm.

8. **a.**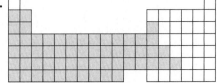

   b. iron, Fe; magnesium, Mg; aluminum, Al; sodium, Na; potassium, K; silver, Ag.

   c.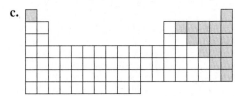

   d. sulfur, S; oxygen, O, carbon, C, chlorine, Cl, fluorine, F (and others).

10. **a.** homogeneous mixture if pulp-free

    b. heterogeneous mixture

    c. heterogeneous mixture

    d. homogeneous mixture

    e. heterogeneous mixture

12. There are several allotropes of sulfur. The most stable and common allotrope consists of eight atoms in a ring. Other common allotropes include rings of 5, 6, 7, 10, and larger number of atoms. Most allotropes are yellow solids, although they can be found in liquid or gaseous forms at appropriate temperatures. Most of these allotropes are created by heating sulfur of the eight-membered ring form.

14. 29 protons, 29 electrons, 35 neutrons

17. **a.** 1 carbon atom, 2 oxygen atoms; all atoms are nonmetals

    b. 2 hydrogen atoms, 1 sulfur atom; all atoms are nonmetals

    c. 1 nitrogen atom, 2 oxygen atoms; all atoms are nonmetals

    d. 1 silicon atom, 2 oxygen atoms; silicon is a metalloid and oxygen is a nonmetal

20. As of 2022, the smallest transistors used in electronics devices have features of 3 nm. This is equivalent to

    $$3 \text{ nm} \times \frac{10^{-9} \text{ m}}{1 \text{ nm}} \times \frac{1 \text{ km}}{10^3 \text{ m}} = 3 \times 10^{-12} \text{ km}.$$

22. There are many steps involved during this process. Try searching "aluminum production and purification" online to find the different methods and steps involved in this process.

24. A thin layer of a transparent conducting material is deposited on the surface of glass. The most common material used for this purpose is indium tin oxide (ITO), which is used in applications ranging from LED displays to solar cells. Do-it-yourself enthusiasts have posted to the Internet a way to coat glass using stannous chloride ($SnCl_2$) to create a layer of conductive tin oxide.

26. Answers will vary depending on device and component choice, but typical components and sizes include length (e.g., 14.5 cm, 145 mm, 145,000 μm, 145,000,000 nm); width (e.g., 7.5 cm, 75 mm, 75,000 μm, 75,000,000 nm); thickness (e.g., 1 cm, 10 mm, 10,000 μm, 10,000,000 nm); camera lens (e.g., 0.3 cm, 3 mm, 3000 μm, 3,000,000 nm); and speaker hole diameter (e.g., 0.03 cm, 0.3 mm, 300 μm, 300,000 nm).

30. The steps involved to convert silica sand, mostly composed of SiO_2, to high-purity Si are outlined in Section 1.7. In contrast, sea sand contains many impurities such as metals (e.g., Fe, Al, Mg, K, Na, Ca, Zn, Ni) and nonmetals (e.g., B, P) that would require extensive pre-processing of the sand via chemical reactions, involving acids and high temperatures. Furthermore, the use of sea sand for

industrial processes would not be sustainable and would cause a variety of environmental consequences. By disturbing sea sand, an area could be changed physically, biologically, and chemically.

32. Although individual cell phones may become smaller and take up less space in, say, a landfill, there may be a much higher impact on the environment, depending on the materials required for making the cell phone and the energy needed for manufacturing the components. To make the devices smaller and less expensive, different materials may be used, which may require more invasive mining or greater waste production in the manufacturing process.

35. The colors of many gemstones come from impurities in the crystal structure. For example, the purple color of amethyst comes from Fe^{3+} ions in a SiO_2 crystal, and the red color of rubies comes from Cr^{3+} in an Al_2O_3 crystal.

37. A thin layer of material is sandwiched between two pieces of glass. When an electrical current is passed through the glass, the material will line up according to the direction of the current, similar to the liquid crystal display (LCD) of common calculators.

39. Apple has removed mercury and arsenic from screens and lead from solder in their electronics.

41. Old electronic devices could be disassembled and parts either could be reused as is or could be mechanically or chemically separated into raw materials to be used for manufacturing new devices.

44. In 2021, there were 240 million iPhones sold worldwide. Although estimates vary, if we assume that each iPhone contains 0.034 g of Au, 0.34 g of Ag, 0.015 g of Pd, 25 g of Al, and 15 g of Cu, that would correspond to 8.2 million grams of Au, 82 million grams of Ag, 3.6 million grams of Pd, 6.0 billion grams of Al, and 3.6 billion grams of Cu! If 25% of iPhones are recycled, the amount of metals recovered would be 0.25 times the above values. To calculate the value of these metals, you would look up the current cost of metals and multiply by the amount of metals recovered. For instance, as of 11/23/22, the current price of gold was $56.38/g, which would equal:

$$8,200,000 \text{ g Au} \times 0.25 \times = \$120,000,000 \frac{\$56.38}{\text{g Au}} \text{ per}$$

year for gold alone! Remember that this is only from iPhones; other major brands such as Samsung, Huawei, etc. also contain precious metals.

46. One consideration includes the use of fossil fuels to power computers and equipment used for synthesis and testing of materials, as well as extensive travel by marketing personnel. Another consideration includes the disposal of waste generated during fabrication and testing of new materials.

48. The increase in demand for rare earth metals amounts to about 40%. Rare earth metals are used in a wide variety of products such as cell phones, computers, rechargeable batteries, wind turbines, speakers, and fluorescent lights. It is doubtful that the U.S. could meet its demand for rare earth elements even with 100% recovery of the metals. The growth of the market for this wide variety of products outpaces the retirement of old products, many of which may not use much, if any, rare earth metals that could be recycled.

Chapter 2

1. a.

$$\frac{0.5 \text{ L}}{1 \text{ breath}} \times \frac{10 \text{ breath}}{1 \text{ min}} \times \frac{60 \text{ min}}{1 \text{ h}} \times 7.5 \text{ h} = 2250 \text{ L}$$

(i.e., 2×10^3 L to one significant figure)

   b. Possibilities include burning less (wood, vegetation, cooking fuels, gasoline, incense), using products that pollute less (low-emission paints), and using motor-less appliances and tools (hand lawnmower, eggbeater, broom, rake).

2. a. $Rn < CO < CO_2 < Ar < O_2 < N_2$

   b. CO and CO_2

   c. CO. By the time this book is in print, CO_2 may be regulated as well.

   d. Rn (radon) and Ar (argon)

4. a. 0.934 part per hundred

$$\times \frac{1,000,000 \text{ parts per million}}{100 \text{ parts per hundred}}$$

= 9340 parts per million
(Move the decimal 4 places to the right.)

   b. $2 \text{ ppm} \times \dfrac{100 \text{ parts per hundred}}{1,000,000 \text{ ppm}}$

= 0.0002 part per hundred, 0.0002%
(Move the decimal 4 places to the left.)
20 ppm is equivalent to 0.0020%. 50 ppm is equivalent to 0.0050%.

   c. $8,500 \text{ ppm} \times \dfrac{100 \text{ parts per hundred}}{1,000,000 \text{ ppm}}$

= 0.85 part per hundred, 0.85%
Be careful not to confuse the absolute humidity calculated in this problem with relative humidity, which is the amount of water vapor in the air compared to the maximum possible amount of water vapor that the air can hold at a particular temperature. For example, in a rainforest, the relative humidity is usually between 75 and 95%.

   c. 8 ppm is 0.0008% (move the decimal 4 places to the left).

5. a. The chemical formula tells the elements present in a compound and the atomic ratio of the elements.

   b. Xe (xenon), N_2O (dinitrogen monoxide or nitrous oxide), CH_4 (methane)

7. Nitrogen is 78.0% of the air, meaning that out of 100 air particles, 78 of them are nitrogen molecules.

$$500 \text{ air particles} \times \frac{78 \text{ nitrogen molecules}}{100 \text{ air particles}}$$
$$= 390 \text{ nitrogen molecules}$$

Oxygen is 21.0% of the air, meaning that out of 100 air particles, 21 of them are oxygen molecules.

$$500 \text{ air particles} \times \frac{21 \text{ oxygen molecules}}{100 \text{ air particles}}$$
$$= 105 \text{ oxygen molecules}$$

Argon is 0.9% of the air, meaning that out of 100 air particles, 0.9 of them are argon atoms.

$$500 \text{ air particles} \times \frac{0.9 \text{ argon atom}}{100 \text{ air particles}}$$
$$= 4.5 \text{ argon atoms (or between 4 and 5 argon atoms)}$$

9. a. Yes, the mass of the reactants equals the mass of the products. The law of conservation of mass applies.

b. No, the numbers of molecules are not the same (four reactant molecules vs. two product molecules).

c. Yes, the numbers of each type of atom present as reactants and products are the same.

11. a. $C_3H_8(g) + 5\,O_2(g) \longrightarrow 3\,CO_2(g) + 4\,H_2O(g)$

b. $2\,C_4H_{10}(g) + 13\,O_2(g) \longrightarrow 8\,CO_2(g) + 10\,H_2O(g)$

c. $2\,C_3H_8(g) + 7\,O_2(g) \longrightarrow 6\,CO(g) + 8\,H_2O(g)$
$2\,C_4H_{10}(g) + 9\,O_2(g) \longrightarrow 8\,CO(g) + 10\,H_2O(g)$

12. a. $2\,C_2H_6(g) + 3\,O_2(g) \longrightarrow 4\,C(s) + 6\,H_2O(g)$

b. $2\,C_2H_6(g) + 5\,O_2(g) \longrightarrow 4\,CO(g) + 6\,H_2O(g)$

c. $2\,C_2H_6(g) + 7\,O_2(g) \longrightarrow 4\,CO_2(g) + 6\,H_2O(g)$

d. The balanced equations show that complete combustion requires the highest ratio of oxygen to ethane (7:2). If a 5:2 ratio is present, carbon monoxide is formed instead of carbon dioxide. If only a 3:2 ratio is available, then carbon (soot and particulate matter) is formed. *Note:* With less oxygen, the products are likely to be mixed, rather than pure CO or pure soot.

14. In respiration, inhaled oxygen reacts with sugar in your body to produce carbon dioxide and water vapor to produce energy. Therefore, exhaled air has a decreased percentage of oxygen and an increased percentage of carbon dioxide. Oxygen is used to metabolize the food we eat.

15. The troposphere is the layer of air closest to Earth—and is the place where we live. It contains 75% of the air, by mass, and is where air currents and storms occur that mix the air in our atmosphere.

17. NO_2 = nitrogen dioxide
N_2O = dinitrogen monoxide
NO = nitrogen monoxide
NCl_3 = nitrogen trichloride
N_2O_4 = dinitrogen tetroxide

19. a. $400 \text{ parts per million} \times \dfrac{100 \text{ parts per hundred}}{1{,}000{,}000 \text{ parts per million}}$
$$= 0.04 \text{ part per hundred} = 4\%$$

b. Carbon monoxide is an air pollutant because when breathed into the lungs, CO can be hazardous to human health.

c. Carbon monoxide interferes with the ability of hemoglobin to carry oxygen throughout your body. Exposure to CO in high enough concentrations can cause a person to die due to lack of oxygen. Shorter-term exposure leads to dizziness or a headache.

22. Carbon monoxide: Mild CO poisoning makes you feel crummy, causing headache, dizziness, or nausea. You will not be able to exert yourself in your normal manner. More severe poisoning may cause unconsciousness.

Particulate matter: Mild PM poisoning will cause lung and cardiovascular distress. Again, you will not have your normal energy level. More severe poisoning can cause a heart attack.

Ozone: Mild ozone poisoning will cause your eyes and throat to burn. It will aggravate your breathing and asthma.

24. In respiration, inhaled oxygen reacts with substances in your body to produce carbon dioxide and water vapor. Therefore, exhaled air has a decreased percentage of oxygen and an increased percentage of carbon dioxide.

26. Here are some possibilities:
- Iron and steel would rust more slowly, prolonging the useful life of many objects made from these materials.
- Fires would burn less vigorously and produce more CO and soot. Logs in your fireplace might last longer, putting out heat more slowly.
- Your body can adapt (just as it does at higher elevations) to lower levels of oxygen. But in this case, the level may be too low for metabolic processes involving oxygen to occur at fast enough rates for life as we currently know it.

28. a. To convert from percent to ppm, move the decimal point 4 places to the right. Alternatively:

$3\% = 3 \text{ pph}$

$$3 \text{ pph} \times \frac{1{,}000{,}000 \text{ ppm}}{100 \text{ pph}} = 30{,}000 \text{ ppm}$$

$$3 \text{ pph} \times \frac{1{,}000{,}000{,}000 \text{ ppb}}{100 \text{ pph}} = 30{,}000{,}000 \text{ ppb}$$

b. The NAAQS for CO in an 8-hour period is 9 ppm. The concentration of CO in cigarette smoke is over three thousand times the 8-hour standard. The NAAQS for CO in a 1-hour period is 35 ppm. The concentration in cigarette smoke is almost nine hundred times the 1-hour standard.

c. Smokers do not die from CO poisoning because they breathe mainly air, not pure cigarette smoke.

30. Reporting the absolute difference, 0.01 ppm, seems to minimize the amount by which the standard is exceeded, at least in the eyes of the general public. Unless the standard is reported as well, there is no way to compare the magnitude of the difference to the magnitude of the standard. Calculating the percentage by comparing the difference (0.01 ppm) to the standard (0.12 ppm) gives 8%, which may give people a better understanding of the amount by which the standard was exceeded.

32. a. The elderly, the young, and people with respiratory diseases such as asthma and emphysema are most affected by PM.

b. December 21–22, December 27, December 31

c. Although PM varies in composition, most of it is less chemically reactive than ozone. It typically is removed from the air by rain or wind.

d. Possibilities include smoke blowing in from a wildfire outside the city, an air inversion, large industrial releases of soot, and a volcanic eruption somewhere in the region that released the ash and soot.

33. a. 15 ppm is 0.0015% and 2% is 20,000 ppm. 20,000 ppm is roughly 1300 times larger than 15.

b. $2\,SO_2 + O_2 \longrightarrow 2\,SO_3$

c. $2\,C_{12}H_{26}(l) + 37\,O_2(g) \longrightarrow 24\,CO_2(g) + 26\,H_2O(g)$

d. Ultimately, burning diesel which is derived from the fossil fuel petroleum is not sustainable. In the short term, diesel engines also are old and have high emissions. So these have a high cost in terms of public health. However, the ultra-low sulfur diesel fuel is definitely a step in the right direction.

36. a. Reducing the number of cars in use will directly and indirectly reduce the concentrations of NO_x, SO_x, CO, CO_2, and ozone in the air.

b. It will reduce ozone, NO_x, SO_x, CO and CO_2 in the air.

39. a. When you are running, you are inhaling air at a greater rate than if you were stationary. Hence, you will inhale a greater concentration of air pollutants relative to others who are not performing physical activities.

b. Running indoors will limit your exposure to air pollutants found outdoors near roadways such as particulate matter, NO_x, SO_x, and tropospheric ozone. However, exercising indoors increases possible exposure to indoor pollutants such as VOCs and carbon monoxide.

42. $6\,m \times 5\,m \times 3\,m = 90\,m^3$. Therefore, the concentration is $3600\,mg/90\,m^3 = 3600000\,mg/90\,m^3$ or $40,000\,mg/m^3$.

44. a. One area listed as "severe" as of 11/24/22 is Los Angeles/San Bernardino counties. The 2015 standard for 8-h ozone is 0.070 ppm.

b. 866,960 people are affected by this ozone level (2010 population). The high population density

and copious amounts of sunshine in this area generate high levels of ground-level ozone.

c. When an area does not meet national air quality standards, the area is classified by EPA as in "nonattainment." When an area returns to meeting national air quality standards, the EPA is asked to re-classify the area as in attainment and in maintenance. Once an area is in maintenance status, the Federal Clean Air Act requires states to implement a Maintenance State Implementation Plan. A maintenance plan includes strategies to keep meeting national air quality standards in that area for 20 years. This plan includes: a) a current, exhaustive inventory of emissions; b) enforceable measures to reduce emissions. These measures may include a combination of state, local, and federal rules, plans, administrative orders, and permits for specific sources; c) a demonstration that the measures will keep the area meeting national air quality standards for 20 years; d) contingency measures that will be applied if the area re-violates national air quality standards.

46. a. The rubber may have come from tires abrading as they roll along the highways. Other sources of PM include soot from incomplete combustion and dirt picked up and blown by the wind.

b. Iron, aluminum, and calcium also are commonly present. Other possibilities include sodium, potassium, magnesium, and sulfur.

c. The edges of the particles appear to be irregular and jagged, thus likely to cause inflammation.

48. a. This graph clearly indicates that exposure to higher carbon monoxide concentrations over longer time periods becomes increasingly life-threatening.

b. CO poses a serious health threat. This gas is colorless and odorless, making it impossible to detect without a monitor or kit. Furthermore, the initial symptoms of carbon monoxide poisoning are not unique, and those suffering from the associated headaches and nausea could easily presume the symptoms are due to a flu-like illness. Untreated, individuals will ultimately lapse into a coma, after which point they will be unable to call for assistance. For reasons such as these, carbon monoxide detectors are lifesaving devices.

50. a. The health hazards associated with isocyanate include irritation of the mucous membrane and skin, tightness in the chest, and difficulty breathing. Isocyanate is also a potential carcinogen for humans and is known to cause cancer in animals.

b. Instead of using nonrenewable, petroleum-based feedstocks to create adhesives, composites, and foams, Professor Wool's processes use feedstocks from bio-based sources. These renewable sources include flax, chicken feathers, and vegetable oils. In addition to being renewable, their production uses less water and energy and is not as toxic as the petroleum-based counterparts.

Chapter 3

1. The chemical formulas of ozone and oxygen are O_3 and O_2, respectively. Both are gases, but they differ in their properties. Oxygen has no odor; ozone has a very sharp odor. Although both are reactive, ozone is much more highly so. Oxygen is necessary for many forms of life; in contrast, ozone is a harmful air pollutant in the troposphere. However, ozone in the stratosphere helps to protect us from the harmful ultraviolet rays of the sun.

3. **a.** The size of the ozone "hole" varies each year, but has been estimated to be as large as 28 million km^2 in area.

$$10 \text{ miles} \times \frac{km}{0.621 \text{ mile}} = 16.1 \text{ km}$$

   b. Yes, the figure is correct, as the stratosphere extends between 15 and 30 km above Earth's surface.

   c. Ozone absorbs UVB and UVC radiation.

6. **a.** The Dobson unit (DU) measures the ozone in a column above a specific location on Earth. If this ozone were compressed at specified conditions of temperature and pressure, it would form a layer. A layer 3-mm thick corresponds to 300 DU. Similarly, a 1-mm layer corresponds to 100 DU.

   b. 320 DU > 275 DU. Thus, 320 DU indicates more total ozone overhead.

7. **a.** Oxygen has 6 valence electrons.

   b. Chlorine has 7 valence electrons.

   c. Nitrogen has 5 valence electrons.

   d. Sulfur has 6 valence electrons.

9. **a.** A group 14 element such as carbon, C

   b. a noble gas such as argon, Ar

   c. a group 16 element such as oxygen, O

10. **a.** ·Ca·   **b.** :C̈l·   **c.** ·N̈·   **d.** He:

12. ·Ö·   Ö=Ö   ·Ö⟨Ö⟩Ö: ·Ö—H

The Lewis structures for the oxygen molecule and the ozone molecule both follow the octet rule. In contrast, the oxygen atom has only six outer electrons and does not follow the octet rule. The hydroxyl radical also does not follow the octet rule and has an unpaired electron. Another resonance structure for the ozone molecule may be drawn; the other molecules do not have resonance structures.

14. **a.** This wavelength is in the microwave region of the spectrum.

   b. This wavelength is in the infrared region of the spectrum.

   c. This wavelength is in the range of violet light in the visible region.

   d. This wavelength is in the UHF/microwave region of the spectrum.

16. *Note:* $c = 3.0 \times 10^8$ m/s and $E = h\nu$ in which $h = 6.63 \times 10^{-34}$ J·s.

   a. $E = (6.63 \times 10^{-34} \text{ J·s})(1.5 \times 10^{10} \text{ s}^{-1}) = 1.0 \times 10^{-24}$ J

   b. $E = (6.63 \times 10^{-34} \text{ J·s})(6 \times 10^{12} \text{ s}^{-1}) = 4 \times 10^{-21}$ J

   c. $E = (6.63 \times 10^{-34} \text{ J·s})(8 \times 10^{14} \text{ s}^{-1}) = 5 \times 10^{-19}$ J

   d. $E = (6.63 \times 10^{-34} \text{ J·s})(2.0 \times 10^{9} \text{ s}^{-1}) = 1.3 \times 10^{-24}$ J

   The most energetic photons correspond to the shortest wavelength, 400 nm.

19. $c = \nu\lambda$ and $\lambda = \frac{c}{\nu}$; $c = 3.0 \times 10^8$ m/s

$$\lambda = \frac{3.0 \times 10^8 \text{ m/s}}{2.45 \times 10^9 \text{/s}} = 1.2 \times 10^{-1} \text{ m}$$

   At 1.2×10^{-1} m, this microwave radiation has a longer wavelength (and lower energy) than X-rays (at $\sim 10^{-1}$ m), but a shorter wavelength (and higher energy) than radio waves (at $\sim 10^3$ m)

23. Answers will vary. To qualify as CFCs, the compounds should contain only carbon, chlorine, and fluorine. Possibilities include

$$
\begin{array}{cc}
\text{:F:} & \text{:F:} \\
| & | \\
\text{:C̈l—C—C̈l:} & \text{:C̈l—C—F̈:} \\
| & | \\
\text{:C̈l:} & \text{:C̈l:} \\
CCl_3F & CCl_2F_2 \\
\text{trichlorofluoromethane} & \text{dichlorodifluoromethane} \\
\text{Freon 11} & \text{Freon 12}
\end{array}
$$

25. **a.** No, a CFC molecule can contain only chlorine, fluorine, and carbon atoms.

   b. HCFC molecules must contain hydrogen, carbon, fluorine, and chlorine atoms, and no other atoms. In order for a molecule to be classified as an HFC, it must contain hydrogen, fluorine, and carbon (but no other atoms).

27. **a.** Cl· has 7 outer electrons. Its Lewis structure is :C̈l·
·NO_2 has $5 + 2(6) = 17$ outer electrons. Its Lewis structure is

Ö::N:Ö̈:   or   Ö=N̈—Ö:

   ClO· has $7 + 6 = 13$ outer electrons. Its Lewis structure is

:C̈l:Ö·   or   :C̈l—Ö·

   ·OH has $6 + 1 = 7$ outer electrons. Its Lewis structure is

·Ö:H   or   ·Ö—H

   b. They all contain an unpaired electron.

29. Graph (a) is a more realistic representation of the relationship between the percent *reduction* in the concentration of ozone and the percent *increase* in UVB radiation. As the ozone layer is depleted, the concentration of UVB that can penetrate into the atmosphere rises. Graph (b) shows a type of inverse relationship, which is not substantiated by experimental facts.

30. The message is that ground-level ozone is a harmful air pollutant. Ozone in the stratosphere, on the other hand, is beneficial because it can absorb harmful UV-B before it reaches the surface of Earth.

32. a. The most energetic fraction of solar UV radiation is the UVC light.

b. Up in the stratosphere where the air is very thin, UVC splits oxygen molecules, O_2, into two oxygen atoms, O. These in turn react with other oxygen molecules to produce ozone, O_3. See Equation 3.5. Without the UVC light (which does not reach the surface of our planet), the ozone layer would not form.

33. a. HFCs are being used to replace HCFCs.

b. HFCs are greenhouse gases!

35. Here are the resonance structures for ozone:

$$:\!\ddot{O}\!=\!\ddot{O}\!-\!\ddot{O}\!:\ \longleftrightarrow\ :\!\ddot{O}\!-\!\ddot{O}\!=\!\ddot{O}$$

Both contain one double bond (expected length of 121 pm) and one single bond (expected length of 132 pm). But, in reality, the bonds are neither single nor double. Rather, the length of each bond is intermediate between the single and double bond lengths. A reasonable prediction would be 126 or 127 pm for both O-to-O bonds, midway between the two lengths

38. With respect to valence electron distribution, the Lewis structures of SO_2 and ozone are identical. This should not be surprising, as sulfur and oxygen are in the same group on the periodic table, and thus have the same number of outer electrons. However, the atoms present in the two Lewis structures differ:

$$\ddot{O}::\ddot{O}:\ddot{O}\ \longleftrightarrow\ :\ddot{O}:\ddot{O}::\ddot{O}\quad\text{and}\quad\ddot{O}::\ddot{S}:\ddot{O}\ \longleftrightarrow\ :\ddot{O}:\ddot{S}::\ddot{O}$$

40. The UV Index, typically a number between 1 and 15, helps people to gauge how intense the sunlight is predicted to be on a particular day. A value of 6.5 (color-coded orange) indicates that there is high risk of harm and that you should protect your eyes and skin. A value of 8–10 indicates a very high risk, and above 11 is an extreme risk.

43. Stratospheric ozone is both formed and broken down in a dynamic system. Unless there are disturbances to this system, the system remains in balance and there is no net change in the concentration of stratospheric ozone.

44. These compounds are useful because they are colorless, odorless, tasteless, and generally inert. However, compounds such as these have long atmospheric lifetimes. They persist in the environment and make their way up to the stratosphere where they cause harm to the ozone layer.

49. a. These compounds once were manufactured as fire suppressants. They are not water-based, so are excellent for specialized uses such as libraries, aircraft, and electronics. However, their production has been halted because they have high ozone-depleting potentials (ODPs).

b. The two halons have different atmospheric lifetimes. According to data from the U.S. EPA, the values are 65 years and 16 years, respectively, for Halon-1301 and Halon-1211. A more interesting question is why the different lifetimes, which is beyond the scope of this text.

c. At the time this graph was drawn, it was thought that no substitutes would be found for some uses of methyl bromide. However, it now is looking more likely that substitutes will be found.

50. $5 + 3(6) + 1 = 24$ electrons

Each atom in all three resonance structures satisfies the octet rule.

53. O_2, O_3, and N_2 all have an even number of valence electrons. In contrast, N_3 would have 15 valence electrons. Molecules with odd numbers of electrons cannot follow the octet rule, making them free radicals and more reactive.

55. Ozonators typically produce ozone either via an electrical discharge or with UV light. The former is similar to the process that produces ozone in a thunderstorm. A bolt of lightning can split O_2 molecules to form O atoms. The latter uses UVC light to split O_2 molecules. In either case, the O atoms then react with another oxygen molecule to produce ozone. The ozone produced works as an effective disinfectant. It can react with many biological molecules, thereby being effective against undesired microbes and viruses. It also can react with many molecules that produce odors.

a. Search the web for examples. Claims include that ozonators can:
- destroy odors from tobacco, smoke, pets, cooking, and chemicals
- kill bacteria and airborne viruses
- remove allergy-causing pollen and microbes
- prevent mold and mildew, the leading cause of Legionnaires' disease
- eliminate toxic fumes from printing, plating processes, and hair and nail salons
- purify water in holding tanks and emergency storage water tanks
- purify drinking water from well sources or city water supplies
- remove undesirable tastes, odors, and colors

b. Ozone can be a harmful pollutant causing damage to both plants and animals. Any device that creates the gas must carefully contain it.

59. **a.** See Figure 3.25. Most months of the year, it is not cold enough in the Arctic for PSCs to form.

b. $HCl + ClONO_2 \longrightarrow Cl_2 + HNO_3$

The nitric acid remains bound to the ice, but the chlorine gas is released into the atmosphere.

c. In the atmosphere in the presence of sunlight, $Cl_2 \longrightarrow 2\ Cl\cdot$

60. **a.** This is a possible Lewis structure:

$$:\!\overset{..}{\underset{..}{Cl}}\!-\!\overset{..}{\underset{..}{O}}\!-\!\overset{..}{\underset{..}{O}}\!-\!\overset{..}{\underset{..}{Cl}}\!:$$

b. If Cl_2O_2 is the actual molecule, then it will have to be broken down by UV photons to $ClO\cdot$ free radicals before it can react with oxygen atoms as shown in Equations 3.8 and 3.9. This would add one additional decomposition reaction in the catalytic destruction of ozone.

Chapter 4

2. These two planets are warmer than would be expected because they have atmospheric gases that produce a "greenhouse effect." Sunlight enters the atmosphere of both Earth and Venus, warming the surfaces of the planet. The atmospheric gases are able to trap some of the heat radiated by the planet surfaces. Without these gases, the planets would be the temperatures expected as a result of their distance from the Sun. *Note:* The high concentration of CO_2 in the Venusian atmosphere (98% CO_2) has led to a "runaway greenhouse effect" and resulted in a surface temperature of about 450 °C!

6. **a.** The rest of the Sun's energy is either absorbed or reflected by the atmosphere. For example, Chapter 3 pointed out that oxygen and stratospheric ozone absorb certain wavelengths of UV light. This chapter points out that clouds may reflect incoming radiation back into space.

b. Under steady-state conditions, 29 MJ/m^2 would leave the atmosphere each day.

7. **a.** As of 2018, the atmospheric concentration of CO_2 was a little above 400 ppm; however, 20,000 years ago the concentration was only about 190 ppm. Looking back to 120,000 years ago, the concentration was about 270 ppm, still ~40% below current levels.

b. The mean atmospheric temperature at present is somewhat above the 1950–1980 mean atmospheric temperature. 20,000 years ago, the mean atmospheric temperature was lower by about 9 °C. However, 120,000 years ago the mean atmospheric temperature was lower than the present temperature by only about 1 °C.

c. Although there appears to be a *correlation* between mean atmospheric temperature and CO_2 concentration, this graph does not prove *causation* of either by the other.

9. **a.** Visible light can enter through the glass, but infrared radiation cannot leave through the glass. There also is little exchange of air with the outside, so the heat cannot be dissipated and the temperature inside the car increases.

b. On clear nights, the heat from Earth can radiate through the atmosphere and out into space. When it is cloudy, the water vapor in the clouds absorbs some of the heat, thus retaining it.

c. In the desert, the temperature swings between night and day tend to be more pronounced. Clouds and humid air make the temperatures more uniform, because they tend to block or scatter incoming solar radiation and trap outgoing heat. *Note:* If the desert contains large urban areas, the pavement and buildings absorb heat during the day. This heat is released at night and thus can keep the temperature high even after the Sun has set.

d. Dark clothing absorbs much of the light that strikes it; in contrast, lighter clothing reflects most of it. The light energy absorbed converts into heat energy, which can increase the risk of heatstroke.

11. **a.** $H\overset{\overset{\displaystyle ..}{S}}{\diagdown}H$ bent

b. $:\!\overset{..}{\underset{..}{Cl}}\overset{\overset{\displaystyle ..}{O}}{\diagdown}\overset{..}{\underset{..}{Cl}}\!:$ bent

c. $:\!N\!=\!N\!=\!\overset{..}{\underset{..}{O}}\!:$ or $:\!N\!\equiv\!N\!-\!\overset{..}{\underset{..}{O}}\!:$ or

$:\!\overset{..}{\underset{..}{N}}\!-\!N\!\equiv\!O\!:$ linear

13. **a.** $3(1) + 4 + 6 + 1 = 14$ outer electrons. This is the Lewis structure:

$$H-\overset{\overset{\displaystyle H}{|}}{\underset{\underset{\displaystyle H}{|}}{C}}-\overset{..}{\underset{..}{O}}-H$$

b. The geometry around the C atom is tetrahedral, and there are no lone pairs. A H—C—H bond angle of about 109.5° is predicted.

c. There are four pairs of electrons around the O atom, two of which are bonding pairs, while the other two are nonbonded pairs. Repulsion between the two nonbonded electron pairs and their repulsion of the bonding pairs is predicted to cause the H—O—C bond angle to be slightly less than 109.5 °C, about 104.5 °C.

15. All can contribute to the greenhouse effect. In each case, the atoms move as the bond stretches or bends, and therefore the charge distribution changes. Unlike the linear CO_2 molecule, the water molecule is bent and so its polarity changes with each of these modes of vibration.

16. a. Use $E = \dfrac{hc}{\lambda}$ to calculate the energies.

$$E = \frac{(6.63 \times 10^{-34} \text{ J} \cdot \text{s}) \times (3.00 \times 10^8 \text{ m/s})}{4.26 \; \mu\text{m} \times \dfrac{1 \text{ m}}{10^6 \; \mu\text{m}}}$$

$$= 4.67 \times 10^{-20} \text{ J}$$

$$E = \frac{(6.63 \times 10^{-34} \text{ J} \cdot \text{s}) \times (3.00 \times 10^8 \text{ m/s})}{15.00 \; \mu\text{m} \times \dfrac{1 \text{ m}}{10^6 \; \mu\text{m}}}$$

$$= 1.33 \times 10^{-20} \text{ J}$$

b. If the vibrating molecule CO_2 collides with another molecule, such as N_2 or O_2, the energy can be transferred to the second molecule. The energy can also be spontaneously emitted back to the atmosphere or into space.

19. a. $C_6H_{12}O_6 \longrightarrow 3 \; CH_4 + 3 \; CO_2$

b. In one day:

$$1.0 \text{ mg glucose} \times \frac{1 \text{ g}}{1000 \text{ mg}} \times \frac{1 \text{ mol glucose}}{180 \text{ g glucose}}$$

$$\times \frac{3 \text{ mol } CO_2}{1 \text{ mol glucose}} \times \frac{44 \text{ g } CO_2}{1 \text{ mol } CO_2} = 7.3 \times 10^{-4} \text{ g}$$

In one year:

$$\frac{7.3 \times 10^{-4} \text{ g } CO_2}{\text{day}} \times \frac{365 \text{ days}}{\text{year}} = 0.27 \text{ g } CO_2/\text{year}$$

21. Let's say that you watch 2 hours of streaming video content each day.

$$\frac{36 \text{ g } CO_2}{\text{h}} \times \frac{2 \text{ h}}{1 \text{ day}} \times \frac{365 \text{ days}}{1 \text{ year}} = \frac{2.6 \times 10^4 \text{ g } CO_2}{\text{year}}$$

22. a. A neutral atom of Rn-222 has 86 protons, 136 neutrons, and 86 electrons.

b. A neutral atom of Rn-220 has 86 protons, 134 neutrons, and 86 electrons. Only the number of neutrons is different.

24. a. $\dfrac{107.87 \text{ g}}{1 \text{ mole}} \times \dfrac{1 \text{ mole}}{6.02 \times 10^{23} \text{ atoms}} = \dfrac{1.79 \times 10^{-22} \text{ g}}{\text{atom}}$

b. $\dfrac{1.79 \times 10^{-22} \text{ g}}{\text{atom}} \times (10 \times 10^{12} \text{ atoms}) = 1.79 \times 10^{-9} \text{ g}$

c. $5.00 \times 10^{45} \text{ atoms} \times \dfrac{1.79 \times 10^{-22} \text{ g}}{\text{atom}} = 8.95 \times 10^{23} \text{ g}$

26. a. The mass percent of Cl in CCl_3F (Freon-11) is

$$\frac{3 \times (35.5 \text{ g/mol})}{12.0 \text{ g/mol} + 3 \times (35.5 \text{ g/mol}) + 19.0 \text{ g/mol}} \times 100$$

$$= 77.5\%$$

b. The mass percent of Cl in CCl_2F_2 is 58.7%.

c. Freon-11: 77.5 g; Freon-12: 58.7 g

d. Freon-11: 77.5 g

$$Cl \times \frac{1 \text{ mol Cl}}{35.5 \text{ g Cl}} \times \frac{6.02 \times 10^{23} \text{ atoms Cl}}{1 \text{ mol Cl}}$$

$$= 1.31 \times 10^{24} \text{ Cl atoms};$$

Freon-12: 9.95×10^{23} Cl atoms.

27. Concentration of carbon atoms =

$$\frac{7.5 \times 10^{17} \text{ g}}{7.5 \times 10^{22} \text{ g}} \times 100 = 0.001\%$$

$$\frac{0.001 \text{ part C in living systems}}{100 \text{ parts C on Earth}}$$

$$= \frac{x \text{ parts C in living systems}}{1,000,000 \text{ parts C on Earth}}$$

$$x = 10 \text{ ppm}$$

32. a. $^{19}_{9}\text{F}$ **b.** $^{56}_{26}\text{Fe}$ **c.** $^{222}_{86}\text{Rn}$

33. a. 4 sig. figs.   **b.** 4 sig. figs.

   c. 4 sig. figs.   **d.** 2 sig. figs.

34. a. 5.12 g   **b.** 2.9 L   **c.** 4.09 g/mL

   d. 5500 nm^2 or $5.5 \times 10^3 \text{ nm}^2$

36. Weather describes the conditions of a region over a short period of time, whereas climate describes the conditions over decades. Weather refers to daily conditions such as high and low temperature, wind, rain, snow, sun, etc., whereas climate refers to average values taken over long periods. Weather can impact farming and wildlife on a short timescale such as what happens in the northern United States when an unexpected frost damages a specific crop. Climate has a much more important role in determining the types of plants and animals that are indigenous to a region or are able to survive and propagate in a region. Global warming is a description of a climate phenomenon that has the potential to destroy ecosystems across the globe.

37. a. Before sophisticated analytical instruments were developed, miners would take caged canaries into the mines to warn them if they encountered any toxic gases (quite prevalent in mine shafts). The canaries were more sensitive to gases like CO. If the canary died, the miners knew they had to get out and into better air quickly.

b. The changes that are occurring in the Arctic may be an early warning sign for the rest of the planet in terms of potential consequences of warmer global temperatures.

c. Significant amounts of methane are trapped in the frozen tundra. If the tundra thawed and released this methane into the atmosphere, this would further accelerate global warming elsewhere because methane is a greenhouse gas.

39. One initial reaction would be that the newspaper reporter has confused "the greenhouse effect" with "global warming." The greenhouse effect is necessary for life on Earth to exist; without it, the average temperature would be −15 °C. Global warming, or the "enhanced greenhouse effect," is what is being blamed for the rise in average global temperatures and the consequences for humans that may result.

42. Substances that absorb visible light have observable colors. For example, if the wavelengths associated with red light are absorbed, the object appears green. Because we cannot see any color associated with either carbon dioxide gas or water vapor, we conclude they do not absorb a significant amount of visible light.

43. The energy required would be smaller for each of the IR-absorbing vibrations if single bonds were present. In general, single bonds between atoms are weaker than double bonds, and therefore less energy will be required to cause stretching and bending.

44. The land surface heats up faster than the ocean, but the ocean has the ability to store energy. For example, in 2020, the average global surface temperature was 0.98 °C above the 20th century average, but it was 1.59 °C over land and only 0.76 °C over the ocean. You will see in Chapter 5 that water has a high specific heat. This means it takes a lot of energy to change its temperature. The other key here is that the oceans are big and mixing can help spread the heat energy. More than 90% of the warming seen on Earth in the past 50 years occurred in the oceans. It is storing the extra energy instead of simply releasing it back into the atmosphere. But, as water warms, it expands. This increase in ocean volume is contributing to rising sea levels.

45. a. $C_2H_5OH + 3\,O_2 \longrightarrow 3\,H_2O + 2\,CO_2$

   b. Two moles of CO_2 are produced for each mole of ethanol burned.

   c. Thirty mol O_2, because for every 1 mole of C_2H_5OH burned, 3 moles of O_2 burn.

47. a. In terms of wavelength, the strongest IR absorbances for water vapor occur around 2.63 μm and 6.67 μm.

   b. Because less energy is needed to bend a molecule than to stretch it, the longer wavelength vibrations at 6.67 μm should represent bending and the shorter wavelength vibrations at 2.63 μm should represent stretching.

48. Model or diagram will vary. The atmosphere acts as the greenhouse windows. The Sun shines on Earth, just as it does on a greenhouse and both the windows and the atmosphere allow sunlight through. Both the windows and the atmosphere do not allow the heat out. The difference between the greenhouse model and the reality of the atmosphere is that the greenhouse has a physical barrier literally trapping heat inside whereas the atmosphere serves as a chemical barrier where the gas molecules absorb and re-emit radiation.

49. The main chemical species involved in ozone depletion are O_3 and CFCs; for climate change, CO_2, CH_4, and N_2O are the main greenhouse gases. Ultraviolet radiation breaks covalent bonds in CFCs, leading to ozone depletion; infrared radiation is absorbed and trapped by atmospheric gases, causing the greenhouse and enhanced greenhouse effects. Predicted consequences of ozone depletion include increased UV exposure at Earth's surface, increased occurrences of skin cancer in humans, and damage to other biological organisms. Climate change consequences include rising sea level, stresses on freshwater resources, loss of biodiversity, and ocean acidification, among others.

51. 73×10^6 metric tons $CH_4 \times \dfrac{12 \text{ metric tons C}}{16 \text{ metric tons CH}_4}$

   $= 5.5 \times 10^7$ metric tons C

54. a. On a per capita basis, the United States would rank first. The population of the United States is smaller than that of China.

   b. The value for metric tons of CO_2 would be higher than that for tons carbon. The former includes the mass of the oxygen; the latter does not.

56. Arrhenius overestimated the temperature increase caused by a doubling of the atmospheric CO_2 concentration by about a factor of two, compared to the IPCC models. The models project a range of 2–4.5 °C increase for doubling of the CO_2 concentration.

57. a. The top 10 listed in 2021 are: United States, China, the Russian Federation, Germany, the United Kingdom, Japan, India, France, Canada, and Ukraine. Countries with large populations, industrial practices, and large supplies of fossil fuels would be predicted to have large amounts of cumulative emissions since 1750 CO_2 emissions. Some may argue that the United States has a responsibility to reduce its per-capita greenhouse emissions and could set a global precedence in doing so.

   b. Without considering legacy pollution, the largest emitter is China. The top 10 nations are then: China, United States, India, the Russian Federation, Japan, Germany, South Korea, Islamic Republic of Iran, Indonesia, and Canada.

   c. Per capita emissions are greater in countries with large industrial practices but with smaller populations such as in the Middle East.

58. a. Burning coal with high sulfur content introduces more SO_x to the atmosphere. This reduces air quality, increases acid precipitation, and speeds the degradation of the environment due to acidification.

   b. Sulfur aerosols reflect incoming solar radiation back toward space. They also serve as nuclei around which water vapor can condense to form sunlight-reflecting cloud particles.

   c. India

59. a. The Quino butterfly has shifted its range to higher, cooler altitudes and has chosen a new plant species to serve as the host to lay its eggs.

b. It is likely that the Quino butterfly will not be able to migrate to even higher/cooler altitudes in the long term. For this species to survive, it will likely need human intervention to help with colonization in other regions.

61. a. The balanced equation for the complete combustion of methane is $CH_4 + 2\,O_2 \longrightarrow CO_2 + 2\,H_2O$ Therefore, burning 1196 mol of methane could produce 1196 mol of CO_2.

b. $1196 \text{ mol} \times \dfrac{44 \text{ g } CO_2}{\text{mol}} = 52{,}624 \text{ g } CO_2$, or 52.62 kg

c. $52.62 \text{ kg} \times \dfrac{1 \text{ metric ton}}{1000 \text{ kg}} = 0.05262$ metric tons

65. In the Maldives, the tourism industry is an example of the tragedy of the commons. As the Maldives is forced to spend a significant portion of its annual budget on energy, the booming tourism industry has become an important source of revenue to the islands' inhabitants. However, rising sea levels threaten these popular vacation spots as beach erosion, saltwater contamination, and flooding are already common occurrences. Additionally, as popularity grows, many of the island resources are becoming depleted. The tourism industry is vital to building up the infrastructure of the islands, but the delicate ecosystems of the islands are becoming damaged in the process. One possible solution is to increase the renewable energy capacity on the islands to free up monetary resources for other important causes.

Chapter 5

1. a. A compound is a pure substance of two or more elements in a fixed, characteristic chemical combination. Water is a compound rather than an element because it contains both the elements H and O in a 2:1 ratio, as evidenced by the chemical formula H_2O.

b. The Lewis structure for water is

$$H \overset{\cdot\cdot}{\underset{}{O}} H$$

The molecule is "bent" because the two lone pairs (nonbonding pairs) of electrons on the oxygen atom occupy space (as do the two bonding pairs of electrons on the O atom). The shape of the water molecule maximizes the space between all these electron pairs.

2. a. N and C, $3.0 - 2.5 = 0.5$

O and S, $3.5 - 2.5 = 1.0$

N and H, $3.0 - 2.1 = 0.9$

F and S, $4.0 - 2.5 = 1.5$

b. N more strongly attracts electrons than C. O more strongly attracts electrons than S. N more strongly attracts electrons than H. F more strongly attracts electrons than S.

c. $N—C < N—H < S—O < S—F$

5. The arrow points to a hydrogen bond, an example of an attractive force *between* water molecules and not *within* each water molecule (as is the case for the polar covalent O—H bond).

6. a. Here is the Lewis structure: $H—\overset{\cdot\cdot}{\underset{\cdot\cdot}{O}}—H$.

b. Lewis structures: $[H]^+$ and $\left[\,\overset{\cdot\cdot}{\underset{\cdot\cdot}{:O}}—H\,\right]^-$.

c. $H^+(aq) + OH^-(aq) \longrightarrow H_2O(l)$

8. a. The volume occupied by 100 g at 1 atm is 50,581 cm^3. The volume at 50 atm is 654 cm^3.

b. The density increases because as pressure increases, the volume contracts. Gases are highly compressible, so their volume becomes smaller as the pressure increases.

9. a. $1 \text{ ppb} = \dfrac{1 \text{ g solute}}{1 \times 10^9 \text{ g } H_2O} \times \dfrac{1 \times 10^6 \text{ µg solute}}{1 \text{ g solute}}$

$\times \dfrac{1000 \text{ g } H_2O}{1 \text{ L } H_2O} = \boxed{\dfrac{1 \times 10^9 \text{ ng solute}}{1 \text{ L } H_2O}}$

$1 \text{ ppt} = \dfrac{1 \text{ g solute}}{1 \times 10^{12} \text{ g } H_2O} \times \dfrac{1 \times 10^9 \text{ ng solute}}{1 \text{ g solute}}$

$\times \dfrac{1000 \text{ g } H_2O}{1 \text{ L } H_2O} = \boxed{\dfrac{1 \text{ ng solute}}{1 \text{ L } H_2O}}$

b. $1 \text{ ppm} = \dfrac{1 \text{ mg solute}}{1 \text{ L } H_2O} \times \dfrac{1 \text{ lb solute}}{453.59237 \text{ g solute}}$

$\times \dfrac{1 \text{ g solute}}{1000 \text{ mg solute}} = \dfrac{2.20 \times 10^{-6} \text{ lb solute}}{1 \text{ L } H_2O} \times \dfrac{1 \text{ lb}}{xL}$

4.54×10^5 L of H_2O needed for 1 lb of solute. Since 1 ppb is 1000 times smaller than 1 ppm, if 4.54×10^5 L of H_2O is required for 1 ppm, then 4.54×10^8 L of H_2O would be required for 1 ppb.

10. a. nitrate = NO_3^-, sulfate = SO_4^{2-}, carbonate = CO_3^{2-}, ammonium = NH_4^+

b. For the nitrate ion, one possibility is nitric acid neutralizing sodium hydroxide.

$H^+(aq) + NO_3^-(aq) + Na^+(aq) + OH^-(aq) \longrightarrow$ $Na^+(aq) + NO_3^-(aq) + H_2O(l)$

For the sulfate ion, one possibility is sulfuric acid neutralizing sodium hydroxide.

$2\,H^+(aq) + SO_4^{2-}(aq) + 2\,Na^+(aq) + 2\,OH^-(aq) \longrightarrow$ $2\,Na^+(aq) + SO_4^{2-}(aq) + 2\,H_2O(l)$

One possibility for carbonate and ammonium ions is carbonic acid neutralizing ammonium hydroxide.

$$H_2CO_3(aq) + 2\,NH_4OH(aq) \rightarrow 2\,NH_4^+(aq) + CO_3^{2-}(aq) + 2\,H_2O(l)$$

Note: Ammonium hydroxide is written in its undissociated form. Similarly, carbonic acid is written in its undissociated form.

13. a. A solution of pH = 6 has 100 times more $[H^+]$ than a solution of pH = 8.

   b. A solution of pH = 5.5 has 10 times more $[H^+]$ than a solution of pH = 6.5.

   c. The solution with $[H^+] = 1 \times 10^{-6}$ M has 100 times more $[H^+]$ than a solution with $[H^+] = 1 \times 10^{-8}$ M.

   d. Using Equation 5.18, the solution with $[OH^-] = 1 \times 10^{-2}$ M has $[H^+] = 1 \times 10^{-12}$ M. The solution with $[OH^-] = 1 \times 10^{-3}$ M has $[H^+] = 1 \times 10^{-11}$ M. Thus, in the second solution, the $[H^+]$ is higher by a factor of 10.

15. All of these compounds are water-soluble.

17. The chemical formula for calcium carbonate is $CaCO_3$. This salt is insoluble in water according to the solubility rules.

20. a. To prepare 2 liters of 1.50-M KOH, weigh 168 g of KOH:

$$2\,L \times \frac{1.50\ mol}{L} = 3.0\ mol\ KOH\quad 3.0\ mol \times 56\ g/mol = 168\ g.$$

   Place the 168 g KOH into a 2-liter volumetric flask. Add distilled (or deionized) water to fill the flask to the mark. *Note:* If you don't have a 2-L volumetric flask, you will need to repeat the procedure twice with a 1-L flask.

   b. To prepare one liter of 0.050-M NaBr, weigh 5.2 g of NaBr and place it into a 1-liter volumetric flask. Add water to fill the flask to the mark.

   c. This should be done with a 100-mL volumetric flask. Weigh 7.0 g of $Mg(OH)_2$ and place it into a 100-mL volumetric flask. Add water to the mark.

21. a. Yes, this is true for hydrocarbons. Distillation towers at a petroleum refinery separate hydrocarbons of different sizes by their boiling point. CH_4 boils at −161.5 °C, C_4H_{10} boils at −1 °C, and C_8H_{18} boils at 125 °C.

   b. Based only on molar mass, you would expect H_2O to have the lowest boiling point because its molar mass is the lowest at 18.0 g/mol.

   c. Water is a polar molecule, while the rest are nonpolar. Both its geometry and its polar covalent bonds contribute to the formation of strong intermolecular forces. Thus, the molar mass is

not the only factor contributing to a substance's boiling point.

22. a. Add the liquids in this order: maple syrup, dishwashing detergent, and then vegetable oil (most to least dense). Three factors must be considered: solubility, density, and the care with which each liquid is poured. Maple syrup will probably slowly dissolve in dishwashing liquid; vegetable oil may be slightly soluble in the detergent. But with careful pouring, these three liquids should not easily mix and probably could be added in any order.

   b. After vigorous mixing, a cloudy emulsion will most likely form. Over time, it will separate into two layers: one with maple syrup and some of the detergent dissolved in water and one with the rest of the detergent dissolved in oil. You may want to try this experiment and observe the results!

24. a. Partially soluble. Orange juice concentrate contains some solids (pulp) that do not dissolve in water. Over time, some of the concentrate may separate from the water. Before drinking, you should give the container a shake or a stir.

   b. Very soluble. Note that ammonia is a gas. Ammonia will dissolve in water in any proportion as it does in cleaning products.

   c. Not soluble. Sometimes you can see chicken fat floating on top of aqueous chicken soup.

   d. Very soluble. When you add laundry detergent to your load of wash, it dissolves in the water.

   e. Partially soluble if the chicken broth contains fat or suspended solids, neither will dissolve in water. Very soluble if the broth is clear and fat-free.

26. a. For Cl and Na, the electronegativity difference is 3.0 − 0.9 = 2.1. For Cl and Si, the electronegativity difference is 3.0 − 1.8 = 1.2.

   b. Larger differences in electronegativity are associated with ionic bonds; smaller differences with covalent bonds.

   c. When electronegativity differences are relatively large, one or more electrons are transferred, forming ions. When electronegativity differences are smaller, neither atom can release its outer electrons to the other, so the outer electrons are shared, resulting in the formation of covalent bonds. In the case of $SiCl_4$, Si and Cl form a polar covalent bond.

28. No, it exceeds the acceptable limit by 35 times. A concentration of 10 ppm is equivalent to 10 mg/L, so 350 mg/L is 350 ppm.

30. a. The solution will conduct electricity, and the bulb will light. Based on Table 5.8, $CaCl_2$ is a soluble salt and therefore releases ions (Ca^{2+} and Cl^-) when it dissolves. These ions carry the current.

b. The solution will not conduct electricity. Although ethanol (C_2H_5OH) is soluble in water, it is a covalent compound and does not form ions.

c. The solution will conduct electricity, and the bulb will light. Sulfuric acid (H_2SO_4) releases ions when it dissolves: H^+ and SO_4^{2-}.

33. a. carbon dioxide, $CO_2(CO_2 + H_2O \longrightarrow H_2CO_3)$

b. sulfur dioxide, $SO_2(SO_2 + H_2O \longrightarrow H_2SO_3)$

35. Possibilities include sodium hydroxide (NaOH), potassium hydroxide (KOH), ammonium hydroxide (NH_4OH), magnesium hydroxide ($Mg(OH)_2$), and calcium hydroxide ($Ca(OH)_2$). In general, bases taste bitter, turn litmus paper blue (and have characteristic color changes with other indicators), have a slippery feel in water, and are caustic to your skin and eyes.

38. Chocolate requires 1700 liters of water to produce a 100-gram bar. This includes the water necessary to grow and process cacao and the sugar used to sweeten it. A 16-ounce glass of beer requires about 140 L of water, mainly for growing and producing the malted barley. Processing cacao and sugar are highly water-intensive compared to processing barley. These are average global values given on The Water Footprint Network.

40. "Pure" drinking water is usually interpreted as water with no dissolved impurities, which is very difficult to achieve. But in reality, water is never pure. For example, if rain has fallen through the atmosphere, it will have picked up carbon dioxide, which explains why all rainwater is slightly acidic. Groundwater can easily pick up water-soluble ions, and ice may contain suspended particulate matter or gases. Even bottled water usually contains dissolved minerals.

42. A mercury concentration of 1.5 ppb means there are 1.5 parts of mercury for every 10^9 parts fish. The caution sign is necessary because mercury is toxic and capable of causing severe neurological effects in humans. The EPA has set the Maximum Contaminant Level for mercury in drinking water at 2 ppb. This is below that limit, but the caution sign is necessary because mercury is a cumulative poison.

44. The diatomic molecule (XY) with a polar covalent bond *must* be polar because the molecule is linear. An example is HCl. The H—Cl bond is polar, and so is the molecule. If the triatomic molecule contains polar bonds, the geometry of the molecule will determine whether the molecule is polar or nonpolar. For example, although CO_2 has polar C=O double bonds, the molecule is linear and, as a result, nonpolar. The H_2O molecule has polar H—O bonds, but is bent. This geometry causes the molecule to be polar.

46. Like water, NH_3 is a polar molecule. It has polar N—H bonds and a trigonal pyramidal geometry. Therefore, despite its low molar mass, considerable energy must be added to liquid NH_3 to overcome the intermolecular forces (hydrogen bonding) among NH_3 molecules.

48. a. The Lewis structure for ethanol is

$$
\begin{array}{ccc}
\text{H} & \text{H} & \\
| & | & \\
\text{H}-\text{C}-\text{C}-\ddot{\underset{\cdot\cdot}{\text{O}}}-\text{H} \\
| & | & \\
\text{H} & \text{H} &
\end{array}
$$

b. The cube sinks because, as is the case for most substances, the density of the solid phase is greater than the density of the liquid phase. Unlike water molecules, ethanol molecules are closer to solid ethanol than liquid ethanol. Therefore, solid ethanol has a greater density than liquid ethanol and sinks.

50. a. Because water has such a high specific heat, it can moderate climate by capturing and absorbing heat from surrounding land and air.

b. If ice were denser than liquid water, it would sink as it forms. As a result, lakes would freeze from the bottom up, killing forms of life that could not tolerate freezing temperatures.

c. Ice is less dense than water, so a given mass will have a greater volume as ice than it does as a liquid. If the pipe is full of water, when it freezes, the volume of ice has to go somewhere.

53. a. $CaCO_3(s) + 2\,H^+(aq) \longrightarrow Ca^{2+}(aq) + CO_2(g) + H_2O(l)$

b. Iron can corrode by rusting, a two-step process:

i) $4\,Fe(s) + 2\,O_2(g) + 8\,H^+(aq) \longrightarrow 4\,Fe^{2+}(aq) + 4\,H_2O(l)$

ii) $4\,Fe^{2+}(aq) + O_2(g) + 4\,H_2O(l) \longrightarrow 2\,Fe_2O_3(s) + 8\,H^+(aq)$

To protect iron from the effects of acid rain, a coating (paint, another metallic film, etc.) can be applied to act as a barrier toward oxidation/rust formation.

54. a. 0.0168, or 1.68%

b. 1.68×10^4 tons S

c. $1.68 \times 10^4 \text{ tons S} \times \dfrac{64 \text{ tons } SO_2}{32 \text{ tons S}} = 3.36 \times 10^4 \text{ tons } SO_2$

d. The SO_3 can react with water to form sulfuric acid: $SO_3(g) + H_2O(l) \longrightarrow H_2SO_4(aq)$

55. The MCLG (a goal) and the MCL (a legal limit) are usually the same for a given contaminant. However, the levels may differ when it is not practical or possible to achieve the health goal set by the MCLG. This sometimes is the case for carcinogens, for which the MCLG is set at zero (assuming that any exposure presents a cancer risk).

57. a. Nitrate ion (NO_3^-) and nitrite ion (NO_2^-).

   b. In the body, oxygen is needed to metabolize ("burn") glucose to produce energy.

   c. The nitrate ion is not volatile. It is a solute that does not evaporate or decompose with heat. Instead, the water in the nitrate-containing solution will evaporate, leaving behind NO_3^-.

59. The two most common desalination techniques are distillation and reverse osmosis. Both of these require energy to remove salts from seawater or brackish water and thus inherently are expensive. This option is used if a less expensive option is unavailable, such as hauling fresh water from a distance.

61. Coffee beans soaked in water are placed in a container where liquid carbon dioxide is injected. The nonpolar solvent, liquid CO_2, attracts caffeine based on the generalization "like dissolves like." This caffeine extract can then be removed from the coffee bean mixture, allowing further processing of the final product.

63. The bond energy of an O—H bond is 467 kJ/mol, or about 10 times greater than the maximum energy of a hydrogen bond. Actually, the hydrogen bonds between water molecules is 20 kJ/mol. This means that the O—H bond energy is about 20 times greater than the bond energy of a hydrogen bond in water.

64. a. Mining waste, gas and oil operations, cement production.

   b. Organic mercury is a carbon compound that contains mercury. These compounds tend to be nonpolar, so they accumulate in fatty tissue comprised of nonpolar molecules ("like dissolves like").

65. a. Glycine contains several polar bonds and has several polar areas in its molecule (everything but the —CH_2 region).

Bond	Electronegativity Difference
O—H	3.5 − 2.1 = 1.4
O—C	3.5 − 2.5 = 1.0
N—H	3.0 − 2.1 = 0.9
N—C	3.0 − 2.5 = 0.5

   b. Yes, hydrogen bonding is possible when O—H and N—H bonds are present, both of which are in glycine.

   c. Because glycine has polar bonds in several areas of the molecule and a relatively small molar mass, glycine should be soluble in water.

67. A complex question! First, you must determine your region's environmental rules and regulations. Most likely, these would apply to releases of chemicals into the soil, air, and water. Then, you would need to monitor what is being released by the industry, in what amounts, and with what occurrence. Compliance with environmental controls, economic factors, and community acceptance of the plant will all affect the success of this plant.

69. a. The PUR Purifier of Water is made by Proctor & Gamble and sold as a packet of chemicals that can be added to a sample of nonpotable water. Each packet contains a powdered flocculent, iron(II) sulfate, and a disinfectant, calcium hypochlorite. The contents are added to 10 liters of nonpotable water, the water is stirred for 5 minutes, and the solids are allowed to settle. The water is then filtered into another container through a cotton cloth. After 20 minutes, the disinfectant inactivates any microbes present (including viruses) and the water is ready for consumption.

   b. Here are a few comparisons. Both systems offer comparable water disinfection, with one catch. The personal Lifestraw does not protect against viruses, but the PUR system does. Both have different uses: one is portable (LifeStraw) and can be used immediately. The other processes a larger volume of water and filters water by gravity instead of by mouth suction. Thus, a larger quantity of water can be purified in a shorter time. Finally, the water filtered with a personal LifeStraw lacks the chemical aftertaste that occurs following treatment with the PUR system.

71. a. Currently, over 90 substances have health-based standards established under the Safe Drinking Water Act. The EPA uses the Unregulated Contaminant Monitoring program every 5 years to establish a list of contaminants and collect data for those contaminants suspected to be present in drinking water.

   b. The EPA uses a CCL to determine what unregulated contaminants could be researched and possibly added as substances to be regulated. This follows the precautionary principle that "stresses the wisdom of acting, even in the absence of full scientific data, before the adverse effects to human health or the environment become significant or irrevocable."

   c. The EPA lists "pesticides, disinfection by-products, chemicals used in commerce, waterborne pathogens, pharmaceuticals, and biological toxins" in its CCL. An example of a specific substance from the CCL-3 list is Halon-1011 (bromochloromethane), used as a fire-extinguishing fluid and as a solvent to make pesticides.

76. 2105 days or 5.7 years

Chapter 6

2. a. CO_2, carbon dioxide.

   b. SO_2 is an air pollutant. Although sulfur is present in low concentrations in coal, large amounts of coal are burned, and collectively large amounts of SO_2 are released.

c. Nitrogen is present in the air (~80% of atmospheric gases). The nitrogen present in air reacts with O_2 (also present in air) at high temperatures to form NO:

$$N_2 + O_2 \xrightarrow{\text{high temperature}} 2\,NO$$

d. Revisit Chapter 2 for the details. From the U.S. EPA website: "Particle exposure [of any size] can lead to a variety of health effects. For example, numerous studies link particle levels to increased hospital admissions and emergency room visits and even to death from heart or lung diseases. Both long- and short-term particle exposures have been linked to health problems. Long-term exposures, such as those experienced by people living for many years in areas with high particle levels, have been associated with problems such as reduced lung function and the development of chronic bronchitis and even premature deaths."

4. a. The fuel in the burner is a source of potential energy. When burned, some of its potential energy is converted into heat through combustion. The heat is converted into kinetic energy of the vaporized water molecules (steam).

b. The kinetic energy of the steam is converted into mechanical energy by spinning a turbine.

c. The mechanical energy generated from the spinning turbine is converted into electrical energy by rotating a wire in a magnetic field.

d. The electrical energy, carried to the city by the power lines, lights bulbs and heats homes.

5. a.
$$\frac{5.00 \times 10^8 \text{ J}}{\text{s}} \times \frac{3600 \text{ s}}{\text{h}} \times \frac{24 \text{ h}}{\text{day}} \times \frac{365 \text{ days}}{\text{year}}$$
$$= 1.58 \times 10^{16} \text{ J of electricity generated per year}$$

$$\frac{1.58 \times 10^{16} \text{ J}}{0.375} = 4.2 \times 10^{16} \text{ J of heat for electricity}$$
generated per year

b. 1.58×10^{16} J produced $\times \dfrac{1 \text{ kJ}}{1000 \text{ J}} \times \dfrac{1 \text{ g}}{30 \text{ kJ}}$

$$= 5.3 \times 10^{11} \text{ g per year}$$

$$5.3 \times 10^{11} \text{ g} \times \frac{1 \text{ metric ton}}{10^6 \text{ g}} = 5.3 \times 10^5 \text{ metric tons}$$

9. A typical power plant burns 1.5 million tons of coal each year. The first calculation is for coal with 50 ppb mercury; the second is for 200 ppb.

$$\frac{x \text{ ton Hg}}{1.5 \times 10^6 \text{ ton coal}} = \frac{50 \text{ ton Hg}}{1 \times 10^9 \text{ ton coal}} \quad x = 0.075 \text{ ton Hg}$$

$$\frac{x \text{ ton Hg}}{1.5 \times 10^6 \text{ ton coal}} = \frac{200 \text{ ton Hg}}{1 \times 10^9 \text{ ton coal}} \quad x = 0.30 \text{ ton Hg}$$

Assuming mercury concentrations in the range of 50–200 ppb, the plant releases between 0.075 and 0.30 ton of Hg per year.

11. a. CH_3CH_3: ethane, $CH_3(CH_2)_2CH_3$: butane

b. C_2H_6 and C_4H_{10}

c. Chemical formulas such as C_4H_{10} are compact and easy to write. The same is true for condensed structural formulas, at least in this particular case. Although structural formulas take longer to draw and take up more space, they clearly reveal the arrangement of all the bonds and atoms.

14. a.

b. There is only one other isomer:

15. Pentane should be a liquid because room temperature (20 °C) is below its boiling point (36 °C) but above its melting point (−130 °C). Triacontane should be solid at room temperature because room temperature is below its melting point (66 °C). Propane should be a gas because room temperature is above its boiling point (−42 °C).

18. a. $C_7H_{16} + 11\,O_2 \longrightarrow 7\,CO_2 + 8\,H_2O$

b. $2.50 \text{ kg} \times \dfrac{10^3 \text{ g}}{\text{kg}} \times \dfrac{1 \text{ mol } C_7H_{16}}{100.2 \text{ g}}$

$$\times \frac{4817 \text{ kJ}}{1 \text{ mol } C_7H_{16}} = 1.2 \times 10^7 \text{ kJ}$$

20. $92 \text{ kcal} \times \dfrac{4.184 \text{ kJ}}{1 \text{ kcal}} = 380 \text{ kJ}$

21. a. exothermic

b. endothermic

c. endothermic

22. a. Bonds broken in the reactants:
1 mol N≡N triple bonds = 1(946 kJ) = 946 kJ
3 mol H—H single bonds = 3(436 kJ) = 1308 kJ
Total energy *absorbed* in breaking bonds = 2254 kJ

Bonds formed in the products:
6 mol N—H single bonds = 6(391 kJ) = 2346 kJ
Total energy *released* in forming bonds = 2346 kJ
Net energy change is (+2254 kJ)
$$+ (-2346 \text{ kJ}) = -92 \text{ kJ}$$
The overall energy change is negative, characteristic of an exothermic reaction.

b. Bonds broken in the reactants:
1 mol H—H single bonds = 1(436 kJ) = 436 kJ
1 mol Cl—Cl single bonds = 1(242 kJ) = 242 kJ
Total energy *absorbed* in breaking bonds = 678 kJ

Bonds formed in the products:
2 mol H—Cl single bonds = 2(431 kJ) = 862 kJ
Total energy *released* in forming bonds = 862 kJ
Net energy change is (+678 kJ)
$$+ (-862 \text{ kJ}) = -184 \text{ kJ}$$

The overall energy change is negative, characteristic of an exothermic reaction.

25. a. $\dfrac{5 \times 10^{12} \text{ J}}{x} = 0.38; \ x = \dfrac{1.32 \times 10^{13} \text{ J}}{\text{day}}$

$= \dfrac{1.32 \times 10^{10} \text{ kJ}}{\text{day}}$

$\dfrac{1.32 \times 10^{10} \text{ kJ}}{\text{day}} \times \dfrac{1 \text{ g}}{30 \text{ kJ}} \times \dfrac{1 \text{ kg}}{1000 \text{ g}} \times \dfrac{\$36.14}{1000 \text{ kg}}$

$= \$13,000/\text{day}$

Coal costs for Plant A = \$13,000/day. Coal costs for Plant B = \$11,000/day.

b. When coal combusts, 12 g of carbon produces 44 g of carbon dioxide. Plant A burns 1.3×10^{10} kJ $\times \frac{1 \text{ g}}{30 \text{ kJ}} = 4.3 \times 10^8$ g of coal per day, creating 1.6×10^9 g CO_2. Plant B burns 3.6×10^8 g of coal per day, creating 1.3×10^9 g of CO_2. Therefore, Plant B emits 3.0×10^8 (30 million) fewer grams of CO_2 each day than Plant A.

c. Energy generated at the power plant × efficiency = total energy required to heat the house
Energy generated at Plant A × 0.38 = 3.5×10^7 kJ of heat

Therefore, energy generated at Plant A = 3.5×10^7 kJ/0.38 = 9.2×10^7 kJ of heat
Since each gram of coal yields 30 kJ of heat, 9.2×10^7 kJ/30 kJ/g = 3.1×10^6 g of coal needed for Plant A

For Plant B:
Energy generated at Plant B (46% efficient) = 3.5×10^7 kJ/0.46 = 7.6×10^7 kJ of heat

Which corresponds to: 7.6×10^7 kJ/30 kJ/g = 2.5×10^6 g of coal needed for Plant B
Hence, 5.3×10^5 more grams (or 530 kg) of coal would need to be burned at Plant A to supply the needed energy to heat the home.

26. a. None of these are isomers. All have different chemical formulas.

b. No, no other isomers are possible for ethene.

c. One other isomer is possible, although it contains a very distinct functional group, called an ether. Its condensed structural formula is CH_3—O—CH_3.

27. a.

b. The second and third are identical.

c. Several other isomers are possible. Here are two:

30. a. $C_{16}H_{34} \longrightarrow C_5H_{12} + C_{11}H_{22}$

The C—C single bond in the center of the molecule and one of the C—H single bonds must be broken. A second C—C single bond must be broken so that a C=C double bond can be formed in its place. A new C—H single bond must form on the shorter product.

b. Bonds broken in the reactants:

c. 2 mol C—C single bonds = 2(365 kJ) = 712 kJ
1 mol C—H single bonds = 1(416 kJ) = 416 kJ
Total energy *absorbed* in breaking bonds = 1128 kJ

Bonds formed in the products:
1 mol C—H single bonds = 1(416 kJ) = 416 kJ
1 mol C=C double bonds = 1(598 kJ) = 598 kJ
Total energy *released* in forming bonds = 1014 kJ
Net energy change is (+1128 kJ)
$$+ (-1014 \text{ kJ}) = 114 \text{ kJ}$$

The overall energy change has a positive sign, characteristic of an endothermic reaction.

32. a. The hydroxyl functional group, —OH. All of the compounds are alcohols.

b. carbon dioxide, CO_2; and water, H_2O.

c. These three compounds are similar in chemical composition and differ only in number of CH_2 groups. Each have hydrogen bonding as their predominant intermolecular force, but methanol would have the lowest boiling point because it has the lowest molecular mass of the group.

d. *n*-propanol, an alcohol with three carbon atoms, is the most similar to glycerol (which also is an alcohol and has three carbon atoms). However, glycerol contains three —OH groups (one on each carbon) in comparison to the one in *n*-propanol.

34. a. $C_6H_{12}O_6 + 6\,O_2 \longrightarrow 6\,CO_2 + 6\,H_2O + energy$

b. Cellulose is one of the primary components of wood. Cellulose is a polymer made up of glucose building blocks. As a result, burning cellulose gives products comparable to burning glucose.

37. Figure 6.5 gives the energy released per gram for the combustion of several fuels. Assuming the densities of octane and ethanol are similar (a good assumption), one gallon of gasoline releases more energy than one gallon of ethanol (44.4 kJ/g of gasoline vs. 26.8 kJ/g of ethanol). This makes sense, because ethanol is an oxygenated fuel; it contains oxygen and thus is already "partially burned."

40. A primary component of wood is cellulose, with a chemical formula that can be approximated with that of glucose, $C_6H_{12}O_6$. Given that the ratio of carbon-to-oxygen in glucose is 1:1, the chemical formula for this soft coal most likely would contain much more oxygen than common types of coal. The same is true for hydrogen, because the ratio of carbon-to-hydrogen in glucose is 1:2.

43. Answers may vary. Here is one example: Wouldn't you rather spill a drop of hot coffee on you than the whole cup at the same temperature? Although the drop and the cup full of coffee may initially have the same temperature, you will receive a bigger burn from the bigger volume of coffee because it has the higher heat content. Heat is a form of energy. In contrast, temperature is a measurement that indicates the direction heat will flow. Heat always flows from an object at high temperature to an object at lower temperature. This means that if hot coffee is added to cold coffee, heat will flow from the hot liquid to the cold liquid, and the final temperature of the mixture will be between the original temperatures of the two individual solutions. Heat depends on the temperature and on how much material is present.

45. $H_2CO(g) + O_2(g) \longrightarrow CO_2(g) + H_2O(g)$

Let x represent the C=O bond energy in H_2CO.

Bonds broken in the reactants:
2 mol C—H single bonds = 2(416 kJ) = 832 kJ
1 mol C=O double bonds = 1(x kJ) = x kJ
1 mol O=O double bonds = 1(498 kJ) = 498 kJ
Total energy *absorbed* by breaking bonds
$$= (1330 + x)\ kJ$$

Bonds formed in the products:
2 mol O—H single bonds = 2(467 kJ) = 934 kJ
2 mol C=O double bonds = 2(803 kJ) = 1606 kJ
Total energy *released* in forming bonds = 2540 kJ
Net energy change:
$$(1330 + x\ kJ) + (2540\ kJ) = -465\ kJ$$
Rearranging the equation:
$$x\ kJ = -465 + 2540 - 330\ kJ \quad x = 745\ kJ$$

This value is less than the bond energy for C=O double bonds in carbon dioxide reported in Table 6.2. The C=O double bonds in carbon dioxide are stronger than the C=O double bond in formaldehyde.

46. CFCs are stable because the bond energies for C—Cl and C—F are large compared to other bond energies. It takes less energy to release Cl atoms from CFCs because the C—Cl bond energy (327 kJ/mol) is lower than the C—F bond energy (485 kJ/mol). HFCs, with their C—F bonds and (no C—Cl bonds), release no Cl atoms.

48. From Figure 6.5, fuels containing oxygen have lower energy content per gram than those without oxygen. For example, ethanol and glucose have proportionately more oxygen than the other fuels listed. In essence, they are already "partially burned" (oxidized) and thus their energy content is lower when you look at the values per gram of fuel.

Here are the values in kilojoules per mole:

methane (CH_4): $\dfrac{50.1\ kJ}{1\ g} \times \dfrac{16.05\ g}{1\ mol} = 802\ kJ/mol$

octane (C_8H_{18}): $\dfrac{44.4\ kJ}{1\ g} \times \dfrac{114\ g}{1\ mol} = 5.06 \times 10^3\ kJ/mol$

coal ($C_{135}H_{96}O_9NS$):

$\dfrac{32.8\ kJ}{1\ g} \times \dfrac{1908\ g}{1\ mol} = 6.26 \times 10^4\ kJ/mol$

ethanol (C_2H_6O):

$\dfrac{26.8\ kJ}{1\ g} \times \dfrac{46\ g}{1\ mol} = 1.23 \times 10^3\ kJ/mol$

glucose: ($C_6H_{12}O_6$):

$\dfrac{14.1\ kJ}{1\ g} \times \dfrac{180\ g}{1\ mol} = 2.54 \times 10^3\ kJ/mol$

With kilojoules per mole, though, the trend observed is different. Here, the fuels with higher numbers of carbon atoms in their chemical formulas (and hence higher molar masses) release more energy when burned. Compare methane and *n*-octane to see the contrast clearly.

50. a. When $n = 1$, the balanced equation is

$$CO + 3 H_2 \longrightarrow CH_4 + H_2O$$

Bonds broken in the reactants:
1 mol C≡O triple bonds = 1(1073 kJ) = 1073 kJ
3 mol H—H single bonds = 3(436 kJ) = 1308 kJ
Total energy *absorbed* in breaking bonds = 2381 kJ

Bonds formed in the products:
4 mol C—H single bonds = 4(416 kJ) = 1664 kJ
2 mol O—H single bonds = 2(467 kJ) = 934 kJ
Total energy *released* in forming bonds = 2598 kJ

Net energy change is (+2381 kJ) + (−2598 kJ)
$$= -217 \text{ kJ}$$

b. Reactions with *n* greater than 1 will release more energy as *n* becomes larger, assuming that we are viewing the energy per mole of the hydrocarbon formed (not per gram). There will always be *n* C≡O triple bonds to break and $(2n + 1)$ H—H single bonds to break. The number of C—H bonds forming will be $(2n + 2)$, the number of O—H bonds forming will be $2n$, and the number of C—C single bonds forming will be $n-1$. Combining these terms shows that as *n* becomes larger, more and more energy will be released.

51. a. The Lewis structure for dimethyl ether is:

$$
\begin{array}{ccc}
\text{H} & & \text{H} \\
\text{H:C:} & \text{Ö:} & \text{C:H} \\
\text{H} & & \text{H}
\end{array}
$$

b. The structural formula for diethyl ether is:

$$
\begin{array}{ccccccc}
& \text{H} & \text{H} & & \text{H} & \text{H} & \\
& | & | & & | & | & \\
\text{H} & - \text{C} & - \text{C} & - \text{O} - & \text{C} & - \text{C} & - \text{H} \\
& | & | & & | & | & \\
& \text{H} & \text{H} & & \text{H} & \text{H} &
\end{array}
$$

c. The common structural feature is an oxygen atom between two carbon atoms.

52. a. The compounds *n*-octane and iso-octane have nearly identical heats of combustion. This makes sense because they have the same number and types of bonds. However, they have very different octane ratings. Therefore, the octane rating is not a measure of the energy content of a substance.

b. Knocking produces an objectionable pinging sound, reduced engine power, overheating, and possible engine damage.

c. The higher octane blends are more expensive to produce because they require more

"processing," including energy-intensive cracking reactions that convert lower octane fuels into higher octane ones.

d. The octane ratings tell you nothing about whether or not oxygenates are present. Although oxygenates are one way to improve the octane rating, they are not the only way.

55.

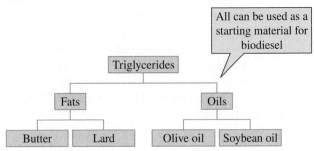

60. The price of electricity varies by locality, but according to the Electric Choice website, the average cost of electricity in the U.S. in 2018 was 13.3 cents per kWh.

Using a 75 W incandescent bulb for 10,000 hours would consume 75,000 Wh of electricity. Using an 18 W compact fluorescent bulb for 10,000 hours would consume 18,000 Wh of electricity.

75 W incandescent bulb:
$$75{,}000 \text{ Wh} \times \frac{1 \text{ kWh}}{1000 \text{ Wh}} \times \frac{\$0.133}{\text{kWh}} = \$9.98$$

18 W compact fluorescent bulb:
$$18{,}000 \text{ Wh} \times \frac{1 \text{ kWh}}{1000 \text{ Wh}} \times \frac{\$0.133}{\text{kWh}} = \$2.39$$

These calculations indicate an electricity savings of $9.98 − $2.39 = $7.59 over the life of a compact fluorescent bulb.

To be most accurate, the lifetime of incandescent bulbs should also be taken into account. Incandescent bulbs burn out after ~750 hours, so 10,000 hours of use would require the use of 10,000/750 = 13.33 incandescent bulbs.

At a popular online retailer, a six-pack of 75 W incandescent bulbs costs $7.97 (or $1.33 each) and a four-pack of 18 W compact fluorescent bulbs costs $5.97 (or $1.49 each). Therefore, the cost of 13.33 incandescent bulbs would be $17.71. When compared to the cost of one compact fluorescent bulb ($1.49), it can be seen that there is an additional savings of $16.22 by using the longer-lasting compact fluorescent bulbs. *Note:* The price of compact fluorescents has been dropping, so it might make sense to recheck the prices.

61. Marisol first educated herself about the emissions from the coal power plant and the trends in health effects in her neighborhood. She realized that the asthma, premature births, and birth defects were disproportionately affecting her community. She then shared her findings with the neighbors. Her map not only showed the physical relationship between emission sources and health impacts, but it also catalyzed community organizing so that members could advocate for change.

63. Consider a natural gas (methane) explosion that releases energy:

$$CH_4 + 2 O_2 \longrightarrow CO_2 + 2 H_2O$$

The bond energies involved are C—H single bonds, 416 kJ/mol; O=O double bonds, 498 kJ/mol; H—O single bonds, 467 kJ/mol; C=O double bonds, 803 kJ/mol. The bond energies of the products are larger than those of the reactants, and thus they release more energy when forming than was necessary to break the bonds of the reactants. This will lead to a large negative net energy change indicating an exothermic reaction.

66. As gasoline additives go, this one ranks high. It is more than double the value of any listed in Table 6.3. The structural formula of TEL shows four ethyl groups (—C_2H_5) around a central lead atom. It is highly branched! Just as iso-octane has a high octane rating because of all of its "branches," so does TEL.

67. a. The diagram shows that the catalyzed pathway requires less activation energy than the uncatalyzed pathway.

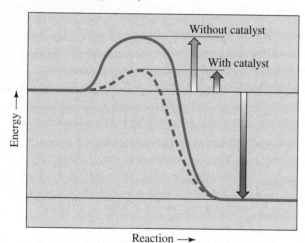

b. In Chapter 2, catalysts were discussed in connection with removing NO from automobile exhaust. Nitrogen oxide can react with oxygen to form NO_2, a criteria pollutant. NO is also involved in forming ozone in the troposphere and contributes to acid rain. To reduce pollution, it is important to reduce NO emissions.

68. Recall that in an endothermic reaction, the potential energy of :N=Ö: the products is greater than the potential energy of the reactants. The opposite is true for an exothermic reaction.

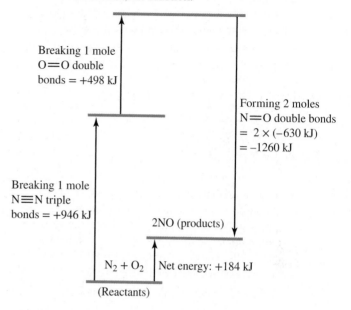

Chapter 7

1. One carbon atom can differ from another in the number of neutrons (such as C-12 and C-13) and in the number of electrons (carbon ions do exist, but we do not discuss them in this text). All carbon atoms differ from all uranium atoms in the number of protons, neutrons, and electrons. Carbon atoms also differ from uranium atoms in their chemical properties.

3. a. 94 protons

b. 93 protons = Np (neptunium), 94 protons = Pu (plutonium)

c. 86 protons

5. E represents energy, m represents mass lost in a nuclear transformation, and c represents the speed of light.

9. a. The alpha particle may have come from the radioactive decay of another radioisotope.

b. $^{1}_{0}n$ represents a neutron.

c. Curium-243 represents an unstable intermediate in the nuclear reaction. This isotope has an extremely short lifetime, decomposing immediately upon formation into Cm-242 with an accompanying neutron.

12. a. $^{1}_{0}n + ^{235}_{92}U \longrightarrow [^{236}_{92}U] \longrightarrow ^{146}_{57}La + ^{87}_{35}Br + 3^{1}_{0}n$

b. $^{1}_{0}n + ^{235}_{92}U \longrightarrow [^{236}_{92}U] \longrightarrow ^{140}_{56}Ba + ^{94}_{36}Kr + 2^{1}_{0}n$

13. **A** represents the control rod assembly, **B** represents the cooling water out of the core, **C** represents the control rods, **D** represents the cooling water into the core, and **E** represents the fuel rods.

15. The primary coolant is the liquid surrounding the fuel bundles and control rods, a liquid that comes in direct contact with the nuclear reactor to carry away heat. The heat from the primary coolant is transferred to the secondary coolant, water in the steam generators that does not come in contact with the reactor. The steam generators are separated from the nuclear reactor, so the secondary coolant is not housed in the containment dome.

16. **a.** $^{1}_{0}n + ^{10}_{5}B \longrightarrow [^{11}_{5}B] \longrightarrow ^{4}_{2}He + ^{7}_{3}Li$

   b. Boron can be used in control rods because it is a good neutron absorber.

17. **a.** $^{239}_{94}Pu \longrightarrow ^{235}_{92}U + ^{4}_{2}He$

   $^{131}_{53}I \longrightarrow ^{131}_{54}Xe + ^{0}_{-1}e + ^{0}_{0}\gamma$

   b. In a particulate form such as a powder or a dust, plutonium can be inhaled. If the plutonium particles become lodged in the lungs, the ionizing radiation they emit (alpha particles) can damage lung cells. The decay products from U-235 also are radioactive and can damage tissue.

   c. Iodine accumulates in the thyroid gland.

   d. After about 10 half-lives, samples have decayed to very low levels. The half-life of Pu-239 is about 24,000 years, so the timescale for a decrease to background level is on the order of hundreds of thousands of years. The half-life of I-131 is 8.5 days, so 10 half-lives is 85 days or about 3 months. A sample of I-131 will decay to low levels on a timescale of months.

19. $^{235}_{92}U \longrightarrow ^{231}_{90}Th + ^{4}_{2}He$

   $^{231}_{90}Th \longrightarrow ^{231}_{91}Pa + ^{0}_{-1}e$

   $^{231}_{91}Pa \longrightarrow ^{227}_{89}Ac + ^{4}_{2}He$

   $^{227}_{89}Ac \longrightarrow ^{227}_{90}Th + ^{0}_{-1}e$

   $^{227}_{90}Th \longrightarrow ^{223}_{88}Ra + ^{4}_{2}He$

   $^{223}_{86}Ra \longrightarrow ^{219}_{86}Rn + ^{4}_{2}He$

20. For this type of question, it is helpful to construct a chart.

# of half-lives	% remaining	% decayed
0	100	0
1	50	50
2	25	75
3	12.5	87.5
4	6.25	93.75
5	3.12	96.88
6	1.56	98.44

22. Perhaps someday it can. However, solar energy is diffuse, unequally distributed over Earth's surface, and still presents us with challenges to economically capture and store it.

25. Alchemists were perhaps the first practical chemists, but they did not have the advantage of knowing anything about atomic structure or nuclear reactions. No chemical reaction can produce gold from another element; a nuclear reaction is required. Even if they had envisioned a nuclear reaction that would produce gold from another isotope, they clearly did not have the means to accomplish this. The situation has indeed changed, and modern-day chemists could design experiments to change lead into gold. The question now is why anyone would want to, as the cost would be prohibitive.

27. **a.** All means of separation depend on the tiny mass difference between U-235 and U-238. For example, it is possible to separate them by converting the uranium sample into gaseous UF_6 and then use gas diffusion. A large high-speed centrifuge also can be used to separate these gas molecules.

   b. The uranium must be enriched to provide a critical mass of U-235 to sustain the chain reaction responsible for energy production in the reactor.

   c. First, the enrichment procedure is both expensive and energy intensive, so the minimum enrichment level capable of sustaining a chain reaction is preferred. Second, reactors using 80–90% fuels would have safety concerns due to the increased possibility of an uncontrolled chain reaction. Third, such reactors would also have significant security issues. The highly enriched fuel can be used directly in nuclear weapons, making the reactors potential terrorist targets.

   d. The difference in the isotopes of uranium is in their nuclear masses. This difference is not enough to significantly affect the chemical reactivity of the two isotopes. For chemical separation, the isotopes of uranium would need to behave differently in a chemical reaction of one sort or another.

29. The Palo Verde power plant produces energy through the process of nuclear fission. Coal- and oil-burning plants generate energy by burning fossil fuels.

31. **a.** The subscript for each element is its atomic number or number of protons, which can be found in the periodic table. The subscript for the neutron is zero, which requires knowing or finding the charge of a neutron in a reference table.

   b. The superscripts cannot be omitted because nuclear equations must specify a specific isotope, and this is something that cannot be determined by looking at the periodic table or another reference.

33. After seven half-lives, 99% of a sample has decayed, which is a reasonable approximation of being "gone." However, actually the radioactivity is not gone, as 0.78% of the radioactive sample still remains. Thus, if you start with a large amount of a radioactive substance (for example, 2000 pounds), after seven half-lives you still have about 10 pounds of radioactive substance left!

34. a. Bananas are rich in potassium (K). One of the naturally occurring isotopes of potassium (K-40) is radioactive, thereby adding to the radioactivity in bananas.

b. Because the natural abundance of K-40 is only 0.01%, the vice president could have stated that bananas are weakly radioactive and that this radioactivity is natural. Potassium-40 has a long half-life (on the order of billions of years), so it is undergoing radioactive decay very slowly.

c. No. Bananas are a good source of potassium, an essential nutrient. The amount of K-40 in bananas is not significant enough to consider eliminating bananas from your diet on the basis of their radioactivity. Furthermore, any potassium (radioactive or not) that you ingest is not retained. Potassium is lost through sweat and in urine.

37. PV devices have demonstrated their practical utility for satellites, highway signs, security and safety lighting, navigational buoys, and automobile recharging stations.

40. a. In the equation $E = mc^2$, the speed of light (c) is 3.00×10^8 meters/second. In order to have joules (J) as the unit of energy (E), the mass (m) must be in kilograms. In addition, use the conversion factor of 1 joule = $1 \text{ kg} \cdot \text{m}^2/\text{s}^2$.

$$50.1 \text{ kJ} \times \frac{10^3 \text{ J}}{1 \text{ kJ}} = m \times \frac{1 \text{ kg}}{10^3 \text{ g}} \times \left(\frac{3.00 \times 10^8 \text{ m}}{\text{s}}\right)^2$$
$$\times \frac{1 \text{ J}}{\text{kg} \cdot \text{m}^2/\text{s}^2}$$

The mass loss is 5.57×10^{-10} g.

b. To produce 50.1 kJ of energy, the ratio of masses is 1.00 g of methane burned to 5.57×10^{-10} g methane converted into energy, or about 1.80×10^9 to 1.

c. In a chemical reaction, mass is conserved, so $E = mc^2$ doesn't apply. The energy released in a chemical reaction is a result of potential energy stored in bonds. In a nuclear reaction, a small amount of mass is lost from reactants to products, which is converted into energy.

41. a. Helium nuclei have more mass, since they have one more neutron than hydrogen nuclei.

b. In the equation $E = mc^2$, the speed of light (c) is 3.00×10^8 meters/second. In order to have joules (J) as the unit of energy (E), the mass (m) must be in kilograms. In addition, use the conversion factor of 1 joule = $1 \text{ kg} \cdot \text{m}^2/\text{s}^2$.

$$E = 0.0265 \text{ g} \times \frac{1 \text{ kg}}{10^3 \text{ g}} \times \left(\frac{3.00 \times 10^8 \text{ m}}{\text{s}}\right)^2 \times \frac{1 \text{ J}}{\text{kg} \cdot \text{m}^2/\text{s}^2}$$

$$E = 2.39 \times 10^{12} \text{ J}$$

44. Tritium, H-3, is a radioisotope of hydrogen. Hydrogen is a gas at room temperature, so it is unlikely that the gas itself is contained in the watches. Several descriptions of this watch mention the stainless-steel screw-on back, again making it unlikely that tritium gas was present inside the watch. Most likely, the tritium is in a compound in the luminous paint. The paint also contains a phosphor; that is, a compound that glows (is phosphorescent) when hit by ionizing radiation such as the beta particles emitted by tritium. Indeed, these watches can glow brightly as claimed by the advertisement.

46. Similarities between a coal-fueled power plant and a nuclear-fueled power plant include:

- Both contain a steam-generating loop where liquid water is turned to steam. The gaseous water turns a turbine to create electricity, and then is re-condensed to form liquid water again.
- The turbine that produces electricity is the same. Each power plant includes a cooling water loop with an external body of water as the cooling source.

Differences between a coal-fueled and nuclear-fueled power plant include:

- The source of the energy to heat the water that turns the turbine is from nuclear fission reactions in the nuclear power plant and from the burning of coal in the coal power plant.
- The nuclear-fueled power plant has an additional cooling loop (the primary coolant) to cool the reactor core. This coolant is a closed loop so that the secondary coolant is not contaminated with radioactive material.

49. Crystalline silicon is used for the production of photovoltaic cells. Two of the common methods for synthesizing crystalline silicon are Czochralski Crystal Growth and Float Zone Crystal Growth, two different processes that involve manipulating silicon in different physical states. Search these two methods online to learn how these techniques are used to synthesize crystalline silicon.

To cope with the shortage of silicon, the photovoltaic industry is undergoing a series of changes. Companies are pursuing business models to increase productivity and secure silicon supply. Researchers are developing and testing a number of different synthetic molecules for the next generation of photovoltaic collectors.

51. a. As of 2019, the largest PV power plant in the United States is the Solar Star power station near Rosamond, California.

b. As of 2019, the Ningxia Hui Autonomous Region of China, and Rajasthan, India, have two of the world's largest PV power plants.

c. Factors include (1) having land available for the array, (2) living in a climate with weather favorable to solar collection, (3) having economic conditions that promote an investment with a long-term payback, (4) having the infrastructure to transmit the electricity to population centers.

Chapter 8

2. **a.** Oxidation. Iron loses two electrons to form the Fe^{2+} ion.

    b. Reduction. The Ni^{4+} ion gains two electrons to form the Ni^{2+} ion.

    c. Oxidation. Each chloride ion loses an electron to form a neutral chlorine atom. These atoms combine to form a chlorine molecule (Cl_2).

3. $Zn(s)$ is oxidized to Zn^{2+} in zinc oxide. $O_2(g)$ is reduced to O^{2-} in zinc oxide.

5. Electric current (an amount of charge per second) is measured in amps (A). In contrast, the electrical potential or the "pressure" behind this current is measured in volts (V).

7. **a.** Oxidation half-reaction: $Li(s) \longrightarrow Li^+(aq) + e^-$

    Reduction half-reaction: $I_2(s) + 2e^- \longrightarrow 2\ I^-(aq)$

    b. Overall cell equation: $2\ Li(s) + I_2(s) \longrightarrow 2\ LiI(s)$

    c. The oxidation half-reaction $Li(s) \longrightarrow Li^+(aq) + e^-$ occurs at the anode. The reduction half-reaction $I_2(s) + 2e^- \longrightarrow 2\ I^-(aq)$ occurs at the cathode.

8. **a.** The voltage from both kinds of cells is the same (1.5 V) because voltage depends on the chemical reaction that is producing the electrical energy and not on the size of the electrodes.

    b. The amount of current (A) produced by a cell depends on the size of the cell. Larger cells contain more materials and can sustain the transfer of electrons over a longer period. For example, the larger D alkaline cell can generate 120 A/h of current, whereas a tiny AAA alkaline cell generates less current at 12 A/h.

10. **a.** Oxidation half-reaction: $Zn(s) \longrightarrow Zn^{2+}(aq) + 2e^-$

    b. Reduction half-reaction: $Hg^{2+}(aq) + 2e^- \longrightarrow Hg(s)$

    c. Mercury was once widely used in batteries. By 1990, an awareness of the dangers of mercury in urban trash had grown. Mercury is a toxic metal and (in some forms) can accumulate in the biosphere. Safer batteries and the need to recycle batteries led to the passage of the Mercury-Containing and Rechargeable Battery Management Act (The Battery Act) in 1996.

12. **a.** $Pb(s) + SO_4^{2-}(aq) \longrightarrow PbSO_4(s) + 2e^-$

    $PbO_2(s) + 4\ H^+(aq) + SO_4^{2-}(aq) + 2e^- \longrightarrow$ $PbSO_4(s) + 2\ H_2O(l)$

    b. The first half-reaction shows electrons being lost, so it represents oxidation. The second half-reaction shows electrons being gained, so it represents reduction.

    c. Lead is oxidized, so lead is the anode. Though not as obvious, the lead dioxide or lead(IV) oxide is reduced (note the electrons on the left side of the half-reaction). Thus, lead(IV) oxide is the cathode.

13. It represents the reduction half-reaction. The conversion of O_2 to H_2O requires a supply of electrons.

15. $R = 12\ V/0.10\ A = 120$ ohms

18. $P = 100\ W/5\ A = 20\ V$

19. The first half-reaction takes place at the anode (hydrogen is oxidized). The second half-reaction takes place at the cathode (oxygen is reduced).

20. PEM stands for proton exchange (or polymer electrolyte) membrane. As in the fuel cell shown in Question 15, $H_2(g)$ is oxidized to form $H^+(aq)$. In the PEM fuel cell $H^+(aq)$ moves through the membrane to react with $O_2(g)$ (which is reduced) to form water. There is no membrane in the fuel cell in Question 15. In addition, the electrodes differ between the two types of fuel cells as do the electrolytes. The PEM has a solid polymer electrolyte membrane coated with a Pt-based catalyst and the other fuel cell uses a $KOH(aq)$ solution as its electrolyte. Finally, the PEM fuel cell operates at room temperature, and the one used for the space mission does not.

22. Oxidation half-reaction:

    $CH_4(g) + 8\ OH^-(aq) \longrightarrow CO_2(g) + 6\ H_2O(l) + 8e^-$

    Reduction half-reaction:

    $2\ O_2(g) + 4\ H_2O(l) + 8e^- \longrightarrow 8\ OH^-(aq)$

    Overall reaction:

    $CH_4(g) + 2\ O_2(g) \longrightarrow CO_2(g) + 2\ H_2O(l)$

24. **a.** $K(s) + \frac{1}{2}\ H_2(g) \longrightarrow KH(s)$ or $2\ K(s) + H_2(g) \longrightarrow 2\ KH(s)$

    b. $KH(s) + H_2O(l) \longrightarrow H_2(g) + KOH(aq)$

    c. Because lithium is less dense than potassium, a given mass will weigh less. This offers an advantage in the handling and transportation of a storage cell. In addition, lithium metal is less reactive than potassium metal, which makes it safer to use in the manufacturing of metal hydride storage systems.

25. Before we can use hydrogen fuel cells more widely, we will need to meet the challenges of safely producing, transporting, and storing large quantities of hydrogen.

29. These batteries derive their voltage from different sets of chemical reactions. A rechargeable battery (such as a Ni–Cd battery) is one in which the oxidation-reduction reaction can be reversed with the input of energy (such as plugging the battery into an electrical outlet.) This recharges the battery. The oxidation–reduction reactions in a nonrechargeable battery, such as an alkaline battery, cannot easily be reversed. Because no simple way exists to recharge alkaline batteries, they are discarded once they stop producing electrical energy.

31. The primary difference is that these produce electricity using different chemical reactions. In addition, a lead–acid storage battery converts chemical energy into electrical energy by means of a reversible reaction. No reactants or products leave the "storage" battery, and the reactants can be reformed during the recharging cycle. A fuel cell also converts chemical energy into electrical energy, but the reaction is not reversible. A fuel cell continues to operate only if fuel and oxidant are continuously added, which is why it is classified as a "flow" battery.

33. a. "ZPower is taking the leading role in launching the next generation of rechargeable, silver-zinc batteries for microbattery applications. This advanced battery offers superior performance over traditional microbattery technologies—delivering 20% more energy than lithium-ion and 100% greater energy than nickel metal-hydride. ZPower batteries also offer a green solution since 95% of the battery elements can be recycled and reused. The mercury-free design is inherently safe due to its water-based electrolyte which is not susceptible to thermal runaway." Visit the ZPower company website for more on the science behind their silver-zinc batteries.

    b. Oxidation half-reaction (zinc is oxidized):

$$Zn(s) \longrightarrow Zn^{2+}(aq) + 2e^-$$

    Reduction half-reaction (silver is reduced):

$$2\,Ag^+(aq) + 2e^- \longrightarrow 2\,Ag(s)$$

37. A Toyota gasoline-battery hybrid car (for example, the Prius) has a gasoline engine sitting side-by-side with a nickel–metal hydride battery and an electric motor. The engine drives the motor to recharge the battery. In addition, regenerative braking recharges the cells in the battery, during which the kinetic energy of the car is stored as electrical energy.

38. All EVs (electric vehicles) share a common technology: the use of electricity to power the vehicle and rechargeable batteries as an energy storage device. If the power goes out, you could drive your vehicle only as far as the energy stored in its batteries would allow. Gasoline-powered vehicles also would be limited. Given that an electrical outage would also render inoperable the pumps at a gasoline filling station, your driving would be limited by the fuel left in your gas tank. Note that in a severe storm, which may topple trees and down power lines, nobody is likely to be driving anywhere!

41. The bond energies listed in Table 6.1 for bonds breaking and forming in gases can be used to calculate the following heats of combustion. These differ somewhat from the values given in the beginning of this chapter, where the product water is given in the *liquid* state.

$$H_2(g) + \tfrac{1}{2}\,O_2(g) \longrightarrow H_2O(g)$$
heat of combustion = 249 kJ/mol

$$CH_4(g) + 2\,O_2(g) \longrightarrow CO_2(g) + 2\,H_2O(g)$$
heat of combustion = 814 kJ/mol

In each case, the units of the calculated heat of combustion can be changed to kJ/gram by dividing by the molar mass of the fuel.

$$\frac{249\ kJ}{mol\ H_2} \times \frac{1\ mol\ H_2}{2.01\ g\ H_2} = \frac{124\ kJ}{g\ H_2}$$

43. a. When water boils, the hydrogen bonds *among* water molecules (intermolecular forces) are disrupted. No bonds are broken within the water molecules.

    b. When water is electrolyzed, the covalent bonds *within* water molecules are broken.

46. An energy input of 249 kJ/mol is required in the electrolysis of water. Most of this energy comes from the burning of fossil fuels in conventional power plants. The inherent inefficiency associated with transforming heat into work limits the usefulness of large-scale electrolysis and makes the process energy-intensive. Although not technically feasible yet, using the power of the sun to produce hydrogen from water could be more thermodynamically efficient and would certainly be a far more sustainable method.

48. a. $1.0\ mol\ H_2 \times \dfrac{2\ mol\ Na}{1\ mol\ H_2} \times \dfrac{23.0\ g\ Na}{1\ mol\ Na} = 46.0\ g\ Na$

    b. $1.1 \times 10^6\ kJ \times \dfrac{1\ mol\ H_2}{249\ kJ} \times \dfrac{2\ mol\ Na}{1\ mol\ H_2} \times \dfrac{23.0\ g\ Na}{1\ mol\ Na}$

    $= 2.10 \times 10^5\ g\ Na$

    c. Using the result from part **a**:

    $46.0\ g\ Na \times \dfrac{1\ kg\ Na}{10^3\ g\ Na} \times \dfrac{\$165}{1\ kg\ Na} = \$7.59$

51. a. Hydrogen is oxidized as it gained an oxygen atom to become water. Oxygen gained hydrogen atoms, so it got reduced.

    b. In the first equation, carbon is oxidized as it gained oxygen atoms to become carbon dioxide. In the second, more complex equation, oxygen is reduced when it gained hydrogen atoms to become water. The carbon in octane lost hydrogen atoms to become a product, which gained an oxygen. The change to both reactants fits this non-electron definition of oxidation and reduction.

54. Here are ways that some of the principles of green chemistry might apply:

    1. *"It is better to prevent waste than to treat or clean up waste after it is formed."* For example, stations are now set up to receive and recycle lead storage batteries. These batteries are no longer sent to landfills or junk yards.

    2. *"It is better to minimize the amount of materials used in the production of a product."* For example, find ways to use less silicon in a PV cell, find ways to reduce the packaging for batteries (for example, "button" batteries often come in a plastic package to prevent point-of-purchase theft). Find another way to prevent theft.

3. *"It is better to use and generate substances that are not toxic."* For example, stop using or minimize the use of batteries that contain toxic metals. This already has been done for mercury batteries. Also develop new batteries that don't contain cadmium, lead, and other toxic materials.

56. Candlelight

Origin	The hot gases that burn and emit light.
Immediate energy source	The hydrocarbon wax, either produced by bees or from petroleum.
Original energy source	Sunlight that drove photosynthesis, which in turn produced the plants from which bees gathered their food (or years ago died and formed fossil fuels).
Products	CO_2, H_2O, and small amounts of soot and CO.
Environmental costs	Primarily the health effects of the particulate matter and soot produced. Also the greenhouse gas produced, CO_2.
Advantages Disadvantages	Convenient, pleasing to view. Produces dirty soot and may cause a fire if unattended.

Light in a battery-powered flashlight

Origin	A wire that glows when it is heated to a high temperature.
Immediate energy source	A chemical reaction in the battery.
Original energy source	Several possibilities, depending on what energy source was used to produce the battery. Could have been fossil fuel consumption (originally solar energy) or nuclear power plant (nuclear fission).
Products	The end products are different chemicals in the battery, while the by-products are those that are produced during the manufacture of the battery, bulb, and flashlight.
Environmental costs	All those associated with the production and disposal of the battery materials, as well as the side products during the combustion of fossil fuels (or nuclear fission) used in its manufacture.
Advantages Disadvantages	Portable, convenient, clean for the user. Somewhat expensive, becomes waste when energy is spent.

Light from an electric lightbulb

Origin	A wire that glows when it is heated to a high temperature.
Immediate energy source	Several possibilities, depending on what energy source was used to produce the electricity (e.g., burning coal or nuclear power plant).
Original energy source	The Sun or the ancient stellar synthesis that produced the uranium and other metals on our planet.
Products	The lightbulb is very clean at the site where it is used, but produces pollutants such as NO_x, SO_2, CO_2, and particulate matter at the power plant (if coal, natural gas, or fuel oil combustion) or spent nuclear fuel (if nuclear).
Environmental costs	All those associated with the production and disposal of the lightbulb, as well as the side products during the combustion of fossil fuels (or nuclear fission) used to provide the electricity.
Advantages Disadvantages	Convenient, safe, inexpensive. Few to the user, except that the energy costs of incandescent lightbulbs are relatively high in comparison to a fluorescent bulb or a light-emitting diode (LED).

Chapter 9

1. Cotton, silk, rubber, wool, and DNA are examples of natural polymers. Synthetic polymers include Kevlar, polyvinyl chloride (PVC), Dacron, polyethylene, polypropylene, and polyethylene terephthalate.

5. a. At the molecular level, increasing the length of the polymer chain would increase its molar mass and the extent of its interactions with neighboring chains. This would be expected to somewhat increase the polymer's rigidity, strength, and melting point.

b. At the molecular level, aligning polyethylene chains with one another means that the structure is more crystalline and highly ordered. This would be expected to give the polymer slightly more density, more rigidity, and more strength. The melting point would also increase.

c. At the molecular level, this would be the opposite of the previous answer. The structure would be less crystalline, less ordered, and possibly somewhat tangled. This would be expected to make the polymer slightly less dense, less rigid, and not as strong. The melting point would decrease.

7. To serve as monomers, hydrocarbons must have a C=C double bond (and contain no elements other than C and H). Here are two possibilities other than ethylene:

Propylene   Styrene

9. Each ethylene monomer has a molar mass of 28 g/mol. To determine the number of monomers in the polymer, divide 40,000 (the molar mass of the polymer) by 28 (the molar mass of the monomer). The result is 1428 monomers, or 1400 to two significant figures. To determine the number of carbon atoms present in the polymer, note that each monomer contains two carbon atoms ($H_2C=CH_2$). Accordingly, the polymer contains 2×1428 carbon atoms, or 2856 carbon atoms. In round numbers, there are roughly 3000 carbon atoms.

11. This is the tail-to-tail, head-to-head arrangement of PVC formed from three monomer units.

12. These are different. The top segment represents the "head-to-tail, head-to-tail" arrangement. The Cl atoms in each case are on alternate carbon atoms. It makes no difference if the atom is on the "top" or on the "bottom" of the chain (these positions are equivalent). In contrast, the bottom segment is "head-to-head, tail-to-tail." The Cl atoms are on adjacent carbon atoms.

14. Note that the question asks "most likely." For plastic use, "most likely" changes over time, so the answers are a moving target. For example, PVC is being phased out in many uses.
    a. PET, polyethylene terephthalate
    b. HDPE, high-density polyethylene
    c. PVC, polyvinyl chloride ("vinyl")
    d. PP, polypropylene
    e. LDPE, low-density polyethylene
    f. PS, polystyrene
    g. HDPE, high-density polyethylene

17. a. phenyl group, alkene
    b. hydroxyl group (or alcohol)
    c. phenyl group, carboxylic acid
    d. amine, carboxylic acid
    e. amine
    f. carboxylic acid

19. Terephthalic acid contains two carboxylic acid groups. Phenylenediamine contains two amine groups. These two monomers react in a condensation polymerization to form amide linkages between the phenyl groups.

22. a. A blowing agent is a gas (or a substance capable of producing a gas) used to manufacture a foamed plastic. For example, a blowing agent is used to produce Styrofoam from PVC.
    b. Carbon dioxide can replace the CFCs or the HCFCs that once were used as blowing agents. Although CO_2 is a greenhouse gas, it still is preferable because CFCs and HCFCs both deplete the ozone layer and are potent greenhouse gases.

24. a. Postconsumer content includes all types of waste: newspapers, cardboard, foam cups, packing peanuts, 2-liter bottles, and plasticware. Preconsumer content includes waste created in the manufacturing process, such as scraps of fabric, plastics, paper, wood, and food.
    b. Not necessarily. For example, most PET soft drink bottles are made from petroleum products and do not include recycled PET.

26. Polypropylene (PP) is a tough plastic and bottle caps need to be tough, standing up to repeated use and not losing their shape or their threads. However, PP melts at a higher temperature than PET and has different properties. So somewhere in the recycling process, PP needs to be separated from PET. In essence, in most cases, the caps need to be removed from the bottles. Beverage companies are currently seeking alternatives.

28. Factors other than the chemical composition of the monomer(s) influence the properties of the polymer. These include length of the chain (the number of monomer units), the three-dimensional arrangement of the chains, the degree of branching in the chain, and orientation of monomer units within the chain (such as head-to-tail).

30. For addition polymerization, the monomer must have a C=C double bond. Although some monomers have benzene rings as part of their structures (styrene, for example), the double bond involved in addition polymerization must not be in the ring—the double bonds in the ring are part of the resonance structure and not a true double bond. An example is the formation of PP from propylene.

    For condensation polymerization, each monomer must have two functional groups that can react and eliminate a small molecule such as water. For example, an alcohol and a carboxylic acid can react to eliminate water. An example is the formation of PET from ethylene glycol and terephthalic acid.

31. In vinyl chloride, each carbon atom has three bonds (two single bonds and one double bond). These three bonds form an equilateral triangle (trigonal) and the Cl—C—H bond angle is about 120°. In the polymer, each carbon atom is connected to other atoms by four single bonds. The double bond is no longer present, and the four bonds point to the corners of a tetrahedron, with a bond angle of about 109.5°.

34. From the bond energies in Table 6.1, it requires 598 kJ/mol to break C=C double bonds. The formation of C—C single bonds releases 365 kJ/mol. If we consider the reaction of two ethylene monomers, two double bonds are broken and replaced with four single bonds (two bonds between the C atoms of the monomers, one between the first monomer and the second, and a bond extending to what would be the third ethylene monomer). Here is the calculation:

$$(2 \times 598 \text{ kJ/mol}) + (4 \times -356 \text{ kJ/mol}) = -228 \text{ kJ/mol}.$$

Thus, the reaction is exothermic.

36. The heat of combustion of polyethylene would be most similar to that of octane. Both are hydrocarbons consisting of carbon–carbon single bonds. The other fuels contain different atoms (and thus different bonds would be broken and formed).

38. **a.** PLA stands for polylactic acid, a polymer.

b. The monomer of PLA is lactic acid. In the United States, lactic acid is produced from corn.

c. Reasons include that (1) corn is a renewable resource, (2) PLA is compostable, and (3) PLA is not a petroleum-based polymer.

d. (1) Although corn is a renewable resource, corn is a crop with its share of controversies. These include the degradation of the land on which it is grown and the runoff of fertilizers and pesticides into nearby waterways. (2) Although PLA is compostable, this is true only in industrial composters that most communities do not have. It degrades slowly if at all in a landfill. (3) Although no oil is used in its production, fuels such as petroleum are nonetheless used in the growing of corn and its transportation.

39. **a.** From the bond energies in Table 6.1, it requires 598 kJ/mol to break C=C double bonds. The formation of C—C single bonds releases 356 kJ/mol. If we consider the reaction of two ethylene monomers, two double bonds are broken and replaced with four single bonds (two bonds between the C atoms of the monomers, one between the first monomer and the second, and a bond extending to what would be the third ethylene monomer). Here is the calculation: $(2 \times 598 \text{ kJ/mol}) + (4 \times -356 \text{ kJ/mol}) = -228 \text{ kJ/mol}.$ With 1000 monomers joining, we multiply the −228 kJ/mol by 500 and the heat released will be 114,000 kJ, or 1.14×10^5 kJ.

b. Overall, heat is released, but some energy must be input to overcome the activation energy. The reaction is so exothermic that, in the early days of polymer manufacture, polymerization vessels exploded. Manufacturers realized that conditions needed to be carefully controlled to avoid this.

44. The Big Six polymers are generally large (in fact huge) molecules with few polar groups. Furthermore, many are hydrocarbons (HDPE, LDPE, PS, PP) and therefore would not be expected to dissolve in polar solvents such as water. "Like dissolves like," so many polymers, including HDPE and LDPE, soften in hydrocarbons or chlorinated hydrocarbons because these nonpolar solvents interact with the nonpolar polymeric chains.

46. **a.** Starch is a polymer of glucose. Many foods are a source of starch, including corn, potatoes, and rice.

b. Advantages of starch packing peanuts: they are lightweight, compostable, and made from a renewable material. Disadvantages: they will degrade if the package gets wet, they are made from what could be eaten as a food, and they are less "springy" than polystyrene foam peanuts.

c. Composting is a good option. Although some can be washed down the drain, this results in the starch needing to be removed later at a water treatment plant.

48. **a.** A plasticizer is a compound added to a hard or rigid plastic in order to soften it.

b. DEHP was added to PVC in order to make soft vinyl products such as boots, shower curtains, clothing items, and flexible tubing.

c. DEHP has been banned for items that infants repeatedly put in their mouths, such as pacifiers. It has also been banned in some medical devices and children's toys. The use of DEHP remains controversial and different restrictions are in place in different countries.

51. **a.** The two main properties are (1) stable over time of intended use and (2) nontoxic. Other factors to consider are low cost, lack of solubility in body fluids, lack of reactivity in body fluids, and ease of implantation.

b. Several different types of contact lenses are on the market, and each uses a different type of polymer. Polymethyl methacrylate (PMMA), one of the earliest polymers used for rigid gas permeable lenses, is structurally similar to Lucite and Plexiglas. Silicone-acrylate materials now are more commonly used under trade names such as Kolfocon. Newer rigid gas permeable (RGP) polymers tend to contain fluorine. Manufacturers' websites are good sources of information.

Desirable properties include being nontoxic, permeable to oxygen, comfortable to wear, and inexpensive. Also desirable is the ability to conform to the shape of the eye and to be easily cleaned (if necessary).

55. **a.** With its high molar mass, this polymer should not dissolve in water. Like many polymers, it would be expected to be an electrical insulator. With its methyl groups, it should be somewhat flammable (but the presence of silicon reduces the flammability and gives it good stability at high temperature).

b. "Silly Putty" bounces, but it breaks when pulled sharply. When it is formed into a shape, it slowly loses this shape over time and flattens out. Ink will stick to Silly Putty and so it can "lift" images from newsprint. These properties are rather unique, and led to the popular acceptance of Silly Putty as a toy long before it was widely used for other products.

c. Silicone rubber is used for flexible bakeware because of its resistance to high temperatures. It also is used in greases, caulking, and tubing.

57. One recent research precedent was reported by the University of Michigan. A link to a summary of this work may be found here: https://phys.org/news/2022-11-recycling-previously-unrecyclable-polyvinyl-chloride.html.

59. As you might suspect from the -ol ending, polyols are alcohols. The "poly" means they have multiple hydroxyl (—OH) groups. Many types of polyols exist, some with only two hydroxyl groups and others with many more. For example, in this chapter, you met ethylene glycol, a polyol with two hydroxyl groups (also called a diol). It served as one of the two monomers for PET. In Section 6.15, you met ethylene glycol and propylene glycol, two polyols. In Chapter 11, in regard to fats and triglycerides, you will read about glycerol (glycerin), a polyol that contains three hydroxyl groups (also called a triol).

Polyols can serve as monomers for any condensation polymers such as polyesters made from a "double acid" and a "double alcohol" (diol). An Internet search for soybean plastics or soybean-based polymers should turn up many examples.

Chapter 10

1. For g/mL, $5\frac{g}{dm^3} \times \left(\frac{1\ dm^3}{1 \times 10^3\ cm^3}\right) = 5 \times 10^{-3}$ g/cm³.

For mg/L, $5\frac{g}{dm^3} \times \left(\frac{1\ dm^3}{1\ L}\right) \times \left(\frac{1 \times 10^3\ mg}{1\ g}\right) = 5 \times 10^3$ mg/L.

For mg/m³, $5\frac{g}{dm^3} \times \left(\frac{1 \times 10^3\ dm^3}{1\ m^3}\right) \times \left(\frac{1 \times 10^3\ mg}{1\ g}\right)$
$$= 5 \times 10^6\ mg/m^3.$$

3. $1\frac{g}{mL} \times \left(\frac{1000\ mg}{1\ g}\right) \times \left(\frac{1000\ mL}{1\ L}\right) = 1 \times 10^6$ mg/L.

In terms of how solutes affect the density of water, their presence increases the density of the liquid water. Because of the nature of solutions, solid solutes in liquid solvents add mass while not adding appreciable volume, and so, density increases.

5. Recall for gases, $\frac{P_1}{T_1} = \frac{P_2}{T_2}$ or $P_1T_2 = P_2T_1$.

Temperature must be expressed in Kelvin, and so, recall K = °C + 273.

(1 atm)(x) = (2 atm)(373 K), x = 746 K, or 473 °C

With a 100 °C increase in temperature, the reaction rate would double 10 times (100/10 = 10). At 2 atm, the rate of reaction would be over 1000 times the original rate at 1 atm (2^{10} = 1024). This emphasizes how pressure cookers cook food so much faster than an open pot of boiling water.

7. How strongly molecules stick together via intermolecular forces determines how easily they will transition from the solid or liquid phase to the gas phase. To be considered volatile, these molecules must make this transition fairly easily and not be held too tightly to other molecules. Weaker intermolecular forces among molecules tend to come from nonpolar features in the molecule, as well as the molecule being compact and not made of *many* branched, but rather single long chains of nonpolar groups. Low molecular masses can also be a factor, but recall nonpolar versus polar molecules tend to be the bigger factor. Numerous molecules that have properties of smells and/or tastes have small rings of nonpolar groups. See examples below:

Lemon   Almond   Thyme

8. a. The browning of cut fruit is the result of a reaction of an enzyme (polyphenol oxidase), which is exposed when the fruit is cut, and oxygen from the air. It is an oxidation process much like the formation of rust on iron structures. These products create an unpleasant sour taste and are arguably a defense mechanism so that the fruit is not eaten by animals or even invaded by germs.

b. This is not the same process as carmelization and the Maillard reaction, which occur at high temperature and with no moisture present.

c. Oxygen prefers to react with acidic vitamin C before reacting with the polyphenol oxidase in fruit, and so, the browning product from the oxidase reaction does not form. Thus, the juice of lemons, limes, or even cranberries, which contain high amounts of vitamin C, when squeezed on fruits is effective at preventing the browning process.

12. All green plants (including vegetables) contain chlorophyll, which also includes a magnesium atom attached to it. This gives them their characteristic color. When exposed to heat (or acid) long enough, the magnesium is replaced by hydrogen atoms. This change in composition alters the color property, and these vegetables turn from bright green to a dull, olive green. In terms of texture, long exposure to heat and acid breaks down the fibrous, crunchy structural molecules of the vegetable, and they become softer.

14. Usually, boiling water is used to make a cup of tea. In order to extract the molecules from the tea leaves, sufficient heating of the water solvent is necessary. Recall that achieving the boiling point of a liquid is actually more related to overcoming the external, atmospheric pressure rather than just merely heating to a particular temperature. Water boils at a lower temperature when the atmospheric pressure is lower, such as at higher altitudes. These lower temperatures may not be sufficient for extracting all the molecules one might enjoy from tea leaves, so, the brewed tea may not be as flavorful.

16. Baking soda and baking powder are both used in baking recipes. They are different mixtures, but both produce the same result. They are leavening agents. Baking soda is a solid basic salt, $NaHCO_3$. Baking powder is a solid mixture of baking soda and a solid acid of some sort, usually cream of tartar, an acidic salt, $KHC_4H_4O_6$. Both leavening agents react to produce carbon dioxide, CO_2, gas which is what causes the airy texture in many baked goods. Baking soda is used with the other ingredients, including some significant acid already, like vinegar or buttermilk. The reaction with vinegar is $HCO_3^-(aq) + HC_2H_3O_2(aq) \longrightarrow C_2H_3O_2^-(aq) + H_2O(l) + CO_2(g)$. In baking powder, once the mixture is in contact with a water solvent, as is the case in many baking recipes, the two acid and base components react: $HCO_3^-(aq) + HC_4H_4O_6^-(aq) \longrightarrow C_4H_4O_6^{2-}(aq) + H_2O(l) + CO_2(g)$. Baking powder is often used in batters that do not have acid in them already.

17. Between 1–2% of alcohol remains in bread. Bread with longer fermentation times will contain higher alcohol concentrations.

18. Microwave radiation causes changing electric fields. Because water is polar, it responds to the changing negative and positive conditions, alternating between repulsions and attractions that ultimately increase the motion of the water molecules. It is this motion (friction) that heats up the food. Other polar molecules are likely to respond in this manner. Among the three molecules listed, only ammonia (NH_3) is polar and is likely to respond like water to microwaves.

19. Capsaicin is the common name for the chemical responsible for the chemical heat in peppers. It has a chemical formula ($C_{18}H_{27}NO_3$), and its molecular structure is

Drinking water is not useful to reduce the burning sensation because capsaicin is insoluble in water; it is mostly a nonpolar molecule. The water merely spreads the capsaicin more around your mouth and across your tongue. Drinking milk or beverages with more fat that can dissolve the capsaicin may be more beneficial for reducing the burning sensation.

22. "Flavor tripping" is a nickname for the concept in the culinary world that deceives the other senses for what you are about to taste. For example, most people see and touch a lemon wedge and are ready for a very sour taste, but after eating a miracle berry, the raw lemon wedge tastes like the sweetest lemonade. Miracle berries contain a particular molecule called miraculin, which is a glycoprotein. It interacts with particular receptors on your tongue, the ones that detect sour, and block them from detecting the sour part, so the lemon tastes sweet.

This concept of blocking flavors might be interesting because it could make things that are perfectly nutritious but taste dreadful more palatable, which could expand diet options for feeding a growing world population.

24. A food mile is a measure of the energy needed just to transport a particular food item from the point of production to the point of consumption. For example, to someone in the midwestern United States, obtaining a pineapple will use more food miles than obtaining an ear of corn.

We can reduce food miles by eating foods that are produced relatively close to where we consume them—the "eat local" movement. Another option is connected to the flavor tripping notion from Question 22. If we can make local, nutritious foods that don't naturally taste good more palatable, we may have more options for eating locally!

Regardless of your opinion of the importance of reducing food miles, the concept of energy use and its ongoing availability is a factor in this analysis.

26. The Maillard reaction takes place only when there is no measurable water in the cooking environment. On a hot

grill, meat will brown because the grill temperatures cause the water on the surface of the meat to evaporate. The meat can then "brown" on the surface. Done properly, this browning sears the surface of the meat and locks moisture inside the meat so it does not dry out. (With moisture, though, the inside will not brown.) Food items do not brown in boiling water because of the abundance of water in the environment, and food cooks in the microwave only if water is present. Recall earlier questions about how a microwave cooks food.

29. **a.** Teflon is a long-chain polymer of repeating units of the monomer C_2F_4. It is mostly nonreactive due to the strong carbon-to-fluorine bonds. Even though the Teflon contains fluorine, a highly electronegative element, the arrangement of the fluorine atoms in the polymer cancel each other and the polymer is nonpolar, making it not want to interact with water. Because nearly all the foods we consume contain some degree of water, cooking with Teflon still allows the heat to transmit but does not cause the food to stick, because the Teflon repels the water in the food.

   b. Ultimately, the side of Teflon that will "stick" to the pan has to be chemically altered to get it to bind to the other pan materials. This is done by breaking many of the carbon-to-fluorine bonds on the side that will stick, making it more chemically conducive to binding to the other pan materials.

30. Gelatin, the primary component in Jell-O, is a protein that provides structure to the liquid mixture, such that once cooled and/or some water evaporates, the web-like structure of the gelatin molecules allows the material to hold a relatively solid shape. (This is similar to how gluten provides structures for many baked breads and goods.)

   Pineapple, kiwi, and papaya contain an enzyme called bromelain, which catalyzes the breakdown of the gelatin protein molecules into its smaller components—its amino acids. These smaller building blocks, when broken apart, cannot provide enough structure to allow the liquid and the Jell-O mixture does not solidify. Ironically, extracts of these fruits are used as digestive aids, to help us when our bodies are not able to do an effective job of breaking down the proteins we consume.

33. A review may be found online here: https://www. frontiersin.org/articles/10.3389/fbioe.2021.612285/ full. Some examples include organic acids (such as lactic acid and acetic acid), vitamins (e.g., folic acid, vitamin C, riboflavin), and a variety of flavor substances in fermented foods.

Chapter 11

1. Malnutrition is caused by a diet lacking the proper mix of nutrients, even though the energy content of the food eaten may be adequate. Undernourishment is caused by the insufficient energy content of the food eaten.

3. **a.** The three different types of macronutrients are fats, carbohydrates, and proteins.

   b. Fats are the highest in energy content, almost a factor of two higher than carbohydrates and proteins, which are similar in energy content.

5. The pie chart indicates more carbohydrate is present than would be found in steak and more protein than would be found in chocolate chip cookies. Of the choices given, the pie chart is likely to represent peanut butter with a high oil content. (See Table 11.1 for confirmation.)

7. A steak is 28% protein, 15% fat, and 57% water.

   18 oz × 0.28 = 5.0 oz of protein

   18 oz × 0.15 = 2.7 oz of fat

   18 oz × 0.57 = 10 oz of water

9. **a.** Here is the structural formula for lactic acid:

   b. Lactic acid would be saturated as a fatty acid because the hydrocarbon chain contains only single bonds between the carbon atoms.

   c. No, lactic acid is not a fatty acid. Although it has the carboxylic acid group characteristic of a fatty acid, it lacks the long hydrocarbon chain (12–24 carbon atoms). Lactic acid also has a hydroxyl group (—OH) that is not found in fatty acids.

11. **a.** flaxseed oil

   b. canola oil

   c. safflower oil

   d. coconut oil

13. **a.** Milk (lactose is "milk sugar")

   b. Many fruits contain fructose ("fruit sugar"). So does honey.

   c. Sucrose ("table sugar") originates from both beets and sugarcane.

   d. Starch is a primary component of potatoes, rice, tapioca, taro root, wheat, and corn.

15. Both starch and cellulose are polymers in which the monomer is glucose. But the glucose units are hooked together differently. Our bodies possess an enzyme that can digest starch. In contrast, we lack the enzyme that would enable us to digest cellulose. Essentially, we can derive nutritional value from a potato but not from a piece of paper.

17. See Figure 11.12 for the chemical structures of fructose and glucose. Observe that the chemical structure of fructose is based on a five-membered ring composed of four C atoms and one O atom. In contrast, the chemical structure of glucose is based on a six-member ring composed of five C atoms and one O atom. Glucose has one —CH_2OH side chain, and fructose has two.

19. The "amino" in *amino acid* indicates that there is an amine functional group present. The "acid" indicates there is an acidic functional group, in this case, a carboxylic acid.

21. Analogous to Equation 11.4a:

$$\text{(chemical structures of two amino acids reacting)} \longrightarrow$$

$$\text{(dipeptide structure)} + H_2O$$

23. Phenylketonuria (PKU) is a disease in which people lack the enzyme necessary to metabolize phenylalanine, an amino acid. Without the enzyme, phenylalanine accumulates in the body and eventually causes problems in brain development. People who are phenylketonurics must carefully limit their intake of foods rich in proteins. Although aspartame is a sweetener, it is a dipeptide of aspartic acid and phenylalanine, as shown in Figure 11.19. Sucralose, with a chemical structure unrelated to amino acids, does not pose any risk to people with this disease.

25. The two methods are increasing crop yields (e.g., by using fertilizers) and devoting more land to agriculture (e.g., deforestation).

27. Several answers are possible. For example, food production may diminish water quality through the amount of (1) water required for irrigation that may deplete aquifers, causing more salty water to seep in; (2) fertilizers used, which may run off into waterways providing excess nutrients that promote algal blooms; (3) herbicides and insecticides used that may contaminate local streams and rivers; and (4) draining of wetlands that serve as natural filters for water.

29. **a.** *Reactive nitrogen* refers to the chemical species of nitrogen that cycle relatively quickly through the biosphere and interconvert via several pathways.

b. Atmospheric nitrogen gas (N_2) is an unreactive form of nitrogen.

c. A natural source of reactive nitrogen is nitrogen-fixing bacteria in soils. An unnatural source of reactive nitrogen is synthetic fertilizers from the Haber–Bosch process.

31. To a chemist, all food is composed of chemical compounds ("chemicals"). These include fats, carbohydrates, proteins, minerals, and water. As a result, it is impossible to go on a "chemical-free" diet. Admittedly, though, the term "chemicals" is commonly taken to mean "added chemicals" or perhaps even "bad chemicals." An "all organic" diet is possible, though, and most organic foods strictly limit what chemicals can be used to raise crops and can be added during food processing.

33. In terms of food chemistry, hydrogenation is the process of "adding hydrogen" to an unsaturated molecule to make it saturated (or more highly saturated). On a molecular level, in hydrogenation a molecule of H_2 is "added" to a C=C double bond to form two new C—H bonds and a C—C bond. Partial hydrogenation is when some of the C=C bonds are hydrogenated but not all of them. The manufacturers must report this because these are two different chemical substances (i.e., partially hydrogenated soybean oil is different from soybean oil). These substances have different properties and different effects on your health.

35. The process of hydrogenating an oil to "add H atoms" converts some of the C=C double bonds in the oil to C—C single bonds. This is desirable in that it improves the product's shelf life (and sometimes the spreadability). However, hydrogenation produces *trans* fats as a side-product, a drawback because *trans* fats both increase the "bad" cholesterol and decrease the "good" cholesterol. Interesterification also reduces the number of C=C double bonds but accomplishes this in a way that does not produce *trans* fat.

37. This is not a good plan based only on the percent of saturated fat in coconut oil relative to the butterfat in cream. Coconut oil is 87% saturated fat, but butterfat is only 63% saturated fat. However, if a person uses a smaller quantity of the nondairy creamer than of cream, his or her total amount of saturated fat consumption may be reduced.

39. According to Table 11.1, peanut butter is a good source of protein but is high in fat/oil, which in this case is peanut oil. On the positive side, unless the peanut butter has been hydrogenated, peanut oil is largely unsaturated, as shown in Figure 11.8.

41. a. $1.5 \text{ g fat} \times \dfrac{9 \text{ Cal}}{\text{g fat}} = 10 \text{ Calories from fat}$

$17 \text{ g carbohydrate} \times \dfrac{4 \text{ Cal}}{\text{g carbohydrate}}$

$= 70 \text{ Calories from carbohydrate}$

$3 \text{ g protein} \times \dfrac{4 \text{ Cal}}{\text{g protein}} = 10 \text{ Calories from protein}$

The total number of Calories is 90 Calories per slice of bread.

b. The percent Calories from fat is

$\dfrac{10 \text{ Cal}}{90 \text{ Cal}} \times 100 = 11\%.$

Note: To one significant figure, this is 10%. This is something worth keeping an eye on, but not something we have emphasized in this textbook.

c. One slice of whole wheat bread qualifies as a nutritious food because it provides a serving of whole grains with few additional Calories from sugars and fats. Remember that if one slice of bread counts for one serving, a sandwich counts as two servings of bread.

43. a. Carbohydrates

b. Possibilities include corn, sugarcane, and tapioca.

c. Digesting the starch to make glucose, fermentation of the glucose to produce ethanol, and distillation of the ethanol. See Section 6.14 for more details.

d. One controversy involves energy costs. The energy gained by burning ethanol in an engine may be less than the energy inputs to produce it. Another controversy involves the environmental costs of growing corn and cutting down forests to produce sugarcane.

45. Answers will vary but could include eating lower-Calorie food, eating a greater variety of food, buying locally grown food, and eating less resource-intensive food such as red meats.

47. Assumptions made in your estimate might include (1) the number of days a year you drink soft drinks (non-diet), (2) the volume you drink or the packets of sugar you add to coffee or tea, and (3) the type of soft drink and grams of sugar it contains.

49. a. There are zero Calories in a packet of Splenda.

b. Splenda is made from sucrose by selectively replacing three of the —OH groups with —Cl groups to produce a molecule that is 600 times sweeter than sucrose. It is made from sugar but is a different chemical compound.

51. a. As you can see from the pie charts, soybeans are higher in protein (~35%) than wheat (~13%). Soybeans can be used to produce many different food products, including soy milk, textured vegetable protein, and tofu.

b. People in some cultures, particularly in Asian countries, are accustomed to obtaining their protein from soy. The taste of soy-based beverages or other foods is familiar and widespread, which would increase soy's appeal and acceptance.

Chapter 12

2. a. Because the $K < 1$, the reactant (here, glucose) is favored over the product (here, fructose).

b. For a system at equilibrium, the ratio of

$K = \dfrac{[\text{fructose}]}{[\text{glucose}]} = 0.74$ Because we know the

concentration of glucose, we can solve for the concentration of fructose using this formula:

$\dfrac{[\text{fructose}]}{0.22 \text{ mM}} = 0.74$

$[\text{fructose}] = 0.74 \times 0.22 \text{ mM}$

$[\text{fructose}] = 0.16 \text{ mM}$

4. In the figure, the AB complex is in the position of the reactants leading to A free and B free as products.

$K = \dfrac{[\text{A free}][\text{B free}]}{[\text{AB complex}]}$

6. Acetic acid since it is closest to our desired pH.

8. There are five different isomers. Here are the structural formulas and line-angle drawings. The hydrogen atoms in the structural formulas have been omitted for clarity.

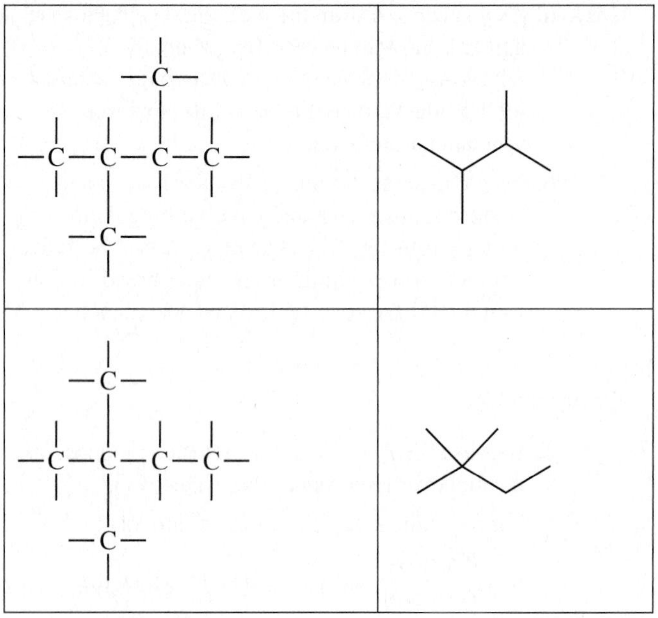

10. **a.** ether

    b. carboxylic acid

    c. ketone

    d. amide

    e. ester

12. **a.** The compound is an alcohol (ethanol). An isomer with a different functional group (an ether) is

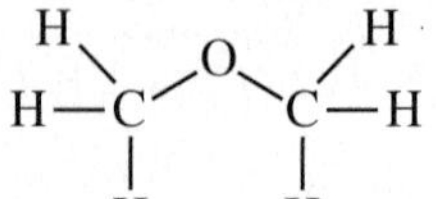

    b. Aldehyde. An isomer with a different functional group (a ketone) is

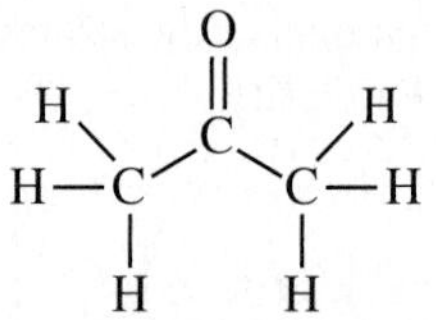

    c. Ester. An isomer with a different functional group (a carboxylic acid) is

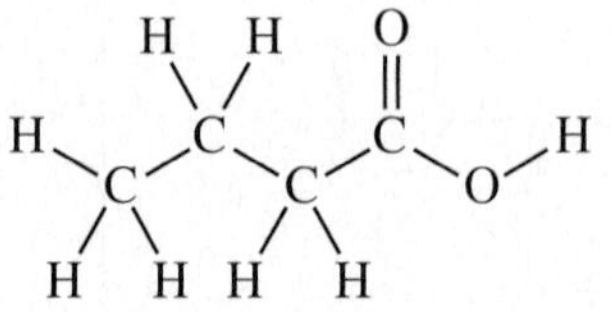

13. **a.** The chemical formula is $C_5H_9N_3$.

    b. The amine functional group, $-NH_2$, is present in histamine.

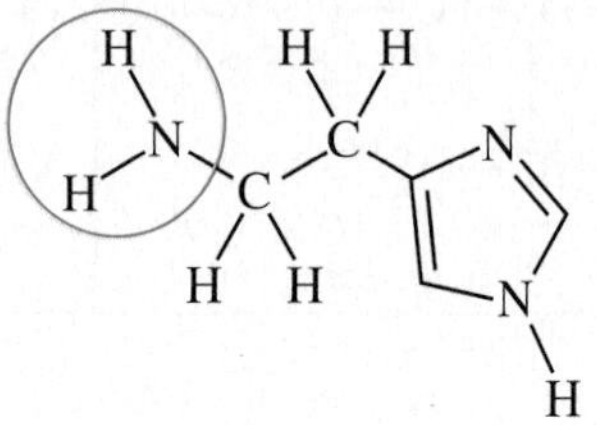

    c. The molecule's most polar region will be around the amine group. This can easily take on a proton in water and (particularly when charged) will be the principal part of the molecule interacting with polar water molecules.

16. **a.** This compound cannot exist in chiral forms. The central carbon atom is bonded to two equivalent $-CH_3$ groups.

    b. This compound can exist in chiral forms. The four groups attached to the central carbon atom are all different.

    c. chiral

    d. not chiral

18. Hormones are chemical signals produced in the body to regulate many physiological events, such as the metabolism of food (insulin), our response to sudden or dangerous events (adrenaline), and many others. Receptors are the proteins that bind to these hormones to transfer the information in that chemical signal. A chemical without a switch to flip will not serve a purpose. Similarly, a protein receptor without a chemical to activate it will be passive.

21.

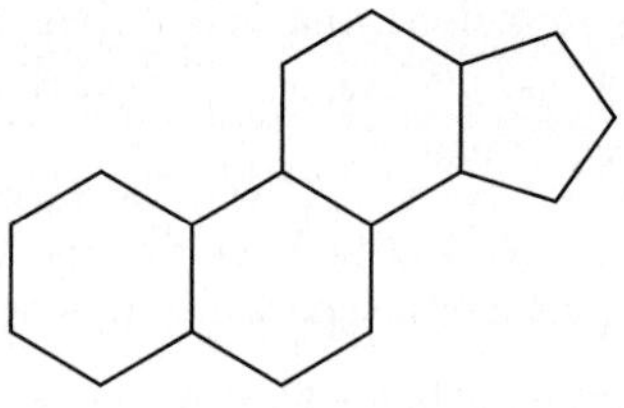

22. Acetic acid is CH_3COOH.

    a.

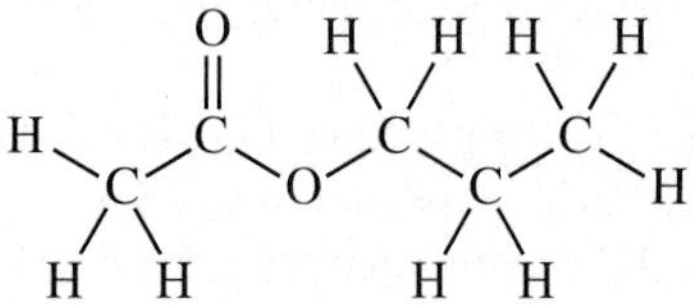

    b.

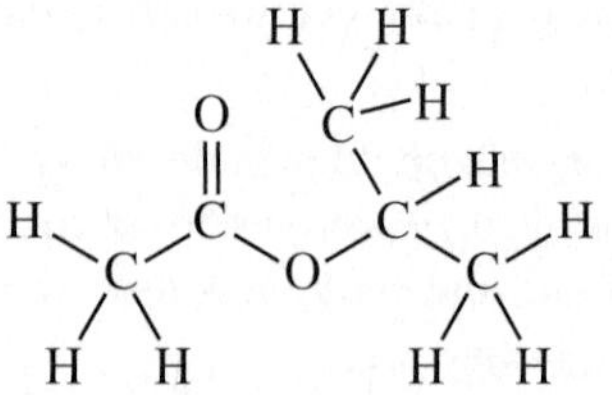

    c.

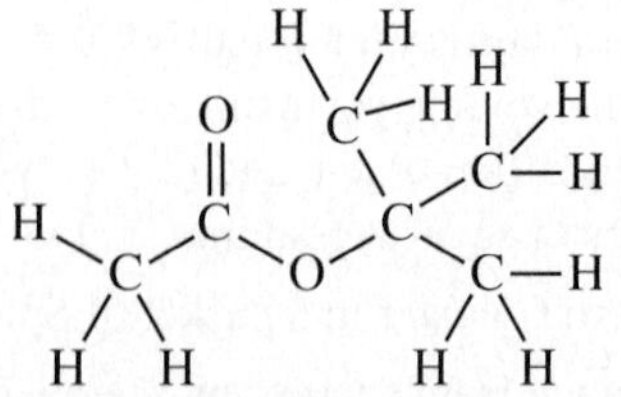

24. A pharmacophore is the three-dimensional arrangement of atoms, or groups of atoms, responsible for the biological activity of a drug molecule.

26. The equilibrium constant shows whether the reactants or products are at a lower energy level. That constant does not provide any information about the barrier between the two. To be more technical, the equilibrium constant provides information about thermodynamics, not kinetics.

28. a. Four single bonds, one triple bond

b. Six single bonds, one double bond

c. 11 single bonds, four double bonds

29. Only three isomers are shown here; some structures are duplicates. #1 and #5 are different representations of the *same* isomer. #2, #3, and #4 represent the *same* isomer. #6 is an isomer *different* from #1 to #5.

31. a. The chemical formula is $C_{16}H_{21}N_3$.

b. Both compounds have nitrogen-containing rings but are not the same size. The major structural similarity is the presence of the

$$-CH_2\ CH_2N\diagdown^{\diagup}$$

group, which is likely the part of each molecule that competes to attach to the receptor site.

34. a. The cellular membrane is primarily composed of lipids.

b. Glycogen is a polymer of glucose, specifically a polysaccharide or, more commonly, a carbohydrate.

c. Enzymes are protein-based catalysts.

37. The analogy compares the receptor site on the surface of a cell to a lock that a unique key can only open. Drug molecules can only bind to receptor sites that match the molecule's geometry.

39. a. Aspirin produces a physiological response in the body.

b. Estrogen causes a physiological response.

c. Antibiotics kill or inhibit the growth of bacteria that cause infections.

d. Penicillin inhibits the growth of bacterial infections.

e. Morphine induces a physiological response.

41. a. The codeine structure has two ether groups, an alcohol group, and an amine group.

Codeine

b. No, a comparison of only two drugs is not enough evidence to draw general conclusions about the role structural changes play in determining drug effectiveness and addictiveness.

44. a. There are 6(4) + 6(1) or 30 electrons available. Here is one possible structural formula of a linear isomer for benzene:

b. Here is the condensed structural formula:
$CH_2{=}C{=}CH{-}CH{=}C{=}CH_2$

c. First, check to see if all of the structures correctly represent C_6H_6 and that each carbon has four bonds. If these conditions are met, the structures with double bonds should differ only in the placement of the lone C—C single bond. However, structures including carbon–carbon triple bonds can also be drawn; these would be distinctly different.

45. Thalidomide's two optical isomers have very different effects in the body. One isomer treats nausea, whereas the other produces mutations in babies born to women who take the drug early in pregnancy. Unfortunately, the body can convert each isomer into the other. When the German maker of thalidomide applied to the FDA for approval to market the drug in the U.S. in the 1960s, its application was rejected repeatedly due to a lack of data proving its safety. In 1998, the FDA approved thalidomide for treating skin lesions caused by leprosy, provided that patients are not pregnant and take precautionary measures to avoid becoming pregnant while on the medication.

48. Information on Presidential Green Chemistry Challenge Awards is collected online. The companies collaborated to develop a novel synthesis incorporating an evolved enzyme. The prior synthetic route required high-pressure, expensive metals, and a costly purification step that can now be eliminated. This matches our understanding of the utility of enzymes as biological replacements for organic synthetic steps. If selected correctly, enzymes can enable the reaction at lower temperatures and with less solvent waste. For more information, visit the U.S. EPA website and search "Presidential Green Chemistry Challenge Winners" for more information.

52. The life of Dorothy Crowfoot Hodgkin (1910–1994) is documented in several biographies. Search the San Diego Supercomputer Center website for a short biography of Dorothy Hodgkin. Dorothy was born in Cairo, Egypt, where her father was in the Ministry of Education and administered archaeological sites. Her mother was a self-taught botanist. Dorothy was educated at Oxford and, with her mentor, J. D. Bernal, she first applied X-ray diffraction to crystals of biological substances, including pepsin, penicillin, cholesterol, and later insulin. Hodgkin was elected a Fellow of the Royal Society in 1947 after publishing the structure of penicillin and was awarded the Nobel Prize in Chemistry in 1964 for solving the structures of important biomolecules, such as vitamin B_{12}. In the words of colleague Max Perutz (Nobelist for his solution of the hemoglobin molecule structure), she was "a great chemist, a saintly, gentle, and tolerant lover of people, and a devoted protagonist of peace."

54. **a.** You will not use the entire bottle before it expires.

    b. Aspirin breaks down over time. It undergoes hydrolysis to form salicylic acid and acetic acid, which don't treat the symptoms you take aspirin for. Salicylic acid, remember, led to some concerning side effects.

55. Search the National Public Radio's (NPR) website for "drug expiration dates" to find an article on why drug expiration dates may be more myth than fact.

Chapter 13

2. Farming and medicine are the most obvious industries that have changed, as seen throughout this chapter, but other options are available. Plastics and recycling are also changing due to new methods for making materials with enzymes. Proposed answers can include the fuel industry with increasing biofuel development.

3. DNA is composed of four nucleotides whereas proteins are built from 20 amino acids. Proteins have primary, secondary, and tertiary structures, whereas DNA is typically only described by its primary and secondary structure. Also, proteins are made on ribosomes and DNA is made in the nucleus of cells.

5. **a.** A nucleotide links a nitrogen-containing base, a sugar, and a phosphate group.

    b. Covalent bonds hold the units together.

8.

9. **a.** Nucleotides polymerize to form DNA when the phosphate group from one nucleotide reacts with the hydroxyl group on another. The nucleotide shown has two hydroxyl groups. The one closer to the phosphate group is the correct site for polymerization, similar to DNA. The second hydroxyl group results in the significant chemical instability seen in this DNA-related polymer.

    b. In the name DNA, the D stands for the deoxyribose in deoxyribonucleic acid. The name for the polymer built with ribose instead is RNA for ribonucleic acid.

12. Sequences are always read in one direction. We read each sequence from left to right on paper to determine which amino acid it codes for. Chemically, the directionality comes from the specific order of bonds connecting the alternating sugars and phosphates in the backbone. Although the second sequence in this question is simply the reverse of the first base sequence, it would code for a different amino acid. (ATG codes for methionine; GTA codes for valine.)

13. **a.** The complementary base sequence is ATAGATC.

    b. Your answer should have two lines between each A and T and three lines between each C and G in the sequence.

15. Each codon consists of a three-nucleotide sequence specific for an amino acid or the start/stop of protein synthesis. All of the codons together make up the code for translating a DNA sequence into a protein's amino acid sequence.

16. **a.** All amino acids must follow the same general structure (see the image below). Only the R group varies between acidic, basic, and neutral amino acids. Correct example R groups for each category include:

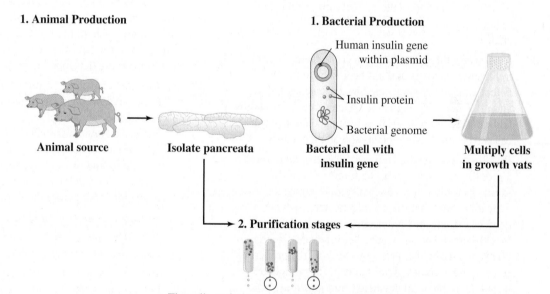

Acidic = glutamic acid ($-CH_2CH_2COOH$)
Basic = lysine ($-CH_2CH_2CH_2CH_2NH_2$)
Polar neutral = serine ($-CH_2OH$)

b. All three categories can make hydrogen bonds, but only acidic and basic amino acids can also make ionic bonds.

c. Acidic amino acids have carboxylic acids within their R groups, whereas basic amino acids often have amine groups. Neutral amino acids may contain hydroxyl groups or amides in their side chains.

18. Only a minor change in the amino acid composition of human hemoglobin leads to sickle-cell anemia. A nonpolar valine in the hemoglobin S chain replaces a specific charged glutamic acid in the sequence. This seemingly innocuous change in the protein's primary structure dramatically affects the shape of the protein. The tertiary structure must change to accommodate the change in the side chain; the nonpolar valine cannot positively interact with water or other polar groups in the same ways as glutamic acid. The change in protein shape leads to sickled red blood cells (under certain conditions) and several health problems.

21. The steps for both actually have several parallels. Both require a source grown in the lab or animal and at least some purification steps before the final medicine. (See the diagram below for more information.)

1. Animal Production   **1. Bacterial Production**

Human insulin gene within plasmid

Insulin protein

Bacterial genome

Animal source   **Isolate pancreata**   **Bacterial cell with insulin gene**   **Multiply cells in growth vats**

2. Purification stages

The cells or tissue must be broken open. Then, the cell contents, a mixture of biomolecules, are separated by size, charge, or other characteristics. Typically, separation occurs by passing through a series of columns.

Injectible purified insulin

23. Adenine-thymine base pairs are less stable than cytosine-guanine base pairs because they have fewer hydrogen bonds holding them together.

24. **a.** X-rays have high energies and wavelengths from 0.01 to 10 nm in the electromagnetic spectrum.

   b. A beam of X-rays is directed at an unknown substance. The nuclei in the substance scatter the X-rays. A detector measures the intensity and pattern of scattered X-rays. If the atoms in the substance are arranged in a regular pattern, the diffracted X-rays can be used to calculate the distance between atoms.

c. One reason is that salts, such as sodium chloride, easily form crystals. In contrast, nucleic acids and proteins are much larger and do not easily form crystals. Another reason is that when nucleic acids and proteins crystallize, their X-ray diffraction patterns are far more complex and difficult to interpret.

26. The two DNA strands are complementary, meaning that the sequence on one strand can be reconstructed from the other. If a single strand breaks, the other strand can be used by repair enzymes to replace any lost nucleotides correctly and reassemble the backbone. If both strands break, the information on how to repair the break is lost, and the correct sequence of DNA will be permanently altered.

28. Many amino acids are represented by more than one DNA codon. For example, GTT, GTC, GTA, or GTG all translate to valine. If an error in the base sequence of DNA did not change the meaning of the codon, the protein would contain the correct amino acid. This redundancy makes the genome more resilient to change.

30. Reasons for the low public concern include (1) insulin treatment is not new and has been seen to be safe before genetically engineered bacteria generated it, and (2) genetically engineered bacteria are kept within laboratories and industrial facilities and are therefore easier to control, while plants spread more easily.

33. Rosalind Franklin was born in 1920, the daughter of a prominent London banking family. All of the family's children were encouraged to pursue an education. Both her undergraduate and graduate degrees were from Cambridge University. During World War II, she suspended her graduate research to contribute to the war effort by studying the properties of coal and graphite. After the war, she completed her Ph.D. in physical chemistry and joined a prominent laboratory in Paris, where she was introduced to the technique of X-ray crystallography. She soon became an expert in the field, moved back to London in 1951 to work at King's College, and in 1952 produced X-ray photographs of DNA. Before publishing the images, she showed them to Maurice Wilkins, another scientist studying DNA. Without Franklin's knowledge, Wilkins shared the photographs with James Watson, a molecular biologist working with Francis Crick to describe the structure of DNA. The X-ray data changed Watson and Crick's hypothesis about the structure, and they published a paper describing DNA as a double helix in 1953. Watson, Crick, and Wilkins shared the Nobel Prize in Physiology or Medicine in 1962, but Franklin did not live to be considered for the prize.

Franklin died of ovarian cancer in 1958. For more information, search the San Diego Supercomputer Center website for Rosalind Franklin or reference *Chemical Achievers: The Human Face of the Chemical Science*, a publication of the Chemical Heritage Foundation, Philadelphia, Pennsylvania, 1997.

34. The most popular explanation for the persistence of the sickle-cell trait is that carrying the trait protects against death from malaria. Certainly, the regions and climates where the trait is historically common overlap closely with regions affected by malaria. The reason that the trait provides resistance to malaria has yet to be fully understood.

36. Patenting GM plants will give distinct advantages to industrialized nations with the funds to conduct further research on GM organisms. The patenting process may be advantageous because (1) it provides a financial incentive to perform the costly research required, and (2) it increases the control and decreases the spread of artificial GM crops. The process may be a disadvantage because it (1) prevents the creative and traditional breeding process by farmers and smaller-scale plant breeders and (2) halts further research by academic scientists into a specific gene. Also, more importantly, patenting may make the technology too expensive for lower-income nations, that many GM crops are being developed to help.

38. **a.** An excellent source of information specific to the EU is the European Commission website. Examples include rapeseed, soybean, carnations, and corn. New traits include herbicide tolerance and insect resistance. Some have been approved for import and processing; very few have been accepted for cultivation. Typically, those allowed were cultivated before the onset of regulations.

b. Examples of biotech timeline:

1973: The first genetically engineered bacteria producing a human protein are created.

1997: Two scientists at Harvard University, Walter Gilbert and Allan Maxam created a method for sequencing DNA using chemicals instead of enzymes. This technique was an important start to identifying DNA sequences efficiently.

1980: The U.S. Supreme Court ruled that genetically engineered organisms can be patented. The ruling allowed Exxon to patent an oil-eating microorganism. More importantly, it opened the door for the commercialization of genetically engineered technologies.

1980: Kary Mullis and others at Cetus Corporation in Berkeley, California, created a method for copying target sequences of DNA outside of cells. The technique is now commonly used in labs across the globe to copy, alter, or create DNA sequences.

1982: Genentech Inc. received FDA approval for the first genetically engineered drug: human insulin protein produced by bacteria.

40. Gene therapy involves introducing normal genes into patients lacking them. It would allow us to treat a specific subset of diseases caused by clear genetic alterations. High-risk clinical trials have seen disappointing results and some promise for a small subset of medical conditions. Other diseases might be amenable to treatment with gene therapy, but developing appropriate protocols is costly, in both time and money. The Human Gene Therapy Subcommittee of the Recombinant DNA Advisory Committee of the National Institutes of Health must approve all proposed uses of gene therapy.

Example histories: The first person treated was a four-year-old girl suffering from severe combined immunodeficiency disease (SCID) in which a genetic defect prevents the formation of a specific enzyme necessary for the health of white blood cells. Children with SCID have extremely weak immune systems and often die before adulthood. Initial results showed promise, but in January 2003, the FDA temporarily halted all gene therapy trials using retroviral vectors in blood stem cells after two children developed a leukemia-like condition. (The first case of leukemia development in a child participating in a clinical trial occurred in 2002.)

Recently, gene therapy has been used more successfully to treat neural degenerative disorders and blindness. Here, the key to this success is that the genes were more carefully controlled and specifically placed. At the time of this writing, studies to assess the long-term effects are ongoing. You can find more details by searching for "gene therapy" on the Science Daily and the U.S. Food and Drug Administration websites.

Chapter 14

1. **a.** Class K (wet chemical, such as aqueous potassium carbonate that includes a detergent)

   b. Class C (carbon dioxide or dry chemical such as ammonium phosphate or sodium bicarbonate)

   c. Class A (water only)

4. Ester, alkene, and nitrile ("cyano") functional groups.

6. Many possible examples, such as accidentally stepping in a pool of blood in a dimly lit house, inadvertently removing or smearing fingerprints when opening doors to secure the scene, or maybe knocking something over as they move through the scene. They may also be tempted to pick up gauze, latex gloves, etc., left by emergency medical personnel or to "tidy up" before more personnel arrive. They may also not wear protective outer clothing and therefore may leave fibers from their own clothing at the crime scene.

9. Many possible examples, such as no sign of forced entry, or the entry is forced beyond what would be required to gain access; only one specific item was stolen; no search for any valuables in an apparent burglary, or no items have been stolen; excessive ransacking, or being too careful of specific items (some items set aside to protect them); survivor of an attack has minor wounds only on the side of the body opposite their own handedness (self-inflicted).

11. The order of places where, and the persons with whom, physical evidence was located from the time it was collected to its submission at trial.

13. Any solvents that have a flash point less than 25 °C would ignite in contact with a match. These include diethyl ether, acetone, tetrahydrofuran, hexanes, ethanol, acetonitrile, and toluene.

15. 2500 nm to 25,000 nm

17. This can be solved by taking the natural log (ln) of both sides of the equation: $2^n = 1,000,000$. That is, $n \ln (2) = \ln (1,000,000)$ or $n = \ln (1,000,000)/\ln (2)$. This will equal 19.9 or 20 cycles. Refer to Appendix 3 for more information about the math behind logarithms.

19. The spectrum shows the presence of alkyl C—H groups (strong C—H stretching peaks around 3000 cm^{-1} and C—H bending peaks around 1380–1400 cm^{-1}), a C=O carbonyl group (peak around 1700 cm^{-1}), and perhaps an ester C—O peak (1170 cm^{-1}). The absence of a medium-strong, broad peak around 3500 cm^{-1} indicates that an alcohol or carboxylic acid —OH is not present. You are not able to determine the concentration of functional groups based on their peak heights; IR spectroscopy is only used to determine whether certain functional groups are present.

20. No. All of these compounds would show the presence of C—H, C=O, C—O and O—H groups. In order to distinguish between these compounds, mass spectrometry or nuclear magnetic resonance (NMR) spectroscopy would need to be performed.

23. Yes. Depending on the alloy comprising the gun, various acid mixtures may be applied to the surface, which will etch the metal and reveal the serial number that lies beneath the scratch marks.

24. In addition to fingerprints, earprints (e.g., left behind in a crime scene by someone listening through a door), or recognizing people by their gait—the way they walk.

25. The mass of an electron is 9.11×10^{-31} kg; by plugging these values into the equation, the best resolution would be 1.2×10^{-11} m. This corresponds to 0.012 nm or 0.12 Å.

27. **a.** Not necessarily. Luminol also exhibits bright chemiluminescence upon contact with other substances such as some paints, varnishes, fruit and vegetable juices, and iron-containing compounds.

b. Although luminol may have an effect on the typing of bloodstains using conventional serological testing, it has been found that luminol *does not interfere* with DNA analyses.

30. Visual inspection of the tungsten filament can be used to determine whether headlights were illuminated. The tungsten filament inside sealed headlights gets extremely hot when headlights are turned on. After a crash with an impact speed of >20 km/h, a hot filament will be significantly deformed, whereas a cold filament will resemble a spring with little/no deformation observed. Often, the glass bulb will shatter during impact, which will allow oxygen to react with the hot filament. This will result in a yellowish-white powder that will coat the surface of the filament, which is readily observable by using scanning electron microscopy. Furthermore, small glass particles and grains will implode into the bulb, which will adhere to the filament as it cools down. In contrast, a cold filament will not react with oxygen and will remain clean and shiny. Glass particulates will not adhere to a cold filament.

Glossary

The numbers at the end of each term indicate the section(s) where the term is defined and explained in the text.

A

acid a compound that releases hydrogen ions, H^+, in aqueous solution (5.8)

acid-neutralizing capacity the capacity of a lake or other body of water to resist a decrease in pH (5.11)

activation energy the energy necessary to initiate a chemical reaction (6.13, 12.1, and 14.4)

active site the catalytic region, often a crevice, in an enzyme that binds only specific reactants and accelerates the desired reaction (12.6 and 13.5)

addition polymerization A type of polymerization in which the monomers add to the growing chain in such a way that the polymer contains all the atoms of the monomer. No other products are formed (9.3)

aerosols tiny solid and/or liquid particles that remain suspended in air (2.11 and 4.8)

albedo a measure of the reflectivity of a surface; the ratio of electromagnetic radiation reflected from a surface relative to the amount of radiation incident on it (4.8)

alcohol a hydrocarbon substituted with one or more —OH groups (hydroxyl groups) bonded to its carbon atoms (6.14)

alkane a hydrocarbon composed of only single bonds between neighboring carbon atoms (6.11)

allotrope different structural forms of the same element (1.6)

alternative energy energy sources that are not reliant on the combustion of fossil fuels (7.1)

ambient air the air surrounding us, usually meaning the outside air (2.7)

amino acid Monomer from which our body builds proteins. Each amino acid molecule contains two functional groups: an amine group (—NH_2) and a carboxylic acid group (—COOH) (9.7)

amino acid residues amino acids which have been incorporated into the peptide chain (11.6 and 13.4)

amorphous disordered material a solid in which the constituent atoms, ions, or molecules are arranged in a random, disordered array (1.8)

amorphous regions a region of a solid that is composed of a random, disordered array of substances (9.5)

amphiphilic a molecule that has both nonpolar and polar groups that make it both lipophilic (fat-soluble) and hydrophilic (water-soluble) (10.10)

anion a negatively charged ion formed by adding an electron(s) to a neutral species (5.4)

anode the electrode at which oxidation takes place (8.1)

anthropogenic caused or produced by human activities, such as industry, transportation, mining, and agriculture (4.4)

aqueous solution a solution in which water is the solvent (5.4)

aquifer an underground permeable rock formation from which groundwater may be extracted using a well (5.3)

aridification the gradual change of a region from wetter to drier climate due primarily from the effects of climate change (5.3)

atom the smallest unit of an element that can exit as a stable, independent entity (1.2)

atomic mass the mass (in grams) of the same number of atoms that are found in exactly 12 g of carbon-12 (4.2)

atomic number the number of protons in the nucleus of an atom (1.5)

autoignition temperature the minimum temperature at which the vapor of a substance spontaneously ignites, even in the absence of an ignition source (14.4)

Avogadro's number equal to 6.02×10^{23}, which is the number of individual species (e.g., atoms, ions, or molecules) in one mole of a substance (4.2)

B

background radiation a measure of the level of ionizing radiation present in the environment, originating from both natural and man-made sources (7.5)

basal metabolic rate (BMR) the minimum amount of energy required per day to support basic body functions (11.8)

base a compound that releases hydroxide ions, OH^-, in aqueous solution (5.8)

battery an energy-storage device that converts the energy released from spontaneous chemical reactions into electrical energy (8.1)

bioaccumulation the increase in concentration of a substance such as a toxin over time in biological creatures (5.7)

biofuel a generic term for a renewable fuel derived from a biological source, such as trees, grasses, animal waste, or agricultural crops (6.14)

biological oxygen demand (BOD) A measure of the amount of dissolved O_2 that microorganisms use up as they decompose the organic wastes found in water. A low BOD is one indicator of good water quality (5.12)

biomagnification the increase in concentration of certain persistent chemicals in successively higher levels of a food chain (5.7)

biomimetic materials materials that try to replicate specific properties of biological materials for use in human applications (9.7)

blowing agent either a gas or a substance capable of producing a gas used to manufacture a foamed plastic (9.5)

boiling point the temperature at which the vapor pressure of a liquid equals the surrounding atmospheric pressure (6.11 and 10.5)

bond dipole The difference in electronegativity between two atoms in a polar covalent bond, which gives rise to partial positive/negative charges on the atoms. A convention to indicate the bond dipole uses an arrow to point in the direction of the more negatively charged end of the covalent bond (5.1)

bond energy the amount of energy that must be absorbed to break a specific chemical bond (6.5)

Brix scale A unit used to quantitatively express the sugar content of a solution, based on measurements provided by a refractometer. One degree Brix (°Bx) is equal to 1 g of sucrose per 100 g of solution (i.e., 1%(w/w)) (10.9)

buffer a solution of a weak acid or base and a conjugate salt that responds only slightly to an external influence to maintain a relatively constant pH (5.11, 12.2)

C

calorie (cal) the amount of heat necessary to raise the temperature of one gram of water by 1 °C (6.4)

calorimeter a device used to experimentally measure the quantity of heat energy released in a combustion reaction (6.4)

capacitance the ability of a material to store an electrical charge (8.8)

capacity the specific energy of an energy storage device given in ampere-hours (Ah). This represents the discharge current a battery can deliver over time (8.3)

carbohydrate a compound that contains carbon, hydrogen, and oxygen, with H and O atoms found in the same 2:1 ratio as in H_2O (11.4)

carbon footprint an estimate of the amount of CO_2 and other greenhouse gas emissions in a given time frame, usually a year (4.8)

carbon neutral a situation in which the CO_2 added to the atmosphere is balanced by the CO_2 removed by photosynthesis, sequestration, a carbon offset, or some other process (6.14)

carcinogen a substance that causes cancer in living tissue (5.7)

carcinogenic capable of causing cancer (2.13)

catalyst a chemical substance that participates in a chemical reaction and influences its rate, without itself undergoing permanent change (2.11 and 3.8)

catalytic cracking a process in which catalysts are used to crack larger hydrocarbon molecules into smaller ones at relatively low temperatures (6.12)

catalytic reforming a process in which the atoms within a molecule are rearranged, usually starting with linear molecules and producing ones with more branches (6.12)

cathode the electrode at which reduction takes place. The cathode receives the electrons produced at the anode (8.1)

cation a positively charged ion formed by removing an electron(s) from a neutral species (5.4)

cellular membrane the dynamic yet protective outer casing of a cell (12.6)

cellulose A naturally occurring compound composed of C, H, and O that provides structural rigidity in plants, shrubs, and trees. Cellulose is a natural polymer of glucose (6.14)

cellulosic ethanol ethanol produced from any plant containing cellulose, typically cornstalks, switchgrass, wood chips, and other materials that are nonedible by humans (6.14)

chain reaction a term that generally refers to any reaction in which one of the products becomes a reactant and thus makes it possible for the reaction to become self-sustaining (7.2)

chemical equation a representation of a chemical reaction using chemical formulas (2.9)

chemical formula a symbolic way to represent the elementary composition of a substance. It reveals both the elements present (by chemical symbols) and the atomic ratio of those elements (by the subscripts) (1.2)

chemical reaction a process whereby substances described as reactants are transformed into different substances called products (2.9)

chemistry the branch of science that focuses on the composition, structure, properties, and changes of matter (1.1)

chiral (optical) isomers compounds with the same chemical formula but different three-dimensional molecular structures and different interaction with plane polarized light (12.5)

chlorofluorocarbons (CFCs) compounds composed of the elements chlorine, fluorine, and carbon (but do not contain the element hydrogen) (3.8)

chromosomes rod-shaped, compact coils of DNA and specialized proteins packed in the nucleus of cells (13.3)

climate A term that describes regional temperatures, humidity, winds, rain, and snowfall over decades, not days. Contrast this with *weather* (4.7)

climate adaptation the ability of a system to adjust to climate change (including climate variability and extremes), to moderate potential damage, to take advantage of opportunities, or to cope with the consequences (4.10)

climate mitigation any action taken to permanently eliminate or reduce the long-term risk and hazards of climate change to human life, property, or the environment (4.10)

codon a sequence of three adjacent nucleotides that either guides the insertion of a specific amino acid or signals the start or end of protein synthesis (13.4)

coenzymes molecules that work in conjunction with enzymes to enhance the enzyme's activity (11.7)

combustion the chemical process of burning; the rapid reaction of fuel with oxygen to release energy in the form of heat and light (2.9)

composition a description of the identity and structure of the subunits that comprise a substance or material (1.1)

compostable under the conditions of either a home composter or an industrial composter, the ability for an item to undergo biological decomposition to form a material (compost) that contains no materials toxic to plant growth (9.10)

compound a pure substance that is comprised of two or more different types of atoms in a fixed, characteristic chemical combination (1.2)

concentrating solar power the use of sunlight to generate electricity through a solar-thermal process (7.9)

condensation polymerization a type of polymerization in which a small molecule such as water is split out (eliminated) when the monomers join to form a polymer (9.6)

condensed structural formula a structural formula in which some bonds are not shown; rather, the structural formula is understood to contain an appropriate number of bonds (6.2)

condenser an apparatus typically composed of recirculating cold water that is used to cool hot vapors, condensing them into liquid for collection (distillation) or extended cycles of reevaporation/condensation (reflux) (10.12)

conjugate acid the species formed by adding a proton to a base (12.2)

conjugate base the species formed by removing a proton from an acid (12.2)

copolymer a polymer formed by the combination of two or more different monomers (9.6)

covalent bond a bond formed when electrons are shared between two atoms (3.6)

cradle-to-cradle a term coined in the 1970s that refers to a regenerative approach to the use of things in which the end of the life cycle of one item dovetails with the beginning of the life cycle of another, so that everything is reused rather than disposed of as waste (1.9 and 9.9)

critical mass the amount of fissionable fuel required to sustain a chain reaction (7.2)

crystal a solid-state material that consists of long-range 3D ordering of its constituent atoms, ions, or molecules (1.8)

crystalline regions in a polymer, a region in which the long polymer molecules are arranged neatly and tightly in a regular pattern (9.5)

crystallization the process of dropping a solid out of a solution in a controllable manner in order to form a solid with an ordered structural array (5.6)

current (electrical) the rate of electron flow through a circuit (8.2)

D

degasification the process of a gas escaping from a liquid or solid (5.6)

denitrification the process of converting nitrates to nitrogen gas (11.12)

density the mass per unit volume (5.2)

deoxyribonucleic acid (DNA) the biological polymer that carries genetic information in all species (13.2)

desalination any process that removes sodium chloride and other minerals from salty water (5.13)

dimensional analysis the conversion of one unit to another using conversion factors (1.4)

dipeptide a compound formed from two amino acids (11.6)

dipole-dipole forces an intermolecular force between polar molecules, which do not contain O—H, N—H, or F—H bonds (5.2)

directed evolution when scientists create an environment where bacteria mimic natural selection by evolving a new trait so they can survive (13.7)

disaccharide a "double sugar" formed by joining two monosaccharide units, such as sucrose (table sugar) (11.4)

distillation a separation process in which a liquid solution is heated to its boiling point and the vapors are condensed and collected (5.13 and 6.11)

distributed generation generating electricity on-site where it is used (i.e., with a fuel cell), thus avoiding the losses of energy that occur over long electric transmission lines (8.10)

doping the process of intentionally adding small amounts of other elements to pure silicon to modify its semiconductor properties (7.10)

double bond a covalent bond consisting of two pairs of shared electrons (3.6)

double helix a spiral consisting of two strands that coil around a central axis (13.3)

double-displacement a common way in which ionic compounds react; the cations and anions are exchanged between reactants to form new ionic products (5.6)

dry cells commercial batteries with an immobilized electrolyte made of paste so there is only enough moisture for ions to flow; they operate in any orientation without spilling because of the lack of free liquid (8.4)

E

electrical energy a form of kinetic energy that results from the flow of electric charge through an electrically conductive material (6.6)

electricity the flow of electrons from one region to another that is driven by a difference in potential energy (8.1)

electrodes the electrical conductors (anode and cathode) in an electrochemical cell that serve as sites for chemical reactions (8.1)

electrolysis the process of passing a direct current of electricity of sufficient voltage to cause a chemical reaction to occur. For example, the electrolysis of water decomposes it into H_2 and O_2 (8.11)

electrolyte a solute that conducts electricity in an aqueous solution (5.6 and 8.4)

electrolytic cell a type of electrochemical cell in which electrical energy is converted to chemical energy (8.11)

electromagnetic (EM) spectrum continuum of waves that ranges from short, high-energy X-rays and gamma rays to long, low-energy radio waves (3.1)

electronegativity a measure of the attraction of an atom for an electron in a chemical bond (4.6 and 5.1)

electrophoresis the movement of charged particles in a fluid or gel under the influence of an applied electric field (14.14)

element one of the 100 or so pure substances in our world from which compounds are formed; Elements contain only one type of atom (1.2)

emulsion A mixture of two or more liquids that are normally immiscible (10.10)

endocrine disrupter a compound that affects the human hormone system, including hormones for reproduction and sexual development (9.11)

endothermic a term applied to any chemical or physical change that absorbs energy (6.4)

energy the ability or capacity of matter to do work, or to produce change (6.3)

energy density the amount of potential energy stored in a given system per unit volume (volumetric energy density) or mass (gravimetric energy density, also referred to as specific energy density) (8.7)

enhanced greenhouse effect the process in which atmospheric gases trap and return *more than* 80% of the heat energy radiated by Earth (4.4)

enzymes proteins that act as biochemical catalysts, influencing the rates of chemical reactions (6.14 and 12.6)

equilibrium the state at which the concentrations of products and reactants are equal during a reversible chemical reaction (12.1)

equilibrium constant the concentration of products divided by the concentration of reactants for a reversible reaction that is at equilibrium (12.1)

essential amino acids those amino acids required for protein synthesis that must be obtained from the diet because the body cannot synthesize them (11.6)

exothermic a term to describe any chemical or physical change accompanied by the release of heat (6.4)

exposure the amount of a substance encountered (2.7)

F

fertilizer natural or synthetic material that is applied to soils or directly to plants to supply plant nutrients and promote plant growth (11.12)

First Law of Thermodynamics Also called the Law of Conservation of Energy, this law states that energy is neither created nor destroyed during any process or transformation (6.3)

Fischer–Tropsch process A method of producing a variety of liquid hydrocarbons from a series of catalyzed reactions using carbon monoxide and hydrogen gases as reactants (6.13)

flash point The temperature at which combustion is possible for a solvent. This occurs when the vapor pressure of a solvent is equal to its lower flammability limit (LFL) (14.4)

flashover The dangerous condition during a fire when the majority of exposed surfaces in a room are heated to their autoignition temperatures. This causes the materials to emit flammable gases, which provide additional fuel to the fire (14.5)

food miles approximate distance that a food item travels from where it was grown to where it was consumed (11.11)

food security maintaining the integrity of availability, access, and use of food (11.13)

foodborne illnesses caused by the presence of bacteria, viruses, parasites, and chemical toxins that can go undetected in food (11.9)

fossil fuel Combustible substances derived from the remnants of prehistoric organisms. The most common examples are coal, petroleum, and natural gas (4.7 and 6.1)

fracking a controversial method of extracting natural gas from underground rock formations through the injection of high-pressure fluids (6.10)

free radical a highly reactive chemical species with one or more unpaired electrons (3.7)

frequency the number of waves passing a fixed point in 1 second (3.1)

fuel any solid, liquid, or gas that may be burned to provide energy in the form of heat or work (6.1)

fuel cell an electrochemical cell that produces electricity by converting the chemical energy of a fuel directly into electricity without burning the fuel (8.9)

functional group a distinctive arrangement of a group of atoms that imparts characteristic properties to the molecules that contain this group (6.14 and 9.6)

G

galvanic cell a type of electrochemical cell that converts the energy released in a spontaneous chemical reaction into electrical energy (8.1)

gas chromatography (GC) a simple and rapid analytical technique commonly used to identify separate components in mixtures of liquids or gases (14.6)

genes short pieces of the genome that code for the production of proteins (13.1)

genetic engineering the direct manipulation of DNA in an organism (13.6)

genome the primary route for inheriting biological information required to build and maintain an organism (13.1)

global warming a popular term used to describe the increase in average global temperatures that results from an enhanced greenhouse effect (4.4)

glycerol a colorless, sweet, viscous liquid formed as a by-product in the production of soaps or biodiesel (6.15 and 11.2)

glycolysis a complex biological process that breaks down glucose in order to provide energy for each living cell (12.6)

green chemistry the design of chemical products and processes that reduce or eliminate the use and generation of hazardous substances (2.14)

greenhouse effect the natural process by which atmospheric gases trap a major portion (about 80%) of the infrared radiation radiated by Earth (4.4)

greenhouse gases Gases capable of absorbing and emitting infrared radiation, thereby warming the atmosphere. Examples include water vapor, carbon dioxide, methane, nitrous oxide, ozone, and chlorofluorocarbons (4.4)

groundwater fresh water found in underground reservoirs also known as aquifers (5.3)

group a column on the periodic table that organizes elements according to the important properties they have in common. Groups are numbered left to right (1.2)

H

Haber–Bosch process an experimental method to catalytically synthesize ammonia (NH_3) from hydrogen and nitrogen gases at elevated temperatures and pressures (11.12)

half-life ($t_{1/2}$) the time required for the level of radioactivity to fall to one half of its initial value (7.6)

half-reactions a type of chemical equation that shows the electrons either lost or gained by the reactants (8.1)

heat the kinetic energy that flows from a hotter object to a colder one (6.3)

heat of combustion the quantity of heat energy given off when a specified amount of a substance burns in oxygen (6.4)

hemoglobin an iron-containing metalloprotein found within red blood cells used for oxygen transport (12.6)

Henry's Law a formula that states that the concentration of dissolved gas in a solution is proportional to its partial pressure in the gas phase (10.10)

heterogeneous mixture a combination of solids, liquids, or gases that are not uniformly distributed throughout the substance (1.2)

homogeneous mixture a single-phase combination of solids, liquids, or gases with a uniform distribution of its constituents throughout the substance (1.2)

hormones chemical messengers produced by the body's endocrine glands (12.6)

hybrid electric vehicle (HEV) a vehicle propelled by a combination of a conventional gasoline engine and an electric motor run by batteries (8.9)

hydrocarbon organic compounds comprised entirely of carbon and hydrogen (6.2)

hydrogen bond an electrostatic attraction between a H atom bonded to a highly electronegative atom (O, N, or F) and a neighboring O, N, or F atom, either in another molecule or in a different part of the same molecule (5.2)

hydrogenation a process in which hydrogen gas, in the presence of a metallic catalyst, adds to a C=C double bond and converts it to a single bond (11.3)

hydrometer a device used to measure the density of a liquid (10.9)

hydronium ion the result of an acid donating a proton (H^+) to a water molecule (5.8)

I

infrared spectrometer an instrument that determines the types of functional groups present in a sample based on preferential absorption of characteristic regions of infrared radiation (4.6)

insulin a hormone produced in the pancreas, which regulates the amount of glucose in the blood. The lack of insulin causes a form of diabetes (11.4 and 13.0)

interesterification any process in which the fatty acids on two or more triglycerides are scrambled to produce a mixture of different triglycerides (11.3)

intermolecular force a force that occurs between molecules (5.2)

ion a positively or negatively charged species that is formed through the addition or subtraction of electrons (5.4)

ionic bond the electrostatic attraction of oppositely charged ions present in a solid ionic compound (5.4)

ionic compound a substance that is composed of oppositely charged ions of metal and nonmetallic elements (5.4)

ionizing radiation radiation with sufficient energy to cause bond breakage and electron removal from atoms or molecules in the medium through which it passes such as air, water, or living tissue (7.5)

isomers molecules with the same chemical formula, but with different structures and properties (6.12)

isotope two or more types of atoms that have equal numbers of protons and different numbers of neutrons (4.2)

J

joule (J) a unit of energy equal to 0.239 cal (6.4)

K

kinetic energy the energy of motion (6.3)

kinetics the branch of science that deals with the rates of reactions (8.7 and 12.1)

L

latent heat a measure of the heat absorbed by a substance to induce a phase change (10.5)

latent prints fingerprints left by the natural oils from one's skin, which result in prints left on a surface that are not visible to the naked eye (14.6)

law of conservation of matter and mass a law stating that in a chemical reaction, matter and mass are conserved (2.9)

Le Châtelier's principle disturbing a system that is in chemical equilibrium will alter the system in a way that opposes the disturbance (12.1)

Lewis structure a representation of an atom or molecule that shows its outer electrons (3.6)

limiting reagent the reactant that is totally consumed during a chemical reaction, hence limiting the amount of product that may be formed (2.9)

line-angle drawing simplified version of a structural formula that is most useful for representing larger molecules (11.2)

lipids a class of compounds that includes not only all triglycerides, but also related compounds such as cholesterol and other steroids (11.2)

London dispersion forces attractive forces between nonpolar molecules such as hydrocarbons (6.11)

lone pair an electron pair in the outermost (valence) shell of an atom that is not shared with another atom, thus not involved in covalent bond formation (3.6)

lower flammability limit (LFL) the lowest concentration of solvent vapor that can ignite in air (14.4)

M

macrominerals elements that are necessary for life (Ca, P, Cl, K, S, Na, and Mg) but not nearly as abundant in the body as O, C, H, and N (11.7)

macronutrient the fats, carbohydrates, and proteins that provide essentially all of the energy and most of the raw material for body repair and synthesis (11.1)

Maillard reaction A reaction that occurs at high temperatures involving functional groups present in sugars and proteins within foods. This reaction results in a browned crust that forms on cooked foods such as eggs, meats, breads, etc. (10.5)

malnutrition a condition caused by a diet lacking in proper nutrients, even though the energy content of the food may be adequate (11.1)

mass number the sum of the number of protons and neutrons in the nucleus of an atom (1.5)

mass spectrometry an analytical technique in which a sample vapor is ionized and the resulting ions are separated according to their mass-to-charge ratios (14.6)

matter any solid, liquid, gas, or plasma that occupies space and has a mass (1.1)

medicinal chemistry the branch of chemistry that deals with the discovery or design of new therapeutic chemicals and their development into useful medicines (12.9)

melanin a dark brown/black pigment occurring in the hair, skin, and eyes in humans and animals. This substance provides some protection against the damaging effects of solar UV radiation (3.4)

metabolism the complex set of chemical processes that are essential in maintaining life (11.1 and 12.6)

methanol a colorless, volatile, and flammable liquid with chemical formula CH_3OH used as a solvent, antifreeze, or denaturant for ethanol. Unlike ethanol, this solvent is poisonous for human consumption, causing blindness or death if ingested (10.12)

microminerals nutrients that the body requires lesser amounts of, such as Fe, Cu, and Zn (11.7)

micronutrients substances such as vitamins and minerals that are needed only in miniscule amounts, but remain essential for the body to produce enzymes, hormones, and other substances needed for proper growth and development (11.7)

minerals ions or ionic compounds that, like vitamins, have a wide range of physiological functions (1.6 and 11.7)

molar mass the mass of Avogadro's number, or one mole, of whatever particles are specified (4.3)

molarity (M) a unit of concentration represented by the number of moles of solute present in one liter of solution (5.5)

mole (mol) an Avogadro's number of objects (4.3)

molecular compound a pure substance that contains two or more atoms from nonmetallic elements (2.5)

molecule two or more atoms held together by chemical bonds in a certain spatial arrangement (1.2)

monomer a small molecule used to synthesize a larger polymer (from *mono* meaning "one" and *meros* meaning "unit") (9.2)

monosaccharide a single sugar, such as fructose or glucose (11.4)

monounsaturated an organic compound, usually a fat, that contains a single multiple bond between adjacent carbon atoms (11.2)

municipal solid waste (MSW) garbage; that is, everything you discard or throw into your trash, including food scraps, grass clippings, and old appliances. MSW does not include all sources, such as waste from industry, agriculture, mining, or construction sites (9.9)

N

n-type semiconductor a semiconductor in which there are freely moving negative charges (electrons) (7.10)

nanotechnology the manipulation of matter with at least one dimension sized between 1 and 100 nanometers (1.4)

neutral solution a solution that is neither acidic nor basic; that is, it has equal concentrations of H^+ and OH^- (5.9)

neutralization reaction a chemical reaction in which the hydrogen ions from an acid combine with the hydroxide ions from a base to form water molecules (5.9)

nitrification the process of converting ammonia in the soil into the nitrate ion (11.12)

nitrogen cycle a set of chemical pathways whereby nitrogen moves through the biosphere (11.12)

nitrogen-fixing bacteria bacteria that remove nitrogen from the air and convert it into ammonia (11.12)

nonelectrolytes a substance that remains intact and is not dissociated into ions in an aqueous solution (5.7)

nonpolar covalent bond a covalent bond in which the electrons are shared equally or nearly equally between atoms (5.1)

normal boiling point the temperature at which the vapor pressure of the liquid is equal to 1 atm (6.11)

nuclear fission the splitting of a large nucleus into smaller ones with the release of energy (7.2)

nuclear fusion when two isotopes are combined under high pressures and temperatures, which releases an enormous amount of energy (7.2)

nuclear radiation radiation emitted by a nucleus, such as alpha, beta, or gamma radiation (7.4)

nucleotide covalently bonded combination of a base, a deoxyribose molecule, and a phosphate group (13.2)

O

ocean acidification the lowering of the ocean pH due to increased atmospheric carbon dioxide (5.11)

octet rule A generalization that electrons are arranged around atoms so that these atoms have a share in eight electrons. Hydrogen is an exception (3.6)

Ohm's law a law stating that electrical current is proportional to voltage, but inversely proportional to resistance (8.2)

oils triglycerides that are liquids at room temperature (6.15 and 11.2)

organic chemistry the branch of chemistry devoted to the study of carbon compounds (4.1 and 12.3)

organic compound a compound that always contains carbon, almost always contains hydrogen, and may contain other elements such as oxygen and nitrogen (2.11)

osmosis the passage of water through a semipermeable membrane from a solution that is less concentrated to a solution that is more concentrated (5.13)

oxidation a process in which a chemical species loses electrons (8.1)

oxidation state a positive or negative number that represents the formal charge of an atom or element. This number indicates the relative likelihood of oxidation (loss of electrons) or reduction (gain of electrons) for the species (8.1)

oxidizing agent a substance that is reduced, and thereby accepts electrons, during a redox reaction (8.10)

oxygenated gasoline a blend of petroleum-derived hydrocarbons with added oxygen-containing compounds such as MTBE, ethanol, or methanol (6.12)

ozone layer a designated region in the stratosphere of maximum ozone concentration (3.5)

P

p-type semiconductor a semiconductor in which there are freely moving positive charges, or holes (7.10)

parts per billion (ppb) one part out of one billion, or 1000 times less concentrated than 1 part per million (5.5)

parts per million (ppm) A concentration of one part out of a million. One ppm is a unit of concentration 10,000 times smaller than 1% (one part per hundred). (5.5)

patent prints fingerprints left by someone who is using a substance such as grease, paint, blood, etc., which results in a visible print on a surface (14.6)

peptide bond the covalent bond that forms when the —COOH group of one amino acid reacts with the —NH_2 group of another, thus joining the two amino acids (9.7)

percent (%) parts per hundred; For example, 15% is 15 parts out of 100 (2.2 and 5.5)

pharmaceuticals therapeutic substances intended to prevent, moderate, or cure illnesses (12.9)

pharmacophore the three-dimensional arrangement of atoms or groups of atoms responsible for the biological activity of a drug molecule (12.10)

photon a way of conceptualizing light as a particle that has energy but no mass (3.2)

photosynthesis the process by which green plants (including algae) and some bacteria capture the energy of sunlight to produce glucose and oxygen from carbon dioxide and water (3.4)

photovoltaic (PV) cell a semiconductor-containing device that converts light into electrical energy (7.9)

plasmids small rings of DNA molecules found in bacteria and some other microscopic organisms (13.6)

plasticizer a compound added in small amounts to a polymer to make the polymer softer and more pliable (9.5)

polar covalent bond a covalent bond in which the electrons are not equally shared, but rather are closer to the more electronegative atom (5.1)

polar stratospheric clouds (PSCs) thin clouds composed of tiny ice crystals formed from the small amount of water vapor present in the stratosphere (3.8)

polyamide a condensation polymer that contains the amide functional group (9.7)

polymer a large molecule built from smaller ones (monomers) that consists of a long chain or chains of atoms covalently bonded together (9.0)

polymerase chain reaction (PCR) a technique that is used to amplify a single copy or a few copies of DNA across several orders of magnitude, generating thousands to millions of copies of a particular DNA sequence (14.14)

polysaccharide a condensation polymer made up of thousands of monosaccharide units. Examples include starch and cellulose (11.4)

polyunsaturated an organic compound, usually a fat, that contains more than one multiple bonds between adjacent carbon atoms (11.2)

postconsumer content material used by a consumer that would otherwise have been discarded as waste (9.9)

potable water water safe for drinking and cooking (5.3)

potential energy Energy of position, or stored energy. In chemistry, we refer to this energy as that stored in the chemical bonds within a molecule that may be released during a chemical reaction (6.3)

power the rate at which electrical energy is transferred through an electrical circuit (7.3 and 8.2)

power density The amount of power (voltage × current) per unit of volume. This refers to the ability of an energy-storage device to take on or deliver power. That is, a battery with a high power density will charge faster than one with a lower power density (8.8)

precautionary principle stresses the wisdom of acting, even in the absence of full scientific data, before the adverse effects on human health or the environment become significant or irrevocable (9.11)

precipitate The solid deposited during a precipitation event, when a solid drops out of a homogeneous solution. This generally refers to an amorphous solid, with no long-range structural order, but it can also be used to describe a crystalline solid deposited slowly from a solution (5.6)

precipitation The process of a solid dropping out of a homogeneous solution. Usually this refers to rapid deposition of a solid, which forms an amorphous solid with a disordered structural array (5.6)

preconsumer content waste left over from the manufacturing process itself, such as scraps and clippings (9.9)

primary structure the unique sequence of the amino acids that make up each protein (13.5)

processed foods foods that have been altered from their natural state by techniques such as canning, cooking, freezing, or adding chemicals such as thickeners or preservatives (11.1)

products the substances listed on the right-hand side of a chemical equation, representing the materials that are formed during a standard chemical reaction (2.9 and 6.2)

protein a polyamide or polypeptide; that is, a polymer built from amino acid monomers (11.6)

protein complementarity combining foods that complement essential amino acid content so that the total diet provides a complete supply of amino acids for protein synthesis (11.6)

proton a subatomic positively charged particle with approximately the same mass as a neutron (5.8)

proton-exchange membrane a polyelectrolyte membrane, which is permeable to H+ ions and coated on both sides with a platinum-based catalyst (8.10)

Q

qualitative information or observations expressed using senses; e.g., changes in texture and firmness (10.9)

quantitative information or observations expressed in numbers (10.9)

quantized an energy distribution that is not continuous, but rather consists of many individual steps (3.2)

R

racemic mixture mixture consisting of equal amounts of each optical isomer of a compound (12.5)

rad a unit of absorbed radiation dose, defined as 1 rad = 0.01 J/kg or 0.01 Gy (Gray) (7.5)

radiation the emission of energy as electromagnetic waves or as moving subatomic particles (3.1)

radiative forcings factors (both natural and anthropogenic) that influence the balance of Earth's incoming and outgoing radiation (4.8)

radioactive decay series a characteristic pathway of radioactive decay that begins with a radioisotope and progresses through a series of steps to eventually produce a stable isotope (7.4)

radioactivity the spontaneous emission of radiation by certain elements (7.4)

radioisotope an isotope that spontaneously emits nuclear radiation (7.4)

radiopharmaceuticals organic molecules that carry radioactive isotopes to specific regions in the body in order to create contrast between tissue areas (12.10)

reactants the substances listed on the left-hand side of a chemical equation, representing the starting materials for a standard chemical reaction (2.9 and 6.2)

reaction quotient the product concentrations divided by the reactant concentrations during a reversible chemical reaction that may or may not be at equilibrium (12.1)

reactive nitrogen the compounds of nitrogen that cycle through the biosphere and interconvert with each other relatively quickly (11.12)

receptor a biomolecule that is typically embedded within the cellular membrane that binds with specific molecules, thus producing some effect in the cell (12.6)

recycled-content products products made from material that otherwise would have been in the waste stream (9.9)

redox reaction an electrochemical reaction consisting of a combination of both reduction and oxidation half-reactions (8.1)

reducing agent a substance that is oxidized, and thereby releases electrons, during a redox reaction (8.10)

reducing agents a substance that is easily oxidized (loses electrons), which results in the reduction of another substance through adding electrons (14.2)

reduction a process in which a chemical species gains electrons (8.1)

reflux the repeated process of heating a solvent to boiling and condensing its vapor (14.2)

reformulated gasoline (RFG) an oxygenated gasoline that also contains a lower percentage of certain more volatile hydrocarbons found in nonoxygenated conventional gasoline (6.12)

refractometry the study of how light propagates through a material or liquid (10.9)

rem referred to as the roentgen equivalent man, an older unit that is a measure of the effects of low levels of ionizing radiation on human health. This unit has largely been replaced by the Sievert, Sv (7.5)

renewable resources those resources that are replenished more quickly over time than they are being consumed (7.1)

replication the process of cell reproduction in which the cell must copy and transmit its genetic information to its progeny (13.3)

residual chlorine The name given to chlorine-containing chemicals that remain in the water after the chlorination step. These include hypochlorous acid (HClO), the hypochlorite ion (ClO^-), and dissolved elemental chlorine (Cl_2) (5.12)

resonance forms Lewis structures that represent hypothetical extremes of electron arrangements in a molecule (3.6)

respiration the process of metabolizing the foods we eat to produce carbon dioxide and water and to release the energy that powers other chemical reactions in our bodies (2.1 and 12.6)

reverse osmosis a process that uses pressure to force the movement of water through a semipermeable membrane from a solution that is more concentrated to a solution that is less concentrated (5.13)

risk assessment the process of evaluating scientific data and making predictions in an organized manner about the probabilities of an outcome (2.7)

rocks heterogeneous solid-state mixtures that contain a variety of ionic compounds (1.6)

S

saturated fatty acid a hydrocarbon containing only single bonds between the carbon atoms (11.2)

scanning electron microscopy (SEM) an analytical technique used to image the surface of a sample by using a high-energy beam of electrons (14.6)

scientific notation a system for writing numbers as the product of a number and 10 raised to the appropriate power (1.4)

secondary batteries rechargeable batteries, which use electroreactions that run in both directions; the transfer of electrons takes place both during the forward (discharging) and the reverse (recharging) processes (8.4)

secondary pollutant a pollutant produced from chemical reactions involving one or more other pollutants (2.12)

secondary structure the folding pattern within a segment of the protein chain (13.5)

self-discharge when an energy storage device loses its charge over time without being connected to an external circuit (8.8)

semiconductor a material that does not normally conduct electricity well, but can do so under certain conditions, such as with exposure to sunlight (7.10)

separator a porous electrical insulator that is placed between the positive and negative electrodes in a battery, which allows the passage of ions (8.4)

shifting baseline the idea that what people expect as "normal" on our planet has changed over time, especially with regard to ecosystems (2.14 and 9.11)

sievert (Sv) the preferred unit of the ionizing radiation dose, which is a measure of the effects of low levels of ionizing radiation on human health. Corresponding to J/kg, this unit represents the biological effect of introducing a joule of radiation energy (100 rad or 1 Gray, Gy) to a kilogram of human tissue (7.5)

single covalent bond a bond formed when two electrons (one pair) are shared between two atoms (3.6)

solubility the degree to which a solute is dissolved in a solvent (5.4)

solute a solid, liquid, or gas that dissolves in a solvent to form a solution (5.4)

solution a homogeneous (of uniform composition) mixture of a solvent and one or more solutes (1.2 and 5.4)

solvent a substance, often a liquid, capable of dissolving one or more pure substances (5.4)

solvent still a laboratory apparatus used to remove oxygen and moisture from organic solvents (14.2)

space-filling model a representation of a molecular compound that shows the volume occupied by the atoms in the molecule (4.5)

specific gravity the ratio of density of a solution to the density of pure solvent (without any dissolved solutes) (10.9)

specific heat the quantity of heat energy that must be absorbed to increase the temperature of one gram of a substance by 1 °C. The standard unit of energy is joules, J (5.2)

speed of light c, the speed at which electromagnetic radiation passes in a vacuum, defined as 3.00×10^8 m/s (3.1)

spherification a culinary process of shaping a liquid into spheres (10.10)

stable isotope analysis a technique that compares the concentration ratio of various isotopes for a sample in order to determine its origin (14.6)

standard temperature and pressure (STP) an ambient temperature of 25 °C and pressure of 1 atm (760 Torr) (5.1)

starch A carbohydrate found in many grains, including corn and wheat. Starch is a natural polymer of glucose (6.14)

steady state a condition in which a dynamic system is in balance so that there is no net change in concentration of the major species involved (3.6)

steroid a class of naturally occurring or synthetic fat-soluble organic compounds that share a common carbon skeleton arranged in four rings (12.8)

strong acid an acid that dissociates completely in water (5.8)

strong base a base that dissociates completely in water (5.8)

strong electrolyte an ionic compound that is 100% dissociated into ions in an aqueous solution (5.6)

structural formula A representation of how the atoms in a molecule are connected. It is a Lewis structure from which the nonbonding electrons have been removed (3.6)

structure–activity relationship (SAR) study a study in which systematic changes are made to a drug molecule followed by an assessment of the resulting changes in activity (12.10)

surface water fresh water found in lakes, rivers, and streams (5.3)

surfactant a molecule that has both polar and nonpolar regions that allow it to help solubilize different classes of molecules (5.7)

sustainability "meeting the needs of the present without compromising the ability of future generations to meet their own needs" (from *Our Common Future,* a 1987 report by the United Nations) (2.14)

sustainable packaging the design and use of packaging materials to reduce their environmental impact and improve the sustainability of all practices (9.8)

T

temperature a measure of the average kinetic energy of the atoms and/or molecules present in a substance (6.3)

tertiary structure the overall molecular shape of the protein defined by the interactions between amino acids far apart in sequence, but close in space (13.5)

thermal cracking a process that breaks large hydrocarbon molecules into smaller ones by heating them to a high temperature (6.12)

thermodynamics the branch of science that deals with the energies and relative spontaneity associated with chemical reactions or processes (12.1)

thermoplastic polymer a plastic that can be melted and reshaped over and over again (9.5)

thermoset plastics a plastic that retains its shape even at elevated temperatures (9.5)

three pillars of sustainability sustainability is meeting the needs of the present without compromising the ability of future generations to meet their own needs. There are three considerations that are equally important for a sustainable practice, which include environmental, social, and economic factors (1.9)

toxicity the intrinsic health hazard of a substance (2.7)

trace mineral an element present in the body, usually at microgram levels, such as I, F, Se, V, Cr, Mn, Co, Ni, Mo, B, Si, and Sn (11.7)

tragedy of the commons the situation in which a resource is common to all and used by many, but has no one in particular who is responsible for it. As a result, the resource may be destroyed by overuse to the detriment of all that use it (2.12 and 5.4)

trans fats triglycerides that are solids at room temperature (6.15 and 11.3)

transgenic an organism resulting from the transfer of genes across species (13.6)

triglycerides a class of compounds that includes both fats and oils. Triglycerides contain three ester functional groups and are formed from a chemical reaction with three fatty acids and the alcohol glycerol (6.15 and 11.2)

trihalomethanes (THMs) compounds such as $CHCl_3$ (chloroform), $CHBr_3$ (bromoform), $CHBrCl_2$ (bromodichloromethane), and $CHBr_2Cl$ (dibromochloromethane) that form from the reaction of chlorine or bromine with organic matter in drinking water (5.12)

triple bond a covalent linkage made up of three pairs of shared electrons (3.6)

Triple Bottom Line a three-way measure of the success of a business based on its benefits to the economy, to society, and to the environment (6.16)

U

undernourishment a condition in which a person's daily caloric intake is insufficient to meet metabolic needs (11.1)

unsaturated fatty acid a fatty acid in which the hydrocarbon chain contains one or more double bonds between carbon atoms (11.2)

V

valence electron(s) an electron(s) in the outermost shell of an atom, which can participate in covalent bonding (3.6)

vapor pressure the pressure exerted by gaseous molecules, as a result of vaporization of a liquid or solid (6.11)

vaporization the process of transferring molecules from the liquid to gaseous state (6.11)

vector a modified plasmid used to carry DNA back into the bacterial host (13.6)

viscosity a measure of a fluid's resistance to flow. For instance, a viscous fluid such as syrup will flow much slower than one with a lower viscosity such as water (10.10)

vitamin an organic compound, with a wide range of physiological functions, that is essential for good health, proper metabolic functioning, and disease prevention (11.7)

volatile organic compounds (VOCs) a substance composed primarily of carbon that is easily vaporized and released into the air (2.11)

volatility the ease at which molecules of a liquid overcome their intermolecular forces to be released into the gaseous phase (6.11)

voltage the difference in electrochemical potential between two electrodes (8.2)

volumetric flask a type of glassware that contains a precise amount of solution when filled to the mark on its neck (5.5)

W

water footprint an estimate of the volume of fresh water used to produce a particular good or to provide a service (5.3)

watt The SI unit of power, equal to 1 J/s (7.3)

wave-particle duality the exhibition of both wave-like (frequency, wavelength) and particle-like (mass, velocity, speed) characteristics by quantum entities such as electrons, atoms, and molecules (3.2)

wavelength the distance between successive peaks in a sequence of waves (3.1)

weak acid an acid that dissociates only to a small extent in aqueous solution (5.8)

weak base a base that dissociates only to a small extent in aqueous solution (5.8)

weak electrolyte an ionic compound that is weakly dissociated (typically <5%) into ions in an aqueous solution (5.6)

weather Conditions that include the daily high and low temperatures, the drizzles and downpours, the blizzards and heat waves, and the fall breezes and hot summer winds, all of which have relatively short durations; Contrast this with *climate* (4.7)

X

X-ray diffraction an analytical technique in which a crystal is hit by a beam of X-rays to generate a pattern that reveals the positions of the atoms in the crystal (13.3)

Index

Periodic Table of the Elements

Key:
24
Cr
52.00

— Atomic number
— Atomic mass

1 1A	2 2A	3 3B	4 4B	5 5B	6 6B	7 7B	8 8B	9 8B	10	11 1B	12 2B	13 3A	14 4A	15 5A	16 6A	17 7A	18 8A
1 H 1.008																	2 He 4.003
3 Li 6.941	4 Be 9.012											5 B 10.81	6 C 12.01	7 N 14.01	8 O 16.00	9 F 19.00	10 Ne 20.18
11 Na 22.99	12 Mg 24.31											13 Al 26.98	14 Si 28.09	15 P 30.97	16 S 32.07	17 Cl 35.45	18 Ar 39.95
19 K 39.10	20 Ca 40.08	21 Sc 44.96	22 Ti 47.88	23 V 50.94	24 Cr 52.00	25 Mn 54.94	26 Fe 55.85	27 Co 58.93	28 Ni 58.69	29 Cu 63.55	30 Zn 65.39	31 Ga 69.72	32 Ge 72.61	33 As 74.92	34 Se 78.96	35 Br 79.90	36 Kr 83.80
37 Rb 85.47	38 Sr 87.62	39 Y 88.91	40 Zr 91.22	41 Nb 92.91	42 Mo 95.94	43 Tc (98)	44 Ru 101.1	45 Rh 102.9	46 Pd 106.4	47 Ag 107.9	48 Cd 112.4	49 In 114.8	50 Sn 118.7	51 Sb 121.8	52 Te 127.6	53 I 126.9	54 Xe 131.3
55 Cs 132.9	56 Ba 137.3	57 La 138.9	72 Hf 178.5	73 Ta 180.9	74 W 183.8	75 Re 186.2	76 Os 190.2	77 Ir 192.2	78 Pt 195.1	79 Au 197.0	80 Hg 200.6	81 Tl 204.4	82 Pb 207.2	83 Bi 209.0	84 Po (209)	85 At (210)	86 Rn (222)
87 Fr (223)	88 Ra (226)	89 Ac (227)	104 Rf (267)	105 Db (268)	106 Sg (269)	107 Bh (270)	108 Hs (277)	109 Mt (278)	110 Ds (281)	111 Rg (282)	112 Cn (285)	113 Nh (286)	114 Fl (289)	115 Mc (289)	116 Lv (293)	117 Ts (294)	118 Og (294)

58 Ce 140.1	59 Pr 140.9	60 Nd 144.2	61 Pm (145)	62 Sm 150.4	63 Eu 152.0	64 Gd 157.3	65 Tb 158.9	66 Dy 162.5	67 Ho 164.9	68 Er 167.3	69 Tm 168.9	70 Yb 173.0	71 Lu 175.0
90 Th 232.0	91 Pa 231.0	92 U 238.0	93 Np (237)	94 Pu (244)	95 Am (243)	96 Cm (247)	97 Bk (247)	98 Cf (251)	99 Es (252)	100 Fm (257)	101 Md (258)	102 No (259)	103 Lr (262)

Metals

Metalloids

Nonmetals

The 1–18 group designation has been recommended by the International Union of Pure and Applied Chemistry (IUPAC).